Handbook of industrial diamonds and diamond films

HANDBOOK OF INDUSTRIAL DIAMONDS AND DIAMOND FILMS

HANDBOOK OF INDUSTRIAL DIAMONDS AND DIAMOND FILMS

edited by

MARK A. PRELAS
University of Missouri–Columbia, Columbia, Missouri

GALINA POPOVICI
University of Missouri–Columbia, Columbia, Missouri

LOUIS K. BIGELOW
Norton Diamond Film, Northboro, Massachusetts

MARCEL DEKKER, INC. NEW YORK • BASEL • HONG KONG

Library of Congress Cataloging-in-Publication Data

Handbook of industrial diamonds and diamond films / [edited by] Mark A. Prelas, Galina Popovici, Louis K. Bigelow.
p. cm.
Includes index.
ISBN 0-8247-9994-1 (hardcover : alk. paper)
1. Diamonds. 2. Diamonds, Industrial. 3. Diamond thin films—Industrial applications. 4. Diamonds, Artificial—Industrial applications. I. Prelas, Mark Antonio. II. Popovici, Galina. III. Bigelow, Louis.
TA455.C3H6 1997
666'.88—dc21

The publisher offers discounts on this book when ordered in bulk quantities. For more information, write to Special Sales/Professional Marketing at the address below.

This book is printed on acid-free paper.

MARCEL DEKKER, INC.
270 Madison Avenue, New York, New York 10016
http://www.dekker.com

Current printing (last digit):
10 9 8 7 6 5 4 3 2 1

PRINTED IN THE UNITED STATES OF AMERICA

Preface

Diamond is one of the oldest engineering materials. It was first used as an abrasive several thousand years ago. In the last 40 years, the development of synthetic diamond in various forms has fueled a revolution in the use of diamond as an engineering material. The process of high-pressure high-temperature diamond synthesis developed by General Electric in the early 1950s was responsible for stunning growth in the abrasives market. During that time, the world's consumption of diamond abrasive materials increased from 5 tonnes to 80 tonnes per year.

In 1954, W. Eversole reported the synthesis of diamond by chemical vapor deposition (CVD). Going strictly by the phase diagram of carbon, it was very difficult for many scientists to accept the proposition that diamond could be obtained by chemical vapor deposition. In 1956, Soviet scientists B. V. Spitsyn and B. V. Deryagin filed a patent disclosure covering the growth of a diamond from carbon tetraiodide that set the stage for substantial chemical vapor deposition work in the USSR. In the 1960s J. Angus in the United States duplicated the earlier work of Eversole. Meanwhile, progress was also made in Deryagin's laboratory in the USSR. The major breakthrough in the early 1970s on the chemical vapor deposition process was the use of atomic hydrogen during the growth phase. The use of atomic hydrogen was independently pursued by J. Angus and by V. Varnin in the USSR. In 1975, Deryagin announced that a high-growth-rate CVD process had been developed. This announcement was greeted with skepticism by scientists worldwide. In the late 1970s and early 1980s, however, N. Setaka's group in Japan also reported high-growth-rate CVD diamond films. These reports motivated the U.S. Department of Defense to increase research funding for CVD diamond in the mid-1980s.

CVD diamond films have since expanded into the markets of electronics, cutting tools, and wear-resistant coatings and have demonstrated other applications in the areas of thermal management, optics, and acoustics. The CVD diamond market is growing broader as new products continue to be developed such as chemical process electrodes and radiation detectors. Additional applications as cold cathode emitters, as high-temperature sensors, and ultimately as semiconductors are also being developed. The impact of diamond on people's lives, while not always immediately obvious, is definitely becoming increasingly significant.

The purpose of this book is to provide a comprehensive view of diamond as an engineering material. Many of the world's leading authorities on diamond have contributed overviews of their specific areas of expertise with an emphasis on diamond's role in modern technology and its various applications as an engineering material. The contributors represent ten countries, demonstrating the international importance of diamond as an engineering material.

The text examines mined diamond as well as synthetic diamond. The scope of the book includes the engineering properties of diamond, methods for its characterization, processes used to produce synthetic diamond, its applications and uses in manufacturing, modeling of diamond-related processes, and economics.

Acknowledgments

The editors are grateful to the authors for their abundant expertise and their enthusiasm. Many of the authors have made key contributions to the science of diamond films. We thank them for their important research accomplishments, without which there would be no industrial technology of diamond films or emerging diamond film industry.

We thank Andrew Benedicktus for his editorial prowess. We also express appreciation for the contributions of Dr. Li-Te Steven Lin and Dr. Salim Khasawinah.

MAP thanks his family, Rosemary S. Roberts, Natalia, and Alexander, for their patience, and his colleagues in the Particulate Systems Research Center.

GP thanks her parents, Valentina and Nikolai, her sons Alexander and Andrei, and her colleagues in the Nuclear Engineering department at the University of Missouri for their continuous support and encouragement.

LKB thanks his wife Helen and daughter Alicia for their encouragement and the management of Saint Gobain Industrial Ceramics Inc. for their support of his participation in this project.

Mark A. Prelas
Galina Popovici
Louis K. Bigelow

Contents

Contributors

John C. Angus Case Western Reserve University, Cleveland, Ohio

Alberto Argoitia Case Western Reserve University, Cleveland, Ohio

Peter K. Bachmann Philips Research Laboratories, Aachen, Germany

S. P. Bozeman North Carolina State University, Raleigh, North Carolina

George R. Brandes Advanced Technology Materials, Inc., Danbury, Connecticut

John V. Busch IBIS Associates, Inc., Wellesley, Massachusetts

James E. Butler Naval Research Laboratory, Washington, D.C.

Mark A. Cappelli Stanford University, Stanford, California

M. Chenevier Université Joseph Fourier de Grenoble, Saint Martin d'Hères, France

Robert E. Clausing* Oak Ridge National Laboratory, Oak Ridge, Tennessee

Alan T. Collins King's College London,Strand, London, England

Mark P. D'Evelyn General Electric Corporate Research and Development, Schenectady, New York, and Rensselaer Polytechnic Institute, Troy, New York

David L. Dreifus Kobe Steel USA Inc., Research Triangle Park, North Carolina

Richard C. Eden Consultant, Northboro, Massachusetts

S. Farhat Université Paris-Nord, Villetaneuse, France

Bradley A. Fox Kobe Steel USA Inc., Research Triangle Park, North Carolina

Naoji Fujimori Sumitomo Electronic Industries, Ltd., Koyakita, Japan

J. M. Galimberti Norton Diamond Film, Northboro, Massachusetts

A. Gicquel Université Paris-Nord, Villetaneuse, France

Peter J. Gielisse Florida State University College of Engineering, Tallahassee, Florida

Steven L. Girshick University of Minnesota, Minneapolis, Minnesota

J. T. Glass Kobe Steel USA Inc., Research Triangle Park, North Carolina

Karen K. Gleason Massachusetts Institute of Technology, Cambridge, Massachusetts

Alexander G. Gontar Institute for Superhard Materials of the National Academy of Sciences of Ukraine, Kiev, Ukraine

---

**Retired.*

David G. Goodwin California Institute of Technology, Pasadena, California

Matt Gordon University of Arkansas, Fayetteville, Arkansas

John E. Graebner Lucent Technologies, Murray Hill, New Jersey

Timothy A. Grotjohn Michigan State University, East Lansing, Michigan

K. Hassouni Université Paris-Nord, Villetaneuse, France

Akimitsu Hatta Osaka University, Suita, Japan

R. A. Hay Norton Diamond Film, Northboro, Massachusetts

Akio Hiraki Osaka University, Suita, Japan

R. Kalish Technion, Haifa, Israel

Christopher S. Kovach Case Western Reserve University, Cleveland, Ohio

M. Lefebvre Office National d'Etudes et de Recherches Aérospatiales, Châtillon, France

Henry O. A. Meyer[†] Purdue University, West Lafayette, Indiana

V. I. Nepsha Almazy Rossii-Sakha Co., Ltd., Moscow, Russia

Sergei M. Pimenov General Physics Institute, Moscow, Russia

S. Prawer University of Melbourne, Parkville, Victoria, Australia

Victor G. Ralchenko General Physics Institute, Moscow, Russia

[†]*Deceased.*

Karen McNamara Rutledge Massachusetts Institute of Technology, Cambridge, Massachusetts

C. D. Scott NASA Lyndon B. Johnson Space Center, Houston, Texas

Michael Seal Sigillum B. V., Amsterdam, The Netherlands

V. R. Shanbhag Norton Company, Worcester, Masssachusetts

Adam T. Singer IBIS Associates, Inc., Wellesley, Massachusetts

B. R. Stoner Kobe Steel USA Inc., Research Triangle Park, North Carolina

K. Subramanian Norton Company, Worcester, Masssachusetts

Alexander M. Zaitsev University of Hagen, Hagen, Germany and Bellarussian State University, Minsk, Belarus

Chapter 1

BAND STRUCTURE

Alan T. Collins
Wheatstone Physics Laboratory, King's College London,
Strand, London WC2R 2LS, UK

Contents

1. Introduction

Diamond is an indirect bandgap material with an energy gap of 5.49 eV. Relatively few impurities can be incorporated into diamond, probably because of the small lattice spacing. The only defects for which the energy levels have been located in the forbidden gap are the boron acceptor, the nitrogen donor centers and the vacancy. In this chapter we will review the experimental measurements which have established the present-day picture of the band structure, and, where appropriate, these results will be compared with the results of theoretical calculations.

2. Indirect Band Gap

2.1 ABSORPTION

Pure diamond (type IIa diamond) exhibits an absorption edge at about 225 nm. Clark [1959] first showed that the dependence of the absorption coefficient on photon energy is of the form expected for allowed indirect electronic transitions, if the Coulomb interaction between the excited electron and the corresponding hole state in the valence band are taken into consideration. Detailed measurements of the features of the absorption edge yield information about certain lattice vibrational energies (phonon energies) and about the band structure, and such an investigation was carried out by Clark et al. [1964]. For an indirect gap semiconductor with a single dominant phonon energy $\hbar\omega$ the phonon absorption process gives rise to an absorption threshold at $E_{gx} - \hbar\omega$ and the phonon emission process gives rise to an absorption threshold at $E_{gx} + \hbar\omega$, where E_{gx} is equal to $E_g - E_x$. Here E_g is the energy separation between the maximum in the valence band and the lowest minimum in the conduction band, and E_x is the exciton binding energy. The electron-phonon interaction processes are temperature dependent, with the phonon absorption process being "frozen out" at low temperatures. By making measurements on a range of diamonds with different thicknesses and over a range of temperatures between 90 K and 600 K Clark et al. established that the low-temperature limit of the energy gap E_g is about 5.49 eV, and that the energy E_x of the indirect exciton is 0.070 ± 0.004 eV. These absorption measurements also suggested that the lowest minima in the conduction band are near the boundaries of the Brillouin zone in the <100> direction.

2.2 LUMINESCENCE

Additional information about the band structure is obtained from luminescence measurements, following excitation of electrons from the valence band to the conduction band. Although this can now be done using an excimer laser, most investigations of the "edge emission" from diamond have used cathodoluminescence. For the intrinsic edge emission, recombination of the electron and hole proceeds first by the formation of an exciton and then a phonon interaction, giving rise to a free-exciton peak which, at low temperatures, has a threshold energy at $E_{gx} - \hbar\omega$. The shape $S(\Delta E)$ of the free-exciton peak as a function of the energy ΔE from the threshold is given [Dean et al. 1965] by $S(\Delta E) = (\Delta E)^{1/2} \exp[-\Delta E/(kT)]$.

At the momentum $\mathbf{K}_c$ of the minima in the conduction band, the transverse acoustic (TA), transverse optical (TO) and longitudinal optical (LO) phonons give rise to three free-exciton peaks A, B and C with thresholds at $E_{gx} - \hbar\omega_{TA}$, $E_{gx} - \hbar\omega_{TO}$ and $E_{gx} - \hbar\omega_{LO}$, respectively [Dean et al. 1965]. Each of these peaks exhibits structure on the high-energy side which is due to the recombination of an indirect exciton associated with the split-off valence band. Cyclotron resonance measurements [Rauch 1962] (see section 5.1) show that this splitting is small for diamond (6 ± 1 meV, compared with 50 meV for Si and 280 meV for Ge), and diamond therefore exhibits this additional structure, while Si and Ge do not [Dean et al. 1965]. Peak B is the strongest free-exciton feature,

Table I. Phonon energies at the momenta of the conduction band minima (**K**$_c$) and the exciton energy gaps and binding energies measured at 100 K.

Parameter	Energy (eV)	Description
$\hbar\omega_{TA}$	0.087 ± 0.002	Transverse acoustic phonon at **K**$_c$
$\hbar\omega_{TO}$	0.141 ± 0.001	Transverse optical phonon at **K**$_c$
$\hbar\omega_{LO}$	0.163 ± 0.001	Longitudinal optical phonon at **K**$_c$
$\hbar\omega_R$	0.167 ± 0.002	Zone center (Raman) phonon
E_{gx}	5.409 ± 0.002	Indirect exciton energy gap associated with the upper valence bands
E_{gx}'	5.416 ± 0.002	Indirect exciton energy gap associated with the lower valence band
E_x	0.080 ± 0.005	Binding energy of the indirect exciton
E_g	5.490 ± 0.005	Indirect energy gap associated with the upper valence bands
E_{4x}	0.053 ± 0.002	Binding energy of the upper-valence-band indirect exciton to the neutral boron acceptor
E_{4x}'	0.048 ± 0.002	Binding energy of the lower-valence-band indirect exciton to the neutral boron acceptor

and two or more replicas of this peak spaced by the energy of the zone-center (Raman) phonon are easily measured.

In addition to the free-exciton peaks, bound-exciton features can be seen in diamonds which contain boron as the major impurity (type IIb diamonds). Here, using the nomenclature of Dean et al. [1965], it is possible to see the zero-phonon lines D_0 at energy E_{gx} - E_{4x} , where E_{4x} is the binding energy of the upper-valence-band indirect exciton to the neutral boron acceptor, and D_0' at energy E_{gx}' - E_{4x}' , where E_{gx}' is the indirect exciton energy gap associated with the lower valence band and E_{4x}' is the binding energy of the lower-valence-band indirect exciton to the neutral boron acceptor. Phonon replicas of these transitions with the TO and LO phonons at **K**$_c$ are also clearly visible and these peaks are further replicated by the zone-center (Raman) phonon.

From the positions of the spectral features Dean et al. [1965] derived the parameters in table I. By comparing the phonon energies at **K**$_c$ with those obtained from neutron scattering data Dean et al. determined that the conduction band minima lie at **K**$_c$ = 0.76 **K**$_{max}$ along the <100> type axes. The energies derived for the Raman, TO and TA phonons are in good agreement with those obtained from the optical absorption measurements [2].

2.3 TEMPERATURE DEPENDENCE

With increasing temperature the energy gap decreases. At room temperature (295 K) E_g is 5.470 ± 0.005 eV, and in the temperature range 135 to 295 K theaverage temperature coefficient is $dE_g/dT = -(5.4 \pm 0.5) \times 10^{-5}$ eV K^{-1} [Clark et al. 1964]. The slope progressively increases to about -1.1×10^{-4} eV K^{-1} at 600 K [Mainwood 1995].

The temperature dependence of the energy gap can be written as [Collins et al. 1990]

$$E_{gx}(T) = E' + \int d\omega\ f(\omega)\ \{n(\omega,T) + \tfrac{1}{2}\} - a(c_{11} + 2c_{12})\ \Delta V(T)/3V \qquad (1)$$

Here $n(\omega,T)$ is the Bose-Einstein occupation number and $f(\omega)d\omega$ is the difference in the electron-phonon coupling for the conduction-band minima and the valence-band maximum for those modes in the frequency range ω to $\omega + d\omega$, the energy gap at 0 K being $E' + \frac{1}{2} \int d\omega\ f(\omega)$. The final term in equation (1) allows for the temperature dependent lattice expansion, $\Delta V(T)/V$ being the fractional volume expansion to temperature T. The terms c_{11} and c_{12} are the elastic constants and a is the change in energy per unit compressional hydrostatic stress (see below). The volume expansion term accounts for only about 6% of the temperature dependence of E_{gx}. Most of the temperature dependence arises from the term $\int d\omega\ f(\omega)\ n(\omega,T)$. The functional form of $f(\omega)$ is not known, but a precise fit can be made to the temperature dependence of E_{gx} using $f(\omega) = c\omega g(\omega)$, where $g(\omega)$ is the density of phonon states for diamond and the constant c of proportionality is given by the fit to E_{gx} [Collins et al. 1990].

2.4 PRESSURE DEPENDENCE

In many semiconductors with an indirect bandgap on a cube axis the bandgap decreases with increasing hydrostatic pressure [Paul 1961]. For diamond, however, the bandgap increases for pressures up to at least 2.3 GPa with a pressure derivative of $dE_g/dP = \sim 6$ meV $(GPa)^{-1}$ [Onodera et al. 1991]. We have noted in section 2.1 that the energy at which a phonon absorption threshold is observed is $h\nu_{obs} = E_g - E_x - \hbar\omega$. Onodera et al. determined that a plot of $h\nu_{obs}$ against pressure had a slope of 6.33 meV $(GPa)^{-1}$. In order to extract the behavior of E_g the pressure dependence of the exciton binding energy and the phonon energy are required. The former is not known, but Onodera et al. estimated the pressure dependence of the phonon energy to be 0.36 meV $(GPa)^{-1}$ from studies of the pressure dependence of the Raman phonon, and subtracted this from the gradient of their graph. The measurements made by Onodera et al. used type IIa diamonds produced by high pressure synthesis, and their value for the pressure derivative agrees well with the 5 ± 1 meV $(GPa)^{-1}$ obtained earlier on natural diamond [Crowther and Dean 1964].

Although the behavior of diamond is unique when compared with other indirect bandgap materials which have a negative pressure coefficient, the calculated values [Fahy et al. 1987] for dE_g/dP agree well with the experimental value in magnitude and sign. Using *ab initio* pseudopotential plane wave and local orbital methods, Fahy et al. obtained pressure coefficients for the indirect bandgap in the range 5.3 to 6.6 meV $(GPa)^{-1}$. These methods frequently underestimate the energy gap by 20 to 50 %, but it has been observed that the pressure derivatives are accurately calculated, and are not sensitive to the use of different functional forms for the exchange-correlation potential [Fahy et al. 1987].

2.5 ISOTOPE SHIFT

Collins et al. [1990] showed that when the isotopic composition is changed from 1.1% ^{13}C (the natural abundance) to 99% ^{13}C the indirect band gap increases by 13.6 ± 0.2 meV. We can infer from this that the shift would be 13.9 ± 0.2 meV on changing from 100% ^{12}C to 100% ^{13}C. Most of this shift is due to the electron-phonon coupling, but there is a small contribution from the decrease in lattice parameter.

Kamo et al. [1988] and Collins et al. [1988] have shown that when diamond is grown from 99% ^{13}C the Raman frequency is decreased by about 50 cm^{-1}. From the small difference between the measured value and the value expected from the average mass Collins et al. inferred that there was a fractional change in the lattice constant of -3.3×10^{-4}. This estimate was critically dependent on the exact isotopic composition of the diamond, as well as the underlying assumption that the Raman frequency varies exactly as the reciprocal of the square root of the average mass. Later measurements of the lattice parameter by Holloway et al. [1991], using precision X-ray diffraction studies, showed that the change is rather smaller at -1.5×10^{-4}.

Collins et al. [1990] showed that the electron phonon-coupling produces the major contribution to the isotope shift of the energy gap. Changing the isotope from ^{12}C to ^{13}C gives a contribution

$$\Delta_1 = \tfrac{1}{2}\,[(12/13)^{1/2} - 1] \int d\omega\, f(\omega) \qquad (2)$$

where $f(\omega)$ is the same function as that used to fit the temperature dependence of the energy gap of ^{12}C diamond. Although the best fit to the temperature dependence is given when $f(\omega) \propto \omega\, g(\omega)$, reasonably good fits are obtained for $f(\omega) \propto \omega^{0.5}\, g(\omega)$ to $f(\omega) \propto \omega^{1.6}\, g(\omega)$. This results in a fairly large uncertainty in the calculated value of Δ_1 , giving Δ_1 = 13.5 ± 2.0 meV.

A second contribution Δ_2 to the isotopic dependence of E_{gx} comes from the volume change produced by changing the isotope. Collins et al. [1990] estimated that $\Delta_2 = 3.0 \pm 1.3$ meV. However, for the fractional volume change they used $\Delta V/V = (1.13 \pm 0.09) \times 10^{-3}$, an average of the values inferred from the shift of the Raman line and that calculated from a typical Grüneisen parameter. The data from Holloway et al. [1991] indicate that $\Delta V/V = 0.45 \times 10^{-3}$, and consequently the value of Δ_2 calculated by Collins et al. should be reduced to 1.2 ± 0.5 meV. The total value calculated for the isotope shift $\Delta_1 + \Delta_2$ of the indirect band gap then becomes 14.7 ± 2.5 meV, in excellent agreement with the experimental value of 13.9 ± 0.2 meV.

Zollner et al. [1992] have used the deformation-potential-type electron-phonon interaction (within the pseudopotential-bond-charge-model framework) to calculate the temperature dependences and isotope shifts of the direct and indirect band gaps in diamond. Their calculations agree quite well with the experimental temperature shift of the indirect band gap found by Clark et al. [1964] and the isotope shift of 17.1 meV calculated on changing from ^{12}C diamond to ^{13}C diamond agrees favorably with the value of Δ_1 , above.

3 Direct Band Gap

Measurements of the energy of the direct band gap, and higher energy inter- and intra-band transitions can only be made using reflectivity measurements since diamond is strongly absorbing in this region.

3.1 REFLECTIVITY MEASUREMENTS

In regions of absorption the refractive index of a material is complex and can be written as $\mathbf{n} = n + ik$, where n is the real part and k is the extinction coefficient. The permittivity is also complex and can be written as $\varepsilon = \varepsilon_1 + i\varepsilon_2$. Since $\varepsilon = \mathbf{n}^2$ it follows that $\varepsilon_1 = n^2 - k^2$ and $\varepsilon_2 = 2nk$.

For light incident normally on a crystal the reflectivity is given by $R = \{(n - 1)^2 + k^2\} / \{(n + 1)^2 + k^2\}$. Reflectance measurements, analysed using the Kramers-Krönig dispersion relationships may then be used to determine the absorption coefficient as a function of energy. Even without a full analysis, prominent absorption transitions are revealed by structure on the reflectivity spectrum [Clark et al. 1964]. Philipp and Taft [1962] measured the optical constants in a type I (nitrogen containing) diamond in the energy range from 5.5 to 23 eV, using a Kramers-Krönig analysis of room-temperature, near-normal-incidence reflectivity data. Structure in ε_2 near 7 eV and 12 eV was observed and associated with transitions at the lowest direct gap at the center of the Brillouin zone and at the <100> zone boundary. (These are referred to as the Γ and X points, respectively).

A later study by Walker and Osantowski [1964] on a type IIa (low nitrogen concentration) diamond at room temperature showed structure in ε_2 at 7, 12, 16, 20 and 24 eV. The new structure at 16, 20 and 24 eV was assigned to transitions near the <111> zone boundary (the L point) of the Brillouin zone. The observation of this structure was attributed to the use of a purer sample than that used by Philipp and Taft [1962].

Philipp and Taft [1964] repeated their earlier measurements over a slightly wider energy range. Structure near 24 eV was observed in these measurements, but otherwise their reflectance spectrum differed only slightly from their earlier investigation. They also re-analysed the data of Walker and Osantowski [1964] and found inconsistencies in their results. Problems can arise when using the Kramers-Krönig relation because it involves computing the phase angle from an integral which extends over all frequencies, whereas the experimental data are taken only over a limited range of frequencies. At energies below the indirect band gap the reflectance may be calculated accurately from the experimental values of the (real) refractive index; at high energies, however, the optical effects are associated with the excitation of core electrons and the behavior of R as a function of frequency ω can be deduced from the asymptotic expression for the permittivity and written in the form $R(\omega) = c\omega^{-4}$ where c is a constant [Philipp and Taft 1964]. Errors in this extrapolation can lead to gross errors in the absorption data extracted by the Kramers-Krönig analysis, and Philipp and Taft concluded that the reflectivity data obtained by Walker and Osantowski were too low for energies above 16 eV.

Clark et al. [1964] recorded the reflectance spectrum between 5 and 14 eV at room temperature and between 5 and 8 eV at 133 K. At 295 K they observed peaks at 7 and 12 eV and an additional weaker peak at 9 eV. At the lower temperature they observed the peak at 7 eV to be sharper, more intense and shifted slightly to higher energy compared with the room-temperature results. In addition the reflectance at low temperature was observed to drop below the room-temperature value at energies above 7.5 eV. Clark et al. associated the 7 eV feature with the threshold for direct transitions, and their data suggested that at 133 K $(E_g)_{direct} = 7.12 \pm 0.01$ eV and at 295 K $(E_g)_{direct} = 7.02 \pm 0.02$ eV. The average temperature dependence in this interval could then be inferred as $d(E_g)_{direct} / dT = -(6.3 \pm 1.8) \times 10^{-4}$ K^{-1}.

The changes in the 7 eV peak at low temperatures observed by Clark et al. [1964] were interpreted by Phillips [1965] as due to a hybrid exciton at the Γ point, and he considered the decrease in the low-temperature reflectance as the start of an anti-resonance. He suggested that the low-temperature curve would display a minimum and rejoin the room-temperature curve near 8.3 eV. On this basis he placed the direct interband gap (the $\Gamma_{25'}$ to Γ_{15} transition) at 8.7 eV, rather than around 7 eV as proposed in the earlier work.

Roberts et al. [1966] and Roberts and Walker [1967] reinvestigated the reflectivity of diamond, at room temperature and at 77 K, taking special care to avoid contamination of the surface of the diamond. They found, for example, that the reflectance in the region 15 to 19 eV of a freshly cleaned specimen decreased from 55% to 32% after being left overnight in the apparatus. They attributed this to contamination by diffusion pump oil, and suggested this type of contamination probably accounted for discrepancies in the reflectance spectra obtained in some of the earlier work. In addition they used a Kramers-Krönig analysis which did not rely on an extrapolation of the high-energy reflectance data.

The room-temperature variation of ε_1 and ε_2 with energy obtained by Roberts and Walker [1967] was in excellent agreement with the data obtained by Philipp and Taft [1964]. In contrast with the measurements by Clark et al. [1964] no structure was observed at 9 eV and the peak at 7.3 eV showed very little temperature dependence. In fact the reflectance data obtained at 77 K were almost the same as those obtained at 300 K, apart from the appearance of a peak at 7.8 eV. The energies of the other features identified from the ε_2 data were at 12.2, 16 and 23 eV; there was evidence that the peak at 23 eV was due to surface contamination.

More recently Logothetidis et al. [1992] have investigated the optical properties of diamond in the energy range 5 - 10 eV using spectroscopic ellipsometry measurements with synchrotron radiation. This technique measures directly both parts of the complex dielectric function and therefore has an advantage over reflectivity measurements which leads to this information indirectly. The position of the direct gap was measured as a

Table II. Parameters for fitting the temperature dependence of the direct gap.

Diamond	Derivative	E_B (eV)	a_B (eV)	θ (K)
IIa	First	7.387 ± 0.180	0.32 ± 0.18	1060 ± 290
	Second	7.590 ± 0.374	0.45 ± 0.37	1300 ± 420
IIb	First	7.271 ± 0.034	0.19 ± 0.03	1000 ± 200
	Second	7.322 ± 0.210	0.18 ± 0.15	970 ± 620

function of temperature between about 100 and 600 K in this work. A number of points emerged from this investigation.

(i) The shape of the dielectric function differed above about 7 eV from that obtained by previous workers, and varied during their experiment after the diamond had been heated to 500 °C.

(ii) The peak at around 7.8 eV observed by Roberts and Walker [1967] in their low-temperature spectrum was absent if the measurement was performed in a vacuum better than 10^{-8} Torr, was just visible if the vacuum was around 10^{-7} Torr and became prominent if the vacuum was worse than 10^{-7} Torr. Logothetidis et al. therefore concluded that this feature is an experimental artefact.

(iii) Logotheditis at al. attempted to fit the first and second derivatives of the experimental $\varepsilon(\omega)$ spectra, instead of the $\varepsilon(\omega)$ itself, so that the features studied are enhanced. The temperature dependence of the energy of the direct gap was fitted using

$$E(T) = E_B - a_B\{1 + 2 / [\exp(\theta/T) - 1]\} \qquad (3)$$

where θ is an average phonon frequency. Data were obtained for a type IIa and a type IIb (semiconducting) diamond by fitting the first derivative and second derivative spectra. The experimental curves indicate that the energy gap for the type IIb diamond appears slightly higher than that of the type IIa diamond and that the energy gaps for both diamonds appear higher when the analysis is based on the second derivative rather than the first. The parameters used to fit the curves are shown in table II and it is clear that the best-fitting parameters a_B and θ differ significantly for the two diamonds and the two methods of analysis. The experimental uncertainties on these parameters are large, however. Although this technique yields important general information about the position of the direct gap and its temperature dependence there are also some significant limitations.

The measured temperature-dependence of the direct gap agrees quite well with that calculated by Zollner et al. [1992]. Between 133 and 295 K the temperature coefficient calculated from the empirical equation (3) and the data in table II is $d(E_g)_{direct} / dT \sim 1 \times 10^{-4}$ K^{-1}. This compares with the value obtained by Clark et al. [1964] of $-(6.3 \pm 1.8) \times 10^{-4}$ K^{-1} which is clearly an over-estimate.

This brief review of the experimental data shows that there are only three features in the reflectivity spectra which are consistently observed, at around 7.3, 12.2 and 16 eV. All workers are agreed that the first represents the direct band gap at the zone center and corresponds to the $\Gamma_{25'}$ to Γ_{15} transition. The 12 eV transition is believed to occur at or near the X zone boundary and corresponds to the transition from the X_4 point in the valence band to the X_1 point in the conduction band. The 16 eV feature is associated with a transition from $\Gamma_{25'}$ in the valence band to $\Gamma_{12'}$ in the conduction band [Saslow et al. 1966].

3.2 THEORETICAL CALCULATIONS

It would be inappropriate for the present author to attempt to review the various theoretical techniques used to calculate phenomena associated with the band structure in diamond. Instead reference will be made briefly to the work of Saslow et al. [1966] to give the flavor of the type of work that has been done where there is a clear connection with experimental measurements. An up-to-date list of the extensive literature in this area is given by Mainwood [1994]. Saslow et al. used the empirical pseudo-potential method to interpret the optical properties. The method involves choosing pseudo-potential form factors which give band structures consistent with the experimental measurements. These form factors are first constrained to give a few of the principal energy gaps in agreement with experiment and then used to determine the electronic band structure at many points within the Brillouin zone. In addition to identifying the transitions listed in the previous paragraph Saslow et al. also calculated that a transition at the L zone boundary, from $L_{3'}$ in the valence band to L_1 in the conduction band occurs at 10.9 eV, apparently in reasonable agreement with the experimental value of about 9 eV found by Clark et al. [1964]. However, the work by Roberts and Walker [1967] casts doubt on whether this feature has actually been observed.

4. Defect Levels - Location of Ground States in the Forbidden Energy Gap

Absorption and luminescence studies reveal well over 100 electronic transitions at defect centers in diamond. Although the energies associated with the transitions can be measured with high accuracy the locations of the ground states within the bandgap are generally unknown. Only for boron, nitrogen and perhaps the vacancy is there evidence to locate the energy levels with respect to the band edges.

4.1 BORON

Boron forms an acceptor in diamond and gives rise to optical absorption, electrical conductivity and photoconductivity. The absorption spectrum of a natural semiconducting (type IIb) diamond, measured at 77 K, exhibits considerable structure in the region 330 to 370 meV due to transitions to excited states of the acceptor and which finally merges with the photoionisation continuum. These data locate the ground state at about 0.37 eV above the valence band. Measurements of the photothermal ionisation in two of the major bands at mean energies of 348 meV and 364 meV yield thermal ionisation energies of 24.5 meV and 8.5 meV, respectively. Summing the two appropriate energies leads to the value 372.5 meV, in each case, for the ionisation energy of the acceptor center [Collins and Lightowlers 1968]. This is in excellent agreement with the value 373 ± 3 meV determined from electrical conductivity measurements in the temperature range 223 to 323 K [Lightowlers and Collins 1966].

A slightly lower ionisation energy of 368.5 ± 1.5 meV was obtained from an analysis of the temperature dependence of the Hall coefficient in the temperature range 160 to 300 K [Collins and Williams 1971].

4.2 NITROGEN

Type Ib diamonds (those that contain nitrogen in isolated substitutional form) have an absorption threshold at about 2.0 eV and Dyer et al. [1965] interpreted this absorption as being due to the ionisation of the substitutional nitrogen. Farrer [1969] examined one natural type Ib diamond and demonstrated that this absorption gave rise to photoconductivity, although he did not determine the sign of the carriers. He also measured the electrical conductivity as a function of temperature, and obtained an ionisation energy of around 1.7 eV.

There is a photoconductivity threshold in most natural diamonds (excluding type IIb in which boron is the major impurity) at 4.05 eV [Denham et al. 1967, Konorova et al. 1965, Vermeulen and Nabarro 1967]. Denham et al. associated this with the isolated substitutional nitrogen. However, natural "type Ib" diamonds are almost always a mixture of type Ib and type IaA material. (A type IaA diamond contains nitrogen in the A-aggregate form; the A aggregate is a pair of nearest-neighbor substitutional nitrogen atoms [Davies 1976].) Vermeulen and Farrer [1975] therefore carried out a series of measurements on synthetic type Ib diamond in which the concentration of aggregated forms of nitrogen is negligible. These diamonds showed no evidence of the sharp threshold in their photoconductivity spectra. The resistances of six diamonds were measured as a function of temperature up to 400 °C, yielding thermal activation energies between 1.6 and 1.7 eV, and from the sign of the thermoelectric coefficient the carriers were shown to be electrons.

Denham et al. [1967] had previously carried out photoHall measurements on a type IIa diamond with photoexcitation at 4.05 eV. The sign of the Hall coefficient showed that the carriers were electrons. The 4.05 eV threshold is now believed to be associated with the A aggregate of nitrogen [Davies 1977a].

These measurements place the ground states of the isolated substitutional nitrogen and the A aggregate of nitrogen respectively 1.7 eV and 4.0 eV below the conduction band.

4.3 THE VACANCY

When diamonds are subjected to radiation damage two dominant absorption bands are produced, the GR1 band with a zero-phonon line at 1.673 eV and the ND1 band with a zero-phonon line at 3.150 eV [Clark et al. 1956, Davies et al. 1992]. The GR1 band predominates in diamonds with a low nitrogen content and the ND1 band is far stronger in type Ib diamonds [Davies 1977b]. The GR1 band is due to transitions at the neutral vacancy and the ND1 band is produced by transitions at the negative vacancy [Davies 1977b]. The shift in the Fermi energy towards the conduction band in type Ib diamonds favours the production of the vacancy in the negative charge state.

Photoconductivity is produced by the absorption of light in the ND1 band [Farrer and Vermeulen 1972], and in some diamonds the intensity of the ND1 band decreases and the intensity of the GR1 band increases [Davies 1977a]. In terms of the model proposed by Davies [1977b] this behavior is consistent with photo-ionisation of the ND1 (V^-) center to produce a GR1 (V^0) center. The charge carrier responsible for the ND1 photoconductivity is therefore expected to be an electron. This observation places the ground state of V^- approximately 3.15 eV below the conduction band.

The neutral vacancy also gives rise to a number of sharp absorption lines labelled GR2 to GR8 between 2.88 and 3.01 eV superimposed on an underlying "ultraviolet continuum." Sharp maxima are observed in the photoconductivity spectrum which correspond exactly to the GR2 - 8 lines in absorption [Farrer and Vermeulen 1972]. There is, however, no photoconductivity associated with the GR1 transition, and the GR1 system is observed, due to competitive absorption, as a series of minima on a background photoconductivity continuum. The threshold energy of the continuum (the R band) is at $\sim$ 1 eV and the p-type response continues up to the indirect absorption edge [Vermeulen and Farrer 1975]. Clark et al. [1976] have demonstrated from photo-magneto-conductivity conductivity measurements that there is no change in the carrier type when the exciting radiation from a tuneable dye laser is scanned from the background continuum through the GR2 line. This shows conclusively that positive holes are produced by photo-excitation of the GR2 level, and it is highly probably that this is also true of GR3 to GR8. Clark and Mitchell [1977] have therefore proposed an energy level diagram for the GR center in which the ground state is $\sim$ 2.4 eV above the top of the valance band and the GR1 - 8 transitions correspond to the excitation of holes. The GR2 - 8 levels then lie inside the valence band and so auto-ionise, producing p-type conductivity.

Lowther [1977] has calculated the ordering and symmetry of the one-electron levels of the neutral vacancy using clusters of 35 and 47 atoms. Here, too, many transitions occur to levels below the top of the valence band. The major objection to the models proposed by Clark and Mitchell [1977] and Lowther [1977] is that neither is likely to produce zero-phonon lines in absorption and photoconductivity which are < 1 meV wide. Stoneham [1977, 1992] has therefore proposed an alternative scheme for the GR defect based on a two-electron transition. In this model the GR2 line is produced by an electronic transition, and the GR3 to GR8 lines correspond to vibronic structure on the GR2 transition. This interpretation also seems unlikely because the line widths of GR3, GR6(a, b) and GR7(a, b) are only slightly greater than that of GR2, and all five components of GR8 are narrower than GR2 [Collins 1978].

We can conclude only that the location of the ground state of the neutral vacancy is still controversial.

5. Effective Masses

The final area of investigation to be considered here is the determination of the effective masses of positive holes which gives some additional information about the structure of the valence bands. Most methods give some type of average value for the three bands, but cyclotron resonance measurements can yield the individual masses for each band. More limited data are available for the effective masses of electrons in the conduction band.

5.1 CYCLOTRON RESONANCE

When a semiconducting crystal is placed in a magnetic flux density of magnitude B the carriers will move in circular paths with angular frequency of rotation given by $\omega = eB/m^*$ where m^* is the effective mass and e is the magnitude of the charge on an electron. This is the cyclotron frequency, and if the crystal is placed in a microwave cavity where the electric field has a frequency f then energy will be absorbed when the magnetic flux density has the value $B = 2\pi f m^*/e$. In order that the resonance is sharp the carrier needs to complete several orbits before it is scattered by collisions with atoms. This can only be achieved in pure materials at low temperatures where scattering by lattice vibrations is small. Under these conditions the thermal excitation of carriers is small or negligible, and it is necessary to shine light on the sample to photo-excite the carriers. With diamond it is possible to excite only holes by using semiconducting crystals (containing boron as an acceptor) and infrared illumination around 0.37 eV.

Rauch [1962] measured the effective hole masses in diamond in the manner described. Natural semiconducting diamonds were cooled to liquid helium temperature and were illuminated with infrared radiation from a monochromator. In this way Rauch was able to differentiate between the two valence bands with masses of $0.70 \pm 0.01\ m_e$ and about $2.1\ m_e$ which are degenerate at $\mathbf{K} = 0$, and the split-off band lying 6 ± 1 meV lower in energy with a mass of $1.06\ m_e$. (Here m_e is the rest mass of a free electron.) This was an absolutely crucial investigation in establishing the valence-band structure, but the results have not been verified by other workers. Several attempts were later made to repeat Rauch's work, but none of them was successful [Hodby, 1971].

Electrons can be excited into the conduction band simultaneously with holes in the valence band by illuminating the diamond with above-bandgap ultraviolet light. Rauch, however, was unable to detect any resonances associated with electrons by using this technique.

5.2 FARADAY ROTATION

When polarised light passes through a optically transparent material placed in a magnetic field the plane of polarisation is rotated by an amount which depends on the average effective mass (and also the optical path length). With easily-achieved magnetic fields the angle of rotation is extremely small for specimens with small linear dimensions. Nevertheless, using natural semiconducting diamond, Prosser [1964] was able to determine $m^* = 0.88 \pm 0.14\ m_e$ at room temperature, which is comparable with the values obtained from cyclotron resonance studies.

5.3 HALL EFFECT

For a partially compensated p-type semiconductor with an acceptor concentration N_A sufficiently small that there is no degeneracy the hole concentration p at temperature T may be written as [Blakemore 1962]

$$p(p + N_D)/(N_A - N_D - p) = (2/g_a)\,[2\pi m^*kT/h^2]^{3/2}\exp(-E_A/kT) \quad (4)$$

where N_D is the donor concentration, g_a is the ground state degeneracy factor for the acceptor and E_A is the acceptor ionisation energy. In principle, by measuring p as a function of T the average effective mass m* can be determined. In practice, as discussed by Collins [1993] this is not straightforward; p can be estimated from the Hall coefficient R using $p = r/Re$ where r is the ratio of the Hall mobility μ_H to the conductivity mobility μ_C. The value of r depends on the scattering processes and most workers set $r = 3\pi/8$ although it almost certainly varies with temperature. Again, most workers have used $g_a = 2$ in equation (4); however, this implies that there is a single, spherically symmetrical valence band with scalar effective mass m*. With a doubly degenerate valence band at $\mathbf{K} = 0$, g_a should be set to 4. In diamond, because the split-off band is only 6 meV below the doubly degenerate band the contribution from the split-off band will vary from 60 to 95 % in the temperature range 150 to 1250 K covered in Hall effect measurements, and at this highest temperature $g_a = 6$ is most appropriate [Collins 1993].

At intermediate temperatures there is a further complication that may make it difficult to fit the data using equation (4), because it may be necessary to take the population of the excited states into account [Blakemore 1962, Collins 1993]. By considering the population of the first few excited states Williams [1970] presented a qualitative argument which showed that the apparent value of m* derived from equation (4) will decrease at temperatures above 400 K.

It is not surprising that the average effective mass derived from Hall-effect measurements is specimen-dependent and appears to vary with temperature [Collins and Lightowlers 1979]. The high-precision measurements by Williams [1970], using $r = 3\pi/8$ and $g_a = 2$, gave a weighted mean in the temperature interval $160\ K < T < 300\ K$ of $m^* = 0.75 \pm 0.1\ m_e$, again comparable with the cyclotron resonance and Faraday rotation data. Effective mass values obtained by other workers using the Hall effect range from 0.16 to 1.1 m_e [Collins and Lightowlers 1979]

5.4 DRIFT VELOCITY

The limiting drift velocity in a material where the carriers are scattered by optical phonons is given by $\sqrt{(8E_{opt}/3\pi m^*)}$ where E_{opt} is the energy of the optical phonon and m* is the density of states effective mass [Moll 1964]. Reggiani et al. [1979] measured the drift velocity of holes in natural diamond using a time-of-flight technique, and determined the variation of the drift mobility with temperature at a fixed electric field of

30 kV cm^{-1}. Because the hole energy was much greater than the 6 meV spin-orbit energy they interpreted their data using a two-band model, neglecting the split-off band. Their analysis yielded heavy and light hole masses of $m_h = 1.1\ m_e$ and $m_\ell \sim 0.3\ m_e$, respectively. These values fitted their data much more closely than the equivalent values of $m_h = 2.30\ m_e$ and $m_\ell = 2.08\ m_e$ derived from the Rauch parameters or $m_h = 0.40\ m_e$ and $m_\ell = 0.28$ which had been derived from a theoretical analysis by Lawaetz [1971]. A similar investigation [Nava et al. 1980] yielded the longitudinal and transverse masses for electrons as $m_L = 1.4\ m_e$ and $m_{T1} = m_{T2} = 0.36\ m_e$. The density of states electron mass can then be evaluated from $m_e^* = (m_L\ m_{T1}\ m_{T2})^{1/3}$ and the density of states hole mass from $m_h^* = (m_h^{3/2} + m_\ell^{3/2})^{2/3}$ [Pan et al 1993], giving $m_e^* = 0.57\ m_e$ and $m_h^* = 1.2\ m_e$. It is interesting to note that if g_a is set to 4 in equation (4) the effective mass value derived by Williams [1970] would need to be increased to $1.2 \pm 0.16\ m_e$.

6. Conclusions

In this chapter I have reviewed some of the experimental work which has established our current picture for the band structure of diamond, and referred briefly to a limited selection of theoretical treatments. It is more usual to make a chapter on Band Structure exclusively theoretical, and I hope that the present work will complement the more traditional approaches. One thing is clear-despite the multitude of theoretical calculations (see Mainwood [1994] for a listing of these) relatively little has been confirmed by experiment. Furthermore, although some of the experimental data and their interpretation are credible, some measurements (for example, spectroscopic studies in the vacuum ultraviolet) suffer from artefacts and in other areas (for example the measurement of the effective masses of holes and electrons) there is considerable disagreement between the techniques used by different workers. The message here is clear: there is no point in theoreticians producing models that cannot be verified experimentally; equally, experimentalists must obtain reliable data. This synergy is essential for further progress to be made in our attempts to understand the band structure of diamond.

7. References

Clark, C. D. (1959), *J. Phys. Chem. Solids*, **8**, 481.

Clark, C. D., Dean, P. J. and Harris, P. V. (1964), *Proc. Roy. Soc. A*, **277,** 312-329.

Clark, C. D. and Mitchell, E. W. J. (1977) *in Radiation Effects in Semiconductors 1976*, Inst. Phys. Conf. Series No. 31, pp 45 - 57.

Clark, C. D., Parsons, B. J. and Vermeulen, L. A. (1976), *Diamond Conference, Bristol* (unpublished).

Collins, A. T. (1993), *Phil. Trans. Roy. Soc. A*, **342**, 233 - 244.

Collins, A. T. and Lightowlers, E. C. (1968), *Phys. Rev.*, **171**, 843-855.

Collins, A. T. and Lightowlers, E. C. (1979), in J. E. Field (ed.) *Properties of Diamond* (Academic Press, London), pp 79 - 105.

Collins, A. T. and Williams, A. W. S. (1971), *J. Phys. C: Solid. St. Phys.*, **4**, 1789-1800.

Collins, A. T., Davies, G, Kanda, H. and Woods, G. S. (1988), *J. Phys. C: Solid. St. Phys.*, **21**, 1363.

Collins, A. T., Lawson, S. C., Davies, G. and Kanda, H. (1990), *Phys. Rev. Lett.*, **65**, 891-894.

Crowther, P.A. and Dean, P. J. (1964) in *Proceedings of the Symposium on Radiative Recombination*, Dunod, Paris, p 104.

Davies, G. (1976), *J. Phys. C: Solid. St. Phys.*, **9**, L537 - L542.

Davies, G. (1977a), *Chem. Phys. Carbon*, **13**, 1-143.

Davies, G. (1977b), *Nature*, **269**, 498 - 500.

Davies, G., Lawson, S. C., Collins, A. T., Mainwood, M. and Sharp, S. J. (1992), *Phys. Rev. B*, **46**, 13157 - 13170.

Dean, P. J., Lightowlers, E. C. and Wight, D. R. (1965), *Phys. Rev.*, **140**, A352 - A358.

Denham, P., Lightowlers, E. C., and Dean, P. J. (1967), *Phys. Rev.*, **161**, 762 -768.

Dyer, H. B., Raal, F. A., du Preez, L. and Loubser, J. H. N. (1965), *Phil. Mag.*, **11**, 763-764.

Fahy, S., Chang, K. J., Louie, S. J. and Cohen, M. L. (1987), *Phys. Rev. B*, **35**, 5856-5859.

Farrer, R. G. (1969), *Solid State Commun.*, **7**, 685.

Farrer, R. G. and Vermeulen L. A. (1972) *J. Phys. C: Solid St. Phys.*, **5**, 2762 - 2768.

Hodby, J. W. (1971), *Diamond Conference, Cambridge* (unpublished).

Holloway, H., Hass, K. C., Tamor, M. A., Anthony, T. R., and Banholzer, W. F. (1991), *Phys. Rev. B*, **44**, 7122 - 7126.

Kamo, M., Yurimoto, H. and Sato, Y. (1988), *Appl. Surf. Sci.*, **33/34**, 553.

Konorova, E. A., Sorokina, L. A. and Shevchenko, S. A., (1965), *Sov. Phys. - Solid State*, 7, 876.

Lightowlers, E. C. and Collins, A. T. (1966), *Phys. Rev.*, **151**, 685-688.

Logothetidis, S., Petalas, J., Polatoglou, H. M. and Fuchs, D. (1992), *Phys. Rev. B*, **46**, 4483-4494.

Lowther, J. E., (1977), *Phil. Mag.*, **36**, 483 - 93.

Mainwood, A. (1994) in Gordon Davies (ed.) *Properties and Growth of Diamond*, Datareview series number 9, INSPEC, The Institution of Electrical Engineers, London, UK, pp 3-8.

Moll, J. L. (1964), *Physics of Semiconductors* (Wiley, New York), Chap. 10.

Nava, F., Canali, C., Jacobini, C. and Reggiani, L. (1980), *Solid State Commun.*, **33**, 475.

Onodera, A., Hasegawa, M., Furuno, K., Kobayashi, M. and Nisida, Y. (1991), *Phys. Rev. B*, **44**, 12176-12179.

Pan, L. S., Han, S., Kania, D. R., Zhao, S., Gan, K. K., Kagan, H., Kass, R., Malchow, R., Morrow, F., Palmer, W. F., White, C., Kim, S. K., Sannes, F., Schnetzer, S., Stone, R., Thomson, G. B., Sugimoto, Y., Fry, A., Kanda, S., Olsen, S., Franklin, M., Ager III, J. W. and Pianetta, P. (1993), *J. Appl. Phys.*, **74**, 1086 - 1095.

Paul, W. (1961), *J. Appl. Phys.*, Supplement to Vol. **32** (8), 2082-2094.

Philipp, H. R. and Taft, E. A. (1962), *Phys. Rev.*, **127**, 159.

Philipp, H. R. and Taft, E. A. (1964), *Phys. Rev.*, **136**, A1445-A1448.

Phillips, J. C. (1965), *Phys. Rev.*, **139**, A1291.

Prosser, V. (1964), *Czech. J. Phys.*, **B15**, 128 - 134.

Reggiani, L., Bosi, S., Canali, C., Nava, F. and Kozlov, S. F. (1979), *Solid State Commun.*, **30**, 333 - 335.

Roberts, R. A., Roessler, D. M. and Walker, W. C. (1966), *Phys. Rev. Lett.*, **17**, 302

Roberts, R. A. and Walker, W. C. (1967), *Phys. Rev.*, **161**, 730-735.

Rauch, C. J. (1962) in A. C. Stickland (ed) *Proc. Int. Conf. Semiconductors, Exeter*, The Institute of Physics and the Physical Society, London, UK, pp 276 - 280.

Saslow, W., Bergstresser, T. K. and Cohen, M. L. (1966), *Phys. Rev. Lett.*, **16**, 354-356.

Stoneham, A. M. (1977), *Diamond Conference, Reading* (unpublished).

Stoneham, A. M. (1992), in J. E. Field (ed.) *The Properties of Natural and Synthetic Diamond* (Academic Press, London), pp. 3 -34.

Vermeulen, L. A. and Farrer, R. G. (1975), *Diamond Research 1975*, pp 18-23.

Vermeulen, L. A. and Nabarro, F. R. N. (1967), *Phil. Trans. Roy. Soc.* A, **262**, 251.

Walker, W. C. and Osantowski, J. (1964), *Phys. Rev.*, **134**, A153.

Williams, A. W. S. (1970), *PhD thesis* (University of London).

Zollner, S., Cardona, M. and Gopolan, S. (1992), *Phys. Rev. B*, **45**, 3376-3385.

Chapter 2

DIAMOND MORPHOLOGY

Robert E. Clausing

Oak Ridge National Laboratory (Retired), Oak Ridge TN 37831-6093, USA

Contents

1. Introduction: "What is Morphology?" and "Why is it Important?"

Morphology as used in this chapter is defined as the internal and external form and structure of a diamond material, either a single crystal or polycrystalline assembly. Morphology is discussed in relation to the processes that created it. The morphology of diamond, natural or man-made, is a consequence of the basic diamond cubic crystal structure and the quality and rate of crystal growth in various crystallographic directions which are, in turn, strongly influenced by the growth conditions. The morphology of diamond materials is especially important because first, so many of the mechanical, physical and chemical properties of diamond single crystals are strongly anisotropic; and second, once the diamond material is created, it is difficult or impossible to significantly modify its structure, defects or impurities by practical processes[i].

The properties of diamond *single crystals* depend on the crystal orientation, and the defects and impurities incorporated in the crystal. The properties of *polycrystalline materials* are not only dependent on the orientation, form, structure, impurities and defects of the individual crystals (often referred to as grains or crystallites) but also the material at or between the individual crystal boundaries. Whether the diamond product grown is a single crystal or is polycrystalline, once the material is grown "What you have is what you get" or more to the point "The structures that you can grow are all that you can get." Heat treatments, mechanical deformation or other ordinary materials processing methods cannot be used in any practical way to significantly alter the bulk structure or properties of the material. This means that if you want to control or engineer the properties of diamond you must control the growth process[ii]

1.1 IMPORTANCE OF MORPHOLOGY

1.1.1 Anisotropic properties of diamond single crystals.

Many of the physical and chemical properties of diamond single crystals vary strongly with crystallographic direction. Some of these anisotropic effects are described in chapter 3 on mechanical properties, in chapter 4 on surface phenomena and in chapter 15, which includes especially the importance of crystallographic orientation on cutting, polishing, oxidation and dissolution. Table I highlights a few of the most notable of these properties to illustrate some of the large differences in properties as a function of crystal orientation. The rate of wear may vary by a factor of 100 depending on the crystal face and direction of polishing [Wilks and Wilks 1979]. Simply reversing the direction of polishing or rotating the crystal ten degrees in the plane of polishing [Wilks and Wilks 1979] can change the material removal rate by a factor of 10. The oxidation rates on the {111} faces are typically about 10 times higher than on the {100} faces [Evans 1979]. The theoretical energy necessary to cleave a diamond on the {111} planes is 10.6 J/cm compared to 18.4 J/cm on the {100} face [Field 1979].

1.1.2 Polycrystalline materials.

Polycrystalline diamond consists of many small crystals grown together to form a film or bulk diamond. The crystallites may be of various degrees of perfection, size and shape as determined by growth conditions. They may have random orientations or they may be more or less aligned with respect to each other and the substrate. The boundaries

---

i It is of course possible to cut, polish or etch its external size or shape or to use it as part of a composite structure or material. It is also possible, strictly speaking, to deform and anneal diamond, but generally this must be done in the regime where diamond is the stable phase, i.e., very high pressures and temperatures (for example, 6 GPa and 1500°C). Dislocations can move and plastic deformation is possible at 1500°C. For a brief review of the subject see [Wilks and Wilks 1991] p. 214-227.

ii Notable exceptions to this are irradiation with high energy particles and ion implantation.

Table I. Some -Anisotropic Properties of Diamond Single Crystals

Property	Crystal Face or Direction	Relative Value	Reference
Abrasion/Wear Rate	{111} face/polishing direction 50o away from <111> toward cube	30	Wilks (1979)
	{111} face/polishing direction 60o away from <111> toward cube	3	Wilks (1979)
Oxidation Rate	{111}	1.5	Evans (1979)
	{110}	0.8	
	{100}	0.2	
Cleavage Energy	{111}	10.6	Field (1979)
	{110}	13.0	
	{100}	18.4	

between the crystals are imperfect because of misalignment of adjacent grains and because they may contain impurities or high concentrations of crystal defects or non-diamond carbon. Even though the grain boundaries are usually mechanically quite strong, they are discontinuities in the structure and will affect bulk properties, especially those which depend on crystal perfection such as thermal conductivity and carrier lifetime in electronic materials. They also affect chemical, mechanical, and other physical properties. A few examples are given below.

Oxidation. Table I in the previous section shows oxidation rates for diamond single crystals. The oxidation rates for polycrystalline diamond films are dependent on the crystallite faces exposed as expected, but are generally very much higher than for single crystal films because of the comparatively rapid oxidation at grain boundaries. Grain boundaries present less perfect material and may include non-diamond carbon. Etch rates for polycrystalline films may be 100 times the rate for single crystal {100} faces, especially if the films consist of small highly defective crystals i.e. microcrystalline diamond films [Clausing]. A polycrystalline <100> textured film with well developed relatively large columnar crystals may resist oxidation of the individual crystallites but fall apart due to rapid oxidation of the grain boundaries between the individual crystals. This example clearly demonstrates the significance of the less perfect diamond at grain boundaries [Sato and Kamo 1989].

Thermal conductivity. The thermal conductivity of diamond is very sensitive to structure. Figure 1 shows that growth conditions change film structure and can produce very large changes in thermal conductivity. The more perfect films i.e. those with higher intensity 1332 cm^{-1} Raman spectroscopy lines offer much better thermal conductivity as shown in (a) of the figure. Gem quality, IIa diamond has twice the thermal conductivity of the best polycrystalline film, i.e., 2000 (W/m K). Isotopically pure IIa diamond made with the carbon 12 isotope has a thermal conductivity of more than 3000 (W/m K) [Ono et al. 1986].

The extraordinarily high thermal conductivity of diamond is due to the rigidity and regularity of the crystal bonds, making phonon conduction very efficient. The difference between the intrinsic conductivity of the individual crystallites and the conductivity of a

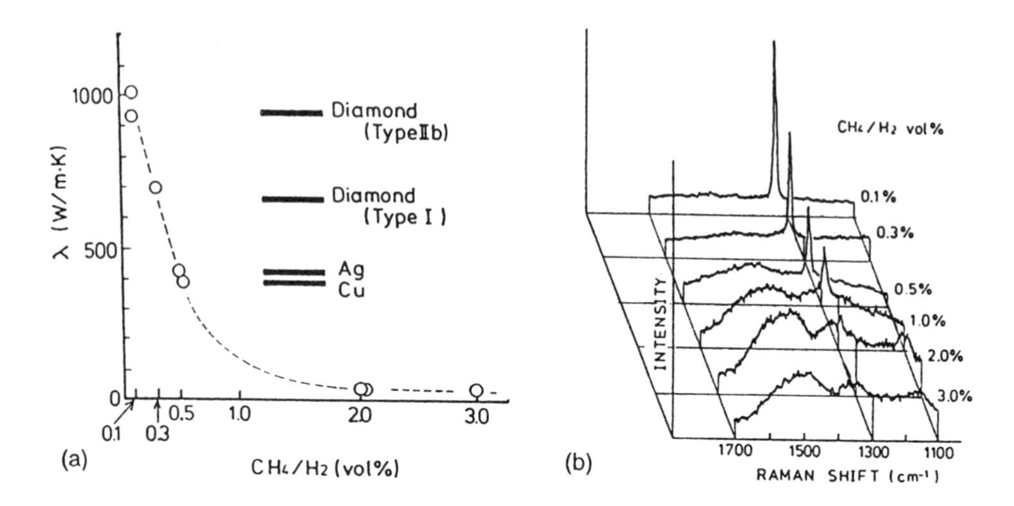

Figure 1. Growth conditions change film structure and thereby the thermal conductivity. (a) Thermal conductivity of CVD diamond films as a function of using different methane concentrations during film depositions. (b) The Raman spectra of the diamond films indicates increasing crystal perfection with decreasing methane concentration [Ono et al. 1986].

polycrystalline film is due to phonon interactions with the grain boundaries. For this reason, fine grain materials i.e. microcrystalline materials, may have thermal conductivities as much as 100 times lower than large grained materials [Graebner et al.].

The thermal conductivity of a diamond film can be directional because of its polycrystalline structure. Thermal conductivity for heat flowing parallel to the plane of the film is often lower than that flowing perpendicular to the because of the columnar nature of the crystallites [Graebner et al. 1994]. The intrinsic conductivity of individual crystallites in a film may equal that of type IIa single crystals, and if the crystallites are long and slender, extending through the film thickness, conductivity through the film can approach that of type IIa single crystal material while being much less in the plane of the film because the heat flow must cross many grain boundaries.

Mechanical Properties. The presence of grain boundaries and highly defective diamond crystal structure can decrease the elastic modulus and hardness while increasing the fracture toughness.

2. Morphology: General Considerations

2.1 FACTORS THAT DETERMINE MORPHOLOGY

The growth tendencies of crystals are determined by the underlying crystal structure (i.e. the symmetrical arrangement of the atoms in a crystal which form a specific regular space lattice) and the growth conditions. It is a cubic lattice belonging to the space group Fd3m. The only active twin plane in practice is {111}. The only active cleavage plane is the {111}. The most common facets on all natural and synthetic crystals are

{111} followed by {100}. Other facets in the <110> zone are found much less frequently and apparently require special growth conditions or dissolution to form. The {111} faces have the lowest free energy while the free energy of{100} faces is nearly 2 times higher. Kinetic factors rather than equilibrium conditions, however are usually the dominant factors in determining growth morphology.

The bounding facets or faces of as grown crystals are a consequence of the differences in the rate of growth in various crystallographic directions. The flatness and other characteristic topographical features of the faces are a consequence of the details of the growth processes. In twinned and polycrystalline materials the morphology is also a function of the nucleation (or renucleation) probability, preferential orientation of nuclei, and in the case of twins, the twin symmetry and competitive growth of twins or other new crystallites.

If we consider initially only one crystal free to grow in all directions, we will find that a well developed crystal will be bounded by the slowest growing crystal faces (i.e., the faces perpendicular to the slowest growing crystallographic directions) because the faster growing directions grow out their corresponding faces to small facets, a point or an edge. In the case of natural diamond this means that the dominant forms are bounded by the family of eight {111} planes, leading to the usual octohedral shape. In synthetic crystals, {111} and {100} facets are usually both present. Figure 2 shows idealized growth morphologies for natural, synthetic HPHT (High Pressure, High Temperature) grown and synthetic CVD grown diamond single crystals. The drawing for the HPHT crystal includes {110} and {112} faces, even though they are not typical, they can be grown in special conditions.

Cubes and octahedrals can be grown by both synthetic processes but usually a combination of {111} and {100} facets form various cubo-octahedral shapes. Figure 3(a) shows the <100> axes in a cubo-octahedron. Note that the edges of all of the facets are <110> directions and that the square {100} facets are not bounded by <100> directions but by <110> directions (at 45 degrees to the crystal axes).

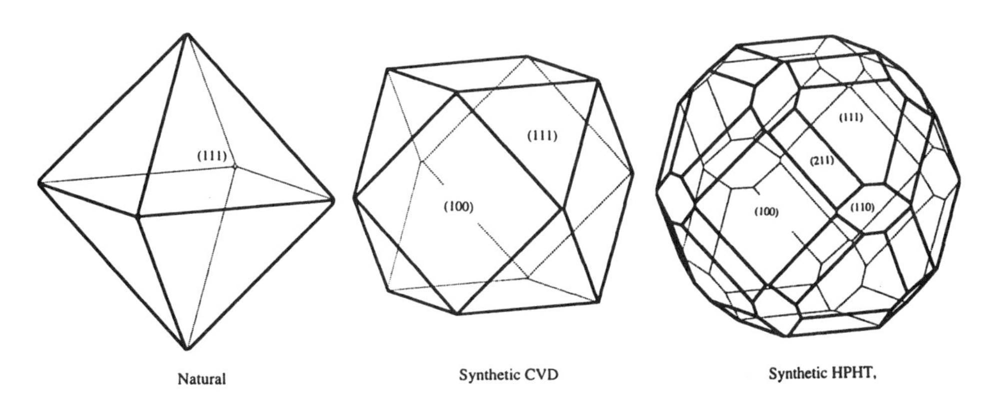

Figure 2. Idealized growth morphologies for natural, synthetic, CVD grown, and synthetic HPHT grown diamond single crystals.

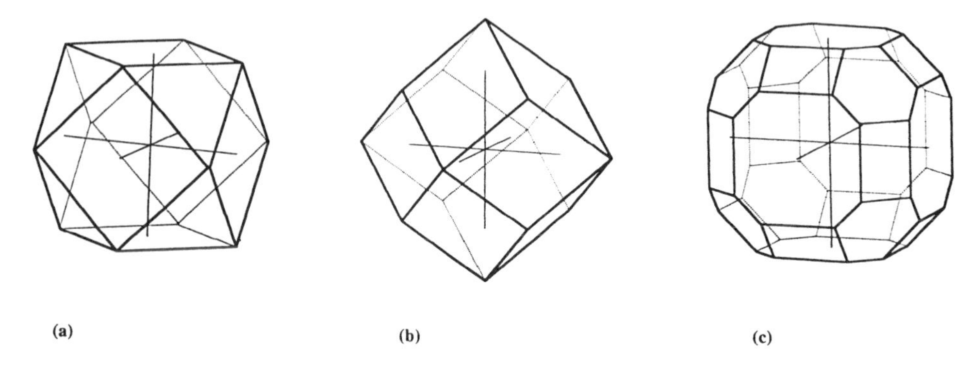

Figure 3. (a) The {100} axes in a cubo-octahedron, (b) a dodecahedron with all {110} faces and (c) a cubo-octahedron with added {110} faces.

The {100} faces, often referred to as cube faces, or 4 point surfaces have four fold symmetry. The {111} faces often referred to as octahedral or 3 point surfaces, have three fold symmetry. In synthetic diamond, {110} and sometimes higher index faces such as {113}, {115}, {117}, {221} or {331} are occasionally seen but they seldom are dominant features. Figure 3(b) shows a dodecahedron with all {110} faces and figure 3(c) shows a cubo-octahedron with added {110} faces. The {110} faces, which have 2 fold symmetry, are often called dodecahedral or 2 point surfaces. Simply put, a {100} cube surface has four equivalent directions at 90 degrees to each other, an {111} octahedral surface has three equivalent directions at 120 degrees to each other, and a {110} dodecahedral surface has only two equivalent directions at 180 degrees to each other.

The shape of cubo-octahedra vary depending on the ratio of the growth velocities in the <100> and <111> directions. The shapes can therefore be used to determine the ratio of growth rates. Figure 4 shows the shapes which result for different growth rate ratios.

Figure 4 shows the crystal shapes for diamond crystals assuming that the limiting faces are {100} and {111} facets but that the crystals grow at different rates on the two types of facets. The distance from the center of the crystal to the center of each of the faces is proportional to the growth velocity in the direction perpendicular to that face, i.e., the direction corresponding to its Miller's indices. The slower growth rates produce the larger corresponding faces. The longest dimension in the crystallite defines the fastest growing direction. This direction is always to a corner and is indicated in the figures by an arrow. The shapes of these cubo-octahedral crystals then reveal the ratio of growth rates in the <100> and the <111> directions. The ratio of growth rates, $R = V<100>/V<111>$, is given under each shape. Note that the longest dimension in the cube is the <111> direction while that in the octahedron is the <100> direction. For a ratio of growth rates equal to 0.87 the longest crystal dimension is in a <110> direction.

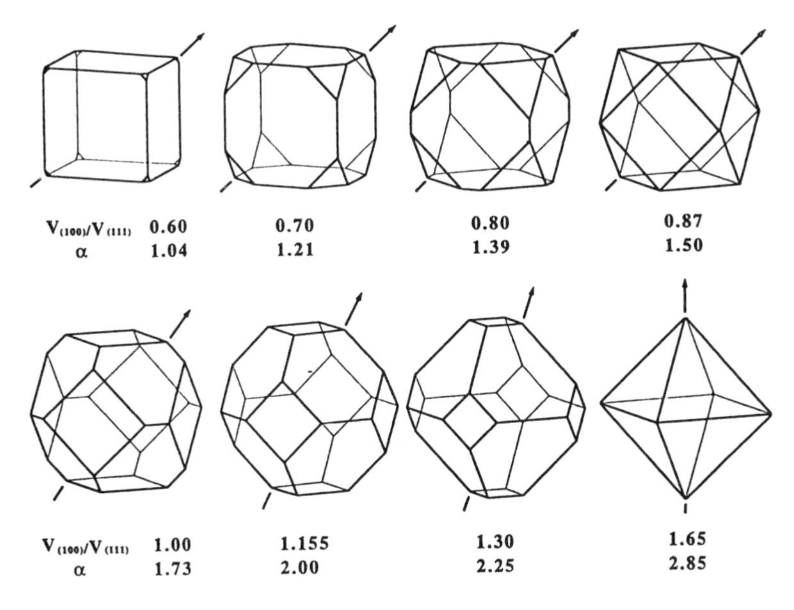

Figure 4. The cubo-octahedral shapes which result as the ratio of growth velocity in the <100> to that in the <111> varies from 0.577 to 1.732.

2.2 THE PERIODIC BOND CHAIN THEORY: THEORETICAL ANALYSIS OF RELATIVE GROWTH RATES

According to the Periodic Bond Chain (PBC) [Hartman and Perdok 1955] and [Hartman] theory, the slowest growing face for diamond is the {111} face but growth in <100> direction should be so fast that the {100} faces grow to extinction. The first prediction seems to be correct but the second is not always true.

The PBC theory relates the growth morphology to crystal structure by considering that the crystal is composed of an interwoven network of chains of strong bonds. The diamond crystal structure contains only one type of PBC, the zigzag chain parallel to the <110> [Hartman 1965]. By definition, F (flat) faces include two or more non-parallel interconnected PBCs within a slice of the crystal equal to the interplanar separation of successive net planes corrected for the extinction conditions of the crystal space group. S (stepped) faces contain one PBC while K (kinked) faces have none. F faces tend to grow by a layer mechanism which works to keep them relatively flat, while S and K faces can grow by random addition at steps and kinks or truly random addition at points and have no reason to remain flat. F faces have a barrier for the nucleation of new growth layers and consequently are normally the slowest growing faces. The {220} has one PBC and the {400} has no complete PBCs within the layer thickness. This may change if the surface is reconstructed. The PBC analysis of diamond is succinctly reviewed elsewhere [VanEnckevort 1994]. The following is a brief summary of the pertinent results.

For diamond, PBC analysis [Hartman 1965] designates the {111} as a slow growing F (Flat) face containing three PBC's. It should therefore be a smooth face which grows slowly, layer by layer, as the result of the difficulty of nucleating new layers. The {220} face and all of the faces in the <110> zone, e.g. the {110}, {112}, {113}, {115}, {117}, {221} and {331}, but excepting the {111} F face, are S (stepped) faces containing one PBC. These faces are predicted to be stepped and to grow more rapidly than {111}.They are of the type {1,1,1±δ). Stepped hillocks or pits with these higher order facets sometimes form on {111} faces. Hillocks are commonly centered on a screw dislocation which provides a perpetual source of new layers as it spins out spiral edges. The {400} and all other faces are theoretically K (kinked) faces and contain no complete PBCs. They are predicted to be atomically rough since nucleation is not necessary and they can grow by random atom addition. Growth on these faces should occur most rapidly since no nucleation is required at all, and they should then grow out of existence to points or edges. Since we do see {100} faces on diamond crystals, they must grow approximately at the same rate as the {111}. The slowed growth of these K faces could be the result of impurities on the surface, or perhaps an adsorbed layer or surface reconstruction, which might provide additional PBCs. Indeed there is evidence that reconstruction of the {100} face to a 2 x 1 structure does occur during CVD growth [Busmann et al. 1992]. This could convert the {100} faces to F type.

Since the diamond {111} is the only F type face, only these slow growing {111} faces should persist during crystal growth. If the {111} are the bounding faces only octahedra should form [Van Enckevort 1994]. Such octahedra need not have equal size faces in real growth conditions if, for example, gradients in nutrients, temperature or other growth conditions exist. Indeed, regular octahedra and distorted octahedra are common in natural diamond crystals. It is now thought that nearly all natural diamond were originally formed under conditions where the octahedral morphologies prevailed. The presence of {100} faces in the growth morphology is not predicted by PBC analysis unless reconstruction of the {100} is allowed. The {110} and <110> zone type S faces are formed only in special growth conditions in agreement with PBC analysis.

3. Morphologies of Natural Diamond

3.1 OBSERVED MORPHOLOGIES

Diamond crystals found in nature exhibit a wide variety of shapes [Bruton 1978]. The shapes often contain inclusions and other flaws. Their exteriors are often rounded perhaps by erosion or imperfect growth conditions. Faceted crystals are usually not perfect and exhibit a wide variety of twins, truncated shapes and distorted octahedra, cubes and dodecahedra. Octahedral forms are by far the most common, followed by dodecahedral and cubic forms. Nicely formed dodecahedra or cubes are rarely found. The octahedral crystal's {111} facets have three fold symmetry, the cubes have square {100} facets with four fold symmetry, and the dodecahedra have rhombic {110} facets with only two fold symmetry. Facets on natural diamond crystals are often domed and edges may be rounded and not straight or sharp. Triangular and square facets often exhibit triangular or square growth terraces, steps or pyramids and/or etch pits. The edges of both the triangular and square facets and the growth and etch features on them are parallel to the <110> directions.

It is thought that, in general, the growth form for diamond at the high temperatures and pressures deep in the earth's silicate mantel is octahedral as is predicted by the PBC theory. Many diamonds recovered from volcanic pipes and alluvial mines have very well formed octagonal shapes. There are, however, also some cubes, usually of low quality, often with rounded or dodecahedral {110} edges. The cubic and dodecahedral crystals may be the result of growth, but it has been proposed that the presence of impurities in the growth environment or subsequent etching, erosion or dissolution may be responsible for most of these shapes.

Sometimes a single crystal develops almost to a ball shape, but often ball shaped diamond material is polycrystalline with crystallites having their <110> crystal axes aligned radially from the center of the ball. The ball shaped cccpolycrystalline diamonds are called ballas or shot boart. The term boart, or crushing boart usually refers to very small, fractured or poor quality single crystals and to randomly oriented polycrystalline masses of minute diamond crystals suitable only to be crushed and used as abrasives.

Because natural crystals are so varied and because the properties of diamond depend so strongly on morphology and impurities they are graded and selected first according to their application. The basic grades are gemstones, industrial and boart. Boart is suitable only for crushing to be used as abrasives. Industrial grades are similar to gemstones except that color, clarity, inclusions or other defects make them unsuitable for gems. Diamonds are also sorted by size, shape, flaws, color and clarity. Bruton has an excellent description of the mining and manufacture of natural diamond, including a lengthy discussion on the morphology of natural diamond with many pictures and illustrations of the grades and appearance of naturally occurring morphologies [Bruton 1978].

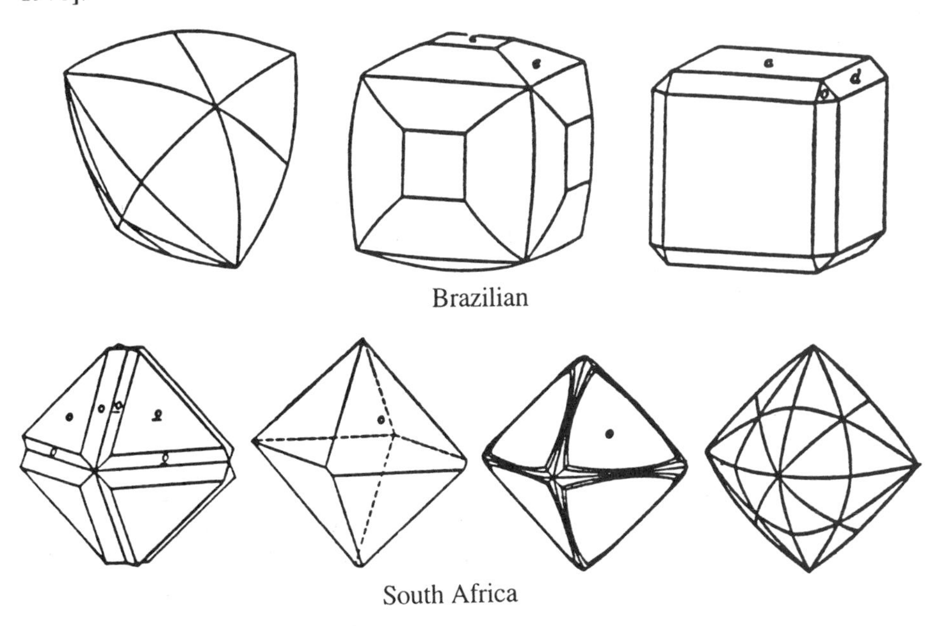

Figure 5. Some of the idealized forms of natural diamond crystals according to [Dana].

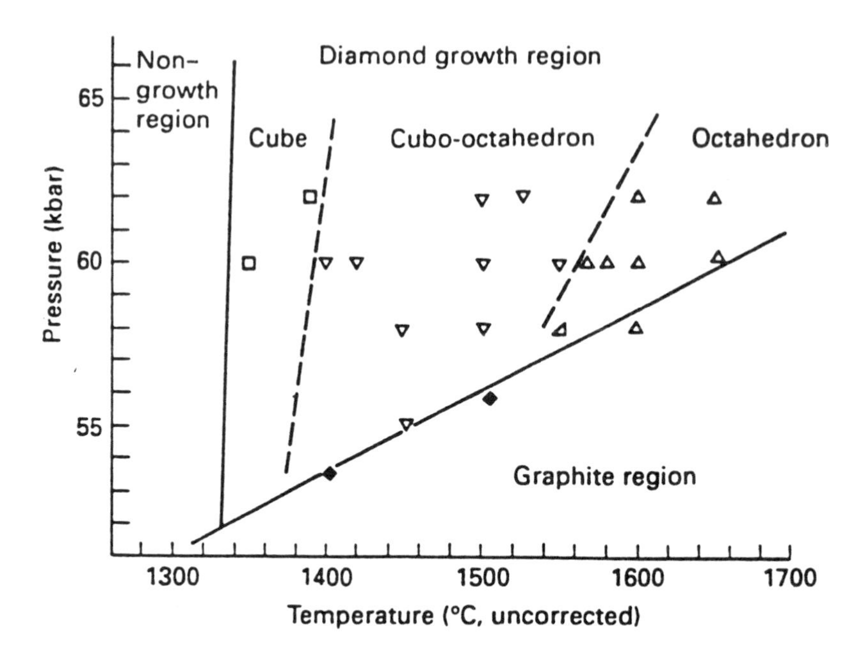

Figure 6. Diagram showing the regions where characteristic crystal morphologies are grown using metal solvents in HPHT diamond synthesis. Higher order faces are grown near the diamond-graphite phase boundary.

Diamonds are classified into type I and type II according to optical absorption by the nitrogen impurity. Type I exhibits pronounced optical absorption due to nitrogen impurities while this optical absorption is too weak to be detected with ordinary methods in type II. Nitrogen is the dominant impurity in type I diamond and may range up to 2500 atomic ppm in type Ia where it is mostly present as aggregates. Type Ib refers to diamond with nitrogen present mostly as single substitutional atoms which exhibit electron paramagnetic resonance. Type Ib is rare in nature but most commercially grown HPHT synthetic diamond belongs to this category with up to 300 atomic ppm of substitutional nitrogen. Type IIa is very low in nitrogen (below 1 or 2 atomic ppm nitrogen) and is often regarded as the most pure form of natural diamond. Type IIb may be even lower in nitrogen, but exhibits semiconducting behavior due to the presence of boron. Both IIa and IIb can exhibit high degrees of crystal perfection.

3.2 RELATIONSHIP OF MORPHOLOGY AND INTERNAL STRUCTURE

Growth of natural diamond crystals often is not uniform. Changing growth conditions can result in layers with different quality. The layers may contain impurities, very fine inclusions, or simply higher densities of crystal defects. The layers are sometimes visible to the eye and some times visible only by x-ray diffraction, optical adsorption or by photo or cathodoluminescence. These layers can be very helpful records of the changing shape of the crystal as growth conditions changed during the growth period. In general, these studies support the hypotheses that the octahedral shape is the usual growth form for natural diamond and that other shapes are often the result of etching or dissolution after the crystals' initial growth. Lang has reviewed a variety of internal growth patterns observed in natural diamond and HPHT synthetic diamond and relates this internal structure to growth morphology [Lang 1979].

Natural and synthetic diamonds also sometimes show inhomogeneous distributions of impurities and internal structures referred to as "growth sectors". These sectors are defined by the crystal face on which the diamond material was formed. There may be pronounced differences in the point defect and impurity concentrations in these growth sectors which can be revealed by color or luminescence. This phenomenon is discussed further in the section on CVD diamond because it is much more common and pronounced in CVD material.

Natural diamonds contain a large variety of crystal faults including twins, stacking faults, voids, inclusions, lathe-like precipitates containing nitrogen (platelets on {100}), dislocations and dislocation loops as well as point defects. These larger defects, except for the dislocations and nitrogen aggregates, are the result of growth conditions. Single crystals also usually contain subgrains or regions which are relatively perfect in themselves but are not perfectly aligned with each other. This is sometimes referred to as a mosaic or subgrain structure. The subgrains are separated by dislocation arrays which form low angle subgrain boundaries.

4. Morphologies of High Temperature High Pressure Synthetic Diamond

4.1 OBSERVED MORPHOLOGIES

In contrast to the morphology of natural diamond which is usually based on octahedral shapes or is derived from such shapes by etching or solution, synthetic diamond is usually cubo-octahedral. All of the cubo-octahedral forms based on {111} and {100} faces (shown in figure 4) can be grown by a suitable choice of parameters [Wilks and Wilks 1991]. Yamaoka et al. have determined a pressure-temperature diagram showing the conditions which produce characteristic growth forms [Yamaoka et al. 1977]. and also described some special conditions under which they were able to grow dodecahedral {110} and {113} faces. Other facets such as the {115} and even {117} facets, have been seen in very pure colorless crystals [Nassau 1993], [Strong and Wentorf 1972]. The diagram showing the regions where characteristic forms are grown is shown in figure 6. An excellent short summary of observed morphologies with macro photographs and additional references are given by [Kanda and Sekine 1994]

4.2 COMMERCIAL DIAMOND SHAPES

Manufactured synthetic HPHT diamond can be optimized for specific applications and thus provide superior properties to natural diamond. Even diamond for sawing or drilling stone has been engineered with a shape optimized to provide up to three times more fracture strength than natural diamond grit. This saw grade diamond is claimed not only to have improved wear resistance but also to require less power, to cut faster and to produce more accurate cuts. The saw grade diamond particles are single crystals without twinning and are of uniform size and shape. The shape appears to have a growth rate ratio R of about 1.1 (α = 2) which should make it quite resistant to fracture on the {111} cleavage plane and free of twins.

In order to develop this saw grade diamond it was essential identify the relevant diamond properties, to quantify the morphology and related characteristics and then to be able to consistently make crystals with these desired characteristics [Banholzer 1995]. The General Electric Superabrasives Division has developed its own method for quantitative classification of crystal morphology [Hackson and Hayden 1993].

5. Morphologies of CVD Synthetic Diamond Crystals and Films

Perhaps the greatest value of the CVD process is that it provides a way to fabricate diamond material into coatings, films and bulk objects with dimensions greatly exceeding those available from natural or HPHT synthetic diamonds. At the present time nearly all CVD diamond is polycrystalline although the growth of single crystal material is a high priority for electronic applications and is important in research on the CVD growth processes.

The morphologies of synthetic diamond crystals and films grown by the activated CVD processes are very diverse and very much a function of the growth conditions. Our knowledge of the relationships between growth conditions and morphology and between morphology and properties is far from complete, but steady progress has been made in the last few years. Substrate temperature, carbon supersaturation and impurities have especially strong effects on morphology. The growth morphology, in turn, is strongly related to the incorporation of film and crystal defects ranging from grain boundaries and voids in polycrystalline films, to twins, point defects and the incorporation of impurities in individual crystallites. The residual stresses, strength, fracture toughness, thermal conductivity, electronic and optical properties of films are all strongly related to growth morphology. As we begin to understand the relationships between growth conditions and morphology it becomes possible to design and grow diamond crystals and films for specific applications.

5.1 CONTROL OF CRYSTAL SHAPES FOR CVD MATERIALS

Because of the ability to grow both single crystals and large area films with controlled morphology, defects and impurities, CVD diamond has great potential for producing engineered diamond material and structures.

It is possible to grow any of the basic single crystal cubo-octahedral shapes based on {111} and {100} facets (shown in figure 4) [Clausing et al. 1992], and many more produced by twinning and renucleation. Under special conditions forms resulting from the stabilization of higher index planes including the {110}, the {221} {331} and {113} can be grown. The parameters chosen by most experimentalists until 1995 tend, however to produce mostly minor variations of the basic cubo-octahedral shapes having growth rate ratios V<100>/V<111> ranging only from 0.8 to 1.1 (growth rate parameter α equal to approximately 1.5 to 2.0). There are many reports in the literature relating morphological development to the deposition parameters [Spitsyn et al. 1981], [Spitsyn 1988], [Kubashi et al. 1988] [Matsumoto and Matsui 1982], [Matsumoto 1983], [Clausing et al. 1989], [Clausing et al. 1991a], [Zhu et al. 1990], [Clausing et al. 1991b] [Locher et al. 1995], [Tamor and Everson 1994], [Locher et al. 1994]or [Chu et al. 1992].

It is very useful to construct maps relating the changes in morphology to changes in growth parameters, particularly changes in temperature and changes in hydrocarbon concentration [Zhu et al. 1990], among others have made such maps labeling the various regions according to the morphology. Some of the attempts to systematize the relationships between morphology and growth conditions are mentioned below. We first review some observed dependencies of morphology on individual parameters and then maps of morphology as functions of temperature and hydrocarbon concentration. Next, in the way of being more quantitative, measurements of growth velocities in specific crystallographic directions. Finally since the ratio of growth rates, R = V<100>/V<111>, (or $\alpha = R\sqrt{3}$) is such a determining factor in morphological development, an improved mapping method which provides isomorphic contours of the growth rate ratio as a function of growth conditions has been developed. Such maps are very useful and can and should be made as a guide for any given reactor system.

5.2 GROWTH RATE RATIOS: V<100>/V<111>

As the growth parameters are changed, the rate of crystal growth on the various facets {111}, {100}, and {110} change, but not all at the same rate. The results reported in the literature vary greatly even in similar systems. Spitsyn first reported determining the comparative rates of growth for CVD diamond in the <100> and <111> directions by measuring the linear dimensions of cubo-octahedra in a microscope. Figure 6 shows the growth rate ratio V<100>/V<111> obtained by [Spitsyn et al 1981]. using a chemical transport reaction and changing the hydrocarbon supersaturation by changing only the substrate (crystallization) temperature. The hydrocarbon concentrations are however, unknown.

These growth rate ratios ranged from about 1.7 to about 0.9 ($\alpha = 3$ to $\alpha = 1.5$) and would grow crystal shapes as shown in figure 4 ranging from octahedra to regular cubo-octahedra as the temperature increased from 800°C to 1100°C. ("the supersaturation decreased"). Spitsyn also noted that growth twins appeared on {111} faces when the growth rate ratios V<100>/V<111> were above 0.9 ($\alpha = 1.5$) and on {100} faces for values below 0.9. He pointed out that the presence of growth twins is an obstacle to growing diamond crystals larger than a few microns. Spitsyn also noted that twined crystals are were more likely to nucleate at higher carbon supersaturations.

Chu et al. 1992, also measured growth of diamond single crystal surfaces using a very sensitive optical interferometery technique for measuring growth rates as a function of crystal direction for a hot filament system. The technique was sensitive enough that even {110} surfaces remained flat during the measurements. Rates obtained for growth on the {111}, {100}, and {110} surfaces are presented in figures 8 and 9. These results were obtained in a more sophisticated system where the hydrocarbon concentration was

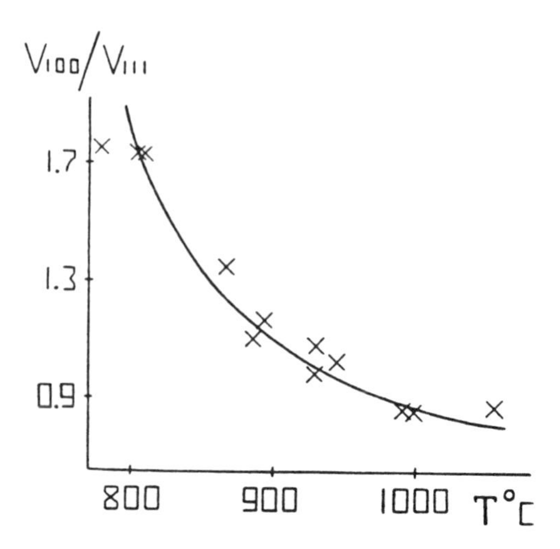

Figure 7. The growth rate ratio V<100>/V<111> obtained at various substrate temperatures by Spitsyn [Spitsyn et al. 1981].

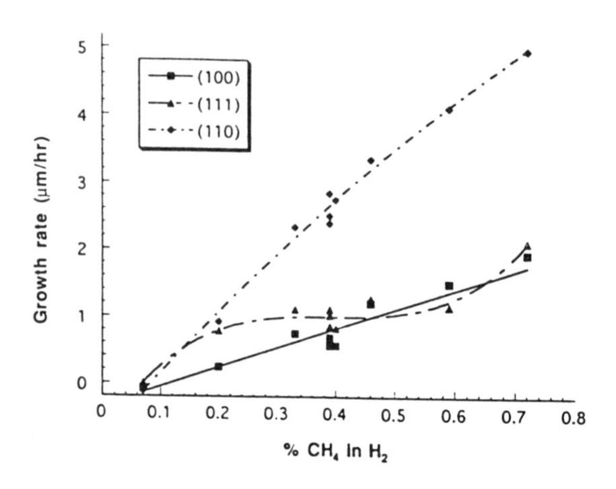

Figure 8. The dependence of homoepitaxial growth rates on methane mole fraction at a substrate temperature of 970°C

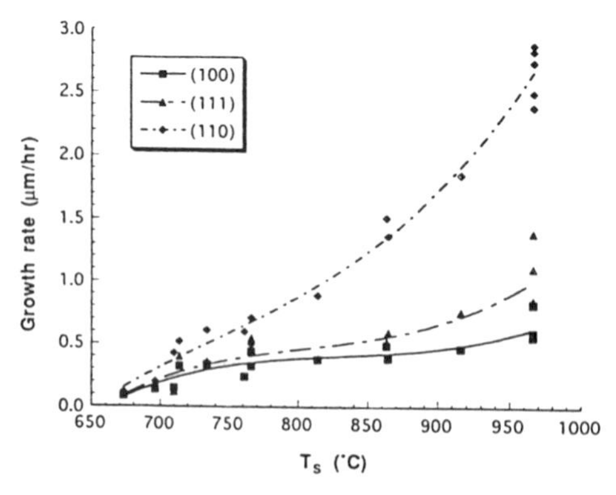

Figure 9. The dependence of homoepitxial growth rates on substrate temperature for a methane mole fraction of 0.4%.

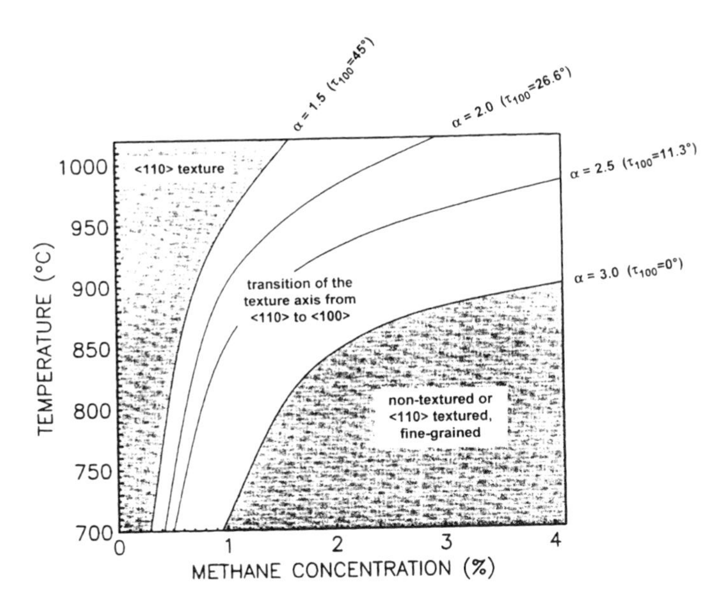

Figure 10. A map of the growth rate ratio parameter α as a function of temperature and methane concentration for a typical microwave plasma assisted CVD growth reactor [Wild et al. 1991].

independently controlled. As mentioned above, such rates are strongly influenced by reactor design and all operating parameters, so they should not be regarded as absolute values but as useful guides. In this system, changing the location of the gas inlet from below the filament to above changed the growth rates by a factor of 1.4. It also changed the nature of the dependence on temperature for growth on the {110}. At 970°C the growth rate ratios V<100>/V<111> ranged from about 0.4 to about 1.5 which would produce nearly the whole range of cubo-octahedral shapes from cubes to octahedra (α = 0.7 to α = 2.6). Cubes are favored at the lowest hydrocarbon levels (0.2% methane). Data as a function of temperature with a methane mole fraction fixed at 0.4% indicates an increase in the growth rate ratio V<100>/V<111> with decreasing temperature i.e. cubes at high temperature changing through a regular cubo-octahedra toward octahedra at lower temperatures.

A variety of diagrams showing the relationship between morphology and growth temperature and gas composition have been presented. The simultaneous development of maps expressing morphology in terms of the growth rate parameter by [Tamor] and [Wild et al. 1993] greatly simplifies and illuminates the relationship between morphology and growth rate parameters (without, however, increasing our fundamental knowledge of the structure sensitive growth mechanisms [Wild et al. 1994] that control the growth rates on individual facets). Figure 10 shows a map of the growth rate ratio parameter α as a function of temperature and methane concentration for a typical microwave plasma assisted CVD growth reactor. The map will be somewhat different for different reactors, different types of CVD reactors, operating parameters, impurities etc. However such maps serve both as general guides and, if done for a specific machine, as an engineering and quality control tool enabling the growth of films optimized for specific applications.

5.3 MORPHOLOGY CONTROL OF POLYCRYSTALLINE CVD FILMS

The morphology of polycrystalline diamond films grown by activated chemical deposition (CVD) varies greatly depending on the deposition conditions. Even small changes in the film nucleation or deposition conditions can cause major changes in the film morphology and crystallite perfection. Areas a few millimeters apart in a film may be quite different unless considerable care is taken to insure uniform deposition conditions. Because the properties of diamond films are so strongly dependent upon the crystallite size, orientation, and perfection, it is important to understand the way in which growth conditions affect film structure in order to design and optimize films for specific applications. Nucleation and growth processes are both important factors in determining the structure and thereby the properties of the films.

The growth conditions for CVD diamond must be carefully chosen to optimize the microstructure for the intended application. Figure 11 shows scanning electron microscope images of the surfaces of films grown at three different temperatures in a hot filament CVD reactor. Only the temperature was changed to grow these three very different films. The films differ strongly in topography, grain size, and crystal perfection as well as the orientation of individual grains. They have large differences in thermal conductivity, wear resistance, chemical reactivity, and other properties as the result of the differences in morphology.

The images on the right are transmission electron micrographs and show the internal crystal perfection of each of these films. Note the wide variation in defect concentration. This illustrates the necessity for carefully controlling the growth conditions to produce

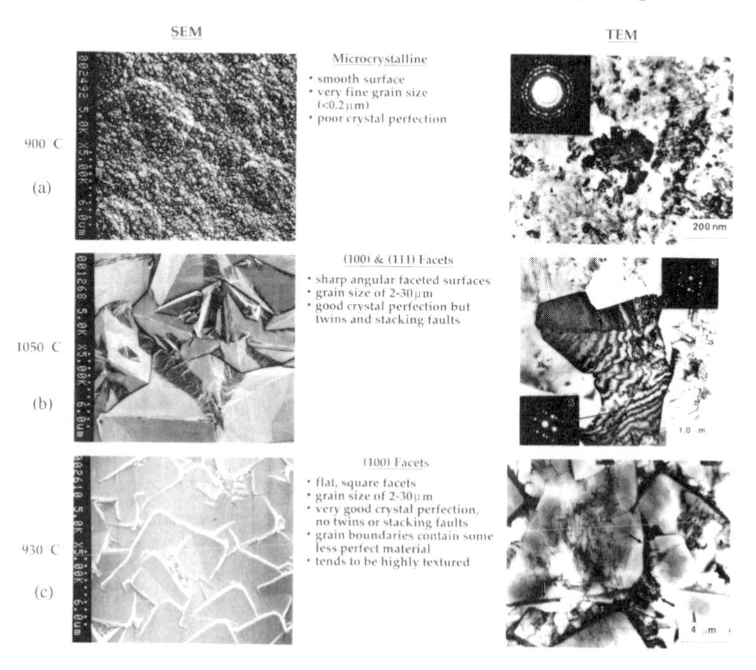

Figure 11. Scanning and transmission electron microscope images of the surfaces of films grown at three different temperatures in a hot filament CVD reactor. Diamond film morphology depends upon the growth conditions. All of these films were grown in a hot filament CVD reactor with 1% methane. Only the substrate temperature was changed.. Film (a) was grown at 1050°C. Film (b) was grown at 900°C. Film (c) was grown at 950°C.

the most desirable properties for a given application. Surface roughness, internal crystal perfection, and grain size and orientation can all be controlled to optimize the product.

5.4 TEXTURE DEVELOPMENT

Most CVD diamond films develop columnar grains in the direction of growth and show strong, preferred orientation as the film thickens, as shown in the micrograph of a fracture surface shown in figure 12. These columnar grains are a natural consequence of the growth processes and the absence of renucleation after film growth begins. This structural evolution has been modeled by [van der Drift 1967]. He called it the principal of evolutionary selection. This model can serve as a guide for growing desired textures.

van der Drift assumed nuclei formed with random orientations on a substrate. The nuclei grow uniformly, perpendicular to each facet of the crystallite, so as to maintain their initial shape. van der Drift assumed that growth occurred on all faces uniformly. If we draw the image of a crystallite its longest dimension will be to a corner where two or more facets meet. This is the fastest growing crystal direction. The crystallites grow until they meet the growing surface of another crystal. The nuclei with their fastest growing crystallographic direction perpendicular to the substrate envelop and overgrow the less favorably oriented ones. This leads to a texture (crystallite orientation) with the fastest growing direction perpendicular to the substrate. (The van der Drift model must be modified somewhat if some face forms grow faster than others or if bulk or surface diffusion is important; but the final conclusion, that the texture is determined by the fastest growing crystallographic direction remains unchanged.)

For real CVD diamond films, growth conditions can be varied to obtain growth rate ratios so as to produce any of the shapes shown in the Figure 4. According to van der Drift's model we should be able to grow textured films with crystallographic directions normal to the substrate which correspond to the longest dimensions of any of the shapes

Figure 12. Scanning electron microscope image of the fracture surface of a typical CVD film.

in the figure. Thus films with nearly pure <100>, <110>,and <111> textures can be grown. Figure 14 a, b, and c show the shapes of individual diamond crystals and the continuous films grown at the same time adjacent to each other on the same substrate. The structure of the films is just what is predicted by the model.

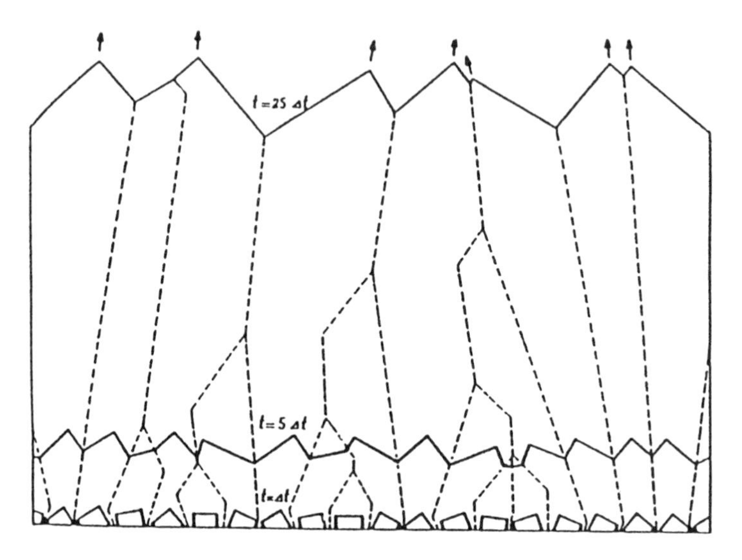

Figure 13. Schematic development of a two dimensional model film which starts from randomly oriented square crystals in a two dimensional space and assumes infinite surface diffusion even along a substrate. The intercrystalline boundaries (dashed lines) as well as the crystal front at different times are shown. Δt is the distance between two neighboring nuclei. Surfaces are shown for several film thicknesses as multiples of Δt. A <11> texture develops.

Figure 14 a, b, and c. The shapes of individual diamond crystals and the continuous films grown at the same time adjacent to each other on the same substrate.

6. Internal Defects Related to Growth Morphology

6.1 GROWTH SECTORS

It has been shown that internal defects and impurity levels of diamond crystals can be related to the crystal facet from which-the section grew, i.e., the growth sector [Wang 1994] and [Shechtman et al. 1993]. The details of the growth processes on {100} and {111} facets can be quite different because of the surface structure, i.e. stereo chemistry. As the crystal grows, the material added to these different facets forms zones within the crystals having different kinds and concentrations of defects and impurities i.e. different crystal quality. These growth zones exist in natural diamonds, and diamonds synthesized by the HPHT process but are especially pronounced in CVD synthesized diamond crystals. Figure 15 shows the development of growth sectors in a crystallite growing as a part of a film with a <001> texture. It has a central (001) facet normal to the direction of growth surrounded by four {111} facets. Figure 16 shows results of scanning and transmission electron microscopic, SEM and TEM, examination of a grain like the one in figure 15. The rear growth sector, i.e., the (-1-11) sector, was removed by fracture allowing transmission diffraction patterns to be obtained from both the (001) and the (-111) growth sectors. The core of this grain, the (001) growth sector, has a much more perfect crystal structure than the {111} growth sectors surrounding it. This is consistent with the commonly observed smoothness of the {100} facets and the three fold surface roughness patterns of the {111} facets. The three fold surface patterns apparently are the result of a growth mechanism involving the creation and growth of microtwins. Although the details of the growth processes are not established, it is certain that under the growth conditions used, {111} growth sectors are filled with microtwins while the material added via {100} facets is not. It has been speculated that the usual growth on {111} is dependent on the nucleation of new growth layers through twinning [Clausing et al. 1991a,b], [Angus and Hayman 1988].

A TEM image of a thinned section perpendicular to the <001> growth direction of a sample similar to the one in Figure 16 is shown in figure 17. The dark areas are microtwinned regions, i.e..{111} growth sectors, surrounding {100} growth sectors which are free of microtwins. This same correlation of higher quality growth from {100} facets and microtwinning in {111} growth sectors is apparently true for other commonly used CVD growth conditions and CVD diamond grown by other techniques [Clausing et al. 1990], [Zhu et al. 1989], [Kaae et al., 1990], [Heatherington et al., 1990], [Badzian et al., 1991]. It is possible to grow CVD diamond without microtwinned {111} growth sectors, for example [Kreutz et al. 1995], very high quality CVD films show little or no microtwinning even in {111} growth sectors but little has been reported to specifically relate decreased microtwinning in {111} growth sectors to growth conditions. Microcrystalline diamond films (such as that in figure 16(b)) have been reported to be highly microtwinned [Clausing et al. 1990].

Growth sectors originating from{100} facets and homoepitaxial layers grown from {100} surfaces tend to be much more perfect than from {111} or {110} surfaces, at least as far as extended defects are concerned. Thus there seems to be a strong incentive to grow <100> textured films, highly oriented <100> polycrystalline films, and single crystal films with only {100} faces where applications require highly perfect crystals.

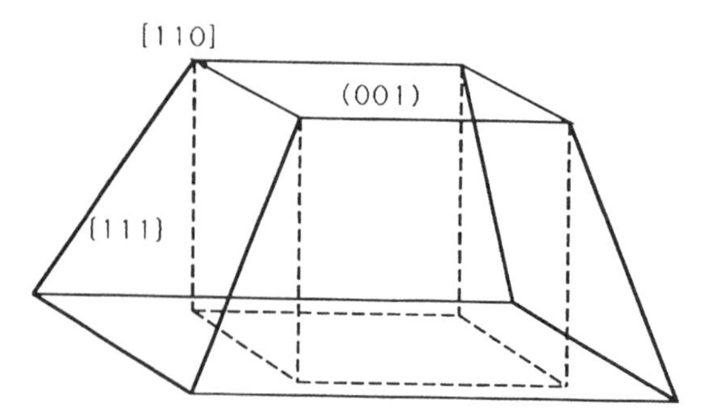

Figure 15. Shows the development of growth zones in a crystallite growing with a <001> texture and having a central (001) surrounded by four {111} facets. The core volume enclosed by the dashed lines grew from the (001) face and is free of microtwins; the surrounding volume was grown on {111} facets and is filled with microtwins. The dotted line indicates the back side of the diamond crystallite.

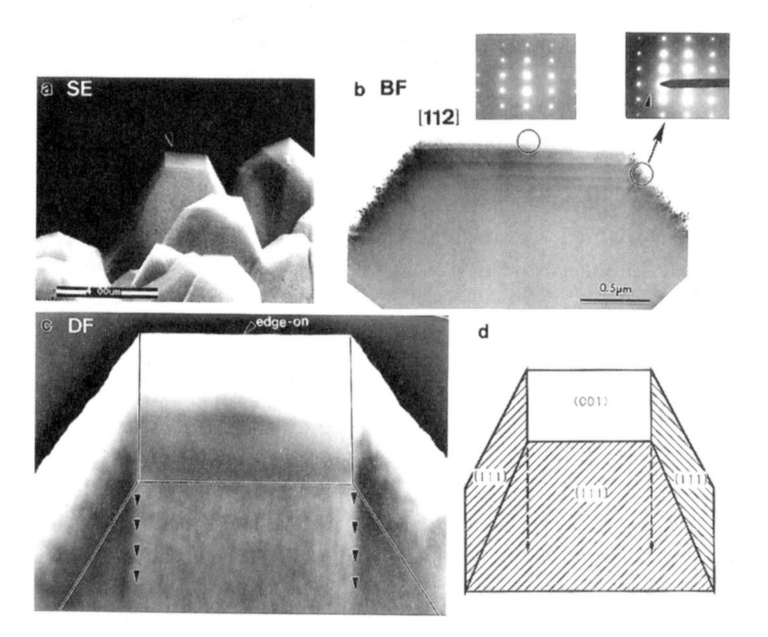

Figure 16. Shows results of scanning and transmission electron microscopic examination of a grain like the one illustrated in figure 14. (a) A SE image and corresponding (b) BF, bright field and (c) DF, dark field images. The geometrical drawing in (d) illustrates the contrast effects seen in (b) and (c). The selected area diffraction patterns shown in the inserts show microtwins only in the (-111) growth sector, i.e. area B.

Figure 17. Shows a TEM image of a thinned section perpendicular to the <001> growth direction of a sample similar to the one in Figure 16. The dark areas are microtwinned regions, {111} growth sectors, surrounding {100} growth sectors free of microtwins.

It is quite apparent that investigation of surface morphology as it relates to the creation of point and extended defects and the growth mechanisms is essential to the control and elimination of crystal defects.

6.2 NUCLEATION

The initial nucleation density, orientation of nuclei, renucleation during growth and the formation of twin nuclei can all greatly influence the morphology of CVD films [Everson 1994], [Tamor and Everson 1994], [Wild et al. 1994a], [Wild et al. 1994b]. Most obviously high nucleation density produces small grain sizes. If the nuclei have preferred orientations or are epitaxial on an oriented substrate they will lead to textured or perhaps in the future to single crystal films.

Nucleation and growth on carbide formers and materials with high carbon solubility is delayed until the surface concentration of carbon can be increased to that necessary to allow nucleation and growth of diamond. For example, there are induction periods on molybdenum and the refractory metals during which a metal carbide layer is formed before it is possible to grow diamond layers [Lux and Haubner 1991]. The high solubility and diffusivity of carbon in iron make diamond coating steels difficult.

Nucleation is difficult on polished silicon or SiC. Ion bombardment to the point of amorphitization can eliminate even nucleation sites created by scratching with diamond grit. This destruction of nucleation sites can be effective enough that patterned growth suitable for semiconductor microelectronics applications can be obtained.

Nucleation can be greatly enhanced by scratching the substrate with diamond grit [Yugo et al. 1990], ultrasonic polishing in a diamond slurry [Yugo et al. 1990], or seeding with fine particles of graphite or diamond is effective in producing high nucleation densities i.e. greater than $10^{10}\,cm^{-2}$.

Textured or even highly oriented pseudo single crystal films can be obtained by controlled seeding [Geis 1990] and [Geis et al. 1991]. Highly oriented polycrystalline films can be obtained by special treatment of silicon single crystal substrates to produce only epitaxially oriented nuclei ("bias enhanced nucleation") [Wolter et al. 1993] and [Jaing et al. 1993] or liquid phase reactions [Yang et al. 1994].

6.3 RENUCLEATION DURING CVD GROWTH

The usual growth processes for polycrystalline CVD films lead to columnar structures dominated by the faster growing crystallites. For the most commonly used growth conditions, renucleation seems to spontaneously occur rather infrequently. This situation allows the development of highly oriented films and can be used to obtain films with properties optimized for special purposes such as thermal or electronic conduction.

Some times it may be desired to obtain fine grained and randomly oriented films for applications such as thin X-ray windows for X-ray lithography. Fracture toughness may also improve in fine grain, randomly oriented material. For this purpose, renucleation can be encouraged by supersaturation of the growth nutrients (lowering the substrate temperature or increasing the hydrocarbons in the gas phase.)

Renucleation can be inhibited by increasing the etching effects of atomic hydrogen and oxygen. The formation and/or survival of small nuclei is less probable when etching rates are increased since grain boundary material and high index planes (which would be exposed on small or randomly oriented nuclei) have relatively high etching rates.

6.4 CONTROL OF TWIN FORMATION

The morphology and quality which develops during the growth of thick diamond films can be influenced by the formation of twins. Twins formed during the nucleation process may survive in thick films only if they are oriented so that the twin plane is perpendicular to the substrate and it is the fastest growing direction. This rarely happens. A more important aspect of twin formation is the formation of penetration twins or contact twins on fast growing facets of polycrystalline or single crystal films. If twins are nucleated on these fast growing surfaces, they will only survive if they grow faster than the face on which they are nucleated. This can happen. Very successful models have been developed [Tamor and Everson 1994] and [Wild et al. 1994] which describe the circumstances under which twins can survive or be prevented from growing on {100} or {111} faces. Unfortunately, the conditions for preventing twin growth on these two different faces do not overlap. This means that the formation and growth of secondary twins cannot be completely eliminated for polycrystalline film growth when both {100} and {111} faces are present unless twin nucleation is completely avoided. Nevertheless, secondary twin formation can be avoided on single crystal homoepitaxial

<100> and <111>films and minimized on highly oriented polycrystalline diamond films if only one type of facet is growing.

The stability of specific crystallographic faces against twin formation is related directly to the ratio of growth rates V<100>/V<111>. Geometrical analysis shows that penetration twins can be avoided on {100} faces when the growth rate ratio is above 1.15 (α values are above 2.0) and that twins on {111} faces are unstable for growth rate ratios less than .886 ($\alpha < 1.5$.) These result are in good agreement Spitsyn's much earlier observations [Spitsyn et al. 1981]. Twins cannot be avoided on {111} when the twin plane is parallel to the facet face.

In summary, the growth of [100] or [111] crystal faces can be stable against twin formation by controlling the growth rate parameter. This is a major consideration in the growth of highly perfect materials for optical and electronic applications.

7. Impurities

7.1 OXYGEN

Oxygen addition to the growth environment for CVD diamond is a common practice. Both improved quality and increased growth rates are usually reported [Spear and Frenklach 1994]. By converting the carbon in the growth atmosphere from hydrocarbons to carbon monoxide the oxygen addition reduces the carbon activity and may change the morphology in the same direction as reducing the hydrocarbon content of the atmosphere without the oxygen addition. The effects of oxygen addition are, however much more complex than a simple reduction of carbon activity. This addition has the effect of increasing the growth rate, suppressing the deposition of non-diamond carbon and improving the 1332 cm^{-1} Raman line intensity. Renucleation seems to be decreased. The effect on the relative growth rates of <100> vs. <111> has not been studied in detail. It has been shown that carefully controlled oxygen additions also permit the growth of well faceted good quality diamond with reduced substrate and filament temperatures [Tolt 1995].

7.2 BORON

Early work on doping to 0.1 at% boron reported a decrease in the lattice parameter of diamond by 0.0009Å [Spitsyn et al. 1981]. The boron atoms were inferred to be mostly in substitutional sites. Increasing the boron to 1.0 at% caused the lattice to expand to match that of undoped diamond, thus it would appeared that the additional boron went into interstitial sites expanding the lattice and compensating the contraction caused by the substitutional boron atoms. More recent work reports that boron additions at all levels expand the lattice.

Heavy boron doping changes the growth morphology of CVD diamond so that a layer by layer growth mode occurs on {111} facets. [Kreutz et al. 1995]. This is in contrast to the growth of undoped films where {111} facets are characterized by the growth of microtwins and stacking faults [Wang et al. 1994].

Boron is reported to be incorporated into {111} growth sectors as much as 100 times more abundantly than into {100} growth sectors [Spitsyn et al. 1981] [Locher et al. 1995]. A very striking example of boron's preference for the {111} growth zone is shown in a crystal grown by the General Electric Company using the HPHT process [Nassau 1993]. The segregation of boron into various growth sectors is {111}>{110}>{100}={113}>{115} for HPHT diamond [Kanda and Sekine 1994].

Light doping with boron is reported to improve the quality of CVD grown diamond while high concentrations promoted graphitization [Nishimara et al. 1989]. Similar results are reported by [Bachmann et al. 1992].

7.3 NITROGEN

The incorporation of substitutional nitrogen into diamond crystals increases the lattice parameter [Lang 1994]. The lattice dilation is about 0.00007Å for 100 ppm atomic nitrogen. The nitrogen can be incorporated into the crystal in many ways; as a single atom substitution (P1 in the electron paramagnetic resonance, EPR spectrum), as N-N pairs (A centers in the IR spectrum) as N-V pairs, N3 centers(3 nitrogen atoms and a vacancy in a (111) plane), W21 centers(3 nitrogen atoms and a vacancy in a (110) plane), B centers (4 nitrogen atoms), H4 centers (a B center with a trapped vacancy), dislocation loops with associated nitrogen, and other aggregates including voidites and platelets [Woods 1994], [Van Enckevort, 1994]. In spite of all these possibilities, which are realized in the case of natural diamond, most nitrogen in synthetic diamond is believed to be single atom substitutional.

The incorporation of modest amounts of nitrogen (up to several hundred ppm) into the atmosphere for diamond film growth by microwave plasma assisted CVD has the effect of increasing both the overall growth rate, and the relative growth rate parameter α [Locher et al. 1994]. The size and smoothness of {100}, cube, faces developed on <100> textured thick films increase as the nitrogen increases up to these levels. Nitrogen is incorporated into the {111} growth sectors as much as 100 times more effectively than into {100} or {110} growth sectors [Yokota 1990], [Lange 1991]. The segregation of nitrogen into various growth sectors is {111}>{100}>{113}>{110} for HPHT diamond synthesized at normal conditions but {100}>{111} at low temperatures [Kanda and Sekine 1994].

8. References

Angus, J. C. and Hayman, C., (1988) *Science*, **241**, 913.

Bachmann, P. K. and Wiechert, D. V., (1992) *Diamond and Related Materials* **1,** 422.

Badzian, A. R., Badzian, T., Wang, X. H., and Hartnett, T., (1991), in *R.* Messier, J.T. Glass, J.E. Buttler, and R. Roy, (eds), *New Diamond Science and Technology*, MRS Int. Conf. Proc. NDST-2, 549.

Banholzer, W. F., (1995), GE Superabrasives develops more wear resistant saw diamond, *Diamond Depositions Science and Technology,* **5 (1),** 3.

Chateigner, D., Brunet, F., Deneuville, A., Germi, P., Pernet, M., Gheeraert, E., Gonon, P., (1995) "Effect of boron incorporation on the lattice parameter and texture of diamond films deposited by chemical vapor deposition," *J. Cryst. Growth*, **148**, 110-115.

Bruton, Eric (1978), *Diamonds*, Chilton Book Co., 366 -395.

Busmann, H. G., Zimmermann-Edling, W., Sprang, H. L., Guntherodt, H. J., and Hertel, I. V.(1992) *Diamond Relat. Mater.* **1,** 979.

Chu, C. J., Hauge, R. H., Margrave, J. L. and D'Evelyn (1992) Growth kinetics of (100), (110), and (111) homoepitaxial diamond films, *Appl. Phys. Lett.* **61 (12)** p 1393-1395.

Clausing R. E., unpublished.

Clausing, R. E., Heatherly, More, K. L., Begun, G. M., (1989) Electron microscopy of the growth features and crystal structures of filament assisted CVD diamond films, *Surface and Coatings Technology*, **39/40**, 199-210.

Clausing, R. E., Heatherly, L. and Specht, E. D., (1991a), Control of texture and defect structure for hot filament assisted CVD diamond films, in R. E. Clausing, L. L. Horton, J. C. Angus, and P. Koidal (eds*), Diamond and Diamond-Like Films and Coatings* (1991), Plenum Press, New York, 611-618.

Clausing, R. E., Heatherly, L., Specht, E. D., and More, K. L., (1991b), Texture development in diamond films grown by hot filament processes *New Diamond Science and Technology* (1991) MRS Int. Conf. Proc., 575-580.

Clausing, R. E., Heatherly, L. L., Horton, L. L., Specht, E. D., Begun, G. M., and Wang, Z. L. (1992) Textures and morphologies of chemical vapor deposited (CVD) diamond, *Diamond and Related Materials* **1** 411-415.

Evans T. (1979) in J. E. Field (ed), *The Properties of Diamond*, Academic Press, 405.

Everson, M. P. et.al. (1994), *Journal of Applied Physics*, **75(1),** 169.

Field J. E. (1979) in J. E. Field (ed), *The Properties of Diamond*, Academic Press, 284.

Geis, M. W., Smith, H. I., Argoitia, A., Angus, J., Ma, G. H. M., Glass, J. T., Butler, J., Robinson, C. R. and Prior, R. (1991) Large-area mosaic diamond films approaching single-crystal quality, *Appl. Phys. Lett.*, **58** 2485-2487.

Graebner, J. E., Jin, S. , Kammlott, G. W., Herb, J. A. and Gardinier, C. F. (1994) Unusually High Thermal Conductivity in Diamond Films, *Appl. Phys. Letters.*

Graebner, J. E., Reiss, M. E., Seibles, L., Harnett, T. M., Miller, R. P., and Robinson, C.J. (1994), *Phys. Rev. B,* **50,** 3702.

Hartman, P. (1973) *Crystal Growth, An Introduction*, North Holland, Amsterdam, 367-402.

Hartman, P. and W. G. Perdok (1955) *Acta Crystallog.* **8**, 49, 521 & 525

Hartman, P., (1965) *Kristallogr.* **121**, 78.

Heatherington, A. V., Wort, C. J. H., and Southworth, P., (1990*), J. Mater. Res.* **5,** 1591.

Jackson, W. E., and Hayden, S. C. (1993) Quantifiable Diamond Characterization Techniques: Shape and Compressive Fracture Strength, General Electric Superabrasives, Worthington, Ohio, Internal Report

Jaing, X., Klanges, C. P., Zachai, R., Hartweg, M. And Fusser, H. J. (1993) *Appl. Phys. Lett.*, **62(26),** 3438.

Kaae, L. J., Grantzel, P. K., Chin, J., and West, W. P., (1990), *J. Mater. Res.* **5**, 1480.

Kanda, H. and Sekine, T., (1994) Characteristics of diamonds grown with metal catalysts in G. Davies, (ed), *Properties and Growth of Diamond*, INSPEC, The Institution of Electrical Engineers, London, 415-426.

Kreutz,T. J., Clausing, R. E., Heatherly, L. Warmak, R. J., Thndat, T., Feigerle, C. S., and Wandelt, K., (1995) Growth mechanisms and defects in boronated CVD diamond as identified by scanning tunneling microscopy, *Physical Review B*, **51**, 14,554 -14,558.

Kubashi, K., Nishimura, K., Kwate, Y. and Horiuchi, T., (1988*), Phys. Rev. B*, **38**, 4067-4084.

Lang, A. R. (1979), Internal structure, in J.E. Field, (ed*), Properties of Diamond,* Academic Press, 425 -469.

Lang, A. R., (1994) Effect of nitrogen on the lattice parameter on diamond in G. Davies (ed), *Properties and Growth of Diamond*, INSPEC, The Institution of Electrical Engineers, London 111-115.

Lang, A. R., Moore, M., Makepeace, A. P. W., Wierzchowski, W., Welbourn, C. W., (1991) *Philos. Trans. R. Soc.* Lound. A (UK) **337,** 497-520.

Locher, R., Wagner, J., Fuchs, F., Mair, M., Gonon, P., Koidl, P., (1994), Optical and electrical characterization of boron-doped diamond films, *Diamond and Related Materials,* **4,** (1995) 678 -683.

Locher, R., Wild, C., Herres, N., Behr, D. and Koidl, P., (1994), Nitrogen stabilized <100> texture in chemical vapor deposited diamond films, *Appl. Phys. Lett.,* **65 (1),** 34-36.

Lux, B. and Haubner, R. (1991) Nucleation and growth of low-pressure diamond, in R. E. Clausing, L. L. Horton, J. C. Angus, and P. Koidal (eds), *Diamond and Diamond-Like Films and Coatings*, Plenum Press, New York, 579-609.

Matsumoto, S., and Matsui, Y., (1982) *Journal of Materials Science* **17,** 3106.

Matsumoto, S., and Matsui, Y., (1983) *Journal of Materials Science* **18**, 1785.

Nassau, Kurt, (1993) Synthesis of bulk diamond: history and present status in R.F> Davis, (ed), *Diamond Films and Coatings,* R. F. Noyes Publications, 31-67.

Nishimara, K., Das, K., Glass, J. T., and Nemanich, R. J., (1989) *Proceedings of NATO Advanced Workshop on the physics and chemistry of carbides, nitrides and borides*, Manchester, UK. Plenum, New York.

Ono, A. ,Baba, T. , Funamoto, H., and Nishikawa, A. (1986) *Japan. J. Appl. Phys.* **vol 25,** L808-810.

Sato, Y., and Kamo, M., (1989) *Surface Coatings Technology*, **39/40**, 183.

Shechtman, D., Hutchison, J.L., Robins, L.H., Farabaugh, E.N., and Feldman, A., (1993) Growth defects in diamond films, *J. of Mater. Res.*, **8**, 473-479.

Spear,K. E., and Frenklach, M., (1994) Mechanisms for CVD diamond growth in K.E. Spear and J.P. Dismukes, (eds), *Synthetic Diamond : Emerging CVD Science and Technology,* John Wiley and Sons, 263.

Spitsyn, B.V., (1988) *Progress in Crystal Growth and Characterization* **17,** 79.

Spitsyn, B. V., Bouilov, L. L., and Derjagin, B. V., (1981), Vapor growth of diamond on diamond and other surfaces, *J. Cryst. Growth* ,**52,** 219-226.

Strong, H. M. and Wentorf, R. H. Jr., (1972) *Naturwiss,* **59,** 1.

Tamor, M. A. and Everson, M. P., 1994 On the role of penetration twins in the morphological development of vapor-grown diamond films, *Journal of Materials Research,* **9,** 1839-1843.

Tolt *Proceedings of the Ectrochemical Society Meeting,* Reno, NV, May 1995.

van der Drift, A., (1967), *Philips research Report* **22**, 267.

Van Enckevort, Willem J. P. (1994) in Karl E. Spear and John P. Dismukes (eds), *Synthetic Diamond: Emerging CVD Science and Technology*, John Wiley & Sons Inc., 307-353.

Van Enckevort, Willem J. P. (1994) in Karl E. Spear and John P. Dismukes (eds), *Synthetic Diamond: Emerging CVD Science and Technology*, John Wiley & Sons Inc., 316.

Wang, Z. L., Bentley, J., Clausing, R. E., Heatherly, L. and Horton, L.L., (1994), Direct correlation of microtwin distribution with growth face morphology of CVD diamond films by a novel TEM technique, *Journal of Materials Research* , **9**, 1552 -1565.

Wild, C., Koidl, P., Mueller-Sebert, W., Walcher, H., Kohl, R., Herres, N., Locher, R., Samelenski, R., Brenn, R., (1993) Chemical vapour deposition and characterization of smooth [100]-faceted diamond films, *Diamond and Related Materials,* **2**, 158-168.

Wild, C., Herres, N., Locher, R., Mulller-Sebert, W.and Koidal, P. (1994) "Control of twin formation: a prerequisite for the growth of thick oriented diamond films" in S. Satio, N. Fujimori, O. Fukunaga, M. Kamo, K. Kobashi, and M. Yoshikawa (eds), *Advances in New Diamond Science and Technology*, MYO, Tokyo, 149-152.

Wild, C., Kohl, R.,Herres, N., R., Mulller-Sebert, W.and Koidal, P., (1994) Oriented CVD diamond films: twin formation, structure and morphology, *Diamond and Related Materials* **3,** 373-381.

Wild, C., Koidl, P., Herres, N., Muller-Sebert, W. and Eckermann, T. (1991) *Electrochem. Soc. Proc.* **91-8** 224.

Wilkes J. and Wilks E. M. (1979) in J. E. Field (ed), *The Properties of Diamond*, Academic Press, 352.

Wilks, John and Wilks Eileen, (1991) *Properties and Applications of Diamond*, Butterworth-Heinemann Ltd, 126-132.

Wolter, S. D.,Stoner, R. B. and Glass, J. T. (1993) *Appl. Phys. Lett.*, **62(11)** 1217.

Woods, G. S., (1994), Chapter 3 Properties of nitrogen in diamond, in G. Davies (ed), *Properties and Growth of Diamond*, INSPEC, The Institution of Electrical Engineers, London, 81-98.

Yakota, Y., Kawarada, H., and Hiraki, A. (1990) *Materials Research Society Symposium Proceedings* **162**, 231.

Yamaoka, S., Komatsu, H., Kanda, H. and Setaka, N. (1977), Growth of diamond with rhombic dodecahedral faces, *Journal of Crystal Growth* **37**, 349-352.

Yang, P. C., Zhu, W. and Glass, J. T. (1994) Diamond nucleation on nickel substrates seeded with non-diamond carbon, *Journal of Materials Research,* **9(5),** 1063-1066.

Yugo, S., Kimura, T., Kanai, H. (1990) in S. Saito, O. Fukunaga and M. Yoshikawa (eds.), *Proc. 1st Int. Conf. on the New Diamond Science and Technology*, KTK Scientific Publishers, Tokyo 119.

Zhu, W., Badzian, A. R., and Messier, R. (1990) *Proceedings of SPIE on Diamond Optics III*, **vol. 1325**, 187-201.

Zhu, W., Badzian, A. R., and Messier, R., (1989), *J. Mater. Res.* **4,** 659.

Chapter 3

MECHANICAL PROPERTIES OF DIAMOND, DIAMOND FILMS, DIAMOND-LIKE CARBON AND LIKE-DIAMOND MATERIALS

Peter J. Gielisse
Florida Agric. & Mech. University, Florida State University
College of Engineering, Tallahassee, FL 32310, USA

Contents

1. General

A treatment of the mechanical properties of diamond requires, first and foremost, a definition of "diamond", along with an indication of what mechanical properties are relevant and significant in actual technical applications. Before the advent of usable diamond thin films, synthesized from the vapor phase at low pressure starting some ten years ago, industrial diamond could be classified into two groups: a naturally occurring mineral subdivided into a few types on the basis of shape, physical and optical properties and "toughness" and the high pressure- high temperature manufactured grit sizes, which today culminate in a world wide production of hundreds of millions of carats (1 gram=5 carats).

The recent major developments in the low pressure vapor synthesis area, no longer requiring high pressure - high temperature (55 kbar, 1500° C) manufacturing capabilities, allow a subdivision of synthesized industrial diamond into two types: single crystals (grit) and polycrystalline thin films, with drastically different application areas.

Table 1. Synthesized diamond types and application areas.

INDUSTRIAL DIAMOND	
SINGLE CRYSTALS	POLYCRYSTALLINE FILMS
High pressure - high temperature 55 kbar; 1500°C	Low pressure - low temperature 0.1-1 mbar; 200-800°C
I ABRASIVE APPLICATIONS	I ABRASIVE APPLICATIONS
▪ Grinding ▪ Cutting ▪ Polishing ▪ Honing	▪ Low friction ▪ High wear protection ▪ Tool coatings
II NON-ABRASIVE APPLICATIONS	II NON-ABRASIVE APPLICATIONS
▪ Electrical insulation ▪ Passive devices (doped) ▪ Chemical resistance ▪ Reinforcement in composites	▪ Electronic devices ▪ Optical windows ▪ Thermal management ▪ Dielectric films

The two categories have been further broken down according to their major distinguishing features in Table 1. A characterization of natural diamond into the traditional four categories based on physical (not mechanical) properties, has been put together in Table 2 for reference purposes. The mechanical properties of these four types do not differ significantly. The two mined products known as carbonado and ballas (see Table 3), constituting only a small portion of mined industrial diamond, have no manufactured equivalents.

Diamond is probably the only material in which size, morphology, shape and structure, influence both type and efficiency of its applications as critically as they do. In other words, the intrinsic physical (mechanical) properties are not sufficient to unambiguously analyze or explain the reasons for its successful applications. A case in point is the highly shape, structure and size differentiated (grit size) industrial diamond, which has virtually completely replaced the mined product in the market place, (see also the section on applications of industrial diamond products). Diamond film products are likely to develop along similar lines, particularly as it concerns the diamond-like carbon films (DLC) which in their properties critically depend on synthesis parameters, allowing customizing or tailoring of the product to the application. A chart indicating the various diamond application categories based on size, covering a range from Angstrom to sub-meter, is given in Fig. 1.

Table 2. Characteristics of single crystal natural diamond.

Type	Ia	Ib	IIa	IIb
Abundance (% total)	98	1-2	very rare	extremely rare
Nitrogen Content (%)	up to 0.1 (platelets) or up to 5000 ppm	up to 0.2 (paramagnetic) or synthetic up to 500 ppm	<5ppm or practically zero	essentially zero
Nitrogen Content (cm^{-3})	10^{19}-10^{21}	-	10^{18}	-
Optical Transparency (nm)	>320	>320	>225	bluish to blue (boron)
Thermal Conductivity (W/m K)	900	900	2600-3200	2600
Electrical Resistivity (Ω cm)	$>10^{16}$	$>10^{16}$	$>10^{16}$	10-10^3 (p-type semiconductor) or up to 10 ppm boron
Dislocation Content	low high mobility	-	high, $>10^5$ low mobility	-
Hardness	possibly not as hard as IIa	-	possibly harder than Ia	

It will be realized that a treatment of mechanical properties of diamond cannot be made along rigorous intrinsic qualifier lines only, but will of necessity have to incorporate other aspects, as demanded by the many types of applications. A case in point would be the comparison of a diamond sphere to a morphologically well developed and shaped crystal with the same intrinsic physical properties and of the same size, which are likely to perform quite differently in grinding operations!

A further problem in defining diamond's mechanical properties is that in most cases diamond is not available in a form required for conventional property testing i.e. as ribbon, plate or fiber. In other cases, particularly in the thin film area, time has been too short to yield desirable results in all characterization areas.

Every attempt has been made in this article to define as closely as possible the processing characteristics of diamond films in the various categories and subcategories, so as to make reasonable quantitative comparisons possible.

Table 3. Diamond nomenclature and definitions.

Designation	Description
Diamond (C-C)	The cubic high density crystalline form of carbon four fold (tetrahedrally) coordinated, forming a covalently bonded aliphatic pure sp^3 hybridized network. Shows a characteristic Raman peak at 1332 cm^{-1}.
Nanodiamond	Diamond of particle size considerably less than one micron, often only fifty to hundreds of Angstrom. It is normally synthesized by shockwave loading of a carbonaceous starting material.
Carbonado	Naturally occurring rare polycrystalline diamond containing graphite and other impurities. Tougher than single crystal diamond, brown to black in color. Major applications: drill crowns in mining and oil (water) exploration. Normally set in tools "as is" since they resist any form of shaping or sizing. Sintered diamond grit (diamond compacts), widely used as cutting tool inserts are often seen as its equivalent, although properties are significantly different.
Ballas	Naturally occurring rare polycrystalline diamond, much like carbonado although normally lighter in color. Roughly ball shaped, the pieces are very tough and impact resistant and therefore used in tooling. They are almost impossible to shape and do not display the well known cleavage characteristics of single crystal diamond. No true synthetic equivalent.
Lonsdaleite 2H-(C-C)	A polymorph of diamond. Sequentially stacked (A,B,A,B,...) hexagonal, high density, four coordinated high pressure carbon lattice (wurtzite structure). Erroneously referred to as "hexagonal diamond" (hexagonal analog of "cubic diamond"). Properties are similar to diamond.
Synthetic Diamond	Diamond as manufactured under high temperature-high pressure conditions (grit sizes) or by shockwave techniques (nanodiamond) or low pressure vapor processes (thin films). Nominally identical to natural diamond. Should not be confused with "faux" diamond (cubic zirconia, yttrium, garnets). Manufactured diamond normally contains impurities not present in natural diamond.
Diamond Films	Thin coatings of single phase, cubic, polycrystalline high density carbon (diamond). Structure is mostly composed of particles of a distinct diamond (cubooctahedral) morphology. Films must be identifiable as diamond by both their X-ray (or electron) diffraction pattern and the characteristic Raman frequency at 1332 cm^{-1}.

Table 3, cont.. Diamond nomenclature and definitions.

Designation	Description
Diamond Like Carbon (DLC)	A thermodynamically metastable amorphous form of carbon, of interest due to its wide range of applications and relative ease of synthesis.
a-C:H	Amorphous diamond-like carbon film (DLC) Amorphous hydrogenated carbon film, normally consists of sp^2 clusters connected by an sp^3 network.
a-C:H:N	Amorphous hydrogenated carbon-nitrogen film, similar to a-C:H with additional C-N bonds.
6H-(C-C)	A chemical vapor deposited (CVD) hexagonal layered form of high density carbon, which comprises a repeat unit along the c-axis of one hexagonal layer followed by two cubic layers. Regarded as a diamond polytype.
a-(C-C)	Amorphic diamond film claimed to be free of hydrogen, normally prepared by a laser plasma discharge.
CN_x / C_3N_4	The product of reactive sputtering of carbon in N_2 discharges. Mostly amorphous with small fractions of crystalline phases. Films of small (<10 nm) crystallites in an amorphous matrix. Material has often been claimed, but not proven, to be harder than diamond.The C_3N_4 form is seen as the analog to the high technology Si_3N_4 ceramic which is widely used in industry.
CVD	Chemical vapor deposition method.
PVD	Physical vapor deposition (sputtering).
HFCVD	Hot filament chemical vapor deposition method.
BPCVD	Bulk polycrystalline chemical vapor deposition method.
FACVD	Filament assisted chemical vapor deposition method.
MPCVD	Microwave plasma enhanced chemical vapor deposition method.
PACVD	Plasma assisted chemical vapor deposition method.

Table 3, cont.. Diamond nomenclature and definitions.

Designation	Description
PECVD	Plasma enhanced chemical vapor deposition method.
C:N	Amorphous carbon-nitride films synthesized from the vapor by various techniques. Can contain micron sized crystallites which X-ray diffraction patterns indicate as having an hexagonal symmetry. Chemical composition, C-80%, N-20%. Raman spectra have two broad peaks with maxima at 1550 cm^{-1} and 1350 cm^{-1}.

Diamond films and DLC films are further described in the next section (2). Nomenclature and definitions are given in Table 3.

The industrial applications of diamond depend for more than 95% on the unique mechanical properties. The three major application categories are stock removal, the scratch and wear resistance area and applications based on (low) frictional characteristics. A broad overview of these subcategories has been put together in Table 4. The listings are not thought of as all inclusive, but are presented as an introduction to present and future diamond applications, in which technological need and ingenuity have been and will be the driving forces.

2. Defining Diamond and Diamond-like Structures

The microlevel arrangements of atoms in the two most important carbon structures, graphite and diamond, are very different. In graphite, six membered aromatic rings of sp^2- hybridized carbon atoms are arranged in layers, with the bond strength between the layers one to two orders of magnitude less than between the atoms within the layer. This imparts the structure with (solid) lubricant characteristics, due to the ease with which the interlayer bonds can be severed in shear. The carbon atoms in diamond, are covalently bonded via a network of tetrahedrally arranged aliphatic sp^3 hybridized carbon atoms. The essential difference lies, therefore, in both the type of hybrization, sp^2 or sp^3 (or as in mixed phases in the sp^3 to sp^2 ratio) and in the structure type. Fig. 2 illustrates the two structure types.

Diamond-like carbon films contain a good portion of sp^2 bonding, often up to 60%, the exact extent of which is determined primarily by the type of deposition technique and the specific process parameters.

The higher the aliphatic sp^3 content the more enhanced the mechanical and physical properties tend to be, although factors such as surface characteristics, foreign phases, impurities, structural inhomogeneities (voids, cracks, filling factors) and grain boundary effects, can make a significant impact on the actual values. DLC properties are

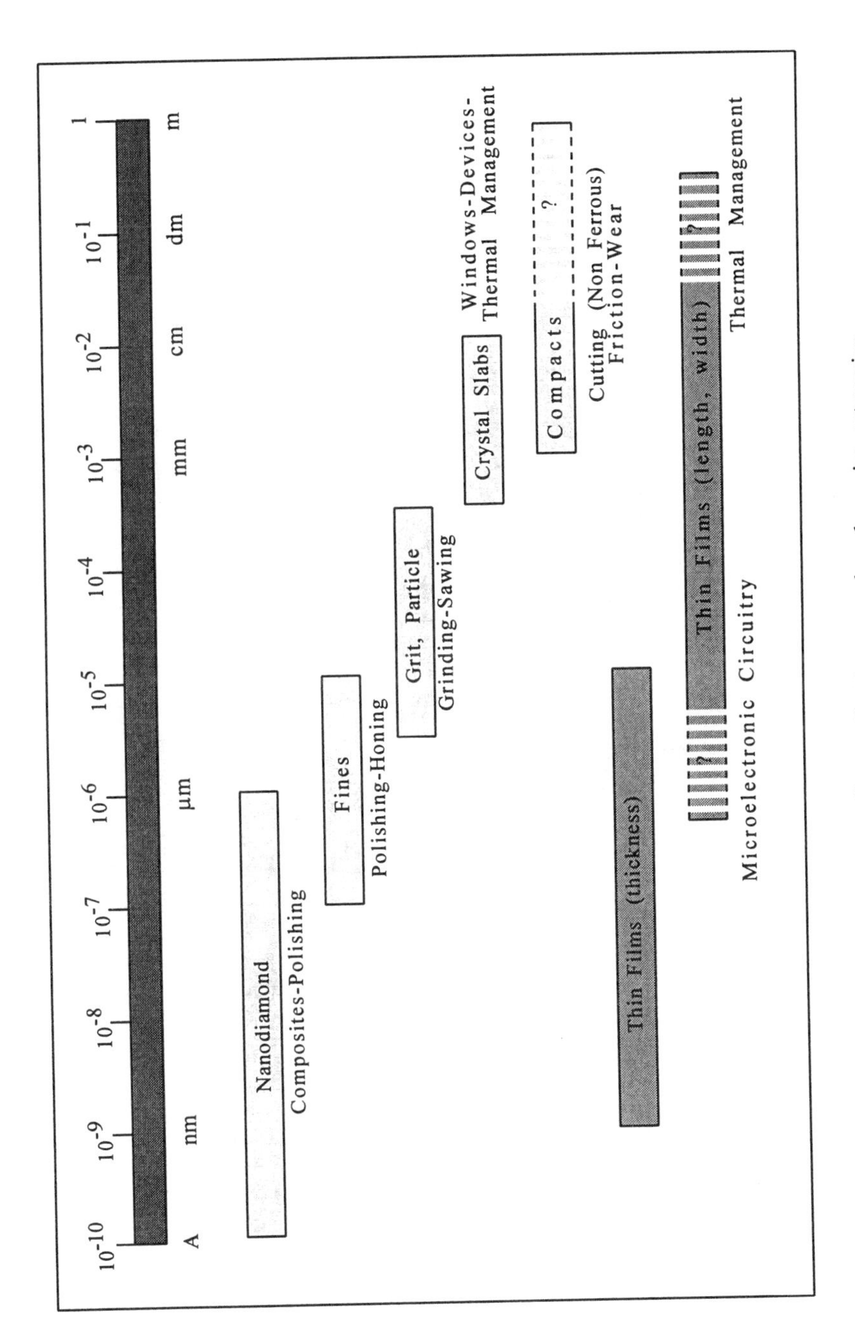

Figure 1. Particle and thin film application areas based on size categories.

Table IV. Diamond applications based on mechanical properties

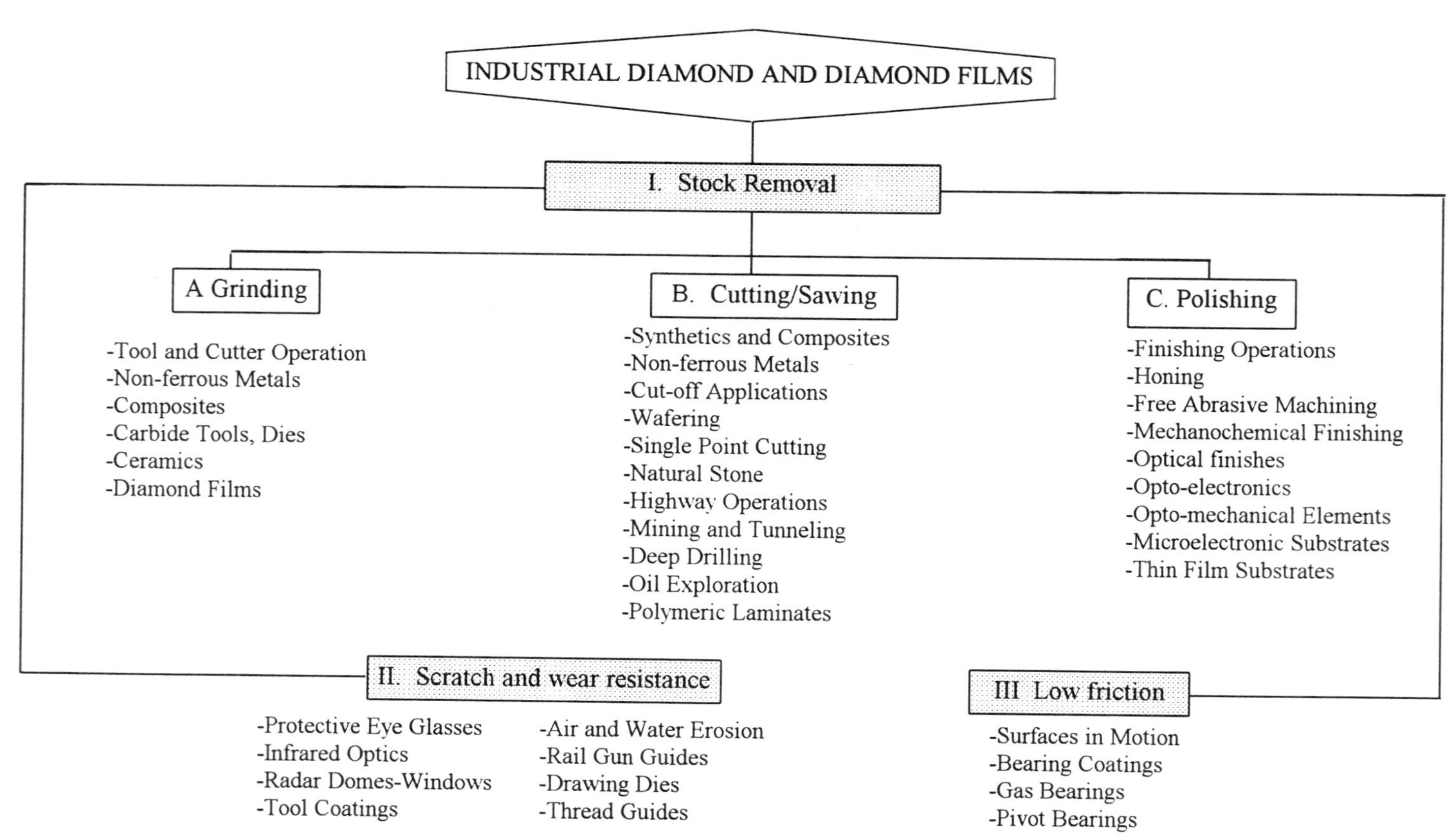

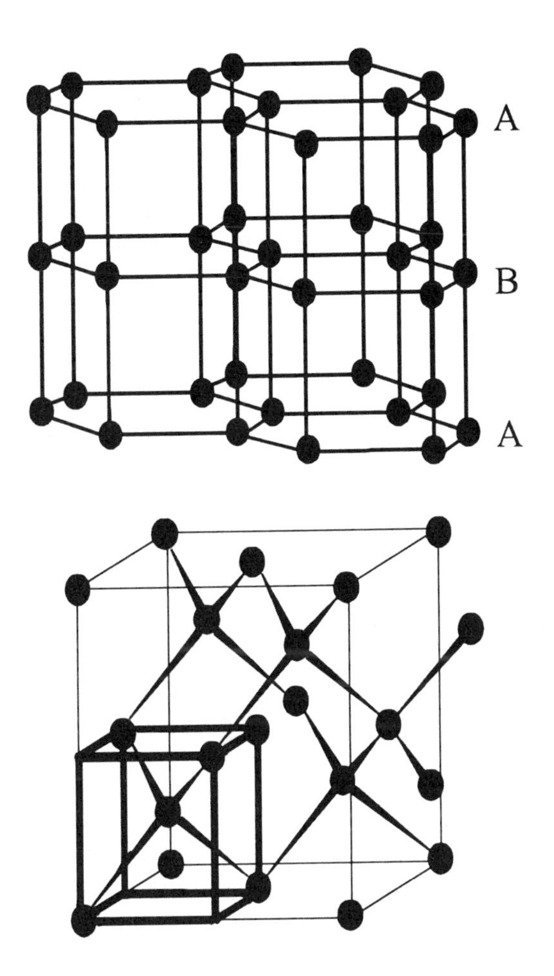

Figure 2. The two main structural types of crystalline carbon, hexagonal graphite (top) and cubic zincblende (diamond).

also influenced by and differentiated on the basis of their (bonded) hydrogen content in hydrogenated films and their hydrogen and nitrogen content in amorphous - C:H:N films.

Diamond films currently are polycrystalline thin coatings consisting of a dense network of high density carbon (diamond) crystallites, generally of cubooctahedral morphology. Depending on purity and particularly on the non-diamond carbon content, free standing films range from essentially opaque to more commonly translucent and are in their best rendition, essentially transparent. Synthesis of single crystal (epitaxial) films on substrates other than diamond, has not yet been accomplished. The films are often characterized with reference to their most common application as being of a tribological, thermal management, or device grade, with transparency (and thermal conductivity values) increasing in the same order.

3. The Strength of Diamond and Diamond Films.

There exist no true tensile strength values for diamond. In the case of films, reliable strength data have normally been derived indirectly from hardness values, from compression tests, or from diaphragm bulge tests. In bulge tests the diamond disks, diaphragms or films, are held in a fixture along their periphery and are exposed to increasing air or gas pressure, till fracture occurs. The actual fracture strength values are then derived by combining strength of materials formulations, such as flat plate theory and bending moment approaches, with the experimental results. In evaluating such diamond film strength data, it should be kept in mind that they are influenced by geometric factors (thickness, grain size, morphology and disk diameter) by growth parameters (substrate treatment, gas type and concentration) as well as by the type and level of the surface roughness, surface crack size (1-10 micron) impurity concentration and residual stress levels.

The fracture strength in polycrystalline brittle materials normally varies as the inverse square root of the grain size, known as the Petch relationship,

$$\sigma = \alpha \; d^{-1/2} \tag{1}$$

where α is an interfacial constraint parameter,
d is the average grain size, and
σ is the strength.

This relationship is often succesfully used in comparing properties of polycrystalline ceramics in which the grain size differs, but the extent of the grain boundary phase remains more or less the same. This is not necessarily the case with vapor phase synthesized diamond films, in which other factors may influence strength. It has been found [Aikawa and Baba 1993] e.g., that crystal quality, often expressed by the development of a microcrystalline film structure as opposed to a more common columnar film structure, tends to promote the highest fracture strength. High quality in diamond films also refers to a low "non-diamond" volume or a high sp^3 (not sp^2 or sp) content. As stated before, the application requirements may in certain cases be more important than the specific diamond physical properties.

The more important (fracture) strength data for diamond single crystals and polycrystalline films, as generated over the last decade or so, are compared in Table 5. In the HFCVD approach it was found that the crystal quality rather than the grain size had the greater effect on fracture strength. The use of a bias current between substrate and filament(s) produced fine agglomerated grains and a higher strength as compared to films with a columnar structure. The latter were more brittle and resulted when no bias was applied. The maximum fracture strength of 2200 MPa was obtained with the highest bias current, 700 mA. Fracture strength tends to be high for films of small grain size and with a low non-diamond content.

Table 5. Fracture strength for diamond single crystals and polycrystalline films.

Material Type	Strength (MPa)	Strength Type and Method	Reference, year
Single crystal Natural Diamond	2800	Fracture Strength	[Field], 1979
Single Crystal Natural Diamond	3500	-	[Field], 1992
Polycrystalline Diamond Films, CVD	500-1400	Tension	[Field], 1993
Polycrystalline Diamond Films, HFCVD	500-2200	Fracture Strength, Bursting pressure nitrogen on diamond membrane	[Aikawa and Baba], 1993
Polycrystalline Diamond Films, BPCVD	746-1138	Disc bursting pressure, simple plate theory	[Susman et al.], 1994
Polycrystalline Diamond Films, FACVD	340-1380	Fracture strength, bursting pressure air on diamond membrane	[Crdinale and Robinson], 1992

The BPCVD film products were evaluated polished and unpolished, with either the nucleation side or the growth side exposed to the pressurization direction. The latter condition did not produce significant differences. The highest bursting stress, 1138 MPa, was obtained for relatively thin (179 m) small diameter (10 mm) discs, while the lowest bursting pressure (746 MPa) was reported for a thicker (296 μm) larger diameter (20 mm) disc.

The FACVD films with a fracture strength ranging from 340 to 1380 MPa, could be differentiated into two groups. Growth on the fine grained substrate (nucleation side) interfaces yielded a high(er) average strength of 880 MPa, while film hardness on the coarser grained diamond growth surfaces averaged 610 MPa.

It will be noted that the maximum fracture strength value for a diamond film of 2200 MPa, Table 5, comes close to that reported for type II natural diamond, (2800-3500) MPa. A more conservative analysis would indicate that the strength value, averaged over all the polycrystalline film data in Table 5, yields approximately 1000 MPa. This is about one third the value for single crystal diamond, in line with similar ratios for other non-metallic materials when comparing single crystal and polycrystalline strength values. The (1000 MPa) average diamond film strength favorably compares to values for other high technology industrial non-metallic materials such as sapphire (430 MPa) silicon nitride (830 MPa) and the other group IV elements, silicon (130 MPa) and germanium (90 MPa).

4. The Hardness of Diamond

4.1 HARDNESS DEFINED

The hardness of a material and particularly that of diamond, is not a rigorously defined physical quantity. The numerical quantity by which hardness is expressed, normally derives from a measurement technique that tests or incorporates more than what can

Table 6. Characteristics of various hardness types

Category	Hardness Types			
	I. Deformation	II. Wear Rate	III. Physical	IV. Indentation
Test Method	Single point stock removal	Multi-point stock removal	Volumetric lattice energy	Stress/strain evaluation
Test Basis	Relative scratch resistance	Relative wear resistance	Resistance to micro-structural breakdown	Resistance to macro-structural breakdown
Scale and Units	1. Moh's scale 2. Modern approaches (J/m^3)	1. Woodddell scale 2. Other approaches (cm^3/min)	1. Plendl-Gielisse 2. Bradt (kcal/mol)	1. Knoop, Vickers 2. Others (kg/mm^2), (GPa)
Diamond Hardness	1. H=10 2. Diamond reference	1. H=45 2. H=?	H=U/V=1350	H=9,000-10,000 (Knoop)
Remarks	-Scratch test type -Could be standardized radily -Suitable for film testing	-Volume wear test -Not easily standardized -suitable for film testing?	-Applicable to single crystals and films -Adaotable to poly-crystalline materials -Sets maximum values	-Very load senstivive -Elastic and plastic contributions -Dissimilar indentors yield incomparable results

truly be considered physical hardness. The results produced by the test techniques are thus sensitive to variables other than those intrinsic to "physical hardness".

In practice, hardness relates to the resistance against a deformation process, which in its measurement methodology and the resultant unit expressions, can differ considerably from one approach to another. Hardness values are not likely to be comparable, even when the same type of units are used. This is best exemplified by the values quoted in the open literature for the hardness of diamond ranging from 500 to 15,000 kg/mm^2 for the Knoop indentor approach and from nominally 10,000 kg/mm^2 to close to 30,000 kg/mm^2 for the Vickers approach.

Attempts at rigorous hardness formulations have over the years led to four fundamentally different approaches: deformation (scratch)-based hardness; wear rate-based hardness; energy- based hardness and the most commonly used, stress-based hardness. The last category can be further subdivided into the tensile non-hydrostatic unrestrained type and the compressive hydrostatic restrained type. Table 6 shows this subdivision, along with characteristics of the various hardness types.

4.1.1 Scratch hardness

The first attempts at a relative hardness expression led to the well known ten point Mohs' scale. It ranked mineral specimen from a value of one for the softest, talc, to a value of ten for to the hardest known, diamond. The arrangement was chosen such that the material in each category could scratch the one at the next level down. The scale shows a remarkable consistency in the doubling of hardness in going from one numerically ranked category to the next, except in the step from nine (corundum, Al_2O_3) to ten (diamond). This range is covered by synthetic materials, of which none were available to Mohs. The scale is currently only of historical significance. Much improved fully instrumented methods, that quantify the energy required for a diamond stylus to scratch a material under standardized conditions, are currently available. The use of such techniques, which in their measurement methodology are likely to come closest to a hardness quantification of a fundamental character, is not yet widespread.

4.1.2 Wear rate hardness.

The wear rate hardness expresses hardness in terms of the extent to which a material is worn away under standardized abrasion conditions. The abrasion during lapping experiments, resulting in the Wooddell hardness numbers and scale, appear to have been the first attempts [Wooddell 1935]. This method has also been implemented through the use of bonded abrasives (i.e. in a grinding wheel configuration) or in free abrasive approaches in which the abrasive is delivered at the interface between a rotating wheel surface and standardized test specimen. Alternative methods in which abrasive jets are impinged on test specimen, have also been used. All these methods tend to be costly and the results much dependent on the specific set-up and test conditions. The method is also not widely practised.

4.1.3 The energy approach

The energy approach was first introduced in the 1960's by Plendl and Gielisse [Plendl and Gielisse 1962]. They proposed the use of the quantity - lattice energy per unit volume -, U/V, i.e. the energy to break down the molecular structure. It was an attempt to provide a fundamental underpinning to hardness, in other words to obtain a strictly physical expression of hardness. They could show that for some thirty solids for which lattice energy values were available, U/V scaled directly with the Woodell hardness number, which is also a volumetric quantity that interprets structural breakdown. The resultant hardness scale, had two attitudes, one for the softer naturally occuring solids from Mohs one (1) to nine (9) and one for all of the technically important synthetic materials representing the high end of the hardness scale between nine (9) and ten (10). Data of this type was used to predict the hardness of c-BN (borazon) shortly after its initial synthesis [Gielisse 1962]. Nothing in this energy approach has been refuted to date. It continues to stand as the only physical and fundamental expression of hardness. It applies in principle to homogeneous high quality single crystals. Polycrystalline materials may require a macro structural modifier, A, so that the hardness expression becomes,

$$H = (U/V)\cdot A \tag{2}$$

where H is the hardness,
U/V is the lattice energy per unit volume, and
A is a macro-structural modifier.

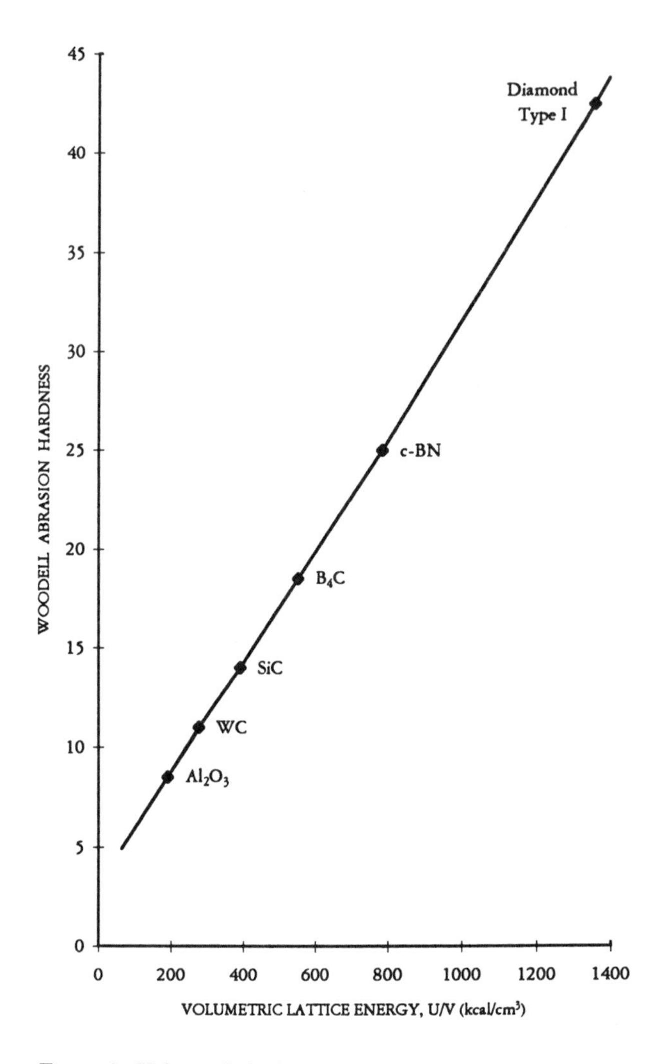

Figure 3. Volumetric lattice energy versus abrasion hardness for hard to superhard materials.

A plot of physical hardness versus abrasion hardness for important industrial solids, is presented in Fig. 3. The values for c-BN were slighty upward corrected [DeVries 1972] from the original values [Plendl and Gielisse 1962].

4.1.4 Indentation Hardness

Indentation hardness may be defined as the resistance of a material against the static (low speed) indentation by an ultra hard indentor of known geometry, into the unrestrained volume of a less compressible body. The hardness determined in this way is related to the plastic work expended in creating the remnant impression on removal of the indentor. Indentation (hardness) measurements are readily and accurately performed on plastically deformable bulk materials (metals). This is most readily shown for the case of spherical (ball) indentors for which it has been shown [Meyer and Disch 1908] that the hardness, H, follows a simple power law,

$$H = L/\pi a^2 = K(a/D)^m \tag{3}$$

where K and m are material constants,
D is the diameter of the sphere,
a is the contact radius, and
L is the maximum force on the indentor.

The indentation geometry and volume remain constant on the release of the indentation force. The exponent, m, has the same value as that in the stress-strain expression for uniaxial tensile loading,

$$\sigma = K \varepsilon^m \tag{4}$$

where ε is the strain and
K is a material's constant.

Diamond and (related) materials which have very large hardness to elastic modulus ratios, show elastic-plastic indentation dynamics. The indentation size and shape relax on the removal of the load i.e. the elastic (recoil) deformation must be taken into account. In the simplest cases the effect is that of a remnant indentation equivalent to one of lesser depth or as if made with a larger diameter sphere. In the above equation the diameter D, will be replaced by one of a greater value, D_p. Similar adjustments will be needed when using diamond shaped conical indentors (Knoop) or square conical indentors (Vickers). Further complications are introduced when lubricants, both intentional and non-intentional, are present. Wide variations in load levels have been used in hardness testing which make the resultant data invariably non-comparable. In polycrystalline materials, the hardness value will tend to reflect an integration over the contributions of the (lower) hardness of the grain boundary component and that of the individual grains. Microindentiation techniques (very low load) are designed to prevent catastrophic brittle failure on identation. Hardness values resulting from such techniques often yield difficult to measure micron sized indentations and thus inaccuracies. They may also reflect the hardness of the individual grains or the grain boundary constituents, rather than a composite value. Very careful microscopic examination is required.

The case of thin (micron-sized) films demands that, in determining their hardness, the contributions of both the substrate and the film are taken into account. Their micron size thickness generally implies that very low load-low penetration depth tests be made. This becomes highly problematic on surfaces with any degree of roughness (relative to film thickness) and for those that are polished but of marginal thickness. In any case the indentation depth cannot be large compared to the film thickness. A rule-of-thumb-value to prevent a substrate contribution, allows a penetration depth of no more than 10% of the film thickness.

It has been proposed that hardness of a coating on a substrate can be determined by a volume law of mixtures hardness model [Burnett and Rickerby 1987a,b; Bull and Rickerby 1989, 1990] for which the composite hardness H_C is given by,

$$H_C = (V_f / V)H_f + (V_s / V)H_s\alpha^3, \quad (H_s < H_f) \qquad (5)$$

or

$$H_C = (V_s / V)H_s + (V_f / V)H_f\alpha^3, \quad (H_s > H_f) \qquad (6)$$

where V_s and V_f are the deforming volumes in the substrate and coating (film) respectively,
V is the total deforming volume,
α is an interfacial constraint parameter and
H V_s and H V_s are the hardnesses of the substrate and the film.

The (fractional) change in the volume of the softer material resulting from the presence of the harder component, is accounted for by α, an empirical interfacial constraint parameter.

The composite hardness expressions incorporate the dependence of hardness on load at small indentation sizes, the so-called "indentation size effect" (ISE) and a "plastic zone size term", to account for (indentor imparted) deviations from the spherical cavity approach. The load bearing profile under an indentor on a stiff film is likely to be non-uniform as well, with further adjustments necessary for substrate materials with a low yield stress to Young's modulus ratio (σ_y/E). All these effects need be incorporated in the value of the interface parameter, α , which is to be experimentally determined for each material.

4.2 THE HARDNESS OF NATURAL DIAMOND SINGLE CRYSTALS

In a quantitative sense, diamond is generally known as the hardest material, with values quoted between 90 and 100 GPa (9,000 -10,000 kg/mm^2 Knoop) and higher. Within present day understanding, diamond is thus about twice as hard as its well known analogue c-BN (cubic boron nitride, borazon) five times as hard as the most widely used technical ceramic, alumina (Al_2O_3, sapphire) and about ten times as hard as the purest single crystal material, silicon.

There are very few reliable and direct measurements of the stress-strain (indentation) type hardness on single crystal diamond. Some of the best and most recent work was carried out by Novikov et.al. [1993] who measured the single crystal hardness of both diamond and c-BN as a function of temperature up to 1200°C under dynamic (300 MN/s) and static (4 N/S) conditions with a Knoop indentor. Their plotted data indicate a hardness of diamond of (70 -75) GPa at room temperature. Static testing values fell to approximately 20 GPa at 1200°C, while the dynamic value remained at 50 GPa at this temperature. The hardness of polycrystalline c-BN at room temperature is about one half

Table 7. Mechancial properties of natural diamond

Material Designation *	Hardness (GPa)	Young's Modulus (GPa)	Poisson's Ratio	Bulk Modulus (GPa)	Density (g/cm^3)	K_{1C}	Reference
Natural Diamond	-	-	-	435	-	-	Knittle et al. [1994]
Natural Diamond	-	1050	0.104	442	-	-	Grimsditch and Ramdas [1975]
Natural Diamond	550-880(?)	1250	-	-	-	-	Hoshino et al. [1989]
c-BN	49	-	-	-	-	-	Gielisse [1964]
c-BN crystal	-	-	-	369±14	-	-	Knittle et al. [1989]
c-BN polycrystalline (static loading)	0.5mm ~40 5mm ~37	-	-	-	-	-	Novikov et al. [1993]
c-BN polycrystalline (dynamic loading)	0.5mm ~42 5mm ~40	-	-	-	-	-	Novikov et al. [1993]
Natural, Ia (static loading)	~70 ~40 (800°C)	-	-	-	3.51	-	Novikov et al. [1993]
Natural, Ia (dynamic loading)	~75 ~60 (800°C)	-	-	-	3.51	-	Novikov et al. [1993]
Natural Diamond, Ia, IIa	~40 (1200°C)	-	-	-	3.51	~8 (Ia) ~5 (IIa)	Novikov et al. [1993]
Natural Diamond	57-104	-	-	-	-	-	Brookes [1970] Loladze et al. [1967]
Natural Diamond, IIa	56-102	1050	-	-	3.51	-	Savvides and Bell [1992]
Natural Diamond	79-102	900-1050	-	-	3.515	-	Bhushan et al. [1992]
Natural Diamond Single Crystal	55-113	1050	0.07	-	3.51	3.4-5.0	Partridge et al [1994]

*Designation as given in applicable reference

Table 8. Mechanical properties of diamond films

Material	Hardness (GPa)	Young's Modulus (GPa)	Poisson's Ratio	Bulk Modulus (GPa)	Density (g/cm^3)	K_{1C} (MPa m$^{1/2}$)	Reference
Diamond Film	80-100	500-533	-	-	-	-	Savvides and Bell [1993]
Diamond Film (MPECVD)	40-75	370-430	-	-	-	-	Bhushan et al. [1992]
Diamond Film (HFCVD)	75-111	-	-	-	-	-	Bull et al. [1992]
Diamond Film (HFCVD)	31-65	-	-	-	-	-	Beetz et al. [1990]
Diamond Film (CVD)	31-90	700-1079	0.07	-	3.5	~6	Partridge et al. [1994]
Diamond Film (CVD) (0-5% CH_4)	-	-	-	380-500	-	-	Sato and Kamo [1989]
Diamond Film (CVD)	-	891 (~820 @ 800°C)	-	-	3.5	-	Seino and Nagai [1993]
Diamond Film Bulk, Polycrystalline (BPCVD)	-	986-1079	0.1	-	-	6	Sussman et al. [1994]
Diamond Film (HFCVD)	-	250-1050	0.29	-	-	-	Aikawa and Baba [1993]

Table 9. Mechanical properties of diamond-like carbon (DLC) films

Material	Hardness (GPa)	Youngs Modulus (GPa)	Poisson's Ratio	Internal Stress (GPa)	Bulk Modulus (GPa)	Density (g/cm^3)	K_{1C} (MPa m$^{1/2}$)	Reference
a-C:H (hydrogenated)	10.5-14	95-130	0.3	1.5-3.5	-	-	-	Dekempeneer et al. [1992]
a-C-N-H	~10 (average)	-	0.2-0.4	1.5-5 (3 average)	-	-	-	Schwan et al. [1994]
a-C:H (diamond like)	-	127 (maximum)	0.2-0.4	0.4-4 (1.5 average)	196	2.0	-	Jiang et al. [1989a]
a-C:H (amophous carbon film), (CH_4)	11-33	90-220	0.22-0.39	-	~135	>12.0	-	Jiang et al. [1990]

Table 9. cont. Mechanical properties of diamond-like carbon (DLC) films

Material	Hardness (GPa)	Youngs Modulus (GPa)	Poisson's Ratio	Internal Stress (GPa)	Bulk Modulus (GPa)	Density (g/cm^3)	K_{IC} ($MPa\ m^{1/2}$)	Reference
a-C:H (amorphous carbon film) (CH_4)	13 (maximum)	140 (maximum)	-	-	-	>20	-	Jiang et al. [1989b]
a-C:H (PECVD sputtered)	7-14	35-136	-	-	-	1.7-1.8	-	Bhushan et al. [1992]
a-C:H (PECVD, r.f.)	33-35	~200	-	-	-	1.6-1.8	-	Bhushan et al. [1992]
a-C:H (PECVD C_2H_2)	5-40	-	-	-	-	-	-	Kleber et al. [1991]
a-C:H (CVD-C_2H_4, C_2H_2)	7-20	-	-	1-6	-	up to 2	-	Dworshak et al. [1990]
DLC	12-30	62-213	-	-	-	-	-	Savvides and Bell [1992]
DLC (hydrogenated amorphous carbon)	200-480 (?)	480-850 (?)	-	-	-	-	-	Hoshino et al. [1989]
DLC (pulsed vacuum arc discharged)	30	300	-	-	-	-	-	Koskinin et al. [1994]
DLC, a-(C) (amorphic diamond)	37	369	-	-	-	-	-	Davanloo et al. [1992]

of that of diamond, 40 GPa (see also Table 7 and Fig.3). The c-BN sample hardness dropped only about 25% at 1000°C but for the 5 m material it was reduced by some 60 %. In all cases the dynamic values were 5-25% higher that those obtained under static conditions.

Nalyetov et.al. [1978] found in 1978 that the hardness and fracture toughness of type Ia diamond are higher than for type II diamond at both room temperature and 1100°C. This was refuted by Brookes et al. [1990] who pointed to a higher dislocation density (see Table 6) in type IIa as the reason for its higher hardness. Novikov et al. [1965] have attempted to settle the issues by revisiting the topic. They remeasured the hardness, fracture toughness and dislocation mobility in both types of diamond at 1200°C. Microhardness measurements in vacuum yielded a hardness of 40 GPa as an average for both types. The K_{IC} values for type IIa (4.2-5.6) MPa·$m^{1/2}$ indicate it to be more brittle than Ia with K_{IC} values of (7.0-8.4) MPa· $m^{1/2}$, which is held to be more tough due to a higher dislocation mobility. Collaborating these findings may be the earlier work of Wilks and Wilks [1965] who found the abrasion resistance of IIa noticeably higher than that of type Ia. Diamond shows a brittle to ductile transition [1993] in vacuum at 1225°C.

The data collected on the hardness of diamond have been tabulated in Table 7.

4.3 THE HARDNESS OF DIAMOND THIN FILMS

Diamond thin films have been grown metastably from the vapor phase in low pressure environments by different deposition techniques. They all more or less involve a plasma growth enviroment. Depositions have been made on a variety of substrates each of which influence the resulting diamond film structure in a different way. As in most other multilayer thin film work, the choice of substrate material proves critical with reference to the desired results. So far, all diamond films have been polycrystalline and are normally columnar in structure. Single crystal epitaxial growth of diamond has not yet been accomplished except on itself or on very small grains of c-BN as substrates [Yoshikava et al. 1990].

It is often stated or possibly assumed, that diamond films have the same properties as single crystal diamond. This is not true. The macrostructure of the films, principally consisting of intergrown crystallites, stressed, with a high defect concentration, a possible void content and the presence of other phases (graphite, carbon) make for a different product. As with other materials, a wide range of diamond film types and qualities are available, resulting in products from opaque to nearly glass clear with thermal conductivities from 500 W/mK to 2000 W/mK and strength variations that cover several orders of magnitude.

Beetz et al. [1990] measured the hardness of diamond films generated with a hot filament CVD method, using an ultra low indentation technique. They obtained hardness values between 31 GPa for low partial pressures of CH_4 and 65 GPa at higher pressures. The latter figure would put their films at a hardness between that of c-BN and

Table 10. Hardness and moduli for diamond, diamond films, and DLC films.*

Types	Hardness H (GPa)	Young's Modulus E, (GPa)	Bulk Modulus, B_0 (GPa)
Diamond-like films	5-49	35-369	100-196
DLC	(21)	(180)	(145)
Diamond films	30-110	250-1050	380-500
	(70)	(730)	(440)
Natural diamond	55-113	900-1250	433-540
	(80)	(1050)	(485)

* Lowest to highest values for each category. Values in brackets represent averages for the data of tables 7, 8 and 9.

diamond. Natural diamond hardness has been reported over a wide range of values from around 50 to over 100 GPa, see Table 7.

Sussman et al. [1994] in work primarily involved with the determination of the bursting strength of very thick (179-300) μm diamond films, felt confident enough to use 10,000 kg/mm^2 for the hardness, in calculating the fracture toughness values of their films.

Partridge et al. [1994] in reviewing the status of diamond fibers tabulated the hardness of CVD films, apparently obtained from the open literature, as (31-90) GPa.

A tabulation of the mechanical properties of diamond films can be found in Table 8.

4.4 THE HARDNESS OF DIAMOND-LIKE CARBON (DLC) FILMS

From a physical property point of view, diamond is a rather well defined substance. This stands in sharp contrast to DLC type films and coatings which are not well defined substances. Their hardness characteristics depend on sp^2 to sp^3 bonding content, the various processing parameters in as many different growth techniques, the presence or absence of bonded hydrogen and nitrogen, the specific structure of the films and the surface character (roughness) to name a few. It is therefore relatively dangerous to evaluate the applicability of DLC films on the basis of one parameter, e.g. their hardness, only. It is, furthermore, well known that not only hardness, but also low wear resistance, low coefficients of friction, a relatively low stiffness and a high elasticity have created unique application areas for DLC films.

The currently available hardness values for a wide varity of DLC type films have been brought together in Table 9. In order to quickly compare DLC film hardness to those of diamond and diamond films, Table 10 is provided. The film descriptions attempt to describe the type of film and its origin, using as much as possible the nomenclature, expressions and descriptive terminology used by the original authors. The reader is referred to the references if more in-depth knowledge is required. Specific information considered essential in the interpretation of the data in Table 9 and 10 is given below.

Jiang et al. [1989] studied the deformation of a-C:H films for a wide variety of deposition parameters and methods, producing polymer-like films with tensile strain and diamond-like films with compressive strain. Their elastic and bulk moduli were calculated from measured shear moduli, G, and microindentation hardness measurements.

The work of Dworschak et al. [1990] showed that low power (20 W) r.f. coupled plasmas with ion energies amounting to only a few electron volts, also led to polymer-like films. Higher power (50-500 W) and biased substrates produced their "hard", a-C:H, films with ion energies up to a maximum of 200 eV.

Further researches by the Jiang et al. team [1990] could show that a high sp^3/sp^2 ratio was not sufficient to produce hard diamond-like films but that a low void content, which could be correlated with the filling factor, was even more important. The filling factor is defined as,

$$f = V_{OCC}/ V_{film} \qquad (7)$$

where f is the filling factor,
V_{film} is the total volume of an a-C:H film, and
V_{OCC} is the volume occupied by carbon and hydrogen atoms in the respective films.

Their hardness data decreased monotonically from 33 GPa to 11 GPa and the E-modulus from 220 to 90 GPa, as the C_2H_2 pressure was decreased. The bulk modulus value, 135 GPa, remained nearly constant.

The films produced by Kleber et al. [1991] in an r.f. discharge (13.56 MHz) discharge at 100 W, yielded the best Knoop hardness values, 40 GPa maximum, at the higher mean ion energies (250 eV) and the lowest hydrogen content (28 at.%). Overall hardness values ranged from 5 GPa to 40 GPa, with values increasing with a decreasing hydrogen content and an increasing mean energy.

The dc magnetron sputtered films and the r.f. PECVD a-C:H films produced by Bhushan et al. [1992], differed considerably in hardness. The plain sputtered films showed a hardness of (7-15) GPa with an average of about 12.5 GPa. The PECVD films were considerable harder at (33-35) GPa, to be compared to their diamond films at (40-75) GPa. As might be expected the sputtered and PECVD films contain sp^2 bonding contributions and a considerable amount of hydrogen. The diamond films contained only a small amount of hydrogen.

A cooperative effort by a Dutch and a Belgian team of Dekempeneer et al. [1992] produced a-C:H films with a capacitively coupled r.f. plasma assisted chemical vapor deposition (PACVD) set up. Nanoindentation measurements, maximum depth of 0.18 m on a 1.7 m thick film, produced a maximum hardness of 13.5 GPa. The primary objective of the work was finding a correlation between hardness and film stress.

Savvides and Bell [1992], have also stated that DLC can have extremely diverse properties. They determined that increasing hardness and modulus values correlate well with increasing energy per condensing carbon atom, to yield a hardness of a-C:H films of (12-30) GPa. This may be compared to the hardness of their diamond films of between 80 GPa and 100 GPa and the "accepted" values of H=56-102 GPa for type IIa diamond. Low energy ion-assisted magnetron sputtering of a graphite target in Ar-H_2 mixtures was used.

Depositions of DLC films at room temperature in a dc plasma of methane and hydrogen gas by Hoshino et al. [1989] showed different properties depending on substrate location. With the anode directly above the substrate cathode, Young's modulus of the film was found to be 480 GPa and with a "sideward" anode this became 850 GPa. The ultra high hardnesses quoted in the article appear out of line with all other values for DLC materials (Table 9).

Hydrogen free DLC coatings were deposited by a pulsed arc discharge method by Koskinen et al. [1994] yielding a layered columnar growth structure. Their hardness analysis resulted in a value of 30 GPa and a modulus of 300 GPa.

Laser plasma discharge by Devanloo et al. [1992] also produced hydrogen free diamond-like material called "amorphic" films. Averaging over twenty separate measurements, the hardness of the 5 m thick films was found to be 37 GPa with a Young's modulus of 369 GPa.

The addition of nitrogen to the CH_4 growth chamber by Schwan et al. [1994] produced amorphous hydrogenated carbon-nitrogen films, a-C:H:N. The presence of nitrile groups was found to reduce hardness as well as stress in the film with increasing nitrogen content. Hardness drops from a value of approximately 17 GPa to about 7 GPa at a 1:1 ratio of N_2 to C_2H_2. Their nitrogen free DLC's show a hardness of (6-17) GPa, depending on gas pressure.

Specific comments with reference to Young's modulus values for DLC films do not appear necessary since most of them were derived from hardness data, more specifically from the slope value of the unloading curve in the load-displacement plot, dP/dh in which P=load (MN) and h the displacement (nm).

5. The Toughness of Diamond

The ability of a material to absorb (impact) energy without failure (fracture) is known as its toughness. It is a characteristic of critical importance in many diamond and diamond system applications. In high speed impact processes such as grinding, the ability to survive high frequency (10-100 kHz) high impact energy pulses at simultaneous particle-workpiece interface temperatures of 1000 -1500°C, is for the greater part determined by diamond's toughness.

Toughness in materials is expressed and quantified in a variety of ways. One approach is through the modulus of rupture (resilience), expressed as,

$$U_r = \int_0^{\varepsilon} \sigma d\varepsilon \tag{8}$$

If one assumes linear elastic behavior, for which $U_r = 1/2\sigma_y\varepsilon_y$ and $\sigma = E\varepsilon$, the modulus of rupture, MOR, becomes,

$$U_r = MOR = \sigma^2/2E \tag{9}$$

where σ is the material's strength,
E is the Young's modulus and
ε_y is the strain.

The MOR expresses the energy stored per unit volume. Numerically it is equivalent to the area under the stress-strain curve. This dimensionless quantity is often used for material comparison purposes. Its value can be determined only after the constitutive properties, Í, E, and/or the strain, ε_y, are known. The MOR values for industrial diamond depend on structure (diamond type) and chemical constitution, but are in any case lower to significantly lower than for gem quality single crystal diamond. Actual values are not generally available, at least not from the open literature.

The fracture toughness parameter, K_{IC}, evaluates a material on the basis of its ability to prevent crack propagation on energy impact. It is expressed, in units of $MPa.m^{1/2}$, as,

$$K_{IC} = y\sigma(\pi a)^{1/2} \tag{10}$$

where a is the half crack size,
y is a dimensionless parameter, with a value near unity, depending on sample and crack geometry.

The K_{IC} values can be determined experimentally or can be derived, directly or indirectly, from a knowledge of other mechanical and physical properties. Only a few determinations of the K_{IC} values for diamond and diamond films have been made, see Tables 7-9. From a numerical point of view these K_{IC} values characterize diamond as a brittle material. Diamond's value may be compared to those for engineering ceramics at an average of 5, for WC-Co at 12, steel at 60 and around 100 for titanium alloys.

The toughness of a material is also expressed as,

$$G_C = K_{IC}^2/E(1-\nu) \tag{11}$$

where G_C is the toughness,
K_{IC} is the fracture toughness parameter,
E is the Young's modulus and
ν is the Poisson ratio.

A more practical and more easily measured quantitative index of diamond's "toughness" is found in the Grinding Ratio (GR) as used in stock removal applications, in which some 75% of all industrial diamond is used. It is defined as

$$GR = \frac{\Delta V_{wp}}{\Delta V_w} = \frac{\textit{volume removed from the workpiece}}{\textit{volume worn away on the grinding wheel}} \tag{12}$$

Table 11. Synthetic single crystal diamond and its applications.*

Type	Morphology shape	Product	Application	Materials
I	Irregular Friable	Grinding wheels Polymer and vitreous bonds	Grinding	Tungsten carbide Non-ferrous metals
IA	As under I Metal coated (Ni)	Grinding wheels Polymer bonds	Wet grinding	Tungsten carbide Metallics
IB	As under I Metal coated (Cu)	Grinding wheels Polymer bonds	Dry grinding	Tungsten carbide
II	Blocky crystals Smooth surface Medium friable	Grinding wheels Metal bonds	Grinding	Non-metallics Metallics
IIA	As under II	Plated tools Metal bonds	Grinding	Non-metallics Metallics
IIB	Slightly elongated crystals Medium friable	Grinding wheels Resin and metallic bonds	Grinding	Metallics
III	Tough, blocky crystals Smooth surface	Saw blades Metal bonds	Sawing Grinding, Shaping	Non-metallics Technical ceramics (Natural) stone
IV	Shape ungraded Very fine size graded products	Loose abrasive polishing wheels Compounded "pastes"	Polishing Lapping Honing	Non-metallics Metallics

* Representatives listing of the most common types only.

Since the ability of a grinding particle to survive high energy impact is not only related to its physical properties, but is also influenced by the properties of the material in which it is held and by certain operational and enviromental conditions, the GR can only be indirectly related to "physical toughness" as defined above. In otherwise comparable situations, it can be regarded as a relative or applied toughness parameter. Values for the GR typically range from 10-100 for demanding (or inefficient) operations and can reach 1000 and up for low energy (high efficiency and low cost) grinding applications. The GR approach defines an applied "dynamic toughness", while the MOR and K_{IC} expressions derive entirely from physical and static premises.

A fourth evaluation of diamond's toughness can be found in the even less available and quantifiable property known as the "friability index". Manufacturers grade their diamond product, in addition to size and shape, on the basis of its ability to withstand impact i.e., its friability. Evaluations of this type are carried out in more or less standardized test procedures, such as in specially designed ball mills. Such methods tend to fracture (comminute) the weaker particles setting a lower impact resistance level for the product. They do, however, fall (very) short of duplicating stress conditions in actual application environments. The operation is, as was in effect initially intended, more an industry internal sorting method and tool than a useful quantifying toughness or friability index. In any case, numerical data are not commonly made available.

Testing of diamond toughness has more recently resulted in other figures of merit such as the toughness index (TI) and the thermal toughness index (TTI). These tests only effect the weak particles and do not provide information on strength or toughness. The General Electric Company recently announced yet another test method, said to be fast and highly accurate, yielding a compressive fracture strength (CFS).

It will be clear that optimal values for the physical properties in Eqs. 10, 11 will lead to high toughness. This is reflected in first instance in the diamond crystal quality. The photomicrographs of Figs. 4 through 9, illustrate the diamond types (categories) described in Table 11. They are representative of the major commercial industrial diamond grades, designed to optimally serve the majority of all cutting and grinding application areas.

Category I, Fig.4, represents a product from a high growth rate synthesis process, resulting in morphologically ill- defined crystals, heavily intergrown and of high void content. It is a low impact resistance, low toughness product. It performs particularly well in the grinding of high MOR non-ferrous materials such as cemented tungsten carbides. The material's performance can be enhanced by (electroless) coating with nickel, Fig.5, to increase the apparent or operational toughness, or with copper to enhance heat transfer in dry grinding operations, see Fig.6.

Category II, Fig.7, represents a medium friability, higher toughness particle, which is morphologically better developed, has a pseudo-cubooctahedral shape and is of higher strength. One of its application areas is the stock removal of medium to high MOR technical ceramics or glasses. Further selection results in sub categories, such as shown in Fig.8.

Category III, illustrated in Fig.9, clearly represents a strong material with highly developed cubooctahedral shape and sharp "cutting edges" (facet intersections). Crystals of this type applied in cutting and sawing operations of natural and synthetic stone, concrete and other high MOR inorganic (ceramic) materials.

6. Tribological Properties of Diamond.

The low frictional and wear properties of a polished natural diamond surface, in contact with itself and with many types of other materials, is well known. The coefficient of friction (COF) of diamond is generally quoted at below 0.1. Many research efforts over the last thirty years [Seal 1979, Bushan and Gusta 1991] have in essence come to the same conclusion. From an application point of view, the developing diamond film and diamond-like film areas are currently of greatest interest in this context.

As-grown diamond films, polycrystalline in nature as a result of random nucleation on various types of substrates, display surface roughnesses ranging from fractions of a micron to tens of microns. Such as grown surfaces are not generally suitable for tribological applications. They display high friction and wear characteristics. Growing smooth diamond films or polishing the rougher surfaces are the solutions that suggest themselves, although one would still have to show low wear and friction characteristics with such altered diamond surfaces.

Fig. 4. Category I, low impact strength diamond.

Fig. 5. Category I, low impact strength diamond particles coated with electroless nickel.

Fig. 6. Category I, low impact strength diamond coated with electroless copper.

Fig. 7. Category II, medium impact strength diamond, blocky morphology.

Fig. 8. Category II, medium impact strength diamond shape selected to render elongated particles.

Fig. 9. Category III, high impact strength diamond. Note blocky, cubooctahedral morphology, displaying many strong cutting edges.

Polishing of diamond films has been attempted by diamond abrasion [Hickey et al. 1991], thermomechanical polishing [Tokura et al. 1992], chemomechanical polishing [Hirata et al. 1992] and technologies which are basically sputtering methods via laser, ion beam or charged particle impacts [Hirata et al. 1992, Bovard et al. 1992, Bouliev et al. 1991]. Abrasion alone is a slow process and, given the nature of the diamond-diamond impact, leads to surfaces reflecting the nature of brittle material failure through spalling and cracking. Chemomechanical polishing is normally resorted to, particularly if elevated temperatures (thermal polishing) can not be tolerated. The photon based techniques are slow-rate processes and may not be suitable in all situations. They also require expensive specialized equipment and are line-of-sight processes. The exact details of the various chemomechanical polishing processes are, like so many aspects of the diamond industry, "proprietary". A few recipes are, however, available [Gielisse 1992].

The careful work of Bushan et al. [1993] who appear to be the first ones to have conducted systematic measurements of friction and wear on polished and unpolished diamond films (not amorphous or DLC films), also indicate that as-deposited, coarse-grained diamond films are not suitable for tribological applications. They need to be polished to produce low friction and wear. Chemomechanical and other polishing processes can reduce film roughness from as high as 650 nm (nominal) to 170 nm and 9.7 nm for chemomechanically polished and laser polished films respectively. In order to obtain reductions of the coefficient of friction from about 0.4 (as- deposited, coarse grain films) to 0.09 (polished, coarse grain films) the rms roughness had to be reduced to below 170 nm, but not necessarily into the single digits. The dominant effects in reducing the coefficient of friction to attain low wear are, the rounding of sharp asperities, the reduction in the asperity slope value and a reduction of asperity height (surface roughness). The surface roughness of the non-diamond member should be less than that of the mating diamond. It was also reported that moisture content and the presence of argon, nitrogen and air do not significantly influence the friction of diamond films. The presence of oxygen does, however, increase the coefficient of friction by about 20%.

M.L. Languell et al., [1994] reported in a recent research summary on the morphological response in linear cyclic dry sliding of polycrystalline diamond films (PDF) on PDF, (cylinder on flat) in a dry nitrogen atmosphere. They concluded that there is a break-in period in tribotesting, in which a very high COF (~0.65) decays to a steady state value (~0.25) within the first twenty cycles (80 cm) of sliding. The asperity tips are altered via a brittle fracture process, to yield a uniform and smoother surface. This break-in process continues till the contact areas are large enough to reduce the stress level to below the fracture limit. Wear track morphology, revealing a modular structure, appears to support the possibility of a tribochemical reaction, although significant graphitization or oxidation is not supported by the results of debris analysis. COF values for PDF on PDF films lower than the "steady state value" of 0.25, were not reported.

Gardos and Sorano [1990] have noted that when diamond film surfaces can be terminated with hydrogen atoms, such as in non-heated and non-oxidizing environments, the COF against bare SiC and diamond-coated SiC pins, is of the order of

0.1. At elevated temperatures (850 °C) oxygen and moisture are absorbed and the value of the coefficient of friction increases to 0.8.

In diamond on diamond film experiments in air, Feng and Field [1990] obtained frictional coefficients of 0.12-0.14, about ten times higher than those obtained in (single crystal) diamond on diamond experiments. Immersion in water decreased the coefficient of friction value to between 0.004 and 0.06. Significant is their finding that surface roughness appreciably influences tribological behavior. The correlation between roughness and COF was found to be linear with the smoothest diamond films giving the lowest values, regardless of the quality of the diamond within the films!

Bull et al. [1992] could also show that the coefficient of friction of diamond films is more or less linearly dependent on surface roughness, varying between 0.1 and 0.6 as roughness, R_a, (arithmetic mean roughness) increased from 0.05 m to 0.35 m.

The frictional behavior of diamond films is essentially independent of film quality, but is very dependent on surface character. The smoother the topology the lower the coefficient of friction and the wear debris volume. Regretfully a definition of "smoothness" or how it is measured is not normally provided, making result comparisons virtually impossible. Reliable and efficient non-contact and on-line surface characterization methods are, however, available [Tu and Gielisse 1995] and should be made part of any quantitative evaluation of the frictional properties of surfaces in relative motion.

7. Materials Potentially Harder than Diamond

The outstanding mechanical properties of diamond have over many years stimulated the search for substances with characteristics surpassing those of diamond. Such searches have been conducted in spite of apparent bond strength and structural limitations.

The first hopes were pinned on polymorphs of hard materials including diamond, such as in the wurtzite type high density carbon, lonsdaleite. This material does, however, display properties very close to those of diamond.

A more recent contender has been a possible analogue to silicon nitride (Si_3N_4) in the form of carbon nitride (C_3N_4). The possible occurence of such a material was apparently first suggested by Sung [1996]. He subsequently interacted with others, whose theoretical considerations and calculations predicted the existence of a nitride of carbon C_3N_4, [Cohen 1985; Liu and Cohen 1989, 1990] held to have a series of excellent physical properties including a hardness and bulk modulus higher than that of diamond [Liu and Cohen 1989; Haller et al. 1992]. Rather extensive experimental investigations [Badzian et al. 1995] have so far failed to produce crystalline C_3N_4, although a turbostratic graphite-like soft phase, intermixed with a small amount of hard (sub) micron sized particles which scratched diamond, did result [Badzian, A. 1995]. There have been numerous attempts by other teams as well, see e.g. the literature references in [Kreider et al. 1995]. It has been possible to generate amorphous carbon nitride coatings with around 40% nitrogen in the structure [Kreider et al. 1995; Chen et al. 1995].

Hardness of these films, measured with an ultra low load indentor (Berkovitch three-sided diamond pyramid indented tip), was given as 10-12 GPa. The films were visually transparent and stable up to 630 °C. The Raman spectrum showed two broad bands with maxima at 1550 and 1350 cm^{-1}, apparently belonging to the =C=C= carbon chain frequency and the =C=N--vibration frequency.

Hardness correlates readily with the elastic moduli, particulary the Young's modulus, $E=\sigma/\varepsilon$, which characterizes and quantifies a material's stiffness. The bulk modulus, K, (the inverse of compressibility, $\kappa=1/V\ dV/dP$), indicating a material's incompressibility can act in a similar way, see Fig 10. Recent pressure-volume measurements [Leger et al. 1994] on high pressure (12-42) GPa phases of ruthenium oxide (RuO_2) and hafnium oxide (HfO_2) indicated bulk moduli of 399 GPa and 553 GPa respectively, close to that of diamond (433 GPa). These metastable phases would therefore be expected to be harder than c-BN while the (third and most dense) high pressure polymorph of HfO_2 would be harder than diamond. Chemical or thermal stabilization of those phases down to room temperature and pressure would be required to verify the anticipated hardness and to put such materials to good use.

As stated earlier, it appears that to-date (1996) and in spite of many attempts by as many groups over some eight years, there is no unambiguous evidence that a harder than diamond C_3N_4 compound or any other phase in the binary system C-N exist. Recent investigations by Badzian [Badzian et al. 1995, Badzian, A. 1995] who attempted the synthesis of the carbon nitride phase in a microwave plasma reactor in the form of thin films deposited on single crystal diamond as the substrate, showed the product to consist of a soft phase and a hard phase. The soft phase was analyzed as having a graphite structure of unknown stacking order, while the hard phase, in the form of particles less than 5 μm and as yet unidentified, did scratch diamond. It should be pointed out that diamond can scratch diamond in certain crystallographic orientations and that even c-BN can scratch diamond.

Recent developments in a materials group known as the "fullerenes" appear to provide evidence of hardness equal to or surpassing that of diamond. These carbonaceous materials are best described as "spherical cage carbon clusters" or as "molecular polyhedral carbon". The predominant phase is C_{60}, although others such as C_{70}, C_{76} and C_{80} also exist. An example of the C_{60} structure is shown in Fig.10. The existence of related molecular and polymer carbon forms as "polyhedral clusters" or "tubular graphite" is currently being explored.

The interest in the mechanical properties of dense carbon fullerene structures lies in the fact that bulk samples of C_{60}, display unusual hardness when submitted to simultaneous pressures (between 6 and 40 GPa), temperatures (up to 1500°C) and shear conditions in various types of high pressure and high temperature apparatus [Blank et al. 1990, 1994; Kuzin 1986] The super hardness of these polycrystalline materials has been described as "...at least above cubic boron nitride hardness" [Blank et al. 1995] and as to the particles as "...easily scratch tungsten carbide alloy, sapphire and even cubic BN " [Blank et al. 1994]. The simultaneous shear deformation and high pressure (up to 6 GPa) on C_{60} led

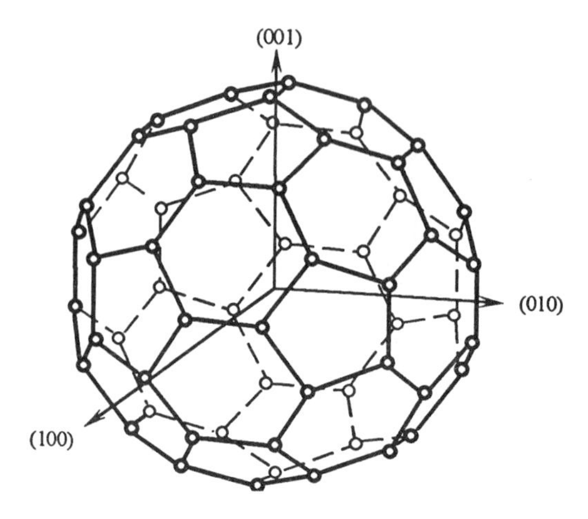

Fig. 10. Geometry and directions in a typical fullerene, C_{60}, cluster [Saito and Oshiyama 1991].

Fig. 11. Diamond crystals synthesized at high pressure and temperature conditions from a fullerene based starting mixture [Vul 1995].

to "high abrasivity" and scratching of the tungsten carbide press anvils, attributed to the formation of phase transformed C-60 into "super hard C-60".

Fullerenes have also been used as the "starting material" in diamond, synthesis, using the standard high pressure-high temperatrure conditions, [Requero et al. 1992, Ruoff and Ruoff 1991]. At about 4 GPa and 1000°C, fullerene-containing soot material (actual fullerene content less than 10%) has yielded diamond crystals varying in size between 100 and 800 micron, as illustrated in figure 11. Note the well developed crystal morphology [Vul 1995]. Although no property data have been reported, it can be assumed that diamond formed from a fullerene source material will have intrinsic mechanical properties very similar or identical to those of single crystal diamond.

There is therefore, at present, no evidence for the existence of a useful crystalline phase harder than diamond.

8. A Hardness Scale for Engineering Materials.

The various hardness testing methods invoke the stress-strain behavior of the material, particularly its tensile character. One should therefore expect a relationship between hardness and Young's modulus in (ultra) hard brittle materials. In turn, the bulk modulus for simple isotropic solids is related to the Young's modulus by the well known equation,

$$K = E/3(1-2\,\nu) \tag{13}$$

where K is the bulk modulus for a simple isotropic solid,
E is the Young's modulus and
ν is the Poisson ratio.

Figure 12 shows plots of hardness versus both moduli. The only solids which are available in single crystal form and for which more or less reliable moduli and hardness figures are available, are diamond, cubic boron nitride and alumina (sapphire) and possibly AlN and SiC. The straight lines are drawn using this data set. The scatter of the points at the lower end is attributable to the polycrystalline nature of the materials and the less-than-reliable data for the hardness or the moduli. A refinement of this data and thus the attitude of these curves, will have to await further input on the physical properties of the binary materials as single crystals.

Leger et al. [1994] have recently used the hardness-versus-bulk modulus relationship in support of materials potentially harder than diamond in cases where polymorphic forms at ultra high pressure indicated a bulk modulus higher than that of diamond, see also section 7.

In the meantime the E-H and possibly the K-H function (Fig.12) can act as a hardness scale, allowing estimation of hardness from a knowledge of the constitutive properties and vice versa. It will be noted that the E-H scale has a convenient 0.1 slope and that the hardness of technically important materials runs from essentialy 10 GPa to 100 GPa. The attitudes of the curves for both moduli below 10 GPa have not yet been fully worked out. The very steep slope of the K-H curve, about four times steeper than the E-H curve, makes it less discriminating, particularly for the important binary solids. It would also exaggerate hardness values estimated from it. A direct relationship between stress-strain type hardness and volumetric lattice hardness already exists [Plendl and Gielisse 1962], which makes the possibility of establishing a fundamentally based practical hardness scale a reality. Present identation hardness testing does not normally involve hydrostatic restrained deformation (the bulk modulus, K, strictly describes a solid's volume change due to hydrostatic pressure only) which makes a workable correlation between indentation hardness and bulk modulus less likely.

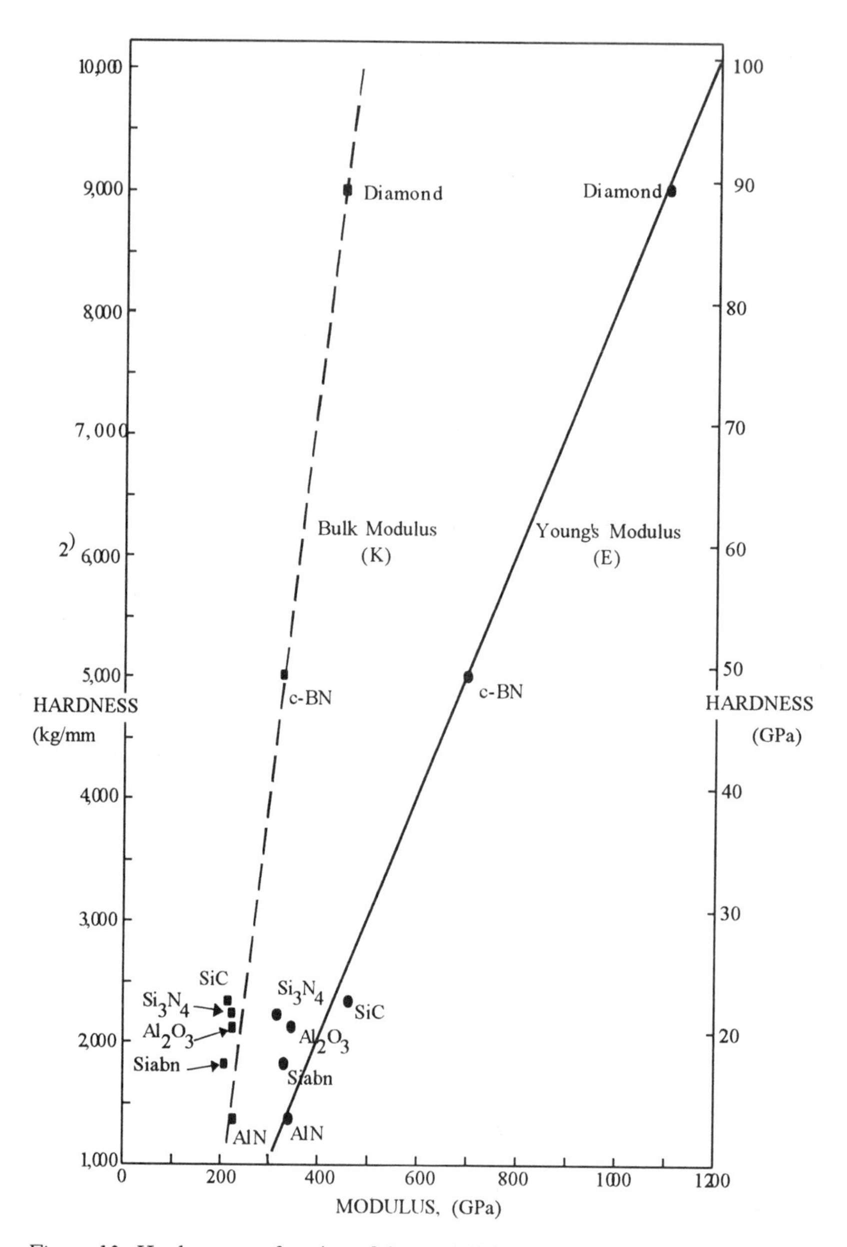

Figure 12. Hardness as a function of the moduli for materials of technological significance.

The benefits of a hardness scale for technically significant materials running from a Young's modulus of (10-100) GPa (essentially1000 to10,000) kg/mm^2 and potentially tied to an energy-volume approach, must be clear. Li and Bradt [1992] have recently argued the need for a fundamental change in hardness testing, away from the stress based presentations and towards an energy per volume approach. Such a scale would also allow for variations in hardness with direction, i.e., elastic and plastic anisotropies which are known to exist in single crystals (diamond, c-BN, sapphire). The Young's modulus in diamond in the <111> was found [McSkinnen and Bond 1957] to be 1196 GPa, about 15% higher than in the <100> at 1050 GPa. The existence of hardness anisotropy in diamond has been the subject of controversy for many years. The author can only suggest seeking the advice of diamond cutters and polishers who will quickly support anisotropy (not just related to the well known easy cleavage directions) and point out the virtually impossible to polish "naats" (nodes) in single crystal diamond.

The problems associated with creating an accurate hardness expression in terms of physical properties, can be exemplified by the recent work of Klein [1992]. He challenged the validity of setting Young's modulus of diamond at E=1050 GPa and its Poisson ratio as ν=2, on the grounds that it ignores the well known anisotropic behavior of E and that it is outright incorrect with reference to the Poisson ratio. The E=1050 GPa value is a minimum and only holds for principal stress. Stress direction, intrinsic anisotropy and crystalline configuration should be taken into account, i.e., directionality equations should determine the actual values of the moduli and the Poisson ratios. Calculations using a knowledge of the elastic constants of diamond yield E=1164 GPa (ν =0.079) in the fully isotropic octahedral plane <111> with peak values of E=1210 GPa along the (111) direction. The very small Poisson's ratio indicates a highly anisotropic volume for diamond.

Similar considerations [Klein 1992] have led to predictions of the values for polycrystalline (CVD-PVD) diamond films, described as "randomly oriented densely packed aggregates of diamond grains", for which it is assumed that the grain boundaries do not impact on the elastic behavior of the aggregate . The use of equations relating E and to K and G,

$$E=9KG/(G+3K) \tag{14}$$

where E is the Young's modulus,
K is the bulk modulus and
G is the shear modulus

and,

$$\nu = (3K/2 - G)/(G+3K) \tag{15}$$

produced values for the shear modulus, G=535 GPa and bulk modulus, K=442 GPa. The calculated value for E resulted in 1143 GPa with ν=0.0691.

For comparison purposes, we can mention results of ultrasonic velocity measurements [Saito and Oshiyama 1991] at 20 MHz (shear wave) and 50 MHz (longitudinal wave) for a total of 50 measurements on six samples of optical grade CVD films (thermal conductivity 13-15 W/cmK) which have yielded E_{ave}=1180 GPa, K_{ave}=558 GPa, G_{ave}=514 GPa and κ_{ave}=0.148, which are quite similar to values obtained by Grimsditch and Ramdas for natural diamond Grimsditch and Ramdas 1975] as far back as 1975.

We can conclude that calculated results do not significantly depart from the experimental results. Reported experimental data for E vary, however, between 1050

GPa and 1210 GPa which would constitute a significant difference in hardness of some 1000 kg/mm^2.

9. References

Aikawa, Y., and Baba, K. (1993) *Jpn. J. Appl. Phys.* 32 (10), 4680

Badzian, A. personal communication, August (1995)

Badzian, A., Badzian, T. and Drawl, W. Synthesis of Nitrides from Microwave Nitrogen Plasma, Applications of Diamond Films and Related Materials: (1995) Third Intern. Conf. Processing, A. Feldman et al. editors, Gaithersburg, MD, USA. 839-843

Beetz Jr, C.P., Cooper, C.V. and Perry, T.A. (1990) *J. Mater. Res.,* 5 (11), 2555

Bhushan, B., Kellock, A.J., Cho, N. and Ager, J. (1992) *J. Mater. Res.*,7 (2), 404

Bhushan,B., Subramaniam, V.V., Malshe, A., Gupta, B.K. and Ruan, J. (1993) *J. Appl. Phys.*, 74 (6), 417

Blank, V.D., Buga, S.G., Dubitskiy, G.A. and Popov, M.Yu. (1994) Physical Properties of Superhard Samples Created from Solid under Pressure up to 12 GPa and Temperatures up to 1000°C, Ibid

Blank, V.D., Buga, S.G., Popov, M.,Yu. and Dubitskiy, G. (1995) Synthesis of the Superhard Materials from C-60 in: Abstract Booklet, Workshop Fullerenes and Atomic Clusters, St. Petersburg, Russia

Blank, V.O. et al. (1990) *Pribory i Technika Experimenta* (in Russian) 2, 186-190

Blank, V.O. et al. (1994) *Jurnal Technicheskoi,* (in Russian) 64, 8, 153

Bouliev, L.L., Chapliev, N.I., Konov, V.I., Pimenov, S.M. ,Smilin, A.A. and Spitsyn, B. (1991) *J. Electrochem Soc.*, 91, 357

Bovard, B.G., Zhao,T. and MacLeod, H.A. (1992) *Appl. Opt.*, 31, 2366

Brookes, C.A. (1970) *Nature*, 228, 660

Brooks, C., Brooks, E.J., Howes, V.R., Roberts, S.G. and Waddington, C.P. (1990) *J. Hard Mater.*, 1, 3

Bull, S.J. and Richarby, D.S. (1990) *Surf. Coat. Technol.*, 42, 149

Bull, S.J. and Rickerby, D.S. (1989) *British Ceram. Soc.Trans J.,* 88, 177

Bull, S.J., Chalker, P.R. and Johnson, C. (1992) *Mater. Sci. and Tech.*, 8, 679

Burnett, P.J. and Rickerby, D.S. (1987) *Thin Solid Films,* 148, 41-50

Burnett, P.J. and Rickerby, D.S. (1987) *Thin Solid Films,* 148, 51

Bushan, B. and Gusta, *B.K.* (1991) *Handbook of Tribology: Materials, Coatings and Surface Treatments.,* McGraw Hill, New York

Cardinale, G.F. and Robinson, C.J. (1992) *J. Mater. Res.*, 7 (6), 1432

Chen, D.G., Lu, F.X., Yang, R., Yu, D.W., Sun, Q.B. and Song, B. (1995) *A Study of Magnetron Sputtered* CN_x *Films,* in: Third Intern. Conf. Propcessing, A. Feldman et al., editors, Gaithesburg, MD, USA. 839-843

Cohen, M.L. (1985) *Phys. Rev. B*, 7988-7991

Davanloo, F., Lee, T.J., Jander, D.R., Park, H., You, J.H. and Collins, C.B. *(1992) J. Appl. Phys.* 71 (3), 1446

Dekempeneer, E.H.A., Jacobs, R., Smeets, J., Meneve, J. and Eersls, L., Blanpain,B., Ross, B.J. and Oostra, D.J. (1992) *ibid.*, 217, 56

DeVries, R. C. (1972) Cubic Boron Nitride: Handbook of Properties, General Electric Internal Report No., 72 CRD 178.

Dworschak, W., Kleber, R., Fuchs, A., Scheppat, B., Keller, G., Jung, K. and Ehrhardt, H. (1990) *Thin Solid Films*, 189, 257

Feng, Z. and Field, J.I. (1990) *Proc. 40th Annual Diamond Conference,* Reading, England

Field, J. E. (1992) *Properties of Natural and Synthetic Diamond* (ed. J.E. Field), London, Academic Press, 474

Field, J.E. ed. (1979) *The Properties of Diamond,* J.E. Field, Academic Press London, 648

Field, J.E., Nicholson, E., Steward, C.R. and Feng, *Z.* (1993) *Philos. Trans. Roy. Soc.*, A,342, 261-275

Gardos, M.N. and Soriano, B.L. (1990) *J.Mater.Res.*, 5 (11), 2599

Gielisse, P.J. (1962). See also Plendl and Gielisse [1962].

Gielisse, P.J. Calculated from the principal infrared (T.O.) frequency and other physical properties (1964)

Grimsditch, M.H. and Ramdas, A.K. (1975) *Phys. Rev.*B, 11, (8), 3139

Gupta, B.K., Malshe, A., Bhushan, B. and Subramaniam, V.V. (1994) *J. of Tribology,* 116, 445-453

Haller, E.E., Cohen, M.L. and Hansen, W.I. (1992) *U-S Patent* 5, 110, 679

Hickey, C.F. Thorpe, T.P. Moorish, A.A. Butler, J. Vold, C. and Snail, K.A. (1991) *SPIE Diamond Optics* IV 1534, 67

Hirata, H., Tokura, H. and Yoshikawa, M. (1992) *ibid*, 212, 43

Hoshino, S., Fujii, K., Shohata, N., Yamaguchi, H., Tsukamoto, Y. and Yanagisawa, M. (1989) *J. Appl. Phys.* 65 (5), 1918

Jiang, X., Reichelt, K. and Stritzker, B. (1989) *J. Appl. Phys.* 66, 5805

Jiang, X., Reichelt, K. and Stritzker, B. (1990) *J. Appl. Phys.*, 68 (3), 1018

Jiang, X., Zou, J.W., Reichelt, K. and Grunberg, P. (1989) *J. Appl. Phys.*, 66 (10), 4729

Kleber, R., Jung, K. and Ehrhard, H., Muhling, I.,Metz, K., Engelke, F. (1991) *Thin Solid Films.*, 205, 274

Klein, C.A. (1992) *Mat. Res. Bull.*, 27,140

Knittle, E., Wentzcovitch., R.M., Jeanliz, R. and Cohen, M.L. (1989) *Nature*, 337, 349

Koskinen, J., Hirvonen, J.P and Hannula, S.P., Pischow, K., Kattelus, H. and Suni, I. (1994) *Diamond and Related Materials*, 3, 1107

Kreider, K.G., Tarlow, M.J., Robins, L.H., Marinenko, R.B. and Smith, D.T. (1995), *Ion Beam Assested Sputtered Carbon Nitride Films,* in Application of Diamond Films and Related Materials: Third Intern. Conf. Processing, A. Feldman et al., editors, Gaithesburg, MD, USA. 839-843 865-868

Kuzin, N.N. et al. *(1986) Doklady Akademii Nauk USSR* (in Russian), 6,1391-1394

Languell, M.L., George, M.A., Wert, J.J. and Davidson, J.L. (July *1994) J. of Metallurgy,* 66

Leger, J.M., Haines, J. and Blanzat, B. (1994) *J. of Mat. Sci. Lett,* 13, 1688

Li, H. and Bradt, R.C. (1992) *Diam Rel. Mat.* 1, 11

Liu, A.Y. and Cohen, M.L. (1989) *Science*, 245, 84

Liu, A.Y. and Cohen, M.L. (1990) *Phys. Rev. B*, 41, 10727

Loladze, T.N. , Bokuchava , G.V. and Davydova, G.A. (1967) *Ind. Lab.*, 33, 1187

McSkinnen, H.J. and Bond, W.L. (1957) *Phys. Rev.*, 105, 1

Meyer, E. and Disch, Z. (1908) *Ing.*, 52, 645-654

Nalyetov, A.N., Klyuer, Yu.A., Grigoryev, O.N., Milman, Yu.V. and Trefilov, V.I. (1978) *Rep. USSR Acad. Sci.* 246, in Russian

Novicov, N.N., Dub, S.N. and Malnev, V.I. (1993) *J. Hard Mater.*, 4, 9

Novikov, N.N., Sirota, Yu.V., Mal'nev, V.I. and Petruha, I.A. (1993) *Diamond and Related Materials,* 2, 1253

Partridge, G.P., May, P.W. and Ashfold, M.N. (1994) ibid., 10, 177

Partridge, P.G., May, P.W., Rego, C.A.and Ashfold. M.N.R. (1994) *Materials Science and Technology,* 10, 505

Plendl, J.N. and Gielisse, P.J. (1962) *Phys. Rev.*, 124, 828-832

Requero, N.N.,Monceau, P.M.,Hodeau, J.L. (1992) *Nature* 355 , 23

Ruoff, R.S. and . Ruoff, A.L. (1991) *Nature* 350, 663

Saito, S. and Oshiyama, A. (1991) Figure adapted from *Phys. Rev. Lett*, 66, 20, 2637

Sato, Y. and Kamo, M. (1989) *Surf. Coat. Technol.*, 39-40, 183

Savvides, N. and Bell, T.J. (1992) *J.Appl. Phys.*, 72 (7), 2791

Savvides, N. and Bell, T.J. (1993) *Thin Solid Films,* 228, 289

Schwan, J., Dworschak, W., Jung, K. and Ehrhardt, H. (1994) *Diamond and Related Materials,* 3, 1034

Seal, M. (1979) *Philos. Mag.* A, 43 ,587

Seino, Y. and Nagai, S. *J.* (1993) *Mater. Sci. Lett.*,12, 324

Sung, C.H. *Mat. Phys. Chem.*, to be published, Jan-Feb (1996)

Sussman, R.S., Brandon, J.R.,Scarsbrook, G.A., Sweeney, C.G., Valentine, T.J.,Whitehead, A.J. and Wort, *C.J.H.* (1994) *Diamond and Related Materials,* 3, 303

Tokura, H., Yang,C.F. and Yoshikawa, M. (1992) *Thin Solid Films,* 212, 49 51. See e.g. reference 2

Tu, M. and Gielisse, P.J. (1995) Application of Optical Scattering in Multichip Module Processing, SPIE's International Symposium on Optical Science, Engineering and Instrumentation, 2541-09, San Diego

Vul, A.Ya. (1995) Joffe Physicotechnical Institute Russian Academy of Sciences, St. Petersburg, Russia, Photo courtesy.

Willks, E.M. and Wilks, J. *R.* (1965) *Physical Properties of Diamond,* in R. Berman (ed.), Clarendon Press. Oxford, 221

Wooddell, C.I., (1935) *J. Electrochem. Soc.*, 68, 111

Yoshikava, M., Ishida,H., Ishitanis, A., Murakami, T., Kaizumi, S. and Inuzuka, T. (1990) *Appl. Phys. Letters*, 57, 428

Chapter 4

SURFACE PROPERTIES OF DIAMOND

Mark P. D'Evelyn

General Electric Corporate Research and Development, P.O. Box 8, Schenectady, NY 12301 USA and Department of Materials Science and Engineering, Rensselaer Polytechnic Institute, Troy, NY 12180-3590 USA

Contents

1. Introduction

The present chapter focuses on the surface properties of diamond, with emphasis on the structure and chemistry of individual crystal faces. Several aspects of the properties of diamond interfaces are treated in other chapters, including the surface band structure and negative electron affinity surfaces in Chapter 1 and metal adsorption and the formation of Schottky barriers and ohmic contacts in Chapter 15. Diamond growth is discussed in Chapters 10 and 11, and the etching of diamond by oxidation has been reviewed by Evans [1979]. A basic familiarity with surface analytical techniques is assumed; various treatments of this material can be found in several standard treatises [Ertl and Küppers 1985; Hudson 1992; Woodruff and Delchar 1986; Zangwill 1988].

The interested reader is referred to earlier reviews of the surface physics and chemistry of diamond by Thomas [1979], Pate [1986], Evans [1992], and Pate [1995]. A considerable number of papers on the surface properties of diamond have appeared during the past 5-10 years, motivated in large part by the burgeoning importance of diamond chemical vapor deposition (CVD).

2. Experimental Issues

Experimental studies of the structure and chemistry of well-defined single crystal diamond surfaces in ultrahigh vacuum (UHV, ca. 10^{-10} Torr) are unfortunately more difficult than comparable studies of many other materials. The experimental issues are outlined briefly below, along with approaches that have proved fruitful.

2.1 SURFACE PREPARATION

The most widely applied method for preparation of clean, well-ordered single crystal surfaces of many materials comprises sputter removal of the contaminated surface region followed by annealing to remove the sputter damage. Unfortunately, this approach is unsuitable for diamond–sputtering produces surface damage that cannot be removed by annealing without extensive graphitization [Marsh and Farnsworth 1964, Lurie and Wilson 1977a; Evans and Thomas 1977, Evans 1978]. Only a few researchers have attempted cleavage to create well-defined (111) surfaces (the cleavage plane) [Cavell et al. 1973; Himpsel et al. 1981], and cleavage in ultrahigh vacuum has not been reported.

2.1.1 Polished surfaces

By far the most widely applied method for preparation of single crystal surfaces of diamond is polishing followed by acid cleaning and heating in ultrahigh vacuum.

Careful polishing of diamond in a direction of easy abrasion produces a relatively smooth surface with parallel grooves and ridges, 20-250 nm wide and 2-12 nm deep, whereas polishing in a direction of difficult abrasion produces a hill-and-valley structure, ca. 30-40 nm wide and a few nm high, suggestive of fracture on the nanometer scale [Couto et al., 1992 and 1994a]. The groove-and-ridge structure appears to result from abrasion of the surface by fine diamond particles, producing locally-high contact pressures and plastic flow [Couto et al. 1994b]. Vigorous polishing on a high-speed cast-iron scaife appears to give the best results [Grodzinski 1953; Tolansky 1954; Wilks and Wilks 1972; Vidali and Frankl 1983; Pate 1986; Wilks and Wilks 1991; Yang 1992], although a mild hand-polishing treatment has sometimes been employed with (100) surfaces [Lurie and Wilson 1977a]. Virtually all polished (111) surfaces are a few degrees off from the exact orientation since the (111) orientation is the hardest.

Polishing of diamond typically leaves surfaces that are predominantly hydrogen-terminated, although oxygen is also present at some level [Derry et al., 1983; Pate 1986]. A mild anneal to several hundred °C typically produces a surface that exhibits a (1×1) low-energy electron diffraction (LEED) pattern, and further heating to ca. 1200°C leads to a clean surface (see Section 3 below) accompanied by a (2×1) LEED pattern in the case of (100) and (111) surfaces [Marsh and Farnsworth 1964; Lander and Morrison 1966; Lurie and Wilson 1977a, Vidali and Frankl 1983; Pate 1986; Derry et al. 1986; Hamza et al. 1988, 1990; Mitsuda et al. 1991; Thomas et al. 1992].

Polished diamond surfaces are sometimes cleaned further by heating in concentrated acid solutions [Marsh and Farnsworth 1964; Lander and Morrison 1966; Maguire 1976; Prins 1989; Mitsuda et al. 1991; Thomas et al. 1992; Struck and D'Evelyn 1993a; Lee and Apai 1993; Baumann et al. 1994]. Acid cleaning is also employed to remove metallic contaminants and/or graphitization suffered during ultrahigh vacuum experiments. Acid solutions that have been used include HCl/HNO_3, CrO_3/H_2SO_4, HNO_3/H_2SO_4, $H_2SO_4/HNO_3/HClO_4$, and HF. Acid treatments typically leave the diamond surface terminated by a mixture of surface oxide species [Ando et al. 1993a, 1993b; Pehrsson 1993; D'Evelyn et al. 1994; Jiang et al. 1996]. Heating an acid-cleaned diamond surface to 900-1100 °C results in disappearance of the surface oxides [Mitsuda et al. 1991; Ando et al. 1993b; D'Evelyn et al. 1994; Baumann et al. 1994; Smentkowski et al. 1995], apparently via desorption as CO and CO_2 [Matsumoto et al. 1977, 1979; Thomas et al. 1992; Jiang et al. 1996].

Related wet chemical cleaning treatments for diamond surfaces include detergents [Derry et al. 1983, 1986; Hansen et al. 1989] and electrochemical etching [Marchywka et al. 1993, Baumann et al. 1994; Humphreys et al. 1994].

Working with polished diamond surfaces has the advantage of enabling virtually any crystallographic orientation to be prepared, but also has several important drawbacks. First, (111) surfaces are slightly off-orientation, as noted above. Second, the residual roughness of the surface will be a function of the polishing technique used, detracting from the reproducibility from sample to sample and between different laboratories. Chang and Lin [1995] estimated that nearly 20% of the sites on as-polished diamond (111)-(1×1) comprise defects, based on infrared spectrosopy of physisorbed CO_2. Annealing is apparently not possible: heating of a polished diamond (100) surface to 1200 °C produced negligible smoothing of the groove-and-ridge structure [D'Evelyn et al. 1994].

An alternative "superpolishing" procedure, using nitrate oxidizing agents applied to a polishing wheel, has demonstrated the reduction of the residual roughness to ca. 2 Å rms, and to be applicable to the (111) as well as (100) orientation [Kühnle and Weis 1995]. A related tribochemical approach utilized diamond powder together with silica nano-particles as a polishing medium [Haisma et al. 1992]. These approaches have not yet been applied to surface chemistry studies, but seem very promising.

2.1.2 CVD-grown surfaces
An alternative surface preparation procedure is to grow a homoepitaxial layer by CVD. Several groups have shown that relatively smooth (100) homoepitaxial surfaces, with a height variation of only a few nanometers over lateral distances of microns, can be grown by CVD [Tsuno et al. 1991; Sutcu et al. 1992a, 1992b; Mori et al. 1993; Sasaki et al. 1993, Lee and Badzian 1995; Stallcup et al. 1995, 1996]. Homoepitaxial growth on an off-axis (100) surface, under appropriate conditions, produces a single-domain (2×1) surface [Tsuno et al. 1994a; Lee and Badzian 1996]. For preparation of low-defect-level surfaces, however, growth conditions must be chosen carefully to avoid the growth of hillocks and penetration twins [Clausing et al. 1992; Sutcu et al. 1992b; Janssen et al. 1992; Posthill et al. 1993; Wild et al. 1994; Tamor and Everson 1994; Everson et al. 1994; Tsuno et al. 1994c; Rawles et al. 1996c; see Chapter 2]. To date, only a handful of groups have reported surface chemistry results on epitaxially-grown diamond (100) [Thomas et al. 1991; Aizawa et al. 1993, 1995; Anzai et al. 1995].

The (111) orientation has received less attention, due to the tendency for (111)-oriented growth surfaces to be highly defective [Hirabayashi 1992; Sutcu et al. 1992b; Nakamura et al. 1992; Kreutz et al. 1995; Rawles et al. 1996c]. However, several groups have demonstrated growth of locally smooth (111) epitaxial layers [Hirabayashi 1992; Sasaki and Kawarada 1993; Tsuno et al. 1994b], and surface chemistry results on epitaxially-grown diamond (111) were reported by Aizawa et al. [1993].

The (110) orientation grows faster than either (100) or (111) under typical CVD conditions [Shiomi et al. 1989; Chu et al. 1992; Janssen et al. 1992]. As a result, the surface morphology is unstable with respect to facetting during growth [Sutcu et al. 1992b; van Enckevort et al. 1995; Szuba et al. 1995] and CVD will not produce a smooth (110) surface. The author is aware of only one report of stable (110)-oriented homoepitaxy [Badzian and Badzian 1993], which was achieved under extreme conditions and produced a highly defective surface.

CVD growth thus represents an attractive method for preparation of (100) or (111) surfaces, particularly the former, but considerable care must be taken that the growth conditions are suitable for generation of smooth morphologies.

2.1.3 Hydrogen plasma-treated surfaces
Recently, several authors have shown that hydrogen plasma treatments can produce a remarkable degree of flattening of diamond (100) [Ravi et al. 1993; Thoms et al. 1994b; Küttel et al. 1995; Lee and Badzian 1996; Stallcup et al. 1996; Hayashi et al. 1996; Rawles et al. 1996b, 1997] and (111) surfaces [Küttel et al. 1995; Chang et al. 1995; Rawles et al. 1996b, 1997]. Treatment of diamond surfaces with atomic hydrogen produced by a hot filament can also produce smoothing [Johansson and Carlsson 1995; Rawles et al. 1997], although to a lesser extent due to the lower atomic hydrogen flux that is produced by a hot filament relative to a high-power microwave plasma. Most authors have attributed the smoothing effect to etching, but Rawles et al. [1996b, 1997] presented evidence that the dominant mechanism is in fact hydrogen-atom-assisted surface diffusion. Under some conditions, however, hydrogen plasma treatment can cause a roughening of the surface on the micron scale [Rawles et al. 1996b]. Several

authors have reported surface chemistry studies on H_2-plasma treated surfaces [Thoms and Butler 1994, 1995; Pehrsson 1995; Mackey et al. 1995a, 1995b, Thoms and Russell 1996].

Diamond surface preparation by hydrogen plasma treatment appears to have the advantages of CVD growth without the complications of morphological instability, and may be applicable to all the low-Miller-index faces. More work will be required, however, to optimize the smoothing process and to better understand the mechanism.

2.1.4 Diamond powder

High-surface area powders offer an alternative approach to achieving surface sensitivity, and a number of diamond surface chemistry studies have been performed on powders. The cleaning treatment that is normally employed comprises heating in concentrated acids [Matsumoto et al. 1977]. A hydrogen plasma pretreatment has also been used [Chang et al. 1995].

2.2 TEMPERATURE MEASUREMENT

Measurement of accurate sample temperatures is a challenge for single-crystal diamond, as direct attachment of a thermocouple is difficult and optical pyrometry is inapplicable. Typically a thermocouple is attached to the sample support, but this can give rise to substantial temperature errors. Several methods for measuring accurate diamond temperatures have been demonstrated. Alexenko and Spitsyn [1991] employed boron-doped CVD diamond as a thermistor. Yang et al. [1992, 1993b] employed optical pyrometry of a sputter-deposited tungsten film on the back side of a diamond sample to calibrate a thermocouple attached to a support. Smentkowski and Yates [1993] obtained accurate temperature readings using a thermocouple embedded in a hole that was laser-drilled into the edge of a diamond crystal. Rawles and D'Evelyn [1995] and Patterson et al. [1995] demonstrated the use of Fizeau laser interferometry as a non-contact temperature probe, performing calibration measurements on diamond samples in a furnace.

To date, pyrometry-calibrated [Yang 1992; Yang et al. 1993b] and edge-inserted [Smentkowski and Yates 1993, 1996a, 1996b] thermocouple measurements have been applied to surface science experiments on single crystal diamond in ultrahigh vacuum, and diamond-thermistor [Spitsyn 1991] and Fizeau-interferometric [Rawles et al. 1996a, 1996c] calibrated temperature measurements have been performed in CVD growth kinetics experiments.

2.3 SURFACE CHARGING

The insulating properties of most diamond samples can create problems for surface science techniques involving electrons. The general experience, however, is that for techniques involving electrons at 100-200 eV and higher energies, including Auger electron spectroscopy, electron energy loss spectroscopy (ELS), x-ray photoelectron spectroscopy (XPS), low energy electron diffraction (LEED), and reflection high energy electron diffraction (RHEED), charging is typically not a major problem [Lander and

Morrison 1966; Maguire 1976; Lurie and Wilson 1977a; Pepper 1982; Pate 1986; Evans and Riley 1986]. Some authors have preferred to work with lightly-B-doped type IIb natural crystals [Marsh and Farnsworth 1964; Himpsel et al. 1979], which are weakly semiconducting. However, for lower energy electrons, as in high-resolution electron energy loss spectroscopy (HREELS) or LEED at lower energies, charging and patchy work function distributions on insulating and also type IIb natural diamond specimens can be crippling. The most-widely applied solutions to this problem are (i) to utilize a synthetic, B-doped diamond specimen or (ii) to grow a conductive, boron-doped epitaxial layer on an insulating diamond substrate by CVD [Aizawa et al. 1993; Thoms and Butler 1994]. The latter approach has also been employed for scanning tunneling microscopy (STM) [Maguire et al. 1992; Sasaki et al. 1993; Kreutz et al. 1995; Kuang et al. 1995; Szuba et al. 1995], although STM can sometimes be performed on undoped, freshly-hydrogenated CVD-grown epilayers [Tsuno et al. 1991; Sasaki and Kawarada 1993; Stallcup et al. 1995].

2.4 SURFACE CARBON: SP^3 VERSUS SP^2 BONDING

Various treatments, especially heating in the presence of adsorbed metal atoms or following ion bombardment, will graphitize diamond surfaces. It is important, therefore, to have suitable diagnostics for sp^3- versus sp^2-bonded surface carbon. Raman spectroscopy is very useful for characterization of the bonding environment of *bulk* carbon, but lacks the sensitivity for surface studies. LEED can provide a clear indication of well-ordered surface diamond, but co-existing patches of disordered, sp^2-bonded surface carbon may go undetected.

The most widely-applied diagnostic, in addition to LEED, is the lineshape of the carbon KVV Auger transition near 250 eV, which is sensitive to the bonding environment because of the involvement of valence orbitals in the electronic transition. Local maxima in the derivative ($E\ dN/dE$) Auger spectra occur at different energies for diamond and graphite [Lurie and Wilson 1977b; Pepper 1981; Khvostov et al. 1985; Pate 1986; Ramaker and Hutson 1987; Mitsuda et al. 1991], and hence can be used to distinguish sp^3- and sp^2-bonded surface carbon.

Graphitic carbon has a $\pi \rightarrow \pi^*$ plasmon transition with an energy near 6 eV, and can be detected even at modest surface concentrations either by electron energy-loss spectroscopy (ELS) [Lurie and Wilson 1977b; Yamazaki and Uchiyama 1993; Beerling and Helms 1994] or as a high-binding energy satellite of the C(1s) XPS peak [McFeely et al. 1974; Evans and Thomas 1977; Smentkowski et al. 1995]. Carbon also exhibits an energy-loss feature near 290 eV associated with ionization [Lurie and Wilson 1977b; Pepper 1981] which can be used as an sp^3/sp^2 diagnostic. Hoffman et al. [1991; 1994] have shown that the fine structure in the secondary electron emission (SEE) spectrum can also be used to differentiate diamond and graphite.

In the case of diamond powders, the only straightforward diagnostic of which the author is aware is the infrared spectrum of adsorbed hydrogen. Surface C-H stretch frequencies lie in the range of 2800-2950 cm^{-1} for sp^3-bonded carbon (see section 4.1), versus 3000-3250 cm^{-1} for sp^2-hybridized carbon.

Diamond surfaces are quite susceptible to degradation and/or graphitization. The deleterious effects of ion bombardment and heating in the presence of surface metallic contamination have already been mentioned. Surface degradation due to electron beam irradiation [Pepper 1981; Pate 1986; Yamazaki and Uchiyama 1993; Hoffman et al. 1996] or repeated annealing or adsorption/desorption cycles [Vidali and Frankl 1983; Thomas et al. 1992; Aizawa et al. 1995; Hoffman et al. 1996] have also been reported. However, other workers have reported that many adsorption/desorption cycles could be performed without significant degradation of the diamond surface [Mitsuda et al. 1991; Smentkowski et al. 1995].

3. Surface Structure

More detailed information about the atomic structure of diamond surfaces is available from theoretical calculations than from experiment. Nonetheless, the main features of the structures of diamond (111) and (100) surfaces have been established experimentally.

3.1 THE (111) CRYSTAL SURFACE

The (111) crystal face, the cleavage surface, has one dangling bond per surface atom in the ideal unreconstructed (1×1) structure. Hydrogenation caps the dangling bonds on the (111) surface and stabilizes the (1×1) structure. The (111)-(1×1) structure is routinely observed by LEED after polishing, cleaving, or growth of an epitaxial layer [Marsh and Farnsworth 1964; Lurie and Wilson 1977a; Himpsel et al. 1981; Pate 1986; Aizawa et al. 1993], and has also been observed in real space by scanning tunneling microscopy [Sasaki and Kawarada 1993; Tsuno et al. 1994b]. Yang et al. [1982] showed by dynamical LEED that the surface C atoms on diamond (111)-(1×1) surfaces, presumably hydrogenated, were located near ideal, bulk-terminated positions. The near-bulk-like structure of the (111)-(1×1):H surface has been confirmed by a number of theoretical calculations, using empirical potentials [Brenner 1990; Harris et al. 1991], semiempirical quantum chemical [Mehandru and Anderson 1990; Zheng and Smith 1991a, 1992a; Frauenheim et al. 1993; Davidson and Pickett 1994a; Sandfort et al. 1995], and first-principles [Zhu and Louie 1992; Stumpf and Marcus 1993; Larsson et al. 1993; Alphonso et al. 1995b; Kern et al. 1996a, 1996b] approaches.

The clean (111) surface exhibits a (2×1) reconstruction [Lurie and Wilson 1977a; Pate 1986; Derry 1986; Mitsuda et al. 1991], and the available evidence supports a π-bonded chain structure proposed originally by Pandey [1982]. Normally, three rotational domains are observed, producing a LEED pattern indistinguishable from a (2×2) structure, but Derry et al. [1986] observed one (2×1) domain to predominate on a 4°-off-axis (111) surface. Kondoh et al. [1993] observed anti-phase domain boundaries by RHEED. Direct evidence for the π-bonded chain reconstruction has been obtained by medium-energy ion scattering [Derry et al. 1986] and by dynamical LEED measurements [Sowa et al. 1988]. In addition, the electronic structure of the π-state surface bands, both occupied [Himpsel et al. 1981; Pate et al. 1981; Pate 1986; Ramaker and Hutson 1987; Namba et al. 1988; Francz and Oelhafen 1995a; Graupner et al. 1995] and unoccupied [Pepper 1982b; Morar et al. 1986b; Kubiak and Kolasinski 1989], are in agreement with predictions based on the π-bonded chain structure.

A number of theoretical papers, employing high-level *ab initio* [Vanderbilt and Louie 1984; Iarlori et al. 1992; Schmidt et al. 1996; Scholze et al. 1996; Kern et al. 1996], non-self-consistent first principles [Alphonso et al. 1995b], semiempirical [Chadi 1984; Dovesi et al. 1987; Badziag and Verwoerd 1988; Zheng and Smith 1991a; Frauenheim et al. 1993; Davidson and Pickett 1994a] and empirical potential [Brenner 1990; Dyson and Smith 1994] calculations, predict that the (111)-(2×1) π-bonded chain structure is lowest in energy of all the clean-surface structural models investigated. Rather than each surface C=C bond being equivalent, many authors predict some degree of dimerization (viz., alternating long and short bonds) within the π-bonded chains [Pandey 1982; Chadi 1984; Dovesi et al. 1987; Badziag and Verwoerd 1988; Zheng and Smith 1991; Iarlori et al. 1992; Kress et al. 1994a]. Other calculations, including some high-level *ab initio* theory, predict negligible dimerization [Chadi 1984; Vanderbilt and Louie 1984; Davidson and Pickett 1994a; Schmidt et al. 1996; Scholze et al. 1996; Kern et al. 1996]. The direct structural studies have not been able to unequivocally determine whether dimerization is present in the π-bonded chains, and the theoretical predictions are still controversial. Alternative surfaces structures, such the single chain configuration [Seiwatz 1964], produced by removal of the topmost layer of carbon atoms from the π-bonded chain structure, are predicted to be somewhat less stable [Zheng and Smith 1991; Scholze et al. 1996; Kern et al. 1996b]. However, such structures might exist in domains on diamond (111) surfaces after certain treatments, e.g., etching reactions. Predictions of graphitization near steps on (111) surfaces have also been made [Davidson and Pickett 1994b], but remain to be observed experimentally. Brink and Verwoerd [1996] estimated step edge and vertex energies on hydrogenated diamond (111).

3.2 THE (100) CRYSTAL SURFACE

The (100) crystal face has two dangling bonds per surface atom in the ideal unreconstructed (1×1) structure. A (100)-(1×1) structure is commonly observed by LEED upon examination of a surface prepared by polishing and mild annealing in UHV [Marsh and Farnsworth 1964; Lurie and Wilson 1977a; Hamza et al. 1990; Thomas et al. 1992]. This surface is terminated by hydrogen and/or oxygen [Derry et al. 1983; Hamza et al. 1990], but at present it appears most likely that the (100)-(1×1):H surface is atomically rough, with surface atoms in a distribution of dihydride and monohydride configurations [Yang and D'Evelyn 1992a; cf. Section 4.1.2] rather than predominantly flat, with two hydrogen atoms on each surface carbon atom (dihydride).

Upon heating, the (100)-(1×1) surface reconstructs to a (2×1) structure, normally with two rotational domains [Lurie and Wilson 1977a; Hamza et al. 1990; Thomas et al. 1992; Lee and Apai 1993; Smentkowski et al. 1995]. By analogy to the Si(100)-(2×1) surface, whose structure is well established [for example, LaFemina 1992; Kolasinski

1995], surface atoms on diamond (100)-(2×1) would be expected to dimerize, reducing the number of dangling bonds per atom from two to one [Verwoerd 1981a]. The surface dimer structure is well established on the hydrogenated, diamond (100)-(2×1):H monohydride surface, which can be prepared by annealing the (100)-(1×1):H surface [Hamza et al. 1990], dosing the clean (100)-(2×1) surface with atomic hydrogen [Thomas et al. 1992; Thoms and Butler 1995; Smentkowski et al. 1995], homoepitaxial CVD growth [Aizawa et al. 1993], or hydrogen plasma treatment [Thoms and Butler 1995]. Direct, real-space observations of monohydride C-C dimers on the (100)-(2×1):H surface have been made by STM [Tsuno et al. 1991, 1994a; Busmann et al. 1992; Maguire et al. 1992; Sasaki et al. 1993a, 1993b, Kawarada et al. 1995; Stallcup et al. 1995, 1996; Kuang et al. 1995] and AFM [Ravi et al. 1993; Hayashi et al. 1996]. Single-domain (2×1) surfaces can be prepared by homoepitaxial growth or hydrogen plasma treatment on a (100) substrate misoriented by several degrees toward $[01\bar{1}]$ [Tsuno et al. 1994a; Lee and Badzian 1996] but, interestingly, the orientation (perpendicular versus parallel) of the dimer rows with respect to the biatomic steps is different in the two cases [Lee and Badzian 1996].

The clean (100)-(2×1) surface, capped by π-bonded C=C dimers, may be prepared by heating the monohydride surface above 1000-1100 °C. An early report that the (100)-(2×1):H monohydride was stable above 1100 °C [Hamza et al. 1990] was questioned by Thomas et al. [1991, 1992] and by Yang and D'Evelyn [1992a]. The nearly-hydrogen-free nature of the (100)-(2×1) surface heated above 1000-1100 °C has been established by HREELS [Lee and Apai 1993; Thoms and Butler 1995; Aizawa et al. 1995; Smentkowski et al. 1995; Thoms and Russell 1996] and is further supported by the observation of occupied surface electronic states within the band gap [Hamza et al. 1990; Wu et al. 1993; Francz and Oelhafen 1995] corresponding to the weak π bonds. The persistence of some surface hydrogen above 1100 °C, particularly as detected by electron-stimulated desorption [Hamza et al. 1990; Nishimori et al. 1995], may be due to defect sites and/or low-level bulk hydrogen [Sellschop et al. 1994]. The dimer structure of the UHV-annealed (100)-(2×1) surface has recently been observed by STM [Tsuno et al. 1994d]. The length of the dimer bond on the clean surface has yet to be determined experimentally.

The dimer-bonded (100)-(2×1) surface structure has been well-established theoretically, although there is still some uncertainty in the dimer bond length and the strength of the π bond. Theoretical investigations have employed *ab initio* cluster [Tsuda et al. 1992; Tsugawa et al. 1993; Jing and Whitten 1994a; Hukka et al. 1994; Weiner et al. 1995], *ab initio* slab [Kress et al. 1994b; Furthmüller et al. 1994; Kobayashi 1994; Krüger and Pollman 1995; Zhang et al. 1995; Ogitsu et al. 1995; Kern et al. 1996a; Scholze et al. 1996a, 1996c], non-self-consistent first principles slab [Yang et al. 1993a; Alphonso et al. 1995a, 1995b], semiempirical cluster [Verwoerd 1981a], and semiempirical slab [Bechstedt and Reichardt 1988; Mehandru and Anderson 1991; Zheng and Smith 1991b; Frauenheim et al. 1993; Skokov et al. 1994a; Davidson and Pickett 1994a; Gavrilenko and Shkrebtii 1995] methods, as well as slab calculations employing empirical potentials [Brenner 1990; Yang and D'Evelyn 1992a, 1992b; Halicioglu 1992; Dyson and Smith 1994]. Nearly all the calculations predict a symmetric C=C dimer, with a bond length in the range of 1.37 - 1.46 Å. The most unambiguous definition of the surface π bond strength, arguably, is the energy required to move a single hydrogen atom from a monohydride dimer to an unoccupied (clean) dimer, breaking the π bond in the latter [Brenner 1990; Hukka et al. 1994 and references therein]. The C-H bond energies calculated via *ab initio* methods by various authors imply values of the π bond strength of 6-28 kcal/mol [Hukka et al. 1994; Tsugawa et al. 1993; Jing and Whitten 1994b; Kobayashi 1994; Weiner et al. 1995], while larger values of 32-35 kcal/mol were calculated with empirical potentials [Brenner 1990; Yang and D'Evelyn 1992a, 1992b]. The predicted dimer bond length and π bond strength are quite sensitive to the treatment of electron correlation and lattice constraints [Jing and Whitten 1994a, 1994b; Hukka et al. 1994; Weiner et al. 1995]. In general, the most recent *ab initio* calculations are probably the most reliable–the bonding structure differs significantly from that in the molecules used to construct the semiempirical and empirical Hamiltonians due to the high amount of strain in the dimer bond. The *ab initio* cluster approaches have the advantage that the treatment of electron correlation can be systematically improved upon, unlike the *ab initio* slab calculations, which typically utilize the local density approximation (LDA), but the former must make artificial assumptions about the constraining effects of the lattice.

3.3 THE (110) CRYSTAL SURFACE

Only limited experimental or theoretical information is available at present on the structure of the (110) surface. The ideal, unreconstructed (110)-(1×1) crystal face has one dangling bond per surface atom, and the zig-zag chains of topmost-layer carbon atoms are rather similar in structure to those in the (111)-(2×1) π-bonded chain. Lurie and Wilson [1977a] observed a (1×1) LEED pattern upon examination of a freshly-polished (110) surface, but did not observe any reconstruction upon heating above ca. 1000 °C, unlike the behavior of (111) and (100) surfaces. In contrast, Fox et al. [1993] observed weak second-order LEED spots, suggesting a (2×1) reconstruction, upon desorption of hydrogen from the (110)-(1×1):H surface. Davidson and Pickett [1994a] predicted, via semiempirical slab calculations, a weak dimerization of the C-C chains (alternating shorter and longer bonds). Alphonso et al. [1995b], using a first-principles approach with a slab geometry, predicted instead a buckled geometry of the surface C-C chains. The clean (110) surface has electronic states in the band gap, qualitatively similar to those on the (111)-(2×1) surface [Pepper 1982a; Davidson and Pickett 1994a, Alphonso et al. 1995b].

4. Chemisorption

A limited number of adsorbates on diamond surfaces have been investigated experimentally and theoretically, with an emphasis on systems that are important in diamond CVD. This section focuses on adsorption and the structure and characterization of chemisorbed species; the energetics of adsorption are discussed together with desorption in Section 5.1.

The kinetics of surface reactions have generally been described using different notations in the diamond CVD and surface science literature. The former has typically used gas-phase-type rate constants (units of cm^3 mol^{-1} s^{-1} for a "bimolecular" reaction), motivated by the reasonable assumption that rates of surface processes should be similar to those of analogous gas-phase reactions [Frenklach and Wang 1991; Harris and Goodwin 1993], although there has been disagreement about the best method for making the estimates. The surface science literature typically defines a sticking or reaction coefficient as the fraction of incident molecules or radicals that react with the surface upon collision, and frequently expresses surface concentrations as fractions of a monolayer. While various definitions of a “monolayer” are possible, the most common choice for low-Miller-index surfaces is the concentration of topmost-layer substrate atoms in a non-reconstructed (bulk-terminated) surface, C_S. For the (111), (100), and (110) surfaces of diamond, C_S = 1.815, 1.572, and 2.223 × 10^{15} atoms cm^{-2}, respectively. Using elementary kinetic theory of gases and the respective definitions, a "gas-phase" bimolecular rate constant k_{AB} for reaction between gas-phase species "A" and surface species "B" may be converted to a gas-surface reaction coefficient S_{AB} by

$$\begin{aligned} S_{AB} &= 4C_S/(N_0 \bar{v}_A)\, k_{AB}\, \theta_B \\ &= 2.636 \times 10^{-14}\, (C_S/10^{15})\, (300\, M_A/T)^{1/2}\, k_{AB}\, \theta_B \end{aligned} \tag{1}$$

where N_0 is Avogadro’s number,

$\bar{v}_A$ is the mean speed of A molecules,

θ_B is the surface coverage of B,

M_A is the molecular weight of A,

T is the temperature,

and the units of C_S, M_A, T, and k are cm^{-2}, g mol^{-1}, K, and cm^3 mol^{-1} s^{-1}, respectively.

In general, sticking/reaction coefficients will depend on adsorbate coverages (e.g., of B and other coadsorbed species), but they must always be less than or equal to unity, which places an upper bound on allowable rate constants. The form of Eq. (1), which follows from the assumption of the applicability of a "bimolecular" rate constant, implies that S is proportional to the surface coverage of “reactant” species, whether vacant sites or adsorbed species, i.e., Langmuirian kinetics. Departures from such behavior can result from either precursor state or steric repulsion effects, of which only the latter has been observed to date on diamond surfaces.

4.1 HYDROGEN

The adsorption of hydrogen on diamond surfaces has received a great deal of attention, due to the critical importance of hydrogen in diamond CVD and its simplicity as an adsorbate. Most single-crystal studies of hydrogen adsorption have been performed in ultrahigh vacuum. The sticking coefficient of molecular hydrogen on clean diamond surfaces has been reported to be very small [Pate 1986; Hamza et al. 1990; Mitsuda et al. 1991; Schulberg et al. 1995], and exposures are generally performed with atomic hydrogen generated by a hot filament. The sticking coefficient for atomic hydrogen on clean diamond has not been measured but is believed to be of order unity [Brenner et al. 1992]. Adsorption follows first-order (Langmuir) kinetics on polycrystalline CVD diamond [Koleske et al. 1995], indicating that chemisorbed hydrogen atoms block one site only and that an incident hydrogen atom must impact the surface near a dangling bond in order for chemisorption to occur. Several workers have reported reaction of hot diamond surfaces with molecular hydrogen in medium vacuum [Marsh and Farnsworth 1964; Lander and Morrison 1966; Aizawa et al. 1995], which is consistent with the rough estimates of the activation energy for H_2 adsorption discussed below. Other researchers have investigated hydrogenated surfaces prepared by CVD growth or by treatment in a hydrogen plasma. Hydrogenated diamond surfaces can also be prepared by heating an oxygen-terminated surface in atmospheric-pressure H_2 to ca. 900 °C, as shown by experiments with powder [Sappok and Boehm 1968a; Ando et al. 1993c]. At coverages between 0 and 1 monolayers, hydrogen adsorption or desorption on the (111), (100), and (110) surfaces is accompanied by significant band bending [Morar et al. 1986b; Wu et al. 1993; Smentkowski et al. 1995; Francz and Oelhafen 1995; Francz et al. 1996].

4.1.1 H/(111)

Hydrogenated diamond (111) has a near-ideal (1×1) structure, as described in the previous section, with no occupied surface electronic states within the band gap observed [Himpsel et al. 1979, 1981; Pate et al. 1980, 1981, 1982, 1984; Pate 1986; Graupner et al. 1995; Francz and Oelhafen 1995] or calculated [Frauenheim et al. 1993; Davidson and Pickett 1994a; Alphonso et al. 1995b; Kern et al. 1996]. The C-H bond is oriented perpendicular to the surface, as determined by electron stimulated desorption ion angular distribution (ESDIAD) [Hamza et al. 1988] and confirmed by the polarization of the stretching mode [Chin et al. 1995a] and by helium scattering [Vidali and Frankl 1983; Vidali et al. 1983]. The C-H bond length has been calculated to be ≈ 1.1 Å by a number of authors [Brenner 1990; Harris et al. 1991; Zheng et al. 1991a, 1992; Zhu and Louie 1992; Stumpf and Marcus 1993; Larsson et al. 1993; Latham et al. 1993; Frauenheim et al. 1993; Davidson and Pickett 1994a; Sandfort et al. 1995; Alphonso et al. 1995b; Kern et al. 1996], but has not yet been measured experimentally. Remarkably, only a small coverage (ca. 5% of a monolayer) of adsorbed hydrogen suffices to reverse the (2×1) clean-surface reconstruction [Mitsuda et al. 1991; Yamada et al. 1992a, 1992b]. Adsorption of hydrogen can be monitored in real time via optical second-harmonic generation [Mitsuda et al. 1991; Seki et al. 1993; Buck and Schaich 1995]. Hydrogen-terminated (111) surfaces prepared by polishing or homoepitaxial growth may contain a substantial concentration of $-CH_3$ groups [Waclawski et al. 1982; Aizawa et al. 1993, 1995]. Dangling-bond sites surrounded by hydrogen on the diamond (111)-(1×1) surface have been predicted to have a substantial electron affinity [Komatsu 1996], and such charged surface species may be important under CVD conditions.

The frequency of the C-H stretching vibrational mode on the (111)-(1×1):H surface, 2838 cm^{-1} at room temperature, has been determined accurately using infrared-visible sum frequency generation [Chin et al. 1992, 1995a; see also Ando et al. 1994b]. Similar values were obtained by HREELS on diamond (111) [Waclawski et al. 1982; Lee and Apai 1993; Aizawa et al. 1995] and by infrared spectroscopy on H_2-plasma-smoothed diamond powder [Chang et al. 1995; Lin et al. 1996]. Good agreement with experiment has been achieved by a number of theoretical calculations [Zhu et al. 1992; Davidson and Pickett 1994a; Pecansky and Waltman 1995; Sandfort et al. 1995; Alphonso et al. 1995b], that showed this surface phonon mode to be nearly dispersionless. The temperature dependence of the CH stretch mode frequency can be accounted for by pure dephasing [Lin et al. 1996]. A metastable mode at 2860-2865 cm^{-1} was also observed [Chin et al. 1992, 1995a], which disappeared upon annealing or at higher hydrogen coverages. This metastable species may comprise hydrogenated 2×1 chains, as suggested by the higher frequency on diamond (110) [McGonigal et al. 1995] and the higher calculated frequency relative to the 1×1 surface [Frauenheim et al. 1996; Köhler et al. 1996]. With adsorbed deuterium the stretching mode frequency decreases to 2115 cm^{-1} [Chin et al. 1995a; Aizawa et al. 1994], and broadens due to coupling with lattice phonons [Chang et al. 1995; Lin et al. 1996]. The first C-H stretching overtone appears at 2728 cm^{-1}, indicating substantial anharmonicity [Chin et al. 1995b; Zhu and Louie 1992], and the lifetime of the $v=1$ state (19 ps) [Chin et al. 1995b] is somewhat smaller than a theoretical prediction [Sun et al. 1994b]. The C-H bending modes couple strongly with the lattice phonons, and several modes with frequecies between 1050 and 1550 cm^{-1} have been predicted [Davidson and Pickett 1994a; Sandfort et al. 1995; Alphonso et al. 1995b; Fraueheim et al. 1996] and observed by HREELS [Waclawski et al. 1982; Lee and Apai 1993; Aizawa et al. 1995]. Only a single deformation mode, at 1331 cm^{-1}, was observed by infrared-visible sum frequency generation [Chin et al. 1995a]. Surface heterogeniety, including the presence of sp^2 carbon, can significantly broaden the C-H vibrational lines [Lee and Apai 1993; Lin et al. 1996]. The Rayleigh-wave dispersion curve of the (1×1):H surface has been investigated by HREELS [Aizawa et al. 1995], inelastic He scattering [Lange and Toennies 1996] and semiempirical calculations [Sandfort et al. 1995].

4.1.2 H/(100)

The behavior of hydrogen on the (100) surface is somewhat more complex than on the (111) surface. The best-understood surface hydride is the (2×1):H monohydride, comprising rows of dimerized carbon atoms with one hydrogen atom per surface carbon atom. The (2×1):H dimer-row structure is well established experimentally, as discussed in Section 3.2. A symmetric C-C dimer, with a bond length in the range of 1.56-1.73 Å and C-H length of 1.1 Å, has been predicted by a number of *ab initio* [Yang et al. 1993a; Latham et al. 1993; Tsugawa et al. 1993; Jing and Whitten 1994b; Hukka et al. 1994; Furthmüller et al. 1994; Kobayashi 1994; Alphonso et al. 1995b; Zhang et al. 1995; Kern et al. 1996a], semiempirical [Verwoerd 1981a, 1981b; Mehandru and Anderson 1991; Zheng and Smith 1991b; Frauenheim et al. 1993; Davidson and Pickett 1994a; Skokov et al. 1994a; Winn and Rassinger 1995; Sandfort et al. 1996], and empirical potential [Brenner 1990; Yang and D'Evelyn 1992a, 1992b] calculations. The behavior of hydrogen at step edges has also been investigated [Harris and Belton 1992; Zhu et al. 1993a, 1993b; Alphonso et al. 1996]. A dynamical LEED study [Yamazaki and Ogio 1994] reported an asymmetric monohydride dimer with a bond length of 1.44 Å. However, it seems likely that this value is erroneous, since lattice strain will force the C-C bond length to be longer than the unstrained value of 1.54 Å, as indicated by the uniformly-larger theoretical predictions cited above.

C-H stretching vibrations on the (100)-(2×1):H surface have been observed near 2910-2930 cm^{-1} by HREELS [Aizawa et al. 1993; Ando et al. 1994a; Thoms and Butler 1995; Pehrsson 1995] and by infrared-visible sum frequency generation [Anzai et al. 1995] and on H_2-plasma-smoothed diamond powder by infrared spectroscopy [Lin et al. 1996]. Theoretical calculations have predicted that symmetric and antisymmetric C-H stretching modes should be split by 0-40 cm^{-1} [Alphonso et al. 1995a, 1995b; Sandfort et al. 1996; Frauenheim et al. 1996] but differ on which mode appears at higher energy. The splitting has not yet been observed experimentally. On Si(100)-(2×1):H the symmetric stretch mode is higher in frequency by 11 cm^{-1} [Chabal and Raghavachari 1984]. Upon deuteration, the stretch mode frequency shifts to 2170-2184 cm^{-1} [Thoms and Butler 1995; Pehrsson 1995; Aizawa et al. 1995; Lin et al. 1996]. Modes assigned to C-H bending vibrations have been observed by HREELS between 1000 and 1450 cm^{-1} [Aizawa et al. 1993, 1995; Ando et al. 1994a; Thoms and Butler 1995; Pehrsson 1995]; while on deuterated surfaces analogous modes at 900 and 1125 cm^{-1} were observed by infrared spectroscopy [Struck and D'Evelyn 1993] and near 725 or 860 and 1120 cm^{-1} by HREELS [Thoms and Butler 1995; Aizawa et al. 1995]. Some of the C-H/C-D modes that have been observed on (100) monohydride surfaces may be due to CH_2, CH_3, step-bonded species, or olefinic species (bonded to sp^2 carbon) but further work will be required for definitive assignments, particularly in the case of the bending modes.

Hydrogen coverages up to approximately two monolayers can be produced on the (100) surface by polishing [Derry et al. 1983], and perhaps by other treatments as well. Theoretical work has shed light on the structure of these species but many aspects remain unclear. A (3×1):1.33H structure, comprising alternating rows of dimer-bonded CH and isolated CH_2 groups, has been predicted to be quite stable [Yang and D'Evelyn 1992a, 1992b; Yang et al. 1993a; Skokov et al. 1994a]. Kuang et al. [1995a, 1995b, 1996] recently observed small concentrations of 3×1 domains on a homoepitaxially-grown surface by STM and assigned them to this species. The (1×1):2H dihydride structure involves a high degree of steric repulsion between the hydrogen atoms in adjacent CH_2 groups, and has been predicted to be metastable with respect to H_2 desorption by a number of calculations [Verwoerd 1981b; Yang and D'Evelyn 1992a, 1992b; Huang and Frenklach 1992; Yang et al. 1993a; Tsugawa et al. 1993; Zhang et al. 1995; Winn and Rassinger 1995]. However, one set of semiempirical calculations predicted a stable (1×1):2H phase [Zheng and Smith 1991], and others reported structures only [Frauenheim et al. 1993; Davidson and Pickett 1994a]. A range of dihydride structures have been predicted, including twisted and canted CH_2 units. The H···H distances in these structures are considerably less than the smallest known value in a molecule, 1.62 Å [Ermer et al. 1985], and the adequacy of the aforementioned theoretical approaches for quantitative description of H-H steric repulsion needs to be established. Several authors have made calculations of structures at hydrogen coverages between 1 and 2 monolayers that involve dangling bonds [Yang and D'Evelyn 1992b; Furthmüller et al. 1994; Winn and Rassinger 1995] and/or electron-deficient bonding of hydrogen [Davidson and Pickett 1994a], but rearrangements of these structures to form dimer bonds and two or more adjacent CH_2 groups appears to lower the surface energy [Yang and D'Evelyn 1992b; Winn and Rassinger 1995].

Experimentally, the structure and stability of (100) hydrides at coverages above one monolayer is as yet unclear. Some authors have reported that exposure of (100)-(2×1) surfaces to atomic hydrogen produces a surface hydride with a (1×1) LEED pattern [Hamza et al. 1990; Lee and Apai 1993], but the maximum hydrogen coverage was not determined. In other work, however, even large exposures to atomic hydrogen did not alter the (2×1) reconstruction, indicating that only the monohydride was formed [Thomas et al. 1991, 1992a, 1992b; Thoms and Butler 1995]. These apparently-conflicting observations may be the consequence of differing degrees of surface roughness [Yang and D'Evelyn 1992a; Struck and D'Evelyn 1993; Thoms and Butler 1995; Aizawa et al. 1995]. Attack of the HC-CH dimer bond by atomic hydrogen is apparently highly inefficient [Thomas et al. 1992b], but the presence of high concentrations of atomic steps may increase the surface reactivity toward hydrogen and will decrease repulsion between surface CH_2 groups. Aizawa et al. [1995] found that 2nd-order LEED spots remained after the initial hydrogenation of a clean (2×1) surface, but repeated cycles of desorption and re-hydrogenation resulted in a (1×1):H surface, presumably disordered. Ravi et al. [1993] and Sasaki et al. [1993b] observed domains of a (1×1) phase on hydrogen-plasma-treated and CVD-grown (100) surfaces, respectively. However, it is not clear whether the latter phase comprised dihydride, oxide or some other species, and most authors have observed predominantly (2×1) phases after either hydrogen plasma treatment or CVD growth.

4.1.3 H/(110), polycrystalline diamond

The behavior of hydrogen on the (110) surface has received less attention. "Herringbone" and (2×1) structures have been observed by STM on B-doped (110) homoepitaxial films [Maguire et al. 1992], presumably hydrogen-terminated, but the connection of these features to an atomic structural model has not been established. The (110) monohydride surface has been predicted to display a bulk-like structure, with normal C-C and C-H bond lengths of approximately 1.54 and 1.1 Å, respectively [Latham et al. 1993; Davidson and Pickett 1994a; Alphonso et al. 1995b]. A broad C-H stretching mode was observed at 2880 cm^{-1} by infrared spectroscopy [McGonigal et al. 1995]. Theoretical calculations [Alphonso et al. 1995] suggest that the greater width of the mode relative to that on (111) may be due to coupling between adjacent CH groups, producing multiple modes, and confirm the higher mode frequency relative to the (111) surface. Hydrogen adsorption also reduces the surface electrical conductivity and increases the infrared reflectance [Mackey et al. 1995].

Hydrogen absorption has also been investigated on polycrystalline CVD diamond and on diamond powder. C-H and C-D vibrational modes similar to those described above have been observed by HREELS on CVD-grown polycrystalline films [Sun et al. 1993a, 1993b, 1994a; Thoms et al. 1994a] and by infrared spectroscopy on high-surface-area diamond powder [Sappok and Boehm 1968; Ando et al. 1993a, 1993b, 1993c, 1995c; Chang et al. 1995; Jiang et al. 1996], and have been assigned to CH, CH_2, and CH_3 species (see next section). Chang et al. [1995] showed that plasma hydrogenation at 1000 °C caused the C-H stretch feature to sharpen into one predominant mode at 2835 cm^{-1}, indicating that most of the surface sites became (111)-like.

4.2 HYDROCARBONS

The adsorption and surface chemistry of hydrocarbons on diamond has a similar level of importance to diamond CVD as that of hydrogen but is more complex and is not as well understood. Many theoretical studies, including detailed diamond growth models, have been published but experimental surface science work to date has been rather limited. Film growth experiments using isotopically-labeled precursors have shown that a one-carbon species, believed to be the methyl radical, is the dominant growth precursor in methane-based diamond CVD under typical hot-filament or microwave plasma conditions [D'Evelyn et al. 1992b; Johnson et al. 1992]. Previous gas-phase diagnostic measurements [Harris et al. 1988; Celii et al. 1989; Hsu 1991] and modeling studies [Harris et al. 1988; Harris 1990; Goodwin and Gavillet 1990; Frenklach and Wang 1991] had identified CH_3 and C_2H_2 as the only two species both sufficiently reactive and present at high enough concentrations to serve as the primary source of diamond carbon. Independent studies have unequivocally demonstrated that diamond growth from CH_3 and atomic hydrogen is feasible [Harris and Martin 1990; Lee et al. 1994; Loh et al. 1996]. Diamond growth from other precursors, including C_2H_2 [Martin 1993; Cappelli and Loh 1994], alkyl halides [Bai et al. 1993; Hukka et al. 1993], atomic carbon [Coltrin and Dandy 1993; Yu and Girshick 1994], and C_2 [Gruen et al. 1994], also appears to be possible under suitable conditions.

A number of theoretical papers have examined the adsorption of CH_3, CH_2, C_2H_2, and C_2H on diamond (111) [Tsuda et al. 1986; Huang et al. 1988; Mehandru and Anderson 1990; Brenner 1990; Valone et al. 1990; Pederson et al. 1991; Mintmire et al. 1991; Harris et al. 1991; Peploski et al. 1992; Zheng and Smith 1992a; Chang et al. 1993a; Larsson et al. 1993, 1995; Frauenheim et al. 1993; Latham et al. 1993; Perry and Raff 1994a, 1994b; Alphonso et al. 1994b; Zheng 1996], (100) [Mehandru and Anderson 1991; Huang and Frenklach 1992; Garrison et al. 1992; Zhu et al. 1993; Frauenheim et al. 1993; Latham et al. 1993; Harris and Goodwin 1993; Alphonso and Ulloa 1993; Alphonso et al. 1994a, 1994b; Feng and Lin 1994; Skokov et al. 1994b, 1995a, 1995b; Pecansky and Waltman 1995; Dawnkaski et al. 1996], and (110) [Belton and Harris 1992; Besler et al. 1992; Latham et al. 1993] surfaces. In general the structures and energetics of the adsorbed species are predicted to be very similar to those in analogous molecules. Stable adsorption of essentially all the species examined was predicted, with the possible exception of C_2H_2 bonded to a single surface site. There is considerable disagreement, however, about the absolute and relative values of the adsorption energies. Adsorption of CH_3 on the (111) surface was predicted to be destabilized at high surface coverages due to steric repulsion [Mehandru and Anderson 1990; Valone et al. 1990; Zheng and Smith 1992a; Larsson et al. 1995], and it is clear that such effects must be accurately accounted for in detailed calculations of adsorbed hydrocarbons at high coverage.

The sticking coefficient of methane on diamond surfaces has been reported to be very low near room temperature [Mitsuda et al. 1991; Yamada et al. 1992b]. Most experimental work has focused on "activated" methane, surface hydrocarbons formed by CVD growth or by reaction with hydrogen, or CH_3 generated by decomposition of a suitable precursor in a hot tube or on a hot filament. Exposure of diamond (111)-(2×1) to CH_4 passed over a hot filament produced shifts in the Auger lineshape and, at higher exposures, reversal of the (2×1) reconstruction [Mitsuda et al. 1991; Yamada et al. 1992a, 1992b; Seki et al. 1993]. C-H vibrational frequencies of 2857 and 2967 cm^{-1} were observed on CH_x-dosed diamond (111) by sum-frequency generation and were assigned to the symmetric and asymmetric stretch modes of adsorbed CH_3 [Seki et al. 1993]. Annealing the CH_x-adsorbed (111)-(1×1) surface above 900 °C caused changes in the Auger lineshape, including the appearance of some graphitic carbon, and the reappearance of second-order LEED spots. The same group later found that cleaner, more reproducible spectra could be obtained by dosing a diamond (111)-(1×1) surface, where the (1×1) structure was stabilized by 5-10% of a monolayer of hydrogen, with CH_3 produced by cracking of $(CH_3)_3COOC(CH_3)_3$ or CH_3I over a 800 °C filament [Chin et al. 1996]. The latter treatment produced new vibrational features at 2896, 2935, and 2975 cm^{-1}, which were assigned to C-H stretch modes of adsorbed CH_3. Exposure of a diamond (111)-(2×1) surface to CH_3 produced by pyrolysis of azomethane produced a surface species with a valence-band photoemission peak near 10 eV and a C(1s) core level at 1.5 eV higher binding energy relative to bulk diamond, which was similarly assigned to adsorbed CH_3 [Klauser et al. 1996a]. Annealing of adsorbed CH_3 to 350 °C and above lead to an increase in the C-H stretch mode of adsorbed hydrogen at 2835-2840 cm^{-1} [Chin et al. 1996], indicating that thermal decomposition of adsorbed hydrocarbon fragments is relatively facile (in the presence of neighboring dangling bond sites, presumably) and is likely to be important under CVD conditions. In contrast to the results summarized above, Nishimori et al. [1995] reported adsorption of unactivated CH_4 on diamond (100) at temperatures between 600 and 800 °C. The temperature dependence of the sticking coefficient, of the order of 10^{-3}, indicated an activation energy for adsorption of 7 kcal/mol and also a puzzling pressure dependence.

The vibrational spectrum of as-grown diamond (111), with C-H stretch modes near 2840 and 2912 cm^{-1} and bending modes near 1081 and 1282 cm^{-1}, was assigned to surface CH_3 groups [Aizawa et al. 1993, 1994; Ando et al. 1994a]. However, the former mode coincides with the frequency of the (111) monohydride, considerable steric hindrance would be expected in a full monolayer of methyl groups (see above), and it seems difficult to exclude spectral contributions from coadsorbed monohydride hydrogen. $(\sqrt{3} \times \sqrt{3})R30°$ domains have been observed on portions of CVD-grown (111) surfaces by STM and were variously attributed to adsorbed, cyclic C_3H_3 species [Busmann et al. 1993] or to CH_3 on every second surface site [Sasaki and Kawarada 1993]. Recently, Köhler et al. [1996] concluded that cyclic C_3H_3 species are relatively unstable.

Ando et al. [1993a, 1993b, 1993c] observed C-H stretch modes near 2840, 2875, and 2935 cm^{-1} on hydrogenated diamond powder, whose amplitudes and frequency depended on the hydrogenation conditions. The modes were assigned to CH_2 (symmetric), a combination of CH and CH_3 (symmetric), and a combination of CH_2 and CH_3 (asymmetric) vibrations, respectively. Jiang et al. [1996] made similar observations and assignments on nanocrystalline diamond powder. Chang et al. [1995]

argued that much of the observed infrared absorption in the C-H region was due to monohydride species on (111) and (100) microfacets rather than exclusively to CH_2 and CH_3. C_2H_2 physisorbed on diamond (111)-(1×1):H, by contrast, exhibits a C-H stretch mode at 3226 cm^{-1} [Chang and Lin 1996].

4.3 OXYGEN AND OXIDES

The adsorption of oxygen and oxygen-containing species has received comparatively little attention to date, either experimental or theoretical, despite the widespread use of oxygen in diamond CVD to improve growth rates and/or film quality. The chemistry of oxygen on diamond is quite rich, but considerable uncertainty remains about the relative stabilities and formation kinetics of different surface oxides and their characterization by infrared and electron spectroscopies.

Early UHV work found little or no reaction of O_2 or H_2O with diamond surfaces except at high temperatures, where degradation of the surface occurred [Marsh and Farnsworth 1964; Lander and Morrison 1966; Lurie and Wilson 1977a]. Exposure of a (100)-(2×1) surface to atomic oxygen and excited molecular oxygen, produced by dosing O_2 over a hot Ir filament, readily converted the surface to a 1×1 state [Thomas et al. 1992a; Pehrsson 1995]. Oxygen-related vibrational modes near 3540, 1720-1800, and 800-1150 cm^{-1} were observed by HREELS, and were assigned to OH stretch, >C=O stretch, and C-O-C and/or C-OH modes, respectively [Pehrsson 1995]. Annealing to 1000 °C restored the 2nd-order LEED spots and eliminated or reduced the vibrational features. Exposure of a (111)-(2×1) surface to atomic oxygen produced by a microwave discharge source produced surface oxides with a 1 eV higher C(1s) binding energy, valence photoemission peaks near 4.2 and 8 eV, and reduced the sharpness but did not eliminate the 2nd-order LEED spots [Klauser et al. 1996b]. Coadsorbed hydrodrogen, however, completely reversed the 2×1 reconstruction.

Struck and D'Evelyn [1993a, 1993b] observed water adsorption, with a sticking coefficient estimated as 10^{-3} to 10^{-2}, and decomposition of surface oxides on hot diamond (100) by infrared spectroscopy. Surface vibrational modes were observed at 1250, 1200, 1125, 1080, and 720 cm^{-1}, and were assigned to C-O-C, C-OH, and >C=O species, based on comparisons with analogous molecules and the results of *ab initio* calculations.

Experiments with diamond powder have demonstrated that clean or hydrogenated diamond surfaces can be oxidized to ca. one monolayer oxygen coverage by heating in oxygen [Sappok and Boehm 1968a, 1968b; Bansal et al. 1972; Matsumoto et al. 1977; Ando et al. 1993d] or water [Matsumoto and Setaka 1979]. Exposure of vacuum-outgassed diamond to oxygen produced sharp infrared absorption features near 1760-1800 cm^{-1} and broader features near 1270-1280 and 1120 cm^{-1} [Sappok and Boehm 1968a, 1968b; Matsumoto et al. 1977], which were assigned to >C=O and C-O-C species, respectively.

Oxidation of initially-hydrogenated diamond powder between 300 and 500 °C gave qualitatively similar results, with the evolution of bands near 1710-1820 cm^{-1}, 1420-1480 cm-1, 1250-1300 cm-1, and 1080-1110 cm^{-1} [Ando et al. 1993e]. The amplitudes and mode frequencies generally increased with oxidation temperature. The former peak was assigned to >C=O moieties in cyclic ketones, lactones, or carboxylic anhydride structures. An increase in the population of the latter with increasing oxidation temperature was supported by the increasing mode frequency and by the results of reaction with aqueous NaOH [Ando et al. 1993e], but contributions of dipole coupling to shifts in the mode frequencies of adsorbed C=O with coverage [Hoffman 1983] also need to be taken into account. The 1250-1300 and 1080-1110 cm^{-1} modes were assigned to C-O-C stretch modes, while the nature of the 1420-1480 cm^{-1} modes was unclear [Ando et al. 1993e]. Surface OH groups prepared by reaction of chlorinated diamond powder with water vapor exhibited O-H stretching and bending modes at 3425-3520 and 1610 cm^{-1}, respectively [Ando et al. 1995a].

Several theoretical studies have examined the structures and stabilities of surface C-O-C ether-type groups, >C=O cyclic ketone groups, and OH groups [Badziag and Verwoerd 1987; Russo 1989; Zheng and Smith 1992b; Frenklach et al. 1993; Whitten et al. 1994; Skokov et al. 1996], and Zheng and Smith [1992b] also examined several surface epoxides (3-membered rings) and peroxides (di-σ-bonded O_2). These studies confirmed the energetically-favorable formation of surface oxides, explaining the facility with which atomic oxygen can reverse the surface reconstruction. However, the predicted energetics differ widely and will require further study.

4.4 HALOGENS

The demonstration of halogen-assisted diamond growth has stimulated significant interest in the interaction of halogens and halocarbons with diamond surfaces. These systems display a rich surface chemistry, including substantial deviations from the behavior of analogous molecular species.

Fluorinated diamond (111) and (100) surfaces have been prepared by a number of methods. Adsorption of F_2 on diamond (100) was much less efficient than that of atomic F [Freedman and Stinespring 1990]. Exposure of the (100) or (111) surface to atomic F or XeF_2 produced saturation coverages of nearly 1 monolayer, with a shoulder in the C(1s) x-ray photoelectron spectrum shifted to 1.8 eV higher binding energy than bulk carbon atoms, indicating formation of a surface monofluoride [Morar et al. 1986a; Freedman and Stinespring 1990; Yamada et al. 1992b; Freedman 1994]. The sticking coefficients for F atoms on the (111) and (100) surfaces were proportional to $(1-\theta_F)$ and $(1-\theta_F)^2$, respectively [Freedman 1994], indicating that reaction with a dangling-bond site is adversely affected by a neighboring F atom in the latter case. Plasma treatment of diamond (111) in SF_6 also produced CF_2 and CF_3 species, as identified by XPS [Cadman et al. 1975], but the latter species may have resulted from ion damage of the surface. Unlike hydrogen, fluorine adsorption did not reverse the (2×1) reconstruction on the (111) surface [Yamada et al. 1992b; Freedman 1994]. X-ray irradiation of physisorbed C_4F_9I on diamond (100) generated adsorbed C_4F_9 and I upon warming to room temperature; the C_4F_9 species decomposed to a surface monofluoride upon heating to 675 °C while the I desorbed just above room temperature [Smentkowski and Yates 1996a, 1996b]. Similar treatment of physisorbed CF_3I produced only surface F and I--adsorbed CF_3 could not be detected [Smentkowski and Yates 1996b]. A partial pressure of F_2 markedly reduced the kinetics of etching of diamond (110) by O_2, presumably due to formation of surface C-F bonds [Chu et al. 1995].

Diamond powders have been fluorinated by exposure of freshly desorbed surfaces to F_2 [Sappok and Boehm 1968a], heating in F_2 [Patterson et al. 1989; Ando et al. 1995b, 1996a], by treatment in a CF_4 plasma [Patterson et al. 1989; Ando et al. 1993d], and by reaction of a chlorinated surface with CHF_3 [Ando et al. 1996]. Under relatively mild fluorination conditions, infrared features were observed near 1085 and 1280 cm^{-1} [Sappok and Boehm 1968a; Patterson et al. 1989; Ando et al. 1995b, 1996a, 1996b]. These features are most likely due to surface monofluoride species, with possible contributions from CF_2 species at step edges. At higher fluorination temperatures [Patterson et al. 1989; Ando et al. 1995b, 1996a], or upon fluorination in a CF_4 plasma [Ando et al. 1993d], an additional infrared peak near 1360 cm^{-1} was observed, most likely due to adsorbed CF_2 and/or CF_3. Fluorination of oxidized diamond surfaces also showed evidence for a COF species [Ando et al. 1993d, 1995b, 1996a].

A handful of theoretical papers have investigated the chemisorption of fluorine and/or fluorocarbons on the low-index surfaces of diamond, including *ab initio* [Pederson and Pickett 1992; Hukka et al. 1995, 1996; Wensell et al. 1995; Larsson 1996] and empirical potential [Harris and Belton 1991; Piekarczyk and Prawer 1993] studies. Surface C-F bonds were predicted to be stronger than surface C-H bonds and similar to those in molecules, with only modest steric repulsion effects. As discussed in Section 5.1, experimental work indicates that the binding energy of halogens at high coverages has been overestimated in the theoretical calculations performed to date. Adsorption of CF_3, which may be useful for diamond growth by atomic layer epitaxy [Hukka et al. 1993, 1996; Lee et al. 1997], was predicted to be quite stable by *ab initio* calculations [Hukka et al. 1996; Larsson 1996] (although not by calculations with an empirical potential [Harris and Belton 1991]), yet adsorbed CF_3 was not observed following photolysis of physisorbed CF_3I [Smentkowski and Yates 1996b].

Less work, either experimental or theoretical, has been performed for the other halogens, although the surface chemistry of chlorine is extremely interesting. Freedman [1994] found that submonolayer coverages of chlorine were produced by exposure of diamond (100) to atomic chlorine, but the surface layer decomposed upon heating to 325 °C. Chlorinated diamond powders and/or polycrystalline CVD diamond films were prepared by heating in Cl_2 [Sappok and Boehm 1968a; Ando et al. 1995a, 1996b], exposure of a freshly-desorbed surface to Cl_2 [Smirnov et al. 1978] or by exposure to ultraviolet irradiation in a Cl_2 atmosphere [Miller and Brown 1995]. Surface C-Cl bonds were not observable by infrared spectroscopy, although weight gains, XPS, or subsequent surface reactions indicated that it was present. Surface chlorine apparently reacts readily with gas phase H_2, CH_4, H_2O or NH_3 to produce H, OH or NH_2 groups, respectively [Smirnov et al. 1979; Gordeev et al. 1980, 1981; Miller and Brown 1995; Ando et al. 1995a, 1996b]. Signficant steric repulsion has been predicted theoretically on Cl-terminated surfaces, and together with the favorable energetics of HCl formation can account at least in part for the high reactivity [Hukka et al. 1995; Wensell et al. 1995; Larsson 1996]. Steric repulsion effects have been predicted to be even greater for adsorbed Br [Larsson 1996], for which only preliminary results have been reported on diamond powder [Smirnov et al. 1978]. Chemisorbed I was produced on diamond (100) by irradiation of physisorbed perfluorinated alkyl iodides at low temperature and decomposed, presumably by desorption, near room temperature [Smenkowski and Yates 1996a, 1996b].

4.5 OTHER SPECIES

The interaction of a handful of additional molecules with diamond surfaces has been reported, but in most cases the results are of a rather preliminary nature. Marsh and Farnsworth [1964] reported that CO_2 reacted with diamond (100) at high temperature in a qualitatively similar way as O_2, suppressing fractional-order diffraction peaks. Lander and Morrison [1966] observed the formation of $(\sqrt{3} \times \sqrt{3})R30°$ structures on diamond (111) upon reaction with phosphorus at about 800 °C. Lurie and Wilson [1977] found that electron beam irradiation enabled the adsorption of H_2S. Hamza and Kubiak [1989] observed the adsorption of NO under the influence of band-gap laser excitation but concluded that only defect sites were active. Surface NH_2 groups were prepared by photochemical or thermal reaction of chlorine adsorbed on polycrystalline CVD diamond or diamond powder with NH_3, exhibiting N-H symmetric and asymmetric stretch modes at 3050 and 3150 cm^{-1}, respectively, and a scissors mode at 1414 cm^{-1} [Miller and Brown 1995; Ando et al. 1995a, 1996b]. Ando et al. [1995a, 1996] suggested that this species may in fact be adsorbed NH_4^+, but the sharp, single H-N-H bending mode is more consistent with NH_2. Amination at 400 °C rather than room temperature caused partial dehydrogenation of the NH_x species, producing >NH and -C≡N groups, as indicated by vibrational modes near 3273 and 3365 cm^{-1} and at 2265 cm^{-1} [Ando et al. 1995a, 1996b], respectively, and a disappearance of the H-N-H bending mode. Dissociative adsorption of submonolayer coverages of HN_3 on clean diamond (100) produced surface N_3 and hydrogen, as indicated by vibrational modes at 2100 and 2950 cm^{-1}, respectively [Thoms and Russell 1996].

5. Surface Reactions

5.1 DESORPTION

The desorption of hydrogen (and deuterium) chemisorbed on each of the low-index faces of diamond, polycrystalline CVD diamond, and diamond powder has been examined in a number of studies. A more limited number of papers have reported on the desorption of CO and CO_2 produced by decomposition of surface oxides and of surface fluorine.

5.1.1 Hydrogen

Hydrogen desorbs from diamond near 800-1050 °C. The only desorption product that has been observed is H_2, which is noteworthy because a concerted elimination of H_2 from adjacent C atoms, forming a C=C double bond, would be expected to be symmetry forbidden. The dynamical mechanism for H_2 desorption has received considerable attention on silicon surfaces [Kolasinski 1995], where the behavior is similar in many ways to that on diamond, but a consensus between experimental and theoretical researchers has not yet emerged. A satisfactory understanding of H_2 desorption from diamond may therefore be expected to yield new insights into the similarities and differences between surface reactions and analogous molecular reactions.

Hydrogen and deuterium desorption from diamond (111) has been investigated by temperature-programmed desorption (TPD) [Mitsuda et al. 1991; Fox et al. 1993] and by time-resolved sum frequency generation [Chin et al. 1995a]. Desorption occurs with approximately first-order kinetics, and the activation energy and preexponential factor were reported as 92±4 kcal/mol and $10^{15\pm2}$ s^{-1}, respectively [Chin et al. 1995a]. These kinetic parameters would predict TPD peak temperatures of 1000°C or 1070°C at heating rates of 5.4 or 40 °C/sec, respectively, in good agreement with the results of Fox et al. [1993] and Mitsuda et al. [1991], respectively; however, Fox et al. [1995] obtained kinetic parameters of 70 kcal/mol and $10^{11.6}$ s^{-1}.

Fox et al. [1993] also investigated hydrogen desorption kinetics on diamond (110) by TPD. They observed first-order kinetics, with an activation energy and preexponential factor of 95±5 kcal/mol and $10^{13.7\pm1}$ sec^{-1}, respectively.

The desorption behavior of hydrogen from diamond (100) has been more controversial and some aspects of the kinetics remain puzzling. Hydrogen desorption with near-first-order kinetics has been observed between 800 and 1050 °C by TPD by Hamza et al. [1990], Thomas et al. [1991, 1992a], Yang et al. [1993b], Hoffman et al. [1996] and McGonigal et al. [1996]. McGonigal et al. [1996] found that the desorption kinetics became zero order at temperatures below 800 °C. The desorption peak was initially attributed to decomposition of the dihydride [Hamza et al. 1990], but the spectroscopic work discussed in Section 3.2 has verified the assignment of the desorption peak to the monohydride [Yang and D'Evelyn 1992a; Thomas et al. 1992a; Yang et al. 1993b]. All of the TPD studies on the (100) surface observed peaks that were significantly broader than those on the (111) or (110) surfaces [Fox et al. 1993], a straightfoward analysis of which implies anomalously low preexponential factors and activation energies. Detailed analysis by Hamza et al. [1990] yielded $10^{5.5}$ s^{-1} and 37 kcal/mol for these quantities, respectively, at low coverages (the high coverage data was compromised by desorption from the sample support). McGonigal et al. [1996] obtained very similar values for isothermal desorption in the first-order kinetic regime. Thomas et al. [1992a] and Yang et al. [1993b] assumed a preexponential factor of 10^{13} s^{-1} and estimated activation energies of 73 and 80 kcal/mol, respectively, but the TPD peak shapes are better fit with lower values of the kinetic parameters [Schulberg et al. 1992]. Even more anomalous kinetic parameters were obtained by Nishimori et al. [1995] by isothermal desorption following CH_4 adsorption, 1.6 s^{-1} and 21±5 kcal/mol. However, these latter results predict TPD behavior very different than that observed by other groups and accordingly must be regarded as suspect.

The simplest explanation for the first-order desorption kinetics, which might have been expected to be second order since two adsorbed hydrogen atoms must associate to form H_2, is that surface π bond formation induces preferential pairing of adsorbed H [Brenner 1990; Yang and D'Evelyn 1992b; Hukka et al. 1994; Koleske et al. 1995; Chin et al. 1995a], as occurs on Si(100)-(2×1) [Boland 1991; D'Evelyn et al. 1992a; Kolasinski 1995, and references therein]. π-bond-induced adsorbate pairing can be anticipated on diamond (111) and (110) as well as on (100)-(2×1) because of the presence of π-bonded chains on the clean surfaces (cf. Sect. 3), unlike the case with Si(111) (which exhibits a (7×7) reconstruction) and Si(110) [LaFemina 1992]. However, the energetics of partially-hydrogenated π-bonded chains on the diamond (111)-(2×1) and (110) surfaces have not yet been investigated, and the possible role of "clustering" of surface hydrogen due to bond strain [Yang and D'Evelyn 1993], which could perhaps explain the onset of zero-order desorption kinetics at low temperatures [McGonigal et al. 1996], should also be considered.

Hydrogen desorption from polycrystalline CVD diamond has been investigated using TPD [Schulberg et al. 1992, 1995; Chua et al. 1994] and time-resolved direct-recoil spectrometry [Koleske et al. 1994, 1995]. No significant isotope effect was observed between desorption of H_2, HD, or D_2 [Schulberg et al. 1992, 1995]. Desorption was shown to be first order, and the kinetic parameters were determined as $10^{7.7}$ s^{-1} and 51 kcal/mol by Schulberg et al. [1992, 1995] or as $10^{10.5\pm0.9}$ s^{-1} and 69±6 kcal/mol by Kokeske et al. [1994, 1995]. Schulberg et al. [1995] also showed that their data could be fit reasonably well by a multiple-site desorption model with a "normal" preexponential factor of 9×10^{12} s^{-1}, where the activation energy for desorption had a Gaussian distribution with mean of 82 kcal/mol and a width of 3 kcal/mol. The desorption behavior of hydrogen from diamond powder [Matsumoto et al. 1973, 1979, 1981; Ando et al. 1993c, 1995c] is similar to that on single-crystal and polycrystalline diamond surfaces, but has not been analyzed quantitatively.

The kinetic parameters for desorption can be interpreted using bond energy considerations and theoretical predictions of desorption energies. The observations that C-H stretch frequencies are essentially equal to those in analogous molecules (the slightly lower frequencies can be explained by a larger reduced mass in the case of surface hydrogen) is a strong indication that surface C-H bonds are approximately as strong as those in molecules, where the typical bond energy of tertiary hydrogen is 96 kcal/mol [Gutman 1990]. The desorption energy of H_2 should therefore be approximately 2×96 - 104 kcal/mol (bond energy of H-H) minus the energy of the surface π bond formed after desorption. Since C-C distances on diamond surfaces are significantly longer than those in normal C=C double bonds and the back bonds deviate significantly from the ideal planar geometry, surface π bonds must be expected to be relatively weak. Therefore, the desorption energy would be expected to be slightly less than 88 kcal/mol. The activation energy for desorption would be expected to be slightly larger than the desorption energy, by an amount equal to the activation energy for dissociative *adsorption* of H_2.

The most reliable predictions of the desorption energy of H_2 from diamond (100) are several recent, high level quantum chemical cluster calculations [Jing and Whitten 1994b; Hukka et al. 1994; Weiner et al. 1995], which obtained values between 74 and 90 kcal/mol, consistent with the bond-energy argument given above. Several other authors have obtained values within or close to this range [Verwoerd 1981a, 1981b; Tsugawa et al. 1993; Kobayashi 1994; Winn and Rassinger 1995]. Brenner [1990] and Yang and D'Evelyn [1992a, 1992b] obtained significantly smaller values, 65 and 47 kcal/mol, respectively, using empirical potentials which apparently overestimate the strength of the π bond on the (100)-(2$\times$1) surface. Other calculations, including some *ab initio* (LDA) and semiempirical slab calculations, yielded widely divergent values [Zheng and Smith 1991b; Yang et al. 1993a; Furthmüller et al. 1994; Alphonso et al. 1995]. Several calculations of the desorption energy of hydrogen from diamond (111) and/or (110) have been reported [Brenner 1990; Zheng and Smith 1992a; Alphonso et al. 1995], but these values differ greatly from one another and none involved high-level quantum chemistry.

The apparently low sticking coefficient for H_2 on diamond surfaces suggests an activation barrier of at least 5-10 kcal/mol. Combined with the theoretical estimates cited above, this suggests that the activation energy for desorption of H_2 should be ≥ 80 kcal/mol. Desorption of atomic hydrogen would be expected to have an activation energy approximately equal to the C-H bond energy (ca. 96 kcal/mol). Observation of H_2 as the only observable desorption product indicates that the activation energy for desorption of H_2 should be significantly less than 96 kcal/mol (else H desorption would compete).

Based on these theoretical considerations, the activation energies reported by Chin et al. [1995a] and Fox et al. [1993] for desorption from the (111) and (110) surfaces appear slightly high, but this could be explained by slight overestimates of the sample temperature (the thermocouples were attached to sample supports). Comparison with the activation energies reported on diamond (100) and polycrystalline diamond indicates a more serious discrepancy. Koleske et al. [1994, 1995] argued that the discrepancy is a consequence of π bond formation after desorption, but the latter has already been accounted for in the calculated desorption energies cited above. The most likely explanation for the anomalously-low apparent activation energies (and preexponential factors) appears to be the existence of sites with a range of effective activation energies for desorption, as suggested by Schulberg et al. [1995]. The microscopic basis for such a phenomenon remains unclear: the similar breadth of TPD peaks (a higher preexponential yields a narrower peak) on polished and homoepitaxial diamond (100) [Thomas et al. 1991] suggests that defect sites are not the explanation. Finally, it will be important to confirm the observation [McGonigal et al. 1996] of zero-order kinetics below 800 °C on the (100) surface, to determine whether similar behavior occurs on (111) and (110), and to establish the microscopic mechanism.

The observation that the activation energy for desorption of H_2 appears to be only slightly larger than the desorption energy suggests that the activation energy for adsorption of molecular H_2 is relatively small, perhaps less than 10-20 kcal/mol. A similar situation occurs on Si(100)-(2×1), where until recently H_2 adsorption had not been observed. However, at elevated gas and surface temperatures the sticking coefficient increases significantly [Kolasinski 1995; Bratu and Höfer 1995], and becomes relatively facile on vicinal Si(100) and Si(111) surfaces with high atomic step densities [Hansen et al. 1996]. If a similar situation exists on diamond surfaces, H_2 adsorption may play a significant role in diamond growth by CVD.

Hydrogen has typically been the only observed desorption product upon heating of hydrogenated diamond surfaces. However, Thomas et al. [1991] observed small quantities of C_2H_2 and CH_3 desorbing near 600 °C and 700 °C, respectively, following very large exposures of homoepitaxial diamond (100)-(2×1) to atomic hydrogen. Matsumoto [1979] observed desorption of C_2H_4 and C_2H_2 from C_2H_6-treated diamond powder.

5.1.2 Oxygen

Adsorption of oxygen on diamond produces etching upon heating, as only carbon-containing desorption products are evolved. Thomas et al. [1992a] observed desorption of CO near 600 °C and smaller quantities of CO_2 near 500 °C from oxygenated diamond (100) by TPD. These TPD peaks were even broader than those of H_2, possibly indicating a range of activation energies for oxygen in different surface configurations. The preexponential factor and activation energy for CO desorption were determined as $10^{11.45}$ s^{-1} and 45 kcal/mol, respectively, from the dependence of the TPD peak temperature on heating rate, and desorption was proposed to occur via β-scission of a C-C=O backbond [Frenklach et al. 1993]. Qualitatively similar behavior was observed with oxidized diamond powders but an additional high-temperature CO peak was observed near 800 °C [Matsumoto et al. 1973, 1977; Ando et al. 1993e; Jiang et al. 1996]. The amplitudes of >C=O and C-O-C vibrational modes decreased in tandem upon heating [Ando et al. 1993e], most likely indicating relatively facile interconversion of these species at elevated temperatures. Oxidation of diamond powder by heating in water vapor rather than O_2 produced a higher CO/CO_2 ratio in the desorption products [Matsumoto and Setaka 1979].

5.1.3 Halogens

Desorption of halogens, particularly fluorine, takes place over a remarkably broad temperature regime, for reasons that are not completely understood. Freedman and Stinespring [1990] and Freedman [1994] observed that the fluorine coverage on diamond (111) and (100), as observed by XPS, began to decay just above room temperature but was not complete until the surface had been annealed to 825 - 925 °C. Atomic F was the only desorption product that could be detected by mass spectrometry [Freedman 1994]. Yamada et al. [1992b] and Smentkowski and Yates [1996b] observed similar behavior on diamond (111) and (100), respectively, but some residual surface fluorine was still present after annealing to 1225 °C. The C-F bond energy on the (100)-(2×1) surface was calculated as 120 kcal/mol [Hukka et al. 1995] and as 115-125 kcal/mol on the (111) surface [Larsson 1996], which suggests that F desorption should be minimal below 1200 °C [Hukka et al. 1995]. These bond energy estimates are consistent with the low-coverage behavior observed by Yamada et al. [1992b] and Smentkowski and Yates [1996b], but not with the behavior at high coverage. It seems possible that the somewhat lower desorption temperatures reported by Freedman [1994] are related to the difficulty of accurate temperature measurements on single crystal diamond. The most likely explanation for the much weaker effective binding energy at high fluorine coverage is steric and/or electronic repulsion effects [Pederson and Pickett 1992; Freedman 1994; Hukka et al. 1995]. Desorption of F and CF_3 from fluorinated diamond powder occurred between 450 and 1300 °C [Ando et al. 1996a].

Desorption of chlorine also takes place over a broad temperature range (ca. -50 to 300°C) on diamond (100) [Freedman 1994]. The C-Cl bond energy was calculated as 87 kcal/mol at low coverage on the (100)-(2×1) surface [Hukka et al. 1995] and as 61-65 kcal/mol at monolayer coverage on the (111) surface [Larsson 1996]. These bond energy estimates predict that surface chlorine should be stable well above 400 °C, in contrast with experiment. However, Ando et al. [1995a, 1996b] found, by TPD, that chlorine on diamond powder desorbed at much higher temperatures, 600-1000 °C, depending on the chlorination conditions. Further work will be necessary to understand and quantify the effects of steric and electronic repulsion and of atomic-scale surface roughness on the thermal stability of chlorine on diamond surfaces.

5.2 ABSTRACTION

Abstraction of surface hydrogen by atomic, gas-phase hydrogen is widely believed to play a critical role in diamond growth by CVD (see Chapters 10 and 11), and this surface reaction has been investigated under both collision-free (high vacuum) and isothermal/collisonal conditions similar to those found in CVD reactors.

In the high and ultrahigh vacuum experiments reported to date, atomic hydrogen was produced by dissociation at a hot filament (1550 to 1800 °C). Since transport to the surface involves few or no gas-phase collisions, the kinetic energy of the hydrogen atoms is approximately Maxwell-Boltzmann at the filament temperature, and the effect of substrate temperature can be investigated independently. Chin et al. [1995a] measured the ratio of the sticking coefficients for abstraction of surface deuterium by atomic hydrogen ($S_{abs,D/H}$) and adsorption of atomic hydrogen ($S_{0,H}$) on clean diamond (111) to be 0.2 at room temperature and T_{fil} = 1800 °C, using time-resolved sum-frequency generation in high vacuum. Results from the converse experiment, abstraction of surface hydrogen by atomic deuterium ($S_{abs,H/D}$), were not presented but were stated to be similar. Thoms et al. [1994c] found, using time-resolved HREELS, that $S_{abs,H/D}/S_{0,D}$ on polycrystalline CVD diamond was 0.05±0.01 at both 80 and 600 °C and T_{fil} = 1800 °C. Koleske et al. [1995] showed that the abstraction and adsorption rates of hydrogen and deuterium on polycrystalline CVD diamond were proportional to the fraction of hydrogen-terminated and open sites, respectively, using time-resolved direct recoil spectrometry. They measured $S_{abs,H/D}/S_{0,D}$ = 0.03±0.01 at 500 °C and T_{fil} = 1560 °C and a weak surface temperature dependence for $S_{abs,H/D}$, with an apparent activation energy of 0.8±0.2 kcal/mol. The much-smaller effective activation energy in the high vacuum studies than that in the isothermal/collisional studies summarized below implies that the kinetic energy of the hydrogen atom is more effective than substrate temperature in surmounting the activation barrier for surface abstraction. Koleske et al. [1995] also observed an isotope effect, with the rate of surface hydrogen abstraction by atomic deuterium being three times higher than the rate of surface deuterium abstraction by atomic hydrogen. The range of the values obtained by different authors are due in part to different filament temperatures [Koleske et al. 1995], reflecting a dependence of reaction rate on gas temperature, but differences in the rates on (100) versus (111) surfaces may also be important.

Several groups have estimated the fraction γ of hydrogen atoms incident on a polycrystalline CVD diamond surface under isothermal, collisional conditions that catalytically recombine to form H_2, which is equal to 2 $S_{abs,H/H}$ in the limit that $\gamma \ll 1$. Gat and Angus [1993], Harris and Weiner [1993], and Proudfit and Cappelli [1993] estimated γ to be 0.3-1, 0.12, and 10^{-4} at substrate temperatures of 1370 °C, 925 °C, and 25 °C, respectively, using thermistor or thermocouple techniques. Krasnoperov et al. [1993] measured γ to be $10^{0.29\pm0.15}$ exp[-(6.7±0.5 kcal/mol)/RT] over the temperature range 25-850°C, using flow tube/photoionization mass spectrometry techniques, in good agreement with the single-temperature measurements. The approximate agreement between the values of $\gamma/2$ summarized above with the values of S_{abs}/S_0 obtained in the high vacuum experiments indicates that the sticking coefficient of atomic hydrogen is of order unity.

Early modeling of hydrogen abstraction from diamond assumed reactivity similar to the tertiary hydrogen atom in isobutane [Harris 1990; Frenklach and Wang 1991], with an activation energy of 7-8 kcal/mol. A number of calculations of the activation energy for surface hydrogen abstraction have been performed using semiempirical quantum chemical methods [Huang et al. 1988; Valone et al. 1990; Besler et al. 1992], but the results differ widely and hence are quantitatively uncertain [Valone et al. 1990; Besler et al. 1992]. A high-level quantum chemical calculation by Page and Brenner [1991] yielded an energy barrier of ≈7.5 kcal/mol, very close to the value assumed in the modeling studies. The Brenner [1990] empirical potential was constructed with an energy barrier of 7.5-9 kcal/mol for hydrogen abstraction, and has been used to estimate rate constants for abstraction and adsorption on both the (111) [Brenner et al. 1992; Brenner and Harrison 1992; Chang et al. 1993b; de Sainte Claire et al. 1994, 1996] and (100) [Dawnkaski et al. 1995a, 1995b] surfaces via classical and semiclassical trajectory simulations and transition-state calculations. These studies derived activation energies for abstraction approximately equal to the energy barrer in the potential, as would be expected, with values of $S_{abs,H/H}$ of 0.02-0.05 at 925 °C. On the (100)-(2×1) surface, H atoms on a dimer are more readily abstracted if the neighboring carbon atom has a dangling bond site rather than another H atom, so that a π bond can be formed, and adsorb more readily on an isolated dangling bond than on a π-bonded dimer [Dawnkaski et al. 1995a, 1995b]. Hase and co-workers showed that the (unactivated) adsorption rate constant for H atoms incident on a single radical site is sensitive to the details of the potential and the dynamics calculation [Accary et al. 1993; Barbarat et al. 1993; de Sainte Claire et al., 1994, 1996; Song et al. 1995].

Less attention has been given to abstraction of surface hydrogen by other species, but Krasnoperov et al. [1993] measured γ for CH_3 radicals colliding with a hydrogen-covered polycrystalline CVD diamond surface to be $10^{0.1\pm0.7}$ exp[-(10.6±3.4 kcal/mol)/*RT*] over the temperature range 460-860°C, using flow tube/photoionization mass spectrometry techniques. Reactive species such as F, Cl, C_2H, and OH should be quite proficient at hydrogen abstraction, given their bond strengths with hydrogen (cf. Section 4.4), and some modeling studies have considered these reactions [Wang and Frenklach 1991; Peploski et al. 1992].

5.3 DIFFUSION

No experimental data is yet available for diffusion of any species on a diamond surface, and until recently surface diffusion was widely believed to be unimportant in diamond growth by CVD. However, calculations using *ab initio* methods [Melnik et al. 1994; Heggie et al. 1996], semiempirical quantum chemistry [Huang and Frenklach 1992; Skokov et al. 1994b], and the Brenner potential [Chang et al. 1994; Dawnkaski et al. 1995b] predict that surface migration of hydrogen has a minimum activation energy in the range of 7-75 kcal/mol, depending on the pathway, implying that migration is likely to occur over at least a few sites on the time scale of diamond growth and that certain hydrogen exchange pathways will be in partial equilibrium [Harris 1990; Skokov et al. 1994b]. Surface diffusion of halogens, particularly chlorine [Melnik et al. 1994], may also be significant. The activation energy for diffusion of hydrogen on silicon surfaces is significantly less than the activation energy for desorption, based on optical second harmonic generation [Reider et al. 1991] and adsorption-kinetic studies [Hansen et al. 1996], and similar behavior might be anticipated on diamond.

The smoothness of CVD-grown diamond films on the nanometer scale is well established (cf. Section 2.1.2), and van Enckevort et al. [1993] have argued that surface migration of carbonaceous species to steps is significant and can be rate-limiting in diamond growth. Mehandru and Anderson [1991] predicted that migration of CH_2 on the (100) surface might be significant, and Frenklach and co-workers [Skokov et al. 1994b, 1995a, 1995b; Frenklach et al. 1994] have proposed detailed models for hydrogen-atom-assisted surface migration of CH_2 and $C{=}CH_2$ on diamond (100)-(2×1). A similar model was investigated by Tsuda et al. [1996], and migration on the clean (100) surface was predicted to be enhanced by photoexcitation [Hata et al. 1994]. While mechanisms other than surface diffusion have been proposed to account for the surface topography of CVD diamond [Harris and Goodwin 1993; Zhu et al. 1993a, 1993b], recent results on H_2-plasma-treated surfaces (cf. Section 2.1.3) suggest that hydrogen-atom-assisted surface diffusion of carbonaceous species is indeed significant. However, considerably more experimental work is needed to quantify the kinetics of surface migration and the role of atomic hydrogen.

6. Concluding Remarks

A great deal has been learned during the past ten years about the structure and reactivity of diamond surfaces, which are analogous in many ways to both silicon surfaces and organic molecules. However, some surface reactions have no satisfactory analogue in gas-phase organic chemistry, e.g., H_2 desorption, and steric and electronic effects have strong effects on the reactivity, particularly with halogens and hydrocarbons at high coverage. Techniques for investigating well-defined diamond surfaces have become much more sophisticated, and considerably more information should be forthcoming within the next decade.

7. Acknowledgments

The author acknowledges partial support from the National Science Foundation (CHE-9214328) for this work and thanks his students, postdoctoral fellows, and collaborators for their many contributions to the author's understanding of the surface properties of diamond.

8. References

Accary, C., Barbarat, P., Hase, W. L., and Hass, K. C. (1993) Importance of energy transfer and lattice properties in H-atom association with the (111) surface of diamond, *J. Phys. Chem.*, **97**, 9934-9941.

Aizawa, T., Ando, T., Kamo, M., and Sato, Y. (1993) High-resolution electron energy-loss spectroscopic study of epitaxially-grown diamond (111) and (100) surfaces, *Phys. Rev. B*, **48**, 18348-18351.

Aizawa, T., Ando, T., Kamo, M., and Sato, Y. (1994) Exchange of hydrogen on diamond (111) surface with gas-phase deuterium, in S. Saito et al. (eds) *Advances in New Diamond Science and Technology* (Proc. ICNDST-4), MYU, Tokyo, pp. 457-460.

Aizawa, T., Ando, T., Yamamoto, K., Kamo, M., and Sato, Y. (1995) Surface vibrational studies of CVD diamond, *Diamond Relat. Mater.*, **4**, 600-606.

Alexenko, A. E. and Spitsyn, B. V. (1991) Some properties of CVD-diamond semiconducting structures, in R. E. Clausing et al. (eds), *Diamond and Diamond-Like Films and Coatings*, NATO-ASI Series, Ser. B: Physics, Vol. 266, Plenum, New York, pp. 789-795.

Alphonso, D. R. and Ulloa, S E. (1993) Molecular-dynamics simulations of methyl-radical deposition on diamond (100) surfaces, *Phys. Rev. B*, **48**, 12235-12239.

Alphonso, D. R., Ulloa, S. E. and Brenner, D. W. (1994a) Hydrocarbon adsorption on a diamond (100) stepped surface, *Phys. Rev. B*, **49**, 4948-4953.

Alphonso, D. R., Yang, S. H., and Drabold, D. A. (1994b) *Ab initio* studies of hydrocarbon adsorption on stepped diamond surfaces, *Phys. Rev. B*, **50**, 15369-15380.

Alphonso, D. R., Drabold, D. A., and Ulloa, S. E. (1995a) Phonon modes of diamond (100) surfaces from ab initio calculations, *Phys. Rev. B*, **51**, 1989-1992.

Alphonso, D. R., Drabold, D. A., and Ulloa, S. E. (1995b) Structural, electronic, and vibrational properties of diamond (100), (111), and (110) surfaces from ab initio calculations, *Phys. Rev. B*, **51**, 14669-14685.

Alphonso, D. R., Drabold, D. A., and Ulloa, S. E. (1996) Structure of diamond(100) stepped surfaces from *ab initio* calculations, *J. Phys. Conden. Matter*, **8**, 641-647.

Ando, T., Inoue, S., Ishii, M., Kamo, M., Sato, Y., Yamada, O. and Nakano, T. (1993a) Fourier-transform infrared photoacoustic studies of hydrogenated diamond surfaces, *J. Chem. Soc.* Faraday Trans., **89**, 749-751.

Ando, T., Ishii, M., Kamo, M. and Sato, Y. (1993b) Diffuse reflectance infrared Fourier-transform study of the plasma hydrogenation of diamond surfaces, *J. Chem. Soc.* Faraday Trans., **89**, 1383-1386.

Ando, T., Ishii, M., Kamo, M. and Sato, Y. (1993c) Thermal hydrogenation of diamond surfaces studied by diffuse reflectance Fourier-transform infrared, temperature-programmed desorption and laser Raman spectroscopy, *J. Chem. Soc.* Faraday Trans., **89**, 1783-1789.

Ando, T., Yamamoto, Y., Ishii, M., Kamo, M. and Sato, Y. (1993d) Diffuse reflectance Fourier-transform infrared study of the plasma-fluorination of diamond surfaces using a microwave discharge in CF_4, *J. Chem. Soc.* Faraday Trans., **89**, 3105-3109.

Ando, T., Yamamoto, Y., Ishii, M., Kamo, M. and Sato, Y. (1993e) Vapour-phase oxidation of diamond surfaces in O_2 studied by diffuse reflectance Fourier-transform infrared and temperature-programmed desorption spectroscopy, *J. Chem. Soc.* Faraday Trans., **89**, 3635-3640.

Ando, T., Aizawa, T., Yamamoto, K., Sato, Y., and Kamo, M. (1994a) The chemisorption of hydrogen on diamond surfaces studied by high resolution electron energy-loss spectroscopy, *Diamond Relat. Mater.*, **3**, 975-979.

Ando, T., Aizawa, T., Kamo, M., Sato, Y., Anzai, T., Yamamoto, H., Wada, A., Domen, K., and Hirose, C. (1994b) A vibrational spectroscopic study of hydrogenated diamond surfaces by infrared-visible sum-frequency generation (SFG), in S. Saito et al. (eds) *Advances in New Diamond Science and Technology* (Proc. ICNDST-4), MYU, Tokyo, pp. 461-464.

Ando, T., Yamamoto, K., Suehara, S., Kamo, M., Sato, Y., Shimosaki, S., and Nishitani-Gamo, M. (1995a) Interaction of chlorine with hydrogenated diamond surfaces, *J. Chin. Chem. Soc.*, **42**, 285-292.

Ando, T., Yamamoto, K., Kamo, M., Sato, Y., Takamatsu, Y., Kawasaki, S., Okino, F., and Touhara, H. (1995b) Diffuse reflectance infrared Fourier-transform study of the direct thermal fluorination of diamond powder surfaces, *J. Chem. Soc.* Faraday Trans., **91**, 3209-3212.

Ando, T., Ishii, M., Kamo, M., and Sato, Y. (1995c) H-D exchange reaction on diamond surfaces studied by diffuse reflectance Fourier transform IR spectroscopy, *Diamond Relat. Mater.* **4**, 607-611.

Ando, T., Yamamoto, K., Matsuzawa, M. Takamatsu, Y., Kawasaki, S., Okino, F., Touhara, H., Kamo, M., and Sato, Y. (1996a) Direct interaction of fluorine with diamond surfaces, *Diamond Relat. Mater.*, **5**, 1021-1025.

Ando, T., Gamo, M. N., Rawles, R. E., Yamamoto, K., Kamo, M., and Sato, Y. (1996b) Chemical modification of diamond surfaces using a chlorinated surface as an intermediate state, *Diamond Relat. Mater.*, **5**, 1136-1142.

Anzai, T., Maeoka, H., Wada, A., Domen, K., Hirose, C., Ando, T., and Sato, Y. (1995) Vibrational sum-frequency generation spectroscopy of a homoepitaxially-grown diamond C(100) surface, *J. Molec. Struct.* **352/353**, 455-463.

Badziag, P. and Verwoerd, W. S. (1987) MNDO analysis of the oxidized diamond (100) surface, *Surf. Sci.* **183**, 469-483.

Badziag, P. and Verwoerd, W. S. (1988) Cluster modeling of the 2×1 chain reconstruction of the diamond (111) surface, *Surf. Sci.* **194**, 535-547.

Badzian, A; Badzian, T; (1993) High temperature epitaxy of diamond, in J. P. Dismukes and K. V. Ravi (eds), *Diamond Materials*, The Electrochemical Society, Pennington, New Jersey, pp. 1060-1066.

Bai, B. J., Chu, C. J., Patterson, D. E., Hauge, R. H., and Margrave, J. L. (1993) Methyl halides as carbon sources in a hot-filament diamond CVD reactor: A new gas phase growth species, *J. Mater. Res.*, **8**, 233-236.

Bansal, R. C., Vastola, F. J., and Walker, P. L., Jr. (1972) Kinetics of chemisorption of oxygen on diamond, *Carbon*, **10**, 443-448.

Barbarat, P., Accary, C., and Hase, W. L. (1993) Comparison of canonical variational transitition state theory rate constants for H atom association with alkyl radicals and with the (111) surface of diamond, *J. Phys. Chem.*, **97**, 11706-11711.

Baumann, P. K., Humphreys, T. P., and Nemanich, R. J. (1994) Comparison of surface cleaning processes for diamond (100), *Mater. Res. Soc. Symp. Proc.*, **339**, 69-74.

Bechstedt, F. and Reichardt, D. (1988) Total energy minimization for surfaces of covalent semiconductors C, Si, Ge, and α-Sn. II. (100)2×1 surfaces, *Surf. Sci.*, **202**, 83-98.

Beerling, T. E. and Helms, C. R. (1994) Effects of oxygen and hydrogen adsorption on the electron energy loss features of diamond surfaces, *Appl. Phys. Lett.*, **65**, 1912-1914.

Belton, D. N. and Harris, S. J. (1992) A mechanism for growth on diamond (110) from acetylene, *J. Chem. Phys.*, **96**, 2371-2377.

Besler, B. H., Hase, W. L., and Hass, K. C. (1992) A theoretical study of growth mechanisms of the (110) surface of diamond from acetylene and hydrogen mixtures, *J. Phys. Chem.*, **96**, 9369-9376.

Boland, J. J. (1991) Evidence of pairing and its role in the recombinative desorption of hydrogen from the Si(100)-2×1 surface, *Phys. Rev. Lett.*, **67**, 1539-1542.

Bratu, P. and Höfer, U. (1995) Phonon-assisted sticking of molecular hydrogen on Si(111)-(7×7), *Phys. Rev. Lett.*, **74**, 1625-1628.

Brenner, D. W. (1990) Empirical potential for hydrocarbons for use in simulating the chemical vapor deposition of diamond films, *Phys. Rev. B*, **42**, 9458-9471; (1992) Erratum, *Phys. Rev. B*, **46**, 1948.

Brenner, D. J., Robertson, D. H., Carty, R. J., Srivastava, D., and Garrison, B. J. (1992) Combining molecular dynamics and Monte Carlo simulations to model chemical vapor deposition: Application to diamond, *Mater. Res. Soc. Symp. Proc.*, **278**, 255-260.

Brenner, D. J. and Harrison, J. A. (1992) Atomistic simulations of diamond films, *Am. Ceram. Soc. Bull.*, **71**, 1821.

Brink, J. H. and Verwoerd, W. S. (1996) Calculation of edge and vertex energies of diamond microcrystals, *Diamond Relat. Mater.*, **5**, 990-994.

Buck, M. and Schaich, Th. (1995) Optical second-harmonic generation on the diamond C(111) surface, *Diamond Relat. Mater.*, **4**, 544-547.

Busmann, H.-G., Zimmermann-Edling, W., Sprang, H., Güntherodt, H.-J., and Hertel, I. V. (1992) Surface views of polycrystalline diamond films: microtwins and flat faces, constricted and free atomic layer motion, *Diamond Rel. Mater.*, **1**, 979-988.

Busmann, H.-G., Lauer, S., Hertel, I. V., Zimmermann-Edling, W., Güntherodt, H.-J., Frauenheim, Th., Blaudeck, and Porezag, D. (1993) Observation of ($\sqrt{3} \times \sqrt{3}$)R30° diamond (111) on vapour-grown polycrystalline films, *Surf. Sci.*, **295**, 340-346.

Cadman, P., Scott, J. D., and Thomas, J. M. (1975) Identification of functional groups on the surface of a fluorinated diamond crystal by photoelectron spectroscopy, *J. Chem. Soc.* Chem. Commun., 654-655.

Cappelli, M. A. and Loh, M. H. (1994) *In-situ* mass spectrometric sampling during supersonic arcjet synthesis of diamond, *Diamond Relat. Mater.*, **3**, 417-421.

Cavell, R. G., Kowalczyk, S. P., Ley, L., Pollak, R. A., Mills, B., Shirley, D. A., and Perry, W. (1973) X-ray photoemission cross-section modulation in diamond, silicon, germanium, methane, silane, and germane, *Phys. Rev. B*, **7**, 5313-5316.

Celii, F. G., Pehrsson, P. E., Wang, H. T., and Butler, J. E. (1989) Infrared detection of gaseous species during the filament-assisted growth of diamond, *Appl. Phys. Lett.*, **52**, 2043-2045.

Chabal, Y. J. and Raghavachari, K. (1984) Surface infrared study of Si(100)-(2×1)H, *Phys. Rev. Lett.*, **53**, 282-285.

Chadi, D. J. (1984) Multiple bonding on C(111)-2×1 surfaces: Surface structural determination from energy minimization, *J. Vac. Sci. Technol.* A, **2**, 948-951.

Chang, H. C. and Lin, J.-C. (1995) Infrared spectroscopy of adsorbed CO_2 as a probe for the surface heterogeneity of diamond (111)-1×1:H, *Appl. Phys. Lett.*, **67**, 2474-2476.

Chang, H. C. and Lin, J.-C. (1996) Structure and bonding of C_2H_2 on diamond C(111)-1×1:H: Infrared spectroscopy and exciton calculations, *J. Phys. Chem.*, **100**, 7018-7025.

Chang, H. C., Lin, J.-C., Wu, J. Y., and Chen, K.-H. (1995) Infrared spectroscopy and vibrational relaxation of CH_x and CD_x stretches on synthetic diamond nanocrystal surfaces, *J. Phys. Chem.*, **99**, 11081-11088.

Chang, X. Y., Thompson, D. L., and Raff, L. M. (1993a) Minimum-energy paths for elementary reactions in low-pressure diamond-film formation, *J. Phys. Chem.*, **97**, 10112-10118.

Chang, X. Y., Perry, M., Peploski, J., Thompson, D. L., and Raff, L. M. (1993b) Theoretical studies of hydrogen-abstraction reactions from diamond and diamond-like surfaces, *J. Chem. Phys.*, **99**, 4748-4758.

Chang, X. Y., Thompson, D. L., and Raff, L. M. (1994) Hydrogen-atom migration on a diamond (111) surface, *J. Chem. Phys.*, **100**, 1765-1766.

Chin, R. P., Huang, J. Y., Shen, Y. R., Chuang, T. J., Seki, H., and Buck, M. (1992) Vibrational spectra of hydrogen on diamond C(111)-(1×1), *Phys. Rev. B*, **45**, 1522-1524.

Chin, R. P., Huang, J. Y., Shen, Y. R., Chuang, T. J., and Seki, H. (1995a) Interaction of atomic hydrogen with the diamond C(111) surface studied by infrared-visible sum-frequency-generation spectroscopy, *Phys. Rev. B*, **52**, 5985-5995.

Chin, R. P., Blase, X., Shen, Y. R., and Louie, S. G. (1995b) Anharmonicity and lifetime of the CH stretch mode on diamond H/C(111)-(1×1), *Europhys. Lett.*, **30**, 399-404.

Chin, R. P., Huang, J. Y., Shen, Y. R., Chuang, T. J., and Seki, H. (1996) Interactions of hydrogen and methyl radicals with diamond C(111) studied by sum-frequency vibrational spectroscopy, *Phys. Rev. B*, **54**, 8243-8251.

Chu, C. J., Hauge, R. H., Margrave, J. L., and D'Evelyn, M. P. (1992) Growth kinetics of (100), (110), and (111) homoepitaxial films, *Appl. Phys. Lett.*, **61**, 1393-1395.

Chu, C. J., Pan, C., Margrave, J. L., and Hauge, R. H. (1995) F_2, H_2O, and O_2 etching rates of diamond and the effects of F_2, HF, and H_2O on the molecular O_2 etching of (110) diamond, *Diamond Relat. Mater.* **4**, 1317-1324.

Chua, L. H., Jackman, R. B., Foord, J. S., Chalker, P. R., Johnston, C. and Romani, S. (1994) Interaction of hydrogen with chemical vapor deposition diamond surfaces: A thermal desorption study, *J. Vac. Sci. Technol.* A, **12**, 3033-3039.

Clausing, R. E., Heatherly, L., Horton, L. L., Specht, E. D., Begun, G. M., and Wang, Z. L. (1992) Textures and morphologies of chemical vapor deposited (CVD) diamond, *Diamond Relat. Mater.*, **1**, 411-415.

Coltrin, M. E. and Dandy, D. S. (1993) Analysis of diamond growth in subatmospheric dc plasma-gun reactors, *J. Appl. Phys. Lett. Phys.*, **74**, 5803-5820.

Couto, M., van Enckevort, W. J. P., Wichman, B., and Seal, M. (1992) Scanning tunneling microscopy of polished diamond surfaces, *Appl. Phys. Lett. Surf. Sci.*, **62**, 263-268.

Couto, M. S., van Enckevort, W. J. P., and Seal, M. (1994a) Diamond polishing mechanisms: An investigation by scanning tunneling microscopy, *Phil. Mag. B*, **69**, 621-641.

Couto, M. S., van Enckevort, W. J. P., and Seal, M. (1994b) On the mechanism of diamond polishing in the soft directions, *J. Hard Mater.*, **5**, 31-47.

Davidson, B. N. and Pickett, W. E. (1994a) Tight-binding study of hydrogen on the C(111), C(100), and C(110) diamond surfaces, *Phys. Rev. B*, **49**, 11253-11267.

Davidson, B. N. and Pickett, W. E. (1994b) Graphite-layer formation at a diamond (111) step, *Phys. Rev. B*, **49**, 14770-14773.

Dawnkaski, E. J., Srivastava, D., and Garrison, B. J. (1995a) Time-dependent Monte Carlo simulations of radical densities and distributions on the diamond (001)(2×1):H surface, *Chem. Phys. Lett.*, **232**, 524-530.

Dawnkaski, E. J., Srivastava, D., and Garrison, B. J. (1995b) Time-dependent Monte Carlo simulations of H reactions on the diamond (001)(2×1) surface under chemical vapor deposition conditions, *J. Chem. Phys.*, **102**, 9401-9411.

Dawnkaski, E. J., Srivastava, D., and Garrison, B. J. (1996) Growth of diamond films on a diamond (001)(2×1):H surface by time-dependent Monte Carlo simulations, *J. Chem. Phys.*, **104**, 5997-6008.

Derry, T. E., Madiba, C. C. P. and Sellschop, J. F. P. (1983) Oxygen and hydrogen on the surface of diamond, *Nucl. Instrum. Methods*, **218**, 559-562.

Derry, T. E., Smit, L., and van der Ween, J. F. (1986) Ion scattering determination of the atomic arrangement at polished diamond (111) surfaces before and after reconstruction, *Surf. Sci.*, **167**, 502-518.

de Sainte Claire, P., Barbarat, P., and Hase, W. L. (1994) *Ab initio* potential and variational transition state theory rate constant for H-atom association with the diamond (111) surface, *J. Chem. Phys.*, **101**, 2476-2488.

de Sainte Claire, P., Song, K., Hase, W. L., and Brenner, D. W. (1996) Comparison of ab initio and empirical potentials for H-atom association with diamond surfaces, *J. Phys. Chem.*, **100**, 1761-1766.

D'Evelyn, M. P., Yang, Y. L., and Sutcu, L. F. (1992a) π-bonded dimers, preferential pairing, and first-order desorption kinetics of hydrogen on Si(100)-(2×1), *J. Chem. Phys.*, **96**, 852-855.

D'Evelyn, M. P., Chu, C. J., Hauge, R. H., and Margrave, J. L. (1992b) Mechanism of diamond growth by hot-filament chemical vapor deposition: Carbon-13 studies, *J. Appl. Phys. Lett. Phys.*, **71**, 1528-1530.

D'Evelyn, M. P., Struck, L. M., and Rawles, R. E. (1994) Surface cleaning, topography, and temperature measurements of single crystal diamond, *Mater. Res. Soc. Symp. Proc.*, **339**, 89-95.

Dovesi, R., Pisani, C., Roetti, C., and Ricart, J. M. (1987) Periodic MINDO/3 study of the reconstructed (111) surface of diamond, *Surf. Sci.* **185**, 120-124.

Dyson, A. J. and Smith, P. V. (1994) Empirical molecular dynamics calculations for the (001) and (111) 2×1 reconstructed surfaces of diamond, *Surf. Sci.*, **316**, 309-316.

Ermer, O., Mason, S. A., Anet, F. A., and Miura, S. S. (1985) Ultrashort nonbonded H···H distance in a half-cage pentacyclododecane, *J. Am. Chem. Soc.*, **107**, 2330-2334.

Ertl, G. and Küppers, J. (1985) *Low Energy Electrons and Surface Chemistry*, 2nd. edition, VCH, Weinheim.

Evans, S. and Thomas, J. M. (1977) The chemical nature of ion-bombarded carbon: A photoelectron spectroscopic study of 'cleaned' surfaces of diamond and graphite, *Proc. Roy. Soc. Lond. A*, **353**, 103-120.

Evans, S. (1978) Depth profiles of ion-induced structural changes in diamond from X-ray photoelectron spectroscopy, *Proc. Roy. Soc. Lond. A*, **360**, 427-443.

Evans, S. and Riley, C. E. (1986) Angular dependence of X-ray excited valence-band photoelectron spectra of diamond, *J. Chem. Soc.*, Faraday Trans. 2, **82**, 541-550.

Evans, S. (1992) Surface properties of diamond, in J. E. Field (ed), *The Properties of Natural and Synthetic Diamond*, Academic Press, London, pp. 181-214.

Evans, T. (1979) Changes produced by high temperature treatment of diamond, in J. E. Field (ed), *The Properties of Diamond*, Academic Press, London, pp. 403-424.

Everson, M. P.; Tamor, M. A.; Scholl, D.; Stoner, B. R.; Sahaida, S. R.; Bade, J. P. (1994) Positive identification of the ubiquitous triangular defect on the (100) faces of vapor-grown diamond, *J. Appl. Phys. Lett. Phys.*, **75**, 169-172.

Feng, K. A. and Lin, Z. D. (1994) Adsorption action of methylene CH_2 during diamond film growth on (100) surface by hot filament chemical vapor deposition, *J. Phys. Chem. Solids*, **55**, 525-529.

Fox, C. A., Kubiak, G. D., Schulberg, M. T., and Hagstrom, S. (1993) Kinetics of hydrogen desorption from the diamond (110) surface, in J. P. Dismukes and K. V. Ravi (eds), *Diamond Materials*, The Electrochemical Society, Pennington, New Jersey, pp. 64-70.

Fox, C. A., Kubiak, G. D., and Schulberg, M. T. (1995) Unpublished work quoted by Schulberg et al. (1995).

Francz, G. and Oelhafen, P. (1995a) Valence band spectroscopy of reconstructed (100) and (111) natural diamond, *Diamond Relat. Mater.*, **4**, 539-543.

Francz, G. and Oelhafen, P. (1995b) Photoelectron spectroscopy of the annealed and deuterium-exposed natural diamond (100) surface, *Surf. Sci.*, **329**, 193-198.

Francz, G., Kania, P., Gantner, G., Stupp, H., and Oelhafen, P. (1996) Photoelectron spectroscopy study natural (100), (110), (111) and CVD diamond surfaces, *Phys. Stat. Solidi* (a), **154**, 91-108.

Frauenheim, Th., Stephan, U., Blaudeck, P., Porezag, D., Busmann, H.-G.; Zimmermann-Edling, W. and Lauer, S. (1993) Stability, reconstruction, and electronic properties of diamond (100) and (111) surfaces, *Phys. Rev. B*, **48**, 18189-18202.

Frauenheim, Th., Köhler, Th., Sternberg, M., Porezag, D., and Pederson, M. R. (1996) Vibrational and electronic signatures of diamond surfaces, *Thin Solid Films*, **272**, 314-330.

Freedman, A. and Stinespring, C. D. (1990) Fluorination of diamond (100) by atomic and molecular beams, *Appl. Phys. Lett.*, **57**, 1194-1196.

Freedman, A. (1994) Halogenation of diamond (100) and (111) surfaces by atomic beams, *J. Appl. Phys.*, **75**, 3112-3120.

Frenklach, M. and Wang, H. (1991) Detailed surface and gas-phase chemical kinetics of diamond deposition, *Phys. Rev. B*, **43**, 1520-1545.

Frenklach, M., Huang, D., Thomas, R. E., Rudder, R. A., and Markunas, R. J. (1993) Activation energy and mechanism of CO desorption from (100) diamond surface, *Appl. Phys. Lett.*, **63**, 3090-3092.

Frenklach, M., Skokov, S., and Weiner, B. (1994) An atomistic model for stepped diamond growth, *Nature*, **372**, 535-537.

Furthmüller, J., Hafner, J. and Kresse, G. (1994) Structural and electronic properties of clean and hydrogenated diamond (100) surfaces, *Europhys. Lett.* **28**, 659-664.

Garrison, B. J., Dawnkaski, E. J., Srivastava, D., and Brenner, D. W. (1992) Molecular dynamics simulations of dimer opening on a diamond (001)-(2$\times$1) surface, Science, **255**, 835-838.

Gat, R. and Angus, J. C. (1993) Hydrogen atom recombination on tungsten and diamond in hot filament assisted deposition of diamond, *J. Appl. Phys.*, **74**, 5981-5989.

Gavrilenko, V. I. and Shkrebtii, A. I. (1995) Anisotropy of optical reflectance of the (001) surface of diamond, *Surf. Sci.*, **324**, 226-232.

Goodwin, D. G. and Gavillet, G. G. (1990) Numerical modeling of the filament-assisted diamond growth environment, *J. Appl. Phys.*, **68**, 6393-6400.

Gordeev, S. K., Smirnov, E. P., Kol'tsov, S. I., and Aleskovskii, V. B. (1980) *Zh. Prikl. Khim.*, **53**, 94-96.

Gordeev, S. K., Smirnov, E. P., and Aleskovskii, V. B. (1981) Mutual influence of functional groups in substitution reactions on diamond surface, *Dokl. Akad. Nauk SSSR*, **261**, 127-130.

Graupner, R., Ristein, J., and Ley, L. (1995) Photoelectron spectroscopy of clean and hydrogen-exposed diamond (111) surfaces, *Surf. Sci.*, **320**, 201-207.

Grodzinski, P. (1953) *Diamond Technology*, 2nd edition, N. A. G. Press, London, Chapter VI.

Gruen, D. M., Liu, S., Krauss, A. R., and Pan, X. (1994) Buckyball microwave plasmas: Fragmentation and diamond film growth, *J. Appl. Phys.*, **75**, 1758-1763.

Gutman, D. (1990) The controversial heat of formation of the *t*-C_4H_9 radical and the tertiary C-H bond energy, *Acc. Chem. Res.*, **23**, 375-380.

Haisma, J., van der Kruis, F. J. H. M., Spierings, B. A. C. M., Oomen, J. M. and Fey, F. M. J. G. (1992) Damage-free tribochemical polishing of diamond at room temperature: a finishing technology, *Precision Engineering*, **14**, 20-27.

Halicioglu, T. (1992) (2×1) Reconstructed patterns of diamond (100) surface, *Diamond Relat. Mater.*, **1**, 963-967.

Hamza, A. V., Kubiak, G. D. and Stulen, R. H. (1988) The role of hydrogen on the diamond C(111)-(2×1) reconstruction, *Surf. Sci.*, **206**, L833-L844.

Hamza, A. V. and Kubiak, G. D. (1989) Adsorption of NO on diamond C(111)-(2×1) by band-gap excitation, *J. Vac. Sci. Technol.* B, 7, 1165-1170.

Hamza, A. V., Kubiak, G. D. and Stulen, R. H. (1990) Hydrogen chemisorption and the structure of the diamond C(100)-(2×1) surface, *Surf. Sci.*, **237**, 35-52; Kubiak, G. D., Schulberg, M. T., and Stulen, R. H. (1992) Erratum, *Surf. Sci.*, **277**, 234.

Hansen, D. A., Halbach, M. R., and Seebauer, E. G. (1996) Experimental measurements of fast adsorption kinetics of H_2 on vicinal Si(100) and (111) surfaces, *J. Chem. Phys.*, **104**, 7338-7343.

Hansen, J. O., Copperthwaite, R. G., Derry, T. E., and Pratt, J. M. (1989) A tensiometric study of diamond (111) and (100) faces, *J. Coll. Interface Sci.* **130**, 347-358.

Harris, S. J., Weiner, A. M., and Perry, T. A. (1988) Measurement of stable species present during filament-assisted diamond growth, *Appl. Phys. Lett.*, **53**, 1605-1607.

Harris, S. J. (1990) Mechanism for diamond growth from methyl radicals, *Appl. Phys. Lett.*, **56**, 2298-2300.

Harris, S. J. and Martin, L. R. (1990) Methyl versus acetylene as diamond growth species, *J. Mater. Res.*, **5**, 2313-2319.

Harris, S. J., Belton, D. N., and Blint, R. J. (1991) Thermochemistry on the hydrogenated diamond (111) surface, *J. Appl. Phys.* **70**, 2654-2659.

Harris, S. J. and Belton, D. N. (1991) Thermochemistry on a fluorinated diamond (111) surface, *Appl. Phys. Lett.*, **59**, 1949-1951.

Harris, S. J. and Belton, D. N. (1992) Diamond growth on a (100)-type step, *Thin Solid Films*, **212**, 193-200.

Harris, S. J. and Goodwin, D. G. (1993) Growth on the reconstructed diamond (100) surface, *J. Phys. Chem.*, **97**, 23-28.

Harris, S. J. and Weiner, A. M. (1993) Reaction kinetics on diamond: Measurement of H atom destruction rates, *J. Appl. Phys.*, **74**, 1022-1026.

Hata, M., Tsuda, M., Fujii, N., and Oikawa, S. (1994) Surface migration enhancement of adatoms in the photoexcited process on reconstructed diamond (001) surfaces, *Surf. Sci.*, **79/80**, 255-263.

Hayashi, K., Yamanaka, S., Okushi, H., and Kajimura, K. (1996) Stepped growth and etching of (001) diamond, *Diamond Relat. Mater.*, **5**, 1002-1005.

Heggie, M. I., Latham, C. D., Jones, B. and Briddon, P. R. (1996) Local density functional modeling of diamond growth and graphitisation, in K. V. Ravi and J. P. Dismukes (eds), *Diamond Materials IV*, The Electrochemical Society, Pennington, New Jersey, pp. 643-648.

Himpsel, F. J.; Knapp, J. A.; Van Vechten, J. A.; Eastman, D. E. (1979) Quantum photoyield of diamond (111)–a stable negative-affinity emitter, *Phys. Rev. B*, **20**, 624-627.

Himpsel, F. J., Eastman, D. E., Heimann, P., and van der Ween, J. F. (1981) Surface states on reconstructed diamond (111), *Phys. Rev. B*, **24**, 7270-7274.

Hiribayashi, K. (1992) Surface microstructures of diamond crystals synthesized in the H_2-CH_4-O_2 system, *J. Appl. Phys.*, **72**, 4083-4087.

Hoffman, A., Prawer, S., and Folman, M. (1991) Secondary electron emission spectroscopy: A sensitive and novel method for the characterization of the near-surface of diamond and diamond films, *Appl. Phys. Lett.*, **58**, 361-363.

Hoffman, A. (1994) Fine structure in the secondary electron emission spectrum as a spectroscopic tool for carbon surface characterization, *Diamond Relat. Mater.*, **3**, 691-695.

Hoffman, A., Bobrov, K., Fisgeer, B., Shechter, H., and Folman, M. (1996) Effects of deuterium adsorption-desorption on the state of diamond: surface degradation and stabilization of sp^3 bonded carbon, *Diamond Relat. Mater.*, **5**, 977-983.

Hoffman, F. M. (1983) Infrared reflection-absorption spectroscopy of adsorbed molecules, *Surf. Sci. Reports*, **3**, 107-192.

Hsu, W. L. (1991) Mole fractions of H, CH_3, and other species during filament-assisted diamond growth, *Appl. Phys. Lett.*, **59**, 1427-1429.

Huang, D., Frenklach, M., and Maroncelli, M. (1988) Energetics of acetylene-addition mechanism of diamond growth, *J. Phys. Chem.*, **92**, 6379-6381.

Huang, D. and Frenklach, M. (1992) Energetics of surface reactions on (100) diamond plane, *J. Phys. Chem.*, **96**, 1868-1875.

Hudson, J. B. (1992) *Surface Science: An Introduction*, Butterworth-Heinemann, Boston.

Hukka, T. I., Rawles, R. E. and D'Evelyn, M. P. (1993) Novel method for chemical vapor deposition and atomic layer epitaxy using radical chemistry, *Thin Solid Films*, **225**, 212-218.

Hukka, T. I., Pakkanen, T. A. and D'Evelyn, M. P. (1994) Chemisorption of hydrogen on the diamond (100)2×1 surface - an ab initio study, *J. Phys. Chem.*, **98**, 12420-12430.

Hukka, T. I., Pakkanen, T. A. and D'Evelyn, M. P. (1995) Chemisorption of fluorine, chlorine, HF, and HCl on the diamond (100)2×1 surface: An *ab initio* study, *J. Phys. Chem.*, **99**, 4710-4719.

Hukka, T. I., Pakkanen, T. A. and D'Evelyn, M. P. (1996) CF_3 radicals as growth precursors and halogen-assisted growth on diamond (100)2×1: An *ab initio* study, *Surf. Sci.* (in press).

Humphreys, T. P., Posthill, J. B., Malta, D. P., Thomas, R. E., Hudson, G. C., and Markunas, R. J. (1994) Surface preparation of single crystal C(001) substrates for homoepitaxial diamond growth, *Mater. Res. Soc. Symp. Proc.* **339**, 51-56.

Iarlori, S., Galli, G., Gygi, F., Parrinello, M., Tosatti, E. (1992) *Phys. Rev. Lett.*, **69**, 2947-2950.

Janssen, G., van Enckevort, W. J. P., Vollenberg, W., and Giling, L. J. (1992) Characterization of single-crystal diamond grown by chemical vapor deposition processes, *Diamond Relat. Mater.*, **1**, 789-800.

Jiang, T., Xu, K., and Ji, S. (1996) FTIR studies on the spectral changes of the surface functional groups of ultradispersed diamond powder synthesized by explosive detonation after treatment in hydrogen, nitrogen, methane, and air at different temperatures, *J. Chem. Soc.* Faraday Trans., **92**, 3401-3406.

Jing, Z. and Whitten, J. L. (1994a) Ab initio studies of diamond (100) surface reconstruction, *Phys. Rev. B*, **50**, 2598-2605.

Jing, Z. and Whitten, J. L. (1994b) Ab initio studies of H chemisorption on C(100) surface, *Surf. Sci.*, **314**, 300-306.

Johansson, E. and Carlsson, J.-O. (1995) Effect of atomic hydrogen on the surface topography of chemically vapour deposited diamond films: an atomic force microscopy study, *Diamond Relat. Mater.*, **4**, 155-163.

Johnson, C. E., Weimer, W. A., and Cerio, F. M. (1992) Efficiency of methane and acetylene in forming diamond by microwave plasma assisted chemical vapor deposition, *J. Mater. Res.*, **7**, 1427-1431.

Kawarada, H., Sasaki, H., and Sato, A. (1995) Scanning-tunneling-microscope observation of the homoepitaxial diamond (001) 2×1 reconstruction observed under atmospheric pressure, *Phys. Rev. B*, **52**, 11351-11358.

Kern, G., Hafner, J., Furthmüller, J. and Kresse, G. (1996a) (2×1) reconstruction and hydrogen-induced de-reconstruction of the diamond (100) and (111) surfaces, *Surf. Sci.*, **352-354**, 745-749.

Kern, G., Hafner, J., Furthmüller, J. and Kresse, G. (1996b) Surface reconstruction and electronic properties of clean and hydrogenated diamond (111) surfaces, *Surf. Sci.*, **357-358**, 422-426.

Khvostov, V. V., Guseva, M. B., Babaev, V. G. and Rylova, O. Yu. (1985) Transformation of diamond and graphite surfaces by ion irradiation, Solid State Commun., **55**, 443-445.

Klauser, R., Chuang, T. J., Smoliar, L. A., and Tzeng, W.-T. (1996a) Adsorption of methyl radicals on diamond C(111) surface studied by synchrotron radiation photoemission, *Chem. Phys. Lett.*, **255**, 32-38.

Klauser, R., Chen, J.-M., Chuang, T. J., Chen, L. M., Shih, M. C., and Lin, J.-C. (1996b) The interaction of oxygen and hydrogen on a diamond C(111) surface: a synchrotron radiation photoemission, LEED and AES study, *Surf. Sci.*, **356**, L410-L416.

Kobayshi, K. (1994) A calculation of diamond C(100)-2×1,2×2 surfaces by using first-principles molecular dynamics, in S. Saito et al. (eds) *Advances in New Diamond Science and Technology* (Proc. ICNDST-4), MYU, Tokyo, pp. 255-258.

Köhler, Th., Sternberg, M., Porezag, D., and Frauenheim, Th. (1996) Surface properties of diamond (111): 1×1, 2×1, and 2×2 reconstructions, Phys. *Stat. Solidi (a),* **154**, 69-89.

Kolasinski, K. W. (1995) Dynamics of hydrogen interactions with Si(100) and Si(111) surfaces, *Int. J. Mod. Phys. B*, **9**, 2753-2809.

Koleske, D. D., Gates, S. M., Thoms, B. D., Russell, J. N., Jr., and Butler, J. E. (1994) Isothermal desorption of hydrogen from polycrystalline diamond films, *Surf. Sci.*, **320**, L105-L111.

Koleske, D. D., Gates, S. M., Thoms, B. D., Russell, J. N., Jr., and Butler, J. E. (1995) Hydrogen on polycrystalline diamond films: Studies of isothermal desorption and atomic deuterium abstraction, *J. Chem. Phys.*, **102**, 992-1002.

Komatsu, S. (1996) Stable anionic site on hydrogenated (111) surface of diamond resulting from hydrogen atom removal under chemical vapor deposition conditions, *J. Appl. Phys.*, **80**, 3319-3326.

Kondoh, E., Tanaka, K. and Ohta, T. (1993) Reflection high-energy electron diffraction observation of anti-phase domain ordering of the 2×1 reconstructed (111) surface of chemical-vapor-deposited diamond, *Jpn. J. Appl. Phys.*, **7A**, L947-L949.

Krasnoperov, L. N., Kalinovski, I. J., Chu, H.-N., and Gutman, D. (1993) Heterogeneous reactions of H atoms and CH_3 radicals with a diamond surface in the 300-1133 K temperature range, *J. Phys. Chem.*, **97**, 11787-11796.

Kress, C., Fiedler, M. and Bechstedt, F. (1994a) Quasi-particle bands for C(111) 2×1 surfaces–Support for the dimerized π-bonded chain model, *Europhys. Lett.*, **28**, 433-438.

Kress, C., Fiedler, M., Schmidt, W. G. and Bechstedt, F. (1994b) Geometrical and electronic structure of the reconstructed diamond (100) surface, *Phys. Rev. B*, **50**, 17697-17700.

Kreutz, T. J., Clausing, R. L., Heatherly, L. Jr., Warmack, R. J., Thundat, T., Feigerle, C. S. and Wandelt, K. (1995) Growth mechanisms and defects in boronated CVD diamond as identified by scanning tunneling microscopy, *Phys. Rev. B*, **51**, 14554-14558.

Krüger, P. and Pollman, J. (1995) Dimer reconstruction of diamond, Si, and Ge (001) surfaces, *Phys. Rev. Lett.*, **74**, 1155-1158.

Kuang, Y., Lee, N., Badzian, A., Tsong, T. T., Badzian, T., and Chen, C. (1995a) Study of antiphase boundaries and local configuration on the (001) surface of homoepitaxial diamond films by scanning tunneling microscopy, *Diamond Relat. Mater.*, **4**, 1371-1375.

Kuang, Y., Wang, Y., Lee, N., Badzian, A., Badzian, T., and Tsong, T. T. (1995b) Surface structure of homoepitaxial diamond (001) films, a scanning tunneling microscopy study, *Appl. Phys. Lett.*, **67**, 3721-3723.

Kuang, Y., Badzian, A., Tsong, T. T., Lee, N., Badzian, T., and Chen, C. (1996) Scanning tunneling microscopy study of antiphase boundaries on the (001) surface of homoepitaxial diamond films, *Thin Solid Films*, **272**, 49-51.

Kubiak, G. D. and Kolasinski, K. W. (1989) Normally unoccupied states on C(111) (diamond) (2×1): Support for a relaxed π-bonded chain model, *Phys. Rev. B*, **39**, 1381-1384.

Kühnle, J. and Weis, O. (1995) Mechanochemical superpolishing of diamond using $NaNO_3$ or KNO_3 as oxidizing agents, *Surf. Sci.*, **340**, 16-22.

Küttel, O. M., Diederich, L., Schaller, E., Carnal, O., and Schalpbach, L. (1995) The preparation and characterization of low surface roughness (111) and (100) natural diamonds by hydrogen plasma, *Surf. Sci.*, **337**, L812-L818.

LaFemina, J. P. (1992) Total-energy calculations of semiconductor surface reconstructions, *Surf. Sci. Reports*, **16**, 133-260.

Lander, J. J. and Morrison, J. (1966) Low energy electron diffraction study of the (111) diamond surface, *Surf. Sci.*, **4**, 241-246.

Lange, G. and Toennies, J. P. (1996) Helium-atom-scattering measurements of surface-phonon dispersion curves of the C(111)-H(1×1) surface, *Phys. Rev. B*, **53**, 9614-9617.

Larsson, K., Lunell, S., and Carlsson, J.-O. (1993) Adsorption of hydrocarbons on a diamond (111) surface: An *ab initio* quantum-mechanical study, *Phys. Rev. B*, **48**, 2666-2674; Erratum (1995) *Phys. Rev. B*, **52**, 8564.

Larsson, K., Carlsson, J.-O., and Lunell, S. (1995) Nearest-neighbor influence on hydrocarbon adsorption on diamond (111) studied by *ab initio* calculations, *Phys. Rev. B*, **51**, 10003-10012-2674.

Larsson, K. (1996) Reaction of diamond surfaces with halogen-containing species, *Mater. Res. Soc. Symp. Proc.*, **416**, 287-291.

Latham, C. D., Heggie, M. I., and Jones, R. (1993) *Ab initio* energetics of CVD growth reactions on the three low-index surfaces of diamond, *Diamond Relat. Mater.*, **2**, 1493-1499.

Lee, J. J., Komarov, S. F., Hudson, J. B., Stokes, E. B. and D'Evelyn, M. P. (1997) Growth of diamond films from a continuous or interrupted CF_4 supply, *Diamond Relat. Mater.* (in press).

Lee, N. and Badzian, A. (1995) Effect of misorientation angles on the surface morphologies of (001) homoepitaxial diamond thin films, *Appl. Phys. Lett.*, **66**, 2203-2205.

Lee, N. and Badzian, A. (1996) H-plasma-annealed and homoepitaxially grown diamond (001) surface structure studied with high-energy electron diffraction, *Phys. Rev. B*, **53**, R1744-R1747.

Lee, S. S., Minsek, D. W., Vestyck, D. J., and Chen, P. (1994) Growth of diamond from atomic hydrogen and a supersonic free jet of methyl radicals, *Science,* **263**, 1596-1598.

Lee, S.-T. and Apai, G. (1993) Surface phonons and CH vibrational modes of diamond (100) and (111) surfaces, *Phys. Rev. B*, **48**, 2684-2693.

Lin, J.-C., Chen, K.-H., Chang, H.-C., Tsai, C.-S., Lin, C.-E., and Wang, J.-K. (1996) The vibrational dephasing and relaxation of CH and CD stretches on diamond surfaces: An anomaly, *J. Chem. Phys.*, **105**, 3975-3983.

Loh, K. P., Foord, J. S., Jackman, R. B., and Singh, N. K. (1996) Growth studies of thin film diamond using molecular beam techniques, *Diamond Relat. Mater.*, **5**, 231-235.

Lurie, P. G. and Wilson, J. M. (1977a) The diamond surface. I. The structure of the clean surface and the interaction with gases and metals, *Surf. Sci.*, **65**, 453-475.

Lurie, P. G. and Wilson, J. M. (1977b) The diamond surface. II. Secondary electron emission, *Surf. Sci.*, **65**, 476-498.

Mackey, B. L.; Russell, J. N. Jr.; Pehrsson, P. E.; Crowell, J. E.; Butler, J. E. (1995a) The effect of hydrogen on the surface electrical conductivity of diamond (110), in K. V. Ravi and J. P. Dismukes (eds), *Diamond Materials IV*, The Electrochemical Society, Pennington, New Jersey, pp. 455-460.

Mackey, B. L.; Russell, J. N. Jr.; Crowell, J. E.; Butler, J. E. (1995b) Effect of surface termination on the electrical conductivity and broad-band internal infrared reflectance of a diamond (110) surface, *Phys. Rev. B*, **52**, R17009-R17012.

Maguire, H. G. (1976) Investigation of the band structure of diamond using low-energy electrons, *Phys. Stat. Solidi (b)*, **76**, 715-726.

Maguire, H. G., Kamo, M., Lang, H. P., Meyer, E., Weissendanger, K., Guntherodt, H.-P. (1992) The structure of conducting and nonconducting homoepitaxial diamond films, *Diamond Relat. Mater.*, **1**, 634-638.

Marchywka, M., Pehrsson, P. E., Binari, S. C., and Moses, D. (1993) Electrochemical patterning of amorphous carbon on diamond, *J. Electrochem. Soc.*, **140**, L19-L22.

Marsh, J. B. and Farnsworth, H. E. (1964) Low-energy electron diffraction studies of (100) and (111) surfaces of semiconducting diamond, *Surf. Sci.*, **1**, 3-21.

Martin, L. R. (1993) High-quality diamonds from an acetylene mechanism, *J. Mater. Sci. Lett.*, **12**, 246-248.

Matsumoto, S., Sato, Y., Setaka, N., and Goto, M. (1973) Mass spectral analysis of thermally desorbed gases from diamond surfaces, *Chem. Lett.*, 1247-1250.

Matsumoto, S., Kanda, H., Sato, Y. and Setaka, N. (1977) Thermal desorption spectra of the oxidized surfaces of diamond powders, *Carbon*, **15**, 299-302.

Matsumoto, S. and Setaka, N. (1979) Thermal desorption spectra of hydrogenated and water treated diamond powders, *Carbon*, **17**, 485-489.

Matsumoto, S. (1979) Evolution of ethylene and acetylene from ethane-treated diamond powders, *Carbon*, **17**, 508-509.

Matsumoto, S., Sato, Y. and Setaka, N. (1981) Effect of the preceding heat treatment on hydrogen chemisorption of diamond powders, *Carbon*, **19**, 232-234.

McFeely, F. R., Kowalczyk, S. P., Ley, L., Cavell, R. G., Pollak, R. A., and Shirley, D. A. (1974) X-ray photoemission studies of diamond, graphite, and glassy *Carbon* valence bands, *Phys. Rev. B*, **9**, 5268-5278.

McGonigal, M., Russell, J. N. Jr., Pehrsson, P. E., Maguire, H. M., and Butler, J. E. (1995) Multiple internal reflection infrared spectroscopy of hydrogen adsorbed on diamond (110), *J. Appl. Phys.* **77**, 4049-4053.

McGonigal, M., Kempel, M. L., Hammond, M. S., and Jamison, K. D. (1996) Isothermal hydrogen desorption from the diamond (100)2$\times$1 surface, *J. Vac. Sci. Technol.* A, **14**, 2308-2314.

Mehandru, S. P. and Anderson, A. B. (1990) Adsorption and bonding of C_1H_x and C_2H_y on unreconstructed diamond(111). Dependence on coverage and coadsorbed hydrogen, *J. Mater. Res.*, **5**, 2286-2295.

Mehandru, S. P. and Anderson, A. B. (1991) Adsorption of H, CH_3, CH_2, and C_2H_2 on 2$\times$1 restructured diamond (100), *Surf. Sci.*, **248**, 369-381.

Melnik, M. S., Goodwin, D. G., and Goddard, W. A. III (1994) *Ab initio* quantum chemical studies of hydrogen and halogen migration on the diamond (110) surface, *Mater. Res. Soc. Symp. Proc.*, **317**, 349-354.

Miller, J. B. and Brown, D. W. (1995) Properties of photochemically modified diamond films, *Diamond Relat. Mater.*, **4**, 435-440.

Mintmire, J. W., Brenner, D. W., Dunlap, B. I., Mowrey, R. C., and White, C. T. (1991) First-principles simulations of diamond surface formation via radical addition, in R. Messier (ed), *New Diamond Science and Technology* (Proc. ICNDST-2), Mater. Res. Soc., Pittsburgh, pp. 57-62.

Mitsuda, Y., Yamada, T., Chuang, T. J., Seki, H., Chin, R. P., Huang, J. Y. and Shen, Y. R. (1991) Interactions of deuterium and hydrocarbon species with the diamond C(111) surface, *Surf. Sci.*, **257**, L633-L641.

Morar, J. F., Himpsel, F. J., Hollinger, G., Jordan, J. L., Hughes, G., and McFeely, F. R. (1986a) C 1*s* excitation studies of diamond (111). I. Surface core levels, *Phys. Rev. B*, **33**, 1340-1345.

Morar, J. F., Himpsel, F. J., Hollinger, G., Jordan, J. L., Hughes, G., and McFeely, F. R. (1986b) C 1*s* excitation studies of diamond (111). II. Unoccupied surface states, *Phys. Rev. B*, **33**, 1346-1349.

Mori, Y., Yagi, H. Deguchi, M., Sogi, T., Yokota, Y., Eimori, N., Yagyu, H., Ohnishi, H., Kitabatake, M., Nishimura, K., Hatta, A., Ito, T., Hirao, T., Sasaki, T., and Hiraki, A. (1993) Characterization of homoepitaxial diamond films grown from carbon monoxide, *Jpn. J. Appl. Phys.*, **32**, 4661-4668.

Nakamura, T., Itoh, H., and Iwahara, H. (1992) Surface topographical variation of single-crystalline diamond grown in CO-H_2 plasma, *J. Mater. Sci.*, **27**, 5745-5750.

Namba, H., Masuda, M., and Kuroda, H. (1988) Electronic states of 2×1 reconstructed surfaces of diamond (111) studied by UPS and EELS, *Appl. Surf. Sci.*, **33/34**, 187-192.

Nishimori, T., Sakamoto, H., Takakuwa, Y. and Kono, S. (1995) Methane adsorption and hydrogen isothermal desorption kinetics on a C(001)-(1×1) surface, *J. Vac. Sci. Technol.* A, **13**, 2781-2786.

Ogitsu, T., Miyazaki, T., Fujita, M. and Okazaki, M. (1995) Role of hydrogen in C and Si (001) homoepitaxy, *Phys. Rev. Lett.*, **75**, 4226-4229.

Page, M. and Brenner, D. W. (1991) Hydrogen abstraction from a diamond surface: Ab initio quantum chemical study with constrained isobutane as a model, *J. Am. Chem. Soc.*, **113**, 3270-3274.

Pandey, K. C. (1982) New dimerized-chain model for the reconstruction of the diamond (111)-(2×1) surface, *Phys. Rev. B*, **25**, 4338-4341.

Pate, B. B. (1986) The diamond surface: Atomic and electronic structure, *Surf. Sci.*, **165**, 83-142.

Pate, B. B. (1995) Surfaces and interfaces of diamond, in L. S. Pan and D. R. Kania (eds), *Diamond: Electronic Properties and Applications*, Kluwer, Boston, pp. 31-60.

Pate, B. B., Spicer, W. E., Ohta, T., and Lindau, I. (1980) Electronic structure of the diamond (111) 1×1 surface: Valence-band structure, band bending, and band gap states, *J. Vac. Sci. Technol.*, **17**, 1087-1093.

Pate, B. B., Stefan, P. M., Binns, C., Jupiter, P. J., Shek, M. L., Lindau, I., and Spicer, W. E. (1981) Formation of surface states on the (111) surface of diamond, *J. Vac. Sci. Technol.*, **19**, 349-354.

Pate, B. B., Hecht, M. H., Binns, C., Lindau, I., and Spicer, W. E. (1982) Photoemission and photon-stimulated desorption studies of diamond(111):hydrogen, *J. Vac. Sci. Technol.*, **21**, 364-367.

Pate, B. B., Oshima, M., Silberman, J. A., Rossi, G., Lindau, I., and Spicer, W. E. (1984) Carbon 1s studies of diamond(111): Surface shifts, hydrogenation, and electron escape depths, *J. Vac. Sci. Technol.* A, **2**, 957-960.

Patterson, D. E., Hauge, R. H., and Margrave, J. L. (1989) Fluorinated diamond films, slabs, and grit, *Mater. Res. Soc. Symp. Proc.*, **140**, 351-356.

Patterson, M. J.; Margrave, J. L.; Hauge, R. H.; Ball, Z.; Sauerbrey, R. (1995) Measurement of the change in the refractive index of diamond with temperature and the etching rate of diamond by the KrF (248 nm) excimer laser, in K. V. Ravi and J. P. Dismukes (eds), *Diamond Materials IV*, The Electrochemical Society, Pennington, New Jersey, pp. 503-508.

Pecansky, J. and Waltman, R. J. (1995) Vibrational spectra of diamond C(111)-(2×1) exposed to hydrogen and methane: Comparison of theory and experiment, *J. Phys. Chem.*, **99**, 3014-3019.

Pederson, M. R., Jackson, K. A., and Pickett, W. E. (1991) Local-density-approximation-based simulations of hydrocarbon interactions with applications to diamond chemical vapor deposition, *Phys. Rev. B*, **44**, 3891-3899.

Pederson, M. R. and Pickett, W. E. (1992) Theoretical investigation of fluorinated and hydrogenated diamond <100> films, *Mater. Res. Soc. Symp. Proc.*, **270**, 389-394.

Pehrsson, P. E. (1993) The acid-cleaned CVD diamond surface, in J. P. Dismukes and K. V. Ravi (eds), *Diamond Materials*, The Electrochemical Society, Pennington, New Jersey, pp. 668-673.

Pehrsson, P. E. (1995) HREELS of the oxidized diamond (100) surface, in K. V. Ravi and J. P. Dismukes (eds), *Diamond Materials IV*, The Electrochemical Society, Pennington, New Jersey, pp. 436-442.

Peploski, J., Thompson, D. L., and Raff, L. M. (1992) Molecular dynamics studies of elementary surface reactions of C_2H_2 and C_2H in low-pressure diamond-film formation, *J. Phys. Chem.*, **96**, 8538-8544.

Pepper, S. V. (1981) Electron spectroscopy of the diamond surface, *Appl. Phys. Lett.*, **38**, 344-346.

Pepper, S. V. (1982a) Transformation of the diamond (110) surface, *J. Vac. Sci. Technol.*, **20**, 213-216.

Pepper, S. V. (1982b) Diamond (111) studied by electron energy loss spectroscopy in the characteristic loss region, *Surf. Sci.*, **123**, 47-60.

Perry, M. D. and Raff, L. M. (1994a) Theoretical studies of elementary chemisorption reactions on an activated diamond ledge surface, *J. Phys. Chem.* **98**, 4375-4381.

Perry, M. D. and Raff, L. M. (1994b) Theoretical studies of elementary chemisorption reactions on an activated diamond (111) terrace, *J. Phys. Chem.* **98**, 8128-8133.

Piekarczyk, W. and Prawer, S. (1993) On the behavior of diamond crystal surfaces during heating in fluorine gas and fluorocarbon-fluorine mixtures, *Diamond Relat. Mater.*, **3**, 66-74.

Posthill, J. B., Malta, D. P., Rudder, R. A., Hudson, G. C., Thomas, R. E., Markunas, R. J., Humphreys, T. P., and Nemanich, R. J. (1993) Homoepitaxial diamond layers grown with different gas mixtures in an rf plasma reactor, in J. P. Dismukes and K. V. Ravi (eds), *Diamond Materials*, The Electrochemical Society, Pennington, New Jersey, pp. 303-309.

Prins, J. F. (1989) Preparation of ohmic contacts to semiconducting diamond, *J. Phys. D*, **22**, 1562-1564.

Proudfit, J. A. and Cappelli, M. A. (1993) Atomic hydrogen recombination on CVD diamond, in J. P. Dismukes and K. V. Ravi (eds), *Diamond Materials*, The Electrochemical Society, Pennington, New Jersey, pp. 290-296.

Ramaker, D. E. and Hutson, F. L. (1987) Experimental evidence for antiferromagnetic spin ordering on the (111)-(2×1) surface of diamond, *Solid State Commun.*, **63**, 335-339.

Ravi, K. V., Oden, P. I., and Yaniv, D. R. (1993) Atomic force microscopy of the {100} surface of flame synthesized diamond, in J. P. Dismukes and K. V. Ravi (eds), *Diamond Materials*, The Electrochemical Society, Pennington, New Jersey, pp. 766-772.

Rawles, R. E. and D'Evelyn, M. P. (1995) Optical properties of diamond at elevated temperatures, in A. Feldman et al. (eds), *Applications of Diamond Films and Related Materials: Third International Conference*, NIST Special Publication 885, Washington, DC, pp. 565-568.

Rawles, R. E., Morris, W. G., and D'Evelyn, M. P. (1996a) Kinetics and morphology of homoepitaxial CVD growth on diamond (100) and (111), *Mater. Res. Soc. Symp. Proc.*, **416**, 13-18.

Rawles, R. E., Gat, R., Morris, W. G., and D'Evelyn, M. P. (1996b) Hydrogen plasma treatment of natural and homoepitaxial diamond, *Mater. Res. Soc. Symp. Proc.*, **416**, 299-304.

Rawles, R. E., Morris, W. G., and D'Evelyn, M. P. (1996c) Effect of growth-rate ratio on surface morphology of homoepitaxial diamond (100) and (111), *Appl. Phys. Lett.* (in press).

Rawles, R. E., Komarov, S. F., Hudson, J. B., Gat, R., Morris, W. G., and D'Evelyn, M. P. (1997) Mechanism of surface smoothing of diamond by a hydrogen plasma, *Diamond Relat. Mater.* (in press).

Reider, G. A., Höfer, U., and Heinz, T. F. (1991) Surface diffusion of hydrogen on Si(111)7×7, *Phys. Rev. Lett.*, **66**, 1994-1997.

Russo, N. (1989) On the reactivity of diamond-like semiconductor surfaces, in C. Morterra et al. (eds) *Structure and Reactivity of Surfaces*, Elsevier, Amsterdam, pp. 809-816.

Sandfort, B., Mazur, A., and Pollmann, J. (1995) Surface phonons of hydrogen-terminated semiconductor surfaces. II. The H:C(111)-(1×1) system, *Phys. Rev. B*, **51**, 7150-7156.

Sandfort, B., Mazur, A., and Pollmann, J. (1996) Surface phonons of hydrogen-terminated semiconductor surfaces. III. Diamond (001) monohydride and dihydride, *Phys. Rev. B*, **54**, 8605-8615.

Sappok, R. and Boehm, H. P. (1968a) Chemie der oberfläche des diamanten–I. Benetzungswärmen, elektronenspinresonanz und infrarotspektren der oberflächen-hydride, -halogenide und -oxide, *Carbon*, **6**, 283-295.

Sappok, R. and Boehm, H. P. (1968b) Chemie der oberfläche des diamanten–II. Bildung, eigenschaften und struktur der oberflächenoxide, *Carbon*, **6**, 573-588.

Sasaki, H. and Kawarada, H. (1993) Structure of chemical vapor deposited diamond (111) surfaces by scanning tunneling microscopy, Jpn. *J. Appl. Phys.*, **32**, L1771-L1774.

Sasaki, H., Aoki, M. and Kawarada, H. (1993a) Reflection electron microscope and scanning tunneling microscope observations of CVD diamond (001) surfaces, *Diamond Relat. Mater.* **2**, 1271-1276.

Sasaki, H., Hasunuma, R., Ohdomari, I., and Kawarada, H. (1993b) STM observations of CVD diamond (001) 2×1/12 reconstructed surfaces, in M. Yoshikawa et al. (eds), *2nd International Conference on the Applications of Diamond Films and Related Materials*, MYU, Tokyo, pp. 805-810.

Schmidt, W. G., Scholze, A. and Bechstedt, F. (1996) Dimerized, buckled, or ideal chains on the diamond (111)2×1 surface?, *Surf. Sci.*, **351**, 183-188.

Scholze, A., Schmidt, W. G., Käckel, P. and Bechstedt, F. (1996a) Diamond (111) and (100) surface: ab initio study of the atomic and electronic structure, *Mater. Sci. Eng. B*, **37**, 158-161.

Scholze, A., Schmidt, W. G. and Bechstedt, F. (1996b) Structure of the diamond (111) surface: Single-dangling-bond versus triple-dangling-bond face, *Phys. Rev. B*, **53**, 13725-13733.

Scholze, A., Schmidt, W. G., and Bechstedt, F. (1996c) Diamond (111) and (100) surface reconstructions, *Thin Solid Films*, **281-282**, 256-259.

Schulberg, M. T., Kubiak, G. D., and Stulen, R. H. (1992) Temperature programmed desorption of hydrogen and deuterium from CVD diamond samples, *Mater. Res. Soc. Symp. Proc.*, **270**, 401-6.

Schulberg, M. T., Fox, C. A., Kubiak, G. D., and Stulen, R. H. (1995) Hydrogen desorption from chemical vapor deposited diamond films, *J. Appl. Phys.*, **77**, 3484-3490.

Seiwatz, R. (1964) Possible structures for clean, annealed surfaces of germanium and silicon, *Surf. Sci.*, **2**, 473-483.

Seki, H., Yamada, T., Chuang, T. J., Chin, R. P., Huang, J. Y., and Shen, Y. R. (1993) Investigation of diamond C(111) (2×1) surface exposed to hydrogen and hydrocarbon species using second-harmonic generation and sum frequency generation, *Diamond Relat. Mater.*, **2**, 567-572.

Sellschop, J. P. F., Connell, S. H., Madiba, C. C. P., Sideras-Haddad, E., Kalbitzer, S., Jans, S., Oberschactsiek, P., Bharuth-Ram, K., Schneider, J. and Kiefl, R. (1994) The nature of the state of hydrogen on the surface and in the bulk of natural and synthetic diamond (using ion beam techniques), *Vacuum*, **45**, 397-402.

Shiomi, H., Nakahata, H., Imai, T., Nishibayashi, Y., and Fujimori, N. (1989) Electrical characteristics of metal contacts to boron-doped diamond epitaxial film, *Jpn. J. Appl. Phys.*, **28**, 758-762.

Skokov, S., Carmer, C. S., Weiner, B. and Frenklach, M. (1994a) Reconstruction of (100) diamond surfaces using molecular dynamics with combined quantum and empirical forces, *Phys. Rev. B*, **49**, 5662-5671.

Skokov, S., Weiner, B., and Frenklach, M. (1994b) Elementary reaction mechanism for growth of diamond (100) surfaces from methyl radicals, *J. Phys. Chem.*, **98**, 7073-7082.

Skokov, S., Weiner, B. and Frenklach, M. (1995a) Chemistry of acetylene on diamond (100) surfaces, *J. Phys. Chem.*, **99**, 5616-5625.

Skokov, S., Weiner, B., Frenklach, M., Frauenheim, T., and Sternberg, M. (1995b) Dimer-row pattern formation in diamond (100) growth, *Phys. Rev. B*, **52**, 5426-5432.

Skokov, S., Weiner, B. and Frenklach, M. (1996) A theoretical study of the energetics and vibrational spectra of oxygenated (100) diamond surfaces, *Mater. Res. Soc. Symp. Proc.*, **416**, 281-286.

Smentkowski, V. S. and Yates, J. T., Jr. (1993) Temperature control and measurement for diamond single crystals in ultrahigh vacuum, *J. Vac. Sci. Technol.* A, **11**, 3002-3006.

Smentkowski, V. S.; Jänsch, H.; Henderson, M. A.; Yates, J. T., Jr. (1995) Deuterium atom interaction with diamond (100) studied by X-ray photoelectron spectroscopy, *Surf. Sci.*, **330**, 207-226.

Smentkowski, V. S. and Yates, J. T., Jr. (1996a) Fluorination of diamond surfaces by irradiation of perfluorinated alkyl iodides, *Science*, **271**, 193-195.

Smentkowski, V. S. and Yates, J. T., Jr. (1996b) Fluoroalkyl iodide decomposition on diamond (100) - an efficient route to the fluorination of diamond surfaces, *Mater. Res. Soc. Symp. Proc.*, **416**, 293-288.

Smirnov, E. P., Gordeev, S. K., Kol'tsov, S. I., and Aleskovskii, V. B. (1978) Synthesis of halide functional groups on the surface of diamond, *Zh. Prikl. Khim.*, **51**, 2572-2577.

Smirnov, E. P., Gordeev, S. K., Kol'tsov, S. I., and Aleskovskii, V. B. (1979) Synthesis of hydride functional groups on the surface of diamond, *Zh. Prikl. Khim.*, **52**, 199-201.

Song, K., de Sainte Claire, P., Hase, W. L., and Hass, K. L. (1995) Comparison of molecular dynamics and variational transition-state-theory calculations of the rate constant for H-atom association with the diamond (111) surface, *Phys. Rev. B*, **52**, 2949.

Sowa, E. C., Kubiak, G. D., Stulen, R. H. and Van Hove, M. A. (1988) Structural analysis of the diamond C(111)-(2×1) reconstructed surface by low-energy electron diffraction, *J. Vac. Sci. Technol.* A, **6**, 832-833.

Spitsyn, B. V. (1991) Origin and evolution of the science and technology of diamond synthesis in the USSR, in R. E. Clausing et al. (eds), *Diamond and Diamond-Like Films and Coatings*, NATO-ASI Series, Ser. B: Physics, Vol. 266, Plenum, New York, pp. 855-873.

Stallcup, R. E., Aviles, A. F., and Perez, J. M. (1995) Atomic resolution ultrahigh vacuum scanning tunneling microscopy of epitaxial diamond (100) films, *Appl. Phys. Lett.*, **66**, 2331-2333.

Stallcup, R. E. II, Villareal, L. M., Lim, S. C., Akwani, A. F., Aviles, A. F., and Perez, J. M. (1996) Atomic structure of the diamond (100) surface studied using scanning tunneling microscopy, *J. Vac. Sci. Technol.* B, **14**, 929-932.

Struck, L. M. (1993) Chemistry of hydrogen and oxygen on the diamond (100) surface, Ph.D. Dissertation, Department of Chemistry, Rice University, (unpublished).

Struck, L. M. and D'Evelyn, M. P. (1993a) Interaction of hydrogen and water with diamond (100): Infrared spectroscopy, *J. Vac. Sci. Technol.* A, **11**, 1992-1997.

Struck, L. M. and D'Evelyn, M. P. (1993b) Infrared spectroscopy of hydrogen and water on diamond (100), in J. P. Dismukes and K. V. Ravi (eds) *Diamond Materials*, The Electrochemical Society, Pennington, New Jersey, pp. 674-681.

Stumpf, R. and Marcus, P. M. (1993) Relaxation of the clean and H-covered C(111) and the clean Si(111)-1×1 surfaces, *Phys. Rev. B*, **47**, 16016-16019.

Sun, B., Zhang, X., Zhang, Q., and Lin, Z. (1993a) Investigation of the growth mechanism of diamond (111) facets using high resolution electron energy loss spectroscopy, *Appl. Phys. Lett.*, **62**, 31-33.

Sun, B., Zhang, X., and Lin, Z. (1993b) Growth mechanism and the order of appearance of diamond (111) and (100) facets, *Phys. Rev. B*, **47**, 9816-9824.

Sun, B., Zhang, X., Zhang, Q., and Lin, Z. (1994a) Investigation of the grown diamond (100) surface using high resolution electron energy loss spectroscopy, *Mater. Res. Bull.*, **28**, 131-135.

Sun, Y.-C., Gai, H., and Voth, G. A. (1994b) Vibrational energy relaxation dynamics of C-H stretching modes on the hydrogen-terminated H/C(111)1×1 surface, *J. Chem. Phys.*, **100**, 3247-3251.

Sutcu, L. F., Thompson, M. S., Chu, C. J., Hauge, R. H., Margrave, J. L. and D'Evelyn, M. P. (1992a) Nanometer-scale morphology of homoepitaxial diamond films by atomic force microscopy, *Appl. Phys. Lett.*, **60**, 1685-1687.

Sutcu, L. F., Chu, C. J., Thompson, M. S., Hauge, R. H., Margrave, J. L. and D'Evelyn, M. P. (1992b) Atomic force microscopy of (100), (110), and (111) homoepitaxial diamond films, *J. Appl. Phys.*, **71**, 5930-5940.

Szuba, S., van Enckevort, W. J. P., and van Kempen, H. (1995) Scanning tunneling microscopy on homo-epitaxial, boron-doped diamond films, *Proc. SPIE*, **2373**, 341-346.

Tamor, M. A. and Everson, M. P. (1994) On the role of penetration twins in the morphological development of vapor-grown diamond films, *J. Mater. Res.*, **9**, 1839-1848.

Thomas, J. M. (1979) Adsorbability of diamond surfaces, in J. E. Field (ed), *The Properties of Diamond*, Academic Press, London, pp. 211-244.

Thomas, R. E., Rudder, R. A. and Markunas, R. J. (1991) Thermal desorption from hydrogenated diamond (100) surfaces, in A. J. Purdes et al. (eds), *Diamond Materials*, The Electrochemical Society, Pennington, New Jersey, pp. 186-192.

Thomas, R. E., Rudder, R. A. and Markunas, R. J. (1992a) Thermal desorption from hydrogenated and oxygenated diamond (100) surfaces, *J. Vac. Sci. Technol.* A, **10**, 2451-2457.

Thomas, R. E., Rudder, R. A., Markunas, R. J., Huang, D., and Frenklach, M. (1992b) Atomic hydrogen adsorption on the reconstructed diamond (100)-(2×1) surface, *J. Chem. Vap. Depos.*, **1**, 6-19.

Thoms, B. D. and Butler, J. E. (1994) HREELS scattering mechanism from diamond surfaces, *Phys. Rev. B*, **50**, 17450-17455.

Thoms, B. D., Pehrsson, P. E., and Butler, J. E. (1994a) A vibrational study of the adsorption and desorption of hydrogen on polycrystalline diamond, *J. Appl. Phys.*, **75**, 1804-1810.

Thoms, B. D., Owens, M. S., Butler, J. E., and Spiro, C. (1994b) Production and characterization of smooth, hydrogen-terminated diamond C(100), *Appl. Phys. Lett.*, **65**, 2957-2959.

Thoms, B. D., Russell, J. N., Jr., Pehrsson, P. E., and Butler, J. E. (1994c) Adsorption and abstraction of hydrogen on polycrystalline diamond, *J. Chem. Phys.*, **100**, 8425-8431.

Thoms, B. D. and Butler, J. E. (1995) HREELS and LEED of H/C(100): the 2×1 monohydride dimer row reconstruction, *Surf. Sci.*, **328**, 291-301.

Thoms, B. D. and Russell, J. N., Jr. (1996) Identification of a surface azide from the reaction of HN_3 with C(100), *Surf. Sci.*, **337**, L807-L811.

Tolanski, S. (1954) *The Microstructures of Diamond Surfaces*, N. A. G. Press, London, p. 58.

Tsuda, M., Nakajima, M. and Oikawa, S. (1986) Epitaxial growth mechanism of diamond crystal in CH_4-H_2 plasma, *J. Am. Chem. Soc.*, **108**, 5780-5783.

Tsuda, M., Oikawa, S., Furukawa, S., Sekine, C., and Hata, M. (1992) Mechanism of the step growth of diamond crystals with carbon atoms, *J. Electrochem. Soc.*, **139**, 1482-1489.

Tsuda, M., Hata, M., Shin-no, Y. and Oikawa, S. (1996) Adatom migration mechanism at S_A steps on (001) surfaces of diamond structure crystals, *Surf. Sci.*, **357-358**, 844-848.

Tsugawa, K., Hoshino, T., Ohdomari, I., and Kawarada, H. (1993) Cluster calculations of diamond (001) surfaces, in M. Yoshikawa et al. (eds), *2nd International Conference on the Applications of Diamond Films and Related Materials*, MYU, Tokyo, pp. 799-804.

Tsuno, T., Imai, T., Nishibayashi, Y., Hamada, K. and Fujimori, N. (1991) Epitaxially grown diamond (100) 2×1/1×2 surface investigated by scanning tunneling microscopy in air, Jpn. *J. Appl. Phys.*, **30**, 1063-1066.

Tsuno, T., Tomikawa, T. Shikata, S., Imai, T., and Fujimori, N. (1994a) Diamond(001) single-domain surface grown by chemical vapor deposition, *Appl. Phys. Lett.*, **64**, 572-574.

Tsuno, T., Tomikawa, T. Shikata, S., and Fujimori, N. (1994b) Diamond homoepitaxial growth on (111) substrate investigated by scanning tunneling microscope, *J. Appl. Phys.*, **75**, 1526-1529.

Tsuno, T., Imai, T. and Fujimori, N. (1994c) Twinning structure and growth hillock on diamond (001) epitaxial film, Jpn. *J. Appl. Phys.*, **33**, 4039-4043

Tsuno, T., Imai, T., Shikata, S., and Fujimori, N. (1994d) Surface structure and morphology of diamond epitaxial films, in S. Saito et al. (eds), Advances in *New Diamond Science and Technology* (Proc. ICNDST-4), MYU, Tokyo, pp. 241-246.

Valone, S. M., Trkula, M., and Laia, J. R. (1990) Possible behavior of a diamond (111) surface in methane/hydrogen systems, *J. Mater. Res.*, **5**, 2296-2304.

van Enckevort, W. J. P., Janssen, G., Vollenberg, W., Schermer, J. J., Giling, L. J., and Seal, M. (1993) CVD diamond growth mechanisms as identified by surface morphology, *Diamond Relat. Mater.* **2**, 997-1003.

van Enckevort, W. J. P., Janssen, G., Vollenberg, W., and Giling, L. J. (1995) Anisotropy in monocrystalline CVD diamond growth. III. Surface morphology of hot filament grown films deposited on planar substrates, *J. Cryst. Growth*, **148**, 365-382.

Vanderbilt, D. and Louie, S. G. (1984) Total energies of diamond (111) surface reconstructions by a linear combination of atomic orbitals method, *Phys. Rev. B*, **30**, 6118-6130.

Verwoerd, W. S. (1981a) A study of the dimer bond on the reconstructed (100) surfaces of diamond and silicon, *Surf. Sci.*, **103**, 404-415.

Verwoerd, W. S. (1981b) The adsorption of hydrogen on the (100) surfaces of silicon and diamond, *Surf. Sci.*, **108**, 153-168.

Vidali, G. and Frankl, D. R. (1983) He-diamond interaction probed by atom beam scattering, *Phys. Rev. B*, **27**, 2480-2487.

Vidali, G., Cole, M. W., Weinberg, W. H., and Steele, W. A. (1983) Helium as a probe of the (111) surface of diamond, *Phys. Rev. Lett.*, **51**, 118-121.

Waclawski, B. J., Pierce, D. T., Swanson, N., and Celotta, R. J. (1982) Direct verification of hydrogen termination of the semiconducting diamond(111) surface, *J. Vac. Sci. Technol.*, **21**, 368-370.

Weiner, B., Skokov, S. and Frenklach, M. (1995) A theoretical analysis of a diamond (100)-(2×1) dimer bond, *J. Chem. Phys.*, **102**, 5486-5491.

Wensell, M. G., Zhang, Z., and Bernholc, J. (1995) Fluorine-based mechanisms for atomic-layer-epitaxial growth on diamond (110), *Phys. Rev. Lett.*, **74**, 4875-4878.

Whitten, J. L., Cremaschi, P., Thomas, R. E., Rudder, R. A., and Markunas, R. J. (1994) Effects of oxygen on surface reconstruction of carbon, Appl. *Surf. Sci.*, **75**, 45-50.

Wild, C., Kohl, R., Herres, N., Müller-Sebert, W., and Koidl, P. (1994) Oriented diamond films: twin formation, structure, and morphology, *Diamond Relat. Mater.* **3**, 373-381.

Wilks, E. M. and Wilks, J. (1972) Resistance of diamond to abrasion, *J. Phys. D*, **5**, 1902-1919.

Wilks, J. and Wilks, E. (1991) *Properties and Applications of Diamond*, Butterworth-Heinemannn, Oxford.

Winn, M. D. and Rassinger, M. (1995) Structure and dynamics of diamond (100) surfaces at high hydrogen coverage, *Europhys. Lett.* **32**, 55-60.

Woodruff, D. P. and Delchar, T. A. (1986) *Modern Techniques of Surface Science*, Cambridge University Press, Cambridge.

Wu, J., Cao, R., Yang, X., Pianetta, P., and Lindau, I. (1993) Photoemission study of diamond (100) surface, *J. Vac. Sci. Technol.* A, **11**, 1048-1051.

Yamada, T., Chuang, T. J., and Seki, H. (1992a) Chemical and structural effects on diamond C(111)-(2×1) surface exposed to F, H, and hydrocarbon species, *Mater. Res. Soc. Symp. Proc.*, **270**, 383-388.

Yamada, T., Chuang, T. J., Seki, H., and Mitsuda, Y. (1992b) Chemisorption of fluorine, hydrogen, and hydrocarbon species on the diamond C(111) surface, *Mol. Phys.*, **76**, 887-908.

Yamazaki, H. and Uchiyama, A. (1993) Graphite formation on natural diamond (111) surfaces by electron irradiation and heat treatment, *Surf. Sci.*, **287/288**, 308-311.

Yamazaki, H. and Ogio, T. (1994) Structure analysis of the diamond (100) surface, in S. Saito et al. (eds), Advances in *New Diamond Science and Technology* (Proc. ICNDST-4), MYU, Tokyo, pp. 439-442.

Yang, S. H., Drabold, D. A. and Adams, J. B. (1993a) Ab initio study of diamond C(100) surfaces, *Phys. Rev. B*, **48**, 5261-5264.

Yang, W. S.; Sokolov, J.; Jona, F.; Marcus, P. M. (1982) Bulk-like structure of diamond (111), *Solid State Commun.*, **41**, 191-193.

Yang, Y. L. (1992) Interaction of hydrogen with group IV semiconductor surfaces, Ph.D. Dissertation, Department of Chemistry, Rice University, (unpublished).

Yang, Y. L. and D'Evelyn, M. P. (1992a) Structure and energetics of clean and hydrogenated diamond (100) surfaces by molecular mechanics, *J. Am. Chem. Soc.*, **114**, 2796-2801.

Yang, Y. L. and D'Evelyn, M. P. (1992b) Theoretical studies of clean and hydrogenated diamond (100) by molecular mechanics, *J. Vac. Sci. Technol.* A, **10**, 978-984.

Yang, Y. L. and D'Evelyn, M. P. (1993) Pairing and clustering of hydrogen on Si(100)2×1: Monte Carlo studies, *J. Vac. Sci. Technol.* A, **11**, 2200-2204.

Yang, Y. L., Struck, L. M., Sutcu, L. F., and D'Evelyn, M. P. (1993b) Chemistry of hydrogen on diamond (100), *Thin Solid Films*, **225**, 203-211.

Yu, B. W. and Girshick, S. L. (1994) Atomic carbon vapor as a diamond growth precursor in thermal plasmas, *J. Appl. Phys.*, **75**, 3914-3923.

Zangwill, A. (1988) *Physics at Surfaces*, Cambridge University Press, Cambridge.

Zhang, Z., Wensell, M. and Bernholc, J. (1995) Surface structures and electron affinities of bare and hydrogenated diamond C(100) surfaces, *Phys. Rev. B.*, **51**, 5291-5296.

Zheng, X. M. and Smith, P. V. (1991a) The structure of the diamond (111) surface – a SLAB-MINDO study, *Surf. Sci.*, **253**, 395-404.

Zheng, X. M. and Smith, P. V. (1991b) The topologies of the clean and hydrogen-terminated C(100) surfaces, *Surf. Sci.*, **256**, 1-8.

Zheng, X. M. and Smith, P. V. (1992a) The chemisorption of hydrogen on the diamond (111) surface, *Surf. Sci.*, **261**, 394-402.

Zheng, X. M. and Smith, P. V. (1992b) The stable configurations for oxygen chemisorption on the diamond (100) and (111) surfaces, *Surf. Sci.*, **262**, 219-234.

Zheng, X. M. (1996) Electronic images of hydrogen-terminated diamond (111) surfaces, *Surf. Sci.*, **364**, 141-150.

Zhu, M., Hauge, R. H., Margrave, J. L., and D'Evelyn, M. P. (1993a) Mechanism for step growth on diamond (100), *Mater. Res. Soc. Symp. Proc.*, **280**, 683-688.

Zhu, M., Hauge, R. H., Margrave, J. L., and D'Evelyn M. P. (1993b) Mechanism for diamond growth on flat and stepped diamond (100) surfaces, in J. P. Dismukes and K. V. Ravi (eds), *Diamond Materials*, The Electrochemical Society, Pennington, New Jersey, pp. 138-145.

Zhu, X. and Louie, S. G. (1992) Anharmonicity of the hydrogen-carbon stretch mode on diamond (111)-1×1, *Phys. Rev. B,* **45**, 3940-3943.

Chapter 5
HEAT CAPACITY, CONDUCTIVITY, AND THE THERMAL COEFFICIENT OF EXPANSION

V. I. Nepsha
Almazy Rossii-Sakha Co., Ltd., Almazny Center, 14 Ul. 1812 Goda
121170, Moscow, Russia

Contents

1. Heat Capacity

Heat capacity is given by the ratio of heat amount $\delta\theta$ absorbed (given back) by the body to the deviation of its temperature, when the last is negligible:

$$C=\delta\theta/\delta T$$

Heat capacity depends not only on the initial and final states (in particular, the body temperature), but also on the path of the transition between them. One distinguishes the heat capacity at a constant volume (C_v) and the heat capacity at a constant pressure (C_p), if in the process of heating, the volume of a body or the pressure remains constant. In dielectric crystals, the heat capacity is determined by the heat capacity of the crystal lattice. At a constant volume, heat is spent only on the change of vibrational energy of the crystal lattice, and the body heat capacity C_v coincides with the phonon heat capacity. If one imagines the crystal lattice as a multitude of independent harmonic oscillators with frequencies corresponding to the normal vibrations of the lattice, then it is possible to calculate the energy of the system using the methods of quantum statistics.

Differentiation of the energy with respect to the temperature gives an expression for heat capacity of a monoatomic crystal lattice:

$$C_v = 3n_o k \int \frac{\left(\hbar\omega/kT\right)^2 e^{\hbar\omega/kT}}{\left(e^{\hbar\omega/kT} - 1\right)^2} D(\omega)d\omega \tag{1}$$

where n_0 is the number of atoms per unit volume, ω is the frequency of normal vibrations of the crystal lattice (i.e. phonons), D(ω) is the spectral density of lattice vibrations, and k and $\hbar$ are the Boltzman and Plank constants respectively.

Since heat capacity is an integral quantity, the exact form of the function D(ω) is not of great importance and Debye theory [1] gives suitable results. According to this theorone takes into account only the acoustic vibrations of the lattice. At the same time, one assumes that independently of the polarization, the phonons are characterized by the same speed of spreading s (the velocity of sound) and by the linear dispersion dependence, $\omega_q = sq$, where q is the wave number. This condition is actually fulfilled only at low temperatures. In addition, in Brillouin zones, where there are allowed values of the vector q, a Debye sphere of the same volume is substituted in inverted space. From these condition, one can imagine that there is a maximum wave number, q_D, and a corresponding maximum frequency of normal vibrations (phonons), ω_D, which is called the Debye frequency.

If one uses cyclic boundary conditions and takes into account that the full number of normal vibrations is equal to the number of atoms in the volume, corresponding to the chosen boundary conditions, then

$q_D = (6\pi^2 n_o)^{1/3}$ and $\omega_D = s(6\pi^2 n_o)^{1/3}$

The spectral density according to the Debye approximation is given by

$D(\omega)d\omega = 3\omega^2/\omega_D^3$

Then from equation (1)

$$C_v = 9n_0 k \int_0^{\omega_D} \frac{(\hbar\omega / kT)^2 e^{\hbar\omega/kT}}{(e^{\hbar\omega/kT} - 1)^2} \frac{\omega^2}{\omega_D^3} d\omega = 9n_0 k (T/\theta)^3 \int_0^{T/\theta} \frac{x^4 e^x}{(e^x - 1)^2} dx \tag{2}$$

Here $x = \hbar\omega/kT$, $\theta = \hbar\omega_D/k$, the Debye temperature. At low temperatures ($T \ll \theta$) one gets from (2) with good accuracy

$C_v = (12\pi^4/5) n_o k (T/\theta)^3$

That is, the heat capacity is proportional to T^3 (Debye relation of T^3 dependence). At high temperatures ($T>>\theta$)

$$C_v \approx 3n_o k$$

That is, the heat capacity does not depend on temperature (the law of Dulong and Pti). In practice, heat capacity temperature dependence is expressed by equation (2) with a single parameter θ for all temperatures, only in general outline which is related to the above mentioned assumptions of Debye theory. The effective Debye temperature $\theta_{ef}(T)$ is the value at which divergence with Debye theory occurs. $\theta_{ef}(T)$ equals the value of θ in equation (2) at which equation (2) agrees with the experimentally determined heat capacity C_v at the given temperature.

While heating a body at a constant pressure some heat is spent on producing the work of its expansion, therefore $C_p>C_v$. The difference between C_p and C_v is related to the coefficient of linear thermal expansion α and compressibility B:

$$C_p - C_v = 9\alpha^2 T/B.$$

For solids $C_p \approx C_v$. This relation is more accurate at lower temperatures. As it follows from paper [2], C_p in diamond exceeds C_v at 300K by not more than 0.15%. This difference increases to 1% near 800K and about 2.5 % at 1100K. Investigations of diamond heat capacity have a long history. In particular, A. Einstein in his publication in 1907 for the first time gave an explanation for the reduction of the heat capacity in solids with reduced temperature, using quantization of the vibrational energy of crystal lattice. He demonstrated the ramifications of the suggested theory using diamond as an example [3].

The most complete measurements of natural diamond heat capacity C_v were carried out in papers [2, 4-6] in the temperature ranges 11-200K; 12.8-277K; 17.4-300K and 300-1100K,respectively. In all of the cited experiments, the investigations were carried out with specimens representing the bulk of natural single-crystal diamond grains with a total mass between 10 and 160 grams. Linear dimensions of separate grains were from 1 to 5 mm. The grains were transparent, contained no inclusions and sometimes had faint yellow, brown or green color. The differentiation of the types by physical classification was not carried out and consequently most of the grains were of nitrogen-containing type Ia diamonds.

The results of C_p cited by different authors are close to each other in overlapping temperature ranges, although they were obtained with somewhat differing methods. The data on heat capacity at temperatures below 75K from the paper [6] are contrary to all other results. The anomalous behavior of C_p between 17.4 and 300 K which was found in this work was not confirmed by later investigations [4,5]. However in works [4,5] it was also noticed that with temperatures below 30-35K, the heat capacity reduced less steeply than it followed from the Debye relation of T^3 dependence. In terms of the Debye effective temperature, it means that reduced T leads to reduced $\theta_{ef}(T)$. The analysis of the measurement technique carried out by authors of paper [4] could not

exclude extra heat capacity below 30-35K as being due to sample properties. The simplest cause of the effect under consideration may be the presence of some micro-inclusions of alien phases in the investigated diamonds. It is also possible that this phenomenon is related to the excitation of one- and two-dimension structural defects (dislocations, plate-like segregations of nitrogen, called "platelets").
In the range from 40 to 100 K the experimental data of natural diamond heat capacity, as well as in the case of other solids (Ge, Si, alkali-halide crystals) are well described up to the temperatures $\theta/20$ by the general expression

$$C_p \approx C_v = aT^3 + bT^5 + cT^7 + ... \quad (3)$$

where $a = 1.48 \cdot 10^{-5}$ J/(kg K^4); $b = 3.07 \cdot 10^{-10}$ J/(kg K^6); $c = 2.79 \cdot 10^{-14}$ J/(kg K^8) [5].

It is obvious that at very low temperatures heat capacity is determined by the first term in expression (3). Then $C_v/T^3 = a$, which gives the ability to determine the Debye effective temperature at 0 K, that is $\theta_{ef}(0)$. The data of paper [5] so determined $\theta_{ef}(0) = (2219 \pm 20)$K. In paper [4], the parameter $\theta_{ef}(0) = (2246 \pm 15)$K was determined from the extrapolation of linear dependence $\theta_{ef}(T) \propto T^2$ in the temperature range near to that pointed out above. As can be seen, the values $\theta_{ef}(0)$ obtained by the different authors are close to each other. It is worth mention once more that in determining $\theta_{ef}(T)$ in both cases the experimental data on heat capacity at $T<40$K were not taken into account.

If $\theta_{ef}(0) = 2219$K, then the mean velocity of acoustic vibrations spreading $s = (1.33 \pm 0.01) \cdot 10^4$ m/s, and if $\theta_{ef}(0) = 2246$K, then $s = (1.345 \pm 0.008) \cdot 10^4$ m/s and the Debye frequencies ω are equal to $(2.90 \pm 0.03) \cdot 10^{14}$ s^{-1} and $(2.94 \pm 0.02) \cdot 10^{14}$ s^{-1}, correspondingly. The values of s found from heat capacity measurements exceed noticeably the value of $s = 1.308 \cdot 10^4$ m/s calculated in paper [7] from the data on diamond elastic constants [8,9]. In this case $\theta_{ef}(0) = 2184$K and $\omega_D = 2.86 \cdot 10^{14}$ s^{-1}.

The data on specific heat and the Debye effective temperature (fig. 1) of natural diamond in the temperature range 10-1000K are listed in Table 1 according to the results of papers: [4] (to 40K); [5] (from 40 to 90K); [6] (from 90 to 300K); [2] (above 300K). The error of measurement of the indicated values makes up: ± 12% at 12K, ± 6% at 20K ± 0.8% at 100K, ±(0.2 to 0.5)% at 200K and at higher temperatures. In spite of the high accuracy of measurements at $T > 100$K, the difference between the results for C_p attained by different investigators in some cases exceeds the error limits and can reach some percents. As was mentioned above, the data in Table 1 was obtained for natural type Ia diamonds. The comparative measurements of heat capacity for type Ia diamonds and type IIa diamonds (nitrogen-free diamonds) give, within the limits of errors, practically the same values both at low temperatures [10] and high temperatures (from 300 to 750K) [11]. From paper [11] it follows that synthetic single crystal type Ib diamonds having a concentration of single substitutional nitrogen atoms from 10^{19} to 10^{20} cm^{-3} also do not differ in their heat capacity from natural diamonds in spite of the fact that they contain some small metallic inclusions arranged along the boundaries of crystal growth pyramids. If one assumes that the Debye temperature does not depend on temperature, then the heat capacity of type Ia and IIa natural single crystalline and

Table 1. Specific heat c_p and effective Debye temperature $\theta_{ef}(T)$ for natural diamond

T,K	c_p, J/(kg K)	$\theta_{ef}(T)$,K	T,K	c_p, J/(kg K)	$\theta_{ef}(T)$,K
10	0.0191	2020	260	378.28	1857
15	0.0590	2100	270	412.07	1858
20	0.1313	2145	280	446.24	1859
25	0.2494	2165	290	480.39	1862
30	0.4135	2195	300	515.74	1861
35	0.6347	2220	320	584.9	
40	0.9835	2192	340	654.2	
50	1.968	2175	360	722.1	
60	3.514	2151	380	788.3	
70	5.823	2121	400	859.1	1869
80	9.169	2083	420	913.0	
90	13.936	2038	440	971.2	
100	20.55	1989	460	1026.6	
110	28.57	1961	480	1079.2	
120	39.02	1928	500	1129.4	1874
130	51.59	1903	550	1242.9	
140	66.15	1886	600	1341.9	1874
150	83.29	1869	650	1427.6	
160	102.43	1858	700	1502.1	1870
170	122.48	1956	750	1566.5	
180	144.78	1854	800	1623.3	1866
190	168.67	1854	850	1673.1	
200	194.52	1854	900	1718.1	1856
210	222.32	1853	950	1759.2	
220	251.62	1852	1000	1798.2	1831
230	281.40	1854	1050	1835.8	
240	311.88	1858	1100	1874.2	1762
250	344.31	1859			

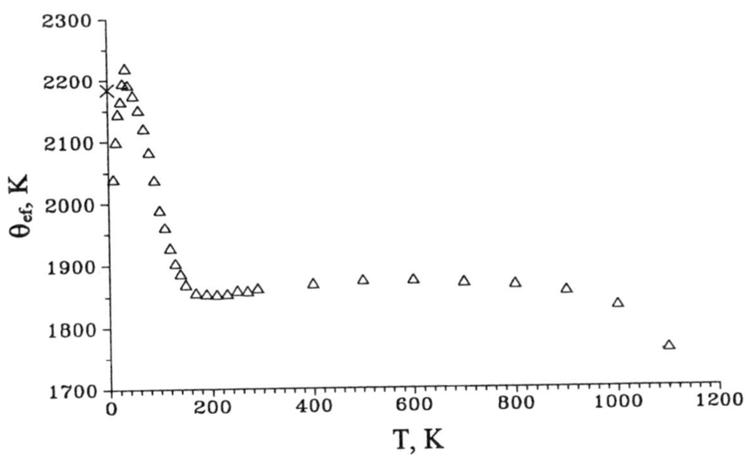

Figure 1. The dependence of the effective Debye temperature θ_{ef} on temperature: Δ - from the measurements of heat capacity; x- from the measurements of the elastic constants.

synthetic type Ib diamonds is best fit by expression (1) with θ =1900K in the temperature range from 300 to 750K [11].

No one has found any differences exceeding the measurement errors for heat capacity of perfect, large type IIa synthetic single crystalline diamonds with different concentrations of ^{13}C (either 1.1% (similar to natural diamonds), and 0.07%) at temperatures 220-740K

[12]. There are some data that the heat capacity of synthetic diamonds can significantly differ at low temperatures. From papers [13,14], the heat capacity of industrial type SAM synthetic diamond powder exceeds the heat capacity of natural diamond by a factor of 2.5 at 50K; a factor of 1.4 at 100K and by 4-5% at 200K. The heat capacity of the powder is approximately that of natural diamond in the 500-700K temperature range. With further temperature growth (up to 1200K), the heat capacity of powder is 2-3% less than that of natural diamond.

Such characteristics of heat capacity basically correspond to the behavior of the heat capacity of two-component mixtures in the case where the components have differing values of the Debye temperature θ and density, as takes place in a diamond-metal "mixture." Obviously the heat capacity of synthetic diamonds with fewer metallic inclusions in it will be closer to that of natural diamond in the whole temperature range. This value of heat capacity is most of all influenced by inclusions at low temperatures. Polycrystalline diamonds produced from sintered diamond powders, where the metal concentration can reach 10%, are well described by the two component heat capacity model. The behavior of the heat capacity of polycrystalline diamonds may be somewhat different because they contain, besides metals, rather noticeable amounts of graphite. Thus the heat capacity of polycrystalline diamond "ballas," synthesized in high pressure chambers where a metal-solvent in the form of a bar situated in the central part of the chamber, exceeds the heat capacity of natural diamond by more than two times at 75K. Such polycrystalline diamond contains up to 5% graphite, 5% nickel and 0.5% chromium. With temperature growth, this difference reduces and becomes less than 1% in the range 350-600K. At temperatures higher than 600K, the heat capacity of "ballas" again exceeds that of natural diamond, and at 1200K this excess reaches 6% [13,14].

The peculiarities of the heat capacity of polycrystalline diamond films produced from the gas phase at low temperatures and pressures are determined basically by the concentration of the non-diamond carbon phase. Films containing practically no non-diamond phase, as determined by their Raman spectra, have heat capacities of 515-543 J/(kg K) at 300K, close to that of natural diamond, or even higher by 5-6% [15].

2. Thermal Expansion

The change of crystal volume V with changing temperature is caused by the temperature dependence of the interatomic forces, and therefore atomic potential energy, leading to changes in the lattice spacing. This is why one has to take into account the terms with higher orders of x in the decomposition of potential energy of atomic interaction, U(x), by their displacement from equilibrium position $x = r - r_0$. If one limits the decomposition to the third degree of x, then

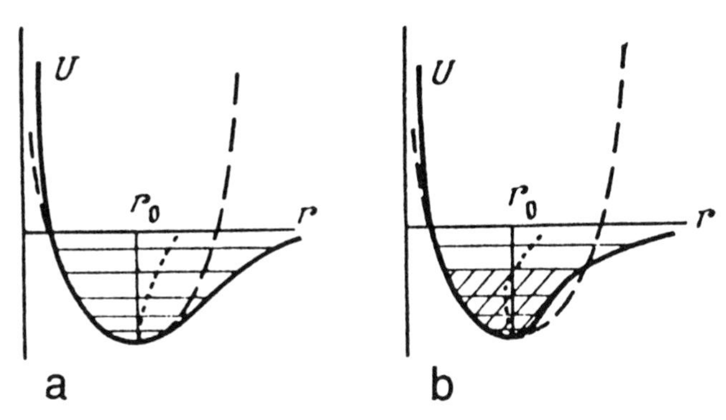

Figure 2. The dependence of the oscillating atom potential energy U on its position r : --- harmonic approach; ... the mean displacement of the oscillating atom; a - the usual form of the dependence U(r); b - the possible form of the dependence U(r) which admits the negative thermal expansion coefficient.

$$U(x) = \frac{1}{2}ax^2 - \frac{1}{3}bx^3$$

and the force, applied to the oscillating atom, will be

$$F = -dU/dx = -ax + bx^2$$

The second term in this expression describes the anharmonicity of oscillations, and increases as the atomic displacement grows larger. The average value of force acting on an atom equals zero. Consequently the mean atom displacement is

$$\bar{x} = b\overline{x^2}/a.$$

In the harmonic approach, (b=0) the mean potential energy $\overline{U}(x) = a\overline{x^2}/2$, and equals the energy of thermal vibrations kT/2. Then $\bar{x} = bkT/a^2$. Thus the mean displacement of oscillating atom relative to the equilibrium position at $b \neq 0$ changes with temperature. Its sign depends on the sign of the anharmonicity coefficient b, and its value depends both on the value of both b and a.

Usually the force of repulsion increases more quickly with atoms approaching each other than the force of attraction with their separation, so the coefficient b is positive and the mean distance between oscillating atoms, $\bar{r} = r_o + \bar{x}$, and consequently the volume V

increase along with temperature. However there may be some cases where U(r) is such that in a determined interval of temperatures, the mean distance between atoms and the volume *V* reduce with the growth of T (Fig.2).

The change of volume with the change of temperature is characterized by the isobaric volume coefficient of thermal expansion

$$\beta = \frac{1}{V}\frac{dV}{dT} \tag{4}$$

For cubic crystals $\beta = 3\alpha$ where

$$\alpha = \frac{1}{l}\frac{dl}{dT} \tag{5}$$

is the linear coefficient of thermal expansion, not depending on crystallographic direction.

From these relations, connecting thermodynamic functions and also assuming the Debye approach on calculations of free and internal energy one gets the Grunaisen relation [16]

$$\beta = \gamma C_v B/V \tag{6}$$

Where C_v and V are the heat capacity and the volume of the body, B is the isothermal compressibility. The value

$$\gamma = -\frac{\partial \ln \theta}{\partial \ln V} \tag{7}$$

is the Grunaisen parameter, characterizing the change of the Debye temperature θ with the change of volume.

As γ in the approximation assumed does not depend on temperature, and B and V depend on T weakly, then the form of the thermal expansion coefficient dependence on temperature will be determined by the form of the heat capacity dependence on temperature. In practice, the frequency spectrum more or less differs from the Debye spectrum, and as was already discussed in the previous section, one can not satisfactorily describe the heat capacity by the Debye expression with θ not dependent on temperature. Because of this, the Grunaisen parameter can not be temperature independent. More over, taking into consideration the contribution of different branches of crystal lattice oscillations with dispersion significantly different from the linear one, one can get negative values of γ and, consequently, β over a definite temperature interval. The above mentioned type of dispersion is especially characteristic for transverse acoustic vibrations at the border of Brillouin zones.

Table 2. The linear coefficient of thermal expansion α for natural diamond

T, K	$\alpha\cdot10^6$, K^{-1}	T, K	$\alpha\cdot10^6$, K^{-1}	T, K	$\alpha\cdot10^6$, K^{-1}
100	0.05	240	0.62	750	3.69
110	0.09	260	0.72	800	3.83
120	0.13	280	0.81	850	3.95
130	0.17	300	1.00	900	4.07
140	0.21	350	1.40	950	4.20
150	0.25	400	1.80	1000	4.32
160	0.29	450	2.18	1050	4.45
170	0.33	500	2.53	1100	4.57
180	0.37	550	2.83	1150	4.70
190	0.41	600	3.09	1200	4.93
200	0.45	650	3.32	1400	5.43
220	0.53	700	3.52	1600	5.87

The authors of papers [17,18,19,29] carried the theoretical analyses of γ and β temperature dependence for crystals with diamond like structure. For germanium and silicon, the Grunaisen parameter is positive at the lowest temperatures. With the growth of temperature as a result of transverse acoustic vibrations, γ decreases and takes on negative values. Further temperature growth leads to an increase in γ values, and it approaches a high temperature limit. Consequently, in the low temperature range, the temperature dependence of β has a negative minimum value. The enumerated relations are in good agreement with experimental data.

The calculations also show that the coefficient of thermal expansion for diamond, unlike germanium and silicon, does not acquire negative values, and must have a more smooth dependence on temperature than T^3 as follows from the Debye approach [19,20]. The results of numerous measurements of the linear expansion coefficient of natural diamond [21-28] are generalized in papers [29,30] and are presented in Table 2 (up to 1100K according to [29]; and at 1120-1600K according to [30]). The data in Table 2 were mostly determined from the precision X-ray measurement of the crystal lattice parameter. Only in paper [25] was the value of α in the range 25-750K was determined from measurements of the sample length (with the use of a quartz dilatometer). The precision of data for α in Table 2 is equal to ±15%. In all of the measurements diamonds were not classified by physical types, therefore one can with great probability attribute them to type Ia.

From the measurements of the crystal lattice parameter for two natural type IIa diamonds, it was found in paper [31] that in the range 1200-2000K, α increases from $5\cdot10^{-6}$ K^{-1} to approximately $5.8\cdot10^{-6}$ K^{-1}. Some differences between absolute values of α and the data in Table 2 scarcely reflects the difference between diamonds of type IIa and Ia, as it may be related to the systematic measurement errors. The differences are more pronounced at higher temperatures, where the rate of increase of α is 1.5 to 2 times higher according to the measurements in the work [31] than in the work [28]. The comparative analysis shows that in the range 320-700K, the coefficient of thermal expansion for synthetic diamond (of type Ib) occasionally exceeds that of natural diamond (of type Ia), but generally does not differ from it noticeably [32]. The difference ($\leq 10\%$) does not exceed the value of usual errors for such measurements. In papers [33, 34] the authors carried out measurements of the crystal lattice parameter for synthetic type Ib diamond in the temperature range 4.2-320K. Nitrogen concentration in this crystal was less than $5 \cdot 10^{19}$ at/cm^{-3}. It was found that from 4.2 to 100K, the linear coefficient of thermal expansion is positive, has insignificant dependence on temperature and is of the order 10^{-8} K^{-1}. In the investigated temperature range the dependence of α on T is expressed by the polynomial

$$\alpha=\alpha_1 T+\alpha_2 T^2+\alpha_3 T^3+\alpha_4 T^4 \qquad (8)$$

where $\alpha_1 = 5.05 \cdot 10^{-10}$ K^{-2}, $\alpha_2 = -5.04 \cdot 10^{-12}$ K^{-3}, $\alpha_3 = 7.79 \cdot 10^{-14}$ K^{-4}, $\alpha_4 = -1.01 \cdot 10^{-11}$ K^{-5}.

The values of α, calculated by means of expression (8) are 1.36 times higher than the values indicated in Table 2 for 100K. Beginning from 110K and up to 240K they are 10-30% lower than the data in Table 2 (maximum difference is at 140-150K). From 260K to 500K the expression (8) gives the values of α different from that in Table 2 by not more than 3%.

In the same works the authors investigated the synthetic crystal irradiated by C^{6+} ions with energy of 100 MeV at liquid nitrogen temperature and then held for some time at room temperature. As a result, lattice parameter increased by $4.32 \cdot 10^{-3}\%$. The irradiated crystal has a somewhat larger temperature interval where the lattice parameter almost does not depend on temperature as compared with non-irradiated crystal, yet in general the experimental results do not give enough reasons to confirm a noticeable difference in α for irradiated and non-irradiated diamond in the range 4.2-320K. The authors of paper [25] reported that α between 40 and 60K has small absolute values with negative sign. This is not confirmed by the above considered recent investigations of synthetic crystals and it is not in agreement with theoretical calculations [19,20].

Obviously diamond differs in the form of its temperature dependence for α from germanium and silicon. Nevertheless there are some general features in anharmonicity discussed in the enumerated materials.

Figure 3 shows the dependence of the Grunaisen parameter, γ, on the reduced temperature $T/\Theta_{ef}(0)$. From the expression (6), $\gamma = 3\alpha/C_v B$. where C_v is the heat capacity of unit volume. In calculations it was assumed than $C_v=C_p$ over the whole temperature range. The compressibility, $B = 2.26 \cdot 10^{-12}$ Pa^{-1} is calculated on the basis of elastic constants ($B = 3/c_{11} + c_{12}$). The values of C_p are taken (with count over to unit volume) from Table 2, and α is taken both from Table 2 and in accordance with expression (8). It is seen from figure 2 that the dependence of γ on T for diamond as well as for germanium and silicon has a minimum at low temperatures although it does not become negative. The minimum value is at $T/\Theta_{ef}(0) \approx 0.1$ to 0.13, which corresponds to the general tendency of displacement to the direction of high reduced temperatures in group IV, i.e.: germanium, silicon and diamond.

The difference between heat expansion of polycrystalline and single-crystal diamonds is basically determined by the contents and distribution of foreign phases. At high

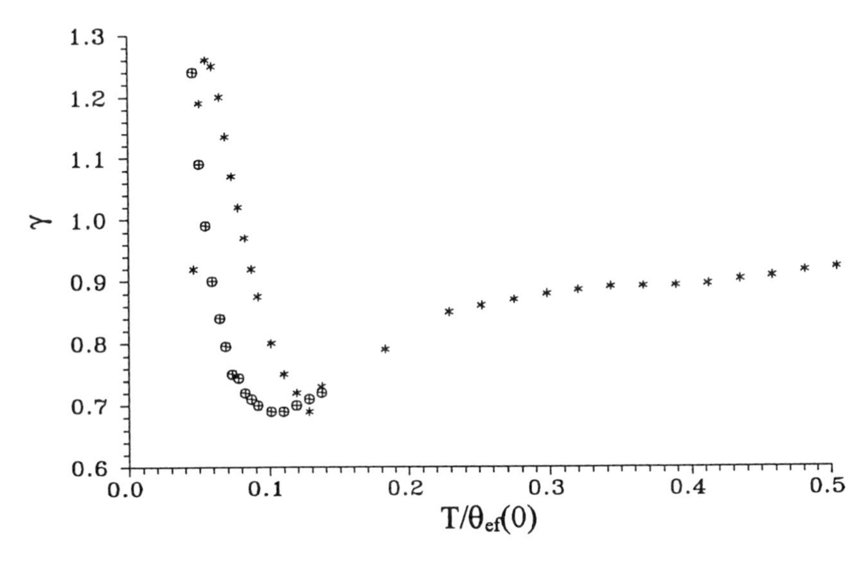

Figure 3. The dependence of the Grunaisen parameter γ on the reduced temperature $T/\theta_{ef}(0)$: * - the results obtained with the use of α values taken from Table 2; ⊕ the results obtained with the use of α values, calculated from the expression (8).

temperatures, polycrystals have values of the coefficient α which is rather close to that of single-crystals. Polycrystalline diamond films, "DIAFILM", with low contents of non-diamond carbon, as determined by the method of Raman spectroscopy, the authors of paper [15] give the data

T,K	300	373-623	873	1023
$\alpha \cdot 10^6$, K^{-1}	0.8-1.2	1.20-1.21	3.84	4.45

3. Thermal Conductivity

3.1 SCATTERING PROCESSES

Heat spreading in an isotropic solid is expressed in stationary conditions by the equation

$$\vec{h} == K\nabla T$$

where $\vec{h}$ is the vector with an absolute value equal to heat flow across the unit section orthogonal to $\vec{h}$, K is the thermal conductivity coefficient or simply the thermal conductivity. The thermal conductivity of diamond as well as other dielectrics is determined by the thermal conductivity of the crystal lattice. In this case, the heat energy is transferred only by phonons and as it follows from the notions developed in the kinetic theory of gases

$$K = c\, s\, l_{ef}/3 \tag{9}$$

Here c is the specific heat of phonons, s is the mean phonon velocity, l_{ef} is the phonon mean free path.

The phonon mean free path is determined by the three kinds of scattering processes. They are the scattering by phonon collisions, crystal lattice defects and specimen boundaries.

The intensity of each of the above processes is characterized by the relaxation time, which is related to the mean free path by the equation

$$\tau_i = l_i/s \tag{10}$$

or the relaxation rate τ_i^{-1}.

The most probable of all phonon interactions are the three-phonon processes in which the collision of two phonons with frequencies ω_1 and ω_2 result in formation of one phonon with frequency ω_3, or conversely one phonon disintegrating into two phonons. The phonon formed as a result of the collision has frequency $\omega_3 = \omega_1 + \omega_2$ and his wave vector $\vec{q}$ (($q = 2\pi/\lambda$), where λ is the phonon wavelength), may be equal to the sum of the wave vectors of phonons interacted before:

$$\vec{q}_3 = \vec{q}_1 + \vec{q}_2$$

if it keeps within the first Brillouin zone, or otherwise

$$\vec{q}_3 = \vec{q}_1 + \vec{q}_2 - 2\pi\vec{b}$$

where b is the vector of reciprocal lattice.

Phonon-phonon interactions of the first type do not change the total quasi-impulse $\hbar\vec{q}$. They are called normal processes of scattering or N-processes. The second type of interactions results in alteration of the total quasi-impulse and they are called Umklapp-processes or U-processes. Obviously normal processes do not make any contribution to thermal resistance, that is, they are non-resistive, but in specific conditions they can play an important role by redistributing phonon frequencies and therefore changing the effectiveness of other scattering processes. The last may be accompanied by the alteration of quasi-impulse and therefore they may be resistive ones.

The expressions for the relaxation rates of N- and U-processes are defined as

$$\tau_N^{-1}=A_N T^m x^n \tag{11a}$$

$$\tau_U^{-1}=A_U T^k x^l \exp(-\alpha/T) \tag{11b}$$

where $x = \hbar\omega/kT$.

Phonon scattering by the defects in the crystal lattice as well as U-processes are of resistive nature and as a rule depend on phonon frequency ω. The scattering by point defects (isotopes, vacancies, single impurities and their aggregations, containing small numbers of atoms) is proportional to ω^4 according to Rayleigh's law. In general form for small point defects concentration the relaxation rate is characterized according to [38] by

$$\tau_{p.d.}^{-1}=nc\frac{V}{\pi\hbar^4 s^3}\left(\frac{k}{\hbar}\right)^4\left(\frac{\Delta M}{2M_o}+\gamma\alpha\right)^2 T^4 x^4 \tag{12}$$

where V is the volume of defect, including the distortion field, c is the relative concentration of impurity, ΔM is the difference between the masses of impurity and substituted atoms, M_o is the mass of atoms forming crystal lattice, α is relative volume change due to scattering formation, n is the number of impurity atoms per formation (for a single impurity atom n = 1), and γ is the Gruneisen constant.

For an impurity creating the field of distortion the volume is taken to be equal to the volume of an elementary lattice cell. For isotopes n=1, $\alpha \approx 0$ in (12) with ΔM equal to the isotopic mass difference and the volume V equal to the volume corresponding to one atom. The relaxation rate for scattering by the segregates having plate-like form and consisting of impurity atoms depends on ω^2. According to [38]

$$\tau_{pl}^{-1}=\frac{8\pi}{s}\rho R^2 h^2\left(\frac{k}{\hbar}\right)^2\left(\frac{\Delta M}{2M_o}+\gamma\alpha\right)^2 I\left(\frac{RkT}{\hbar s}\right)T^2 x^2 \tag{13}$$

where ρ is the segregates content per unit volume, R and h are the radius and thickness of platelets (2R>>h), I $(RkT/\hbar s)$ is a multiplier expressed by cylindrical Bessel functions of zero and first order.

The next type of scattering process which is considered while investigating diamond thermal conductivity is the scattering by segregates of isoteric form with sizes on the order of some tens of Angstrom. In this case the expression for relaxation rate depends on the correlation between phonon wavelength and the segregate diameter a. Phonons with wavelength λ much longer than the size of the segregates are scattered by them exactly like point defects according to Rayleigh's law ($\tau_{seg}^{-1}\sim\omega^4$). When the phonons' wavelength is much shorter than a, then so called geometric type scattering occurs and the relaxation rate τ_{seg}^{-1} does not depend on frequency. For phonons with λ on the order of a it is possible either to have a smooth transition from Rayleigh to geometric scattering with shortening λ, or τ_{seg}^{-1} has oscillations with decreasing amplitudes. In this case the character of the process depends on the correlation between the densities and elastic properties of crystal matrix and segregate. The simplest expression for the relaxation rate of phonon scattering by isometric segregates is proposed in paper [39]:

$$\tau_{seg}^{-1} = \begin{cases} \dfrac{\pi}{4(1.5)^4}\left(\dfrac{k}{\hbar}\right)^4 \dfrac{\rho a^6}{s^3} T^4 x^4 & for\ x < \dfrac{1.5\hbar s}{kaT} \\ \dfrac{\pi}{4} s\rho a^2 & for\ x \geq \dfrac{1.5\hbar s}{kaT} \end{cases} \tag{14}$$

Here, ρ, exactly like in equation (13), is the segregate content per unit volume.

The most complex problem is the theoretical description of phonon scattering by a dislocation stress field. It is caused by dislocations that have complex and, as a rule, loop-like form and are rarely simply screw or edge dislocations. Besides that, when studying even linear dislocations the choice of model and the assumptions used in calculations are not single valued, which leads to rather different results. But these differences are of quantity character and concern only the coefficient of ω in the expression for relaxation rate. As an example it is possible to deduce the result of one of the calculations of τ_{disl}^{-1} for linear edge dislocations [40]:

$$\tau_{disl}^{-1} \approx \frac{200\sigma b^2 k}{\pi\hbar}\left(\frac{1-2\nu}{1-\nu}\right) Tx \tag{15}$$

Here σ is dislocation density, b is the value of Burgers vector and ν is the Poisson coefficient. The problem becomes even more complicated if the dislocation stress fields interact with each other. In this case the relaxation rate in different ranges of ω can be proportional to ω^3 or frequency independent.

Phonon scattering by sample boundaries was studied in the works [35, 41-43]. For a cylindrical sample with length L and diameter d, the relaxation rate for phonon scattering by boundaries is given by

$$\tau_b^{-1} \approx s\left(\frac{1}{d}\frac{1-P}{1+P} + \frac{1}{L}\right) \tag{16}$$

where P is the portion of phonons speculary reflected by specimen surface, which depends on the correlation between the effective value of its roughness H_{ef} and phonon wavelength λ

$$P = \exp\left[-\left(2H_{ef} \frac{kTx}{\pi s} \right)^2 \right] \tag{17}$$

In the case of a sample with a rectangular cross-section with sides d_1 and d_2 the value d in expression (16) is replaced with $d = 1.12\sqrt{d_1 d_2}$. If $\lambda \ll H_{ef}$ then the portion of specularly reflected phonons equals zero (phonons are scattered diffusely) and the expression (16) transfers for L>>d, to the frequency-independent expression of Kasimir for absolutely rough surfaces [41]:

$$\tau_b^{-1} = s/d \tag{18}$$

For $\lambda \geq H_{ef}$ the portion of specularly reflected phonons becomes considerable and τ_b^{-1} depends on ω. With this τ_b^{-1} decreases as if it were the result of increasing d by (1+P)/(1-P) times. The above expressions for τ_b^{-1} can be used only in the case when τ_b^{-1} or relaxation rate for other resistive processes τ_R^{-1} is greater than the relaxation rate for non-resistive normal processes τ_N^{-1}, which is not always the case.

Callaway [44] applied the relaxation method to the kinetic Boltzman equation, and for crystal lattice thermal conductivity, formed the next expression:

$$K = K_1 + K_2 \tag{19}$$

$$K_1 = \frac{k}{2\pi^2 s}\left(\frac{k}{\hbar}\right)^3 T^3 \int_0^{\theta/T} \frac{\tau_R \tau_N}{\tau_R + \tau_N} \frac{x^4 e^x}{(e^x - 1)^2} dx \tag{19a}$$

$$K_2 = \frac{k}{2\pi^2 s}\left(\frac{k}{\hbar}\right)^3 T^3 \frac{\left[\int_0^{\theta/T} \frac{\tau_R}{\tau_R + \tau_N} x^4 e^x \left(e^x - 1\right)^{-2} dx \right]^2}{\int_0^{\theta/T} \frac{1}{\tau_R + \tau_N} x^4 e^x (e^x - 1)^{-2} dx} \tag{19b}$$

Here τ_R is the total relaxation time for all resistive processes and it is connected with the relaxation time τ_i of each of these processes by the relation $\tau_R^{-1} = \sum_i \tau_i^{-1}$; τ_N is the relaxation time for normal processes.

Callaway's expression in the above form assumes the Debye spectrum for the frequencies of crystal lattice vibrations and the linear dependence of vibration frequency on the wave vector. The above dependence is not influenced by vibration polarization and the Callaway expression completely corresponds to the limits of the Debye approximation.

If resistive processes prevail, that is $\tau_R \ll \tau_N$ or $\tau_R^{-1} \gg \tau_N^{-1}$, then

$$K = K_1 = \frac{k}{2\pi^2 s}\left(\frac{k}{\hbar}\right)^3 T^3 \int_0^{\theta/T} \tau_R \frac{x^4 e^x}{(e^x - 1)^2} dx \tag{20}$$

From comparison of expressions (2), (9), (20) and taking into account (10)

$$l_{ef} = s\langle\tau_R\rangle = s\langle(\Sigma\tau_i^{-1})^{-1}\rangle = \langle(\Sigma l_i^{-1})^{-1}\rangle \tag{21}$$

where < > means the average over the frequency spectrum (in this case over the Debye spectrum). According to this approximation, one takes into account phonon scattering by the specimen boundaries in accordance with the expressions (16)-(18) along with other resistive processes.

If normal processes prevail, then $\tau_N \ll \tau_R$ ($\tau_N^{-1} \gg \tau_R^{-1}$). It means that before resistive scattering takes place, a lot of normal phonon-phonon collisions occur. In these collisions the total quasi-impulse and consequently the heat flow does not change. In this case

$$K = K_2 = \frac{k}{2\pi^2 s}\left(\frac{k}{\hbar}\right)^3 T^3 \frac{\left[\int_0^{\theta/T} x^4 e^x \left(e^x - 1\right)^{-2} dx\right]^2}{\int_0^{\theta/T} \tau_R^{-1} x^4 e^x (e^x - 1)^{-2} dx} \tag{22}$$

The expression (22) describes the thermal conductivity in the Ziman limit [35], which is characterized by each of the resistive processes (besides boundary scattering) giving maximum effectivity, and the total crystal lattice thermal resistivity, W=1/K, is equal to the sum of thermal resistivities W_i, determined by different scattering processes. The effective free path in the Ziman limit is

$$l_{ef} = s\langle\tau_R^{-1}\rangle^{-1} = s(\Sigma\langle\tau_i^{-1}\rangle)^{-1} = \Sigma\langle l_i^{-1}\rangle)^{-1} \tag{23}$$

As one can see from (21) and (23) the two limiting cases considered above differ from each other by the method of averaging the relaxation rates over the frequency spectrum.

Taking into consideration phonon interactions with the boundaries of a specimen by means of expressions (16)-(18) in the Ziman Limit does not always lead to correct results and a special analysis is demanded. Thermal conductivity of diamond as well as of most of dielectric crystals is usually analyzed on the basis of expression (20), that is, when one assumes that the normal phonon-phonon collisions are rare and can be neglected. However, the recent experimental data on thermal conductivity of synthetic single-crystals with different contents of the ^{13}C isotope [45] gave some reasons to suppose that the role of normal processes is of great importance [46, 47]. Apparently the use of expression (20) is most justified for polycrystalline diamonds because in this case, the resistive processes are significantly intensified by phonon scattering on microcrystallite boundaries.

3.2 TYPE IIa AND TYPE IIb DIAMONDS

Type IIa diamonds mostly correspond to the notions of ideal crystals, and differ from crystals of other types in that the concentration of nitrogen, which is the most characteristic for diamonds, does not exceed 10^{17} cm^{-3}. They do not absorb radiation in the visible and ultraviolet ranges of spectrum (up to 225 nm) and also in the range of single-phonon absorption (7-9 mm).

The temperature dependence of thermal conductivity of type IIa natural diamonds in the temperature range 3-300K was investigated in several works [48-50]. The measurements of the thermal conductivity of synthetic type IIa diamonds (with nitrogen concentrations $\leq 10^{16}$ cm^{-3}) were carried out in a somewhat lesser temperature interval [51]. Temperature dependencies of thermal conductivity for two specimens of natural diamond with cross section about 1 mm^2 are shown in figure 4. These diamonds have maximum and minimum values of thermal conductivity. As can be seen from the figure, the maximum of thermal conductivity for different specimens of natural type IIa diamonds can be found at somewhat different temperatures, and its value can differ from the mean value by 10-15%. The thermal conductivity of synthetic type IIa diamond exceeds by 1.2-1.6 times, that of natural diamonds at maximum and reaches 1750 W m^{-1} K^{-1}. The thermal conductivity maximum is in the low temperature range, near 65K. Apparently these peculiarities are the consequence of the somewhat larger cross section of the synthetic diamond specimen.

At 300K the thermal conductivity value for synthetic diamond approaches to the lowest value for natural crystals (2000 W m^{-1} K^{-1}) so that the total scatter of K for type IIa diamonds is within the limits 2000-3000 W m^{-1} K^{-1} at 300K.

The most interesting are the data of thermal conductivity in the high temperature region listed in Table 3. The idea about the scatter of K values is given by the results of measurements in twenty seven specimens, made from carefully selected natural type IIa diamonds [52]. The mean values are

$$K_{IIa}(320) = 1930 \text{ W m}^{-1} \text{ K}^{-1}; \quad K_{IIa}(450) = 1250 \text{ W m}^{-1} \text{ K}^{-1}$$

The thermal conductivity measurements in synthetic type IIa diamonds in the range 150-400K [55] give results rather close to that indicated in figure 3 and in Table 3.

The authors of the works [48, 54, 56] used expression (20) to describe the temperature dependence of type IIa diamond thermal conductivity. Recall, this expression assumes phonon-phonon interactions to be unimportant.

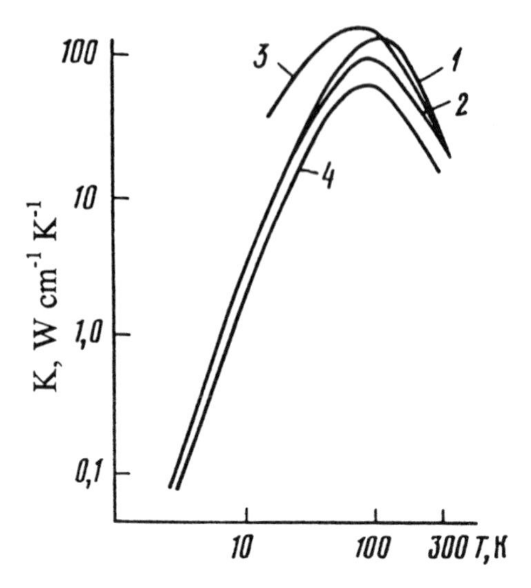

Figure 4. The thermal conductivity of type II diamonds: (1): the natural type IIa crystal [50], (2): the natural type IIa crystal [48]; (3): the synthetic type IIa crystal [51]; (4): the natural type IIb crystal [49].

Table 3. Thermal conductivity of type IIa diamonds with T>300K

T, K	K, W m^{-1} K^{-1}
300	2000-2500 [48-51]
230	1800-2020 [52]
450	1180-1350 [52]
500	1170 [53]
700	800 [53]
1000	530 [53]

Table 4. Parameters B, A_U, k, α in expression for relaxation rate of resistive processes of phonon scattering in type IIa diamonds

B	A_U	k	α	temperature interval	source
$3.41 \cdot 10^{-2}$	1.3	4	270	up to 300K	[48]
$3.41 \cdot 10^{-2}$	0.85	4	160	up to 500K	[56]
$4.2 \cdot 10^{-2}$	830	3	540	up to 1000K	[54]

In all of the works mentioned above, their authors basically took into consideration three resistive mechanisms of scattering, that is the scattering by specimen boundaries, by point defects and also the scattering caused by phonon-phonon Umklapp processes. Consequently the summed relaxation rate is

$$\tau_R^{-1}=\tau_b^{-1} + BT^4x^4 + A_U T^k x^l \exp(-\alpha/T) \quad (24)$$

Parameters B, A_U, k, α suggested by different authors for the description of temperature dependence of type IIa diamond thermal conductivity are listed in Table 4. In all cases concerning Table 4, l was taken to be equal to 2.

For the whole temperature interval the best results can be obtained with the use of parameters suggested in the work [54], as they were determined from fitting the results of equation (20) to the experimental data, including the high temperature data.

For all three variants there is a general feature, that the value of parameter B is 5-6 times higher than in the case of only point defects in type IIa diamonds with a carbon isotope content of 1.1%. The assumption was made that type IIa diamonds contain up to $2\cdot10^{19}cm^{-3}$ impurity nitrogen atoms, or a somewhat lesser quantity of other heavier impurities [48], or up to $10^{19}cm^{-3}$ of vacancies [54]. However, two arguments discount these assumptions. Firstly, such concentrations of vacancies and impurities are evident under optical spectroscopy. Secondly, both synthetic and natural diamonds are unlikely to have so nearly equal concentrations of non-controlled surplus point defects, as the small differences in the thermal conductivity suggests. It is possible to take into account some point defects of unknown character in a somewhat artificial method which enables one to reach a formal correlation between the experimental and calculated data, however, this approach is not well grounded. In the case considered above this approach assumed complete prevalence of resistive phonon scattering over normal phonon-phonon processes.

This conclusion is confirmed in that the experimental data on type IIa diamond thermal conductivity can be successfully described in another extreme approach with the supposition that normal processes prevail over resistive processes ($\tau_N^{-1}>>\tau_R^{-1}$), i.e. in the Ziman limit [57] with separate consideration of longitudinal and transverse oscillations in expression (22). It turns out that such an approach does not require taking into consideration any point defects except ^{13}C isotopes. The expression for the relaxation rate of phonon-phonon Umklapp-processes suitable for this case is

$$\tau_U^{-1} = 151.2\, T^3x^2 \exp(-730/T).$$

The relaxation rate with all scattering mechanisms present including normal processes, decreases with decreasing phonon frequency. Scattering at specimen boundaries with absolutely rough surfaces is the only exception. That is why at rather low temperatures it becomes prevalent. Then phonon free length l is determined by the specimen dimensions and does not depend on temperature. With this the thermal conductivity changes with temperature in accordance with (9) as well as thermal capacity. As in the Debye approach at low temperature $C(T) \sim T^3$, therefore $K \sim T^3$. However, the experimental measurements [48] testify that at low temperatures the phonon free path length is 2-3 times longer than the diametric size of the specimen. Similar results were obtained in the work [7] while investigating the thermal conductivity of some type IIa diamonds in the temperature range 0.5-20K. It was ascertained that the phonon free path length almost does not depend on T only in temperature range 1-10K ($K \sim T^n$ where n = 2.8 to 2.9) and, as was reported in the work [48], it exceeds the value corresponding to the dimension of specimen. With the temperature below 1K, l increases quite sharply.

It was also established that specimen surface treatment by powders with different grain sizes (0-0.5 μm and 50μm) has a faint influence on the character of the K dependence on T. Such a treatment does not allow the attainment of any correlation between the values of l determined from thermal conductivity measurements and the values determined by the real dimensions of the specimen. This study was a success only after a long anneal (10 hours) at 1100°C. The results obtained can be explained if one supposes that after polishing, the diamond surface consists of two kinds of sections. Some sections have a rather large roughness size $H_1^{ef} \geq 1500$Å. At T ≥ 1K, the phonon wavelength λ is less than H_1^{ef} and they are scattered by such surface sections diffusely (P_l=0). At T≤ 1K the value of λ at first becomes comparable with H_1^{ef} and then exceeds it, which leads to the increase of phonon specular reflection ($P_1 \neq 0$) and to the rapid increase of l with decreasing T. Other sections of the surface behave with such a small roughness size that $P_2 \approx 1$ and is almost independent of the phonon wavelength λ in the range from 1K to at least 10K. In order for this to be observed the value of H_2^{ef} must be less then 100-150 Å. It can be shown that a surface consists of two kinds of sections, each of which has its own roughness size H_1^{ef} and H_2^{ef}, so one gets in expression (16)

$$P = P_1 s_1 + P_2 s_2,$$

where s_1 and s_2 are the area portions for each kind of sections and P_1 and P_2 are the portions of phonons speculary reflected by the sections. In the case considered above $P_1 = 0$, $P_2 = 1$ and $P = s_2$ in the temperature range 1-10K. According to (16) it corresponds to the increase of the diametric specimen size D by $(1+s_2)/(1-s_2)$ times. As the value $(1+s_2)/(1-s_2)$ changes in the limits from 2 to 3 [7, 48], the sections of the second kind occupy from 30 to 50% of total surface area.

It was supposed in the work [7] that in the temperature range from 1 to 10K the phonons are scattered not by the external surface, but by some under-surface layer containing defects created in the polishing process. If one assumes that the last concerns only the surface sections of the second kind, then the surface model suggested above conforms with the ideas about fragile materials destruction while processing with free abrasive grains. In accordance with these ideas, in the initial passage of the abrasive grain some trace is formed in the near surface layer of the treated material. This trace consists of a set of circular cracks (sections of the second kind) and only after the second passage over these sections does final distribution and removal of the material take place (sections of the first kind). Under these conditions the surface state is determined in both cases not only by the size of treating grains but on the whole by their configuration.

While interpreting the experimentally found dependence of K on T at low temperatures one is obliged to take into account the influence of dislocations. In work [58] the expression for relaxation rate in type IIa diamonds (24) was supplemented by the term

$\tau_{disl}^{-1} = 3.01 \cdot 10^4$ Tx, which corresponds to the dislocation density $\sigma \approx 7.7 \cdot 10^6$ cm^{-2} according to (15). Type IIb diamonds (semiconductive) have a rather high thermal conductivity. This is caused by the absence of acceptor - compensating donor nitrogen impurities. The dependence of K on T for type IIb diamond is shown in figure 4 (according to the data of paper [49]). In this paper it is pointed out that at low temperatures, thermal conductivity values are lower than it follows from the cross-section dimensions of the specimen. As a possible reason for this, the authors suggest phonon scattering by electrons.

Three type IIb diamonds were investigated in the temperature interval 0.5-20K [59]. It was found that the mean phonon free path length l reduces at temperatures below 14K. At the same time the dependence of l on T has two minimum values, one of which is displayed at 5.5K and another at either 0.75, 1.2 or 1.35K for different specimens. It was supposed that these minimum values are the consequence of resonance phonon scattering by acceptor centers. The most specimens of type IIb diamonds (18 pieces) were investigated in the work [52] at 320 and 450K. In all of these specimens the absorption coefficient α at wavelength 3.56 μm (2810 cm^{-1}) and specific resistance ρ, correlating to each other were measured. The coefficient $\alpha_{3.56}$ changed from 0 to 6.1 cm^{-1}, and ρ changed from 10^5 to 1 $\Omega \cdot$m. The thermal conductivity of type IIb diamonds does not markedly differ from that of type IIa diamonds and changes from 1840 to 2020 W m^{-1} K^{-1} and from 1140 to 1350 W m^{-1} K^{-1} at 450K. With this there is no noticeable correlation between the thermal conductivity and the values of $\alpha_{3.56}$ and ρ. The estimates found in the work [52] with the use of expression (12) testify that such a high thermal conductivity of type IIb diamonds may only be in the case of boron concentrations below $3.5 \cdot 10^{18}$ cm^{-3}. Since the field of distortion created by the boron impurity was not taken into account in the use of equation (12), this value may be even lower. All results cited above were obtained in natural type IIb diamonds.

3.3 TYPE Ib DIAMONDS

Most natural diamonds are type Ia according to the physical classification. They are characterized by a high concentration (up to $5 \cdot 10^{20}$ cm^{-3}) of nitrogen in different aggregated forms. A and B 1-defects are the most common types of defects. Paper [61] discusses the structure and optical spectra of these and other defects. Crystals containing only A or B1-defects are almost as rare as type IIa diamonds. Most diamonds simultaneously contain A and B1-defects and a whole series of other defects, which are of a minor importance.

Measurements of the thermal conductivity of type Ia diamonds in the temperature range 3 -300K are discussed in the works [48-50]. These diamonds do not differ from type IIa diamonds in their K dependence on T at all but the lowest temperatures, at which the phonon free path length is limited by the specimen dimensions, leading to type Ia diamonds have a significantly lower thermal conductivity.

The most general investigations of type Ia diamond thermal conductivity were carried out at temperatures 320 and 450K [52, 60-62]. The measurement of a great number of specimens (about 100 pieces) with a different content of A, B1 and B2 defects, as determined by IR absorption spectra determined the influence of each of these defects on the value of the thermal conductivity. From the analysis of the experimental data it follows that B2-defects (or "platelets") which are aggregates in the plate-like form, oriented in the cubic planes of the crystal lattice do not exert any influence on the thermal conductivity at the temperatures of interest. Apparently the cause of this is their low concentration ($\leq 10^{15}$ cm^{-3}). The value of thermal conductivity is determined by the content of A and B1-defects. Each of them makes an independent contribution to the crystal thermal resistance W = 1/K. The experimental results can be expressed by

$$W(320) = 1/K(320) = W_{IIa}(320) + 2.6\cdot10^{-5}\,\alpha_{B1} + 2.1\cdot10^{-5}\,\alpha_A^{4/5} \quad (25)$$

$$W(450) = 1/K(450) = W_{IIa}(450) + 2.5\cdot10^{-5}\,\alpha_{B1} + 2.7\cdot10^{-5}\,\alpha_A^{4/5} \quad (26)$$

where $W_{IIa}(320)$ and $W_{IIa}(450)$ are the thermal resistance of type IIa diamonds at 320 and 450K, corresponding to the mean values of thermal conductivity $K_{IIa}(320)$=1930 W m^{-1} K^{-1} and $K_{IIa}(450)$=1250 W m^{-1} K^{-1}, α_{B1} and α_A are the absorption coefficients (in cm^{-1})at the maxima (8.5 and 7.8 μm respectively) of the most intensive bands of B1 and A systems of IR spectra. As A and B1 systems overlap and each system gives its own contribution to the absorption at 7.8 and 8.5 μm, α_A and α_{B1} are determined by the expressions [61]

$$\begin{aligned} \alpha_A &= 1.2\,\alpha_{7.8} - 0.49\,\alpha_{8.5} \\ \alpha_{B1} &= 1.2\,\alpha_{8.5} - 0.51\,\alpha_{7.8} \end{aligned} \quad (27)$$

Here $\alpha_{7.8}$, $\alpha_{8.5}$ are the absorption coefficients, determined at 7.8 and 8.5 μm. Expressions (25)-(27) rather accurately enable the determination of the thermal conductivity from optical measurements. Comparison between such calculated values and experimentally determined K values disagree on average by ±5%, and only in some cases reach ±10%.

With somewhat less accuracy one can evaluate the value of K at 320 and 450 K without separating the contributions from A and B1-defects by the additional thermal resistance W_{ad} [60]:

$$W_{ad} = 2.6\cdot10^{-5}\,\alpha_{8.5}\ (W^{-1}\ m\ K) \quad (28)$$

The range of diamond thermal conductivity at 320K varies from 500 to about 1800 W m^{-1} K^{-1}, and at 420K from about 400 to 1200 W m^{-1} K^{-1}.

The temperature dependence of K becomes logarithmic in the range 200-600K [48, 52, 63, 64]. Work [63] suggested the following expression for K in the 200-600K region.

$$K(T) = K(320)\left[\frac{K(450)}{K(320)}\right]^{\frac{\ln(T/320)}{\ln(450/320)}} \quad (29)$$

Thus on the basis of data received from IR absorption spectra of type Ia diamonds, one can get rather complete information about their thermal conductivity.

As the additional thermal resistance determined by A and B1-defects enters into the total thermal resistance of type Ia diamonds additively, one can consider as correct the supposition that the normal phonon-phonon processes play a significant role, and the approximation of $\tau_N^{-1} >> \tau_R^{-1}$ becomes more acceptable than $\tau_N^{-1} << \tau_R^{-1}$. Nevertheless the authors of the paper [62] made an attempt to interpret the expressions (25) and (26) based on the later assumption.
From the comparison of (25) and (26) with calculations based on equations (20), (24) and (12), it follows that A-defects scatter phonons as point defects consisting of one [62] or two [65] nitrogen atoms. The experimental data on the influence of B1-defects on the thermal resistance correspond to the expression for the phonon relaxation rate in the form (13), that is, B1-defects scatter phonons as flattened volume aggregates. Another conclusion was made on the basis of low temperature measurements of the thermal conductivity of type I diamonds containing B1-defects [58]. Satisfactory correlation between the calculated and experimental data was achieved with the assumption that B1-defects scatter phonons as isometric aggregates with dimensions of 44-65 Å, using expression (14) for the relaxation rate.

Type Ib diamonds differ by a higher content of single substitutional nitrogen atoms (C-defects), which reveal themselves in optical absorption spectra such as EPR. The thermal conductivity of a rather limited number of natural diamonds with an increased C-defect content was measured [48, 66]. C-defects contents in these crystals amounted from 10^{17} to $1.5 \cdot 10^{18}$ cm^{-3}. From the results concerning these crystals it is impossible to make definite conclusions about the C-defects' influence on thermal conductivity because natural crystals contain both C and A-defects, the concentrations of where are unknown. Additionally, crystals with an increased C-defect content have a fibrous structure [61], which can also influence the value of K. This supposition is in agreement with the results of measurements on three natural diamonds in which nitrogen is contained in small concentrations simultaneously in the form of C and A-defects. The additional thermal resistance in these crystals at 320K is higher than at 450K, which is not in agreement with the Rayleigh mechanism of phonon scattering, and may only be in the case when scattering by extended defects plays a significant role.

Unlike natural diamonds, synthetic diamonds contain nitrogen mainly in the form of single substitutional nitrogen atoms (C-defects), and consequently relate to type Ib. But until now the thermal conductivity of only perfect synthetic single-crystalline diamonds with a rather high nitrogen concentration (10^{19} cm^{-3}) has been measured [51]. Near its maximum, the thermal conductivity exceeds that of type IIa diamonds. At temperatures higher than 200K, the thermal conductivity reduces with temperature more abruptly, and becomes somewhat less than type IIa diamonds at 300K. The thermal conductivity of synthetic single-crystal diamonds grown in conditions of spontaneous crystallization is determined not by C-defect content, but by the perfection of their structure, and reduces from about 900 W m^{-1} K^{-1} at 300K to about 700 W m^{-1} K^{-1} at 600K[52,64].

The influence of C-defects on diamond thermal conductivity was calculated in the work [67] in an approach with $\tau_R^{-1} >> \tau_N^{-1}$. If ones uses in (12) $\alpha = 0.65$ [68] (for nitrogen in A-defects $\alpha = 0.15$ [69]), then (in W^{-1} m K)

$$W(320) = 1/K(320) \approx W_{IIa}(320) + 8.5\cdot10^{-20} N_c^{4/5}$$
$$W(450) = 1/K(450) \approx W_{IIa}(450) + 12.4\cdot10^{-20} N_c^{4/5} \quad (30)$$

Where N_c is the contents of single nitrogen atoms (C-defects) in cm^{-3}.
As nitrogen concentration in A-form is related to α_A by the correlation $N_A = 5.8\cdot10^{18}\ \alpha_A\ cm^{-3}$ [69] then it follows from (25) and (26) that for diamonds containing only A-defects

$$W(320) = 1/K(320) \approx W_{IIa}(320) + 2.05\cdot10^{-20} N_A^{4/5}$$
$$W(450) = 1/K(450) \approx W_{IIa}(450) + 2.63\cdot10^{-20} N_A^{4/5} \quad (31)$$

Thus it follows from the calculations that the single nitrogen atoms (C-defects) must contribute to the thermal resistance at temperatures from concentrations from 4 to 5 times higher than that of nitrogen atoms in A-defects. It is a consequence of a larger crystal lattice distortion by single nitrogen atoms in comparison to aggregated atoms.

3.4 THERMAL CONDUCTIVITY IN DIAMONDS WITH VARIABLE ISOTOPE CONTENT

The data on thermal conductivity of perfectly nitrogen free synthetic diamonds with various concentration of ^{13}C are listed in the works [45, 54, 57, 70]. The dependence of diamond thermal conductivity on ^{13}C concentration at 150 and 300K is shown in figure 5. Figure 6 illustrates the K dependence on T for diamonds with a usual natural content (1.1%) of ^{13}C and a decreased content (~0.1%) of this isotope. Both figures are taken from the work [55], and include almost all the material on this phenomenon accumulated to date.

Undoubtedly one must consider the high thermal conductivity of crystals with 0.07% content of ^{13}C isotope as a sensation, because at 300K it reaches 3320 W m^{-1} K^{-1}, approximately 1.5 times higher than the value of thermal conductivity for type II synthetic or natural diamonds with natural isotopic content [45]. It is even more surprising because it is difficult to explain such a high isotopic influence on diamond thermal conductivity in the limits of the usually used expression (20). The calculations show that the reduction of ^{13}C concentration from 1 to 0.1% can increase the thermal conductivity at 300K only by a few percent.

This difficulty can be removed if one assumes that the normal phonon-phonon interations play a significant role in the processes of heat transfer [46, 47]. Using expression (22), that is considering that $\tau_N^{-1} >> \tau_R^{-1}$, one can get a rather good agreement between the calculated results and the experimental data, at least for low concentrations of the ^{13}C isotope.

The above mentioned works served as a basis for later work on the thermal conductivity of diamonds with variable isotope content, analyzed with the use of the general

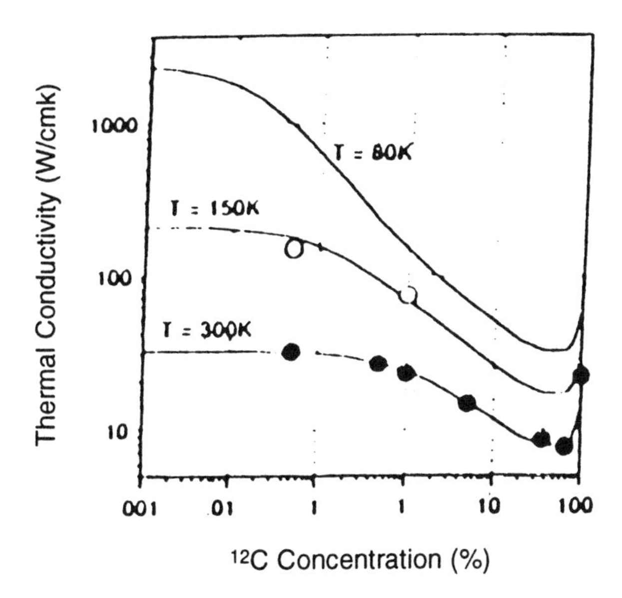

Figure 5. The thermal conductivity dependence on the ^{13}C isotope content. The solid lines are the calculated data, the symbols are the experimental results (taken from the work [55]).

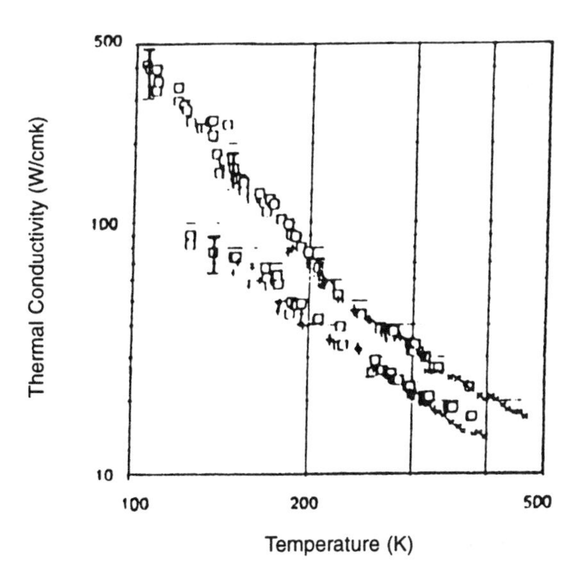

Figure 6. The temperature dependence of diamond thermal conductivity with the 1.1% ^{13}C isotope content (corresponding to the natural content) and 0.1% content (taken from the work [55]): □ - the data from the work [55]; + - the data from the work [57]; x - the data from the work [54].

Callaway expression (19). According to the general forms (11a) and (11b) the next expressions for τ_N^{-1} and τ_U^{-1} were suggested in the works [71, 72]:

$$\tau_N^{-1} = 2.2\, T^4 x$$
$$\tau_U^{-1} = T^4 x^2 \exp(-500/T) \tag{32}$$

and

$$\tau_N^{-1} = 4.7 \cdot 10^{-2}\, T^5 x^2$$
$$\tau_U^{-1} = 9.1 \cdot 10^{-4}\, T^5 x^2 \exp(-100/T) \tag{33}$$

in the work [73]

$$\tau_N^{-1} = 0.011\, T^5 x$$
$$\tau_U^{-1} = 0.9\, T^4 x^4 \exp(-500/T) \tag{34}$$

in the work [55]

$$\tau_N^{-1} = 1.5\, T^4 x$$
$$\tau_U^{-1} = 162\, T^3 x^2 \exp(-670/T) \tag{35}$$

and in the work [57]

$$\tau_N^{-1} >> \tau_U^{-1}$$
$$\tau_U^{-1} = 151.2\, T^3 x^2 \exp(-730/T) \tag{36}$$

In the last two cases the transverse and longitudinal phonon modes were taken into account separately. From the variety of suggested expressions it is seen that the use of only experimental data on thermal conductivity does not allow one to determine definitely the parameters m, n, k, l and α. Therefore it is useful to take into account the most general notions about the ratio between the relaxation times τ_N and τ_U, which avoids incorrect estimates or determination of the temperature range where they have only a formal right to exist.

So in all temperature ranges the ratio l_U/l_N must decrease with the growth of T [35, 36]. From (11 a,b) and (10) it follows that

$$l_U/l_N = \tau_U/\tau_N \approx T^{k-m} x^{l-n} \exp(-\alpha/T).$$

At low temperatures phonon frequency $\omega \sim T$ (x = const) and

$$l_U/l_N \approx T^{n-k} \exp(-\alpha/T).$$

Then at m-k>0 the ratio l_U/l_N has a minimum at $T = \alpha(m-k)^{-1}$

With the growth of T, the phonon frequency becomes more independent of T and approaches the Debye frequency ω_D. Consequently

$$l_U/l_N \approx T^{m-k-n+l} \exp(-\alpha/T).$$

and if m-k-n+l>0 then the ratio l_U/l_N will have minimum at $T=\alpha(m-k-n+l)^{-1}$. It follows from this that the expressions (32) can be used at least up to 500K. The expressions (33) formally can be used at any temperature. The expressions (34) give a minimum ratio l_U/l_N in the range 250-500K. The direct calculations show, that in this case l_U/l_N rapidly grows beginning fıom 300K. Therefore they are suitable for application only at T<300K. The application of expressions (35) in the temperature range up to 1000K as it was made in the work [55] is not quite correct because according to them, the ratio l_U/l_N has a minimum in the range 335-670K. One ought to remember that the general form of τ_N^{-1} (11a) was obtained with the low-temperature approximation; with this series of assumptions it is difficult to hope it would be useful for application in the wider temperature range.

3.5 HYDRODYNAMIC BEHAVIOR OF PHONONS (POISEUILLE FLOW, THE SECOND SOUND)

The results of calculations of phonon mean free path lengths l_U and l_N on the basis of expression (19) are shown In figure 7, where τ_N^{-1} and τ_U^{-1} were used in accordance with (32) and (34). Phonon scattering by the specimen boundaries was not taken into account so the crystal was considered to be infinite. If the crystal lattice consists of only ^{12}C (100% ^{12}C) and has no defects, then the only resistive process is the U-process and $l_R \equiv l_U$. As it is seen in figure 7 in a relatively wide temperature range $l_R >> l_N$ in a single-isotope defectless diamond. The peculiarities of phonon scattering by the specimen boundaries at $l_R >> l_N$ were analyzed for the first time in the work [74] where the criterion for the existence of Poiseuille flow was formulated. In this process, the phonon free path length l_{ef} exceeds the characteristic specimen dimension d by approximately d/l_N times. Later on, the authors of the works [75, 76] analyzed in detail, on the basis of Monte-Carlo calculations, the behavior of l at different correlations between l_R, l_N and the diameter d of a specimen with infinite length in the process of phonon diffuse scattering by a boundary surface.

The following situations can be realized at $l_R > l_N$.

1. If l_N > d then the Knudsen limit takes place and l_{ef} = d. For any d there is a low temperature at which it is fulfilled.

2. When, with the growth of temperature, l_N and d become equal, and at the same time $l_R/l_N \geq 30$, then the situation of Knudsen minimum takes place. At these conditions $l_{ef} \approx 0.8$d.

3. If with the further growth of T the conditions $d/l_N \geq 10$ and $l_R/l_N \geq 100$ are simultaneously fulfilled, then beginning from this point l_{ef} exceeds d by not less than 1.3-1.5 times.

4. If there is a temperature range where $d/l_N \geq 30$ and $l_R/l_N \geq 3 \cdot 10^3$ then Poiseulle phonon flow takes place for $l_{ef} \approx 0.1\, d^2/l_N$.

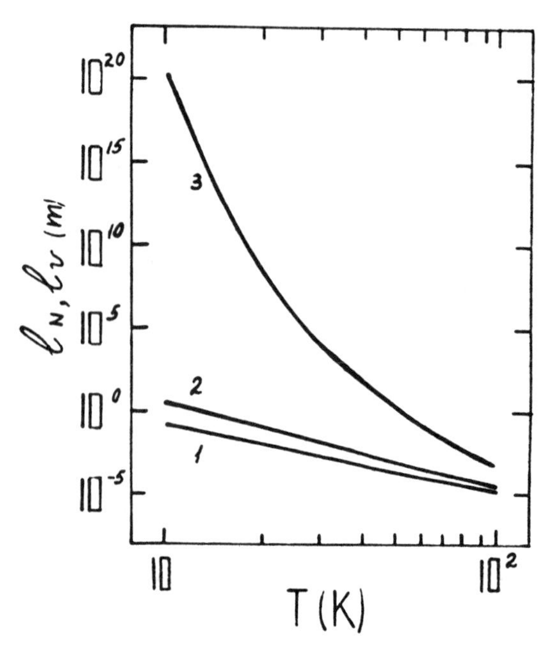

Figure 7. The temperature dependence of l_N and l_u for isotopically pure (100% ^{12}C) diamond (taken from the work [77]): 1 - l_N in accordance with (32); 2⁻ l_N in accordance with (34); 3- l_U in accordance with (32) and (34).

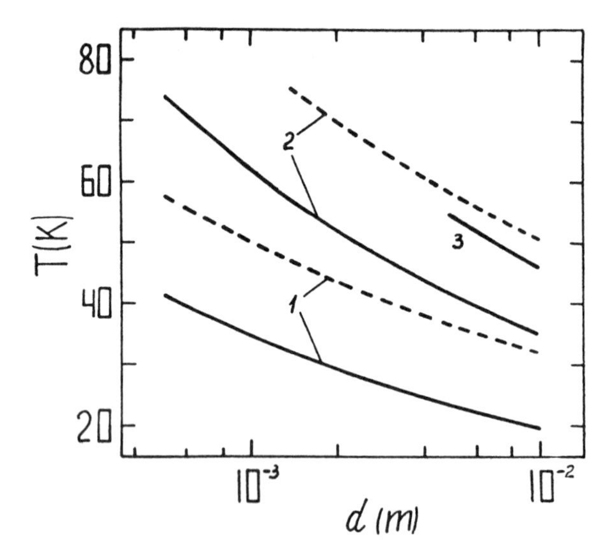

Figure 8. The temperatures at which the Knudsen minimum exists (1); l_{ef} exceeds d not less than 1.3-1.5 times (2); Poiseuille phonon flow (3) for the specimens with different diameters d. The solid lines correspond to (32), the dotted lines correspond to (34). (The figure is taken from the work [77]).

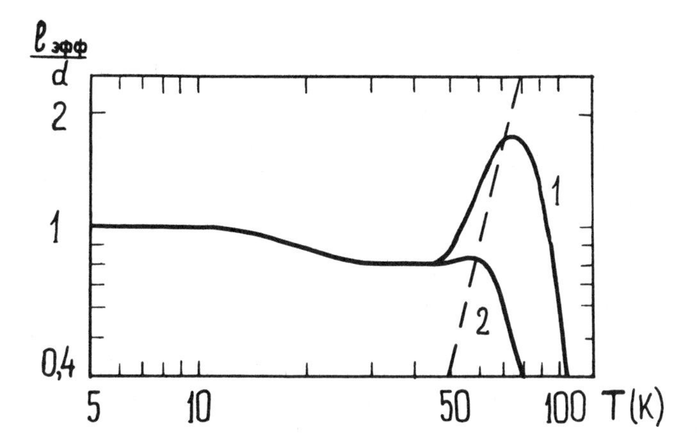

Figure 9. l_{ef} /d dependence on temperature, calculated for the infinite specimen with the diameter 1 mm: 1 - without ^{13}C isotope; 2 - ^{13}C isotope content is 0.07%; --- l_{ef} =0.1 d^2/l_{ef}.

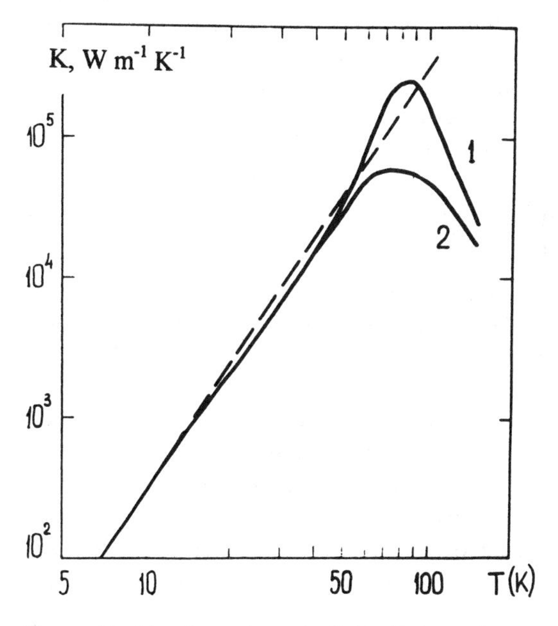

Figure 10. The thermal conductivity K dependence on temperature, calculated for the infinite specimen with the diameter 1 mm (taken from the work [72]): 1 - without ^{13}C isotope; 2 - for the 0.07% content of ^{13}C isotope the cubic K dependence T.

Figure 8 shows the temperatures at which, according to the calculations, some effects related to the specific phonon scattering by the specimen boundaries are displayed. With this, N-processes play a significant role for specimens of single-isotope defectless diamond having different diameters d [77]. One can expect Poiseuille flow when $l_{ef} \approx 0.1\ d^2/l_N$ in a single-isotope diamond only if d>5 mm. l_{ef} can noticeably exceed d and the Knudsen minimum in specimens with a diameter of about 1 mm.

The dependence of l_{ef}/d on T for a single-isotope diamond and a diamond with 0.07% of ^{13}C at d = 1 mm is shown in figure 9 [72]. Isotope content is seen to have a strong influence on the character of phonon scattering by the specimen surface. Figure 10 shows how the peculiarities of phonon scattering by the specimen boundaries influences the thermal conductivity at the conditions of a significant role for N-processes. There is a simpler method of taking into account N-processes for phonons interacting with boundaries. When one uses τ_b^{-1} in calculations of K according to the general Callaway expression (19), τ_b^{-1} takes a form like the one suggested in the work [71]:

$$\tau_b^{-1} = s^2\tau_N / (0.1\ d^2 + Sd\tau_N) \tag{37}$$

According to (37), and if the corresponding conditions are fulfilled, then with the growth of T after the Knudsen limit instead of l_{ef} decreasing to 9.8d (the Knudsen minimum) it begins to grow to $l_{ef} \approx 0.1\ d^2/l_N$. Nevertheless, the expression (37) gives more correct results than (18) if $l_R > d > l_N$.

The second sound is another hydrodynamic effect. On the conditions that phonon flow keeps its total wave vector, any excitations in the phonon gas (local changes in energy or temperature) can be transferred along the crystal, which leads to the rise of harmonic temperature waves analogous to the elastic waves in an ordinary gas or crystal. Such waves (the second sound) are spread with a speed of about $s/\sqrt{3}$. The conditions for second sound spreading in solids were discussed in the works [74, 78, 79] and are reduced to the demands $l_N \ll l_R$ and $\lambda/l_N \gg 1$ (λ is the wavelength of oscillations spreading in the phonon gas). Under these conditions the relaxation is proportional to $(l_N/\lambda + \lambda/l_R)$. It follows from this that the relaxation has minimum when $\lambda = \sqrt{l_N l_R}$. The calculations show that in a single-isotope perfect diamond with dimensions of about 1 cm^3, oscillations with frequency (1 to 7)$\cdot 10^6$ Hz must be spread with a minimum relaxation in the temperature range 40-70K [71, 72].

Presently it is impossible to state with definite confidence that one can observe the above hydrodynamic effects in diamond. But even the presence of the serious reasons for discussion of this problem is of great interest. For an ideal single-isotope crystal lattice the main demand, $l_N \ll l_R$, reduces to the condition $l_N \ll l_U$ ($\tau_N^{-1} \gg \tau_U^{-1}$), which is always fulfilled at rather low temperatures; the correlation improves as T decreases with respect to 0.5 θ. From this point of view, diamond as a material with a high Debye temperature θ possesses an advantage over other crystals.

In the case of real crystals it is necessary to fulfill the condition $l_U \ll l_r$ ($\tau_U^{-1} \gg \tau_r^{-1}$) where the index r means the phonon scattering processes at all kinds of defects in a crystal. It is fulfilled more simply the higher the degree of anharmonicity of crystal lattice oscillations. From this point of view, diamond is a rather unsuitable material which in particular is confirmed by low values of its thermal coefficient of expansion and the character of its temperature dependence. The coefficient, as well as phonon-phonon interactions is determined by the value of crystal anharmonicity.

But in the limits of the simplest analysis stated in section 2 of this chapter, one can imagine the situation where the potential energy of the oscillating atom, U, is well described in general by a quadratic dependence on its displacement. But at the same time the details of potential energy dependence on atom displacement are such that in each mode the oscillations are noticeably different from anharmonic. These circumstances will not be in contradiction if the mean atom displacement from the equilibrium position for neighboring modes is minorly different in value but having opposite signs. Then the effectiveness of phonon-phonon interactions can be considerably higher than it follows from the information on anharmoticity degree, obtained by the measurements of the thermal expansion coefficient.

3.6 THE THERMAL CONDUCTIVITY OF DIAMOND FILMS

The peculiarities of diamond film structure makes it worthwhile to separate them into two conventional classes: the thin films and the thick films. One can attribute films with a thickness not exceeding 30 μm to the first class. In such films, grain dimensions do not exceed several microns and a regular change of grain dimensions in the direction of growth front is rather small. The thermal conductivity of thin films changes over a wide range and can exceed 1000W m^{-1} K^{-1} at temperatures 370-400K. Figure 11 shows the dependence of the thermal conductivity of films with a thickness not exceeding 30 μm, on the methane content in the feeding gas [80]. The methane content reduction in the feed gas results in thermal conductivity growth which is simultaneously accompanied by an increase in the ratio between the intensity of two spectral bands - the band at 1332 cm^{-1}, characteristic of the diamond phase in the Raman spectrum and the band at 1550 cm^{-1}. There is also a correlation between the value of K and the full width at half-maximum (FWHM) of the band 1332 cm^{-1} (Fig. 12). Thus according to the characteristics of the Raman spectrum of the film one can judge the value of its thermal conductivity. The film thermal conductivity is higher the more its Raman spectrum corresponds to the spectrum of a pure diamond phase.

The thermal conductivity of diamond films was investigated for the first time in a wide temperature range (from 10 to 300K) in the work [84]. The results obtained from measuring one film (with thickness 14 μm) are shown in figure 13. The authors of the work [84] managed to describe the K dependence on T in the range 30-200K by the expression (20) taking into account, besides the mechanisms of phonon scattering considered for a single-crystal diamond, the scattering by grain boundaries. Such an approach to the analysis of the thermal conductivity of polycrystalline materials was previously considered by R. Berman [36] and the quantity estimates of the value K for a diamond with grain dimensions from 1 to 1000 μm, obtained in the work [85], showed that phonon scattering by the grain boundaries may be a factor in determining the thermal conductivity of a polycrystalline diamond.

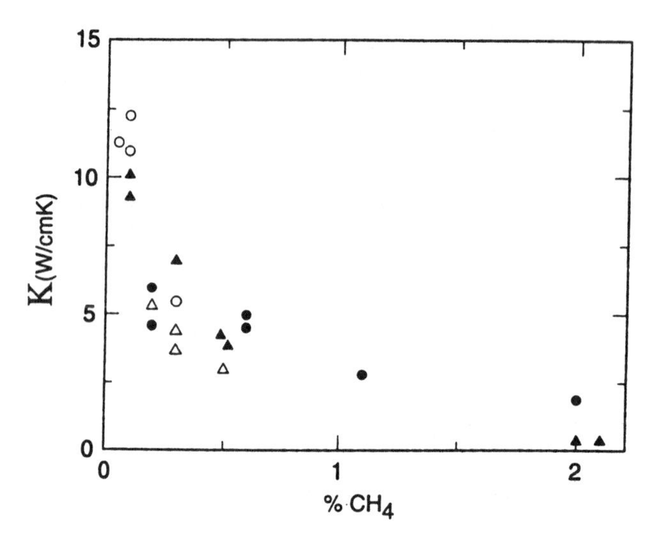

Figure 11. The thermal conductivity dependence on methane content in the feed gas used in the process of thin diamond film production (taken from the work [80]): Δ - according to the work [81]; o - according to the work [82]; • - according to the work [83].

For the description of the dependence of diamond films thermal conductivity on temperature in a wide range of temperatures the authors of the works [56, 86, 87] suggested two approaches. The first one takes into account the possibility of phonon specular reflection during scattering by the grain boundaries (expression (16) for the relaxation rate). The experimental data [84] and the calculated results are in good agreement if in the limits of the first approach the value of boundary roughness H_{ef} is 9-10Å for the grain dimensions (~1.6 μm, close to the ones determined from electron-microscopic measurements).

From the analysis carried out with the use of the second approach, it follows that a diamond film contains aggregates of non-diamond phase with diameters a = 12 Å and concentrations $\rho \approx 1.1 \cdot 10^{18}$ cm^{-3}, which makes up ~0.1% from the film value. As the values H_{ef} and a are close, both approaches considered above are not in contradiction to each other if one assumes that the aggregates are near the inter-granular boundaries. As it follows from the detailed survey of the works devoted to the investigations of diamond film thermal conductivity [80], the dimensions of non-diamond phase aggregates, found from the measurements of K dependence on T, are varied from 10 to 17 Å, and their volume contents is varied from 0.01 to 0.1%. According to expression (14), the thermal conductivity is determined by the value ρa^2 or by the ratio between the volume contents of aggregates and their dimension, V/a, at temperatures at which phonon scattering by the aggregates is of a geometric character. The less is ρa^2 or V/a, the higher K is.

If a film has a thickness of more than 30μm, then one cannot neglect the change of its structure with the removal of the growth front from the substrate, as this leads to a series of peculiarities of thermal conductivity behavior which are analyzed in details in the

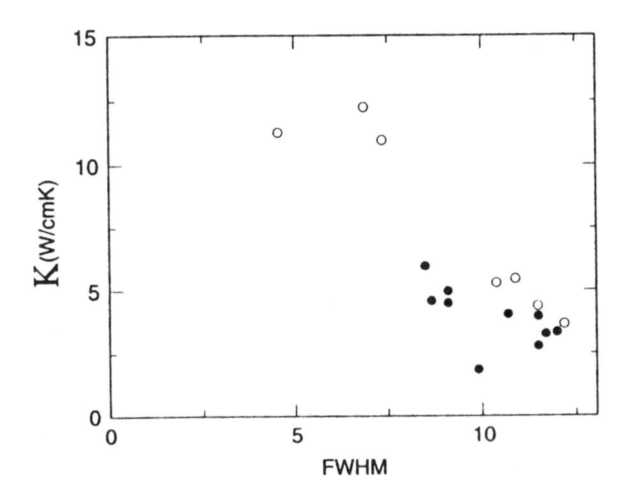

Figure 12. The thermal conductivity of thin diamond films as a function of the full width at half-maximum (FWHM, in cm^{-1}) of the Raman line at 1332 cm^{-1} (taken from the work [80]): o - according to the work [82]; • - according to the work [83];

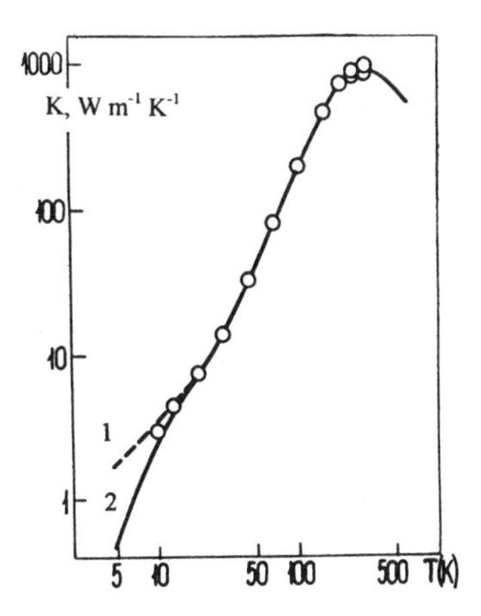

Figure 13. The temperature dependence of the thin diamond film thermal conductivity: o- the experimental data from the work [84]; 1 - the results calculated on the basis of phonon scattering by the grain boundaries with taking into account the specular reflection [56, 86, 87]; 2 - the results calculated on the basis of phonon scattering by non-diamond phase aggregates [56, 86, 87].

works [80, 88, 89]. Diamond film structure is represented schematically in figure 14, taken from the work [88]. As shown in this figure, the film is a non-homogeneous material in which the mean grain size increases with the increase in distance from the substrate. In addition, due to the column-like form of the grains, the number of inter-grain boundaries in the direction parallel to the film surface is greater than in the orthogonal direction. Therefore, as the distance from the initial growth surface increases, the value of the local thermal conductivity, K_{loc}, increases and displays anisotropic behavior, with the heat flow directed along the film surface, $K_{loc\parallel}$, smaller than $K_{loc\perp}$ for the heat flow spreading in the orthogonal direction.

The investigations of thick films thermal conductivity in a wide temperature range [88-91] showed that the general form of K dependence on T is like that which is observed in the case of thin films. However, so far as most of such films contain grains of greater size, their thermal conductivity can markedly exceed the thermal conductivity of thin films.

There is some difference between the measured thermal conductivity values in the directions along the film surface $K_{\parallel}$ and perpendicular to it $K_{\perp}$, as shown in figure 14. The measured values $K_{\parallel}$ and $K_{\perp}$ at room temperature and the estimates of $K_{loc\parallel}$ and $K_{loc\perp}$ for the films of different thickness are shown in figure 15 [80,89]. It follows from this figure that the local thermal conductivity in the sections far the substrate is comparable with that of type IIa single-crystals. High thermal conductivity of the thick films gives a broad opportunity for their use in microelectronics as heat sink elements. In order to achieve their maximum efficiency of use one must take into account the factors of film inhomogeneity while designing different applications.

3.7 MATERIALS SINTERED FROM DIAMOND POWDERS

A great number of experiments on diamond powders sintering at high pressures and temperatures were carried out in the work [92], where it was shown that one can select such thermodynamic conditions, sintering time, diamond grains size and additives such that the thermal conductivity of sintered material obtained could run up to 920 W m^{-1} K^{-1} at T=300K. This value corresponds to the average value of natural single-crystal thermal conductivity. Until recently only the thermal conductivity of sintered material like Syndite have been investigated rather well [93, 94]. This material is used as a tool-making material, although it was shown that after a special chemical treatment it can serve as a heat sink substrate for electronic devices. Sintered Syndite-type materials are made from diamond powders with grain sizes from 10 to 100 μm with the addition of about 5 volume percent cobalt. The investigations of polished specimens show that during the sintering process, the grains do not crush and their size in the sintered material correspond to the initial volume. As it follows from the measurements carried out in works [93, 94], K dependence on T has a maximum near 380K and its position does not depend on grain sized. At the same time the thermal conductivity value increases with the growth of d (Fig. 16). More reliably this dependence was determined for grain sizes from 10 to 50 μm. The number of measured specimens with grain size from 75 to 100 μm is small and there are some grounds to consider that this material has a rather great scatter in K values for d = 100 μm. Treatment with some chemicals enables one to remove from the sintered material a considerable part of the cobalt which results in the reduction of the thermal conductivity by about 10%.

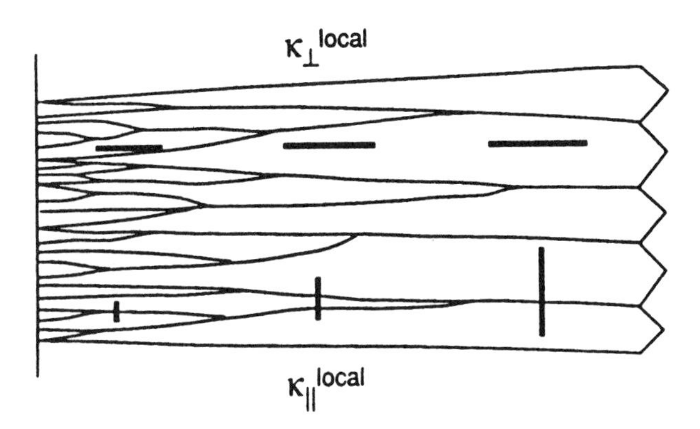

Figure 14. Schematic view of the typical cone-shaped grain structure of diamond films. The lengths of the bars are roughly proportional to local conductivity for heat flow perpendicular and parallel to the plane of the sample. (From reference 88.)

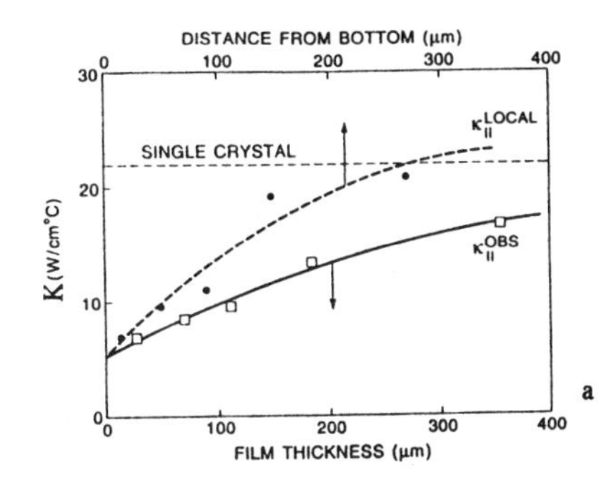

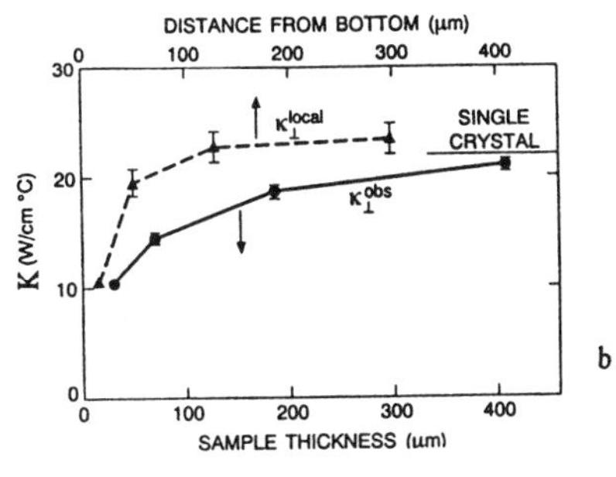

Figure 15. The dependence of the measured thermal conductivity K^{obs} and calculated local thermal conductivity K^{loc} at room temperature on the film thickness (taken from the work [88] and [89]): a) for heat flow along the film surface; b) for heat flow perpendicular to the film surface.

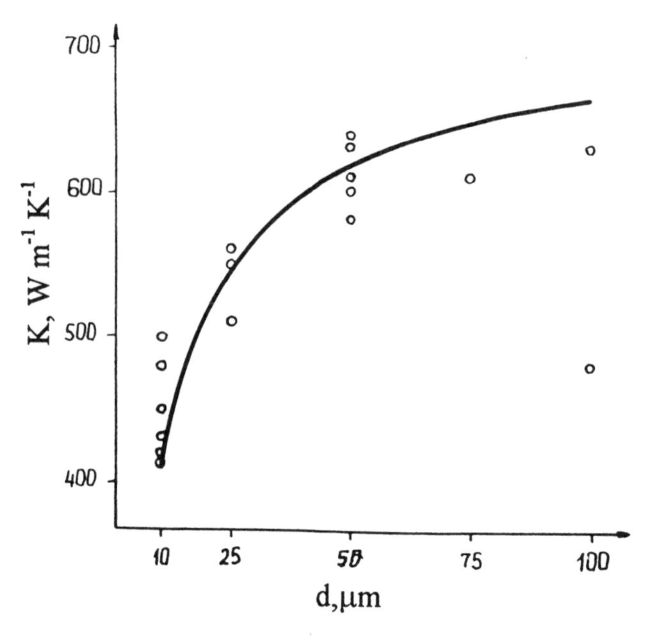

Figure 16. The dependence of the thermal conductivity of sintered diamond material of Syndite type on grain size at 380K: o - the experimental data [93], --- - the calculated results obtained according to the model of the structure with interpenetrating components.

The estimates based on expression (20) with taking into account phonon scattering by grain boundaries show that a polycrystal consisting of about 0.5 μm grains has K values (about 500W m^{-1} K^{-1} at 380K) almost the same as those peculiar to the sintered materials discussed, as well as a maximum in K dependence on T in the range 350-400K [56]. Such a model assumes that sintered materials produced from diamond powders with different sizes have the same structure and, consequently, the same thermal conductivity. The totality of experimental data for sintered material of Syndite-type can be explained if one uses the model suggested in the works [56, 86].

The possibility of metal phase removal from the sintered material by chemical etching gives some reasons to consider it as a material consisting of two inter-penetrating components with an elementary cell shown in figure 17. A diamond grain is considered to consist of two regions. The inner region has little difference from the initial grain in structure and properties. The outer region with a thickness of h is considerably modified as a result of such processes as a near-surface grain crushing, plastic deformation and possibly secondary synthesis. If one assumes that $h \ll 2d$ then in the limits of the theory of generalized conductivity the effective thermal conductivity K_{ef} of such a material may be expressed

$$K_{ef} = \frac{c^2 K^d_{diam} K_h d}{K_h d + 2K^d_{diam} h} + \frac{(1-c)K_{ph}[c(1-c)K_{ph} + (1+c^2)K^d_{diam}]}{cK_{ph} + (1-c)K^d_{diam}} \qquad (38)$$

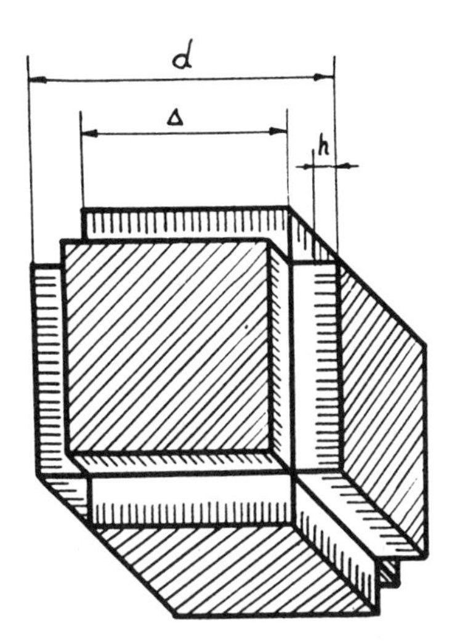

Figure 17. The elementary cell of diamond sintered material according to the model of a structure with interpreting components.

Here K^d_{diam} is the inner region thermal conductivity which equals the thermal conductivity of the initial grain (in general it depends on d), K_h is the thermal conductivity of the outer modified grain region, K_{ph} is the thermal conductivity of the second (metallic) phase, $c = \Delta/d$. The value c is connected with the volume content of the second phase V_{ph}. For $0 \leq V_{ph} \leq 0.5$, $c = 0.5 - \cos(\beta/3)$, where $\beta = \arccos(1 - 2V_{ph})$ in the region $3\pi/2 \leq \beta \leq 2\pi$.

All the results of the measurements of the sintered materials thermal conductivity in the range 320-459K are well described by expression (38) at $V_{ph} = 0.05$; $K_{ph} \approx 50$ W m^{-1} K^{-1}, which is close to the thermal conductivity of cobalt, about 69 W m^{-1} K^{-1}. $K^d_{diam} = 870$, 860, 750 W m^{-1} K^{-1} at 320, 380, 450K,respectively. The last values of thermal conductivity are well correlated with the thermal conductivity of synthetic diamonds produced by the method of spontaneous crystallization [64].

The thickness h of the outer region is the same for all grain sizes, about 2μm. In all cases, the structure of the region is characterized by phonon scattering on the boundaries of the grains with a size about 0.2μm and corresponding thermal conductivity K_h. The dependence of sintered materials thermal conductivity and calculated on the basis of expression (38) is shown in figure 17. The assumption $2h >> d$ for d = 10μm is less acceptable than in other cases owing to which the calculated values of K are somewhat lower. At $d \leq 5$μm the sintered material thermal conductivity is entirely determined by the thermal conductivity of the modified region. Then the sintered materials from any diamonds with $K^d_{diam} \geq 500$ W m^{-1} K^{-1} must have thermal conductivity not higher than 360 W m^{-1} K^{-1}. The use of type IIa diamonds with high thermal conductivity can lead to increased thermal conductivity in comparison to that peculiar to Syndite only if the grain size will exceed 10 μm. The estimates show that in the case $d < 50$ μm, the thermal conductivity of sintered material produced from type IIa diamonds using technology analogous to the one used for Syndite can reach 900-1000 W m^{-1} K^{-1} at room temperature.

3.8 TWO-COMPONENT DIAMOND COMPOSITES

One can analyze the effective thermal conductivity K_{ef} of a composite consisting of a continuous medium and diamond particles dispersed in it. In general the calculations of such thermal conductivity K_{ef} are carried out on the basis of the general theory of conductivity [95]. With this one assumes regular particle distribution throughout the volume, the absence of convective and radiative mechanisms of heat transfer, and the absence of heat resistance at the boundary between a particle a continuous medium. From this consideration it follows that the thermal conductivity of a composite varies between the value of the thermal conductivity of the constituent components depending on the concentration of dispersed particles. Consequently the above thermal conductivity can not be lower than the thermal conductivity of the worse heat conducting component. But while investigating diamond-containing materials on the basis of copper and nickel it was found that K_{ef} of both composites was markedly lower than the thermal conductivity of the binder (copper and nickel) [96]. This can be explained if the thermal resistance of the boundary between the forming components is considerable. The influence of this factor on the composite thermal conductivity was investigated in the limits of the theory of generalized conductivity in the works [97, 98].

An elementary cell of a composite is shown in figure 18 for the case when the dispersed particles have the form of a rectangular parallelepiped. While calculating K_{ef}, two ways

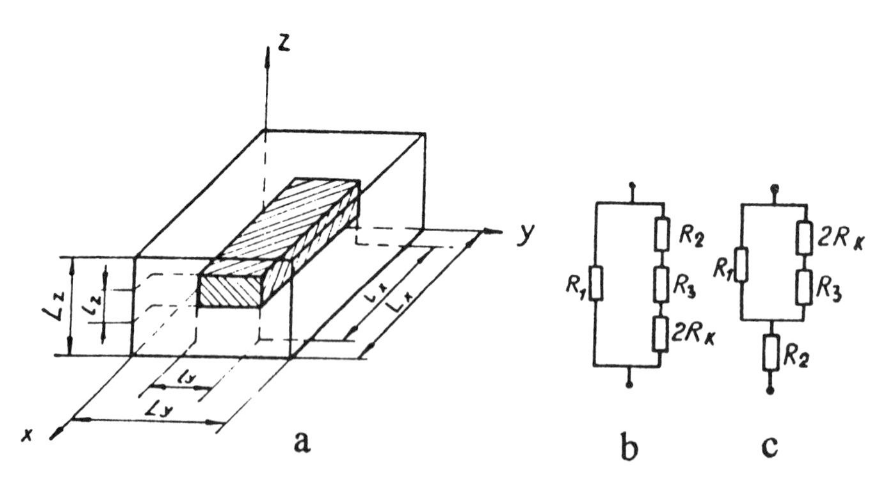

Figure 18. The elementary cell of a composite material with dispersed particles in the form of rectangular parallelepipeds (a); the equivalent schemes of the thermal resistances connections while dividing the cell by the adiabatic planes (b) and by the isothermal planes (c).

dividing the cell into parts were used. According to the first method, the cell is divided by the planes which are perpendicular to the heat flow and are considered to be isothermal (the isothermal approach). According to the second way the cell is divided by the planes, parallel to the flow while assuming the absence of heat exchange through them (the adiabatic approach). The corresponding equivalent schemes for the calculation of elementary cell thermal resistance are shown in figure 18 b,c (R_1, R_2 are the thermal resistances of the region of binder, R_3 is the thermal resistance of the particle, $R_k = (\alpha_k l_i l_j)^{-1}$ is the thermal resistance of the boundary between the particle and the binder expressed via the thermal conductivity coefficient α_k of the boundary). It is supposed that the thermal resistance of the particle-medium and medium-particle junctions are the same.

According to the equivalent schemes in the adiabatic approach

$$\frac{K_{ef}}{K_m} = 1 + \frac{V_d\left(\frac{K_d}{K_m} - 2\frac{K_d}{l_x\alpha_k} - 1\right)}{\frac{K_d}{K_m} - \frac{V_d}{s}\left(\frac{K_d}{K_m} - 2\frac{K_d}{l_x\alpha_k} - 1\right)} \tag{39a}$$

and in the isothermal approach

$$\frac{K_{ef}}{K_m} = \frac{1 - \frac{V_d}{k_x}\left(1 - \frac{K_d}{K_m}\right) + 2\frac{K_d}{l_x\alpha_k}\left(1 - \frac{V_d}{k_x}\right)}{1 - \frac{V_d}{k_x}(1 - k_x)\left(1 - \frac{K_d}{K_m}\right) + 2\frac{K_d}{l_x\alpha_k}\left(1 + V_d - \frac{V_d}{k_x}\right)} \tag{39b}$$

Here V_d and K_d are the volume content and the thermal conductivity coefficient of dispersed material, K_m is the thermal conductivity coefficient of the binder, s is the ratio of the cross sections of the particle and the elementary cell in the plane perpendicular to the heat flow, l_x is the particle size in the heat flow direction, $k_x = l_x/L$ is the ratio of the linear dimensions of a particle and a cell in the same direction. The best approach is given by the mean value of K_{ef} calculated by the two methods above, but a relatively small difference between allows the use of any of these variants.

The important consequence of taking into account the finite value of α_k is that $K_{ef}=K_m$ even for $K_d > K_m$ if

$$\frac{K_d}{K_m} - 2\frac{K_d}{l_x\alpha_k} = 1$$

Moreover if this difference is less than one then $K_{ef} < K_m$. In connection with this while considering composites with $K_d > K_m$, it is useful to introduce the value of the critical size of the dispersed particles

$$l_{cr} = \frac{2K_d}{\left(\frac{K_d}{K_m} - 1\right)\alpha_k} \quad (40)$$

For the particle size in the heat flow direction $l_x > l_{cr}$ one gets $K_{ef} > K_m$, and at $l_x < l_{cr}$ one gets $K_{ef} < K_m$ for any particle concentration.

The isothermal and adiabatic approaches give the same value l_{cr} not depending on V_d. If one investigates the material having $K_d >> K_m$, then

$$l_{cr} \approx 2K_m/\alpha_k \quad (41)$$

The calculated values l_{cr} for composites based on copper and nickel with dispersed particles of type IIa diamond (natural or synthetic) and the mean value of thermal conductivity are listed in Table 5. The values α_k correspond to the maximum thermal conductivity of the contact regions between the polished copper surface, that is 10^5 W m^{-2} K^{-1} [99], and the thermal conductivity $1.67\cdot10^7$ W m^{-2} K^{-1} of the boundary between the polished diamond surface and titanium vacuum evaporated onto it [100]. From the measurements of the effective thermal conductivity of the composites on the basis of the galvanically precipitated copper and nickel [96] it follows that for the *Table 5.* The critical particle size l_{cr} for diamond containing composites on the basis of copper and nickel

K_d, W m^{-1} K^{-1}	K_m, W m^{-1} K^{-1}	l_{cr}, m ($\alpha_k = 10^5$ W m^{-2} K^{-1})	l_{cr}, m ($\alpha_k = 1.67\cdot10^7$ W m^{-2} K^{-1})
2000	400 (Cu)	10^{-2}	$6\cdot10^{-5}$
(type IIa diamond)	100 (Ni)	$2.1\cdot10^{-3}$	$1.26\cdot10^{-5}$
1000	400 (Cu)	$1.3\cdot10^{-2}$	$8\cdot10^{-5}$
(type Ia, Ib diamond)	100 (Ni)	$2.2\cdot10^{-3}$	$1.3\cdot10^{-5}$

diamond-binder boundary $\alpha_k \approx 1.2\cdot10^6$ W m^{-2} K^{-1} as in the case of copper and nickel. The use of natural or synthetic diamonds also does not lead to the change of α_k. For the galvanically produced composite on the basis of copper $l_{cr} \approx 1$ mm, and on the basis of nickel $l_{cr} \approx 200$ μm. Thus the galvanically produced composite will have the effective thermal conductivity higher than the thermal conductivity of a binder only in a case when the diamond particles size is greater than 1 mm in the case of copper and greater than 200 μm in case of nickel.

The authors of the work [101] investigated the effective thermal conductivity of the zinc sulfide-diamond composite produced by pressing at 200 MPa and temperature 1200-1300K. The analysis of the results was carried out in particular on the basis of the expression for K_{ef} considered in the works [102, 103]. Here it is supposed that the particles have spherical form with radius a.

$$\frac{K_{ef}}{K_m} = \frac{\left[K_d\left(1+2\frac{K_m}{a\cdot\alpha_k}\right)+2K_m\right]+2V_d\left[K_d\left(1-2\frac{K_m}{a\cdot\alpha_k}\right)-K_m\right]}{\left[K_d\left(1+2\frac{K_m}{a\cdot\alpha_k}\right)+2K_m\right]-V_d\left[K_d\left(1-2\frac{K_m}{a\cdot\alpha_k}\right)-K_m\right]}$$

As the thermal conductivity of ZnS is 17.4 W m^{-1} K^{-1}, then the ratio $K_d/K_m \ll 1$ and the critical particle size l_{cr}=2a is determined by expression (41). The measurements of K_{ef} showed that for ZnS/diamond composite the critical size $l_{cr} \approx 3\mu m$ and $\alpha_k \approx 1.16 \cdot 10^7$ W m^{-2} K^{-1}. The theoretical estimates of the thermal conductivity value for the boundary ZnS/diamond give the values of α_k from 1.7 to $3.7 \cdot 10^7$ W m^{-2} K^{-1} which corresponds to $l_{cr} \leq 2$ μm. Such a difference between the experimental and theoretical results may be considered acceptable. As a matter of fact, l_{cr} is apparently even smaller because the value of α_k obtained from the results of the measurements of the composite thermal conductivity is determined by including the real square of the two components boundary, which depends on the technology of the material production and the character of particle surface roughness. In particular, the low value of α_k for galvanically produced composites can be explained first of all in that most of the diamond grain surface does not have direct contact with the binder. The effective thermal conductivity of composites has a weak dependence on particles form if they are isometric and change considerably with the change to flattened and stretched forms. For the same volume concentration, the distorted particles give higher values of K_{ef} in comparison to the isometric particles if their linear size is greater than l_{cr}, and they give lower values if the indicated size is less than l_{cr}. With this the transfer from the isometrical particles to the plate-like particles gives a greater change of K_{ef} than the transfer to the needle-like particles. The influence of the particles form and arrangement on the effective thermal conductivity of composites with taking into account the thermal resistance of boundary between a particle and the binder is analyzed in detail in works [97, 98].

4. References

1. Debye P.: *Ann. Phys.* (1912)v.**39,** p. 789.

2. Victor A.C.: *J.Chem.Phys.* (1962) v.**36** N 7 p.1903

3. Einstein A.: *Ann.Phys.* (1907) v.**22** p.180

4. Burk D.L. and S.A.Friedberg: *Phys.Rev.* (1958) v.**111** N 5 p 1275.

5. Desnoyers I.E. and Morrison J.A.: *Phil.Mag.* 1958 v.**3** N 25 p.42.

6. De Sorbo W.: *J. Chem. Phys.* (1953) v.**21** N 5 p.876.

7. Vandersande J.W.: *Phys. Rev. B* (1976) v.**13** N 10 p.4560.

8. McSkimin H.J., Bond W.L.: *Phys. Rev.* (1957) v.**105** N 1 p.116.

9. McSkimin H.J., Andreatch P., Glyrn P.: *J. Appl. Phys.* (1972) v. N 7 p.2944.

10. Berman R., Poulter I.: I. *Chem. Phys.* (1953) v.**21** N 7 p.1906.

11. Stukalenko V.V., Ositinskaya T.D., Podoba A.P.: In the book *Davleniem pri vysokich temperaturach*". Kiev (1987) p.22.

12.. Morelli Donald M, Smith C.W., Heremans I., Banholzer W.F. and. Anthony T.R: *New Diamond Science and Technology 1991 MRS Int. Conf Proc.* p.869.

13. Volga V.I., Buchnev L.I., Markelov N.V., Dymov B.K.: *Sinteticheskie Almazy* (1976)N3 p.9.

14. Zhdanov V.M., Turdakin V.A., Buchnev A.M. et.al.: *Neorganicheskie Materialy* (1976)v. **12**N lip. 2078.

15. Sassmann, R.S.: *Ind. Diam. Rev.* (1993)v. **53N555** p.63.

16. Gruneisen E.: *Ann. Physic* (1912) v.**39** p.257.

17. Oskotskii V.S.: *Fizika tverdogo tela* (1964) v.**6** N 5 p.1294.

18. Bienenstock A.: *Phil. Mag.* (1964) v.**9** p.755.

19. Dolling G., Cowley R.A.: *Proc. Phys. Soc.* (1966) v.**88** p.463.

20. Xu C.H., Wang C.Z., Chan C.T. and Ho K.M.: *Phys. Rev. B* (1991) v.. **43** p.5024.

21. Krishnan R.S.: *Proc. Indian Acad. Sci.* (1946) v. **A-24** p.33.

22. Straumanis B.Z., Aka I.: *Amer. Chem. Soc.* (1951) v.**73** p.5643.

23. Berman R., Thewlis I.: *Nature* (1955) v.**176** p.834.

24. Thewlis J., Davey A.R.: *Phil. Mag.* (1956) v.**1** p.409.

25. Novikova S.I.: *Fizika tverdogo tela* (1960) v.**2** N 7 p.1617.

26. Kozhina I.I., Tolkachev S.S.: *Vestnik Leningradskogo universiteta* (1965) N **10** p.91.

27. Skinner B.J.: *Am. Mineralogist* (1957) v.**42** p.39.

28. Mauer F.A., Bolz L.H.: *Wright Air Development Center Techical Report* (1957) N **55** p.473, Supplement (unpublished).

29. Novikova S.I.: *Teplovoe rasshirenie tverdych tel. Moskva, Nauka* (1974) 292 p. 30.

30. Slack G.A, Bartram S.F.: *J. Appl. Phys.* (1975) v.**46** N 1 p.89.

31. Wright A.C.J.: *Diamond Res.. Suppl. Industr. Diamond Rev.* (1965) p.10.

32. Sochor M.I., Vitol V.D.: *Kristallografiya (*1969) v.**14** N 4 p.734.

33. Haruna K., Maeta H., Ohashi, K. and Koike T.: *Jpn J. Appl. Phys.* (1992) v.**31** p. 2527.

34. Haruna K., Maeta H.: *Diamond and Relat. Mater.* (1993) v.**2** p.859.

35. Ziman J.M.: *Electrons and Phonons.* Clarendon Press, Oxford (1960).

36. Berman R.: *Thermal Conduction in Solids.* Oxford University Press, Oxford (1976).

37 Mogilevskii B.M., Chudnovskii A.F.: *Teploprovodnost poluprovodnikov.* Moskva, Nauka (1972) 536 p.

38. Turk L.A., Klemens P.O.: *Phys. Rev* (1974) v.**9** p.4422.

39. Neumaier K: *J. Low. Temp. Phys.* (1969) v.**1** p.77.

40. Oskotskii V.S., Smirnov I.A.: *Defecty v kristallach i teploprovodnost.* Leningrad Nauka(1972) 160 p.

41. Casimir H.B.G.: *Physica* 1938 v.**5** N 6 p.495.

42. Berman R., Simon F.E., Ziman J.M.: *Proc. Roy. Soc.* (1953) v. **A220** p.171.

43. Campisi G.I., Frankl D.K.: *Phys. Rev.* (1974) v. **l0B** N 6 p.2644

44. Callaway: *Phys. Rev.* (1959) v.**113** p.1046.

45. Anthony T.R., Banholzer W.F., Fleischer J.F., Lanhua Wei, Kno P.K., Thomas R.L., Pryor R.W.: *Phys. Rev. B* (1990) v.**42** p.1104.

46. Nepsha V.I., Grinberg V.R., Klyuev Yu.A., Naletov A.M., Bokiy O.B. : *Sov. Phys. Dokl.* (1991) v.**36** p.228.

47. Berman R.: *Phys. Rev. B* (1992) v.**45** N 10 p.5726.

48. Berman R, Martinez M.: *Diamond Res. Suppi. Industr. Diamond Rev.* (1976) p. 7.

49. Berman R., Foster E.L., Ziman J.M.: *Proc. Roy. Soc.* (1956) v.**237** N 1210 p.

50. Berman R., Hadson P.R.W., Martinez M.: *J. Phys. C: Solid State Phys.* (1975) v. **8** p. L430.

51. Slack GA..: *J. Phys. and Chem. Solids* (1973) v. **34** N. 2 p. 321.

52. Burgemeister E.A.: *Physica B* (1978) v. **93B** N 2 p.165.

53. Vandersande J.W. et. al: The discussed results are considered in [54].

54. Onn D.O., Witek A., Qiu Y.Z., Anthony T.R., Banholzer W.F.: *Phys. Rev. Lett.* (1992) v.**68** N 18 p.2806.

55. Lanhua Wei, Kuo P.K., Thomas R.I., Anthony T.R., Banholzer W.F.: *Phys. Rev. Lett.* (1993)v. **70** N 24 p. 3764.

56. Nepsha V.I., Orinberg V.R., Naletov A.M., Klyuev Yu.A.: *Sverchtverdye Materialy* (1990) N 6 p.21.

57. Olson J.R. et. al: The discussed results are considered in [55].

58. Vandersande J.W.:*J. Phys. C: Solid State Phys.* (1980) v.**13** N 5 p.759.

59. Vandersande J.W.: *Diamond Res. Suppl. Industr. Diamond Rev* (1973) p.21.

60. Burgemeister E.A.: *J. Phys.C: Solid. State Phys.* (1980) v.**13**, N 33 p.963.

61. Bokiy G.B., Bezrukov O.N., Klyuev Yu.A., Naletov A.M., Nepsha V.I.: *Prirodnye i sinteticheskie almazy.* Moskva, Nauka (1986) p. 221.

62. Nepsha V.I., Klyuev Yu.A., Reshetnikov N.F., Naletov A.M., Dudenkov Yu.A., Kulakov B.M.: *Sverchtverdye Materialy* (1982) N 4 p.28.

63. Reshetnikov N.F.,. Nepsha V.I., Klyuev Yu.A., Pavlov Yu.A: *Sverchtverdye Materialy* (1984) N 2 p.23.

64. Ositinskaya T.D.: *Sverchtverdye Materialy* (1980) N 4 p.13.

65. Ositinskaya T.D., Podova A.P., Shinegera S.V.: *Sverchtverdye Materialy* (1991) N 2 p. 19.

66. Hadson P.R.W., Phakey P.P: *Nature* (1977) v.**269** p.227.

67. Nepsha V.I., Naletov A.M., Reshetnikov N.F., Klyuev Yu.A., Bokiy O.B.: *Doklady Akademii Nauk SSSR* (1985) v.. **284** N 4 p.843.

68. Lisoivan V.I., Nadolinnyi V.A.: *Doklady Akademii Nauk SSSR* (1984), **274** N 1,72.

69. Kaiser W., Bond W.L.: *Phys. Rev.* (1959) v.**115** p.857.

70. Anthony T.R., Banhobzer W.F.: *Diamond and Relat. Mater.* (1992) v.**1** p.717.

71. Nepsha V.I., Orinberg V.R., Klyuev Yu.A., Kolchemanov NA., Naletov A.M.: *Sverchtverdye Materialy* (1992) N 2 p.19.

72. Nepsha V.I., Orinberg V.R., Klyuev Yu.A., Naletov A.M.: *Diamond and Relat. Mat.* (1993) v.**2** p.862.

73. Hass K.C., Tamor M.A., Anthony T.R., Banholzer W.F.: *Phys. Rev. B* (1992) v. **45** N.13 p.7171.

74. Gurzhi R.N.: *Zhurnal eksperimentalnoi i teoreticheskoi fiziki* (1964) v.**46** p.720.

75. Mezhov-Deglin L.P., Kopylov V.M., Medvedev E.S.: *Zhurnal eksperimentalnoi teoreticheskoi fiziki* (1974) v.**67** p.1123.

76. Mezhov-Deglin L.P.: *Fizika tverdogo tela* (1980) v.**22** p.1748.

77. Nepsha V.I., Orinberg V.R., Klyuev Yu.A., Naletov A.M., Bokiy G.B: *Phys. Dokl.* (1993) v.**38** (9) p.380.

78. Prohofsky E.W., Krumhansl, J.A.: *Phys. Rev.* (1964) v.**133** N 5A, p. 1403.

79. Ouyer R.A., Krumhansl, J.A.: *Phys. Rev.* (1964) v.**133** N 5A, p.1411.

80. Graebner J.E.: *Diamond Films and Technology* (1993) v.**3** N 2 p.77.

81. Ono A., Baba T., Funamoto H., Nishikawa A.: *Jpn. J. Appl. Phys.* (1986) v.**25** p. L808.

82. Herb J.A., Bailey C., Ravi K.V., Denning P.A.: *Electrochem. Soc. Proc. First International Symposium on Diamond and Diamond-like Films* (1989) p.366.

83. Graebner J.E., Mucha J.A., Seibles L., Kammlott OW.: *J. Appl. Phys.* (1992) v.**71** p.3143.

84. Morelli D.T., Beetz C.P., Perry T.A.: *J. Appl. Phys.* (1988) v.**64** N 6 p.3063.

85. Nepsha V.I., Reshetnikov N.F., Klyuev Yu.A., Bokiy G.B., Pavlov Yu.A: *Sov. Phys. Dokl.* (1985) v.**30** p.547.

86. Nepsha V.I., Orinberg V.R., Klyuev Yu.A., Kolchemanov N.A., Naletov A.M.: *Proc. of the Second International Conference on New Diamond Science and*

Tehnology. Ed. R. Messier, J.T. Glass, J.E. Butler and R. Roy. Materials Research Society, Pittsburg (1990) p.887.

87. Nepsha V.I., Grinberg V.R., Klyuev Yu.A., Kolchemanov N,A., Naletov A.M.: *Surface and Coatings Technology* (1991) v.**47** p.388.

88. Graebner J.E., Jin S., Kamrnlott G.W., Herb J.A., Gardinier C.F.: *Nature* (1992) v.**359** p.401.

89. Graebner J.E., Jin S., Kammlott G.W., Herb J.A., Gardinier C.F.: *Appl. Phys. Lett.*(1992)v. **60** p. 1576

90. Morelli D.T., Hartnett, T.M., Robinson C.J.: *Appl. Phys. Lett.* (1991) v.**59** p.2112.

91. Graebner J.E., Herb, J.A.: *Diamond Films Technol.* (1992) v.**1** p.155.

92. Pope B.J., Norton M.D., Hall H.T., Divita S., Bowman L.S., Adaniya H.N.: *Proc. of the 4-th Intern. Conf on High Pressure* (1975) p.404.

93. Burgemeister E.A.: *Ind. Diamond Rev.* (1980) N 3 p.87.

94. Burgemeister E.A., Rosenberg H.M.: *J. of Materials Science* (1981), v. **16** p.1730.

95. Maxwell J. C.: *A Treatise on Electricity and Magnetism* v.1. Oxford University Press (1904) p.205.

96. Grinberg V.R., Zhuravlev V.V., Klyuev Yu.A., Nepsha V.I.: *Trudy VNIIALMAZ* (1987) p.92.

97. Nepsha V.I., Grinberg V.R., Klyuev Yu.A., Kolchemanov NA., Zhuravlev V.V.: *Sverkhtverdye Materialy* (1989) N 6 p.18.

98. Nepsha V.I., Grinberg V.R., Klyuev Yu.A., Kolchemanov N. A., Zhuravlev V. V.: *Sverkhtverdye Materialy* (1990) N 1 p.21.

99. Shlykov Yu.P., Ganin E.A.: *Kontaktnyi teploobmen.* Moskva, Gosenergoizdat (1963) p.144.

100. Burgemeister E.A.: *J. Phys. D: Appl. Phys* (1977) v.**10** N 14 p.1923.

101. Every AG., Tzou Y., Hasselman D.P.H. and Ray R.: *Acta metall. mater* (1992) v. **40** N.1 p.123.

102. Benvensite Y.: *J. Ann. Phys.* (1987) v.**61** p.2840.

103. Hasselman, D.P.H. and Johnson, L.F., *J. Comp. Mater.* (1987) **21**, 508.

Chapter 6
THERMAL MEASUREMENT TECHNIQUES

John E. Graebner
Bell Laboratories, Lucent Technologies, Murray Hill, NJ 07974, U.S.A.

Contents

1. Introduction

Accurate measurement of the thermal conductivity κ or thermal diffusivity D of diamond is made difficult by a number of characteristics of this remarkable material. The very high values of κ and D for the best material make the temperature gradients (which are needed in any measurement) small and/or the required heating frequency high. In addition, the samples of CVD diamond available at the present time span a wide range of κ (a factor of 10), depending on growth conditions. Perhaps the most serious problems for CVD diamond are the large gradient (with respect to distance z from the substrate surface) and the anisotropy in both κ and D that have been found for even the best samples [Graebner 1993a, 1995a], despite the fact that the crystal structure of diamond is cubic. These inhomogeneities have been correlated with the microstructure usually observed, in which oriented columnar grains develop as growth proceeds from the initial random nucleation stage to the thick plate stage. The gradient and anisotropy in κ and D mean that a single number is not sufficient to fully describe the transfer of heat, as it would be for a homogeneous material with a cubic crystal structure. Consequently, it is useful to have a variety of experimental techniques available to measure κ and/or D for heat flowing in any of several different directions and for measurements over various length scales.

The thermal conductivity κ appears in the heat conduction equation [Berman 1976], which is basically a statement of the conservation of energy,

$$C\delta T / \delta t - \kappa \nabla^2 T = J(x, y, z, t) \tag{1}$$

where C is the heat capacity per unit volume and J is the source of heat (energy per unit volume per unit time). A standard reference [Carslaw and Jaeger 1959] provides solutions of Eq. (1) under a wide range of boundary conditions in space and time. One case of special interest for the measurement of κ is loss or gain of energy by exchange of thermal (blackbody) radiation with the environment or by exchange of thermal energy with the surrounding gas if the sample is not in a vacuum. This loss or gain through the surface causes serious complication for the measurement of κ which has not always been taken into account in the literature.

We begin our discussion of measurement techniques with a few general precautions. While the basic equations of electrical and thermal conduction are similar, in practice the measurement of thermal conductivity is much more difficult because it is not easy to measure temperature and heat flow as accurately as one can measure voltage and electrical current. Furthermore it is difficult to control stray thermal currents. The heat that is applied to one portion of a sample may have a number of paths to thermal ground in addition to the intended path through the sample. Thermal conduction along thermometer and heater lead wires, for example, must be minimized. The stray paths that are most common, by far, at temperatures above approximately 100K are through radiation and convection. The heat lost by radiation varies as the fourth power of the absolute temperature, so that it rapidly becomes more important at room temperature and above. Errors of a factor of two or more in the deduced conductivity can easily be made under conditions which at first do not appear to be unusual. One can calculate the *maximum* amount of heat that can be lost by radiation, using the Stefan-Boltzmann law, but the emissivity of the surface is rarely that of an ideal blackbody and is not easy to measure. One must therefore either use a measurement technique that allows one to detect and correct for any significant amounts of radiative loss during the measurement, or adjust the experimental conditions to reduce the maximum possible amount of radiation to acceptable levels. Convective loses (including simple conduction through the surrounding gas as well as by convection (movement) of the gas) can be eliminated by placing the sample in a vacuum, but this solution is not always feasible. An important quantity is the ratio $R = k_{surf}/k_{diff}$, where k_{surf} is the conductance (watts per degree of temperature difference) from the surface of the sample to thermal ground, due to both radiation and convection, and k_{diff} is the conductance due to diffusion of heat through the sample to thermal ground. This ratio depends sensitively on geometry and absolute temperature. The portion of the sample that is important for calculating R is different for steady-state and non-steady-state techniques. For the former, it is the whole sample. For the case of periodic application of heat (see below), it is only that portion of the sample extending one diffusion length from the heater in the direction of the heat flow. By operating at a sufficiently high frequency, one can reduce the thermal diffusion length enough to reduce R by several orders of magnitude, as discussed below.

The techniques that are important for thermal conductivity measurements in CVD diamond are discussed here in order of increasing complexity of time dependence (steady-state, periodic, and pulsed heating) and increasing complexity of geometry (increasing dimensionality). We include all the methods that have been applied to diamond so far, and we suggest potentially useful variations in several cases.

2. Steady-State Heating

For the case of steady-state heating, Eq. (1) reduces to Poisson's equation,

$$\kappa \nabla^2 T = -J \tag{2}$$

Solutions to Eq. (2) can be found for many different geometries [Carslaw and Jaeger 1959].

2.1 HEATED BAR

Perhaps the most traditional method for measuring the conductivity of a solid is to use a long sample of uniform cross section, $A = ZW$ (Z and W are thickness and width, respectively), which is thermally grounded at one end and fitted with a heater supplying steady power P at the other end (inset, Fig. 1). In this case, Eq. 2 reduces easily to an expression for the conductivity in terms of P, A, and the measured gradient $\Delta T/\Delta x$:

$$\kappa_{\parallel} = P/(A\ \Delta T/\Delta x) \tag{3}$$

Equation (3) is valid under the condition that all the heat is conducted through the sample. However, if the surfaces of the sample (with total area pL, where L = sample length and perimeter $p = 2(W + Z)$) lose energy by radiating at a rate P_{rad}, an effective conductance k_{rad} between the sample and its surroundings at temperature T_o can be defined [Graebner and Herb 1992] as

$$k_{rad} \approx \frac{P_{rad}}{(\Delta T)_{avg}} \approx \frac{\sigma \varepsilon p L (T_{avg}^4 - T_o^4)}{(\Delta T)_{avg}}, \tag{4}$$

where σ is the Stefan-Boltzmann constant (σ=5.67$\times$10^{-12} W cm^{-2} K^{-4}), ε is the emissivity which lies between 0 and 1.0 depending on surface conditions, and the average temperature is $T_{avg} \approx T_o + (\Delta T)_{avg}$. Expanding $T_{avg}{}^4$ in a power series, we find

$$k_{rad} \approx 4\sigma\varepsilon T_o^3 pL = H_{rad} pL \tag{5}$$

where H_{rad} is defined as the net radiant flux per unit surface area per degree of temperature difference. At room temperature with maximum emissivity ε = 1.0, H_{rad} = 0.61 mW cm^{-2} °C^{-1}. By analogy, we can define an effective conductance k_{conv} due to convective losses as $k_{conv} = H_{conv}pL$, although it is not so easy to derive an analytical expression for H_{conv} as for H_{rad}. It should be noted that k_{rad} and k_{conv} are expected to be roughly linear in the temperature difference $(\Delta T)_{avg}$ between sample and surroundings only for small values of $(\Delta T)_{avg}$. It is often found that, for samples in one atmosphere of air at room temperature, H_{rad} and H_{conv} are comparable unless the emissivity is particularly high or low. The total surface conductance $k_{surf} = k_{rad} + k_{conv}$ can be compared with $k_{diff} = \kappa WZ/L$ through the sample, so that the ratio of the two channels for heat flow is $R = k_{rad}/k_{diff} + k_{conv}/k_{diff} = R_{rad} + R_{conv}$, where

$$R_{rad} = k_{rad} / k_{diff} = \frac{H_{rad} p}{\kappa A} L^2 T_o^3 \tag{6}$$

Radiation loss is therefore most severe for long, thin plate-like samples with low conductivity at high temperatures. For a thin plate of CVD diamond with Z = 0.0010 cm, W = 0.5 cm, L = 2 cm, and κ = 5 Wcm^{-1}K^{-1}, Eq. (6) yields R = 1.0 at room temperature, indicating that approximately half of the heat generated by the heater is radiated from the surface of the sample if ε = 1.0. The measured temperature drop ΔT, if approximated by a straight line, would therefore be too small, and the conductivity calculated from Eq. (3) would be approximately twice the correct conductivity. For materials of lower conductivity, the error would be correspondingly larger.

In the case of non-negligible surface loss from the heated bar, an extra term can be included in the steady-state heat balance equation:

$$\kappa \frac{\partial^2 T}{\partial x^2} - \frac{Hp}{A} T = 0 \tag{7}$$

where $H = H_{rad} + H_{conv}$ and the source of heat at one end is incorporated as a boundary condition on the heat flux P/A there. One can define $\mu = \sqrt{Hp / \kappa A}$ as a measure of the importance of surface loss compared to conduction. The solution to Eq. (7) is then [Carslaw and Jaeger 1959]

$$T(x) = \frac{T_1 \sinh(\mu(L - x)) + T_2 \sinh(\mu x)}{\sinh(\mu L)} \tag{8}$$

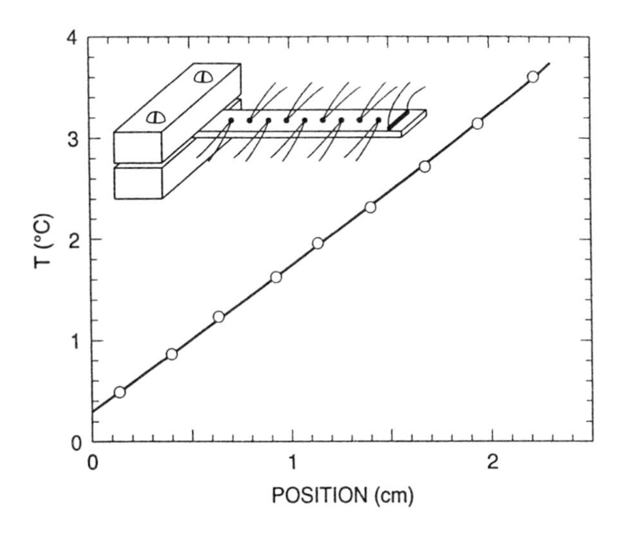

Figure 1. Typical data for a sample measured with the heated bar technique. The bar is equipped with a heater at one end, a row of thermocouples, and a clamp to thermal ground at the other end. The temperature T is measured relative to thermal ground. With the heater energized, the slight upward curvature of the data, superimposed on the overall positive slope, indicates non-negligible loss of heat from the surface of the sample. The solid line through the points is a fit of Eq. (8) to the data, yielding the surface heat loss coefficient as well as the thermal conductivity $\kappa_{\parallel}$ for heat flowing parallel to the long axis of the sample.

where L is the distance between points with temperatures T_1 and T_2. One can fit Eq. (8) to the data for T(x) by varying T_1, T_2, and μ for a best fit. Then from the boundary condition on the heat flux at the heater, $\kappa[\partial T/\partial x]_{x=L} = P/A$, the conductivity can be calculated as

$$\kappa_{\parallel} = \frac{P\sinh(\mu L)}{\mu A\left[T_2\cosh(\mu L) - T_1\right]}. \tag{9}$$

Equation (8) is thus used to model the heat flow in the bar as a function of x and Eq. (9) is used, in effect, to extrapolate the slope to its value at $x = L$ (the heater position), where negligible heat flow has been lost to radiation. Figure 1 shows an example of a measurement under conditions of non-negligible heat loss. It is important to measure the temperature at many positions, especially near the heater, to get a good measure of the curvature. For the sample in Fig. 1, a straight-line analysis of $\partial T/\partial x$ and use of Eq. (3) would yield a conductivity value that is 15% too high, compared with the use of Eqs. (8) and (9).

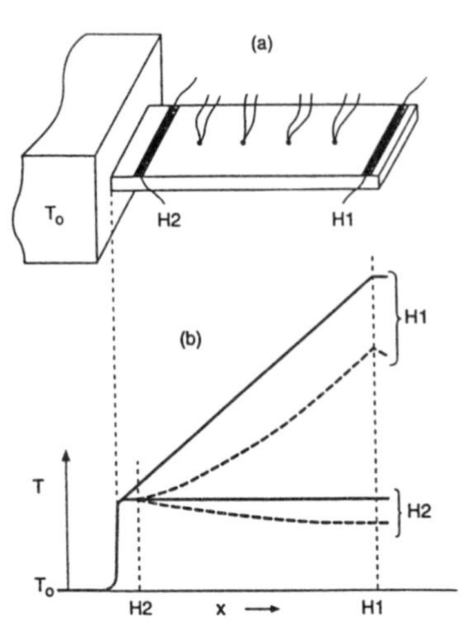

Figure 2. (a) Experimental arrangement for the two-heater heated bar technique for measuring thermal conductivity $\kappa_{\|}$. Heater H2 is used to test for surface heat loss. (b) In the absence of surface heat loss, one expects straight-line data as indicated schematically by the solid lines for either H1 or H2 energized. In the presence of significant surface heat loss, one expects the curved (dashed) lines, as described in the text. (Graebner et al. 1994a)

2.2 TWO-HEATER HEATED BAR

A method with high sensitivity to detect and correct for the presence of surface loss without using a large number of thermocouples involves [Graebner and Herb 1992, Graebner et al. 1992a, 1994a] the use of a second heater (H2 in Fig. 2(a)) placed on the sample near thermal ground. With power dissipated only in H2, the temperature profile in thermal equilibrium is flat in the absence of radiative heat loss but is described by Eq. (8) if loss takes place from all points on the surface (Fig. 2(b)). The surface loss with power applied to H2 can be fit by means of Eq. (8), modified for zero flux at $x = L$, in order to determine μ. This value of μ can then be used with Eqs. (8) and (9) to obtain $\kappa_{\|}$ when power is applied to H1. The emissivity of surfaces depends very sensitively on surface treatment, so that even the order of magnitude of the radiation loss is difficult to estimate without the use of H2. For typical thick diamond samples (e.g., $0.01 \times 0.5 \times 1.0$ cm^3) of reasonably high conductivity (≈ 10 Wcm^{-1}K^{-1}), the maximum radiation correction is typically 1% at room temperature and roughly 3% at 100°C. These corrections increase rapidly as the sample length is increased or the thickness is decreased.

Discrete heaters and resistance thermometers could, in principle, be used for these measurements. The advantage of using heaters evaporated directly onto the electrically insulating sample, and very small thermocouples, is that the surface area of such elements is small and usually of low emissivity in the infrared, reducing the risk of radiative loss. Furthermore, they afford precise placement on the sample, enabling the measurement of small samples ($L = W = 0.5$ cm). The distance between either heater and the nearest thermometer, however, should be at least 5 times the thickness Z to ensure a uniform distribution of the thermal current across the cross section of the sample in the vicinity of the thermometers. With the sample in vacuum and with good

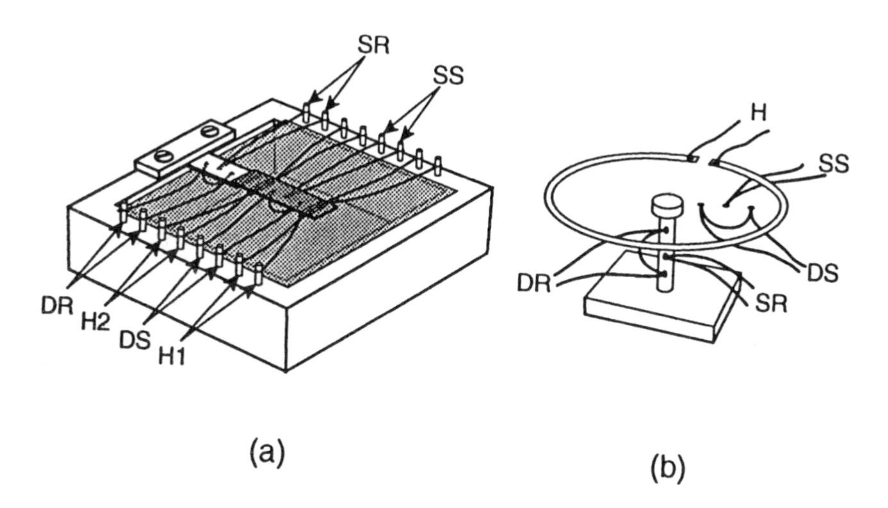

Figure 3. (a) Apparatus for measuring the thermal conductivity of a long thin plate. The sample is placed on top of the heaters, H1 and H2, and on single (SS) and differential (DS) thermocouples. It also overlaps a reference strip of known conductivity which serves as a path to thermal ground, i.e., the large thermal mass of the supporting block, typically copper. The reference strip is equipped with thermocouples labeled SR and DR for measuring the flow of heat through it. The heaters and sample thermocouples are glued permanently to a thin film of plastic (Kapton) which is stretched over insulating material (typically plastic foam) filling a well in the supporting block. The sample is covered with insulating material and a weight (not shown). The entire assembly is housed in a vacuum chamber. (b) Circular analog of the apparatus in (a), showing the arrangement to be used for a large-area sample of irregular outline. The thermal geometry is defined by a ring heater supported by insulating material (not shown) and a thermal ground at the center. The vertical post serves as the thermal path to ground, with thermocouples to measure the amount of heat flowing through it. As in (a), the sample is simply placed on top of the heater and thermocouples SS and DS, and covered by insulating material and a weight. (Graebner 1993b)

temperature control of the thermal ground, a resolution of 0.001°C can be obtained at room temperature and above. With a thermal gradient of only 0.2°C/cm, for example, the overall accuracy in the measurement of κ is usually limited by the accuracy of the sample dimensions. For polished, laser-cut samples, an accuracy of 1-2% is possible, while for as-grown CVD diamond samples with their typical faceted rough growth surface and variations in total thickness of a few percent over distances of 1 cm, an absolute accuracy of 3-5% is typical. The effective thickness of rough samples can be estimated [Graebner et al. 1994a] by $Z_{eff} = m / \rho WL$, where m is the sample mass and $\rho = 3.51$ g/cm^3 is the density of bulk diamond. This usually results in $Z_{eff} = 0.95 \times Z_{max}$, where Z_{max} is the distance from the substrate side to the tops of the pyramids on the growth side, as measured by a standard thickness gauge.

2.3 INSULATED PLATE

The effects of radiation loss may be minimized by use of a technique [Graebner 1993b] which is aimed at rapid, accurate measurement of the thermal conductivity of a flat plate with minimal requirements on sample preparation. To avoid the time-consuming task of attaching thermocouples and heaters to a sample, they are mounted permanently on a bed of highly thermally insulating material, as shown in Fig. 3(a). The sample is then pressed gently onto these elements with a small amount of grease to improve thermal contact, and covered with more insulating and thermally reflective material to reduce loss of heat by radiation. The apparatus is located in a vacuum chamber to eliminate loss of heat by conduction through the gas. One end of the sample is connected to thermal ground through a standard conductor (such as copper) equipped with thermocouples to determine the amount of heat flowing through it from the sample. This heat flow can be compared with the heat generated in heater H1. A second heater H2 can be used as in Fig. 2 as a check for the presence of radiative or convective loss. If a *series* of single thermocouples is used, and one finds $\partial^2T/\partial x^2 \neq 0$, Eqs. (8) and (9) can be used to analyze the data. Measurements on copper and diamond yield an accuracy of 5-10%.

For samples in the shape of a thin flat plate of large surface area and with perhaps an irregular outline, one way of establishing a simple pattern of heat flow is to use a heater in the shape of a circle with a connection to thermal ground at the center, as illustrated in Fig. 3(b). The post between the center of the circle and the heat sink serves as the reference conductor. The two-dimensional temperature distribution in the sample is constant outside the ring in the absence of surface heat loss by radiation, etc. The conductivity is determined from the temperature distribution [Carslaw and Jaeger 1959] within the ring:

$$\kappa_{\parallel} = \frac{P\ln(r_2 / r_1)}{2\pi Z(T_2 - T_1)}, \tag{10}$$

where T_1 and T_2 are the temperatures at radial positions r_1 and r_2, respectively, and Z is the plate thickness. A second ring heater of radius just larger than the center contact can be used as H2 in Fig. 2 to check for heat loss.

2.4 RADIATING BAR

The radiative loss of heat from a sample is used to *advantage* in a steady-state technique [Ono et al. 1986] in which the dominant mechanism for loss of heat is radiation. A long, thin sample is held at its ends by two posts (Fig. 4) which are maintained at some temperature above the temperature T_s of the environment. Heat is lost by radiation from both surfaces of the film and is replaced by thermal conduction along the film. The thermal conductivity $\kappa_{\parallel}$ can be obtained by measuring the distribution of temperature T as a function of x, the distance along the film:

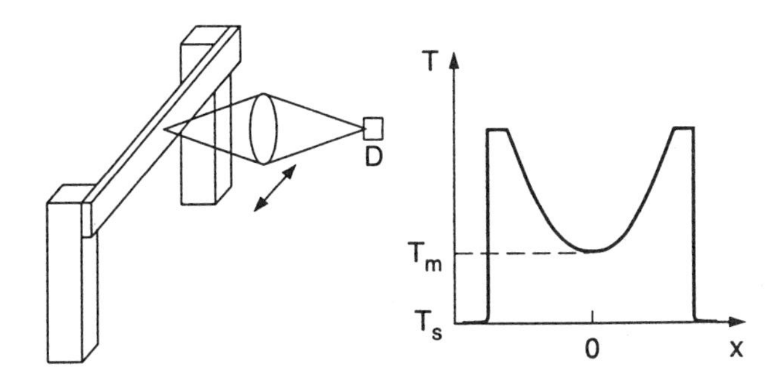

Figure 4. Schematic diagram of the radiating-bar technique for measuring $\kappa_{\parallel}$. An infrared detector D is used to measure the temperature profile (right) of a sample attached at its ends to heated posts (left). $\kappa_{\parallel}$ is determined by fitting the profile to Eq. (11). (After Ono et al. 1986)

$$\kappa_{\parallel} = \frac{8\varepsilon\sigma T_m^3 x^2}{Z\{\operatorname{arccos h}(4T_m^3(T-T_m)/(T_m^4 - T_s^4)+1)\}^2}. \tag{11}$$

Here, T_m is the minimum temperature along the film, and x = 0 at the center of the film. To maximize the radiative loss, the film is usually coated with black paint whose emissivity must be accurately known or measured. The technique is especially useful for thin films of relatively poor thermal conductivity. It can be used with both ends of the sample attached to the hot posts or, more easily, only one end attached to a single post. In the latter case, the thermal distribution should be identical to that of the two-heater heated bar technique with power applied to H2 (Fig. 2(b) and Eq. (8)) for the case of severe radiation loss. The accuracy of the technique is 10-15%, as determined by measurements of Cu and Ag foils. Undoubtedly one of the most crucial aspects of this technique is obtaining a thin coating of known and reproducible emissivity [Herb et al. 1989]. A detailed study of the effect of different coatings has not appeared, but one expects that the coating should be kept thermally thin so that there is little or no thermal gradient within the coating in a direction perpendicular to the film. Measurements have been reported for the temperature range of 70 - 140 °C, with a temperature drop along the film of, typically, 15°C. Presumably, the sensitivity of the thermometry is seriously degraded at room temperature. The large thermal gradient would also lead to inaccuracies for materials, such as high-quality diamond, which exhibit a temperature-dependent conductivity.

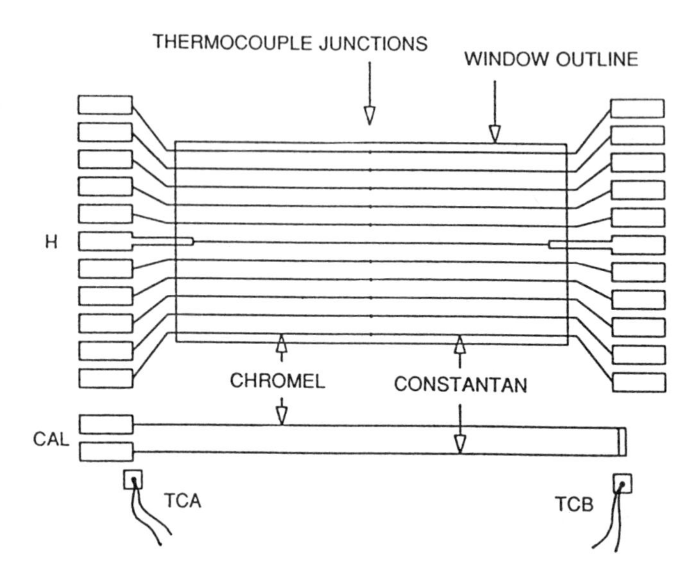

Figure 5. The pattern of heaters (H), thermocouples, and gold pads (closed rectangular outlines) deposited on the surface of a diamond film on a silicon substrate, in the vicinity of a window in the substrate. Also shown is a thermocouple used for calibration. Wire thermocouples TCA and TCB are used to measure the temperature gradient imposed from right to left by an external heater while the voltage at CAL is measured. (Graebner et al. 1992b)

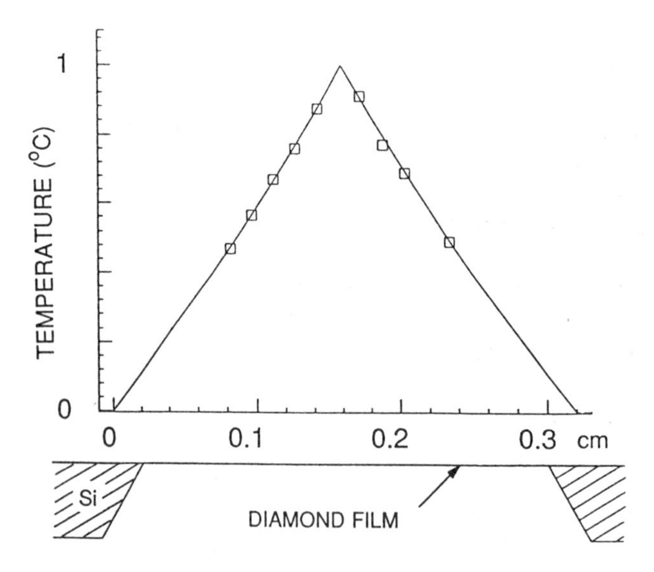

Figure 6. Temperature profile of the diamond film shown in Fig. 5, relative to the temperature of the silicon substrate. The data (squares) are obtained with the thin-film thermocouples. The solid curve represents a numerical simulation of the heat flow which has been fit to the data. The fact that the temperature difference does not extrapolate to zero at the edge of the window indicates a thermal resistance between film and substrate. (Graebner et al. 1992b)

2.5 LINE-HEATED PLATE

Another way to minimize radiation effects is to decrease the length L of the sample by using thin-film thermocouples and thin-film heaters deposited onto the sample using shadow masks to precisely define the geometry [Graebner et al. 1992b]. For example, very thin samples of diamond deposited on silicon can be made free-standing by etching a window entirely through the silicon, as shown in Fig. 5. The remaining substrate then serves as a convenient support as well as heat sink for the film. A line heater at the center of the window can be used to generate a two-dimensional temperature gradient, measurable with thin-film thermocouples or by infrared imaging. By solving Poisson's equation for the particular geometry and heater power, one can fit the data (Fig. 6) by adjusting only a single parameter, $\kappa_{\|}$. Certain deviations from the simple model can be interpreted as evidence for a significant thermal resistance at the boundary between diamond and substrate, possibly indicating poor adhesion.

3. Periodic Heating

The analysis of thermal waves generated by periodic heating is a technique pioneered by Ångström in the 19th century [Ångström 1861]. It has appeared in many modifications. By studying the dynamics of thermal wave propagation, one can determine the thermal diffusion length $\delta_T = \sqrt{2D/\omega}$ for a heating frequency ω measured in radians per sec, thus determining the thermal diffusivity D, defined as κ/C where C is the heat capacity per unit volume. If the absolute power input is also known, both the specific heat and diffusivity may be obtained, allowing a determination of the thermal conductivity. If, as in most non-contact heating techniques, C cannot be determined simultaneously, a value must be assumed for C in order to convert the measured D into κ. It was realized [Salamon et al. 1974] that the problem of surface heat loss can be largely avoided by heating at a sufficiently high frequency.

3.1 END-HEATED BAR

Most treatments of this experimental technique assume a long bar which is heated at one end and which loses heat by radiation or conduction through a surrounding gas, and possibly through direct contact with thermal ground at the other end. The heat balance equation has an extra term [Salamon et al. 1974] to account for the loss of heat through the surface:

$$C\frac{\partial T}{\partial t} = \kappa\frac{\partial^2 T}{\partial x^2} - \frac{Hp}{A}T. \tag{12}$$

Again, the heat applied at one end, P(t), appears as a boundary condition; C is the heat capacity per unit volume; and H ($Wcm^{-2}{}^{\circ}C^{-1}$) is the conductance per unit area between sample and thermal bath. One can define a relaxation time to the bath, $\tau_b = AC/pH \approx ZC/2H$, where the approximation is for a thin sample ($Z \ll W$). It is convenient to choose a sinusoidal heat input at one end of the sample: $P(0,t) = P_o \sin \omega t$. The general solution to Eq. (12) can be obtained in closed form for several simple cases. Choosing the operating frequency ω sets the length scale, δ_T, for the measurement. We distinguish three frequency regimes of interest corresponding to different relationships between δ_T and either the total sample length or the length for which heat loss to the bath is competitive with conduction along the sample.

a. $\omega << 2D/L^2$, or equivalently $\delta_T >> L$

For low frequencies, the sample is very close to thermal equilibrium at all times. If the geometry and thermal conductivity are such as to make radiation a problem in the steady-state solution described above, it will be a problem here as well. The only advantage over the steady-state methods described above is that the signal/noise ratio can be improved somewhat by reducing the effect of 1/f noise.

b. $2D/L^2 < \omega < 2H/ZC = \tau^{-1}_b$

The first inequality indicates that the diffusion length δ_T is shorter than the sample. Bearing in mind the general forms for the diffusive conductance $k_{diff} = \kappa A/L$ and surface conductance $k_{surf} = HpL$, the second inequality, which can be rewritten $\kappa ZW/\delta_T < HW\delta_T$, indicates that in this regime the conductance through a length δ_T of the sample is less than the conductance out through the surface of the same length of sample. In this crossover regime, the solution is complicated by the appearance of both D and H [Salamon et al. 1974, Hatta et al. 1985].

c. $\tau_b^{-1} << \omega$, or equivalently $HW\delta_T << \kappa ZW/\delta_T$

In this regime, a length δ_T of the sample has a much higher thermal conductance along its length than through the surface to thermal ground. Under these conditions, the effects of conduction via the gas or radiation can be ignored. The solution is given approximately by the simple expression:

$$T(x,t) = T_o e^{-x/\delta_T} \sin(\omega t - x/\delta_T - \theta) + T_{mean}(x) \tag{13}$$

where $T_o = P_o/2\omega ZC$. The constant phase shift θ is a result of the boundary condition being a sinusoidal heat flux, not a sinusoidal temperature. The thermal wave of Eq. (13) has a period $2\pi/\omega$ in time and $2\pi\delta_T$ in distance, and is very strongly attenuated as a function of position. At a distance of one wavelength from the heater, $x = 2\pi\delta_T$, the prefactor attenuates the sinusoid by a factor of $e^{-2\pi} = 0.002$. By measuring either the phase or the amplitude of T(x) near the heater, one can obtain δ_T and therefore D. In practice, it is usually easier and more accurate to measure the phase. This amounts to measuring the phase velocity $v_p = \omega\delta_T = (2\omega D)^{1/2}$. If absolute measurements of heat input P_o and temperature T_o are made, the specific heat C is also determined. However, many of the most commonly used versions of this approach apply heat by radiation and therefore do not measure the power input accurately; hence, only D is measured.

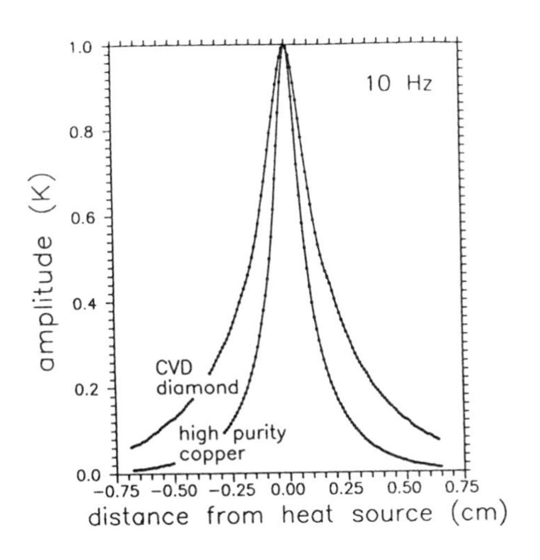

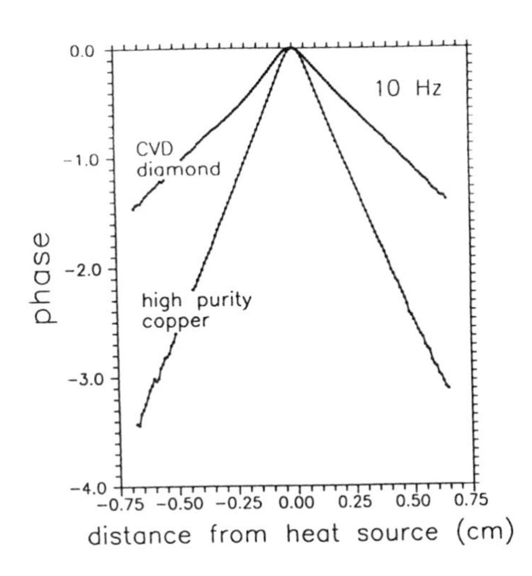

Figure 7. Amplitude and phase of the thermal wave as a function of radial distance from a heated spot, for a 150-μm-thick copper sheet. Analytical expressions have been compared to these data to obtain the in-plane diffusivity $D_{\|}$. The amplitudes are normalized to 1K at the location of the laser spot. (Visser et al. 1992)

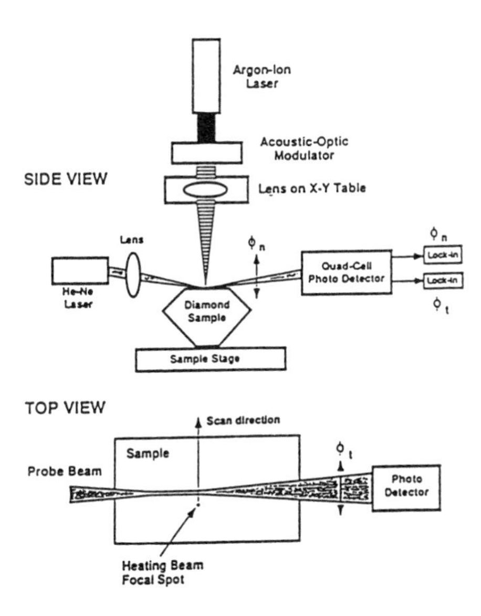

Figure 8. Schematic diagram of the apparatus used in the mirage detection of temperature changes at the surface of a sample. The argon laser beam is modulated and focused to a spot on the sample, producing thermal waves traveling hemispherically into the sample. The temperature of the surface is mapped out by the He-Ne probe beam which is deflected by the gradients in index of refraction of the air just above the sample. (Anthony et al. 1990)

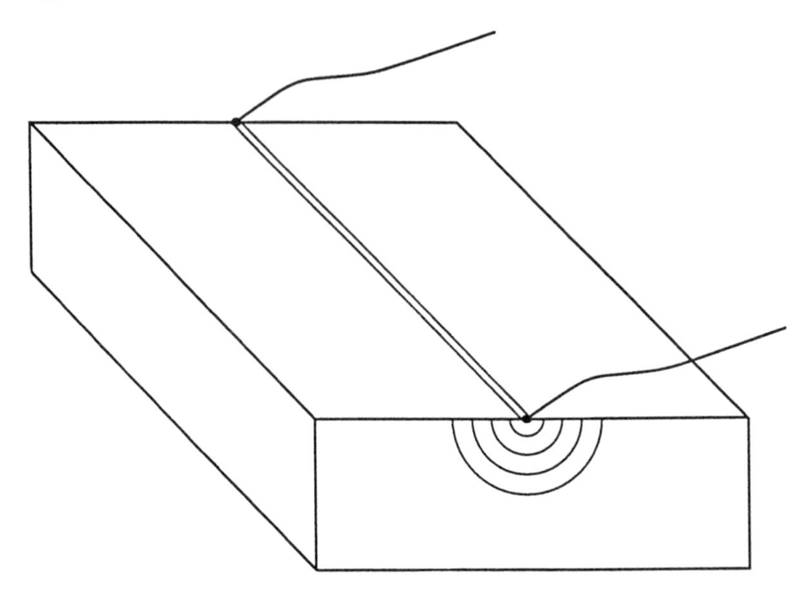

Figure 9. Schematic diagram of a sample equipped with a line heater for the measurement of κ using the 3ω method. The heater also serves as a thermometer to detect the amplitude and phase of thermal waves traveling into the sample, indicated schematically on the end of the sample. (Graebner 1993a)

We note that the commonly used term "thermal wave" is somewhat misleading. The absence of a second derivative with respect to time in Eq. (12) precludes the possibility of having truly propagating thermal waves. The time-independent mean temperature added to the highly damped expression in Eq. (13) assures that the temperature never becomes negative and that the slope $\partial T/\partial x$ never changes sign (which would imply that heat flows *toward* the source). Only for certain conditions involving the phonon mean free path due to normal processes [Berman 1976] can wavelike propagation occur (second sound), but this has not been observed in diamond.

Various techniques have been used to measure the x-dependence of the ac temperature. Discrete thermometers such as thermocouples can be attached at a number of locations (Fig. 1), or an infrared detector can be scanned [Kosky 1993] along x. Another variation [Hatta et al. 1986, 1987] exposes the sample to optical radiation with a mask which can be moved in the x direction, allowing the use of a single thermometer at a fixed position on the sample. In any case, the operating frequency must be chosen carefully; it must be high enough to satisfy the criterion (c) above, but not so high that δ_T becomes comparable to or less than Z (in order to maintain simple one-dimensional heat flow near the surface where the heat is applied).

We also mention an approach [Ohta et al. 1989] which uses the same one-dimensional geometry as the above techniques but with a pulsed-laser source of heat. The temperature measured at some distance from the heated zone can be compared with the calculated response to a pulse to derive a value for $D_{\|}$. For long, thin samples, one must ensure that the measuring point is close enough to the source of heat that radiation loss is not a problem.

3.2 SPOT HEATING ON A PLATE

Radial heat flow in a plate has been studied by using a modulated laser beam to heat a spot of either small diameter ($\approx 10\ \mu$ m) [Albin et al. 1990, Visser et al. 1992] or large diameter (up to several millimeters) [Frederikse et al. 1992]. The radial heat flow is most conveniently measured by infrared thermometry on either the same or the opposite side from the heated spot. The measurements have been done with the sample in air, so that surface losses by gas conduction as well as by thermal radiation were expected to be important and were included in two analyses [Feldman et al. 1990, Visser et al. 1992]. The solution to the two-dimensional heat balance equation in cylindrical coordinates includes a sinusoidal time dependence and a Bessel-function radial dependence. A fit of the theoretical solution to data (Fig. 7) is used to obtain both $D_{\|}$ and the surface heat loss. The dependence of the phase shift on frequency has been shown to be particularly useful. Again, if the absolute power absorbed is not known, the heat capacity cannot be measured, so that the standard heat capacity for the material must be assumed in order to convert $D_{\|}$ to $\kappa_{\|}$. A three-dimensional analysis has been given [Feldman et al. 1990], which is especially important for a small spot size on a thick sample with detection on the opposite face. In this case, the measurement is sensitive , in principle, to both $D_{\|}$ and $D_{\perp}$. A study of the black coating to absorb the heat and emit thermal radiation was reported [Visser et al. 1992]. For the low frequencies used (0.5 - 50 Hz), a graphite spray coating was found not to distort the measurements, though some problems were observed with thicker coatings. The overall accuracy of this technique seems to be about 5-10%, as determined by measuring the diffusivity of copper foils. Analyses for multi-layered samples have also been given [Albin et al. 1990, Visser et al. 1992, and Feldman et al. 1990]. The effect of thermal radiation reflected internally in an infrared-transparent sample was discussed [Visser et al. 1992], although this phenomenon is likely to be important only at high temperatures [Vandersande et al. 1991].

3.3 SPOT HEATING ON A SOLID

Periodic heating of a spot on the surface of the material under investigation has been used to measure the diffusivity of thick samples using the mirage technique, Fig. 8, for thermal sensing [Anthony et al. 1990, Charbonnier and Fournier 1986, Plamann et al. 1994, and Wei et al. 1993]. The thermal distribution along the surface as a function of time is monitored by the deflection of a probe laser beam as it traverses the air just above the surface (mirage effect). To within a thermal diffusion length of air at the modulation frequency, the air follows the temperature of the surface and causes a small deflection of the beam due to its temperature-dependent index of refraction. The theoretical deflection is a complicated function of the index of refraction integrated along the beam path. The modulated deflection is measured as a function of the radial distance from the spot and fitted with multi-parameter expressions to obtain a value for the diffusivity. The measured diffusivity is a hemispherical average. The technique does not distinguish between $D_{\|}$ and $D_{\perp}$, but at sufficiently high modulation frequency the penetration depth δ_T can be substantially less than the thickness of the sample. For example, for high-quality diamond with $D = 10\ cm^2/sec$ and $\omega = 2\pi \times 30$ kHz, $\delta_T = 100$ μm; hence, the measurement samples the material to a depth of only approximately 100 μm. For shallower penetration, higher frequencies must be used. At high frequencies, however, the mirage (heated air) can not respond quickly enough and some faster form of thermometry must be used, such as blackbody radiation or temperature-dependent reflectivity. Radiation losses are not always taken into account in the analysis but are probably not important except for material of very low thermal diffusivity and low measuring frequency. We not that the mirage technique can also be applied to spot heating of a plate by using the appropriate mathematical model to interpret the data.

3.4 PLANE AND LINE HEATING

A metal heater in contact with a sample can be used also as a thermometer if it has a non-zero temperature coefficient of resistance. This technique was first used to measure the thermal properties of liquids [Sandberg et al. 1977, Atalla et al. 1981, Birge 1986, and Birge and Nagel 1987]. If the heater is excited with an ac electrical current $I = I_o \sin \omega t$, heat is generated in the heater resistance R_o at a rate $P = I_o^2 R \sin^2 \omega t$. The oscillating temperature $T = T_o \sin^2 \omega t = (T_o/2)[1 - \cos(2\omega t)]$ causes a variation in resistance $R = R_o(1 + \alpha T)$ due to a non-zero temperature coefficient of resistance α. The voltage across the heater, $V = IR$, then has a component at a frequency 3ω : $V_{3\omega} = (\alpha I_o R_o T_o/4) \sin(3\omega t + \varphi)$, and both T_o and φ are related to the propagation of thermal waves into the material. By measuring the voltage and phase at a frequency of 3ω and comparing with a theoretical expression, one can deduce the thermal characteristics such as κ, D, and C. For this reason, the method is sometimes referred to as the 3ω technique [Cahill 1990].

The particular combination of thermal properties deduced from the data depends on the geometry. With a plane heater on the surface of the sample, one measures the product $\kappa_\perp C$, where $\kappa_\perp$ indicates the direction perpendicular to the plane of the heater. For a line heater on the surface, with cylindrical waves propagating into the material adjacent to the heater (Fig. 9), the measured quantity is simply κ. In this case, κ is an average of $\kappa_\perp$ and the component of $\kappa_\parallel$ that is perpendicular to the axis of the heater. The fact that different quantities are measured with different heater geometries can be understood qualitatively by considering the temperature variation ΔT per cycle at the heater caused by absorption of the heater energy in one cycle, P/ω. If c is the heat capacity $C \times \Delta V$ of a volume ΔV determined by the thermal length δ_T, setting $\Delta T = P/(\omega c)$ yields, for a plane heater on a thick sample, $\Delta T = P(2\omega C\kappa_\perp)^{-1/2}$, and for a line heater on a thick sample, $\Delta T = P(2\pi\kappa_{avg})^{-1}$. The same considerations yield $\Delta T = PZ^{-1}(8\omega C\kappa_\parallel)^{-1/2}$ for a line heater on a thin film of thickness Z ($Z << \delta_T$).

For a line heater on a thick sample, it was realized [Salamon et al. 1974, Cahill 1990] that the relevant distance for calculating the loss of heat by radiation is the thermal diffusion length δ_T along the surface and that by operating at sufficiently high frequency, the effects of radiation can be made insignificantly small compared to conduction. The 3ω method can thus be used reliably to measure the direction-averaged conductivity up to high temperatures [Cahill and Pohl 1987]. As with the mirage method, operation at high frequency allows one to measure only to a depth δ_T below the surface.

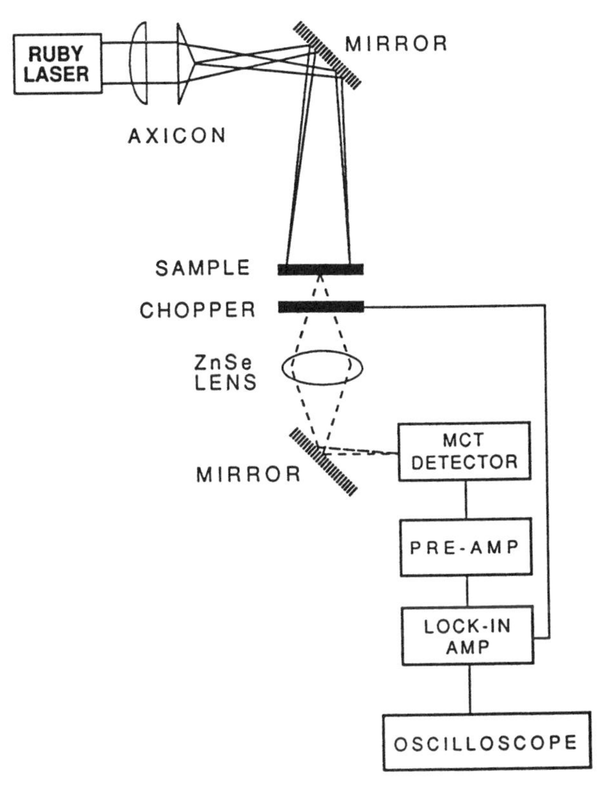

Figure 10. Schematic view of apparatus for the converging wave technique. The axicon is used to focus the beam of a laser into an annular shape on the sample. An infrared detector monitors the temperature of the center of the ring as the thermal wave propagates radially inward. The chopper is used to improve the signal-to-noise ratio of the infrared thermometry. (Lu and Swann 1991)

4. Pulsed Heating

4.1 ANNULAR HEATING

In the converging-wave technique [Cielo et al. 1986, Enguehard et al. 1990, Lu and Swann 1991], the surface of the sample is heated in a narrow annular pattern and the temperature at the center of the annulus is monitored. The time of arrival of the thermal wave at the center is a direct measure of the in-plane thermal diffusivity, $D_{\|}$. The heat can be applied by a pulsed laser projected through a shadow mask or, more efficiently, focused into an annular pattern by means of an axicon (lens of conical shape). The temperature at the center can be monitored by a simple infrared detector (Fig. 10). To establish heat flow parallel to the plane of the sample, the radius of the annulus should be much larger than the sample thickness. However, a large annulus is much more sensitive to the problem of radiation loss by conduction through a gas or by radiation. Such heat loss has been included [Enguehard et al. 1990] in the analysis and can be measured by fitting the full theoretical expression to the data. The effect of a thick sample has also been included. The analysis was originally given for a detection point at the center of the heated annulus and on the same side of the sample as the annulus. Detecting the temperature on the opposite side of a free-standing sample [Lu and Swann 1991] has the advantage of a more convenient geometry. We point out that it also has the potential advantage of measuring the perpendicular diffusivity, $D_{\perp}$, which would become dominant as the annular radius is decreased. In principle, one should be able to detect any difference in the parallel and perpendicular diffusivities by studying the shape of the arriving pulse as a function of annular radius.

It is worth noting that an advantage of annular heating over spot heating is that the absorbed power in the former is distributed over a larger area than in the latter, avoiding the generation of very high local temperatures. On the other hand, the pulse is localized in time, offsetting this advantage. Another advantage is that, because the geometry is defined by the annulus, a plate of arbitrary outline can be measured, as long as the distance from the annular center to any point on the outline is at least 1.5 times the radius of the annulus [Lu and Swann 1991]. The geometrical requirements on the sample are thus easier to satisfy than those in the one-dimensional version of this technique [Ohta et al. 1989].

4.2 THERMAL GRATING ON A SURFACE

Measuring thermal or electrical diffusivity by creating a grating pattern and following its disappearance is a well-established optical technique [Eichler et al. 1986]. One can create a standing wave pattern by interference between two nearly-collinear coherent beams. If this pattern is absorbed on (within) a sample, a grating of hotter and cooler parallel lines (planes) is established during the time the beams are pulsed. A continuous-wave third beam can be used to probe the amplitude of the thermal grating as a function of time. The probe beam can be focused to examine the individual ridges and troughs, or it can be diffracted from the thermal grating to provide an average over the area covered by both the grating and the probe beam. With either method of detection, the rate of disappearance of the grating is directly related to the thermal diffusivity.

The first use [Käding et al. 1993] of a thermal grating in diamond used a split beam (wavelength λ = 532 nm) interfering at a total angle θ to produce a grating with wave vector $q = 2\pi/\Lambda$, where Λ is the period of the grating (Fig. 11). The period is simply related to θ by $\Lambda = \lambda/(2 \sin(\theta/2))$. The intense interference pattern was absorbed by a Cr coating on the surface of the sample, producing a series of ridges because of the thermal expansion of the material below each hot line of the grating. The probe beam from a He-Ne cw laser was focused to approximately a 10 μm spot on the surface of the sample. The small angular deflection of the reflected beam measured the local shape of the expansion grating as a function of time (Fig. 12). The measured decay of the grating amplitude was exponential, $A = A_o \exp(-t/\tau_g)$. The decay constant τ_g was interpreted as $1/(Dq^2)$ from which D was obtained. It was shown that the diffusivity measured in this experiment is $D_{\parallel}$; the only effect of $D_{\perp}$ is to determine the depth L_z to which the measurement probes, where $L_z = (\Lambda/\pi)(D_{\perp}/D_{\parallel})^{1/2}$. Accurate measurements of the profile in $D_{\parallel}(z)$ therefore require independent knowledge of the profile in $D_{\perp}(z)$.

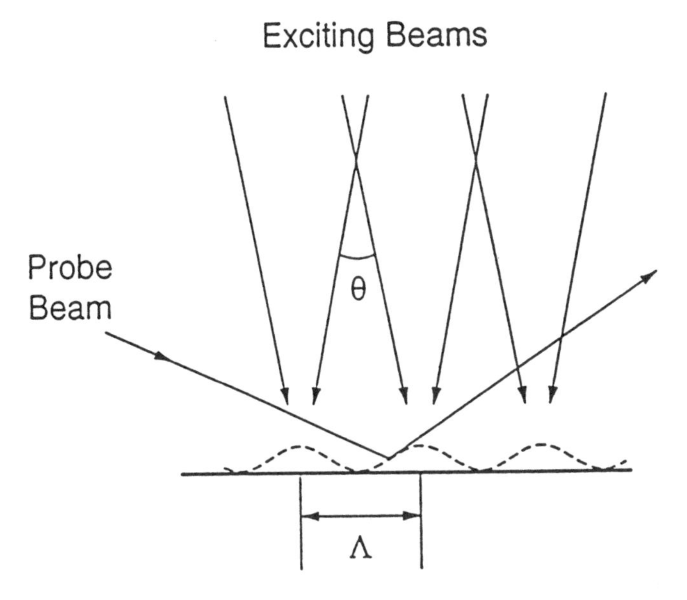

Figure 11. Schematic diagram of the interference pattern generated on the surface of a sample, resulting in a temporary series of ridges due to thermal expansion of the heated areas. The probe beam is used to map out the sinusoidal surface displacement, typically 0.1 nm, which is highly exaggerated compared to the grating spacing $\Lambda \approx 50$ μm.

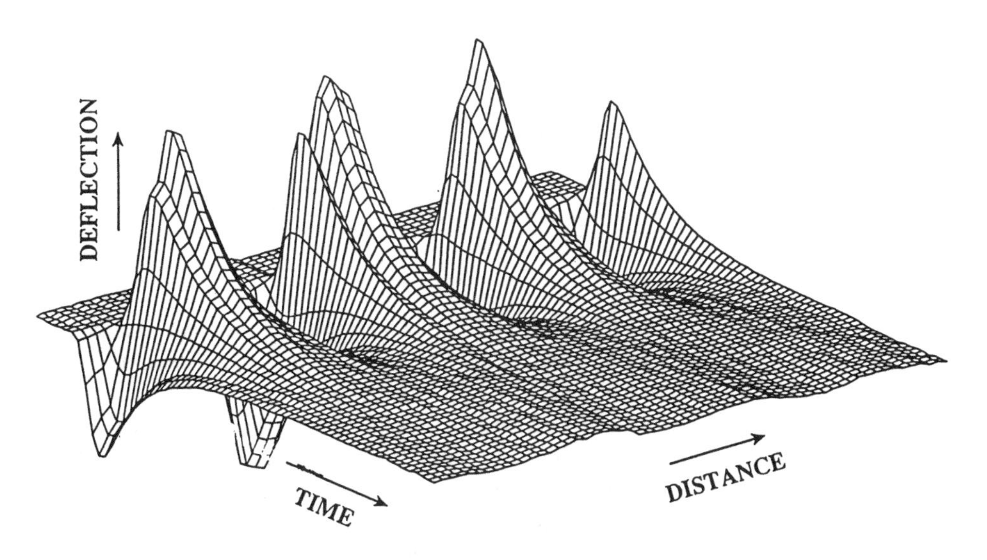

Figure 12. Time decay of a thermal grating on a silicon sample as a function of the position of the probe beam along the grating axis. The size of the mesh is 5.9 μm × 0.3 μs. (Kading et al. 1993)

Local detection of the decay of a TTG can also be accomplished [Graebner 1995b] by direct detection of the local temperature using infrared thermometry of a spot (approximately 30 μm in diameter) on the surface (Fig. 13). For a thin film ($Z << \Lambda$), the decay is exponential, and the method has been applied to roughly 11-μm-thick CVD diamond films. Infrared detection is generally less sensitive at room temperature than the determination of the local slope using a probe laser beam, but the sensitivity grows rapidly with increasing temperature.

4.3 THERMAL GRATING IN THE BULK

Another application of the thermal grating technique [Tokmakoff et al. 1993] made use of the interference of infrared beams ($\lambda = 4.1$ μm) to produce a thermal grating *within* the diamond by the intrinsic absorption in the two-phonon band. The grating was in the form of parallel planes of higher temperature, with the spacing between the planes equal to Λ. Values of Λ in the range 12-18 μm were used. A time-delayed probe pulse was diffracted from the grating at the Bragg angle due to the spatial modulation of the real part of the index of refraction. In this case, the decay time constant τ_g was assumed to be equal to $1/(2Dq^2)$, where the factor of 2 appears because the signal was expected to be proportional to the square of the induced polarization. The data could be tested for a linear dependence of the inverse decay time on q^2. The deduced diffusivity was that which governs heat flow in a direction perpendicular to the planes of the grating, which in the present geometry was essentially $D_{\parallel}$ for heat flow parallel to the surface of the sample.

4.4 LARGE-AREA SURFACE HEATING

The flash method for measuring $D_{\perp}$ was introduced over 30 years ago [Parker et al. 1961]. It consists of heating one side of a plate with a pulse of radiation and monitoring the temperature of the opposite face for the arrival of the heat wave. Because of the simple geometry of the problem, the solution of the one-dimensional diffusion equation usually represents the data well (Fig. 14). If the radiation pulse delivers an amount of energy Q per unit area of the front surface, the temperature rise of the rear surface is

$$T(t) = \frac{Q}{ZC}\left\{1 + 2\sum_{n=1}^{\infty}(-1)^n \exp\left(\frac{-n^2\pi^2 D_{\perp}}{Z^2}t\right)\right\}, \tag{14}$$

where Z is the sample thickness and C is the heat capacity per unit volume. As pointed out in the original treatment, the time scale and the thickness determine $D_{\perp}$. For example, the time $t_{1/2}$ for which $T(t) = T_{final}/2$ can be used to evaluate $D_{\perp}$ by the following relation: $D_{\perp}=1.38\ Z^2/\pi^2 t_{1/2}$. The heat capacity can be determined if Q can be measured absolutely, but in practice this is not easy. By using a well-characterized and reproducible surface coating to absorb the radiant energy and by maintaining a constant geometrical arrangement of radiant source and sample [Vandersande et al. 1989, Vining et al. 1989], the heat capacity can be obtained by calibration with a material of known specific heat.

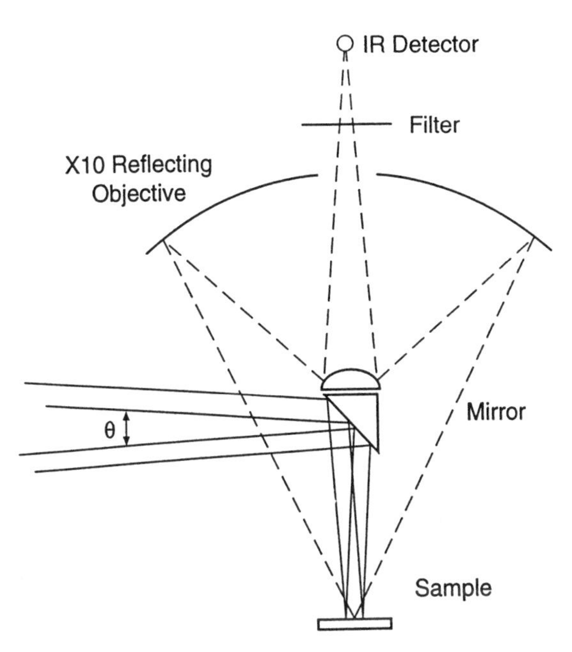

Figure 13. Experimental arrangement for infrared detection of temperature variations at a small spot on the sample where a transient thermal grating is generated. (after Graebner 1995b)

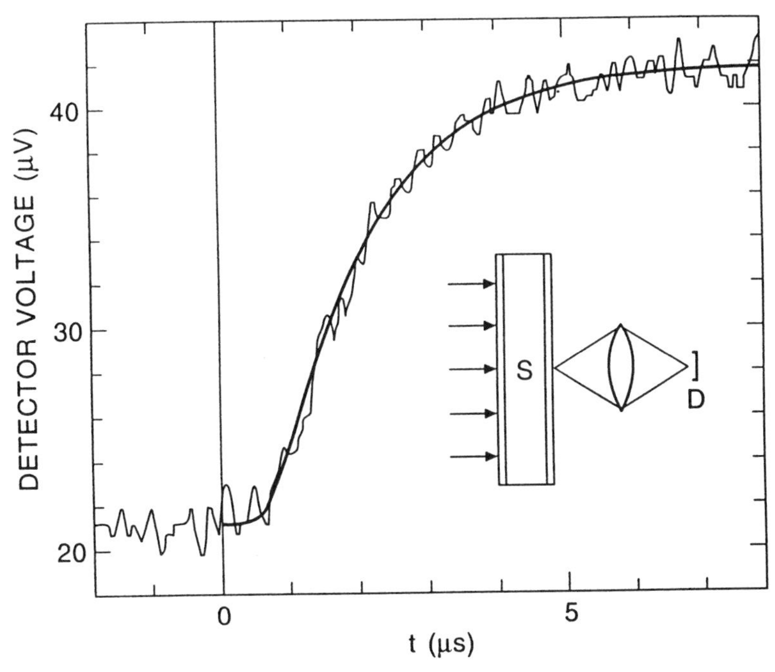

Figure 14. Typical data for the measurement of $D_\perp$ of a 300-μm-thick sample of CVD diamond by the flash diffusivity method. The laser pulse occurs at the zero of time, and the total rise in temperature is approximately 0.3°C. The solid curve through the data is a least-squares fit of Eq. (14) with a value of $D_\perp$ = 3.0 cm^2/sec. The inset shows the sample S coated on both sides with a thin film of titanium. The thermal radiation from the right side of the sample is focused onto an infrared detector D. (Graebner et al. 1994a)

A number of precautions must be observed to obtain accurate diffusivity data. The length of the pulse must be kept small compared to the transit time in order to approximate the assumption of a short pulse. In practice, this means that the pulse should be no longer than a few percent of $t_{1/2}$. For thick ($\approx$ 1 mm) samples of low or medium diffusivity, a flash lamp with pulse length of 10 - 1000 μs is adequate [Log and Jackson 1991], hence the characteristic name for this method. For thin samples of high-conductivity material such as CVD diamond, it is more convenient [Graebner et al. 1992c] to use a Q-switched laser with a typical pulse length of 10 ns, corresponding to a minimum sample thickness of approximately 30 μm for diamond. With such a short pulse, higher demands are placed on the optically dense coating. A thermally thin but optically thick film of high absorbtivity and high adhesion to the surface of the sample is required. However, yet another problem is encountered with thick samples ($Z \approx 1$ mm) for which, even in the case of diamond, ten nanoseconds is only $10^{-4} \times t_{1/2}$. In this case, during the time τ of the pulse, the heat wave travels into the sample a distance $z_\tau = (7D\tau)^{1/2} \approx 10$ μm, or 1% of Z. By conservation of energy, the temperature rise in this front portion of the sample reaches 100 times the highest (final) temperature rise of the rear face, T_{final}. The total input power must be kept low enough to keep T_{final} as small as possible, consistent with a reasonable signal-to-noise ratio.

The *average* temperature of the sample in the region where thermal diffusion is taking place, however, is much less than the maximum temperature of the front face. It has been suggested [Graebner et al. 1994a] that $T_{avg} \approx T_{final} \times \ln(Z/z_\tau)$. Thus, for a 1-mm-thick sample with the input power adjusted to give $T_{final} \approx 0.5$ °C, T_{avg} is only 2.3°C higher than the temperature just before the pulse.

The thermal diffusivity can be obtained from the time $t_{1/2}$, as described above. It is much more accurate to use the full data curve by performing a least-squares fit of the theoretical expression, Eq. (14), to the data, which is possible with only modest computing power. Typically, ten terms in the summation are more than adequate to represent Eq. (14) at times longer than approximately $t_{1/2}/20$.

To improve the temperature resolution, signal averaging techniques can be used. Even with the wide bandwidth of a fast (10-ns rise time) cooled HgCdTe infrared detector monitoring the rear face, a temperature resolution of 0.02°C at room temperature can be obtained after averaging 1000 pulses [Graebner et al. 1994a].

In cases where the sample is not free-standing, it is desirable to measure the thermal properties from one side only. The laser flash technique has been applied in this way to ceramic materials [Kehoe et al. 1995], but it was pointed out that accurate results depend on having precise knowledge of the optical absorption at the surface and the heat penetration during the time of the laser pulse.

5. Discussion of Techniques

With the above techniques to choose from, it is not obvious which method is best suited to a particular application. To aid in making comparisons, we have listed [Graebner

1993a] some of the features of each method in Table 1. There is a degree of arbitrariness in some of the entries, and many considerations are difficult to summarize in a word or two. The entries for typical uncertainty indicate the accuracy possible with a reasonable degree of effort. The accuracy of the volume thermal grating technique with its very short length scale has not yet been established for high-conductivity diamond. The temperature range indicated is that attainable without the use of very special apparatus. The temperature limits are not well defined; rather, they indicate convenient limits of currently used techniques such as infrared thermometry, or epoxy for attachment of thermocouples or heater wires. However, upper limits for several techniques (including all the dc techniques except for the radiating bar) are temperatures above which thermal radiation corrections become very important for CVD diamond. The component of κ or D that is measured is straightforward except for the mirage and 3ω techniques, which measure a combination of the two directions depending on the exact geometry of the heater, the thermal penetration depth δ_T, and the sample thickness. (As noted above, some of the other techniques require care in order to assure measurement of only the parallel or perpendicular component.) Some techniques have special requirements for the sample geometry and some need rather laborious attachment of heaters and thermometers, while others simply require black coatings or nothing at all. Most methods can be used with only moderate skill on the part of the operator, with the exceptions perhaps of the mirage and thermal grating techniques. Finally, there is a wide range of cost. "Low cost" in the table refers to the cost of a system with apparatus such as voltmeters, a simple evaporation chamber, a vacuum chamber for measurement, and a computer for data acquisition. Adding a pulsed laser and/or infrared scanning camera brings the cost up to "intermediate," while more elaborate optical setups are indicated as "high cost." The most appropriate technique for a particular application will usually depend on all the above considerations.

Some of the techniques can be combined to measure several different properties on the same sample. The heated bar approach could be used with either dc ($\kappa_{\|}$) or ac ($D_{\|}$) heating, thereby measuring the specific heat. The TTG approach for measuring $D_{\|}$, especially with thermal detection, Fig. 13, could be used with very little modification to measure $D_{\perp}$. That is, guiding the full laser beam around the interferometer and onto the rear surface of the sample makes it into a laser flash apparatus. As pointed out above, the annular heating technique for $D_{\|}$ could also be converted to a laser flash method ($D_{\perp}$) by simply removing the axicon lens and using a wider-bandwidth detection system.

As noted above, it is sometimes inconvenient to remove a sample from its substrate for measurement, precluding use of several of the above techniques. The remaining choices are ac techniques which use only one surface, such as spot-, line-, or plane-heating, or a surface transient thermal grating. These approaches all probe into the material a depth related to the measuring frequency or the grating wavelength. If the depth is kept less than the film thickness, the properties of only the film are measured. For larger depths, a theoretical model of a multi-layer assembly can be constructed to analyze the thermal properties of the various layers and interfaces. This approach borders on the rather large field of non-destructive testing of layered materials [see, for example, Heuret et al. 1990]. The main difficulties with many-parameter models are, of course, the questions of uniqueness and the degree of confidence one should attach to the fitting parameters.

Table 1. Comparison of different methods for measuring thermal conductivity or diffusivity.

Technique	Quantity Measured	Typical Uncertainty	Temperature Range	Sample Geometry	Sample Preparation	Operator Skill	Relative Cost
DC							
Heated Bar (Rad. Corr.)	$\kappa_{\parallel}$	3-20%	0-600K	uniform bar	detailed	moderate	low
Two-heater Heated Bar	$\kappa_{\parallel}$	3-5%	0-600K	uniform bar	detailed	moderate	low
Insulated Plate	$\kappa_{\parallel}$	10%	300-400K	plate or bar	none	moderate	low
Radiating Bar	$\kappa_{\parallel}$	10%	400-600K	uniform bar	black coating	moderate	low - medium
Line-heated Plate	$\kappa_{\parallel}$	5%	0-600K	plate	detailed	moderate	low
PERIODIC							
End-heated Bar	$D_{\parallel}$	5%	0-600K	uniform bar	varied	moderate	low
Spot-heated Plate	$D_{\parallel}$	5%	200-1000K	plate	black coating	moderate	medium
Spot-heated Solid (mirage)	D_{avg}	5%	100-400K	flat surface	black coating	high	high
Line-heated Plate (3ω)	κ_{avg}	3-5%	0-1000K	flat surface	narrow heater	moderate	medium
Plane-heated Plate (3ω)	$\kappa_{\perp}C$	3-5%	0-1000K	flat surface	broad heater	moderate	medium
PULSED							
Annular Heating	$D_{\parallel}$	10%	300-1000K	plate	black coating	moderate	medium
Thermal Grating on Surface	$D_{\parallel}$	10%	100-1000K	flat surface	black coating	high	medium
Thermal Grating in Bulk	$D_{\parallel}$	10%?	100-1000K	flat surface	none	high	high
Laser Flash	$D_{\perp}$	3-5%	200-1000K	plate	black coatings	moderate	medium

The depth-dependent techniques are potentially very useful for mapping out the depth dependence of the local diffusivity. The results of a measurement of, say, $D_{\parallel}$ to a depth L_z yield the average over that depth. As L_z is varied, the average results follow any gradient of the local $D_{\parallel}^{local}$ with respect to z that might be present. If $D_{\parallel}^{local}$ is a linear function of z, then the value $D_{\parallel}(L_z) = D_{\parallel}^{local}(L_z/2)$. If $D_{\parallel}^{local}$ is a non-linear function of z, then a more elaborate approach is needed, as described below.

The gradient with respect to z of the local conductivity $\kappa_{\parallel}^{local}$ of CVD diamond was first observed by another technique [Graebner et al. 1992a, 1994b]. A series of five samples with thickness Z ranging from 27 to 355 μm was prepared with the growth conditions

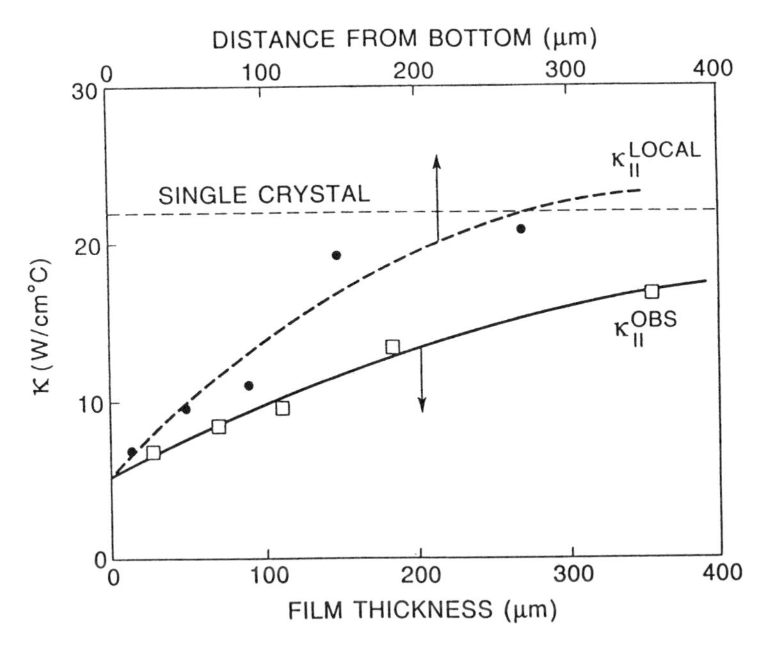

Figure 15. The in-plane thermal conductivity $\kappa_{\|}$ of a series of CVD diamond films of various thickness (squares). The rise of the measured conductance with increasing thickness indicates that the *local* conductivity increases with distance from the bottom. The local conductivity is derived from the data by either a discrete (solid circles) or continuous (dashed line) differentiation procedure as described in the text. (Graebner et al. 1992a)

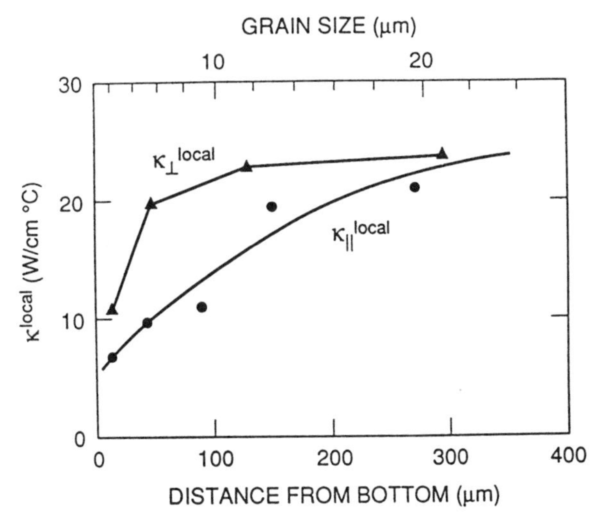

Figure 16. Local conductivity for heat flow parallel to the film $\kappa_{\|}^{local}$ (from Fig. 15) compared with that for heat flow perpendicular to the film, $\kappa_{\perp}^{local}$ Large anisotropy, up to a factor of 2, is observed in the range 20-200 μm from the bottom. The nonlinear top scale gives the grain size as a function of distance from the bottom of the samples (bottom scale) and indicates the inhomogeneous microstructure which is thought to be responsible for the large gradient and anisotropy in the local conductivity. (Graebner et al. 1992d)

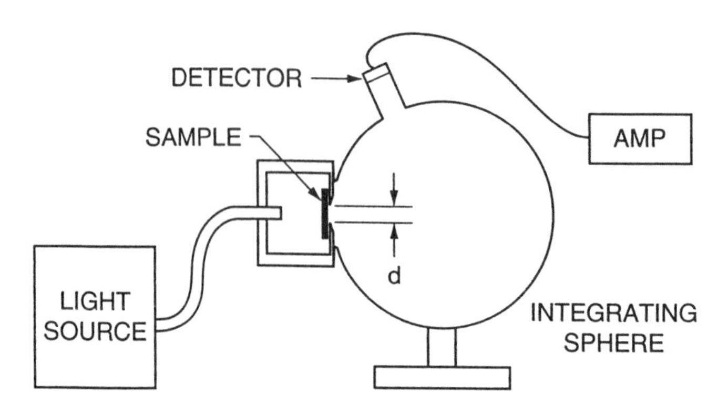

Figure 17. Apparatus for measuring the optical absorption of a sample. The light source (100W halogen bulb) illuminates the sample which transmits some of the light to an integrating sphere where it is measured by a detector. (Graebner 1995c)

held the same for all five samples, the only difference being the growth time. The samples therefore all shared a common microstructural profile. The heated-bar measurements of $\kappa_{\|}$, yielding an average over the thickness of each sample, exhibited a positive dependence of $\kappa_{\|}^{obs}$ on thickness (Fig. 15). A local conductivity $\kappa_{\|}^{local}$ was derived by a differentiation performed on the slab *conductance* $Z\kappa_{\|}^{obs}$, i.e., $\kappa_{\|}^{local}(Z) = \partial(Z\kappa_{\|}^{obs})/\partial Z$. This was most easily done by fitting the observed data with a polynomial and then using the polynomial for $\kappa_{\|}^{obs}$ in the partial derivative. A similar approach was used [Graebner et al. 1992d] with laser flash data for $\kappa_{\perp}^{obs}$ on a similar series of four samples, taking into account that the slabs of uniform conductivity at different heights were now encountered in series by the diffusing heat, rather than in parallel. The derived $\kappa_{\perp}^{local}$ agreed with $\kappa_{\|}^{local}$ at small z (near zero) and large z (above approximately 300 μm) but was up to a factor of two higher at intermediate values of z (Fig. 16). This observation of considerable anisotropy for $20 < z < 200$ μm is discussed above. The flash method measures the diffusivity, of course, while the dc heated bar measures conductivity. The good agreement between $\kappa_{\|}^{local}$ and $\kappa_{\perp}^{local}$ (the latter derived from $D_{\perp}^{obs}$ assuming bulk heat capacity) is a strong self-consistency check and supports the idea that the conductivity and not the heat capacity is the property that is strongly z-dependent. Recently, direct measurements of heat capacity for both poor-quality and high-quality CVD diamond show that the heat capacity and mass density are each within 1% of the bulk-diamond values [Graebner 1996].

A gradient in $D_{\|}^{local}$ similar to that discussed above has recently been reported [Käding et al. 1994] with the TTG technique. The results are qualitatively similar [Graebner et al. 1996], even though the samples are quite different (<100> and quasi-heteroepitaxial <100>, vs. <110> in [Graebner et al. 1992a]).

6. Alternative Measurements

Because of the difficulty in making reliable measurements of thermal conductivity, it would often be convenient to have a simpler measurement of another property that scales with conductivity. Several observations of a correlation between thermal resistivity and infrared absorption in the 8μm wavelength range have been reported [Burgemeister 1978, Vandersande 1980, Morelli and Uher 1993, Bachmann et al. 1995]. Recently, however, a correlation of the thermal resistivity with the optical absorption integrated over the *visible* spectrum has been observed [Graebner 1995c]. Such a quantity is easily and quickly measured with the simple and inexpensive apparatus shown schematically in Fig. 17. The integrating sphere allows measurement of samples with rough surfaces. The absorption coefficient α is obtained from the measured thickness Z and the detector output voltage V (or V_o) obtained with (or without) the sample present:

$$\alpha = Z^{-1} \log(0.706\, V_o / V) \tag{15}$$

The numerical constant takes into account reflections which occur at the surfaces because the index of refraction of diamond is different from 1.0 Figure 18 illustrates the correlation between α and $\kappa_{\|}$ for many samples grown by various techniques in ten laboratories. The solid line is a fit to the data using a simple function:

$$\kappa = A\left(\frac{\alpha}{B} + 1\right)^{-1} \tag{16}$$

with $A = 20.21\ \mathrm{Wcm^{-1}K^{-1}}$ and $B = 136.7\ \mathrm{cm^{-1}}$. A similar correlation and fit is made for $\kappa_{\perp}$, with slightly larger fitting parameters. The data for the isotopically purified samples (solid circles) are excluded from the fit. The data lie within approximately $\pm 3\ \mathrm{Wcm^{-1}K^{-1}}$ of the curve for $\kappa_{\|}$. The smaller scatter in the $\kappa_{\|}$ data for the high-conductivity samples (low α) may be due in part to the much greater effort spent in obtaining reliable measurements of κ for these remarkable specimens.

Equation (15) can be rearranged to obtain a direct dependence of optical absorption on thermal resistivity $W = \kappa^{-1}$:

$$\alpha = B\left(\frac{W}{W_o} - 1\right) = B\left(\frac{\Delta W}{W_o}\right) \tag{17}$$

where $W_o = A^{-1}$ and $\Delta W = W - W_o$. For perfect diamond, one expects $\alpha \rightarrow 0$ as $W \rightarrow W_o$, where W_o is the thermal resistivity due only to intrinsic phonon-scattering processes such as Umklapp scattering. Equation (17) is intuitively appealing, as one expects at least some types of impurities or defects to give rise to both optical absorption and thermal resistance. It has been suggested [Robins et al. 1991] that the dark color of

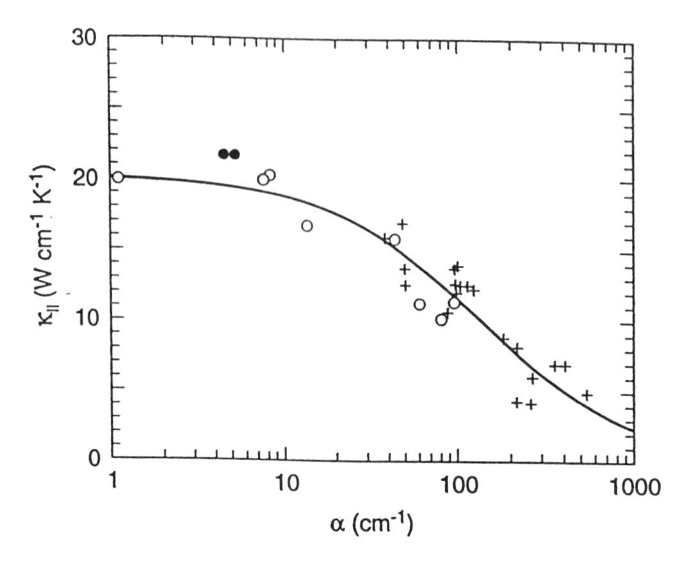

Figure 18. Parallel thermal conductivity $\kappa_{\|}$ at 25°C as a function of the optical absorption coefficient α for 30 samples of CVD diamond covering a wide range of $\kappa_{\|}$ and α. The two filled circles are data for two nominally identical samples of CVD diamond with isotopic enrichment of ^{12}C. The full line is a least-squares fit of Eq. (16) to the data, with the exception of the isotopically enriched samples. (Graebner 1995c)

poor quality CVD diamond is due to light absorption by amorphous carbon, which would also be expected to scatter phonons.

7. Conclusion

CVD diamond is clearly one of the most difficult materials to characterize thermally. This is due not only to its extremely high conductivity/diffusivity but also its columnar microstructure and the resulting inhomogeneity, which creates a gradient in the local conductivity. The gradient depends on the direction of heat flow, i.e., the gradient is different for $\kappa_{\|}^{local}$ and $\kappa_{\perp}^{local}$, resulting in a pronounced anisotropy in the local conductivity. Some measurement techniques are more sensitive to this gradient and anisotropy than others.

Many different techniques have been used to measure κ and/or D for CVD diamond. Choosing the most appropriate technique for a particular application depends on many considerations: the desired accuracy, the direction of heat flow that is considered important, the required speed of measurement, the degree of expertise available, whether κ or D is preferred, and the length scale of interest. We have summarized the relevant features of a dozen or so measurement techniques.

As an example of the difficulty in obtaining reliable results for CVD diamond, we point out a round-robin series of measurements conducted under the auspices of the National Institute of Standards and Technology (NIST) in the United States [Feldman et al. 1994]. Ten specimens with conductivities roughly spanning the range from 4 to 20

W/cm K were measured at ten laboratories using eight different techniques. The spread in values obtained for any particular specimen was usually outside the error limits estimated by each individual lab, and the highest and lowest values for a given specimen typically differed by a factor of two. Most of this large variation was attributed to specimen inhomogeneity and to the fact that different techniques are sensitive in different ways to the inhomogeneity. A second round robin with much more uniform CVD diamond specimens and twice as many laboratories is currently being conducted and the results are expected to be more closely clustered. However, the first round-robin series illustrates the need for considerable caution when measuring the thermal properties of CVD diamond.

8. References

Albin, S., Winfree, W. P. and Crews, B. S. (1990) Thermal diffusivity of diamond films using a laser pulse technique, *J. Electrochem. Soc.*, **137**, 1973-1976.

Anthony, T. R., Banholzer, W. R., Fleischer, J. F., Wei, L., Kuo, P. K., Thomas, R. L. and Pryor, R. W. (1 990) Thermal diffusivity of isotopically enriched ^{12}C diamond, *Phys. Rev. B,* **42**, 1104 - 1111.

Atalla, S. R., El-Sharkawy, A. A. and Gasser, F. A. (1981) Measurement of thermal properties of liquids with an ac heated-wire technique, *Int. J. of Thermophysics,* **2**, 155-162.

Ångström, A. J. (1861) *Annln. Phys., Lpz.,* **114**, 513.

Bachmann, P. K., Hagemann, H. J., Lade, H., Leers, D., Wiechert, D. U., Wilson, H., Fournier, D. and Plamann, K. (1995) Thermal properties of C/H-, C/H/O-, and C/H/X-grown polycrystalline CVD diamond, to appear in *Diamond and Related Materials.*

Berman, R. (1976) *Thermal Conduction in Solids,* Oxford University Press.

Birge, N. O. (1986) Specific-heat spectroscopy of glycerol and propylene glycol near the glass transition, *Phys. Rev. B,* **34**, 1631-1642.

Birge, N. O. and Nagel, S. R. (1987) Wide-frequency specific heat spectrometer, *Rev. Sci. Instrum.,* **58**, 1464-1470.

Burgemeister, E. A. (1978) Thermal conductivity of natural diamond between 320 and 450 K, *Physica,* **93B**, 165-179.

Cahill, D. G. (1990) Thermal conductivity measurement from 30 to 750 K: the 3ω method, *Rev. Sci. Instrum.*, **61**, 802-808.

Cahill, D. G., and Pohl, R. O. (1987) Thermal conductivity of amorphous solids above the plateau, *Phys. Rev. B,* **35**, 4067-4073.

Carslaw, H. S. and Jaeger, J. C. (1959) *Conduction of Heat in Solids,* Second Edition, Clarendon Press.

Charbonnier, F. and Fournier, D. (1986) Compact design for photothermal deflection (mirage): Spectroscopy and imaging, *Rev. Sci. Instrum.,* **57**, 1126-1128.

Cielo, P., Utracki, L. A. and Lamontagne, M. (1986) Thermal diffusivity measurements by the converging-thermal-wave technique, *Can. J. Phys.,* **64**, 1172.

Eichler, H. J., Günter, P. and Pohl, D. W. (1986) *Laser-Induced Dynamic Gratings,* Springer-Verlag.

Enguehard, R., Boscher, D., Deom, A. and Balageas, D. (1990) Measurement of the thermal radial diffusivity of anisotropic materials by the converging thermal wave technique, *Materials Science and Engineering,* **B5**, 127-134.

Feldman, A., Frederikse, H. P. R. and Norton, S. J. (1990) Analysis of thermal wave propagation in diamond films, *Diamond Optics III,* SPIE vol. **1325**, 304.

Feldman, A., Holly, S., Klein, C. A. and Lu, G. (1994) Conference Report for the Workshop on Characterizing Diamond Films III, *J. Res. Natl. Inst. Stand. Technol.,* **99**, 287-293.

Frederikse, H. P. R., Fields, R. J. and Feldman, A. (1992) Thermal and electrical properties of copper-tin and nickel-tin intermetallics, *J. Appl. Phys.* **72**, 2879-2882.

Graebner, J. E. (1993a) Thermal conductivity of CVD diamond: techniques and results, *Diamond Films and Tech.,* **3**, 77-130.

Graebner, J. E. (1993b) Simple method for measuring the thermal conductivity of a thin plate, *Rev. Sci. Instrum.* **64**, 3245-3247.

Graebner, J. E. (1995a) "Thermal conductivity of diamond" in L. S. Pan and D. R. Kania (eds), *Diamond: Electronic Properties and Applications,* Kluwer Academic Publishers, pp. 285-318.

Graebner, J. E. (1995b) Measurement of thermal diffusivity by optical excitation and infrared detection of a transient thermal grating, *Rev. Sci. Instrum.,* **66**, 3903-3906.

Graebner, J. E. (1995c) Simple correlation between optical absorption and thermal conductivity of CVD diamond, *Diamond and Relat. Mater.,* **4**, 1196-1199.

Graebner, J. E. (1996) Measurements of specific heat and mass density in CVD diamond, *Diamond and Relat. Mater.*, **5**, 1366-1370.

Graebner, J. E. and Herb, J. A. (1992) Dominance of intrinsic phonon scattering in CVD diamond, *Diamond Films and Tech.*, **1**, 155-164.

Graebner, J. E., Jin, S., Kammlott, G. W., Herb, J. A. and Gardinier, C. F. (1992a) Unusually high thermal conductivity in diamond films, *Appl. Phys. Lett.*, **60**, 1576-1578.

Graebner, J. E., Mucha, J. A. and Baiocchi, F. A. (1996) Sources of thermal resistance in chemically vapor deposited diamond, *Diamond and Relat. Mater.*, **5**, 682-687.

Graebner, J. E., Mucha, J. A., Seibles, L. and Kammlott, G. W. (1992b) The thermal conductivity of chemical-vapor-deposited diamond films on silicon, *J. Appl. Phys.*, **71**, 3143-3146.

Graebner, J. E., Jin, S., Kammlott, G. W., Bacon, B., Seibles, L. and Banholzer, W. R. (1992c) Anisotropic thermal conductivity in CVD diamond, *J. Appl. Phys.*, **71**, 5353-5356.

Graebner, J. E., Jin, S., Kammlott, G. W., Herb, J. A. and Gardinier, C. F. (1992d) Large anisotropic thermal conductivity in synthetic diamond films, *Nature*, **359**, 401-403.

Graebner, J. E., Reiss, M. E., Seibles, L., Hartnett, T. M., Miller, R. P. and Robinson, C. J. (1994a) Phonon scattering in chemical-vapor-deposited diamond, *Phys. Rev. B*, **50**, 3702-3713.

Graebner, J. E., Jin, S., Herb, J. A. and Gardinier, C. F. (1994b) Local thermal conductivity in chemical-vapor-deposited diamond, *J. Appl. Phys.*, **76**, 1552-1556.

Hatta, I., Sasuga, Y., Kato, R. and Maesono, A. (1985) Thermal diffusivity measurement of thin films by means of an ac calorimetric method, *Rev. Sci. Instrum.*, **56**, 1643-1647.

Hatta, I., Kato, R. and Maesono, A. (1986) Development of ac calorimetric method for thermal diffusivity measurement. I. Contribution of thermocouple attachment in a thin sample, *Jpn. J. Appl. Phys.*, **25**, L493-L495.

Hatta, I., Kato, R. and Maesono, A. (1987) Development of ac calorimetric method for thermal diffusivity measurement. II. Sample dimension required for the measurement, *Jpn. J. Appl. Phys.*, **26**, 475-478.

Herb, J. A., Bailey, C., Ravi, K. V. and Dennig, P. A. (1989) The impact of deposition parameters on the thermal conductivity of CVD thin diamond films, *J. Electrochem. Soc.*, **89-12**, 366.

Heuret, M., van Schel, E., Egee, M. and Danjoux, R. (1990) Bonding analysis of layered materials by photothermal radiometry, *Mat. Sci. and Eng.*, **B5**, 119.

Käding, O. W., Matthias, E., Zachai, R., Füsser, H.-J. and Münzinger, P. (1993) Thermal diffusivities of thin diamond films on silicon, *Diamond Relat. Mater.*, **2**, 1185.

Käding, O. W., Rössler, M., Zachai, R., Füsser, H.-J. and Matthias, E. (1994) Lateral thermal diffusivity of epitaxial diamond films, *Diamond Relat. Mater.*, **3**, 1178-1182.

Kehoe, L., Kelly, P. V., O'Connor, G. M., O'Reilly, M. and Crean, G. M. (1995) A measurement methodology for laser-based thermal diffusivity measurement of advanced multichip module ceramic materials, *IEEE Semitherm-95*, San Jose, CA, pp. 92-101.

Kosky, P. G. (1993) A method of measurement of thermal conductivity: application to free-standing CVD diamond sheet, *Rev. Sci. Instrum.*, **64**, 1071-1075.

Log, T. and Jackson, T. B. (1991) Simple and inexpensive flash technique for determining thermal diffusivity of ceramics, *J. Am. Ceram. Soc.*, **74**, 941-944.

Lu, G. and Swann, W. T. (1991) Measurement of thermal diffusivity of polycrystalline diamond film by the converging thermal wave technique, *Appl. Phys. Lett.*, **59**, 1556-1558.

Morelli, D. T. and Uher, C. (1993) Correlating optical absorption and thermal conductivity in diamond, *Appl. Phys. Lett.*, **63**, 165-167.

Ohta, H., Shibata, H. and Waseda, Y. (1989) New attempt for measuring thermal diffusivity of thin films by means of a laser flash method, *Rev. Sci. Instrum.*, **60**, 317-321.

Ono, A., Baba, T., Funamoto, H. and Nishikawa, A. (1986) Thermal conductivity of diamond films synthesized by microwave plasma CVD, *Jpn. J. Appl. Phys.*, **25**, L808-L810.

Parker, W. J., Jenkins, R. J., Butler, C. P. and Abbott, G. L. (1961) Flash method of determining thermal diffusivity, heat capacity, and thermal conductivity, *J. Appl. Phys.*, **32**, 1679-1684.

Plamann, K., Fournier, D., Anger, E. and Gicquel, A. (1994) Photothermal examination of the heat diffusion inhomogeneity in diamond films of sub-micron thickness, *Diamond and Relat. Mater.*, **3**, 752-756.

Robins, L. H., Farabaugh, E. N. and Feldman, A. (1991) Determination of the optical constants of thin chemical-vapor-deposited diamond windows from 0.5 to 6.5 eV, *Diamond Optics IV,* SPIE vol. **1534**, 105-116.

Salamon, M. B., Garnier, P. R., Golding, B. and Buehler, E. (1974) Simultaneous measurement of the thermal diffusivity and specific heat near phase transitions, *J. Phys. Chem. Solids,* **35**, 851-859.

Sandberg, O., Andersson, P. and Bäckström, B. (1977) Heat capacity and thermal conductivity from pulsed wire probe measurement under pressure, *J. Phys. E, Scientific Instruments,* **10**, 474-477.

Tokmakoff, A., Banholzer, W. F. and Fayer, M. D. (1993) Thermal diffusivity measurements of natural and isotopically enriched diamond by picosecond infrared transient grating experiments, *Appl. Phys. A,,***56**, 87-90.

Vandersande, J. W. (1980) A correlation between the infrared absorption features and the low temperature thermal conductivity of different types of natural diamonds, *J. Phys. C: Solid St. Phys.,* **13**, 759-764.

Vandersande, J. W., Vining, C. B. and Zoltan, A. (1991) Thermal conductivity of natural type IIa diamond between 500K and 1250K, *J. Electrochem. Soc.,* **91-8**, 443.

Vandersande, J. W., Zoltan, A. and Wood, C. (1989) Accurate determination of specific heat at high temperatures using the flash diffusivity method, *International J. of Thermophysics,* **10**, 251-257.

Vining, C. B., Zoltan, A. and Vandersande, J. W. (1989) Determination of the thermal diffusivity and specific heat using an exponential heat pulse, including heat-loss effect, *International J. of Thermophysics,* **10**, 259-268.

Visser, E. P., Versteegen, E. H. and van Enckevort, W. J. P. (1992) Measurement of thermal diffusion in thin films using a modulated laser technique: application to CVD diamond films, *J. Appl. Phys.,* **71**, 3238-3248.

Wei, L., Kuo, P. K., Thomas, R. L., Anthony, T. R. and Banholzer, W. F. (1993) Thermal conductivity of isotopically modified single crystal diamond, *Phys. Rev. Lett.,* **70**, 3764.

Chapter 7
OPTICAL PROPERTIES

Alexander M. Zaitsev
University of Hagen, 58084 Germany
Belarussian State University, Minsk 220080, Belarus

Contents

Abbreviations and Designations

A	absorption;
CL	cathodoluminescence (excitation by external electron beam);
CVD	chemical vapor deposition;
E_g	indirect energy band gap;
E_{dg}	direct energy band gap;
EA	electro-absorption;
EL	electroluminescence (excitation by internal electrical current);

Abbreviations and Designations, cont.

EPL	excitation of photoluminescence;
FWHM	full width at half maximum;
k	wave vector;
L	luminescence;
LA	longitudinal acoustic;
LO	longitudinal optical;
LVM	local vibrational mode;
MPCVD	microwave plasma assisted CVD;
n	refractive index;
p_{ij}	non-zero elements of elasto-optic tensor;
P	pressure;
PCCVD	polycrystalline chemical vapor deposited;
PL	photoluminescence;
R	reflection;
S	Huang-Rhys factor;
SFG	sum-frequency generation;
T	temperature;
TA	transversal acoustic;
TO	transversal optical;
XL	X-ray luminescence;
ZPL	zero-phonon line;
λ	wave length;
σ	stress;
μ	absorption coefficient;
ε	dielectric constant.

1. Introduction

Many outstanding optical properties of diamond make it an attractive material for optical and optoelectronic applications. Diamond has a very wide optical transparency band ranging from 0.22 μm (fundamental absorption edge) to far infrared. Only a moderate intrinsic absorption band between 2.5 to 7 μm disturbs the perfection of diamond transparency in the infrared region. Being transparent in the ultraviolet, visible and infrared spectral regions diamond offers a great opportunity for lattice defects to reveal optical activity of their electronic and vibrational transitions. The large energy band gap (5.49 eV) is a favorable condition particularly in the case of luminescence, because effective radiative electronic transitions require the ground and excited electronic states to be lying within the band gap. By now more than 100 vibrational and more than 400 electronic optically active centers have been known in diamond in a spectral range from 20 to 0.17 μm, that is from the vacuum ultraviolet till the middle infrared region. To fill up this large spectral range diamond possesses many optically active defects of different origin, including intrinsic and impurity related, point and extended ones. Both types of the intrinsic point defects (vacancy and interstitial related) in diamond are optically active. Many impurities are known to form optically active point defects in diamond, they are: H, He, Li, B, N, O, Ne, Si, Ti, Cr, Ni, Zn, Zr, Ag, Xe and Tl.

Nitrogen is the impurity of special importance for optics of diamond. First, nitrogen is responsible for the majority of the optical centers. Second, many of the most intense and the most interesting optical centers are known as nitrogen related. Nitrogen forms optically active defects in many ways: single isolated atoms, multi-atom complexes, complexes with intrinsic lattice defects and with other impurities. So far the presence of nitrogen almost in any form in the diamond lattice immediately changes its optical properties. One of the consequences of such a high optical activity of nitrogen is the physical classification of diamond primarily based on the nitrogen related optical absorption.

An important optically related feature of diamond is its high Debye temperature (almost 2000 K). Actually this value is the highest one among all known solids. Owing to the high Debye temperature, a substantial excitation of phonons in diamond and, consequently, electron-phonon coupling in optically active defects, occurs at relatively high temperatures. As a result many optical centers in diamond retain their spectral structure and transition probabilities to relatively high temperatures. For instance, some luminescence centers remain active even above 500°C.

Physical backgrounds of the optical properties of diamond have already been discussed in detail in a number of books and review articles. The aim of the present chapter is to present only the experimental facts related to the optical properties of diamond without their detailed explanations. Some qualitative features of the optical processes in diamond are given too.

2. Refraction

2.1 NATURAL DIAMOND

Table I. Refractive index n of natural diamond at different wavelengths [1, 2, 3, 7, 74, 130, 341]

λ, [nm]	n		
0.225	2.729	0.48613	2.43554
0.226	2.7151	0.5	2.432; 2.4324
0.2265	2.7151	0.5270	2.4269
0.25	2.627; 2.6333	0.5350	2.4278
0.3	2.545; 2.5407	0.5460	2.4237
0.35	2.490; 2.4928	0.5461	2.42309
0.3969	2.4653	0.55	2.4230
0.397	2.4648	0.5641	2.4237
0.4	2.463; 2.4641	0.5890	2.4173
0.4102	2.4592	0.5892	2.4176
0.4308	2.4512	0.58929	2.41726
0.4358	2.44902	0.5893	2.4195; 2.41681
0.4410	2.4482	0.6	2.415; 2.4159
0.45	2.4454	0.65	2.4105
0.4860	2.4354	0.65428	2.40990
0.4861	2.4354; 2.43488	0.6560	2.4099

Table I., cont, Refractive index n of natural diamond at different wavelengths [1, 2, 3, 7, 74, 130, 341]

λ, [nm]	n	λ, [nm]	n
0.65628	2.40990	3.0	2.382; 2.3782
0.6563	2.4103; 2.40967	4.0	2.3773
0.6708	2.4135	5.0	2.381; 2.3767
0.6876	2.4077	6.0	2.3763
0.7	2.405; 2.4062	7.0	2.3761
0.75	2.4028	8.0	2.3759
0.7590	2.4024	9.0	2.3758
0.7628	2.4024	10.0	2.380; 2.3756
0.8	2.400	12.0	2.3755
0.9	2.396	14.0	2.3753
1.0	2.394	15.0	2.380
1.2	2.390	16.0	2.3752
1.5	2.386	18.0	2.3751
2.0	2.383	20.0	2.380; 2.3750
2.5	2.3786	25.0	2.380; 2.3749

The data of this table can be approximated by the expression:

$$n^2 = 1 + \frac{0.3306\ \lambda^2}{\lambda^2 - (175.0)^2} + \frac{4.3356\ \lambda^2}{\lambda^2 - (106.0)^2}, \tag{1}$$

where λ is given in [nm] units [11, 311].

Spectral dependence of the refractive index in vacuum ultraviolet is given in fig. 1 [322, 494]:

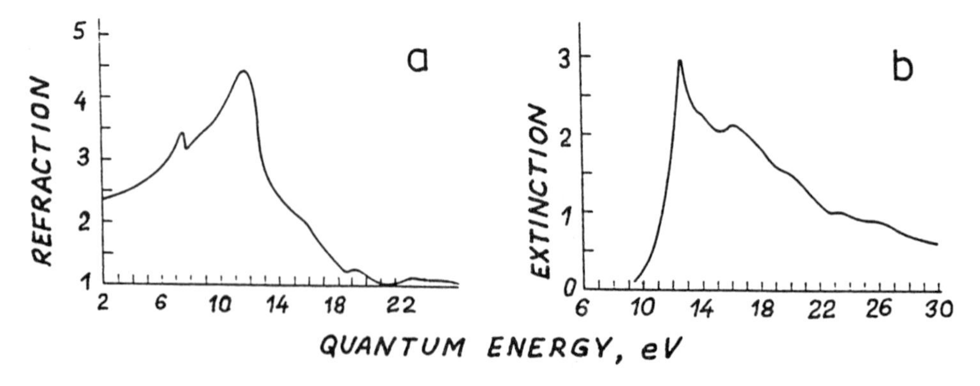

Figure 1. Spectral dependence of the real (a) and imaginary (extinction coefficient) (b) of the refractive index in type IIa diamond [494, fig. 2, 3].

The mean values of n for natural diamond at a wavelength of 547 nm is 2.4236 [3].

Temperature dependence at normal pressure and pressure dependence at room temperature of the refractive index n are given by [12]:

$$(1/n)(dn/dT)_P = +4.04\times10^{-6}\ K^{-1}. \tag{2}$$

$$(1/n)(dn/dP)_T = -0.36\times10^{-12}\ Pa^{-1}. \tag{3}$$

2.2 SYNTHETIC DIAMOND

Refractive index for synthetic diamonds at a wavelength of 580 nm (experimental error is ±0.0004) [7]:

octahedral shaped crystals	2.4183 to 2.4216
cubo-octahedral with dominant cubic facets	2.4182 to 2.4238
cubo-octahedral with dominant octahedral facets	2.4167 to 2.4192

The mean values of n for diamond at a wavelength of 547 nm are [3]:

for octahedral shaped	2.4259;
for cubo-octahedral shaped	2.4243.

2.3 PIEZO-OPTIC AND ELASTO-OPTIC CONSTANTS [130, 459, 460].

Uniaxial stress induced birefringence for (100) and (111) directions:

$$n_{II} - n_{\perp} = -0.5\,n^3 \frac{(p_{11} - p_{12})}{(c_{11} - c_{12})}\sigma_{100}, \tag{4}$$

$$n_{II} - n_{\perp} = -0.5\,n^3 \frac{p_{44}}{c_{44}}\sigma_{111}, \tag{5}$$

Hydrostatic stress induced change of the refractive index:

$$\partial n / \partial P = 0.5\,n^3 \frac{(p_{11} + 2p_{12})}{(c_{11} + 2c_{12})}\sigma. \tag{6}$$

Table II. Values of components of elasto-optic tensor at different wavelengths [459, 460].

λ[nm]	250 nm	500 nm	700 nm
p11 -p12	-0.365	-0.31	- 0.31
p44	-0.2	-0.17 (- 0.18)	-0.17

3. Reflection

Table III. Reflection coefficient R of natural diamond at different wavelengths [1, 2, 3, 7, 130, 341]

λ, [nm]	R	λ, [nm]	R
0.225	0.2150	1.0	0.1687
0.25	0.2012	1.2	0.1681
0.3	0.1899	1.5	0.1676
0.35	0.1823	2.0	0.1671
0.4	0.1785	3.0	0.1670
0.5	0.1741	5.0	0.1669
0.6	0.1717	10.0	0.1668
0.7	0.1703	15.0	0.1668
0.8	0.1696	20.0	0.1668
0.9	0.1690	25.0	0.1668

The reflection spectrum of type IIa natural diamond is presented in fig. 2.

4. Infrared Absorption

The ideal diamond lattice does not absorb light in the one-phonon spectral region. Intrinsic absorption of perfect diamonds is observed only in the two and three-phonon regions [3, 324, 338, 353, 355, 461, 462] (fig. 3, 4). The main peaks of the absorption are:

2.79 μm (3580 cm^{-1}), absorption coefficient of 1.7±0.3 to 3±2 cm^{-1};
3.07 μm (3260 cm^{-1});
3.91 μm (2560 cm^{-1}), 2TO, absorption coefficient of 4.6±0.3 cm^{-1};
4.11 μm (2430 cm^{-1}), TO+LO(L), absorption coefficient of 4.6±0.3 cm^{-1};
4.62 μm (2170 cm^{-1}), LO+LA(L), absorption coefficient of 12.8±0.3 cm^{-1};
4.92 μm (2030 cm^{-1}), TO+TA, absorption coefficient of 12.8±0.3 cm^{-1};
5.04 μm (1980 cm^{-1}), LO+TA(X), absorption coefficient of 12.8±0.3 cm^{-1}.

If disturbed by intrinsic or impurity defects, diamond lattice can show a weak intrinsic absorption in the one-phonon spectral region. For instance, irradiation with fast electrons or neutrons induces a one phonon absorption, which anneals out at temperatures above 1100°C presumably due to vacancy-interstitial annihilation [74] (fig. 4).

4.1 DEFECT INDUCED VIBRATIONAL BANDS

1.161 μm (8615 cm^{-1}), FWHM of 150 cm^{-1}, observed in natural hydrogen rich diamonds of light gray color, hydrogen related line [284] (fig. 5).

1.211 μm (8255 cm^{-1}), FWHM of 150 cm^{-1}, observed in natural hydrogen rich diamonds of light gray color, hydrogen related line [284] (fig. 5).

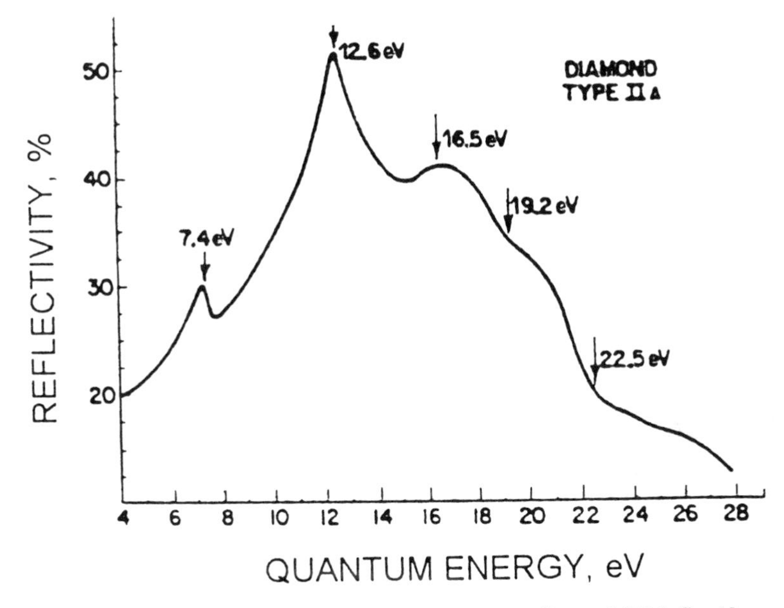

Figure 2. Reflection spectrum of a polished type IIa diamond [494, fig. 1].

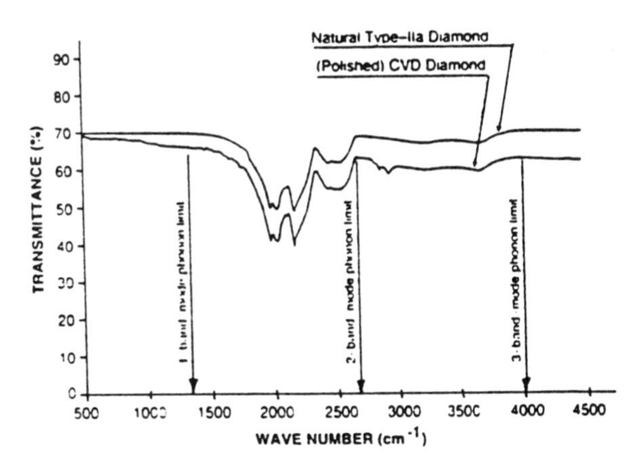

Figure 3. Spectral regions of one-, two-, and three-phonon absorption in diamond [353, fig. 3].

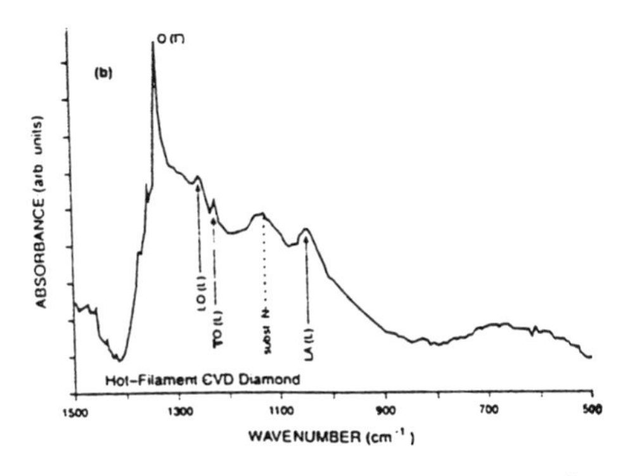

Figure 4. One-phonon absorption of CVD diamond films [353, fig. 4].

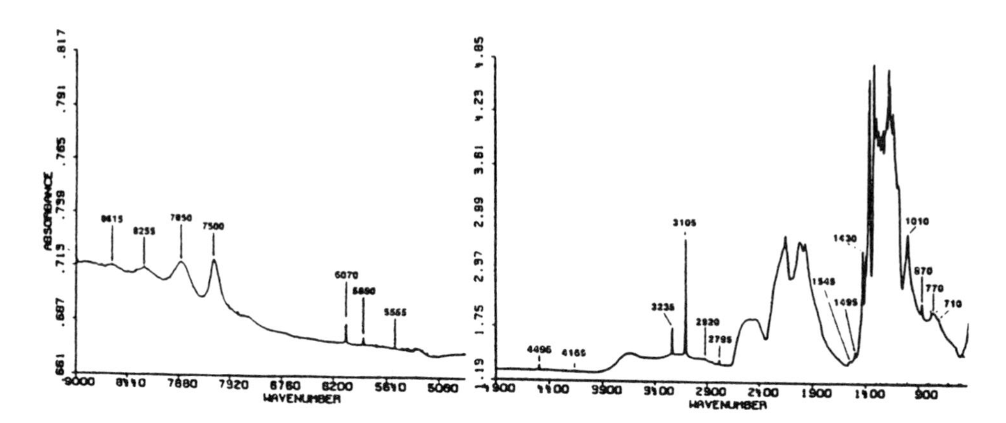

Figure 5. Infrared absorption of hydrogen-rich natural diamond of a light gray color [284, fig. 1, 2].

1.274 μm (7850 cm^{-1}), FWHM is of 180 cm^{-1}, observed in natural hydrogen-rich diamonds of light gray color, hydrogen related line [284] (fig. 5).

1.333 μm (7500 cm^{-1}), FWHM is of 120 cm^{-1}, observed in natural hydrogen-rich diamonds of light gray color, hydrogen related line [284] (fig. 5).

1.647 μm (6070 cm^{-1}), see the 3.219 μm (3107 cm^{-1}) line.

1.701 μm (5880 cm^{-1}), see the 3.219 μm (3107 cm^{-1}) line.

1.800 μm (5555 cm^{-1}), see the 3.219 μm (3107 cm^{-1}) line.

2.028 μm (4932 cm^{-1}), a weak narrow line observed in ^{13}C CVD films [326] (fig. 6).

2.070 μm (4830 cm^{-1}), a weak narrow line observed in ^{13}C CVD films [326] (fig. 6).

2.126 μm (4703 cm^{-1}), a narrow line observed in natural hydrogen-rich-diamonds of light gray color [284] (fig. 5).

2.223 μm (4498 cm^{-1}), see the 3.219 μm (3107 cm^{-1}) line.

2.399 μm (4168 cm^{-1}), see the 3.219 μm (3107 cm^{-1}) line.

2.401 μm (4165 cm^{-1}), a narrow line observed in natural hydrogen-rich-diamonds of light gray color [284] (fig. 5).

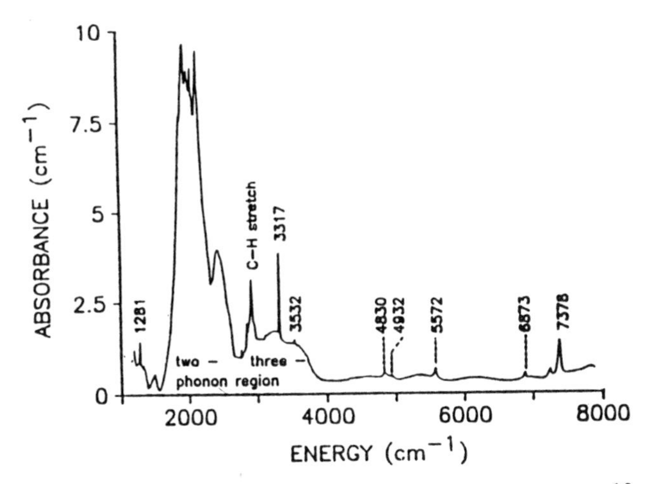

Figure 6. Absorption spectrum of a homo-epitaxially grown ^{13}C CVD diamond film [326, fig. 1].

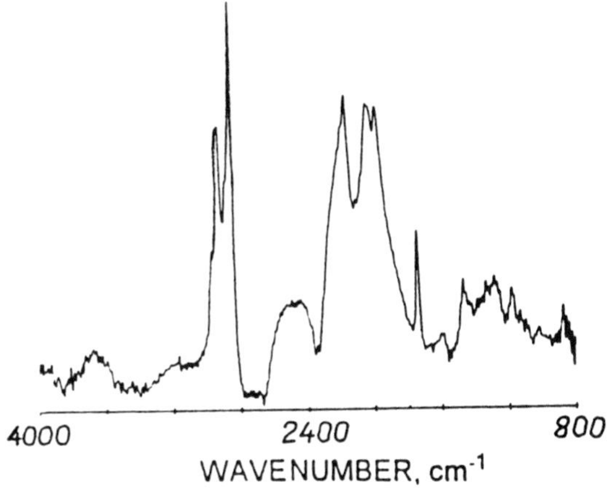

Figure 7. Absorption spectrum of a free-standing flame-grown CVD diamond film [254, fig. 5].

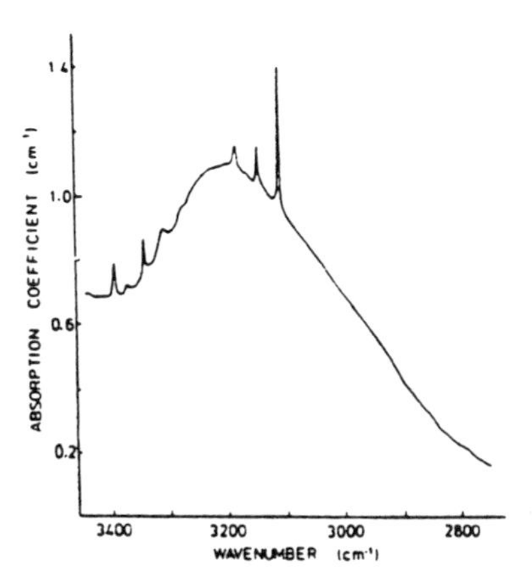

Figure 8. Absorption spectrum of type Ib diamond [183, fig. 2].

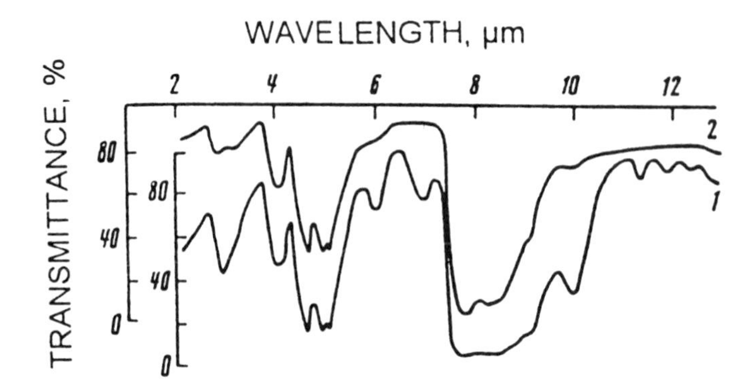

Figure 9. IR absorption spectrum of type Ia natural diamond of a gray color: 1 - of the whole crystal; 2 - of its transparent central area [3, fig. 45a].

2.740 μm (3650 cm^{-1}), a broad band with FWHM of 200 cm^{-1}, observed in single-crystal films grown in an acetylene combustion flame, attributed to a part of the three-phonon absorption of diamond lattice [254] (fig. 7).

2.831 μm (3532 cm^{-1}), a weak sharp line observed in ^{13}C hydrogen containing CVD films; in fully deuterated films the line disappears, ascribed to a hydrogen related vibration, nitrogen is not involved in the vibration [326] (fig. 6).

2.9 to 3.2 μm (3100 - 3500 cm^{-1}), several sharp lines at 3107, ~3137, 3145, ~3181, 3310, 3343, 3372, 3394 cm^{-1}, in natural diamonds showing Ib type character; the lines in the 2.9 to 3.0 μm area are probably due to N-H bond stretching vibrations [74, 183] (fig. 8).

2.9 μm (3400 cm^{-1}) and 6.1 μm (1640 cm^{-1}), observed in type Ia diamonds, ascribed to inclusions of water [3, 302] (fig. 9).

3.009 μm (3323 cm^{-1}), in CVD films, the line shifts to 3.015 μm (3317 cm^{-1}) in ^{13}C films, in fully deuterated films the line disappears, ascribed to a hydrogen related vibration, nitrogen is not involved in the vibration [326] (fig. 6).

3.030 μm (3300 cm^{-1}), triple bond configuration sp^1 C-H, observed only in a-C:H films deposited below 100°C [48].

3.091 μm (3235 cm^{-1}), a hydrogen related line, intense in the gray-violet hydrogen rich diamonds [284] (fig. 5).

3.219 μm (3107 cm^{-1}), a narrow line (fig. 10) observed in natural and CVD hydrogen-rich diamonds, may be very intense in diamonds of a light gray color, absorption strength in type Ia diamonds has been recorded up to 13 cm^{-1} [183, 284]; the line is absent in PCCVD films [48]; in ^{13}C:^{14}N diamonds the line shifts to 3.228 μm

(3098 cm^{-1}); ascribed to a carbon-hydrogen vibration in the cis-form of the di-substituted ethylene group–CH=CH– [74, 252] or in the vinilidene group >C=CH_2 (vibration of sp^2 bonds) [74, 183]; the vibration is localized probably at interfacial surfaces forming C-H bonds [130]; this vibration is interpreted as the stretch one (s), a bend vibration (b) giving a line at 1405 cm^{-1} (7.117 μm); the s and b vibrations produce the combinations 2b (first overtone at 3.589 μm (2786 cm^{-1})), 3b (second overtone at 2.399 μm (4168 cm^{-1})) and s+b at 2.223 μm (4498 cm^{-1}) [52, 130, 183, 75, 177, 284]; the line is always accompanied by the 3.228 μm (3098 cm^{-1}) weak line of 1% intensity of the 3.219 μm line, this satellite is ascribed to ^{13}C-H vibration [183, 326, 327]; there are correlated lines at 1.647 μm (6070 cm^{-1}) tentatively ascribed to the first overtone of the 3.219 μm line, at 1.701 μm (5880 cm^{-1}) ascribed to a combination band, and at 1.800 μm (5555 cm^{-1}) ascribed to the third overtone of the 7.117 μm line [74, 284].

3.202 μm (3123.6 cm^{-1}), a sharp line observed in homo-epitaxial CVD films, is proposed to be an analog of the 3.219 μm vibration occurring in natural diamonds; in ^{13}C films the spectral position is at 3.211 μm (3114.5 cm^{-1}), in fully deuterated films the line disappears, is attributed to a hydrogen related vibration involving only one carbon atom, nitrogen is not involved in the vibration [326] (fig. 6).

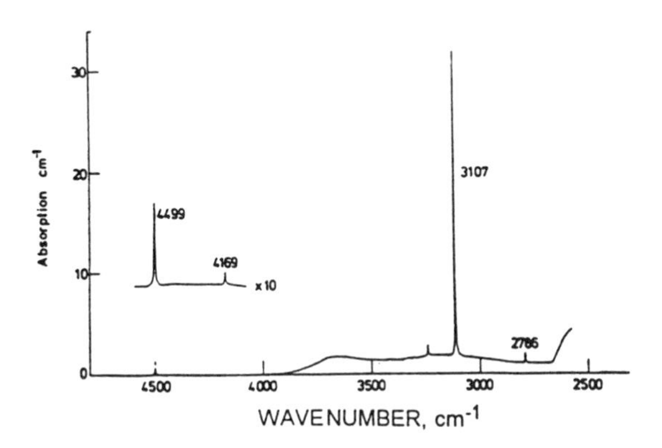

Figure 10. Absorption spectrum of a natural diamond [52, fig. 1b].

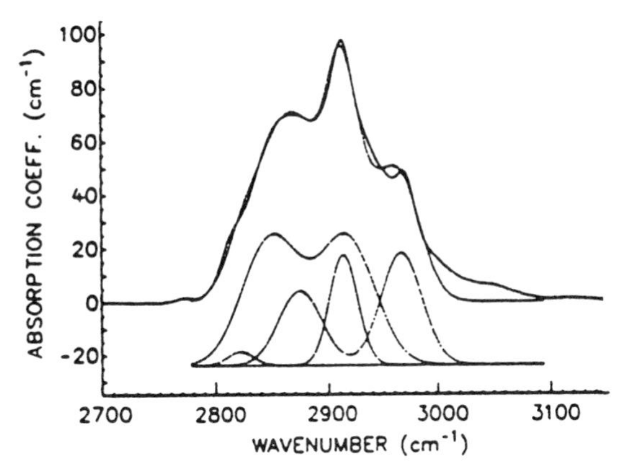

Figure 11. Absorption spectrum of a PCCVD diamond film [48, fig. 2].

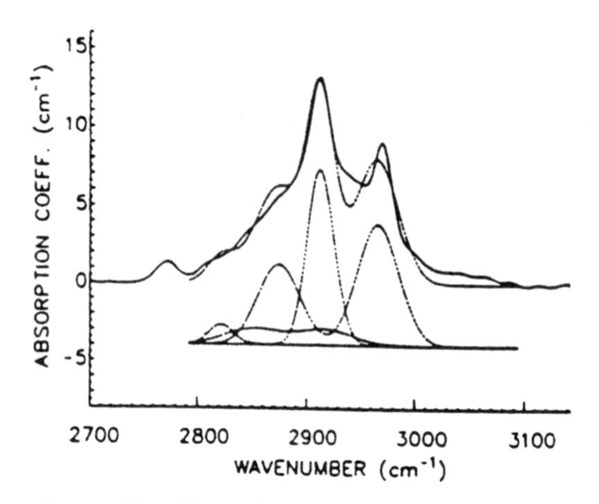

Figure 12. Absorption spectrum of a hot filament CVD diamond film grown on silicon substrate [48, fig. 2].

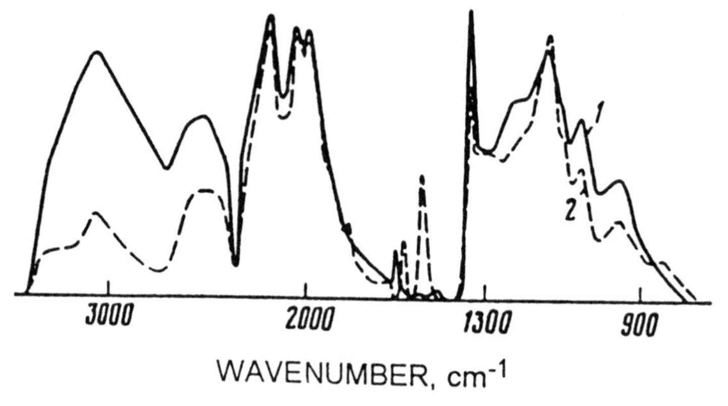

Figure 13. Absorption spectrum of a synthetic diamond irradiated with neutrons at a dose of 10^{18} cm^{-2}: as irradiated (full line), after annealing at 900°C (broken line) [3, fig. 11b].

3.228 μm (3098 cm^{-1}), see the 3.219 μm (3107 cm^{-1}) line.

3.279 μm (3050 cm^{-1}), is observed in PCCVD films, FWHM of 35 to 70 cm^{-1}, ascribed to amorphous sp^2 C-H stretch vibrations of a symmetry Cs:A [48, 63] (fig. 11).

3.3 μm (3000 cm^{-1}), a broad band from ~3300 to 2800 cm^{-1} observed in synthetic diamonds irradiated with neutrons, the band partially anneals after heating at 800°C [3] (fig. 13).

3.306 μm (2972 cm^{-1}) and 3.365 μm (3025 cm^{-1}) doublet, width of the lines of 30 cm^{-1}, observed in PCCVD films, ascribed to olef. sp^2 CH_2 configuration of the C-H stretch vibrations, symmetry C_{2V}:A_1, B_1 [48] (fig. 11).

3.333 μm (3000 cm^{-1}), FWHM of 30 cm^{-1}, observed in PCCVD films, ascribed to olef. sp^2 C–H stretch vibrations, symmetry Cs:A [48, 63].

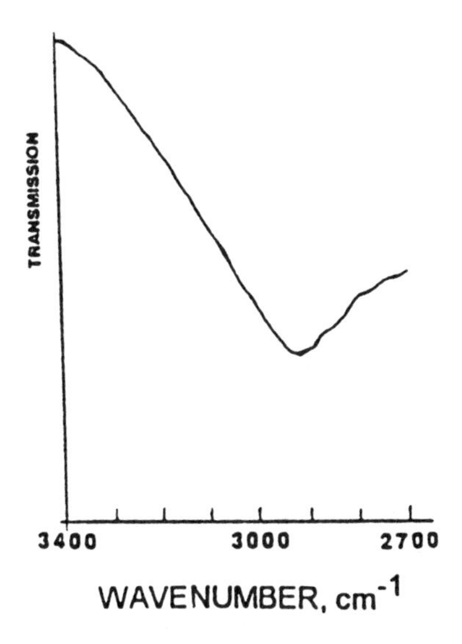

Figure 14. Transmission spectrum of a-C:H film [433, fig. 6].

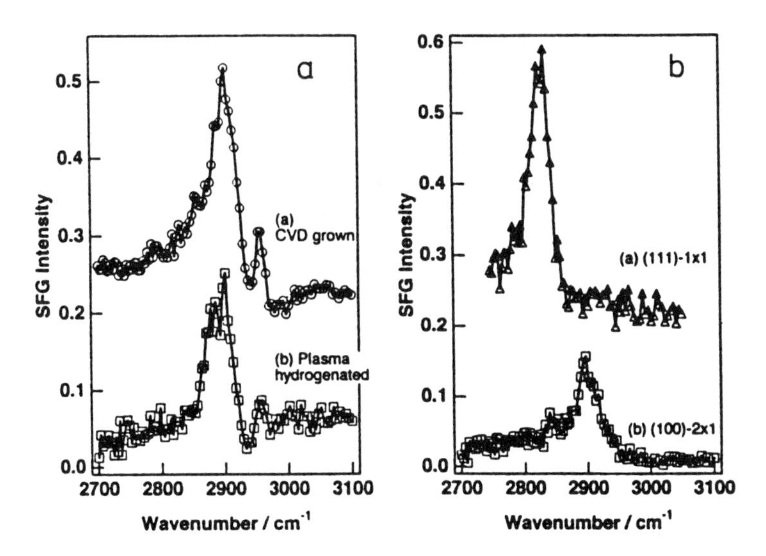

Figure 15. SFG spectra of natural diamond surfaces: (I) (100) surface; (a) homoepitaxially grown surface, and (b) plasma-hydrogenated surface; (II) plasma-hydrogenated surfaces; (a) (111)-1 ×1 surface, and (b) (100)-2×1 surface [402, fig. 2, 3].

3.425 μm (2920 cm^{-1}), a weak narrow line observed in natural hydrogen-rich diamonds of light gray color [284] (fig. 4).

3.41 μm (2930 cm^{-1}), a broad band from 2820 to 2950 cm^{-1}, observed in a-C:H films, ascribed to hydrogen bonded to sp^3 carbon (as in methane) [433, 434] (fig. 14).

3.368 to 3.382 μm (2957 to 2969 cm^{-1}), width of 25 to 85 cm^{-1}, observed in CVD films, ascribed to stretching vibrations of sp^3-bonded CH_3 asymmetric configuration [63, 256, 257].

3.384 μm (2955 cm^{-1}), FWHM of 20 cm^{-1}, observed by SFG method from the (100)-2×1 surface of epitaxially grown diamond, ascribed to C–H stretching vibrations of sp^3-hybridized bonds [402] (fig. 15).

3.421 μm (2923 cm^{-1}) and 3.504 μm (2854 cm^{-1}), FWHM of each line of 30 to 50 cm^{-1}, a doublet in a range from 2800 to 3000 cm^{-1}, the spectral position of the lines may change within ranges 2840 to 2862 cm^{-1} and 2904 to 2927 cm^{-1} respectively depending on the origin and structure of diamond, observed in CVD diamond films, the lines are ascribed to the asymmetric and symmetric stretching modes, respectively, of sp^3-bonded methylene groups $-CH_2-$, symmetry $C_{2V}:A_1$, B_1 [48, 63, 74, 254, 255 471, 473] (fig. 7); the doublet characterizes the type IIc diamond; concentration of hydrogen in type IIc diamonds can be found by [254, 457]:

$$N_H[ppm] = 10^3\ \mu_{2840cm^{-1}}[cm^{-1}]. \qquad (7)$$

3.432 μm (2914 cm^{-1}), width of 30 cm^{-1}, observed in PCCVD diamond, ascribed to sp^3 CH configuration of the C–H stretch vibrations, symmetry $C_{3V}:A_1$ [48] (fig. 11).

3.450 μm (2900 cm^{-1}), FWHM of 40 cm^{-1}, observed by SFG method on (100) diamond surface and plasma-hydrogenated (100)-2×1 diamond surface, ascribed to C–H stretching vibrations of sp^3-hybridized bonds [402] (fig. 15).

3.478 μm (2875 cm^{-1}, may change from 2862 to 2884 cm^{-1}) and 3.370 μm (2967 cm^{-1}), width of 35 to 45 cm^{-1}, observed in CVD diamond, ascribed to the C–H stretch vibrations of sp^3-bonded methyl groups $-CH_3$, symmetry $C_{3V}:A_1$, E [48, 50, 63, 74, 256, 257] (fig. 11).

3.480 μm (2870 cm^{-1}), a broad band in a range from 2820 - 2950 cm^{-1}, observed in CVD films grown at methane/hydrogen ratio above 6%, ascribed to vibrations of hydrogen bonded to sp^3 configured C [433] (fig. 16).

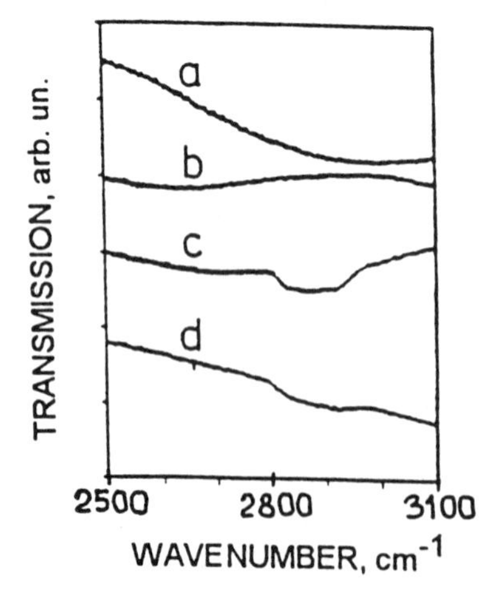

Figure 16. Transmission spectrum of CVD films grown in methane/hydrogen mixture at methane content of: (a) 1.5%, (b) 3%, (c) 6% and (d) 8% [433, fig. 2].

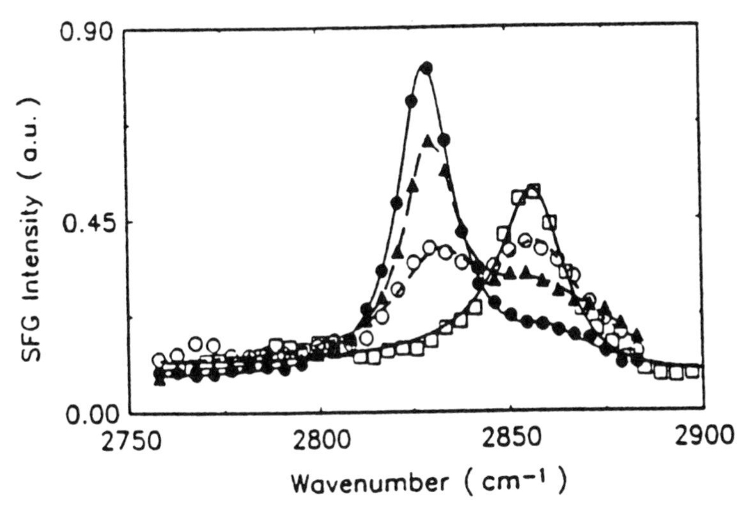

Figure 17. SFG spectra from a freshly formed (111) surface of type IIa natural diamond with different surface hydrogen coverage: 42% (open squares), 51% (open circles), 68% (solid triangles), 77% of monolayer hydrogen coverage (solid circles) [49, fig. 1].

3.497 μm (2860 cm^{-1}), FWHM ~ 25 cm^{-1}, observed by SFG method from clean (111) diamond surface dosed to a hydrogen coverage of ~ 5% when the reconstruction from (2×1) to (1×1) begins [49] (fig. 17).

3.504 μm (2854 cm^{-1}), see 3.421 μm (2923 cm^{-1}).

3.521 μm (2840 cm^{-1}), FWHM of 40 cm^{-1}, observed by SFG method from plasma-hydrogenated (111)-1×1 diamond surfaces, ascribed to C–H symmetric stretching vibrations of CH_3 species [402] (fig. 15).

3.527 μm (2835 cm^{-1}), a line of a width of 24 cm^{-1}, observed in PCCVD diamond films, proposed to be hydrogen associated with crystalline grain boundaries [63].

3.534 μm (2830 cm^{-1}), FWHM of 8 to 15 cm^{-1}, observed by SFG method from (111) surfaces of natural type IIa diamonds after covering with atomic hydrogen, ascribed to the C–H stretch mode on the ideal (111) diamond surface [49, 402] (fig. 17).

3.544 μm (2822 cm^{-1}), FWHM of 27 cm^{-1}, observed in PCCVD films, diamond C(111)-H stretch vibrations, symmetry C_{3V}:A_1, the line is interpreted as hydrogen directly bonded to the crystalline diamond at sites like dislocations, the center is formed on H-terminated unreconstructed (111) planes [48] (fig. 11).

3.589 μm (2786 cm^{-1}) see the 3.219 μm (3107 cm^{-1}) line.

3.606 μm (2773 cm^{-1}), a weak line observed in PCCVD films [48] (fig. 12).

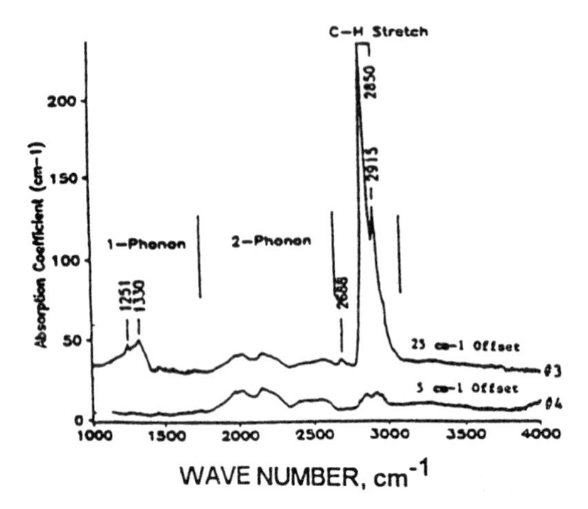

Figure 18. Absorption spectrum of a CVD diamond film grown from oxygen acetylene mixture with increased acetylene concentration at a relatively low temperature [474, fig. 2].

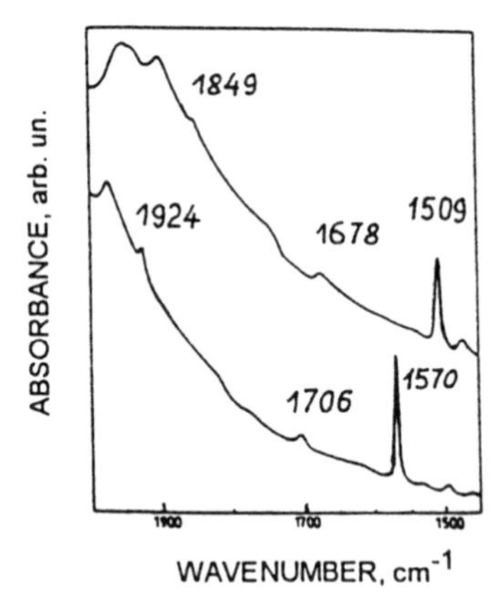

Figure 19. Absorption spectra of ^{13}C (A) and ^{12}C (B) synthetic diamonds after 2 MeV electron irradiation and annealing at 450°C [75, fig. 2].

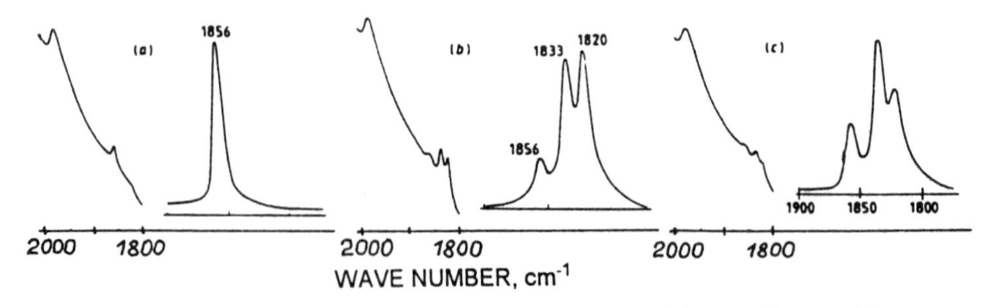

Figure 20. Absorption spectra of synthetic diamonds doped with ^{14}N (a), ^{15}N and ^{14}N in the ratio 70:30 (b), and ^{15}N and ^{14}N in the ratio 55:45 (c), all after neutron irradiation at a dose of 5×10^{17} cm^{-2} and annealing for 3 h at 1400°C [29, fig. 3].

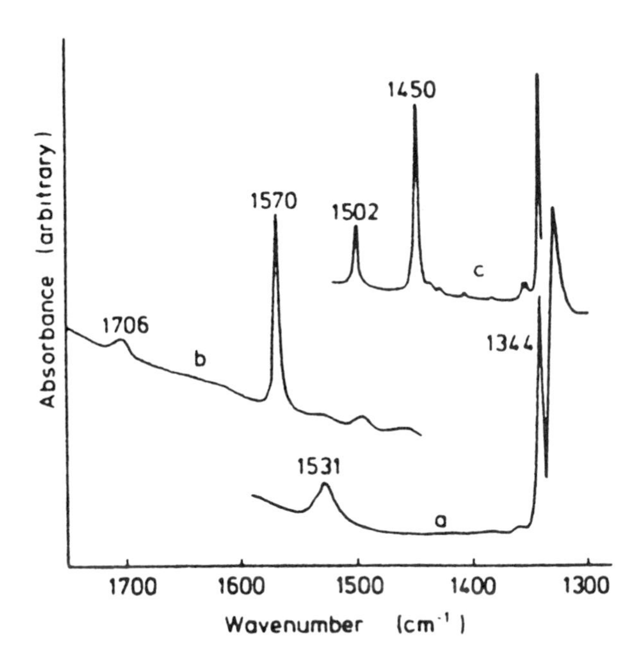

Figure 21. Absorption spectra taken from type !b diamonds after irradiation followed by annealing: (a) as irradiated, (b) annealing at 450°C, (c) further annealing at 800°C [443, fig. 8].

3.720 μm (2688 cm^{-1}), a weak narrow line observed in flame grown CVD films showing strong hydrogen related absorption [474] (fig. 18).

3.92 μm (2550 cm^{-1}), a broad band within a range from 2400 to 2700 cm^{-1} observed in synthetic diamonds, enhanced by irradiation [3] (fig. 13).

5.198 μm (1924.0 cm^{-1}), observed in heavily neutron irradiated natural diamonds after annealing at 300-400°C, the same spectral position in ^{12}C:^{15}N diamonds, shifts to 5.408 μm (1849.0 cm^{-1}) in ^{13}C:^{14}N diamonds, a vibration involving only carbon atoms, very tentatively a carbon interstitial center, this vibration (239 meV) interacts with the 5RL center [74, 75, 77, 130, 177]; ascribed to vibrations of the groups >C=C=C< [74, 130] (fig. 19).

5.388 μm (1856 cm^{-1}), FWHM of 10 cm^{-1}, observed in type IaA natural and synthetic nitrogen containing diamonds after irradiation and annealing above 500°C [74, 130, 29]; stable at temperatures above 1400°C, in ^{12}C:^{14}N+^{15}N diamonds two additional lines appear at 5.456 μm (1833 cm^{-1}) and 5.495 μm (1820 cm^{-1}); ascribed to a radiation defect involving two nitrogen atoms [74] (fig. 20).

5.750 μm (1739 cm^{-1}), a sharp peak, observed in combustion flame grown CVD films, might originate from a C=O stretching vibration [254] (fig. 7).

5.861 μm (1706.1 cm^{-1}), observed in irradiated type Ib diamonds (including synthetic) after annealing at 300 to 450°C, can be observed in diamonds exposed to very heavy neutron irradiation and annealed at 600°C; the line anneals out at 800°C [443]; the line shifts to 5.961 μm (1677.7 cm^{-1}) in ^{12}C:^{15}N diamonds and to 5.974 μm (1674.0 cm^{-1}) in ^{13}C:^{14}N diamonds, ascribed to a double bonded groups >C=N – vibration [74, 130, 177] (fig. 21).

6.369 μm (1570.3 cm^{-1}), a narrow line appears in types Ia, Ib and IIa diamonds immediately after irradiation (electrons and neutrons), is increased after annealing at temperatures 300-400°C as the 1531 cm^{-1} center anneals out, anneals out in the region 650-750°C, the line does not change the spectral position in ^{12}C:^{15}N diamonds, in ^{13}C:^{14}N diamonds the line is observed at 6.630 μm (1508.8 cm^{-1}), the line is very tentatively ascribed to a split interstitial carbon center (two carbon atoms) [3, 74, 75, 130, 176, 177, 443] (fig. 21).

6.534 μm (1531 cm^{-1}), the line appears in type Ia, Ib and IIa diamonds directly after electron or neutron irradiation, particularly strong in type Ib diamonds, the center is stable to 400°C; the line has the same spectral position in ^{12}C:^{15}N diamonds and shifts to 6.799 μm (1470.7 cm^{-1}) in ^{13}C:^{14}N diamonds, carbons only are involved in the vibration, very tentatively a nitrogen interstitial atom surrounded by carbon atoms [3, 77, 130, 176, 177, 443]; (fig. 21).

6.623 μm (1510 cm^{-1}), a narrow line appearing in synthetic and nitrogen containing natural diamonds after irradiation and annealing at 800°C, a possible origin is a complex combining interstitial nitrogen atoms bound to vacancies [3] (fig. 14).

6.658 μm (1502 cm^{-1}), a narrow line, observed only in type Ib diamonds after irradiation and subsequent annealing at 750°C, the line anneals out at 1000°C [443]; the line changes its spectral position to 6.761 μm (1479.1 cm^{-1}) in ^{12}C:^{15}N diamonds and to 6.789 μm (1473.0 cm^{-1}) in ^{13}C:^{14}N diamonds, ascribed to C–N vibration, the center is possibly another charge state of the 6.896 μm center [74, 75, 77, 130, 177] (fig. 21).

A group of lines in spectral regions from 6.752 μm (1481 cm^{-1}) to 6.944 μm (1440 cm^{-1}), from 8.834 (1132 cm^{-1}) to 10 μm (1000 cm^{-1}) and at 14 μm (700 cm^{-1}), is observed in CVD films, the lines have been ascribed to deformation modes of sp^2-hybridized C–H bonds [74, 256, 257].

6.826 μm (1465 cm^{-1}), is observed in CVD diamond monocrystalline films, has been ascribed to an H–C–H bending mode of CH_2 groups [74, 254, 255] (fig. 7).

6.895 μm (1450.3 cm^{-1}), possibly the H1a center [259]; observed in both types Ia and Ib diamonds (including synthetic) after irradiation and annealing at 750°C, intensity of the line is enhanced by 1000°C annealing [443]; the line changes its spectral position to 7.0126 μm (1426.0 cm^{-1}) in ^{12}C:^{15}N diamonds and to 7.025 μm (1423.4 cm^{-1}) in ^{13}C:^{14}N diamonds, the line is ascribed to a C–N vibration, the nitrogen atom occupies possibly an interstitial position [74, 130, 75, 176, 177]; there is a model comprising interstitial nitrogen atoms bound to vacancies [3] (fig. 21).

6.9 μm (1450 cm^{-1}), a band in a spectral regions from 6.8 μm (1470 cm^{-1}) to 7 μm (1430 cm^{-1}) accompanied by features at 9.132 μm (1095 cm^{-1}) and 11.4 μm (880 cm^{-1}), observed in type Ia diamonds, a possible origin is inclusions of carbonates [3, 302] (fig. 9).

6.993 μm (1430 cm^{-1}), a weak line observed in low nitrogen type Ib natural diamonds, naturally occurring center [3, 176].

7.042 μm (1420 cm^{-1}), a narrow line appearing in synthetic diamonds direct after irradiation, ascribed to an intrinsic interstitial type defect [3] (fig. 13).

7.117 μm (1405 cm^{-1}), see 3.219 μm (3107 cm^{-1}).

7.3 μm (1370 cm^{-1}), a sharp line the spectral position of which may vary within a range from 7.273 to 7.366 μm (1358 to 1380 cm^{-1}), the B' (or B2) aggregate (platelets); observed in nitrogen containing diamonds (fig. 22); the line shows no spectral shift in ^{12}C:^{15}N diamonds, the B' aggregates probably do not involve nitrogen [130, 166, 167, 168, 178, 183]; the precise position of the line depends on the size of the platelets: the larger the platelets, the greater the wave length; the line is accompanied by a peak at 7.0 μm (1430 cm^{-1}) and a weak broad band at 30.3 μm (330 cm^{-1}) [3]; concentration of platelets N_P can be found by [168]:

$$N_P\left[\text{Platelet area in } \mu m^2 \text{ / volume in } \mu m^3\right] = (9 \pm 2) \times 10^{-3} \int Peak\ B'\left[cm^{-2}\right] \tag{8}$$

7.353 μm (1360 cm^{-1}), a weak line naturally occurring in low nitrogen Ib natural diamonds [3].

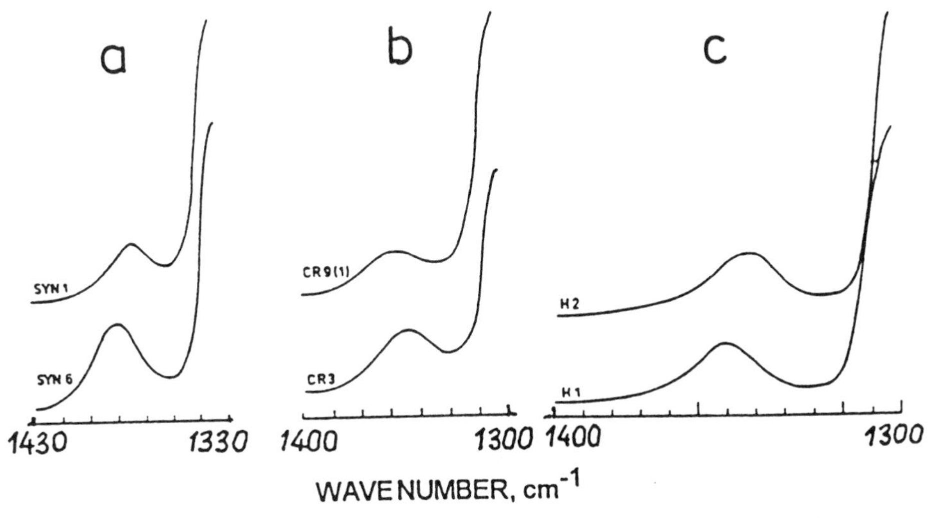

Figure 22. The B' absorption in synthetic diamond: (a) - ^{12}C:^{14}N; (b) - ^{12}C:^{14}N+^{15}N; (c) - ^{13}C:^{14}N [178, fig. 1, 2, 3].

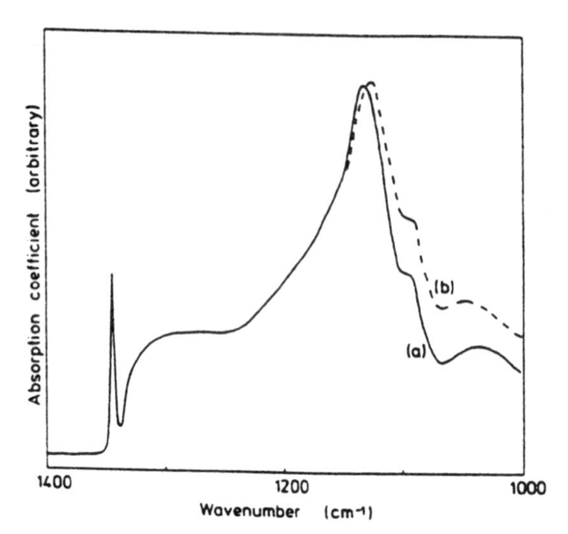

Figure 23. Absorption spectra of synthetic diamond: (a) - a ^{14}N containing diamond; (b) - a diamond containing roughly equal concentrations of ^{14}N and ^{15}N [173, fig. 1].

7.440 μm (1344 cm^{-1}), the C defect, a sharp peak associated with a local vibration at single substitutional nitrogen, the peak correlates with lines at 8.850 μm (1130 cm^{-1}) [74, 75, 130, 134, 172, 173, 177] and 9.1 μm (1100 cm^{-1}); the spectral position of the peak shifts to 7.740 μm (1292 cm^{-1}) in ^{13}C diamonds, no shift of the peak occurs in ^{12}C:^{15}N diamonds, ascribed to a vibration of only carbon atoms [76], nitrogen, if involved, is virtually in a stationary position [74] (fig. 23, 24).

7.508 μm (1332 cm^{-1}), a sharp line at the Raman frequency; induced by the B-aggregate of nitrogen, by neutron irradiation (dose of 10^{18} cm^{-2}) [3]; enhanced in synthetic diamonds grown in presence of nickel [130, 89, 112,] (fig. 169).

A continuum starting at 7.51 μm (1332 cm^{-1}), one-phonon absorption of the A and B aggregates of nitrogen [130, 140-143, 146] (fig. 168, 169).

7.553 μm (1324 cm^{-1}), the D center, a complicated spectrum in a range from 1100 to 1330 cm^{-1} with the most intense peak at 1324 cm^{-1}, observed only in diamonds containing the B defects, is tentatively attributed to nitrogen containing defects [142] (fig. 24).

7.752 μm (1290 cm^{-1}), a broad band observed in boron doped diamonds, ascribed to a vibration of two neighboring B and N atoms; the center considerably stimulates the lattice vibration at 1332 cm^{-1} due to polarization of the B–N bond [3, 298] (fig. 25).

7.800 μm (1282 cm^{-1}), the most intense peak of the one-phonon absorption of the A aggregate of nitrogen within a range from 1050 to 1330 cm^{-1} [130, 140-142] (fig. 24); naturally occurring in type Ia natural diamonds; can be produced in type Ib diamonds by radiation damage and subsequent heating to 1500°C [267, 442, 443], or by sole heating at temperatures 1700 to 2100°C [441], or during growth at temperatures between 1350 and 1500°C [443, 446, 447]. The peak shifts to 7.886

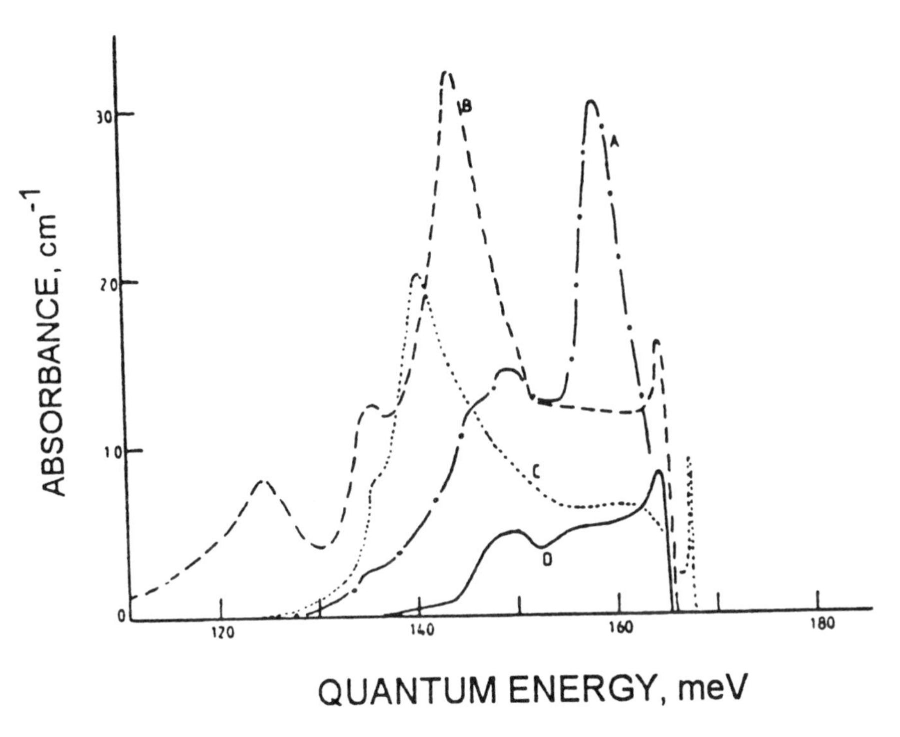

Figure 24. Absorption spectra of the A, B, C and D defects in a natural diamond with nitrogen concentration of 2×10^{20} cm^{-1} [142, fig. 3].

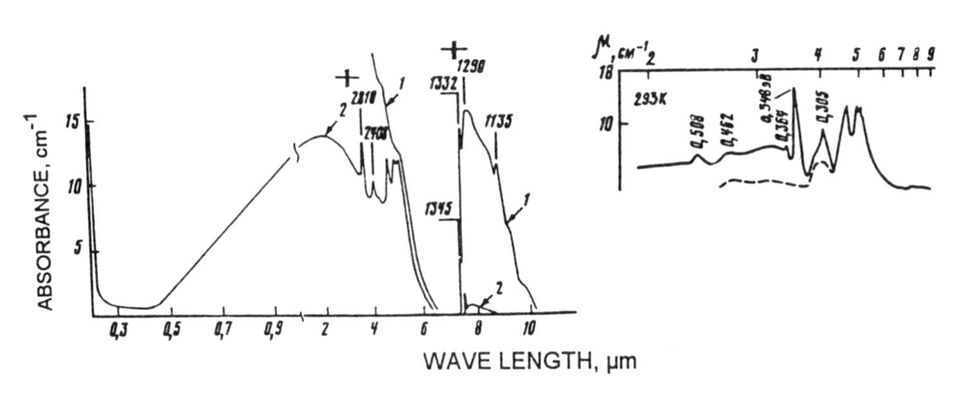

Figure 25. Absorption of a synthetic p-type diamond [3, fig. 17b].

μm (1268 cm^{-1}) in ^{12}C:^{15}N diamonds [130, 174]. Absorption coefficient μ_A of the band can reach a value of 85 cm^{-1} [3]; Nitrogen concentration in the A aggregates are evaluated using the following expressions:

$$N_A[\text{ppm}] = 17.5\ \mu_A[\text{cm}^{-1}]\ [74, 264];$$
$$N_A[\text{ppm}] = 33.3\ \mu_A[\text{cm}^{-1}]\ [268]; \qquad (9)$$
$$N_A[\text{ppm}] = 16.2\ \mu_A[\text{cm}^{-1}]\ [363];$$

The conversion coefficient between the nitrogen concentration and the strength of the 7.8 μm absorption band ranges from 10.9 to 33 atomic ppm/cm^{-1} for the A-centers [329]. The B defects also contribute in the absorption at 7.8 μm with a conversion coefficient ranging from 64 to 103.8 atomic ppm of nitrogen per 1 cm^{-1} [264, 329].

8.000 μm (1250 cm^{-1}), the F center, a complicated band in a range of 1100 to 1335 cm^{-1} with maxima at 1155, 1250 and 1332 cm^{-1}, observed in nitrogen containing diamonds [142] (fig. 26).

8.065 μm (1240 cm^{-1}), in hydrogen-rich diamonds showing type Ib character [114, 183].

8.197 μm (1220 cm^{-1}), a broad band within a range from 1550 to 900 cm^{-1} with features at 7.506 μm (1332 cm^{-1}), 5.090 μm (1100 cm^{-1}), 9.804 μm (1020 cm^{-1}) and 10.8 μm (926 cm^{-1}); observed in polycrystalline natural and synthetic diamonds containing lonsdaleite and graphite, ascribed to absorption of diamond lattice distorted by inclusions of lonsdaleite [3, 303, 304] (fig. 27).

8.251 μm (1215 cm^{-1}), a feature of the one-phonon absorption of the A aggregate of nitrogen [130, 140-142] (fig. 24).

8.3 μm (1200 cm^{-1}), a very complicated structured band in a range from 1550 to 500 cm^{-1} consisting of a number of narrow lines (the most intense are 1545, 1495, 1430, 1010, 870, 770, 710 cm^{-1} lines), observed in natural hydrogen-rich diamonds of light gray color [284] (fig. 4).

8.4 μm (1190 cm^{-1}), absorption induced by radiation damage in low nitrogen diamonds irradiated with neutrons (irradiation dose of 10^{18} cm^{-2}) [3] (fig. 28).

8.511 μm (1175 cm^{-1}). The most intense line of the B-aggregate absorption within a range from 850 to 1330 cm^{-1}. In some publications the center is labeled as B1 or B_1. The spectrum shows a shoulder at 9.124 μm (1100 cm^{-1}) and a subsidiary peak at 9.901 μm (1010 cm^{-1}) (fig. 24). The most widely used model of the B aggregate is a cluster of four nitrogen atoms surrounding a vacancy [130, 140-143, 146]. Absorption coefficient of the 8.511 μm line (μ_B) can reach a value of 40 cm^{-1} [3]. There are the following relations for evaluation of the nitrogen content in the B aggregates:

$$N_B[cm^{-3}] = 7.6\times10^{18}\ \mu_B[cm^{-1}],$$
$$N_B[cm^{-3}] = 2.4\times10^{18}\ \mu_B[cm^{-1}], \qquad (10)$$

There is a good correlation between the B aggregate absorption and the N9 and N10 centers [3]:

$$\mu_B = 0.16\ \mu_{236\ nm} = 2.66\ \mu_{240\ nm}. \qquad (11)$$

The B aggregates are naturally occurring defects in type Ia natural diamonds. These defects can be formed also in type Ib diamonds by radiation damage and subsequent heating to 2200°C [444]. The A-aggregates can be converted into the B-aggregates without radiation damage at temperatures around 2600°C [445].

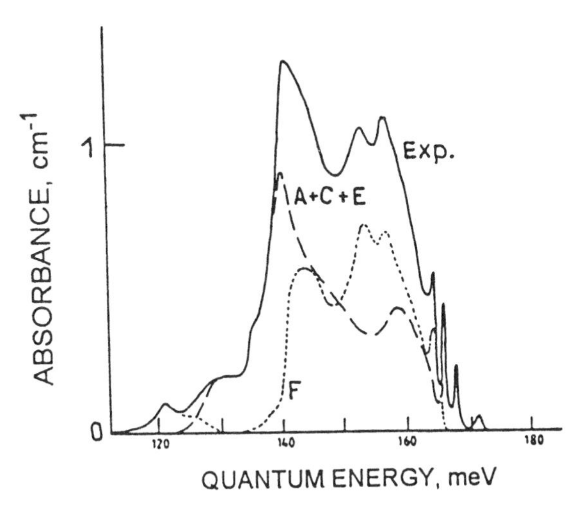

Figure 26. A decomposition of the one-phonon absorption spectrum of a diamond into the components A, C, E, and F [142, fig. 9].

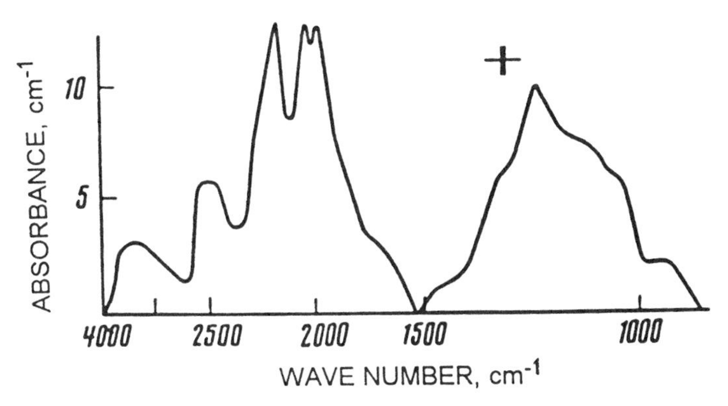

Figure 27. Absorption spectrum of a diamond containing inclusions of graphite and lonsdaleite [3, fig. 50].

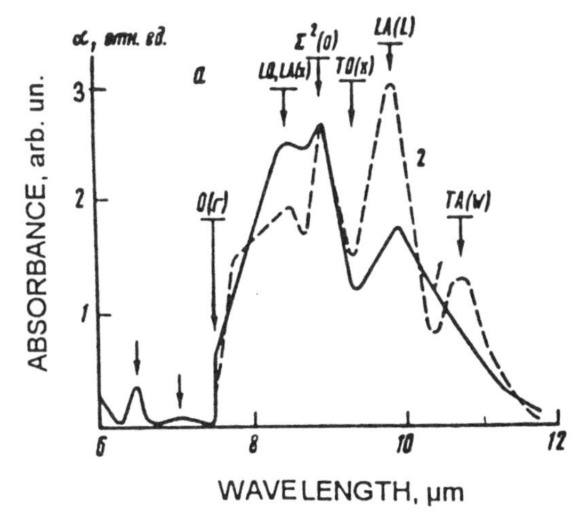

Figure 28. Absorption of type IIa (1) and type Ia (2) natural diamonds irradiated with neutrons at a dose of 10^{18} cm^{-2} [3, fig. 11a].

8.850 μm (1130 cm^{-1}), The C center. The main absorption line due to single substitutional nitrogen atoms [130, 134]. The spectrum extends from 1000 to 1330 cm^{-1}. The center is labeled as P1 in electronic paramagnetic resonance. The maximum of the band shifts to 8.969 μm (1115 cm^{-1}) in ^{12}C:^{15}N diamonds [130, 173]. The substitutional nitrogen concentration N_C can be evaluated from the absorption strength μ_C at wave length 8.850 μm by the following expressions:

$$
\begin{aligned}
N_C[ppm] &= (22 \pm 1.1)\ \mu_C[cm^{-1}]\ [74, 265], \\
N_C[ppm] &= 25\ \mu_C[cm^{-1}]\ [266, 363], \\
N_C[ppm] &= 45\ \mu_C[cm^{-1}]\ [267], \\
N_C\ [cm^{-3}] &= (\text{from } 0.7 \text{ to } 4.1)\times 10^{18}\ \mu_C\ [cm^{-1}]\ [3]
\end{aligned} \tag{12}
$$

There are the following correlations between the IR absorption and UV absorption continuum ascribed to substitutional nitrogen [3, 266, 505]:

$$
\begin{aligned}
\mu_{477\,nm} &= 1.4\ \mu_C, \\
\mu_{270\,nm} &= 45\ \mu_C, \\
\mu_{270\,nm} &= 21\ \mu_C.
\end{aligned} \tag{13}
$$

9.0 μm (1115 cm^{-1}), absorption induced by radiation damage in low nitrogen diamonds irradiated with neutrons (irradiation dose of 10^{18} cm^{-2}) [3] (fig. 28).

9.1 μm (1090 cm^{-1}). A feature characteristic of the A and B aggregates (fig. 24).

9.132 μm (1095 cm^{-1}), see 6.9 μm (1450 cm^{-1})

9.4 μm (1065 cm^{-1}), intrinsic radiation damage center appearing in neutron irradiated diamonds (irradiation dose of 10^{18} cm^{-2}) [3].

9.524 μm (1050 cm^{-1}), the E center, observed in nitrogen containing diamonds [142] (fig. 26).

9.9 μm (1010 cm^{-1}), absorption induced by radiation damage in nitrogen low diamonds irradiated with neutrons (irradiation dose of 10^{18} cm^{-2}) [3, **355**] (fig. 28).

10.42 μm (960 cm^{-1}). A complicated absorption band accompanied by features in spectral regions from 17.3 to 17.7 μm (565 - 578 cm^{-1}), 10.9 μm (920 cm^{-1}), 11.63 (860 cm^{-1}), 12.2 μm (822 cm^{-1}), 19.6 μm (510 cm^{-1}) and 20.6 μm (485 cm^{-1}). The spectrum is observed in synthetic diamonds grown with Ni-Mn, Co-Mn, Co-Fe and Ni-Fe metal-catalysts. It is ascribed very tentatively to defect complexes incorporating metal atoms [3] (fig. 29).

10.85 μm (920 cm^{-1}), radiation damage center in nitrogen containing diamonds irradiated with neutrons (irradiation dose of 10^{18} cm^{-2}) [3] (fig. 28).

11 μm (910 cm^{-1}), a band of FWHM 25 cm^{-1}, observed in synthetic diamonds grown with As; possibly relates to As containing defects [3, 298] (fig. 47).

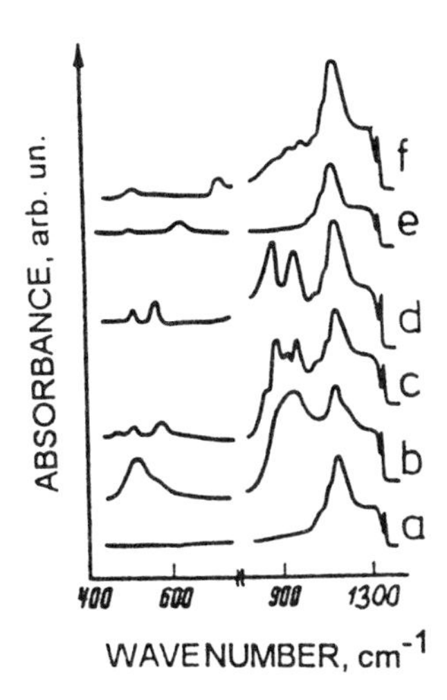

Figure 29. Absorption spectra of synthetic diamonds grown with: (a, b, c) Ni+Mn; (d) Co+Mn; (e) Ni+Fe; (f) Co+Fe [3, fig. 16].

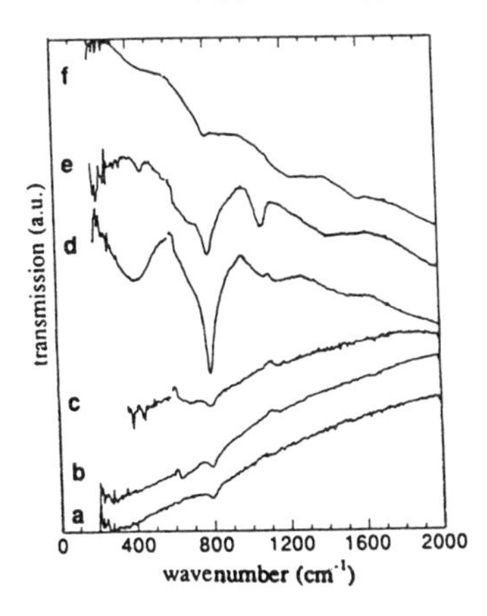

Figure 30. Transmission spectra of CVD diamond film deposited from a 0.5%CH4/H2 mixture on (a) an unscratched Si substrate, and of films deposited from (b) 0.1%, (c) 0.3%, (d) 0.5%, (e) 1.2%, and (f) 2% mixtures on scratched Si substrates [121, fig. 1].

11.4 μm (880 cm^{-1}), see 6.9 μm (1450 cm^{-1}).

12.1 μm (830 cm^{-1}) and 13.7 μm (732 cm^{-1}), observed in some type Ia diamonds. A possible origin is inclusions of nitrites [3, 302] (fig. 22).

12.5 μm (800 cm^{-1}), the most intense peak with FWHM of 60 cm^{-1} in a complicated band in a range from 600 to 1000 cm^{-1}, observed in CVD films, ascribed to stretching vibration mode of the Si–C bonds [121, 122] (fig. 30).

12.8 μm (780 cm^{-1}). A feature due to the B-aggregate of nitrogen [3] (fig. 31).

20.6 μm (484 cm^{-1}). A feature due to the A-aggregate of nitrogen [3] (fig. 29, 31).

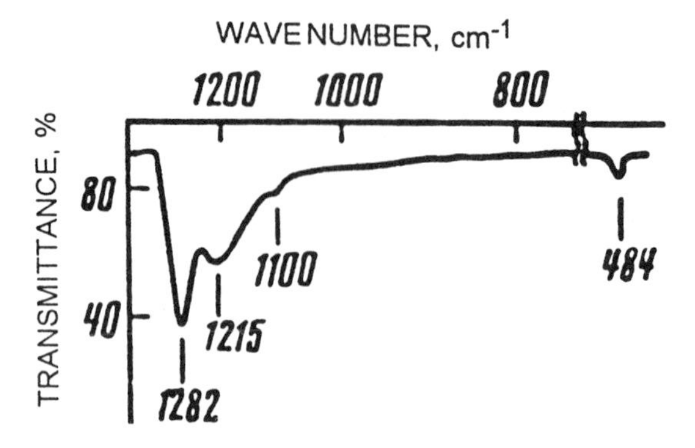

Figure 31. Absorption spectrum of type IaB natural diamond [3, fig. 2].

4.2 LATTICE PHONONS

One phonon dispersion curves and one phonon density of states for the perfect diamond crystal are shown in fig. 32, 33b [8, 9].

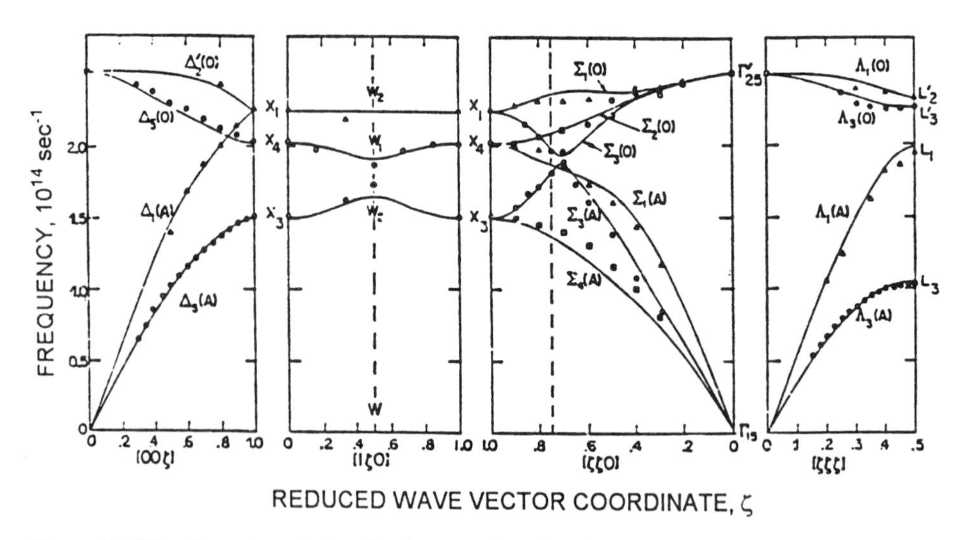

Figure 32. The dispersion relation for the normal modes of vibration of diamond in the principal symmetry directions at 296 K[8, fig. 1].

Table IV. Critical-point phonon frequencies in diamond as measured in optical absorption [4, 353, 354, 355]:

Symmetry point	Phonon branch	Phonon energy in cm^{-1} (meV)
Γ	Δ'2(O), Δ5(O)	1332.5 (165)
X	Δ'2(O), Δ1(A)	1185 (147), 1191 (148)
X	Δ5(O)	1069 (133), 1072 (133)
X	Δ5(A)	807 (100), 829 (103)
L	Λ1(O)	1242±37, 1252 (155), 1256 (156)
L	Λ3(O)	1206 (149), 1220 (151)
L	Λ1(A)	1006 (125), 1033 (128)
L	Λ3(A)	563 (70), 553 (69)
K	Σ1(O)	1230 (152), 1239 (154)
K	Σ2(O)	1109 (137), 1111 (138)
K	Σ3(O)	1045 (130), 1042 (129)
K	Σ1(A)	988 (122), 992 (123)
K	Σ3(A)	980 (121), 978 (121)
K	Σ4(A)	764 (95)
W	Z(U)	1179 (146), 1146 (142)
W	Z(M)	999 (124), 1019 (126)
W	Z(L)	908 (113), 918 (114)

Table V. Critical-point two-phonon absorption observed in diamond [353 - 356]:

Energy [meV]	Assignment
225	LO(L)+TA(L)
232	TO(X)+TA(X), Σ2O+Σ4A
244	Σ1A+Σ3A
247	L(X)+TA(X), 2Σ1A
251	Σ3O+Σ3A
253	Σ3O+Σ1A, 2TO(W)
258	Σ2O+Σ3A, 2Σ3O
262	Σ2O+Σ1A
267	Σ2O+Σ3O
270	L(W)+TO(W)
274	Σ1O+Σ3A
281	LO(L)+LA(L), L(X)+TO(X)
292	Σ1O+Σ2O
302	2TO(L)
315	2LO(L)
330	2O(Γ) (IR inactive)

5. Raman scattering

5.1 ONE PHONON RAMAN BANDS

The following Raman bands can be observed in diamond related materials (the first figures show spectral positions of the band maxima):

321 cm^{-1}, FWHM of 10 cm^{-1}. A feature observed in meteoritic carbon [453] (fig. 33a).

500 cm^{-1}, FWHM of 550 cm^{-1}. A feature of carbon films [47].

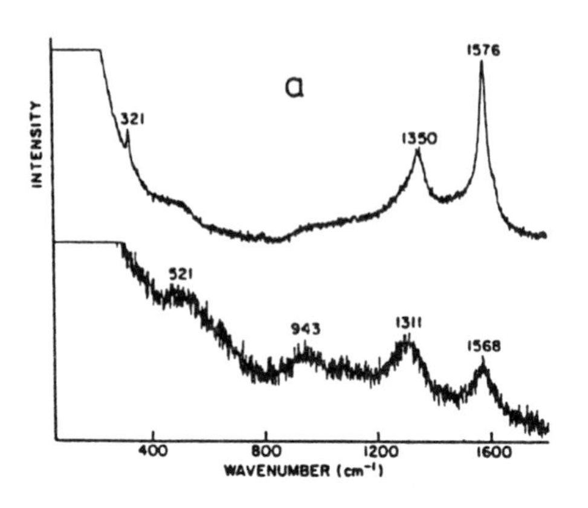

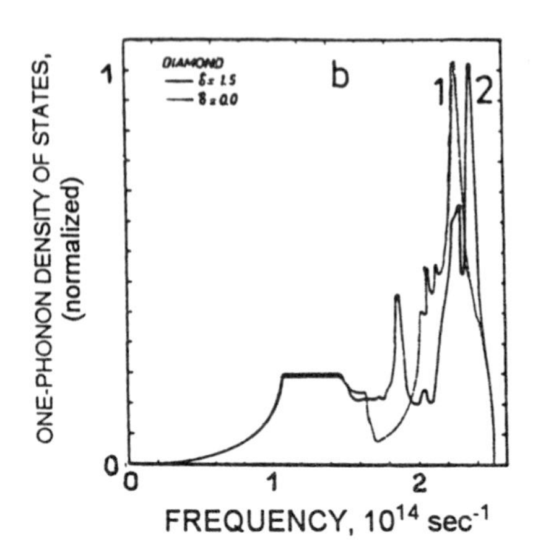

Figure 33. (a) - Raman spectra of meteoritic carbon [453, fig. 5]; (b) - one-phonon density of states of diamond calculated with different 2nd neighbor force constant parameter δ: δ = 0 (1) and δ = 1.5 (2) [9, fig. 3].

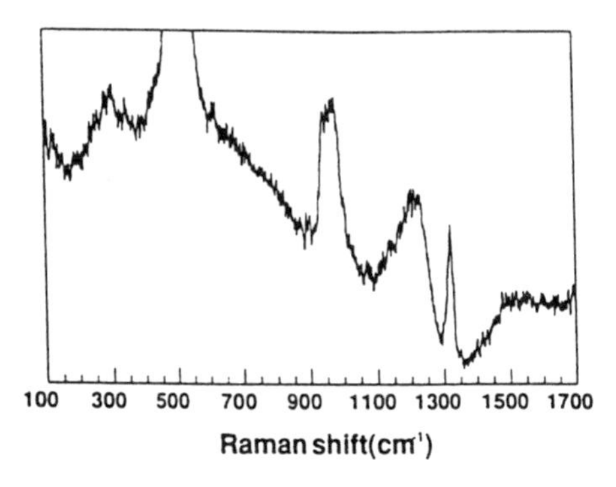

Figure 34. Raman spectrum of a MPCVD diamond film deposited on Si substrate with B:C ratio in the gas phase of 10000 ppm [276, fig. 3c].

521 cm^{-1}, FWHM of 100-200 cm^{-1}. A feature observed in meteoritic carbon [453]) (fig. 33).

550 cm^{-1}, FWHM of 70 cm^{-1}. A band observed in microwave plasma deposited CVD diamond films doped with boron at 10000 ppm. The feature is attributed to a one phonon diamond band activated due to a relaxation of the selection rules by very high boron concentration [276] (fig. 34).

943 cm^{-1}, FWHM of 150 cm^{-1}. A feature observed in meteoritic carbon [453] (fig. 33).

1132 to 1145 cm^{-1}, spectral position may vary, FWHM of 30 to 80 cm^{-1}. The width of the band increases with the increase of energy of the band maximum. The feature is observed in diamond films and is attributed to nanocrystalline diamond [46, 47, 65, 67, 68, 71, 72, 453, 454]. It is also interpreted as a disordered sp^3 bonded carbon, which can be considered as precursor of diamond formation [74, 263, 433]. (fig. 35).

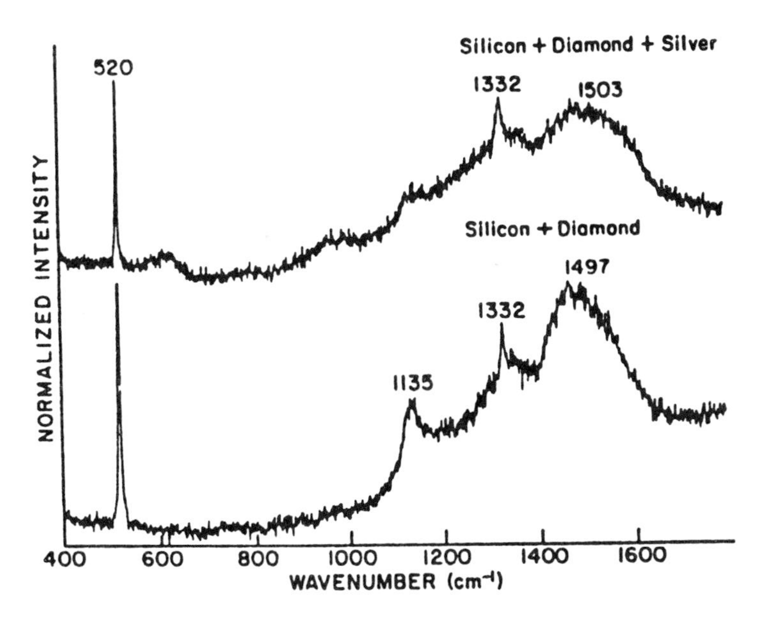

Figure 35. Raman spectrum of a thick CVD diamond film grown by hot filament method [453, fig. 3].

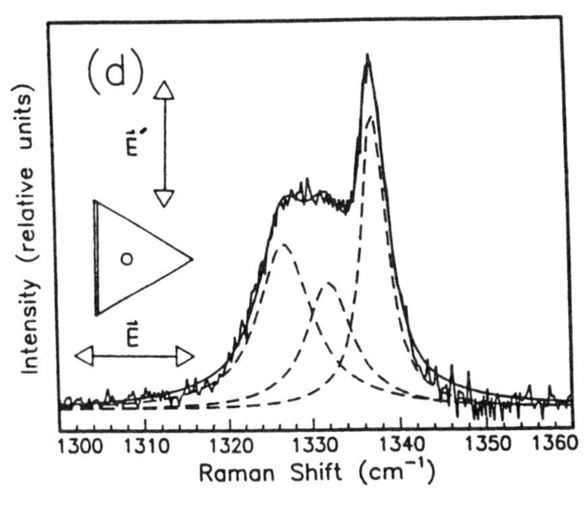

Figure 36. Raman spectrum taken from a large isolated CVD diamond single crystal with electric field vector of exciting light **E** parallel to [112] and orientation of analyzer **E**' perpendicular to [112] [54, fig. 3d].

1200 to 1300 cm^{-1}. A small peak observed in ion implanted diamond. It is ascribed to a non-diamond disordered phase [487].

1220 cm^{-1}, FWHM of 150 cm^{-1}. A band observed in microwave plasma deposited CVD diamond films doped with boron at around 10000 ppm. The feature is attributed to a one phonon intrinsic band activated due to a relaxation of the selection rules by very high boron concentration [276] (fig. 34).

1318 cm^{-1}, FWHM of 40 cm^{-1}. A band is observed in meteoritic carbon. Position of the maximum may range from 1311 to 1350 cm^{-1} and its width may increase up to 100 cm^{-1}. The feature is ascribed tentatively to lonsdaleite inclusions [453] (fig. 33).

1319 cm^{-1}. A small peak is observed from surface of cut diamonds. The feature is ascribed very tentatively to a lonsdaleite formation [74, 64].

1326 cm^{-1}, FWHM of 8 cm^{-1}. A line is observed in CVD diamond films. The feature is ascribed to stacking faults on (111) planes [54, 319, 321] (fig. 36).

1330 cm^{-1}, FWHM of 100 cm^{-1}. A band is observed in CVD diamond films. It is ascribed to intermediate carbon defects in diamond crystallites controlling the confinement length of diamond phonons [123] (fig. 37).

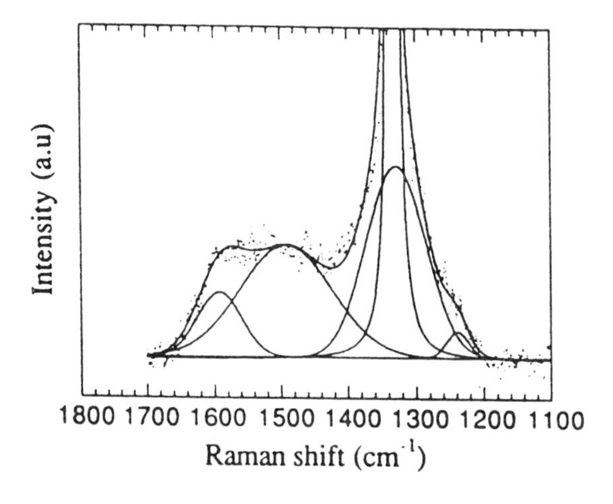

Figure 37. Decomposition of the Raman spectrum of a CVD diamond film deposited on Si substrate from a CH4/H2 mixture containing 0.1% of CH4 [123, fig. 2].

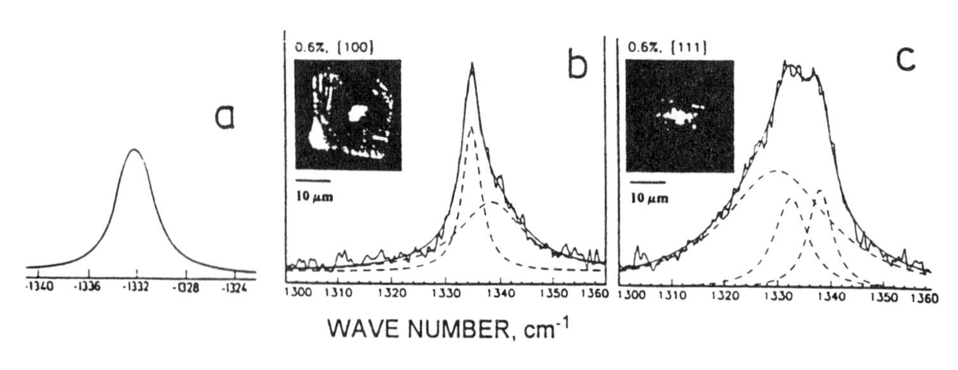

Figure 38. Raman spectra taken from a perfect natural diamond (a) and from [100] (b) and [111] (c) facets of single crystals of CVD diamond film [476, fig. 4; 319, fig. 1c, 1d].

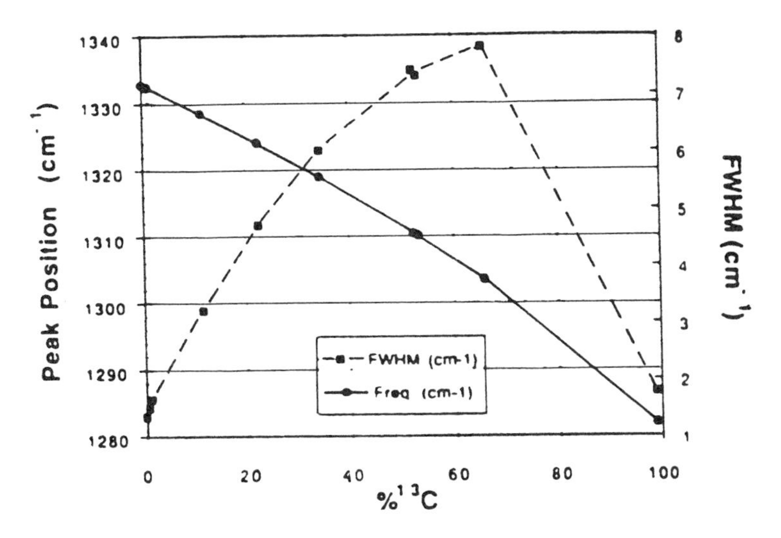

Figure 39. Peak position and FWHM for first-order Raman line as a function of isotopic composition [350, fig. 8].

1332 cm^{-1}. First order Raman spectrum of undisturbed diamond lattice (fig. 38). FWHM of the line in natural perfect diamond is of 1.5 cm^{-1}, typical FWHM in CVD diamond films is of 7 cm^{-1}. In disordered synthesized diamonds the width of the line can exceed 14 cm^{-1} [348]. In CVD isolated single crystals the line can be split into two components with separation up to 7 cm^{-1} as a result of directional strain fields [319, 321]. In highly boron doped CVD films the line shifts down to 1322 cm^{-1} [276]. The position of the line may change from 1315 to 1326 cm^{-1} in diamond powders prepared by shock-loading technique, which is explained by the presence of lonsdaleite inclusions [74, 64]. Shift of the line position with change of the ^{12}C:^{13}C isotopic composition is not linear with average mass, but departs from linearity by about 5 cm^{-1} at the middle of the range [130, 317, 350, 369] (fig. 39).

In pure ^{13}C diamond the line is found at 1280 cm^{-1} [119]. The line shifts linearly with strain: $\Delta\omega/\omega \cong -(1.0)S$, where 1.0 is the approximate mode Gruneisen parameter [14]. The splitting of the line as function of uniaxial stress along <111> and <001> directions is [130, 477, 478]:

$$\Delta\omega_{<111>} = 2.2\pm0.02 \text{ cm}^{-1}/\text{GPa},$$
$$\Delta\omega_{<001>} = 0.73\pm0.010 \text{ cm}^{-1}/\text{GPa}; \qquad (14)$$

and the hydrostatic component of the uniaxial stress gives a contribution [130]:

$$\Delta\omega_H = 3.2\pm0.2 \text{ cm}^{-1}/\text{GPa}. \qquad (15)$$

The hydrostatic shift rate of the line at pressures up to 40 GPa has a value within a range +(2.7 to 3.6)±0.3 cm^{-1}/GPa [14, 58, 130, 477-482]) and it has been measured as +1.693 cm^{-1}/GPa at pressures up to 200 GPa [406]. The energy of the line reduces with temperature [4, 61, 472] (fig. 40). The temperature dependence of the line spectral position within a range of 300 to 2000 K can be given by the expression [472]:

$$\omega[\text{cm}^{-1}] = -1.124\times10^{-5}\,(T[K])^2 + -6.71\times10^{-3}\,(T[K]) + 1334.5. \qquad (16)$$

Features observed immediately above the 1332 cm^{-1} peak are ascribed to sp^2 containing carbon materials [119, 120].

1340 to 1347 cm^{-1}, FWHM may reach a value of 330 cm^{-1}, ascribed to nanocrystalline graphite, disordered graphite or to sp^2-hybridized carbon phases [47, 65, 70].

1350 to 1360 cm^{-1}, attributed to inclusions of amorphous diamond-like carbon [58, 59, 68], disordered graphite phases [64, 234], or microcrystalline defective graphite [119].

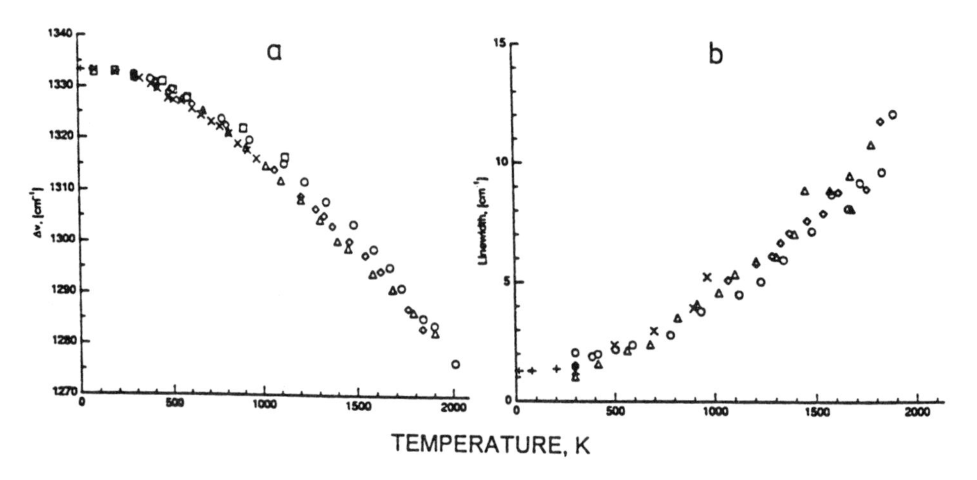

Figure 40. Peak position (a) and FWHM (b) of the first-order Raman line in natural diamond versus temperature [472, fig. 5, 8].

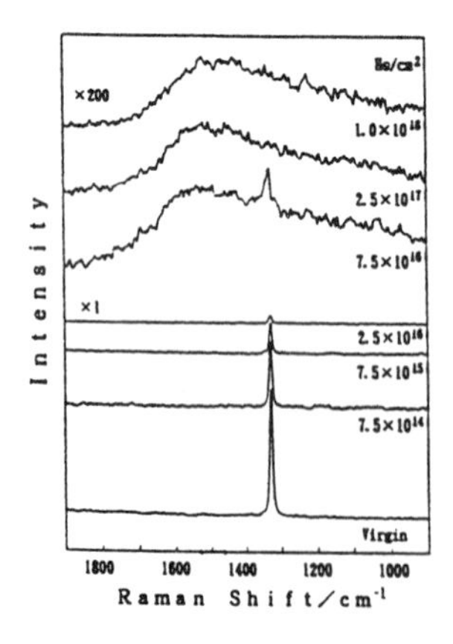

Figure 41. Raman spectra of type Ib synthetic diamonds irradiated with 25 keV He ions to different doses at room temperature [367, fig. 1].

1367 to 1470 cm^{-1}, FWHM of 70 to 80 cm^{-1}, is proposed to be a diamond precursor [47, 65, 58, 59].

1390-1420 cm^{-1}, ascribed to micro-crystalline graphite [35].

1450 cm^{-1}, amorphous carbon phase with sp^2 bonding [62].

1450 to 1480 cm^{-1}, FWHM of 35 to 60 cm^{-1}, is attributed to vibrations of polyacetylene molecules [66, 68].

1495 to 1564 cm^{-1}, FWHM of 150 to 220 cm^{-1}, observed in diamond, in diamond-like carbon films, in heavy ion implanted type IIa diamonds [123, 453, 487]; ascribed to inclusions of disordered sp^2-bonded carbon, intermediate carbon defects in diamond crystallites controlling the confinement length of diamond phonons, or amorphous diamond-like carbon [35, 39-42, 58, 59, 123, 453] (fig. 35, 37).

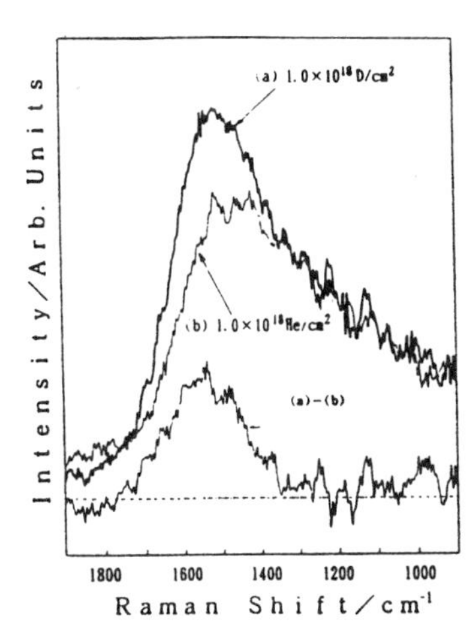

Figure 42. Raman spectra of type Ib synthetic diamonds irradiated with D and He ions both to a dose of 10^{18} cm^{-2} at room temperature and a difference between the two spectra [367, fig. 6].

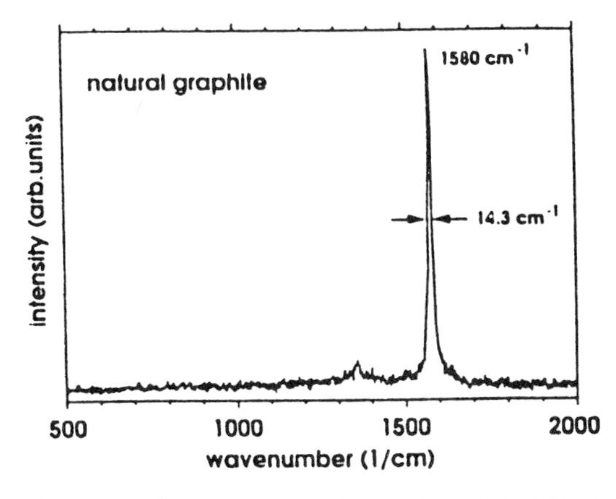

Figure 43. Raman spectra of natural graphite [47, fig. 6a].

1500 cm^{-1}, FWHM of 350 to 500 cm^{-1}, a broad asymmetric band emerging in a range from 1000 to 1700 cm^{-1}, appears after high dose implantation of light ions [367] (fig. 41).

1550 cm^{-1}, FWHM of 250 cm^{-1}, appears after high dose implantation of D^+ ions in type Ib synthetic diamonds, attributed to some disordered structure containing C–D bonds [367] (fig. 42).

1560 cm^{-1}, FWHM of 110 cm^{-1}, graphite-like inclusions consisting sp^2-hybridized carbon atoms [31, 32, 47, 65].

1568 to 1576 cm^{-1}, FWHM 35 to 80 cm^{-1}, observed in meteoritic carbon [453]; ascribed to crystalline graphite [64, 68] (fig. 33).

1580 cm^{-1}, FWHM of 14 cm^{-1}, carbon-carbon stretching mode in single crystal graphite [47] (fig. 43).

1580 to 1600 cm^{-1}, FWHM of 100 cm^{-1}, observed in as-grown and oxidized CVD diamond films and in explosive diamond nano-crystallites [219]; ascribed to defective or micro-crystalline graphite structures [35, 119, 120, 234, 262, 455] (fig. 44).

1590 cm^{-1}, disordered graphite [60, 64].

1600 cm^{-1} FWHM of 20 cm^{-1}, amorphous carbon [68];

1630 cm^{-1}, FWHM of 60 cm^{-1}, observed in ion implanted type IIa natural diamonds, ascribed to radiation point defects [487] (fig. 45).

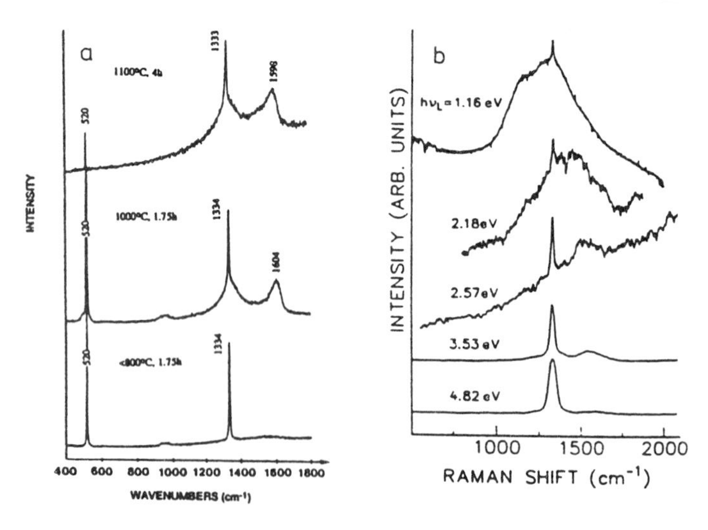

Figure 44. (a) Raman spectra of CVD diamond films deposited on Si substrate and oxidized at different temperatures in 0.25%O2 + Ar atmosphere [455, fig. 2]; (b) Raman spectra of a PC CVD diamond film excited at different photon energies indicated in the figure [344, fig. 3]

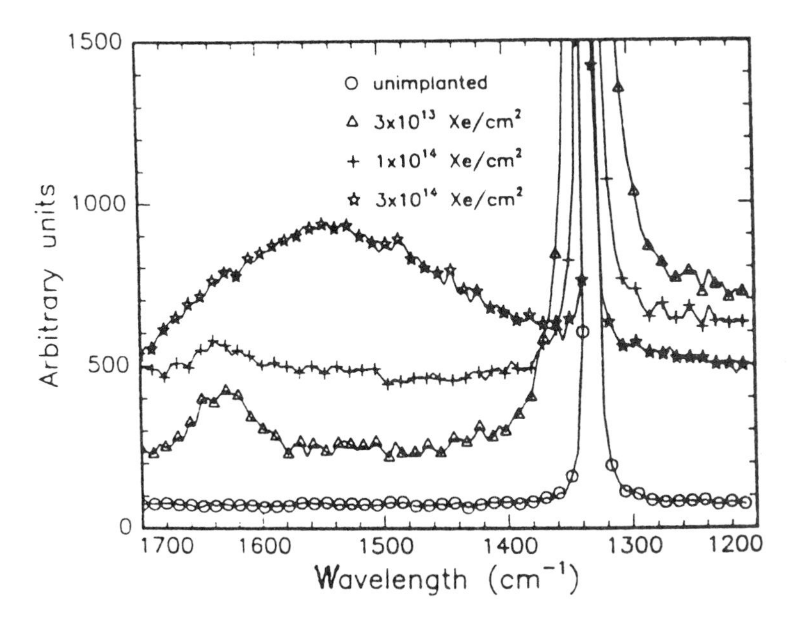

Figure 45. Raman spectra of type IIa diamond unimplanted and implanted with 320 MeV Xe ions to different doses [487, fig. 1].

5.2. TWO-PHONON RAMAN SCATTERING

Typical two phonon Raman spectrum of undisturbed diamond lattice is presented in fig. 46.

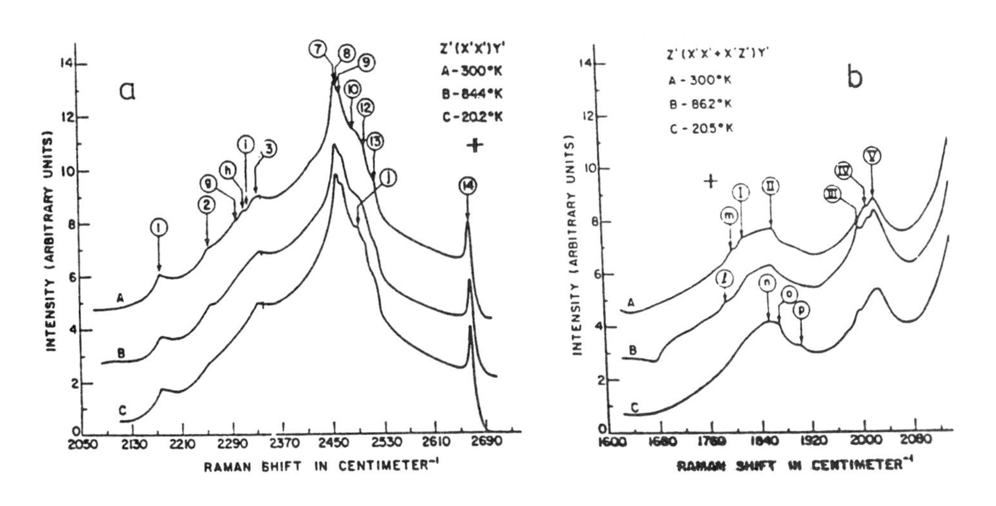

Figure 46. Two-phonon Raman spectra of diamond: (a) Z'(X'X')Y' scattering; (b) Z'(X'X'+X'Z')Y' scattering [4, fig. 3, 6, 8].

Table VI. Allowed Two Phonon Transitions in Raman Scattering [6, 7].

Symmetry point	Phonons
Γ	2O
L	2TO
L	2LO
L	2LA
L	2TA
L	TO+LA
L	LO+TA
X	2TO
X	2L
X	2TA
X	TO+L
X	TO+TA
X	L+TA
W	2TO
W	2L
W	2TA
W	TO+L
W	TO+TA
W	L+TA

Table VII. Pressure induced shifts of the two-phonon bands [130, 480].

Spectral position, cm^{-1}	Assignment	Shift rate, cm^{-1}/GPa
1864	$TA(X^3)+TO(X^4)$	5.6±0.8
2178	$LO(W^2)+TO(W^1)$	8.7±0.5
2256	$LO(X^1)+TO(X^4)$	8.3±0.5
2333		8.1±0.5
2360		6.2±0.8
2370	$2LO(X^1)$	6.9±0.8
2460	$LO(L^{2-})+TO(L^{3-})$ or $2O(\Sigma^1)$	7.4±0.5
2467		8.5±0.5
2491		6.6±0.6
2501	$2LO(L^{2-})$	7.2±0.5
2519		8.6±0.6
2667	$2O(\Gamma^{25+})$	7.1±0.4

5.3 MISCELLANEOUS

In resonant Raman scattering two components A (the main line at 1510 cm^{-1}) and B (the main line at 1430 cm^{-1}) have been found, the B component is strongly excited in a wavelength range 476.5 to 514.5 nm (2.41-2.6 eV), there is a splash of both centers after annealing at 1300°C, centers anneal out at 2200°C [342, 343].

Resonant Raman scattering in CVD films of a different origin is presented in [344]. Amorphous carbon inclusions are detected most sensitively at low quantum energy excitation, e. g., 1.16 eV. In contrast the diamond Raman line (1332 cm^{-1}) is better resolved for high quantum energy excitation, e. g., 4.82 eV (fig. 44b).

The defects composed of sp^3 bonded carbon atoms are revealed in diamond Raman scattering at frequencies < 1332 cm^{-1}. In contrast, the sp^2 (graphitic) bonded species are revealed at > 1332 cm^{-1}. The reason is that the sp^2 carbon bonds are stronger than the sp^3 carbon bonds [451].

The relative Raman cross-section for diamond to graphite is of 1:50 [452]. The relative intensity for diamond(sp^3) to graphite(sp^2) bonded carbon in CVD films is 1:75, that is the scattering due to graphite is 75 times stronger than that due to diamond [120].

A sharp peak at 520 cm^{-1} observed in Raman spectra of diamond films deposited on silicon substrates is due to the substrate (see fig. 44).

6. Optical Electronic Transitions

6.1 OPTICAL BANDS

The Mössbauer-type optical centers are given by energies of their zero-phonon transitions. The broad bands are given by spectral positions of their maxima or by the spectral position of their onset at longer wave lengths. If not indicated, all the spectral positions are given for liquid nitrogen temperature. The technique used for observation (A, CL, etc.) is stated after the energy of the transition.

0.30 eV (4132 nm), 0.351 eV (3534 nm), 0.507 eV (2448 nm), A, in synthetic diamonds grown with As impurity, possibly an As-related donor center [3,298].

0.37 eV (3350 nm), A, EA; in some papers the center is called the B center; a continuum absorption starting at about 0.370 eV and extending to about 2.2 eV (fig. 25, 48, 49); observed in natural type IIb diamonds, HTHP synthetic diamonds and diamond films doped with boron, and B+ ion implanted diamonds [40, 483]. There are maxima at 0.304, 0.348, 0.363, 0.462, 0.508, 0.625 0.670 and 0.830 eV.

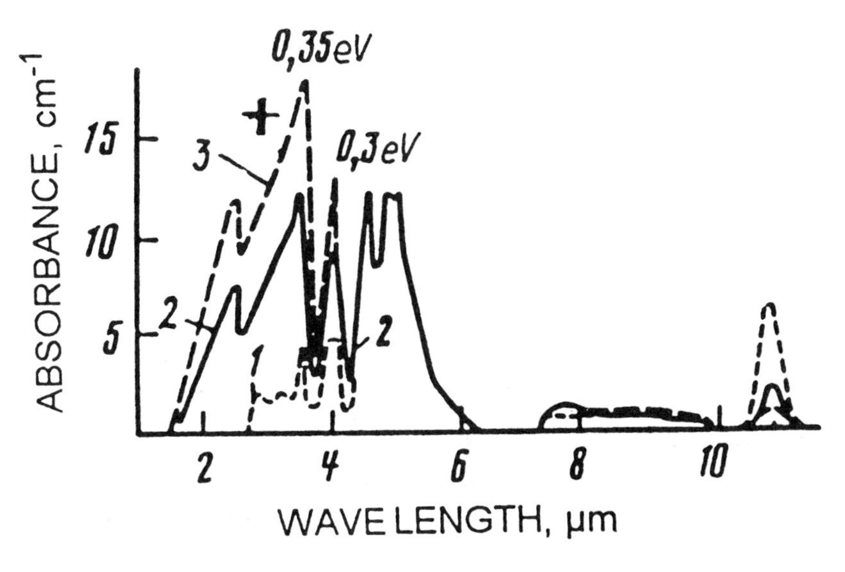

Figure 47. Absorption spectra of n-type synthetic diamonds grown with As [3, fig.17].

Table VIII. Optical transition from the ground to excited states of boron acceptor showing spin-orbital splitting [464, 466].

Electronic transition*	Energy, eV [466]	Energy, eV [464]
II-2[Γ8(g)-Γ8(4)] and I-2[Γ7(g)-Γ6(4)]	0.3042±2	0.3046
II - 3[Γ7(g) - Γ8(3)]	0.3356±1	0.3360
I - 3[Γ8(g) - Γ8(3)]	0.3373±1	0.3377
-	0.3404	0.3408
II - 4[Γ7(g) - Γ8(5)]	0.34148	0.3418
-	0.34210	0.3425
I - 4[Γ8(g) - Γ8(5)]	0.34353	0.3439
-	0.3451	0.3456
-	0.34637	0.3473
II - 5[Γ8(g) - Γ6(1)]	0.34710	0.3497
I - 5	0.34913	0.3528
II - 6	0.3524	0.3552
I - 6	0.35456	0.3562
II - 7	0.35579	0.3585
I - 7	0.35789	0.3628
II - 8	0.36273	0.3653
I - 8	0.3646	

*where I and II denote the ground states and the figures 2 to 8 denote the excited states.

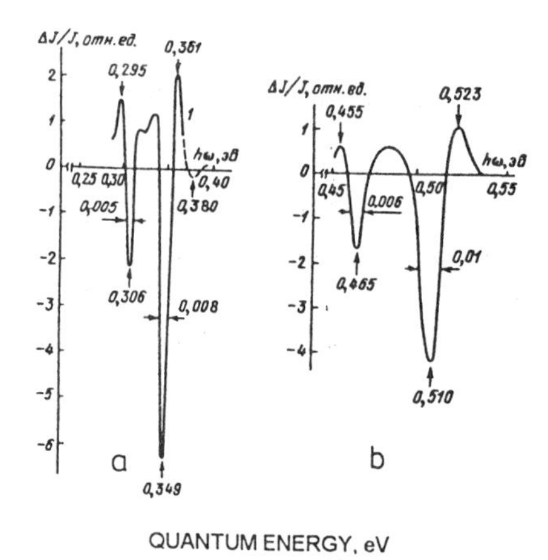

Figure 48. EA spectra of a natural diamond implanted with boron ions: (a) - zero-phonon electronic transitions; (b) - electronic transitions accompanied by emission of 160 meV TO phonons [483, fig. 3.3].

The bands at 0.348 and 0.363 eV exhibit fine structure at low temperature. All the bands in a range of 0.300 to 0.365 eV exhibit doublet structures with an equal splitting (probably spin-orbital splitting) of 0.002 eV, the components being thermolized [466, 467] (see Table VIII). The center is ascribed to boron acceptors. The features of energies less than 0.37 eV are due to transitions of the bound holes between the ground state and various excited states of the acceptor center [246]. The more energetic peaks are transitions of the bound holes with emission of one or more lattice phonons of energy 159 meV. The 0.830 eV peak is the three phonon replica; Huang-Rhys factor of the electron-phonon coupling is S = 0.18±0.02 [130, 246, 247]. A step at 0.530 eV is believed to be due to transitions assisted by the $LO_{k=0}$ phonon [130, 246]. More than 30 peaks in an interval from 0.330 to 0.371 eV are interpreted as transitions to excited states, some of the lines having stress induced origin [248-250]. There is boron related absorption in the one-phonon spectral region (below 0.165 eV), the integrated absorption of which is about 0.02 of the integrated absorption of the 0.348 eV peak [130, 246]. There are weak peaks at 0.266, 0.270 and 0.290 eV assigned to excited state transitions [130, 247, 251]. The lines at 0.304, 0.348, 0.363, 0.462 and 0.508 eV are especially well detected by EA technique showing simultaneously Stark-effect on the 0.304 eV line [483, 484]. Concentration of uncompensated boron acceptors can be calculated by the formulae [3, 299, 300, 457]:

$$\left(N_A - N_D\right)[ppm] = 4.45x10^{-3} \int_{0.325meV}^{0.360meV} A(E) \cdot dE[meV / cm] \quad (17)$$

$$\left(N_A - N_D\right)[cm^{-3}] = 0.63x10^{16} I_{0.348eV} = 0.54x10^{14} \mu_{0.348eV} \quad (18)$$

where I is the integrated line intensity.

0.6124 eV (2024 nm), A, ZPL of the H1b center, observed in type Ia diamonds after radiation damage and annealing at above 1000°C; formed by trapping the 2.085 eV center on the A aggregates of nitrogen; symmetry monoclinic-I; very weak vibronic side band, S ~ 0.1 [74, 84, 259, 485]; no isotopic shift of the ZPL in ^{13}C diamonds [130] (fig. 50).

0.6408 eV (1934 nm), A, ZPL of the H1c center; appearance as for the H1b center except that the 2.085 eV center is trapped at the B aggregates of nitrogen [74, 84, 485] (fig. 50).

0.6906 eV (1795 nm, 5572 cm^{-1}), A, FWHM is around 50 cm^{-1}, observed in homoepitaxial CVD diamond films, no shift of the line in ^{13}C diamonds, in ^{12}C:D diamond films the line shifts to 0.6754 eV (5449 cm^{-1}), in ^{12}C:50%H:50%D diamond the line splits in three components at 0.6796, 0.6835 and 0.6871 eV (5483, 5514, 5544 cm^{-1}); ascribed to an electronic transition at a complicated hydrogen containing center [326] (fig. 6).

0.77 eV (1610 nm), the amber center, A, most readily occurs in natural type Ib diamonds; interacts with vibrations of energy 83 and 89 meV, the center shows a continuous absorption up to at least 3.5 eV superimposed by two bands at 2.2 and 3.3 eV, the center shows also a peak at 0.52 eV [104].

0.8515 eV (1456 nm, 6870 cm^{-1}), A, FWHM is around 50 cm^{-1}, observed in homoepitaxial CVD diamond films, in ^{12}C:[50%H:50%D] diamond the line splits in two components at 0.8470 and 0.8524 eV (6833 and 6877 cm^{-1}), in ^{13}C diamond the line shifts to 0.8517 eV (6873 cm^{-1}), ascribed to electronic transition at a hydrogen related center [326] (fig. 6).

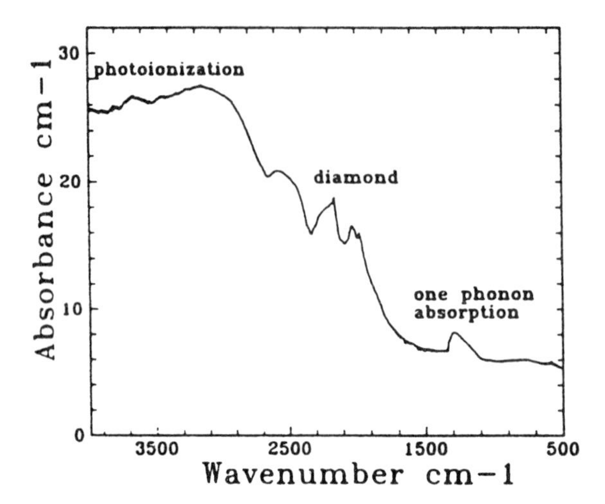

Figure 49. Absorption spectrum of a heavily boron doped homoepitaxial CVD diamond [457, fig. 9.13].

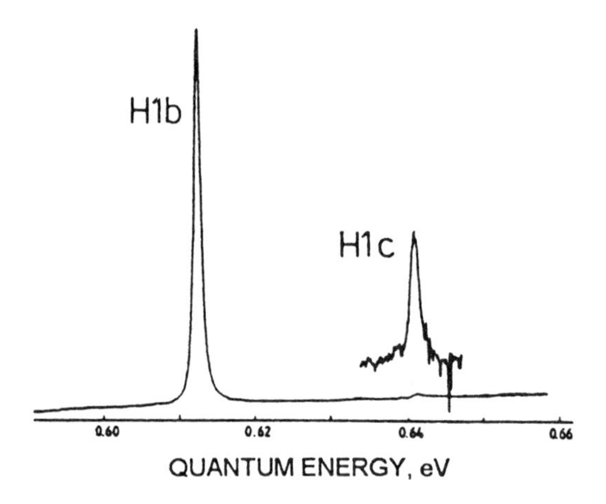

Figure 50. Absorption of a synthetic diamond heat treated at 2150°C and then irradiated with 2 MeV electrons and annealed at 1100°C [84, fig. 6].

0.8972 eV (1382 nm, 7238 cm^{-1}), A, FWHM is around 50 cm^{-1}, observed in homoepitaxial CVD films, in ^{13}C diamond the spectral position of the line is 0.8974 eV (7240 cm^{-1}), in ^{12}C:D films the spectral position is 0.8913 eV (7191 cm^{-1}), in ^{12}C:[50%H:50%D] diamond the line splits in two components at 0.8913 and 0.8972 eV (7191 and 7238 cm^{-1}), ascribed to electronic transition at a hydrogen related center [326] (fig. 6).

0.9130 eV (1358 nm, 7366 cm^{-1}), A, observed in homoepitaxial CVD films, FWHM is around 50 cm^{-1}, in ^{12}C:D diamond the spectral position is 0.9078 eV (7324 cm^{-1}), in ^{13}C diamond the line shifts to 0.9145 eV (7378 cm^{-1}), in ^{12}C:[50%H:50%D] diamond the spectral position is 0.9105 eV (7346 cm^{-1}), ascribed to electronic transition at an hydrogen related center [326] (fig. 6).

1.053 eV (1177 nm), PL, ZPL, observed at low temperatures in type Ia diamond [438].

1.2 eV (1030 nm), CL, a broad band, FWHM of 0.33 eV, observed in type IaA diamonds [13, 169].

1.2 eV (1030 nm), CL, in all types of natural diamonds, the center shows a strong electron-phonon coupling, S ~ 9 at 50 K, interacts with vibration of energy 53 meV [13, 169].

1.22 eV (1016 nm), A, ZPL, the line consists of up to eight overlapping components: 1.21148 eV (FWHM 1.46 meV), 1.21327 eV (FWHM 1.9 meV), 1.21554 eV (FWHM 1.13 meV), 1.21753 eV (FWHM 3 meV), 1.2192 eV (FWHM 2.67 meV) - the main line, 1.2255 eV (FWHM 3.82 meV), 1.2280 eV (FWHM 5.22 meV) and 1.23018 eV (FWHM 1.11 meV), observed in synthetic diamonds grown by the temperature-gradient method using nickel catalyst and nitrogen getter, the center interacts with vibrations of energies 64, 113, 152 and 164 meV, the center is dichroic, the unique peculiarity - two mutually perpendicular dipole moments exist at the center, there is a correlation between the decay of the 1.40 eV nickel related center and strengthening of the 1.22 eV center, the center is exclusively confined to the {111} growth sectors, the center is proposed to be a different charge state of the 1.40 eV nickel center [130, 89, 112, 179] (fig. 51).

1.250 eV (991.8 nm), CL, ZPL, appears in natural diamonds implanted with Ti ions [288, 289] (fig. 52).

1.25 eV (992 nm), CL, PL, FWHM of 0.3 eV, a band possibly related to the giant B' platelets; luminescence is strongly polarized with the electric vector lying in the plane of the platelets [74, 130, 169, 438] (fig. 53).

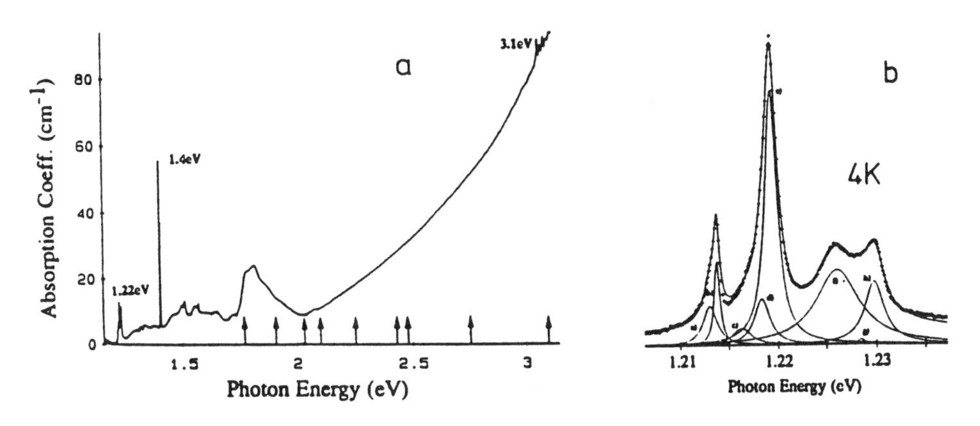

Figure 51. Absorption of a typical brown synthetic diamond grown using Ni with a nitrogen getter (a) and the fine structure of the 1.22 eV ZPL recorded at4 K (b) [179, fig. 1, 2].

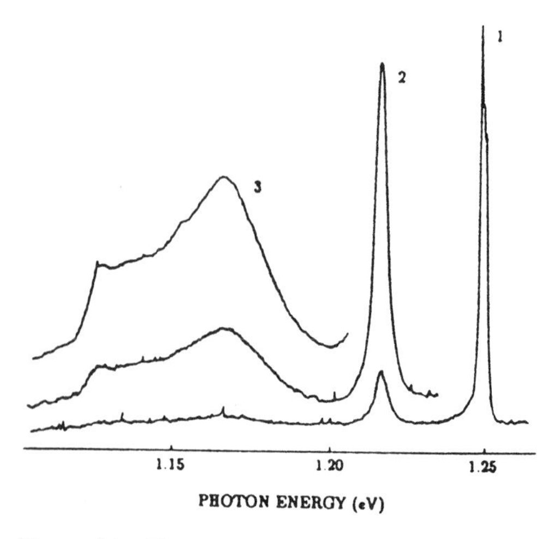

Figure 52. CL spectra of a Ti-implanted natural diamond subsequently annealed at above 1400°C [288, fig. 1].

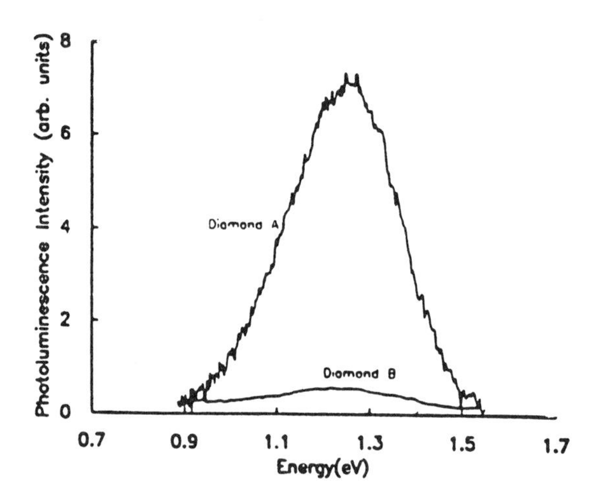

Figure 53. Ar laser excited luminescence at 300 K in type Ia diamonds [438, fig. 8].

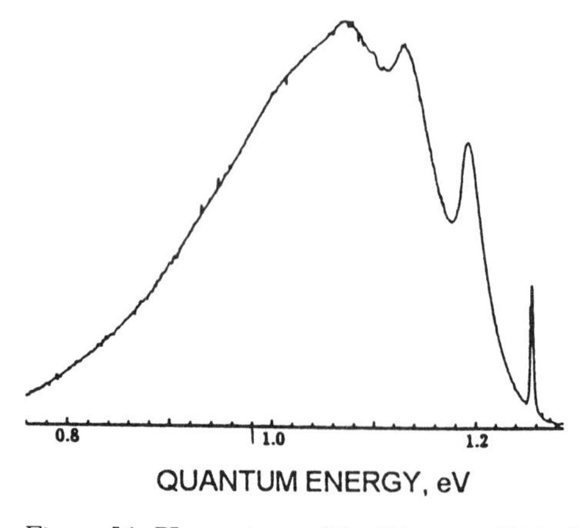

Figure 54. PL spectrum of the H2 center [253, fig. 1].

1.256 eV (986.3 nm), A, PL, ZPL, the H2 center, is not excited in CL [443]. The center appears in synthetic diamonds after radiation damage and 1700°C annealing. It can be also observed in some natural type I diamonds after irradiation and 500 to 600°C annealing, lower or higher annealing temperatures do not result in the center formation. In natural diamonds the center is stable up to 900°C. A modulation of intensities of the H2 and H3 centers is achieved by optical bleaching [224]. Symmetry of the H2 center is rhombic-I. In absorption the center interacts with LVM of an energy 167.1 meV ascribed to a carbon-carbon vibration [253,450]. The ZPL shifts in ^{13}C diamonds by +0.93 meV [130]. The center is accompanied by a LVM transition at 1.424 eV [253] and it interacts with a vibration of energy 70 meV, S ~ 4 [13]. A possible model is two nitrogen atoms sandwiching a vacancy along (110) direction (the negative charge state of the H3 center) [130, 177, 224] (fig. 54).

1.264 eV, CL, ZPL, observed in natural diamonds, interacts with 60 meV vibrations, S ~ 1 [13].

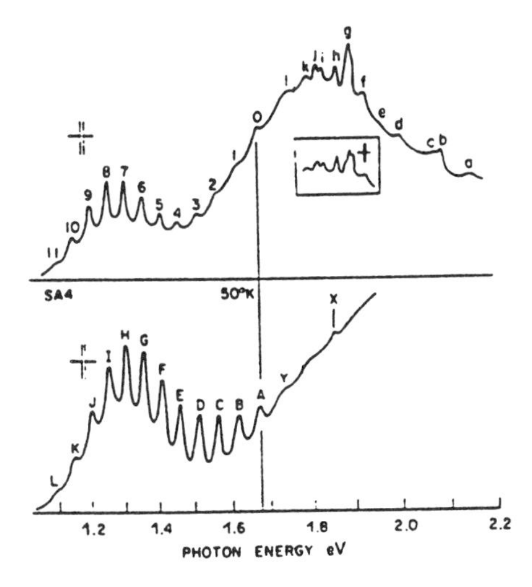

Figure 55. CL spectrum of the C and B bands in type IIa diamond [169, fig. 11].

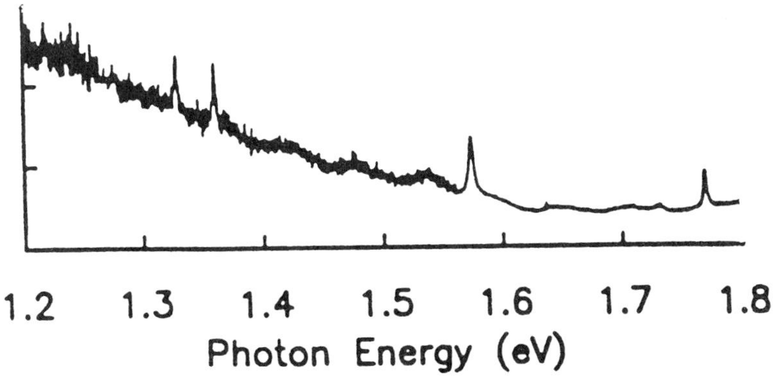

Figure 56. Ar laser excited luminescence in type Ia diamond at 17 K [438, fig. 9].

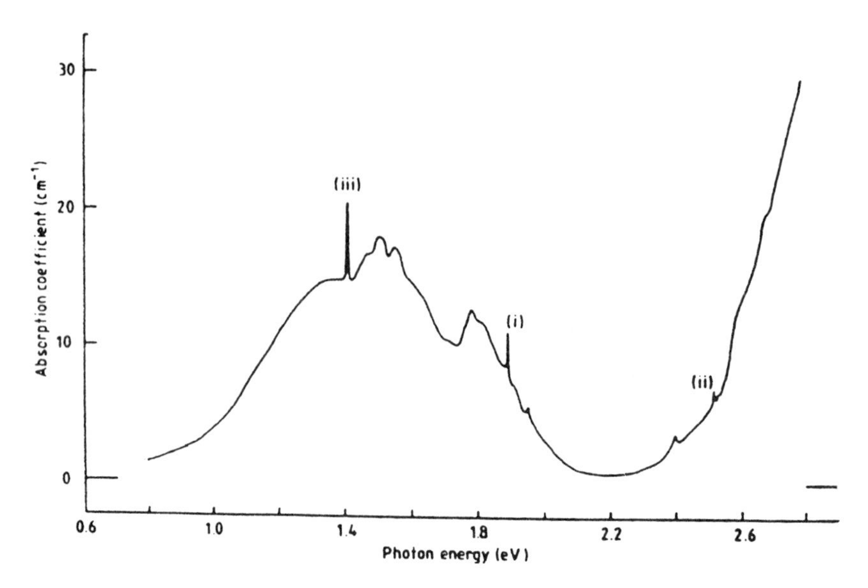

Figure 57. Absorption of a synthetic diamond, grown using a nickel catalyst-solvent [112, fig. 2].

1.277 eV (970.5 nm), CL, ZPL, observed in natural diamonds implanted with Ti ions and subsequently annealed at 800°C [289].

1.3 eV (950 nm), CL, the C band, A broad structured band naturally occurring in all except type Ib diamonds. The center interacts with 53 meV vibration, S ~ 7 to 8. The ZPL is expected to be at around 1.8 eV [104, 169] (fig. 55).

1.328 eV (933.4 nm), PL, ZPL, observed at low temperatures in type Ia diamonds [438, 440] (fig. 56).

1.360 eV (911.4 nm), PL, observed at low temperatures in type Ia diamonds [438, 440] (fig. 56).

1.4 eV (890 nm), A, a broad band, FWHM of 0.5 eV, observed in synthetic diamonds grown in presence of nickel and adding nitrogen getters [130, 89, 112, 179, 282] (fig. 57).

1.4008, 1.4035 eV (884.85, 883.15 nm), the 1.40 eV center, ZPL doublet, A, PL, CL, observed in synthetic diamonds grown in nickel containing ambient, rarely in type Ib natural diamonds, in natural type IIa diamonds implanted with Ni ions [81]. The doublet structure of the ZPL is a result of spin-orbital interaction splitting the ground state by 2.7 meV [89, 112]. The fine structure of each ZPL line consists of at least 4 equispaced components associated with the stable Ni isotopes (1.40376, 1.40096 eV - ^{58}Ni; 1.40359, 1.40079 eV - ^{60}Ni; 1.40343, 1.40063 eV - ^{61}Ni+^{62}Ni; 1.40327 eV - ^{64}Ni). The excited state is a non degenerate orbital state. The center has a trigonal symmetry. Nitrogen suppresses the absorption intensity. The center segregates in {111} growth sectors of synthetic diamonds [83]. Decay time in CL is of 20 to 33 ns [13, 130, 180]. In ^{13}C diamonds ZPL shifts by -0.5 meV [75, 130].

The center anneals out at 2150°C [357]. The 1.40 eV center and the 2.56 nickel related center luminescence intensities relate as $I_{1.4} \sim (I_{2.56})^{0.5}$ [112]. The center interacts with 60 and 165 meV vibrations; S ~ 1.6. The center emission is strongly polarized in synthetic diamonds with the **E** vector preferentially aligned with the {111} growth planes [448]. The center is tentatively associated with the NIRIM-2 paramagnetic center [364]. The common model of the center is substitutional Ni^+ ions relaxed along ⟨111⟩ direction with the electronic configuration ($3d^9$, S=1/2), the electronic transition occurring between a Γ_4 - Γ_5 and Γ_4 - Γ_4 [82, 180, 351, 365]. An alternative model is a defect containing two nickel atoms [486] (fig. 58).

1.428 eV (868.0 nm), CL, ZPL; observed in type Ib synthetic diamonds grown using nickel and annealed to 1700°C; anneals out on heating to 2200°C [74, 84]; formed only in the {111} growth sectors; ascribed to a nickel related defect [336].

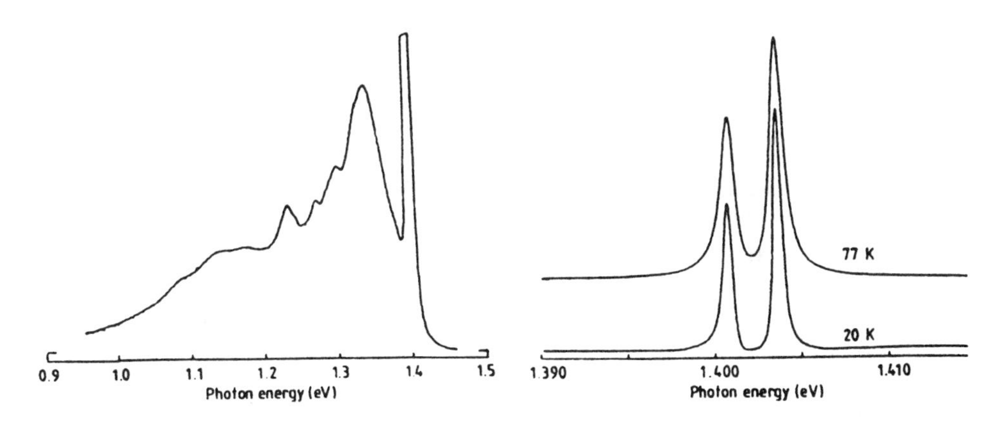

Figure 58. CL spectra of the 1.4 eV center taken from synthetic diamonds, grown using nickel catalyst-solvent [112].

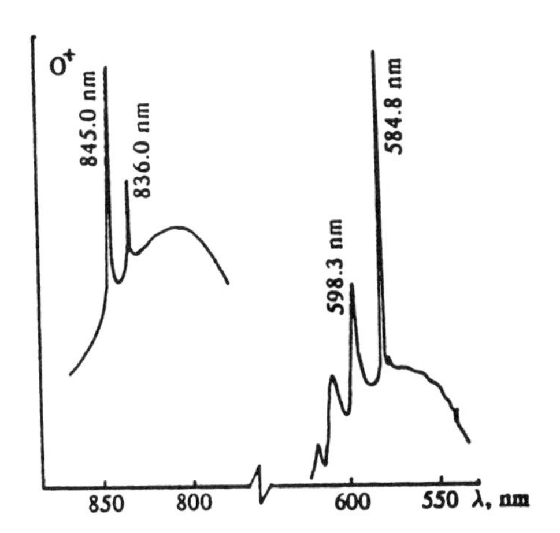

Figure 59. CL spectra of oxygen implanted low nitrogen natural diamonds annealed at 1650°C [368, fig. 3].

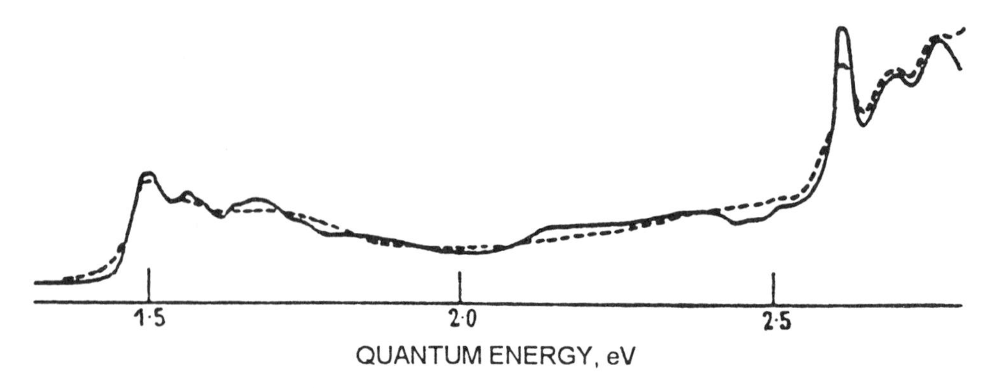

Figure 60. Absorption spectrum of a type I natural diamond [144, fig. 2a].

1.448 eV (856.0 nm), CL, ZPL; observed in type Ib synthetic diamonds grown using nickel and annealed to 1700°C; anneals out on heating to 2200°C [74, 84]; formed only in the {111} growth sectors; ascribed to a nickel related defect [336].

1.466 eV (845 nm), CL, ZPL, created in oxygen ion implanted natural diamonds; the center appears only after 1500°C annealing; ascribed to an oxygen containing defect, which does not contain nitrogen [330, 368] (fig. 59).

1.482 eV (836 nm), CL, ZPL, created in oxygen ion implanted natural diamonds; the center appears only after 1500°C annealing; ascribed to an oxygen containing defect, which does not contain nitrogen [330, 368] (fig. 59).

1.50 eV (826 nm), the N1 center, A, naturally occurring in type Ia diamonds, little temperature dependence; dominant interaction with 60 meV vibrations; similar in appearance to the N2 center [13, 104, 144] (fig. 60).

1.521 eV (814.9 nm), A, ZPL, observed in type IIb diamonds after radiation damage. The center is strongly photo-chromic when irradiated with light of quantum energy > 2 eV [3, 74].

1.525 eV (813 nm), PL, ZPL, observed in type IaB natural diamonds irradiated by neutrons with a dose of 10^{19} cm^{-2} and subsequently annealed at 950°C; in the ZPL spectral holes of a width of 22 cm^{-1} can be burned at temperatures to about 300 K [26, 69, 405] (fig. 61).

1.527 eV (811.5 nm), CL, the most intense ZPL of a center formed in diamonds of any type by Xe^+ ion implantation and subsequent annealing at temperatures above 700°C. The center interacts with vibrations of energy 69 meV. The luminescence intensity strongly increases with the annealing temperature up to 1500°C, ascribed to a Xe containing defect [376, 377] (fig. 62).

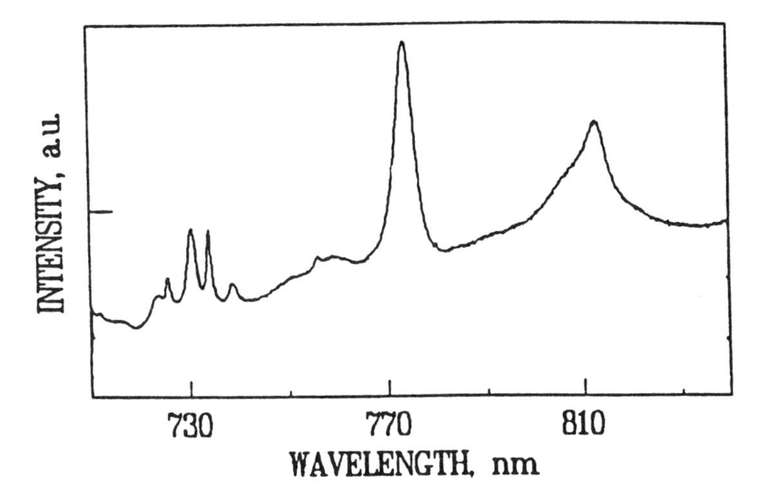

Figure 61. PL spectrum of a neutron irradiated type IaB diamond at a temperature of 5 K [405].

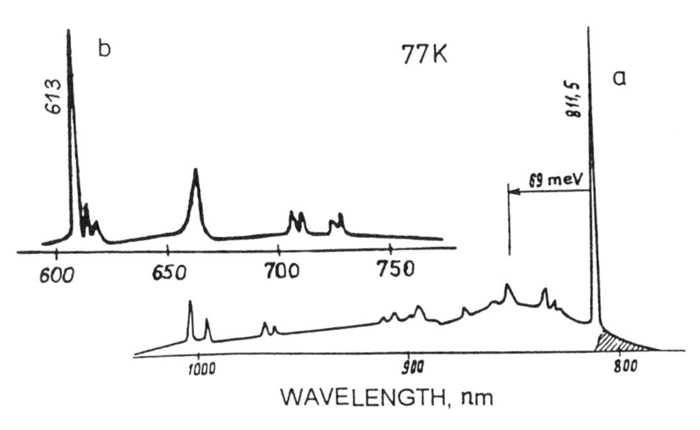

Figure 62. CL spectra of type IIa diamond implanted with Xe (a) and Tl (c) ions and subsequently annealed at 1400°C [376, fig. 2.36, 2.43].

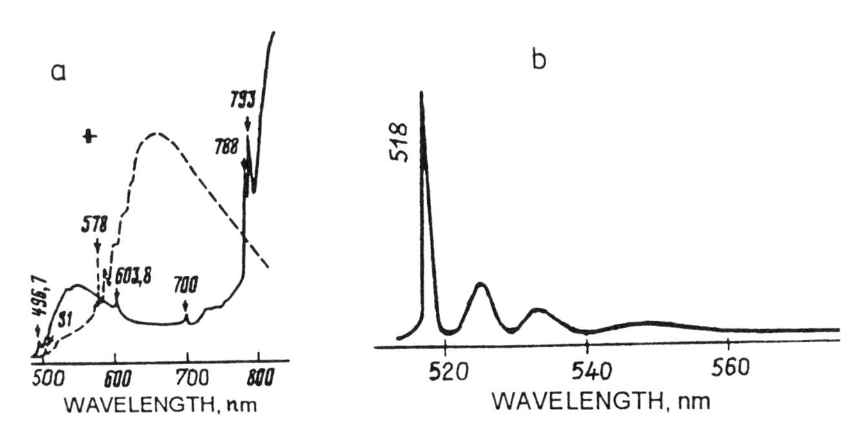

Figure 63. (a) - PL spectrum of a natural diamond [3, fig. 9c]; (b) - CL spectrum of type IIa diamond implanted with Zn ions and annealed at 1400°C [376, fig. 2.43].

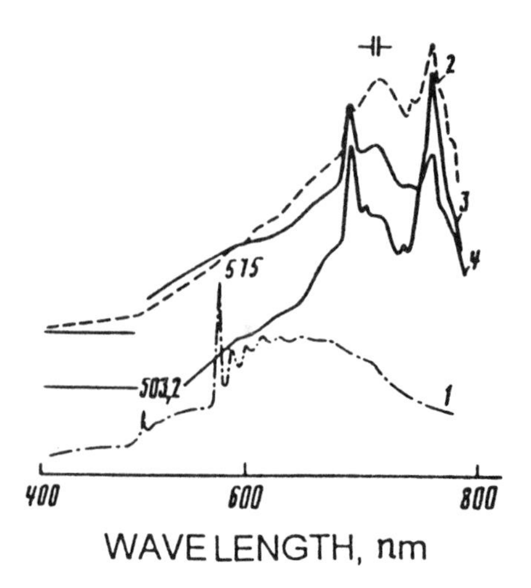

Figure 64. PL spectra of natural polycrystalline diamonds [3, fig. 48].

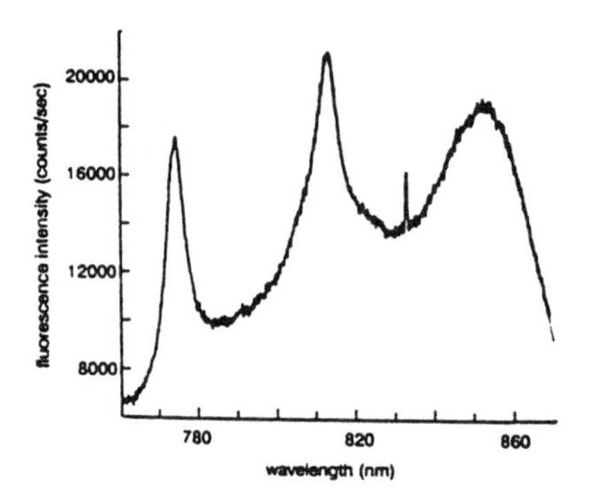

Figure 65. PL spectrum of a type IaB diamond irradiated with neutrons and annealed at 950°C [26, fig. 1].

1.559 eV (795 nm), PL, in low nitrogen Ib natural diamonds, naturally occurring [3].

1.562 eV (793.5 nm), A, PL, CL ZPL, naturally occurring center, particularly intense in type IaB diamonds, also observed in as-grown synthetic diamonds grown by the temperature gradient method with nickel containing catalyst, is induced in synthetic diamonds by heating above 1700°C, is stable above 2200°C [74, 84, 292, 357]; PL intensity correlates with the intensity of the S2 center; interacts with vibrations of energies 40 and 70 meV; PL excited in two broad bands centered at 2.88 eV and 3.65 eV [3, 74, 85]; present only in the {111} growth sectors of synthetic diamonds; ascribed to a nickel related defect [336, 361] (fig. 63).

1.573 eV (788 nm), PL, ZPL, naturally occurring in type Ia diamonds, in type Ia yellow diamonds and diamonds of a mixed cubo-octahedral habit; correlates with the S2 center intensity; interacts with vibrations of energy 38, 47 and 56 meV, excited in two broad bands centered at 2.88 eV and 3.65 eV, coexists with the N2 and N3 centers, attributed to an aggregated nitrogen defect [3, 74, 85, 86, 295, 438] (fig. 55, 63).

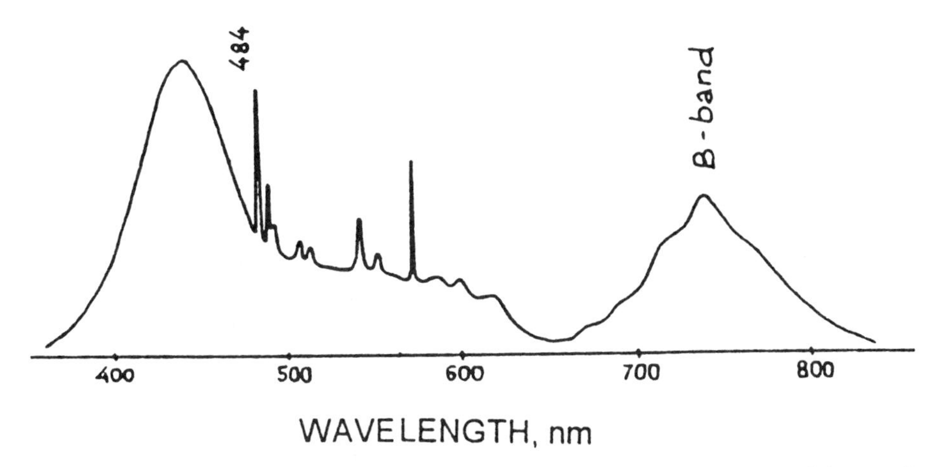

Figure 66. CL spectrum of a type IIa diamond irradiated with 69 MeV Ni ions and annealed at 1700°C [375, fig. 2].

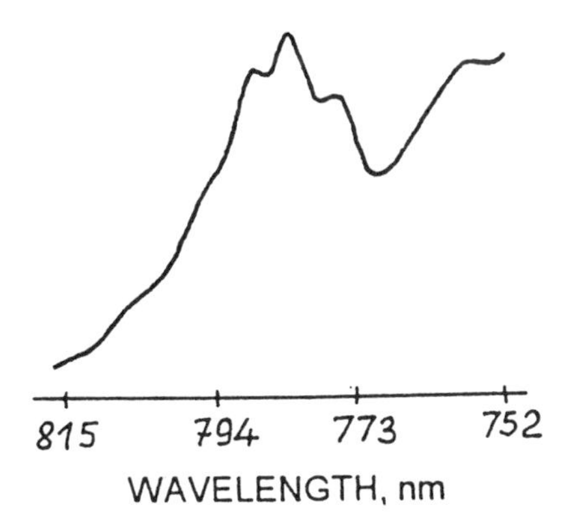

Figure 67. PL spectrum of a MPCVD diamond film [407, fig. 5].

1.593 eV (778 nm), PL, naturally occurring in lonsdaleite containing diamonds [3] (fig. 64).

1.602 eV (774 nm), is observed in IaB natural diamonds irradiated with neutrons (10^{19} cm^{-2}) and subsequently annealed at 950°C, spectral holes of a width 22 cm^{-1} can be burned in this line up to room temperature, nitrogen aggregates are proposed to be involved in formation of the corresponding defect [26, 69] (fig. 65).

1.6 eV (770 nm), the B-band (the same name as for the band at 1.85 eV), CL, a broad band of a width of 0.5 eV, appears in ion irradiated low nitrogen diamonds after annealing above 600°C, stable at temperatures above 2200°C, the electronic transition interacts with a 65 meV vibration, the band intensity grows linearly with ion dose, is strongly quenched by nitrogen [88, 116, 375, 376, 386] (fig. 66).

1.635 eV (758 nm), PL, naturally occurring in lonsdaleite containing diamonds [3].

1.637 eV (757 nm), PL, ZPL, in CVD diamond films grown at high temperature, a set of lines 1.598, 1.586, 1.575, 1.568 and 1.540 eV is also ascribed to this center; S ~ 3.0 [407, 408] (fig. 67).

1.65 eV (750 nm), PL, a broad band with the maximum at 700 to 800 nm, observed in natural polycrystalline diamonds containing lonsdaleite inclusions [3] (fig. 64).

1.660 eV (746.7 nm), CL, observed in synthetic diamond containing aggregated nitrogen after annealing to 1700°C, only in diamonds grown using nickel, is present only in the {111} growth sectors, anneals out on heating to 2200°C [74, 84]; ascribed to a nickel related center [336].

1.666 eV (744 nm), A, PL, ZPL, observed in type IaB natural diamond after 10^{19} neutron/cm^2 irradiation and subsequent 950°C annealing [405]; transition probably interacts with 136 meV vibration (fig. 61).

1.67 eV (741 nm), CL, ZPL, the most intense line of a set of lines in a range of 730 to 780 nm appearing in type IIa diamonds implanted with Cr^+ ions and annealed above 1000°C, a Cr related center [92, 376] (fig. 143b).

1.673 and 1.665 eV (740.9 and 744.4 nm) doublet, ZPL, the GR1 center (abbreviation of General Radiation), A, PL, CL, in absorption accompanied by a broad absorption band extending to about 2.4 eV and a series of sharp lines GR2-GR8 (2.881-3.00 eV) superimposed on a broad band extending into the UV region; anneals out at around 650°C; introduced by 2 MeV electrons at room temperature with a rate of 0.075 cm^{-1} in type IIa diamonds and 0.105 cm^{-1} in type IaA diamonds [188]; The GR1 electronic transition occurs between a doubly degenerate ground orbital state ^1E and triply degenerate orbital excited state $^1T_2(C_{3V})$ in a T_d point symmetry defect, excited state g-value being of -0.08±0.02. The GR1 and GR2-8 centers have the same ground state. The GR1 center shows no inversion symmetry in Stark-effect investigations (linear Stark-effect) [221, 222]. The splitting of the GR1 ZPL (8 meV) results from a Jahn-Teller distortion of the ground state [211, 212]; the GR1 center interacts with vibrations of 41 meV (e mode) and 93 meV (t mode) in absorption [213] but 36 meV in luminescence [104, 169, 427]. The GR2-8 lines arise from ^1E to $^1T_1(C_{3V})$ transitions in a defect of T_d point symmetry, their exact spectral positions and excited state g-values are: GR2 2.881 eV (-0.50±0.10), GR3 2.888 eV (-0.40±0.10), GR4 2.902 eV ($|g|<0.1$), GR5 2.940 eV ($|g|<0.1$), GR6a 2.958 eV (-0.60±0.10, effective g value is -2.0), GR6b 2.960 eV (-0.95± 0.10, effective g value is of -2.0), GR7a 2.976 eV (-0.55±0.10), GR7b 2.982 eV (-0.60±0.10), GR8a 2.996 eV (-0.65±0.10, effective g value is of -1.8), GR8b 2.997 eV (-0.05±0.05, effective g value is of -1.8), GR8c 3.001 eV (-0.30±0.10), GR8d 3.004 eV (-1.00±0.10), GR8e 3.006 eV ($|g|<0.2$) [502]. Lifetime in PL drops to 2.5 ns by temperature about 0 K and is around 0.4 ns at 473 K and around 3.1 ns at 77 K; the luminescence efficiency is about 0.014; absorption has a highly non-radiative recombination character [182]. Radiative lifetime is 182 ns. Activation energy of non-radiative processes is 0.15 eV [204, 242]. The 1.673 eV transition has a non-radiative decay channel [182]. No luminescence is observed from the GR2-8 lines. The GR1 luminescence can be excited through the GR2-8 lines [205].

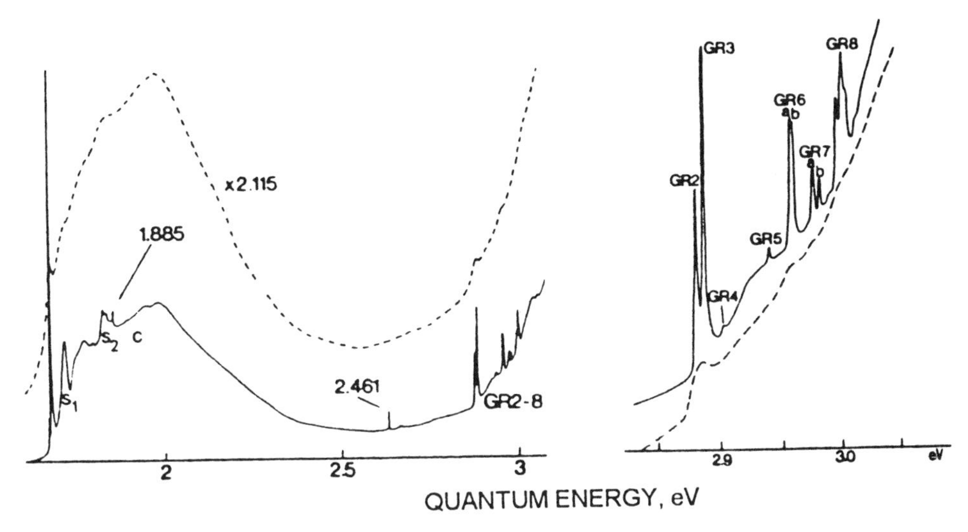

Figure 68. Absorption spectra of a type IIa diamond irradiated with 2 MeV electrons [13, fig. 27, 28].

No polarization of the GR1 luminescence when excited with polarized light [206]. In CL the GR1 ZPL has a set of broader low-energy satellites resulting from perturbed vacancies [207, 208]. No photoconductivity on the GR1 center. The GR2-8 lines show positive (hole) photoconductivity [209, 210]. A hole burning in the GR1 ZPL has been shown, inhomogeneous line width is about 1000 GHz, hole width is of 195 MHz, lifetime 3 ns, the hole lives for several minutes [227]. The GR1 ZPL in ^{13}C diamonds shifts by +2.9 meV [75, 130]. The GR1 luminescence is strongly quenched by ion implanted nitrogen [88, 379]. The GR center is ascribed to neutral single vacancy V^o. There is a model of the GR center as a donor defect [130, 188-199]. There is an opinion that the GR2-8 centers relate more to V^- (negatively charged vacancy) than V^o [393] (fig. 68).

1.681 eV (737.5 nm), ZPL, Si center, A, PL, CL, EL [398], observed in CVD films, high PT synthetic diamonds doped with Si and in Si^+ ion implanted diamonds. Probably is predominantly incorporated in {111} growth sectors in synthetic diamonds [331]. PL spectrum of the center consists of the following lines: 1.6810, 1.6390, 1.7241, 1.6164, 1.7484, 1.5557, 1.8096, 1.5274, 1.8381, 1.7686, 1.7849, 1.8590 and 1.8784 eV [16]. Several weak narrow lines are observed in CL in a range between the 1.681 eV and 1.62 eV lines [316]. PL intensity at room temperature is about 6 times lower than at 100 K. The center interacts with 33 meV vibration in absorption; the Huang-Rhys factor in absorption is found to be 0.24±0.02. In PL the center interacts with 63, 124 and 155 meV [331]. The excited state is split by 0.8 meV causing ZPL doublet 1.6822 and 1.6831 eV; below 60 K each component of the doublet splits by two sub-components separated by 0.2 meV (ZPL spectrum) [27, 314]. In PL spectra the lines at 765, 767, 797, 814 (triplet) nm are considered as vibrational replica of energies 42, 64, 125 and (148, 155, 163) meV respectively, the 64 meV replica being ascribed to a Si-Si quasi local vibration, the 125, 148, 155 and 163 meV replica is ascribed to the LA, TO, LO phonons at the L point and the O phonon in the center of the Brillouin zone

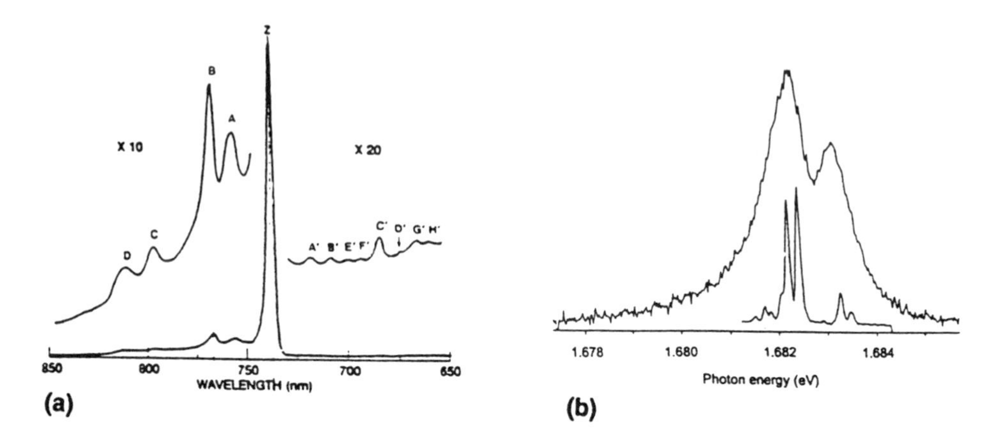

Figure 69. (a) PL spectra of a CVD diamond film recorded at 10 K; peaks A-D and A'-H' are vibronic lines related to the ZPL by vibration emission and absorption, respectively [16, fig. 1]; (b) fine structure of the 1.681 eV center ZPL at 4.2 K [314, fig. 1].

respectively [415, 421]. The center is severely suppressed in heavily boron doped synthetic diamonds; is quenched by the presence of oxygen [45, 130, 182, 186]. The luminescence efficiency is very high (about 80%); the radiative efficiency of absorption is relatively high [182]. The center is characterized by two radiative decay times of 2.3 ns and 100 µs, the long time decay being explained by interaction with a reservoir [312]; there is a non-radiative decay channel [182]. Annealing increases the PL intensity in CVD films above 900°C, there is an intensity splash at 1300°C; the center is stable above 2200°C; FWHM of the ZPL is stable till 1100°C and decreases with further temperature down to about 40 cm^{-1} at 2200°C [44, 343]. There are the following models of the center: (i) substitutional silicon atom + vacancy [44]; (ii) a center containing silicon and nitrogen [90]; (iii) two Si atoms located along <111> axis and bound with vacancies and a two silicon atom model [45, 415, 421]. (fig. 69).

1.685 eV, A, ZPL, observed in all types irradiated diamonds, correlates with the R2 paramagnetic center, anneals out at about 400°C, is proposed to be a <001> distorted vacancy, possibly, carbon atom in <001> split interstitial state trapped by the vacancy along that axis [335]. The center is related to the 1.860 eV center; it freezes out on lowering temperature at the same rate as the 1.665 eV component of the GR1 center and is not observed below 80 K indicating that it is a transition from a level 6 meV above the ground state [13, 104, 422].

1.689 eV (734 nm), PL, observed in type IaB natural diamonds irradiated by 10^{19} neutron/cm^2 and subsequently annealed at 950°C [69].

1.693 eV (732.1 nm) and a complex series of sharp lines between 2.2 and 2.7 eV (560-460 nm), A, observed in synthetic type Ib diamonds after annealing to 1700°C. The center may be seen in diamonds grown at high temperatures [74, 84]; observed

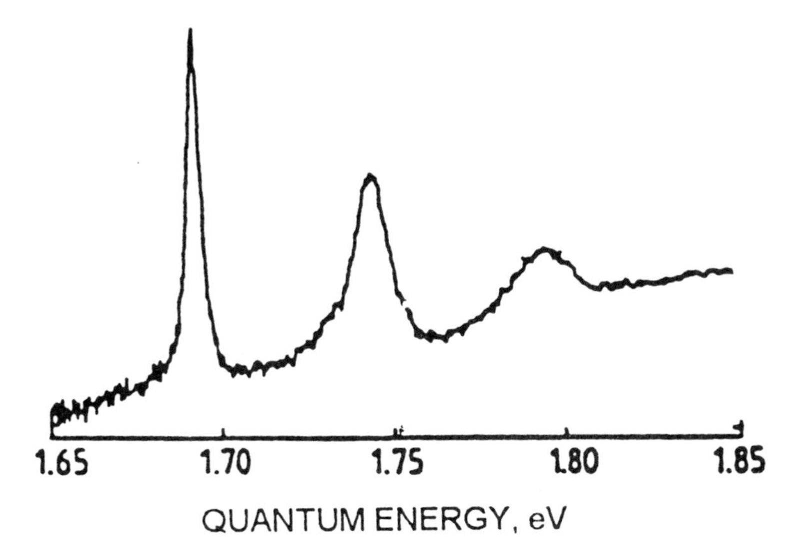

Figure 70. Absorption spectrum of a synthetic diamond treated at 1700°C [84, fig.1b].

only in diamonds grown using nickel [74]; is present only in the {111} growth sectors [336]. The center anneals out on heating to 2000°C, interacts predominantly with vibrations of 52 meV [84, 357, 361]; nickel related center [336] (fig. 70).

1.694 eV (732 nm), A, observed in type IaB natural diamonds irradiated by 10^{19} neutron/cm^2 and subsequently annealed at 950°C [69].

1.696 eV (730.7 nm), A, PL, ZPL, accompanied by a set of narrow lines at 739, 734, 726, 728 and 724 nm, observed in IaB natural diamonds irradiated by 10^{19} neutron/cm^2 and subsequently annealed 950°C annealing. Spectral holes can be burned in the line at temperatures up to 200 K [69, 405] (fig. 61).

1.7 eV (730 nm), PL, XL, a broad band with the maximum in a range 700 to 730 nm, observed in smoky-brown diamonds subject to plastic deformation and in natural polycrystalline diamonds containing lonsdaleite inclusions, observed also in ballas, possible origin - dislocations [3, 295] (fig. 64).

1.705 eV (727.0 nm), CL, ZPL, observed in type Ib synthetic diamonds after annealing to 1700°C; anneals out on heating to 2200°C [74, 84]. The center appears only in diamonds grown using nickel and is present only in the {111} growth sectors; ascribed to a nickel related defect [336].

1.707 eV (726.1 nm), PL, ZPL, observed in brown diamonds which do not show yellow luminescence when excited at 365 nm. The center interacts with vibrations of 51 meV, S=3.6; slow luminescent center, decay time within a range of 1 to 15 ms [74, 87]; the low temperature lifetime is 5.7 ms after excitation at temperatures 50 to 220 K [130].

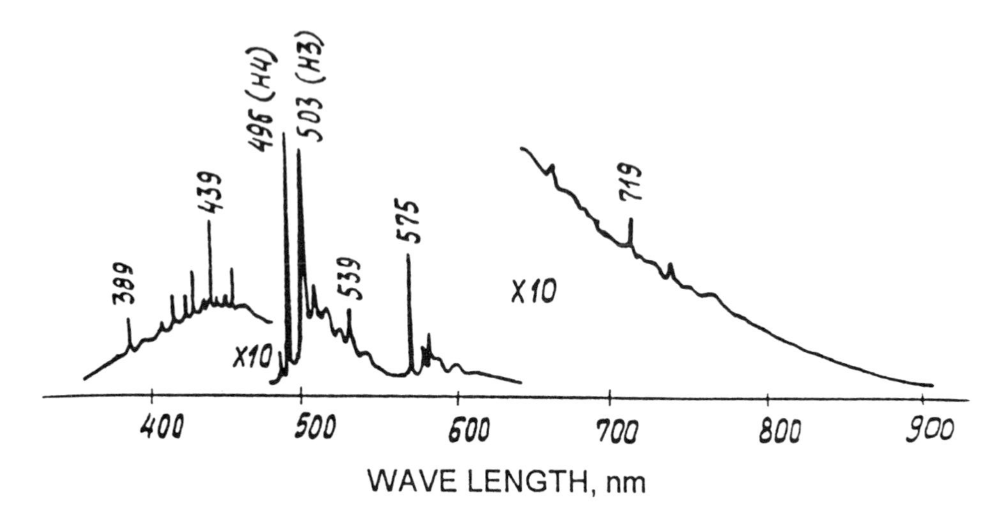

Figure 71. CL spectrum of a type Ia diamond after ion irradiation and subsequent annealing at 1400°C [88, fig.3].

1.72 eV (719 nm), CL, ZPL, observed in type Ia natural diamonds after ion implantation and subsequent annealing at 800°C [88] (fig. 71).

1.723 and 1.731 eV (716 and 719.5 nm), the NG-center, CL, ZPL doublet. The center is created in low nitrogen IIa diamonds by Ne^+ ion implantation, is strongly quenched by annealing at 700°C, is stable at temperatures above 1400°C, interacts predominantly with 25 meV vibration, S ~ 5, tentatively ascribed to a vacancy related defect trapping one Ne atom [376, 381, 382] (fig. 72a)

1.74 eV (714 nm), PL, naturally occurring in natural lonsdaleite containing diamonds [3] (fig. 64).

1.746 eV (710.2 nm), A, ZPL, observed in type Ib synthetic diamonds grown by the temperature gradient method using nickel containing catalyst [361].

1.762 eV (703.6 nm), CL, ZPL, observed in Ib diamonds, interacts with 33 meV vibrations [13, 423].

1.767 eV (701.5 nm), PL, a small peak observed in microwave plasma deposited CVD films [498] (fig. 73).

1.770 eV (700.3 nm), PL, ZPL, naturally occurring center in type Ia yellow diamonds and diamonds of a mixed cubo-octahedral habit. The center interacts with vibration of 63 meV, coexists with the N2 and N3 centers, attributed to an aggregated nitrogen defect [74, 85, 86, 438, 439] (fig. 55).

1.77 eV (700 nm), A, a broad band, FWHM of about 0.25 eV, observed in hydrogen rich brown-grayish yellow, gray-violet and chameleon diamonds [284] (fig. 74).

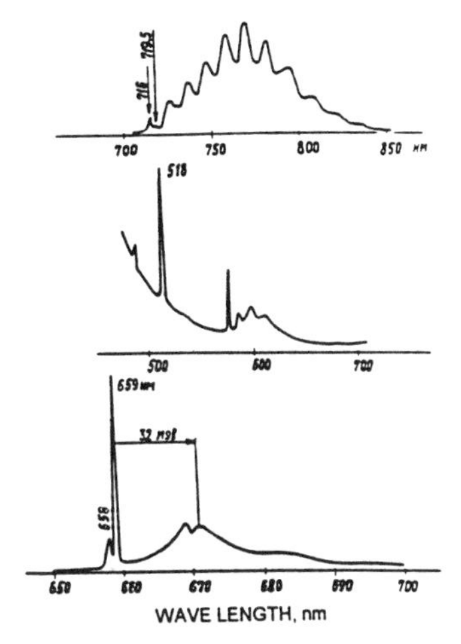

Figure 72. CL spectra of a type IIa natural diamond implanted with 60 keV Ne ions [376, fig. 2.41 a, b, c].

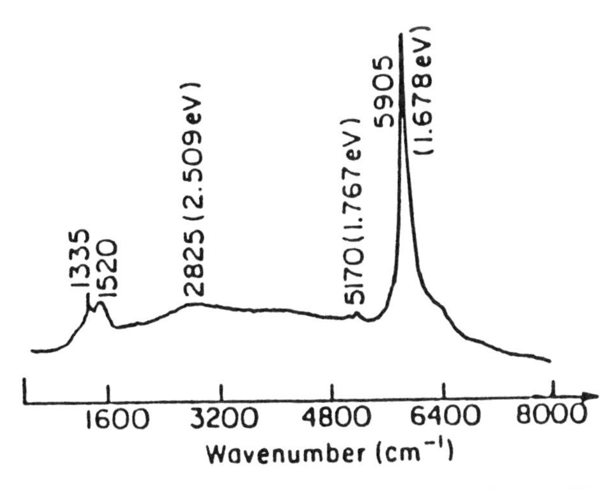

Figure 73. PL spectrum of a MPCVD diamond film on silicon substrate [498, fig. 27b].

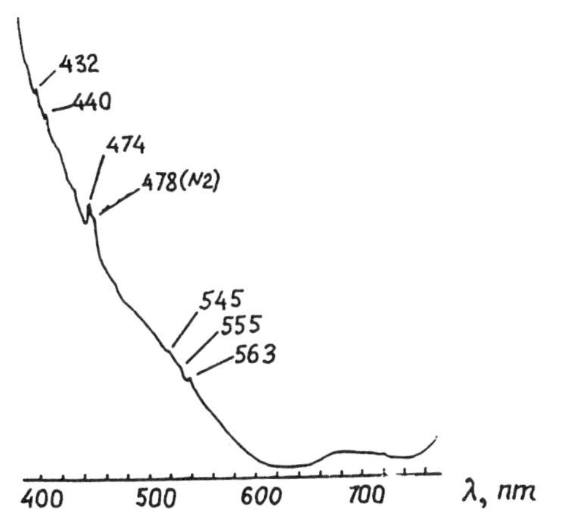

Figure 74. Absorption spectrum of a brownish-yellow natural diamond [284, fig. 3].

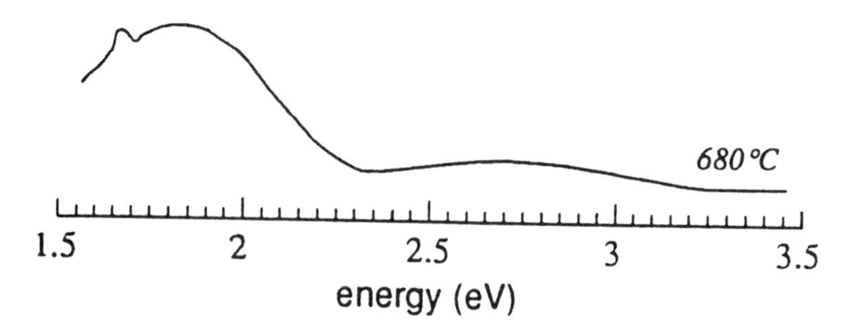

Figure 75. CL spectrum of a MPCVD diamond film grown on silicon substrate at deposition temperature 680°C [280, fig. 1].

1.77 eV (700 nm), PL, ZPL, a weak system occasionally occurring in type Ia gray diamonds, intensity correlates with the S2 center intensity. The center interacts with vibrations of 40 and 70 meV; excited in two broad bands at 2.88 eV and 3.65 eV [3, 295].

1.77 to 1.78 eV (697 to 701 nm), PL, a narrow band naturally occurring in lonsdaleite containing diamonds [3] (fig. 64).

1.79 eV (690 nm), CL, a broad band of a width 0.5 eV, observed in microwave plasma deposited CVD and high temperature-pressure synthetic diamonds, attributed to radiative recombination between remote (about 5th neighbor) donor-acceptor [280] (fig. 75).

1.795 eV (690.5 nm), A, ZPL, observed in type Ib synthetic diamonds grown by the temperature gradient method using nickel containing catalyst [361].

1.8 eV (690 nm), A, PL, CL, a broad band, FWHM of 0.35 eV, observed in brown diamonds, possibly dislocations [1, 3, 13, 114, 169, 287, 297].

1.819 eV (681.4 nm), PL, ZPL, observed in brown diamonds which do not show yellow luminescence when excited at 365 nm: interacts with vibrations of 49 meV, S=3.8, slow luminescent center, the low temperature lifetime is 17.5 ms after excitation at temperatures in a range of 50-120 K [74, 87, 88, 130].

1.821 eV (681 nm), A, PL, observed in type IaB natural diamonds irradiated by 10^{19} neutron/cm^2 and subsequently annealed at 950°C. Spectral holes can be burned up in this line at temperatures up to 200 K [69, 405].

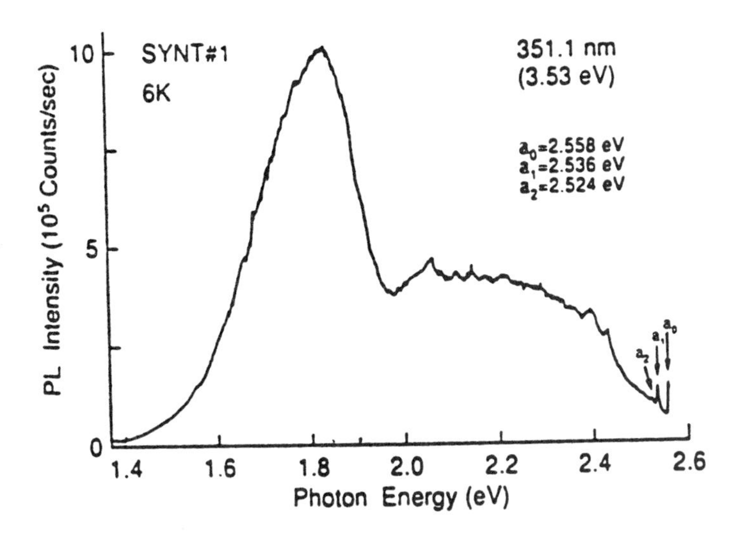

Figure 76. PL spectrum of a synthetic type IIb diamond grown using Ni as solvent-catalyst [309, fig. 3].

1.84 eV (675 nm), PL, a band in synthetic boron doped diamonds, FWHM > 0.25 eV [309]; ascribed to donor-acceptor recombination from distant pairs (boron + nitrogen complexes) [318, 409] (fig. 76).

1.85 to 1.86 eV (667 to 670 nm), PL, a narrow band naturally occurring in lonsdaleite containing diamonds [3] (fig. 64).

1.85 eV, CL, the B band, a band with FWHM of 0.3 eV, naturally occurring in IIa diamonds, obscured in type I crystals [104, 169] (fig. 54).

1.859 eV (667 nm), A, observed in irradiated diamonds of all types, correlates with the R2 paramagnetic center, possibly interstitial carbon related center [3, 104, 296]; destroyed by heating at 400°C [13] (fig. 68)

1.88 eV (658 nm), PL, a narrow band naturally occurring in lonsdaleite containing diamonds [3].

1.880 and 1.884 eV (659 and 658 nm), CL, ZPL doublet of a center created in low nitrogen IIa diamonds by Ne^+ ion implantation, interacts with a vibration of energy 32 meV, S ~ 1, the center anneals out at temperatures below 550°C, the center is tentatively ascribed to an interstitial type defect containing Ne [376, 391] (fig. 72).

1.883 eV (658.3 nm), A, PL, ZPL, observed in type Ib synthetic diamonds grown by the temperature gradient method in a nickel containing ambient. Radiation damage product [13]; anneals out at 2150°C [357]. In the phonon sideband there are features at 1.906, 1.914, 1.943 eV. The absorption is most intense in high nitrogen diamonds; segregated in the {111} growth sectors [74, 89, 90, 112, 179]. In ^{13}C diamonds ZPL shifts by +3.6 meV [130]. The center interacts with 61 meV vibrations in absorption and 65 meV vibrations in luminescence [332]. The

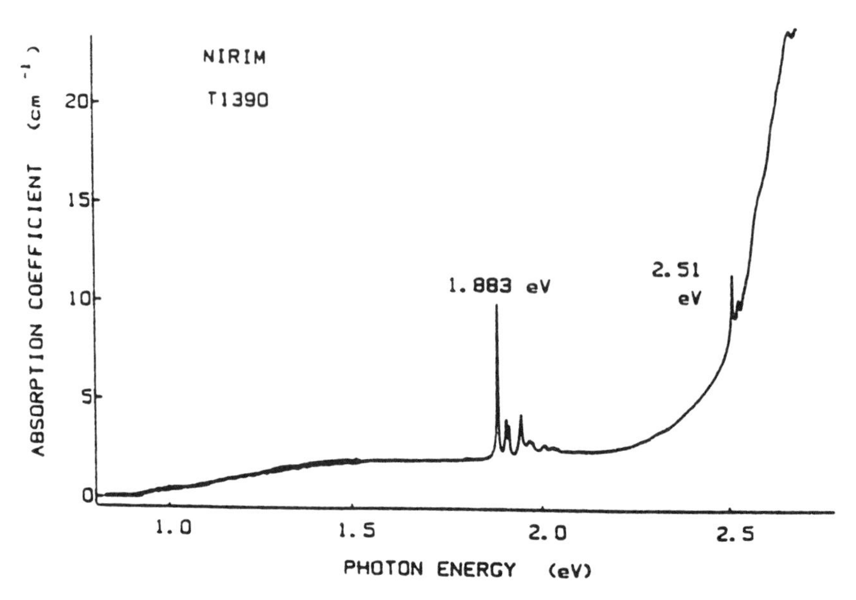

Figure 77. Absorption spectrum of a yellow-brown synthetic diamond grown using Ni and without nitrogen getters [282, fig. 1a].

electronic transition occurs between E (excited) and T_2 (ground) states at a tetragonal defect. The center is proposed to be related to nickel [282]; probably substitutional Ni^- ions [333] (fig. 77).

1.893 eV (655 nm), A, observed in type IaB natural diamonds irradiated by 10^{19} neutron/cm^2 and subsequently annealed at 950°C. Spectral holes can be burned up in this line at temperatures up to 200 K [69, 405].

1.9 eV (650 nm), A, a broad band, FWHM of 0.5 eV. The band provides a green color in some synthetic diamonds grown at temperatures close to the melting point of the solvent-catalyst. ZPL is expected to be at around 1 eV [74, 258] (fig. 78).

1.9 eV, PL, CL, a broad band of a width from 0.36 to 0.86 eV, observed in CVD films, for instance in films deposited at low temperatures without oxygen. The band might be related to the 2.156 eV center; the behavior of the band is very similar to that of the luminescence from hydrogenated Si_xC_{1-x} alloys [275, 342, 343] (fig. 79).

1.903 eV (651.5 nm), A, ZPL, observed in type IaB natural diamonds irradiated by 10^{19} neutron/cm^2 and subsequently annealed at 950°C [69].

1.908 eV (649.5 nm) and 1.925 eV (644 nm), A, PL, ZPL doublet, observed in type IaB natural diamonds irradiated by 10^{19} neutron/cm^2 and subsequently annealed at 950°C. Spectral holes can be burned in this line at temperatures up to 200 K [69, 405]. The doublet is caused by splitting in the excited state. The center relates to the ZPLs at 2.087, 2.321 and 2.341 (doublet caused by splitting in the excited state), 2.712 and 2.718 eV [13, 424].

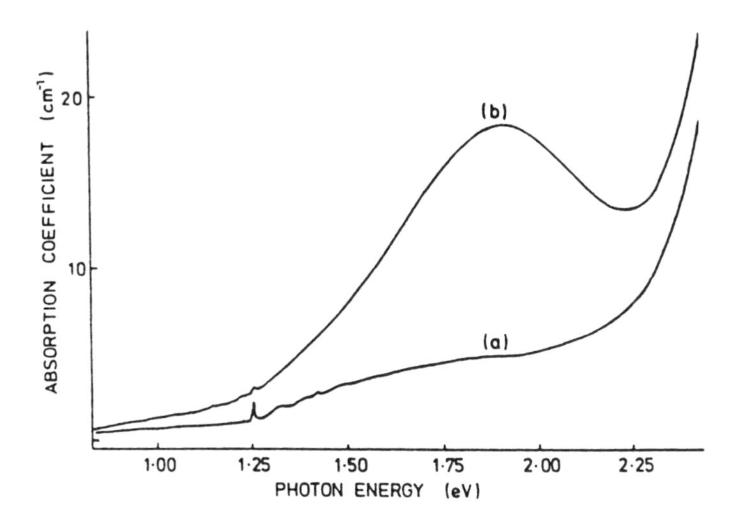

Figure 78. Absorption spectrum of a synthetic diamond grown using Co-Fe or Ni-Fe as solvent-catalyst: as measured at 77 K (a) and 250 K (b) [258, fig. 1].

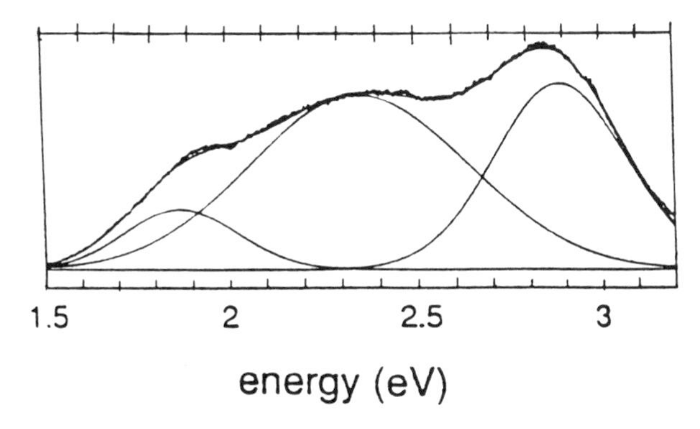

Figure 79. CL spectrum of a MPCVD diamond film deposited on silicon substrate at 630°C. The three Gaussian decompositions are superimposed on the spectra [275, fig. 1].

1.923 eV (644.6 nm), PL, probably one phonon replica of the 1.967 eV center [343].

1.945 eV (638 nm), A, PL, PL excitation, the NV center, CL in nitrogen containing diamonds irradiated with high energy ions [380, 384, 385]; naturally occurring in nitrogen containing diamonds of any origin; produced in these diamonds by any irradiation and annealing above 550°C; produced in type IIa diamonds by N^+ ion implantation and annealing above 700°C; anneals out in irradiated and ion implanted diamonds at about 1500°C. The center produces an intense purple coloration of synthetic diamonds subjected to a heavy radiation damage and 800°C annealing. ZPL shifts in ^{13}C diamonds by +2.1 meV [75, 130]. $^3A(A1)$–$^3E(C_{3V})$ electronic transition in a defect of C_{3V} symmetry. Excited state g-values are g_{orb} = 0.1 and g_e = 2.0 [202]. Linear Stark effect (no inversion symmetry) [222]. The center interacts with 65 meV vibrations, S = 3.65. Oscillator strength relatively to the GR1 center ZPL is NV/GR1 = 1.15 [188]. Inhomogeneous width of the ZPL is around 750 GHz. A hole of width 25 MHz can be burned in the line, the hole lives for several minutes, photochemical process of the hole burning, the hole is stable to temperatures of 120 K [130, 227-231, 241]. Lifetime in PL is 11.6 ns at temperatures 77 to 700 K. The quantum efficiency is about 1. The ground state parameters are: ^{3}A symmetry, g=2.0028, D=2.88 GHz, $A_{\parallel}$ = 2.32 MHz (in ^{13}C diamond 205 MHz), $A_{\perp}$ = 2.10 MHz (in ^{13}C diamond 123 MHz), P = 5.04 MHz [130, 232, 233]. The ground state is a spin triplet. The excited state g-value is 0.1 and the spin-orbit splitting is 1 cm^{-1} [202, 503]. There is energy transfer between the centers and nitrogen [130, 241]. The center originates from nitrogen-vacancy pairs. (fig. 80, 138).

1.946 eV (636.9 nm), PL, ZPL, observed in brown diamonds which do not show yellow luminescence when excited at 365 nm; interacts preferentially with 42 meV vibrations; decay time within a range 5 to 15 ms [74, 87].

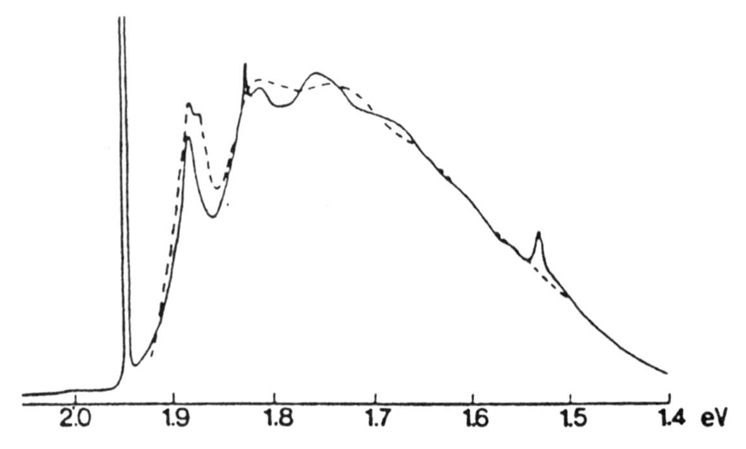

Figure 80. PL spectrum of an irradiated and annealed type Ib diamond [13, fig. 32].

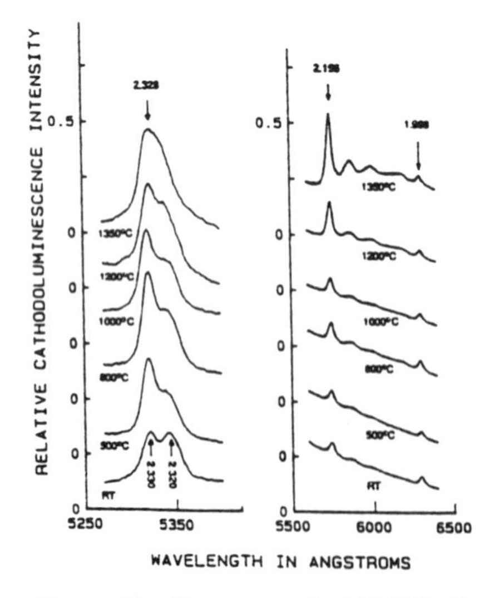

Figure 81. CL spectra of a MPCVD diamond film deposited on silicon and annealed at different temperatures to 1350°C [20, fig. 2b].

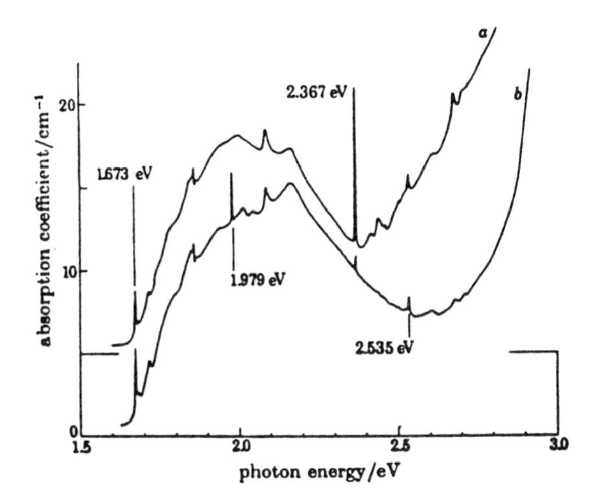

Figure 82. Absorption spectra of an electron irradiated natural type Ib diamond (a) before, and (b) after bleaching of the 2.367 eV band [91, fig. 1].

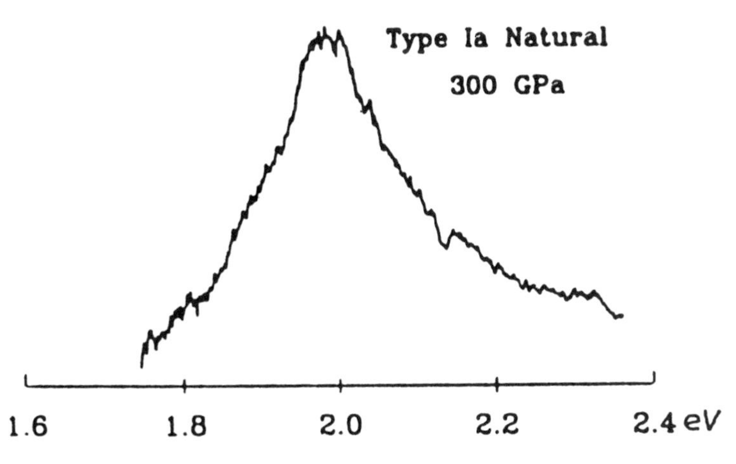

Figure 83. PL spectrum of a type Ia natural diamond at a pressure of 300 GPa at 300 K [438, fig. 12].

1.95 eV (636 nm), PL, XL, ZPL, naturally occurring in ballas [3]; probably the 1.945 eV center [279].

1.967 eV (630.3 nm), PL, ZPL, observed in as grown CVD diamond films; intensity is almost unchanged by annealing to 1000 - 1200°C, a sharp splash of the intensity after annealing at 1300 - 1350°C, stable at temperatures above 2200°C, predominant interaction with 58 meV vibrations [20, 74, 343] (fig. 81).

1.979 eV (626.3 nm), A, ZPL, observed in electron-irradiated type Ib diamonds, intensity increases as the 2.367 center is bleached, annealed out at 200°C, ascribed to a radiation damage center [13, 74, 91] (fig. 82).

1.98 eV (626 nm), PL, ZPL, naturally occurring in ballas [3].

2.0 eV (630 nm), A, a broad band observed in smoky-brown diamonds. Possible origin - dislocations [1, 3, 297].

2.0 eV, PL, a broad band of FWHM 0.2 eV, appears in type Ia diamonds under pressures above 100 - 250 GPa [438] (fig. 83).

A broad band with maximum located at 2 to 2.3 eV (540 to 620 nm), PL, especially effectively excited at 284 nm. Possible origin - the B' center [3].

2.001 eV (619.4 nm), PL, observed at 80 K in type I diamonds, interacts with 30 meV vibrations, possibly a transition at the H3 center [130, 162, 163].

2.01 eV (617 nm), PL, XL, ZPL, naturally occurring in ballas [3]

2.012 (616.2) 2.059 (602), 2.161 (573.5), 2.177 (569.3), 2.215 (559.7), 2.239 (553.5), 2.632 (471), 2.903 (427) eV (nm), CL, the most intense ZPLs of a set of lines in a spectral range of 400 to 650 nm appearing in ion implanted low nitrogen diamonds subsequently annealed at about 1300°C; particularly intense after carbon ion implantation and in diamonds irradiated with high energy ions. Most of the centers possess relatively weak phonon assisted bands; temperature broadening of the ZPLs in the temperature range 80 to 300 K is much lower than that of the vacancy related centers (for instance the GR1 center). Proposed tentatively to be related to intrinsic interstitial type defects [376, 377, 378, 387] (fig. 84).

2.016 (614.7), 2.293 (540.6), 2.538 (488.3), 2.965 (418) eV (nm), CL, the most intense ZPLs of a set of lines in a spectral range of 400 to 650 nm appearing in ion implanted low nitrogen diamonds subsequently annealed at about 1400°C; particularly intense after carbon ion implantation and in diamonds irradiated with high energy ions. Most of the centers possess relatively weak phonon assisted bands. Temperature broadening of all mentioned ZPLs in the temperature range 80 to 300 K is much lower than that of the vacancy related centers (for instance the GR1 center). Proposed tentatively to be related to intrinsic interstitial type defects [376, 377, 378] (fig. 84).

2.02 eV (613 nm), CL, ZPL, the most intense line of a set of lines within a range of 600 to 730 nm appearing in type IIa diamonds after Tl^+ ion implantation and subsequent annealing above 1000°C, ascribed to a Tl containing center [92, 376] (fig. 62).

2.051 eV (604.3 nm), PL, ZPL, observed in diamonds showing yellow luminescence when excited at 365 nm. The center is excited only by a relatively long wavelength (525 nm) light. Decay time is several milliseconds [74, 93, 94, 95].

2.052 eV (604.0 nm), PL, ZPL, in natural brown diamonds; slow luminescent center; the low temperature lifetime is 0.17 ms after excitation and 0.35 ms upon thermalization of the spin sub levels; decay time drops to 0.15 ms on temperature increase to 150 K; the temperature range of non-exponential decay is 15-30 K;

quartet-doublet spin-forbidden transition; interaction with 33 meV vibration, S~3.4 [130, 247, 345] (fig. 85).

2.053 eV (603.8 nm), PL, ZPL, naturally occurring in Ia diamonds, particularly those with mixed cubo-octahedral growth [74, 85]; correlates with the S2 center, interacts with vibrations of 40 and 70 meV, excited in two broad bands at 2.88 eV and 3.65 eV [3].

2.055 eV (603.2 nm), CL, ZPL, observed in type Ib synthetic diamonds after heating at temperatures above 1700°C when most of the nitrogen aggregates in the A centers [74, 84] (fig. 86).

2.057 eV (602.7 nm), CL, ZPL, in some synthetic Ib diamonds after high energy C^+ ion irradiation and subsequent annealing at 1600°C, interacts with 95 meV vibrations, S ~ 1.5 [10] (fig. 85).

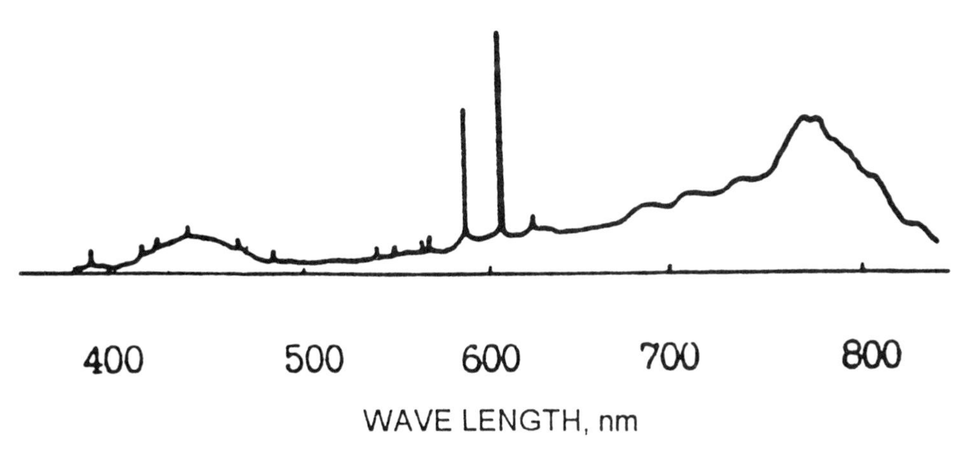

Figure 84. CL spectrum of a type IIa natural diamond irradiated with 82 MeV carbon ions and annealed at 1400°C [378, fig. 1d].

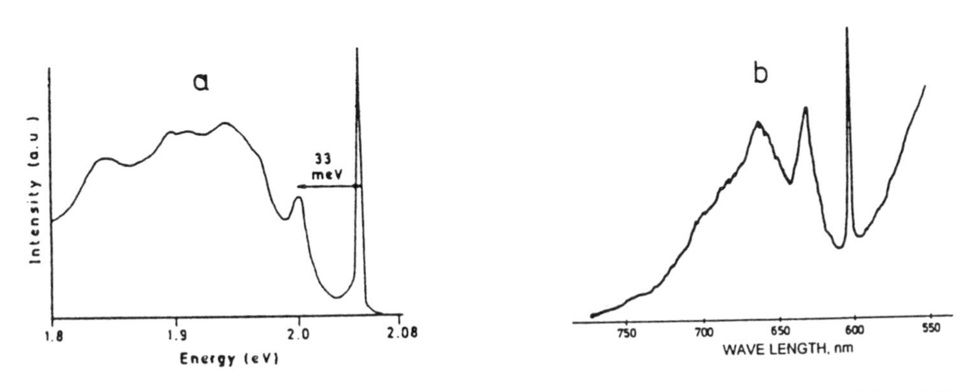

Figure 85. (a) - PL spectrum of the 2.052 eV center [345, fig. 2]; (b) - CL spectrum of the 2.057 eV center [10].

2.072 eV (598.3 nm), CL, ZPL, in oxygen ~~ion implanted~~ types IIa and Ia diamonds after subsequent annealing at above 1500°C; possibly interacts with 47 meV vibrations, an oxygen containing center [368] (fig. 59).

2.082 eV (595.3 nm), PL, ZPL, observed in natural brown diamonds showing yellow luminescence when excited at 365 nm. The center is excited only by a relatively long wavelength (525 nm) light. Decay time is several milliseconds [74, 93, 94, 95]; slow luminescent center; the low temperature lifetime is 0.32 ms after excitation at temperatures above 10 K [247, 130]; lifetime drops down to about 0.05 ms at 170 K. The center interacts with vibrations of energy less than 40 meV [345] (fig. 87).

2.086 eV (594.4 nm), A, not seen in luminescence, ZPL, observed in any nitrogen containing diamonds subjected to primary radiation damage (or after subsequent annealing at temperatures above 275°C), maximum is reached after 800°C annealing, anneals out at 1000-1100°C. E to E electric-dipole transition at a trigonal center of D_{3d} symmetry [74, 95, 13, 96]; quadratic Stark effect (inversion

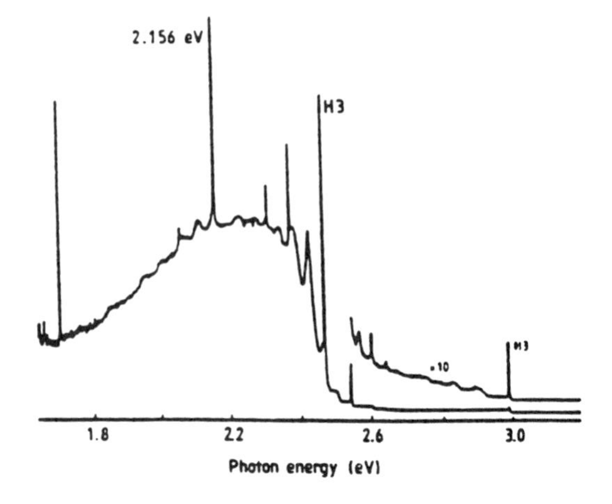

Figure 86. CL spectrum of a synthetic diamond heat treated at 1700°C [84, fig. 4].

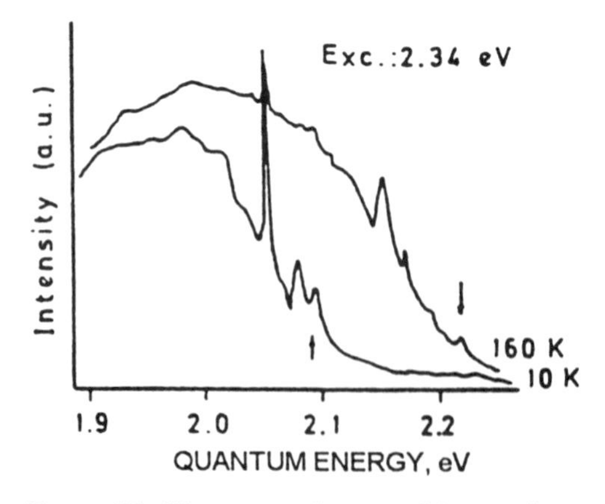

Figure 87. PL spectra of a natural brown diamond recorded between 0.1 and 1 ms after the exciting light pulse at 10 and 160 K [345, fig. 1a].

symmetry) [222, 223]. ZPL very rapidly shifts with temperature and splits under stress. Dominant interaction with 75 and 165 meV, S = 2.1. Strong correlation in intensity with 2.916 eV absorption line. Intensity reduction during annealing correlates with the growth of the H1b and H1c centers [74] (fig. 88).

2.087 eV (594 nm), A, observed in type IaB natural diamonds irradiated by 10^{19} neutron/cm^2 and subsequently annealed at 950°C [69], see 1.908 eV, could relate to the 2.806 eV center [279].

2.088 eV, PL, ZPL, observed in natural brown diamonds, a weak line [345].

2.099 eV (590.5 nm), PL, ZPL, in low nitrogen Ib natural diamonds, naturally occurring [3].

2.1 eV (590 nm), A, a broad complicated band in a range 500 to 600 nm found in hydrogen rich grayish violet diamonds [284] (fig. 89).

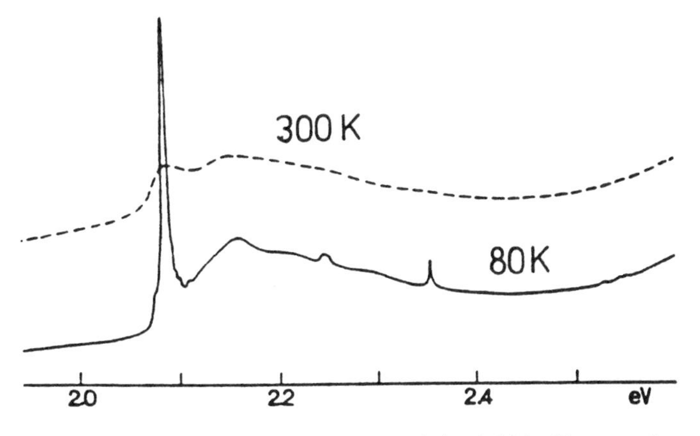

Figure 88. Absorption spectrum of the 2.086 eV center in a natural type Ib diamond subjected to radiation damage and subsequent annealing at 800 K [13, fig. A2].

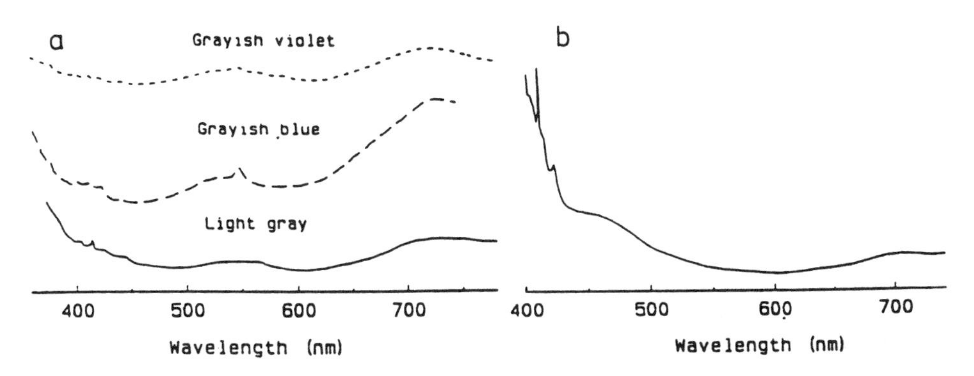

Figure 89. Typical absorption spectra of three hydrogen rich diamonds from the "gray to violet family" (a) and of a "chameleon" diamond (b) [284, fig. 4].

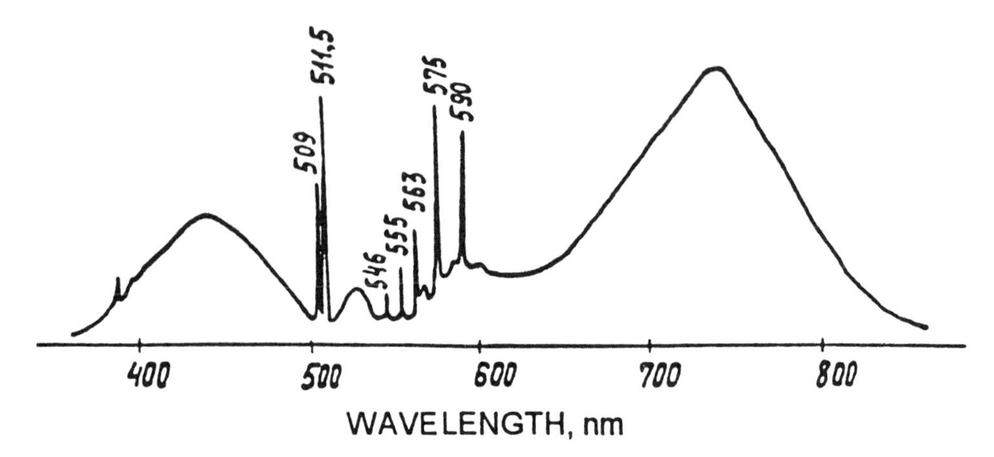

Figure 90. CL spectra of type IIa diamond implanted with 82 MeV carbon ions and annealing at 1000°C [376, fig. 2.7b].

2.1 eV (580 nm), CL, a broad band, most clearly appears in type IIa diamonds after Zr^+ ion implantation and annealing at temperatures above 1000°C [92, 376].

2.101 (590), 2.202 (563), 2.423 (511.5), 2.435 (509) eV (nm), CL, the most intense ZPLs of a set of lines in a spectral range of 500 to 600 nm appearing in ion implanted low nitrogen diamonds subsequently annealed above 700°C; annealed out at about 1300°C, particularly intense after carbon ion implantation and in diamonds irradiated with high energy ions. The centers 2.101 and 2.423 eV can be created in natural diamonds by low energy electrons (subthreshold defect production) [388, 389]. Most of the centers possess relatively weak phonon assisted bands. Proposed tentatively to be related to intrinsic interstitial type defects [376, 377, 378] (fig. 90).

2.114 eV (586.3 nm), PL, ZPL, brown diamonds which do not show yellow luminescence when excited at 365 nm. vibration 46 meV, S=3.8, Decay time within a range 5 - 15 ms [74, 87].

2.119 eV (584.8 nm), CL, ZPL, observed in oxygen ion implanted natural type Ia and IIa diamonds after subsequent annealing at temperatures above 1500°C; ascribed to an oxygen containing center, possibly not containing nitrogen [330, 368]; possibly relates to the 2.072 eV center (fig. 59).

2.133 eV (581.0 nm), PL, ZPL, in diamonds showing yellow luminescence when excited at 365 nm. Predominantly interacts with vibrations of 31 meV, S ~ 10. Decay time is 6 ns [74, 93, 94, 95].

2.14 to 2.25 eV (550 to 580 nm), CL, a broad band of FWHM 0.4 to 54 eV, possibly relates to the B' platelets [74, 130, 171, 154] (fig. 91).

2.145 eV (577.8 nm), PL, CL, ZPL, in natural brown diamonds showing yellow luminescence when excited at 365 nm. The center dominates under excitation at

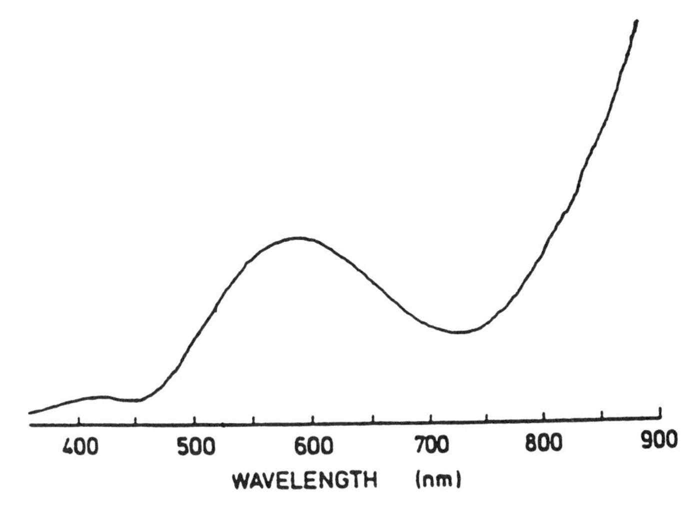

Figure 91. CL spectrum at 160°C of a natural diamond showing yellow emission of giant platelets [154, fig. 6].

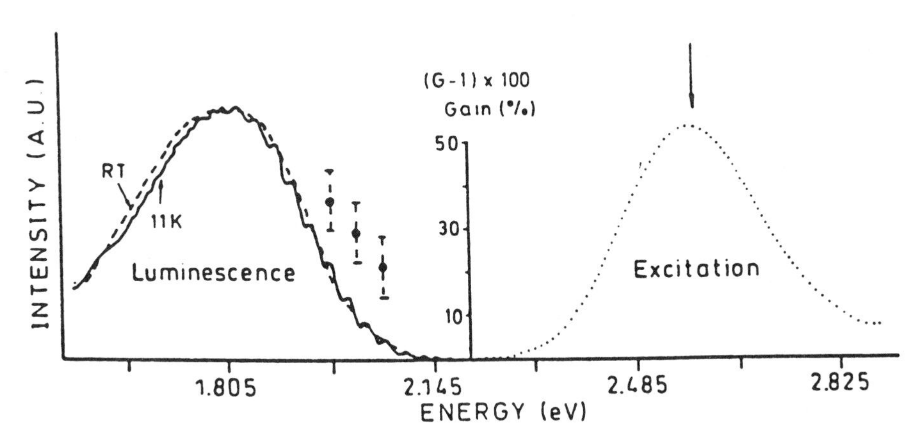

Figure 92. PL spectra of a natural diamond recorded at 11K and room temperature [435, fig. 2].

460 nm. In polycrystalline natural diamonds ZPL can be split into two components at 578.7 and 581.1 nm [3]. Electronic transition at a defect of monoclinic-I symmetry [74, 93, 94, 95, 97]. Predominant interaction with vibrations of 30 meV [435]; very strong electron-phonon coupling, S ~ 9.9 [3, 295, 345]. Decay time is 6 ns in a temperature range of 77 to 600 K [93]. The center is destroyed upon 2 MeV electron irradiation followed by annealing at 700°C. The center presents a tunable stimulated emission at room temperature [435]. A vacancy related center. Possibly relates to an intrinsic defect [3] (fig. 92).

2.156 eV (574.8 nm), PL, ZPL, in brown diamonds showing yellow luminescence when excited at 365 nm [74, 93, 94, 95].

2.156 eV (574.9 nm), CL, in regions of indentations on (001) surfaces of synthetic and type IaA natural diamonds [74, 261].

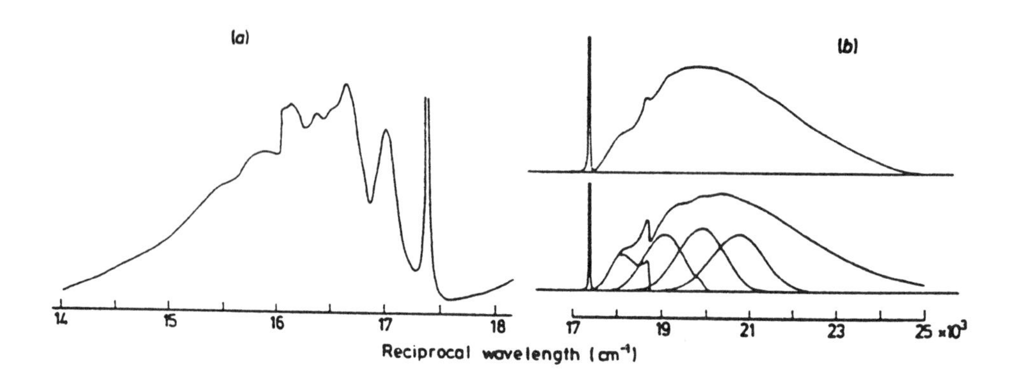

Figure 93. CL (a) and absorption (b) spectra of the 2.156 eV center [30, fig. 1].

2.156 eV (575.5 nm), A, PL, CL, Excit. of PL, EL [80, 398, 399], ZPL, the 575 nm center; in some publications labeled as the T_1 center [3]; naturally occurring in nitrogen containing natural and synthetic diamonds including diamond films, created in type IIa diamonds by N^+ implantation and annealing above 500°C; the center is activated in irradiated nitrogen containing diamonds by annealing above 500°C or by an electron beam of subthreshold energy (< 300 keV) [30]; can be produced by plastic deformation of diamond crystal, e. g. by polishing [337]; can be produced by high energy C^+ ion irradiation as a result of nuclear reactions [245]; very intense in fault-free single crystal diamond films [41]. The center shows linear Stark effect (no inversion symmetry) [222]. In ^{13}C diamonds ZPL shifts by +3 meV [75, 130]. Interacts with vibrations of an energy 46±2 meV, S = 3.3. Radiative lifetime ~ 29 ns [13,]. Predominantly formed in {111} growth sectors of both HTHP and CVD diamonds [432]. The center can be or of trigonal symmetry [30, 105] or C_{2V} symmetry [379, 390]. Nitrogen-vacancy model [19, 390]; proposed to be a single interstitial nitrogen atom bound to a nearest vacancy along the <001> axis [116, 379, 383] (fig. 93).

2.163 eV (573.0 nm), PL, ZPL, transition from a higher energy level of the 2.052 eV center [345].

2.167 eV (572.1 nm), PL, ZPL, in diamonds showing yellow luminescence when excited at 365 nm [74, 93, 94, 95].

2.175 eV (569.9 nm), PL, ZPL, transition from a higher energy of the 2.082 eV center [345].

2.20 eV (563 nm), A, ZPL, in natural diamonds having a high hydrogen content [74, 284]; accompanied by 555 and 545 nm bands (possibly phonon side bands) [284] (fig. 74).

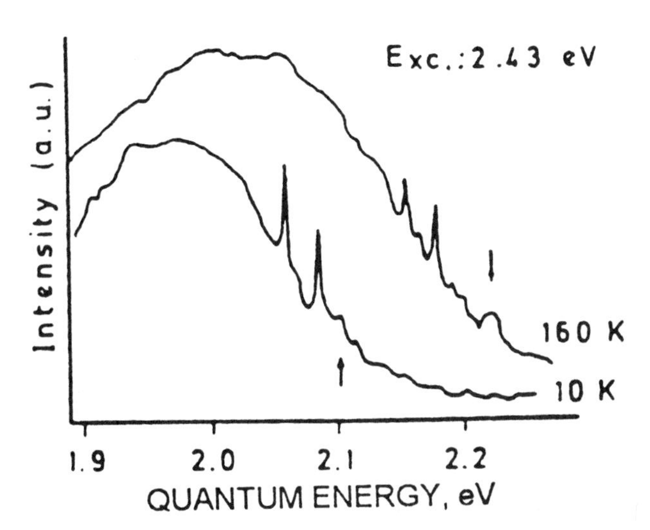

Figure 94. Time-resolved spectra of a natural brown diamond seen between 0.1 and 1 ms after the light pulse at 10 and 160 K [345, fig. 1b].

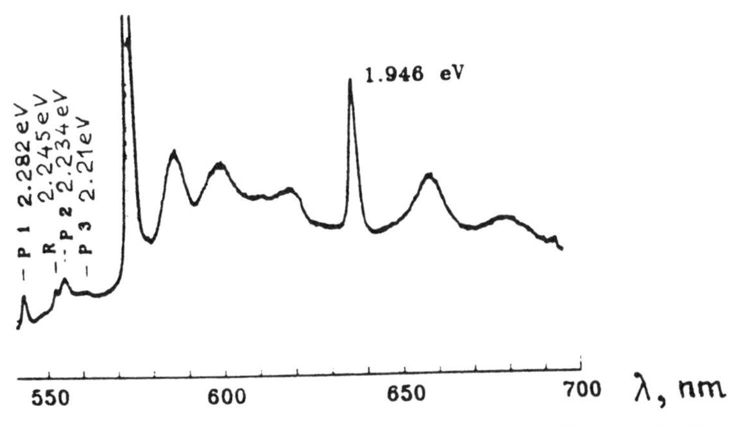

Figure 95. PL spectrum of a free standing flame-grown diamond film [254, fig. 6].

2.200 eV (563.4 nm), CL, ZPL, observed in CVD films [74, 340] (fig. 129a).

2.205 eV, PL, in natural brown diamonds, a weak line [345].

2.208 eV (561.3 nm), CL, ZPL, in synthetic Ib diamond grown using pure cobalt as the solvent-catalyst [74].

2.21 eV (561 nm), CL, in type IIa diamonds after H^+ and C^+ ion implantation and 1000°C annealing [88].

2.21 eV (561 nm), PL, in combustion flame films, possibly ZPL line [254] (fig. 95).

2.210 eV (560.9 nm), CL, ZPL, in synthetic Ib diamond grown using pure cobalt as the solvent-catalyst [74].

2.225 (557), 2.330 (532), 2.426 (510.8), 2.440 (508), 2.558 (484.5), 3.00 (412) eV (nm), CL, the most intense ZPLs of a set of lines in a spectral range of 400 to 600 nm appearing in ion implanted low nitrogen diamonds subsequently annealed at temperatures above 500°C; annealed out simultaneously with the GR1, TR12 and 3H centers at about 700°C; particularly intense after carbon ion implantation and in diamonds irradiated with high energy ions. Most of the centers possess relatively weak phonon assisted bands. Proposed tentatively to be related to intrinsic interstitial type defects [88, 375, 376, 377, 378, 387] (fig. 96).

2.23 eV (555 nm), CL, ZPL, IIa natural diamonds after H ion implantation [74, 88].

2.23 eV (556 nm), A, a broad band of FWHM 0.35 eV, observed in natural diamonds, rarely observed, insensitive to temperature [13].

2.234 eV (554.8 nm), PL, a narrow line, possibly ZPL, observed in combustion flame grown diamond films [254] (fig. 95).

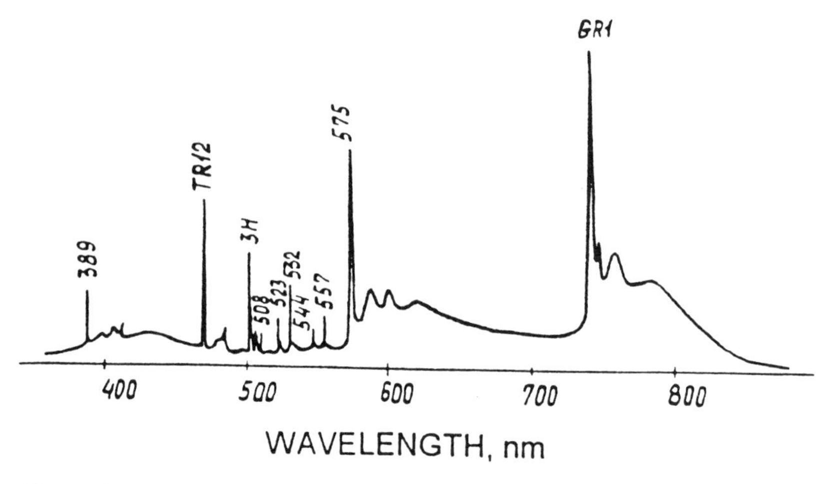

Figure 96. CL spectra of type IIa diamond implanted with 82 MeV carbon ions and annealing at 500°C[376, fig. 2.7a].

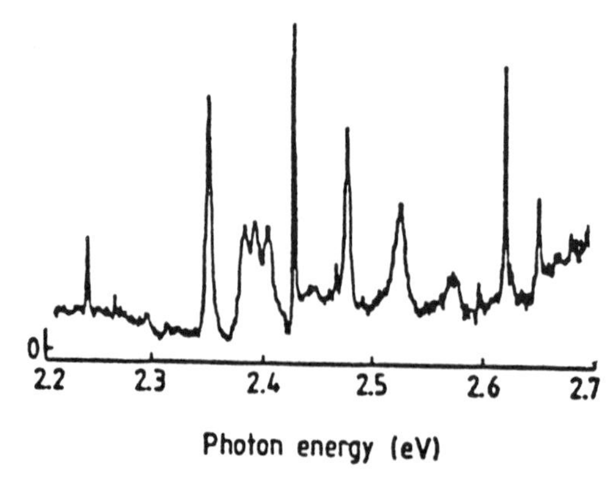

Figure 97. Absorption spectrum of a synthetic diamond heat treated at 1700°C [84, fig. 1c].

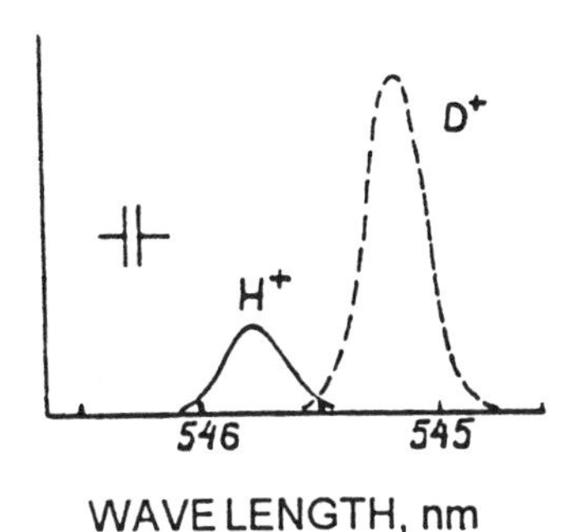

Figure 98. CL spectrum of the 2.271 eV ZPL in a natural diamond implanted by H^+ and D^+ ions and annealed at about 1000°C [184, fig. 12].

2.241 eV (553.1 nm), A, the lowest energy component of a complex series of weak lines at 2.241, 2.352, 2.385, 2.395, 2.406, 2.430, 2.480, 2.553, 2.624 and 2.654 eV observed in synthetic type Ib diamond after partial aggregation of the nitrogen by heating to e.g. 1700°C. All the lines anneal out on heating to 2000°C, except those at 2.624 and 2.654 eV which persist until at least 2400°C [74, 84] (fig. 97).

2.25 eV (550 nm), A, a broad band observed in smoky-brown diamonds, possibly relates to dislocations [1, 3, 297].

2.25 eV (550 nm), A, a rare broad band of FWHM 0.2 eV observed in natural diamonds at temperatures below 300 K [13].

2.271 eV (545.8 nm), CL, ZPL, in type IIa diamonds after H^+ ion implantation; after D^+ ion implantation the line shifts to 2.273 eV (545.2 nm); the center persists after annealing at 1000°C; possibly hydrogen related defect [88, 130, 184] (fig. 98).

2.278 eV (544.1 nm), CL, ZPL, observed in type Ib synthetic diamonds grown using pure cobalt as the solvent-catalyst [74].

2.28 eV (563 nm), long time phosphorescence, ZPL, the N3b center; observed in natural diamonds containing the N3 and B' centers; the phonon assisted band structure is similar to that of the N3 center; ascribed to the N3 center, the excited state of which is shifted down in the vicinity of the B' defects [3, 291, 379] (fig. 99).

2.282 eV (543.2 nm), PL, a narrow line, possibly ZPL, observed in combustion flame grown diamond films [254] (fig. 95).

2.295 eV (540.1 nm), A, ZPL, in type Ib synthetic diamonds grown by the temperature gradient method using nickel containing catalyst [361].

2.30 eV (539 nm), CL, ZPL, in Ia diamonds after ion irradiation and annealing above 800°C, in N^+ implanted IIa diamonds followed by 1400°C annealing; a center of tetragonal symmetry [88, 99].

2.30 eV (540 nm), the N3b center, PL, ZPL, observed at 4.2 K in natural diamonds containing the B' aggregates; the spectral structure of the center is very similar to that of the N3 center; excited in the N3 absorption spectrum; the spectral position of the center shifts to the red with increasing the energy of the B' center transition; tentatively ascribed to the N3 centers localized in the vicinity of the B' aggregates [370] (fig. 99).

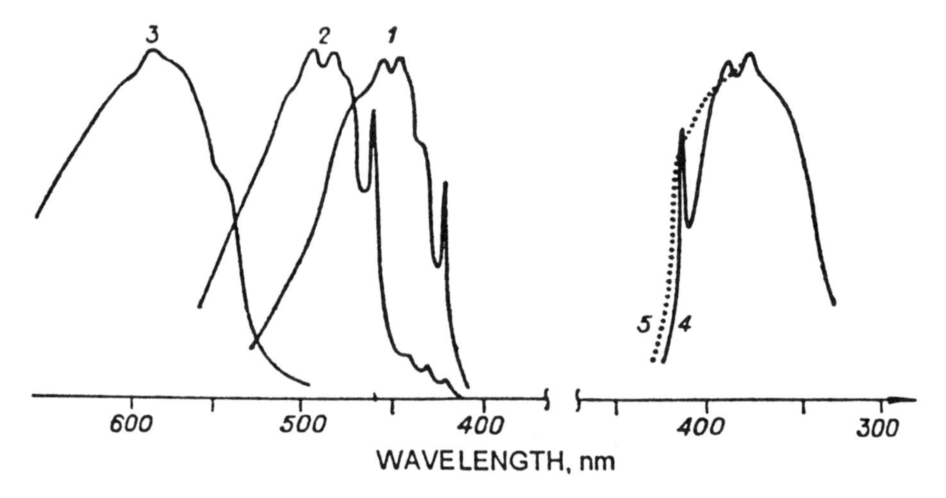

Figure 99. PL spectra of the N3, N3a and N3b centers (1, 2, 3) as well as the excitation spectra of the N3 and N3a (4) and N3b (5) centers in nitrogen containing natural diamonds [370, fig. 2].

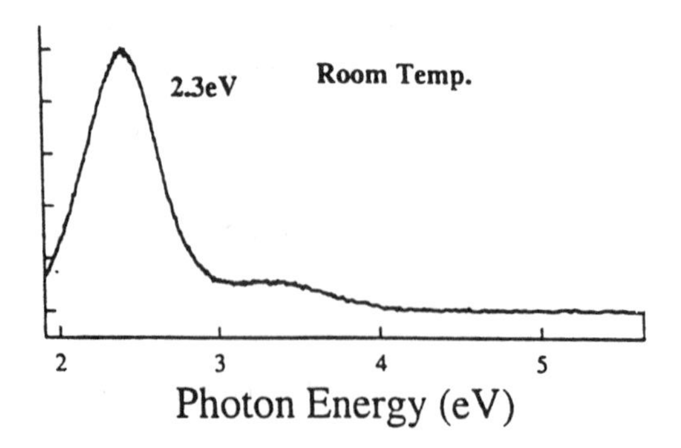

Figure 100. CL spectrum from a boron doped CVD diamond film at room temperature [500, fig. 4b].

2.1 to 2.73 eV (590 to 450 nm), CL, PL, a broad band of FWHM 0.4 to 0.8 eV, a common feature in CVD and HTHP synthetic diamonds, pronounced in boron doped diamonds [38], observed in polycrystalline films after B ion implantation [37]; in CVD films the spectral position shifts to the red as the deposition temperature decreases [275]; relatively insensitive to ion irradiation damage, is localized within {100} growth sectors in crystallites of polycrystalline CVD films and synthetic HTHP diamonds [419, 420]; dispersed nitrogen does not affect strongly on this band [413, 414, 500]; aggregated nitrogen and nickel can quench the band [270]; rapid annealing at 1000°C quenches the band [499]; the band is also quenched by electron irradiation of 1 keV energy [499]. The band ascribed to boron related centers [38, 130, 270, 413, 418], for instance, to substitutional boron [36, 37, 38, 124], or to hydrogen [499] (fig. 100).

2.302 eV (538.4 nm), PL, observed in CVD films after electron irradiation [74].

2.313 eV (535.9 nm), A, CL, in natural diamonds showing signs of radiation damage; may be also produced in type IaB diamonds by irradiation and annealing at temperatures above 1300°C [100, 101, 74] (fig. 101).

2.321 and 2.341 eV, see 1.908 eV.

2.322 and 2.335 eV (533.8 and 530.8 nm), CL, ZPLs, observed in diamond films, frequently a dominating line alone with the 575 nm center, both centers persist after heating at 1400°C; decay time of the broader line (2.322 eV) is of 1.0±0.1 μs and the narrower line (2.335 eV) of 9.6±0.8 ns [312]; the centers are localized on rod-like macro defects possibly caused by strain [328]; related to nitrogen, possibly nitrogen-vacancy complex [20-22, 36, 74, 124, 339, 340] (fig. 81, 102).

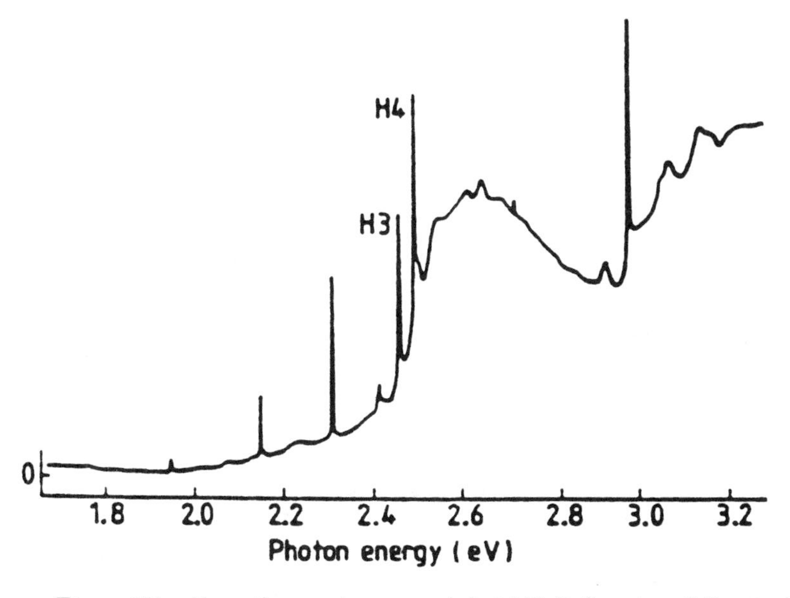

Figure 101. Absorption spectrum recorded at 100 K for a type IaB natural diamond after electron irradiation and annealing at 1500°C [101, fig. 3b].

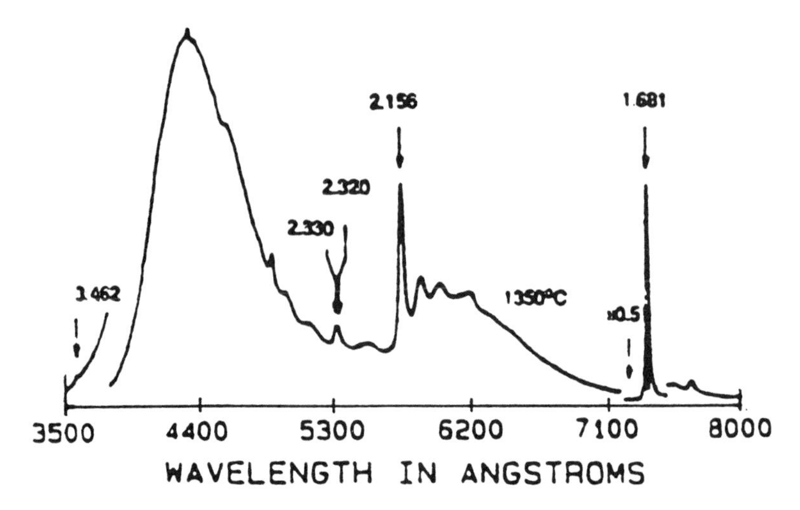

Figure 102. CL spectrum of a MPCVD diamond film deposited on silicon substrate and annealed to 1350°C [20, fig. 1].

2.350 eV (527.4 nm), A, ZPL, in type Ib synthetic diamonds grown by the temperature gradient method using nickel containing catalyst and synthetic diamonds after high PT treatment (1950°C) [361].

2.36 eV (525 nm), CL, a band of a spectral shape similar to that of a broadened H3 center, observed in brown diamonds, associated with H3 center strain-broadened near straight dislocations and dislocation triangles [395].

2.363 eV (524.6 nm), PL, ZPL, observed in brown diamonds 1 ms after excitation [74].

2.367 eV (523.6 nm), A, ZPL, observed in type Ib diamonds, radiation damage product, particularly intense after electron irradiation at low temperatures (<250 K), anneals out at 200°C apparently producing the 2.535 eV center [443, 91], coupled with the 1.979 eV center by optical bleaching and heating. A possible model: nitrogen bound to one or more carbon interstitials [74, 91, 13] (fig. 82).

2.367 eV (523.6 nm), CL, ZPL, created in type Ib synthetic diamonds after heating at 1700°C when partial aggregation of the nitrogen occurs [84], observed also in diamonds grown using pure cobalt as the solvent catalyst [74]. Not the same center as that seen in absorption.

2.369 eV (523.3 nm), the S2 center, the A-line, A [104], PL, PL excitation, inactive in CL, naturally occurring center; particularly intense in type IaB diamonds in which the luminescence intensity correlates with that of the 1.56 eV system; observed also in absorption in irradiated type Ia diamonds; induced in synthetic diamonds by 2150 and 2200 K annealing [292, 357]. The lines at 2.534, 2.594, 2.623, 2.634 eV (B-E lines) are possibly associated with the S2 center [103, 357]; the B line is an unresolved triplet or quartet [104, 74]. In PL excitation spectra of the center in as-grown synthetic diamonds there is a fine structure: 3.560 eV (348.1 nm), 3.502 eV (354.0 nm), 3.474 eV (356.0 nm), 3.443 eV (360.1 nm), 3.412 eV (363.4 nm), 3.380 eV (366.7 nm), 3.341 eV (371.1 nm), 2,527 eV (471.9 nm), 2.598 eV (477.2

nm) [357]. Slow luminescent center; the A transition decay is 0.24 ms (does not depend on temperature), the B and C transition decays are 3.5 μs at 80 K [3]; the low temperature lifetime is 0.5 ms after excitation, 1.5 ms upon thermalization of the spin sub-levels, the temperature range of non-exponential decay is 20-30 K, triplet-singlet spin-forbidden transition [130]. The A and C transitions interact with 40(32) and 72 meV vibrations, the B transition interacts with 45 meV vibrations, S ~ 0.4 for the A transition [3, 104, 103]. The S2 center intensity does not change considerably up to 500°C [3]. The center is paramagnetic - the NE2 center [357]. A possible model - Ni^+ ion bound to divacancy surrounded by two or three nitrogen atoms in the nearest co-ordination sphere [357] (fig. 103).

2.369 eV (523.2 nm), A, PL, in natural diamonds, in synthetic nitrogen containing diamonds grown with nickel containing catalyst after high PT treatment; in PL excitation and absorption in as-grown synthetic diamonds there is a fine structure related to the center: 3.689 eV (336.1 nm), 3.607 eV (343.7 nm), 3.572 eV (347.1 nm), 2.598 eV (477.2 nm). The center is paramagnetic - the NE3 center [357]. A possible model: - Ni^+ ion bound to divacancy surrounded by two or three nitrogen atoms in the nearest co-ordination sphere [357] (fig. 104).

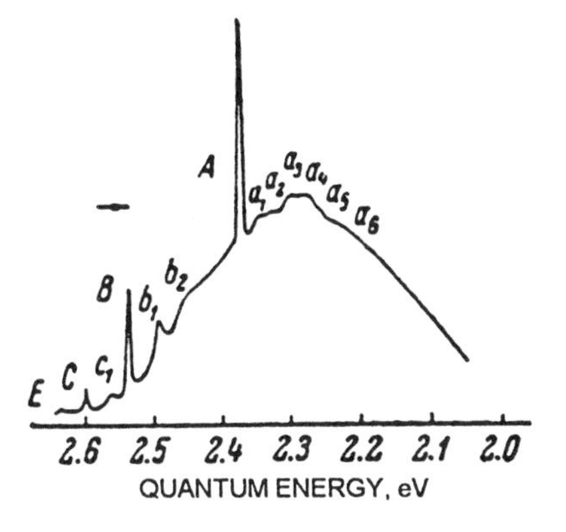

Figure 103. PL spectrum of a natural diamond showing the S2 center emission [103, fig. 1].

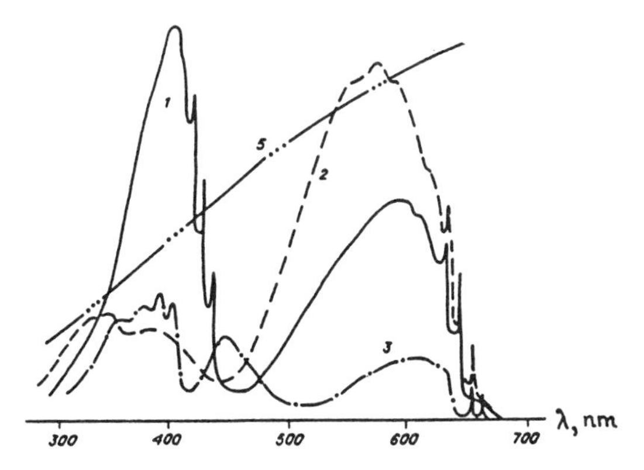

Figure 104. PL excitation spectra of S2 (1), S3 (2) and 523.2 nm center (3) in synthetic diamonds grown by the temperature gradient method;(5) - spectrum of the exciting lamp [361, fig. 6].

2.372 (522.5), 2.33 (532), 2.315 (535.5), 2.223 (556), 2.213 (560), 2.15 (577) eV (nm), CL, the most intense ZPLs in a very complicated spectra (more than 50 ZPLs) appearing within a spectral range from 2.1 to 2.5 eV in diamond of any type after He^+ ion implantation and subsequent annealing at temperatures above 450°C; set of the lines may differ considerably depending on nitrogen content. Most of the lines anneal out at temperatures from 600 to 700°C. The 2.372 eV line is the strongest in type IIa natural diamonds and it anneals out at 800°C. All the transitions reveal a weak electron-phonon coupling, Debye-Waller factor is less than 0.02. The transitions 2.213 and 2.315 eV appear to occur at one center, which is the most temperature resistant and it anneals out at 1200°C; this center has a very complicated vibration assisted side-band where vibrations of energies 38, 59, 89 and 130 meV dominate; both lines have identical spectral shape each revealing at least three components with spectral separation of 0.5 meV. A very tentative model for the centers the $^3D_3$ – $^3P_3$ electronic transitions in neutral individual He atoms disturbed by different radiation defects [376, 381, 382] (fig. 105).

2.379 eV (521.0 nm), CL, ZPL, in type Ib synthetic diamonds grown using pure cobalt as the solvent-catalyst [74].

2.385 eV (519.7 nm), CL, ZPL, in type Ib synthetic diamonds grown using pure cobalt as the solvent-catalyst [74].

2.39 eV (518 nm), CL, the most intense ZPL of a set of lines within a range of 510 to 560 nm appearing in type IIa diamonds after Zn^+ ion implantation and annealing above 1000°C; ascribed to a Zn related center [92, 376] (fig. 63).

2.39 eV (518 nm), CL, ZPL, observed in type IIa diamonds implanted with Ne^+ ions and subsequently annealed at temperature above 600°C; tentatively ascribed to a neon containing defect [376] (fig. 72).

2.391 eV (518.4 nm), PL, ZPL, observed in brown diamonds which do not show yellow luminescence when excited at 365 nm; dominant interaction with 29 and 47 meV vibrations, S=4.0 [74, 87].

2.395 eV (517.5 nm), PL, ZPL, in synthetic boron doped diamonds, interacts with 50 meV vibrations [309]; exponential decay with a time constant of 228 ns; a pure intracenter transition exhibiting no recombination pathways; tentatively ascribed to a boron involving complex [315] (fig. 106).

2.400 eV (516.5 nm), the M1 center, A, ZPL, observed in diamonds of all types, radiation induced center, stable at temperatures above 800°C, interacts with 21 meV vibrations, S < 1 [13] (fig. 107).

2.403 eV (515.8 nm), CL, ZPL, in natural brown diamonds showing yellow PL; decay time in CL at a temperature of 77 K is 32.5 ns [74, 93, 105].

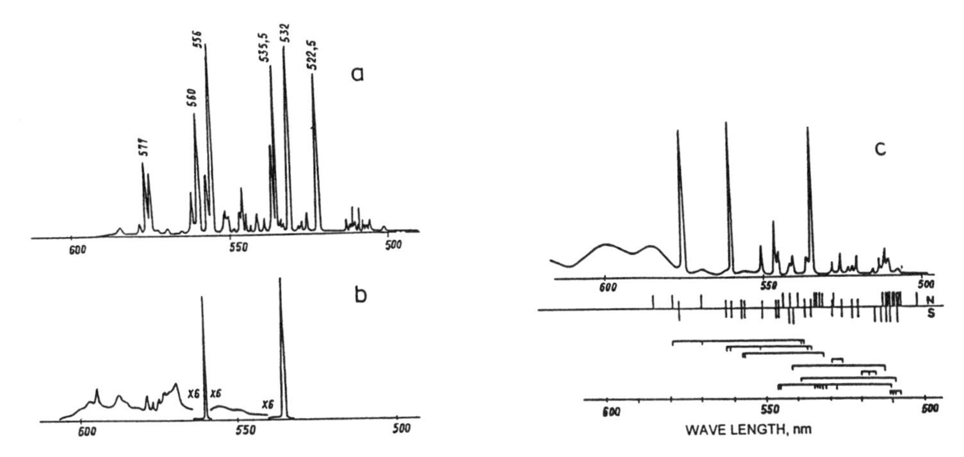

Figure 105. CL spectra of natural type IIa diamond (a, b) and synthetic nitrogen containing diamond (b) after implantation of He^{+} ions and subsequent annealing at 550°C (a), 900°C (b) and 700°C (c) [376, fig. 2.38].

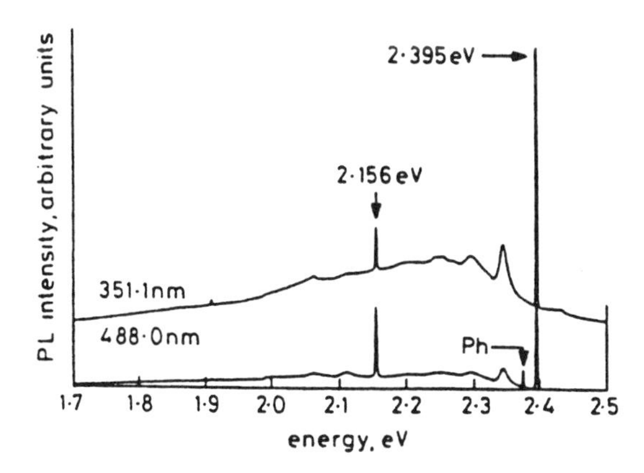

Figure 106. Continuous wave PL spectra from a boron doped HTHP synthetic diamond grown using Fe-Al as a solvent-catalyst [315, fig. 1].

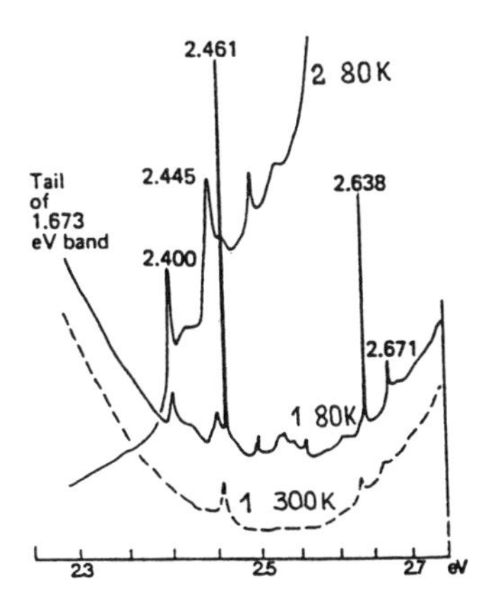

Figure 107. Absorption spectra of type IIa diamonds after electron irradiation (1) and after neutron irradiation and annealing at about 800°C (2) [13, fig. A3].

2.417 eV (513 nm), A, occasionally in PL, ZPL, in irradiated and subsequently annealed at 600°C type Ia diamonds, possibly a component of the H4 center, the line is 10 times weaker than the H4 center ZPL [3]; a center of C_{1h} symmetry with σ_h plane being perpendicular to <110> axis; linear Stark effect (no inversion symmetry) [13, 220, 222] (fig. 108).

2.42 eV (512 nm), CL, a line observed in some virgin good quality CVD films [36, 124] (fig. 109).

2.424 eV (511.3 nm), PL, ZPL, in brown diamonds which do not show yellow luminescence when excited at 365 nm; dominant interaction with 46 and 58 meV vibrations, S = 2.5 [74, 87].

2.426 eV (510.9 nm), PL, ZPL, in CVD films after electron irradiation; the center anneals out at temperatures above 400°C, complete annealing after 700°C [74].

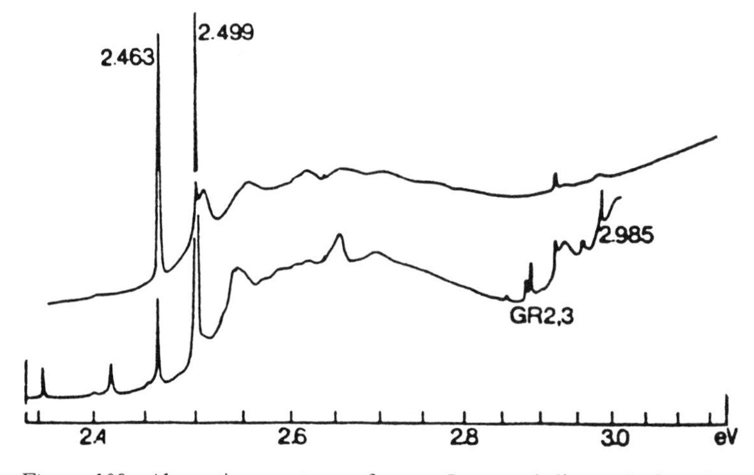

Figure 108. Absorption spectrum of a type Ia natural diamond after electron irradiation and annealing [13, fig. 36].

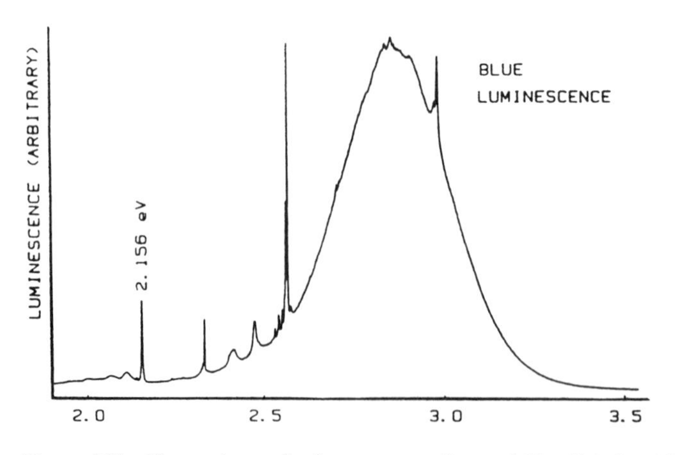

Figure 109. CL spectrum of a flame grown diamond film [36, fig. 15].

2.429 eV (510.3 nm) and 2.463 eV (503.2 nm, the main line), PL, no CL, the S1 center, ZPLs; the doublet is caused by a splitting of the excited state; the ratio of the transition probabilities of the second and the first ZPL is 20; transition from the higher excited state is not detectable at temperatures below 40 K. Observed in brown natural diamonds which do not show yellow luminescence when excited at 365 nm; created in synthetic type Ib diamonds by heating at 2200 K [3, 292]. Concentration of the centers does not exceed 2×10^{17} cm^{-3}; the concentration is reduced with increase of the A aggregate content, energy transfer occurs between the S1 centers and the H3 centers [3, 130, 252]. Five dominant phonons interact with the center the most effective of which are of energy 45, 57 and 69 meV, S = 5.9 - 7. The center of trigonal symmetry [74, 87, 106, 107, 108, 109]. Linear Stark effect (no inversion symmetry) [221]. The ground state lies by 2.8 eV below the conduction band [3]. Slow luminescent center; decay time within a range of 4 ms; the low temperature lifetime is 6 ms immediately after excitation and 10 ms upon thermalization of the spin sub-levels; the temperature range of non-exponential decay is 20-50 K [107, 130]. The center is paramagnetic (C_{2h} or C_s symmetry - contradiction with optically estimated symmetry); a possible model - single nitrogen atom bound to vacancy [3, 293, 294, 504] (fig. 110).

2.43 eV (511 nm), CL, ZPL, observed in type IIa diamonds after H^+ or C^+ ion implantation followed by annealing at 1000°C [88].

2.445 eV (507.0 nm), the M2-center, A, a narrow line observed in all diamonds; radiation induced center; persists to temperatures above 800°C; dominant interaction with 21 meV vibration, S < 1 [13] (fig. 107).

2.461 eV (503.6 nm), the AH3 center, ZPL, PL, observed in synthetic type Ib diamonds after electron irradiation. Anneals out like the 3H center at 500°C; but, unlike the 3H center, is not reactivated by an ionization dose of electrons. Dominant interaction with 74 meV vibrations [74].

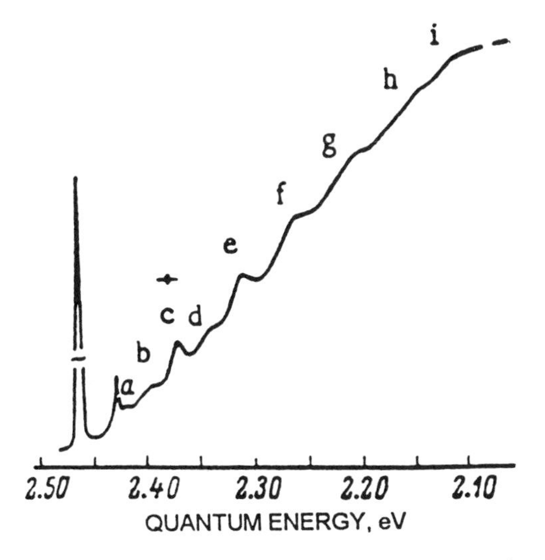

Figure 110. PL spectrum of the S1 center taken from a natural diamond [108, fig. 3].

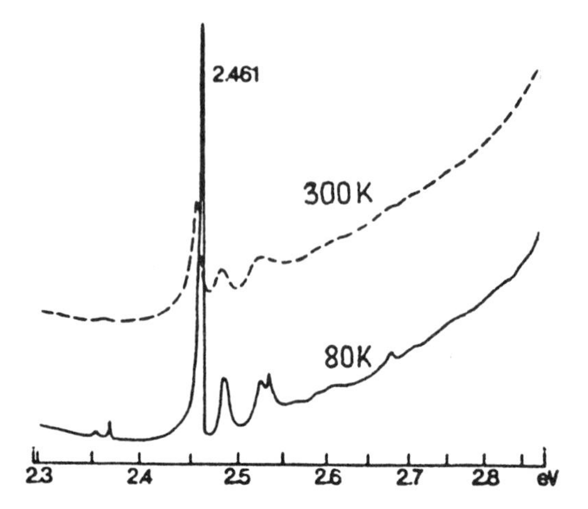

Figure 111. Absorption spectrum of a natural diamond after 2 MeV electron irradiation at room temperature [13, fig. A4].

2.462 eV (503.4 nm), the 3H center, A, PL, CL, PL excitation [74]; ZPL, observed in diamonds of all types after irradiation (for instance in CVD films [90]); quenched by UV illumination; enhanced by heating at about 300°C (effect is particularly strong in diamonds implanted by heavy ions); anneals out at temperatures above 500°C in nitrogen rich diamonds; stable to higher temperatures in type IIb diamonds; reactivated by ionization dose of electrons; bleached by an intense electron beam [104, 429]. Dominant interaction with 67 meV vibration, $S \sim 0.8$; rhombic I symmetry [13, 104, 422, 428]. Correlation with the A1 paramagnetic center. Two tentative models: (i) V-C-C-V complex (two carbon atoms in (110)-plane) and (ii) oxygen-vacancy pair [104] (fig. 111).

2.463 eV (503.2 nm), the H3 center, A, PL, CL, PL excitation, EL [398, 399], ZPL; naturally occurring in synthetic and nitrogen containing natural diamonds; created in type Ia diamonds after any irradiation and subsequent annealing at temperatures above 500°C; created in CVD films by electron irradiation and 800°C annealing [281]; created in natural type I diamonds by electrons of a few keV energy followed by annealing [388, 389]; created in natural type IIa diamonds after N^+ ion implantation and annealing above 800°C [28, 381]; created by strong mechanical deformation (in the regions of indentations, around slip traces, dislocation lines) [74, 261, 397]. The center is destroyed by heating above 1400°C for several hours [100]. In CL accompanied by a line at 2.305 eV (537.8 nm). A line at 3.368 eV seen in A and PL excitation is ascribed to a transition from the ground state of the H3 center [13, 160, 426]. Electronic transition occurs between bonding and anti bonding states localized on carbon atoms forming vacancy [334]. Isotope shift of the ZPL in ^{13}C diamonds is +5.0 meV [75, 130]. Dominant interaction with 41 and 152 meV vibrations, $S \sim 3$ [13]. Decay time in PL is 16 ns, radiative decay time of 17 ns in a temperature range 77 to 700 K; quantum efficiency is 0.95, $g_e/g_g = 1$ [130, 237, 238]; reversible population of two triplet levels results in a delayed emission with a lifetime of 3.7 ms at temperatures below 90 K and 7.6 ms at temperatures above 160 K, above 220 K the delayed emission is quenched by non-radiative processes to the ground state [130, 251]. The H3 center luminescence is efficient up to ~ 500°C [310]. The center is quenched by energy transfer to the A nitrogen aggregates, the probability of the transfer for one center to one aggregate at one lattice spacing in unit time is of 0.5×10^{15} s^{-1} [130, 237, 239]. Oscillator strength relatively to the GR1 ZPL line is H3/GR1 = 0.82 [188]. The center is formed of an A nitrogen aggregate and vacancies. There are two models: (i) N-V-N complex (pair of substitutional nitrogen atoms separated by vacancy), symmetry rhombic-I (C_{2V}) [130, 145, 217]; (ii) V-N-N-V complex (A nitrogen aggregate bound to two vacancies), symmetry monoclinic-I (C_{2h}), no inversion symmetry [28, 116, 221, 371, 372, 28, 383, 390] (fig. 112).

2.464 eV (503 nm), CL, ZPL, observed in some dilute CO CVD-grown diamond [218, 340].

2.464 and 2.487 eV (498.2 and 502.8 nm), CL, ZPLs, in natural type IaB diamonds appears following oxygen ion implantation and subsequent 1000°C annealing; intensity sharply increases by annealing at 1400-1500°C; the center contains oxygen and, possibly, a fragment of the nitrogen B aggregate [330].

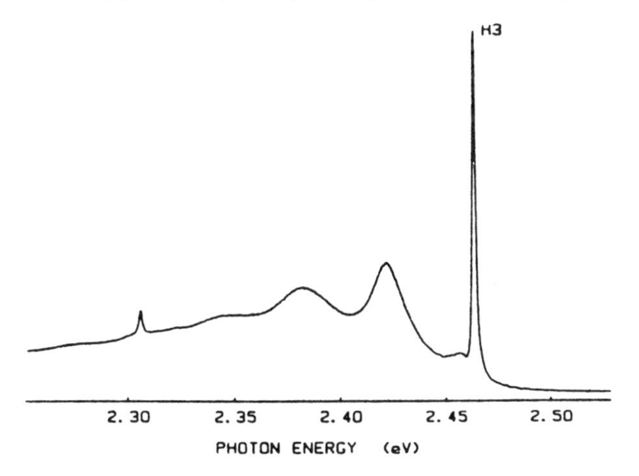

Figure 112. CL spectrum of the H3 center taken from [100] growth sector of a synthetic diamond containing about 200 ppm of nitrogen [36, fig. 8].

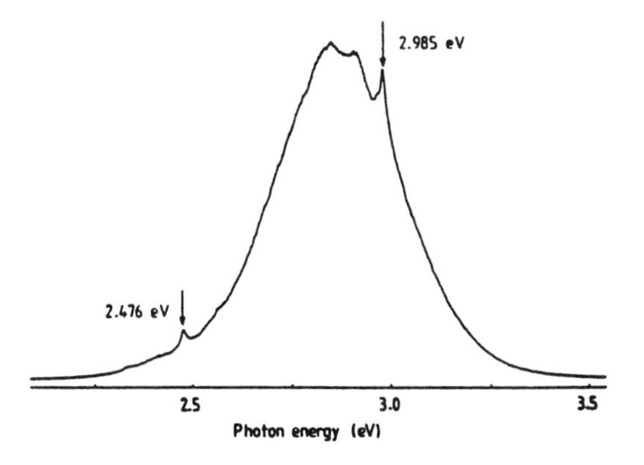

Figure 113. CL of a single crystal CVD diamond film [98, fig. 2a].

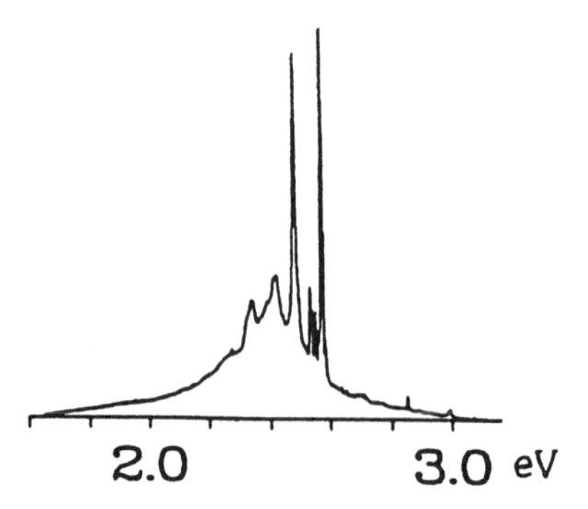

Figure 114. Time resolved CL spectra of a MPCVD diamond film (scan with gate delay at 0.2 μs [312, fig. 1iii].

2.47 eV (501.8 nm), CL, a line observed in some virgin good quality CVD films [36, 124, 340] (fig. 109, 129).

2.476 eV (500.6 nm), CL, a narrow line observed in some CVD films [74, 90, 98] (fig. 113).

2.477 eV (500.5 nm), A, ZPL, in type Ib synthetic diamonds grown by the temperature gradient method using a nickel containing catalyst and in synthetic diamonds after high PT treatment (1950°C) [361].

2.478 eV (500.2 nm), CL, ZPL, observed in natural brown diamonds showing yellow PL; a center of triclinic symmetry ?; decay time in CL at 77 K is 40 ns [74, 93, 105].

2.479 eV (500.0 nm), L, in diamond films [340].

2.48 eV (499.8 nm), CL, ZPL, observed in CVD diamond films; decay time 1.0±0.1 μs [312] (fig. 114).

2.490 eV (497.8 nm), PL, ZPL, observed in brown diamonds 1 ms after excitation, strong electron-phonon coupling, S = 6.6 [74, 102] (fig. 134b).

2.495 eV (496.7 nm) and 2.620 eV (473.3 nm), the S3 center, PL, PL excitation, ZPLs, observed in natural diamonds particularly in those of mixed cubo-octahedral growth habit; observed in synthetic diamonds grown in the presence of Ni after annealing at 2150 and 2200 K [345, 292, 357, 361]. In PL excitation spectra from as-grown synthetic diamonds there is related fine structure: 2.658 eV (466.5 nm), 2.652 eV (467.6 nm), 2.623 eV (472.8 nm), 2.598 eV (477.2 nm), 2.592 eV (478.4 nm) [357]; the center interacts with 40 and 70 meV vibrations [361]. Slow luminescence center, the low temperature lifetime is 0.18 ms after excitation, 0.36 ms upon thermalization of the spin sub-levels, the temperature range of non-exponential decay is 20-30 K; quartet-doublet spin-forbidden transition. Ratio of the transition probabilities of the second and the first ZPL is 10^4 [246, 130]. The center correlates with the cloud-like defects responsible for the "hydrogen" absorption lines at 1405 and 3107 cm^{-1} [74, 110]. The center is paramagnetic - the NE1 center [357]. A possible model - Ni^+ ion in divacancy surrounded by two or three nitrogen atoms in the nearest co-ordination sphere [357] (fig. 115).

2.497 eV (496.4 nm), CL, ZPL, observed in some CVD diamonds [340] (fig. 129).

2.498 eV (496.2 nm), the H4 center, CL, PL, A, PL excitation, ZPL, naturally occurring in type IaB diamonds; induced by radiation and subsequent annealing at temperatures above 600°C; formed in type Ib synthetic diamonds after heating at about 2400°C [84]. A center of monoclinic-I (C_{1h}) symmetry with σ_h plane being perpendicular to <110> axis [147, 220, 130]. Linear Stark effect (no inversion symmetry) [222]. Oscillator strength relative to the GR1 center ZPL is H4/GR1 = 0.86 [188]. The H4 center, the H3 center, the A and B aggregates relate in absorption as: H4/H3 = 1.5 [μ_B/μ_A] [130]. Inhomogenous width of ZPL is of 800 GHz, hole burning in the ZPL has been shown [227]. Decay time is 19 ns in CL at a temperature of 21 K, quantum efficiency is close to unity [130, 240]. The electronic transition occurs between bonding and anti bonding states localized on carbon atoms surrounded vacancy [334]. Dominant interaction with 40 and 154 meV vibrations [13]. A possible model: - the B nitrogen aggregate bound to a vacancy (fig. 116).

2.501 eV (495.6 nm), PL, ZPL, naturally occurring in ballas [3].

2.500 eV (495.8 nm), A, ZPL doublet, radiation damage product in type IIa diamonds, stable to annealing above 800°C [13] (fig. 107).

2.507 eV (494.5 nm), CL, ZPL, observed in natural diamonds implanted with Ti^+ ions and annealed at temperatures above 1300°C [289].

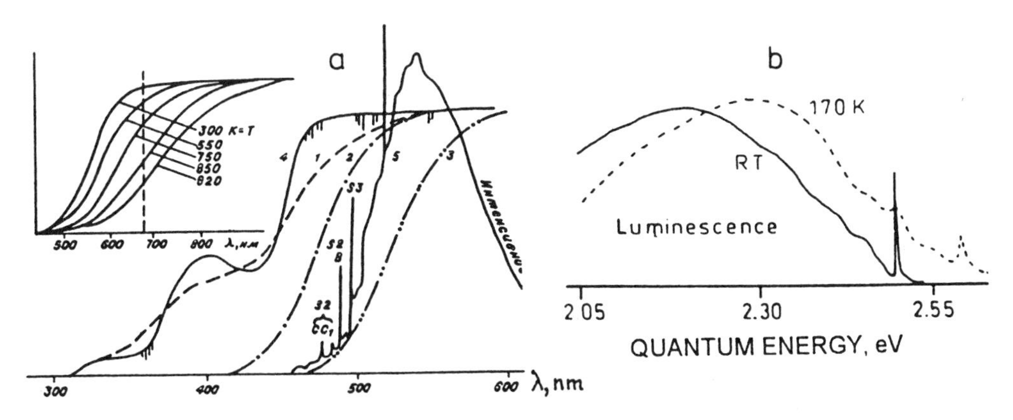

Figure 115. (a) - Transmission and PL spectra of synthetic diamonds grown by the temperature gradient method: as-grown (1, 2, 3, 5), and annealed at 2150 K (4) in colorless (1) and yellow (2) sectors [361, fig. 3]; (b) - PL spectra of the S3 center [433, fig. 3].

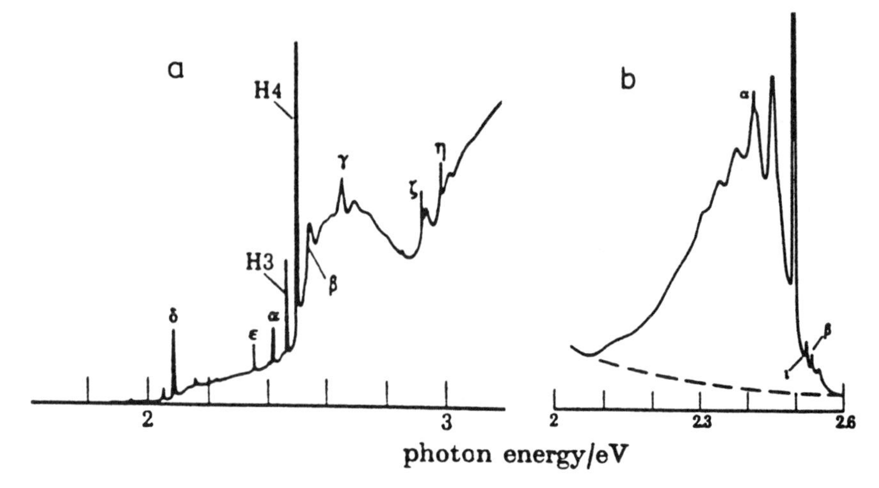

Figure 116. Absorption (a) and photoluminescence (b) spectra of natural diamonds after electron irradiation and annealing at about 700°C [220, fig. 1].

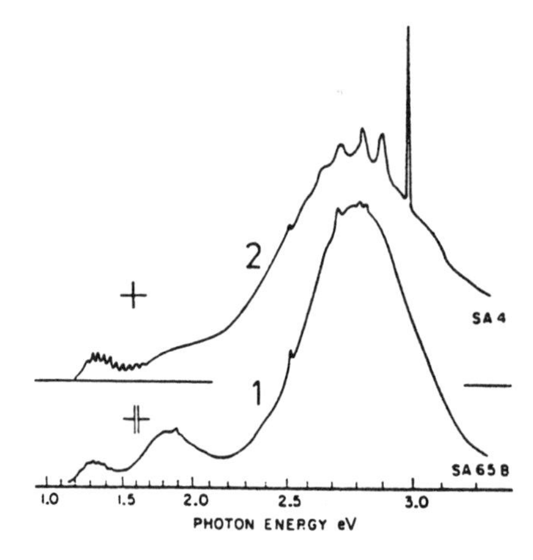

Figure 117. CL spectra from type IIb (1) and intermediate type (2) natural diamonds [169, fig. 10].

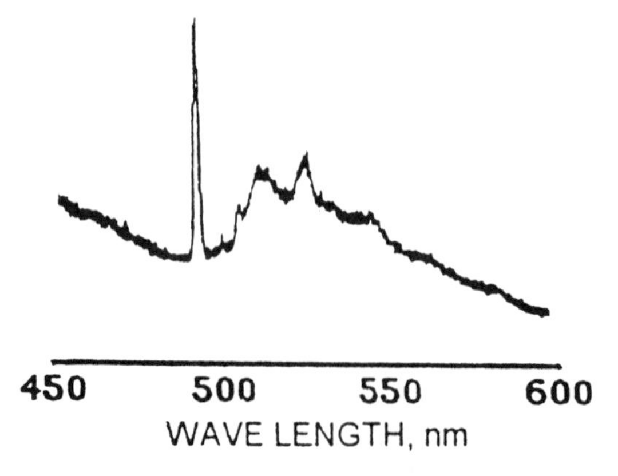

Figure 118. CL spectrum of a natural brown diamond [395, fig. 1b].

2.51 eV (494 nm), A, ZPL, in type Ib synthetic diamonds grown by the temperature gradient method using nickel; intense in diamonds with a high nitrogen concentration; segregated in the (111) growth sectors; tentatively ascribed to a nickel related center [74, 89, 83, 112, 179, 282]. The center correlates with an EPR center (g = 2.0319) which has been identified as Ni^- with $3d^7$ electronic configuration [180] (fig. 77).

2.520 eV (492.0 nm), CL, a narrow line naturally occurring in type IIb diamonds both synthetic and natural [13, 169] (fig. 117).

2.523 eV (491.3 nm), CL, ZPL, observed in fault-free areas of brown diamonds, is proposed to be connected with N-C-C-N defect [396]; the phonon assisted band consist of peaks at 63, 97, 157 and 243 meV; probably related to the 490.7 nm center [395] (fig. 118).

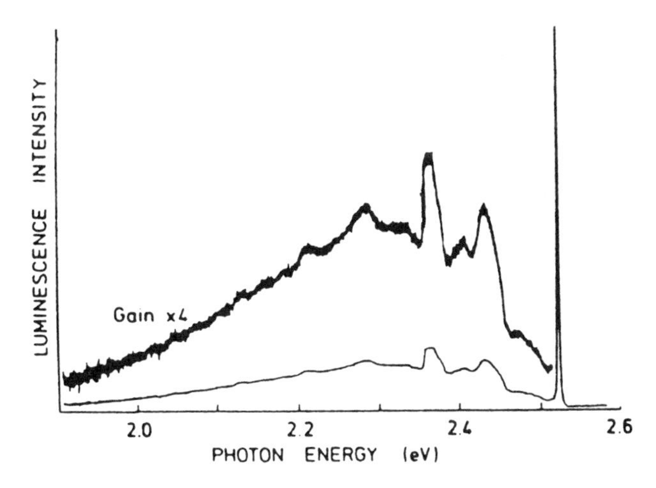

Figure 119. CL spectrum of the 2.526 eV center in natural diamond [154, fig. 3].

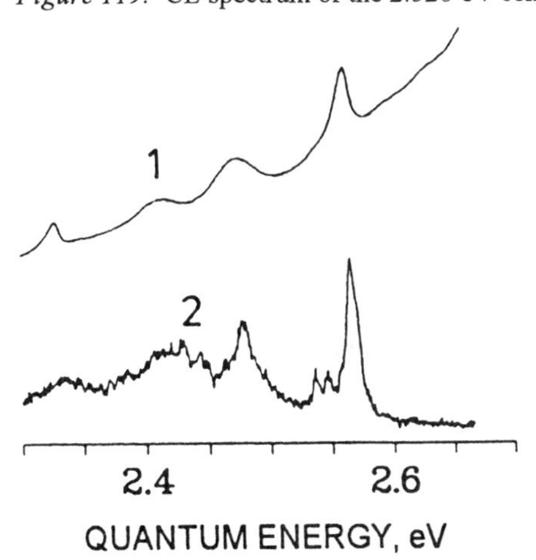

Figure 120. Conventional (1) and time-resolved (2) spectra of a MP CVD diamond film [312, fig. 4].

2.526 eV (490.7 nm), A, CL, PL, ZPL, observed in type IaB diamonds; associated with defects decorating slip traces; naturally occurring and generated by plastic deformation; naturally occurring in carbonado [3]; dominant interaction with vibrations 80, 106, 143 meV, S = 2.8; a center of monoclinic I symmetry; decay time in CL at 77 K is 70 ns [74, 102, 29, 111, 93, 154, 290]; a possible model: N-V-N complex [3] (fig. 119).

2.535 eV (489.0 nm) (I), A, no CL; ZPL, radiation damage product in type Ib diamonds, enhanced by annealing at 200°C simultaneously with annealing of the 2.367 eV center, anneals out at 350°C; dominant interaction with 37 and 74 meV vibrations, S ~ 1 [13]; probably relates to nitrogen bound to carbon interstitials [74, 91, 443] (fig. 82).

2.535 eV (489.0 nm) (II), CL, ZPL, observed in type Ib synthetic diamonds after heating at 1700°C when a partial nitrogen aggregation is produced; no relation to the 2.535 eV (I) absorption system [74, 84] (fig. 86).

2.536 eV (489 nm), A, a narrow line; possibly a component of the H4 center, the line is 50 times weaker than the H4 ZPL [3]; C_{1h} symmetry with σ_h plane being perpendicular to <110> axis [220] (fig. 116).

2.537 eV (488.5 nm), PL, the main component of a group of narrow lines naturally occurring in carbonado [3].

2.537 eV (488.6 nm), CL, in CVD films, decay time 1.8±0.1 μs [312] (fig. 120).

2.543 eV (487.5 nm), the TH5-center, A, ZPL, in type IIa diamonds after irradiation and subsequent annealing at 600°C; the related lines are: 2.552, 2.563, 2.572, 2.583, 2.590, 2.602, 2.610 and 2.618 eV; proposed to be divacancy related defect [13, 130, 236, 259] (fig. 121).

2.547 eV, CL, ZPL, in CVD diamond films; decay time 1.3±0.1 μs [312] (fig. 114, 120).

2.556 eV (484.9 nm), PL, ZPL, observed in brown diamonds 1 ms after excitation [74].

2.556 eV (484.9 nm), CL, ZPL, in type Ib synthetic diamonds grown using pure cobalt as the solvent-catalyst [74].

2.556 eV (484.9 nm), PL, ZPL, in type Ia irradiated diamonds, in IaB diamonds the ZPL has four components at 2.5541, 2.5555, 2.5570 and 2.5584 eV resembling the 484 nm nickel line; probably associated with slip traces, some correlation in intensity with the 490.7 nm system [74].

2.56 eV (484 nm), CL, PL, EL [400], ZPL, observed in (111) growth sectors of synthetic Ib diamonds grown using nickel; in natural diamonds after Ni^+ ion implantation; in some CVD diamond films [51, 181]. ZPL consists of four thermalizing components at 2.5587, 2.5602, 2.5618, 2.5634 eV resulted from splitting of the excited state; further fine structure is seen at low temperature [314].

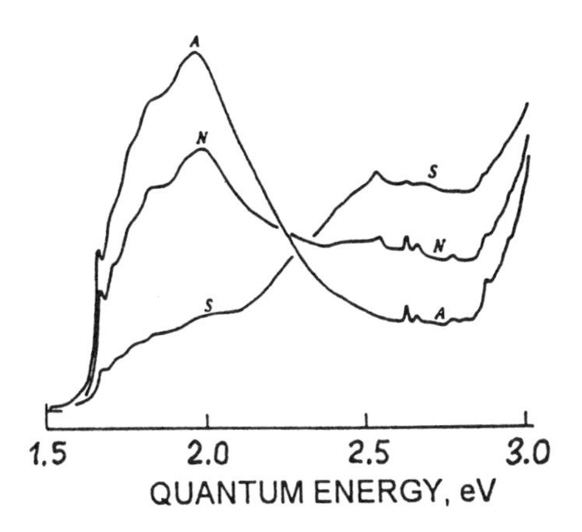

Figure 121. Room temperature absorption spectra of a natural diamond irradiated with 1 MeV electrons (A) and subsequently annealed at 600°C for 4 h (N) and 28 h (S) [259, fig. 4].

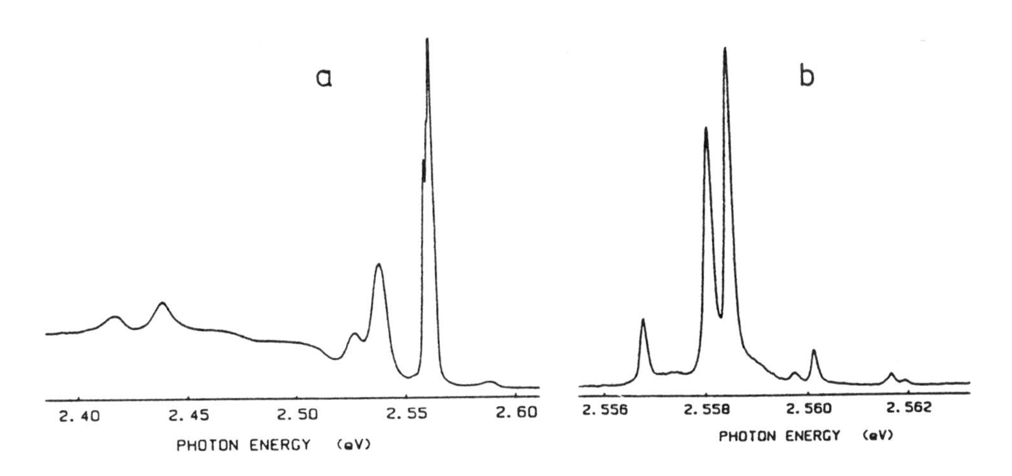

Figure 122. CL spectrum of the 2.56 eV center taken from a [111] facet of a synthetic diamond at 80 K (a) and fine structure in the ZPL region recorded at 8 K [36, fig. 9; 83, fig. 8].

ZPL is thermolized with a broad line at 2.588 eV (two or more overlapping states) which is believed to be a second excited state of the center. No absorption sideband but strong PL is excited by long wave UV light; CL decay time is 10.5 ns [74, 81, 112, 130, 89, 179, 130]; anneals out at 2150°C [357]; nickel related center probably incorporating one Ni atom [486] (fig. 122).

2.56 eV (484 nm), CL, a broad band with FWHM of about 0.2 eV, observed in randomly oriented polycrystalline diamond films; in undoped (100) highly oriented diamond films the band shift to a position 2.61 eV (474 nm) [416] (fig. 123).

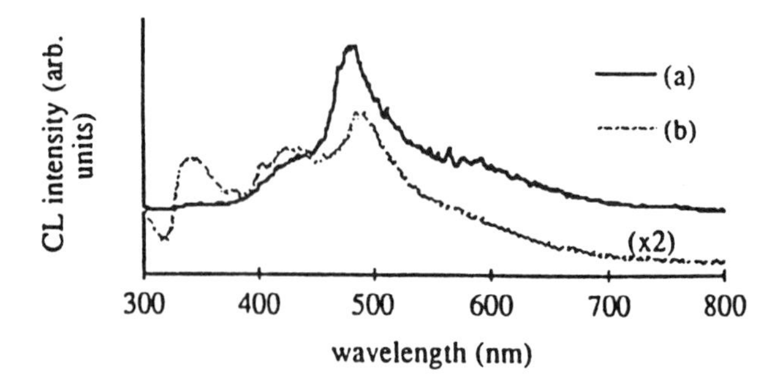

Figure 123. CL spectra of CVD diamond films: (a) - (100) oriented, and (b) - random polycrystalline [416, fig. 1].

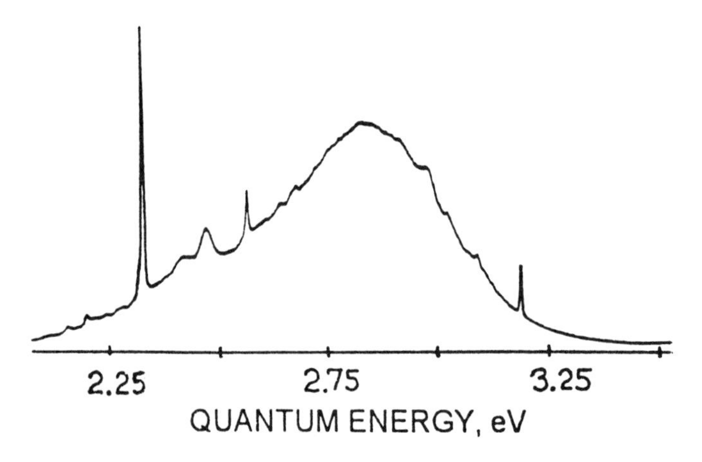

Figure 124. CL spectrum of a MP CVD diamond film [340, fig. 2b].

2.557 eV, CL, ZPL, observed in CVD diamond films, decay time 1.2±0.1 μs [312] (fig. 114, 120).

2.561 eV (484.0 nm), CL, a narrow line observed in MPCVD films on Si substrate; proposed to be an oxygen-related center [130, 187] (fig. 131).

2.567 eV (482.9 nm), CL, ZPL, observed in some CVD diamond films [36, 74, 98, 124, 187, 340]; decay time 1.5±0.1 μs [312] (fig. 114, 124).

2.57 eV (482 nm), CL, the most intense line of a set of very narrow ZPLs in a range from 2.52 to 2.58 eV; observed in some virgin good quality CVD films [36, 124] (fig. 109).

2.572 eV, CL, ZPL, in CVD films, decay time 1.3±0.1 μs [124, 312] (fig. 114).

2.590 eV (478.6 nm), CL, ZPL, observed in type Ib synthetic diamonds grown using pure cobalt as the solvent-catalyst [74].

2.596 eV (477.6 nm), the N2 center, A, naturally occurring in type Ia diamonds; ascribed to a vibronic transition at the N3 center in which the electronic excitation is forbidden [3, 130, 144, 161, 284]; interacts with 89 meV vibrations; the lines of the center are always relatively broad [13] (fig. 60, 74).

2.599 eV (476.9 nm), CL, ZPL, observed in some CVD diamond films [74, 340] (fig. 129).

2.6 eV (480 nm) a broad absorption band, causing the amber color of some diamonds, FWHM of 0.25 eV; observed in yellow luminescent diamonds of a mixed type Ia and Ib diamonds; intense in hydrogen rich chameleon diamonds [284]; there is a corresponding mirror image in PL, very large Huang-Rhys factor and the ZPL is undetectable [74, 114] (fig. 125).

2.61 eV (474 nm), A, observed in natural diamonds having a high hydrogen content [74, 284] (fig. 74).

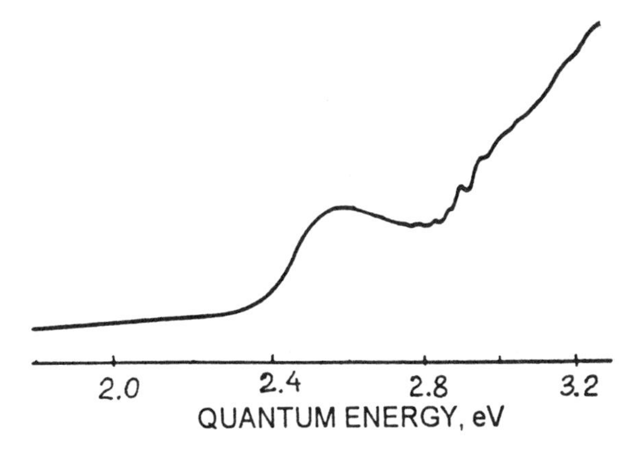

Figure 125. Absorption spectrum of a typical yellow luminescent diamond [114, fig. 1a].

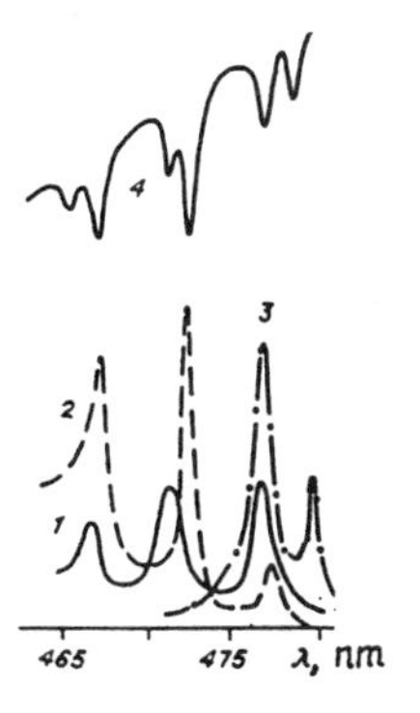

Figure 126. PL excitation (1-3) and transmission (4) spectra of synthetic diamonds grown by the temperature gradient method [361, fig. 6].

2.624 eV (472.8 nm), A, PL excitation of the S3 center, in Ib synthetic diamonds grown by the temperature gradient method and in synthetic diamonds after high PT treatment [83, 361] (fig. 126).

2.638 eV (469.9 nm), the TR12-center, A, PL, PL excit., CL, ZPL, observed in all types of diamond after any irradiation treatment [373]; also in CVD films [90]. One-phonon sideband at 2.671 eV in absorption is relatively sharp and is called TR13 [74, 113]; there are lines possibly related to the TR12 center: 2.645 (TR12a), 2.68, 2.70, 2.777 (TR14), 2.788 (TR15), 2.817 (TR16) and 2.831 (TR17) eV [144]. The center interacts with quasi local vibrations of 34 (dominant), 67 and 140 meV in absorption and 70 meV in luminescence, as well as with local vibration of 200 meV; the 200 meV vibration is carbon related and involves several carbon atoms [177]; S ~ 0.6 - 0.8 in absorption [13, 104, 113]. The symmetry of the center is probably monoclinic-I with a symmetry plane (110) [104, 113, 374]. The center enhanced by heating at temperatures 400-600°C in type II diamonds, this effect is particularly strong in diamonds implanted with high energy heavy ions; in type Ia diamonds the enhancement does not occur; the center anneals out at temperatures above 700°C [104, 431, 376]. Strongly quenched by nitrogen including ion implanted nitrogen [88, 104, 379]. The center is tentatively ascribed to a defect involving vacancies and interstitial carbon atoms [88] (fig. 127).

2.642 eV (469.2 nm), CL, ZPL, observed in some CVD diamond films [74, 340] (fig. 129).

2.649 eV (467.9 nm), CL, ZPL, observed in natural brown diamonds which show yellow luminescence when excited at 365 nm; closely related to the 2.699, 2.721 and 2.748 eV lines; dominant interaction with vibrations of 36 meV; possible symmetry of the center is rhombic I [74, 105, 205, 93] (fig. 132).

2.65 eV (468 nm), CL, ZPL, observed in type IIa diamonds after H^+ ion implantation [74, 88].

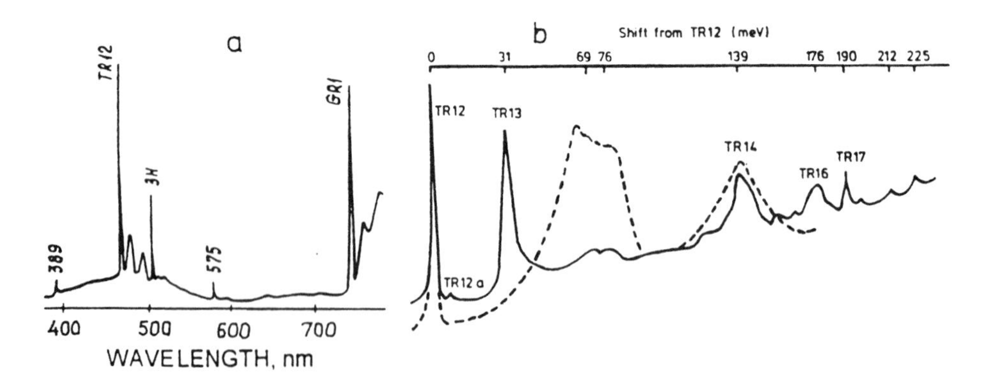

Figure 127. CL spectrum of a type IIa natural diamond as-irradiated with carbon ions at a dose of 10^{14} cm^{-2} (a), and a comparison of the TR absorption (full line) and PL (broken line) spectra (b) [376, fig. 2.2a; 104, fig. 24].

2.65 eV (468 nm) and a set of accompanying lines at 2.62, 2.59, 2.51, 2.47, 2.40, 2.39, 2.38, 2.35, 2.30 and 2.24 eV, A, due to Ni in HTHP synthetic diamond [84, 325]

2.651 eV (467.5 nm), PL, ZPL, observed in CVD diamond films grown by direct current plasma-jet method; the center interacts with vibrations of 73 meV, S = 3.2 [74].

2.652 eV (467 nm), A, possibly a component of the H4 system [3, 220] (fig. 116).

2.653 eV (467.2 nm), A, PL excitation of the S3 center, observed in type Ib synthetic diamonds grown by the temperature gradient method and in synthetic diamonds after high PT treatment [361] (fig. 126).

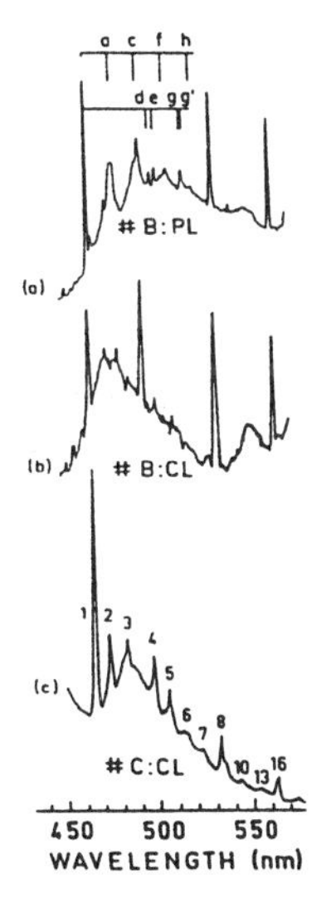

Figure 128. PL (a) and CL (b) spectra of an undoped [100] epitaxial CVD diamond film, and CL from a boron doped PCCVD diamond (c) [73, fig. 2].

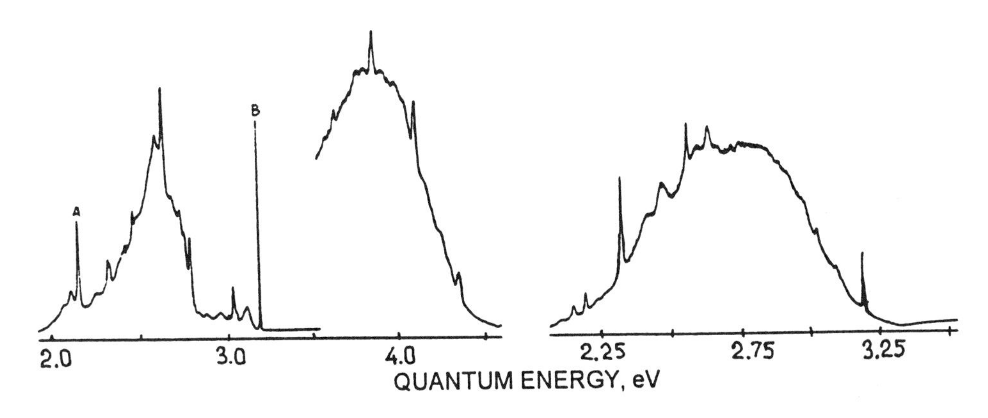

Figure 129. CL spectra of MPCVD diamond films [340, fig. 2a, 3].

2.67 eV (464 nm), CL [218, 340]; accompanied by a set of narrow lines at 2.62, 2.57, 2.50, 2.46, 2.41, 2.37, 2.33, 2.28, 2.24, 2.20, 2.11, 2.07 eV; observed in some CVD diamond films; ZPLs; possibly relates to the 2.67 eV band reported in [187]; ascribed to donor-acceptor pair recombination (possibly nitrogen and boron) [73] (fig. 128).

2.670 eV (464 nm), CL, a complicated band in a range from 2.2 to 2.8 eV, observed in some CVD diamonds [340] (fig. 129).

2.68 eV (461 nm), long time phosphorescence, ZPL, the N3a center, observed at a temperature of 4.2 K in natural diamonds containing the N3 centers and B' aggregates; the phonon assisted band structure is similar to that of the N3 center; the center is excited in the N3 absorption spectrum; ascribed to the N3 center, the excited state of which is shifted down due to interaction with the B' defects [3, 291, 370] (fig. 99).

2.694 eV (460.1 nm), PL, ZPL, observed in synthetic diamonds grown in the presence of Al and Ti; a slow luminescence center [130].

2.70 eV (459 nm), CL, ZPL, observed in epitaxial undoped CVD diamond films grown on [100] natural diamond surface at 77 K [73] (fig. 130).

2.7 eV (460 nm), a complicated band in a region of 2.75 to 2.65 eV consisting of several narrow lines the most intense of which are at 2.676, 2.681, 2.692, 2.696, 2.702 and 2.725; CL, observed in MPCVD films grown from oxygen containing gas mixture on Si substrate; are proposed to be oxygen-related centers [130, 187] (fig. 131).

2.712 eV, see 1.908 eV.

2.718 eV, see 1.908 eV.

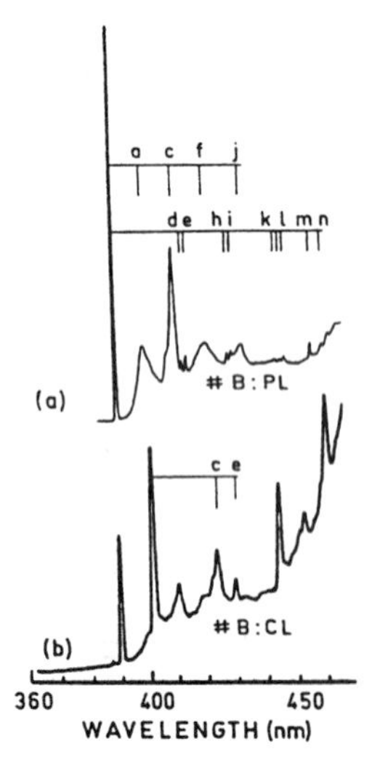

Figure 130. PL (a) and CL (b) spectra of a [100] homoepitaxial CVD diamond film [73, fig. 5].

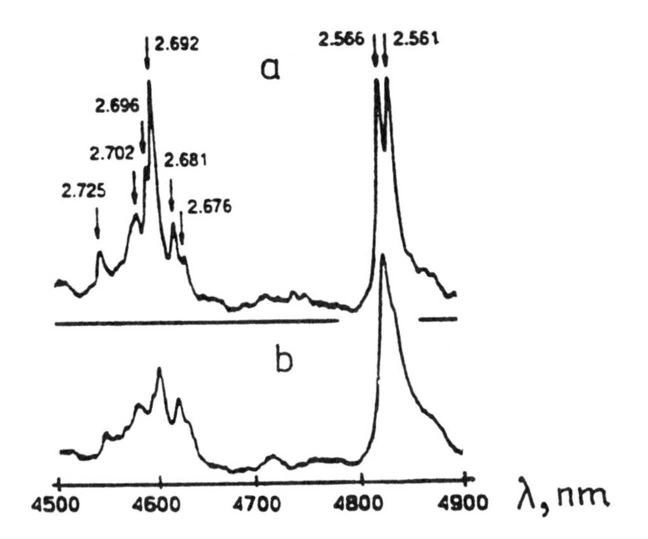

Figure 131. CL spectra of CVD diamond films grown with oxygen and boron (a) and oxygen (b) [187, fig. 2].

2.721 eV (455.5 nm), PL, PL excit., CL; in natural brown diamonds showing yellow luminescence under 365 nm excitation, vibration 34 meV, S = 7.5, monoclinic I, decay time in PL at temperatures $\leq$ 300 K is 5.8 ns, activation energy for non-radiative transitions is 0.127 eV [36, 74, 93, 114, 205, 287] (fig. 132).

2.725 eV, CL, a narrow line naturally occurring in type IIb diamonds both synthetic and natural [13, 169] (fig. 117).

2.748 eV (451.0 nm), CL, ZPL, in natural brown diamonds showing yellow luminescence under 365 nm excitation; closely related to the 2.649, 2.699, 2.721 eV lines, possibly rhombic I symmetry [74, 205, 93] (fig. 132).

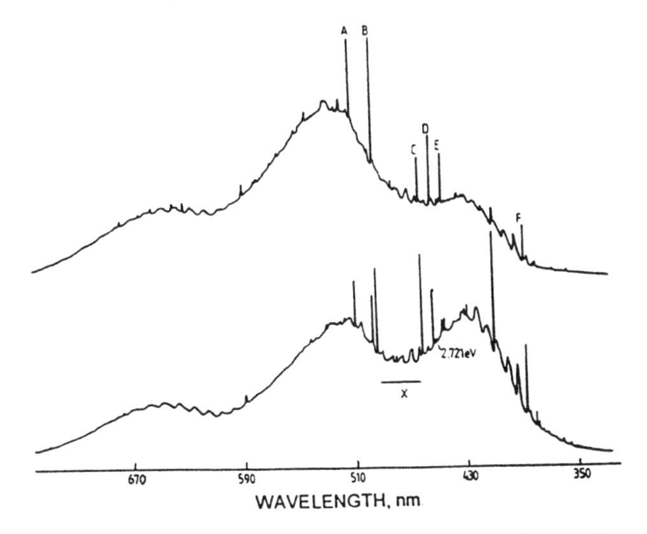

Figure 132. CL spectra of yellow-photoluminescent brown diamonds [205, fig. 1].

2.756 eV (449.7 nm), CL, ZPL, in some CVD diamonds [74, 340] (fig. 129).

2.767 eV (448 nm), A, a sharp line observed in hydrogen rich gray-violet diamonds [284].

2.792 eV (444 nm), A, a sharp line observed in hydrogen rich gray-violet diamonds [284].

2.786 eV (444.9 nm), CL, ZPL, observed in some CVD diamonds [74, 340] (fig. 129).2.8 eV (440 nm), A, a broad band of FWHM 0.25 eV appearing in nitrogen containing synthetic diamonds after high PT (~1900°C, 5.5 GPa) treatment [357, 361] (fig. 115).

2.807 eV (441.6 nm), CL, ZPL, radiation damage product, the strongest in type Ib diamonds, post irradiation annealing at 600°C increases the center intensity strongly; the center can be activated athermally by electron beam; naturally occurring in CVD diamonds [73, 90, 340]; created in type IIa diamonds by N^+ ion implantation; in ion implanted diamond anneals out at 1200°C; ZPL shifts in ^{13}C diamonds by +3.4 meV [130]. A set of sharp lines between 2.60 and 2.64 eV is associated with local vibrational modes connected to carbon atoms (for instance the 180 meV mode). A model involving a single nitrogen + a few carbon interstitials [74, 76, 115]. In many respects analogous to the 3.188 eV center (fig. 133); probably observed in epitaxial undoped CVD diamond films grown on [100] natural diamond surface as lines at 2.80 eV (443 nm) and 2.81 eV (441 nm) [73, 75, 78] (fig. 130).

2.815 eV (440.3 nm), PL, ZPL, paramagnetic, possibly relates to NV_2 complex [293, 294].

2.817 eV (440 nm), A, a narrow line observed in hydrogen rich diamonds [284] (fig. 89).

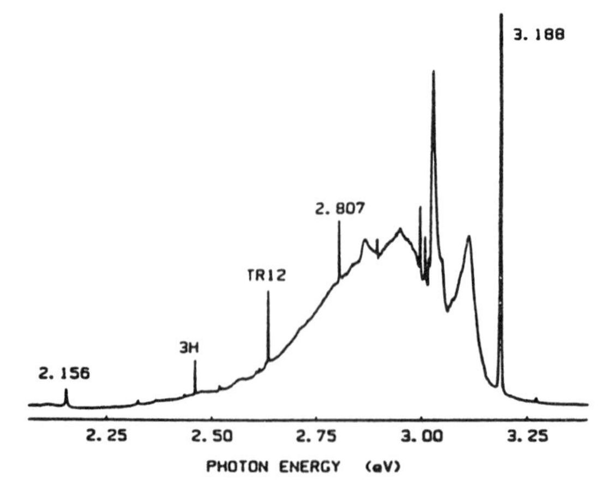

Figure 133. M CL spectrum of a single crystal CVD diamond after electron irradiation [90, fig. 6].

2.818 eV (439.9 nm), PL, observed in brown diamonds; slow luminescent center, the low temperature lifetime is 1.04 ms after excitation and 1.7 ms upon thermalization of the spin sub-levels, the temperature range of non-exponential decay is 20-60 K; triplet-singlet spin-forbidden transition [102, 130]; dominant vibrations 50 and 140 meV, S = 6.6 [74]. The center is paramagnetic; the triplet sublevel lifetimes are 0.5 ms.(T_x), 1.8 ms (T_y) and 23 ms (T_z); cross-relaxation effects are observed as abrupt changes in the luminescence intensity [359, 360]; tentatively the center involves a vacancy or oxygen atoms as, for instance, an O–C–O defect in the (110) plane [358, 359] (fig. 134).

2.82 eV (439 nm), CL, in ion implanted type Ia diamonds after annealing above 800°C, in type IIa diamonds after N^+ ion implantation followed by 1400°C annealing; a center of monoclinic-I or trigonal symmetry with electronic transitions between A and E states; probably related to a vacancy type nitrogen containing defect [74, 88, 99].

2.833 eV (437.6 nm), CL, a weak line naturally occurring in type IIb diamonds [13].

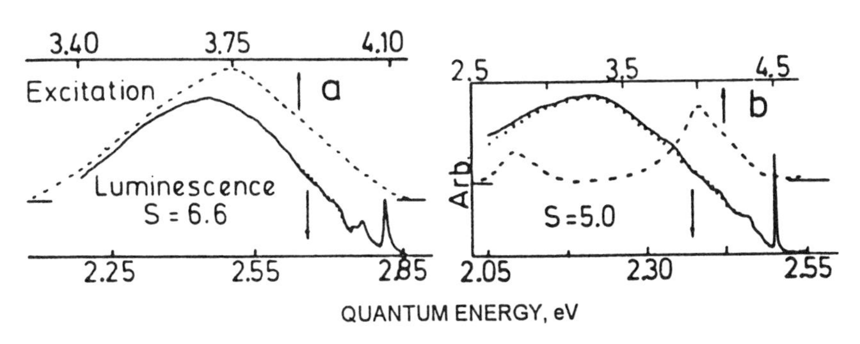

Figure 134. spectra (full line) and excitation spectra (broken line) of the 2.818 eV center (a) and 2.490 eV center (b) recorded in brown natural diamond [102, fig. 1, 2].

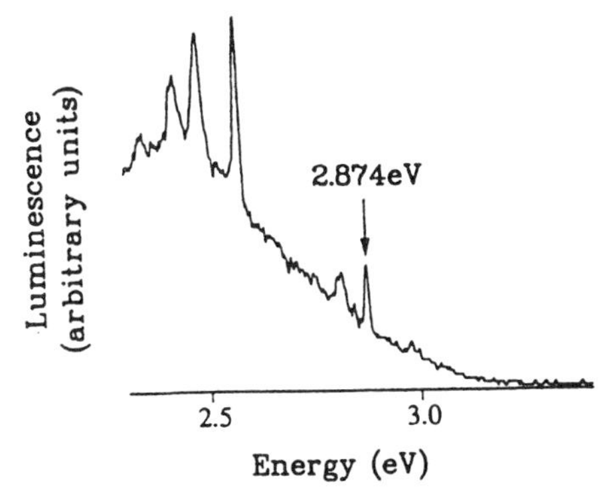

Figure 135. Time resolved spectrum of a CVD diamond film with a gate delay of 0.5 μs [312, fig. 3a].

2.87 eV (432 nm), A, a narrow line observed in natural diamonds with a high hydrogen content [74, 284] (fig. 89).

2.874 eV and 2.802 eV, CL, doublet of narrow lines observed in CVD diamond films, decay time 0.6±0.1 μs [312] (fig. 135).

2.88 eV (435 nm), the A-band, CL, EL [74, 80], FWHM of 0.45 eV, the spectral position of the maximum may range from 2.8 to 2.99 eV; observed in type II natural diamonds, CVD films and HPHT synthetic diamonds [24]; the band widens in CVD films as the deposition temperature decreases [275]; in polycrystalline CVD films is particularly strong around grain boundaries [416]; the band is quenched by ion implantation but recovers with subsequent annealing [37]; nitrogen quenches the band [270]; Non-exponential decay, the decay time may change from less than 1 ms to greater than 50 ms as the intensity drops to 1/e of the initial value [130, 249], in CVD films there are two decay times depending on the band width according to the relations: 45±10 ns/eV and 625±195 ns/eV [312]. A model of donor-acceptor recombination at dislocations [25]. A model of luminescence on pure dislocations (excluding D-A recombination) [55]. Electronic transitions from deep lying acceptor centers to the valence band [79]. (fig. 136).

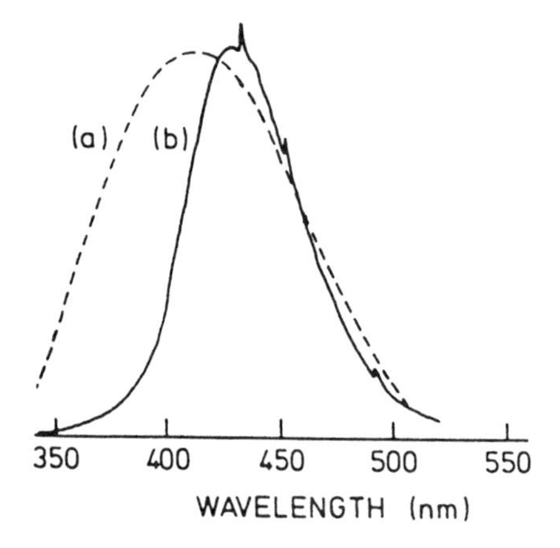

Figure 136. CL of the A band from natural diamonds: (a) type Ia, (b) type IIa and IIb [36, fig. 4].

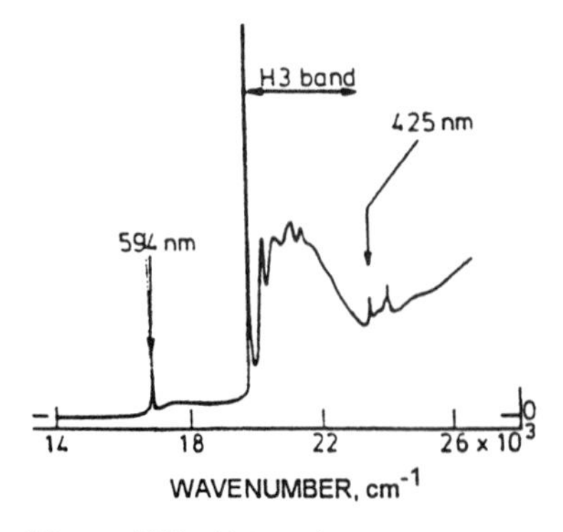

Figure 137. Absorption spectrum of a nitrogen containing diamond after electron irradiation and annealing at about 700°C [96, fig. 1b].

2.88 eV (430 nm), a broad band in a range from 390 to 480 nm, the α-band, observed in PL excitation of the S2 and S3 centers; at 80 K the band is assisted with ZPLs at 478.9, 477.6, 472.3, 467 nm [3].

2.889 eV (429.0 nm), CL, ZPL, in type Ib synthetic diamonds grown using pure cobalt as the solvent-catalyst [74].

2.916 eV (425.1 nm), A, ZPL, in type I diamonds, possibly electronic transition to a higher excited state of the 595 nm center; a center of D_{3d} symmetry; quadratic Stark effect (inversion symmetry) [74, 96, 130, 222] (fig. 137).

2.916 eV (425 nm), A, a narrow band observed in hydrogen rich gray-violet and chameleon diamonds [284].

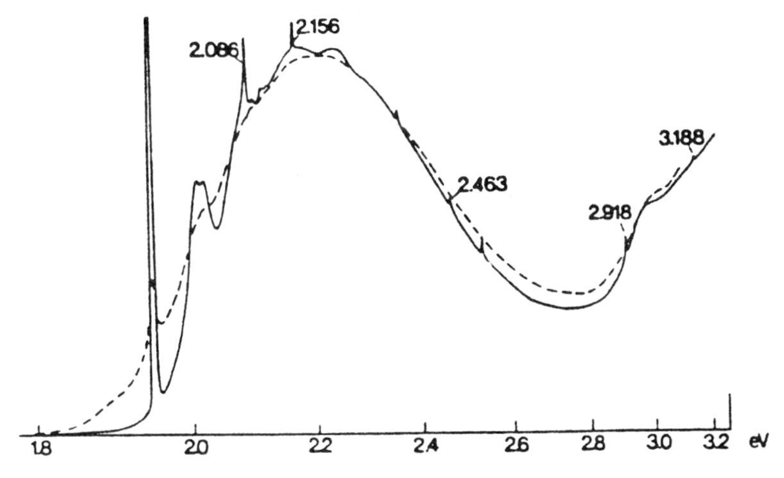

Figure 138. Absorption spectrum of an electron irradiated and annealed type Ib diamond recorded at 80 K (solid line) and 300 K (broken line) [13, fig. 31].

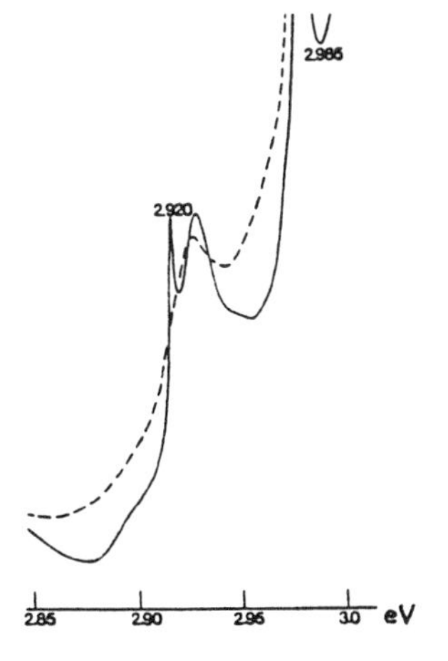

Figure 139. Absorption spectrum of a type IaB diamond after 2 MeV electron irradiation and annealing at 900 K [13, fig. A7].

2.918 eV (424.9 nm), A, ZPL, in type Ib diamonds, observed after irradiation and annealing at about 600°C; dominant interaction with 55 meV vibrations [13] (fig. 138).

2.920 eV (424.6 nm), A, in type IaB diamonds, after irradiation and annealing at about 600°C, interacts with 12 meV vibration, probably related to the H4 center [13] (fig. 139).

2.925 eV (423.8 nm). A, the H6 center; in type I diamonds after irradiation and annealing at temperatures above 500°C [259] (fig. 147).

2.964 eV (418.2 nm) and 2.974 eV (416.8 nm), PL, doublet of ZPLs, observed in brown natural and in synthetic diamonds grown in the presence of Ti-Al; electronic transitions from two excited states to a common ground state, the center is excited in a broad band at about 3.8 - 3.9 eV [102, 248]; slow luminescent center, the low temperature lifetime (below 15 K) is 40 μs after excitation and 130 μs upon thermalization of the spin sub-levels in a temperature range of 30 to 150 K, the temperature range of non-exponential decay is 20-40 K; triplet-singlet spin-forbidden transition, triplet level is not radiative but feeds the singlet emitting levels; non-radiative processes are not observed up to room temperature; the activation energy of non-radiative decay is 0.155 eV; the ratio of the transition probabilities of the second and the first ZPLs is 65; [248, 130]; the 2.974 eV transition interacts with 92 (the most intense), 38 and 130 meV vibrations, S = 4.0; the 2.964 eV transition interacts with 79 (the most intense) and 38 meV vibrations, S = 3.7 [calculated from spectra in 366]; excitation spectrum of the center consists of a broad band with maximum at 3.75 eV (330 nm) [366] (fig. 140).

2.971 eV (417.2 nm), PL, ZPL, in type Ib synthetic low nitrogen diamonds grown using iron-nickel solvent-catalyst; excited with light of wave length > 360 nm [74].

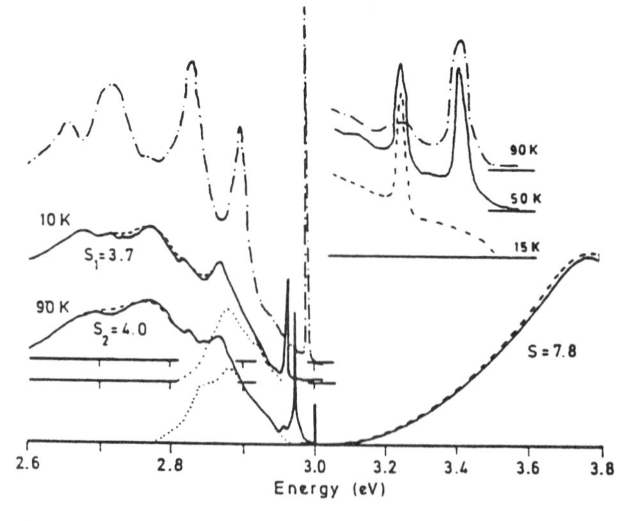

Figure 140. PL and PL excitation spectra of the 2.964 eV center [248, fig. 1].

2.975 eV (416.8 nm). A, the H7 center; in type I diamonds after irradiation and annealed at temperatures above 500°C [259] (fig. 147).

2.985 eV (415.2 nm), CL, ZPL, observed in some virgin CVD diamond films [36, 124, 340]; dominant interaction with 70 meV vibrations; the ZPL width is very sensitive to mechanical stress; the center appears unrelated to the N3 center; (fig. 109, 129, 141b).

2.985 eV (415.2 nm), the N3 center, A, CL, PL, PL excitation; ZPL, a common feature in most type Ia diamonds containing the B aggregates, observed in some synthetic diamonds; in CVD films ? [90]; created in type IIa diamonds by nitrogen ion implantation and subsequent annealing at temperatures above 1200°C; created in synthetic diamonds by heating at temperatures above 1700°C [84, 441]; arises within stacking faults on {111} planes in brown diamonds [395]. ZPL is split by 0.59 meV (splitting of the excited state); excited electronic state is coupled predominantly to totally symmetric vibrational modes; excited state g-value is positive [203]; A to E(C_{3V}) electronic transition at a defect of C_{3V} symmetry; very strong linear Stark effect (no inversion symmetry) [221, 225]; dominant interaction with 93 and 165 meV vibrations, S = 3.45 [13]; no Jahn-Teller effects. Intrinsic decay time at low temperature in PL is about 40 ns in a temperature range of 77 to 400 K and it is reduced by interaction with the A-aggregates; the radiative decay time is 150 ns; reversible population of a quartet level leads to delayed emission at above 90 K with a decay time of 7 ms that decreases with temperature increase [130, 250]. Quantum efficiency is 0.29. Activation energy of the non-radiative processes is 0.566 eV [239]. The paramagnetic P2 center is ascribed to the N3 center. Inhomogeneous ZPL width is of 350 GHz; hole burning in the ZPL has been demonstrated [130, 227]; Emission is quenched by energy transfer to the A aggregates, the probability of the transfer for one N3 center to one A aggregate at one lattice spacing in unit time is 0.3×10^{16} s^{-1} [130, 237, 239]. The center is excited at temperatures up to 500°C [3]. Model of the N3 center: trio of substitutional nitrogen atoms in a (111) plane bonded to a common vacancy [108, 116, 130, 141, 144, 147, 149, 154-159, 161, 164, 165, 203, 226, 284, 394] (fig. 141).

2.99 eV (414.5 nm), CL, in CVD films, interacts with vibrations of 70 meV; the ZPL is a doublet split by about 6 meV; decay time 63±5 ns [73, 312] (fig. 114).

3.004 eV (412.7 nm). A, the H8 center; in type I diamonds after irradiation and annealing at temperatures above 500°C [259].

3.04 eV (407.8 nm), the R9 center, A, PC, ZPL, radiation induced center in type Ia and IIa diamonds; a line at 3.09 eV possibly relates to the R9 center [13, 144, 425].

3.062 eV (405.0 nm), A, the H9 center; in type I diamonds after irradiation and annealing at temperatures above 500°C [259] (fig. 147).

3.065 and 3.076 eV (404.4 and 403.0 nm), A, Luminescence, ZPL doublet observed in synthetic diamonds grown in presence of nickel; the center is confined to the {111}

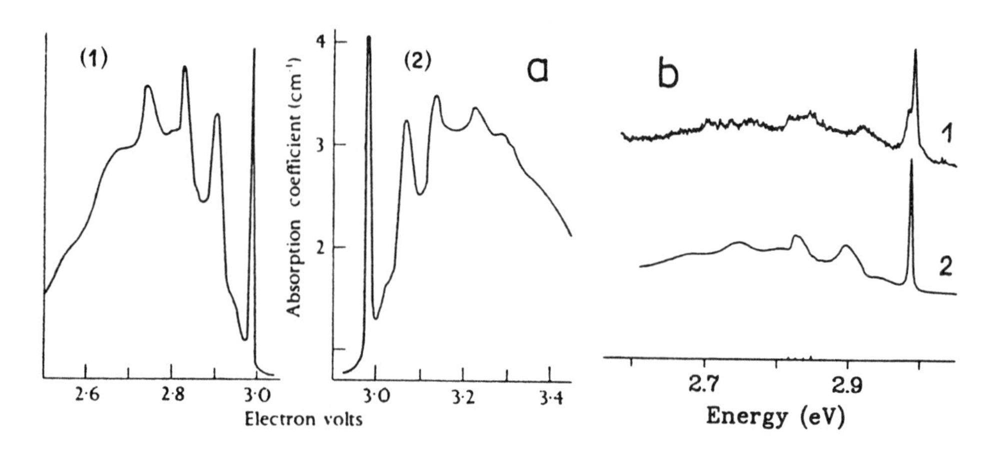

Figure 141. (a) - comparison between the emission (1) and absorption (2) spectra of the N3 center [163, fig. 237]; (b) - time resolved spectrum of the 2.985 eV center in CVD film (1) in comparison with the N3 center from a natural type IIa diamond (2) [312, fig. 2].

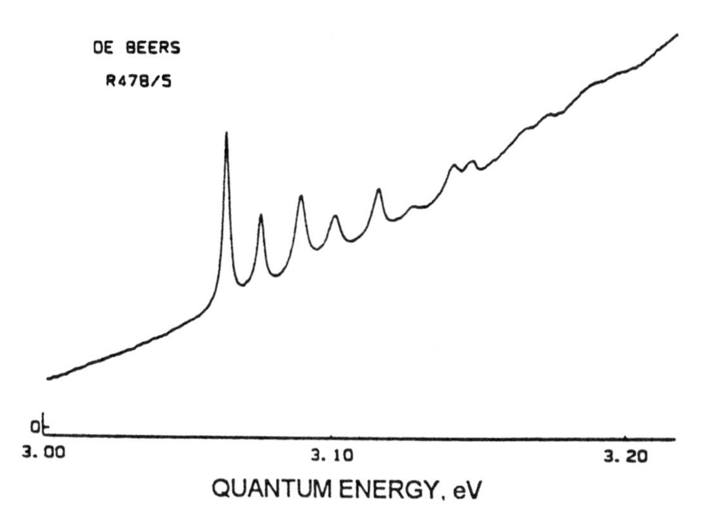

Figure 142. Absorption spectrum of a pale green synthetic diamond grown by the temperature gradient method [282, fig. 2b].

sectors [74, 130, 89, 112, 179]; dominant interaction with 26 meV vibrations; tentatively ascribed to a nickel related center (fig. 142).

3.07 eV (404 nm), A, a narrow line observed in hydrogen rich gray-violet diamonds [284].

3.09 and 3.11 eV (398.5 and 401 nm), CL, ZPL doublet, the most intense of a set of lines within a range 385 to 430 nm observed in type IIa diamonds after Ag^+ ion implantation and subsequent annealing at temperatures above 1000°C [74, 92, 376] (fig. 143).

3.092 eV (400.9 nm), CL, ZPL, observed at 77 K in some CVD films, for instance in epitaxial undoped CVD diamond films grown on [100] natural diamond surface, interacts with quasi local 154 meV and local (connected with carbon) 216 meV vibrations [73, 74, 340] (fig. 129, 130).

3.12 eV, CL, a broad band of FWHM 0.67 eV, observed only in boron-doped HPHT diamonds containing very small concentration of uncompensated nitrogen; dispersed nitrogen and uncompensated boron quench the band; tentatively ascribed to compensated boron [270, 500] (fig. 144).

3.149 eV (393.6 nm), ZPL, the ND1 center (or the R10 center [144]), A, PC, observed in all types of diamond as a result of radiation damage at room temperature; anneals out at temperatures above 500°C. Transition between A (ground) to T (excited) electronic levels at a defect of Td symmetry; excited state g-value is

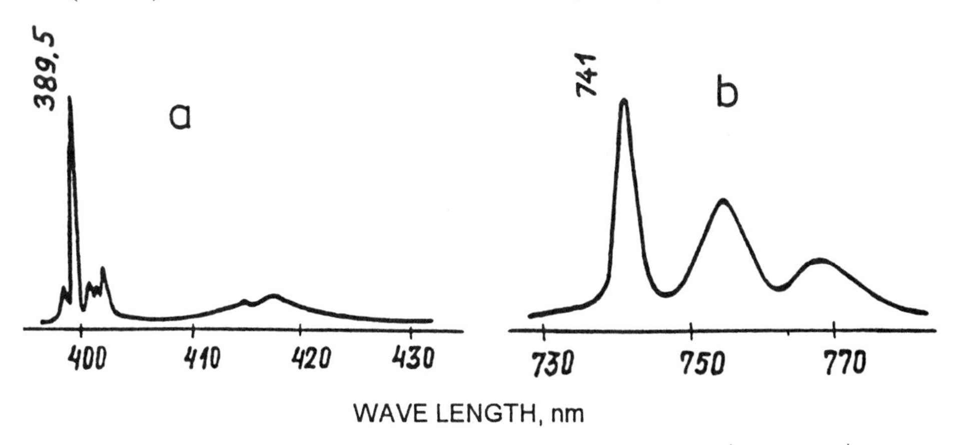

Figure 143. CL spectrum of type IIa natural diamonds implanted with Ag^+ (a) and Cr^+ (b) ions and subsequently annealed at 1400°C [376, fig. 2.36].

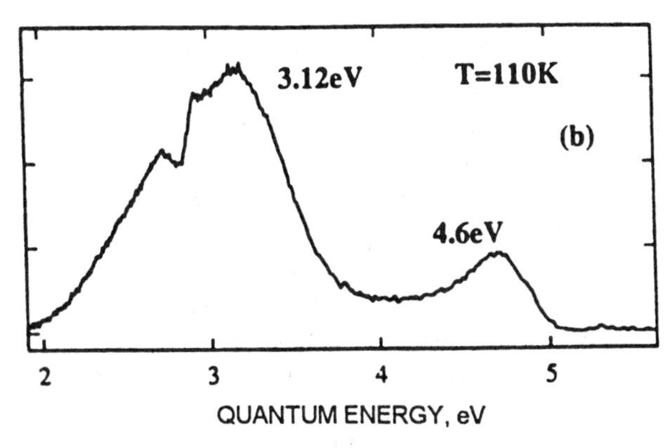

Figure 144. CL spectrum of a nitrogen gettered synthetic HPHT boron doped diamond [500, fig. 1b].

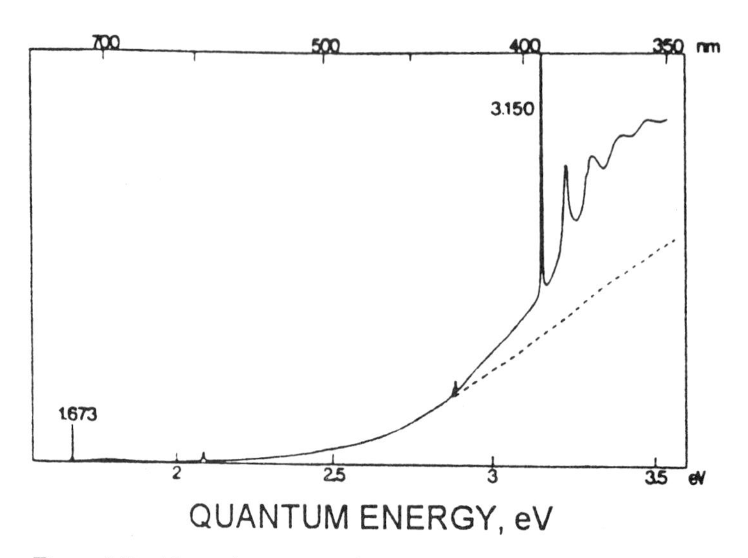

Figure 145. Absorption spectra of a type Ib diamond before (broken line) and after (full line) irradiation by 2 MeV electrons [13, fig. 20].

positive [203]. In photoconductivity the center acts as a donor. Normally the oscillator strength relative to the GR1 transition is about 4.0 [188]; in type IIb diamond the ND1 absorption is much weaker [200]. Linear Stark effect (no inversion symmetry) [221]. Dominant interaction with 80 meV vibrations, S = 3.18 [13]. The center ascribed to negatively charged mono-vacancy V^- [13, 104, 130, 188, 190, 200, 201, 203, 214-216, 371, 372] (fig. 145).

3.150 eV (393.5 nm), ZPL, CL, in natural brown diamonds showing yellow PL; decay time in CL at 77 K is 45 ns; similar behavior as for the 3.204 and 3.224 eV centers; dominant interaction with 54 meV vibrations; the center is not related to the ND1 center [74, 93].

3.170 eV (391.0 nm), ZPL, CL, in natural brown diamonds showing yellow PL [74, 93].

3.188 eV (389 nm), ZPL, A, CL, PL, radiation damage product in all types of diamond; a common feature of CVD films; in PL of some CVD films can be shifted to 3.20 eV [73]; intensity strongly increases in electron irradiated diamonds after 600°C annealing; strongest in type Ib diamonds; absorption is very weak and observed only in type Ib diamonds; strongly produced in CVD films by N^+ ion implantation and subsequent plasma hydrogenation at 900°C [418]. Athermally activated with a keV energy electron beam [22]. Sharp structure between 2.99 and 3.02 eV is associated with transitions involving localized vibrational modes. Dominant interaction with 75 meV acoustic phonon and the 165 meV zone-center LO phonon; S = 1.82 [13]; all the vibrational sidebands occur due to carbon vibrations except a mode 178.8 meV attributed to a C-N vibration [29, 115, 130]. ZPL shifts in ^{13}C diamonds by +3.3 meV [75, 130]. Possible models of the center are: (i) defect containing interstitial nitrogen atom [28, 116 340], (ii) defect containing substitutional nitrogen atom bound to interstitial carbon atoms [29, 76], (iii) single

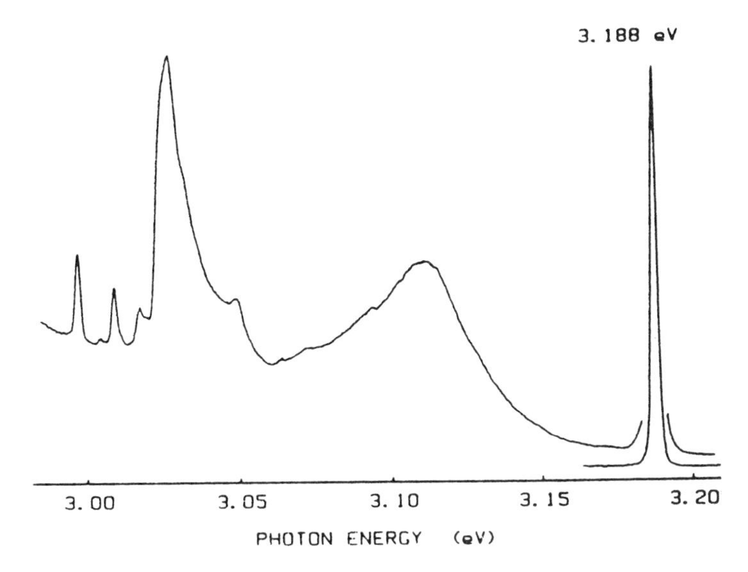

Figure 146. CL of the 3.188 eV center from a single crystal MPCVD diamond film after radiation damage and annealing at 400°C [36, fig. 7].

substitutional nitrogen shifted along <111> axis forming an elongated C–N pseudo-molecule, [73, 74, 29, 75, 76, 77] (fig. 146).

3.204 eV (386.9 nm), ZPL, CL, in natural brown diamonds showing yellow PL; similar behavior to the 3.105 and 3.224 eV centers; dominant interaction with vibrations 54 meV [74, 93, 105].

3.216 eV (385.4 nm), ZPL, CL, in natural brown diamonds showing yellow PL when excited with 365 nm light [74, 93, 105].

3.224 eV (384.5 nm), ZPL, CL, in natural brown diamonds showing yellow PL when excited with 365 nm light; similar behavior to the 3.150 and 3.204 eV lines; dominant interaction with vibrations of 54 meV [74, 93, 105].

3.26 eV (380 nm), A, a narrow line in hydrogen rich gray-violet diamonds [284].

3.270 eV (379.2 nm), A, PL excitation of the H3 center; the H10 center; a relatively broad line observed in type I diamonds after irradiation and annealing at temperatures above 500°C [259]; electronic transition to an excited state of the H3 center [260] (fig. 147).

3.272 eV (378.8 nm), ZPL, CL, in some CVD films [74, 340].

3.3 eV (380 nm), A, a broad band of FWHM 0.2 eV; in type Ib diamonds; absorption is polarized parallel with respect to (111) axis; ascribed to transition from the ground to the first excited state of the trigonally distorted substitutional nitrogen donor [135] (fig. 154).

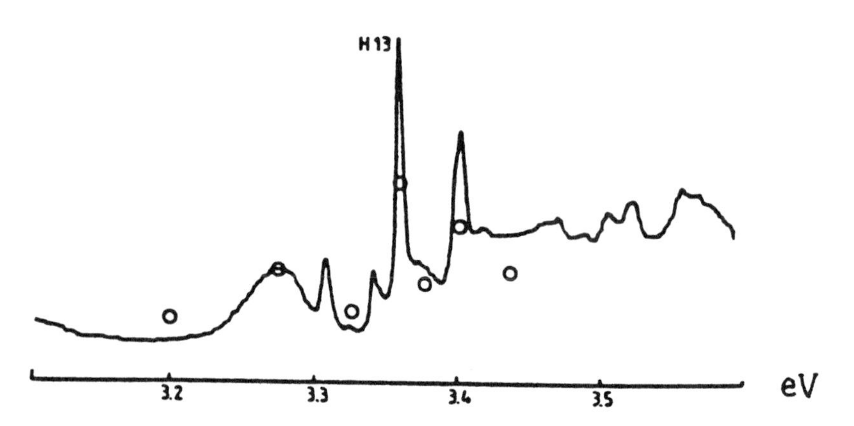

Figure 147. (a) - Absorption spectrum of high excited states of the H3 center (solid curve) and PL excitation spectrum (points)[260, fig. 2].(b) - Absorption spectra of a type I diamond after irradiation with 1.17 and 1.33 MeV-rays (broken line) and after subsequent heating at 450°C (solid line) recorded at 80 K [259].

3.311 eV (374.4 nm), A, PL excitation of the H3 center; the H11 center; a narrow line observed in type I diamonds after irradiation and annealing at temperatures above 500°C [259]; electronic transition to an excited state of the H3 center [260] (fig. 147).

3.343 eV (371.0 nm), A, PL excitation of the H3 center; the H12 center; a narrow line observed in type I diamonds after irradiation and annealing at temperatures above 500°C [259]; electronic transition to an excited state of the H3 center [260] (fig. 147).

3.361 eV (368.9 nm), A, PL excitation of the H3 center; the H13 center; a narrow line observed in type I diamonds after irradiation and annealing at temperatures above 500°C [259]; electronic transition to an excited state of the H3 center [260] (fig. 147).

3.380 eV (366.7 nm), A, PL excitation of the S2 center, a narrow line observed in type Ib synthetic diamonds grown by the temperature gradient method and in synthetic Ib diamonds by annealing at temperatures above 2200°C [74, 84, 361] (fig. 148).

3.403 eV (364.3 nm), A, PL excitation of the H3 center; the H14 center; a narrow line observed in type I diamonds after irradiation and annealing at temperatures above 500°C [259]; electronic transition to an excited state of the H3 center [260] (fig. 147).

3.412 eV (363.4 nm), A, PL excitation of the S2 center, a narrow line observed in type Ib synthetic diamonds grown by the temperature gradient method and in synthetic diamonds after high PT treatment [84, 361] (fig. 148).

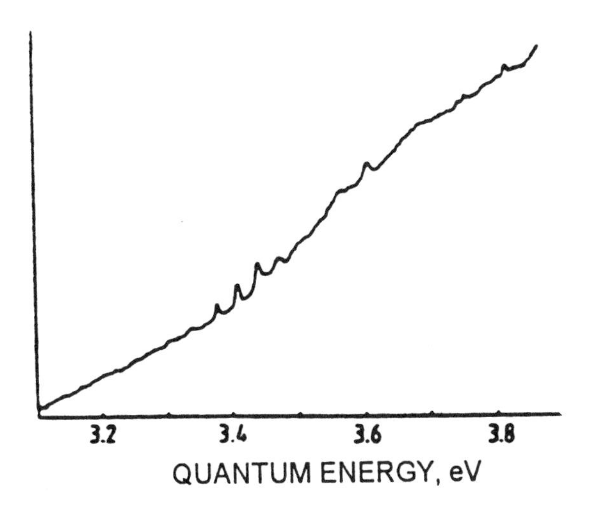

Figure 148. Absorption spectrum of a synthetic diamond heat treated at 2400°C. The absorption coefficient has been arbitrarily set to zero at 3.0 eV [84, fig. 2].

3.443 eV (360.1 nm), A, PL excitation of the S2 center, a narrow line observed in type Ib synthetic diamonds grown by the temperature gradient method and in synthetic diamonds after high PT treatment [84, 361] (fig. 148).

3.462 eV (358.0 nm), CL, a narrow line observed in CVD diamond films after annealing at 1200°C [20].

3.463 eV (358.0 nm), A, PL excitation of the H3 center; the H15 center; a narrow line observed in type I diamonds after irradiation and annealing at temperatures above 500°C [259]; electronic transition to an excited state of the H3 center [260] (fig. 147).

3.47 eV (357.2 nm), CL, a band of a width 0.35 to 0.6 eV; the band may change its spectral position from 3.5 to 3.35 eV [413, 416, 421, 500]; produced in type IIb diamonds by cold ion implantation and subsequent rapid 1200°C annealing; weak intensity in type IIa diamonds; observed in as-grown boron doped synthetic HTHP and CVD diamonds; observed in undoped CVD diamond films [421]; generated by boron ion implantation. The band correlates in intensity with the 4.6 eV band. Possible origin: uncompensated boron acceptors [306, 307, 404] (fig. 149).

3.474 eV (356.0 nm), A, PL excitation of the S2 center, in Ib synthetic diamonds grown by the temperature gradient method and in synthetic diamonds after high PT treatment [84, 361] (fig. 148).

3.504 eV (353.8 nm), A, PL excitation of the H3 center; the H16 center; a narrow line observed in type I diamonds after irradiation and annealing at temperatures above 500°C [259]; electronic transition to an excited state of the H3 center [260] (fig. 147).

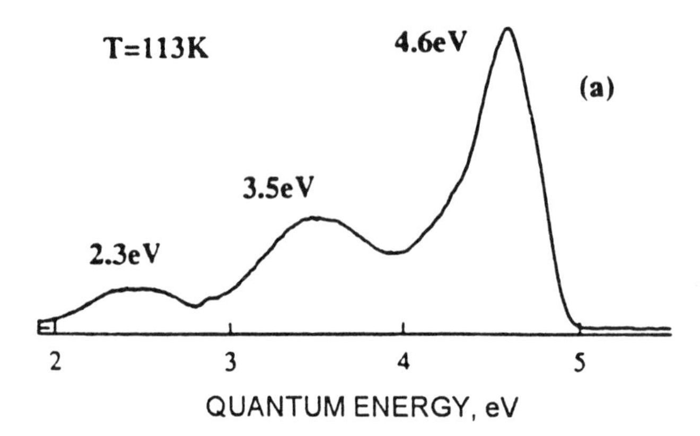

Figure 149. CL spectrum from a boron doped CVD diamond film recorded at 113 K [500, fig. 4a].

3.543 eV (349.9 nm), A, PL excitation of the H3 center; the H17 center; a narrow line observed in type I diamonds after irradiation and annealing at temperatures above 500°C [259]; electronic transition to an excited state of the H3 center [260] (fig. 147).

3.560 eV (348.2 nm), A, PL excitation of the H3 center; the H18 center; a narrow line observed in type I diamonds after irradiation and annealing at temperatures above 500°C [259]; electronic transition to an excited state of the H3 center [260] (fig. 147).

3.57 eV (347 nm), CL, a narrow line appearing in CVD films after 1200°C annealing [20] (fig. 150).

3.570 eV (347.2 nm) or 3.560 eV (348.2 nm), A, PL excitation of the S2 center, a narrow line observed in type Ib synthetic diamonds grown by the temperature gradient method and in synthetic diamonds after high PT treatment [84, 361].

3.6 eV (340 nm), A, a broad band of FWHM 0.35 eV appearing in nitrogen containing synthetic diamonds after high PT (~1900°C, 5.5 GPa) treatment [357, 361] (fig. 115).

3.603 eV (344.2 nm), the N4 center, A, PL excitation, naturally occurring, vibronic transition at the N3 center (between the ground state and nondegenerate excited state) in which the electronic excitation is forbidden; there is a related line at 3.58 eV; dominant interaction with 79 meV vibrations [3, 130, 144, 160, 161] (fig. 151).

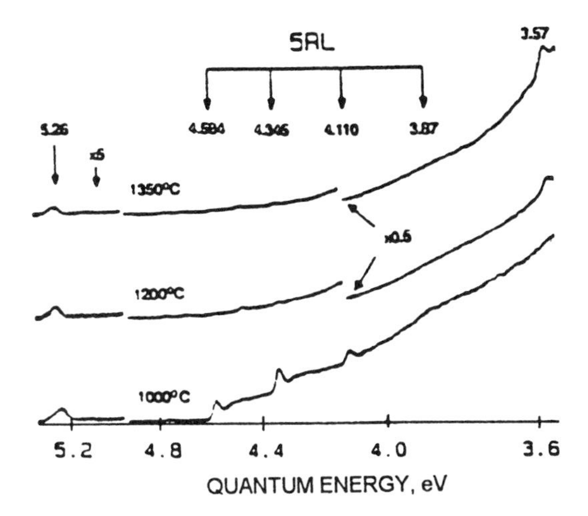

Figure 150. CL spectra of a MPCVD diamond film subjected to annealing at different temperatures [20, fig. 3].

3.607 eV (343.6 nm) or 3.612 eV, A, PL excitation of the S2 center, a narrow line in type Ib synthetic diamonds grown by the temperature gradient method and in synthetic diamonds after high PT treatment [84, 361].

3.65 eV (340 nm), a broad band from 310 to 370 nm, the β-band, observed in PL excitation of the S2 and S3 centers; at 80 K the band is assisted with ZPLs at 367.0, 363.6, 360.4 and 356.7 nm [3].

3.75 eV (330 nm), CL, a broad band of FWHM 0.7 eV; in CVD films; originated from oxygen diffused from saturated CrO_3+H_2SO_4 solution at 200°C followed by an H plasma treatment; ascribed to oxygen containing defects [185] (fig. 155).

3.762 eV (329.6 nm), the N5 center, ZPL, A, in type IaA diamonds naturally occurring center; electronic transition A_1 to A_2 at a defect of trigonal symmetry [104]; the center interacts with vibrations of 113 and 159 meV; ascribed to an internal transitions at the A center [130, 144, 145].

3.853 eV (321.7 nm), CL, a sharp line observed in some CVD diamond films [74].

3.9 eV (320 nm), A, a broad band of FWHM 0.25 eV; in type Ib diamonds; absorption is polarized perpendicular with respect to (111) axis; ascribed to transition from the ground to the first excited state of the trigonally distorted substitutional nitrogen donor [135] (fig. 154).

3.901 eV (317.7 nm), A, ZPL; in type IaA diamonds naturally occurring center; electronic A - E transition at a defect of trigonal symmetry [104]; interacts with vibrations of energy 113 meV; internal transition at the A center [130, 144, 145].

3.928 eV (315.6 nm), A, the N6 center, ZPL; naturally occurring in type IaA diamonds; electronic transition A - E at a defect of trigonal symmetry [104]; the center

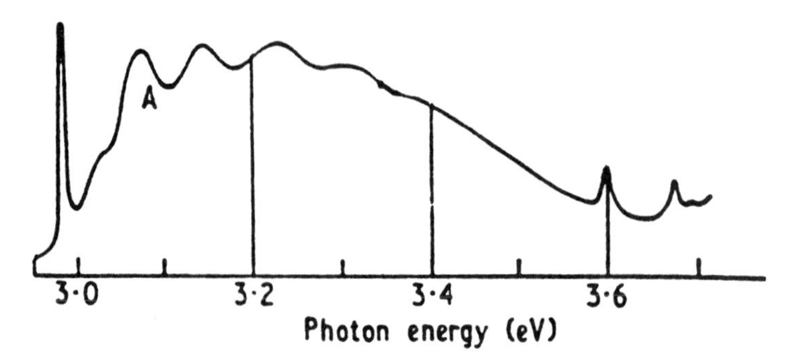

Figure 151. PL excitation spectrum of the N3 center [160, fig. 1].

interacts with vibrations of 113 and 159 meV; the vibronic transition coupled with 113 meV vibration at 4.042 eV (306.5 nm) is labeled as the N7 center [3, 13, 104]; ascribed to an internal transitions at the A center [130, 144, 145]; concentration of the A-aggregates can be evaluated by the expression: $N_A = 11.6\times10^{18}\ \mu_{306.5}$ [3, 13].

3.988 eV (310.9 nm), ZPL, the R11 center, A, in type IIa diamonds, radiation induced center; lines at 4.04, 4.11, 4.33, 4.63 eV possibly relate to the center [13, 144].

4 eV (310 nm), CL, EL, a band of a width of ~ 0.7 eV; produced in natural IIa diamonds at liquid nitrogen temperature by C^+ ion implantation and subsequent rapid annealing at 1200°C; emission intensity rapidly drops by 335 K; tentatively assigned to an electron transitions from the conduction band to deep lying donor centers [79, 305, 308] (fig. 152).

4.042 eV (306.5 nm), see 3.926 eV (the N6 center).

4.059 eV (305.4 nm), ZPL, A, in type Ib natural and synthetic diamonds; naturally occurring center; dominant interaction with vibrations of 61 and 120 meV, S = 6; electronic transition at a defect of trigonal symmetry, A_1 level; ascribed to single substitutional nitrogen [74, 117, 130].

4.137, 4.27, 4.316, 4.355 eV (299.7, 290, 287.2, 284.7 nm), A, narrow lines naturally occurring in type Ib diamonds; ascribed to electronic transitions at isolated substitutional nitrogen atoms [13].

4.184 eV (296.2 nm), A, the first line of a group of three naturally-occurring zero-phonon lines (the others are at 4.191 and 4.197 eV); observed in natural Ib diamonds, not in synthetic Ib diamonds; all the centers interact with vibrations of energy 36 meV; ascribed to isolated substitutional nitrogen [13, 74, 90, 117, 130].

4.19 eV (296 nm), the N8 center, A, a narrow line naturally occurring in type IaA diamonds; ascribed to intracenter transition at the A-aggregate [3, 13, 104, 144];

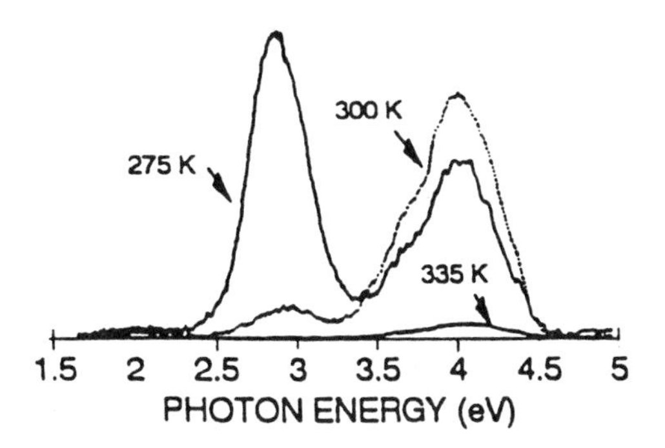

Figure 152. EL spectra of a diode made on a type IIb natural diamond by low temperature carbon ion implantation and subsequent rapid annealing [308, fig. 2].

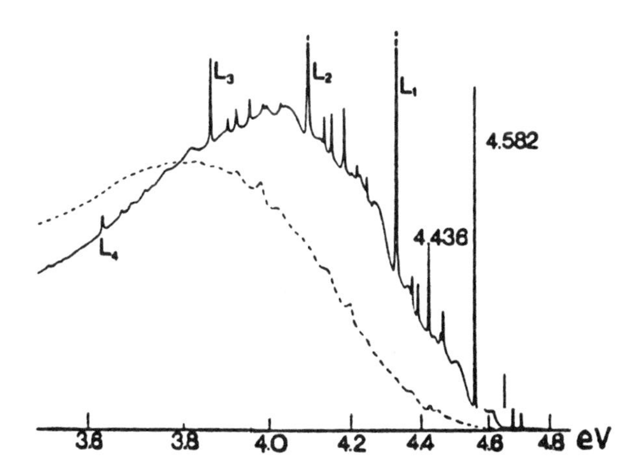

Figure 153. CL spectrum of a type II natural diamond after 2 MeV electron irradiation recorded at 90 K (full line) and 300 K (broken line) [13, fig. A9].

4.328 eV (286.4 nm), A, observed in type Ib diamonds after radiation damage and annealing at 800°C; intensity of the absorption system and its stress splitting parameters correlate with those of the 1.945 eV center; electronic A–E transition at a defect of trigonal symmetry [74].

4.374 eV (283.4 nm), A, a weak narrow peak, tentatively ascribed to the B' center [3].

4.43 eV (280 nm), A, PL, a weak band relating possibly to the B' center [3, 130, 170].

4.436 eV (279.5 nm), CL, observed in type IIa and IIb diamonds, radiation induced center; dominant interaction with 237 meV vibration; the center is not related to the 5RL center [13] (fig. 153).

4.468 eV (277.4 nm), A, in type Ib diamonds, naturally occurring center [74, 117].

4.567 eV (271.4 nm), ZPL, A, in type Ib natural and synthetic diamonds, naturally occurring center; dominant interaction with vibrations of energies 59 meV and 106 meV, S = 1.8; electronic transition at a defect of trigonal symmetry, A_1 level; ascribed to single substitutional nitrogen [13, 74, 117].

4.582 eV (270.5 nm), ZPL, the 5RL center; A, CL, absorption observed only in type II; radiation damage center; observed in type IIa, IIb and Ib diamonds of various origin; annealing at 500°C optimizes the absorption intensity; in diamond films the center is gradually annealed within a range of 800 to 1200°C [20, 74]. The center interacts in CL with local modes of energy 237 (dominant), 193 and 175 meV; in absorption with vibrations of 202 (dominant) and 167 meV. Huang-Rhys factor for the dominant CL mode is of 1.6. Interaction with a vibration of 58 meV. Decay time in CL is of 7 ns at temperatures <150 K and 13 ns at 180 K, activation energy for non-radiative transitions is around 0.32 eV [243]. ZPL shifts in ^{13}C diamonds by +8 meV [75, 130]. The center of rhombic-I symmetry. The center is proposed to contain a light atom (boron) [23]. The center may involve the carbon interstitials [75]. The mode 237 meV is carbon related [75]. The center is possibly produced by <100>-split interstitial [76] (fig. 153).

4.59 eV (270 nm), a band, A, possibly related to the B' platelets [130, 170].

4.6 eV (270 nm), a band of FWHM 0.4 eV, CL, observed in synthetic HPHT and CVD boron doped diamonds; maximum intensity at 130 K, completely quenched above 200 K, strongly quenched by nitrogen defects and particularly by atomically dispersed nitrogen [269, 413, 500]; produced by low temperature ion implantation and subsequent rapid annealing at 1200°C in type IIb diamonds; weak intensity in type IIa diamonds; observed in as-grown synthetic type IIb diamonds; created by boron ion implantation; ascribed to boron related defects; possibly, uncompensated boron acceptors [306, 307, 404] (fig. 149).

4.6 eV (270 nm), A, a band of FWHM 0.55 eV; in type Ib diamonds; absorption is polarized parallel to (111) axis; assigned to a transition from the valence band to a substitutional nitrogen donor level [130, 134, 135] (fig. 154).

4.639 eV (267.3 nm), ZPL, CL, in type IIa and IIb diamonds, probably a radiation induced center [13] (fig. 153).

4.64 eV (267 nm), CL, a band of FWHM 0.2 eV; in CVD films; originated from oxygen diffused from saturated CrO_3+H_2SO_4 solution at 200°C followed by an H plasma treatment; possibly a complex of oxygen and carbon interstitial atoms [130, 185] (fig. 155).

4.646 eV (266.8 nm), A, in type Ia diamonds; a weak line, possibly the B' center [3].

4.676 eV (265.1 nm), ZPL, CL, in type IIa and IIb diamonds, probably a radiation induced center [13] (fig. 153).

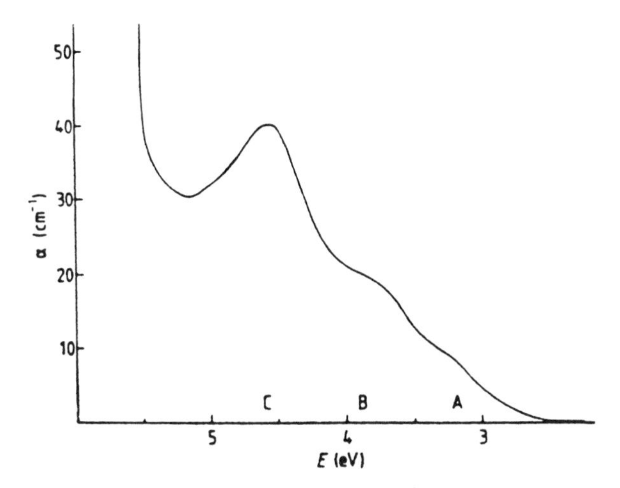

Figure 154. Absorption spectrum of a nitrogen doped synthetic diamond [135, fig. 1].

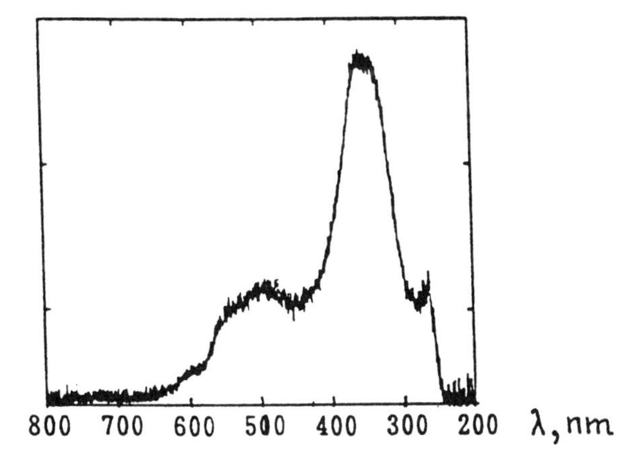

Figure 155. CL spectrum of a MPCVD diamond film subjected to the following treatment: boiling in a saturated solution of CrO3 in H2SO4 at 200°C, followed by a rinse in a solution of H2O2 and NH4OH, followed by hydrogen plasma at 800°C [185, fig. 3a].

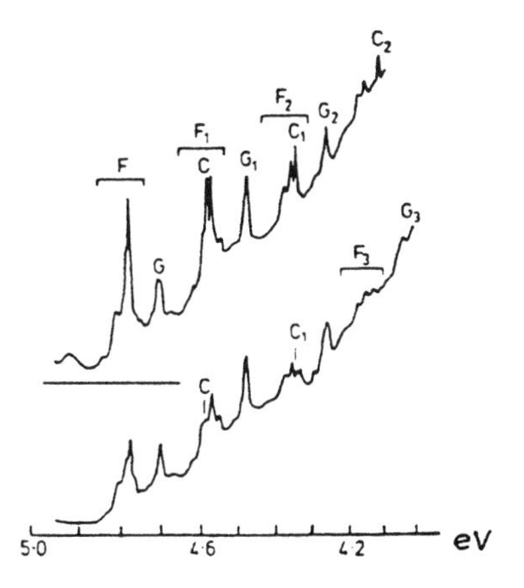

Figure 156. CL spectra of the 2BD center recorded at 50 (upper) and 100 K (lower) [104, fig. 26].

4.698 eV (263.9 nm), the 2BD(G) center; A, CL, ZPL consists of four components, observed in type IIb irradiated diamonds; dominant interaction with 220 meV vibrations, S ~ 2 [13]; possibly an interstitial related boron containing defect [3, 104] (fig. 156).

4.69 eV (270 nm), A, PL, a weak band relating possibly to the B' center [3, 170].

4.755 eV (260.7 nm), CL, at 100 K, low energy threshold of the D4' line, recombination of excitons bound to acceptor with emission of 150 meV phonons [362] (fig. 157).

4.757 eV (260.6 nm), 4.832 eV (256.5 nm), 4.950 eV (250.4 nm), CL, narrow lines observed in undoped CVD diamond films at 4.2 K; ascribed to excitons bound to dislocations [34] (fig. 158).

4.76 eV (260 nm), a narrow line observed in CVD diamond films, stable at 1200°C [20].

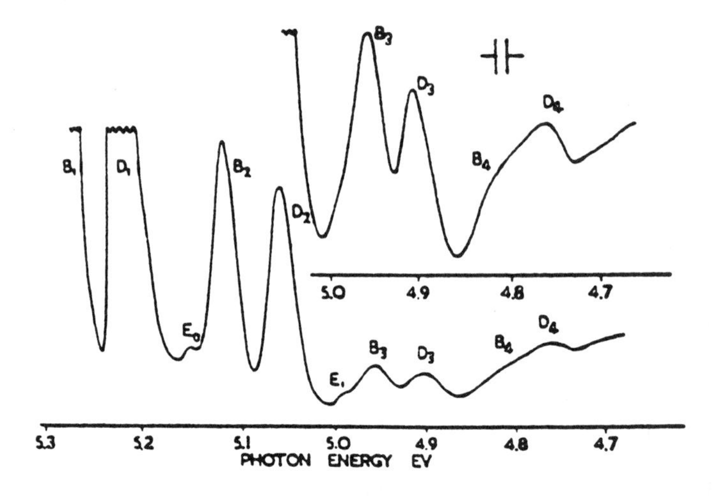

Figure 157. CL spectrum of a natural type IIb diamond at 80 K [362, fig. 6].

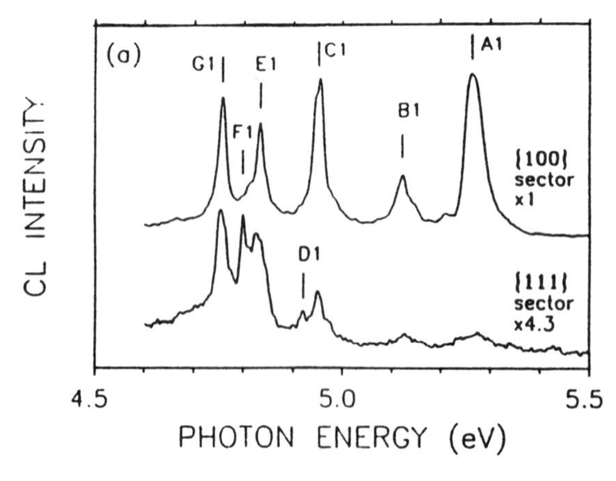

Figure 158. CL spectra from [111] and [111] growth sectors of nominally undoped diamond single crystals grown by CVD deposition method [34, fig. 2a].

4.777 eV (259.5 nm), the 2BD(F) center, A, CL; in type IIb diamonds; ZPL consists of four components at 4.777, 4.781, 4.803 and 4.830 eV; dominant interaction with 210 meV vibrations, S ~ 1, [13]; possibly a center related to interstitial boron atoms [3, 104] (fig. 156).

4.83 eV (257 nm), CL, a narrow line occasionally observed in CVD diamond films, after 1000°C anneal [20].

4.832 eV, see 4.757 eV.

4.846, 4.855 eV (255.8, 255.4 nm), the T and T' centers, CL, naturally occurring in type IaB diamonds; the doublet structure is caused probably by the excited state splitting; interaction with 148 meV vibration, S ~ 1 [13].

4.890 eV (253.5 nm), CL, at 100 K, low energy threshold of the D3 line, indirect exciton bound to neutral acceptor associated with upper valence band and emitting 141 meV TO and two 167 meV Raman phonons [362] (fig. 159).

4.903 eV (252.8 nm), CL, at 100 K, low energy threshold of the D3' line, indirect exciton bound to neutral acceptor associated with lower valence band and emitting 141 meV TO and two 167 Raman phonons [362] (fig. 159).

4.93 eV (251.4 nm), CL, low energy threshold of the B3 line, indirect exciton associated with upper valence band emitting 141 meV TO and two 167 meV Raman phonon [362] (fig. 159).

4.950 eV, see 4.757 eV.

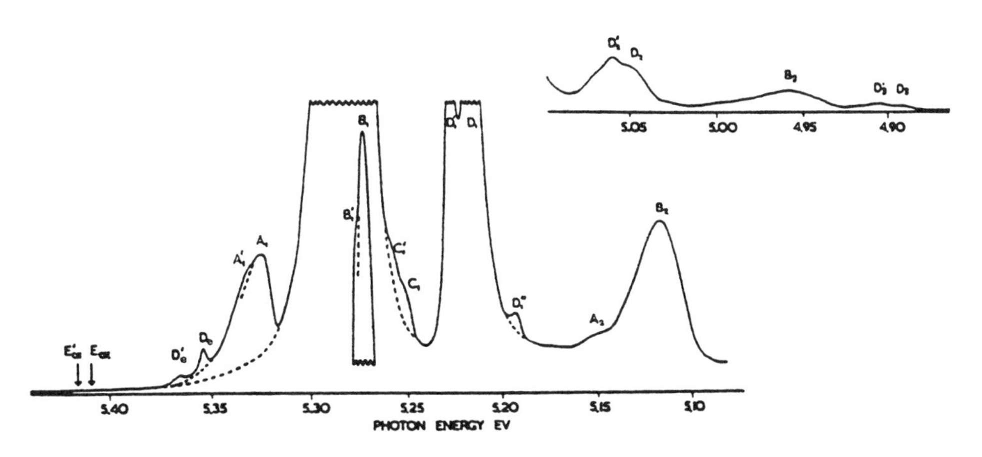

Figure 159. CL spectra of a relatively strain-free natural type IIb diamond recorded at 100 K [362, fig. 1].

4.986 eV (248.7 nm), the N center, CL, naturally occurring in type IaB diamonds, S << 1 [13].

4.99 eV (248 nm), the N10 center, A, PL excitation of the A-band; naturally occurring in type IaB diamonds with low content of the A nitrogen aggregates ($\mu_A < 1\text{-}2\ cm^{-1}$); relates to the 5.17 eV center; ascribed to a defect related to the B nitrogen aggregates [3, 153] (fig. 160).

4.999 eV (248.0 nm), the M center, CL, naturally occurring in type IaB diamonds, S << 1 [13].

5.032, 5.042 eV (246.4, 245.9 nm), the L center, CL, a doublet naturally occurring in type IaB diamonds, possibly interacts with 141 meV vibration, S ~ 1 [13].

5.048 eV (245.5 nm), CL, at 100 K, low energy threshold of the D_2 line, indirect exciton bound to neutral acceptor associated with upper valence band and emitting 141 meV TO and 167 meV Raman phonons [362] (fig. 159).

5.060 eV (245.0 nm), CL, at 100 K, low energy threshold of the D_2' line, indirect exciton bound to neutral acceptor associated with lower valence band and emitting 141 meV TO and 167 meV Raman phonons [362] (159).

5.092 eV (243.5 nm), the K center, CL, naturally occurring in type IaB diamonds [13].

5.10 eV (243.0 nm), CL, low energy threshold of the B_2 line, indirect exciton associated with upper valence band emitting 141 meV TO and 167 meV Raman phonon [362] (fig. 159).

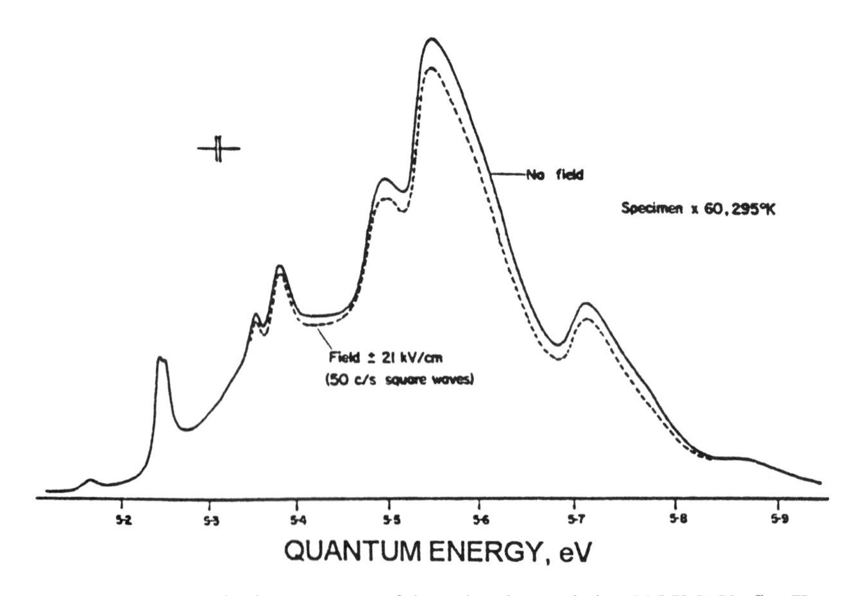

Figure 160. PL excitation spectrum of the A-band recorded at 295 K [153, fig. 7].

5.145 eV (240.9 nm), CL, low energy threshold of the A2 line, indirect exciton associated with upper valence band emitting 100 meV TA and an optical phonon [362] (fig. 159).

5.17 eV (240 nm), A, PL excitation of the N10 center and the A-band; naturally occurring in type IaB diamonds with low content of the A nitrogen aggregates (μ_A < 1-2 cm^{-1}); relates to the N10 center; a possible origin is the B aggregate of nitrogen [3, 153] (fig. 160).

5.18 eV (239 nm), the E_o center, CL, naturally occurring or radiation induced center in type IIa and IIb diamonds, low energy threshold of the line at 5.135 eV at 80 K; interaction with 165 meV vibration, S ~ 0.5 [13, 133].

5.193 eV (238.7 nm), CL, at 100 K, low energy threshold of the D_1'' line, indirect exciton bound to neutral acceptor associated with upper valence band and emitting 163 meV LO phonon [362]; shifts to 5.215 eV in ^{13}C diamond at 77 K [463] (fig. 159).

5.215 eV (237.8 nm), CL, at 100 K, low energy threshold of the D_1 line, indirect exciton bound to neutral acceptor associated with upper valence band and emitting 141 meV TO phonon [362]; peak energy at a temperature of 80 K is at 5.21 eV [53]; shifts to 5.2346 eV at 77 K in ^{13}C diamond [463] (fig. 159).

5.227 eV (237.1 nm), CL, at 100 K, low energy threshold of the D_1' line, indirect exciton bound to neutral acceptor associated with lower valence band emitting LO phonon of an energy 163 meV [362]; shifts to 5.2465 eV at 77 K in ^{13}C diamond [463] (fig. 159).

5.246 eV (236.3 nm), low energy threshold of the C_1 line, CL, indirect exciton associated with upper valence band emitting 163 meV LO phonon [362]; the threshold shifts to 5.266 eV in ^{13}C diamond at 77 K [463] (fig. 159).

5.253 eV (236.0 nm), low energy threshold of the C_1' line, CL, indirect exciton associated with lower valence band emitting 163 meV LO phonon [362] (fig. 159).

5.254, 5.264 and 5.279 eV (235.9, 235.5 and 234.8 nm), ZPL triplet, the N9 center, A, PL, PL excitation of the A-band [153], CL, PC; naturally occurring in type IaB diamonds; doublet splitting of the ZPL occurs in the excited state; interaction with vibrations of energy 76, 141 and 163 meV, S = 2.16 [13]; a possible origin is the B aggregate of nitrogen [33, 147-153, 130] (fig. 161).

5.258 eV (235.8 nm), the S_o center, ZPL, CL, naturally occurring in type IIb synthetic diamonds [13, 133].

5.2636 eV (235.49 nm), CL, a narrow line observed in ^{12}C synthetic diamonds at a temperature of 15 K [316] (fig. 162).

5.268 eV (235.3 nm), CL, low energy threshold of the B_1 line, indirect exciton associated with upper valence band emitting 141 meV TO phonon; the binding

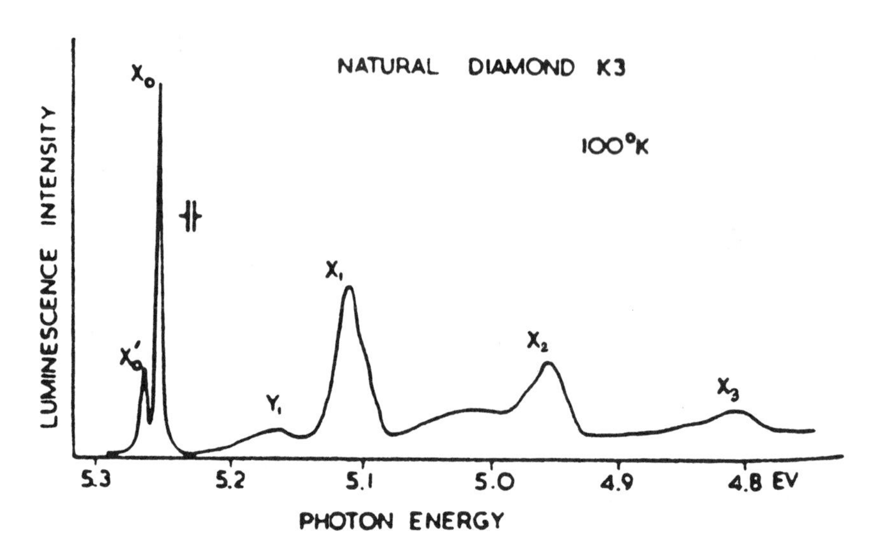

Figure 161. CL spectrum of the N9 center from natural type I diamond [150, fig. 1].

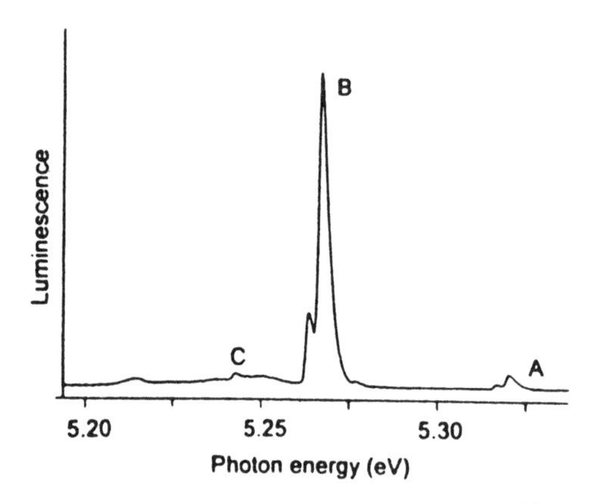

Figure 162. Free-exciton CL spectrum of a ^{12}C HTHP synthetic diamond [316, fig. 1].

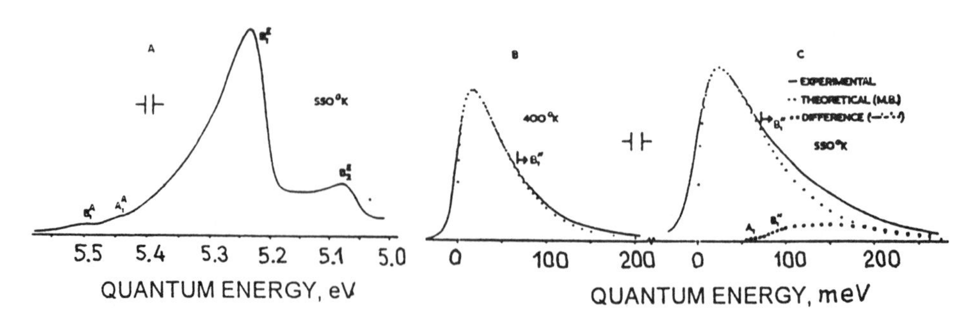

Figure 163. CL spectra of a natural type IIb diamond recorded at elevated temperatures [362, fig. 2].

energy of the free excitons is found to be 80 meV [362]; peak position at a temperature of 80 K is at 5.27 eV (235 nm) [53]; at temperature 550 K the threshold shifts to 5.212 eV [362]; the intensity in luminescence is almost constant up to a temperature 120 K [411]; the threshold shifts to 5.288 eV in ^{13}C diamond at 77 K [463] (fig. 159).

5.28 eV (235 nm), CL, at a temperature of 550 K, low energy threshold of the B_1'' line, ascribed to the recombination of free unbound electrons and holes [362] (fig. 163).

5.322 eV (232.9 nm), low energy threshold of the A_1 line, L, indirect exciton associated with upper valence band emitting 87 meV TA phonon [362]; the threshold shifts to 5.341 eV in ^{13}C diamond at 77 K [463] (fig. 159).

5.329 eV (232.6 nm), low energy threshold of the A_1' line, CL, indirect exciton associated with lower valence band emitting 87 meV TA phonon [362] (fig. 159).

5.35 eV (232 nm), in PL excitation of the A-band at room temperature [153] (fig. 160).

5.356 eV (231.4 nm), CL, at 100 K, ZPL, the D_o line, indirect exciton bound to neutral acceptor and associated with upper valence band [362]; 5.370 eV at 77 K in ^{13}C diamond [463]; S ~ 60 [13] (fig. 159).

5.368 eV (230.9 nm), CL, at 100 K, ZPL, the D_o' line, indirect exciton bound to neutral acceptor and associated with lower valence band [362]; 5.382 eV at 77 K in ^{13}C diamond [463]; S ~ 60 [13] (fig. 159).

5.38 eV (230 nm), in PL excitation of the A-band at room temperature [153] (fig. 160).

5.49 eV (226 nm), CL, at 550 K, low energy threshold of the B_1^A line, indirect exciton associated with upper valence band absorbing 141 meV TO phonon [362] (fig. 163).

5.50 eV (225 nm), a band observed at room temperature in PL excitation of the A-band [153] (fig. 160).

5.5 eV (225 nm), a structure observed in PL excitation spectra; tentatively ascribed to the N9 center [130, 149, 150, 138].

5.56 eV (223 nm), a band observed at room temperature in PL excitation of the A-band [153] (fig. 160).

5.715 eV (216.9 nm), a band observed at room temperature in PL excitation of the A-band [153] (fig. 160).

5.87 eV (211 nm), a band observed at room temperature in PL excitation of the A-band [153] (fig. 160).

12.5 eV (99.2 nm), A, a band of a width ~ 1 eV; electronic transitions in the maximum of joint conduction-valence band density of states [322] (fig. 164).

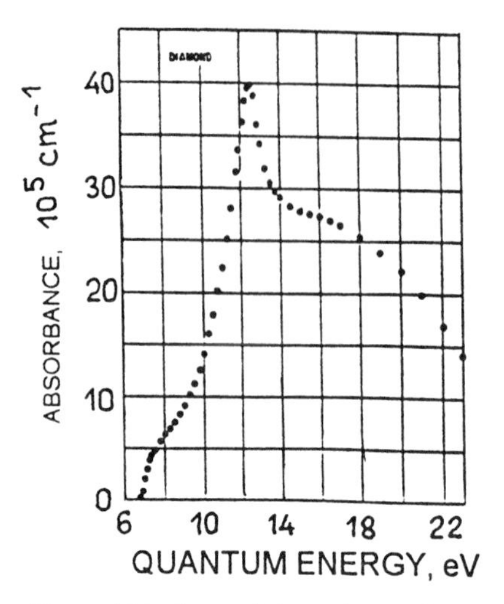

Figure 164. Spectral dependence of the absorption coefficient of diamond [322, fig. 4].

6.2 OPTICAL CONTINUA

Absorption over the whole visible region, strength of which monotonically increases with quantum energy, superimposed by three broad maxima at 2.25, 2.0 and 1.8 eV; observed in smoky-brown diamonds subject to plastic deformation, possible origin - dislocations [1, 3, 297].

A continuum from 0.35 to 4.1 eV (3500 to 300 nm) with a weak band at about 2 eV (620 nm), A, observed in type IIa natural and PCCVD diamonds after ion implantation [273, 274].

A continuum from 0.62 eV (2000 nm) to 6.2 eV (200 nm) observed in type IIb diamonds implanted at 250°C [272].

Absorption starts at about 1.7 eV and rises continuously with quantum energy; in type Ib diamonds; this continuum is composed at room temperature of three broad bands peaking at 3.3 eV (the A-feature), 3.9 eV (the B-feature) and 4.6 eV (the C-feature) (see above) [104, 130, 134, 135, 352]; ascribed to electronic transition at substitutional nitrogen donors; optical ionization energy of the nitrogen donor is found to be about 2.2 eV [104] or somewhat lower than 2.1 eV [352].

Absorption to higher energy than a threshold at 1.7 eV, in synthetic diamonds grown in presence of nickel [89, 112, 179] (fig. 51).

Absorption starting from 3.9 eV (320 nm) and rising to shorter wave length; the so called secondary absorption edge; ascribed to electronic transitions from the A aggregates of nitrogen to the conduction band [3, 13, 143] (fig. 168).

4.97 eV (249 nm), CL, at 100 K, the E_1 threshold, electronic transitions from the conduction band to the boron acceptor level with emission of 167 meV Raman phonons [362].

5.135 eV (241 nm), CL, at 100 K, the E_0 threshold, electronic transitions from the conduction band to the boron acceptor level [362].

5.49 eV (225 nm), at room temperature, A, onset of a strong absorption, indirect electronic transitions from the maximum of the valence band (**k** = 0) to the conduction band minimum (**k** = 0, 0, 0.76); indirect energy band gap E_g [133]. The transitions are accompanied with the absorption or emission of TA (87±2 meV), TO (141±2 meV) and LO (163±1 meV) phonons. In ^{13}C diamonds TO phonon is 135.4±0.6 meV and LO phonon is 155±0.7 meV [130, 133, 463]. E_g is 13.6±0.2 meV higher for ^{13}C diamond [463] (fig. 165).

Temperature dependence of E_g is given by the following expression [17, 18] (fig. 166):

$$E_g(T) = E_g(0) + \frac{1.979 \times 10^{-4}\, T^2}{T - 1437}. \tag{19}$$

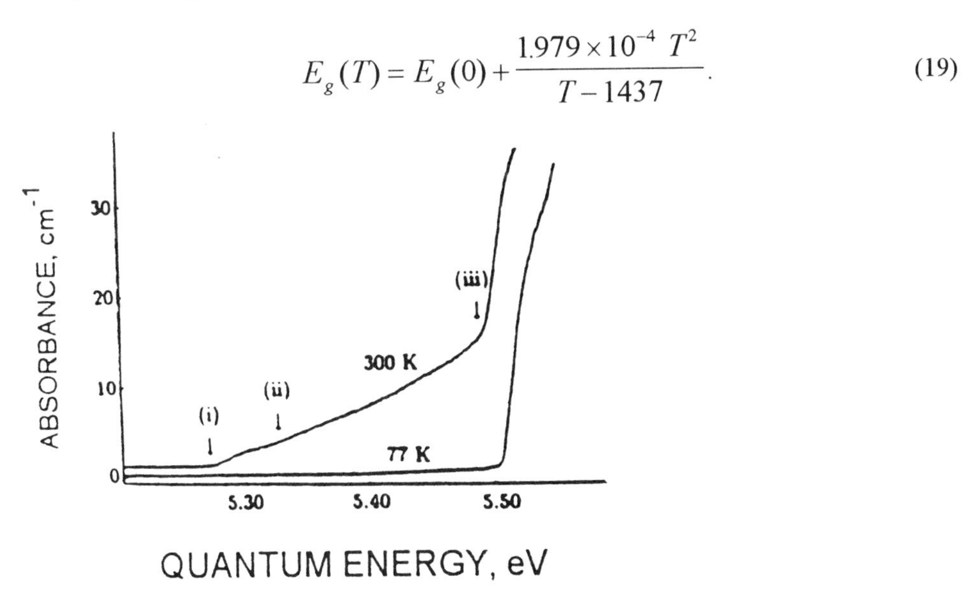

Figure 165. Absorption spectra of ^{13}C diamond measured at 77 and 300 K. Processes with thresholds i and ii involve the destruction of TO and TA phonons, and threshold iii involves the creation of a TA phonon [463, fig. 3].

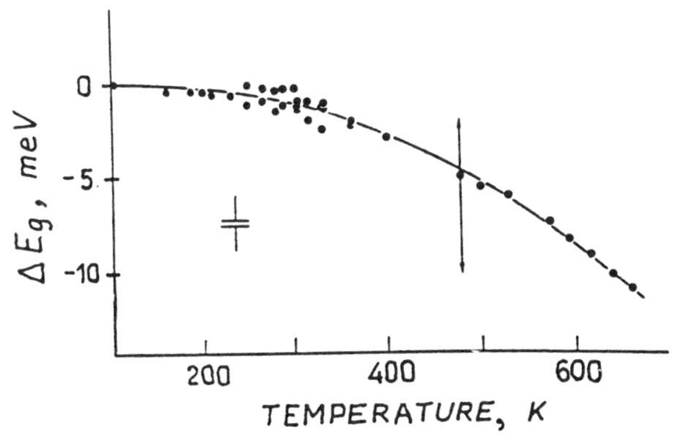

Figure 166. Temperature dependence of the indirect energy gap of diamond [132, fig. 8].

Table IX. dEg/dT [eV/K] values of diamond for different temperatures [130, 132].

Temperature, K	K 200 K	300 K	600 K
dE_g/dT	$-5.4\pm0.5\times10^{-5}$	-1.1×10^{-4}	-2.6×10^{-4}

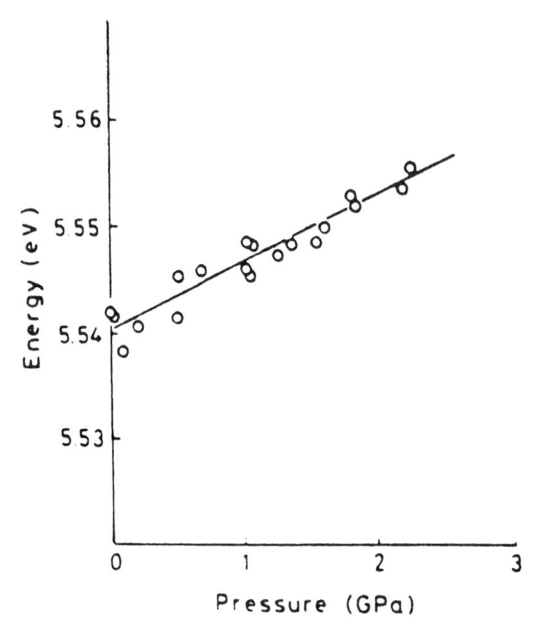

Figure 167. Indirect energy band gap versus pressure [131, fig. 4].

Dependence of Eg on hydrostatic pressure up to 2.3 GPa is given by [130, 131]:

$$dE_g/dP = +\ 5.97\ \ meV/GPa. \quad (20)$$

Dependence of Eg on nonhydrostatic pressure up to 4.5 GPa is given by [349]:

$$dE_g/dP = +\ 5\ meV/GPa. \quad (21)$$

Anisotropic stress reduces the band gap of diamond. At very large pressures around 150 - 300 GPa the Eg value drops down to 1.5 eV [130, 491].

7.3 eV (170 nm), A, a threshold ascribed to the direct electronic transitions from the valance band maximum (**k** = 0) to the conduction band local minimum at **k** = 0 ($\Gamma'_{25} \rightarrow \Gamma_{15}$ transitions) [322, 492]. Temperature dependence of the direct energy gap E_{dg} in a temperature range of 133 to 295 K is given by the expression [130, 132]:

$$dE_{dg}/dT = -(6.3\pm1.8)\times10^{-5}\ eV/K. \quad (22)$$

Pressure dependence of E_{dg} is presented by theoretical data [130, 488-490]:

$$dE_{dg}/dP = +\ 6.1,\ 6.2,\ 6.3,\ or\ 7.0\ meV/GPa. \quad (23)$$

6.3 ELECTRON-PHONON COUPLING

Table. X. Phonons interacting with optical electronic transitions in absorption and/or luminescence

Energy, meV	Optical center
12	2.920 eV
21	2.445 eV (M2 center)
	2.400 eV(M1 center)
25	1.723 eV (NG center)
	3.065 eV
29	2.391 eV
30	2.145 eV
32	1.880 eV (Ne center)
33	1.681 eV (Si center)
	1.762 eV
	2.052 eV
34	2.638 eV (TR12 center)
	2.721 eV
36	GR1 center
	2.699 eV
	2.649 eV
	4.184 eV
	4.191 eV
	4.197 eV
37	2.533 eV
38	1.573 eV
	2.466 eV
	2.964 eV
	2.974 eV
40	1.57 eV
	1.56 eV
	1.562 eV
	1.77 eV
	2.495 eV (S3 center)
41	GR1 center
	2.463 eV (H3 center)
42	1.681 eV (Si center)
	1.946 eV
46	2.424 eV
47	1.573 eV
	2.072 eV (O center)
	2.391 eV
50	2.395 eV
	2.818 eV
52	1.693 eV
53	1.3 eV (C band)
	1.2 eV (band)
54	3.150 eV
	3.204 eV
	3.224 eV
55	2.918 eV
56	1.573 eV
57	2.429 eV (S1 center)
58	1.967 eV
	2.424 eV
	4.582 eV (5RL center)
59	4.567 eV
60	1.264 eV
	1.40 eV nickel center
	1.50 eV (N1 center)
61	4.059 eV

Table X., cont. Phonons interacting with optical electronic transitions in absorption and/or luminescence

Energy, meV	Optical center
63	1.681 eV (Si center)
	1.770 eV
	2.523 eV
64	1.22 eV
	1.681 eV (Si center)
65	1.6 eV (the B-band)
	1.945 eV (NV center)
67	2.462 eV (3H center)
	2.638 eV (TR12 center)
69	1.527 eV (Xe-center)
70	1.57 eV
	1.56 eV
	H2 center
	1.562 eV
	1.77 eV
	2.086 eV
	2.495 eV (S3 center)
	2.638 eV (TR12 center)
	2.985 eV
	2.99 eV
73	2.651 eV
74	2.461 eV (AH3 center)
	2.533 eV
75	3.188 eV
79	3.603 eV (N4 center)
	2.964 eV
	3.603 eV (N4 center)
80	2.526 eV
	3.149 eV (ND1 center)
83	0.77 eV
87	TA at kph = kc
89	0.77 eV
	2.596 eV (N2 center)
92	2.974 eV
93	GR1 center
	2.985 eV (N3 center)
97	2.523 eV
106	2.526 eV
	4.567 eV
113	1.22 eV
	3.762 eV (N5 center)
	3.901 eV
	3.928 eV (N6 center)
120	4.059 eV
125	1.681 eV (Si center); LA (L)
130	2.974 eV
140	2.638 eV (TR12 center)
	2.818 eV

Table X., cont. Phonons interacting with optical electronic transitions in absorption and/or luminescence

Energy, meV	Optical Center
141	TO at kph = kc; 135.4 meV in ^{13}C diamond
	TA at kph = kmax <100>
	5.032 eV (L-center)
143	2.526 eV
148	1.681 eV (Si center); TO (L)
	4.846 eV (T center)
152	1.22 eV
	2.463 eV (H3 center)
154	3.10 eV
155	1.681 eV (Si center); LO (L)
159	3.762 eV (N5 center)
	3.928 eV (N6 center)
164	1.22 eV
163	LO phonon at kph = kc; 155 meV in ^{13}C diamond
	1.681 eV (Si center); O (k=0)
165	1.40 eV nickel center
	2.086 eV
	2.985 eV (N3 center)
	3.188 eV
167 (165)	Raman phonon at k = 0
	4.582 eV (5RL center)
	1.256 eV (H2 center)
175	4.582 eV (5RL center)
193	4.582 eV (5RL center)
200	2.638 eV (TR12 center)
202	4.582 eV (5RL center)
210	4.777 eV (2BD center)
216	3.10 eV
220	4.698 eV (2BD center)
237	4.436 eV
	4.582 eV (5RL center)

7. Coloration of diamond

Colorless diamonds are rather rare. There are mostly of type IIa or pure type IaA [412]. Usually diamonds are colored. The following reasons for coloration are known:

Red.

Upon change of temperature from 77 to 900 K the color of type Ib diamonds changes reversibly from yellow to red and further to dark reddish brown due to the "red" broadening of the nitrogen absorption edge [352, 361];

Yellow synthetic diamonds can be given a deep red color by ion irradiation and subsequent annealing at about 500°C. The effect is particularly pronounced when using high energy ions [475].

Yellow.

The yellow color of most synthetic diamonds and natural type Ib diamonds is due to absorption by single substitutional nitrogen [89, 179].

The color of most pale yellow natural diamonds is due to the N3 center absorption [412].

Nickel impurity gives nitrogen rich synthetic diamonds a yellow-green color [83, 179].

Green.

Light apple-green color of nitrogen containing diamonds can be caused by a few hour high temperature treatment under pressure as a result of the A-aggregate formation [357].

Some natural diamonds have a green "skin" resulting from α-particle damage producing the GR1 absorption [412].

Natural colorless diamonds get grayish-green color after light ion irradiation [475].

Nickel impurity can give a green color to synthetic diamonds with a low nitrogen concentration [83, 179].

Blue.

The blue color of type IIb diamonds is due to absorption by boron acceptors [89, 179].

Dark blue color of irradiated natural diamonds may be caused by the GR1 center absorption [449].

Brown.

Brown coloration of natural diamonds is very common. There is a grayish brown color due to an absorption over the whole visible spectral range monotonically rising to shorter wavelengths [412].

A reddish brown color of natural diamonds originates from an absorption over the whole visible spectral range and two additional broad bands located at about 550 and 380 nm [412].

Nickel impurity gives a brown color to synthetic diamonds with a very low nitrogen concentration. Deep brown color of synthetic diamonds grown by the temperature-gradient method using a nickel catalyst and nitrogen getter is caused by a continuous absorption with a threshold at around 1.7 eV [83, 179].

White.

White diamonds are usually hydrogen rich diamonds. They are generally pure type IaB diamonds [284].

Chameleon diamonds typically change their coloration reversibly from a gray-green to a brighter yellow when heated in the flame [285, 286]. These diamonds are hydrogen-rich diamonds of pure IaA type [284].

8. Physical Classification of Diamond.

The physical classification of diamond is based on the optical absorption of nitrogen, boron and hydrogen related defects and paramagnetic absorption of single substitutional nitrogen.

Type I comprises the diamonds, impurity related optical and paramagnetic absorption of which are dominated by nitrogen defects.

Type Ia comprises type I diamonds which do not show the absorption due to single substitutional nitrogen. Nitrogen is contained in type Ia diamonds as nonparamagnetic aggregates.

Type IaA (type Ia in some publications) comprises type Ia diamonds containing predominantly the A aggregates of nitrogen (fig. 168).

Type IaB (type III in some publications) comprises type Ia diamonds containing predominantly the B aggregates of nitrogen (fig. 169).

Type Ib comprises type I diamonds which contain paramagnetic active nitrogen as the dominating defects (fig. 170).

Type II comprises the diamonds showing no optical and paramagnetic absorption due to the nitrogen related defects.

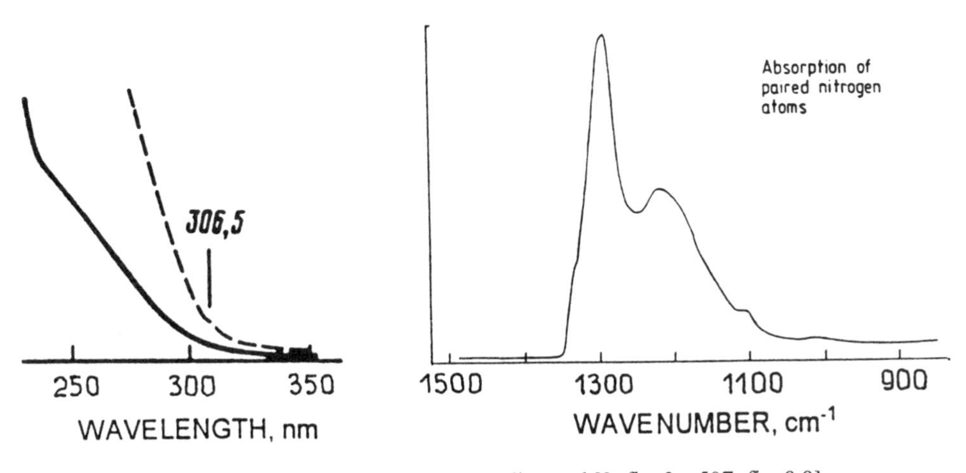

Figure 168. Absorption spectrum of pure type IaA diamond [3, fig. 2a; 507, fig. 8.8].

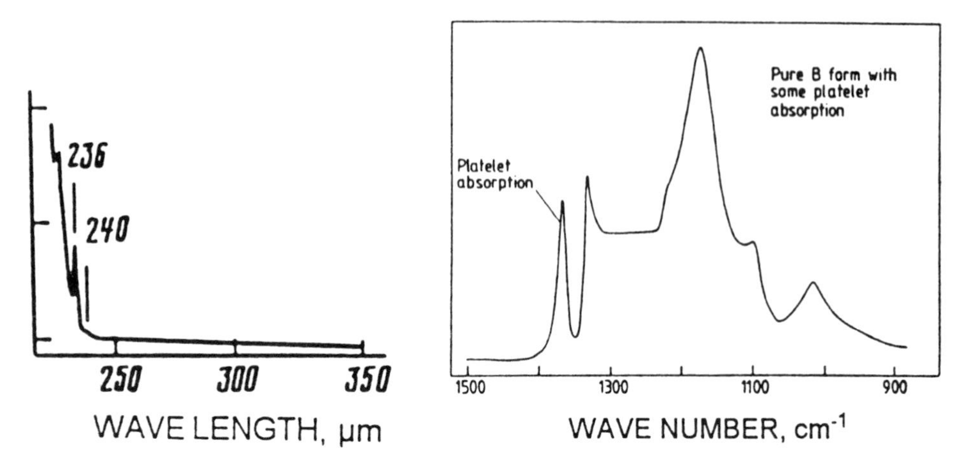

Figure 169. Absorption spectrum of pure type IaB diamond [3, fig. 2b; 313, fig. 8.8].

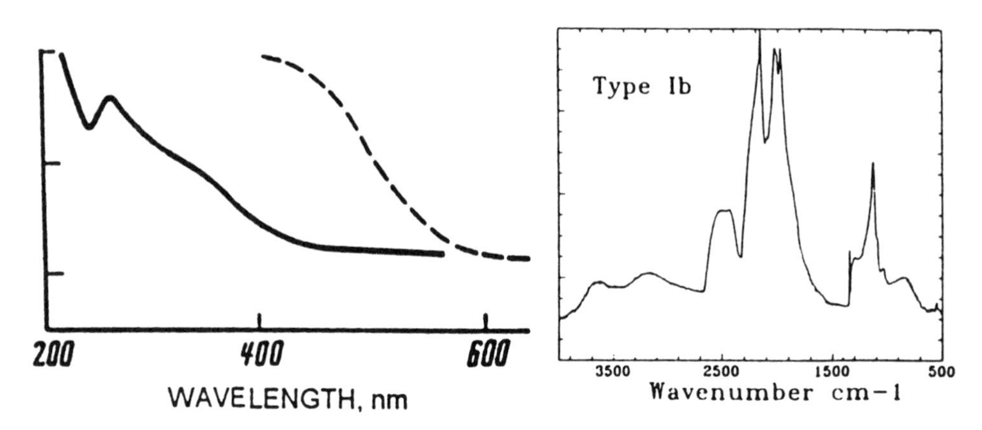

Figure 170. Absorption spectrum of pure type Ib diamond [3, fig. 2c; 457, fig. 9.9e].

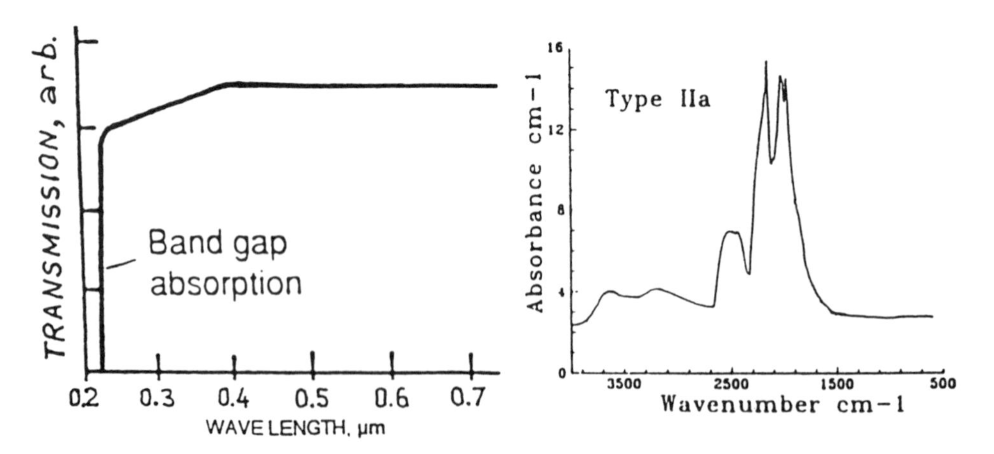

Figure 171. Absorption spectrum of pure type IIa diamond [457, fig. 9.9a; 458, fig. 11.4].

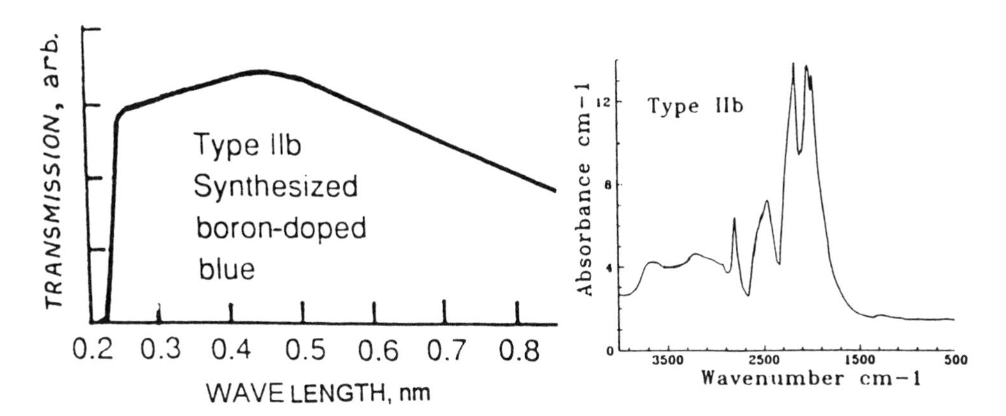

Figure 172. Absorption spectrum of pure type IIb boron containing diamond [457, fig. 9.9b; 458, fig. 11.4].

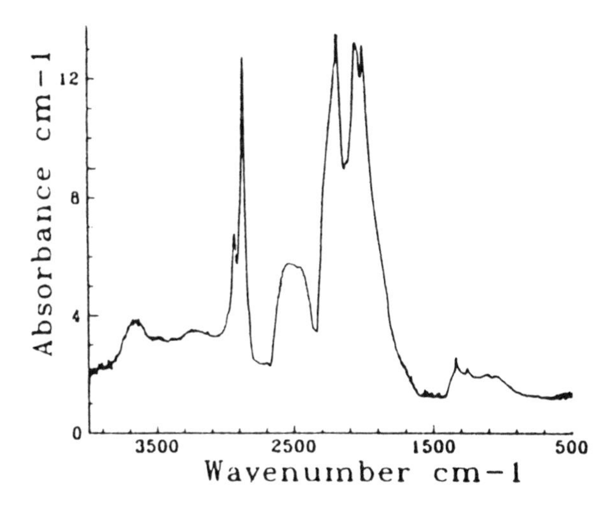

Figure 173. Absorption spectrum of a type IIc diamond [457, fig. 9.9c]

Type IIa comprises type II diamonds which do not show optical absorption due to boron and hydrogen impurity. They are the most optically transparent diamonds (fig. 171).

Type IIb comprises type II diamonds showing optical absorption due to boron impurity (fig. 172). In some publications type IIb is mentioned for any semiconducting diamonds containing uncompensated acceptors or donors.

Type IIc comprises type II diamonds with dominating hydrogen related absorption at around 2900 cm^{-1} [254] (fig. 173).

Table. XI. Threshold of laser ablation in [J/cm^2] units [401].

Type of diamond	ArF laser, λ = 193 nm	KrF laser, λ = 248 nm
Ib	3	3
IIa	4	25

9. Optically Induced Damage

Diamond crystal structure can be changed by absorption of powerful light fluxes.

The laser damage threshold for 193 nm light increases by an order of magnitude when concentration of ^{13}C isotope decreases from 1.07% to 0.07%. In the former case laser energy flux of 300 MJ/cm^2 is sufficient to damage diamond. In isotopically pure diamond 3000 MJ/cm^2 flux can not cause damage [350, 456].

Effect of annealing of 2.8 MeV C^+ ion implanted diamond by 530 nm Nd-silicate glass laser, pulse duration of 15-40 ns [320]:

10 to 23 J/cm^2 - nearly complete regrowth of the implanted layer for doses up to 2 $\times 10^{15}$ cm^{-2};

23 to 28 J/cm^2 - a buried graphite layer is formed for doses greater than 2×10^{15} cm^{-2};

> 30 J/cm^2 - the ion implanted layer is completely graphitized.

10. Miscellaneous

In this paragraph different data related to the optical properties of diamond are collected.

A broad band in A and PL in a spectral range of 700 to 850 nm in diamonds irradiated with neutrons is polarized, the polarization direction being connected with the neutron irradiation direction [405].

Quasi hydrostatic pressure up to 253 GPa does not produce in synthetic diamond any PL centers within a spectral range of 500 to 900 nm [406].

Free-exciton recombination may be very strong in diamond films grown by the combustion flame method [411]. The intensity becomes comparable to that in direct band gap semiconductors at the same excitation conditions [410].

Penetration depth L of electrons of energy E from 5 to 70 keV exciting CL can be found from the expression [74]:

$$L = 0.018\, E^{1.825}. \quad (24)$$

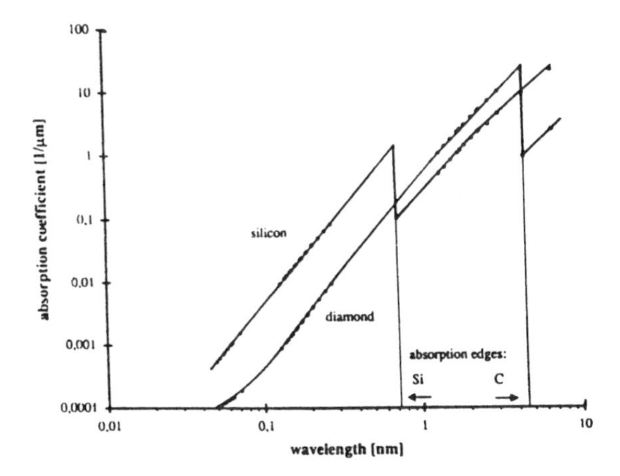

Figure 174. X-ray absorption spectrum of diamond in a wave length range of 0.01 to 10 nm, vertical line indicate Kα absorption edge [324, fig. 4].

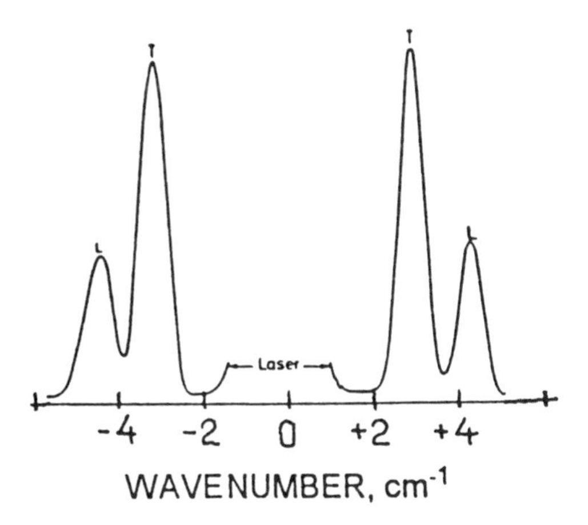

Figure 175. Brillouin scattering spectrum of diamond recorded with exciting light 488.0 nm at 300 K. Incident light along ki | | [100]. Scattered light along ks | | [010] [476, fig. 4].

Brewster angle of diamond at a wavelength of 589.29 nm is 67.53° [74].

Dielectric constant at a temperature of 300 K is $\varepsilon = 5.70 \pm 0.05$ [12, 130]. Temperature and pressure dependencies of the dielectric constant is given by:

$$\varepsilon = 5.70111 - 5.35167\times10^{-5}\,T + 1.6603\times10^{-7}\,T^2, \quad (25)$$

$$(1/\varepsilon)(d\varepsilon/dT)_P = 8.09\times10^{-6}\ K^{-1}, \quad (26)$$

$$(1/\varepsilon)(d\varepsilon/dP)_T = -0.72\times10^{-12}\ Pa^{-1}. \quad (27)$$

Internal nonhomogenous stress P (in GPa) in diamond lattice can be evaluated from FWHM values (in meV) of some optical centers [15, 506]:

P = 0.04 FWHM - using the GR1 center;
P = 0.15 FWHM - using the 2.156 eV nitrogen related center;
P = 0.32 FWHM - using the 3.188 eV nitrogen related center; (28)
P = 0.29 FWHM - using the 1.681 eV silicon related center.

Absorption strength in the one-phonon range (at a wave number of 1200 cm^{-1}) $\mu_{1200cm^{-1}}$ can be used for evaluation of room temperature thermal resistivity (with respect to that of the defect free crystal) T_W of diamond [264, 346, 347] (fig. 176).

Absorption strength in the one-phonon range (at a wave number of 1200 cm^{-1}) $\mu_{1200\ cm^{-1}}$ can be used for evaluation of the total defect concentration in diamond [346] (fig. 177).

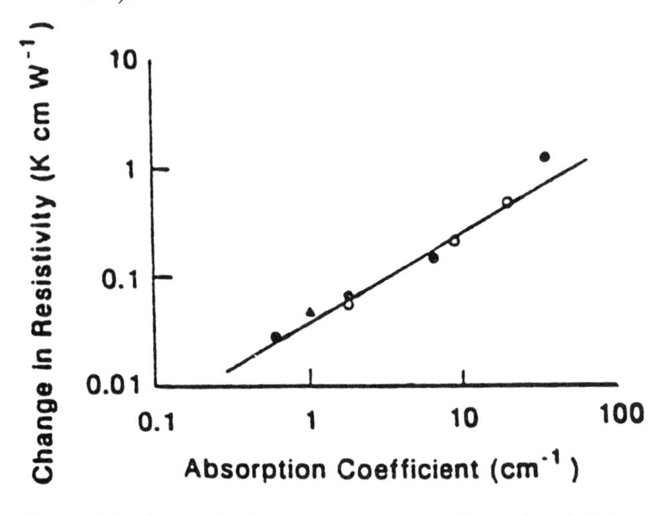

Figure 176. Change in the room temperature thermal resistivity versus absorption coefficient at 1200 cm^{-1} [346, fig. 4a]

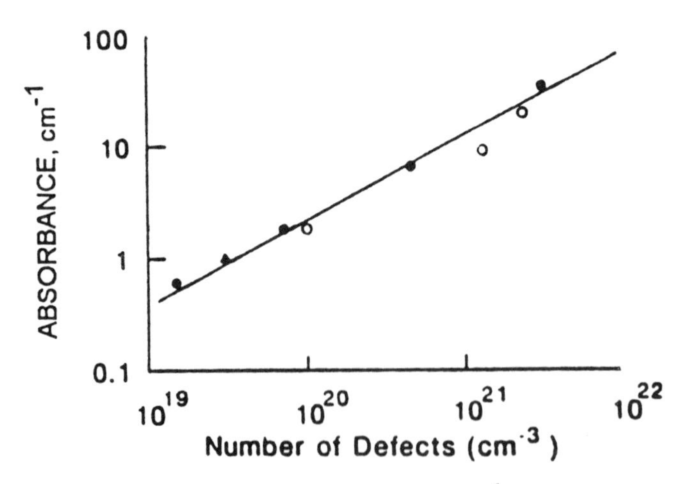

Figure 177. Absorption coefficient at 1200 cm^{-1} versus total defect concentration determined from thermal conductivity [346, fig. 4b]

Thermal conductivity of diamonds at 320 and 450 K is well described by the following empirical expressions:

$$\Lambda_{320} = (5.18\times10^{-4}[\text{m K/W}] + 2.6\times10^{-5}\,\mu_B[\text{cm}^{-1}] + 2.1\times10^{-5}\,\mu_A[\text{cm}^{-1}]^{0.8})^{-1}, \quad (29)$$

$$\Lambda_{450} = (8\times10^{-4}[\text{m K/W}] + 2.5\times10^{-5}\,\mu_B[\text{cm}^{-1}] + 2.7\times10^{-5}\,\mu_A[\text{cm}^{-1}]^{0.8})^{-1}, \quad (30)$$

where μ_A and μ_B are absorption coefficients for the principal A and B bands [436].

Relation between the magnitude of lattice dilatation and the strength of infrared absorption due to substitutional nitrogen:

$$\Delta a_0/a_0 = (2.75\pm0.14)\times10^{-6}\,\mu_C[\text{cm}^{-1}], \quad (31)$$

where μ_C is the absorption coefficient at wave number 1130 cm^{-1} [437].

The secondary absorption edge in type Ia diamonds is shifted from 3.7 to 2.5 eV at a pressure of 364 GPa [438].

The absorption and luminescence features attributed to Ni impurity in synthetic diamonds are segregated exclusively in the {111} growth sectors [83].

Most type IIa diamonds have a weak continuous absorption wing starting in the visible and gradually increasing up to the band gap absorption edge. The wavelength dependence of this continuum can be expressed by: $\mu \sim \lambda^{-3.6}$ [163].

Emission intensity ratio of the D1 (5.21 eV) bound exciton to the B1 (5.27 eV) free exciton in boron doped PC CVD films is given as a function of boron concentration and electrical conductivity [53] (fig. 178).

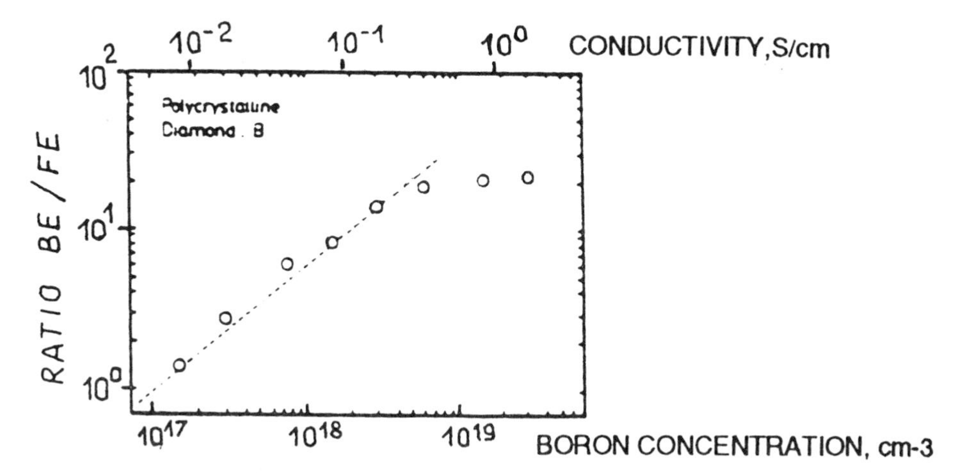

Figure 178. Emission intensity ratio IBE/IFE at 80 K in boron doped polycrystalline diamond films as a function of boron concentration and conductivity [53, fig. 5].

11. Acknowledgments

The author thanks the Alexander von Humboldt Foundation and Deutsche Forschungsgemeinschaft (Germany) for provision of the fellowships and opportunity to carry out this work. The author gratefully acknowledges all the authors, the editors, and publishers of the *Journal of Luminescence*, the *Journal of Physics*, *Diamond and Related Materials*, *Philosophical Magazine*, the *Journal of Materials Science*, *Solid State Communications*, the *Journal of Applied Physics*, *Applied Physics Letters*, the *Physical Review*, the *Journal of Physics and Chemistry of Solids*, the *Electronic Letters*, *Surface Coatings and Technology*, *Nuclear Instruments and Methods in Physics Research*, *Thin Solid Films*, the *Physica*; MYU K. K., Materials Research Society (USA), KTK Scientific Publishers/Terra Scientific Publishing Company, NIRIM (Tsukuba, Japan), Adam Hilger Ltd., Marcel Dekker Inc., the Royal Society, INSPEC (UK), IOP Publishing Ltd., Oxford University Press, Trans Tech Publications Ltd., John Wiley & Sons, Inc., Noyes Publications (USA) for permission to reproduce published figures. Unceasing support and assistance of Dr. V. S. Varichenko, Dr. A. A. Melnikov and Prof. W. R. Fahrner are highly appreciated too. Finally, many thanks go to Dr. A. T. Collins for constructively reviewing this manuscript.

12. References

1. Yu. L. Orlov, *Mineralogy of Diamond*, Moscow, Izd. Acad. of Sci. USSR, 1973, 222 p, (in Russian).
2. D. V. Fedoseev, N. V. Novikov, A. S. Vishnevskii, I. G. Teremetskaja, (1981). *Diamonds*, Handbook, Kiev, Naukova Dumka, (in Russian).
3. G. B. Bokii, G. N. Bezrukov, Yu. A. Klyuev, A. M. Naletov, V. I. Nepsha, (1986). *Natural and Synthetic Diamonds*, Moscow, Nauka.
4. S. A. Solin, A. E. Ramdas, *Phys. Rev. B*, **1** (1970) 1687-1698.
5. R. Loudson, *Proc. Phys. Soc. London*, **84**, pt. 3 (1964) 379-388.
6. F. A. Johnson, Lattice bands in diamond and zincblende crystals, in *Progress in Semiconductors*, London: Hevwood book Temple press, vol. 9, 1965, p. 179-235.
7. N. V. Novikov, Ju. A. Kocherzhinskii, L. A. Shulman, T. D. Ositinskaja, V. G. Malogolovets, A. V. Lysenko, V. I. Malnev, G. F. Nevstruev, E. A. Pugatch, G. P. Bogatyreva, A. S. Vishnevskii, (1987). *Physical Properties of Diamond*, Handbook, Kiev, Naukova Dumka.
8. J. L. Warren, J. L. Jarnell, G. Dolling, R. A. Cowley, *Phys. Rev.*, **158** (1967) 805-808.
9. R. Wehrner, H. Borik, W. Kress et al., *Solid State Communic.*, **5** (1967) 307-309.
10. V. S. Varichenko, A. M. Zaitsev, CL Investigations of Synthetic Diamonds Irradiated with 82 MeV Carbon Ions, 1985, unpublished.
11. F. Peter, *Z. Phys.*, **15** (1923) 358-368.
12. J. Fontanella, R. L. Johnston, J. H. Colwell, C. Andeen, *Appl. Optics*, **16** (1977) 2949-2951.
13. G. Davies, The optical properties of diamond, in *Chemistry and Physics of Carbon*, Marcel Dekker, New York, vol. **13**, (1977)1-143.
14. H. Boppart, J. van Straaten, I. F. Silvera, *Phys. Rev. B*, **32** (1985) 1423-1429.
15. A. T. Collins, S. H. Robertson, *J. Mater. Sci. Letters*, **4** (1985) 681-683.
16. T. Feng, B. D. Schwartz, *J. Appl. Phys.*, **73** (1993) 1415-1425.

17. J. I. Pankove, *Optical Properties of Semiconductors*, Dover Ps, New York, 1971, p. 27.
18. Y. P. Yarshni, *Physica*, **34** (1967) 149-154.
19. L. H. Robins, L. P. Cook, E. N. Farabaugh, Proc. SPIE, vol. **969**, (1989) 86.
20. J. Ruan, W. J. Choyke, W. D. Partlow, *J. Appl. Phys.*, **69** (1991) 6632-6636.
21. V. S. Vavilov, A. A. Gippius, A. M. Zaitsev, B. V. Deryagin, B. V. Spitsyn, A. E. Alexenko, *Sov. Phys. Semicond.*, **14** (1980) 1078-1079.
22. H. Robins, L. P. Cook, E. N. Farabaugh, A. Feldman, *Phys. Rev. B*, **39** (1989) 13367-13377.
23. J. Mazzaschi, J. C. Brabant, M. Brousseau, F. Viollot, *Phys. Rev. Appl.*, **15** (1980) 9.
24. W. D. Partlow, J. Ruan, R. E. Witkowski, W. J. Choyke, D. S. Knight, *J. Appl. Phys.*, **67** (1990) 7019-7025.
25. N. Yamomoto, J. C. H. Spencer, D. Fathy, *Philos. Mag. B*, **49** (1984) 609-629.
26. R. Bauer, A. Osvet, I. Sildos, U. Bogner, *J. of Luminescence*, **56** (1993) 57-60.
27. A. T. Collins, Lars Allers, *Diamond and Related Mater.*, **3** (1994) 932-935.
28. A. M. Zaitsev, A. A. Gippius, V. S. Vavilov, *Sov. Phys. Semicond.*, **16** (1982) 252-256.
29. A. T. Collins, G. S. Woods, *J. Phys. C*, **20** (1987) L797-L801.
30. G. Davies, *J. Phys. C*, **12** (1979) 2551-2566.
31. K. Lobashi, K. Nishimura, Y. Kawate, T. Horiuchi, *Phys. Rev. B*, **38** (1988) 4067-4084.
32. Y. Sato, M. Kamo, *Surf. Coat. Technol.*, **39/40** (1989) 183-188.
33. D. R. Wight, P. J. Dean, *Phys. Rev.*, **154** (1967) 689-696.
34. L. H. Robins, E. N. Farabaugh, A. Feldmann, *Phys. Rev. B*, **48** (1993) 14167-14181.
35. E. Gheeraert, A. Deneuville, A. M. Bonnot, L. Abello, *Diamond and Related Mater.*, **1** (1992) 525-531.
36. A. T. Collins, *Diamond and Related Mater.*, **1** (1992) 457-469.
37. Gheeraert, F. Fontaine, A. Deneuville, Y. L. Khong, A. T. Collins, *Diamond and Related Mater.*, **3** (1994) 737-740.
38. J. Ruan, K. Kobashi, W. J. Choyke, *Appl. Phys. Lett.*, **60** (1992) 3138-3140.
39. S. Matsumoto, Y. Sato, M. Kamo, N. Setaka, *Jap. J. Appl. Phys.*, **21** (1982) L183-L185.
40. R. Locher, J. Wagner, F. Fuchs, M. Maier, P. Gonon, P. Koidl, *Diamond and Related Mater.*, **4** (1995) 678-683.
41. R. J. Graham, K. V. Ravi, *Appl. Phys. Letters*, **60** (1992) 1310-1312.
42. A. M. Bonnot, *Phys. Rev. B*, **41** (1990) 6040-6049.
43. C. V. Cooper, C. P. Beetz, *J. Surface and Coatings Technology*, **47** (1991) 375-387.
44. C. D. Clark, C. B. Dickerson, *Surface and Coatings Technology*, **47** (1991) 336-343.
45. A. M. Zaitsev, V. S. Vavilov, A. A. Gippius, *Sov. Phys. Leb. Inst. Rep.*, **10** (1981) 15-17.
46. W. A. Yarbrough, R. Roy, in A. Badzian, M. Geis and G. Johnson (eds), *Diamond and Diamond-Like Materials*, Mater. Res. Soc., Pittsburgh, PA, Extended Abstracts, Vol. **EA-15**, (1988) 33.
47. P. K. Bachmann, D. U. Wiechert, "Characterization and properties of artificially grown diamond", in R. E. Clausing, L. L. Horton, J. C. Angus and P. Koidl (eds), *Diamond and Diamond-Like Films and Coatings*, Plenum Press, New York, 1991, 677-713.

48. B. Dischler, C. Wild, W. Müller-Sebert, P. Koidl, *Physica B,* **185** (1993) 217-221.
49. R. P. Chin, J. Y. Huang, Y. R. Shen, T. J. Chuang, H. Seki, M. Buck, *Phys. Rev. B*, **45** (1992) 1522-1524.
50. E. Gheeraert, A. Deneuville, *Diamond and Related Mater.*, **1** (1992) 584-593.
51. R. J. Graham, F. Shaapur, Y. Kamo, B. R. Stoner, *Appl. Phys. Letters.* **65** (1994) 292-294.
52. G. Davies, A. T. Collins, P. Spear, *Solid State Communic.*, **49** (1984) 433-437.
53. H. Kawarada, H. Matsuyama, Y. Yokota, T. Sogi, A. Yamaguchi, A. Hiraki, *Phys. Rev. B*, **47** (1993) 3633-3637.
54. S-A. Stuart, S. Prawer, P. S. Weiser, *Diamond and Related Mater.*, **2** (1993) 753-757.
55. J. Ruan, Koji Kobashi, W. J. Choyke, *Appl. Phys. Letters*, **60** (1992) 3138-3140.
56. S. Han, S. G. Prussin, J. W. Ager III, L. S. Pan, D. R. Kania, S. M. Lane, R. S. Wagner, "Radiation Damage Study of Polycrystalline CVD and Natural Type IIa Diamonds Using Raman and Photoluminescence Spectroscopies", Report at *the 8th Int. Conf. IBMM-92*, Sept. 7-11, 1992, Heidelberg, Germany.
57. R. S. Sussmann, J. R. Brandon, G. A. Scarsbrook, C. G. Sweeney, T. J. Valentine, A. J. Whitehead, C. J. H. Wort, *Diamond and Related Materials*, **3** (1994) 303-312.
58. S. K. Sharma, H. K. Mao, P. M. Bell, J. A. Xu, *J. Raman Spectroscopy*, **16**, (1985) 350.
59. M. Mermoux, F. Roy, B. Marcus, L. Abello, G. Lucazeau, *Diamond and Related Mater.*, **1** (1992) 519.
60. L. Fayette, B. Marcus, M. Mermoux, L. Abello, G. Lucazeau, *Diamond and Related Mater.*, **3** (1994) 438-442.
61. H. Herchen, M. A. Cappelli, *Phys. Rev. B*, **43** (1991) 11740-11744.
62. Chia-Fu Chen, Sheng-Hsiung Chen, Hsien-Wen Ko, S. E. Hsu, *Diamond and Related Mater.*, **3** (1994) 443-447.
63. P. John, D. K. Milne, I. C. Drummond, M. G. Jubber, J. I. B. Wilson, J. A. Savage, *Diamond and Related Mater.*, **3** (1994) 486-491.
64. D. S. Knight, B. White, *J. Mater. Res.*, **4** (1989) 385.
65. J. Gerber, M. Weiler, O. Sohr, K. Jung, H. Ehrhardt, *Diamond and Related Mater.*, **3** (1994) 506-509.
66. Y. Muranaka, H. Yamashita, H. Miyadera, *J. Vac. Sci. Technol. A*, **9** (1991) 76-80.
67. D. S. Knight, W. B. White, SPIE, vol. **1055** (1989) 144.
68. J. Beckman, R. B. Jackman, J. S. Foord, *Diamond and Related Mater.*, **3** (1994) 602-607.
69. I. Sildos, A. Osvet, *Diamond and Related Mater.*, **3** (1994) 725-727.
70. P. K. Bachmann, H. Lade, D. Leers, D. U. Wiechert, G. S. A. M. Theunissen, *Diamond and Related Mater.*, **3** (1994) 799-804.
71. H. Eto, Y. Tamou, Y. Ohsawa, N. Kikuchi, *Diamond and Related Mater.*, **1** (1992) 373.
72. A. K. Mehlmann, S. Berger, A. Fayer, S. F. Dirnfeld, M. Bamberger, Y. Avigal, A. Hoffman, R. Porath, *Diamond and Related Mater.*, **3** (1994) 805-809.
73. B. Dischler, W. Rothemund, K. Maier, C. Wild, H. Biebl, P. Koidl, *Diamond and Related Mater.*, **3** (1994) 825-830.
74. J. E. Field (ed), *The Properties of Natural and Synthetic Diamond*, , Academic Press, London, 1992.

75. A. T. Collins, G. Davies, H. Kanda, G. S. Woods, *J. Phys. C*, **21** (1988) 1363-1376.
76. A. T. Collins, P. J. Woad, G. S. Woods, H. Kanda, *Diamond and Related Mater.*, **2** (1993) 136-141.
77. P. R. Briddon, R. Jones, *Physica B,* **185** (1993) 179-189.
78. Y. L. Khong, A. T. Collins, *Diamond and Related Mater.*, **2** (1993) 1-9.
79. J. F. Prins, *Diamond and Related Mater.*, **3** (1994) 922-925.
80. B. Burchard, A. M. Zaitsev, W. R. Fahrner, A. A. Melnikov, A. V. Denisenko, V. S. Varichenko, *Diamond and Related Mater.*, **3** (1994) 947-950.
81. A. A. Gippius, V. S. Vavilov, A. M. Zaitsev, B. S. Zhakupbekov, *Physica B,* **116** (1983) 187-192.
82. M. H. Nazare, A. J. Neves, G. Davies, *Mater. Res. Soc., Symp. Proc.*, **162** (1990) 249-254.
83. A. T. Collins, H. Kanda, R. C. Burns, *Phil. Mag. B*, **61** (1990) 797-810.
84. A. T. Collins, M. Stanley, *J. Phys. D*, **18** (1985) 2537-2546.
85. S. P. Plotnikova, Yu. A. Klyuev, I. A. Parfianovich, *Mineral. Zh.*, **2** (1980) 75-78 (in Russian).
86. Y. K. Vohra, C. A. Vanderborgh, S. Desgreniers, A. L. Ruolff, *Phys. Rev. B*, **39** (1989) 5464-5467.
87. M. N. Nazare, M. F. Thomas, M. I. B. Jorge, *Solid State Communic.*, **55** (1985) 577-581.
88. A. A. Gippius, A. M. Zaitsev, V. S. Vavilov, *Sov. Phys. Semicond.*, **16** (1982) 256-261.
89. A. T. Collins, P. M. Spear, *J. Phys. D*, **15** (1982) L183-L187.
90. A. T. Collins, M. Kamo, Y. Sato, *J. Mater. Res.*, **5** (1990) 2507-2514.
91. A. T. Collins, S. Rafique, *Proc. R. Soc. London A,* **367** (1979) 81-97.
92. V. S. Vavilov, A. A. Gippius, V. A. Dravin, A. M. Zaitsev, B. S. Zhakupbekov, *Sov. Phys. Semicond.*, **16** (1982) 1288-1290.
93. M. I. B. Jorge, M. E. Pereira, M. F. Thomaz, G. Davies, A. T. Collins, *Portugal Phys.*, **14** (1983) 195-210.
94. M. H. Nazare, M. I. B. Jorge, M. F. Thomaz, *J. Phys. C*, **18** (1985) 2371-2379.
95. A. T. Collins, *Nature*, **273** (1978) 654-655.
96. G. Davies, M. H. Nazare, *J. Phys. C*, **13** (1980) 4127-4136.
97. M. E. Pereira, M. I. B. Jorge, *Solid State Communic.*, **61** (1987) 75-78.
98. A. T. Collins, M. Kamo, Y. Sato, *J. Phys.: Condensed Matter*, **1** (1989) 4029-4033.
99. A. M. Zaitsev, A. A. Gippius, V. S. Vavilov, *Sov. Phys. Semicond.*, **16** (1982) 252-256.
100. A: T. Collins, *Inst. Phys. Conf.* Ser. No. **46**, (1979) 327-333.
101. A. T. Collins, *J. Phys. D*, **15** (1982) 1431-1438.
102. E. Pereira, L. Santos, *J. Luminescence*, **40/41** (1988) 139-140.
103. V. E. Ilin, E. V. Sobolev, O. P. Yureva, *Sov. Phys. Solid State*, **12** (1971) 1721-1722.
104. J. Walker *Rep. Prog. Phys.*, **42** (1979) 1605-1659.
105. K. Mohammed, G. Davies, A. T. Collins, *J. Phys. C,* **15** (1982) 2779-2788.
106. M. E. Pereira, M. I. B. Jorge, M. F. Thomaz, *J. Luminesc.*, **31/32** (1984) 179-181.
107. M. E. Pereira, M. I. Barradas, M. F. Thomaz, *J. Phys. C,* **20** (1987) 4923-4932.
108. E. V. Sobolev, V. E. Ilin, O. P. Yureva, *Sov. Phys. Solid State*, **11** (1969) 938-943.
109. D. S. Nedzvetskii, V. A. Gaisin, *Sov. Phys. Solid State,* **15** (1973) 427-428.

110. C. M. Welbourn, M. L. Rooney, D. J. F. Evans, *J. Cryst. Growth*, **94** (1989) 229-252.
111. M. H. Nazare, G. S. Woods, M. C. Assuncao, *Mater. Sci. Engin. B*, **11** (1991) 341-345.
112. A. T. Collins, P. M. Spear, *J. Phys. C*, **16** (1983) 963-973.
113. G. Davies, C. Foy, K. O'Donnell, *J. Phys. C*, **14** (1981) 4153-4165.
114. A. T. Collins, K. Mohammed, *J. Phys. C*, **15** (1982) 147-158.
115. A. T. Collins, S. C. Lawson, *J. Phys: Condens. Matter.*, **1** (1989) 6929-6937.
116. A. M. Zaitsev, *Cathodoluminescence of single crystals, epitaxial films and ion implanted layers of diamond*, Ph.D. Dissertation, P.N.Lebedev Physical Institute of Acad. Sci. USSR, 1980.
117. M. N. Nazare, A. J. Neves, *J. Phys. C*, **20** (1987) 2713-2722.
118. K. Doverspike, J. E. Butler, J. A. Freitas, *Diamond and Related Mater.* **2** (1993) 1078-1082.
119. K. M. McNamara, K. K. Gleason, D. J. Vestyck, J. E. Butler, *Diamond and Related Mater.* **1** (1992) 1145-1155.
120. R. E. Shroder, R. J. Nemanich, J. T. Glass, *Phys. Rev. B*, **41** (1990) 3738.
121. E. Gheeraert, A. Deneuville, *Diamond and Related Mater.*, **1** (1992) 584-587.
122. W. G. Spitzer, D. A. Kleinman, C. J. Frosch, *Phys. Rev.*, **113** (1959) 133.
123. E. Gheeraert, A. Deneuville, A. M. Bonnot, L. Abello, *Diamond and Related Mater.* **1** (1992) 525-528.
124. E. Gheeraert, F. Fontaine, A. Deneuville, Y. L. Khong, A. T. Collins, *Diamond and Related Mater.* **3** (1994) 737-740.
125. K. Okumura, J. Mort, M. A. Machonkin, *Philos. Mag. Letters*, **65** (1992) 105.
126. R. Vaitkus, T. Inushima, S. Yamazaki, *Appl. Phys. Letters*, **62** (1993) 2384.
127. P. Gonon, A. Deneuville, E. Gheeraert, F. Fontaine, *Diamond and Related Mater.* **3** (1994) 836-839.
128. M. D. Bell, W. Levio, *Phys. Rev.*, **111** (1958) 1227-1235.
129. L. A. Vermeulen, A. Halperin, *J. Phys. Chem.*, **45** (1984) 771-778.
130. G. Davies (ed), (1994). *Properties and Growth of Diamond*, INSPEC, London, UK.
131. A. Onodera, M. Hasegawa, K. Furuno, Kobayashi, Y. Nisida, *Phys. Rev. B*, **44** (1991) 12176-12179.
132. C. D. Clark, P. J. Dean, P. V. Harris, *Proc. R. Soc. Lond. A*, **277** (1964) 312-329.
133. P. J. Dean, E. C. Lightowlers, D. R. Wight, *Phys. Rev. A*, **140** (1965) 352-365.
134. H. B. Dyer, F. A. Raal, L. du Preez, J. H. N. Loubser, *Philos. Mag.*, **11** (1965) 763-774.
135. J. Koppitz, O. F. Schimmer, M. Seal, *J. Phys. C*, **19** (1986) 1123-1133.
136. R. G. Farrer, *Solid St. Communic.*, **7** (1969) 685-.
137. L. A. Vermeulen, R. G. Farrer, *Diamond Research,* (1975) 62.
138. P. Denham, E. C. Lightowlers, P. J. Dean, *Phys. Rev.*, **161** (1967) 762-768.
139. P. J. Dean, *Phys. Rev.*, **139** (1967) 588.
140. G. Davies, in M. A. Nusimovici (ed), *Proc. Int. Conf. on Phonons*, Flammarion, Paris, (1971) 382-386.
141. G. S. Woods, *Proc. R. Soc. London A*, **407** (1986) 219-238.
142. C. D. Clark, S. T. Davey, *J. Phys. C.* **17** (1984) 1127-1140.
143. G. B. B. M. Sutherland, Blackwell, W. G. Simeral, *Nature,* **174** (1954) 901-904.

144. C. D. Clark, R. W. Ditchburn, H. B. Dyer, *Proc. R. Soc. London A*, **234** (1956) 363-381.
145. G. Davies, *J. Phys. C*, **9** (1976) L537-L542.
146. R. Robertson, J. J. Fox, A. E. Martin, *Philos. Trans. R. Soc. Lond. A*, **232** (1934) 463-535.
147. G. Davies, I. Summersgill, *Diamond Research*, (1973), 6-15.
148. F. A. Raal, *Proc. Phys. Soc. Lond.*, **74** (1959) 647-649.
149. P. A. Crowther, P. J. Dean, *J. Phys. Chem. Solids*, **28** (1967) 1115-1136.
150. D. R. Wight, P. J. Dean, *Phys. Rev.*, **154** (1967) 689-696.
151. J. Nahum, A. Halperin, *J. Phys. Chem. Solids*, **23** (1962) 345-358.
152. P. J. Dean, J. C. Male, *Proc. R. Soc. Lond.*, **277** (1964) 330-347.
153. P. J. Dean, J. C. Male, *J. Phys. Chem. Solids*, **25** (1964) 1369-1383.
154. A. T. Collins, G. S. Woods, *Philos. Mag. B*, **45** (1982) 385-397.
155. D. S. Nedzvetskii, N. Dymke, *Opt. Spectrosc.*, **28** (1970) 41-45 (in Russian).
156. D. S. Nedzvetskii, V. A. Gaisin, *Opt. Spectrosc.*, **36** (1974) 123-125 (in Russian).
157. A. A. Kaplyanskii, V. I. Kolyshkin, V. N. Medvedev, *Sov. Phys. Solid State*, **12** (1970) 1193-1195.
158. G. Davies, *J. Phys. C*, **7** (1974) 3797-3809.
159. M. F. Thomaz, C. L. Braga, *J. Phys. C*, **5** (1972) L1-L4.
160. C. D. Clark, C. A. Norris, *J. Phys. C*, **3** (1970) 651-658.
161. G. Davies, C. M. Welbourn, J. H. N. Loubser, *Diamond Research*, (1978) 23-30.
162. H. B. Dyer, I. G. Matthews, *Proc. R. Soc. London A*, **243** (1957) 320-335.
163. C. D. Clark, in R. Berman (ed), *The Physical Properties of Diamond*, Clarendon, Oxford, England, 1965, 295-324.
164. J. A. van Wyk, *J. Phys. C*, **15** (1982) L981-983.
165. J. E. Lowther, *J. Phys. Chem. Solids*, **45** (1984) 127-131.
166. E. V. Sobolev, S. E. Lenskaya, V. I. Lisoivan, *J. Struct. Chem.*, **9** (1968) 917-920.
167. T. Evans, P. Rainey, *Proc. R. Soc. London A*, **344** (1975) 111-130.
168. N. Sumida, A. R. Lang, *Proc. R. Soc. London A*, **419** (1988) 235-257.
169. D. R. Wight, P. J. Dean, E. C. Lightowlers, C. D. Mobsby, *J. Luminesc.*, **4** (1971)169-193.
170. E. V. Sobolev, V. E. Il'in, S. V. Lenskaya, O. P. Yureva, *J. Appl. Spectroscopy*, **9** (1968) 1108-1110 (in Russian).
171. A. S. Vishnevskii, O. N. Grigoriev, V. G. Malogolovets, V. I. Trefilov, in *Proc. Conf. on Synthetic Diamonds - Key to Technical Progress*, vol.1, Kiev, USSR, August 1974, Naukova Dumka, 1974, 30-36 (in Russian).
172. C. D. Clark, A. T. Collins, G. S. Woods, in J. E. Field (ed), *The Properties of Natural and Synthetic Diamond*, Academic Press, London, 1992, 35-79.
173. A. T. Collins, G. S. Woods, *Philos. Mag.*, **46** (1982) 77-83.
174. A. T. Collins, M. Stanley, G. S. Woods, *J. Phys. C*, **20** (1987) 969-974.
175. G. S. Woods, A. T. Collins, *J. Phys. C*, **15** (1982) L949-L952.
176. G. S. Woods, *Philos. Mag. B*, **50** (1984) 673.
177. A. Mainwood, A. T. Collins, P. Woad, Mater. Sci. Forum, **143/147** (1994) 29-34.
178. G. S. Woods, I. Kiflawi, H. Kanda, T. Evans, *Philos. Mag. B*, **67** (1993) 651-658.
179. S. C. Lawson, H. Kanda, M. Sekita, *Philos. Mag. B*, **68** (1993) 39.
180. M. H. Nazare, A. J. Neves, G. Davies, *Phys. Rev. B*, **43** (1991) 14196.

181. A. V. Denisenko, V. S. Varichenko, A. M. Zaitsev, A. R. Filipp, V. A. Laptev, S. A. Martynov, V. P. Varnin, I. G. Teremetskaja, Proc. of the 1st Int. Seminar on Diamond Films, June 30 - July 6, Ulan-Ude, Russia, (1992) 8-9.
182. T. G. Bilodeau, K. Doverspike, U. Strom, J. A. Freitas, R. Rameshan, *Diamond and Related Mater.*, **2** (1993) 699-703.
183. G. S. Woods, A. T. Collins, *J. Phys. Chem. Solids*, **44** (1983) 471-475.
184. A. A. Gippius, V. S. Vavilov, A. M. Zaitsev, B. S. Zhakupbekov, *Physica B*, **116** (1983) 187-194.
185. Y. Mori, N. Eimori, H. Kozuka, Y. Yokota, J. Moon, J. S. Ma, T. Ito, A. Hiraki, *Appl. Phys. Letters*, **60** (1992) 47-49.
186. J. Ruan, K. Kobashi, W. J. Choyke, *Appl. Phys. Letters*, **60** (1992) 1884-1886.
187. J. Ruan, W. J. Choyke, K. Kobashi, *Appl. Phys. Letters*, **62** (1993) 1379-1381.
188. G. Davies, S. C. Lawson, A. T. Collins, A. Mainwood, S. J. Sharp, *Phys. Rev. B*, **46** (1992) 13157-13170.
189. A. Halperin, L. A. Vermeulen, *J. Phys. Chem. Solids*, **44** (1983) 77-78.
190. C. D. Clark, E. W. J. Mitchell, B. J. Parsons, in J. E. Field (ed), *The Properties of Diamond*, Acad. Press, 1979, 23-77.
191. C. D. Clark, E. W. J. Mitchell, *Inst. Phys. Conf.* Ser. **31** (1977) 45-57.
192. J. Walker, L. A. Vermeulen, C. D. Clark, *Proc. R. Soc. London A*, **341** (1974) 253-266.
193. C. D. Clark, J. Walker, *Proc. R Soc. London A,* **334** (1973) 241-257.
194. L. A. Vermeulen, C. D. Clark, J. Walker, *Inst. Phys. Conf.* Ser. **23** (1975) 294-300.
195. J. C. Bourgoin, M. Lannoo, *Point Defects in Semiconductors*, vol.I, "Theoretical Aspects", (1981) and vol. II, "Experimental Aspects", (1983), Springer Verlag.
196. G. Davies, C M. Penchina, *Proc. R. Soc. London A*, **338** (1973) 359-374.
197. A. T. Collins, A. W. S. Williams, *J. Phys. C*, **4** (1971) 1789-1800.
198. A. T. Collins, E. C. Lightowlers, in J. E. Field (ed), *The Properties of Diamond*, Acad. Press, 1979, 79-105.
199. A. T. Collins, *Philos. Trans. R. Soc. London A*, **342** (1993) 233-244.
200. G. Davies, *Nature*, **269** (1977) 498-500.
201. C. D. Clark, E. W. J. Mitchell, *Radiat. Effects*, **9** (1971) 219-234.
202. N. R. S. Reddy, N. B. Manson, E. R. Krausz, *J. Luminescence*, **38** (1987) 46-50.
203. I. N. Douglas, W. A. Runciman, *Phys. Chem. Mineral.* **1** (1977) 129.
204. G. Davies, M. F. Thomaz, M. H. Nazare, M. M. Martin, D. Shaw, *J. Phys. C*, **20** (1987) L13-L18.
205. K. Mohammed, G. Davies, A. T, Collins, *J. Phys. C*, **15** (1982) 2789-2800.
206. C. D. Clark, C. A. Norris, *J. Phys. C*, **4** (1971) 2223.
207. A. T. Collins, *J. Phys. C*. **11** (1978) 2453-2463.
208. J. Mazzaschi, J. Barrau, J. C. Brabant, M. Brousseau, F. Voillot, *J. Luminesc.*, **24/25** (1981) 343-346.
209. R. G. Farrer, L. A. Vermeulen, *J. Phys. C.*, **5** (1972) 2762-2768.
210. C. D. Clark, B. J. Parsons, L. A. Vermeulen, *J. Phys. C.*, **12** (1979) 2597.
211. J. E. Lowther, *Solid State Communic.*, **20** (1976) 933-937.
212. A. T. Collins, *J. Phys. C*, **11** (1978) 1957.
213. G. Davies, *J. Phys. C,* **15** (1982) L149.
214. G. Davies, E. C. Lightowlers, *J. Phys. C*, **3** (1970) 638.
215. H. B. Dyer, L. du Preez, *J. Chem. Phys.*, **42** (1965) 1898.

216. G. Davies, *Nature,* **269** (1977) 498.
217. G. Davies, M. H. Nazare, M. F. Hamer, *Proc. R. Soc. London A*, **351** (1976) 245.
218. R. J. Graham, J. B. Posthill, R. A. Rudder, R. J. Narkunas, *Appl. Phys. Letters*, **59** (1991) 2463-2465.
219. M. Yoshikawa, Y. Mori, H. Obata, M. Maegawa, G. Katagiri, H. Ishida, A. Ishitani, *Appl. Phys. Letters*, **67** (1995) 694-696.
220. E. S. de Sa, G. Davies, *Proc. R. Soc. London A*, **357** (1977) 231-251.
221. A. A. Kaplyanskii, V. I. Kolyshkin, V. N. Medvedev, A. P. Skvortsov, *Sov. Phys. Solid State*, **12** (1971) 2867-2872.
222. G. Davies, N. B. Manson, *Ind. Diamond Rev*., (1980) 50.
223. G. Davies, M. H. Nazare, *J. Phys. C*, **13** (1980) 4127-4136.
224. Y. Mita, Y. Nisida, K. Suito, A. Onodera, S. Yazu, *J. Phys: Condens Matter.*, **2** (1990) 8567.
225. A. A. Kaplyanskii, V. I. Kolyshkin, V. N. Medvedev, *Sov. Phys. Solid State*, **12** (1970) 1193-1195.
226. G. Davies, *Rep. Prog. Phys*., **44** (1981) 787.
227. R. T. Harley, M. J. Henderson, R. M. Macfarlane, *J. Phys. C.*, **17** (1984) L233-L236.
228. Y. Nisida, Y. Mita, *Diamond Optics III*, SPIE, vol. **1325** (1990) 296-303.
229. Y. Nisida, et al., in S. Saito, O. Fukunaga, M. Yoshikawa (eds), *Science and Technology of New Diamond*, KTK/Terra, Japan, 1990, 363-367.
230. D. A. Redman, S. Brown, S. C. Rand, *J. Opt. Soc. Am. B*, **9** (1992) 768-774.
231. N. R. S. Reddy, N. B. Manson, E. R. Krausz, *J. Luminesc*., **38** (1987) 46-47.
232. J. H. N. Loubser, J. A. van Wyk, *Diamond Research*, (1977) 4-8.
233. N. B. Manson, X.-F. He, P. T. H. Fisk, *Opt. Letters*, **15** (1990) 1094-1096.
234. S. Sato, H. Watanabe, T. Takahashi, Y. Abe, M. Iwaki, *Nucl. Instrum. Methods Phys. Res. B*, **59/60** (1991) 1391-1394.
235. J. F. Prins, F. A. Raal, Radiat. *Effects Express*, **1** (1987) 1.
236. C. D. Clark, R. W. Ditchburn, H. B. Dyer, *Proc. R. Soc. London A*, **235** (1956) 305.
237. M. D. Crossfield, G. Davies, A. T. Collins, E. C. Lightowlers, *J. Phys. C*, **7** (1974) 1909-1917.
238. G. Davies, M. N. Nazare, M. F. Hamer, *Proc. R. Soc. London A*, **351** (1976) 245.
239. M. F. Thomaz, G. Davies, *Proc. R. Soc. London A*, **362** (1978) 405.
240. A. T. Collins, M. F. Thomaz, M. I. B. Jorge, *J. Phys. C,* **16** (1983) 5417-5425.
241. A. T. Collins, M. F. Thomaz, M. I. B. Jorge, *J. Phys. C*, **16** (1983) 2177.
242. A. T. Collins, *J. Phys. C,* **20** (1987) 2027-2033.
243. J. Mazzaschi, I. Barrau, J. C. Brabant, M. Brousseau, F. Voillot, *Rev. Phys. Appl.*, **15** (1980) 9.
244. A. T. Collins, P. M. Spear, *J. Phys. C,* **19** (1986) 6845-6858.
245. V. S. Varichenko, V. G. Gordeev, A. M. Zaitsev, V. A. Nikolaenko, V. F. Stelmakh, Method of doping of diamond with nitrogen, *USSR Patent*, 1989, Int. Cl. C01B 31/06.
246. E. Pereira, L. Santos, *J. Luminesc*., **45** (1990) 454.
247. E. Pereira, L. Santos, *J. Luminesc*., **38** (1987) 181-183.
248. E. Pereira, L. Santos, *Physica B*, **185** (1993) 222-227.
249. P. J. Dean, *Phys. Rev. A*, **139** (1965) 588.

250. E. Pereira, T. Monteiro, *J. Luminesc.*, **54** (1990) 443.
251. E. Pereira, T. Monteiro, *J. Luminesc.*, **48/49** (1991) 811.
252. E. Pereira, T. Rodrigues, L. Pereira, Report at *the Diamond Conference*, Oxford, UK, 1987.
253. S. C. Lawson, G. Davies, A. T. Collins, A. Mainwood, *J. Phys.: Condens. Matter*, **4** (1992) 3439-3452.
254. G. Janssen, W. Vollenberg, J. Giling, W. J. P. van Enckevort, J. J. D. Schaminee, M. Seal, *Surf. Coatings Technol.*, **47** (1991) 113-126.
255. G. Janssen, W. J. P. van Enckevort, J. J. D. Schaminee, W. Vollenberg, J. Giling, M. Seal, *J. Crystal Growth*, **104** (1990) 752-757.
256. A. Dehbi-Alaoui, A. S. James, A. Matthews, *Surf. Coatings Technol.,* **43/44** (1990) 88-98.
257. A. Dehbi-Alaoui, P. Holiday, A. Matthews, *Surf. Coatings Technol.,* **47** (1991) 327-335.
258. A. T. Collins, S. C. Lawson, *Phil. Mag. Letters*, **60** (1989) 117-122.
259. C. D. Clark, R. W. Ditchburn, H. B. Dyer, *Proc. R. Soc. London A*, **237** (1956) 75-89.
260. A. T. Collins, *J. Phys. C,* 16 (1983) 6691-6694.
261. A. T. Collins, S. H. Robertson, *J. Mater. Sci. Letters*, **4** (1985) 681-684.
262. M. Nakamizo, H. Honda, M. Inagaki, *Carbon,* **16** (1978) 281-283.
263. R. J. Nemanich, J. T. Glass, G. Lucovsky, R. E, Shroder, *J. Vac. Sci. Technol.*, **A6** (1988) 1783-1787.
264. G. S. Woods, G. C. Purser, A. S. S. Mtimkulu, A. T. Collins, *J. Phys. Chem. Solids*, **51** (1990) 1191-1197.
265. G. S. Woods, J. A. van Wyk, A. T. Collins, *Phil. Magazine B*, **62** (1990) 589-595.
266. R. M. Chrenko, H. M. Strong, R. E. Tuft, *Phil. Mag.*, **23** (1971) 313-318.
267. A. T. Collins, *J. Phys. C,* **13** (1980) 2641-2650.
268. W. Kaiser, W. L. Bond, *Phys. Rev.*, **115** (1959) 857-863.
269. S. C. Lawson, H. Kanda, H. Kiyota, T. Tsutsumi, H. Kawarada, J. Appl. Phys., 77(1995) 1729-1734.
270. S. C. Lawson, H. Kanda, K. Era, Y. Sato, A study of broad band cathodoluminescence from boron-doped high-pressure synthetic and chemical vapor deposited diamond, (1994) unpublished paper, NIRIM, Tsukuba, Japan.
271. P. R. Brosious, J. W. Corbett, J. C. Bourgoin, *Phys. Stat. Sol.*, **21** (1974) 677-683.
272. J. P. F. Sellschop, in J. E. Field (ed), *The Properties of Natural and Synthetic Diamond,* Acad. Press, London, (1992) 81-179.
273. J. F. Prins, *Phys. Rev. B.* **38** (1988) 5576-5584.
274. E. Gheeraert, A. Deneuville, E. Bustarret, F. Fontaine, "Determination of Weak Optical Absorption Coefficient in Polycrystalline Diamond Thin Films by Photothermal Deflection Spectroscopy", Report at the Int. Conference Diamond Films'94, Sept. 1994, Lucca, Italy.
275. A. Deneuville, P. Gonon, E. Gheeraert, A. T. Collins, Y. L. Khong, *Diamond and Related Materials*, **2** (1993) 737-741.
276. E. Gheeraert, P. Gonon, A. Deneuville, L. Abello, G. Lucazeau, *Diamond and Related Mater.*, **2** (1993) 742-745.
277. P. V. Huong, *Diamond and Related Mater.*, **1** (1991) 33.
278. P. Bou, L. Vandenbulcke, *J. Electrochem. Soc.*, **138** (1991) 2991-3000.

279. A. T. Collins, personal communication, 1995.
280. A. Deneuville, A. C. Papadopoulo, E. Gheeraert, P. Gonon, Suppl. **261** (1992) 277, Le vide, les couches minces.
281. A. T. Collins, M. Kamo, Y. Sato, *Mat. Res. Soc. Symp. Proc.*, **162** (1990) 225-230.
282. A. T. Collins, in S. Saito, O. Fukunaga, M. Yoshikawa (ed), *Science and Technology of New Diamond*, KTK Scientific Publishers/Terra Scientific Publishing Company, 1990, 273-278.
283. A. T. Collins, S. C. Lawson, G. Davies, H. Kanda, *Materials Science Forum*, **65/66** (1990) 199-204.
284. E. Fritsch, K. Scarratt, A. T. Collins, in *New Diamond Science and Technology*, MRS Int. Conf. Proc., 1991, 671-676.
285. F. A. Raal, *Indiaqua*, **31** (1982) 127-131.
286. F. A. Raal, *Am. Miner.*, **54** (1969) 292-296.
287. A. T. Collins, in A. J. Purdes (ed), *Proc. of the 2nd Int. Symp. on Diamond Materials*, Electrochem. Soc. Inc. 1992, 408-419.
288. A. A. Gippius, A. T. Collins, *Solid State Communic.*, **88** (1993) 637-638.
289. A. A. Gippius, A. T. Collins, S. A. Kazarian, *Mater. Sci. Forum*, **143-147** (1994) 41-47.
290. E. J. Brookes, A. T. Collins, G. S. Woods, *J. Hard Mater.*, **4** (1993) 97-105.
291. A. P. Eliseev, E. V. Sobolev, *Superhard Materials*, **1** (1979) 19-25 (in Russian).
292. Yu. A. Klyuev, A. M. Naletov, V. I. Nepsha, et al., *J. Phys. Chem.*, 56 (1982) 524-531 (in Russian).
293. M. I. Samoilovitch, V. P. Butuzov, Ju. P. Solodova, *Diamonds*, 7 (1972) 1-7 (in Russian).
294. M. Ja. Scherbakova, E. V. Sobolev, V. A. Nadolinny, *Rept. of the Acad. of Sci. of the USSR*, **204** (1972) 851-854 (in Russian).
295. S. P. Plotnikova, Yu. A. Klyuev, I. A. Parfianovich, *Mineral. Journal,* **4** (1980) 75-80 (in Russian).
296. E. V. Sobolev, O. P. Yurjeva, *Phys. And Techn. of Semicond.*, **16** (1982) 1513-1515 (in Russian).
297. G. O. Gomon, *Diamonds,* Leningrad, Izd. Mashinostroenie, 1966 (in Russian).
298. Yu. A. Klyuev, V. I. Nepsha, G. N. Bezrukov, *Phys. Thechn. of Semicond.*, **8** (1974) 1619-1622 (in Russian).
299. R. M. Chrenko, *Phys. Rev. B*, **7** (1973) 4560-4567.
300. E. C. Lightowlers, A. T. Collins, *Diamond Research*, 1976, 14-26.
301. V. S. Vavilov, E. A. Konorova, E. B. Stepanova, E. M. Truchan, *Phys. Techn. Semicond.*, **13** (1979) 1033-1036 (in Russian).
302. E. M. Galimov, Yu. A. Klyuev, I. N. Ivanovskaja, *Rept. of Acad. of Sci. Of the USSR*, **249** (1979) 958-962 (in Russian).
303. Yu. A. Klyuev, V. I. Nepsha, N. I. Epishina, *Rept. of Acad. Sci. USSR*, **240** (1978) 1104-1107 (in Russian).
304. E. M. Galimov, I. N. Ivanovskaja, Yu. A. Klyuev, *Geochemistry*, **4** (1980) 533-538 (in Russian).
305. J. F. Prins, Cathodoluminescence as a probe to study residual radiation damage in ion-implanted diamond, IUMRS Conf., Tokyo, 31 Aug. - 4 Sept. 1993.
306. J. F. Prins, Optical studies on ion beam-doped diamond layers, ICNDST-4 Conf. 17-22 July, 1994, Kobe, Japan.

307. J. F. Prins, Cathodoluminescence bands in virgin and ion-beam-doped natural type II diamonds, Diamond Conference, Reading, 1994.
308. J. F. Prins, *Diamond and Related Mater.*, **3** (1994) 922-925.
309. J. A. Freitas, Jr, U. Strom, A. T. Collins, *Diamond and Related Mater.*, **2** (1993) 87-91.
310. A. A. Melnikov, A. V. Denisenko, A. M. Zaitsev, W. R. Fahrner, V. S. Varichenko, High Temperature Light Emitting Diodes on Insulating Diamond, Report on the 3rd Int. Conf. on Appl. of Diam. Films and Related Mater., 21-24 August, 1995, Gaithersburg, Maryland, USA.
311. A. T. Collins, *Physica B,* **185** (1993) 284-296.
312. Y. L. Khong, A. T. Collins, L. Allers, (1984). *Diamond and Related Mater.*, **3** (1994) 1023-1027.
313. G. Davies, *Diamond,* Adam Hilger Ltd., Techno House, Bristol.
314. A. T. Collins, Spectroscopic Studies of Point Defects in CVD Diamonds and High Pressure Synthetic Diamonds, Proc. of the NIRIM Int. Symp. On Advanced Mater. 94, Tsukuba, Japan, 13-17 March, 1994.
315. J. A. Freitas, Jr. P. B. Klein, A. T. Collins, *Electronic Letters*, **29** (1993) 1727-1727.
316. A. T. Collins, E. C. Lightowlers, V. Higgs, L. Allers, S. J. Sharp, in S. Saito, N. Fujimori, O. Fukunaga, M. Kamo, K. Kobashi, M. Yoshikawa (eds), *Advances in New Diamond Science and Technology*, MYU, Tokyo, 1994, 307-310.
317. K. C. Hass, M. A. Tamor, T. R. Anthony, W. F. Banholzer, *Phys. Rev. B*, **44** (1991) 7123.
318. J. A. Freitas, Jr. P. B. Klein, A. T. Collins, *Appl. Phys. Letters*, **64** (1994) 2136-2138.
319. S.-A. Stuart, S. Prawer, P. S. Weiser, *Appl. Phys. Letters*, **62** (1993) 1227-1229.
320. S. Prawer, D. N. Jamieson, R. Kalish, *Phys. Rev. Letters*, **69** (1992) 2991-2994.
321. S. -A. Stuart, S. Prawer, P. S. Weiser, *Diamond and Related Materials*, **2** (1993) 753-757.
322. H. R. Phillip, E. A. Taft, *Phys. Rev.*, **127** (1962) 159.
323. F. J. Himpsel, J. A. Knapp, J. A. van Vechten, D. E. Eastman, *Phys. Rev.*, **20** (1979) 624.
324. P. Koidl, C. -P. Klages, *Diamond and Related Materials*, **1** (1992) 1065-1074.
325. S. C. Lawson, H. Kanda, *Diamond and Related Mater.*, **2** (1993) 130.
326. F. Fuchs, C. Wild, K. Schwarz, W. Müller-Sebert, P. Koidl, *Appl. Phys. Letters,* **66** (1995) 177-179.
327. E. Fritsch, K. V. G. Scarratt, SPIE, vol. **1146** (1989) 201.
328. N. C. Burton, G. M. Meaden, Cathodoluminescence studies of CVD diamond films in the scanning and transmission electron microscopes, *Proc. of the Diamond Conference*, 1993, Bristol, p. 5.1.
329. I. Kiflawi, S. R. Boyd, G. S. Woods, Infrared determination of the nitrogen concentration in pure type Ia diamonds, *Proc. of the Diamond Conference*, 1993, Bristol, p. 8.1.
330. A. A. Gippius, S. A. Kazarian, Oxygen luminescent centers in Diamond, *Proc. of the Diamond Conference*, 1993, Bristol, p. 16.1.
331. C. D. Clark, H. Kanda, I. Kiflawi, G. Sittas, Silicon in Diamond, *Proc. of the Diamond Conference*, 1993, Bristol, p. 17.1.

332. M. H. Nazare, L. M. Rino, H. Kanda, Synthetic diamond: the optical band at 1.885 eV, *Proc. of the Diamond Conference*, 1993, Bristol, p. 24.1.
333. M. H. Nazare, J. C. Azevedo, L. M. Rino, H. Kanda, Optically active nickel in diamond: the 1.885 eV defect, *Proc. of the Diamond Conference*, 1993, Bristol, p. 25.1.
334. R. Jones, V. J. B. Torres, P. R. Briddon, S. Öberg, Theory of the H3 and H4 defects in irradiated type I diamond, *Proc. of the Diamond Conference*, 1993, Bristol, p. 42.1.
335. A. Mainwood, J. E. Lowther, J. Van Wyk, The distorted vacancy - a model for the R2 EPR and the 1.685 eV optical center, *Proc. of the Diamond Conference*, 1993, Bristol, p. 43.1.
336. A. T. Collins, P. J. Woad, Nickel-related cathodoluminescence features in synthetic diamonds annealed at 1700 to 1900°C, *Proc. of the Diamond Conference*, 1993, Bristol, p. 15.1.
337. B. L. Jones, Cathodoluminescence studies of synthetic diamond after deformation at 1100°C, *Proc. of the Diamond Conference*, 1993, Bristol, P23.1.
338. Willingham, T. Hartnett, C. Robinson, C. Klein, in Y. Tzeng, M. Yoshikawa, M. Murakawa, A. Feldman (eds), *Application of Diamond Films and Related Materials*, Elsevier Science Publishers B. V., (1991) 157-162.
339. P. R. Chalker, in R. E. Clausing, L. L. Hirton, J. C. Angus, P. Koidl (eds), *Diamond and Diamond-like Films and Coatings*, Plenum Press, NATO ASI Series, B: Physics, vol. **266** (1991) 127-150.
340. A. T. Collins, M. Kamo, Y. Sato, *J. Phys. D*, **22** (1989) 1402-1405.
341. A. Feldman, L. H. Robins, in Y. Tzeng, M. Yoshikawa, M. Murakawa, A. Feldman (eds), *Application of Diamond Films and Related Materials*, Elsevier Science Publishers B. V., (1991) 181-188.
342. C. D. Clark, C. D. Dickerson, *J. Phys.: Condensed Matter*, **4** (1992) 869-878.
343. C. D. Clark, C. D. Dickerson, in A. Lettington and J. W. Steeds (eds), *Thin Film Diamond*, Chapman & Hall, London, (1994) 83-90.
344. J. Wagner, C. Wild, P. Koidl, *Appl. Phys. Letters*, **59** (1991) 779-781.
345. L. Santos, E. Pereira, *J. of Luminescence*, **60/61** (1994) 614-617.
346. D. T. Morelli, C. Uher, *Appl. Phys. Letters*, **63** (1993) 165-167.
347. E. A. Burgemeister, *Physica B*, **93** (1978) 165-179.
348. P. V. Huong, *Mater. Sci. and Engineering B*, **11** (1992) 235-242.
349. O. Madelung, M. Schulz and H. Weiss (eds), *Landoldt-Börnstein Tables*, Springer, Berlin, vol. **17a** (1982) 36ff.
350. W. F. Bahnholzer, T. R. Anthony, *Thin Solid Films*, **212** (1992) 1-10.
351. G. Davies, A. J. Neves, M. H. Nazare, *Europhysics Letters*, **9** (1989) 47-52.
352. W. J. P. Enckevort, E. H. Versteegen, *J. Phys. C: Solid State Physics*, **4** (1992) 2361-2313.
353. C. A. Klein, T. M. Hartnett, C. J. Robinson, *Phys. Rev. B*, **45** (1992) 12854-12863.
354. S. A. Solin, A. K. Ramdas, *Phys. Rev. B*, **1** (1970) 1687-1695.
355. J. R. Hardy, S. D. Smith, *Philos. Mag.*, **6** (1961) 1163-1170.
356. R. Wehner, H. Borik, W. Kress, A. A. Goodwin, S. D. Smith, *Solid State Communications*, **5** (1967) 307-311.
357. V. A. Nadolinny, A. P. Yelisseyev, *Diamond and Related Materials*, **3** (1993) 17-21.

358. M. Glasbeek, I. Hiromitsu, J. Westra, *J. Electrochem. Soc.*, **140** (1993) 1399-1402.
359. J. Westra, S. Sitters, M. Glasbeek, *Phys. Rev. B*, **45** (1992) 5699-5702.
360. I. Hiromitsu, J. Westra, M. Glasbeek, *Phys. Rev. B*, **46** (1992) 5303-5310.
361. A. P. Eliseev, G. M. Rylov, E. N. Fedorova, V. G. Vins, B. N. Feigelson, A. P. Sharapov, N. E. Ulanov, A. N. Zhiltsov, A. I. Kungurov, Peculiarities of Defect Content and Mechanical Treatment of Large Synthetic Single Crystals, *Preprint of the Joint Inst. of Geology, Geophysics and Mineralogy*, Novosibirsk, **5**, (1992) 28 p. (in Russian).
362. P. J. Dean, E. C. Lightowlers, D. R. Wight, *Phys. Rev. A*, **140** (1965) 352-368.
363. I. Kiflawi, A. E. Mayer, P. M. Spear, J. A. van Wyk, G. S. Woods, *Philos. Mag. B*, **69** (1994) 1141-1147.
364. J. Isoya, H. Kanda, Y. Uchida, *Phys. Rev. B*, **42** (1990) 9843-9850.
365. L. Paslovsky, J. E. Lowther, *J. of Luminescence*, **50** (1992) 353-359.
366. E. Pereira, L. Santos, *Physica B*, **185** (1993) 222-227.
367. K. Niwase, Y. Kakimoto, I. Tanaka, T. Tanabe, *Nuclear Instrum. and Meth. in Phys. Res., B*, **91** (1994) 78-82.
368. A. A. Gippius, *Mater. Sci. Forum*, **83-87** (1992) 1219-1224.
369. K. C. Hass, M. A. Tamor, T. R. Anthony, W. F. Banholzer, *Phys. Rev. B*, **45** (1992) 7171-7182.
370. E. V. Sobolev, A. P. Eliseev, *J. of Struct. Chemistry*, **17** (1976) 933-935 (in Russian).
371. E. V. Sobolev, *Proc. of the III Symp. on Growth and Synthesis of Semiconducting Crystals and Films*, Novosibirsk, USSR, (1972) 71 (in Russian).
372. E. V. Sobolev, *Proc. of the III Symp. on Growth and Synthesis of Semiconducting Crystals and Films*, Novosibirsk, USSR, (1975) 175 (in Russian).
373. A. A. Gippius, A. M. Zaitsev, V. S. Vavilov, *Proc. of the Int. Conf. on Radiat. Effects in Semicond. and Related Mater.*, USSR, Tbilisi, (1979) 419-423.
374. J. Walker, *Inst. Phys. Conf.* Ser. **23**, Chapter 4 (1975) 317-324.
375. A. R. Filipp, V. V. Tkachev, V. S. Varichenko, A. M. Zaitsev, A. R. Chelyadinskii, Yu. A. Klyuev, *Diamond and Related Mater.*, **1** (1992) 271-276.
376. A. M. Zaitsev, *Luminescence of Ion-Implanted Diamond and Cubic Boron Nitride*, Dissertation for Doctor's degree, the University of Minsk, 1992, 367 p. (in Russian).
377. A. M. Zaitsev, *Materials Science and Engineering B*, **11** (1992) 179-190.
378. V. S. Varichenko, A. Yu. Didyk, A. M. Zaitsev, V. I. Kuznetsov, V. M. Kulakov, A. A. Melnikov, S. P. Plotnikova, V. A. Skuratov, V. F. Stelmakh, V. D. Shestakov, Defect Production in Diamond by High Energy Ion Implantation, *Preprint JINR* No **R14-88-44**, (1988) 12 p. (in Russian).
379. A. M. Zaitsev, A. G. Uliyashin, H. Ali Noor, *Superhard Materials*, **1** (1991) 18-24 (in Russian).
380. V. S. Varichenko, E. D. Vorobjev, A. M. Zaitsev, V. A. Laptev, M. I. Samoilovich, V. A. Skuratov, V. F. Stelmakh, Defect Production in Synthetic Diamonds by High Energy Ion Implantation, Preprint JINR No 14-85-893, Dubna, 1985, 8 p. (in Russian).
381. V. D. Tkachev, A. M. Zaitsev, V. V. Tkachev, *Phys. Stat. Solidi b*, **129** (1985) 129-133.
382. V. D. Tkachev, A. M. Zaitsev, V. V. Tkachev, *Sov. Phys. Semiconductors*, **19** (1985) 585-587.

383. A. G. Uliyashin, A. M. Zaitsev, H. Ali Noor, *Mater. Sci. and Engineering B*, **11** (1992) 359-362.
384. V. S. Varichenko, E. D. Vorobjev, A. M. Zaitsev, V. A. Laptev, M. I. Samoilovich, V. A. Skuratov, V. F. Stelmakh, *Sov. Phys. Semicond.*, **21** (1987) 668-671.
385. V. S. Varichenko, A. M. Zaitsev, V. F. Stelmakh, *Phys. Stat. Solidi a*, **95** (1986) K25-K28.
386. V. S. Varichenko, A. M. Zaitsev, V. F. Stelmakh, *Phys. Stat. Solidi a*, **95** (1986) K123-K126.
387. V. S. Varichenko, I. A. Dobrinets, V. A. Dravin, A. M. Zaitsev, *J. of Appl. Spectroscopy*, Minsk, Belarus, **51** (1989) 218-224 (in Russian).
388. A. M. Zaitsev, A. A. Gippius, V. S. Vavilov, *J. of Experim. and Theor. Phys. Letters*, **31** (1980) 181-184.
389. A. A. Gippius, A. M. Zaitsev, V. S. Vavilov, *Proc. II Int. Conf. on Defects and Radiat. Effects in Semiconductors*, Japan, Kyoto, 1980, 121-125.
390. A. A. Gippius, V. S. Vavilov, A. M. Zaitsev, V. V. Ushakov, *Proc. of the 2nd Soviet-American Workshop on Ion Implantation*, USSR, Pustshino, 1979, 11-23.
391. V. D. Tkachev, A. M. Zaitsev, V. V. Tkachev, *Proc. of the Working Meeting on Ion Implantation in Semicond. and Other Materials and Ion Beam Devices*, Hungary, Balatonaliga, 1985, 87-88.
392. E. D. Palik, *Handbook of Optical Constant of Solids*, Academic, New York, 1985.
393. J. E. Lowther, *J. of Luminesc.*, **60/61** (1994) 531-534.
394. M. Ja Shcherbakova, V. A. Nadolinnyi, E. V. Sobolev, *J. of Struct. Chemistry*, **19** (1978) 305-309 (in Russian).
395. R. J. Graham, P. R. Buseck, , *Philos. Mag. B*, **70** (1994) 1177-1185.
396. M. Ja Shcherbakova, E. V. Sobolev, V. A. Nadolinnyi, V. A. Aksenov, *Soviet Phys. Doklady*, **20** (1976) 725-728.
397. W. J. P. van Enckevort, E. P. Visser, *Philos. Mag. B*, **62** (1990) 597.
398. A. A. Melnikov, A, V. Denisenko, A. M. Zaitsev, V. S. Varichenko, V. P. Varnin, I. G. Teremetskaja, V. A. Laptev, Ju. A. Palyanov, W. R. Fahrner, B. Burchard, in S. Saito, N. Fujimory, O. Fukunaga, M. Kamo, K. Kobashi, M. Yoshikawa (eds), *Advances in New Diamond Science and Technology*, MYU, Tokyo, 1994, 263-266.
399. W. R. Fahrner, B. Burchard, A. M. Zaitsev, A. A. Melnikov, A, V. Denisenko, V. S. Varichenko, in S. Saito, N. Fujimory, O. Fukunaga, M. Kamo, K. Kobashi, M. Yoshikawa (eds), *Advances in New Diamond Science and Technology*, MYU, Tokyo, 1994, 737-740
400. A. A. Melnikov, A. M. Zaitsev, V. S. Varichenko, H. Kanda, W. R. Fahrner, P-i-p and p-i-n electronic structures on synthetic diamond (1995), to be published.
401. K. Harano, N. Ota, N. Fujimori, in S. Saito, N. Fujimori, O. Fukunaga, M. Kamo, K. Kobashi, M. Yoshikawa (eds), *Advances in New Diamond Science and Technology*, MYU, Tokyo, 1994, 497-500.
402. T. Ando, T. Aizawa, M. Kamo, Y. Sato, T. Anzai, H. Yamomoto, A. Wada, K. Domen, C. Hirose, in S. Saito, N. Fujimori, O. Fukunaga, M. Kamo, K. Kobashi, M. Yoshikawa (eds), *Advances in New Diamond Science and Technology*, MYU, Tokyo, 1994, 461-464.
403. T. Aizawa, T. Ando, M. Kamo, Y. Sato, *Phys. Rev. B*, **48** (1993) 18348.

404. J. F. Prins, in S. Saito, N. Fujimori, O. Fukunaga, M. Kamo, K. Kobashi, M. Yoshikawa (eds), *Advances in New Diamond Science and Technology*, MYU, Tokyo, 1994, 443-448.
405. I. Sildos, A. Osvet, in S. Saito, N. Fujimori, O. Fukunaga, M. Kamo, K. Kobashi, M. Yoshikawa (eds), *Advances in New Diamond Science and Technology*, MYU, Tokyo, 1994, 395-398.
406. Y. K. Vohra, T. S. McCauley, S. S. Vagarali, in S. Saito, N. Fujimori, O. Fukunaga, M. Kamo, K. Kobashi, M. Yoshikawa (eds), *Advances in New Diamond Science and Technology*, MYU, Tokyo, 1994, 391-394.
407. T. S. McCauley, Y. K. Vohra, in S. Saito, N. Fujimori, O. Fukunaga, M. Kamo, K. Kobashi, M. Yoshikawa (eds), *Advances in New Diamond Science and Technology*, MYU, Tokyo, 1994, 371-374.
408. T. S. McCauley, Y. K. Vohra, *Phys. Rev. B*, **49** (1994) 5046-5049.
409. J. A. Freitas, P. B. Klein, A. T. Collins, in S. Saito, N. Fujimori, O. Fukunaga, M. Kamo, K. Kobashi, M. Yoshikawa (eds), *Advances in New Diamond Science and Technology*, MYU, Tokyo, 1994, 321-326.
410. H. Kawarada, T. Tsutsumi, Jun-ichi- Imamura, in S. Saito, N. Fujimori, O. Fukunaga, M. Kamo, K. Kobashi, M. Yoshikawa (eds), *Advances in New Diamond Science and Technology*, MYU, Tokyo, 1994, 301-306.
411. T. Tsutsumi, J. Imamura, M. Suzuki, S. Takeuchi, M. Murakawa, H. Kawarada, in S. Saito, N. Fujimori, O. Fukunaga, M. Kamo, K. Kobashi, M. Yoshikawa (eds), *Advances in New Diamond Science and Technology*, MYU, Tokyo, 1994, 363-366.
412. C. M. Welbourn, in S. Saito, N. Fujimori, O. Fukunaga, M. Kamo, K. Kobashi, M. Yoshikawa (eds), *Advances in New Diamond Science and Technology*, MYU, Tokyo, 1994, 327-332.

413. S. C. Lawson, H. Kanda, H. Kiyota, T. Tsutsumi, H. Kawarada, in S. Saito, N. Fujimori, O. Fukunaga, M. Kamo, K. Kobashi, M. Yoshikawa (eds), *Advances in New Diamond Science and Technology*, MYU, Tokyo, 1994, 315-320.
414. R. C. Burns, V. Cvetkovic, C. N. Dodge, D. J. F. Evans, M. T. Rooney, P. M. Spear, C. M. Welbourn, *J. of Crystal Growth*, **104** (1990) 257.
415. A. A. Gorokhovsky, A. V. Turukhin, R. R. Alfano, W. Phillips, *Appl. Phys. Letters*, **66** (1995) 43-45.
416. D. Buhaenko, C. Beer, P. Ellis, L. Walker, B. Stoner, G. Tessmer, R. Fauber, in S. Saito, N. Fujimori, O. Fukunaga, M. Kamo, K. Kobashi, M. Yoshikawa (eds), *Advances in New Diamond Science and Technology*, MYU, Tokyo, 1994, 311-314.
417. H. Kawarada, Y. Yokota, Y. Mori, K. Nishimura, A. Hiraki, *J. Appl. Phys.* **67** (1990) 983.
418. Y. Mori, M. Deguchi, N. Eimori, J. S. Ma, K. Nishimura, M. Kitabatake, T. Ito, T. Hirao, A. Hiraki, *Jap. J. Appl. Phys.*, **31** (1992) L1191-L1194.
419. H. Kawarada, K. Nishimura, T. Ito, J.-I. Suzuki, K.-S. Mar, Y. Yokota, A. Hiraki, *Jap. J. Appl. Phys.*, **27** (1988) L683-L686.
420. G. S. Woods, A. R. Lang, *J. Crystal Growth*, **28** (1975) 215.
421. R. J. Graham, T. D. Moustakas, M. M. Disko, *J. Appl. Phys.*, **69** (1991) 3212-3218.
422. G. Davies, *Proc. Roy. Soc. London A*, *336* (1974) 507.
423. O. Reinkober, *Ann. Phys.*, **34** (1911) 342.
424. G. Davies, *J. Phys. C*, **5** (1972) 2534.

425. R. G. Farrer, L. A. Vermeulen, *J. Phys. C*, **5** (1972) 2762.
426. R. J. Elliott, I. G. Matthews, and E. W. J. Mitchell, *Phil. Mag.*, **3** (1958) 360.
427. D. S. Nedzvetskii, V. A. Gaisin, *Sov. Phys. Solid State,* **14** (1973) 2535-2538.
428. D. S. Nedzvetskii, V. A. Gaisin, *Sov. Phys. Solid State,* **16** (1973) 1600-1601.
429. P. L. Hanley, I. Kiflawi, A. R. Lang, *Phil. Trans. R. Soc. A,* **284** (1977) 329-368.
430. J. Walker, *Rad. Eff. in Semicond.*, Inst. Phys. Conf. Ser. No. **31**, Bristol, 1976, 510-511.
431. J. Walker, *J. Phys. C*, **10** (1977) 3031-3037.
432. Y. Yokota, H. Kawarada, A. Hiraki, in J. T. Glass, R. Messier, N. Fujimori, *Diamond, SiC and Related Wide Gap Semiconductors*, Proc. Mater. Res. Soc., **162** (1990) 231-236.
433. Y. M. LeGrice, E. C. Buehler, R. J. Nemanich, J. T. Glass, K. Kobashi, F. Jansen, M. A. Machonkin, C. C. Tsai, in J. T. Glass, R. Messier, N. Fujimori, *Diamond, SiC and Related Wide Gap Semiconductors*, Proc. Mater. Res. Soc., **162** (1990) 267-272.
434. G. Herzberg, *Molecular Spectra and Molecular Structure II. Infrared and Raman Spectra of Polyatomic Molecules*, D. van Nostrand Co. Inc., Princeton NJ, (1962).
435. L. Santos, E. Pereira, in J. T. Glass, R. Messier, N. Fujimori, *Diamond, SiC and Related Wide Gap Semiconductors*, Proc. Mater. Res. Soc., **162** (1990) 291-294.
436. Yu. A. Klyuev, A. M. Naletov, V. I. Nepsha, in R. Messier, J. T. Glass, J. E. Butler, R. Roy, *New Diamond Science and Technology*, Materials Research Society, Pittsburgh, Pennsylvania, (1991) 149-153.
437. A. R. Lang, A. P. W. Makepeace, M. Moore, W. Wierzchowski, in R. Messier, J. T. Glass, J. E. Butler, R. Roy, *New Diamond Science and Technology*, Materials Research Society, Pittsburgh, Pennsylvania, (1991) 557-560.
438. A. L. Ruoff, Y. K. Vohra, S. Desgreniers, in R. Messier, J. T. Glass, J. E. Butler, R. Roy, *New Diamond Science and Technology*, Materials Research Society, Pittsburgh, Pennsylvania, (1991) 645-657.
439. Y. K. Vohra, C. A. Vanderborgh, S. Desgreniers, A. L. Ruoff, *Phys. Rev. B*, **39** (1989) 5464.
440. S. Desgreniers, Y. K. Vohra, A. L. Ruoff, *Solid State Communic.*, **70** (1989) 705.
441. R. M. Chrenko, R. M. Tuft, H. M. Strong, *Nature*, **270** (1977) 141.
442. A. T. Collins, *J. Phys. C*, **11** (1978) L417.
443. A. T. Collins, in R. Messier, J. T. Glass, J. E. Butler, R. Roy, *New Diamond Science and Technology*, Materials Research Society, Pittsburgh, Pennsylvania, (1991) 659-670.
444. B. P. Allen, T. Evans, *Proc. R. Soc. London A*, **375** (1981) 93.
445. T. Evans, Z. Qi, J. Maguire, *J. Phys. C*, **14** (1981) L379.
446. Yu. A. Klyuev, V. I. Nepsha, A. M. Naletov, *Sov. Phys. Solid State*, **16** (1975)2118.
447. H. Kanda, T. Ohsawa, S. Yamaoka, Paper 3-13 at the 1st Int. Conf. on New Diam. Sci. and Techn., Tokyo, 1988.
448. A. T. Collins, *J. Phys.: Condense. Matter*, **1** (1989)439.
449. E. Fritsch, J. E. Shigley, in R. Messier, J. T. Glass, J. E. Butler, R. Roy, *New Diamond Science and Technology*, Materials Research Society, Pittsburgh, Pennsylvania, (1991) 677-681.

450. S. C. Lawson, A. T. Collins, S. Satoh, H. Kanda, in R. Messier, J. T. Glass, J. E. Butler, R. Roy, *New Diamond Science and Technology*, Materials Research Society, Pittsburgh, Pennsylvania, (1991) 709-714.
451. R. J. Nemanich, L. Bergman, Y. M. LeGrice, R. E. Shroder, in R. Messier, J. T. Glass, J. E. Butler, R. Roy, *New Diamond Science and Technology*, Materials Research Society, Pittsburgh, Pennsylvania, (1991) 741-752.
452. N. Wada, S. A. Solin, *Physica B+C*, **105** (1981) 353.
453. D. S. Knight, L. Pilione, W. B. White, in R. Messier, J. T. Glass, J. E. Butler, R. Roy, *New Diamond Science and Technology*, Materials Research Society, Pittsburgh, Pennsylvania, (1991) 753-758.
454. M. Green, R. C. Hyer, S. C. Sharma, K. K. Mishra, T. D. Black, A. R. Chourasia, D. R. Chopra, in R. Messier, J. T. Glass, J. E. Butler, R. Roy, *New Diamond Science and Technology*, Materials Research Society, Pittsburgh, Pennsylvania, (1991) 759-764.
455. Wei Zhu, Xiao Hong Wang, A. R. Badzian, R. Messier, in R. Messier, J. T. Glass, J. E. Butler, R. Roy, *New Diamond Science and Technology*, Materials Research Society, Pittsburgh, Pennsylvania, (1991) 821-826.
456. J. W. Bray, T. R. Anthony, in R. Messier, J. T. Glass, J. E. Butler, R. Roy, *New Diamond Science and Technology*, Materials Research Society, Pittsburgh, Pennsylvania, (1991) 839-850.
457. W. J. P. van Enckevort, in K. E. Spear and J. P. Dismukes (eds), *Synthetic Diamond: Emerging CVD Science and Technology*, John Wiley & Sons, Inc., (1994) 307-353.
458. J. I. Pankove, C. -H. Qiu, in K. E. Spear and J. P. Dismukes (eds), *Synthetic Diamond: Emerging CVD Science and Technology*, John Wiley & Sons, Inc., (1994) 401-418.
459. M. H. Grimsditch, E. Anastassakis, M. Cardona, *Phys. Rev. B*, **19** (1979) 3240.
460. M. H. Grimsditch, A. K. Ramdas, *Phys. Rev.*, **11** (1975) 3139.
461. J. J. Charette, *Ind. Diam. Review*, **26** No 305 (1966) 144-148.
462. Yu. A. Klyuev, *Diamonds*, USSR, **6** (1971) 9-12 (in Russian).
463. A. T. Collins, S. C. Lawson, G. Davies, H. Kanda, *Phys. Rev. Letters*, **65** (1990) 891-894.
464. S. D. Smith, W. Taylor, *Proc. Phys. Soc. London*, **79** (1962) 1142-1143.
465. G. Davies, R. Stedman, *J. Phys. C*, **20** (1987) 2119-2127.
466. P. A. Crowther, P. J. Dean, W. F. Sherman, *Phys. Rev.*, **154** (1967) 772-785.
467. J. J. Charette, *Physica*, **27** (1961) 1061.
468. A. T. Collins, E. C. Lightowlers, *Phys. Rev.*, **171** (1968) 843-855.
469. A. T. Collins, P. J. Dean, E. C. Lightowlers, W. F. Sherman, *Phys. Rev. A*, **140** (1965) 1272-1274.
470. E. V. Sobolev, Lenskaja, in Yu. L. Orlov, *The Mineralogy of the Diamond*, Wiley, New York, (1973) 47-58.
471. X. X. Bi, P. C. Eklund, J. G. Zhang, A. M. Rao, T. A. Perry, C. P. Beetz, *J. Mater. Res.*, **5** (1990) 811-817.
472. H. Herchen, M. A. Cappelli, M. Landstrass, M. A. Plano, D. Moyer, *Thin Solid Films*, **212** (1992) 206-215.

473. C. Wild, N. Heres, J. Wagner, P. Koidl, T. Anthony, in Proc. Optical and structural characterization of CVD diamond, 1st Int. Symp. on Diamond and Diamond-Like Films, The Electrochem. Soc. Pennington, NJ, (1989).
474. R. Phillips, J. Wei, Y. Tzeng, *Thin Solid Films*, **212** (1992) 30-34.
475. A. M. Zaitsev, V. S. Varichenko, Coloration of diamonds irradiated with high energy ions, unpublished.
476. A. K. Ramdas, Vibrational band structure of diamond, in G. Davies (ed), *Properties and Growth of Diamond*, INSPEC, London, UK, (1994) 13-22.
477. M. H. Grimsditch, E. Anastassakis, M. Cardona, *Phys. Rev.*, **18** (1978) 901.
478. E. Anastassakis, A. Cantarero, M. Cardona, *Phys. Rev.*, **41** (1990) 7529.
479. S. S. Mitra, O. Brafman, W. B. Daniels, R. K. Crawford, *Phys. Rev.*, **186** (1969) 942.
480. B. J. Parsons, *Proc. R. Soc. London A*, **352** (1976) 397.
481. M. Hanfland, K. Syassen, S. Fahy, S. G. Louie, M. L. Cohen, *Phys. Rev. B*, **31** (1985) 6896.
482. A. F. Goncharov, I. N. Makarenko, S. M. Stishov, *JETP Letters*, **41** (1985) 184 (in Russian).
483. V. S. Vavilov, A. A. Gippius, E. A. Konorova, (1985). *Electronic and Optical Processes in Diamond*, Moscow, NAUKA, (in Russian).
484. T. A. Karatygina, E. A. Konorova, *Phys. and Techn. of Semicond.*, **10** (1976) 80 (in Russian).
485. A. T. Collins, G. Davies, G. S. Woods, *J.Phys. C: Solid State Physics*, **19** (1986) 3933-3944.
486. A. M. Zaitsev, A. R. Filipp, Nickel containing centers in diamond (1992), unpublished.
487. P. F. Lai, S. Prawer, A. E. C. Spargo, L. A. Bursill, Studies of defects on ion irradiated diamond, presented at the Int. Conf. IBMM 95, Canberra, Australia, 1995, No 07.046.
488. R. W. Godby, M. Schulter, L. J. Sham, *Phys. Rev. Letters*, **56** (1986) 2415.
489. R. W. Godby, M. Schulter, L. J. Sham, *Phys. Rev. B*, **36** (1987) 6497.
490. R. Orlando, R. Dovesi, C. Roetti, V. R. Saunders, *J. Phys. C: Condens. Matter*, **2** (1990) 7769.
491. H. K. Mao, R. J. Hemley, *Nature*, **351** (1991) 721.
492. R. A. Roberts, W. C. Walker, *Phys. Rev.*, **161** (1967) 730-735.
493. G. S. Painter, D. E. Ellis, A. R. Lubinsky, *Phys. Rev.*, **4** (1971) 3610-3622.
494. W. C. Walker, J. Osantowski, *Phys. Rev. A*, **134** (1964) 153-157.
495. F. J. Himpsel, J. F. van der Veen, D. E. Eastman, *Phys. Rev. B*, **22** (1980) 1967-1971.
496. T. Gora, R. Staley, J. D. Rimstidt, J. Sharma, *Phys. Rev. B*, **5** (1972) 2309-2314.
497. F. R. McFeely, S. P. Kowalczyk, L. Ley, R. G. Cavell, R. A. Pollak, D. A. Shirley, *Phys. Rev. B*, **9** (1974) 5268-5278.
498. Wei Zhu, Hua-Shuang Kong, J. Glass, Characterization of Diamond Films, in R. F. Davis (ed), *Diamond Films and Coatings*, Noyes Publications, Park Ridge, New Jersey, USA, (1993) 244-338.
499. Xinxing Yang, A. V. Barnes, M. M. Albert, R. G. Albridge, J. T. McKinley, N. H. Tolk, J. L. Davidson, *J. Appl. Phys.*, **77** (1995) 1758-1761.

500. S. C. Lawson, H. Kanda, H. Kiyota, T. Tsutsumi, H. Kawarada, *J. Appl. Phys.*, **77** (1995) 1729-1734.
501. J. P. F. Sellshop, H. J. Annegarn, R. J. Keddy, C. C. P. Madib, M. J. Renan, *Nucl. Instr. Meth.*, **149** (1978) 321-328.
502. N. B. Manson, J. Porsch, W. A. Runciman, *J. Phys. C*, **13** (1980) L1005.
503. E. van Oort, N. B. Manson, M. Glasbeek, *J. Phys. C*, **21** (1988) 4385.
504. P. E. Klingsporn, M. D. Bell, W. J. Leivo, *J. Appl. Phys.*, **41** (1970) 2977-2980.
505. E. V. Sobolev, Yu. A. Litvin, N. D. Samsonenko, V. E. Illin, S. V. Lenskaya, V. P. Butuzov, *Sov. Phys. Solid State*, **10** (1969) 1789.
506. A. M. Zaitsev, B. Burchard, A. V. Denisenko, W. R. Fahrner, A. A. Melnikov, V. S. Varichenko, Luminescence of Diamond and Diamond Based Light Emitting Devices, (1994), unpublished paper, Belarussian State University, Belarus, and University of Hagen, Germany.

Chapter 8
ELECTRICAL AND ELECTRONIC PROPERTIES

Alexander G. Gontar
Institute for Superhard Materials of the National Academy of Sciences of Ukraine
Kiev 254074, Ukraine

Contents

1. Mobility

The first data on carrier mobilities were published in 1924 [Lenz 1924] followed by additional work, quite a bit later in 1951 [Klick and Maurer 1951]. Up to the early seventies only a few papers had been published on carrier transport properties. Since natural semiconducting diamonds do not show electron conductivity, electron mobility was measured on crystal insulators, the carriers being excited by light [Redfield 1954] and [Konorova and Shevchenko 1967], α-particles [Pearlstein and Sutton 1950] or electrons [Konorova et al. 1966]. Depending on the sample, the mobility values at T=300 K ranged from 720 to 2000 $cm^2V^{-1}s^{-1}$. In all cases, at near-room and lower temperatures, mobility follows the $\mu \sim T^{-1.5}$ law, which is indicative of the predominance of phonon scattering [Redfield 1954], [Austin and Wolfe 1956], [Jonson 1964], and [Konorova and Shevchenko 1967]. Several investigators [Konorova and Shevchenko 1967] have established that the spread in mobility values is caused by the nonuniform distribution of excited carrier concentration, which in turn depends on both the (nonuniform) distribution of impurities and the carrier life time variations [Konorova and Kozlov 1966].

Hall mobility of holes was determined on rare semiconducting natural diamonds with a hole conductivity (type IIb diamond) [Austin and Wolfe 1956], [Wedepohl 1957], [Bate and Willardson 1959], [Jonson et al. 1964], [Dean et al.1965], and [Williams 1970]. Measurements on carefully selected uniform specimen showed values at 290 K of 1550 ± 150 $cm^2V^{-1}s^{-1}$ [Dean et al.1965] and 1600±400 $cm^2V^{-1}s^{-1}$ [Williams 1970]. The results of these precision measurements were discussed in [Williams 1970], [Collins and

Lightowlers 1979], and [Collins 1989] and are given in Table 1 and Fig.1. For temperatures up to about 350K, the temperature dependence of hole conductivity obeys the $\mu \sim T^{-1.5}$ law. Above 400K the temperature dependence is sharper and follows a $\mu \sim T^{-2.8}$ law, which is assumed to be governed by the optical mode carrier scattering [Wedepohl 1957] and [Dean et al.1965]. As can be seen from Fig. 1, the mobility has a tendency to be temperature independent below 200 K. This is attributed to the increased contribution of the impurity scattering and the measured mobility described by the expression $1/\mu = T^S/\mu_0 + 1/\mu_I$, where μ_I is a temperature independent mobility [Collins 1989]. Standard scattering theories, however, suggest that at low temperatures, μ_I should vary as $T^{+1.5}$ [Smith 1961].

Based on Hall measurements, the effective mass of holes was found to be $m_h/m_o = 1.1 \pm 0.2$ [Dean et al.1965] and $m_h/m_o = 0.75 \pm 0.05$ [Williams 1970]. These values do not agree with the results from effective mass calculations on cyclotron resonance data, in which the contributions from all three zones were allowed for: heavy hole band m_h/m_o=2.1, light hole band m_h/m_o= 0.7 and split-of band m_h/m_o=1.06 [Rauch 1961] and [Rauch 1963]. The disagreement between cyclotron resonance data and the electrical measurements in assessing carrier effective masses is not understood up to now [Vavilov et al. 1985].

Table 1. Hall measurement data for natural semiconducting diamond.

Sample	Impurity concentration, (cm^{-3})		Hole mobility at 290K, (cm^2/V·s)	Mobility index, *s*, in $\mu \sim T^S$		Effective mass, m_h/m_o
	$N_A \times 10^{-15}$	$N_D \times 10^{-15}$		250K	650K	
A100[1]	32 ±3.0	5.6 ±1.2	1550 ± .150	-1.5	-3.0	1.1 ± 0.2
E3[1]	71 ± 7.0	3.0 ± 0.6	1550 ± 150	-1.5	-3.0	1.1 ± 0.2
E4[1]	33 ± 3.0	1.1 ± 0.2	800 ± 80	-1.5	-3.0	0.4 ± 0.1
D114[2]	55 ± 2.5	1.8 ± 0.1	1870 ± 60	-1.3	-2.9	0.75 ± 0.5
D119[2]	83 ± 3.0	4.2 ± 0.2	1200 ± 40	-1.5	-2.9	0.75 ± 0.5
SA65C[2]	50 ± 2.5	0.9 ± 60	2010 ± 60	-1.95	-3.1	0.65 ± 0.5
SA65E[2]	78 ± 3.0	2.8 ± 0.2	1230 ± 40	-1.4	-2.8	0.75 ±.0.5
mean values[3]			1600 ± 400			0.75 ± 0.5

[1] [Dean et al.1965]
[2] [Collins and Lightowlers 1979]
[3] [Williams 1970]

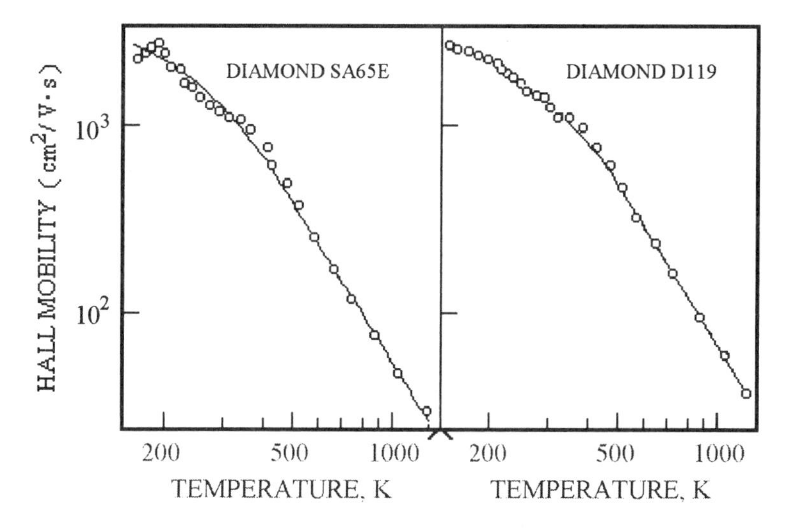

Figure 1. Hall mobility of diamond as a function of temperature [Williams 1970]. The continuous curve refers to the expression $1/\mu = T^s/\mu_0 + 1/\mu_I$, s=3,0 and μ=3000 $cm^2V^{-1}s^{-1}$ [Collins 1989].

Values for charge carrier mobility in strong electric fields were first obtained in 1966-1967 [Konorova et al. 1966] and [Konorova and Shevchenko 1967]. At 190 K the mobility of electrons and holes in fields above 500 V cm^{-1} decreased with an increase in electrical field according to $E^{-1.5}$. In fields above 6×10^3 V cm^{-1}, the electron drift velocity tends to saturate as earlier attributed to interaction with optical phonons. The experimentally defined saturated carrier drift velocity is reported as 10^7 cm s^{-1} [Konorova et al. 1966]. This result was later supported by measurements of drift velocities of holes and electrons in diamond irradiated with α -particles with energy of 5.4 MeV [Kozlov et al. 1974a] and [Kozlov et al. 1974b].

During 1979-1983 Italian researchers published results on drift mobility of charge carriers in diamond [Reggiani et al. 1979], [Nava et al. 1979], [Nava et al. 1980], [Reggiani et al. 1981], and [Reggiani et al. 1983]. The drift mobility was obtained from the ratio drift velocity to electric field, υ_d/E, at the lowest field strength. Drift velocity measurements have also been performed with the time-of-flight technique [Martini et al. 1972] and [Canali et al. 1975] in the temperature range between 85 and 700 K. The total experimental error of υ_d was within ± 5 % [Canali et al. 1975]. The samples were prepared from high purity natural type IIa diamond crystals from the Yakutia (Russia) deposits and carefully selected by photoconductivity and absorption coefficient measurements, to ensure that the concentration of nitrogen, acting as a deep donor, does not exceed 10^{19} cm^{-3}. The crystals were cut along <100> ± 1° and <110> ± 15°, into plates between 0.06-0.4 mm thick. The plates were annealed at 1000-1300 °C to remove stress.

The experimental results show that at T < 400 K the ohmic drift mobility of holes depends on temperature according to $\mu \sim T^{-1.5}$ (Fig. 2). Above 400 K, the temperature

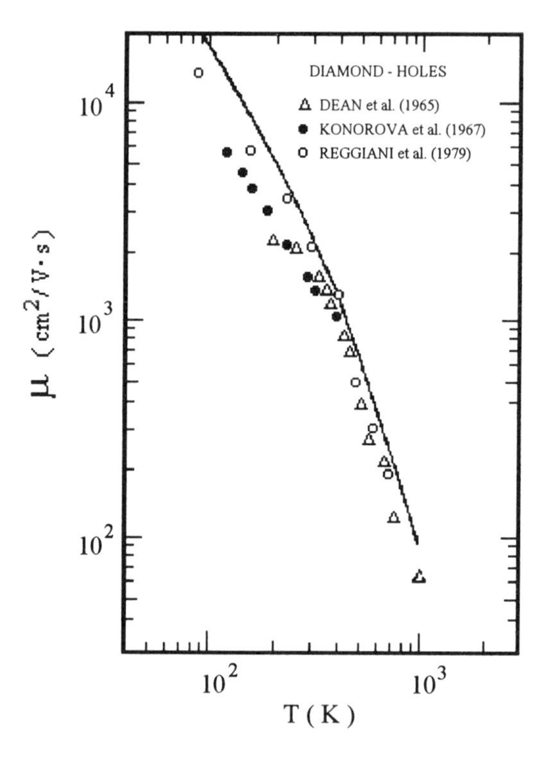

Figure 2. Hole mobility as a function of temperature in natural diamond. The continuous curve refers to theoretical calculations [Reggiani et al.1979].

dependence becomes more critical and follows a $T^{-2.8}$ law. At room temperature, the hole mobility becomes 2100 $cm^2V^{-1}s^{-1}$. Over the entire temperature range under study, the drift mobility value is higher than the Hall mobility. At T<400 K, the difference is about 40 %, i.e. the Hall factor value in diamond is less than unity. By analogy with silicon [Nakagawa and Zukotynski 1978] and [Lin et al. 1981], such a difference can be observed in high purity specimen having negligible impurity scattering. Analysis of the Hall factor for holes in diamond has shown, that in the 100-1000 K temperature range, its value is lower than unity, resulting from the warped and nonparabolic features of the valence band [Reggiani et al. 1983].

The electron mobility at room temperature is 2400 $cm^2V^{-1}s^{-1}$. Below 400K the temperature dependence follows $\mu \sim T^{-1.5}$ and above 400 K the $T^{-2.8}$ law, controlled mainly by optical phonon scattering (Fig. 3). The drift mobility for electrons as well as for holes, is higher than the Hall mobility.

At 300 K, the drift velocity of electrons and holes in fields exceeding 10^3 Vcm^{-1} becomes dependent on the field [Konorova et al. 1966], see figures 4 and 5. Below 300 K, the drift velocity of holes and electrons becomes saturated in fields higher than 30 kV cm^{-1}. At 300 K, the saturated drift velocities of electrons and holes are $1.5 \pm 0.1 \times 10^7$ cm s^{-1}

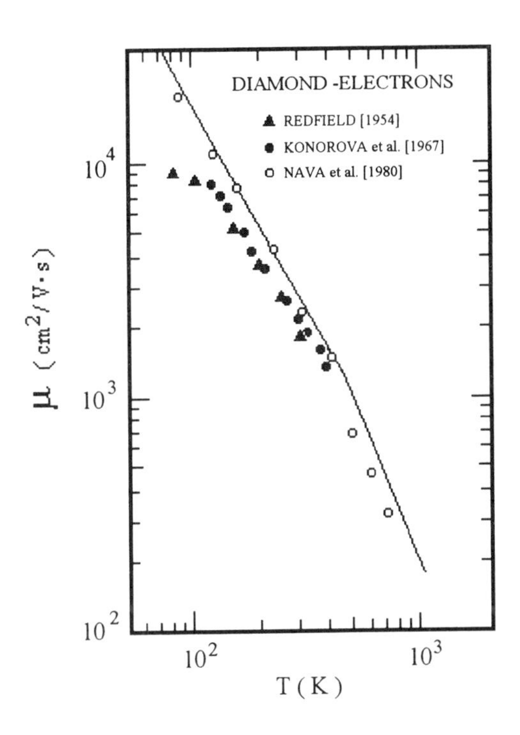

Figure 3. Electron mobility as a function of temperature in natural diamond. The continuous curve refers to theoretical calculations [Nava et al. 1980].

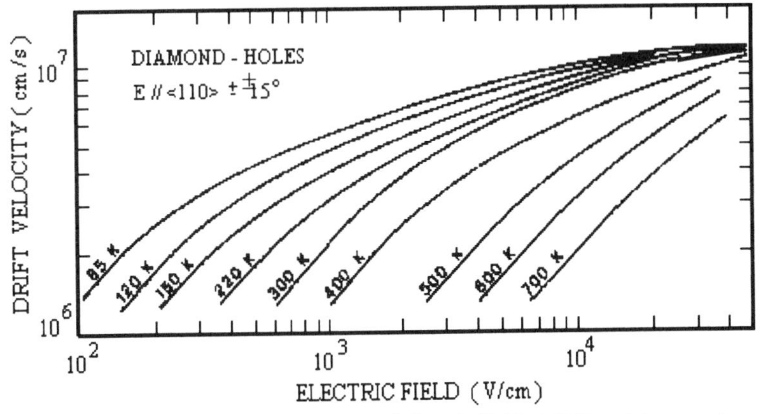

Figure 4. Hole drift velocity as a function of electric field at different temperatures [Nava et al. 1979].

[Rauch 1963] and $1.1 \pm 0.1\times10^7$ cm s^{-1} [Reggiani et al. 1979] and [Reggiani et al. 1981], respectively. At temperatures higher than 300 K, there is no saturation of carrier drift velocities, even at the highest field. The temperature independence of the saturation value of the drift velocity at low temperatures (Figs. 4 and 5) points to the predominant role of carrier-optical phonon interactions [Nava et al. 1979].

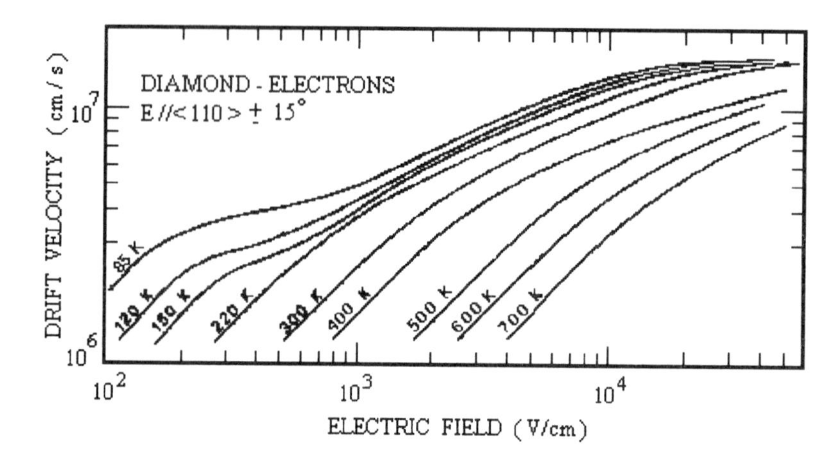

Figure 5. Electron drift velocity as a function of electric field at different temperatures [Nava et al.1979].

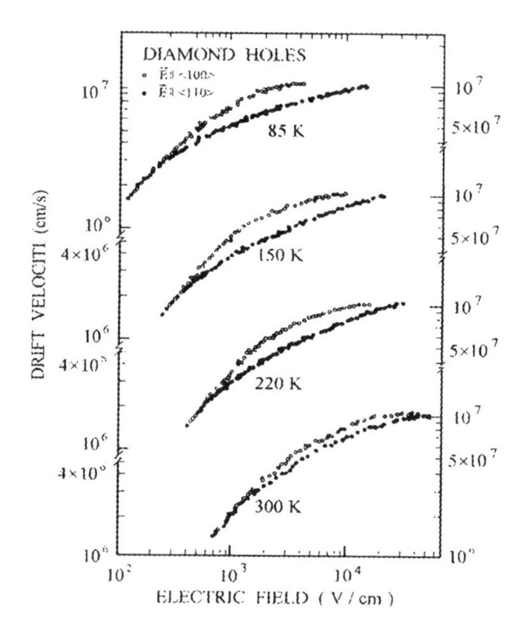

Figure 6. Hole-drift velocity as a function of electric field at different temperatures [Reggiani 1981].

Measurements in strong fields have revealed anisotropy of the carrier drift velocity (Figs. 6,7), along the <100> and <110>, with $\upsilon_{d110} > \upsilon_{d100}$ for electrons and $\upsilon_{d100} > \upsilon_{d110}$ for holes. With a temperature decrease, anisotropy increases and attains the highest values $[(\upsilon_{d100} / \upsilon_{d110}) - 1]$ of ~ 40 % at 85K with an electric field, E=2.5 kV cm^{-1} [Nava et al. 1980] for holes and $[(\upsilon_{d_{110}}/\upsilon_{d_{100}})^{-1}]$ of~50% at 85 K with E=1.8 kV cm^{-1} [Reggiani et al. 1981] for electrons. The drift velocity limiting values depend only slightly on field direction about the crystallographic axes at temperatures of 300 K and below.

With reference to the theoretical interpretation of the results, Boltzmann's equation was solved by the Monte-Carlo technique [Reggiani et al. 1977]. Due to the high degree of specimen purity and the temperature at which they were studied, only lattice scattering mechanisms such as acoustical and nonpolar optical phonon scattering, were considered. The microscopic model for holes makes allowance only for two zones of heavy, m_h, and light, m_l, hole effective masses. The zones were thought to be parabolic and of spherical symmetry. This does not appear to contradict the results of cyclotron resonance studies [Rauch 1961]. A zone, split off due to spin orbital interaction was disregarded, because of very small effective hole mass and their low concentration in the zone, under the experimental conditions. Transport theory for electrons allowed calculation of the diamond zone structure [Herman 1952] and [Leite et al. 1975] with anisotropy similar to that of silicon, with a minimum of the conduction-bands at (0.76; 0; 0) k_{max}. Electron scattering has been studied for acoustic and three inter-valleys phonons: longitudinal optical (LO)-*g*-scattering, longitudinal acoustic (LA) and transverse optical (TO) phonons - *f*-scattering .

Calculated and experimental data for three types of dependencies for holes [Reggiani et al. 1979], [Reggiani et al. 1981] and [Reggiani et al. 1983] and electrons [Nava et al. 1980] and [Nava et al. 1979] separately, have been compared: the temperature dependence of the ohmic drift mobility; the anisotropy of the drift velocity obtained with the electric field applied along <110> and <100> and the temperature dependence of the drift velocity at high electric fields.

The best agreement between theoretical and experimental results for hole transport properties were obtained for the effective masses of heavy and light holes, the deformation potential in acoustic phonon scattering E_1 and the deformation potential in optical phonon scattering $(D_tK)_{opt}$. In the analysis of electron transport properties, effective mass and deformation potential values were determined from the temperature dependence of the ohmic mobility. Deformation potentials for all three types of intervalley scattering were assumed to be equal. From anisotropy of the drift velocity, a ratio of longitudinal and traverse electron effective masses was defined. The electron effective masses themselves were assessed independently from the drift velocity values at the highest fields [Nava et al. 1980]. The numerical results of electron and hole effective mass calculations as well as band structure values and other diamond properties used in the calculations, are listed in Table 2.

Measurements of transport properties of natural diamond using the time-of-flight technique have shown that in the ohmic region, the temperature dependence of the product $\mu\tau$ for holes and electrons reflects the thermal behaviour of the carrier mobility [Nava et al. 1979]. At the same time, at field strength between 10 and 30 kVcm^{-1} and in the 85-700 K temperature range (Fig.8 and Fig.9), $\mu\tau$ is practically constant. Carrier

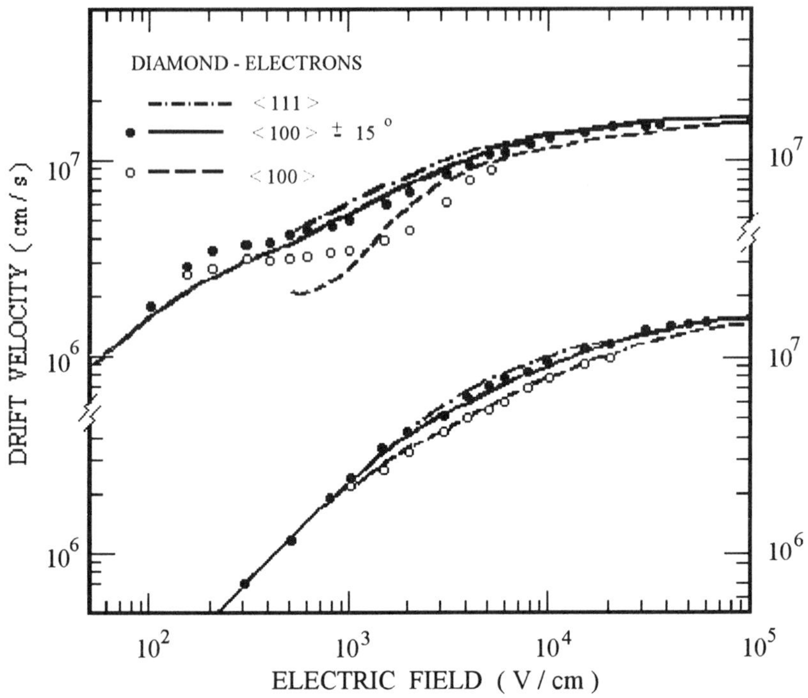

Figure 7. Electron drift velocity as a function of electric field in natural diamond at 85 and 300K. The dot - dash curve refers to theoretical results along <111> [Reggiani 1981].

energy increases with the field intensity. Figure 10 shows the results of calculations made for holes [Reggiani et al. 1981] using the data of Table 2. At first the hole mean energy grows slowly due to interaction with optical phonons. In fields above 2×10^5 Vcm^{-1}, it increases according to $E\sim E^{3/2}$. At $E=2\times10^6$ Vcm^{-1}, hole energy becomes equal to that of a forbidden zone, i.e., one can expect shock ionization in such fields. Shock ionization in diamond was observed with pulsed laser radiation ($\lambda=1.06$ μm [Lin et al. 1978]) and in static electric fields with p^+-n-i-n structures, produced by lithium and boron implantation [Konorova et al. 1983]. In the first case, the threshold field value for breakdown was estimated at 2.7×10^7 V cm^{-1}. The calculation results of the highest field of an electrical breakdown in semiconducting diamond structures are given in Table 3.

The fact that the electric field breakdown values obtained in static electric fields [Konorova et al. 1983] are lower by an order of magnitude as compared to laser radiation [Lin et al. 1978] can have two reasons. Firstly, the optical frequency is 10^{14} s^{-1} and according to theory [Das and Ferry 1976], mobility at such frequencies and in strong fields, becomes a complex value and its real part decreases with increase in frequency. Secondly, at a pulse width of 30 ps and owing to a low concentration of free electrons, breakdown progress may well take stronger fields.

Table 2. Physical property values of diamond.

Parameters	Values	Reference
Lattice constant, (cm)	$a_o = 3.57 x 10^{-8}$	[Yarnell et al. 1964]
Crystal density, (g cm^{-3})	$\rho = 3.51$	[Weast 1974]
Static dielectric constant	$C_o = 5.7$	[Dean et al.1965]
Longitudinal sound velocity[a], (cm sec^{-1})	$S_l = 18.21 \times 10^5$	
Transverse sound velocity[a], (cm sec^{-1})	$S_t = 12.30 \times 10^5$	
Optical phonon equivalent temp., K	$\theta_{opt} = 1938$	
Energy gap, (eV)	$E_g = 5.49$	[Clark et al. 1964]
Spin orbit energy, (eV)	$\Delta = 0.006$	[Rauch 1962]
	ELECTRONS	
Intervalley phonon equivalent temp., K	θ_g (LO) =1900	[Dean et al.1965
Intervalley phonon equivalent temp., K	θ_{f_1} (LA) =1560	[Dean et al.1965
Intervalley phonon equivalent temp., K	θ_{f_2} (TO)=1720	[Dean et al.1965
Acoustic deformation potential, (eV)	$E_l = 8.70$	[Nava et al. 1980]
Longitudinal electron effective mass	$m_l = 1.4$	[Nava et al. 1980]
Transverse electron effective mass	$m_t = 0.36$	[Nava et al. 1980]
Intervalley deformation potential, (eV cm^{-1})	$(D_tK)_g = 8 \times 10^8$	[Nava et al. 1980]
Intervalley deformation potential, (eV cm^{-1})	$(D_tK)_{f_1} = 8 \times 10^8$	[Nava et al. 1980]
Intervalley deformation potential, (eV cm^{-1})	$(D_tK)_{f_2} = 8 \times 10^8$	[Nava et al. 1980]
Drift mobility (T=300K), ($cm^2V^{-1}s^{-1}$)	2400	[Nava et al. 1980]
Saturation of the drtift velocity (cm s^{-1}) ($T \leq 300K$, $E \geq 10^4 Vcm^{-1}$)	1.50×10^7	[Nava et al. 1980]
	HOLES	
Heavy-hole effective mass	$m_h = 1.1$	[Reggiani et al. 1979]
Light-hole effective mass	$m_l = 0.3$	[Reggiani et al. 1979]
Acoustic deformation potential, (eV)	$E_l = 5.5$	[Reggiani et al. 1979]
Optical deformation potential, (eV cm^{-1})	$(D_tK)_{opt} = 21 \times 10^8$	[Reggiani et al. 1979]
Drift mobility (T=300K), ($cm^2V^{-1}s^{-1}$)	2100	[Reggiani et al. 1979]
Saturation of the drtift velocity (cm s^{-1}) ($T \leq 300K$, $E \geq 10^4 Vcm^{-1}$)	1.05×10^7	[Reggiani et al. 1979]

[a] Derived from the values of the elastic constant reported by C.Kittel, (1966) *Introduction to Solid St. Phys.* Wiley, New York.

The above data on mobility apply to the purest and structurally most perfect natural insulators [Reggiani et al. 1979] and [Nava et al. 1980] and lightly-doped semiconducting type IIb crystals [Dean et al.1965], [Williams 1970], and [Collins and Lightowlers 1979]. These values are the highest for diamond. Good agreement between experimental data and the results from calculations, which allowed for trace impurities could be obtained [Reggiani et al. 1983]. Data on the change in free carrier mobility with increase in concentration were obtained by increasing the dopant level in synthetic diamonds see, e.g. [Williams et al.1970] and [Grot et al. 1991] or from carrier density in type IIa single crystals by pulse laser excitation (6.11 eV photons, 5 ps in length,

Table 3. Maximum values of the electrical breakdown strength in diamond [Konorova et al. 1983].

Sample	Calculated donors concentration N_o (cm^{-3})	Thickness of the layer with N_o concentration (μ m)	Electric field E_{max}, ($V\ cm^{-1}$)
79-93	1×10^{17}	0.25	1.4×10^{6}
78-61	8×10^{16}	0.3	1.1×10^{6}
78-12	2×10^{16}	0.6	1.3×10^{6}

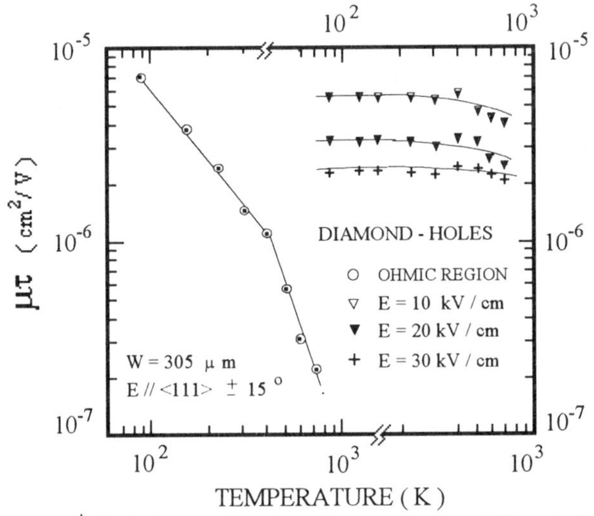

Figure 8. The product $\mu\tau^{+}$ for holes as a function of temperature at different electric fields [Nava et al. 1979].

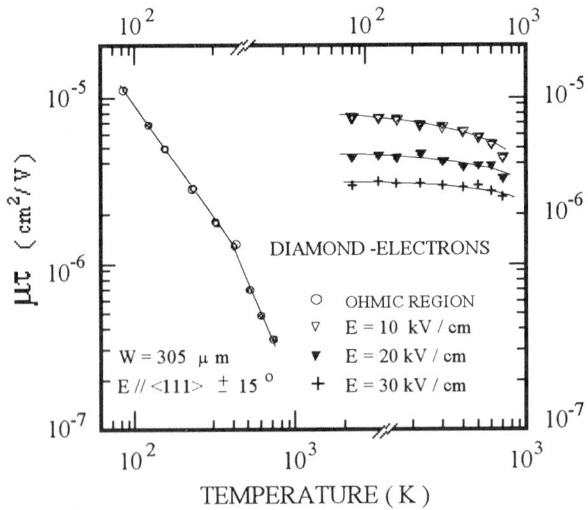

Figure 9. The product $\mu\tau^{+}$ for electrons as a function of temperature at different electric fields [Nava et al. 1979].

maximum energy of 50 μJ) [Kania et al. 1990], [Pan et al. 1990], and [Pan et al.1993c]. Figures 11 and 12 present mobility as a function of carrier concentration for different types of diamond.

In the case of photoexcited carriers and under conditions of low free carrier densities (less than 10^{15} cm^{-3}), an applied electric field of 200 Vcm^{-1} and the 120-410K temperature range, the combined electron and hole mobilities (μ_{eh}) are limited by acoustic phonon scattering, $\mu \sim T^{-1.5}$. At high carrier densities, mobility decreases because of increased electron-hole scattering and mobility is determined by the Mathiesson rule [Pan et al. 1993c] as follows:

$$\langle \mu \rangle^{-1} = \mu_{eh}^{-1} + \mu_{o}^{-1} \quad (1)$$

$$\mu_{eh}^{-1} = 3.11 \times 10^{16} \times T_c^{-1.5} (np)^{-0.5} \left(\ln\{1 + 1.77 \times 10^{8} [T_c^{2} (np)^{-1/3}]\} \right)^{-1}$$

in which: μ_o = 3200 cm $cm^2V^{-1}s^{-1}$, as extracted from the low density regime,
T_c is the carrier temperature (K), and
n and p are the electron and hole densities (cm^{-3}).

A fixed electron to hole mobility ratio (μ_n=800 $cm^2V^{-1}s^{-1}$ and μ_p=2400 $cm^2V^{-1}s^{-1}$) has been reported in [Pan et al.1993c]. For high carrier densities (>$10^{16}cm^{-3}$), the electron-hole scattering results in a mobility temperature dependence according to $\mu \sim T^{1.5}$.

There is little data on hole mobility in synthetic semiconducting diamond single crystals. Synthetic diamonds are characterized by a distinct sectorial-zone structure [Woods and Lang 1975] and [Vishnevsky, 1975], uneven impurity distribution over the volume [Vishnevsky et al. 1974], [Burns 1990], [Rooney 1992], and [Vishnevsky et al. 1992] and a considerable distortion of the crystalline lattice [Tkach 1993]. At T = 300K, carrier mobility was found to be lower than in natural crystals with values ranging from 88 up to1300 $cm^2V^{-1}s^{-1}$ [Collins 1989], [Revin and Slesarev 1982], [Baranskii et al. 1987], and [Smirnova and Gontar 1993]. Data on hole mobility in boron-doped synthetic diamond are given in Table 4.

Undoped diamond films have characteristics similar to those of semi-insulating materials with a high concentration of electrical defects. Their transport properties are characterized by space-charge limited currents in the presence of traps [Srikanth and Ashok 1988] or electrical conduction, with activation energies ranging from 0.8 to 1.8 eV between 200 and 1200 °C [Spitsyn, 1990], [Vandersande and Zoltan 1991], and [Stoner et al. 1992]. The possibility of obtaining p-type semiconducting diamond films by in-situ boron-doping during growth has been recorded by [Proferl et al. 1973], [Aleksenko et al. 1977], [Spitsyn et al. 1981], [Fujimori et al. 1986], [Sato et al. 1987], [Kobashi et al. 1988], and [Okano et al. 1988]. These results gave rise to the possibility that semiconducting diamond films could be used as semiconducting devices which, in turn, has stimulated researches in this direction [Gildenblat et al. 1988], [Gildenblat et al. 1990b], [Shiomi et al.1989a], and [Shiomi et al. 1989b]. At the same time the

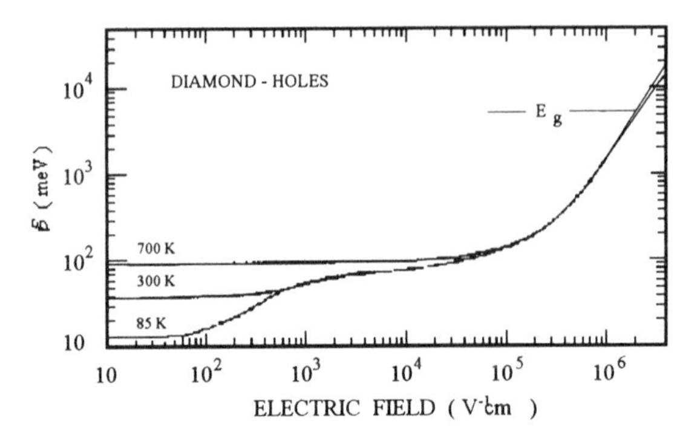

Figure 10. Mean hole energy as a function of electric field at temperatures as indicated. E_g indicates the diamond energy gap [Reggiani, 1981].

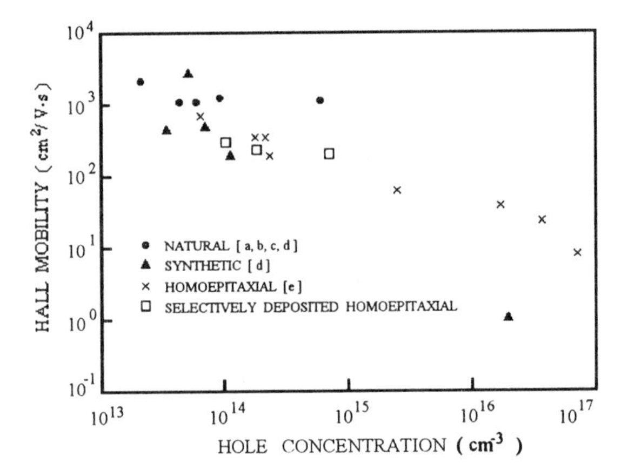

Figure 11. Hall mobility as a function of hole concentration for different types of semiconducting diamond [Grot et al. 1990] : a - [Klick et al. 1951], b - [Austin et al. 1956], c - [Wedepohl 1957] , d - [Williams et al. 1970], e - [Fujimori et al. 1990].

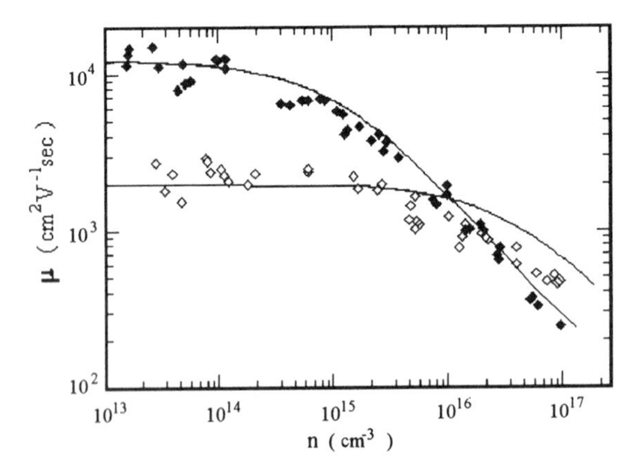

Figure 12. Combined electron and hole mobility as a function of density, at T = 50K (solid points) and T=400K (open points). The curves were calculated using Eq. (1), with corrections for high field effects [Pan et al 1993c].

Table 4. Mobility in synthetic diamond.

Sample	Mobility, $(cm^2V^{-1}s^{-1})$	Impurity concentration, cm^{-3}			Mobility index, s, at 300K
		$N_A \times 10^{-17}$	$N_D \times 10^{-17}$	$(N_A-N_D) \times 10^{-16}$	
A2[1]	800 ± 60	1.26	1.13	1.3	-1.5
B2[1]	700 ± 80	3.14	3.02	1.2	-1.5
C2[1]	600 ± 70	1.57	1.45	1.2	-1.5
ISM-4[2]	88 ± 5	-	-	3.5	-0.7
ISM-5[2]	163 ± 8	-	-	1.5	-1.5
ISM-8[2]	290 ± 14	-	-	55.0	-1.5
6[3]	1300 ± 100	5.10	4.08	10.2	-1.5
8[3]	1300 ± 100	20.0	4.60	154.0	-1.5
11[3]	1200 ± 100	26.8	9.91	168.9	-1.5
2[3]	800 ± 90	2.60	2.03	5.7	-1.5

[1][Revin and Slesarev 1982]
[2][Baranskii et al. 1987]
[3][Smirnova and Gontar 1993]

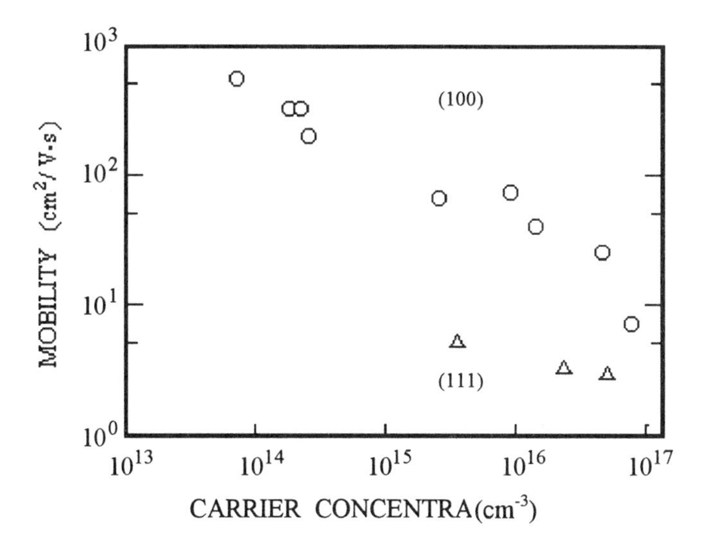

Figure 13. Relationship between carrier concentration and carrier mobility for boron-doped epitaxial films deposited on (100) and (110) silicon substrates [Fujimori et al. 1990].

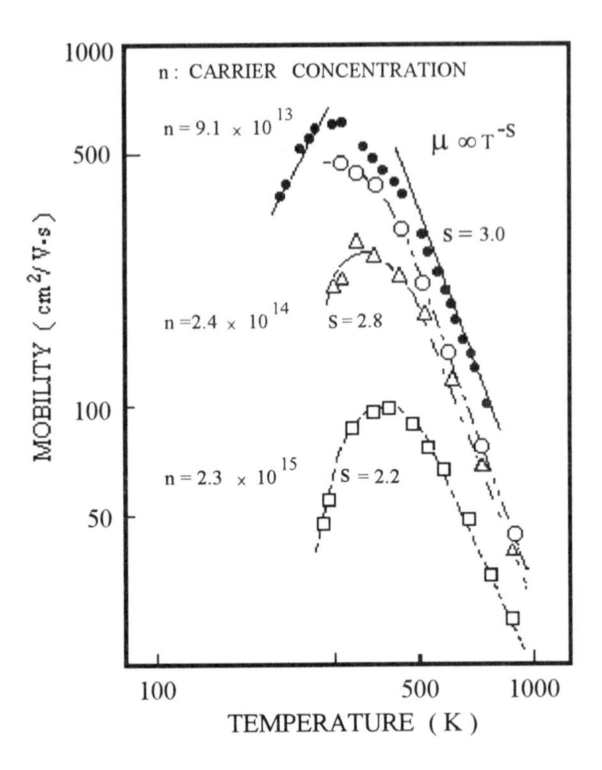

Figure 14. Carrier mobility as a function of temperature for boron-doped epitaxial films. □, Δ, O - data of [Fujimori et al. 1990]; ● - data of [Visser et al. 1992].

technique of selective deposition of polycrystalline [Davidson et al. 1989] and homoepitaxial films is progressing rapidly [Gildenblat et al. 1990a], [Grot et al. 1991], and [Gildenblat et al. 1991].

The first Hall effect measurements on boron-doped epitaxial diamond films indicated hole mobilities of 470 $cm^2V^{-1}s^{-1}$ at room temperature and 70 $cm^2V^{-1}s^{-1}$ at T=500 ^0C [Fujimori et al. 1990]. Further results are shown in Figures 13 and 14 [Fujimori et al. 1990] and [Visser et al. 1992].

The hole mobility in homoepitaxial diamond films is much lower than that in high quality diamond single crystals (Fig.11). The data of Fig.13 and in Table 5 clearly bring out that with an increase in hole concentration from 1.3×10^{14} cm^{-3} to 1.0×10^{17} cm^{-3}, the mobility in all homoepitaxial diamond films deposited under different conditions, decreases from 590 $cm^2V^{-1}s^{-1}$ to 7.8 $cm^2V^{-1}s^{-1}$ [Fujimori et al. 1990], [Visser et al. 1992], and [Fox et al. 1993].

At T > 400 K, the temperature dependence of the carrier mobility follows $\mu \sim T^{\,S}$, in which the mobility index, s, takes on the values -3.0, -2.8, -2.9 and -2.2 for samples with the carrier concentration $9{,}1\times10^{13}$, 1.3×10^{14}, 2.4×10^{14}, 2.3×10^{14} cm^{-3}, respectively [Fujimori et al. 1990] and [Visser et al. 1992]. The mobility maxima lie in the 290-350 K range, see Fig.14. In the low-temperature range, the mobility index for impurity scattering (s=2.2) deviates from that obtained from theory. Other scattering mechanisms are felt to be responsible [Visser et al. 1992].

Homoepitaxial diamond films with a high carrier mobility have been reported by [Landstrass et al.1993]. The combined electron and hole mobility at 300K was 3500 $cm^2V^{-1}s^{-1}$. Based on the earlier measurements of hole and electron mobilities [Reggiani et al.1979] and [Nava et al.1980], one can state the following values, μ_h=1635 $cm^2V^{-1}s^{-1}$ and μ_e=1865 $cm^2V^{-1}s^{-1}$. These undoped films are very pure and have a perfect structure. Raman spectroscopy of these films indicate a FWHM of only 2.6 cm^{-1} as compared to 2.4 cm^{-1} for type IIa natural diamond single crystals. With an increase in photoexcited carrier concentrations above $10^{15}cm^{-3}$, the combined mobility, degraded due to electron-hole scattering as is the case in type IIa single crystals [Kania et al. 1990], [Pan et al 1990], and [Pan et al 1993c]. Attempts at growing such films with boron and phosphorus result in mobility decrease down to 500 and 200 $cm^2V^{-1}s^{-1}$, respectively (see Table 5) [Landstrass et al.1993].

Data on Hall mobility in heavily boron-doped polycrystalline diamond films grown on Al_2O_3 substrates, are given in [Nishimura et al.1991]. At a carrier concentration of 3×10^{18} cm^{-3}, the mobility was 0.6 $cm^2V^{-1}s^{-1}$. Lower concentration boron-doped films exhibited higher mobility values. At 300 K, the mobility ranged from 32 to 2 $cm^2V^{-1}s^{-1}$ with Hall concentration in the 2.3×10^{15} to 1.72×10^{17} cm^{-3} range and boron concentrations between 10^{17} and 10^{21} cm^{-3} [Aslam et al.1992b]. Table 6 summarizes deposition parameters and the electrical properties for polycrystalline diamond films. In polycrystalline as well as homoepitaxial films on single crystals, mobility decreases with an increase in carrier concentration (heavier doping) but is much lower than that in homoepitaxial films. Grain boundaries, carrier trapping centers and intergranular defects are all detrimental to mobility. A comparison of simultaneously grown homoepitaxial and polycrystalline CVD diamond films confirms the lower mobility in polycrystalline films [Fox et al. 1993].

An increase in boron concentration from 10^{16} cm^{-3} to 10^{20} cm^{-3} decreases hole mobility, and high impurity concentrations dictate hopping or impurity band conduction with low mobility [Fox et al. 1993]. It has been suggested that carrier scattering effects on neutral and ionized impurities and heavy effective mass in impurity bands have a decisive impact on transport properties in heavily doped films [Nishimura et al. 1991]. The structure of doped films could be improved by deposition on undoped highly-oriented diamond films with low angle grain boundaries (less than 5°) [Stoner et al. 1993]. In these films the hole mobility at T=300K became 165 $cm^2V^{-1}s^{-1}$ approximating the values obtained for homoepitaxial films grown on single crystal diamond. The data of Tables 5 and 6 clearly show that mobility in homoepitaxial and polycrystalline films is much lower than that in diamond single crystals. In recent years, however, it has become possible to obtain CVD diamond with characteristics close to that of type IIa natural diamond [Pan et al.1992], [Pan et al.1993a], [Pan et al.1993b], and [Plano et al. 1993]. The deposition parameters of such films and their characteristics are given in Table 7.

Table 5. Summary of the deposition parameters and electrical properties of the homoepitaxial diamond films.

Deposition parameters	Sample	Carrier concentration (cm^{-3})	Mobility ($cm^2V^{-1}s^{-1}$)	Mobility index, s, in $\mu \sim T^{-s}$ at T > 400K	Comments	References
1	2	3	4	5	6	7
MPA CVD; MP at 2.45 GHz	-	9.1×10^{13}	470.0	-3.0	Hall measurement;	[Fujimori et al. 1990]
Gases: CH_4, H_2, B_2H_6	7328	2.2×10^{14}	310.0	- 2.8	Growth Rates:	
B/C ratio : 100-500 ppm	-	2.3×10^{15}	50.0	-2.2	0.8 μm / H (100)	
Pressure: 40 Torr, T_S = 950 °C	73207	2.9×10^{15}	4.9	-	0.3 μm / H (110)	
Substrates:diamond (110) and	72904	1.7×10^{16}	3.0	-		
(100) planes	72302	1.0×10^{17}	7.8	-		
	70712	1.1×10^{17}	85.0			
MPA CVD; MP at 2.45 GHz						
Gases: CH_4 in H_2	HED1	2.1×10^{14}	519.0	- 2.43	Hall measurement;	[Fox et al. 1993]
C/H ratio: 75 or 2310 ppm						
Substrates: type-IIa diamond	HED2	6.8×10^{16}	13.9			
(100) planes						
MPA CVD; MP at 2.45 GHz					Hall measurement;	[Grot et al. 1991]
Gases: 1.0% CH_4 in H_2					Films thickness:	
Dopant: boron powder	S205	1.0×10^{14}	290	-	0.15 μm.	
Pressure: 80Torr, T_S = 910 °C	S214	1.8×10^{14}	260	-	0.08 μm.	
Substrates: type-Ia diamond	S212	6.9×10^{14}	210	-	0.16 μm.	
(100) plane & SiO_2 mask						
HFACVD;	HF-3	1.32×10^{14}	590	-2.8	Hall measurement;	[Visser et al. 1992]
Gases: 0.67% CH_4 in H_2				2.2 (T<300K)	Films thickness:	
Dopant: hexagonal BN					1.5μm	
Pressure: 66 torr						
T_s =735…915 °C						
Substrates: Type IIa diamond						
(100) (110) planes						

Table 5. Summary of the deposition parameters and electrical properties of the homoepitaxial diamond films.

Deposition parameters	Sample	Carrier concentration (cm^{-3})	Mobility ($cm^2V^{-1}s^{-1}$)	Mobility index, s, in $\mu \sim T^{-s}$ at T > 400K	Comments	References
MPACVD: MP at 2.45 GHz Gases: 1.0% CH_4 in H_2 Dopant: diborane or phosphine gases Pressure: 60 Torr T_s=800 °C Substrates: type Ia and IIa diamond (100) planes	-	undoped $1x10^{17}$ (boron) $1x10^{17}$ (phosphan)	3500 500 200		Pulse photoconductivity measurement. Combined electron and hole mobilities Growth rates: 2-10 μm/hour Raman FWHM: 2.6 cm^{-1}	Landstrass et al. 1993]

MPACVD - Microwave Plasma Assisted Chemical Vapor Deposition ; HFACVD-Hot Filament Assisted Chemical Vapor Deposition ; FWHM - Full Width at Half-Maximum diamond peak in Raman spectra; T_S - Substrate temperature; MP -Microwave Power

Table 7. Microwave deposition parameters and electrical properties for high quality diamond films [Plano et al. 1993].

Sample	Substrate temperature (^oC)	Pressure (Torr)	Microwave power (W)	Methane concentration (%)	Mobility at 200 V cm^{-1} ($cm^2V^{-1}s^{-1}$)	Raman FWHM (cm^{-1})
Film A	650	50	1000	0.1	50	7.1
Film B	700	90	1900	3.0	1000	4.8
Film C	950	100	1900	1.0	4000	3.3
Diamond type -IIa	-	-	-	-	4000	2.4

Table 6. Summary of the deposition parameters and electrical properties of the polycrystalline diamond films.

Deposition parameters	Sample	Carrier concentration (cm^{-3})	Mobility ($cm^2V^{-1}s^{-1}$)	Mobility index, s, in $\mu \sim T^{-s}$ at $T > 400K$	Comments	References
MPA CVD,		3.0×10^{18}	0.6	-	Hall measurement;	[Nishimura et al.
Gases: 0.5% CH_4 in H_2,					Film thickness:	1991]
Dopant: B_2H_6		2.0×10^{20}	0.28	-	1.4 μm.	
B/C ratio: 0-400 ppm					Raman peak	
Pressure: 31.5 Torr, T_S=800 ^Oc		2.25×10^{20}	0.22	-	at 1337 -1341cm^{-1}	
Substrate: polycrystalline Al_2O_3						
HFACVD;		2.30×10^{15}	32		Hall mesurement;	[Aslam et al. 1992b]
Gases:		1.19×10^{16}	-		Film thickness:	
Dopant: amorphous boron powder		1.03×10^{16}	-		from 1 to 3 μm.	
(5N8) purity		1.72×10^{16}	2		RP - at 1332 cm^{-1}	
Pressure: unknown;						
T_S = unknown						
Substrates: SiO_2 on silicon						
MPA CVD,	highly	5.9×10^{13}	165 ± 9	-	Hall measurement;	[Stoner et al. 1993]
Gases: CH_4 in H_2	oriented				Film thickness:	
Dopant: B_2H_6;'		-	2.0		1.5 μm.	
B/C ratio: 44 ppm	non-			-		
Substrate: oriented diamond	oriented					
films; non-oriented diamond films						
MPA CVD,	PCD1	2.98×10^{16}	33.3	-2.85	Hall measurement;	[Fox et al. 1993]
Gases: CH_4 in H_2					Grain size : 10 μm	
Dopant: B_2H_6 '	PCD2	3.73×10^{19}	0.89	3.95	$N_A=2\times10^{18}$ cm^{-3}	
B/C ratio: 75 and 2310 ppm						
Substrate: intrinsic diamond films					$N_A=2\times10^{20}$ cm^{-3}	

MPACVD - Microwave Plasma Assisted Chemical Vapor Deposition ; HFACVD-Hot Filament Assisted Chemical Vapor Deposition ;
FWHM - Full Width at Half-Maximum diamond peak in Raman spectra; T_S - Substrate temperature; MP - Microwave Power
RP- diamond peak in Raman spectra.

Film properties improve with an increase in grain size, growth temperature, purity and structural perfection. Grain size increases with film thickness [Plano et al. 1993] and [Yarbrough and Messier 1990], but temperature and chemistry also play a very important role. Transport properties of the films cited in Table 7 were determined by pulse transient photoconductive method [Kania et al. 1990]. The absorption depth of 6.1 eV photons is less than 2 μm and free carriers are emitted from it. Therefore, the data obtained characterizes a surface layer on the grown side of the films [Pan et al.1993b].

For the best films (film C), the combined electron and hole mobility at 300 K at an applied electric field of 200 Vcm^{-1} is 4000 $cm^2V^{-1}s^{-1}$ and is similar in value to that for natural type IIa diamond Fig.15). At low concentrations photoexcited carriers, the mobility saturates and becomes independent of the carrier densities. For the best samples, the mobility decreases according to $\mu \sim T^{-1.5}$, limited by acoustic-phonon scattering. In other samples, mobility at low densities has been found to be rather insensitive to temperature [Pan et al.1993a].

In all cases in which carrier densities exceed 10^{16} cm^{-3}, mobility drops, because electron-hole scattering becomes predominant [Pan et al. 1993]. Mobility at low carrier densities (about $10^{16}cm^{-3}$) and decay times (at $\sim10^{18}$ cm^{-3}) increased with film thickness towards the growth side. Figure 16 shows these variations in material quality [Pan et al.1993b]. The phenomenon of improved properties with an increase in film thickness was first noted when measuring thermal conductivity of CVD diamond [Graebner et al. 1992] and mobility in polycrystalline silicon [Kamins 1971]. While investigating CVD diamond film characteristics, it was observed that a "gradient in electrical properties has implications for electronic uses of CVD diamond" [Plano et al. 1994].

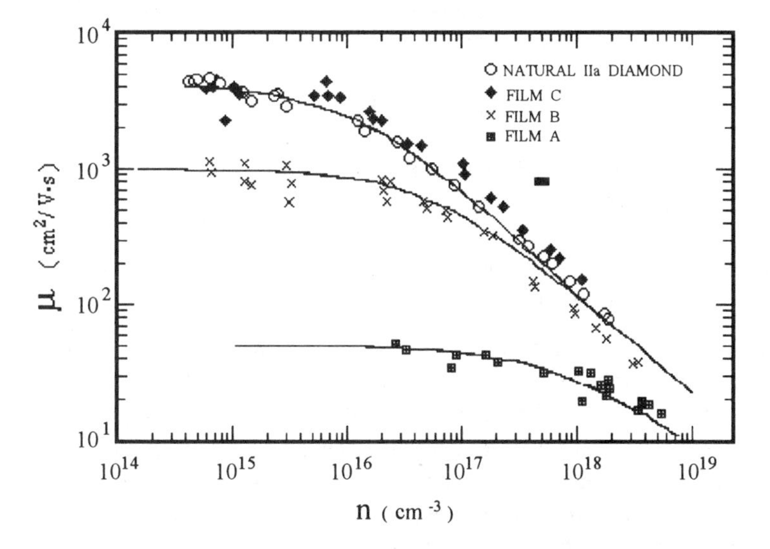

Figure 15. Combined electron and hole mobility measured with intrinsic photoconductivity for films A, B and C, and a natural type IIa diamond. The mobility falls off at high densities due to electron-hole scattering. The solid lines are calculated mobilities based on electron - hole scattering [Pan et al. 1993].

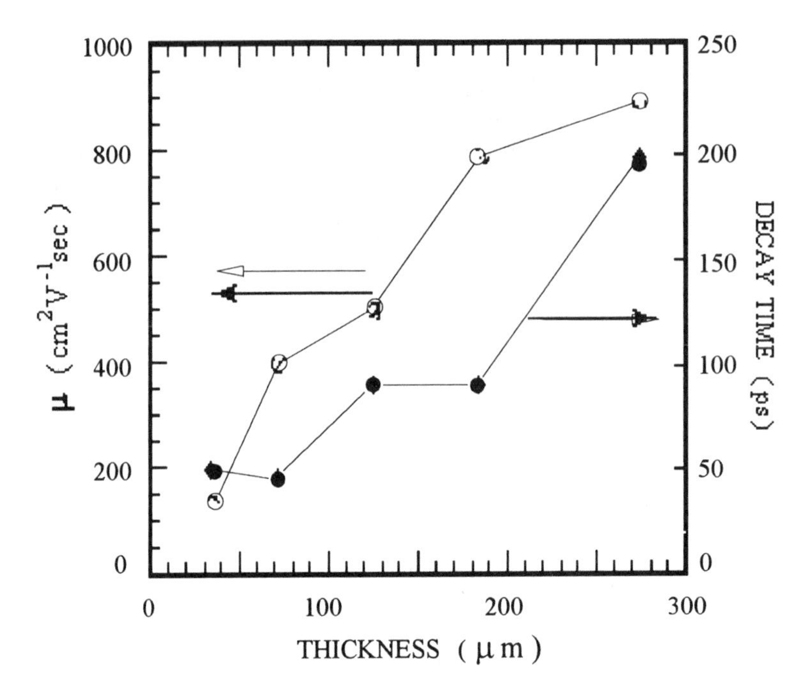

Figure 16. Combined electron and hole mobility at low densities (10^{16} m^{-3}) (open circles) and decay time at 10^{18} cm^{-3} (solid circles) [Pan et al. 1993].

2. Magnetoresistance

Resistance change in a magnetic field is defined by the charge carrier mobility and magnetic field strength, according to,

$$\Delta\rho/\rho = M\,\mu^2\,H^2 \tag{2}$$

in which: M is the kinetic coefficient of magnetoresistance and
H is the magnetic field strength.

The value of the coefficient of magnetoresistance is, in turn, determined by the energy, E, and the dependence of the carier relaxation time, τ, in the valence band. In most cases, this can be expressed as $\tau \sim E^{-\lambda}$, where λ is a function of the type of charge carrier scattering. If the carrier relaxation times are energy dependent, then carriers of different energy drift with different velocity. The Lorentz force is counterbalanced by the Hall field only for carriers having some average energy value E. The paths of other carriers are twisted by the magnetic field, resulting in magnetoresistance. The effect of the Hall field on the experimental magnetoresistance values is effected by sample geometry and shape. The coefficient of magnetoresistance varies for different mechanisms of scattering and valence band structure. Its value is not generally known in advance, and is normally derived through model calculation or from applicable theory.

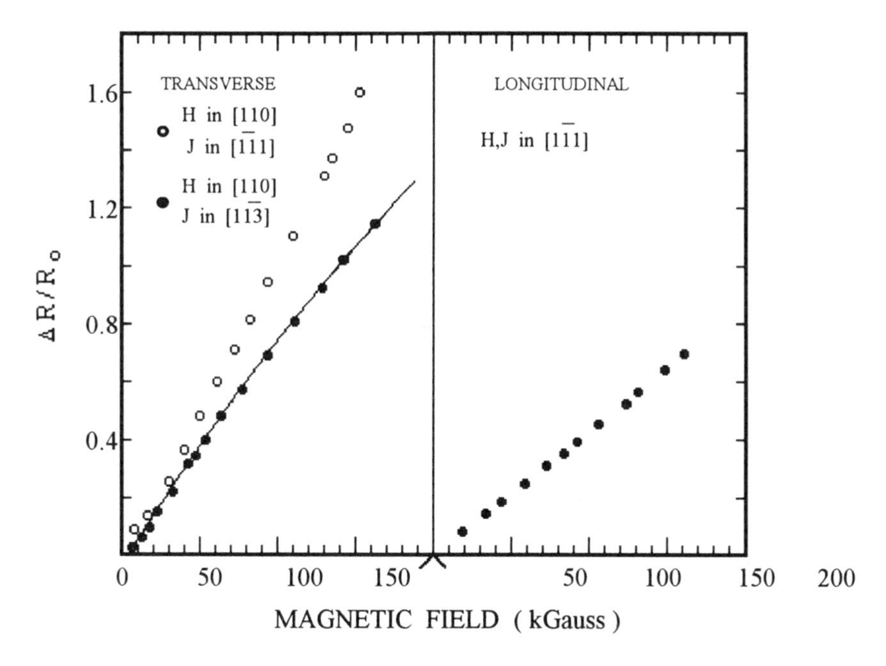

Figure 17. Transverse and longitudinal magnetoresistance of semiconducting diamond at lattice temperature T=307.7K [Russell and Leivo 1972].

Problems in making magnetoresistance measurements and in the interpretation of the data were discussed in Collins and Lightowlers [1979]. It was initially reported that the Hall coefficient was independent of magnetic field up to 1T, while others [Bate and Willardson 1959] found that it increased by 10% when the magnetic field strength varied from 0.05 to 1 T. Magnetoresistance has been analyzed using a three valence band model [Kemmey and Mitchell 1961] assuming that the carrier relaxation time could be expressed as $\tau=\tau_0 E^{-\lambda}$. The authors used theoretically derived band parameters rather than those from experiments and assumed that τ and λ had the same values for all three bands. Their results were in rough agreement with theory for the transverse magnetoresistance coefficients, but the longitudinal effect was much larger than predicted by theory.

The magnetoresistance of p-type diamond at room temperature has been determined in high magnetic fields up to 20T [Russel and Leivo 1972]. The longitudinal magnetoresistance was measured in the [1$\bar{1}$3] and [$\bar{1}$11] directions. Transverse magnetoresistance was measured for the sample current along [110] and the magnetic field in the [$\bar{1}$11] and [110] directions as well as for sample current along [$\bar{1}$11] and magnetic fields in the [1$\bar{1}$3] and [110] directions. The ratios of longitudinal magnetoresistance to the transverse magnetoresistance for various directions of magnetic field and current are plotted in figure 17.

Table 8. Lattice mobilities and parameter β values [Russel and Leivo 1972].

Parameters	Band 1	Band 2	Band 3
μ_L (in $cm^2V^{-1}s^{-1}$)	3900	195	1365
β	0.001	0.0001	0.0003

Theoretical calculations have been made for transverse magnetoresistance only. The assumptions were hole conduction from three uncoupled valence bands, spherical energy surfaces and mixed scattering by lattice vibrations and ionized impurities. The relaxation time for each band and for mixed scattering was expressed as,

$$\tau = \frac{3\sqrt{\pi}}{4}\frac{m^*}{e}\mu_L\frac{\alpha^{3/2}}{\beta+\alpha^2}, \quad \alpha = \frac{E_v - IE}{kT} \qquad (3)$$

where: μ_L is the lattice mobility for a spherical band,
β is a slowly varying function of energy,
E is the carrier energy and
E_v the energy at the top of the valence band.

The value of β gives an indication of the degree of impurity scattering, for pure lattice scattering β=0. In calculations, the values of the effective masses of holes in the three valence bands from Rauch [1963] have been used. The calculated parameters μ_L and β and the parameters that give the best fit for the transverse magnetoresistance are given in Table 8.

From these data the conduction mobility was determined to be 900 $cm^2V^{-1}s^{-1}$. It was also concluded that the lattice scattering predominates over ionized impurity scattering at room temperature. Mobility calculations from measurements of magnetoresistance in diamond-like semiconductors are given in [Sokolov and Stepanov 1974] and [Sokolov and Stepanov 1975]. Using these data, the values of the effective drift mobility for synthetic diamond single crystals of octahedral [Revin and Slesarev 1982] and cubic [Smirnova and Gontar 1993] habit could be determined.

3. Piezoresistance

The piezoresistive effect of p-type semiconductors, such as boron-doped diamond, is defined by the structural characteristics of the valence band [Smith 1954] and [Pikus and Bir 1959]. For diamond the energy maximum is in the center of the Brillouin zone at k=0 for all the three bands $P_{3/2}^{3/2}$, $P_{3/2}^{1/2}$, $P_{1/2}$. Spin-orbital interaction results in the partial removal of the degeneracy of the $P_{1/2}$ band and the band lowers to 0.006 eV. The $P_{3/2}^{3/2}$ bands degeneracies at k=0 and at k ≠ 0 have different curvatures, thus forming two energy branches for light and heavy holes. Superposition of deformation disturbs the symmetry of the lattice field. Because of this the degeneracy is removed. The top of the light and heavy hole zones become displaced by different amounts in opposite directions. This changes the concentration of light and heavy holes, which have different mobilities and causes variations in the electrical resistance even, when the total concentration of holes remains unchanged.

The piezoresistive effect has been evaluated in boron-doped polycrystalline diamond films [Aslam et al. 1992a]. Piezoresistive properties of polycrystalline films are determined not only by the crystalline substance itself, but also by the transport properties of the grain boundaries [French and Evans 1989] and [Zhao and Bao 1989]. The extent of grain-orientation (texture) is also of great importance [Ohmura 1992].The relationship between the electrical resistance variation $\Delta R/R$ and the piezoresistive strain $\Delta L/L$ due to the mechanical loading and the elastic deformation, and known as the gauge factor, K, is expressed as,

$$\mathrm{K} = \frac{\Delta R / R}{\Delta L / L} = \frac{\Delta R / R}{\varepsilon} \qquad (4)$$

Using different material parameters, the gauge factor can also be expressed as,

$$\mathrm{K} = 1 + 2\nu + \frac{\Delta\rho}{\rho\varepsilon} \qquad (5)$$

for which : ε is the mechanical strain,

ρ the resistivity and

ν Poisson's ratio.

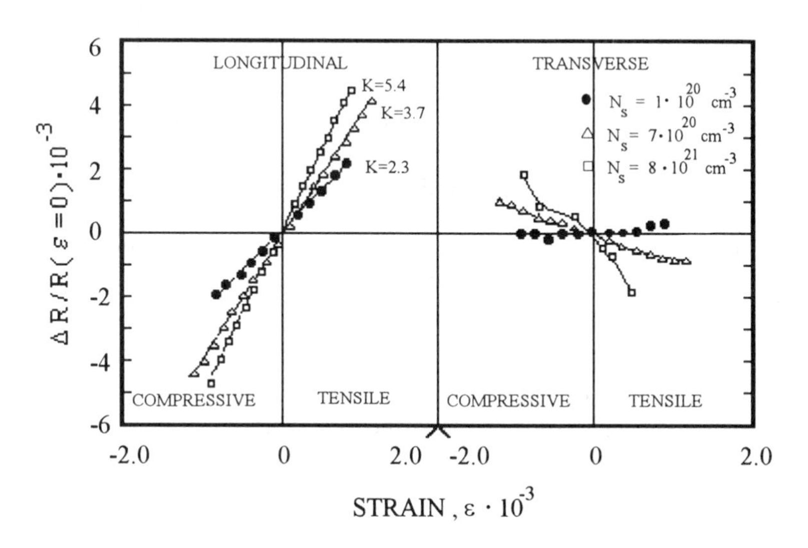

Figure 18. Longitudinal and transverse piezoresistive effect of polycrystalline diamond films. N_S= boron concentration, K= gauge factor [Dorsch et al. 1993].

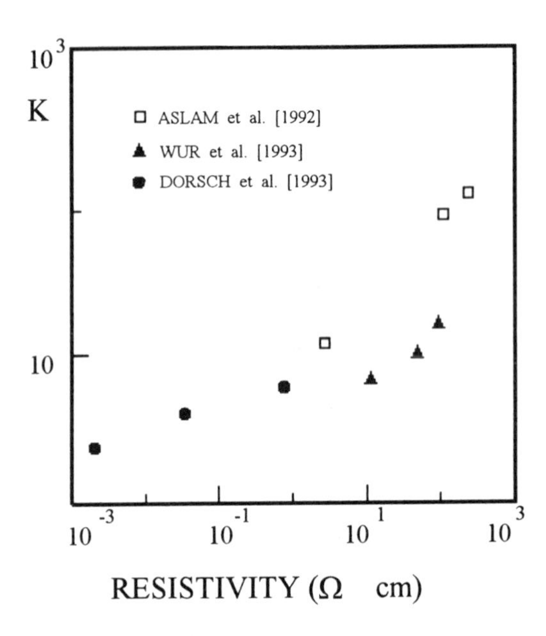

Figure 19. Gauge factor versus resistivity for boron - doped diamond films [Dorsch et al.1993].

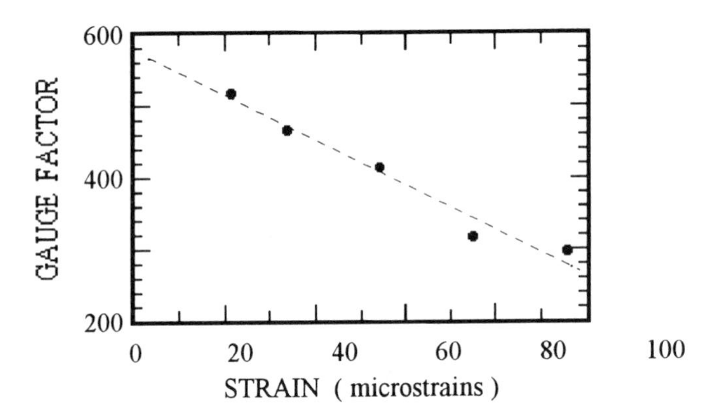

Figure 20. Strain dependence of the piezoresistive gauge factor of a (100) - oriented homoepitaxial diamond film. Both strain and current flow are parallel to the [100] direction [Aslam et al.1992]

When the strain direction coincides with the direction of the electrical field and the current density, the gauge factor measures the longitudinal piezoresistive effect. When the strain is perpendicular to the field and current density, it measures the transverse piezoresistive effect.

According to Dorsch et al. [1993], the longitudinal piezoresistive effect for Microwave Plasma Assisted Chemical Vapor Deposited (MPACVD) polycrystalline diamond films

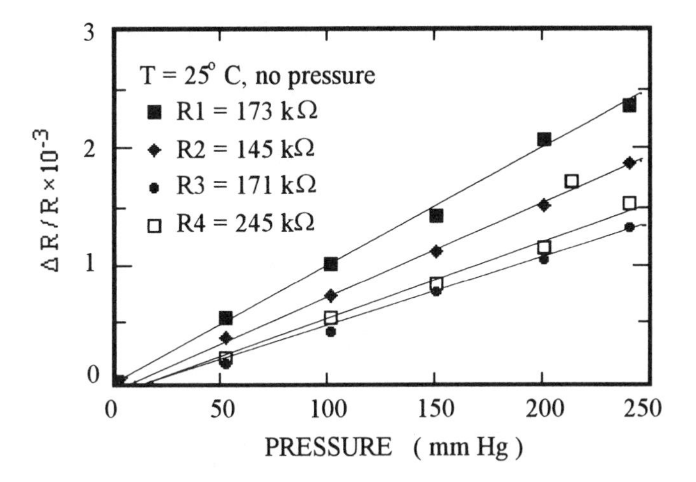

Figure 21. Resistance change versus pressure of polycrystalline diamond piezoresistors [Wur et al. 1993]. (■, ♦) - longitudinal effect, (●, □) - transverse effect.

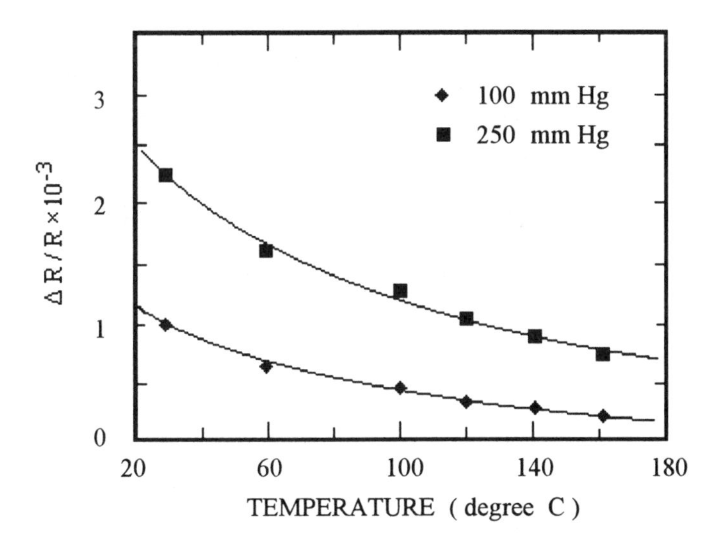

Figure 22. Resistance change versus temperature at two pressures [Wur et al. 1993]. ♦ - 100mm Hg, ■ - 250mm Hg.

is greater than the transverse one (Fig.18). Figure 19 shows gauge factor values of boron-doped polycrystalline diamond films [Aslam et al. 1992], [Dorsch et al. 1993], and [Wur et al. 1993]. The gauge factor increases with an increase in the film resistivity (decrease in the level of doping). With a boron concentration of 10^{18}cm^{-3}, the gauge factor attains a value of 300 at room temperature [Wang and Liano 1994]. Figure 20 displays the strain dependence of the piezoresistive gauge factor for homoepitaxial diamond films [Aslam et al. 1992].

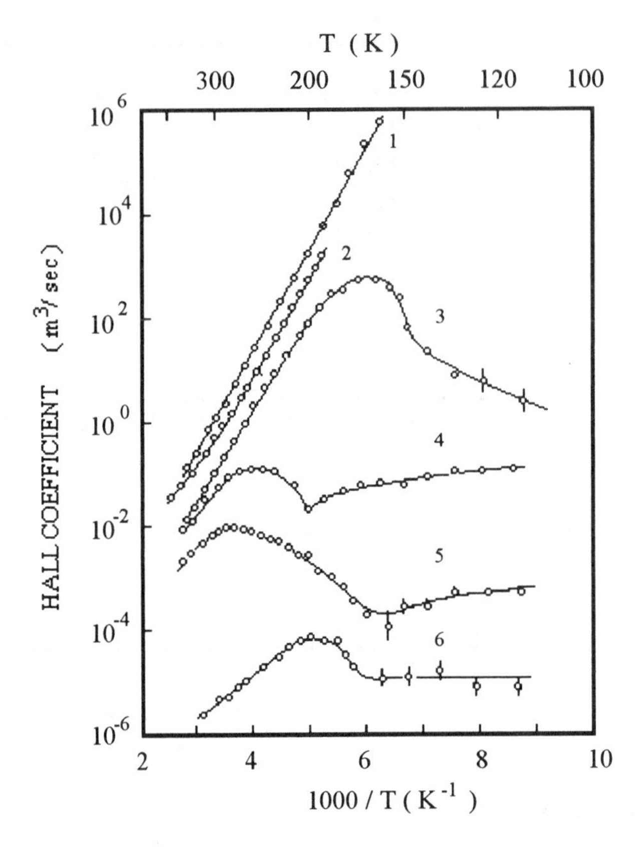

Figure 23. Hall coefficient R versus temperature for one natural (1) and five synthetic (2-6) semiconducting diamonds. The absolute values for the synthetic diamonds have only order of magnitude significance because of their irregular shapes and the curves have been scaled as indicated below to avoid intersection. (1) R x 10, (2) R x 3, (3) R x 1, (4) R x 1, (5) R x 10^{-1}, (6) R x 10^{-2} [Williams et al. 1970].

At room temperature the piezoresistance change ΔR/R is linearly related to the applied strain (Fig. 21) and at constant strain decreases in value with a temperature increase (Fig. 22).The piezoresistive effect of polycrystalline diamond films depends heavily on doping level, grain size, defect density and film thickness [Wang and Liano 1994], [Davidson et al. 1993], and [Werner et al. 1994].

4. Hall Effect Measurements

High precision Hall coefficient measurements have been performed on natural type IIb and synthetic semiconducting diamond [Dean et al.1965] and [Williams et al. 1970]. Temperature dependence of the Hall coefficient and conductivity are shown in figures 23 and 24. These data can be satisfactorily explained by assuming that diamond is a partially compensated p-type semiconductor.

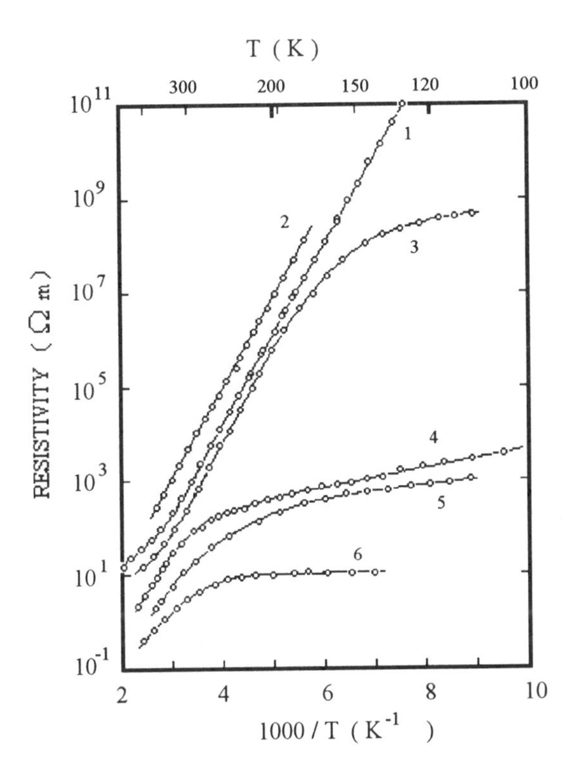

Figure 24. Resistivity ρ plotted versus temperature for one natural (1) and five synthetic (2-6) semiconducting diamonds. The absolute values for the synthetic diamonds have only order of magnitude significance because of their irregular shapes and the curves have been scaled as indicated below to avoid intersection. (1) $\rho \times 10^3$, (2) $\rho \times 10^3$, (3) $\rho \times 10^3$, (4) $\rho \times 30$, (5) $\rho \times 10$, (6) $\rho \times 1$ [Williams et al. 1970].

For the temperature dependence of free hole concentration p, most researchers have used the expression

$$\frac{p(p + N_D)}{N_A - N_D - p} = \left(\frac{2\pi m^* kT}{h^2} \right)^{3/2} exp\left(- E_A /kT\right) \tag{6}$$

where: N_A and N_D are the acceptor and donor concentrations;
E_A is the acceptor ionization energy;
m^* is the effective mass of the holes;
h and k are the Planck and Boltzmann constants.

The degeneracy factor was assumed to be 2 [Visser et al. 1992]. The values of p were determined from Hall coefficient, R_H, values according to

$$R_H = r\,(1/pe),\quad r = \mu_H / \mu_C \tag{7}$$

where: r is the Hall-coefficient factor and μ_H and μ_C are the Hall and conductivity mobilities.

As with any kinetic coefficient, the Hall-coefficient factor is defined by the band structure and the dependence of relaxation time on energy, i.e. on the scattering mechanism. In most cases the Hall factor is assumed to be equal to $3\pi/8$ [Dean et al.1965] and [Williams et al. 1970] or 1 [Visser et al. 1992] and [Wynands et al.1994] and temperature-independent. This assumption is justified for nondegenerate carriers in a single spherical band for which scattering by long-wavelength acoustical phonons is predominant. Dean et al. [1965] concluded that under these conditions the departure of the Hall-factor from $3\pi/8$ does not, at any temperature, exceed 30%. In analyzing Hall-effect and electrical conductivity measurements, the majority of researchers have used equations (6) and (7) and adjusted m^* to fit the experimental data. Values obtained for N_A and N_D using Lee's method [Lee 1957] were inserted together with the value of E_A obtained from a plot of $k(\ln p - 3/2\ln T)$ vs T^{-1} in equation (6), while m^* was calculated over the range of temperatures employed. The parameters of a semiconducting diamond obtained from Hall measurements are given in Table 9.

More detailed analyses of the band parameters and scattering mechanisms which determine the appropriate values for m^* and the variation of Hall-coefficient factor with temperature were made by Reggiani et al. [1983]. Theoretical calculations of the Hall-coefficient factor were made for different sets of the valence-band parameters as given in Table 9. Calculations based on data of the sets I, II and III, result in r-values that systematically exceed 1 between 100 and 1000K, while data of the sets IV, V and VI, exhibit values very close to each other and consistent with the experimental r-values at T=300K, as calculated from the ratio between Hall and drift mobility experimental data. The best fit between theory and experiment was achieved when the data of set V were used in the calculations: $|A|=3.61$; $|B|=0.18$; $|C|=3.76$.

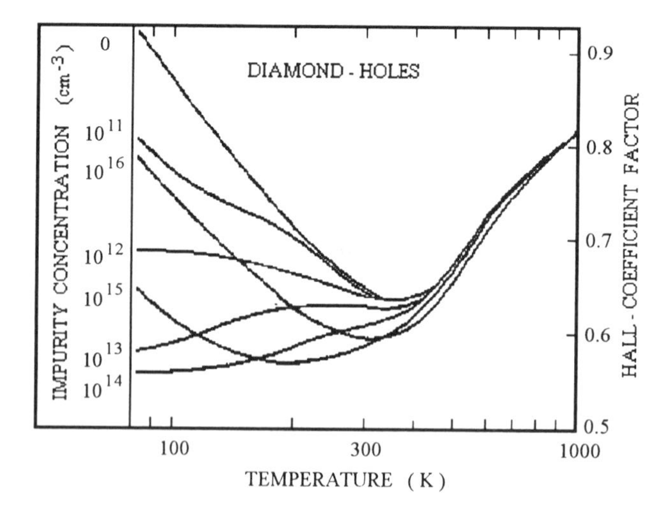

Figure 25. Theoretical calculations of the Hall-coefficient factor (right scale) for the impurity concentrations as indicated (values on the left give the impurity concentration for each curve) [Reggiani et al. 1983].

Table 9. Inverse valence-band parameters and related quantities in diamond.

	Band parameters					
	I Hall (1958)	II Rauch (1963)	III Lewaetz (1971)	IV Saslow et al. (1966)	V van Haeringen and Junginger (1969)	VI van Vechten (1972)
/A/	1.61	0.94	4.62	3.63	3.61	3.72
/B/	0.11	0.44	0.84	0.44	0.18	0.42
/C/	0.78	0.40	3.14	3.80	3.67	3.88
m_h	0.89	2.30	0.40	1.02	1.08	0.92
m_l	0.70	2.08	0.28	0.38	0.36	0.36
k_{110}/k_{100}	1.25	1.10	1.86	2.60	2.60	2.40
k_{111}/k_{100}	1.14	1.06	1.20	1.52	1.52	1.49

A,B and C are given in units of $\hbar/2m$, and the effective masses in terms of the free-electron mass:

$$\frac{k_{110}}{k_{100}} = \left[\frac{2(1-b)}{2+b-3[(c^2/3)+b^2]^{1/2}}\right]^{1/2}, \quad \frac{k_{111}}{k_{100}} = \left[\frac{(1-b)}{1-[(c^2/3)+b^2]^{1/2}}\right]^{1/2}$$

where: $b=|B|/|A|$; $c=|C|/|A|$.

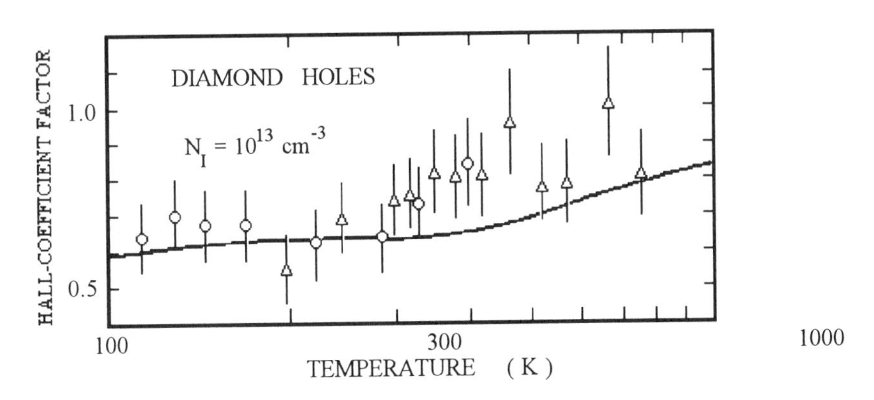

Figure 26. Comparison between theoretical and experimental Hall-coefficient factor. Theory uses one value of V and includes ionized impurity scattering with concentration $N_I=10^{13}$ cm^{-3} [Reggiani et al. 1983]. O and Δ refer to data obtained from Hall-mobility measurements [Konorova et al. 1967] and [Dean and al. 1965] .

Calculations have shown that the effect of ionized impurity scattering becomes important below about 300K for concentrations up to $N_I=10^{16}$cm^{-3}. From figure 25 it can be seen that above about 500K optical-phonon scattering tends to increase the value of *r* over the entire range of ionized impurity concentrations. A comparison between theoretical and experimental Hall-factors is given in figure 26. There is agreement between theory and experiment below 300K. At about 300K the theoretical curve lies consistently below the experimental points. Reggiani et al. (1983) suggest that this is because the split-off band is neglected in their theory and the inclusion of the split-offband in the calculations should increase the value of the Hall-coefficient factor by 10-15% at temperatures above approximately 293K.

5. Acknowledgement

The author thanks Prof. Peter Gielisse for his valuable help in discussing the manuscript.

6. References

Aleksenko, A.E., Vavilov, V.S., Derjaguin, B.V., Gukasyan, M.A.,Karatygina, T.A., Konorova, E.A., Sergienko,V.F., Spitsyn, B.V. and Tkachenko, S.D.(1977) Charge transfer and the nature of acceptor in semiconductor epitaxial diamond layers, *Sov. Phys. Dokl.* **22,** 166-170.

Aslam, M., Taher, I., Masood, A., Tamor, M.A. and Potter, T.I. (1992a) Piezoresistivity in vapor-deposited diamond films, *Appl. Phys.Lett.*, **60**(23), 2923-2925.

Aslam, M.M., Tamor, M.A. and Potter, T.F. (1992b) Synthesis and electrical characterization of boron-doped thin diamond films, *Appl. Phys. Lett.*, **61**(15), 1832-1834.

Austin, J.G. and Wolfe, R. (1956) Electrical and optical properties of a semiconducting diamond, *Proc.Phys.Soc.*, **B69**(3), 329-338.

Baranskii, P.I., Malogolovets, V.G., Torishnii, V.I. and Chipenko, G.V. (1987) Temperature dependence of hole mobility in a semiconductor synthetic diamond, *Sov. Phys. Semicond.*, **21**, 45-50.

Bate, R.T. and Willardson, R.K (1959) Hall coefficient and magnetoresistance in semiconducting diamond, *Proc. Phys. Soc.*, **B74,** Pt 3(477), 363-367.

Brophy, J.J.(1955) Preliminary study of the electrical properties of a semiconducting diamond, *Phys.Rev.,* **99**(4), 1336-1337

Burns, R.C., Cvetkovic,V., Dodge, C.N., Evans, D.J.F., Rooney, M-L.T., Spear, P.M., and Welbourn, C.M. (1990) Growth-sector dependence of optical features in large synthetic diamonds, *J. Cryst. Growth.,* **104**(2), 257-279.

Canali, C., Jacoboni, C., Nava, F., Ottaviani, G. and Alberigi-Quaranta, A. (1975) Electron drift velocity in silicon, *Phys.Rev. B,* **12**(4), 2265-2284.

Clark, C.D., Dean, P.J. and Harris, P.V. (1964) Intrinsic edge absorption in diamond, *Proc. R. Soc. Lond. A,* **277,** 312-329

Collins, A.T. (1989) Diamond electronic devices - a critical appraisal, *Semicond. Sci.Technol.*, **4**, 605-611.

Collins, A.T. and Lightowlers, E.C. (1979) Electrical Properties, In I.E.Field (ed.), *The Propertes of Diamond*, Academic Press, London, pp.79-105.

Das, P., Ferry, D.K (1976) Hot electron microwave conductivity of wide bandgap semiconductors, *Sol. St. Electron.,* **19**, p.851-855.

Davidson, J.L., Ellis, C. and Ramesham, R. (1989) Selective deposition of diamond films, *J. Electronic Mat.* **18**(6), 711-715.
Davidson, J.L., Wur, D., Kang, W.P. (1993) The Piezoresistance of boron-doped diamond on an undoped diamond membrane, in J.P.Dismukes and K.V.Ravi (eds), *Proceeding of the Third International Symposium on Diamond Materials,* Peninghton, NJ, **97**(13), pp.1048-1053.
Dean, P.J., Lightowlers, E.C. and Wight, D.R.(1965) Intrinsic and extrinsic recombination radiation from natural and synthetic aluminum-doped diamond, *Phys. Rev.*, **140**(1A), A352-A368.
Dorsch, O., Holzner, K., Werner, M., Obermeier, E., Harper, R.E., Johnston, C., Chalker, P.R. and Buckley-Golder, I.M. (1993) Piezoresistive effect of boron-doped diamond thin films, *Diamond and Related Materials*, 2, 1096-1099.
French, P.J. and Evans, A.G.R. (1989) Piezoresistance in polysilicon and its applications to strain gauges, *Solid-St.Electr.*, **32**(1), 1-10.
Fujimori, N., Imai, T. and Doi, A. (1986) Characterization of conducting diamond films, *Vacuum*, **36**(1-3), 99-102.
Fujimori, N., Nakahata,H. and Imai, T. (1990) Properties of Boron-Doped Epitaxial Diamond Films, *Japan. J. of Appl.Phys.*, **29**, 824-827.
Fox, B.A., Malta, D.M., Wynands,H.A. and van Windhein, J.A. (1993) Electronic properties of boron doped homoepitaxial and polycrystalline diamond thin films and natural single crystals, in J.P.Dismukes and K.V.Ravi (eds), *Proceedings of the Third International Symposium on Diamond Materials* Peninghton, NJ, **97**(13) pp.759-765.
Gildenblat, G.Sh., Grot, S.A., and Badzian, A.R.(1990a) Device applications of homoepitaxial and polycrystalline diamond films, invited paper, I-th Proc. of International Conf. on Electronic Materials, 1990, Newark, NJ, USA.
Gildenblat, G.Sh., Grot, S.A., Hatfield, C.W., Badzian, A.R., Badzian, T. (1990b) High-temperature Schottky diodes with thin-film diamond base, *IEEE Electron Dev. Lett.*, **EDL-11**(9), 371-372.
Gildenblat, G.Sh., Grot,S.A. and Badzian,A.R. (1991)The electrical properties and device applications of homoepitaxial and polycrystalline diamond films, *Pros. IEEE*, **79**, 647-668.
Gildenblat, G.Sh., Grot, S.A., Wronski, C.R., Badzian, A.R., Badzian, T. and Messier, R. (1988) Electrical characteristics of Schotky diodes fabricated using plasma assisted chemical vapor deposited diamond films, *Appl. Phys. Lett.*, **53**(7), 586-588.
Graebner, J.E., Jin, S., Kammlott, G.W, Herb, J.A. and Gardinier, C.F. (1992) Unusually high thermal conductivity in diamond films, *Appl.Phys.Lett.*, **60**(13), 1576-1578.
Grot, S.A., Hatfild, C.W., Gildenblat, G.Sh., Badzian, A.R., and Badzian, T. (1991) Electrical properties of selectively grown homoepitaxial diamond films, *Appl. Phys.,Lett.*, **58**(14), 1542-1544.
Hall, G.G. (1958) The electronic structure of diamond, silicon and germanium, *Philos.Magazine,* **3**(29), 429.
Herman, F. (1952) Electronic structure of the diamond crystal, *Phys.Rev.,* **88**(5), 1210-1211.

Jonson, C., Stein, H., Young, T., Wayland, J. and Leivo, W. (1964) Photoeffects and related properties of semiconducting diamonds, *J.Phys.Chem.Solids*, **25**(8), 827-836.

Kamins, T.L. (1971) Hall mobility in chemically deposited polycrystalline silicon, *J.Appl.Phys.* **42**(11), 4357-4365.

Kania, D.R, Pan, L.S, Bell, P., Landen, O.L., Kornblum, H., Pianetta, P. and Perry, M.D. (1990) Absolute x-ray power measurements with subnanosecond time using type IIa diamond photoconductors, *J. Appl. Phys.*, **68**(1), 124 -130.

Kemmey, P.J. and Mitchell, E.V.J.(1961) The magneto-resistance of *p*-type semiconducting diamond, *Pros. R. Sos.Lond.*, **263**(1314),420-432.

Klick, C.C.and Maurer, R.J. (1951) The mobility of electrons in diamond, *Phys. Rev.*, **81**(1), 124-130.

Kobashi, K., Kawate, Y., Horiuchi, T. and Nishimura, K. (1988) Synthesis of diamonds by use of microwave plasma chemical-vapor deposition: Morphology and growth of diamond films, *Phis.Rev. B,* **38**, 4067-4084.

Konorova, E.A., Kozlov, S.F. and Vavilov,V.S. (1966) Ionization currents in diamonds in case of 500-1000keV electron bombardment, *Fizika Tverdogo Tela*, **8**(1), 1-8.

Konorova, E.A., Kuznetsov, U.A., and Sergienko, V.F. (1983) Impact ionization in semiconductor structures based on ion-implanted diamond, *Sov. Phys. Semicond.*, **17**(1), 108-112.

Konorova, E.A. and Shevchenko, S.A., (1967) Study of current carrier mobility in diamonds, *Sov. Phys. Semicond.*, **1**, 364-370.

Kozlov, S.F., Belcarz, E., Hage-Ali, M., Stuck, R. and Siffert, P. (1974a) Diamond nuclear radiation detectors, *Nucl. Inst. Meth.*, **117**(1), 277-283.

Kozlov, S.F., Belcars, E., .Hage-Ali, M., Stuck, R. and Siffert, P. (1974b) CRN/PN 74-1, Strasburg, France.

Landstrass, M.I., Plano, M.A., Moreno, M.A., Williams, S.Mc., Pan, L.S., Kania, D.R. and Han, S. (1993) Device properties of homoepitaxially grown diamond, *Diamond and Related Materials*, **2**, 1033-1037.

Lee, P.A.(1957) Determination of the impurity concentrations in a semiconductor from Hall coefficient measurements, *Br. J. Appl. Phys.*, **8**(8), 340-343.

Leite, J.R., Bennett, B.I. and Herman, F. (1975) Electronic structure of the diamond crystal based on an improved cellular calculation, *Phys. Rev., B*, **12**(4), 1466-1481.

Lenz, H. (1924) Uber den Hall-effekt des lichtelek-trischen primarstromes bei isolierenden kristallen, *Physik Z.*, **25**(17), 435-439.

Lewaetz, P. (1971) Valence-band parameters in cubic semiconductors, *Phys.Rev.,* **B4**( ), 3460-3467.

Lin, J.F., Li, S.S., Linares, L.C. and Teng, K.W. (1981) Theoretical analysis of Hall factor and Hall mobility in p-type silicon, *Solid St.Electron.*, **24**(9)**,** 827- 833.

Lin, P., Yen, R. and Bloembergen, N. (1978) Dielectric breakdown threshold, two-photon absorption, and other optical damage mechanisms in diamond, *JEEE of Quantum Electronics*, **14**(8), 574-576.

Martini, M., Mayer J.W. and Zonio K.R. (1972) Drift velocity and trapping in semiconductors transient charge technique, in R.Wolf (ed), *Appl. Solid St. Sci.*, **3**, Academic Press, New York, pp.181-261.

Nakagawa, H. and Zukotynski, S. (1978) Drift mobility and Hall coefficient factor of holes in germanium and silicon, *Canadian J. Phys.*, **56**(3), 364-372.

Nava, F., Canali, C., Artuso, M., Gatti, E., Manfredi, P.F. and Kozlov, S.F. (1979) Transport properties of natural diamond used as nuclear particle detectors for a wide temperature range, *IEEE Trans. Nucl. Sci.*, **NS-26**, 308-315.

Nava, F., Canali, C., Jacoboni, C., Reggiani, L. and Kozlov, S.F. (1980) Electron effective masses and lattice scattering in natural diamond, *Solid.St.Comm.*, **33**(4), 475-477.

Nishimura, K., Das, K. and Glass, I.T.(1991) Material and electrical characterization of polycrystalline boron-doped diamond films grown by microwave plasma chemical vapor deposition, *J. Appl. Phys.*, **69**(5), 3142-3148.

Ohmura, J. (1992) *Jpn.J.Phys.Soc.* , **61** (1), 217-226.

Okano, K., Naruki, H., Akiba, Y., Kurosu, T., Jida, M. and Hirose, Y. (1988) Synthesis of diamond thin films having semiconductive properties, *Jpn J. Appl. Phys.*, **27**(2), L173-L175.

Pan L.S., Han S., Kania D.R., Plano M.A.and M.J.Landstrass. (1993a) Electrical properties of high quality diamond films, *Diamond and Related Materials*, **2,** 820-824.

Pan, L.S., Han, S., Kania, D.R., Plano, M.A., Landstrass, M.I., Zhao, S. and Kagan, H. (1993b) Comparison of high electrical quality CVD diamond and natural single-crystal IIa diamond, in J.D.Dismukes and K.V.Ravi (eds), *Proceedings of the Third International Symposium on Diamond Materials,* Penington, NJ, **97-13**, pp.735-739, Presented at the 3 rd Int'l Symp. on Diamond Materials, Honolulu, Hawaii, May, (1993), p.16-21.

Pan, L.S., Kania, D.R., Han,S.,Ager III,J.W., Landstrass,M., Landen,O.L. and Pianetta,P. (1992) Electrical transport properties of undoped CVD diamond films, *Science,* **255**(5046) 830-833.

Pan, L.S., Kania, D.R., Pianetta, P., Ager III, J.W, Landstrass, M.I. and Han, S. (1993c) Temperature dependent mobility in single-crystal and chemical vapor-deposited diamond, *J. Appl.Phys.*, **73**(6), 2888-2894.

Pan, L.S., Kania, D.R., Pianetta, P. and Landen, O.L. (1990) Carrier density - dependent photoconductivity in diamond, *Appl. Phys. Lett.*, **57**(6), 623-625.

Pearlstein, E.A. and Sutton, R.B. (1950) Mobility of electrons and holes in diamond, *Phys.Rev.*, **79**(5), 907-911.

Pikus, G.E. and Bir, G.L. (1959) The effect of deformation on the electric properties of hole germanium and silicon, *Fizika Tverdogo Tela*, **1**(12), 1828-1840.

Plano, M.A., Landstrass, M.I., Pan, L.S., Han, S., Kania, D.R, Mc Williams S. and Ager III, J.W. (1993) Polycrystalline CVD Diamond Films with High Electrical Mobility, *Science,* **260**, 1310-1312.

Plano, M.A., Zhao, S., Gardinier, C.F., Landstrass, M.J. and Kania, D.R. (1994) Thickness dependence of the electrical characteristics of chemical vapor deposited diamond films, *Appl. Phys. Lett.*, **64**(2), 193 -195.

Proferl, D.J., Gardner, N.C. and Angus, I.C. (1973) Growth of boron-doped diamond seed crystals by vapor deposition, *J. Appl. Phys.*, **44**(4), 1428-1434.

Rauch, C.J. (1961) Millimeter cyclotron resonance experiments in diamond, *Phys. Rev Lett.,* 7(3), 83-84.

Rauch, C.J. (1962) Millimeter cyclotron resonance in diamond, in A.C.Stickland (ed.), *Proc. Int.Conf. on Physics of Semiconductors*, Exeter, The Institute of Physics and the Physical Society, London, pp. 276-280.

Redfield, A.G. (1954) Electronic Hall effect in diamond, *Phys.Rev.*, **94**(3), 526-537.
Reggiani, L., Bosi,S., Canali, C., Nava F. and Kozlov, S.F. (1979) On the scattering and effective mass of holes in natural diamond, *Solid St. Comm.*, **30**(6), 333-335.
Reggiani, L., Bosi, S., Canali, C., Nava, F. and Kozlov, S.F. (1981) Hole-drift velocity in natural diamond, *Phys. Rev.B*, **23**, 3050-3057.
Reggiani, L., Canali, C., Nava, F. and Ottaviani, G. (1977) Hole drift velocity in germanium, *Phys. Rev.*, **16**(6), 2781-2791.
Reggiani, L., Waechter, D. and Zukotynski, S. (1983) Hall-coefficient factor and valence-band parameters of holes in natural diamond, *Phys. Rev.B*, **28**(6), 3550-3555.
Revin, O.G., Slesarev, V.N. (1982) Hole concentration and degree of compensation in artificial p-type diamond, *Fiz. and Techn. Semiconductor*, **16**, 2219-2222.
Rooney, Marie-Line T. (1992) Growth in boron-doped synthetic diamonds, *J. Cryst. Growth.*, **116**(1/2), 15-21.
Russel, K.J. and Leivo,W.I. (1972) High-field magnetoresistance of semiconducting diamond, *Phys. Rev.*, **B6**(12), 4588-4596.
Saslow, W., Bergstresser, T.K. and Cohen, M.L. (1966) Band structure and optical properties of diamond, *Phys. Rev. Lett.*, **16**(9), 354 -356.
Sato,Y., Kamo, M. and Setaka, N. (1987) in P.Vincenzini (ed) *High Tech. Ceramics*, Elsevier, New.York, pp. 1719 - 1724.
Shiomi, H., Nakahata, H., Imai,T., Nishibayashi, Y. and Fujimori, N. (1989a) Electrical characteristics of metal contacts to boron-doped diamond epitaxial film, *Japan. J. Appl. Phys.*, **28**(5), 758-762.
Shiomi, H., Nishibayashi, Y. and Fujimori, N. (1989b) Field-effect transistors using boron-doped diamond epitaxial films, *Japan. J. Appl. Phys.*, **28**(12), L2153 - L2154.
Smirnova, O.I., Gontar, A.G. (1993) Study of hopping conductivity in boron-doped semiconducting diamond, *J. of Superhard. Mat.*, **1**, , Alerton Press. Inc., New-York, pp.10 -18.
Smith, C.S. (1954) Piezoresistance effect in germanium and silicon, *Phys. Rev.*, **94**(1), 42-49.
Smith, R.A. (1961) *Semiconductors* (Cambrige,University Press), Pt.5.4.
Sokolov, Ya.F., Stepanov, B.G. (1974) Physical principles of using magnetoresistance effect for measuring mobility and concentration of current carriers, *Mikroelectronika* (Russian), **3**(2), 142-153.
Sokolov, Ya.F., Stepanov, B.G. (1975) Physical principles of using magnetoresistance effect for measuring mobility and concentration of current carriers, *Mikroelectronika* (Russian), **4**(5), 414-421.
Spitsyn, B.V. (1990) Chemical crystallization of diamond from the activated vapor phase, *J. Crys. Growth.* Part 2, **99**(1-4), 1162-1167.
Spitsyn, B.V, Bouilov, L.L. and Derjaguin, B.V. (1981) Vapor growth of diamond on diamond and other surfaces, *J. Cryst. Growth.*, **52** (1), 219-226.
Srikanth, K. and Ashok, S. (1988) Electrical conduction in thin film diamond, *Thin Solid Films*, **164**, 187-190.
Stoner, B.R., Chien-ten Kao, Malta, D.M. and Glass, R.C. (1993) Hall effect measurements on boron-doped, highly oriented diamond films grown on silicon, via

microwave plasma chemical vapor deposition, *Appl. Phys. Lett.*, **62**(19), 2347-2349.

Stoner, B.R., Glass, J.T., Bergaman, L., Nemanich, R.I., Zoltan, L.D. and Vandersande, J.W. (1992) Electrical conductivity and photoluminescence of diamond films grown by downstream microwave plasma CVD, *J. Electron. Mater.*, **21**, 629-634.

Tkach, V.N. (1993) Precision studies of synthetic diamonds using Kossel's method, *Diamond and Related Materials,* **3**, 112 -115.

Vandersande, J.W. and Zoltan, L.D. (1991) High temperature electrical conductivity measurements of natural diamond and diamond films, *Surf. Coatings Technol.*, **47**(1-3), 392-400.

van Haeringen, W. and Junginger, H.G. (1969) Empirical pseudopotential approach to the band structures of diamond and silicon carbide, *Solid State Commun.*, **7**, 1135-1137.

van Vechten, J.A. (1972) Band structure of diamond from a local, Heine-Abarenkov, pseudopotential, *Bull.of the American Phys.Soc.,* **17**(3), 237.

Vavilov, V.S., Gippius, A.A., Konorova, E.A. (1985) *Electronic and optical processes in diamond* (in Russian), *Nauka*, Moscow, Pt.2, pp. 18-29.

Vavilov, V.S. and Konorova, E.A. (1976) Semiconducting diamond, *Sov. Phys. Usp.* **19**, 301-316.

Wang, W.L. and Liano, K.I.(1994) Piezoresistivity studies of polycrystalline p-type diamond films, in Abstr.book *The 5-th European Conf. on Diamond, Diamond-like and Related Materials,* Tuscany, Italy, 25-30 Sept.,1994.

Vishnevsky, A.S. (1975) Sectorial structure and laminar growth of synthetic diamond crystals, *J. Cryst. Growth.*, **29**, 296 -300.

Vishnevsky, A.S., Gontar, A.G., Romanko, L.A., Zelentsova, T.I., Radin, I.V. and Rusov, V.D. (1992) Use of track autoradiography in crystal and the physical study of synthetic diamond monocrystals, *Nucl. Track. Radiat. Meas.*, **20**, 309 -313.

Vishnevsky, A.S., Prikhna, A.I., Ositinskaya, T.D. and Gontar.A.G. (1974) Internal structure and electrical conductivity of synthetic diamond crystals doped with boron, *Sintheticheskiye Almazy*, **2**, 5-8.

Visser, E.P., Bauhuis, G.T., Jaussen, G., Vollenberg, W., van Enekevort, Willen J.P. and Giling, L.J. (1992) Electrical conduction in homoepitaxial, boron-doped diamond films, *J. Phys.: Condens. Matter.*, **4**, 7365-7376.

Weast, R.C.(ed.) (1971) *Handbook of Chemistry and Physics*, Chemical Rubber Co.,Ohio.

Wedepohl, P.T. (1957) Electrical and optical properties of type IIb diamond, *Proc. Phys. Soc.* **B70,** Pt 2(446), 177-185.

Werner, M., Dorch, O., Obermeier, E. (1994) Properties and applications of polycrystalline diamond double layers as a basic material for high temperature sensors, *Electron Technology*, Institute of Electron technology, Warszawa, **27**(3/4), 83-95,

Williams, A.W.S., Thesis, Ph.D. (1970) University of London.

Williams, A., Lightowlers E.C. and Collins, A.T. (1970) Impurity conduction in synthetic semiconducting diamond, *J. Phys. C.: Solid St. Phys.,* **3**(8), 1727-1735.

Woods, G.S. and Lang, A.R. (1975) Cathodoluminescense, optical absorption and X-ray topographic studies of synthetic diamonds, *J. Cryst. Growth.*, **28**, 215 - 226.

Wur, D., Davidson, I.L., Kang, W.P. and Kinser, D. (1993) A polycrystalline diamond film piezoresistive microsensor, *The 7-th International Conf. on Solid-State Sensor and Actuator's Digest of Technical Paper. Transducers.* Yokohama, Japan, July, p.722.

Wynands, H.A., Malta, D.M., Fox, B.A., von Vindheim, J.A., Fleurial, J.P., Irvine D. and Vandersande, J.(1994) Impurity-characterization agreement in type-IIb single-crystal diamond by high-temperature Hall-effect, capacitance-voltage, and secondary-ion mass-spectroscopy measurement, *Phys.Rev.*, **B49**(8), 5745-5748.

Yarbrough, A. and Messier, R. (1990) Current issues and problems in the chemical vapor deposition of diamond, *Science*, **247**(4943), 688-696.

Yarnell, J.L., Warren, J.L. and .Wenzel, R.G. (1964) Lattice vibrations in diamond, *Phys. Rev. Lett.*, **13**(1), 13-15.

Zhao, G. and Bao, M. (1989) *Chinese J. Semicond.*, **10**(9), 692-701.

Chapter 9
CHARACTERIZATION METHODS

Karen McNamara Rutledge and Karen K. Gleason
Massachusetts Institute of Technology, Cambridge MA 02139, USA

Contents

1. Overview

The industrial use of diamond stems directly from its unique properties. Characterization is the key link for understanding how these properties are related to the microstucture of diamond and to its growth conditions. Both bulk and surface defects influence the performance of diamond in a wide range of applications. For example, adhesion is critical for metal-bonded diamond in both cutting tools and electronic heat sinks and this adhesion is expected to be dramatically influenced by surface defects. However, even if good adhesion is achieved, the material must possess the desired hardness and thermal conductivity in order to be useful. These properties are often controlled by impurities, lattice defects and vacancies in the bulk of the crystal.

The techniques summarized in this chapter have been successful in elucidating the structure of diamond. Table 1 is intended only as a rough guide to information that can be provided by these different characterization methods. Nevertheless, it is clear that no single technique can provide all of the desired information. However, for all techniques,

Table 1. Comparison of Characterization Methods

section	techniques	depth probed (nm)	width probed (nm)	imaging	depth profiling	in situ	sp2 carbon	hydrogen	metals	other impurities	electronic states	vacancies	texture	disorder	usage
2.1	Raman	sample de-pendant	10^3		3	2	1							1	1
2.2	IR	bulk	10^4		3	3	2	1		No					1
3.1	ESR	bulk	bulk	3					1	many		1		2	3
3.2	NMR	bulk	bulk				1	1		F				1	3
4.1	RBS	10^3	10^6		1				1	heavy				1	3
4.2	ERS	10^3	10^6		1			1							2
4.3	SIMS	2	10^3		1				1	many					3
5.3	AES	2	100	3	1		1			many				1	1
5.1	XPS	3	10^5		2		1			many				1	1
5.2	UPS	1	10^6		2	3	2				3				3
6.1	SEM	10^3	10	1									1	2	1
6.1	EMPA	10^3	10^3	2					1	many					3
6.2	TEM	200	5	1									2	1	2
6.4	STM	1	10	1							1				3
6.5	AFM	1	1	1											3
7.1	XRD	10^4	10^6										1	1	1
7.2	ED	10	10	3									1	1	3
7.3	RHEED	1	10^5											1	3
8.1	Ellips-ometry	10	10			1	1							1	3
8.2	Mössbauer	bulk	bulk						1			1		1	3
8.3	Positron Annihilation	bulk	bulk									1		1	3

1= often 2=occasionally 3=seldom

sample care and preparation is critical. Since many impurities which effect the performance of diamond are found at very low levels, surface contamination can interfere with bulk measurements as well as with surface sensitive techniques.

More detail on the strengths and limitations of individual methods appear in the sections which follow. These sections outline the required experimental methodology and nature of the information which can be derived. While this chapter is illustrated with specific examples from the diamond literature, please refer to other chapters in this handbook for detailed discussions of the relationship between a particular type of defect and a specific property. Many of the examples represent novel uses of a characterization tool, demonstrating the potential to further exploit these techniques for extending our knowledge and control over diamond's properties and performance.

2. Vibrational Spectroscopy

Both infrared (IR) and Raman spectroscopies examine the vibrational interactions between atoms. Since these interactions occur at similar energies, the range of frequencies probed by Raman and IR are often the same. However, the nature of these two techniques is more complementary than similar. In IR spectroscopy, the incident photons are absorbed directly by the bond as phonons. In Raman spectroscopy, however, the process is one of scattering, where the incident photon contributes or receives only a small amount of energy and is scattered to a higher or lower frequency. The IR absorption process, then, is related to a change in the electric dipole moment, while the Raman interaction results in changes in polarizability. As a result of symmetry considerations, the selection rules for the two interactions differ, allowing transitions in the IR which do not occur in the Raman and vice versa. Thus, when combined, Raman and IR offer a great deal of information about the bonding and structure of materials.

2.1 INFRARED SPECTROSCOPY (IR)

2.1.1 Introduction

Fourier transform (FTIR) spectroscopy is an attractive technique for analyzing impurities in diamond since it is non-destructive and measurement times are relatively short. FTIR can be used to obtain information on hydrogen contents, carbon bonding environments, and bulk impurities. This information may be quantitative when absorption coefficients are known or can be obtained for similar materials. For example, similar absorption coefficients have been reported in the literature for alkanes, paraffinic hydrocarbons, and a-Si:C:H and may provide a basis for comparison to the an effective absorption coefficient of hydrogen-carbon bonds in diamond thin films.

The high optical transmission of natural diamond, particularly in the 8-12 μm region of the IR, makes it an attractive material for free standing IR windows and optical coatings. Polycrystalline films are particularly suited to these applications, although grain boundaries and other defects not found in single crystal diamond may limit their transparency. FTIR spectroscopy provides methods for both measuring a property (transmission) as well as for gaining insight into the defects and impurities which effect that property.

Infrared spectra are typically obtained on commercial instruments under atmospheric conditions. In the simplest experiment, the incident beam of irradiation enters the sample at an angle of 90° with respect to the surface. The transmitted energy is collected on the opposite side. Such measurements are made over a range of frequencies, typically 900 to 4000 cm^{-1} in order to obtain an IR spectrum. Samples with smooth, polished surfaces yield the most accurate results. Scattering must be accounted for when the sample has a rough surface. Contributions due to surface and internal scattering processes are often estimated using an appropriate polynomial baseline fit. The measured transmittance, T, is then converted to absorbance, A:

$$A = \ln(T_o / T) \quad (1)$$

where T_o refers to the polynomial fit.

The area probed in IR spectroscopy varies from a few microns to a few millimeters, depending on whether a microscope is used to focus the incident beam or not. Direct transmission measurements can be made on freestanding diamond samples in a matter of minutes. Although the thickness which can be examined varies with absorption coefficient, single crystals and most CVD diamond have sufficiently low absorption that thickness does not become an issue. In order to examine films which remain attached to a foreign substrate, reflection techniques must be used.

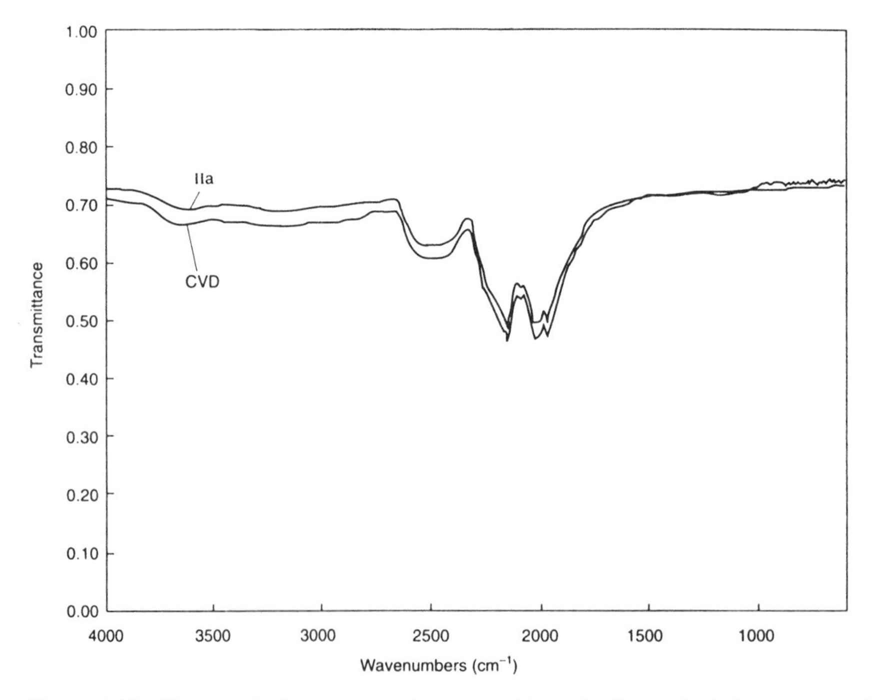

Figure 1. The IR transmission spectrum for a natural type IIa diamond window compared to that of a state-of-the-art polycrystalline CVD sample. [McNamara 1994a]

2.1.2 Applications to Diamond

A typical IR spectrum for a type IIa natural diamond is shown in Figure 1 [McNamara 1994a]. Note the intrinsic absorption in the two-phonon range between 1333 and 2666 cm^{-1} which is present in all diamond. In figure 1, the transmission in the 8-12 μm region (850-1250 cm^{-1}), is ~70%, approaching the theoretical maximum for diamond. However, diamonds containing defects often show additional absorption in the symmetry forbidden one-phonon region below 1333 cm^{-1} and in the region between 2750 and 3300 cm^{-1}, known as the CH-stretch region. The CH-stretch absorptions are particularly prevalent in CVD diamond, while the nitrogen and vacancy related absorptions of type I natural and synthetic diamonds are observed in the one-phonon region.

Unlike the spectrum of a high-quality CVD diamond film of figure 1, which compares favorably with that of natural diamond, most CVD spectra show additional features. The full infrared spectra of a representative CVD film is shown in figure 2 [McNamara 1994a]. It shows absorptions in the one, two, and three phonon regions. While the two phonon absorption is intrinsic to pure diamond, it is the absorptions outside of this region which yield information about the impurities and defects in diamond. While varying in intensity with respect to the two phonon absorption, these one and three phonon absorptions appear at the same characteristic frequencies in the spectra of many diamond samples.

The CH-stretch region. A large number of the recent infrared studies on diamond, particularly CVD diamond, have concentrated on the absorption in the CH stretch region of the IR [Hartnett and Miller 1990, Xin Bi 1990, Zhang 1992, McNamara 1992c, 1992d, 1994a and 1994b, Dischler 1993]. The stretching vibrations of carbon-hydrogen bond occur in this region (2750-3300 cm^{-1}), providing information about hydrogen which is covalently bonded to carbon. Note that free hydrogen atoms or H_2 molecules do not give rise to absorptions in the CH-stretch region.

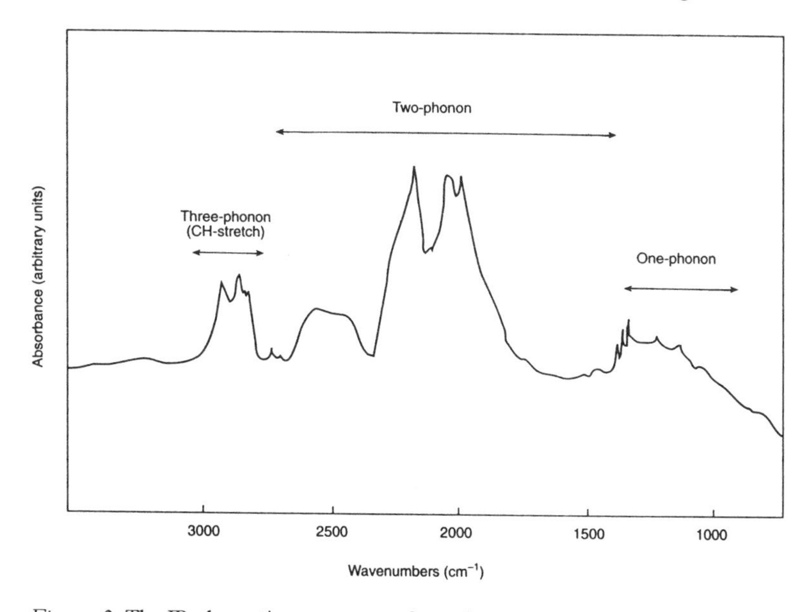

Figure 2. The IR absorption spectrum of a typical CVD diamond system. [McNamara 1994a]

The frequency of the C-H bond vibrations shift slightly in response to differences in local environments. For example, hydrogen bonded to sp^2 bonded carbon appears above 2950 cm^{-1}, while that associated with sp^3 bonded carbon below 3000 cm^{-1} [Bellamy 1980, Socrates 1980]. Thus, the CH-stretch region of the IR spectrum yields information about the carbon bonding environment as well as that of hydrogen. Indeed, absorptions at ~3025 cm^{-1}, related to hydrogen bonded to sp^2 carbon, have been observed in diamond films described as being of "poor quality" [Zhang 1992, Dischler 1993, McNamara 1994a].

The number of hydrogen atoms bonded to a single carbon will also cause variation in the frequency of the stretch vibration (Table 1) [McNamara 1992c, Dischler 1993]. The dominant absorptions in the CH-stretch region typically appear near 2850 cm^{-1} and 2920 cm^{-1} [Hartnett and Miller 1990] and are indicative of the symmetric and asymmetric stretching of sp^3-CH_2 groups. In many diamond films, the absorption intensity for the symmetric and asymmetric absorptions of sp^3-CH_2 groups are clearly unequal [McNamara 1994a]. The somewhat less intense absorptions present in the spectrum at 2880 cm^{-1} and 2960 cm^{-1} are similarly related to the symmetric and asymmetric stretching vibrations of sp^3-CH_3 groups.

Two other absorptions have also been observed in CVD diamond in a number of studies, but their origin has remained the subject of debate[McNamara 1992c, 1994a, 1994b, Zhang 1992, Dischler 1993]. These absorptions appear at frequencies below those normally observed for CH stretching vibrations in alkanes and amorphous carbon, specifically 2820 and 2833 cm^{-1}. They have been observed in films produced by microwave, hot-filament, and DC arc-jet deposition. One possibility is that the peak at 2820 cm^{-1} is related to the hydrogen terminated <111> surface of diamond [Dischler 1993]. Others suggest that the absorptions at 2820 and 2833 cm^{-1} can be attributed to nitrogen and oxygen related defects [Zhang 1992, McNamara 1994a and 1994b]. These peaks may be the CH stretch vibrations of the carbon-hydrogen bond in an N-CH_3 and O-CH_3 group, respectively. Observation of these groups is well documented in the organic chemistry literature and is considered positive identification of such a group, as no other characteristic CH-stretch absorptions appear at such low frequencies [McNamara 1994a and 1994b].

Least squares fits to the experimentally measured CH-stretch absorption, such as shown in figure 3, can determine the CH_x species present [Dischler 1993, McNamara 1994a]. However, it should be noted that there are a large number of fitted parameters, and the fits do not provide quantitative information. It is possible to decrease the number of fitted parameters by fixing positions, widths and/or intensity ratios of the individual components [Dischler 1993]. However, the ratio of peak intensities for the symmetric and asymmetric line pairs should not be assumed, since the diamond environment is ill-defined, and it is possible for this ratio to vary considerably [McNamara 1994a].

Hydrogen content. The integrated intensity of the C-H stretch region correlates linearly with hydrogen concentrations obtained by nuclear reaction analysis [Hartnett and Miller 1990] and by nuclear magnetic resonance (NMR) [McNamara 1992c and 1994a]. These comparisons determined quantitative absorption coefficients for CH bonds in diamond, allowing the calculation of the total hydrogen content in these films from the integrated

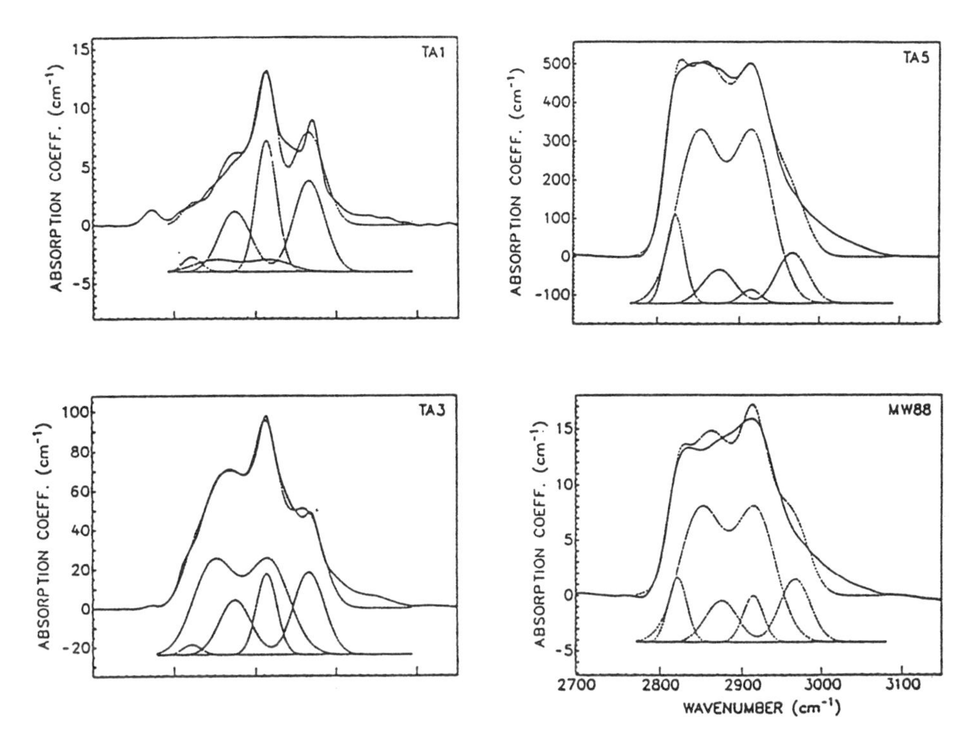

Figure 3. Least squares fit to the experimentally observed CH-stretch absorption assuming constant peak widths, central frequencies, and intensity ratios [Dischler 1993].

area under the CH stretch absorption. The calculated absorption coefficient, $4.3 \pm 0.8 \times 10^3$ l mol^{-1} cm^{-2} represents an effective absorption coefficient for the entire CH-stretch region and is consistent with values for C-H bonds in long-chain alkanes and paraffinic hydrocarbons. This agreement between the IR, which detects C-H_x bonds, and methods which measure total hydrogen content suggests that the majority of hydrogen in CVD diamond is covalently bonded to the lattice.

The C-H stretch absorption coefficient is no longer accurate for films with greater than 0.2 at. % hydrogen [McNamara 1992a and 1994a]. Such films often containing sp^2 C-H groups (absorptions above 2950 cm^{-1}), which may have different absorption coefficients from the sp^3 C-H bonds [McNamara 1994a].

One Phonon Region. The presence of oxygen and nitrogen is important in CVD diamond because these impurities are known to cause absorptions in the 8-12 μm wavelength region, where transparency is critical for many applications. Additional evidence of both oxygen and nitrogen incorporation in the films can be obtained on examination of this normally symmetry-disallowed one-phonon region. An expansion of the IR spectrum for a CVD diamond film, between 1000 cm^{-1} and 1600 cm^{-1}, is shown in figure 4 [Klein 1991]. There are a number of absorptions present, many of which can be related to oxygen and nitrogen containing groups [McNamara 1994a] as well as to vacancies and other point defects [Klein 1991]. Klein [1991] gives a comprehensive examination of these defects in CVD diamond. The basis for this work

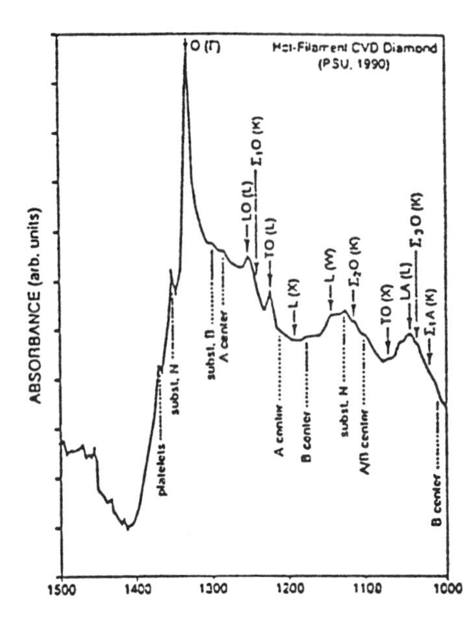

Figure 4. Absorption in the symmetry disallowed 1-phonon region for a CVD diamond film. [Klein 1991]

are the absorption studies for this region in both natural and synthetic diamonds [Woods 1983, Chrenko 1975].

IR Emission Spectroscopy.
Both transmission and reflection FTIR spectroscopies are performed using an incident beam of IR irradiation which interacts with the sample, being absorbed, transmitted, reflected, or since these techniques require a directed beam and specific geometries, they are typically performed ex-situ. However, infrared irradiation will be emitted from a diamond sample upon sufficient heating without any incident radiation. Infrared emission spectroscopy takes advantage of this fact in both the ex-situ and in-situ study of externally heated diamond films. For example, Spiberg et al. [1993] used the technique to correlate ex-situ IR emission of heated diamond films with their room temperature thermal conductivity. Assuming high thermal conductivity as the property, they were then able to use the in-situ emission of films grown by DC arc-jet CVD as an on-line monitor of film quality. IR emission is particularly attractive as an in-situ technique in that it does not perturb the growth environment and requires only detection equipment external to the growth environment.

2.2 RAMAN SPECTROSCOPY

2.2.1 Introduction

Many of the potential applications for polycrystalline diamond films require high quality, defect-free diamond. Consistent reproduction of such high quality material requires an understanding and definition of "high quality". While many techniques have been used to evaluate diamond film quality, few have been applied so extensively as Raman spectroscopy. This section explores the basics of Raman spectroscopy and its applications in diamond.

Raman spectroscopy is one of the most common methods of diamond film characterization because of the ability to distinguish between different forms of carbon [Knight and White 1985, Celii and Butler 1991, Bachmann and Wiechert 1992]. Raman spectroscopy is non- destructive and is readily available to many laboratories. The observed 'Raman' spectra are often the sum of contributions from the Raman scattering of the excitation photons, as well as fluorescence/photoluminescence from states excited in the material, and resonance Raman contributions due to nearby electronic states [Clark and Dickerson 1992]. This latter effect is particularly important when experiments are performed with photon energies exceeding the band gap of components in the material analyzed [Wagner 1991]. Care must be taken in assigning bands in any single spectrum because of these interfering effects.

A simple, although seldom explored, method for distinguishing Raman scattering from fluorescence/photoluminescence is to change the energy of the incident photon source and observe changes in the inelastic energy loss shift or intensity of the band. An example of this technique is found in figure 5 from McNamara [1992b] where the Raman spectra of a diamond film at two different incident energies are shown. Only the features with a constant peak position are attributed to Raman effects.

Raman spectroscopy is used to investigate quantized molecular resonances by observing transitions between vibrational energy levels caused by inelastic scattering of high energy visible photons. When a photon is scattered inelastically in an interaction with lattice phonons, it may gain or lose energy causing it to be scattered to a higher or lower frequency, respectively. Those energy levels found below the original frequency are traditionally called Stokes frequencies, while those scattered above the original frequency are called anti-Stokes. Since the original population of molecular energy levels follows a Boltzmann distribution, the intensity of the Stokes frequencies is expected to be much greater than that of the anti-Stokes. In practice, observed Raman spectra are almost always the Stokes lines.

2.2.2 Raman Spectra of Carbon

For diamond, the single phonon (first order) Raman spectra produced using visible photons (e.g. 514.5 nm from the Ar+ ion laser) gives a single band at 1332 cm^{-1} [Solin & Ramdas 1970; Knight and White 1985; Schroder 1990]. Generally, Raman scattering immediately above the diamond one-phonon band is assigned to sp^2 carbon containing materials [Schroder 1990]. Three generally distinguishable types of sp^2 carbon include: crystalline graphite (a single band at 1580 cm^{-1}) [Tuinstra and Koenig 1969, Knight and White 1985, Schroder 1990], defective or microcrystalline graphite (two broad bands at ca. 1580 and 1350 cm^{-1}) [Wright 1976, Knight and White 1985], and amorphous carbon (a broad asymmetric band peaking at around 1500±40 cm^{-1})[Wright 1976, Knight and White 1985; Dines 1991]. An additional band is occasionally observed at ca. 1120 to 1140 cm^{-1} [Schroder 1990, Liou 1990, Dines 1991], which has not been definitely assigned. The effective scattering efficiencies vary considerably, due in part to resonant Raman effects. [Wagner 1991]. For example, crystalline graphite has a scattering efficiency ~50 times greater than that of single crystal diamond [Knight and White 1985]. A typical multi-component Raman spectrum from a diamond film is shown in figure 6 [Knight and White 1985].

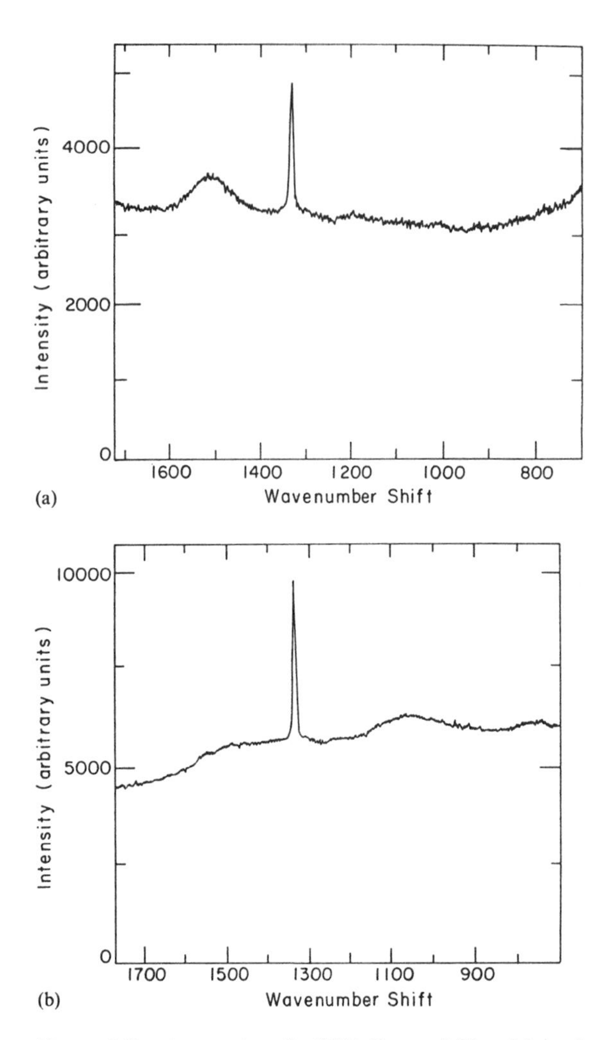

Figure 5. Raman spectra of a CVD diamond film obtained with an incident source at 514.5 and 488 nm, respectively. [McNamara 1992b]

2.2.3 Peak Position

Isotope Effects. Since the Raman shift is related to the inverse of the reduced mass, the position of the Raman band for diamond, as well as that of the other forms of carbon, is dependent on the isotopic composition [Chrenko 1988]. At the natural abundance of the ^{13}C isotope, 1.1 %, the diamond band appears at its characteristic 1332 cm-1 position. However, in pure ^{13}C diamond, this diamond band shifts to 1280 cm^{-1}. Raman lines of all of the carbon bonding environments in CVD diamond are expected to shift to lower wave numbers as the ^{13}C concentration increases. The isotopic shift of the Raman peaks has been exploited in number of studies on the kinetics of CVD diamond growth [Chu 1991; McNamara and Gleason 1992a and 1993].

Temperature. The Raman shift observed for diamond can vary with sample temperature. Bachmann and Wiechert [1992] found this shift to be linear with temperature for CVD diamond between 300 and 700 K, while Zouboulis and Grimsditch

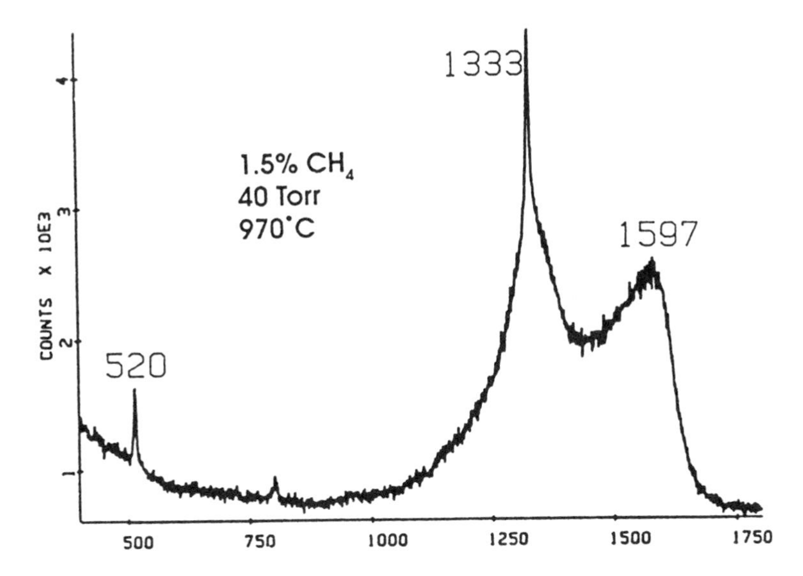

Figure 6. Raman spectrum from a CVD diamond film containing multiple carbon bonding environments. [Knight and White 1985].

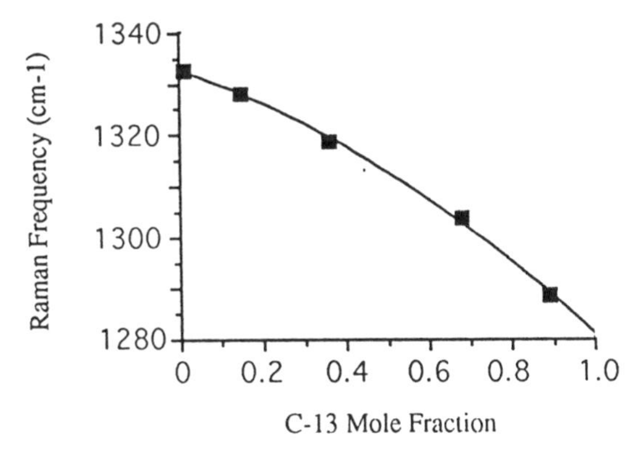

Figure 7. Peak position of the diamond 1-phonon Raman line as a function of isotopic ^{13}C enrichment. Plot of data from Chrenko [1988].

[1991] found non-linear heating effects below 1000 K and a linear change in shift above that temperature for single crystal diamond. The temperature dependence of the Raman shift is particularly important in microRaman spectroscopy where high power densities may result from the relatively small sampling area and cause local sample temperatures to increase during spectral acquisition [Hannsen 1991].

Stress. Raman peak position may also vary as a function of film stress [Gheeraert 1992, Gupta 1989, Windischmann and Epps 1992]. For the 1332 cm^{-1} diamond line, both positive and negative wavenumber shifts have been observed, corresponding to tensile stress and compressive stress, respectively. A uniaxial stress on single crystal diamond is expected to split the triply degenerate optical phonon into a singlet and a doublet with different linear stress coefficients, however such splitting has not been observed in CVD diamond and some natural diamond [Gupta 1989]. The absolute value of the total stress in diamond films is typically below 0.5 GPa, corresponding to Raman shifts of a few wavenumbers [Gheeraert 1992, Windischmann and Epps 1992]. In addition, Windischmann and Epps [1992] have suggested that the stress measured by Raman shift is the result of a combined intrinsic tensile stress and a thermal compressive stress. Raman spectroscopy has also been used to monitor diamond growth and stress development in situ [Bernardez and McCarty 1993].

Other. Disorder within a bonding environment may also contribute to Raman peak shift [McNamara 1992]. In addition, the position of the 1350 cm^{-1} band of disordered graphite has also been observed to shift with changes in the excitation photon energy [Wagner 1991].

2.2.4 Linewidth

The width of the diamond Raman peak can be affected by the properties of the material and has been related to disorder in the sp^3 bonding environment. For example, Fabisiak [1993] observed an increase in the linewidth of the 1332 cm^{-1} line width with increasing ESR signal. The full-width at half-maximum of the 1332 cm^{-1} peak in type IIa diamonds is on the order of 2 cm^{-1}, while polycrystalline diamond films can demonstrate linewidths between 2 and 14 cm^{-1}. The width of the Raman can also be affected by grain size and sample temperature as well [Bachmann and Wiechert 1992].

2.2.5 Resolution and Limitations

With standard optics, the lateral resolution of most Raman instruments is of the order of 0.1 to 1 mm. However, using microscope optics, resolution of less than 1 μm can be achieved and a significant increase in signal collection also results from using high numerical aperture optics [Oakes 1991]. As mentioned, care must be taken in micro-Raman spectroscopy to avoid local heating or burning of the samples due to high laser power density [Hannsen 1991]. Micron-resolution allows Raman spectra to be obtained from individual crystals within a polycrystalline CVD diamond film. Recent studies have demonstrated that such Raman spectra may vary with both crystallographic orientation and polarization of the incident irradiation in CVD diamond [Stuart 1993 and 1994]. In each case, changes in both linewidth and position of the diamond Raman line were observed.

Quantitative analysis of inhomogeneous or multicomponent carbon materials with Raman spectroscopy is difficult because of the varying contributions of each component to the penetration of the exciting photon beam into the sample and the escape of the scattered photons [McNamara 1992b, Stuart 1993 and 1994]. However, the qualitative use of Raman spectroscopy has been invaluable to the rapid development of diamond chemical vapor deposition (CVD) technology [Oakes 1991].

Raman has the advantage of giving fast, spatially resolved information and requiring limited sample size. Care must be taken not to underestimate the sp^3 nature of films showing strong amorphous carbon absorptions (~1350 cm^{-1}). Since the Raman scattering efficiency of sp^2 bonded carbon greatly exceeds that of sp^3 bonded material [Knight and White 1985], it may be that the sp^3 signal is simply obscured by the strong amorphous carbon peak. It is also possible that a disordered sp^3 phase does not allow

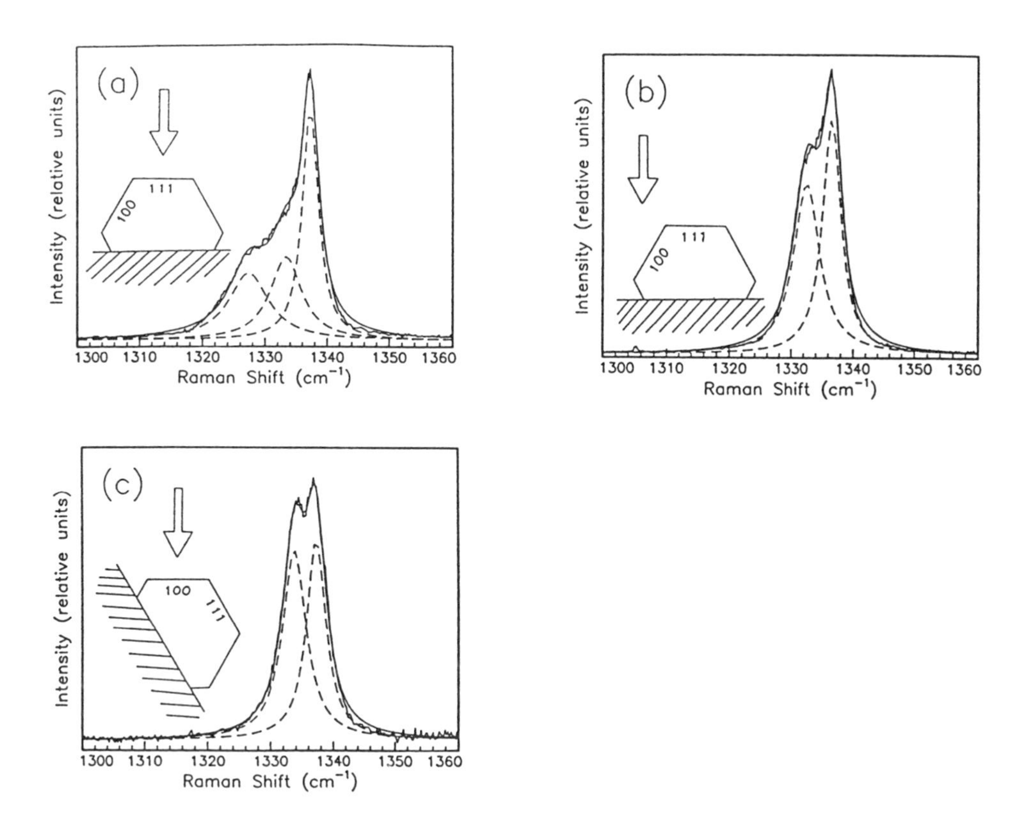

Figure 8. Raman spectra of diamond as a function of crystallographic orientation [Stuart 1993].

the 1332 cm^{-1} transition, leaving disordered sp^3 bonded carbon undetected in the Raman spectrum [Sato 1986]. This has serious implications for quantitative evaluation of film quality using Raman spectroscopy. In some cases, the sp^3 content of the film may be much higher than that anticipated from the Raman spectrum [McNamara 1992b]. In addition to differences in scattering behavior, the sp^2 and sp^3 bonded carbon may have different absorptivities which can lead to non-uniform sampling as a function of depth in the film. These and other sampling difficulties have been discussed previously [McNamara 1992b, Nemanich 1991].

2.2.6 *In-situ Techniques*

Raman spectroscopy has recently been used as an in-situ monitor of diamond film growth in both hot-filament (HF) and microwave (MW) CVD environments [Bernardez 1993; Fayette 1994a and 1994b]. A gated multichannel detection system is used to differentiate Raman signals from background emissions in the reactors. Such systems have been used to monitor changes in the linewidth of the 1332 cm^{-1} diamond peak, the evolution of sp^2 carbon and stress development in diamond films as a function of deposition conditions [Bernardez 1993; Fayette 1994a and 1994b].

3. Magnetic Resonance

3.1 ELECTRON SPIN RESONANCE (ESR)

3.1.1 Introduction

ESR (also called electron paramagnetic resonance, (EPR))detects defects which have an unpaired electron. In diamond, unpaired electrons are often localized at impurity atom sites or at lattice defects such as vacancies. Detailed information can be obtained about the identities of the atoms and symmetry which surrounds the unpaired electron. Paramagnetic defects in diamond are often also optically active. Loubser and van Wyk [1978] present an excellent review of the basic principles of ESR spectroscopy and its application to the study of natural and HPHT diamonds. More recently, ESR has been used to characterize polycrystalline CVD diamond films.

An ESR peak is observed when microwaves are applied at the resonance frequency of the unpaired electron in an external magnetic field. This resonance frequency is proportional to g, the spectroscopic splitting factor. For a free electron, g=2.0023. Shifts from this value are indicative of the orbital contribution to magnetic momentum of the defect. Anisotropy in the g-value, evidenced by the orientational dependence of the ESR spectra, provides information on the symmetry of a defect and thus can be used to probe the local distortion in lattice geometry at the defect site.

Nuclei having angular momentum (^{1}H, ^{13}C, ^{14}N, ^{31}P, etc., see also NMR section) which are in the proximity of the paramagnetic defect can often be identified as a result of hyperfine interactions. The strength and orientation dependence of hyperfine couplings depend on the location of the nuclear spin (or spins) with respect to the unpaired electron.

The ability to resolve g values and hyperfine coupling constants depends on the linewidths in the ESR spectra. Extremely small ESR linewidths (0.04 G) have been obtained from an isopure ^{12}C diamond single crystal with a low nitrogen content (Fig. 9)[Zhang 1994]. The identical defect in a CVD diamond displayed an ESR linewidth of 0.21G [Hoinkis 1991]. Higher concentrations of nuclear spins (^{13}C, ^{1}H etc.) and paramagnetic defects in the film lead to the broader linewidth. For the same defect in HPHT diamond, linewidths of greater than10 G can result due to the high concentration of magnetically active defects and impurities [Singh 1990].

Paramagnetic defect concentrations can be determined from the integrated intensity of the corresponding ESR peak. Sensitivity limits of $\sim 10^{10}$ cm^{-3} have been achieved for room temperature ESR measurements and improve inversely with temperature.

3.1.2 Impurity Related Defects

Defects are named by a letter followed by a number which denote the laboratory and order of discovery. An isolated impurity nitrogen atom, substituted for carbon in the diamond lattice, is known as a P1 center. The resonance frequency of a P1 center

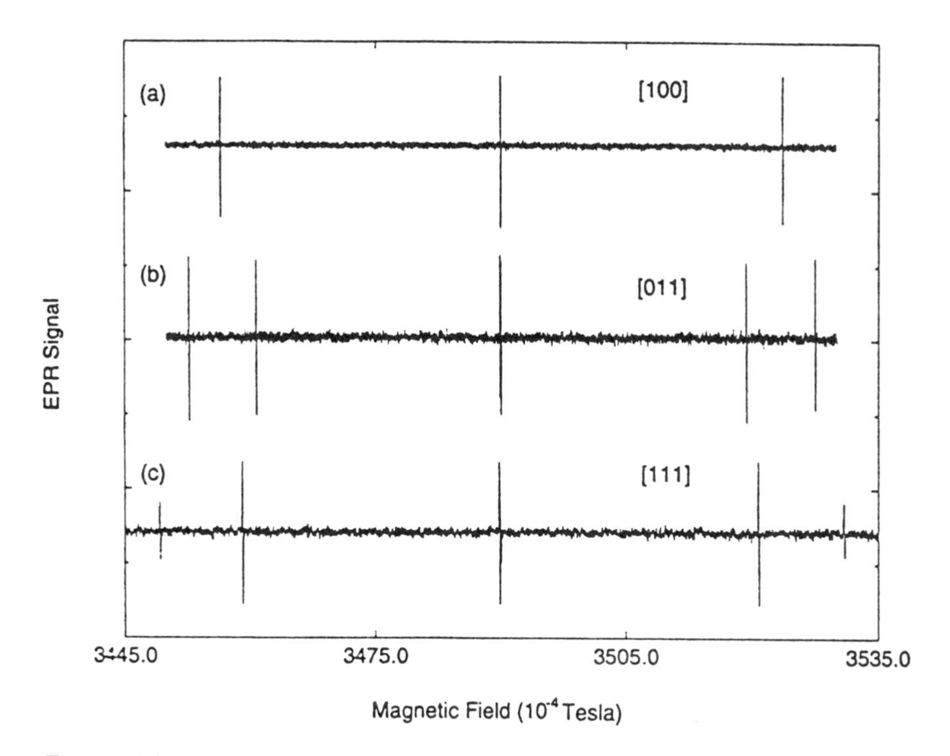

Figure 9. High-resolution EPR spectra obtained for each of the three high symmetry directions of a ^{12}C isopure single crystal diamond aligned parallel to the external magnetic field. The differences between the three spectra reflect the hyperfine anisotropy of the P1 center. Expansion of the center peak (not shown here) also revealed g-tensor anisotropy [Zhang 1994].

corresponds to g=2.0024, close to the free electron value. The hyperfine interaction of the unpaired electron with ^{14}N, a spin-1 nucleus, produces a characteristic splitting in the ESR spectra (Fig.9). Analysis of the axially symmetric hyperfine couplings reveals that the unpaired electron occupies an antibonding orbital pointed along the C-N bond direction and is localized primarily on the carbon atom. A variety of other ESR active defects involving groupings of two or three nitrogen atoms have also been catalogued [Loubser and van Wyk 1978].

The P1 center has been observed in various types of diamond at different concentration levels: type IIa, the purest natural diamond ($\sim 10^{18}$ cm^{-3}), HPHT grit ($\sim 10^{20}$ cm^{-3}) and microwave CVD films ($\sim 10^{17}$ cm^{-3}) [Hoinkis 1991]. No additional P1 centers are formed by a 1 hour anneal in N_2 at 1000°C.

Like nitrogen, boron has a high solubility in diamond. Broad boron-related ESR lineshapes have been observed in natural and HPHT diamonds [Loubser and van Wyk 1978] and in polycrystalline films [Jin 1994]. Phosphorus-related ESR active defects have also been observed in HPHT diamond [Samonenko 1992].

ESR has also been used to study metallic impurities such as iron, nickel and cobalt in HPHT diamond [Loubser and van Wyk 1978]. Copper has also been observed [Singh 1990]. These defects are generally not observed in natural diamonds. Ferromagnetic impurities from HPHT catalysts can significantly broaden the ESR lineshape, making it difficult to resolve structural features [Loubser and van Wyk 1978].

Variation in substitutional nitrogen and nickel concentrations within a HPHT crystal have been imaged by microwave scanning ESR [Furusawa and Ikeya 1990]. Lateral resolution of 0.5 mm was achieved, and defects within 0.2 mm thickness of the crystal surface are observed. Of the (111) facets, some were nickel-rich while all had high nitrogen contents. ESR imaging, which is analogous in principle to NMR imaging (known as MRI), was used to map out the two dimensional spatial distribution of nitrogen defects in diamond crystals with 100 to 400 μm resolution [Zommerfelds and Hoch 1986].

3.1.3 Damage Centers

Irradiation with energetic particles can induce vacancy and interstitial formation in diamond [Loubser and van Wyk 1978]. Subsequent atomic rearrangement can result in multivacancy formation. ESR active centers also appear after high temperature annealing. Increased thermal energy allows vacancies to move away from impurity traps. The trapping of mobile vacancies can give rise to give vacancy chains of various length and may also produce vacancy loops.

Mechanical deformation can also alter the ESR spectra of diamond [Loubser and van Wyk 1978]. Plastic deformation can cause dislocations to slip such that a nitrogen atom is displaced, producing a W7 center which gives the diamond a brown color. Grinding diamond can produce vacancy rows (O1 centers), remove nickel (W8 centers), and/or produce dangling bond centers($g=2.0026 \pm 0.0002$). The dangling bond center intensity increases with decreasing particle size, presumably indicating this defect was forming on the external surface of the diamond crystallites. A similar dangling bond center in crystalline silicon and germanium is associated with internal/external surfaces or amorphous regions [Erchak 1993]. In fact, the ESR measured silicon dangling bond concentration of substrates was shown not to affect the rate of diamond nucleation [Dennig 1992]. The silicon dangling bond concentrations were varied by abrasion and etching of the wafer.

Ion implantation of diamond by either boron, carbon, or nitrogen gives ESR spectra which represent the superposition of several defect types [Loubser and van Wyk 1978]. At large ion doses, dangling bond defects are dominant, indicating the formation of amorphous regions.

3.1.4 Applications to Polycrystalline CVD Films

The spectra of CVD films represent the superposition of lines from a variety of defects. Some may arise from nitrogen [Watanabe and Sugata 1988, Hoinkis, 1991] and metal impurities incorporated during deposition. Dangling bond defects, like those produced by grinding and ion implantation of natural and HPHT diamond (see also damage centers), are observed in as-deposited CVD diamond films. The dangling bonds and nitrogen-related P1 centers have similar g-values but have been distinguished by performing ESR at different microwave power levels. The P1 centers saturate at low microwave powers while dangling bond defects saturate at high power [Jin 1994].

Dangling bond defects are often assumed to dominate the ESR spectra of CVD diamond films. Their concentration, N_s, ranges from 10^{17} to 10^{19} cm^{-3} [Watanabe and Sugata 1988, Fabisiask 1993, Fanciulli and Moustakas 1993]. Decreased N_s correlates with improved Raman spectra [Wantanabe and Sugata 1988, Fabisiask 1993] and increased grain size [Fabisiask 1993]. Hydrogen plasma treatment has been noted to decrease N_s [Erchak 1992]. Annealing in nitrogen below ~600°C reduces N_s indicating structural relaxation, while higher temperature anneals increase N_s, perhaps due the evolution and/or rearrangement of hydrogen [Zhang 1992].

Linewidths of dangling bond defects in polycrystalline diamond range from 1.9 to 12 gauss [Watanabe and Sugata 1988, Erchak 1992, Zhang 1992, Fabisiask 1993 and Fanciulli and Moustakis 1993]. The smaller linewidths are observed in films with lower dangling bond densities, consistent with the reduction of dipole-dipole interactions between randomly distributed paramagnetic centers [Erchak 1992].

Shoulders are often observed on the dangling bond center lineshape (Fig.10), varying in intensity from film to film [Watanabe and Sugata 1988, Fanciulli and Moustakis 1993, Holder 1994, Jin 1994]. Similar lineshapes have also been seen in amorphous hydrogenated carbon films [Miller and McKenzie 1983]. The shoulders are absent in polycrystalline diamond grown from deuterated reactants [Jin 1994]. Experiments at different external magnetic fields (Fig.10) show these shoulders do not represent the superposition of two different lines. Rather, they are satellite transitions from the simultaneous microwave flipping of an unpaired electron and a proton which are coupled through dipole-dipole interaction [Holder 1994]. No first order hyperfine splitting from ^1H is observed. Thus, the unpaired electron is not localized at a proton, but is most likely localized at a carbon vacancy with one or more nearby hydrogen atoms [Holder 1994].

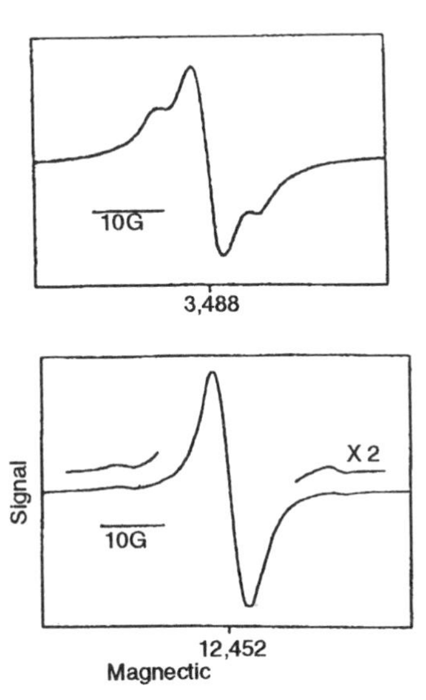

Figure 10. Magnetic field dependence of the ESR spectrum of a dangling bond defect in a polycrystalline film. The shoulders on the central feature which are quite strong at 9.5 GHz (top), are barely resolved at 35 GHz (bottom) [Holder 1994].

3.2 NUCLEAR MAGNETIC RESONANCE (NMR)

3.2.1 Introduction

Solid-state NMR is sensitive to the concentration and local bonding environment of nuclei possessing a magnetic moment. Thus, while ^{12}C is undetectable by NMR, quantitative sp^2 to sp^3 bonded carbon ratios have been determined by ^{13}C NMR of natural abundance and isotopically enriched diamond. Quantitative measurements of impurities such as ^{1}H and ^{19}F, averaged over the entire sample volume, can also be obtained. Proton NMR is particularly valuable for measuring hydrogen concentrations, since this mobile, low Z element is difficult to quantify by other spectroscopies. In addition, NMR is able to distinguish uniform versus segregated distributions of an element. For instance, NMR has shown that hydrogen is clustered at internal surfaces, such as grain boundaries and voids, in polycrystalline diamond. Temperature dependent measurement can be used to determine the mobility of various bonding configurations.

In the NMR experiment, an external magnetic field is used to produce an energy difference in the states of the nuclear spins. Radio-frequency waves are applied to measure the resonance frequencies corresponding to the energy splitting. This low-energy irradiation is unlikely to result in chemical alteration of the diamond samples, unlike some higher energy spectroscopies. More detailed reviews concerning the spectroscopic aspects of NMR as applied to materials science [Brundle 1992] and to diamond [Gleason 1995] are available.

Bonding environments are generally differentiated through isotropic chemical shift differences. The resulting frequency shift of a peak with respect to a compound is linearly proportional to the magnetic field strength in which the sample resides. Thus, by using higher magnetic fields, chemical shift resolution of different bonding environments is improved. For ease of comparing spectra obtained at different magnetic fields, peak positions are generally expressed in parts per million rather than absolute frequencies.

In contrast to peak positions, the NMR linewidths are often reported directly in frequency units. These linewidths reflect a number of NMR interactions. Dipolar broadening increases with the density of NMR active nuclei and is independent of the applied magnetic field. Broad lines can also result from anisotropic bonding configurations (e.g. sp^2 carbon). Magic-angle spinning (MAS) removes broadening by both of these mechanisms provided the linewidth is smaller than the attainable sample rotation rate of several kHz. However, broadening which reflects the disorder within a given bonding environment cannot be removed by MAS. Line broadening can also result from interactions of the NMR active nuclei with paramagnetic and ferromagnetic defects. Spatial variations in the magnetic susceptibility within the sample can also lead to increased linewidths.

Diamond single crystals have been studied by NMR [Henrichs 1984, Hoch and Reynhardt 1988]. Macroscopic sections of diamond films, either free-standing or intact on their substrates, can be used for either static [McNamara 1992a] or MAS [Merwin 1994] NMR experiments. Powdered samples, formed by crushing or removal of a film from its substrate, necessitate the use of a sample holder, a potential source of background signals. Typically greater than 5 x 10^{18} ^{13}C nuclei are required [McNamara 1992b], corresponding to ~10 mg of a natural abundance diamond, equivalent to a 10 μm thick film over an area of 2.8 cm^2. ^{13}C enrichment can be used to decrease sample size requirements of HPHT [Hoch and Reynardt 1988] or CVD films [McNamara 1992a, 1992b, 1993, Lock 1992]. However, enrichments of greater than 25% will limit spectral resolution of sp^2 versus sp^3 carbon due to homonuclear dipolar line broadening. Due to their higher magnetic moments, ^1H and ^{19}F detection requires only greater than 10^{17} total nuclei [Levy and Gleason 1993]. At these sample sizes, care must be taken to minimize background nuclear signals arising from the NMR probe head and sample.

For quantitative measurements, it is important to insure that a sufficient spin lattice relaxation (see section 3.2.6) occurs between signal acquisitions and that the effects of broadening by paramagnetic and ferromagnetic impurities are below the NMR detection limits [McNamara 1992b]. Note that the range of paramagnetic defect concentrations (see section 3.1.4) corresponds to only ~0.0001 to 0.01 at.%, significantly lower than the concentrations of nuclear environments usually being measured by NMR.

3.2.2 Carbon-13

Bulk diamond is characterized by a single resonance at 36±2 ppm relative to tetramethylsilane [Retcofsky and Friedel 1973, Wilkie 1983]. This sp^3 bonded carbon peak is easily resolved from peaks from sp^2 carbon which appear between 120 to 200 ppm. Lower values are observed for amorphous carbon films [Petrich 1989] while soot gives a peak centered at 189 ppm [McNamara 1992b].

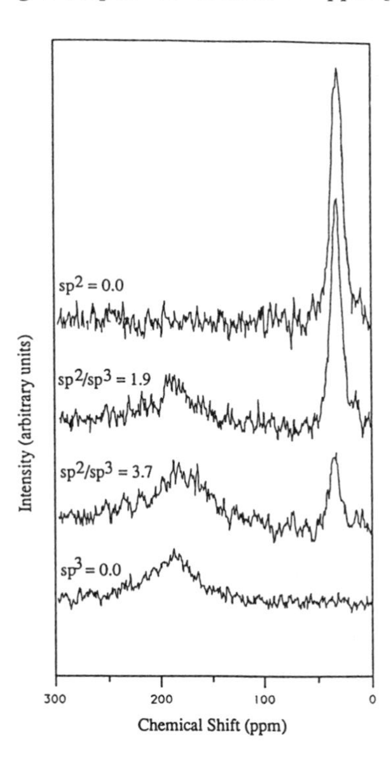

Figure 11. ^{13}C NMR spectra for four mixtures of natural diamond with soot. The ratio of the two NMR peak areas agrees within experimental error with the known w/w ratio of the two components [McNamara, 1992b].

Figure 11 shows that the ratio of sp^2/sp^3 bonded carbon can be quantitatively determined from the ratio of the integrated peak areas of the direct polarization ^{13}C NMR spectra [McNamara 1992b].

Proton decoupling is not required for crystalline diamond since the hydrogen content is low. The low proton concentration also means that neither quantitation nor enhanced sensitivity is obtained using cross-polarization of ^{13}C by ^1H, a technique which is commonly employed for other solids.

The sp^2 bonded carbon in visibly faceted films produced via hot-filament, DC arc-jet, or microwave plasma CVD is generally below the ^{13}C NMR detection limit [McNamara 1994a] An sp^2 resonance has been observed in a film of poor morphology and no detectable diamond one-phonon Raman absorption. The intensities of this peak relative to the sp^3 resonance indicate that 11% of the carbon in this film is sp^2 bonded. Thus, Raman spectroscopy may be so sensitive to sp^2 bonded carbon that sp^3 bonded carbon in films containing as much as 90% sp^3 bonded carbon, this tetragonal configuration may remain undetected (Fig. 12).

This is undesirable when new deposition conditions are being explored, since conditions which yield a majority of sp^3 bonded sites can be overlooked by the Raman analysis. However, Raman spectroscopy can identify small amounts of sp^2 carbon which are well below NMR detection limits.

The isotopic abundance of ^{13}C can be determined via the homonuclear dipolar contribution to the sp^3 linewidth. These measurements are in excellent agreement with ^{13}C enrichments determined through the isotopic shift of the diamond one-phonon absorption in the Raman spectra (see also section 2.1.3) [McNamara 1993].

For natural abundance CVD films, the static spectra shows broadening in addition to homonuclear dipole interactions. MAS at sufficient speeds can narrow the linewidths due to dipolar broadening, chemical shift anisotropy and variations in magnetic susceptibility [McNamara 1992a]. Thus, any remaining broadening can be attributed to the heterogeneity of bond angles and bond lengths in the film. Variations in MAS linewidths from 6 to 55 kHz in natural abundance samples have been linearly correlated to increases in Raman one-phonon linewidth, which is also associated with increasing disorder of the diamond phase [Merwin 1994].

Dynamic nuclear polarization (DNP) has also been used to characterize natural and HPHT diamond [Duijvestijn 1985, Wind 1985] and CVD films [Loch 1992 and 1993]. In DNP, polarization transfer from paramagnetic defects (see also section 3.1.4) is used to enhance the NMR signal of nearby ^{13}C nuclei. This advantage is somewhat offset by the fact that DNP is generally performed at lower magnetic fields than conventional NMR. In addition, the sp^2/sp^3 ratio cannot be quantified by DNP because the polarization dynamics of these samples is not well known.

DNP has identified a small broad shoulder on the primary sp^3 peak, at ~45 ppm as hydrogenated sp^3 bonded carbon and a sharp peak at 28 ppm as freely rotating methyl groups. Only one sample gave a DNP spectra with an sp^2 resonance (~140 ppm). The

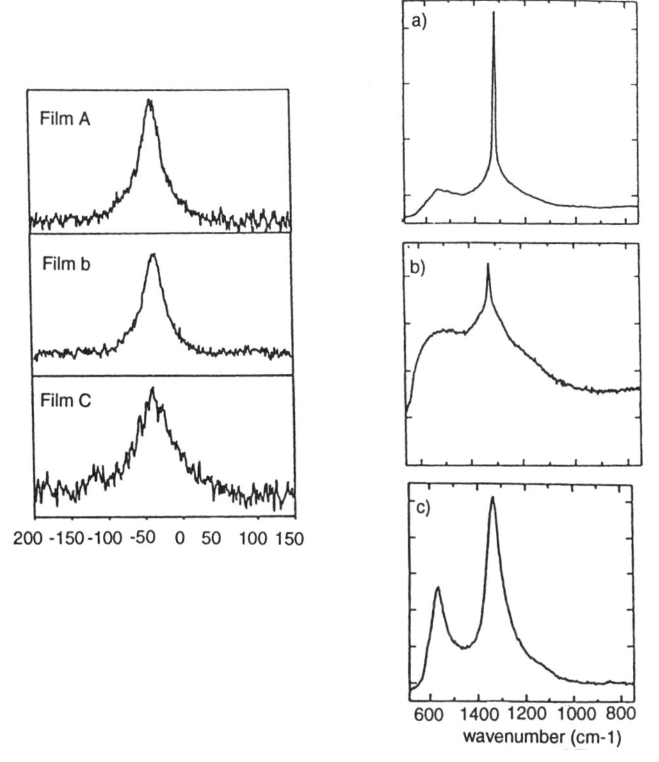

Figure 12. On the right, ^{13}C NMR spectra of three CVD diamond films, with the corresponding Raman spectra shown on the left. While only a small sp^2 peak is seen in the bottom NMR spectra, the corresponding Raman spectra show no evidence of a one phonon diamond absorption. [McNamara 1992b]

polarization dynamics suggested this sp^2 carbon is localized near hydrogenated sites [Loch 1993].

3.2.3 Proton NMR

The location and distribution of hydrogen provides insight into both the diamond deposition chemistry and structure-property relationships in these films. The ^1H NMR spectra yield concentrations ranging from less than 0.017 to ~1 at. % H for CVD films. These concentrations show a linear relationship to the CH_x stretching absorption in the infrared (IR) spectra for both hot-filament (see also section 2.2.2) [McNamara 1992c] and microwave films [McNamara 1994a]. Thus, hydrogen detected via NMR is predominately covalently bound to carbon in the film. This correlation also shows the IR measurements are quantitative for less than 0.2 at. % H.

The room-temperature ^1H NMR spectra of CVD diamond films (Fig. 13) often contain two components, a narrow Lorentzian and a broad Gaussian, indicating at least two different bonding configurations for hydrogen [McNamara 1992d].

A "hole-burning" NMR experiment shows these two hydrogen environments are within 5Å of each other. At 100 K, the Lorentzian component is broadened. This is indicative of reduced molecular motion, potentially the slowing of the rotation in $-OCH_3$ and

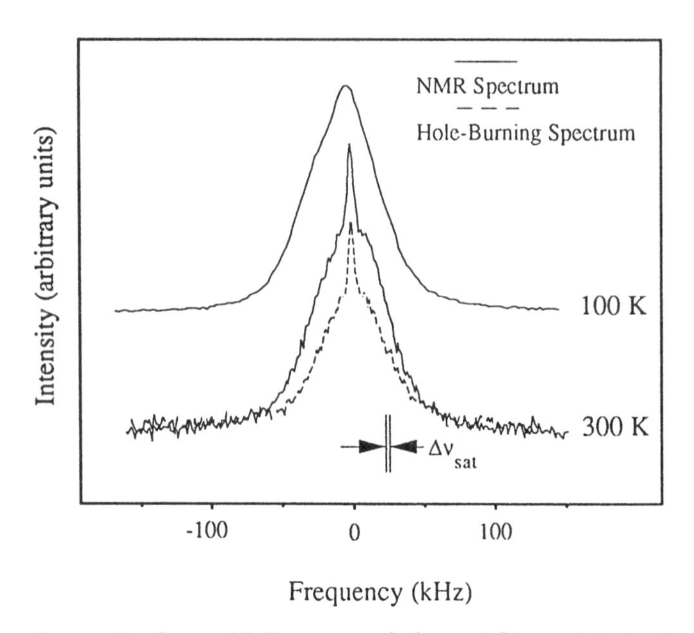

Figure 13. Proton NMR spectra of diamond film at 300 K shows two components. Upon lowering the temperature to 100 K, the broad Gaussian remains the same while the narrow Lorentzian broadens. Selective saturation of the spectra in the wings of the broad Gaussian (hole-burning) produces attenuation in both line shapes. (McNamara 1992d]

-NCH_3 groups [McNamara 1994a and 1994b]. The Gaussian component is unchanged at 100 K, indicating rigidly held hydrogen [McNamara 1992d]. As the hydrogen content increases, it can sometimes become difficult to resolve the narrow feature from the dominant broad line [Pruski 1994].

Although the average hydrogen concentration in the CVD films is low, the proton homonuclear dipolar line broadening is large (~60 kHz). This discrepancy indicates locally high hydrogen concentrations exist, requiring significant segregation of hydrogen in polycrystalline diamond [McNamara 1992c and 1992d]. Randomly dispersed CH_2 groups would provide too little homonuclear broadening. However, the broadening is consistent with the areal densities for hydrogen passivated diamond surfaces. Both internal surface area (for example grain boundaries) as well as that of the top growth surface are potential sites for such passivation. Internal voids may also contribute to the surface area [Pruski 1994] and influence the thermal conductivity of diamond films [McNamara 1995].

The assumption of a surface distribution of hydrogen in polycrystalline diamond films is also supported by ^1H multiple quantum (MQ) NMR studies [Levy and Gleason 1992]. Up to 20 hydrogen atoms within a 10Å radius were correlated with CVD diamond films despite the very low (< 0.1 at.% H) bulk concentrations, confirming the large scale clustering of hydrogen. The initial MQ growth for the films is slower than that for known bulk hydrogen distributions, such as CaH_2. However, the MQ growth for the films is coincident with a known two-dimensional proton distribution, intentionally hydrogenated diamond powder. Thus, the MQ results are also consistent with hydrogen clustering on surfaces in CVD diamond.

3.2.4 Fluorine-19 NMR

The surfaces of diamond powder exposed to CF_4 containing plasmas have been examined using solid-state ^{19}F NMR [Scruggs and Gleason 1993]. Surface coverages between 5% and 50% of the available surface bond density were observed. The ^{19}F MQ NMR experiment shows that this fluorine is not uniformly distributed on these surfaces. High-speed (15 kHz) magic-angle-spinning (MAS) was used to average the effects of ^{19}F chemical shift anisotropy. The isotropic chemical shifts were resolved and assigned relative to $CFCl_3$ as follows: CF =148±1 ppm; CF_2 = 106±2 and 123 ppm; CF_3 = 78±1 ppm. The major species was monofluoride and only 5 to 10% was bonded as CF_3. These CF_x functionalities were the only fluorine-containing species observed.

3.2.5 Other Nuclei

The nuclei discussed in this section: ^{13}C, ^{1}H and ^{19}F, all have a spin of 1/2. Subject to the constraint of sensitivity limits, other spin-1/2 nuclei, such as ^{31}P could also be studied by NMR. Quadrupolar nuclei, such as ^{2}H and ^{11}B, which have spin greater than 1/2 could also be examined. For these elements, the interaction of the quadrupolar nucleus with the symmetry and orientation of the surrounding electronic field gradient will often dominate the spectra and thus determine the information that can be obtained [Brundle 1992].

3.2.6 Spin-Lattice Relaxation

Paramagnetic defects facilitate spin-lattice relaxation in both natural and HPHT diamond. Spin-lattice relaxation can be dominated either by direct ^{13}C relaxation to paramagnetic defects (see ESR section) or by spin diffusion between dipolar-coupled pairs of ^{13}C nuclei [Henrichs 1984, and Hoch and Reynhardt 1988]. Spin diffusion is slow when no ^{13}C enrichment is employed and MAS is used, resulting in non-exponential relaxation. In contrast, pure exponential spin lattice relaxation behavior, having a time constant T_1, is observed in static spectra of highly ^{13}C enriched CVD films, suggesting spin diffusion is dominant [McNamara 1992a]. Analysis of the spin-lattice relaxation in polycrystalline films is difficult because of the higher complexity of possible defect distributions. A study of ^{13}C spin-diffusion rates as a function of isotopic enrichment suggests that paramagnetic centers are distributed uniformly in polycrystalline CVD diamond films, rather than being localized at grain boundaries [Pruski 1994].

At a fixed degree of ^{13}C enrichment, an increase in the ^{13}C T_1 qualitatively indicates a lower relaxation center density, hence an increase in crystal quality. The T_1's of commercial natural diamond powder are on the order of tens of seconds, while significantly longer time constants, on the order of a few hours, can be observed for large, gem-quality natural diamonds [Henrichs 1984]. An increase in T_1 has also been observed within a series of CVD diamond films as the morphology observed by scanning electron microscopy improved [McNamara 1992a]. Experimentally, T_1 is important because it determines the delay time required between NMR signal acquisitions.

Relatively fast spin-lattice relaxation has been observed for both ^{1}H and ^{19}F bonded to diamond. For ^{1}H, T_1 is between 5 ms and 1 s at room temperature [Levy and Gleason 1992, Lock 1993]. These values increase after annealing CVD diamond at 500 to 750 °C, suggesting passivation of paramagnetic centers, consistent with direct ESR measurement [Mitra and Gleason 1992]. For ^{19}F, anisotropic spin-lattice relaxation is observed, yielding non-exponential behavior with an associated time constant of ~0.5 s.

4. Ion Scattering Techniques

4.1 RUTHERFORD BACKSCATTERING (RBS)

4.1.1 Introduction

For Rutherford backscattering spectrometry (RBS), the surface of a solid sample acts as a target for a beam of monoenergetic particles. A small fraction of the incident particles are backscattered as a result of elastic collisions with atoms in the sample. The energy of a particle scattered by such a collision allows the concentration profiles of heavy elements to be determined as a function of distance from the sample's surface. This depth profiling is nondestructive in the sense that the surface is not sputtered away. Note that RBS does not yield information about chemical bonding since the kinematics of the collision are not sensitive to the electronic configuration of the target species. When the rows of atoms in a crystal (especially a single-crystal) are precisely aligned parallel to particle beam, the incident ions can channel through the sample. This channeling can be used gain additional structural information as well as improved depth resolution. More information on this technique can be found in Brundle [1992].

4.1.2 Experimental

An accelerator creates a collimated beam of particles required for RBS. In most cases, high energy (0.5 to 5.0 MeV) He ions are used as the incident species. These He ions can penetrate greater than 1 μm into the sample. The vast majority of He ions are not scattered and instead become implanted in the sample. When an elastic collision takes place between an incident ion and an atom in the solid target, some energy is transferred to the sample and the energy of the He ion is reduced. An analysis based on the conservation of energy and momentum determines the mass of the target atom. In addition, the He ion gradually loses energy as it travels into the solid prior to the collision and as it reemerges after scattering. Since the energy loss of light particles is well behaved in the MeV range, this phenomenon allows concentration depth profiles to be obtained.

The energy of the He which is backscattered is analyzed with a semiconductor nuclear particle detector. The optimal detection angle for discriminating between the masses of different elements is 180°, but is usually limited to less than 170° as a result of spatial constraints. The scattering cross section is typically determined experimentally for quantitative measurement of specific elements. The depth resolution of such experiments may be improved by the use of grazing angle techniques.

4.1.3 Sensitivity

The scattering cross section causes RBS to be preferentially sensitive to heavy elements. The ultimate detection limit is set by the onset of sputtering. Sputtering occurs when a significant fraction of secondary species acquire sufficient energy to escape the solid as a result of the collisional cascade caused by the impinging ion. This leads to erosion of the material and places a limit on the sensitivity of RBS. The best sensitivity obtainable by RBS for heavy scatterers, such as gold, is approximately 5×10^{11} atoms/cm^2. RBS measurements have significant uncertainty on an absolute scale and are more useful for relative comparisons.

4.1.4 Applications to Diamond

Impurities. Heavy atom impurities in both CVD films and diamond single crystals have been measured by RBS. In polycrystalline films grown by microwave CVD, RBS has been used to measure oxygen content [Ingram and Lake 1992]. Oxygen impurities in single crystal diamond have also been measured by RBS [Baumann 1994]. In addition, contamination of hot filament CVD diamond films by filament materials, such as tungsten and tantalum, has been observed by RBS [Hinneberg 1992]. The tungsten concentrations were between 1 and 300 ppm and are a strong function of filament temperature. The tungsten concentration increased by two orders of magnitude when the filament temperature was raised from 1800 to 2200°C. Tantalum contamination levels at comparable temperatures were about an order of magnitude greater than those from tungsten [Hinneberg 1992]. These studies serve to accent the usefulness of RBS measurements for comparative analysis.

Crystal Order and Orientation. While channeling is generally performed on single crystals, a channeling study of polycrystalline diamond was used to determine the preferential growth orientation and the growth parameter, α, of these films [Samlenski 1994]. The density of scattering centers was found to have a very strong dependence on the {111} character of the films. Ion channeling was possible on these polycrystalline samples only because of the very high degree of preferential orientation.

In addition, the crystalline perfection of substrates for diamond growth and layers of other materials grown on diamond have both been investigated by RBS. In order to determine the effect of substrate crystallinity on diamond nucleation, single crystal silicon wafers were treated by ion implantation and then examined for crystallinity using RBS with ion channeling [Kobashi 1993]. Channeling was also used to study copper films deposited epitaxially on diamond [Baumann 1994].

4.2 ELASTIC RECOIL SPECTROMETRY (ERS)

4.2.1 Introduction

Forward or elastic recoil spectrometry (ERS) is a modification of the RBS technique (see section above) used to obtain information on light nuclei within a sample, such as protons (^{1}H) and deuterons (^{2}H). Thus, ERS complements RBS, which provides information on heavy nuclei. In an elastic collision where the mass of the incident particle is greater than that of the target, the incident energy is transferred to the light particle in a recoil collision. Here, the incident particle is not scattered backward, nor is its energy detected. Instead, the energy of recoiled particles is measured by placing the detector at a forward angle and the sample at a glancing angle with respect to the incident beam. Incident and forward angles are usually 5° and 30° with respect to the target, respectively. Similar to RBS, elastic recoil may be used to determine depth profiles in solids. Unlike RBS, however, ERS is complicated by the fact that both hydrogen (or deuterium) and the incident helium ion are scattered in the forward direction. In order to prevent helium ions from reaching the detector, a mylar film with a thickness of ~10 μm is placed between the target and the detector. Since the stopping power of hydrogen with respect to helium is sufficiently low, the mylar is effective in stopping the helium while allowing the passage of hydrogen to the detector. Depth profiles can be determined from the energy loss of the two particles on the inward and outward paths, respectively.

Since ERS detects atoms lighter than the incident helium ions, it has been primarily used to study ^{1}H or ^{2}H in diamond films [Ingram 1992 and 1993]. Hydrogen concentrations of over 1.0 at. % have been measured by this technique, in disagreement with techniques such as NMR (see section 3.2.3) and IR spectroscopy (see section 2.2.2) which give lower hydrogen concentrations by about an order of magnitude. This may be due, in part, to the uncertainty of absolute concentrations or background hydrogen levels in the ERS method. Another factor is a thin (~10 nm) layer of hydrocarbon contamination on the surfaces of CVD diamond films. This contamination has been imaged with TEM, and the ERS measured hydrogen content was 10 at.% [Engel 1994]. Cleaning, for instance with an isopropanol rinse, can remove the contaminate layer, and is required for accurate results. Cleaning is also important for other techniques to measure hydrogen, such as nuclear reaction analysis (see section 8.4). Depth profiling was used to study the effects of annealing on both oxygen (RBS) and hydrogen (ERS) concentrations within the films, and showed that although surface impurities could be removed by annealing, concentrations in the bulk remained unaffected by the process [Ingram 1992]. ERS has also been used to study the mechanisms of proton and deuterium incorporation into CVD diamond [Ingram 1993].

4.3 SECONDARY ION MASS SPECTROSCOPY (SIMS)

4.3.1 Introduction

Sputtering is a process by which atoms in a solid are ejected as a result of bombardment by a beam of energetic particles. In contrast to backscattering techniques (see previous section on RBS and ERS), the incoming species used for sputtering are typically low energy (< 50 keV), heavy particles, such as Ar+. These particles undergo elastic collisions, transferring energy to the target atoms, and eventually cause ejection of some atoms from the sample. The number of direct recoils, or ejection events, is low. However, a complex series of collisions leads to secondary recoil events from ions, excited states, and neutrals. The ratio of ions to neutrals in such sputtering will be a highly sensitive function of surface conditions. The collection of these secondary ions and their separation as a function of mass is known as secondary ion mass spectroscopy (SIMS). It is one of the most sensitive techniques available in surface analysis and is used to measure trace impurities in solids. A more detailed description can be found in Brundle [1992].

4.3.2 Resolution

SIMS analysis is complicated by the fact that both positive and negative ion scattering may occur simultaneously. In practice, either positively or negatively charged species are detected. In addition, not only singly ionized species, but also multiply ionized atoms and clusters of atoms appear in the spectrum. In most cases, the singly ionized species will predominate, but quantitative concentrations may be difficult to obtain. Interfering mass effects from charged atom clusters may pose serious limits on mass resolution in some SIMS experiments, especially when hydrocarbon fragments are involved. SIMS is often used for depth profiling of impurities, however, care must be taken in the experimental setup to avoid sputtering from the surface layers as the depth of sputtering increases, since this reduces depth resolution. Sputtering from the outer surface layers can be minimized by rastering across a length of sample and gating the detector to accept only sputtering from the central portion of the etch pit. This insures minimization of side wall effects. Some of the difficulties in depth profiling hydrogen with SIMS are discussed by Magee and Botnick [1981].

4.3.3 Applications to Diamond

SIMS has been used to detect a number of impurities in diamond [Cifre 1992, Cook 1992, Baker 1993]. In single crystal HPHT diamond, SIMS has been used to determine boron dopant levels as well as metal contamination [Baker 1993]. The concentration depth profile of boron implanted into single crystal diamond to depths of less than0.5 μm was quantitated by SIMS [Maby 1981]. Tungsten incorporation in hot-filament CVD diamond has been monitored as a function of process conditions by SIMS and found to increase with decreasing methane fraction in the gas phase [Cifre 1992]. Depth profiles for both the unintentionally incorporated tungsten and intentionally incorporated boron were also measured in this study. No attempt was made, however, to quantify the relative contents [Cifre 1992]. Finally, SIMS has been used to monitor hydrogen and oxygen incorporation in CVD diamond and to screen for other trace impurities [Cook 1992].

5. Electron Emission Spectroscopies

5.1 X-RAY PHOTOELECTRON SPECTROSCOPY (XPS)

5.1.1 Introduction

The main use of XPS in materials analysis is to determine the chemical binding of atoms in the surface region (top ~5 to ~50Å) of the sample. Unlike RBS and ERS, which depend on the interaction of a charged species with the nucleus of the atom via scattering, XPS depends on the interaction of X-rays with atomic electrons via photon absorption. Photoelectron spectroscopy is also known by the acronym, ESCA (Electron Spectroscopy for Chemical Analysis), which serves to highlight the importance of chemical bonding. For more detail, see Brundle [1992].

The process of interest in XPS is the absorption of a quantum of energy, hν, which causes the ejection of an electron from an inner shell, the photoelectron, whose kinetic energy is related to the binding state of the atom of origin. In XPS, a beam of monochromatic X-rays impinges on a surface. X-rays are used because these photons can penetrate the solid and interact with inner shell electrons. In practice, the energy range of interest is approximately 10 eV to 0.1 MeV. The most common X-ray sources are the $K\alpha$ transitions of magnesium (1486.6 eV) and aluminum (1356.6 eV). Unlike lower energy spectroscopies, such as those in the visible range, the outermost electrons are not the object of study. It is the innermost electrons, associated with a specific atom, which are important. Escape depths for these electrons are typically only 10-20Å. These give rise to surface sensitivity of XPS which requires careful cleaning procedures and ultra-high vacuum equipment. Typical XPS measurement times are from 1 to 5 hours.

The photoelectrons released by X-ray bombardment are then detected by an electron energy analyzer, which uses the deflection of the electron in an electrostatic or magnetic field to determine the electron energy. The relevant quantity for atom identification is the binding energy. The binding energy can be determined from the known incident energy and the measured kinetic energy of the photoelectron (after correction for spectrometer and sample charging effects). In reality, such energies are expressed relative to a reference level, typically the Fermi level in solids. For insulators, such as diamond, care is required because of the uncertainty of the Fermi level within the gap and because of sample charging effects. Coating with a thin layer of gold is sometimes used to overcome such difficulties.

Typical XPS spectra show sharp peaks at the energies of the excited photoelectrons. These peaks correspond to electrons which escape without energy loss and will often display higher energy tails which correspond to electrons which have undergone inelastic scattering or energy loss along the outward path. The energy of the photoelectron will be typical of the element from which it originates, but its exact location will depend on the chemical environment of that element as well. Any change in the chemical environment of the element will redistribute the valance electrons, causing the inner electrons to experience a change in potential, shifting the binding energy. Typical ranges for such chemical shifts are about 8 eV, increasing with the greater electronegativity of the surrounding atoms.

Line intensities or peak areas can be used to derive quantitative information from photoelectron spectra, however, ratios are more reliable than absolute values as a number of assumptions are involved in quantifying spectra. These include homogeneity of the sample, clean, smooth surfaces, and isotropic photoelectron emission. All of these are potential concerns, particularly in examining polycrystalline diamond films.

5.1.2 Applications to Diamond

XPS has been used by a number of investigators to probe the surface of bulk diamond films [Cook 1992, Ababou 1992, Dowling 1992] as well as the interface between early diamond growth and the substrate [Haq 1993, Williams 1991]. The most common feature studied in the XPS of diamond films is the carbon 1s core photoelectron (C 1s) peak. This peak is often used to identify the carbon bonding environment within the sample. The accepted peak position for diamond is 284.4 eV, while for graphite, it is 284.3 eV [Cook 1992, Ababou 1992, Wagal 1992]. Often absolute peak position are calibrated using a reference, such as the Au $4f_{7/2}$ peak [Cook 1992] or the Cu $2p_{3/2}$ or Cu

3p peaks[Ababou 1992]. Still, energy resolution may not be sufficient to resolve the graphite and diamond lines. In fact, information about other carbon species, such as carbides, is often obtained only by deconvoluting overlapping peaks [Ababou 1992].

The carbon valence band spectrum may provide information on carbon bonding as well. For example, Cook [1992] was able to demonstrate that cleaning diamond samples by argon ion etching was actually damaging the sample surface. Before etching took place, peaks in the valance band appeared at 12 eV and 17 eV. These can be attributed to sp^3 bonding electrons and the valance band s-electron distribution, respectively. After argon ion etching, however, the sp^3 peak appears substantially attenuated and an additional peak, characteristic of graphite, appears at 8 eV. Considerations of potential damage to the diamond surface are important for spectroscopies which rely on high energy beams for cleaning and sputtering.

XPS has been used to identify other elements in diamond films, including oxygen, silicon, and tantalum [Pehrsson 1993, Williams 1991, Jubber 1993]. Both the C 1s and O 1s spectra were examined to determine the effects of acid cleaning on the diamond surface, and the formation of a surface oxygen layer was identified [Pehrsson 1993]. Another study made use of simultaneous Si 2p and C 1s, and Ta 4f and C 1s, data to trace nucleation events on silicon and tantalum substrates, respectively [Williams 1991]. Similar studies have been carried out by a number of other authors as well [Haq 1993, Ababou 1992]. Finally, XPS has been used to determine quantitative levels of these elements in diamond [Oelhafen 1993].

5.2 ULTRAVIOLET PHOTOELECTRON SPECTROSCOPY (UPS)

UPS is similar to XPS in all but its energy source. In this case, He I and/or He II radiation from a resonance discharge lamp provides the incident energy. This is typically 21.2 eV and 40.8 eV for He I and He II, respectively. Unlike XPS, ultraviolet radiation has only sufficient energy to excite photoemission of electrons from the valance band and shallow core levels. Thus, it may provide insight about electronic bonding configurations. XPS and UPS are often used as complimentary valence band spectroscopy techniques [Oelhafen 1993, Cinti 1992]. UPS measurements can typically be obtained in under half an hour.

5.3 AUGER ELECTRON SPECTROSCOPY (AES)

5.3.1 Introduction

Auger electron spectroscopy (AES) is primarily used to examine low concentrations of impurities near a solid surface. Like other electron emission spectroscopies, it is carried out under ultra-high vacuum conditions. While XPS uses X-ray bombardment of the sample to produce electron emission via a single electron process, Auger transition requires electron beam excitation of an atom in the sample having at least three electrons. Three letters, such as KLL, signify the three electronic levels in the atom which were involved in a given Auger transition.

Sampling depths for AES range between 5 and 100Å. AES is particularly useful for detecting low Z elements, such as carbon, nitrogen, and oxygen. However, Auger cannot detect hydrogen since Auger transitions involve three electrons. Although chemical shifts are evident in both XPS and AES spectra, they are far more difficult to interpret for the Auger electron process. In addition, Auger linewidths tend to be broader than those observed in XPS spectra. Thus, XPS is typically used to examine chemical shifts and chemical binding.

When combined with sputtering, Auger can be used to examine impurity gradients and interfacial concentration. In such depth profiling, care must be taken to avoid intermixing due to the sputtering. It is quite useful to combine AES with RBS. RBS gives depth profile information on heavy elements without the added complications of sputtering, and AES gives profile information on both heavy and light masses. Both AES and RBS techniques can be used to examine layers as thin as 1000Å.

5.3.2 Experimental

For AES, an electron gun generates a beam which is focused on the sample. Typically, the energy of this primary electron beam is between 0.1 and 15.0 keV. The electron bombardment of the sample can produce a vacancy in the inner electronic shell of the atom. An outer electron from the same atom fills the hole left by an inner electron vacancy. The energy released by this electronic rearrangement can eject another electron from the atom. The ejected Auger electron must pass through an aperture on the cylindrical mirror analyzer (CMA) prior to detection by an electron multiplier. The voltage applied to the CMA determines the energy of Auger electrons which can pass through the aperture.

Auger electron transitions typically appear as small features on a large secondary electron background, and as a result, most Auger spectrometers operate in differential mode. Backscattering of incident electrons, angular yield dependencies, and variations in surface roughness of the sample can further complicate spectral interpretation. Together these effects contribute to the uncertainty of quantitative Auger analysis, however the sensitivity of the technique to impurities is still relatively high at about 1000 ppm or 0.1 atomic percent.

5.3.3 Applications to Diamond

Because the Auger lineshape is sensitive to electronic structure, diamond, having only sp^3 bonds, shows a fine structure distinct from that of either graphite or amorphous carbon. In figure 14, the Auger spectrum of natural diamond shows higher intensity in the peak closest to the primary carbon KLL transition and slightly lower intensity for the lower energy peak [Hoffman 1994]. For graphite, the reverse is true. Thus, Auger has been widely used to identify the diamond "signature" in CVD films [Cook 1992, Shih 1992, Skytt 1993, Wagal 1992, and Davis 1994]. It has been used to determine changes in the film surface structure as a function of gas phase methane concentration, etching, and annealing as well [Skytt 1993; Cook 1992].

One of the most significant concerns when performing AES on diamond samples, is the possibility of surface damage to the sample as a result of the incident electron beam or the ion beam used for sputtering in the case of depth profiling or material removal. This can lead to misinterpretation of results if not considered carefully. Also note that the incident electron beam may cause nonconductive samples to charge. These effects have

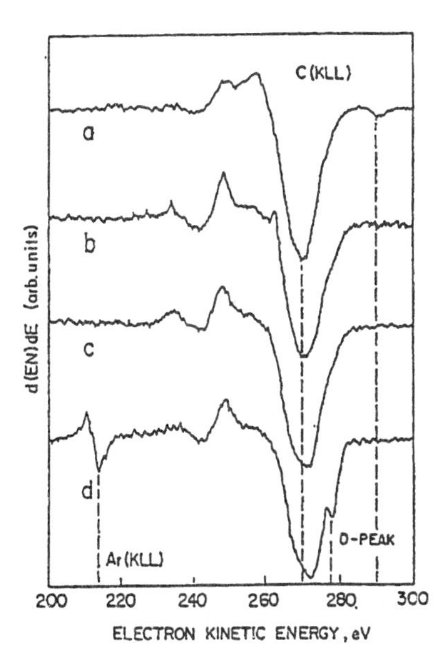

Figure 14. Comparison of C(KLL) Auger line shapes for (a) type IIa diamond; (b) HDPG; (c) graphitic carbon; and (d) graphitic carbon after Ar^+ irradiation. [Hoffman 1994].

been observed in both CVD diamond [Shih 1992, Cook 1992, Ababou 1992] and single crystal high pressure high temperature diamond [Bugaets 1994]. Figure 15 shows the change in the Auger spectra obtained from a single crystal diamond as a function of the incident electron beam energy as well the development of argon ion beam damage [Bugaets 1994]. Ababou [1992] attempted to circumvent this issue by use of a scanning Auger microscope which allows the use of a low energy, low intensity electron beam to image the surface. This study identified the type of carbon deposits present on various substrates.

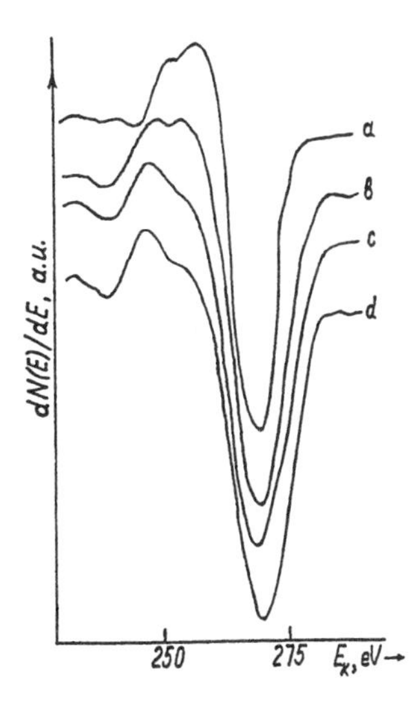

Figure 15. Auger line shapes obtained for (a) diamond; (b) diamond after 3 keV electron irradiation; (c) diamond after 6 keV electron irradiation, and (d) diamond after 3 keV Ar^+ irradiation. [Bugaets 1994].

Finally, AES has been used to identify and quantify impurities in CVD diamond [Hoffman 1994, Mearini 1994]. Hoffman [1994] was able to use Auger to determine differences in impurity incorporation between the growth and substrate side of CVD diamond deposited on silicon. While both silicon and oxygen were detected on the substrate side, neither was observed on the growth surface. Similarly, Mearini [1994] was able to explain the unusual emission behavior of a CVD diamond sample by detecting both sodium and sulfur in the near surface region with AES and throughout the material using AES with depth profiling. While one of the strongest features of AES is its ability to detect such trace impurities, great care must be taken to insure that surface contamination does not bias results.

6. Microscopy

6.1 SCANNING ELECTRON MICROSCOPY (SEM)

This technique is one of the most widely used to examine crystallite size and facetting in CVD diamond films. Magnification of sample topology ranges from 10 x to 300,000 x. Besides providing higher magnification than optical microscopy, SEM also had a greater depth of field. Electron Microprobe Analysis (EMPA), typically performed as Energy-Dispersive X-ray Spectroscopy (EDS), is often available on SEM instruments, providing complementary identification of non-trace elements with $Z>3$ present in the samples. More information on both SEM and EMPA can be found in Brundle [1992].

The primary electron beam impinged upon the sample has an energy between 0.5 and 50 keV, with 20 to 30 keV being the most commonly used range. Images are formed using either backscattered electrons (BSEs) or secondary electrons (SEs) emitted from the sample. A BSE has an energy greater than 50 eV and hence a relatively large escape depth from the sample. Since most BSEs results from elastic scattering events, signals from regions containing high Z elements appear brighter, thus providing some elemental contrast. The limited escape depth of the lower energy SEs provides higher surface sensitivity images (Fig. 16).

In addition to producing BSEs and SEs, the primary electron beam causes X-ray emission from sample depths between 100 nm to 5 μm. Based on the energy of these emitted X-rays, the EMPA attachment available on many SEM instruments identifies the elemental composition of the sample. While EMPA is not sensitive to trace elements, this technique is useful for rapidly identifying the presence of non-carbon crystallites.

Sample preparation can be an important factor for SEM. Insulating diamond surfaces may require gold coating prior to SEM analysis. Alternatively, low primary beam voltages (<2.0 keV) can be employed in order to avoid charging effects. Gold coating can also be avoided by using environmental SEM (ESEM, in which the presence of water vapor in the analysis chamber ameliorates surface charging. Images of cross-sections of CVD films have also been used to study diamond morphology. For example, a cross-section with polishing (Fig. 17) shows a large void [McNamara 1995].

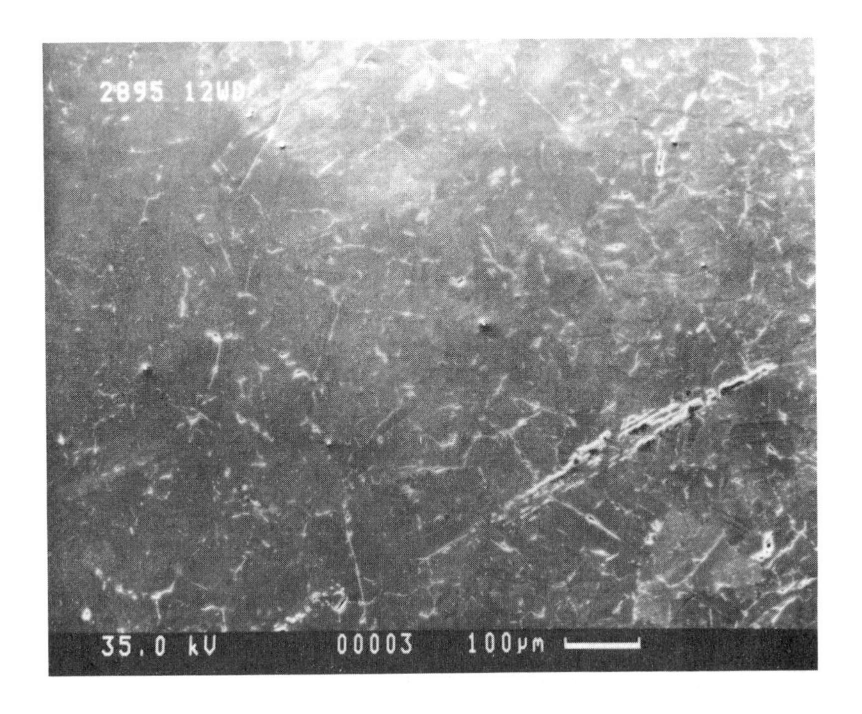

Figure 16. Two SEMs of the same region of a polished CVD diamond surface. The top image shows surface sensitivity provided by SEs. The bottom image was made with BSE's. (Courtesy of Dr. John Sutliff, Corporate Research and Development, General Electric Company.)

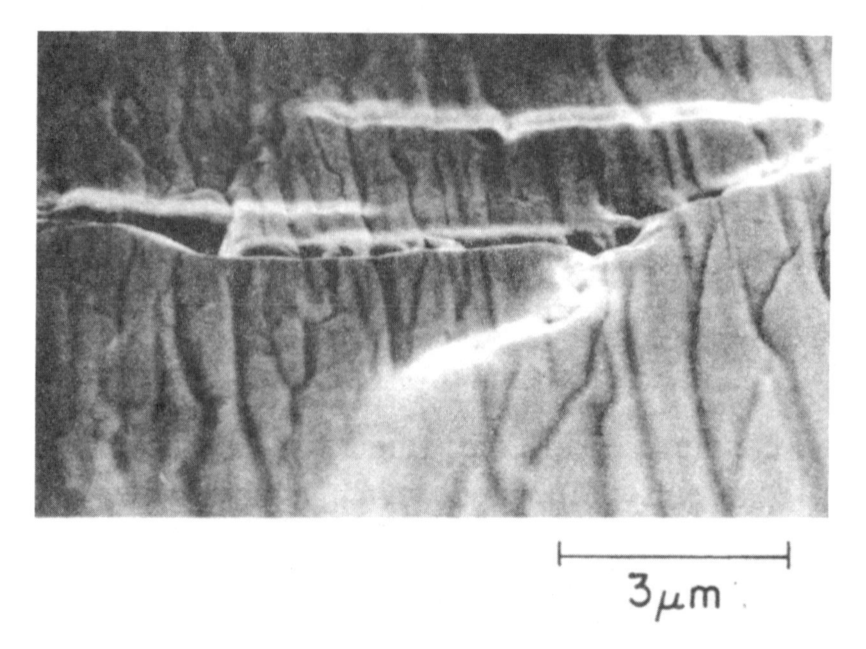

Figure 17. A void, greater than 10 μm in length, oriented along the direction of film growth, is observed in a SEM of polished diamond cross-section [McNamara 1995].

6.2 TRANSMISSION ELECTRON MICROSCOPY (TEM)

A transmission electron microscope can provide very high magnification (>10^6 x) images and electron diffraction patterns from the same sample. Complementary chemical characterization is often available on TEM instruments in the form of electron energy loss spectroscopy (EELS). More detail on TEM and EELS can be found in Brundle [1992]. Details of using TEM for understanding grain boundaries, growth interfaces, defects and facetting in diamond are discussed in detail in Chapter 7 of this handbook.

As compared to SEM, samples for TEM are very thin (<200 nm) and thus, more difficult to prepare. However, since electrons are transmitted through the sample, surface charging does not occur during TEM. Also, the energy of the primary beam for TEM (100 to 400 keV) is higher than that for SEM. This high energy corresponds to a short electron wavelength, making high resolution (<0.2 nm) images possible. At the same time, lattice parameters can be measured to high accuracy via electron diffraction.

6.3 LOW-ENERGY ELECTRON MICROSCOPY (LEEM) & PHOTOELECTRON ELECTRON MICROSCOPY (PEEM)

These two types of electron microscopies provide real-times surface images of the dynamics of monolayer formation. These images can be combined with in situ sample heating and gas absorption. Low-Energy Electron Microscopy is the real space imaging partner to Low-Energy Electron Diffraction (LEED) [Tromp 1994]. In PEEM, variation in the emission intensity of photoelectrons (see also XPS) is used to form an image. The growth of an ordered carbon monolayer on a molybdenum substrate has been monitored by PEEM [Garcia 1992]. During this study, the molybdenum surface was exposed to a 5% methane in H_2 gas mixture, which was thermally excited by a hot-filament.

6.4 SCANNING TUNNELING MICROSCOPY (STM)

6.4.1 Introduction

At maximum resolution, STM can be used to examine surface reconstruction and atomic steps, features crucial for providing insight on the mechanism of diamond growth. Increasing the scan size allows profilometry of larger scale features such as surface roughness, grain boundaries and crystallite shapes.

In the STM experiment, electrons tunnel between a sharp tip and a surface, requiring the tip-to-sample separation to be just a few angstroms. The magnitude of the tunnel current depends on the tip-to-sample distance, allowing the 3-D topography of the surface to be profiled. The tunnel current also depends on the local density of electronic states at the surface. The bias voltage used selects which of these electronic states are involved in tunneling and the direction of current flow. This electronic aspect complicates the interpretation of an STM image as a purely profilometric measurement but also provides a means to measure local variations in electronic properties and to perform spectroscopic imaging. However, in general, the atomic composition of the surface is not obtained via STM.

Both vertical and horizontal resolution of STM can approach 1Å for a square area ~100Å on a side. The small lattice constant of diamond makes atomic scale resolution inherently challenging. Images over larger regions (~1 mm per side) can be made at lower resolution. Typically, recording an image requires only several minutes.

The STM experiment can be performed under atmospheric conditions using an electrically conductive surface. However, the use of ultrahigh vacuum (UHV) for STM experiments generally improves the maximum resolution obtainable by providing a clean surface for observation.

The tip is a crucial factor in performing STM. A single atom tip is required to prevent multiple imaging. Since the image represents a convolution of the observed surface with the tip profile, the macroscopic size of the tip can limit the resolution of sharp vertical drops and narrow features. Finally, extremely rough topography may result in the tip physically contacting the surface. For more information on the STM technique see Hansma [1988] and Brundle [1992].

6.4.2 Experimental Requirements

Stable STM images of diamond surfaces have been obtained using tunnel currents in the range of 0.1 to 1.0 nA [Turner 1991a, Baker 1993]. Stable images of diamond surfaces have been made with bias between -3 V and +7 V [Turner 1991a]. Little change was seen in the image of a (100) homoepitaxial film as the bias was varied from -2 to +2 V [Tsuno 1991]. However, at very large negative biases, unstable images have been reported [Turner 1991a], suggesting localized electronic states exist in this energy range. Stable images are easier to obtain for relatively large sample to tip gaps where there is low density of states [Baker 1993]. However, using a large gap limits the ability to achieve atomic scale resolution.

For the STM experiment, the resistance of the sample should be less than that of the tunnel gap (~4 x 10^{+10} ohms) [Baker 1993]. Thus, the high resistivity of intrinsic diamond would be expected to be problematic and as a result several STM studies employ boron doped diamond [Turner 1991a, Everson 1991, Baker 1993]. However, higher conductivities have been observed for CVD films [Landstrass 1989]. In fact, STM studies on films without intentional doping have also been successful [Tsuno 1991, Tsuno 1994, Vazquez 1994b, Frauenheim 1994]. These studies are not subject to the possible influence of boron incorporation on diamond morphology [Wang 1992]. Illumination by a xenon arc lamp to generate photo-induced carriers has also been used to improve the conductivity of diamond surfaces during the STM measurements [Mercer 1994]. The electronic conductivity of a nominally intrinsic diamond surface is likely to be affected by defects including impurities [Baker 1993], adsorbates [Everson 1991] and non-diamond bonded carbon [Busman 1991, Tamor 1993]. The decreased defect density with film thickness has been suggested as one reason that it is difficult to perform STM on thick films [Turner 1991b]. Indeed, "notches" resulting from low conductivity patches have been observed on an initial STM scan of a diamond surface which are absent upon subsequent traces [Everson 1991]. Presumably the low conductivity area is due to an adsorbate which is removed by the first pass of the STM tip. Cleaning with an oxygen plasma also eliminated observation of the notches, suggesting hydrocarbon contamination could be responsible for this phenomenon [Everson 1991].

6.4.3 Applications to Diamond

Single crystal STM studies have been performed on the (111) trigonal face of a large single crystal heavily boron doped diamond [Baker 1993]. This deep blue crystal was synthesized at high pressures and temperatures and contained few visible defects. Clean surfaces were exposed by etching with acid, after which the surfaces were maintained in ethanol until insertion into the ultrahigh vacuum chamber (10^{-10} torr) for STM analysis.

As a result of the p-type doping and raising the sample temperature to 500°C, the surface conductivity was sufficient for generating stable STM image in many regions [Baker 1993]. However, stable images could not be obtained at some spots indicating heterogeneity in surface conductivity. Impurities such as metals, hydrogen or oxygen may be responsible for this phenomenon. The length scale of the conductivity variation is small, as it appears uniform to XPS using a 300 μm diameter spot size. Levels of greater than 10% oxygen surface coverage were observed even on cleaned surface via XPS [Baker 1993].

Step heights corresponding to a single interplane spacing on the (111) surface (2.1 Å) were observed on the trigonal crystal face. Small angular features termed "vees" were also seen. These vees were mostly of similar size (10-12 Å in length and base width) and crystallographic orientation. The latter observation suggests a preferred direction of etching or growth. Tip imaging effects prevent accurate measurement of the depth of these narrow vees. Despite the care given to surface preparation, atomic scale resolution of the surface reconstruction was not reported [Baker 1993].

Homoepitaxial Films. Homoepitaxial films grown by microwave plasma enhanced CVD have been studied by STM in air [Tsuno 1991 and 1994, Maguire 1992]. The stability of the (111) surface was confirmed by LEED [Tsuno 1994], while the (100) surface was shown to be stable using RHEED [Tsuno 1991].

The surface of an epitaxial (100) film grown with no intentional dopant consisted primarily of dimers resulting from 2 x l and l x 2 reconstructions (Fig. 18), resulting in lines of atoms parallel to the (110) and $(1\bar{1}0)$ directions [Tsuno 1991, Frauenheim 1994]. The separation between parallel rows was consistent with twice the lattice constant (5.04Å) on the (100) surface [Tsuno 1991]. The STM image of a boron doped (100) epitaxial layer showed only unidirectional striations [Maguire 1992].

An epitaxial (111) surface without intentional doping maintained a lxl periodicity [Tsuno 1994]. Most of the steps occurred in the (112) direction, corresponding to a bilayer change (2.06Å height). The intersection of three such steps was found to result in triangular islands and spiral-like features. In addition, some $(11\bar{2})$ bilayer steps and (112) double bilayer steps were seen. Typically, the lateral between steps were 10 to l00Å wide. Atomic kinks, which are potential sites for carbon addition, were seen in the (112) single bilayer step every 50- 100Å [Tsuno 1994].

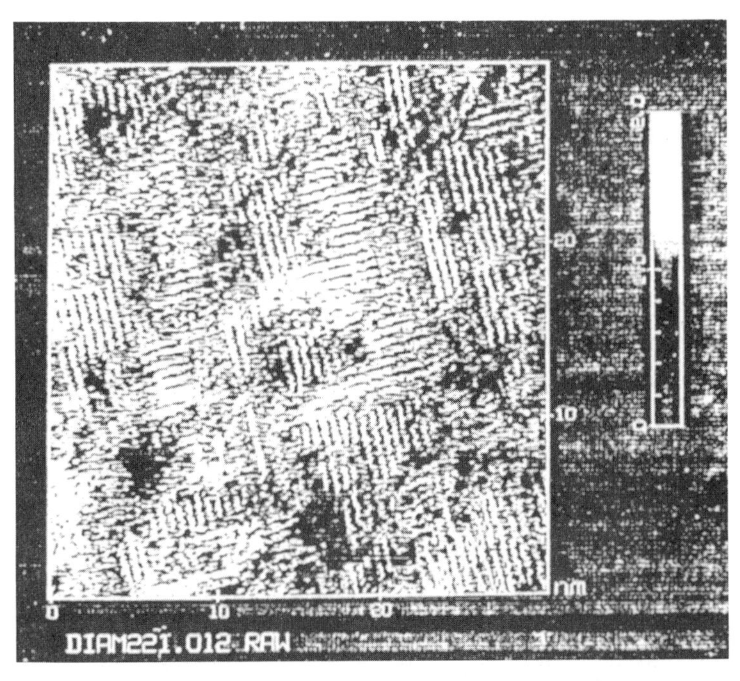

Figure 18. STM image of perpendicular atomic rows resulting from 2 x l and l x 2 reconstruction on the (100) surface of an epitaxially grown CVD film. [Frauenheim 1994]

The STM image of a boron-doped (110) epitaxial layer resolves a striated herringbone having both single and double atomic steps (Fig. 19). Reasons cited for the improved resolution in this case were either that the doped surface has a lower hydrophilicity and/or an increased electronic charge-density at surface [Maguire 1992].

Polycrystalline Materials. On a (100) facet, clear perpendicular rows were seen resulting from the 2 x l and l x 2 reconstruction and step heights of 1.1Å were measured [Frauenheim 1994]. Images of the (111) facets showed some regions of lxl periodicity and others of (√3x√3)R30°. Note that both the 2 x 1 / 1 x 2 reconstruction of the (100) surface and the 1 x 1 periodicity of the (111) surface on this polycrystalline hot-filament CVD film were also seen for the microwave CVD epitaxial film, suggesting some similarity in growth mechanism.

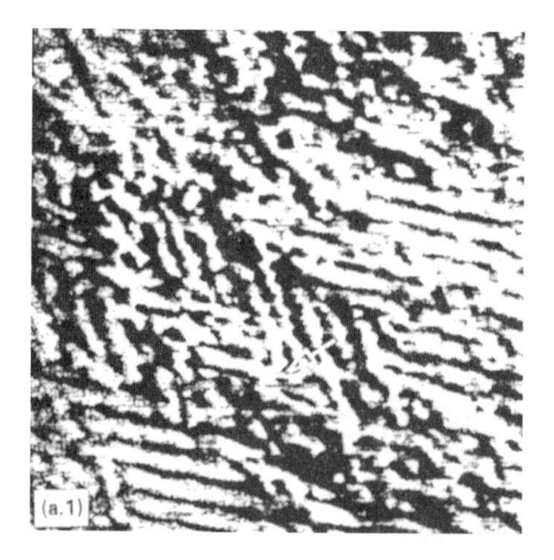

Figure 19. Atom rows on an STM image of a (110) CVD diamond surface [Maguire 1992].

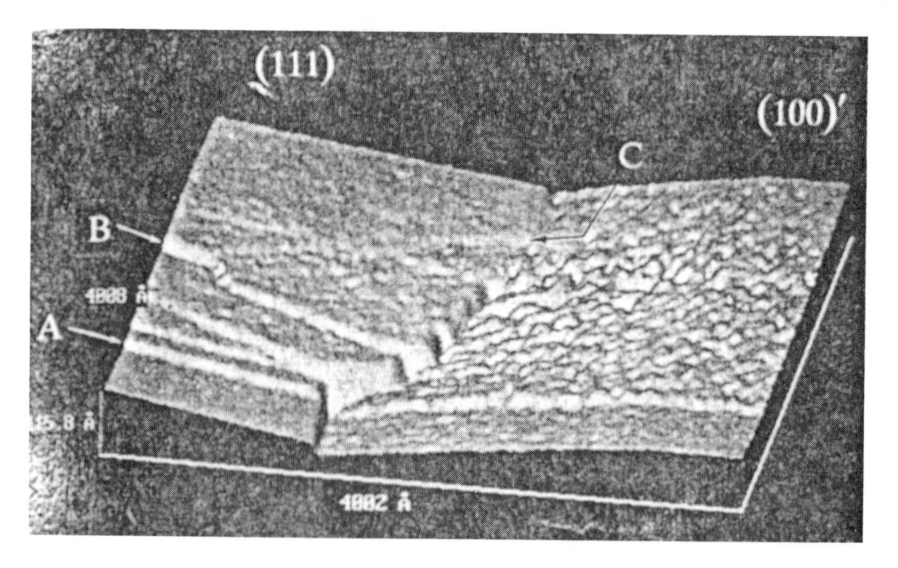

Figure 20. Surface roughness on a (111) versus a (100) surfaces as observed by STM [Everson 1991].

Surface roughness can be quantitated via STM. During the initial stages of microwave growth the rms roughness of (111) surfaces was found to be greater than that for (100) faces [Vazquez 1994a and 1994b]. The opposite situation has also been imaged (Fig. 20) [Everson 1991].

Angles between crystal planes can be quantitated. This is important for studying mechanisms of twin [Everson 1994] and nm scale defect [Turner 1993] formation. The discrete variation in slope along the pyramid structures on a (100) textured film suggests terrace widths between the steps in this feature corresponding to an integral number of dimers [Turner 1993].

Small (400 nm wide, 100-150 nm high), isolated diamond crystallites have been imaged, having ill-defined triangular or hexagonal geometries [Vazquez 1994a]. Improved facetting was observed upon formation of contiguous film. The spatial relationship between diamond nuclei and scratches on the silicon substrate and increasing surface roughness developed between nuclei has been examined [Turner 1991b].

More limited use has been made of the STM for probing the local electronic structure of surfaces resulting from the diamond CVD process. Local I-V curves have been obtained from the regions between the nuclei of a non-contiguous deposition onto a silicon wafer, which were consistent with silicon carbide formation [Turner 1991b]. However, no I-V measurements were obtained from the diamond nuclei themselves [Turner 1991b]. The local variations in electronic properties have been used to perform spectroscopic imaging [Everson 1991]. Both dI/dV and dI/dZ, where Z is the tip-sample distance, measurements were used to produce contrast between the diamond nuclei and a non-diamond substrate.

6.5 ATOMIC FORCE MICROSCOPY (AFM)

6.5.1 Introduction

Like STM, AFM (also called scanning force microscopy (SFM)) is a scanning probe microscopy. AFM provides high magnification three-dimensional images of solid surfaces, similar in appearance to the STM images described in the section above. However, no sample preparation is required for AFM [Chernoff 1992]. The key to the technique is the precise translation of a sharp tip and measurement of vertical displacement due to attractive and repulsive forces between the tip and the surface of the sample. Thus, unlike STM, AFM does not require electrically conductive surfaces. More detailed discussions of the AFM technique can be found elsewhere [Burnham 1991, Brundle 1992].

Like STM, maximum resolution requires a clean surface, suggesting measurements should be performed under ultrahigh vacuum. However, most AFM studies of diamond surfaces to date have been performed in air. Thus, the absorption of water, hydrocarbons and other contaminants from ambient has the potential to affect small scale imaging particularly if surface energy of the sample is altered by the adsorbate [Burnham 1991]. Bumps, which might be due to adsorbed contaminants have been seen [Sutcu 1992a]. However, the (100) diamond surface is apparently remarkably stable and no differences in image were observed before and after a surface cleaning in boiling acid [Sutcu 1992a].

Tip convolution effects are more prevalent, since the AFM cantilevers are generally higher than STM tips[George 1994]. This affects the measurement of narrow features and sharp angles. In addition, AFM images can show rounding of edges and apexes as compared to SEM micrographs due to these tip convolution effects [Baranauskas 1992]. It is also possible for the AFM tip to damage the surface.

For diamond films, repulsive forces of 10^{-6} to 10^{-8} N are often used to ensure intimate contact over the entire scan area [Sutcu 1992a and 1992b , Chernoff 1992, Perry 1993]. Horizontal scan dimensions range between 0.5 to 65 μm [Chernoff 1992, George 1994].

6.5.2 Applications to Diamond

Homoepitaxial Films. The morphology of the (100) surface is influenced by the carbon concentration used in the gaseous reactants. Using 0.3% CH_4, μm scale roughness was observed for both microwave [Maguire 1992] and hot-filament [Sutcu 1992a and 1992b] growth. Thus, the AFM images on these samples have limited resolution. However, larger features such as pyramids [Sutcu 1992a and 1992b], penetration twins [Sutcu 1992a and 1992b], and pits [Maguire 1992, Karasawa 1993] can be observed. The AFM images show these features am often oriented with respect to the underlying substrate [Sutcu 1992a and 1992b, Karasawa 1993].

In contrast, using 1.6 % CH_4 gives atomically (100) smooth surfaces although 1 to 2 nm roughness is present over a 1000 nm length scale [Sutcu 1992a and 1992b]. Over part of the surface, orthogonal domains containing parallel lines 5.04 ± 0.04Å apart can be resolved, indicative of 2xl/lx2 reconstruction. However, the atoms within the dimers were not resolved. The upper rows are shown to be 0.9Å above and rotated by 90° with respect to underlying layer, analogous to type A mono-atomic steps on the Si (100) and Ge (100) surfaces. Under these conditions, the flat (100) surfaces imply that growth at steps, ledges or kinks occurs much more rapidly than nucleation of a new layer.

A hot-filament CVD (111) epitaxial layer had 50-500 nm roughness, .limiting AFM resolution to ~10 nm [Sutcu 1992a and 1992b]. Tensile stress, evidenced by a shift in the Raman spectra, produced spontaneous fractures. Cleavage occurred along (111) planes and thus the crack pattern formed equilateral triangles and 60° parallelograms of various sizes. Some delamination was also noted.

For a (110) homoepitaxial hot-filament film, atomic scale smoothness was noted on the 0.5-5 nm scale but roughness was present on the 1 μm scale [Sutcu 1992b]. No faceting was observed on hills with slopes as high as 40° suggestive of the reason this face is not normally seen in polycrystalline films [Sutcu 1992a].

Polycrystalline Films. For polycrystalline films grown on silicon, AFM has been used to give quantitative measurements of roughness and grain size [Baranauskas 1992, Chernoff 1992, George 1994]. Height variations of less than 10 nm to greater than0.1 μm have been measured. Grain sizes measured by AFM are generally larger than those determined by XRD [Chernoff 1992]. This suggests either that the surface grains which AFM detects are larger or that XRD is more sensitive to twins and other defects.

For non-contiguous diamond growth on silicon, nucleation on scratches has been seen with AFM [Chernoff 1992]. AFM was also used to monitor the initial nucleation and oriented growth of diamond nuclei grown via microwave CVD while a negative dc bias was applied to the (100) silicon substrate [Jiang 1993a]. Formation of an intermediate layer prior to diamond nucleation on silicon has been suggested by AFM studies of the growth surface [Chernoff 1992, George 1994]. This intermediate region has also been examined by AFM in cross-section [George 1994].

Near atomic scale resolution on (111) and (100) surfaces of a polycrystalline film were obtained using AFM under ambient conditions and employing no special surface treatment. The (111) surface showed a spacing of 0.25 nm while a periodicity length of 0.35 nm was seen on the (100) surface [Baranauskas 1992]. In contrast, another high-resolution AFM study of a (100) diamond surface deposited on silicon, gave a 0.25 nm periodicity [Badzian and Badzian 1993].

Mechanical Measurements. In an in situ mechanical measurement, the effect of load on the deformation of a step in cleaved (111) diamond surface was examined using a 300 nm diamond AFM tip. The interaction of a diamond tip indenting a diamond surface has been studied by molecular dynamics simulations [Harrison 1992]. Ex situ AFM also can be used to monitor tribological measurements. AFM also revealed that unidirectional polishing leaves undulations in the orthogonal direction in a epitaxial diamond film [Maguire 1992]. AFM measurements inside a wear track show a dramatic (~ factor of 10) decrease in surface roughness, as compared to adjacent regions (Fig. 21). Depressions created by grain pull out were also identified [Perry 1993].

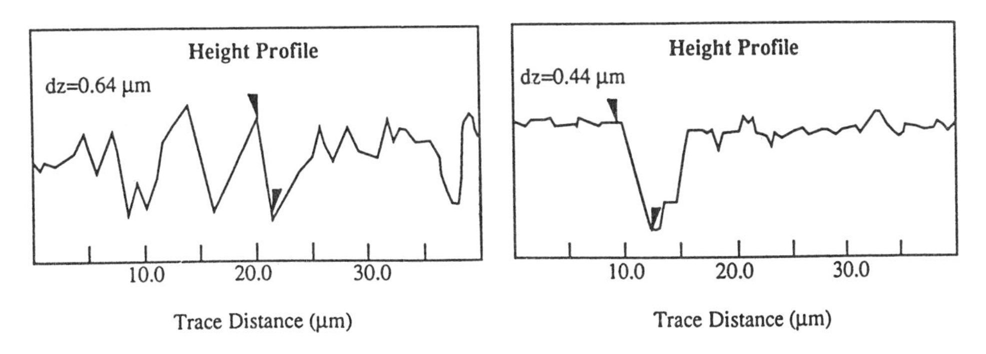

Figure 21. Graphs of the vertical tip displacement across the center of an AFM image, allowing the RMS roughness of polycrystalline diamond to be quantified before (right) and after (left) wear measurements. [Perry 1993]

7. Diffraction

7.1 X-RAY DIFFRACTION (XRD)

7.1.1 Introduction

X-ray diffraction (XRD) is a powerful and commonly available technique for identifying the presence of crystalline phases. In addition to Raman spectroscopy and scanning electron microscopy, XRD is one of the more common characterization tools used to confirm that diamond synthesis has indeed been achieved. The quantitative, high-accuracy measurements of interatomic spacings provided by XRD have also motivated some detailed studies. The XRD patterns also provide information on strain, grain size, preferential orientation and epitaxy. In addition, XRD is non-destructive and can sometimes be used in situ.

Relative to electron diffraction, diffraction intensities for XRD are low. Also, diffraction intensity is inversely proportional to sample volume. Thus, XRD patterns typically represent a spatial average over several mm^2 of area and a few μm of depth. Less commonly, the brightness of a synchrotron irradiation source allows XRD to be performed on smaller samples. The diffracted intensities are largest for high Z elements. Generally, hydrogen is difficult to detect by XRD.

X-rays with wavelength, λ, between 0.5 and 2Å are impinged upon a sample. The diffracted X-rays are measured at 2θ, the angle between the X-ray source and detector. The diffraction can then be described by Bragg's law ($\lambda = 2d \sin\theta$), where d is the spacing between atomic planes within the sample. Three Miller indices, h, k and l, are used to specify the orientation of each set of parallel atomic planes in the crystal. The relative intensity of a diffraction peak from a given set of planes can be determined by using the symmetry of the lattice to calculate a structure factor. For more detailed information on XRD, see Brundle [1992].

7.1.2 Phase Identification

The lattice parameter, a, for cubic diamond at 27°C and 1 atm was found to be 3.5667Å [Joint Committee on Powder Diffraction Standards 1974]. The positions and relative intensities of the strongest XRD peaks having $2\theta < 130°$ for powdered diamond are given in Table 11.

The XRD patterns for four hexagonal and two rhombohedral diamond polytypes have also been calculated [Ownby 1992]. These polytypes differ from cubic diamond only in stacking sequence. Of course, amorphous carbon will not give rise to a XRD pattern.

Table 2. XRD Patterns for Diamond

hkl	2θ*	Relative intensity
111	41.9	100
220	75.3	2.5
311	91.5	16
400	119.5	8

*for CuKα1 radiation (λ=1.5405Å)

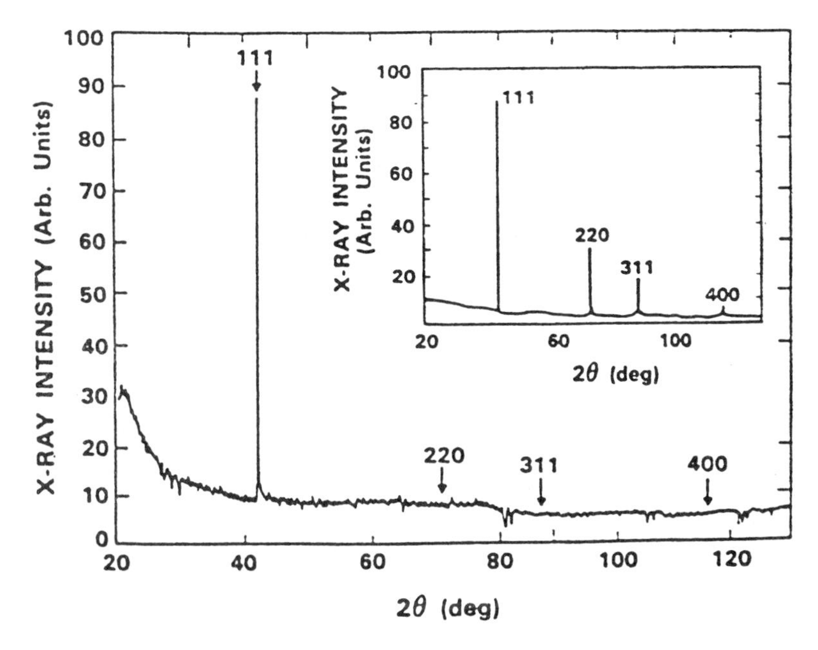

Figure 22. An XRD pattern from diamond seed crystals having a (111) texture. The inset shows the same data for diamond seed crystals with random orientation. [Geis 1989]

Small quantities of other crystalline materials which may have formed simultaneously with diamond can also be identified using XRD. After HPHT growth using nickel, invar and monel as catalyst solvents, Ni_xC (x>4), Cu and SiO_2 (α-quartz) were all observed using XRD [Singh 1990]. A W_2C phase has been identified using low-angle XRD on a laser-ion-deposited carbon film on a tungsten substrate [Wagal 1992].

7.1.3 Crystallite Orientation

The relative intensities of the XRD peaks can be used to determine the orientation of diamond crystallites [Geis 1989]. Preferential orientation, referred to as texture, occurs under some growth conditions. The preferred crystallite orientation parallel to the direction of the incident the X-ray will have the dominant diffraction peak. For example, figure 22 shows the XRD pattern from diamond seed crystallites preferentially attached to a silicon substrate with a strong (111) texture. In contrast, the insert in figure 22 shows the typical intensity distribution reflecting the structure factors for a randomly oriented powder (Table II).

The distribution of crystalline orientations (i.e., the degree of texture) can be quantitated through the variation in diffraction intensity with sample rotation. For these measurements, the X-ray detector angle remains fixed at the $2\theta_{hkl}$ value of the dominant diffraction peak [Geis 1989]. Pole figures can be produced by sample rotation about an axis normal (azimuthal angle) or parallel (polar angle) to the substrate [Wild 1990]. Diffraction from the underlying substrate can be used to determine the degree of epitaxy between the film and its substrate [Jiang 1993b; Kohl 1993].

7.1.4 Strain, Crystallite Size and Defects

The precise measurements of interatomic spacings by XRD allow strain to be studied. Homogeneous strain shifts the diffraction peak position. In some CVD diamond films, the observed 2θ values were higher than those in Table II, indicative of smaller lattice spacings [Windischmann and Epps 1990]. Volume expansion of synthetic single crystals after irradiation with 100 MeV carbon ions and 3 MeV electrons have also been measured by XRD [Vance 1971]. Fractional lattice parameter increases of up to one part in 10^4 could be observed [Maeta 1993]. This irradiation also produced a dramatic (up to 500%) increase in the integrated intensity of the XRD peaks. The increased intensity was attributed to diffraction from strained planes containing ion induced point defects and dislocation loops [Maeta 1993].

Defects such as stacking faults can broaden the diffraction peaks. The linewidths of these peaks are also effected by inhomogeneous strain and crystal size. These latter two effects can sometimes be resolved by analyzing several peaks since broadening resulting inhomogeneous strain increases with $\sin\theta$ while that due to finite crystal size is independent of $\sin\theta$. Assuming other broadening mechanisms are absent, the Scherrer formula:

$$L \sim \lambda / \left((\Delta 2\theta)(\cos\theta)\right) \tag{2}$$

can be used to estimate the crystallite size L. The grain size of polycrystalline diamond films has been estimated in this manner [Windischmann and Epps 1990].

7.1.5 Detailed Applications to Diamond

Using synchrotron radiation, the Laue method of X-ray diffraction was performed on an icosohedral CVD diamond only 5 μm in dimension [Ohsumi 1992]. Within this crystallite, seventeen domains were identified and shown to be interrelated predominately by spinel twins. A synchrotron source was also used to perform X-ray topography on natural and synthetic single crystals [Black 1993]. These measurements are sensitive to variations in strain and defect density within the crystal.

The asymmetric broadening of the (111) diamond diffraction peak has been studied by in situ XRD at high pressures (10 GPa) and temperatures (1550°C). These experiments provided insight into the processes responsible for yield stress of diamond [Weidner 1994] . Under even higher pressure (14 GPa) and employing synchrotron radiation, the phase transformation of graphite to hexagonal diamond was followed by in situ XRD [Yagi 1992].

Structure factors were measured using Pendellösung-fringes from a synthetic diamond single crystal having parallel faces [Takama 1990]. These values are important for evaluating deformation charge densities and the Debye-Waller B factor. The original determination of the structure factors for diamond relied on absolute integrated diffraction intensities observed from a fine-powder sample [Göttlicher and Wölfel 1959]. However, such absolute intensity measurements are inherently subject to many experimental factors. This difficulty is eliminated by using Pendllösung beats, since only extrema in position need to be determined.

7.2 ELECTRON DIFFRACTION

Electron diffraction of three-dimensional solids and transmission electron microscopy (TEM) are typically performed in the same apparatus. Thus, the same sample requirements apply (see TEM section). The information content of an electron diffraction experiment is similar to that of XRD. One major difference is the smaller sample volume required for electron diffraction. Electron diffraction can map out spatial variations in the structure and orientation within the sample. In addition, the electron wavelength is shorter than the X-ray wavelengths used for XDR. At an incident electron energy of ~100 keV, the wavelength of the electron is only 0.037Å. Thus, electron diffraction can determine lattice parameters to four significant figures. General information on electron diffraction can be found in Brundle [1992]. Application of electron diffraction for elucidating diamond structure and defects in detail can be found elsewhere in this handbook.

In a scanning electron microscope (SEM), backscatter Kikuchi diffraction patterns have been obtained from polycrystalline diamond films [Geier 1994]. A mapping of preferential crystallite orientation over a 60 μm x 60 μm area with submicron resolution identified twin boundaries. Thus, the information content is similar to that obtained by X-ray pole figures (see section 7.1.3). However, the backscatter Kikuchi diffraction method is only sensitive to the top 40 μm depth of the exposed surface, while X-rays from which pole figures represent an average orientation over the larger X-ray penetration depth.

7.3 LOW-ENERGY ELECTRON DIFFRACTION (LEED) AND REFLECTION HIGH-ENERGY ELECTRON DIFFRACTION (RHEED)

7.3.1 Introduction

These experiments are two-dimensional forms of electron diffraction. Both LEED and RHEED probe the periodic arrangement of atoms on a surface by electron diffraction. Both techniques determine lattice parameters, symmetry, orientation and reconstruction of clean crystalline surfaces. Background in both techniques is an indication of surface defects. The orientation of adsorbates with respect to the underlying surface arrangement can also be monitored. Also, in-situ RHEED examinations of the quality of layer by layer growth are possible. To maintain well-defined surfaces, both LEED and RHEED require ultrahigh vacuum.

LEED uses low energy electrons (10-1000 eV) which have limited penetration depth and thus provide the surface sensitivity of this technique. RHEED uses higher energy electrons (5 keV to 50 keV) striking the surface at a grazing angle (1-5°). The detector monitors elastically scattered electrons diffracted from the surface. More information on both LEED and RHEED can be found in Brundle [1992].

In most cases, it is desirable to have samples with a single crystalline orientation over the area exposed to the electron beam. Typical beam diameters for LEED are 100 μm, but can be made as small as 10 μm, allowing the experiment to be performed on individual crystallites of at least this size. Amorphous surfaces, which can result from depth profile sputtering, give no diffraction pattern. Sample charging on insulating surfaces can be problematic, but can be reduced by using lower beam currents [Sowa 1988]. Possible damage from the electron beam should also be considered.

7.3.2 Applications of LEED to Diamond

A 2xl reconstructed (111) surface provided a high quality LEED pattern. For this work, a Type IIa natural diamond was cleaved in air, mechanically polished and annealed at 1275 K for one minute [Sowa 1988]. Hydrogen passivated surfaces results in lxl reconstruction of both the (100) and (111) diamond surfaces [Küttel 1994]. The (100) surface shows four spots in a square pattern while the (111) surface displays six LEED spots in a hexagonal arrangement. These authors also reported X-ray photoelectron and Auger electron diffraction patterns from diamond surfaces. A homoepitaxial film grown on a (111) substrate also displayed a lxl LEED pattern [Tsuno]. A change from a lxl to a 2x1 LEED pattern has been observed as both (111) and (100) surfaces of natural diamond reconstruct upon hydrogen desorption [Kubiak 1991]. The loss of chemisorbed hydrogen was induced by increasing the diamond's temperature to over 1000°C. When pits and surface pucking in the diamond surface are visible, a higher background was seen in the LEED pattern [Maguire 1992].

7.3.3 Applications of RHEED to Diamond

RHEED has been performed on homoepitaxial diamond on (100) substrates. Tsuno [1991] observed lx2 and 2xl reconstruction with a 2xl unit cell size of 5.14 x 2.52Å. This surface was found to be stable in air. Differences in the quality of the epitaxial layers grown with between 2% and 8% methane in H_2 on (100) surfaces were also monitored by RHEED [Shiomi 1990]. Higher quality layers produced higher contrast between the diffraction lines and the background and gave fewer extra spots and halos.

High-resolution RHEED was used to study the early stages of diamond growth on Si (100) [Jiang 1993a]. The small crystallites had the lattice spacing of diamond. The experiment also revealed a preferential (100) orientation of the diamond nuclei with respect to the silicon substrate. A RHEED study of silicon wafers after ion bombardment monitored the effect of substrate crystal quality on diamond nucleation [Kobayashi 1993].

8. Additional Methods

8.1 ELLIPSOMETRY

8.1.1 Introduction

Ellipsometry uses the polarization properties of light reflected from a thin sample to infer information about the structure and/or composition of that material. In diamond films, it is used most often to study the film-substrate interface [Bachmann and Wiechert 1992, Cifre 1993, Hong 1994]. This is appropriate since ellipsometry is highly surface sensitive, having monolayer resolution. In addition, because ellipsometry depends on reflected light, it is greatly affected by surface roughness, limiting the application to regions in which the grain size is smaller than the wavelength used for observation. Typically, a wavelength of 280 nm is used for ellipsometric measurements.

8.1.2 Applications to Diamond

Ellipsometry is a nondestructive technique and has potential as an in-situ diagnostic technique [Hong 1994]. Typically the measurement provides a pseudodielectric function which can be compared to that predicted by various models of the surface. For diamond, these models usually include several layers consisting of silicon carbide, amorphous carbon, voids and polycrystalline diamond. These transition or nucleation layers are less than 200Å in thickness [Bachmann and Wiechert 1992, Cifre 1993, Hong 1994]. The technique has been applied to films up to 700 nm in thickness by Windischmann and Epps [1991] to determine the sp^2 carbon fraction in microwave diamond films.

8.1.3 Related Techniques

Other measurements of elastic light scattering and reflectivity have also been used to examine diamond film growth in-situ [Mathis and Bonnot 1993, Bonnot 1994 and John 1993]. Based on observations of both scattered and reflected light, Mathis and Bonnot [1993] were able to determine the index of refraction of HFCVD diamond and propose a model for early diamond growth [Mathis and Bonnot 1993, Bonnot 1994]. In another study, the interference of reflected light was used to monitor growth rates for thin microwave CVD diamond based on an assumed index of refraction [John 1993]. Finally, Locher [1993] has applied reflection and scattering techniques to thick (~50 μm) highly-oriented microwave CVD diamond films. It is the high degree of orientation and optical smoothness which allows observation of these unusually thick samples by light scattering techniques.

8.1.4 Limitations

The extreme surface sensitivity makes ellipsometry attractive for in-situ analysis of the early stages of diamond growth. The primary drawbacks of ellipsometric measurements are their sensitivity to surface roughness and that the interpretation of the data is limited to simple model structures. An excellent review of ellipsometric techniques is provided by Collins [1989].

8.2 MÖSSBAUER SPECTROSCOPY

8.2.1 Introduction

Mössbauer spectroscopy has been used to study ^{57}Fe and ^{133}Cs incorporated into diamond. For these nuclei, the lineshape reflects the symmetry of the surrounding electronic orbitals and neighboring ions. In addition, the intensity of the resonance is sensitive to vibrational characteristics of the impurity. Both of these characteristics allow substitutional atoms to be distinguished from interstitials. Also, ion-implanted Mössbauer active impurities can serve as a probe of damage and amorphization processes in diamond. Finally, Mössbauer spectra of iron/carbon mixtures at high pressures and temperatures has been used to study HPHT catalysis.

In Mössbauer spectroscopy, γ-irradiation produces a resonant transition from the ground state to an excited state of a nucleus. The probability of observing the Mössbauer effect for light elements such as carbon, nitrogen or oxygen is very small, but it has been seen in many of the heavier elements. However, ^{57}Fe is the most commonly studied nucleus. For this isotope, the γ-ray source is the 14.4 keV nuclear excited state of ^{57}Fe, produced from the radioactive decay of ^{57}Co. For details on the theory and practice of Mössbauer spectroscopy see Goldanskii [1968].

The precise energy at which a Mössbauer resonance occurs can differentiate between various bonding environments and is reported as an isomer shift. Since the Doppler effect is used to vary the energy of the γ-ray source, the isomer shift is reported in units of mm/s. Large values of the isomer shift indicate a high density of s-orbital electrons at the nuclei participating in the resonance.

A hyperfine structure in the Mössbauer spectra results when the quadrupole moment of the excited state of the nucleus interacts with a noncubic electric field. The symmetry of this electronic field depends on the both the electronic configuration of the atom and the locations of the surrounding atoms.

The area of a Mössbauer peak depends on the mean square vibrational displacement of the nucleus participating in the resonance. The Debye temperature associated with vibrational modes of a Mössbauer resolved site can be calculated from the experimentally observed variation in Mössbauer intensity with temperature. For a substitutional site, the Debye temperature can also be predicted based on the atomic mass of the impurity and the vibrational force constants of its bonds to the diamond lattice. If the force constants for iron-carbon and carbon-carbon bonds were identical, the Debye temperature for substitutional ^{57}Fe would be 1023 K, as compared to 2230 K for the diamond lattice itself [Sawicki and Sawicka, 1990].

8.2.2 Applications to Diamond

Iron. In order to study ^{57}Fe incorporation, the radioactive parent ion ^{57}Co was implanted into natural diamond at energies between 50 and 100 keV [Sawicka 1981, Sawicki and Sawicka 1982, de Potter 1983, Sawicki and Sawicka 1990]. Implantation of nanophase diamond has also been studied and yielded qualitatively similar results [Sinor 1994]. Typically, between one and three days was needed to acquire each spectrum.

The atomic radius of iron (1.26Å) is large compared to that of carbon (0.77Å) [Sawicka 1981]. Thus, the introduction of substitutional atoms without damage is expected to be difficult. In addition, a substitutional impurity would need to satisfy the electronic constraints imposed by the diamond lattice. For a tetrahedral lattice site, the quadrupole splitting should be absent due to the symmetry of the surrounding electronic field and thus, a narrow singlet line should be observed [Van Rossum 1979]. In addition, the small lattice constant of diamond induces a strong compression of electronic shells of substitutional atoms, resulting in large isomer shifts.

At low ion implantation doses ($<10^{14}/cm^2$), several features of the Mössbauer spectra (Fig. 23) suggest that up to 20% of the ^{57}Fe can be incorporated in substitutional sites [Sawicki and Sawicka 1990]. First, one component of the deconvoluted lineshape is a

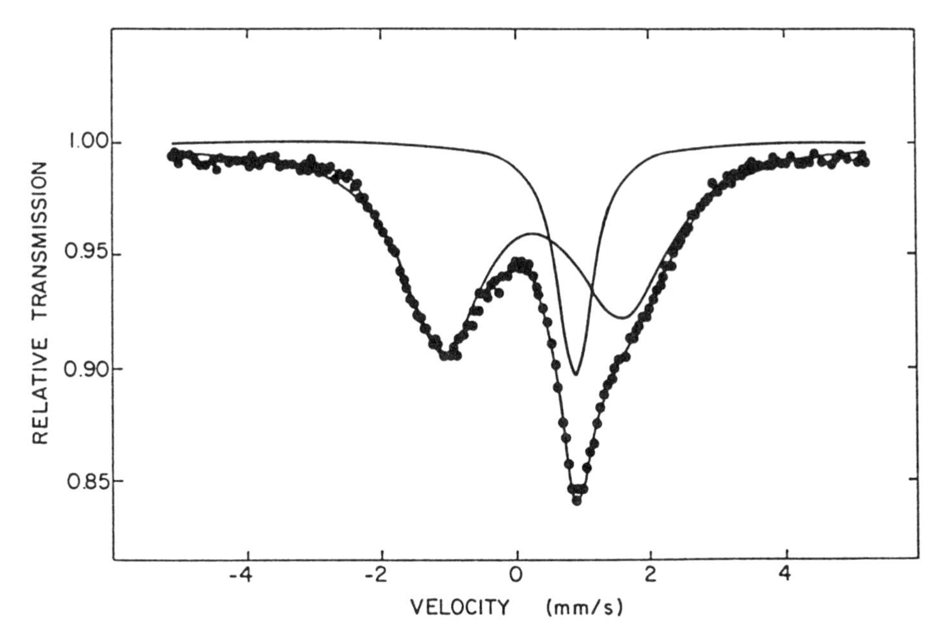

Figure 23. Mössbauer spectrum of ^{57}Fe from a diamond after hot-implantation with ^{57}Co ions showing the deconvolution into a Lorentzian line and an asymmetric quadrupole doublet [Sawicki 1990].

Lorentzian singlet. Although this singlet has a symmetric lineshape, its large linewidth suggests heterogeneity exists among the iron sites produced by ion implantation [Sawicki and Sawicka 1990]. Also as expected for a substitutional impurity, the isomer shift is very high, equivalent to compression by ~ 1500 kbar of pressure. The measured Debye temperature of ~1300 K also suggests a substitutional site for which the force constants of the impurity are greater than those for the diamond matrix. Thus, the singlet component has been assigned as substitutional Fe, with four hybridized sp^3 electrons tetragonally oriented towards the nearest neighbor carbon atoms [Sawicki and Sawicka 1990].

The second lineshape component in figure 23 is an asymmetric doublet with a large quadrupolar splitting. The large electronic field gradient responsible for this splitting cannot arise solely from the 3d electrons of iron. Thus, a non- symmetric distribution of neighboring carbon must exist about these ^{57}Fe nuclei. Further analysis suggests this iron is likely to reside in hexagonal or off -center interstitial sites or may jump between these two types of positions [Sawicki and Sawicka 1990]. This components assignment as to interstitial ^{57}Fe is also supported by its low isomer shift and low Debye temperature (550 K ± 50) relative to the singlet.

The relative proportion of singlet and doublet components seen after exposure of diamond to low ion doses can be changed by thermal treatment. Sawicka [1981] found only a 3% contribution of singlet to the as-implanted lineshape which increased to 20% upon annealing at 600 K. This indicates the healing of disrupted carbon bonds and the ability to create substitutional impurity sites. Alternatively, "hot implantation", where ions strike a diamond heated to ~600 K, can be used to minimize damage for low ion doses [de Potter 1983].

The partial reversibility of implantation damage by thermal treatments suggests that an amorphous region is not formed initially at low ion doses [de Potter 1983]. In addition, the formation of an amorphous phase would allow the release of volume compression and thus produce a sharp decrease in the isomer shift [Sawicki and Sawicka 1982]. This effect is not seen at low ion implantation doses. Similarly, a reduction in non-zero quadrupole splittings would be expected but is not observed. The direct amorphization of diamond may be inhibited by the significant volume expansion required to form an amorphous carbon or a graphitic phase. Thus, retaining a tetrahedral lattice would be favored in order to avoid large compressive forces. Instead, the damage at low ion doses has been attributed to the formation of isolated or clustered carbon interstitials and possibly some graphitized zones [de Potter 1983].

At higher doses of 10^{14}-10^{15} ions/cm^2, the Mössbauer spectra of diamond do indicate the formation of amorphous carbon [de Potter 1983]. Although after low ion doses, the Mössbauer spectra of graphite and diamond are different, the line shapes become comparable after high ion doses, indicating similar electronic environments have been formed in both materials [de Potter 1983]. However, the recoil-free fraction remains higher for diamond after implantation, indicating its vibrational characteristics remain different from the ion damaged graphite [de Potter 1983]. In some cases, a black layer formed on the top surface of the diamond. Acid cleaning removed this layer with less than 3% of ^{57}Co being lost, indicating this amorphized layer contains only a small fraction of the implanted impurity. In addition, no change in the Mössbauer spectra was observed after the acid treatment [Sawicki and Sawicka 1990]. At very high doses ($>10^{17}$ ions/cm^2), the shape consists of several quadrupole doublets, indicative of numerous bonding environments for ^{57}Fe. In addition, the formation of other Co-C phases has been noted [de Potter 1983].

Cesium. A two component lineshape similar to that of figure 23 was also seen for ^{133}Cs incorporated into diamond via implantation of the parent ^{133}Xe ions at doses less than 5×10^{14} ions/cm^2 and an energy of 70 keV [Van Rossum 1979]. The singlet displays a high isomer shift and is assigned as substitutional cesium. The doublet at a lower isomer shift is taken to be cesium at an interstitial location. Also, in analogy to the ^{57}Fe studies, the proportion of the singlet line in the ^{133}Cs Mössbauer spectra could be increased by a 600 K anneal or by performing hot implantation. At higher ion doses, growth of amorphous layer resulted from the ^{133}Cs incorporation.

HPHT Synthesis. Mössbauer spectroscopy has also been used to study iron/graphite mixtures containing less than 0.3 at.% ^{57}Fe subjected to 55 kbar of pressure at 1500°C, similar to the conditions utilized for HPHT diamond synthesis [de Oliveira 1993]. Although iron containing compounds were found on the graphite surface, no ^{57}Fe was either dissolved or intercalated into the graphite. If water was present, hydrated compounds such as $Fe(OH)_2$, iron oxalate ($FeC_2O_4 \cdot (H_2O)_2$) or iron squarate ($FeC_4O_4 \cdot (H_2O)_2$) could be identified in the Mössbauer spectrum. This was suggested as a reason why water is detrimental to HPHT diamond growth [de Oliveira 1993].

8.3 POSITRON ANNIHILATION SPECTROSCOPY (PAS)

8.3.1 Introduction

Room-temperature PAS is predominantly sensitive to vacancy type defects although impurities can also act as positron traps at low temperatures [Dannefaer 1994]. However, PAS is not sensitive to planar defects such as twin planes or stacking faults [Dannefaer 1993]. More detailed information on this technique is available elsewhere [Brandt and Dupasquier 1983]. PAS of diamond has been performed on single crystals [Dannefaer 1994] and free standing films greater than10 μm in thickness [Dannefaer 1993]. Measurements have also been made on thinner films intact on silicon wafers, although this requires substrate effects to be accounted for [Uedono 1990, Suzuki 1992].

PAS is a non-destructive technique in which a positron beam is impinged upon a sample under UHV conditions [Sharma 1990]. A pair of gamma rays is produced for each positron annihilated by an electron from the sample. The positron lifetime, τ, is a quantitative measurement of the electron density at the annihilation site. Neutral vacancies give rise to longer lifetimes since their electron densities are lower than in the bulk [Dannefaer 1994]. Longer lifetimes indicate larger vacancy sizes [Suzuki 1992]. The number of positrons trapped with a given τ can be used to calculate the corresponding vacancy concentration [Dannefaer 1993]. The lower limit of vacancy concentration detectable is ~0.2 ppm while at ~ 20 ppm, all positrons will be annihilated. If sufficient empty space is present, the positron can form a bound state with an electron (positronium) before annihilation, yielding lifetimes greater than 800 ps [Suzuki 1992, Dannefaer 1993].

8.3.2 Application to Diamond

Single Crystals. In single crystal diamonds, positron annihilation occurs predominantly in the bulk (τ ~100 ps) as typical vacancy concentrations are low (~10^{16} cm^{-3}) [Dannefaer 1994]. No positronium formation is observed in single crystal diamond [Suzuki 1992]. Irradiation of type Ia diamond with 3.5 MeV electrons has produced neutral vacancies (τ ~155 ps) and divacancies (τ ~180 ps) [Dannefaer 1992]. In contrast, undoped CVD films can have a significant fraction (>30%) of a 300 to 470 ps lifetime component resulting from multivacancies [Dannefaer 1993]. The corresponding void geometry has been modeled as a 2 to 5Å diameter sphere [Sharma 1990].

CVD Films. Vacancy cluster concentrations in CVD films are estimated at 10^{17}-10^{18} cm^{-3} [Dannefer 1993]. Some monovacancies and divacancies may also be present, but are difficult to resolve. In addition, ~ 5% lifetime component in some CVD films can be ascribed to positronium formation, which may occur in large crystallite vacancies or in void regions between crystallites.

The types of vacancies present and their concentrations are influenced by film deposition conditions. The mean positron lifetime derived from figure 24 was found to increase as the methane percentage in the growth chamber increased. Thus, longer positron lifetime is a qualitative indication of reduced film quality [Suzuki 1992].

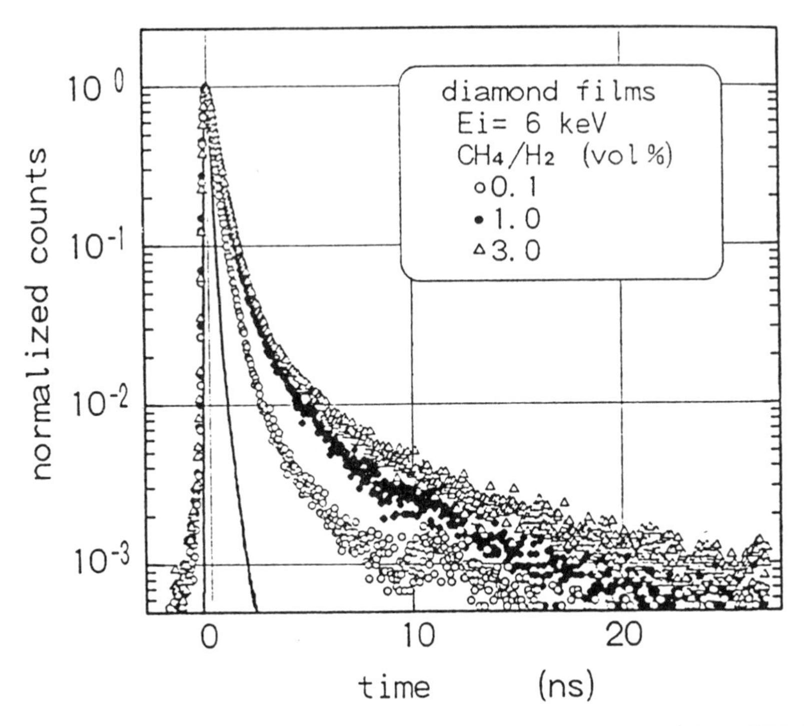

Figure 24. Positron lifetime spectra of bulk diamond (solid line) and three CVD diamond films (data points as shown in legend). Longer positron lifetimes qualitatively correspond to reduced diamond quality [Suzuki 1992].

For boron doping, PAS shows a substantially reduced vacancy concentration [Dannefaer 1993], consistent with morphological trends observed by other techniques [Wang 1992]. Annealing of a boron doped film at greater than 700°C resulted in lowering vacancy concentration while increasing their average size [Dannefaer 1993]. Little change in an undoped film resulted from the identical thermal treatment.

In addition to lifetime measurements, Doppler broadening spectra [Uedono 1990] and angular correlation function measurements [Dannefaer 1994] can be used to probe the momentum distribution of annihilating electrons. Compared to bulk crystals, the electron momentum distribution in diamond films is sharp, as would be expected for formation of positronium. The Doppler spectra has also been used to calculate mean free paths of ~30 nm for positions in CVD diamond. A decrease in mean free path was observed as the concentration of methane used to deposit the film increased [Uedono 1990].

8.4 NUCLEAR METHODS

Several different categories of techniques involving induced nuclear reactions provide quantitative concentrations of many trace impurities in diamond [Sellschop 1974]. Neutron activation analysis detects both light elements such as nitrogen, oxygen and boron and heavier impurities, including iron, cobalt and nickel. Measurements of 61 isotopes, representing 44 elements, have been performed in natural diamonds [Sellschop

1975]. For some heavy elements, neutron activation analysis can detect concentrations as low as one part in 10^{12}. Note that these measurements do not identify differences in chemical bonding states. In addition, these measurements may cause significant radiation damage to the diamond lattice, sometimes resulting in radioactive samples. However, these techniques are often termed as non-destructive in that no sputtering of the sample is required. Thus, it has been use to examine expensive single crystal natural diamonds. Cleaning diamond prior to analysis is crucial for obtaining accurate measurements of the trace elements [Sellschop 1774].

In both natural and synthetic diamonds, hydrogen concentrations have been examined by nuclear reaction analysis (NRA) [Sellschop 1992; Sellschop 1994, Windischmann and Epps 1991]. Hydrogen is well suited to such analysis since there are nuclear reactions, using ^{15}N and ^{19}F, which give interference-free hydrogen signatures. These signatures are gamma rays of unique energy which are detected when the incident species undergoes nuclear reaction with hydrogen at surfaces or in the bulk of the diamond. By varying the energy of the incident beam, depth profiling can be achieved with this technique as well [Sellschop 1992]. With few exceptions [Windischmann and Epps 1991], NRA yields higher ^{1}H contents than those measured by other techniques. This may be related to uncertainty in absolute concentrations and/or sample cleanliness (see also section 4.2).

9. Acknowledgments

The authors gratefully acknowledge the support of the Office of Naval Research and the National Science Foundation (CTS-9057119) in preparing this chapter.

10. References

Ababou, A., Carriere, B., Goetz, G., Guille, J., Marcus, B., Mermoux, M., Mosser, A., Romeo, M., and LeNormand, F. (1992) Surface characterization of microwave-assisted chemically vapor deposited carbon deposits on silicon and transition metal substrates, *Diamond and Related Materials, 1,* 875-881.

Badzian, A. and Badzian, T. (1993) Diamond homoepitaxy by chemical vapor deposition, *Diamond and Related Materials,* **2,** 147-157.

Bachmann, P.K. and Wiechert, D.U. (1992) Optical characterization of diamond, *Diamond and Related Materials,* **1,** 422-433.

Baker, S.M., Rossman, G.R. and Baldeschwieler, J.D. (1993) Topology of synthetic, boron-doped diamond by scanning tunneling microscopy, *Diamond and Related Materials,* **3**, 94-97.

Baranauskas, V., Fukai, M., Rodigues, C.R., Parizotto, N., and Trava-Airoldi, VJ. (1992) Direct observation of chemical vapor deposited diamond films by atomic form microscopy, *Appl. Phys. Lett.,* **60,** 1567-1569.

Baumann, P.K., Humphreys, T.P., Nemanich, R.J., Ishibashi, K., Parik, N.R., Porter, L.M., and Davies, R.F. (1994) Epitaxial Cu contacts on semiconducting diamond, *Diamond and Related Materials,* **3,** 883-886.

Bellamy, LJ. (1980) *The Infrared Spectra of Complex Molecules,* vol. 2, Chapman and Hall.

Bernardez, LJ. and McCarty, K.F. (1993) Determination of diamond film quality during growth using in-situ Raman spectroscopy, *Diamond and Related Materials,* **3,** 22-29.

Black, D.R., Burdette, H.E., and Banholzer, W. (1993) X-ray diffraction imaging of man-made and natural diamond, *Diamond and Related Materials, 2,* 121-125.

Bonnot, M., Mathis, B.S., and Moulin , S. (1994) Growth kinetic analysis of diamond films by in-situ elastic scattering of light and reflectivity, *Diamond and Related Materials,* **3,** 426-430.

Brundle, C.R., Evans, C.A. and Wilson, S. (eds) (1992) *Encyclopedia of Materials Characterization,* Butterworth-Heinemann.

Brandt, W. and Dupasquier, A. (eds) (1983) *Positron Solid-State Physics, North-*Holland.

Bugaets, O.P., Smekhnov, A.A., and Kuzenkov, S.P. (1994) Electron spectroscopies of the Surfaces of Diamond and Cubic Boron Nitride, *J. Electron Spectroscopy and Related Phenomena,* **68,** 713-718.

Burnham, N.A., Colton, RJ., and Pollock, H.M. (1991) Interpretation issues in force microscopy, *J. Vac. Sci. Technol. A,* **9,** 2548-2556.

Busmann, H.-G., Sprang, H., Hertel, I.V., Zimmermann-Edling, W., and Guntherodt, H.-J. (1991) Scanning tunneling microscopy on chemical vapor deposited diamond films, *Appl. Phys. Lett.,* **59,** 295-297.

Butler, J.E., Celii, F.G., Oakes, D.B., Hanssen, L.M., Carrington, W.A., and Snail, K.A. (1990) *High Temp. Sci.,* **27,** 183-197.

Celii, F.G. and Butler, J.E. (1991) *Ann. Rev. Phys. Chem.,* **42,** 643-684.

Chernoff, D.A. and Windischmann, H. (1992) Atomic force microscope imaging for process characterization in diamond film deposition, *J. Vac. Sci. Technol. A* 10, 2126-2130.

Chrenko, R.M. and Strong, H.M. (1975) Physical properties of diamond, *Technical Information Series,* General Electric Corporate Research and Development, 75CRD089.

Chrenko, R.M. (1988) ^{13}C-doped Diamond: Raman Spectra, *J. Appl. Phys.,* **63,** 5873-5875.

Chu, C.J., D'Evelyn, M.P., Hauge, R.H., and Margrave, J.L., (1991) Mechanism of diamond growth by chemical vapor deposition on diamond (100), (111). and (110) surfaces: carbon-13 studies, *J. Appl. Phys.,* **70**,1695-1705.

Cifre, J., Morenza, J.L., Lopez, F., and Esteve, J., (1992) Analysis of contamination in diamond films by secondary non mass spectroscopy, *Diamond and Related Materials,* **1,** 500-503.

Cifre, J., Campmany, J., Bertan, E., and Esteve, J. (1993) Spectroscopic ellipsometry measurements of diamond-crystalline Si interface in chemically vapour-deposited polycrystalline diamond films, *Diamond and Related Materials,* **2,** 728-731.

Cinti, R.C., Mathis, B.S., and Bonnot, A.M. (1992) Surface sensitive electron spectroscopy study of diamond films prepared by hot filament-assisted CVD, *Surface Science,* **279**, 265-271.

Clark, C.D. and Dickerson, C.B. (1992) Raman and photoluminescence spectra of as-grown CVD diamond films, *J. Phys. Condens. Matter,* **4,** 869-878.

Collins, R.W., Cong, Y., Kim, Y.T., Vedam, K., Liou, Y., Inspektor, A., and Messier, R. (1989) Real-time spectroscopic ellipsometry characterization of diamond and diamond-like carbon, *Thin Solid Films,* **8,** 565-578.

Cong, Y., An, I., Nguyen, H.V., Vedam, K., Messier, R., and Collins, R.W. (1991) Real time monitoring of filament-assisted chemically vapor deposited diamond by spectroscopic ellipsometry, *Surface and Coatings Technology,* **49,** 381-386.

Cook, A., Fitzgerald, A.G., Storey, B.E., Wilson, J.I.B., John, P., Jubber, M.G., Milne, D., Drummond, I., Savage, J.A., and Haq, S. (1992) A microbeam analytical characterization of diamond films, *Diamond and Related Materials,* **1**, 478-485.

Dannefaer, S., Mascher, P., and Kerr, D. (1992) Defect characterization in diamonds by means of positron annihilation, *Diamond and Related Materials,* **1**, 407-410.

Dannefaer, S., Bretagnon, T., and Kerr, D. (1993) Positron lifetime investigations of diamond films, *Diamond and Related Materials,* **2**, 1479-1482.

Dannefaer, S. (1994) Positron annihilation in diamond, in G. Davies (ed) *Properties and Growth of Diamond, INSPEC.*

Davis, R.F. (1994) Deposition and characterization of diamond, silicon carbide, and gallium nitride thin films, *J. Crystal Growth,* **137,** 161-169.

de Oliveira, L.S. and da Jornada, J.A.H. (1993) Study of high pressure diamond synthesis by Mössbauer spectroscopy, *Diamond and Related Materials,* **2**, 1322-1326.

de Potter, M. and Langouche, G.(1983) Mössbauer study of the amorphous process in diamond, *J. Phys. B: Condensed Matter*, **53**, 89-93.

Dennig, P.A., Shiomi, H., Stevenson, D.A., and Johnson, N.M. (1992) Influence of substrate treatments on diamond thin film nucleation, *Thin Solid Films* **212,** 63-67.

Dines, T.J., Tither, D., Dehbi, **A.,** and Matthews, A. (1991) Raman spectra of hard carbon films and hard carbon films containing secondary elements, *Carbon,* **29,** 225-231.

Dischler, B., Wild, C., Müller-Serbet, W., and Koidl, P. (1993) Hydrogen in polycrystalline diamond, *Physica B,* **185,** 217-221.

Dowling, D.P., Ahern, M.J., Kelly, T.C., Meenan, B.J., Brown, N.M.D., O'Connor, G.M., and Glynn, T.J. (1992) Characterization Study of Diamond and Diamond-like Carbon, *Surface and Coatings Technology,* **53,** 177-183.

Duijvestijn, M.J., van der Lugt, C., Smidt, J., Wind, R.A., Zilm, K.W.. and Staplin, D.C. (1983) ^{13}C NMR spectroscopy in diamonds using dynamic nuclear polarization, *Chem. Phys. Lett.,* **102,** 25-28.

Engel, W., Ingrain, D.C., Keay, J.C., and Kordesch, M.E. (1994) Removal of nondiamond carbon from the surface of CVD diamond films, *Diamond and Related Materials,* **3,** 1227-1229.

Erchak, D.P., Ulyashin, A.G., Glefand, R.B., Penian, N.M., Zaitsev, A.M., Varichenko, V.S., Efimov, V.G., and Stelmakh, V.F. (1992) Hydrogen passivation of the paramagnetic centers in amorphous region of ion implanted diamond, *Nuclear Instruments and Methods in Physics Research,* **B69,** 271-276.

Erchak, D.P., Efimov, V.G., Azarko, I.I., Denisenko, A.B., Penina, N.M., Stelmakh, V.F., Varichenko, V.S., Zaitsev, A.M., Melnikov, A.A., Ulyashin, A.G., Shlopak, N.V., Bouilov, L.L., Varnin, V.P., Botev, A.A., Sokolina, G.A., and Teremetskaya, I.G. (1993) Electron paramagnetic resonance of boron-implanted natural diamonds and epitaxial diamond films, *Diamond and Related Materials* **2**, 1164-1167.

Everson, M.P. and Turner , M. A. (1991) Studies of nucleation and growth morphology of boron-doped diamond microcrystals by scanning tunneling microscopy, *J. Vac. Sci. Technol. B,* **9,** 1570-1576.

Everson, M.P., Turner, M.A., Scholl, D., Stoner, B.R., Sahaida, S.R., and Bade, J.P. (1994) Positive identification of the ubiquitous triangular defect on the (100) faces of vapor-grown diamond, *J. Appl. Phys.* **75,** 169-172.

Fabisiak, K., Maar-Stumm, M., and Blank, E. (1993) Defects in chemically vapor-deposited diamond films studied by electron spin resonance and Raman spectroscopy, *Diamond and Related Materials,* **2,** 722-727.

Fanciulli, M. and Moustakas, T.D. (1993) Defects in diamond thin films, *Phys. Rev. B* **48,** 14982-14988.

Fayette, L., Marcus, B., Mermoux, M., Abello, L., and Lucazeau, G. (1994a) In-situ Raman investigation of diamond films during growth and etching processes, *Diamond and Related Materials,* **3,** 438-442.

Fayette, L., Marcus, B., Mermoux, M., Rosman, N., Abello, L., and Lucazeau, G. (1994b) In-situ Raman spectroscopy during diamond growth in a microwave plasma reactor, *J. Appl. Phys.,* **76,** 1604-1608.

Frauenheim, Th., Stephan, U., Blaudeck, P., Porezag, D., Busman, H.-G., and Zimmermann-Edling, W. (1994) Stability and reconstruction of diamond (100) and (111) surfaces, *Diamond and Related Materials,* **3,** 966-974.

Furusawa, M. and Ikeya, M. (1990) Distribution of nitrogen and nicel in a synthetic diamond crystal observed with scanning ESR imaging, *J. Physical Society of Japan,* **59**, 2340-2343.

Garcia, **A.,** Wang, C.. and Kordesch, M.E. (1992) Controlled deposition and lateral growth of an ordered monolayer of carbon on Mo(100) observed in situ, *Appl. Phys. Lett.,* **61,** 2984-2986.

Geier, S., Schreck, M., Hessmer, R., Rauschenbach, B., Stritzker, B., Kunze, K., and Adams, B.L. (1994) Characterization of the near-interface region of chemical vapor deposited diamond films on silicon by backscatter Kikuchi diffraction, *Appl. Phys. Lett.,* **65,** 1781-1783.

Geis, M.W. (1989) Growth of textured diamond films on foreign substrates from attached seed crystals, *Appl. Phys. Lett.,* **55**, 550-552.

George, M.A., Burger, A., Collins, W.E., Davidson, J.L., Barnes, A.V., and Tolk, N.H. (1994) Investigation of nucleation and growth processes of diamond films by atomic force microscopy, *J. Appl. Phys.,* **76,** 4099 4106.

Gheeraert, E., Deneuville, A., Bonnet, A.M., and Abello, L. (1992) Defects and stress analysis of the Raman spectrum of diamond films, *Diamond ad Related Materials,* **1**, 525-528.

Gleason, K.K. Diamond thin films, in D.M. Grant and R.K. Harris (eds.) *Encyclopedia of Nuclear Magnetic Resonance* (Wiley, to be published 1995).

Goldanski, V.I. and Herber, R.H. (1968), *Chemical Applications of Mosssbauer Spectroscopy,* Academic Press.

Gottlicher, S., and Wölfel, E. (1959) Röntgenographische Bestimmung der Elektronenverteilung in Kristallen: VII Die Elektronendicten im Diamantgitter und im Gitter des Silicium, Z. *Electrochem.,* **63,** 891 - 901.

Gupta, Y.M., Horn, P.D., and Yoo, C.S. (1989) Time-resolved Raman spectrum of shock-compressed diamond," *Appl. Phys. Lett..* **55** 33-35.

Hansma, P.K., Elings, V.B., Marti, O., and Bracker, C.E. (1988) Scanning tunneling microscopy and atomic force microscopy: application to biology and technology, *Science,* **242,** 209-216.

Hanssen, L.M., Snail, K.A., Carrington, W.A., Butler, J.E., Kellogg, S., and Oakes, D.B. (1991) Diamond and non-diamond synthesis is an oxygen acetylene flame, *Thin Solid Films,* **196,** 271-281.

Haq, S., Somerton, C., Tunnicliffe, D., Savage, J.A., John, P., Milne, D.K., Jubber, M.G., Wilson, J.I.B. (1993) An X-ray photoelectron spectroscopy study of the surface layers between diamond crystallites and silicon substrate deposited by microwave-plasma chemical vapour deposition, *Diamond and Related Materials,* **2,** 558-561.

Harrison, J.A., White, C.T., Colton, R.J., and Brenner, D.W. (1992) Nanoscale investigation of indentation, adhesion and fracture of diamond (111) surfaces *Surf. Sci.,* **271,** 57-67.

Hartnett, T. and Miller, R.P. (1990) Potential limitations for using CVD diamond as LWIR windows, *SPIE Proceedings,* **1307,** 60-69.

Henrichs, P.M., Cofield, M.L., Young, R.H. and Hewitt, J.M. (1984) Nuclear spin-lattice relaxation via paramagnetic centers in solids: ^{13}C NMR of diamonds, *J. Magn. Reson.,* **58,** 85-94.

Hinneberg, J., Eck, M., and Schmidt, K. (1992) Hot-filament grown diamond films on Si: characterization of impurities, *Diamond and Related Materials,* **1**. 810-813.

Hoch, M.J.R. and Reynhardt, E.C., (1988) Nuclear spin-lattice relaxation of dilute spins in semiconducting diamond, *Phys. Rev. B,* **37,** 9222-9226.

Hoffman, A., (1994) Fine structure in the secondary electron emission spectrum as a spectroscopic tool for carbon surface characterization, *Diamond ad Related Materials,* **3,** 691-695.

Hoinkis, M., Weber, E.R., Landstrass, M.I., Plano, M.A., Han, S., and Kania, D.R. (1991) Paramagnetic nitrogen in chemical vapor deposition diamond films, *Appl. Phys. Lett.,* **59,** 1870-1871.

Holder, S.L., Rowan, L.G. and Krebs, JJ. (1994) Electron Paramagnetic resonance forbidden transitions from hydrogen in polycrystalline diamond films, *Appl. Phys. Lett.,* **64,** 1091-1093.

Hong, B., Wakagi, M., Collins, R.W., An, I., Engdahl, N.C., Drawl, W. and Messier, R. (1994) Real-time spectroscopic ellipsometry studies of diamond film growth by microwave plasma-enhanced chemical vapour deposition, *Diamond and Related Materials,* **3.** 431-434.

Ingram, C. and Lake, M.L. (1992) Hydrogen content of chemical vapour deposited diamond, *Surface Modification Technologies,* 227-234.

Ingram, C., Keay, J.C., Tang, C., Lake, M.L., and Ting, J-M (1993) Trapping of hydrogen in diamond, *Diamond and Related Materials,* **2**, 1414-1419.

Jia, H., Shinar, J., Lang, D.P., and Pruski, M. (1993). Nature of the native defect ESR and hydrogen-dangling-bond centers in thin diamond films, *Phys. Rev. B,* **48**, 17595-17598.

Jiang, X., Schiffmann, K., Westphal, A., and Klages, C.-P. (1993a) Atomic force-microscopic study of heteroepitaxial diamond nucleation on (100) silicon, *Appl. Phys. Lett.,* **63,** 1203-1205.

Jiang, X., Klages, C.-P., Zachai, R., Hartweg, M., and Füsser, H.-J. (1993b) Epitaxial diamond thin films on (001) silicon substrates, *Appl. Phys. Lett.,* **62,** 3438-3440.

Jin, S., Fanciulli, M., Moustakas, T.D. and Robins, L.H. (1994) Electronic characterization of diamond films prepared by electron cyclotron resonance microwave plasma, *Diamond and Related Materials,* **3**, 878-882

John, P., Drummond, I.C., Milne, D.K., Jubber, M.G., and Wilson, J.I.B. (1993) Fundamental limits to growth rates in a methane-hydrogen microwave plasma, *Diamond and Related Materials,* **3**, 56-60.

Joint Committee on Powder Diffraction Standards *(1974), Selected Powder Diffraction Data for Minerals: Data Book,* 1st ed., Joint Committee on Powder Diffraction Standards , pp. 91, 190, 610, 804.

Jubber, M.G., Wilson, J.I.B., Drummond, I.C., John, P., Milne, D.K. (1993) Microwave plasma chemical vapour deposition of high purity diamond films, *Diamond and Related Materials,* **2,** 402-406.

Karasawa, S., Mitsuhashi, M., Ohio, S., Kobayashi, K., Watanabe, T., Hirai, K., Hirai, K., Horiguchi, K., and Togashi, F. (1993) Crystal growth of epitaxial CVD diamond using ^{13}C isotope and characterization of dislocation by Raman spectroscopy , *J. Cryst. Growth,* **128,** 403 - 407.

Klein, C., Hartnett, T., Miller, R., and Robinson, C.J. "Lattice Vibrational Modes and Defect-Activated IR Absorptions in CVD Diamond", Proceedings of the 2nd Intentional Conference on Diamond Materials (ECS, Pennington, NJ 1991), 435-442.

Knight, D.S. and White, W.B. (1985) Characterization of diamond films by Raman spectroscopy, *J. Mater. Res.,* **4,** 385-393.

Kobayashi, K., Kumagai, M., Karasawa, S. and Watanabe, T. (1993) Effect of ion implantation and surface structure of silicon on diamond film nucleation, *J. Cryst. Growth,* **128,** 408 - 412.

Kohl, R., Wild, C., Herres, N., Koidl, P., Stoner, B.R., and Glass, J.T. (1993) Oriented nucleation and growth of diamond films on β-SiC and Si, *Appl. Phys. Lett.*, **63**, 1792-1794.

Küttel, O.M., Agostino, R.G., Fasel, R., Osterwalder, J., and Schlapbach, L. (1994) X-ray photoelectron and Auger electron diffraction study of diamond and graphite surfaces, *Surf. Sci.,* **312,** 131-142.

Landstrass, M.I. and Ravi, K.V. (1989) Hydrogen passivation of electrically active defects in diamond, *Appl. Phys. Lett.,* **55**, 1391-1393.

Levy, D.H. and Gleason, K.K. (1992) Multiple-quantum NMR as a probe for the dimensionality of hydrogen in polycrystalline powders and diamond films, *J. Phys. Chem.,* **31**, 8125-8131.

Levy, D.H. and Gleason, K.K. (1993) An NMR probe for low-level ^{1}H detection in solids, *J. Vac. Sci. Technol.,* **A 11**, 195-198.

Liou, Y., Weimer, R., Knight, D., Messier, R. (1990) *Appl. Phys. Lett.,* **56**, 437-439.

Locher, R., Wild, C., Muller-Serbert, W., Kohl, R., Koidl, P. (1993) Optical reflection and angle-resolved light scattering from polycrystalline diamond films, *Diamond and Related Materials,* **2,** 11248-1252.

Lock, H. and Maciel, G.E. (1992) Natural-abundance ^{13}C dynamic nuclear polarization experiments on chemical vapor deposited diamond film, *J. Mater. Res.,* **7,** 2791-2797.

Lock, H., Wind, R.A., Maciel, G.E., and Johnson, C.E. (1993) A study of ^{13}C enriched chemical vapor deposited diamond films by means of ^{13}C nuclear magnetic resonance, electron paramagnetic resonance and dynamic nuclear polarization, *J. Chem. Phys.,* **99**, 3363-3373.

Loubser, J.H.N. and van Wyk, J.A. (1978) Electron spin resonance in the study of diamond, *Rep. Prog. Phys.,* **41,** 1201-1248.

Maby, E.W., Magee, C.W., and Morewood, J.H. (1981) Volume expansion of ion-implanted diamond, *Appl. Phys. Lett.,* **39,** 157-158.

Maeta, H., Haruna, K., Bang, L., and Ono, F. (1993) X-ray diffraction studies of self-ion irradiated synthetic single crystal diamond, *Nuclear Instruments and Methods in Physics Research,* **B80/81,** 1477-1479.

Magee, C.W. and Botnick, E.M. (1981) Hydrogen depth profiling using SIMS-problems and their solutions, *J. Vac. Sci. Technol.,* **19,** 47-52.

Maguire, H.G., Kamo, M., Lang, H.P. and Guntherodt H.-J. (1992) Surface morphology determination of LPCVD homoepitaxial diamond using scanning tunneling and atomic force microscopy, *Appl. Surface Sci.,* **60/61,** 301-307.

Mathis, B.S. and Bonnot, A.M. (1993) Elastic scattering of light and reflectivity development during low pressure diamond growth, *Diamond and Related Materials,* **2,** 718-721.

McNamara, K.M. and Gleason, K.K. (1992a) Selectively ^{13}C enriched CVD diamond films studied by NMR, *J. Appl. Phys.,* **71**, 2884-2889.

McNamara, K.M., Gleason, K.K., and Butler, J., and Vestyck, DJ. (1992b) Evaluation of diamond films by NMR and Raman spectroscopy, *Diamond and Related Materials,* **19** 1145-1155.

McNamara, K.M., Gleason, K.K., and Robinson, C.J. (1992c) Quantitative correlation of infrared absorption with nuclear magnetic resonance measured hydrogen content in diamond films, *J. Vac. Sci. Technol. A.* **10**, 3143-3148.

McNamara, K.M., Levy, D.H., Gleason, K.K. and Robinson, C.J. (1992d) Nuclear Magnetic Resonance and absorption studies of polycrystalline diamond, *Appl. Phys. Lett.*, **60,** 580-582.

McNamara, K.M. and Gleason, K.K. (1993) Comparison of tantalum and rhenium filaments in diamond chemical vapor deposition, J. *Electrochem. Soc.*, **140,** L22-L24.

McNamara, K.M., Williams, B.E., Gleason, K.K. and Scruggs, B.E. (1994a) Identification of defects and impurities in chemical vapor deposited diamond through infrared spectroscopy, J. *Appl. Phys.*, **76**, 2466-2472.

McNamara, K.M. and Gleason, K.K. (1994b) Radial distribution of hydrogen in CVD diamond, *Chem. of Materials,* **6,** 39-43.

McNamara, K.M., Gleason, K.K. and Scruggs, B.E. (1994c) The effect of impurities on the IR absorption of chemically vapor deposited diamond, *Thin Solid Films,* **253,** 157-161.

McNamara, K.M., Scruggs, B.E. and Gleason, K.K. (1995) The influence of hydrogen incorporation on the thermal conductivity of polycrystalline diamond, *J. Appl. Phys.*, **77,** 1459-1462.

Mearini, G.T., Krainsky, I.L., Dayton, J.A. Jr., Wang, Y., Zorman, C.A., Angus, J.C., and Hoffman, R.W. (1994) Stable secondary electron emission observations from chemical vapor deposited diamond, Appl. *Phys. Lett.*, **65**, 2702-2704.

Mercer, T.W., Carroll, D.L., Liang, Y., DiNardo, N.J., and Bonnell, D.A. (1994) Photoinduced scanning tunneling microscopy of insulating diamond films", *J. Appl. Phys.*, **75,** 8225-8227.

Merwin, L.H., Johnson, C.E. and Weimer, W.A. (1994) ^{13}C NMR investigations of CVD diamond: correlation of NMR and Raman spectral linewidths, *J. Mater. Res.*, **9,** 631-635.

Miller, D.J. and McKenzie, D.R. Electron spin resonance study of amorphous hydrogenated carbon films, *Thin Solid Films,* **108**, 257-264.

Mitra, S. and Gleason, K.K. (1992) IH NMR Studies on the Effects of Annealing on Chemical Vapor Deposition (CVD) Diamond, *Diamond and Related Materials,* **2,** 126-129.

Neimanich, RJ., Bergman, L., LeGrice, Y.M., and Shroder, R.E. (1991) Raman characterization of diamond film growth, in T.S. Sudarshan and J.F. Braza *(eds), Surface Modification Technologies* VI, The Minerals, Metals and Materials Society, pp. 741-752.

Oakes, D.B., Butler, J.E., Snail, K.A., Carrington, W.A., and Hanssen, L. M. (1991) Diamond synthesis in oxygen-acetylene flames: inhomogeneities in the effects of hydrogen addition, *J. Appl. Phys.*, **69**, 2602-2610.

Oelhafen, P., Francz, G., and Kania, P. (1993) Characterization of diamond films by photoelectron spectroscopy , *Proc. Electrochem. Soc.*, **93-17**, 799-807.

Ohsumi, K., Takase, T., Hagiya, K., Shimizugawa, Y., Miyamoto, M., Mitsuda, Y., and M. Ohmasa (1992) Characterization of 5-pm-sized icosahedral chemical vapor deposited by synchrotron X-ray diffraction with Laue method, *Rev. Sci. Instrum.*, **63**, 1181-1184.

Ownby, P.D., Yang, X., and Liu, J. (1992) Calculated X-ray diffraction data for diamond polytypes, *J. Am. Ceram. Soc.*, **75**, 1876-1883.

Pehrsson, P.E. (1993). The Acid-cleaned CVD Diamond Surface, *Proc. Electrochem. Soc.*, **93-17**, 668-673.

Perry, S.S., Ager, J.W., III, and Somorjai, G. A. (1993) Combined surface characterization and tribological (friction and wear) studies of CVD diamond films, *J. Mater. Res.*, **8**, 2577-2586.

Petrich, M.A. (1989) Nuclear magnetic resonance studies of amorphous hydrogenated carbon, *Mat. Sci. Forum*, **52-53**, 387 - 406.

Pruski, M., Lang, D.P., Hwang, S.-J., Jia, H., and Shinar, J., (1994) On the structure of thin diamond films: a ^{1}H and ^{13}C nuclear magnetic resonance study, *Phys. Rev. B*, **49**, 10635-10642.

Retcofsky, H.L. and Friedel, R.A. (1973) Carbon-13 magnetic resonance in diamonds, coals, and graphite, *J. Phys. Chem.*, **77**, 68-71.

Samlenski, G., Flemig, R., Brenn, C., Wild, W., Muller-Sebert, and Koidl, P. (1994) Characterization of structure and defects in textured diamond films by ion channeling, *Diamond and Related Materials*, **3**, 1091-1096.

Samonenko, N.D., Tokil, V.V., and Gorban, S.V. (1992) Electron paramagnetic resonance of phosphorous in diamond, *Sov. Phys. Solid State*, **33**, 14091410.

Sato, Y., Kamo, M., and Setaka N. (1986) Characterization of diamond and diamond-like carbon films, *Proc. Ninth Symposium ISIAT86*, 223-228.

Sawicka, B.D., Sawicki, J.A., and de Waard, H. (1981). Mössbauer effect evidence of high internal pressure at iron atoms implanted in diamond, *Phys. Lett.*, **85A**, 303-307

Sawicki, J.A. and Sawicka, B.D, (1982) Implantation induced diamond to amorphous-carbon transition studied by conversion electron Mössbauer spectroscopy, *Nuclear Instruments and Methods,* **194,** 465-469.

Sawicki, J.A. and Sawicka, B.D. (1990) Properties of ^{57}Fe hot-implanted into diamond crystals studied by Mössbauer emission spectroscopy between 4 and 300K. *Nuclear Instruments and Methods in Physics Research ,* **B 46,** 38-45.

Scruggs, B.E. and Gleason, K.K. (1993) Analysis of Fluorocarbon Treated Diamond Powders by Solid-State F-19 Nuclear Magnetic Resonance, *J. Phys. Chem.,* **97,** 9187-9195.

Sellschop, J.P.F., Bibby, D.M., Erasmus, C.S., and Mingay, D.W. (1974) Determination of impurities in diamond by nuclear methods, *Diamond Research,* **11,** 43-50.

Sellschop, J.P.F., Bibby, D.M., Erasmus, C.S., Fesq, H.W., and Kable E.J.D. (1975) *The Determination of Trace Elements in Natural Diamonds by Instrumental Neutron-Activation Analysis,* National Institute for Metallurgy.

Sellschop, J.P.F., Connell, S.H., Madiba, C.C.P., Sideras- Haddad, E., Stemmet, M.C., Bharuth-Ram, K., Appel, H., Kundig, W., Patterson, B., and Holzschuh, E. (1992) Hydrogen in and on natural and synthetic diamond, *Nuclear Instruments and Methods in Physics Research.,* **B68,** 133-140.

Sellschop, J.P.F., Connell, S.H., Madiba, C.C.P., Haddad, E.-S., Kalbitzer, S., Jans, S., Oberschactsiek, P., Baruth-Ram, K., Schneider, J., and Kiefl, R. (1994) The nature of the state of hydrogen on the surface and in the bulk of natural and synthetic diamond, *Vacuum,* **45,** 396-402.

Sharma, S.C., Dark, C.A., Hyer, R.C., Green, M., Black, T.D., Chourasia, A.R., Chopra, D.R. and Misha, K.K. (1990) Deposition of diamond films at low pressures and their characterization by positron annihilation, Raman, scanning electron microscopy, and X-ray photoelectron spectroscopy, *Appl. Phys. Lett.,* **56**, 1781-1783.

Shih, H.C., Sung, C.P., Tang, Y.S., and Chen, J.G. (1992) Microstructure and characterization of diamond films grown on various substrates, *Surface and Coatings Technology,* **52,** 105-114.

Shiomi, H., Tanabe, K., Nishibayashi, Y., and Fujimori, N. (1990) Epitaxial growth of high quality film by the microwave plasma-assisted chemical-vapor-deposition method, *Japan. J. Appl. Phys.,* **29**, 34-40.

Shroder, R.E., Nemanich, R.J., and Glass, J.T. (1990) Raman Analysis of the Composite Structures in Diamond Thin Films, *Phys. Rev. B,* **41,** 3738 3745.

Singh, B.P., Gupta, S.K., Dhawan, U., and Lal, K., (1990) Characterization of synthetic diamonds by EPR and X-ray diffraction techniques, *J. Mater. Sci.,* **25,** 1487-1490.

Sinor, T.W., Standifird, J.D., Davanloo, F., Taylor, K.N., Hong, C., Carroll, J.J., and Collins, C.B. (1994) Mössbauer effect measurement of the recoil-free fraction for ^{57}Fe implanted in nanophase diamond film, *Appl. Phys. Lett.,* **64,** 1221-1223.

Skytt, P., Johansson, E., Wassdahl, N., Wiell, T., Guo, J.H., Carlsson, J.O., and Nordgren, J. (1993) A soft X-ray emission study of EFCVD diamond *films, Diamond and Related Materials,* **3**, 1-6.

Socrates, G., (1980), *Infrared Characteristic Group Frequencies,* John Wiley & Sons.

Solin, S.A. and Ramdas, A.K. (1970) Raman spectrum of diamond, *Phys. Rev. B,* **19** 1688-1698.

Sowa, E.C., Kubiak, G.D., Stulen, R.H., and Van Hove, M.A. (1988) Summary abstract: structural analysis of the diamond C(111)-(2xl) reconstructed surface by low-energy electron diffraction, *J. Vac. Sci. Technol.* A, **6**, 832-833.

Spiberg, P., Woodin, R.L., Butler, J.E., and Dahr, L. (1993) *In-situ* Fourier transform IR emission spectroscopy of diamond chemical vapor deposition, *Diamond and Related Materials,* **2,** 708-712.

Stuart, S.A., S. Prawer, and P.S. Weiser, (1993) Variation of the Raman diamond lineshape with crystallographic orientation of isolated chemical vapour-deposited diamond crystals, *Diamond and Related Materials,* **2**. 753-757.

Stuart, S.A., Prawer, S., and Weiser, P.S. (1994) Polarized Raman spectroscopy of chemically vapour deposited diamond films, *Appl. Phys. Lett.,* **65,** 2248-2250.

Sutcu, L.F., Chu, C.J., Thompson, M.S., Hauge, R.H., Margrave, J.L. and D'Evelyn, M.P. (1992a) Atomic force microscopy of (100), (110), and (111) homoepitaxial diamond *films. Appl. Phys.,* **71,** 5930-5940.

Sutcu, L.F., Thompson, M.S., Chu, C.J., Hauge, R.H., Margrave, J.L., and D'Evelyn, M.P. (1992b) Nanometer-scale morphology of homoepitaxial diamond films by atomic force microscopy, *Appl. Phys. Lett.,* **60**, 1685 1687.

Suzuki, R., Kobayashi, Y., Mikado, T., Ohgaki, H., Chiwaki, M., Yamazaki, T., Uedono, A., Tanigawa, S. and Funamoto, H. (1992) Characterization of diamond films by means of a pulsed positron beam, *Jpn. J. Appl. Phys.,* **31,** 2237-2240.

Takama, T., Tsuchiya, K., Kobayashi, K., and Sato, S. (1990) Measurement of the Structure Factors of Diamond, *Acta Cryst.,* **A46,** 514-517.

Tromp, R.M. (1994) Low-energy electron microscopy, *MRS Bulletin,* June 1994, 44-46.

Tuinstra, F. and Koenig, J.L. (1970) Raman Spectrum of Graphite, *J. Chem. Phys.*, **53**, 1126-1130.

Turner, K.F., LeGrice, Y.M., Stoner, B.R., Glass, J.T., and Nemanich, R.J. (1991a) Topography and nucleation of chemical vapor deposition of diamond on silicon by scanning tunneling microscopy, *J. Vac. Sci. Technol. B,* **9,** 914-919.

Turner, K.F., Stoner, B.R., Bergman, L., Glass, J.T., and Nemanich, R.J., (1991b) Observation of surface modification and nucleation during deposition of diamond on silicon by scanning tunneling microscopy, J. *Appl. Phys.,* **69,** 6400-6405.

Turner, M.A. and Everson, M.P. (1993) Speculation on the reconstruction of vicinal (100) faces of vapor-deposited diamond, *J. Mater. Res.,* **8,** 1770 1772.

Tsuno, T., Imai, T., Nishibayashi, Y., Hamada, K. and Fujimori, N. (1991) Epitaxial grown diamond (001) 2xl/lx2 surface investigated by scanning tunneling microscopy in air, *Jap. J. Appl. Phys.,* **30**, 1063-1066.

Tsuno, T., Tomikawa, T., Shikata, S.-I., and Fujimori, N. (1994). Diamond homoepitaxial growth on (111) substrate investigated by scanning tunneling microscope, *J. Appl. Phys.,* **75,** 1526-1529.

Uedono, A., Tanigawa, S., Funamoto, H., Nishikawa, A., Takahashi, K. (1990) Characterization of diamond films synthesized on Si from a gas phase in microwave plasma by slow positrons, *Jpn. J. Appl. Phys.,* **29,** 555-559.

Vance, E.R. (I 97 1) X-ray study of neutron irradiated diamonds, *J. Phys. C: Solid St. Phys.,* **4**, 257-262.

Van Rossum, M., De Bruyn, J., Langouche, G., de Potter, M., and Coussement, R. (1979) Observation of the amorphization process in diamond by Mössbauer spectroscopy, *Phys. Lett.,* **73A**, 127-128.

Vazquez, L., Sanchez, O., Messequer, F.,and Albella, J.M. (1994a) STM nanometric study of the initial stages of diamond film growth: quantitative measurement of the (111) and (100) surface roughness, *Diamond and Related Materials,* **3,** 715-719.

Vazquez, L., Sanchez, O., and Albella, J.M. (1994b) Scanning tunneling microscopy morphological study of the first stages of growth of microwave chemical vapor deposited thin diamond films, *J. Vac. Sci. Technol. B,* **12**, 1-7.

Wagal, S.S., Adhi, K.P., Joag, D.S., Sharma, A.K., Abhyankar, N., and Kulkarni, S.K. (1992) A study of laser-ion-deposited carbon films on tungsten by X-ray diffraction, field ion microscopy, and electron spectroscopy, *J. Appl. Phys.,* **71**, 1052-1054.

Wagner, J., Wild, C., and Koidl, P. (199 1) Resonance effects in Raman scattering from polycrystalline diamond films, *Appl. Phys. Lett.,* **59**, 779-781.

Wang, X.H., Ma, G.-H. M., Zhu, W., Glass, J.T., Bergman, L., Turner, K.F., and Nemanich, RJ. (1992) Effects of boron doping on the surface morphology and structural imperfections of diamond films, *Diamond and Related Materials,* **1**, 828-835.

Watanabe, I. and Sugata, K. (1988) ESR in diamond thin films synthesized by microwave plasma chemical vapor deposition, *Jap. J. Appl. Phys.,* **27,** 1808-1811.

Weidner, D.J., Wang, Y., and Vaughan, M.T. (1994) Strength of diamond, *Science,* **266,** 419 - 422.

Wild, C., Herres, N., and Koidl, P. (1990) Texture formation in polycrystalline diamond films, *J. Appl. Phys.,* **68**, 973-978.

Wilkie, C.A., Ehlert, T.C., and Haworth, D.T. (1978)J. *Inorg. Nucl. Chem., A* solid-state ^{13}C-NMR study of diamonds and graphites, **40,** 1983-1987.

Williams, B.E., Stoner, B.R., Asbury-y, D.A. and Glass, J.T. (1991) In-vacuo surface analysis of diamond nucleation and growth on Si (111) W polycrystalline tantalum, in R.E. Clausing (ed), *Diamond and Diamond like Film and Coatings,* Plenum Press.

Wind, R.A., Duijvestijn, M.J., van der Lugt, C., Manenschijn, A., and Vriend, J. (1985) Application of dynamic nuclear polarization in ^{13}C NMR in solids, *Prog. NMR Spectrosc.,* **17,** 33-67.

Windischmann, H., Epps, G.F., Cong, Y., and Collins, R.W. (1991) Intrinsic stress in diamond films prepared by microwave plasma CVD, *J. Appl. Phys.,* **69,** 2231-2237.

Windischmann, Hand Epps, G.F. (1990) Properties of diamond membranes for X-ray lithography, *J. Appl. Phys.,* **68,** 5665-5673.

Windischmann, H and Epps, G.F. (1992) Free-standing diamond membranes: optical, morphological, and mechanical properties, *Diamond and Related Materials,* **1.** 656-664.

Woods, G.S. and Collins, A.T. (1983) absorption spectra of hydrogen complexes in Type I diamonds, *J. Phys. Chem. Solids,* **44,** 471-475.

Wright, R.B., Varma, R., and Gruen, D.M. (1976) Raman scattering and SEM studies of graphite and silicon carbide surfaces bombarded with energetic protons, deuterons, and helium ions," *J. Nuc. Mater.,* 63, 415-421.

Xin Bi, X., Eklund, P.C., Zhang, J.G., Rao, A.M., Perry, T.A., and Beetz, C.P. Jr., (1990) Optical properties of chemically vapor deposited diamond, *J. Mater. Res.,* **5**, 811-817.

Yagi, T., Utsumi, W., Yamakata, M., Kikegawa, T., and Shimomura, O. (1992) High-pressure in situ X-ray diffraction study of the phase transformation from graphite to hexagonal diamond at room temperature, *Physical Review B,* **46**, 6031-6039.

Zhang, S., Ke, S.C., Zvanut, M.E., Tohver, H.T. and Yohra, Y.K. (1994) g tensor for substitutional nitrogen in diamond *Phys. Rev. B,* **49,** 15392 - 15395.

Zhang,W., Zhang F., Wu, Q. and Chen, G. (1992) Study of influence of annealing on defects in diamond films with ESR and IR measurements, *Materials Letters* **15,** 292-297.

Zouboulis, E.S. and Grimsditch, M. (1991) Raman scattering in diamond up to 1900K, *Phys. Rev. B,* **43,** 12490-12493.

Zommerfelds, W. and Hoch, M.J.R. (1986) Imaging of paramagnetic defect centers in solids, *J. Magn. Reson.,* **67,** 177-188.

Chapter 10

NATURAL DIAMOND

Henry O. A. Meyer*
Purdue University, West Lafayette IN 47907, U.S.A.

Michael Seal
Sigillum B. V., P.O.B. 7129, Amsterdam, The Netherlands

Contents

Deceased.

1. Diamond in Nature

1.1 HISTORICAL

It is unknown when the mineral diamond was first discovered but by the fourth century B.C. a Sanskrit manuscript, the Artha-Sastra of Kautilya, describes an active trade of diamonds in India. Although there is reference to diamond in the Bible, some scholars suggest that this reference is incorrect and arises from misinterpretation of the original Hebrew text, particularly the word yahalom and its later Greek synonym adamas. Both these words in later centuries refer to diamond, but in early times their use was much broader and more or less meant "a hard substance." Adamas for example was used in early Grecian times to refer solely to iron.

Diamond, albeit for industrial purposes, is mentioned in a Chinese second century B.C. manuscript and this suggests trade between China and India. Certainly by the first century A.D. there is record of diamond trade between India and the Mediterranean. Pliny the Elder, at this time, because of mystical properties ascribed by the Romans to diamond considered it to be of the highest value among precious stones. Later in Europe diamond, even into the sixteenth century, seems to have been relegated to a lower standing after emerald and ruby. However, with the advent of cutting and polishing techniques, which allowed the brilliance and fire of diamond to be revealed, diamond returned to its current premier status among gemstones.

1.2 INDIA

Although diamond was undoubtedly mined in India by the fourth century B.C. it is not until much later in the fifth century A.D. in Indian (Sanskrit) literature and only in the sixteenth and seventeenth centuries in European literature that the locations and methods of diamond mining in India are revealed. The deposits in India, apart from that at Panna, are all secondary or tertiary and the diamonds were obtained from ancient alluvial conglomerates and related deposits as well as from recent river gravels. Tavernier, a gemstone merchant who traveled to India during the 1600's, mentions seeing 60,000 persons, including women and children, toiling for diamonds. Stones above a certain weight were reserved solely for the ruling class and several of these have become the classic historic diamonds of India, for example, the Koh-i-Noor, (originally thought to be 600 cts when rough but now cut to 108.93 cts and part of the British Crown jewels), the Regent (cut to 140 cts and part of the French Crown jewels now in the Louvre) and the Hope (45.5 cts when cut and now in the Smithsonian Museum, Washington, D.C.).

The primary sources, apart from the Panna mine, of the Indian diamonds have not yet been discovered in spite of extensive exploration, which is currently ongoing. The present production from India, including that of Panna, is relatively small, probably less than 20,000 ct/yr, and mostly from small alluvial deposits of river terraces and modern river gravels.

Although most people consider only India as a source of historic diamonds there are suggestions that diamond was also mined in Borneo (now Kalimantan) by 600 A.D. However, the production, relative to that of India at the time, would have been small and the major markets were probably China, S.E. Asia, and India itself.

1.3 BRAZIL

By the early 1700's the diamond fields of India were in decline but in 1725 diamond was discovered in the Diamantina region of eastern Minas Gerais State in Portuguese Brazil. The diamonds here were also from alluvial deposits. Exploration for diamond in other parts of Brazil was undertaken and new diamond fields in western Minas Gerais, Mato Grosso and Bahia States were soon discovered. For almost 150 years Brazil was the major producer of diamonds and in fact within five to ten years of the Brazilian discovery of diamond the prices fell sharply in Europe because of an oversupply. A similar problem occurred in 1773 but was resolved by the Portuguese Crown effectively forming a monopoly by limiting the areas of diamond production and raising taxes on diamond.

The methods of mining were somewhat similar to those of India in as much as the diamond-bearing gravels or soft rock were washed and the resulting material sieved and sorted for diamond. Currently, diamond has been found in almost all the States of

Brazil but always in secondary or tertiary alluvial deposits. To date no economic primary source of diamond has been discovered in Brazil in spite of extensive modern exploration. One or two major mining operations based on large dredges on the Jequitinhonha River in Minas Gerais are in operation. These together with other mining companies produce about 4% of the official Brazilian production. Virtually all other deposits are worked by either very small mining companies or local miners known as garimpeiros. As a consequence the actual production of diamonds in Brazil is difficult to assess as only a portion is declared to Government.

It was in Brazil that carbonado (black, polycrystalline aggregates of diamond) were first found. The largest to date weighed 3,167 ct (the Sergio Carbonado) and was found in 1905 in Bahia State, an area that has historically produced a high proportion of boart, balas and carbonado. In contrast, large gem diamonds (up to 1,680 ct) occur mainly in the alluvial deposits of Western Minas Gerais.

1.4 SOUTH AFRICA

In 1866 a diamond was discovered on the banks of the Vaal River in the northern Cape Province of South Africa. Subsequent discoveries led to a major diamond rush centered on the Vaal and Orange rivers in the region and were referred to as the River Diggings. As with the discoveries in India, Borneo and Brazil these diamonds were all associated with alluvial deposits. In 1870 a diamond was found on a farm (Koffiefontein) many kilometers away from the main River Diggings. Similar discoveries were made on the farm Jagersfontein, also distant from the River Diggings. Though it was not known at the time the diamonds on these two farms were derived directly from primary source rocks. In the same year other Dry Diggings, as they were called, were started at a place called Colesburg Kopje. The richness of the deposit and the ensuing diamond rush resulted in the name being changed eventually to Kimberley. The non-alluvial primary volcanic source rock in which the diamonds occurred was subsequently called Kimberlite. Originally Colesburg Kopje (Kimberley) was in the Boer republic of the Orange Free State, but the British maintained that the boundary between the Orange Free State and the Cape Province (British) had been incorrectly drawn and effectively annexed the town and the diamond mining area.

There were four major eroded volcanic vents, usually referred to as kimberlite pipes because of their pipe-like shape at depth. These became the major mines of Dutoitspan, Bultfontein, DeBeers, and Kimberley. All are situated within a radius of 4 km. A fifth kimberlite pipe, Wesselton, was discovered about 20 years later in 1890. During the first rush to the Kimberley diggings numerous miners were able to work small individual claims on each pipe. As the mining became deeper the difficulties of working individual claims became apparent and soon miners combined into small cooperatives or were bought outright by other companies. Eventually in 1888 DeBeers Consolidated Mines Ltd was formed and took over the assets of the DeBeers and Kimberley mines to become the world's major diamond producer.

In 1902 a new kimberlite was discovered near Pretoria in the Transvaal and proved to be diamond-bearing. This was named the Premier Mine by its owner, Thomas Cullinan, and in five years equaled the combined output of the five Kimberley mines. In 1905 the largest gem diamond in the world, the Cullinan (3,106 ct) was found at the Premier Mine.

With the discovery in South Africa of kimberlite, the primary source of diamond at the Earth's surface, major exploration around the world for similar rock was undertaken. Within South Africa (including the Boer republics of the Orange Free State and the Transvaal) numerous kimberlites were discovered, but not all were diamond-bearing or economic. Most of these kimberlites were owned by private individuals or small companies and some were profitably mined for a time and the diamonds sold independently. Similarly, today there are several small diamond mines, based on kimberlite, that are operated by individual companies outside of the DeBeers organization. However, their production may or may not be sold through the Central Selling Organization.

1.5 NAMIBIA AND MARINE DIAMONDS

In 1908 diamonds were found in the desert sands and gravels of the Luderitz area of what was then German South West Africa. In places the diamonds could be simply picked from the desert surface as a result of the wind having blown away the less dense and finer material - a geological process in the formation of what is referred to as lag gravel, or desert pavement. The production was controlled by Germany until the First World War. After the war control of the diamond fields passed to a mining company known as Anglo-American formed in 1917 by Ernest Oppenheimer with backing from Herbert Hoover (a mining engineer and future U.S. President), the U.S. Newmont Mining Corporation, and the J.P. Morgan bank. The new diamond mining operation in South West Africa was called the Consolidated Diamond Mines of South West Africa (CDM). A consequence of Oppenheimer's control of Anglo-American Corporation with its extensive gold and diamond holdings was that he eventually became Chairman of DeBeers in 1926.

The origin of the diamonds in South West Africa (now Namibia) was debated but the general consensus of geologists was they were associated with marine sands and gravels. Similar deposits were found on the coast of South Africa, just south of the Orange River and today the South African Government controls production of a mine at Alexander Bay, (the State Alluvial Diggings) whereas DeBeers produces from two other mines on the coast of South Africa.

The idea of possibly mining diamonds from the sea-bed off the coast of South Africa and Namibia was first mooted in 1957 by two small South African Companies but due to legal problems their production was modest. However, the stage had been set and others began to explore and recover diamonds from the sea-bed. Currently, a number of large companies, including DeBeers, are exploring and developing techniques for recovering diamonds.

The source of the marine diamonds, 95% of which are gem, is believed to be the kimberlites of the interior part of southern Africa. These kimberlites intruded the continent 90 million years ago and have been eroded extensively. The diamonds released through the geological agents of weathering and erosion have been carried by rivers, predominantly the Orange river and its tributaries, to the coast. Fluctuations of the land/sea interface, changes in the position of the mouth of the Orange river, and normal marine processes have resulted in redistribution of the diamonds in marine terraces above and below the present sea level along the coast. It is roughly estimated that possibly 3 billion carats of diamond have been eroded from the kimberlites in the interior.

1.6 RUSSIA

Although alluvial diamonds have been known to occur in the Ural Mountains since 1859 it was not until 1953, following years of intense exploration, that diamonds were found in Siberia. These were alluvial diamonds and occurred in a tributary of the River Lena. The following year the first kimberlite, named Zarnitsa, was discovered. Although diamond-bearing it was of relatively low grade and was not initially mined. As with southern Africa, there are presently at least 300 kimberlites known in Siberia but the majority are either not diamond-bearing or are uneconomic to mine.

The towns of Mirny and Udachnaya, formed as a result of the diamond mining and exploration, are the major diamond-mining centers in what is now known as Sakha State (previously Yakutia). Because of the adverse cold temperatures the Russians had to modify and develop new techniques for mining and diamond extraction from the ore. Currently, all producing mines are still open-pit but plans for underground extraction have been proposed. The major mine is the Udachnaya mine and is responsible for about 85% of the production from Sakha.

The grade (i.e. carats per 100 tonnes of kimberlite) of the ore in several kimberlites is relatively high and the quality of the diamonds produced is very good. Several mines are being refurbished to extend their useful life (e.g. the Mir and International mines) and a number of new ones (e.g. Jubilanya) are being brought on line. The overall production from Sakha is high and Siberia thus ranks among the major world producers.

Recently, kimberlites and diamonds have been discovered in the region of Arkhangel, on the southern shore of the White Sea, about 700 km north east of St. Petersburg. Further exploration and possible mine development are underway.

1.7 AUSTRALIA

Alluvial diamonds were mined in the Copeton/Bingara area of New South Wales during the latter half of the nineteenth century. The source of these diamonds is still a mystery. In the early 1970's renewed interest in diamond exploration in Australia occurred and led to the discovery in Western Australia of the Ellendale pipes which contained diamonds but in sub-economic quantities. This discovery revolutionized diamond exploration because the pipes were not kimberlite but a completely different type of rock known as lamproite. The rock, similar to kimberlite, is volcanic in origin but differs in mineralogy and geochemistry. Subsequently, the Argyle lamproite was discovered in the Kimberley region of Western Australia. This lamproite is believed to have the worlds highest grade of ore (up to 400 ct per 100 tonnes in places) and thus makes the Argyle mine, which came into production in about 1982, the worlds top producer.

Unfortunately, the production consists of small diamonds of which about 90% are industrial or near gem. Pink colored diamonds are relatively common.

1.8 OTHER DIAMOND PRODUCING COUNTRIES

Diamonds have been found in mineable quantities in several other countries in Africa, and America. Some of the deposits are alluvial and some are kimberlite. For example alluvial diamonds were found in the Kasai region of the Belgian Congo (now Zaire) in 1908. This led to further exploration in both Angola and Zaire. In Zaire the rich diamond deposits of the Mbuji Mayi area, based on kimberlites, are mined. Until the advent of Australia (and ignoring the Siberian production which was secret for many years) Zaire was the world's major producer of diamonds, albeit mostly industrial.

Large alluvial deposits were found in Angola by 1916 but it was only in the 1950's that kimberlite pipes were discovered. Two major and distinct regions of alluvial diamond production occur in the Central African Republic but most of the activity is based on artisanal workings. Diamonds were first found here in 1930. Diamonds have also been mined in the neighboring countries of Gabon and Congo-Brazzaville.

In West Africa diamond occurs in Ghana (1929), Sierra Leone (1930), Guinea (1930), Mali (1950) and Cote d'Ivoire. In this last country diamonds were mined for a number of years in the 1970's from a deposit which was not recognized at the time to be a lamproitic source.

The Williamson diamond mine was discovered in Tanzania in 1940 and this mine (now referred to as the Mwadui mine) has continued production to this day, although this is now much reduced. Alluvial diamonds were also found in small quantities in Southern Rhodesia (now Zimbabwe) and currently there is renewed exploration and production from one kimberlite.

After about thirteen years of exploration DeBeers geologists discovered the rich Orapa kimberlite in Botswana in the late 1960's. This mine occurs in the Letlhakane field of about 35 kimberlites, of which not all are economic. Further to the south DeBeers geologists also discovered the Jwaneng kimberlite beneath 150 ft of Kalahari sands and this too is a major producer of diamonds. DeBeers and the Government of Botswana have a joint company which exploits these mines.

Diamonds are also produced from kimberlites in Lesotho and Swaziland in southern Africa. In the case of Lesotho, DeBeers extracted diamonds from the worlds highest diamond mine, that of Letseng-le-Terai. Here the grade was extremely low, about 3 ct per 100 tonnes, but based on exploration statistics it was believed the mine would produce individual diamonds in excess of 100 carats. This actually occurred and thus a small number of large diamonds made the mine economic in spite of the low grade.

In South America diamonds were discovered in Guyana and Venezuela by the turn of the century but not until about 1920 did the production amount to anything. Diamond mining activities continue in both these countries. In North America, particularly the U.S.A., isolated diamond finds have been recorded since the mid 1800's, particularly in the south eastern States of South Carolina and Georgia, the mid-western States of Wisconsin, Indiana and Ohio, and in California. In 1905 diamond was discovered in a volcanic rock in Murfreesboro, western Arkansas. Diamonds were mined from this locality until the 1930's, and until the 19-70's this was the only primary source and diamond mine in the U.S. Because of the existence of diamond the Arkansas rock was referred to kimberlite but it has since been shown to be a lamproite.

In the 1970's diamond-bearing kimberlites were found to occur in the Rocky Mountains in the States of Colorado and Wyoming. However, in spite of intensive exploration only one possible small kimberlite is economic.

In 1991 kimberlite was discovered in the North West Territories of Canada. This discovery led to intense exploration throughout much of Canada and the position today is that there is a possibility of three to five diamond-bearing kimberlites being mined by the turn of the century.

In China diamond was found in alluvial deposits at the turn of the century in Shandong Province and also known in the Yuan river in Hunan Province. In 1963, geological exploration teams found kimberlites in Shandong Province and later in Liaoning Province. One kimberlite (Changma mine) in Shandong was mined for a period but is now more or less abandoned apart from some desultory work. The main production is from the Pipe #50 mine in Liaoning. Most of the production is used within China.

Diamond-bearing kimberlites have also been recently discovered in Finland but few details are available. Diamond has also been recorded associated with tin gravels in Thailand, and in Borneo.

1.9 OTHER OCCURRENCES OF DIAMOND

Diamond has been found in rocks other than kimberlite and lamproite, but for the most part the discovery is of scientific interest only. In Morocco and Spain evidence for diamond in oceanic rocks is evidenced by the occurrence of graphite pseudomorphs after diamond. Small diamonds are also reported by the Chinese in oceanic rocks in Tibet.

Recently, the discovery of microscopic diamond in garnet in metamorphic rocks in Khazakstan has caused rethinking of crustal processes among some geologists. A similar occurrence appears to be in metamorphic rocks of Dabie Shan in China. Neither of the occurrences are of economic importance.

Tiny black diamonds have been known to occur since the late 1800's in certain carbonaceous meteorites and more recently the occurrence of diamond as interstellar dust has been postulated. Lonsdaleite (hexagonal diamond) has also been recorded from meteorites, and in northern Siberia polycrystalline masses of diamond and lonsdaleite are referred to as yakutite, after the region in which they are found.

2. Diamond Production and Trade

2.1 DIAMOND PRODUCTION

Since the late 1980's diamond production worldwide has been around 100 million carats per year estimated to be worth over U.S.$5 billion. This figure includes both industrial and gem quality diamonds from alluvial and primary sources. It is also an estimate as a proportion is the unknown non-declared illicit diamonds.

Australia, from the Argyle mine, is foremost in production with a figure of about 40 million ct/yr but less than about 5% of the stones are good quality gems. Zaire ranks second with an annual output in the region of 20 million ct, but again well over 90% industrial. Zaire is unique in that a high proportion of the diamonds have a clear diamond core that is overgrown by a coat of fibrous steel gray diamond. These are referred to as Congo coats, or coated diamond.

The major good quality diamond producing countries are Botswana (17 million ct/yr), Russia (estimated to be of the order of 15 million ct/yr), South Africa (11 million ct/yr), Angola (2 million ct/yr) and Namibia (1 million ct/yr). Good quality diamonds are also produced from other countries in Africa, South America and elsewhere but the productions are all generally less than half a million ct/yr, except for Brazil whose production is estimated to be about 1.5 million ct/yr.

Polycrystalline aggregates of diamond (e.g. framesite, ballas, boart) are found in the production of many countries and are invariably classified as industrial material. Carbonado is generally found only in Brazil, the Central African Republic and Russia.

2.2 DIAMOND TRADE

All mines owned and operated by DeBeers and subsidiaries supply their production to the Central Selling Organisation (CSO). Mines in other countries in which DeBeers has a percentage interest, e.g. Botswana, Australia, Ghana, etc. sell a percentage of their production to the CSO based on renewable contracts which run for a number of years. The remaining percentage may be sold individually by the company concerned. The CSO has sold an agreed percentage of Russian diamonds since the late 1950's whereas Russia itself was allowed under the agreement to sell a specific amount of rough. Currently (1995) Russia has indicated they may want to sell some of their rough diamonds privately and serious negotiations are underway between the CSO and Russia.

Diamonds sold outside of the CSO, other than those mentioned above, are mainly produced from alluvial deposits in South America, and Africa. Individual companies will have diamond buyers in the various countries often with specific instructions, from time to time, to buy certain types of goods. These goods are usually declared to the

Government concerned, export taxes paid and the diamonds shipped to Europe, New York, Tel Aviv or elsewhere.

The diamond buyers have an excellent unofficial intelligence network and within hours the price ranges for various goods are more or less similar from buyer to buyer worldwide. Nevertheless, due to stock requirements, or other reasons, a diamond company may inform its buyers in various countries to pay an extra premium for diamonds in a particular size or quality range. In this case the other buyers may hold off buying such goods until the normal price is re-established. The CSO also participates in this market and has its own buyers who compete with everyone else.

A somewhat anomalous country for diamond buying is Liberia. Although kimberlites and alluvial diamonds are mined in Liberia the total production is probably only a few thousand carats. However, one will read in some tabulations of world production that Liberia produced in some years the order of 1 million ct, although it has now decreased to about 100,000 ct due to political instability and other causes. This so-called production is incorrect and is in fact the export of what were once illicit, or smuggled diamonds from elsewhere. These diamonds travel from surrounding and often distant countries in Africa to Liberia and once sold to a Liberian registered diamond buyer become legal. These diamonds, of course, were not registered with the Government in the country in which they were mined and thus in total worldwide compilations of diamond production one has to take into account the Liberian exports.

From time to time other countries also fulfil this role. For example at the time of Angolan independence Lisbon in Portugal became a buyers market. Historically Lisbon was only reverting to a role it played during the 17th and 18th centuries when it was a clearing house for Indian goods which arrived via its trading colony of Goa in India. In the 19th century it was the main market for diamonds from its Brazilian colony and also exerted strict control over production in order to keep prices stable. In the 1980's diamond buyers could be found in Havana purchasing goods smuggled out of Angola, and for a short time in the early 1980's Burundi was also a potential market in spite of no diamonds being mined there.

3. Primary and Secondary Sources of Diamond

The major primary sources of diamond at the Earth's surface are the relatively rare volcanic rocks kimberlite and lamproite, and these only occur in continental regions. However, in spite of the number of occurrences of these rocks worldwide only a minority of them contain diamonds, and even fewer contain diamond in sufficient quantities to be economically viable to mine. Prolonged weathering and erosion of these rocks results in the hard and chemically nonreactive diamond entering into the sedimentary geological cycle and being concentrated in secondary, or even tertiary, alluvial (river or marine) deposits. These alluvial (or placer) deposits may be hundreds of kilometers from the primary source and in many cases the geographical position and geological age of the original source rock is unknown.

3.1 KIMBERLITE

Kimberlite is a rare type of igneous volcanic rock and was named in 1880 after the town of Kimberley, South Africa, where it was first discovered to be the primary host for diamond. It is formed at depths probably in excess of 200 km within the Earth and is believed to travel rapidly to the Earth's surface where it erupts as a volcano. No modern equivalent is known and the last kimberlite eruption probably occurred about 25 million years ago, although kimberlite has erupted at the Earth's surface periodically throughout geological time. The majority of kimberlites form cone-shaped bodies, with the narrower end of the cone at depth. These types of bodies are known as diatremes but are commonly referred to as pipes and have a surface expression often less than 100's of meters in diameter. The largest kimberlite, the Mwadui (Williamson mine) pipe in Tanzania has a surface area of 146 hectares (361 acres) whereas the rich diamond-bearing Mir pipe in Sakha was only 6.9 hectares (17 acres) in size. At depth most kimberlite pipes terminate as narrow dikes. Kimberlite dikes may also reach the surface where they occur as thin, often a few meters wide, vertical sheet-like bodies that may extend for several miles. Some of these kimberlite dikes are diamond-bearing in sufficient grade to warrant mining and in Southern Africa are often referred to as fissure mines. Rare occurrences of kimberlite sills (horizontal sheet-like bodies) are also known.

Kimberlites often occur in groups or clusters of similar age (kimberlite fields) but within a field not all, or none, may be diamond-bearing nor are the kimberlites identical in texture and mineralogy. Further a single kimberlite pipe may consist of several varieties of kimberlitic rock with different textures, mineral content and diamond grade. This indicates that a single volcanic vent has been subjected over a period of time to multiple kimberlite eruptions.

Kimberlite is a hybrid igneous volcanic volatile-rich potassic ultramafic rock. The major constituent minerals (some of which may not be present) include olivine, phlogopite mica, calcite, monticellite, apatite, ilmenite and spinel. These minerals usually form a fine grained matrix in which larger minerals (referred to as megacrysts) such as olivine, garnet, ilmenite, pyroxene, mica and diamond occur. Some of these megacryst minerals are foreign to kimberlite and represent minerals from disaggregated deep mantle rocks probably torn from the sides of the conduit through which the kimberlite passed on its way to the surface. These specific minerals are known as xenocrysts and include garnet, ilmenite, pyroxene and diamond. Thus it is important to realize that diamond does not form in kimberlite but rather that kimberlite is the transporting medium that carries diamond from its place of formation (about 200km depth) to the Earth's surface. Lamproite also has a similar role as a transporting agent for diamond. The other xenocryst minerals (garnet, ilmenite, pyroxene) are important as they are used extensively as diamond indicator minerals in geochemical exploration for kimberlites and diamond.

A further complication in the study of kimberlite is the variation in lithology, or texture, of kimberlite within the same diatreme or pipe. At the surface within the crater part of the pipe the kimberlite may be fine grained ash or tuff type material (crater facies), some of which may have been reworked by weathering and erosion. With depth and within the pipe proper the kimberlite may be more tuffaceous and brecciated (diatreme facies), and in the deeper levels the kimberlite may be massive with minimum brecciation (root zone facies). What type of kimberlite is exposed in a pipe depends entirely of the level of erosion. Crater facies kimberlites are rare (e.g. Orapa, Botswana; Mwadui, Tanzania) whereas diatreme facies (the five Kimberley mines) and the deeper root-zone facies are more common. In general, a diamond-bearing kimberlite exposed at the crater level has a much greater economic potential than a deeply eroded diatreme level kimberlite which in turn has more potential than a root zone facies kimberlite. Thus in exploration for a primary source of diamond the best possible target is a crater level kimberlite.

Another complication with regard to kimberlite is that, based on mineral and whole-rock isotopic evidence, kimberlite was subdivided into two groups - Group I and Group II. Whereas Group I kimberlites occur worldwide, only Group II is known to occur in Southern Africa. Following a suggestion by an early researcher of kimberlites in South Africa the name Orangeite has been resurrected for Group II kimberlites and there is some evidence to suggest that orangeites (or Group II kimberlites) are a distinct rock type and are thus not kimberlite sensu stricto.

3.2 LAMPROITE

Although lamproite as a rock has been known since the late 1930's it was not known to contain diamond until the 1970's. Exploration for diamond in Western Australia, where lamproites are relatively common, in the mid 1970's resulted in detailed sampling of the Ellendale lamproites and the subsequent discovery that they were diamond-bearing. This completely unexpected discovery caused major reassessment of exploration methods and programs worldwide. However, compared to the many known diamond-bearing kimberlites only four lamproites worldwide (Ellendale and Argyle in Australia; Mufreesboro, Arkansas in the U.S.A.; and the Bobi dike in the Cote d'Ivoire) are known to contain diamond and have warranted exploitation.

Similar to kimberlite, lamproite is a volcanic igneous rock that has formed at depth in the Earth and has erupted with sufficient velocity to carry diamond and other xenocrysts to the surface. It occurs as pipes, dikes and sills but differs in overall shape from kimberlite in that the pipe often has a more champagne-glass shape near the surface than the narrower cone-shaped kimberlite. Similar to kimberlites, lamproites have a small surface area. Whereas no kimberlitic lava is known, some lamproites occur as lava.

Lamproite forms a family of rock types which differ in mineral content and geochemical composition. The lamproite family in general consists of ultramafic magnesium-rich rocks that contain varying proportions of the minerals olivine, pyroxene, leucite, sanidine, spinel, mica, amphibole and exotic Ti-Ba minerals. Depending upon the major minerals present the lamproites are subdivided into different named types (e.g olivine lamproite, leucite lamproite). The diamond-bearing lamproites in Australia are mainly olivine lamproites. Exploration methods for lamproite require the search for diamond-indicator minerals, other than diamond, different from those used in the search for kimberlites.

3.3 DISTRIBUTION OF DIAMOND BY GEOLOGICAL PROCESSES

In spite of its apparent rarity diamond occurs throughout all continents, except Antarctica in which no exploration has been undertaken. Diamonds do not normally occur in oceanic rocks. It has been recognized for many years that diamond-bearing kimberlites are mostly confined to old continental (cratonic) nuclei. Thus continental regions composed of rocks older than 2.5 billion years are prime target areas for the exploration of primary sources (kimberlite) of diamond. In contrast, diamond-bearing lamproites do not follow this pattern and may occur in regions of the continental crust that have been deformed by older mountain building events.

Subsequent to erosion of the kimberlite or lamproite, diamond, because of its hardness and chemical resistance to all but highly oxidizing agents, is released from the primary rock and enters the sedimentary environment. The diamonds are initially spread on the ground surface at or close to the primary source rock and form colluvial or elluvial deposits. With time these diamonds, due to weathering processes, enter local rivers and may be transported many hundreds of kilometers. The processes of river deposition and erosion result in some diamonds being deposited in river gravels thus forming an alluvial (Placer) deposit, which in turn may be later eroded and the diamonds moved further downstream and redeposited again. Generally with progression downstream the average size of diamond decreases. Some of these alluvial deposits may be isolated from the river as a result of various geological processes and eventually the deposit becomes a part of the stratigraphic succession only to be later eroded and the diamonds released into a new, maybe modern, river. Thus diamonds may be found in secondary or tertiary alluvial deposits that are remote from the original primary source and furthermore these deposits may occur in any region of a continent and are not confined to the regions in which older rocks occur.

Examples of ancient alluvial deposits which contain diamond are the gold-bearing Witwatersrand conglomerate (2.6 billion years age) in South Africa, the Birimian rocks (approx. 2.1 billion years old) of Ghana, and the sandstones and conglomerates of the Roraima Group (1.7 billion years age) in Venezuela, Guyana and northern Brazil. The primary sources of the diamonds in these rocks are unknown but the distribution provides evidence at least of the antiquity of diamond (and kimberlite) on the Earth's surface. The diamonds held within the Roraima rocks are being released through the agents of weathering and erosion. These diamonds are now found associated with the modern river gravels and sands (e.g. the Rio Caroni) of Venezuela from which they are recovered by local miners. In Sierra Leone and Guinea, for example, both old and modern river deposits have been exploited for diamond; only later were the primary source kimberlites found and in most cases are uneconomic to mine. Thus nature through its geological processes has concentrated the diamonds in river deposits which are then economically viable.

4. Diamond Exploration

The techniques used in diamond exploration depend upon the size, nature and specific target. In an unexplored region one may initially use geochemical rather than geophysical methods, as the former are generally less costly. In contrast, if kimberlites (or lamproites) are known to occur in an area, geophysical methods may be used in an attempt to isolate rapidly specific targets which will then be geochemically sampled.

4.1 GEOPHYSICAL TECHNIQUES

Because of the small area of most kimberlites they are often less than the size of a pixel in remote sensing data and thus most kimberlites are not visible by satellite remote imagery. However, satellite remote sensing methods, including false color and radar, are used to provide regional geological coverage from which possible structural controls (e.g. faults, fracture zones, etc.) may be determined. Such structural features may be zones which have controlled the intrusion of igneous bodies, possibly kimberlitic or lamproitic.

Aerial photography (normal or false color, and radar) may also be used and because of the higher resolution some kimberlites can be clearly identified. However, as with satellite data, specific training techniques based on known spectral characteristics of kimberlites, overlying soils and vegetation have to be evaluated first in order to make the techniques useful.

Normally, the geophysical techniques used are aerial gravimetric, electrical and magnetic (EM). It must be remembered that the response to any method depends upon the difference in property between the target and the surrounding country rock into which it is emplaced. If there is no difference the method will reveal no anomaly and will thus be unsuccessful. Thus a kimberlite pipe containing a high proportion of iron-bearing minerals (e.g. garnet, ilmenite, chromite) intruded into sedimentary rocks may give a clear magnetic anomaly as compared to one intruded into basaltic rocks.

The aim of whichever geophysical method is used is to isolate small geophysical anomalies which help limit the area of search. Such anomalies may or may not be kimberlitic and thus require further evaluation, usually by geochemical methods or drilling. The major drawback of aerial geophysics is the cost per line km flown. Because kimberlites (and lamproites) are small it is easy to miss one if the geophysical line spacing is too wide. Thus a choice has to be made between flight line spacing, potential target size and cost. Once a kimberlite has been identified by geochemical sampling, ground based geophysical methods are usually used to delineate the surface size and the projected shape at depth.

4.2 GEOCHEMICAL TECHNIQUES

Under the broad title of geochemical techniques are included geochemical analysis of rocks and soil, mineralogical and petrological studies of rock, soil, sand, gravel, glacial deposits (e.g. till) and river sediment samples, plus the chemical analysis of selected minerals from the above samples.

In the initial search for diamond from primary source rocks (kimberlite and lamproite) or alluvial deposits the usual method is a regional geochemical sampling program. The samples, depending upon the geographical terrain, may be taken from rivers, river terraces, glacial deposits, desert sands, or soil. The weight of the sample taken depends upon a number of factors, but generally most parties in the field attempt to sieve the sample and collect material less than about 2mm in size. A minimum weight of 3 kg is considered satisfactory but depending upon the nature of the sample material 20 kg or more may be considered necessary. The samples may be further processed in the field camp or be sent directly to a central processing laboratory. The light fraction is removed and the heavy mineral concentrate cleaned and examined for diamond and diamond indicator minerals such as garnet, ilmenite, chromite and pyroxene.

Extensive research on minerals included in diamond (e.g. garnet, olivine, ilmenite, chromite, pyroxene) and on diamond indicator minerals has shown them to have specific chemical compositions which are unique and appear to be related to crystallization in the same geochemical and physical environments as diamond. For example certain claret colored garnets with low contents of calcium and relatively high concentrations of chromium occur in kimberlite and have compositions very similar to garnets included inside diamonds. The diamond indicator minerals observed in the samples are chemically analyzed and the resulting data processed to determine whether or not any of the minerals satisfy the compositional range of the similar minerals which occur as inclusions in diamond. If this is the case the field party may be directed to sample more closely a specific area in order to isolate the source.

This geochemical technique of mineral chemical analysis is rapid and very cost effective and can also be used to indicate which kimberlite(s) should be first bulk sampled (almost equivalent to a small mining operation) when several kimberlites have been discovered in an area.

5. Diamond Mining and Processing

At the present time the only deep underground diamond mines are all in South Africa and operated either by DeBeers or private companies. This situation may change in the future as open-pit surface mines in Siberia, Zaire and elsewhere change to underground activities. The majority of diamonds are currently produced from open-pit or surface workings, such as the Argyle mine in Western Australia; Orapa mine, Botswana; and the Udachanya mine in Siberia. In fact, the whole diamond production from South America is from surface alluvial mining operations.

5.1 ALLUVIAL MINING

Surface mining of diamonds may range from primitive one man operations through to extremely sophisticated and complex mines. In the simplest operation, invariably associated with alluvial diamond mining, the diamond-bearing gravel is exposed by manual digging, washed and sieved and the resulting material carefully searched for diamond. The next step-up in mining utilizes hydraulic methods. In this instance the sterile overburden is removed by washing away using high powered water hoses. When the diamond-bearing gravel is reached the material (including diamond) is hosed into a sump and pumped into a mechanical jig. The jig operates in such a way that the light material is carried away by the water flow and the heavy minerals, including diamond, sink into the bed of the jig and are later recovered. This type of mining using a jig is popular in Venezuela, Guyana and Brazil.

In parts of West Africa and in Kalimantan (Borneo) the local people use a technique of shallow underground mining to obtain the diamond-bearing gravel. Shafts up to 20m deep are dug manually, and when the diamond-bearing gravel is reached it is removed from a series of tunnels which radiate out from the end of the shaft and may eventually coalesce.

When water is not in abundant supply, or due to environmental considerations, the overburden may be stripped off using bulldozers and the diamond-bearing gravel transported to a central washing and processing facility. Depending upon the nature and size of the fragments constituting the gravel, the gravel may be passed over a vibrating sieve or through a rotating trommel and the oversize material removed. The selected finer sized material is then further processed to remove the light fraction and the heavy mineral concentrate collected, washed and sorted for diamond. In several African countries, particularly South Africa, the preferred equipment for separating the light fraction from the concentrate is the rotary pan, which is also used in the processing of kimberlite.

5.2 OPEN-PIT MINING OF KIMBERLITE AND LAMPROITE

The initial development of the kimberlite pipes in South Africa at the time of their discovery was by surface or open-pit methods. Numerous claims over the surface of the pipe were worked by individuals or small groups of men. As the workings became deeper, greater difficulty was encountered in removing the diamond-bearing ore and control of the workings passed to major companies which held large numbers of claims. Eventually one company, DeBeers Consolidated Mines, gained complete control at Kimberley.

Open-pit mining is generally more cost effective and profitable than underground mining. For this reason all kimberlite and lamproite pipes are initially developed as open-pit mines and only later, often after several years, is underground development started. The usual method in developing an open-pit mine is first to cut back a series of stepped benches in the country rock immediately surrounding the pipe. This is all waste material but is necessary in order to avoid collapse of the pit walls. The ore is removed by blasting and trucked to the surface via spiral ramps which utilize the benches constituting the walls. As the pit progressively becomes deeper so the benches in the pit walls have to be cut back in order to retain slope stability. Nevertheless, there comes a point when the cost of cutting back the waste benches is more than the economics of underground mining and at this point the open-pit has run its useful life. Prior to this, however, mine management will have been planning for underground extraction and will have already sunk a shaft at the edge of the pipe and may have undertaken initial development work for the underground activity. Thus the changeover from open-pit to underground mine is a progressive change and designed to minimize interruption of production.

During the last few years the Finsch mine in South Africa has been changing from open-pit to underground, and a number of mines in Siberia which are presently open-pit are being redesigned as possible underground ones. The Orapa and Jwaneng mines in Botswana, and the Argyle mine in Western Australia are all open-pit mines.

5.3 UNDERGROUND MINING

In most underground mines two vertical shafts, for safety purposes, are sunk into the rocks surrounding the kimberlite and at a specific level a horizontal tunnel is driven from the shaft towards the kimberlite pipe. This tunnel is the major haulage and access passage to the kimberlite, and from it and into the pipe are driven a number of subsidiary tunnels or drifts. The removal of the kimberlite ore was originally done by a method known as chambering. The chamber levels were developed from sub-vertical shafts about 12 m apart. Alternate and successive cuts, 8m wide, extend across the pipe and vertically to the footwall (floor) of the level above. Half of the cut width is left as a support pillar. These pillars are positioned so that they lie above cuts which are being chambered below. Thus as the chamber reaches the level above the now unsupported pillar at that level starts to collapse into the chamber below. Ore is removed from the lower level of the chamber through drifts spaced about 7m apart. As chambering continues there may be several different levels from which the ore is being removed. The ore is usually loaded manually into ore trucks and the trucks pushed to an ore-pass into which the broken kimberlite is tipped. The ore-pass also contains a large sieve (a grizzly), and material that would not pass through the grizzly is broken down. Beneath the grizzly is the main haulage adit and the material is then transported to the shaft and hoisted to the surface.

Chambering has been superseded by block-caving, a method used throughout the world in many massive ore bodies that have nothing to do with diamonds. In this method a number of drifts on the same level are developed into the kimberlite ore. The drifts, at 14 m intervals, are often lined with concrete or steel and along the drifts (known as scraper drifts) conical openings (drawpoints) are raised vertically into the ore. These raises extend up into the footwall of the level above, and a whole section of the kimberlite is mined out. Eventually the kimberlite above the footwall starts to break-up and collapses into the conical draw points in the scraper drifts. The broken ore is removed from these draw points, passed through an ore shoot and hoisted to the surface. In some instances the ore prior to hoisting passes through a primary crushing circuit within the mine itself. Eventually when all the kimberlite ore has been removed from the drawpoints only overburden or sunlight remains. Prior to this, another series of scraper drifts, drawpoints, etc. will have been prepared at a level maybe 300 m below, and the whole process repeated.

At the Premier mine a method known as slot mining was used for a time. In this method benches of kimberlite, as developed in open-pit mining, were successively blasted and the kimberlite rock fell into large conical depressions constructed in the floor of the pit. The ore was caused to gravitate through the cones by drawing off underground the material at the base of the cone, scraping it over grizzlies and moving it through ore passes to the main haulage level.

In fissure (kimberlite dike) mines the kimberlite which may be a meter or two wide, but extend for thousands of meters, is removed at successive levels by a method known as overhead shrinkage mining. A main haulage and access tunnel is either developed in the kimberlite or close to it in the surrounding country rock. Vertical raises are developed in the kimberlite and the hanging wall blasted successively until the footwall of the mined out level above is reached. The broken ore is removed through the vertical raises (draw-points) in the access drifts and hoisted to the surface.

5.4 MARINE DIAMOND MINING

Herein the term marine diamonds is used for those diamonds which have been deposited by marine processes. Thus the diamond-bearing marine terraces, now above sea-level, on and behind the shoreline in southern Namibia and in Namaqualand on the west coast of South Africa are included. The major deposits in Namibia are those of CDM and in South Africa are the State Alluvial Diggings, owned by the Government of South Africa, and the mines Kleinzee, and Koingnaas owned by the Namaqualand division of DeBeers.

For many years the method of mining these deposits was essentially similar to that used in mining any alluvial deposit. The overburden sands were stripped off and the exposed diamond-bearing gravel trucked to a central plant for processing. However, when the gravel lay on bedrock laborers were used to sweep the gullies, pot-holes, and other depressions in the exposed bed-rock surface with brooms for diamonds. As the onshore marine terraces were exploited CDM began to mine the foreshore area and out into the surf zone. The method used was first to form two large embankments of concrete blocks and sand at right angles to the shore and a kilometer or so apart. The ends of these embankments may reach tens of meters into the surf zone and were joined by a similar one that was built in and parallel to the surf zone. The rectangular area thus enclosed is referred to as a paddock and in this mining area the overburden sand is removed and the diamond-bearing gravel mined to the bedrock which itself is cleaned by hand.

In the late 1970's small groups of people started to recover diamonds from the surf zone off the west coast of South Africa. Two methods are used. In one method, divers dragging suction hoses walk out into the surf and probe the gullies and fissures on the seabed for gravel. The suction hoses pump the gravel onto the shore where it is processed and the diamonds recovered. Air is supplied to the divers from compressors, sometimes mounted on tractors which are driven to the waters edge. The second method utilizes the same principle of suction hoses controlled by divers, but here the equipment is aboard small, often converted fishing, vessels. These sail close in-shore and the divers on the seabed search for potentially diamond-bearing gravel and pump it via the hoses onto the vessel where it may be initially sized in a rotating screen. The coarse material and fine sand are returned to the sea and the gravel collected and processed on-shore.

More recently large ships have been converted or designed specifically for diamond exploration and mining in waters too deep for prolonged activity by divers. These vessels use various techniques, some still in the experimental stage, to lift the gravel from the seabed and process it onboard for the recovery of diamonds. The most successful technique to-date, designed by DeBeers Marine division, has been the use of a very large diameter drill to strip the sand and gravel from the seabed. Another technique, but of limited use as it can only operate in up to certain water depths, is the airlift. Here compressed air is forced down a pipe which rests on the seabed. As the air rises it returns inside a second pipe and carries with it sand and gravel to the recovery vessel. However, the only successful mining operation for these deep water diamonds was that undertaken by DeBeers in about 1991, now producing about 400,000 ct/yr. All other deep-water operations off the coast of southern Africa are all diamond exploration projects designed to prove the presence and number of diamonds prior to active mining.

5.5 PROCESSING OF DIAMOND-BEARING ORE

In the early days of the Kimberley mines the kimberlite ore was hoisted from the workings in the pit to the surface and then spread over areas known as "floors" to weather and break-up into softer material. This material was later scraped up from the floors and processed in rotary pans to provide a heavy mineral concentrate which was then sorted for diamond. At the present time kimberlite (or lamproite) from either an open-pit or underground mine is first crushed and screened to a particular size. The initial size is carefully selected so as to minimize breakage of any possible large diamonds. Following the first processing the material may be recrushed to a finer size, reprocessed and recrushed again in an attempt to release all the potential diamonds down to possibly 1 mm size or less.

A rotary pan is a shallow (sides about 1 m or less) flat bottomed vessel up to roughly 5 m in diameter. The material to be processed, either crushed kimberlite or sandy gravel, is mixed with water to produce a slurry (the puddle) of density such that the light material will float and diamond and other heavy minerals (e.g garnet, ilmenite) will settle to the bottom. A set of arms with vertical metal prongs rotates over the pan and the prongs rake the material to keep it stirred. Periodically the concentrate which has gathered at the bottom of the pan is removed and sorted for diamond. In small alluvial operations the sorting for diamond is often done by first sieving the concentrate in water and then sorting manually through each sieve size on a table covered with an old piece of felt-type carpeting. Large commercial alluvial operations (e.g CDM in Namibia) use more sophisticated equipment, similar to that used in the processing of kimberlite and lamproite, for the actual separation and sorting of the diamond-bearing concentrate.

Most modern mines have more or less abandoned rotary pans for obtaining a heavy mineral concentrate and instead use what are called Heavy Media Separation (HMS) units. These HMS units can vary in construction but may be either cone or cyclone types. Both rely on the use of finely ground ferro-silicon powder (sometimes with the addition of magnetite) mixed with water to produce a medium of about 2.6 gm/cm^3 density. The crushed diamond-bearing ore is fed into the unit at the top in the case of the cone HMS and tangentially in the cyclone HMS. The denser material or concentrate, including diamond, settles towards the bottom of the cone, or cyclone, and is removed whereas the lighter material which floats is removed from the top. Both the concentrate and float are passed over electromagnetic drums in order to recover the ferro-silicon which is then recycled for use. In some instances the initial concentrate may be further reprocessed in another HMS unit with a higher density medium in order to concentrate further the heavy minerals.

A major breakthrough in sorting was the discovery that diamond would stick to grease. This discovery led to the development of the grease table and to an increase in mechanization of processing. The simplest grease table consists of a vibrating grease covered platform over which flows water carrying the diamond-bearing heavy mineral concentrate from the rotary pan. Most of the minerals in the concentrate pass over the grease but the diamonds remain behind stuck to the grease. Periodically the table would be stopped and the diamonds removed from the grease, boiled in acid to remove any adhering grease, and washed. Variations on this simple table were designed, including moving grease covered conveyor belts and automatic scrapers to remove the diamonds. The development of these latter grease tables was an attempt to minimize operator handling of diamonds which is a major source of potential loss in any processing operation where diamonds are concerned.

Although grease tables are still in use in some mines most modern processing plants use machines which differentiate between diamond and other minerals based on the florescent properties of diamond when subject to X-radiation. The heavy mineral concentrate, including diamond, passes in a uniform and constant stream through an X-ray beam. Diamond fluoresces, in contrast to most other minerals, and the fluorescent emitted light is recorded by a photomultiplier tube which sends a message to an air jet which is timed to shoot the diamond from the concentrate stream into a separate container. The whole process operates within sealed equipment and thus there is no operator handling of the diamonds. Periodically either security officials will remove the container in which the diamonds are collected for later sorting, or the diamonds themselves will fall onto a sealed conveyor which passes through sealed glove boxes for immediate sorting.

The flow plans for processing plants vary considerably from mine to mine worldwide. A number of factors have to be considered in the design, for example, the nature of the primary ore, the maximum and minimum size of diamonds in the ore, the economics of recovery of the smallest diamonds, the percentage of the heavy minerals in the ore, water and power consumption, etc. Furthermore, a processing plant may combine several recovery techniques, for example, HMS and rotary pans, X-ray sorters and grease tables. Small diamonds may be recovered by skin-flotation, or caustic fusion methods, whereas in other mines the concentrate may be further processed by electrostatic or magnetic separation methods.

6. Mineralogy and Geochemistry of Diamond

A considerable amount of research has been undertaken on the mineralogy and geochemistry of diamond and its associated mineral inclusions. These studies led to an understanding of the genesis of diamond in nature and also of the physical and chemical changes that diamond has undergone since its formation. Some of the evidence is derived from investigations of diamond itself but much of the knowledge has been obtained from the minerals that have grown with diamond and become inclusions in the diamond host.

6.1 CHARACTERIZATION OF NATURAL DIAMONDS

The characterization considered here is based on the physical features of diamond, for example, crystal form, crystal angularity, color, surface features, as a function of diamond size and geographic locality. The most important physical characters however, appear to be crystal form and color. Based on the above physical features diamonds have been characterized from southern Africa, Western Australia, U.S.A., China and in part Siberia. A result of such studies, for example, suggests on the basis of color and morphology that diamonds from kimberlite or lamproite are not significantly different.

It is important to remember when discussing the results of mineralogical and geochemical studies of diamond that the data are often skewed to occurrences in southern Africa. This is undoubtedly due to the ready availability of diamonds from this region. However, one must always question whether geological models of diamond genesis erected on data from southern Africa are applicable to diamonds worldwide.

The morphological features recorded for diamonds from kimberlites in southern Africa appear to be independent of depth within the kimberlite pipe and also kimberlite type. It has been suggested that the morphology of diamonds throughout southern Africa is fairly similar and that octahedra, twinned octahedra (maccles) and polycrystalline aggregates are predominant. These shapes have been modified by resorption processes and result in dodecahedra and related forms. It is generally accepted for diamonds worldwide that octahedra reflect growth conditions whereas dodecahedra are the result of resorption, often of pre-existing octahedral forms. Comparable morphological data for other localities in the world are limited but it appears that the proportions of morphological shapes vary from place to place.

In terms of color the majority of diamonds are colorless, yellow or brown but distinct differences in proportions of these colors and the addition of other colors may occur between localities worldwide. For example, a small percentage of blue stones (Type IIb) is a regular feature of the production from the Premier mine in South Africa, whereas diamonds with green transparent surfaces occur in the upper oxidized zones of kimberlites (e.g. Finsch mine, South Africa). It is reported that in southern Africa as diamond size increases so does the proportion of yellow stones relative to colorless and brown ones. Furthermore, diamonds that exhibit plastic deformation, observed as single or multiple lines on the diamond surfaces, are predominantly brown in color.

Surface features have been used extensively to characterize diamonds and an attempt has been made to place these surface features into a chronological order. The features include ones due to crystal growth, plastic deformation, resorption, frictional damage, etching and abrasion. The results of this type of study have been useful in exploration programs to determine the sources of various alluvial diamonds. In Swaziland alluvial diamonds were recognized to have similar surface features to diamonds from the Dokolwayo kimberlite, and in Shandong, China, alluvial diamonds in the Tancheng field were found on the basis of surface features to have been derived from the Changli kimberlites.

6.2 MINERAL INCLUSIONS IN DIAMOND

Because of the unreactive chemical nature of diamond the mineral inclusions within diamond have been armored from chemical re-equilibration, not only with the diamond host, but also with the geological and geochemical environment surrounding the diamond during its growth and subsequent history. Thus mineral inclusions provide unique information, not only about diamond genesis but also about the Earth's mantle wherein diamond has crystallized. Much of the mineralogical, geochemical, isotopic and geological knowledge about diamond and models of diamond occurrence are based on the results of mineral inclusion research.

An important conclusion derived from inclusion studies is that there is no direct genetic link between diamond and kimberlite or lamproite. Diamond was in existence long before these rocks came into being and the only relationship is that kimberlite and lamproite transported diamond from depth to the surface. Another important result, especially for exploration, is the presence in kimberlite or alluvial deposits of chemically equivalent minerals to those occurring as inclusions in diamond. The proven existence of such minerals in a deposit makes it a prime target for further detailed exploration.

Inclusions have been examined in diamonds from localities worldwide and have been found to be generally similar in overall species and chemistry. This conclusion suggests that inclusions cannot be used to determine provenance of a particular diamond, unlike several other gemstones whose inclusions often characterize a gem from a particular locality. A possible exception is the suite of inclusions which occurs in alluvial deposits at Copeton in New South Wales, Australia. These inclusions are generally less than 100 microns in size, are monomineralic and often display good crystal faces. Others may be irregular, platy or have bulbous terminations. Some mineral inclusions although belonging to the orthorhombic or monoclinic crystal systems may display cubo-octahedral morphology. Such inclusions are considered to have grown simultaneously with the host and are referred to as syngenetic inclusions. Other inclusions may pre-date the diamond (protogenetic) and have been incorporated into the host during diamond growth, whereas some can be definitely identified to have formed after diamond growth (epigenetic) and may have entered the diamond along open cleavage planes or cracks. Several of these epigenetic inclusions are the results of alteration of pre-existing inclusions. Large diamonds are more likely to contain large inclusions, but otherwise no relationship has been demonstrated between diamond and inclusion size. Several inclusions may occur in one diamond and they may be all the same or of more than one species. In some instances a single inclusion may consist of two or more different minerals (polymineralic). Some inclusions are crystallographically oriented relative to the diamond host, particularly olivine inclusions (orthorhombic) whose b-axis is often parallel to the [111] axis of diamond.

The first major discovery based on the chemical analysis of mineral inclusions was that the inclusions could be divided into two, mutually distinct suites, an ultramafic (peridotitic) suite and an eclogitic suite. This division is applicable to diamonds worldwide, and over 40 different minerals have been identified as inclusions. Ultramafic suite inclusions (and their host diamond) have formed in a similar geochemical environment to that of ultramafic rocks, and eclogitic inclusions in an environment similar to that of eclogite rocks. Thus natural diamonds crystallized in two geochemically distinct environments. The only exceptions to this rule are the diamonds from Copeton, NSW, SE Australia. Here a unique suite of inclusions occurs, referred to as the calc-silicate suite, and the origin of these inclusions and host diamonds is a subject of debate.

Within a kimberlite pipe rock fragments derived from the Earth's crust and mantle often occur. These are unrelated to the genesis of kimberlite and are referred to as xenoliths. The two main types of mantle xenoliths are ultramafic and eclogitic. Furthermore, some xenoliths may contain diamond, particularly eclogitic xenoliths. Some kimberlites contain both ultramafic and eclogitic xenoliths whereas others may contain a predominance of one type (e.g. eclogite xenoliths, Roberts Victor kimberlite, South Africa; ultramafic xenoliths, Kimberley mines, South Africa). A similar situation exists with regard to diamonds and inclusion suites. Some kimberlites contain predominantly eclogitic suite inclusions (diamonds), e.g. Premier Mine, South Africa, whereas others may contain predominantly ultramafic inclusions, e.g. Pipe 50, Liaoning Province, China. Furthermore, some kimberlites that contain an abundance of eclogitic inclusions may contain predominantly ultramafic suite diamonds. The reason for this is presently unknown. However, on a worldwide basis ultramafic inclusions are predominant over eclogitic suite inclusions. Generally olivine is the most common inclusion followed by garnet, sulfides, spinel, and pyroxenes; however, this relative abundance may vary for individual localities or even regions. Thus in Southern Africa garnet is the major inclusion in three kimberlites followed by olivine, pyroxene, spinel and other minerals. To date no macroscopic fluid inclusions have been observed in diamond.

6.3 TEMPERATURES AND PRESSURES OF DIAMOND FORMATION

The temperatures (geothermometry) and pressures (geobarometry) of diamond formation can be calculated from the chemistry of specific coexisting mineral inclusions in diamond. Suitable inclusions are, for example, clinopyroxene (diopside) coexisting with orthopyroxene (enstatite), and garnet/pyroxene pairs. Several geothermometers and barometers have been calibrated by experimental phase equilibria studies and compared with naturally occurring mineral systems. Ideally the inclusion should be bi- or polyminerallic as this allows diffusion of elements between phases in response to changing physical conditions of temperature and pressure. The temperatures and pressures derived from such inclusions are considered to be final equilibration values and will be in general lower than the original temperature of formation. In contrast, use of monominerallic inclusions in the same host diamond but separated from each other can be used with certain caveats. One assumption is that the two inclusions grew more of less simultaneously together with the diamond. This may not be the case as evidenced by the stratigraphy of diamond growth, and thus the inclusions may have grown at different times and not have been in equilibrium with each other or the same geochemical environment. Temperatures and pressures (assuming simultaneous growth of the two inclusions) should reflect original T and P of diamond formation. One further problem is that different geothermometers and barometers can give different T's and P's for the same inclusion pair. Thus one has carefully to interpret the data related to geothermobarometry. Furthermore some multiple pairs of inclusions in the same diamond give different T's and P's suggesting disequilibrium.

In the case of ultramafic suite inclusions suitable pairs for determination of both T and P are often available. Unfortunately this is not the case for the eclogitic suite inclusions and there is no suitable geobarometer for the composition of garnet/clinopyroxene that predominates in the eclogitic suite. As a consequence it is necessary to assume a pressure in order to calculate a temperature. A value of 50 kilobars (5Gpa) is often assumed.

In spite of the drawbacks mentioned above the results for diamonds worldwide give an overall temperature range for both suites of between about 1400 and 900°C. However, the eclogitic suite inclusions (and presumably host diamonds) have equilibrated at higher temperatures than the ultramafic suite inclusions which appear to have formed in the sub-solidus field. Ultramafic suite inclusions generally have formed at pressures in excess of 50 Kilobars (5 GPa) and this is one reason for the assumption that eclogitic suite inclusions have also equilibrated at these pressures. These T's and P's are equivalent to conditions at depths of approximately 180 to 200 km in the lithospheric mantle of the Earth.

A unique set of inclusions has been observed in diamonds from Monastery Mine, South Africa, and Juina alluvial deposit, Brazil. These consist of garnet inclusions which contain pyroxene in solid solution. The amount of pyroxene in solid solution is a function of pressure and experimental investigations of the garnet/pyroxene system have demonstrated that these inclusions (and host diamonds) may have formed at depths equivalent to about 350 kilometers in the asthenospheric region of the Earth's mantle. These inclusions are the deepest pristine samples of the mantle available and are important for understanding the geochemical environment at these depths but also have implications regarding the early formation of the Earth. One other important point is that the occurrence of these asthenospheric diamonds indicates that the transporting kimberlite must have originated at least at these depths or beyond.

6.4 RADIOGENIC ISOTOPE AGES OF INCLUSIONS

There were early suggestions, based on the chemistry and nature of the inclusions, that diamonds were unrelated to the host kimberlite rock and thus were not phenocrysts in kimberlite but rather xenocrysts. These were further strengthened in the mid and late 1970's by two studies of K, Rb, Sr, Pb and U isotopes in selected parcels of diamonds with inclusions from the Premier, Finsch and Kimberley kimberlite mines. The results indicated that the diamonds were much older than the intrusion of the host kimberlites. However, for various reasons these results were challenged and it was not until 1984 that an investigation of Nd-Sm isotopes on selected syngenetic ultramafic garnet inclusions demonstrated unequivocally that the inclusion (and hence host diamonds) were formed 3 billion years ago whereas the kimberlite erupted only 90 million years ago.

Further isotopic studies of both ultramafic and eclogitic suite inclusions have corroborated this result and confirmed that diamond is in fact a xenocryst in kimberlite. Thus kimberlite (and lamproite) as they pass through the mantle on their way to the surface simply pick up diamonds that have sat in the mantle for billions of years and transport them upwards. However, it appears based on the present isotopic results that ultramafic inclusions (and diamonds) are generally older than eclogitic suite inclusions (and diamonds). The youngest diamond so far determined is one from Zaire and has an age of 622 million years compared to the kimberlite intrusion age of about 90 million years. It thus appears that diamonds have formed continuously through geologic time, certainly from 3.3 billion years ago.

6.5 TRACE ELEMENT IMPURITIES IN DIAMOND

At least 62 elements have been reported to occur in diamond, but most are at the ppb and ppt level. Nitrogen is an exception and is the major impurity in most natural diamonds, often present at levels of several parts per thousand. Most of these elements have been identified by Instrumental Neutron Activation Analysis (INAA). In many instances the trace elements are believed to be contained in sub-microscopic inclusions within the diamond host and the more common, or abundant, elements are those associated with rock forming minerals, especially those associated with diamond. It has thus been suggested that the trace elements, and the sub-microscopic inclusions, represent the composition of the original magma in which the diamonds formed. These magmas have picritic or komatiitic compositions.

Recent studies have demonstrated that particular trace elements, (e.g. Au, Ir, Cr, Sc) can be used to determine the suite (ultramafic or eclogitic) to which the diamond may be assigned, in spite of the absence of macroscopic inclusions. Furthermore, the relative abundances of specific trace elements in diamonds from a particular locality appear to lie within specific ranges. These ranges differ for localities worldwide. Thus it may be possible to use such abundance ranges as indicators of provenance for diamonds.

It is generally believed that most of the impurity elements in diamond, apart from nitrogen and boron, are non-substitutional. Substitutional nitrogen and boron impart various levels of yellow or blue color, respectively, to the diamond host. Nitrogen can be present at up to about 5000ppm (0.5 wt %). The occurrence or absence of nitrogen is used as a basis for dividing diamonds into two types; Type I (nitrogen present) and Type II (nitrogen absent). This division is further sub-divided into Type Ia, Type Ib, Type IIa and Type IIb, predominantly on the basis of infra-red spectra. Type IIb diamonds contain substitutional boron (<0.25 ppm), have a blue color and display semiconducting properties. Although rare in nature a well known source for such diamonds is the Premier Mine in South Africa. This classification is discussed further in Section 8.4 below.

Type I diamonds are the most common and account for about 98% of all diamonds. At the time of formation of diamond the nitrogen enters the lattice as isolated substitutional atoms (Type Ib diamond - very rare). Over a period of time at the elevated temperatures and pressures in the mantle the nitrogen diffuses and aggregates first to pairs of nitrogen atoms and then to more complex arrangements. Because the process of aggregation depends upon the length of time the diamond has experienced high temperatures in the mantle, if the residence time is known then one can calculate the temperature, or conversely, if the temperature is known one can determine the residence time in the mantle. The nitrogen content and the aggregation state can be determined from infra-red spectra. Assuming diamond ages in the range obtained from radiogenic isotope studies (3 to 1 billion years) temperatures calculated from the infra-red nitrogen aggregation data lie in the region of 1050 to 1200°C, similar to the range obtained from mineral inclusion thermometry. However, diamonds from a single source may show a range of temperatures, and even one or more populations based on a limited temperature range.

6.6 STABLE ISOTOPE STUDIES OF DIAMOND

Stable isotope studies of diamond include ^{13}C, ^{15}N, Argon, Neon, Krypton, and Helium. However, by far the greatest study has been confined to ^{13}C. In the early 1950's the average for ^{13}C was believed to lie in the range -5 to -9 per mil. With continued investigations of diamonds worldwide it became obvious that this range extended from +3 to -30 per mil. A major peak occurs at about -7 per mil and this generally reflects diamonds of the ultramafic suite. The broader range, reaching to beyond -30 per mil is considered to include eclogitic suite diamonds, as well as those consisting of polycrystalline aggregates. There is a suggestion that Type II diamonds may have a peak around -15 per mil.

Much of the data on carbon isotopes has been collected on bulk samples, but it is well documented that there occurs variation of ^{13}C within single diamonds. A group of unique diamonds with positive ^{13}C (+3 per mil) occurs in the alluvial deposits of Copeton/Bingara, New South Wales, Australia.

Fewer studies have been undertaken on ^{15}N of diamonds. The general range is between +3.2 and -11 per mil, at total nitrogen contents between 100 and 2100 ppm.

$^{3}He/^{4}He$ investigations were initially undertaken on alluvial diamonds and the results were ambiguous; one result providing an age older than the age of the Earth. Recent studies tend to support a cosmogenic component and thus the best investigations need to be done on diamonds immediately extracted from the deeper parts of an underground mine.

The early results of $^{40}Ar/^{36}Ar$ have always been questioned but more recent controlled experiments show a range of $^{40}Ar/^{36}Ar$ from 2473 to 259.

7. Geological Summary

Several important facts have been noted in the above discussion regarding the genesis of natural diamond. In summary they may be listed as:

a) Diamonds occur in primary deposits in the ancient cores (cratons) of continents, especially in rocks older than 2.8 Ga that have been intruded by kimberlite.

b) Multiple intrusions of diamond-bearing kimberlites have occurred throughout these cratons over successive geological time (e.g. the Kaapvaal craton, Southern Africa; the Siberian craton, Russia).

c) These cratons, constituting lithospheric rock have been detached from the underlying asthenosphere.

d) Diamonds are often ancient megacrysts in kimberlite and lamproite and thus are unrelated to the genesis of these two rock types. Both appear to be transporting mechanisms which carry diamonds from depths of the order of 200 Km, or even deeper, to the surface.

e) The majority of diamonds have formed in the range of 1200 to 1000°C and pressures approximately in the region of 6.5 GPa. These are generally termed lithospheric diamonds to differentiate them from some rare diamonds in South Africa and Brazil, which on the basis of inclusion chemistry appear to have formed in the region of 350 km depth and are referred to as asthenospheric diamonds.

f) On the basis of inclusions diamonds are subdivided into two main suites: Ultramafic suite, and Eclogitic suite. A unique suite, the Calc-silicate suite, occurs in alluvial diamonds in the Copeton/Bingara area of New South Wales, Australia. The significance of this latter suite is still an open question.

g) The majority of diamonds belong to the ultramafic suite and have ^{13}C values in the range of -5 to -7 per mil. Eclogitic diamonds span a larger range, to beyond -30 per mil, and include polycrystalline aggregates of diamond. Others range upwards to +3 per mil.

h) Ultramafic suite diamonds also occur in ultramafic (lherzolite, harzburgite, dunite) xenoliths in kimberlite and are believed to represent the original rocks in which these diamond formed. In contrast eclogitic suite diamonds are thought to have signatures of basaltic rock which has been subducted into the mantle and the carbon inverted from graphite to diamond.

In order to explain many of the above features it has been suggested, particularly as a model for the Kaapvaal Craton, South Africa, that the majority of diamonds form over a period of time in a cool, thick cratonic keel associated with ultramafic rocks. The kimberlite (or lamproite) which transports the diamonds to the Earth's surface probably derives from greater depths within the asthenosphere and at this stage may be considered a proto-kimberlite magma, possibly of picritic or komatitic nature. Some of this asthenospheric magma may be the transporting mechanism for the asthenospheric diamonds.

The eclogitic suite diamonds start their life as non-diamond carbon, either in marine sediments or associated with the basaltic rocks which are major features of the oceanic crust. With time these rocks are subducted and the transformation of these basaltic rocks to eclogitic occurs accompanied by the change of graphite to diamond. With continued subduction these rocks, including diamond are underplated to the lower surface and interior of the cool thickened cratonic root zone, where they too are available for transport to the surface. At the edges of the craton the thermal regime is generally warmer and not conducive to the preservation of diamond. Thus as one moves away from the central and older part of a craton the less likely is one to find diamond-bearing kimberlites. Finally the model also explains why repeated eruptions over time of diamond-bearing kimberlite have occurred; in essence the lithospheric keel stores the diamonds as a cargo awaiting the puncture of the ascending kimberlite to move the cargo to the surface.

The main discourse above has been on what are referred to as "macro" diamonds, those that range from say a few tenths up to several hundreds of carats. They are believed to be megacrysts in the kimberlite and lamproite. However, much interest has been focussed on so-called "micro-diamonds." These are small stones variably defined with an upper size limit of 1mm or 0.5mm or some similar number. These micro-diamonds are important and recent work has suggested they may represent diamond that has grown in proto-kimberlite or lamproite. This is especially convincing in view of the fact that most micro-diamonds are growth form octahedra and have suffered very little resorption, which is often a major feature of the larger macro-diamonds. In essence micro-diamonds could be phenocrysts in kimberlite and lamproite, and are thus related directly to the formation of these rocks.

8. Classifications

8.1 GEM V. INDUSTRIAL

Though diamond is or rather should be pure carbon, there are many variations in quality depending largely on the impurity content, but also on the shapes of the crystals and the perfection of crystallization. Impurities can be in dispersed form showing up most significantly as color, or as discrete inclusions which may be present singly or in clouds. Commercially the most important distinction is between gem and industrial. Gem or "cuttable" diamonds are, as the name indicates, those destined to be cut into jewelry. They are in general single crystals, though there may sometimes be regions of twinned material (known in the trade as "knaats") which add to the difficulty of cutting, but may be acceptable if the boundaries are more or less invisible. Ideally there should be no inclusions, but this is a counsel of perfection. Flawless diamonds are rare, highly prized and highly priced. Color is equally important, white being the norm of the trade. White means transparent or water-clear, free from any tint, though gem appraisers recognize more shades of white than most people can imagine. Tints shading into the more common browns, dull yellows, and grays reduce the value, though (perhaps perversely) clear coloration in one of the rare shades of pink, blue, green or clear yellow may increase it. Gem diamonds, then, are those which can be cut into an attractive shape with facets to provide refraction and internal reflection of incident light, and sufficient clarity and an appropriate color to allow the dispersion of the light within the diamond and the play of spectral colors or "fire" which is one of the distinguishing features of this gemstone. Industrial diamonds are all others.

Industrial diamonds cover a gamut of qualities ranging from "near gem" at the expensive end to "crushing boart" at the cheaper end. Near-gem diamonds usually have the clarity and crystal perfection of gem, but a color which is unacceptable or minimally acceptable for jewelry. They are used for production of precisely shaped industrial diamond components such as turning tools or surgical knives. Crushing boart is diamond of such low quality (usually occasioned by multitudinous inclusions or cracks) that it is only fit to be crushed to powder for abrasive use. In between come qualities which generally are named for their specific use.

Diestones, for example, are good quality diamonds often of rather flattened dodecahedral shape, and clear enough to permit viewing of the interior through facets polished for that purpose. After drilling of shaped holes, diestones are used for the drawing of wire. The diestone material needs to be good enough that no crack or inclusion marches on the hole and strong enough to withstand the quite large forces involved in wire drawing. Of lower quality are dresser stones used to make relatively rough tools for the shaping of aluminum oxide or silicon carbide grinding wheels or to open up the texture of their surfaces and make them free-cutting.

Size is also an important factor in determining use. The so-called "drilling" eponymous from the use, ranges from a few pieces per carat weight down to about 100 or sometimes more stones per carat. The larger sizes are used in surface-set drills for the oil and gas industry. Closed, i.e. generally equidimensional diamonds without re-entrant angles are in demand for reasons of strength and the absence of fissures or sharp corners which could snag in the often very inhomogeneous rock being drilled. Since these are sizes which are also appropriate for the production of the smaller gems, commercial pressures dictate that the material be sorted to remove the white stones. At the other end of the size range the diamonds are too small for surface setting and are used in sintered metal drill crowns, incorporated in depth in the abrasive layer. They are also too small for most gem production and accordingly are not sorted to the same extent. The inherent material quality may thus be better.

Industrial diamonds may also occur with radially fibrous growth. The result is usually a "coated diamond" in which the fibrous coat overlays a monocrystalline center. The inherent stresses associated with this are usually sufficient to produce gross cracking of the central material, but sometimes the stone presents a lottery in which the prize of a central gem is hidden below an opaque coat. Normally, however, coated diamonds are only sorted for use as dressers or boart. They are common in the production of the Congo area.

Another growth variant is carbonado. Traditionally from Brazil, the material is polycrystalline throughout. This makes it impossible to shape by polishing, but also very tough. It is commonly shaped by breaking into pieces of the desired size, and used directly in appropriate tools and drills.

Industrial qualities make up more than 90% of the world natural diamond production by weight, but less than 10% by value. There is, however, insufficient natural diamond to satisfy world demand for diamond abrasives and some other products, and high-pressure, high-temperature synthetic diamond dominates some industrial markets. With the tremendous variation between natural diamonds, classification of the industrial qualities in a general scheme is difficult, and is largely a matter of the practice of individual companies or sections of the industry. The classification of gem diamonds has reached a more advanced stage, probably because of the nature and value of the final product.

Table 1. Clarity grades in polished gemstones.

Grade	Description	Inclusion size (maximum)
clean	loupe clean 10×	5 μm
vvs 1	very very small 1	12 μm
vvs 2	very very small 2	25 μm
vs 1	very small 1	40 μm
vs 2	very small 2	70 μm
si	small inclusions	150 μm
p 1	first piqué	500 μm
p 2	second piqué	1.5 mm
p 3	third piqué	3.0 mm

Table 2. Simplified comparison of color grading scales [after Bruton, 1970].

G.I.A.	Scan.D.N. (> ½ ct.)	U.K.
D	river	finest white
E	river	finest white
F	top Wesselton	fine white
G	top Wesselton	fine white
H	Wesselton	white
I	top crystal	commercial white
J	crystal	top silver cape
K	crystal/top cape	top silver cape/top cape
L	top cape	silver cape
M	cape	light cape
N	cape	light cape
O-Q	light yellow	cape
R-X	yellow	dark cape

8.2 GEM CLASSIFICATION

The assessment of polished gems is usually done on the basis of the "4 C's." These stand for carat weight, clarity, color, and cut. Carat weight needs no further explanation except to note that the internationally used metric carat is defined as 0.2 g exactly. Clarity refers to the presence or absence of inclusions. Normally these are looked for with a 10× loupe. The descriptions have been adopted by most diamond grading laboratories and are based on the "International Rules for Grading Polished Diamonds" of the World Federation of Diamond Bourses. They range from loupe-clean to 3rd piqué (Table 1).

As regards color, there are various systems in use, but that of the Gemological Institute of America (G.I.A.) is perhaps the most common. Colors range from D (finest white) to X (yellow). The white grades are very finely spaced. Experts can disagree by one grade in assessing a stone, and without training it is difficult even to judge within three grades

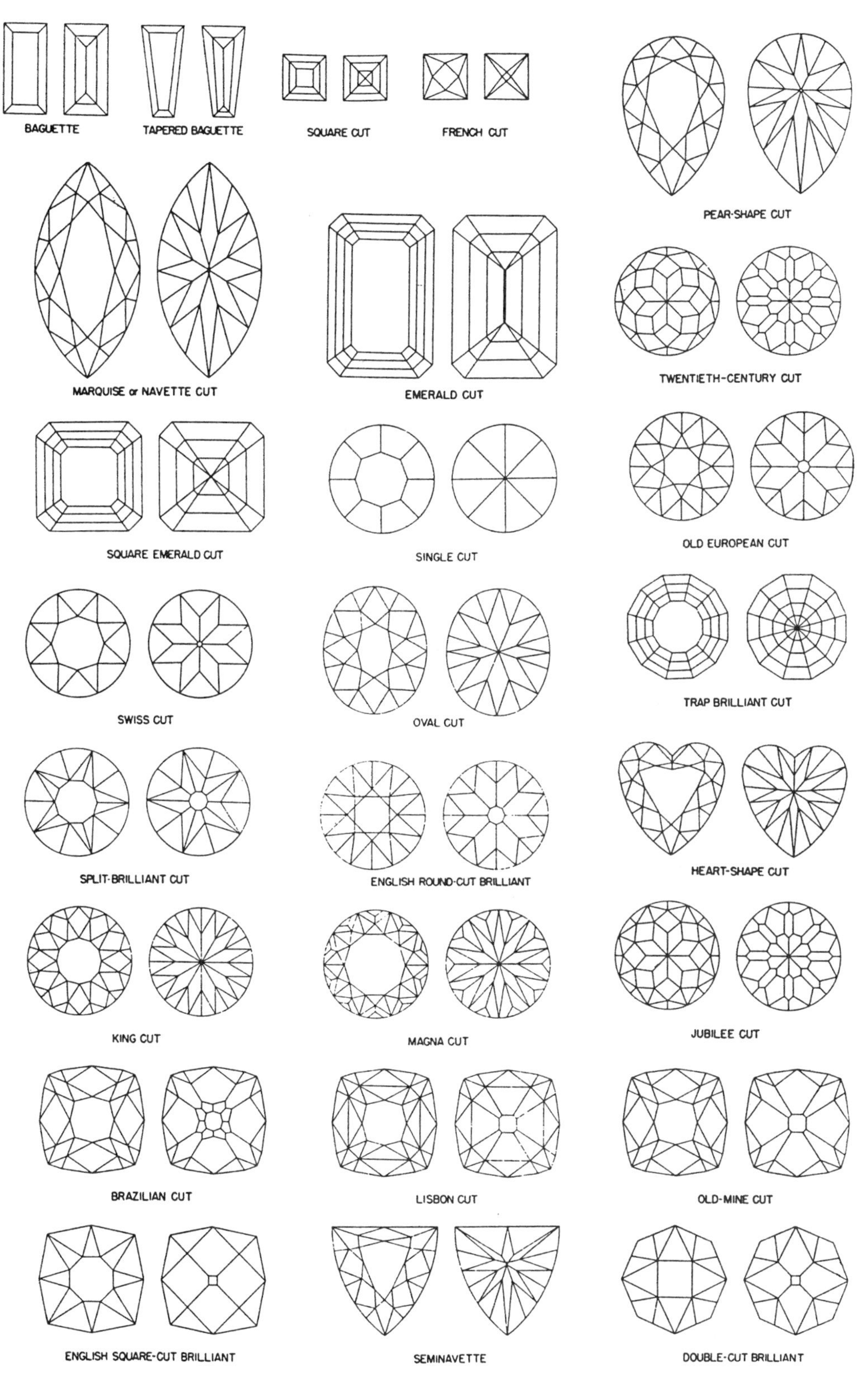

Figure 1. Diamond fancy cuts

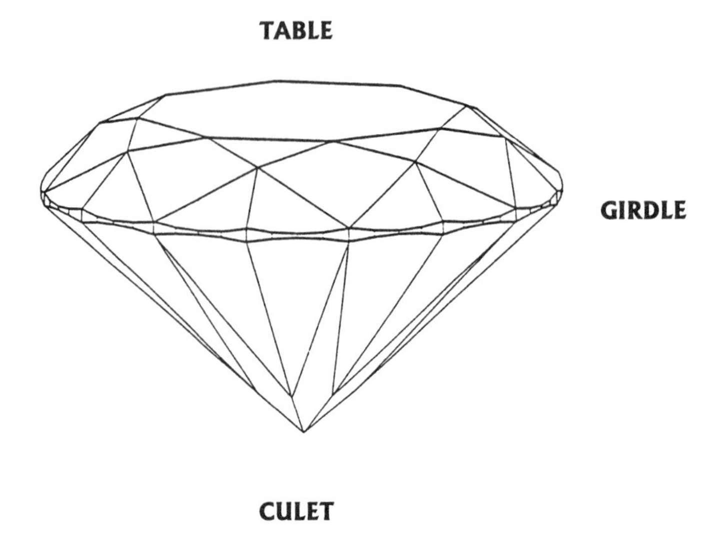

The ideal (Scan. D. N.) cut has:

- **table diameter 57.5%,**
- **table to girdle 14.6%,**
- **girdle to culet 43.1%, of the girdle diameter.**

Figure 2. Brilliant-cut diamond

on either side of the correct one. Other systems are also in common use. The Scandinavian Diamond Nomenclature and Grading Standards (Scan.D.N.) were introduced in 1970 by the four Scandinavian countries and have found wide acceptance. A more traditional system is in common use in the U.K., and the American Gem Society has a system based on numbers from 0 to 10. One step on this system is roughly equivalent to two on the G.I.A scale, with A.G.S. zero roughly equivalent to G.I.A. D&E, though there are non-linearities lower in the comparison. Scan.D.N. imposes different scales for small and for large diamonds, as is optically logical. Table 2 shows a much simplified comparison of the more commonly used scales.

The last "C", cut, is a subject in itself. The brilliant shape with 57 facets is the norm, but there are many fancy cuts (Figure 1). Further, the proportions are always important, diamonds which are too deep or too shallow lacking fire and appearing dark in the center, or with a "fish-eye" effect. An ideal proportion for a brilliant-cut diamond is shown in Figure 2. There are also considerations of symmetry and a requirement that four facets coming together should apparently meet at a point when the stone is viewed with a 10× loupe, i.e. there should be no visible closure line.

8.3 INDUSTRIAL CLASSIFICATION

As indicated in section 4.1, industrial diamonds do not lend themselves to a precise general classification. Baumgold [1979] published an excellent overview, which necessarily was somewhat oversimplified. He described eight qualities of "drilling" and

Table 3. Comparison of Christensen and U.S. standard mesh sieves.

	Christensen sieves		U.S. standard mesh	
stones per carat	no.	opening (round), mm	no.	opening (square), mm.
4.2-4.5	12	2.90	8	2.39
7	10	2.50	10	2.01
11-12	8	2.10	12	1.68
20	6	1.80	14	1.40
30-33	4	1.50	16	1.19
58	1	1.25	18	1.00

ten of toolstones with color photographs of each. He also reproduced part of an American National Standard [1975] which lists and describes some twenty two qualities of dressing tool stones. He included in his article a brief account of the technology of diamond abrasives.

8.3.1 Drilling

An early account of this material was given by Long [1960]. A detailed more formal system of classification followed in a joint project of the Industrial Diamond Association of America and the U.S. Army Corps of Engineers to specify qualities of drilling and set up reference samples. These samples were made available for inspection at the Corps of Engineers Southwest Division Laboratory in Dallas.

The 1960's and 1970's saw a great expansion in the drilling business and many companies adapted their product specifications to the exigencies of their own processes, markets, and raw material supplies. In particular, the supply of the best and strongest qualities of mined drilling was supplemented with processed diamond derived from lower quality material. This was subjected to a variety of processes including impacting, tumbling and abrasion with diamond slurry to produce rounded particles of high strength. The weaker or cracked particles had been broken and thus removed. Surfaces were sometimes polished to improve the appearance and allegedly the performance of the product.

Another quality known as "casting" was used in impregnated (sintered) bits. This was derived from Congo or West African whole diamonds, rounded or cuboid in shape and often of the coated type. It was, however, not of the strength needed for surface-setting.

Today much of the high quality end of the drilling market has gone to synthetic polycrystalline diamond (PCD) segments. These can be supplied as segments or full rounds which are ideal for brazing to the drill forms in a surface-set pattern.

For the sizing of drilling Christensen sieves are in most common use. These comprise metal plates with round holes, fitting in a frame which is shaken by hand. Woven wire mesh sieves are also used. The dimensions of the square openings for the U.S. mesh sizes are specified in an ASTM Standard [1961]. The mesh numbers are the approximate numbers of openings per inch. Because of the different shapes and tolerances of the sieve openings, the products from Christensen sieving and wire mesh sieving are not directly comparable. A very rough guide for some of the middle sizes follows in Table 3 above.

8.3.2 Abrasives

Natural diamond abrasives are produced by crushing low grade mined diamonds. By now they account for only a small proportion of the total diamond abrasive market, the greater part of which is served by synthetic diamond grits. Particle size and shape distributions are important for all diamond grits. They affect the performance of the grinding wheels made from them. Shape is controlled by the type of crushing and by processes which separate rounded from flat particles, e.g. by sorting on inclined vibrating tables. Size separation is effected by sieving in the coarser sizes or by Stokes law methods (settling, centrifuging, and elutriation) for the finer sizes. Sieving is a non-trivial process in that square mesh sieves can pass flatter particles on the diagonal of the opening or elongated particles lengthwise, yet hold back the same particles if they are presented differently by a variation of the vibratory motion of the sieve shaker. The problem is compounded by variations in the openings of nominally identical sieves. These can vary quite widely due to lack of uniformity of the weaving of wire sieves or of the plating rate for electroformed sieves (due for instance to different electrical paths to center and edge of a sheet of sieve material affecting the current density and rate of metal deposition across the sheet during production).

The sizing of diamond grits and powders is the subject of several national and international standards. For the grit sizes there is good standardization, the international (ISO 6106), U.S. (ANSI B74-16), and European (FEPA) test sieving standards being in almost perfect agreement. The procedure for checking sizes involves four sieves for each size: an oversize limiting sieve through which 99.9% must pass; an upper control sieve which determines allowable oversize (maximum 8 to 15% depending on size); a lower control sieve which determines the on size fraction (minimum 80 to 90%, again depending on size) and allowable undersize (maximum again 8 to 15%); an undersize limiting sieve through which no more than 2% may pass. The sieve dimensions are tabulated on the following page (Table 4).

8.3.3 Polishing Powders

Though strictly speaking also abrasives, the finest sizes of diamond powder are distinguished from the grits both by the method of production and by the applications. Sieving on electroformed sieves may now be used for the sizes immediately finer than D46 = 325/400 mesh (Table 4), but the tolerances on sieve openings are in general absolute (not better than ± 1 μm at present) rather than proportional to the aperture sizes. This precludes their use for the finest sizes. Historically the solution has been to grade the powders by settling, centrifuging, or elutriation in liquids of appropriate viscosity. The sizes were and are measured by examination under microscopes fitted with calibrated graticules, and expressed in ranges of microns (μm). Since the coarser sizes were designated by mesh numbers and called "mesh sizes," the finer sizes acquired the name of "micron powders." The applications are also different, mesh sizes in general being used for stock removal and micron powders for surface finishing.

Table 4. ISO/ANSI/FEPA test sieving standards.

ISO/ FEPA size	U.S. mesh size	Oversize limiting seive (μm)	upper control sieve (μm)	lower control sieve (μm)	undersize limiting sieve (μm)
D1181	16/18	1700	1180	1000	710
D1001	18/20	1400	1000	850	600
D851	20/25	1180	850	710	500
D711	25/30	1000	710	600	425
D601	30/35	850	600	500	355
D501	35/40	710	500	425	300
D426	40/45	600	455	360	255
D356	45/50	500	384	302	213
D301	50/60	455	322	255	181
D251	60/70	384	271	213	151
D213	70/80	322	227	181	127
D181	80/100	271	197	151	107
D151	100/120	227	165	127	90
D126	120/140	197	139	107	75
D107	140/170	165	116	90	65
D91	170/200	139	97	75	57
D76	200/230	116	85	65	49
D64	230/270	97	75	57	41
D54	270/325	85	65	49	~
D46	325/400	75	57	41	~

Notes: 1. Calibrated woven wire mesh sieves are specified down to 500 μm opening. Below that electroformed sieves are specified. This results in an apparent discontinuity in the progression of numbers in the table above.

2. The ISO and FEPA designations differ very slightly. Those in the table are the FEPA designations. The ISO ones are sufficiently similar for the correspondence to be obvious.

The measurement in μm does bring problems. One is trying to use a single linear dimension to define the "size" of an irregular three-dimensional particle. Conceptually it is impossible unless one makes assumptions. The first of these is inherent in the method of measurement by microscope. One is looking at a two-dimensional projection in which usually the flat particles will be lying on their largest faces and the needle-shaped ones with their lengths in the plane of projection. The next problem is to define the one-dimensional size of the two-dimensional image of an individual particle in the microscope. Three definitions have been used, namely the diameters of the inscribed, equal-area, and circumscribed circles. Of these the most commonly used is the last, equivalent to the longest dimension of the particle.

Standards using the circumscribed circle as the measure of particle size include the FEPA [1977] and IDA [1985]. Both define a nominal range for each size, with tolerance for undersize and oversize and a maximum permissible particle size. Typical ranges for the IDA standard are 0-½μm, 0-1μm, 0-2μm, 1-3μm, 2-4μm through a dozen intermediate sizes to 36-54μm, 40-60μm, 54-80μm, The FEPA standard prefers to define mean sizes and maximum standard deviations for the distribution of particles. It also takes account of the Mach effect [1865 and later] in which an image boundary of moderate contrast is seen in sharpened form due to interaction between neighboring areas of the retina. The amount and apparent position of the sharpening is a function of individual vision. It has been suggested that the effect is an evolutionary consequence of the need to distinguish predators against a background into which they blend. Be that as it may, the effect is that different operators see the same particle in a microscope differently and measure it differently. The FEPA standard attempts to allow for this by introducing a calibration factor for each operator. It is probably more scientific than the other diamond micron powder standards, but less convenient in use.

The ANSI [1974] standard uses as a measure of size the diameter of the circle of area equal to that of the projected image of the particle. At the time this was thought to facilitate automated image analysis, but advances in that technology have made it now possible to use virtually any desired measure of particle size. The ranges specified in the ANSI standard (0-1μm, 0-2μm, 2-4μm, 4-8μm, 8-12μm, 12-22μm, 22-36μm, 36-54μm, 54-80μm) are similar to those of the IDA standard though in coarser steps. However, because of the different definitions of particle size there is no simple equivalence between nominally identical sizes measured according to different standards.

8.4 SCIENTIFIC CLASSIFICATION

The current scientific classification of diamonds derives from observations in the 1920's and 1930's that different diamonds differed often quite markedly in some of their physical properties. The first definitive classification based on multiple properties was that of Robertson, Fox, and Martin [1934] who recognized two types on the basis of the properties listed in Table 5.

It was noted that color was not a function of the type, and there were no systematic differences as regards specific gravity, refractive index, dielectric constant, triboluminescence, the Raman difference, or electron diffraction patterns.

Since 1934 there has been a great deal published on the spectral classification of diamonds. The types have been subdivided and the impurities and optical centers responsible for the various spectral features identified. The essentials of the main sub-classification are summarized in Table 6. The state of aggregation of nitrogen impurity affects the vibrational (infrared) absorption spectrum and has led to a further sub-classification (Table 7). One can take the details further, but the intricacies of the spectra become almost overwhelming for a non-specialist. Over 100 vibronic bands have been identified in absorption, cathodoluminescence, and photoluminescence spectroscopy. The matter is complicated further by the fact that natural diamonds are inhomogeneous, sometimes having adjoining regions of different spectral type, and often being of mixed or intermediate type where combinations or varying concentrations of different defects occur in the same crystal region. Some of the boundaries can be revealed by etching diamonds with fused potassium nitrate [Seal, 1965]. An example is shown here (Figure 3).

Table 5. Diamond types proposed in 1934

	Type I	Type II
Occurrence:	Common.	Rarer.
Form:	Derivatives of cubic system.	Same, but with fine, parallel laminations.
Isotropy in polarized light:	Considerable anisotropy.	Nearly isotropic.
IR absorption at -170°C:	At 3, 4.1, 4.8 and 8 μm	Same, except no band at 8μm.
UV absorption:	Below 300nm.	Below 225nm.
Photoelectric conductivity:	Small, even at high volts.	Present even at low volts.
X-ray pattern:	111/222 small.	111/222 large.

Table 6. Sub-classification of diamond types.

	Type IA	Type IB	Type IIA	Type IIB
Main impurity:	nitrogen	nitrogen	-	boron
Nitrogen concentration (p.p.m.):	500-5000	5-500	<100	<100
Typical color:	white/yellow	yellow	brown/white	blue/gray
Electrical conductivity:	insulator	insulator	insulator	p-type semiconductor
Thermal conductivity at 320K (W/m K)*:	530-1760	1750	1900	1950
Electron Spin Resonance:	no	yes	no	no

* Thermal conductivity data from Burgemeister [1978].

Table 7. Nitrogen aggregation and diamond type.

Type	Main infrared band (μm)	Description of aggregate or optical center
IB	8.85	Isolated substitutional N.
IAa	7.8	Pair of N atoms.
IAb	8.4	Larger N groups (even number of N atoms: 4, 6, or 8?)
Platelet	7.3	Platelets of large extent and low thickness, probably a structure of C & N atoms.

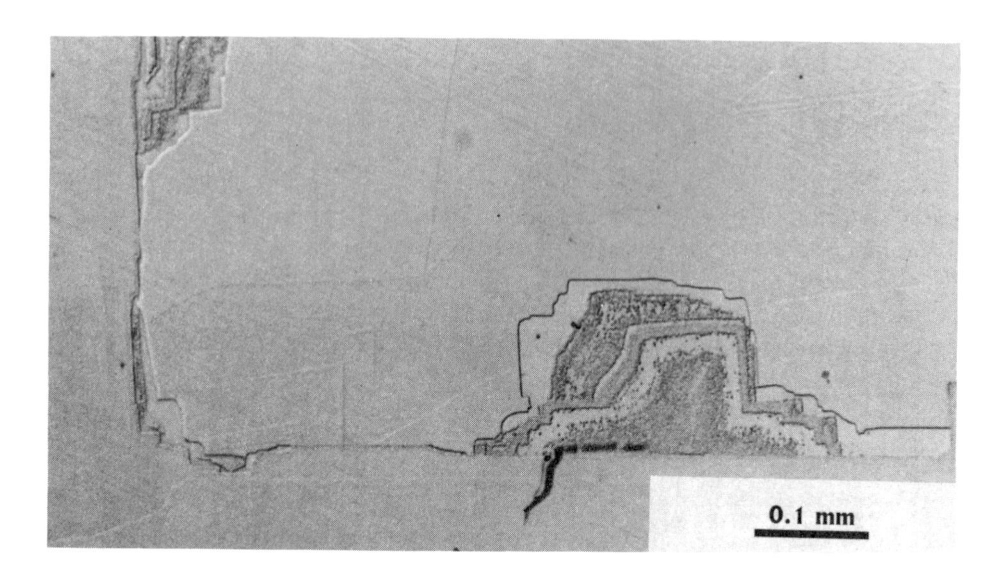

Figure 3. Diamond etched with fused potassium nitrate.

8.5 BOOKS ON DIAMOND CLASSIFICATION

A great deal of the background to the classification of natural diamonds is surveyed in three books published at roughly 40-year intervals. The first of these, entitled simply "Der Diamant" by von Fersmann and Goldschmidt [1911] is a compilation of detailed descriptions of some 131 diamond crystals, with an atlas of meticulously executed drawings showing all the principal forms of diamond crystals. The second, Grodzinski's "Diamond Technology" in its definitive [1953] edition covers not only the processes of diamond shaping and use, but includes lists and tables of the grading standards of the time. The most modern of the three is Field's "The Properties of Natural and Synthetic Diamond" [1992]. It gives a detailed account of current knowledge of the physical science of diamond, and in chapter 2 by Clark et al. the background to the optical classification of diamonds (summarized also in tables in the Appendix by Field). The three books were written respectively from the standpoints of classical descriptive mineralogy, industrial technology, and modern physics. They thus cover very different topics.

As regards diamond gems and jewelry, one of the best books is Bruton's "Diamonds" [1970 & 1978]. Another useful publication is the G.I.A. "Diamond Dictionary" [1993]. On the industrial side the I.D.A. book "The Industrial Diamond" [1979] is good, but beginning to show its age. Similarly, the major field of diamond abrasive use and the classification of abrasives was covered in a book "Diamond Grinding of Metals" by Hughes [1978]. This does, however, concentrate on the products of one manufacturer and there is really no book which surveys the whole field of modern diamond abrasives. For this, one is best advised to read and compare the trade literature.

9. Fabrication Methods

9.1 INFLUENCE OF CRYSTAL ORIENTATION

The major industrial techniques for shaping diamonds comprise cleaving, sawing, and laser cutting for dividing individual diamonds in two or more parts, and bruting and polishing for material removal and surface preparation. The use of these techniques is constrained by the effects of the atomic lattice of the diamond. Thus the major cleavage is on {111} which are the planes of greatest spacing of the atomic layers. Cleaving as an industrial technique is restricted to the splitting of diamond on those planes. A {110} cleavage does exist and diamonds under suitably oriented high stress distributions (as in high pressure anvil use) do often fracture on {110} but this cleavage is too unreliable for industrial use. The {111} cleavage is more perfect in type II diamonds than type I, which may sometimes fracture conchoidally, an ever present risk for the cleaver.

If one wants to split a diamond in two parts along a {111} plane, cleaving is the method of choice. For division along other planes, e.g. {110} or {100}, one is obliged to use another method, usually sawing with a thin disc charged with diamond powder. This method in turn is constrained as regards orientation. The technique is basically abrasion and there is a strong anisotropy in the resistance of diamond to abrasion. On {100} and {110} planes, <001> directions are those of easy abrasion, whilst others such as <011> may be several hundredfold more abrasion-resistant. Thus in sawing a well-shaped octahedral diamond across the middle, one saws from point to point and not along an edge.

Flat polishing, which is also abrasion, is subject to the same constraints on the planes and directions of polishing. Bruting, a rough turning operation, removes material by small-scale chipping and is not particularly orientation-sensitive. It is only used to generate cylindrical surfaces such as the girdles of brilliant-cut gemstones which have a matte, rough appearance. Laser-cutting and the newer hot-metal techniques remove material by carbon phase transformation or chemically and are not crystal orientation-sensitive.

9.2 CLEAVING

Traditional cleaving started with study of the diamond shape and internal flaws to determine the best position and direction for splitting. A notch was then formed to give a key for the cleaving blade. This was usually done by rubbing the edge of an industrial diamond backwards and forwards over the diamond to be cleaved. Both diamonds were abraded, but the desired result was a V-shaped notch in the diamond to be split. A steel blade was then inserted in the notch, oriented along the cleavage plane and given a blow with a round mallet in that direction. If skillfully done, the diamond splits in just two pieces as planned.

Part of the basis of the art of cleaving lies in the way of holding the diamond. This is partially embedded in "cleaver's cement," a sealing-wax-like compound but of rather harder consistency. The material is a mixture of shellac, rosin, brick-dust and glass-dust, sometimes with proprietary additives. It softens rather abruptly at about 80°C and the diamond can be pushed into position in it by hand. A good composition will have sufficient acoustic match to the diamond to permit passage of the shock wave from the cleaving blow across the diamond-cement interface with minimal reflection. Reflected shock waves are the cause of shattering and a function of the cement is to prevent that.

The cleaver's knife is usually of a quenched carbon steel, with an angle of 20° to 25° over the edge. The mallet is of steel or hard wood, of round cross-section and somewhat truncheon-like shape. The blow must be of controlled force, sharp but not hard. Therein lies the skill.

9.3 SAWING

The discs used to saw diamonds are generally of bronze or phosphor bronze and about 7 cm diameter. They are clamped between flanges with only the edge protruding. This prevents undue flexing of the blade which could widen or allow deviation of the cut with consequent wastage of the expensive diamond workpiece material. Obviously the amount of edge protrusion has to be at least equal to the thickness of diamond to be cut. In turn this puts constraints on the thickness of the sawing blade, thicker diamonds requiring more edge protrusion and therefore stiffer, i.e. thicker, blades. Commonly blade thicknesses range from about 0.05 mm for sawing the smallest diamonds to 0.15 mm for large diamonds. The diamond is normally fed by gravity onto the saw which rotates at speeds up to 10,000 r.p.m. A lever and fixed weight arrangement controls the force. The abrasive is diamond powder applied to the edge of the disc as a suspension in oil (commonly olive oil). Peripheral speeds are about 35 m/s.

Cutting rates are quite low, never more than a few mm per hour and more usually measured in tenths of a millimeter per hour. A one carat diamond can take a day to saw across the middle and large diamonds can take many days or weeks. Consequently diamond sawyeries handling significant production will have hundreds of sawing machines. Figures 4 and 5 show respectively an individual machine and a sawyery.

9.4 POLISHING

Diamond polishing is normally done on a flat, cast-iron lap charged with diamond powder. The cast iron is of a porous type. The powder is applied either dry or as a suspension in oil and rubbed into the pores of the iron, traditionally with the thumb. Powder retention is aided by previously scoring the iron surface with a silicon carbide stick. Laps up to about 35 cm diameter are in common use. They are mounted on vertical spindles and rotate at about 2,500 r.p.m. giving peripheral speeds of 45 m/s and about 35 m/s in the polishing zone of the flat surface. The laps are commonly referred

Figure 4. Diamond sawing machine.

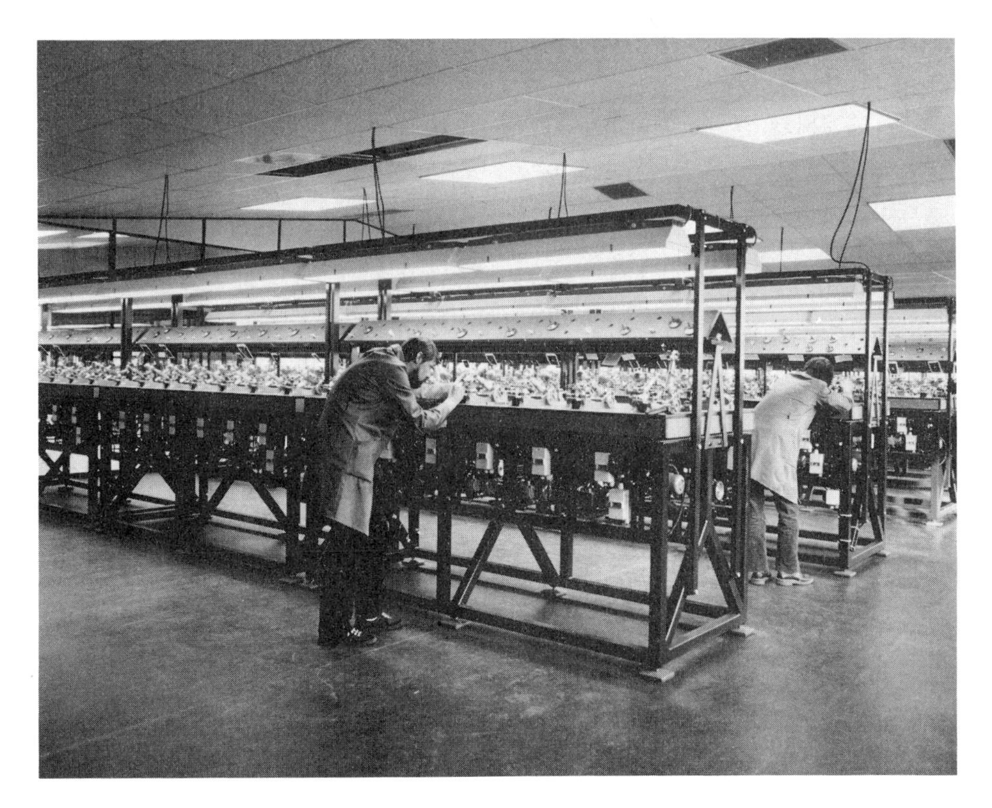

Figure 5. A diamond sawyery.

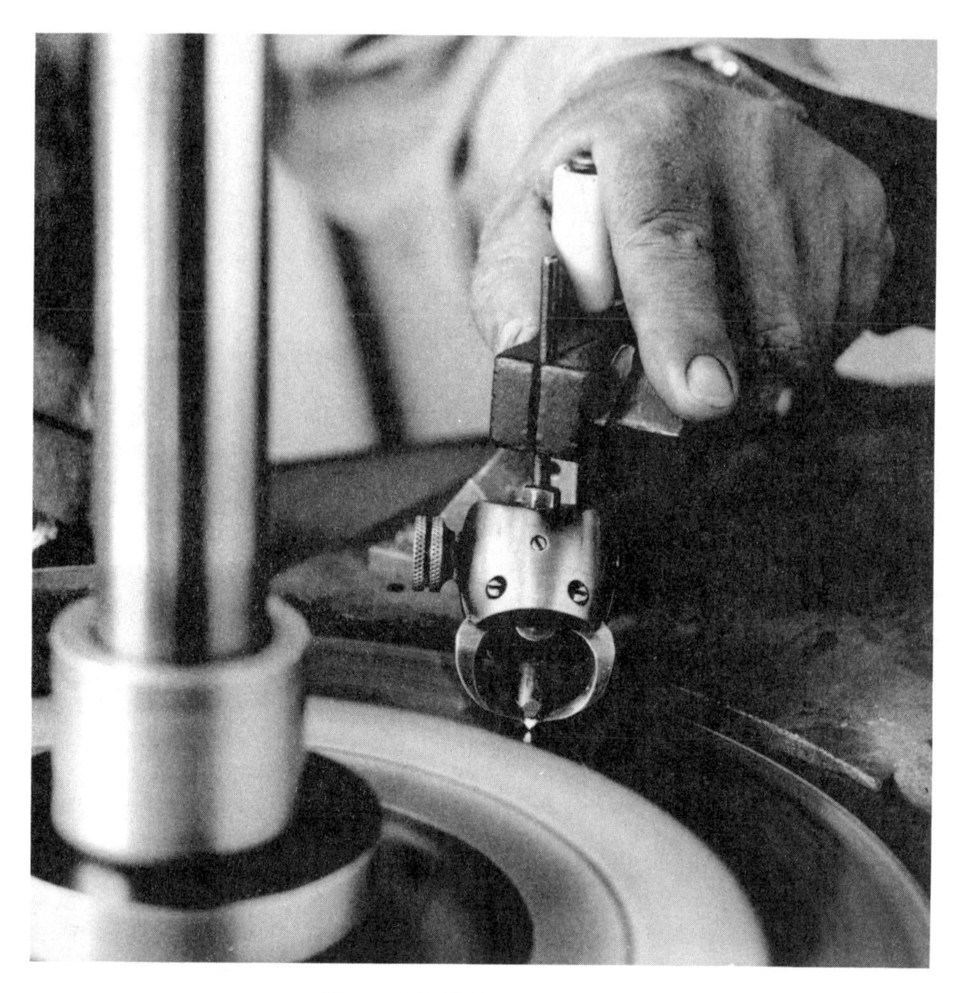

Figure 6. Polishing of a diamond.

to in the trade as “scaifes.” They are large, heavy and made of an inherently weak metal. They thus need to be tested by manufacturers at over-speed before use to avoid the risk of centrifugal bursting, and manufacturer’s recommendations regarding rotational speeds and balancing strictly observed at the factory. Figure 6 shows a scaife in use.

The diamond to be polished is either held mechanically using an arrangement of claws, or more traditionally in a mass of lead into which it was pushed at elevated temperature near the softening point. Positioning was formerly done by bending a copper rod in the mount, but nowadays through use of a goniometer head. In most factories, the arrangement holding the diamond (“dop”) and its support (“tang”) are held in the hand and the diamond thus brought manually into position on the polishing wheel. The required force (depending on the size of diamond and its orientation and state of polish) is applied by hand.

9.5 NON-TRADITIONAL METHODS

Non-traditional methods of working diamond are in some cases quite old. By the early 1950's methods involving the application of heat or electric current (spark erosion and electrolytic methods) or chemical interaction (hot oxygen, silica, zirconium carbide, etc.) had all been tried. Sometimes success was claimed, but the methods never really became adopted on a wide scale. Details of many of the trials are given in Godzenski's book "Diamond Technology."
More recently the difficulty of polishing polycrystalline diamond has renewed interest in chemical (hot iron, nickel, manganese, etc.) or ion and laser milling techniques. The most widely used has been laser cutting as an alternative to mechanical sawing and in the gem trade for drilling out inclusions in natural diamonds. Neodymium-YAG lasers are the most widely used, though diamond is transparent at their operating wavelength (just above 1μm). The problem of coupling laser energy into the diamond is usually solved by coating the diamond with an opaque layer. Once cutting starts the black carbon formed by the high temperatures under the laser beam provides a continuously generated opaque layer and ensures continuation of the cut. Excimer lasers have also been used recently to advantage.

10. Markets

10.1 SUPPLY

According to figures published by the United States Department of the Interior Bureau of Mines, worldwide natural diamond production in 1992 was 103 million carats. The major producing countries were (by weight): Australia (39%), Botswana (16%), Zaire (15%), Russia (11%), and South Africa (10%). Other significant supply came from Angola, Namibia, Brazil, and West Africa. Figures by weight are, however, somewhat misleading as the quality of the production varies greatly from country to country. Namibia, for instance, produces very high quality diamonds, substantially more than 80% of the production being of gem or near gem quality. At the other extreme, Zaire produces largely industrial qualities, less than 20% of the total being gem or near gem. Overall the Bureau of Mines figures indicate that about 50% of the worldwide total of 103 million carats was gem or near gem. The term "gem or near gem" takes account of the fact that the poorer gem qualities overlap the better industrial qualities. Depending on market demand the borderline qualities may end up being polished to gems or to industrial components.

Figures from the De Beers Central Selling Organization indicate that total sales by them of rough diamonds were slightly over $3.4 billion in 1992, but sales were unusually low in that year. In the following year 1993, they reached a record high of over $4.3 billion. Of course these are the figures for one company only, albeit the world's largest diamond supplier. Figures for the remainder of the market are difficult to come by, especially as significant quantities of diamonds are smuggled and figures for the proportions of Russian production coming on the market through De Beers and otherwise are quite obscure since the break-up of the Soviet Union. Reportedly the Russian production is worth well over $1 billion per year in total.

10.2 CONSUMPTION

The values of industrial and of gem diamonds are very different. They range from one or two dollars per carat for the lowest grade industrial boart, to well over $10,000 per carat for the very rare, very large, perfect white diamonds that make the news headlines. The borderline is about $300 to $500 per carat. This means that in money terms the market is dominated by the gem business. Of the De Beers $4.3 billion sales in 1993, probably more than 90% was gem. The 10% industrial is of course still important, but less so than it was.

The main industrial diamond application is as an abrasive grit. Boart is crushed and sieved to size for that purpose. However, even if all the 50 million carats or so of natural industrial diamond mined each year were crushing boart (which it is not), this would represent less than 10% of the world total consumption of diamond grit which by 1992 was in excess of 500 million carat annually, the overwhelming majority of which was produced by high temperature, high pressure synthesis.

Other applications of diamond, both natural and synthetic, include rock and concrete drilling, sawing of all kinds of hard materials, wire drawing, turning and milling, as well as a host of niche applications. The inexorable trend, though, is towards high-pressure, high-temperature synthetic diamond if only because 50 million carats or so per year of natural industrial diamond is not sufficient to supply the world's industries. The use of natural industrial diamond survives most stubbornly in applications where large perfect crystals polished to high precision are needed, as in turning tools for computer hard discs, surgical knives or high pressure anvils. It also survives in applications where the shape, strength or surface roughness of natural diamonds is important as in rock drilling, or where large pieces of low quality diamond are needed, as in dresser tools. Nevertheless, the future for industrial diamond seems to lie ever more with synthetic as the costs of this come down and better tailoring of the material to specific applications becomes possible.

11. Acknowledgements

Figures 1, 2, 4, 5, and 6 are reproduced here by kind permission of Drukker International B.V., Cuijk, The Netherlands.

12. References

12.1 GENERAL BIBLIOGRAPHY (BOOKS)

Berman, R. (ed.) (1965), *Physical Properties of Diamond,* Clarendon Press, Oxford.

Bruton, E. (1970 & 1978), *Diamonds,* 1st & 2nd eds., N.A.G. Press Ltd., London.

Davies, G. (1984), *Diamond,* Adam Hilger Ltd., Bristol.

Dawson, J.B. (1980), *Kimberlites and their Xenoliths,* Springer-Verlag, Berlin.

Deutsche Diamanten-Gesellschaft m.b.H. (1914), *Die deutschen Diamanten und ihre Gewinnung,* Dietrich Reimer (Ernst Bohsen), Berlin.

von Fersmann, A. and Goldschmidt, V. (1911), *Der Diamant,* Carl Winter, Heidelberg.

Field, J.E. (ed.) (1979), *The Properties of Diamond,* Academic Press, London.

Field, J.E. (ed.) (1992), *The Properties of Natural and Synthetic Diamond,* Academic Press, London.

Grodzinski, P. (1953), *Diamond Technology,* 2nd ed., N.A.G. Press Ltd., London.

Gurney, J.J. (1989), *Kimberlites and Related Rocks,* Special Publication No. 14, Vol. 2, Geological Society of Australia.

Hughes, F. (1978), *Diamond Grinding of Metals,* Industrial Diamond Information Bureau, Ascot, Berkshire U.K.

Industrial Diamond Association of America (publ.) (1964), *The Industrial Diamond,* Skyland NC.

Liddicoat, R.T. (ed.) (1993), *The G.I.A. Diamond Dictionary,* 3rd ed., Gemological Institute of America, Santa Monica CA.

Orlov, Yu. L. (1973), *The Mineralogy of the Diamond,* John Wiley, New York NY.

Raman, C.V. (republ. 1988) "Optics of Minerals and Diamond", S. Ramaseshan (ed.), *Scientific Papers of C.V. Raman, Vol.4,* Indian Academy of Sciences, Bangalore, and Oxford University Press.

Tolansky, S. (1962), *The History and Use of Diamond,* Methuen & Co., London.

Tolansky, S. (1968), *The Strategic Diamond,* Oliver and Boyd, Edinburgh.

Williams, A.F. (1932), *The Genesis of the Diamond,* 2 vols., E. Benn & Sons, London.

Wilson, A.N. (1981), *Diamonds: From Birth to Eternity,* Gemological Institute of America, Santa Monica CA.

12.2 COMMERCIAL STANDARDS

American National Standards Institute (ANSI), New York NY:

- (1971) *Checking the Size of Diamond Abrasive Grain,* no. B74.16-1971.
- (1973) *Test for Bulk Density of Diamond Abrasive Grains,* no. B74.17-1973.
- (1974) *Grading of Diamond Powder in Sub-sieve Sizes,* no. B74.20-1974.

American Society for Testing and Materials (ASTM), Philadelphia PA:

- (1981) *Standard Specifications for Sieves for Testing Purposes,* no. E11-81.

Fédération Européenne des Fabricants de Produits Abrasifs (FEPA), Paris:
- (1972) *FEPA Standard for Diamond Grain Sizes.*
- (1977) *FEPA Standard for Diamond Micron Powder Sizes.*

- Industrial Diamond Association of America, Inc., Skyland NC:
- (1963) *Grading of Diamond Powder in Sub-sieve Sizes,* no. CS261-63.
- (1984) *IDA Graded Powder Standard.*
- (1993) *A Review of Diamond and CBN Sizing and Standards.*

- International Organization for Standardization, Geneva:
- *Standard for Diamond Grain Sizes,* no. 6106.

12.3 SPECIFIC REFERENCES

Baumgold, C. (1964) Diamond Forms and Technology in *The Industrial Diamond,* Industrial Diamond Association of America, Skyland NC.

Harris, J.W. (1987) Recent Physical, Chemical, and Isotopic Research of Diamond, in P.H. Nixon (ed.), *Mantle Xenoliths,* J. Wiley & Sons, Chichester, pp. 477-500.

Long, A.E. (1960), *Glossary of the Diamond Drilling Industry,* U.S. Bureau of Mines Bulletin no. 583, U.S. Government Printing Office, Washington DC.

Mach, E. (1865 onwards) various papers reviewed and summarised in F. Ratliff (1965) *Mach Bands: Quantitative Studies on Neural Networks in the Retina,* Holden-Day, San Francisco CA, and in R.W. Ditchburn (1973) *Eye Movements and Visual Perception,* Clarendon Press, Oxford, pp. 284-289.

Meyer, H.O.A. (1987) Inclusions in Diamonds, in P.H. Nixon (ed.), *Mantle Xenoliths,* J. Wiley and Sons, Chichester, pp. 501-523.

Robertson, R., Fox, J.J. and Martin, A.E. (1934) Two Types of Diamond, *Phil. Trans. Roy. Soc. A,* 232, 463-535.

Seal, M. (1965) Structure in Diamonds as Revealed by Etching, *Amer. Mineral.,* 50, 105-123.

Chapter 11
THEORY OF DIAMOND CHEMICAL VAPOR DEPOSITION

David G. Goodwin
California Institute of Technology, Pasadena, CA 91125, USA

James E. Butler
Naval Research Laboratory, Washington, D. C. 20375, USA

1. Introduction

The complex chemical processes occurring during the chemical vapor deposition (CVD) of diamond are fascinating and exciting from many perspectives. First, how does one understand the process of growing a material under conditions in which it is metastable, e.g. what are the critical details and limitations? What are the complex gas phase, surface and bulk chemical processes which lead to diamond vs. graphite or amorphous carbon?

Secondly, this process provides a model for understanding many other chemical vapor deposition processes. It is comprised of all the relevant features of CVD, such as gas phase chemistry, complex heat and mass transport, nucleation, surface chemistry, bulk chemistry and diffusion, and temporal dynamics (Figure 1). And finally, the technological impact and applications enabled by diamond materials are important in many ways for society and the economy. Industry is already using CVD diamond in applications such as cutting tools,

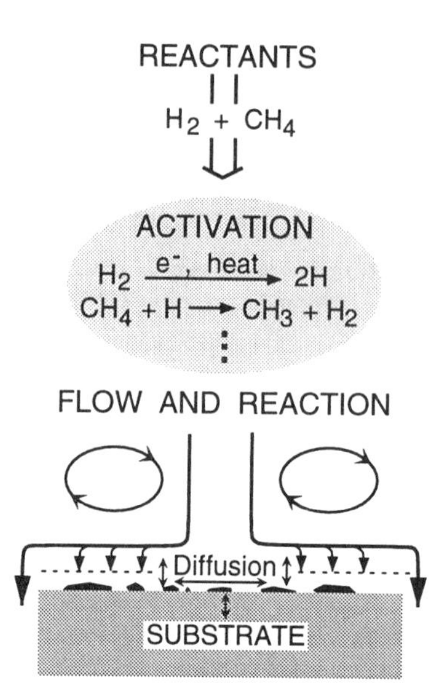

Figure 1. Schematic of processes occurring during diamond CVD [Butler and Woodin 1993].

electronic thermal management, optical windows, and radiation detectors.

Diamond CVD has been researched since the early 1950's. Initially the research was conducted at a modest level by a few innovative scientists, with only a few publications appearing before the mid-1980's. In the 1980's, several groups created deposition processes which produced good quality diamond at technologically relevant deposition rates. This led to a major increase in the number of researchers and in the resources used to study diamond CVD, and to a dramatic increase in the scientific understanding (and number of publications) from 1985 to 1995. Previous reviews and book chapters have covered the motivation for the scientific and technological interest in diamond materials, the variety of techniques employed for diamond growth, and addressed many of the issues important to diamond CVD. (See, for example DeVries [1987], Angus and Hayman [1988], Spitsyn et al. [1988], Spear [1989], Angus et al. [1989], Yarbrough and Messier [1990], Nemanich [1991], Celii and Butler [1991], Yarbrough [1992], Pehrsson et al. [1993], Angus et al. [1991], Butler and Woodin [1993], Spear and Frenklach [1994], Liu and Dandy [1995].)

The goal of this chapter is to review the important features of the growth environment, and to present the current understanding of critical aspects of the growth process. The growth of a solid material from gaseous reactants is fundamentally a surface chemical process which depends on the flux of reactant species to the surface, and products from it, as well as the surface structure and temperature. There is general agreement that hydrogen, particularly atomic hydrogen, plays an important role in the gaseous and surface chemistry critical to diamond CVD. Hence, we begin this chapter with a discussion of the gaseous chemistry of both the hydrogen and relevant hydrocarbon species, followed by an examination of the evidence regarding which hydrocarbon species contribute to diamond growth. Then we delve into the diamond growth chemistry, first with relevant experimental observations, then a discussion of some aspects of the surface chemistry, followed by a review of models proposed for elements of the growth mechanisms. Finally, we conclude with a discussion of the chemistry affecting the quality of the CVD diamond and the incorporation of defects.

2. The Gas-Phase Chemical Environment

To understand diamond CVD, it is first necessary to characterize the chemical environment the diamond film is exposed to during growth. In many other CVD processes this is straightforward, since the pressure is low enough that gas phase chemistry is negligible (low-pressure CVD of polysilicon) or is limited to a few precursor decomposition reactions (organometalic CVD of GaAs). However, in diamond CVD, free radicals – particularly atomic hydrogen – play crucial roles. The presence of radicals insures that gas-phase chemistry is an integral part of diamond CVD. A mixture of radicals, molecules, and in some cases ions, impinges on the substrate, even though the feedstock gas composition may be a simple methane/hydrogen mixture.

There have been many studies of the gas-phase chemistry during diamond chemical vapor deposition [Celii et al. 1988, Harris et al. 1988, Celii and Butler 1989, 1992, Celii et al. 1990, Harris et al. 1989, Harris and Weiner 1989, 1990, Goodwin and Gavillet 1990, 1991, Hsu 1991a, 1992, Corat and Goodwin 1993, McMaster et al. 1994, 1995].Most studies, both experimental and computational, have focused on low-pressure, benchtop hot-filament or microwave systems with a typical pressure of 20 – 30 Torr, a substrate temperature of 800 – 900 °C, and a gas composition of 1% or less methane in hydrogen. These process parameters result in a growth rate of order 1 μm/hr. While this choice of conditions is somewhat arbitrary, it is one which reliably produces diamond films and it allows comparison of results obtained by different laboratories. Most of the discussion in this section will consider low-pressure, low-growth-rate systems of this type.

Only a few gas-phase measurements have been carried out in higher growth rate environments, such as arcjets [Raiche and Jeffries 1993, Cappelli and Loh 1994, Reeve and Weimer 1994, 1995] or atmospheric-pressure thermal plasmas [Green et al. 1993]. This is largely due to the difficulty of making measurements in these luminous, high-speed, high-heat-flux environments. A principal objective of the work with low-pressure hot-filament or microwave systems has been to validate numerical models, which can then be used to simulate the more energetic high-growth-rate processes. Models of high-rate processes developed in this way have generally been quite successful [Goodwin 1991, Coltrin and Dandy 1993, Girshick, Li, Yu and Han 1993].

In this section, the characteristics of the gas phase in diamond CVD reactors are summarized, and the controlling chemistry discussed. Since atomic hydrogen plays such an important role in diamond CVD, it is discussed first, followed by a discussion of hydrocarbon chemistry.

2.1 ATOMIC HYDROGEN

Atomic hydrogen is perhaps the most critical determinant of diamond film quality and growth rate (see section 5). Its production mechanisms, loss mechanisms, and concentration profiles have been the subject of many studies. The picture which emerges from this work is summarized here.

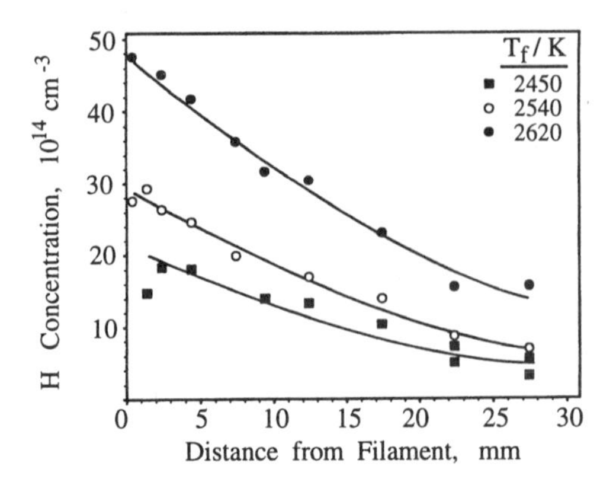

Figure 2. Atomic hydrogen profile measured by two-photon fluorescence vs. distance from a 2 mm diameter Ta filament [Schäfer et al. 1991].

2.1.1 *H Production Mechanisms*

In hot-filament systems, atomic hydrogen is produced heterogeneously by thermal decomposition of H_2 on the hot filament surface. The atomic hydrogen produced diffuses rapidly away from the filament, resulting in a concentration profile near the filament such as shown in Figure 2.

As the hydrocarbon content of the gas is increased beyond a critical value, the measured H concentration drops suddenly [Celii and Butler 1989]. Sommer and Smith [1990] have pointed out that this critical hydrocarbon fraction (which depends on filament temperature) corresponds closely to the solubility limit of carbon in hydrogen. The most likely explanation for this effect is that a graphite layer covers the filament when the critical hydrocarbon fraction is exceeded, poisoning its catalytic activity for H production.

The H concentration scales with filament temperature in the manner expected for thermal equilibrium [Celii and Butler 1989, Schäfer et al. 1991, Hsu 1991b, Connell et al. 1995]. However, the absolute concentration level at the filament is always observed to be less than the chemical equilibrium value at the filament temperature, varying anywhere from 12% [Schäfer et al. 1991] to 60% [Connell et al. 1995] of equilibrium depending on pressure, filament diameter, filament material, and the hydrocarbon content of the gas. The sub-equilibrium concentration is a result of the competition between H production and diffusion of H away from the filament. Of course, once H has diffused to cooler regions away from the filament, the *equilibrium* H concentration at the *local* gas temperature is negligible, and H is therefore present in superequilibrium concentrations.

In plasma-enhanced systems such as microwave, RF, or DC arcjet reactors, H is produced homogeneously in the plasma. The external energy input couples directly to the free electrons in the plasma. These energetic electrons may have several eV of energy, and may directly produce H through

$$H_2 + e^- \rightarrow H + H + e^-. \tag{1}$$

This reaction actually proceeds through successive vibrational excitation of H_2 by electron impact, leading finally to dissociation. Calculating the H_2 dissociation rate due to electron impact requires a complete understanding of the electron energy distribution, which may be non-Maxwellian in low-pressure systems.

In higher-pressure thermal plasmas (> 100 Torr), the electrons rapidly transfer energy to the heavy particles through elastic collisions, resulting in heating of the gas. In this case, the electrons and heavy particles come to the same temperature (which may be 3000 – 5000 K) and H_2 dissociation by impact with heavy particles,

$$H_2 + M \rightleftharpoons H + H + M, \tag{2}$$

is also significant. At these temperatures, the kinetics of H production are fast enough that this reaction runs in both directions and H approaches its equilibrium value at the local gas temperature. In many cases, the H concentration may remain in equilibrium up to the edge of the boundary layer above the substrate, and only attain superequilibrium as H diffuses across the boundary layer to the substrate.

2.1.2 *H Loss Mechanisms*

The steady state level of atomic hydrogen in the reactor is determined by a balance of the H atom production rate and the destruction rate due to homogeneous chemistry and wall recombination (assuming no atomic hydrogen enters or leaves the reactor).

Homogeneous Recombination. For typical diamond CVD conditions, homogeneous recombination of H is a slow process, and H atoms are able to diffuse to the walls or to the substrate before recombining in the gas. The rate of the direct recombination reaction

$$H + H + M \rightarrow H_2 + M \tag{3}$$

is pressure-dependent, due to the need for a third body (M) to carry away the excess heat of recombination. At 20 Torr, the characteristic time for this reaction is of order 1 s [Harris et al. 1988, Goodwin and Gavillet 1990]. In time t an atom may diffuse a distance of order $\sqrt{Dt}$, where D is the diffusion coefficient. Using $D = 0.12$ m^2/s for H in H_2 at 20 Torr implies a diffusion distance of 35 cm in 1 s. Therefore, since the substrate is usually at most 1 – 2 cm from the location of H production, H atoms in an H_2 background are able to freely diffuse to the substrate without homogeneously recombining.

In the presence of a small amount of hydrocarbon, a second path competes with reaction (3) and in many cases dominates the homogeneous recombination rate. This path is due to the two reactions

$$H + CH_3 + M \rightarrow CH_4 + M \tag{4}$$

and

$$H + CH_4 \rightleftharpoons CH_3 + H_2. \tag{5}$$

Accounting for these and similar reactions, Goodwin and Gavillet [1990] have show from numerical simulations that the homogeneous recombination time is reduced to about 50 ms for a gas composition of 0.5% CH_4 in H_2. However, this still implies a diffusion distance of 8 cm. We are left with the conclusion that homogeneous recombination of H may be neglected under typical low pressure (20 Torr) CVD conditions. Note however, that at hydrocarbon concentrations of several percent (higher than typically used for diamond growth), gas-phase H profiles are affected by the presence of the hydrocarbons, indicating a contribution from homogeneous recombination [Schäfer et al. 1991].

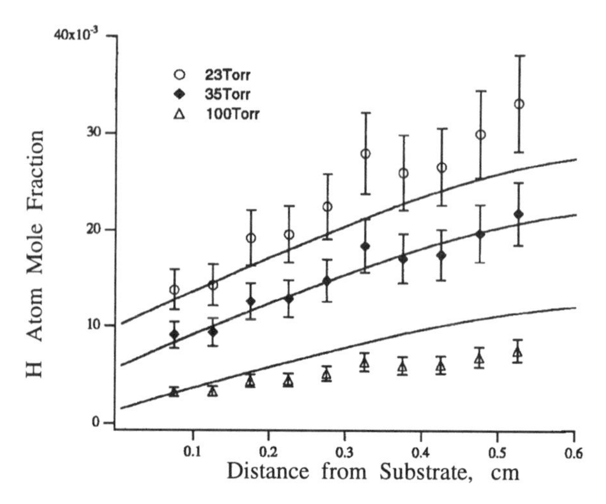

Figure 3. Atomic hydrogen profiles near a diamond substrate measured by third-harmonic generation [Connell et al. 1995].

In thermal plasma and combustion systems, H is nearly in chemical equilibrium outside the boundary layer. In these systems, homogeneous recombination is only of concern within the boundary layer. Numerical simulations have shown that it is usually unimportant in high-speed arcjet or combustion systems [Goodwin 1991], although it is predicted to become significant at pressures above about 2 atm [Goodwin 1993b]. Homogeneous recombination is predicted to significantly deplete H in the boundary layer in some atmospheric-pressure plasma torches with relatively low velocity [Owano et al. 1991].

Heterogeneous Recombination. Since homogeneous recombination of H atoms is usually negligible, the loss of H atoms must occur primarily on reactor walls and on the diamond surface itself. Measurements of the H concentration profile near a diamond substrate clearly show that diamond is a sink for H at typical substrate temperatures (Figure 3).

The loss of atomic hydrogen on diamond may be quantified by the recombination coefficient γ_H, defined as the H loss rate on the surface divided by the H collision rate with the surface. The recombination coefficient has been measured by Harris and Weiner [1993], Krasnoperov et al. [1993], Gat and Angus [1993], Thoms et al. [1994], and Proudfit and Cappelli [1993]. Most of these measurements are in reasonable agreement with one another and with theoretical predictions [Goodwin 1993a]. The measurements of Krasnoperov et al. [1993] were made over the widest temperature range (300 to 1119 K). Their results may be represented by the expression

$$\gamma_H = 4.0 \times 10^{-4} + 1.95 \exp(-3025/T), \tag{6}$$

with T in Kelvin. At a typical substrate temperature of 1200 K, this expression gives $\gamma_H = 0.16$, which compares well with the value of 0.12 measured by Harris and Weiner [1993] at this temperature.

This recombination coefficient is high, and shows that the diamond surface is a strong sink for H atoms. Also, since the reaction $2H \rightarrow H_2$ is exothermic by 104 kcal/mol, recombination of atomic hydrogen on diamond heats the substrate, and in fact is the primary means of substrate heating in many systems. In high-growth-rate systems such as arcjets, a large flux of H to the substrate is employed; this necessarily results in a large substrate heat flux.

2.2 HYDROCARBON CHEMISTRY

The first hydrocarbon concentration measurements during diamond growth were made by Celii et al. [1988], who used infrared diode laser absorption spectroscopy to detect acetylene, the methyl radical, and ethylene in a 25 Torr hot-filament reactor with an input gas of 0.5% methane in hydrogen. The measured acetylene concentration represented conversion of 10 – 20% of the initial methane. Since the gas-phase conversion of methane to acetylene requires several sequential reactions with atomic hydrogen, this observation clearly showed that significant gas-phase chemistry was occurring.

Typical mole fractions of species at the substrate under low-pressure diamond CVD conditions are shown in Figure 4, as a function of the carbon content of the feed gas. These results are for a 20 Torr microwave system, and were measured using molecular beam mass spectrometry, extracting gas through a small pinhole in the substrate [McMaster et al. 1995].

A notable feature of these results is that the mole fractions are almost independent of whether carbon is introduced into the reactor as methane or acetylene. In either case, both are found at the substrate, and these two species account for the majority of the gas-phase carbon. Ethylene (C_2H_4) is present at much lower levels, and ethane (C_2H_6) is not detected.

The only two radical species which are detectable under these conditions are atomic hydrogen and the methyl radical CH_3. The atomic hydrogen mole fraction is seen to be nearly independent of carbon content in the gas, at a level slightly below 10^{-3}. The methyl radical mole fraction rises with increasing carbon content, and is typically about an order of magnitude lower than methane.

The mole fractions shown in Figure 4 are very similar to those found in hot-filament systems [Harris et al. 1988, Celii et al. 1988, Harris and Weiner 1990, Hsu 1991a,b, McMaster et al. 1994].At first this is surprising, since plasma chemistry would be expected to be important in the microwave system, and negligible in the hot-filament system. However, Hsu [1992] has pointed out that the ion density in a 2.45 GHz microwave system is limited to $\sim 10^{11}$ cm^{-3} by the requirement that the microwave beam be able to penetrate the plasma. Comparing reaction rates for ion-neutral and neutral-neutral reactions using this upper-bound ion density, Hsu finds the neutral-neutral reactions are dominant in typical 20 Torr microwave systems. Therefore, the gas-phase chemistry in both hot-filament and microwave systems is primarily neutral chemistry. This may not be true, however, at much lower pressure (for example, at 10^{-3} Torr in an Electron Cyclotron Resonance microwave system).

To fully understand the reasons for the results presented above would require a detailed analysis, accounting for chemical kinetics in the gas and on the surface and species and

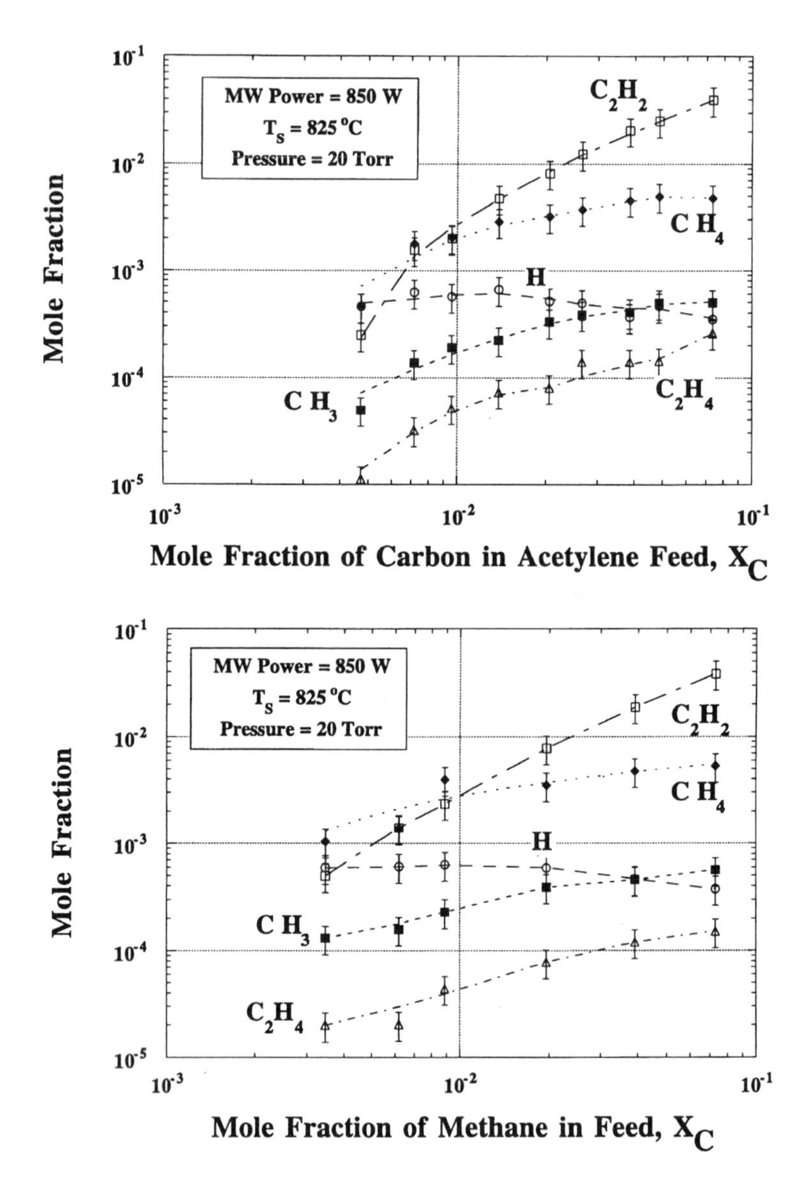

Figure 4. Mole fractions measured at the substrate in a microwave plasma reactor as a function of carbon mole fraction in the feed gas X_C. In (a), the carbon is introduced as acetylene, and in (b) the carbon is introduced as methane [McMaster et al. 1995].

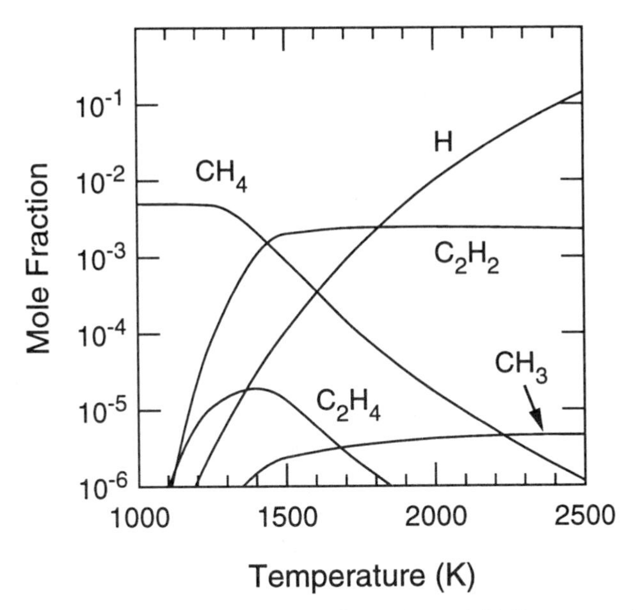

Figure 5. Homogeneous equilibrium mole fractions for a mixture initially composed of 0.5% CH_4 in H_2 at 20 Torr.

energy transport by diffusion and convection. Several models of this type have been presented, although in most cases highly idealized reactors are considered, in order to make the simulation tractable [Goodwin and Gavillet 1990, 1991, Goodwin 1991, Kim and Cappelli 1992, Coltrin and Dandy 1993, Meeks et al. 1993, Yu et al. 1993].

Fortunately, relatively simple arguments based on gas-phase chemistry can account at least qualitatively for much of the hydrocarbon distribution in the gas phase. Below, some of the factors controlling the gas-phase hydrocarbon chemistry are discussed, beginning with the simplest model (equilibrium), followed by more detailed discussion of the chemistry of the C_1, C_2, and heavier hydrocarbons.

2.2.1 *Chemical Equilibrium*

Although temperature gradients and finite residence time prevent the gas from attaining chemical equilibrium, the equilibrium composition provides a rough first approximation to the actual composition [Butler and Celii 1989, Janssen et al. 1989]. The homogeneous (gas-phase only) equilibrium composition for a mixture of 0.5% CH_4 in H_2 at 20 Torr is shown in Figure 5 for temperatures from 1000 K to 2500 K. All species with concentration greater than 1 ppm at 2000 K are shown. (The species C_3, C_4H, C, and CH_2 are predicted to have concentrations of approximately 1 – 2 ppm at 2500 K, but their concentrations fall off rapidly at lower temperature.)

Under these conditions, the dominant hydrocarbon is methane below about 1500 K, while above this temperature it is acetylene. The reason for this becomes apparent if we consider the reaction

$$2CH_4 \rightleftharpoons C_2H_2 + 3H_2. \tag{7}$$

This reaction is endothermic by 97 kcal/mol, but the change in standard-state entropy ΔS^0

is +65 cal/mol/K, due largely to the translational entropy of the 3 H_2 molecules produced. Since the equilibrium state is the one which minimizes the Gibbs free energy $G = H - TS$, at low temperature this reaction proceeds to the left to minimize the enthalpy H, while at high temperature maximizing the entropy is more important and it proceeds to the right. For this carbon/hydrogen ratio, the shift from CH_4 to C_2H_2 occurs at about 1440 K at 20 Torr, while at 1 atm it does not occur until approximately 1820 K.

The equilibrium calculation is, of course, a drastic oversimplification of the true situation during diamond CVD. Typical reactors contain both high temperature regions where acetylene may be produced, and cool regions near the walls and substrate where methane is favored. In most cases, these regions are only millimeters apart, and rapid diffusion results in a mixed gas composition containing both methane and acetylene. For these reasons, the CH_4/C_2H_2 ratio at the substrate depends on the details of the reactor geometry, residence times, rate constants, etc., and must in general be measured.

Another reason the gas is not in chemical equilibrium is that H near the substrate is present in superequilibrium amounts. This induces chemistry which otherwise wouldn't occur (for example, methane to acetylene conversion at low temperature). To get a more accurate description of the gas-phase chemical environment, models which go beyond the equilibrium assumption are needed.

2.2.2 *C_1 Chemistry*

The qualitative distribution of species within the C_1 system (C through CH_4) is governed by two simple observations: (a) the bimolecular hydrogen shift reactions

$$\mathrm{CH_m + H \rightleftharpoons CH_{m-1} + H_2}, \tag{8}$$

for $m = 1-4$ occur rapidly in both the forward and reverse directions; and (b) the pressure-dependent recombination reactions

$$\mathrm{CH_{m-1} + H + M \rightarrow CH_m + M} \tag{9}$$

are slow, and usually may be neglected.

The hydrogen shift reactions are rapid since the activation barriers are typically low (a few kcal/mol) in both directions. Therefore, at steady state the species concentrations adjust so that the forward and backward rates are equal, in which case the reaction is said to be in partial equilibrium. Equating the forward and reverse reaction rates results in

$$\frac{[\mathrm{CH_{m-1}}][\mathrm{H_2}]}{[\mathrm{CH_m}][\mathrm{H}]} = K_m(T) \tag{10}$$

where $K_m(T)$ is the equilibrium constant for the mth hydrogen shift reaction, which depends only on temperature.

Numerical simulations show that the reaction

$$\mathrm{CH_4 + H \rightleftharpoons CH_3 + H_2} \tag{11}$$

is in partial equilibrium in hot-filament systems at positions 1 mm or more away from the substrate [Goodwin and Gavillet 1991]. Closer than 1 mm, the fall-off in H due to

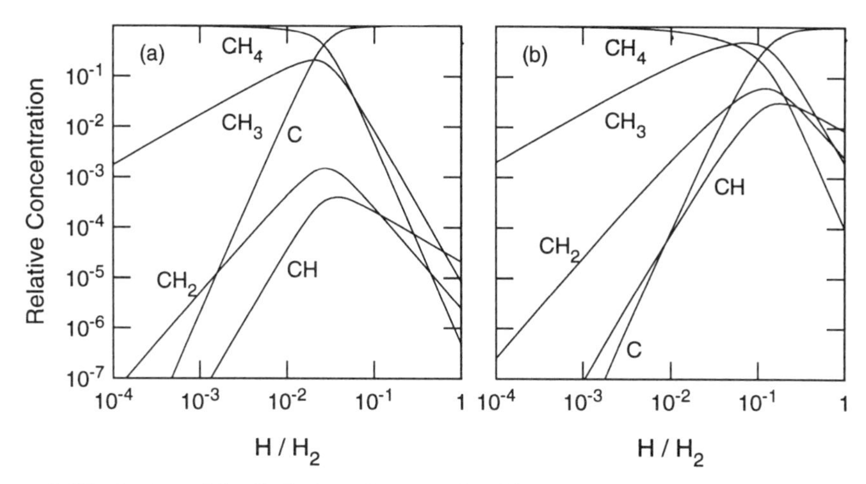

Figure 6. Distribution of the C_1 hydrocarbons predicted by Equation (12). (a) $T = 1000$ K; (b) $T = 2000$ K.

recombination on the surface is too rapid for the CH_3 and CH_4 concentrations to follow, and partial equilibrium breaks down. In this case, the CH_3 concentration at the substrate is effectively frozen at the value obtained by partial equilibrium $\sim$ 1 mm away from the substrate. For the other C_1 hydrogen shift reactions ($m < 4$), equation 10 is predicted to remain valid closer to the substrate than for the methyl/methane shift reaction.

If we apply equation 10 sequentially, we find that

$$\frac{[\mathrm{CH_m}]}{[\mathrm{CH_4}]} = \left(\frac{[\mathrm{H}]}{[\mathrm{H_2}]}\right)^{4-m} \prod_{j=m+1}^{4} K_j(T). \tag{12}$$

Equation 12 provides an algebraic relationship for the distribution of the C_1 species as a function of only the H/H_2 ratio and the local temperature. This relationship is plotted in Figure 6, normalized to the sum of all C_1 species, for the two temperatures $T = 1000$ K and $T = 2000$ K.

From this figure, it is seen that the most abundant C_1 radical is CH_3 if the H/H_2 ratio is below a few percent, while above a few percent it is atomic carbon. For this reason, CH_3 and (in highly activated environments) C are the C_1 radicals most often postulated to be important for diamond growth.

Due to its possible role in diamond growth, the methyl radical has been the subject of numerous diagnostic studies [Celii et al. 1988, Celii and Butler 1992, Corat and Goodwin 1993, Harris and Weiner 1990, Menningen et al. 1993, 1995, Zalicki et al. 1995, Wahl et al. 1996] A quantitative spatial profile of CH_3 measured recently in a hot-filament reactor is shown in Figure 7. The CH_3 depletion seen near the filament may have several causes. Thermal diffusion will cause all hydrocarbons to be depleted near the hot filament [Harris et al. 1989, DebRoy et al. 1990]. Also, the H concentration near the filament may be above a few percent, which would cause atomic carbon to dominates the C_1 distribution (Figure 6); as H falls off, the distribution shifts to CH_3, and then further away to CH_4.

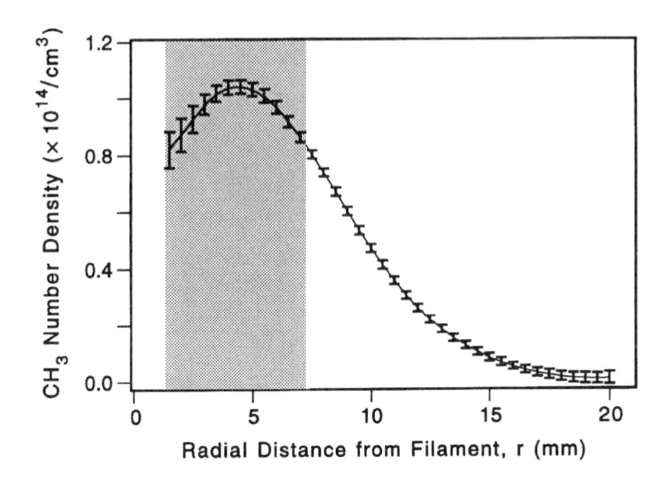

Figure 7. Radial distribution of the CH_3 concentration near a 2400 K, 0.2 mm diameter tungsten filament without a substrate present, measured using cavity ring-down spectroscopy. The inlet gas composition is 0.6% CH_4 in H_2 and the pressure is 20 Torr. The shaded region indicates the region where the gas temperature is between 1250 and 2000 K [Zalicki et al. 1995].

As was the case above for atomic hydrogen, introducing a diamond substrate perturbs the CH_3 concentration near the substrate. Measurements of CH_3 very close to the substrate [Corat and Goodwin 1993, McMaster et al. 1994, Wahl et al. 1996] show that its concentration depends on substrate temperature, even though McMaster et al. have shown that the H concentration is relatively temperature-independent (Figure 4). Below 1000 K the substrate-temperature dependence of the CH_3 mole fraction can be described by an activation energy of 3 – 4 kcal/mole. One plausible explanation for this effect is recombination of CH_3 with H to form CH_4 or with another CH_3 to form C_2H_6 in the cool gas layer near the substrate [Corat and Goodwin 1993].

2.2.3 *C_2 Chemistry*

The distribution of species within the C_2 system (C_2 through C_2H_6) cannot be fully explained by simple partial equilibria of hydrogen shift reactions, unlike the case for the C_1 system. This results from the fact that the species C_2H_2, C_2H_4, and C_2H_6 are all stable molecules, and thus reactions such as

$$C_2H_2 + H_2 \rightarrow C_2H_3 + H \tag{13}$$

have high activation energies and are consequently very slow.

Both C_2H [Cappelli and Paul 1990, Rau and Picht 1992, Komaki et al. 1993] and C_2 [Gruen et al. 1995] have been mentioned as possible diamond growth species. The concentration of C_2H may be approximately related to that of acetylene, since the reaction

$$C_2H_2 + H \rightleftharpoons C_2H + H_2 \tag{14}$$

is typically close to partial equilibrium. This reaction is endothermic by 27 kcal/mole, which leads to a low concentration of C_2H. For example, assuming reaction (14) is in partial equilibrium and taking C_2H_2 and H mole fractions of roughly 1% each yields a C_2H mole fraction of 10^{-6} at 2000 K and 10^{-9} at 1000 K. These numbers suggest that C_2H is unlikely to be important for diamond CVD.

The C_2 present in the CVD growth environment will be comprised of a mixture of the C_2(X $^1\Sigma_g^+$) electronic state and a low-lying (610 cm^{-1}) C_2(a $^3\Pi_u$) state, which will have very different reaction mechanisms and reactivities. The singlet state reacts roughly ten times faster than the triplet with H_2 and CH_4, while the reverse is the case for reactions with unsaturated compounds such as C_2H_2 and C_2H_4 [Donnelly and Pasternack 1979, Pitts et al. 1982].

Singlet C_2 reacts rapidly via a low-barrier pathway with H_2 [Pitts et al. 1982]. The rate constant measured by Pitts et al. ($A = 10^{14}\,\text{cm}^3\,\text{mol}^{-1}\,\text{s}^{-1}$; $E_a = 2.92$ kcal/mol) indicates that singlet C_2 will be consumed after only a few collisions with H_2. Since triplet C_2 can react with H_2 by H abstraction, and in addition collisions will tend to equilibrate singlet and triplet C_2, it is expected that the C_2 concentration will be negligibly small under typical CVD conditions with H_2 as the carrier gas.

The concentrations of all C_2H_n species with $n > 2$ are generally low, since these species are less stable thermodynamically at high temperature and in the presence of atomic hydrogen than acetylene. Ethane is rapidly converted to ethylene through

$$C_2H_6 + H \rightarrow C_2H_5 + H_2 \tag{15}$$

followed by

$$C_2H_5 + H \rightarrow C_2H_4 + H_2 \tag{16}$$

or

$$C_2H_5 + M \rightarrow C_2H_4 + H + M. \tag{17}$$

The ethylene formed is, in turn, converted to acetylene by

$$C_2H_4 + H \rightarrow C_2H_3 + H_2, \tag{18}$$

followed by

$$C_2H_3 + H \rightarrow C_2H_2 + H_2 \tag{19}$$

or

$$C_2H_3 + M \rightarrow C_2H_2 + H + M. \tag{20}$$

2.2.4 *$C_1 \rightarrow C_2$ Conversion*

If a C_1 species is introduced into the reactor (for example, methane), the conversion to heavier hydrocarbons is initiated by rapid H abstraction, which, as discussed above, produces primarily methyl radicals for H/H_2 below a few percent. These methyl radicals may recombine through the reactions

$$CH_3 + CH_3 \rightarrow C_2H_6, \tag{21}$$

$$CH_3 + CH_3 \rightarrow C_2H_5 + H, \tag{22}$$

or

$$CH_3 + CH_3 \rightarrow C_2H_4 + H_2. \tag{23}$$

Once a C_2 species is created, it is rapidly converted to C_2H_2 through the reactions discussed above.

2.2.5 *$C_2 \rightarrow C_1$ Conversion*

Although the gas-phase process discussed above which converts C_1 species to acetylene is rapid and well-understood, no analogous gas-phase processes are known which create C_1 species rapidly from acetylene. Nevertheless, the species distribution at the substrate is very similar in hot-filament and microwave systems whether carbon is introduced to the system as methane or acetylene (Figure 4). This observation indicates that a mechanism must be operating to effectively convert acetylene to methane and other C_1 species under diamond CVD conditions.

The most likely possibility is that conversion occurs heterogeneously on reactor walls or even on the diamond surface. One scenario would involve pyrolysis of acetylene, forming an sp^2 (graphitic) carbon film on the walls, which is then attacked by atomic hydrogen. Since it is known that the etch product of H attack on graphite is CH_3 [Rye 1977], this provides a possible mechanism for C_1 production from acetylene.

2.2.6 *C_3 and Higher Hydrocarbons*

Species with three or more carbon atoms are generally believed to be unimportant for diamond growth. Harris et al. [1989] used sampling mass spectroscopy to measure mole fractions of species with up to 10 carbon atoms in a 20 Torr hot-filament system. With an input gas of 0.3% methane in hydrogen, hydrocarbons with 3 carbon atoms had a total mole fraction of slightly less than 10^{-5} at a distance of 2 mm from the filament. The concentration of heavier species dropped off rapidly with increasing number of carbons. The C_6, C_9, and C_{10} signals were dominated by the aromatic species benzene, indene, and naphthalene, with mole fractions of 7×10^{-7}, 7×10^{-8}, and 5×10^{-8}, respectively.

Frenklach [1989] has proposed that condensation of aromatic species on the diamond surface leads to incorporation of non-diamond carbon. This proposal was tested by Martin and Harris [1991], in which benzene was purposely added in large amounts to a low-pressure flowtube reactor. No deleterious effect on the deposited diamond was observed in this experiment due to benzene addition.

2.3 EFFECTS OF OXYGEN ADDITION

Although most diamond CVD is carried out with hydrocarbon/hydrogen gas mixtures, it is also common to add a small amount of oxygen, or an oxygen-containing compound. Several studies have reported enhanced growth rates or quality due to oxygen addition [Kawato and Kondo 1987, Liou et al. 1990, Chang et al. 1988, Mucha et al. 1989, Kapoor et al. 1995].

Due to rapid gas-phase chemistry, once again the equilibrium composition provides a first estimate of the effects of oxygen addition on the gas composition. Since acetylene is typically the most abundant hydrocarbon, let us consider the oxidation reaction

$$C_2H_2 + O_2 \rightleftharpoons 2CO + H_2. \tag{24}$$

This reaction is highly exothermic, and at equilibrium, it shifts strongly to the right, proceeding until essentially all of one reactant is depleted. For example, at 1000 K, the free energy of reaction is $\Delta G^0 = -136.6$ kcal/mol, resulting in an equilibrium constant of $K_p = 7 \times 10^{29}$.

Harris and Weiner [1989] have reported mass spectral measurements during filament-assisted diamond growth with various mixtures of CH_4, O_2, and H_2. In all cases, they found that adding O_2 reduced the hydrocarbon mole fractions due to conversion to CO and H_2. No O_2 was detected experimentally at the substrate, indicating all injected O_2 was rapidly consumed.

These results are also consistent with the empirical observations of Bachmann et al. [1991]. They observed that essentially all C/H/O gas compositions used successfully for diamond growth – whether in a flame, plasma, or filament system – have an overall C:O ratio close to 1. For C:O greater than 1, poor quality diamond or non-diamond carbon is deposited, while for C:O less than 1 no film can be grown. These observations are consistent with the idea that the oxygen rapidly oxidizes any available hydrocarbon, forming CO and H_2. For C:O slightly greater than 1, this results in an environment consisting of a small amount of residual hydrocarbon in a background gas primarily of H_2 and CO. Assuming CO is not chemically active for diamond growth, this environment would be expected to be very similar to a standard oxygen-free diamond CVD environment. On the other hand, for C:O slightly less than 1, the hydrocarbon is fully oxidized and oxygen remains, resulting in no possible film growth.

Even though the primary effect of oxygen addition is to oxidize some or all of the hydrocarbon to form CO, there are other effects which may be significant for film growth. For example, Harris and Weiner [1989] calculated that oxygen addition to a filament-assisted system should result in a slight increase in the H level, and an OH level of approximately 10^{-6} mole fraction. They estimate that this OH concentration could oxidize pyrolytic, non-diamond carbon at a rate of 0.25 μm/hr. This estimate suggests that OH may play a role similar to H, and thus oxygen addition may aid diamond growth through creation of OH radicals.

2.4 GAS-PHASE SPECIES TRANSPORT

To accurately describe the chemical environment at the substrate, it is important to understand not only gas-phase chemistry but also how species are transported to the substrate. In particular, atomic hydrogen transport is important, since it drives much of the hydrocarbon chemistry and its concentration will vary greatly through the reactor, due to recombination on walls and on the substrate. In this section, some of the important transport issues in diamond CVD will be discussed, focusing on H transport. This discussion is taken largely from the analysis of Goodwin [1993b].

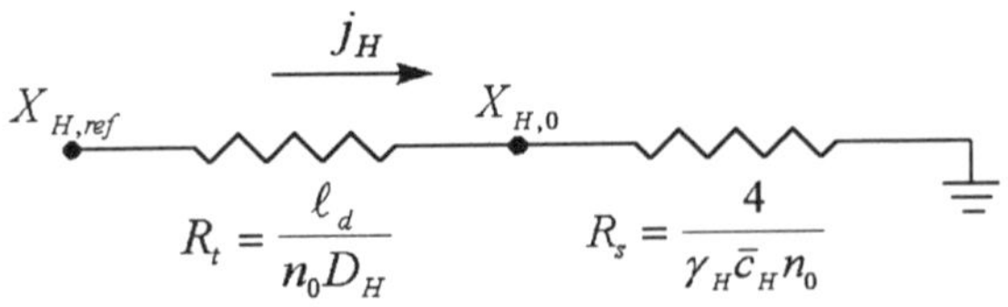

j_H

$X_{H,ref}$ $X_{H,0}$

$R_t = \frac{\ell_d}{n_0 D_H}$ $R_s = \frac{4}{\gamma_H \bar{c}_H n_0}$

Figure 8. Electrical circuit analogy for H transport to the substrate.

2.4.1 *Diffusive vs. Convective Transport*

Transport of species to the substrate may occur either by diffusion (transport by random molecular motion) or convection (transport by bulk gas motion). Which mode is dominant is determined by the value of the Peclet number, defined as

$$\mathrm{Pe} = \frac{uL}{D}, \tag{25}$$

where L is a characteristic length scale of the reactor, u a characteristic velocity, and D the diffusion coefficient, which will in general be different for each species.

If Pe $\ll 1$, then transport occurs primarily by diffusion, and the flux of species to the substrate is essentially independent of the bulk gas velocity. In the other limit, Pe $\gg 1$, species transport is primarily by convection. In this case, diffusive transport is limited to the thin boundary layer which forms over the substrate, in which the gas velocity, temperature, and species concentrations all vary sharply. In the convective case, the transport of species to the substrate is highly dependent on the velocity field in the reactor. Most low-pressure hot-filament and microwave reactors operate in the low-Pe diffusive mode, while all combustion systems and arcjets operate in the high-Pe convective mode.

2.4.2 *Atomic Hydrogen Transport*

Since atomic hydrogen recombines rapidly on the substrate at diamond growth temperatures (see section 2.1.2), there will always be a net loss of H on the substrate which must be balanced by a flux of H to the surface. For this reason, there will be an H concentration gradient near the substrate (Figure 3). Simple transport theory may be used to estimate the magnitude of this gradient and the net flux of H to the substrate.

A characteristic length scale for diffusion ℓ_d may be defined locally at the substrate by

$$\left(\frac{dX_H}{dz}\right)_{z=0} = \frac{X_{H,ref} - X_{H,0}}{\ell_d}. \tag{26}$$

The left-hand side of equation 26 is the H mole fraction gradient at the substrate. The quantity $X_{H,0}$ is the H mole fraction at the substrate, and $X_{H,ref}$ is the H mole fraction at some reference point in the reactor, which is assumed to be known. For example, in a hot-filament reactor, the H mole fraction at the filament is approximately given by the equilibrium value (see section 2.1.1), as it is in thermal plasma or combustion systems outside the boundary layer.

Applying Fick's law for the flux j_H of H to the substrate, and equating this flux to the recombination rate on the substrate, expressions for j_H and $X_{H,0}$ are obtained. These are

TABLE I. Numerical values of transport quantities for dilute H in H_2. T is in Kelvin and p is in Torr.

Quantity	Expression	Units
D_H	$0.143T^{1.65}p^{-1}$	cm^2/s
$\bar{c}_H$	$1.45 \times 10^4 T^{0.5}$	cm/s
n_0	$1.6 \times 10^{-5} pT^{-1}$	mol/cm^3
R_t/ℓ_d	$4.37 \times 10^5 T^{-0.65}$	cm-s/mol
R_s	$8.84T^{0.5}p^{-1}e^{3025/T}$	cm-s/mol

most simply understood in terms of the electrical circuit analogy shown in Figure 8. The H flux and H mole fraction are analogous to an electrical current and electrical potential, respectively. The current flows to ground through two resistors in series. The first is a transport resistance

$$R_t = \frac{\ell_d}{n_0 D_H} \tag{27}$$

and the second is a surface chemistry resistance

$$R_s = \frac{4}{\gamma_H \bar{c}_H n_0}. \tag{28}$$

Here n_0 is the total gas molar density p/RT at the substrate, D_H is the diffusion coefficient of H through the background gas, and $\bar{c}_H$ is the mean thermal speed of an H atom at the substrate temperature. Note that $1/R_s$ is simply the H recombination rate on the surface which would be obtained if the gas were pure atomic hydrogen. Numerical values of these quantities for a dilute mixture of H in H_2 are given in Table I. (Adding 1% hydrocarbon to the gas does not appreciably change these values.)

From the circuit in Figure 8, it is clear that

$$j_H = \frac{X_{H,ref}}{R_t + R_s} \tag{29}$$

and

$$\frac{X_{H,0}}{X_{H,ref}} = \frac{1}{1 + R_t/R_s}. \tag{30}$$

To achieve the highest possible diamond growth rate without sacrificing quality, it is often desired to maximize the flux of H to the substrate. Equation 29 shows that this can be done in one of three ways. First of all, the driving potential $X_{H,ref}$ can be increased, for example by increasing the filament or plasma temperature. Once $X_{H,ref}$ is maximized to the extent possible, further increase of the H flux requires lowering the transport resistance, either by increasing $n_0 D_H$ or decreasing ℓ_d. The product $n_0 D_H$ cannot be varied significantly, since

TABLE II. Diffusion lengths at surfaces 1 and 2 in the low-Pe limit for simple 1D geometries, assuming no gas-phase chemistry and constant properties.

Geometry	$\ell_{d,1}$	$\ell_{d,2}$
Parallel planes separated by L	L	L
Concentric cylinders, $R_1 < R_2$	$R_1 \ln(R_2/R_1)$	$R_2 \ln(R_2/R_1)$
Concentric spheres, $R_1 < R_2$	$R_1(1 - R_1/R_2)$	$R_2(R_2/R_1 - 1)$

it is independent of pressure and only weakly dependent on temperature. Thus, the only practical way to increase j_H further is to decrease the diffusion length ℓ_d.

2.4.3 *The Diffusion Length For Low Pe*

If Pe $\ll$ 1 and gas-phase chemistry and transport property variations are not significant, then ℓ_d is a function only of reactor geometry [Goodwin 1993a]. Analytical values for some simple one-dimensional geometries are listed in Table II.

As an example of how the expressions in Table II may be used, consider an idealized model of a hot-filament reactor, consisting of 2 long concentric cylinders. The inner cylinder of radius R_1 represents the filament, and the outer cylinder of radius R_2 represents the substrate and walls. Let us take the experimental conditions of McMaster et al. [1994], who have reported a measured H mole fraction at the substrate in an experiment with a single, uncoiled filament of diameter 0.25 mm and length 5 cm, a filament-substrate separation of 1.3 cm, a substrate temperature of 1098 K, a filament temperature of 2600 K, and a pressure of 20 Torr.

From Table II, taking $R_1 = 0.0125$ mm and $R_2 = 1.3$ cm results in $\ell_d \approx 6$ cm. (Note that ℓ_d may be larger than R_2.) Evaluating R_t and R_s from Table I yields $R_t = 2.76 \times 10^4$ cm-s/mol and $R_s = 230$ cm-s/mol. Thus, from equation 30, $X_{H,0}/X_{H,fil} \approx 8.3 \times 10^{-3}$. If we assume the equilibrium value for $X_{H,fil}$ (0.21), then we estimate $X_{H,0} \approx 1.7 \times 10^{-3}$. The experimental value McMaster et al. report for $X_{H,0}$ is between 1×10^{-3} and 2×10^{-3} for carbon addition of less than 1%.

2.4.4 *The Diffusion Length For High Pe*

For high-Pe convection-dominated reactors, equations 29 and 30 still apply but now the length scale ℓ_d depends on flow conditions as well as geometry. In this case, ℓ_d is identical to the H boundary layer thickness δ_H, defined by

$$\delta_H = \frac{X_{H,\infty} - X_{H,0}}{(dX_H/dz)_{z=0}}. \tag{31}$$

Here $X_{H,\infty}$ is the H mole fraction immediately outside the boundary layer.

Many convective diamond growth reactors employ either a high-speed jet which impinges at normal incidence on the substrate (arcjets), or the substrate is immersed in a large-area gas flow (RF torches). In axisymmetric flows of this type, the thickness of the boundary layer is fairly uniform near the center of the jet or substrate (whichever is smaller) and is given by [Goodwin 1993a, Kays and Crawford 1980]

$$\delta_H = 1.32\,(\nu/a)^{1/2}\,(D_H/\nu)^{0.4}. \tag{32}$$

Here ν is the kinematic viscosity of the gas and a is the stagnation point strain rate, given by

$$a = C\frac{U_\infty}{d}, \tag{33}$$

where U_∞ is the gas velocity far upstream of the substrate, and d is the diameter of the substrate or gas jet, whichever is smaller. The parameter C is a non-dimensional constant which depends on the details of the flow but is of order 1 [Goodwin 1993a].

Using property values appropriate for a dilute hydrocarbon mixture in H_2 at 1200 K, eq. 32 simplifies to

$$\delta_H \approx \frac{160}{\sqrt{ap}}\ \text{cm}, \tag{34}$$

with a in s^{-1} and p in Torr. Although this expression has been derived assuming no gas-phase chemistry and constant transport properties, comparison to more detailed numerical simulations shows that it is accurate to $\pm 20\%$ even including gas-phase chemistry and property variations encountered for typical arcjet conditions [Goodwin 1993b].

2.4.5 *Optimal Conditions for H Transport in Convective Reactors*

For a convective reactor, maximizing the delivery of H to the substrate requires maximizing $X_{H,\infty}$ and minimizing δ_H. Achieving high dissociation of H_2 ($X_{H,\infty}$) is the most direct and effective way to increase j_H. Beyond this, δ_H can be decreased either by increasing the flow speed U_∞, or increasing the pressure.

The flow speed cannot be increased indefinitely, since it is typically produced by accelerating a hot, high-pressure gas through a nozzle. The maximum U_∞ which can be achieved with a nozzle is of order the sound speed in the hot gas upstream. For example, for fully-dissociated hydrogen at 5000 K, the maximum velocity achievable is approximately 8 km/s.

If the substrate is a disk of diameter d_s immersed in this 8 km/s flow, the a parameter is of order

$$a_{max} \approx \frac{6 \times 10^5}{d_s}\ \text{s}^{-1}, \tag{35}$$

with d_s in cm. This value is asymptotically approached as the flow Mach number exceeds 1.0. For Mach numbers much in excess of 1.0, the stagnation pressure drop through the shock wave thickens the boundary layer and is detrimental for H transport to the surface. This leads to the conclusion that the optimal flow has Mach number between 1 and 2 [Goodwin 1993b].

Using the a_{max} value above, it is possible to calculate the minimum achievable boundary layer thickness at a given pressure:

$$\delta_{H,min} \approx 0.2\sqrt{d_s/p}\ \text{cm}, \tag{36}$$

with d_s in cm and p in Torr. For example, for a substrate diameter of 10 cm and a pressure of 100 Torr, a boundary layer (and diffusion length) as small as 0.6 mm is achievable. Note that this value is 2 orders of magnitude smaller than the value of ℓ_d calculated above for a "typical" hot-filament system. The ability to achieve very small diffusion lengths for H transport uniformly over the substrate is a primary advantage of high-speed convective systems such as arcjets for high-rate diamond growth.

The analysis presented above neglects gas-phase chemistry. In particular, homogeneous recombination in the boundary layer is neglected. Goodwin [1993b] has analyzed this problem numerically, and finds that homogeneous recombination only begins to alter these results at a pressure of 1 atm or greater, for typical flow parameters. Since most convective reactors operate below this pressure, homogeneous recombination in the boundary layer does not usually play a significant role.

3. The Growth Species

The question of which carbon-bearing gas-phase species is most responsible for bringing carbon to the growing diamond film has attracted much interest. In addition to its fundamental nature, this question is important for reactor design, since once the "growth species" is identified, a reactor could be designed to maximize the concentration of this species.

There have been many suggestions for the growth species, including small radicals (C, CH, C_2, C_2H, CH_3), ions (CH_3^+), and large hydrocarbons which have a similar structure to diamond (adamantane). Some of these were suggested based on observation of characteristic emission spectra in plasma or flame environments during diamond growth. However, some species, such as C_2 and CH, may produce intense visible emission due to electron-impact excitation (in plasmas) or chemiluminescence (in flames) even at concentrations far too low to account for measured growth rates, while other abundant species (CH_3 and C_2H_2, for example) have no prominent visible emission bands. For this reason, it is difficult to draw conclusions about the significance of a given species from its detection in the plasma or flame emission spectrum (even if, as often reported, the emission intensity from a particular species correlates with film growth rate or quality).

The observation that diamond may be readily grown in hot-filament reactors at rates comparable to plasma systems operating at the same pressure and flow velocity indicates that species which might be found in a plasma (ions, electrons, electronically-excited neutrals) but not in a "thermal" environment are probably not important for diamond CVD. For this reason, most work on determining the growth species has focused on neutral species. Of course, it is possible that additional species become important in some plasma or flame environments. However, as will be discussed below, numerical simulations have shown that the growth rates measured in many different environments (hot-filament, microwave plasma, RF plasma, oxyacetylene flame, DC arcjet) spanning two orders of magnitude can all be accounted for (to within a factor of 2–3) by a single growth species and a single surface mechanism. Thus, while contributions from additional species may occur in these environments, they do not appear to be *necessary* to explain measured growth rates.

Harris et al. [1988] have pointed out that any potential growth species must have a collision

TABLE III. Summary of results for ^{13}C mole fractions (%) measured in the diamond film, CH_4, and C_2H_2 for hot-filament [D'Evelyn et al. 1992] and microwave [Johnson et al. 1992] experiments.

		^{13}C Mole Fraction		
CVD method	Film type	Film	CH_4	C_2H_2
Hot-filament	polycrystalline	58.2 ± 3.6	61.6 ± 5.5	32.4 ± 5.6
Hot-filament	homoepitaxial	56.8 ± 1.2	58.6 ± 5.4	34.9 ± 5.3
Microwave	polycrystalline	77	83	29

frequency with the surface at least as great as the rate at which carbon is incorporating into the film (corrected for the number of carbons n_c in the growth species). For typical hot-filament growth rates (~ 1 μm/hr), this implies a growth species concentration at the surface of at least $3 \times 10^{10}/\sqrt{n_c}$ cm^{-3}. Furthermore, the gas-phase production rate at steady state must be equal to the incorporation rate in order to sustain this concentration.

Harris et al. [1988] measured stable species concentrations in a hot-filament system, and used a zero-dimensional kinetic model to estimate unmeasured species concentrations at the substrate. They concluded that only CH_4, CH_3, C_2H_2, and C_2H_4 are present in sufficient quantities to account for measured growth rates. When the low reactivity of methane is taken into account, only CH_3 and C_2H_2 are left as likely growth species under typical hot-filament conditions.

A variety of experiments have been conducted to distinguish between diamond growth from the methyl radical and from acetylene. In a series of experiments at Rice University [Chu et al. 1990, 1991, D'Evelyn et al. 1992], isotope labeling was used differentiate carbon incorporated from CH_3 and from C_2H_2. In these experiments, a 2:1 mixture of ^{13}C methane and ^{12}C acetylene was introduced into a hot-filament reactor, and the isotope ratio of the resulting film was measured from the peak position of the diamond Raman line. It was assumed that CH_4 and CH_3 were in isotopic equilibrium, due to rapid interconversion via $CH_4 + H \rightleftharpoons CH_3 + H_2$. Some isotope scrambling between CH_4 and C_2H_2 also occurs, due to $C_1 \rightleftharpoons C_2$ interconversion chemistry. However, by injecting the gas close to the substrate it was possible to avoid complete scrambling.

The isotope ratios of methane and acetylene within 1 mm of the substrate were measured using matrix-isolation infrared spectroscopy. The results, summarized in Table , clearly show that the diamond film isotope ratio matches that of methane to within experimental error, and does not match that of acetylene. From these results, it was concluded that a species in isotopic equilibrium with methane (presumably CH_3) is the dominant growth species under typical hot-filament conditions.

Johnson et al. [1992] carried out similar isotope-labeling experiments using an Evenson cavity microwave plasma reactor rather than a hot-filament system. The gas used was 1% $^{13}CH_4$ and 0.5% $^{12}C_2H_2$ in H_2 at a pressure of 21 Torr. Fast gas flow rates were used to avoid complete isotope scrambling in the plasma. The measured ^{13}C mole fractions are also given in Table III. Again, the results suggest that a species in isotopic equilibrium with methane is responsible for diamond growth.

Another class of experiments has focused on using very low pressure, high-velocity reactors which limit the residence time for gas-phase chemistry. Martin and co-workers at the Aerospace Corporation have used a low-pressure, remote plasma flowtube to study diamond growth with minimal $C_1 \rightleftharpoons C_2$ interconversion. In these experiments, a hydrocarbon (either methane or acetylene) is injected into an argon/hydrogen mixture downstream of a microwave discharge at a pressure of 3 Torr. The residence time from the hydrocarbon injection to the substrate is of order a few milliseconds, which effectively suppresses $C_1 \rightleftharpoons C_2$ interconversion. Diamond is found to grow with either methane or acetylene injection, but the rate is faster by an order of magnitude and the quality better with methane injection [Martin and Hill 1990]. A numerical simulation of this experiment shows that only CH_3 can account for the growth when methane is injected, and only acetylene can account for the growth when acetylene is injected [Harris and Martin 1990].

In a subsequent study, Martin [1993] found that good-quality diamond could be grown using acetylene if the flowtube reactor was first run on methane for a short period; it was suggested that the methane may be necessary in the initial stages of nucleation and growth. In these experiments, the reactivity of acetylene for diamond growth was estimated to be a factor of 10 lower than that of the methyl radical; this acetylene reactivity is approximately an order of magnitude larger than the previous estimate of Harris and Martin [1990].

A similar study was carried out by Loh and Cappelli [1993], who used a supersonic DC arcjet with downstream hydrocarbon injection. The pressure at the substrate was 400 Pa, and the hydrocarbon residence time was estimated to be 7 μs. Diamond growth occurred with either methane or acetylene injection. Gases sampled through a small hole in the substrate show the presence of a small amount of C_1 species when acetylene is injected. However, Loh and Cappelli argue that the amount is insufficient to account for the measured diamond growth rates. They conclude that both acetylene and methyl are growth species.

In subsequent work, Loh and Cappelli measured the methyl radical concentration at the substrate [Loh and Cappelli 1997] with methane injection. This measurement showed that only a fraction of the methane was converted to CH_3 during the short reactor residence time. Using this measured concentration, they concluded that the methyl radical reactivity for diamond growth was two orders of magnitude greater than that of acetylene.

A particularly clean experiment is that of Lee et al. [1994]. In this work, a substrate containing small (0.1 – 0.25 μm) diamond seed crystals was exposed simultaneously to a supersonic free jet of methyl radicals (created by pyrolysis of azomethane in helium) and to an effusive glow-discharge jet of atomic hydrogen in helium. The design of the experiment suppressed or eliminated gas-phase chemistry, and there was no detectable interconversion of CH_3 to C_2H_2. With the substrate exposed simultaneously to CH_3 and H, good quality 10 μm diamond particles were grown on the seed crystals in the temperature range 650–850 °C. No diamond was grown if the pyrolysis of azomethane was turned off (azomethane striking the substrate rather than CH_3), or if the glow discharge was turned off (H_2 instead of H). Substituting acetylene for azomethane (without pyrolysis) gave primarily graphitic carbon.

Finally, studies have been carried out to measure the growth rate as a function of pressure and input gas composition. Harris and Weiner [1991] used a microbalance to measure the growth rate of polycrystalline diamond on platinum as a function of methane fraction R in a hot-filament system. The measurements were repeated for pressures from 20 Torr to 200 Torr, under conditions where the substrate temperature was essentially constant. They found that the growth rate G followed approximately a power law in R:

$$G \propto R^{\alpha}. \tag{37}$$

The exponent α is found to be pressure-dependent, decreasing from $\alpha = 1$ at 20 Torr to $\alpha = 0.5$ at 200 Torr. This behavior tracked very well the CH_3 concentration as a function of R and pressure calculated from a detailed kinetic model. In a later paper, the acetylene concentration was measured for the same conditions as the diamond growth experiments [Harris and Weiner 1992]; no correlation was found between the growth rate and the acetylene concentration.

Wang and Angus [1993] and Evans and Angus [1996] have conducted experiments in which a microbalance was used to measure the growth rate as a function of gas-phase hydrocarbon content in a hot-filament reactor. The hydrocarbon was introduced as either methane, acetylene, ethylene, or ethane. For methane addition, the growth rate was found to be first-order in CH_4 up to 1% methane, and then transitioned to zero-order at higher methane fractions. For the two-carbon gases, the growth rate is half-order up to a C_2H_x fraction of 0.64%. This is the behavior expected if the growth species is a C_1 hydrocarbon, assuming the residence time is long enough for approximate equilibration of C_1 and C_2 species.

Modeling studies have also considered the question of the growth species. Goodwin [1991] showed that a methyl radical growth mechanism could explain the high growth rates observed in flame and arcjet experiments, when account is taken of the chemistry occurring in the boundary layer above the substrate. The results of growth rate predictions from numerical simulations for a variety of environments are compared to measured rates in Figure 9. These calculations were all made using a particular CH_3 growth mechanism [Harris 1990, Harris and Goodwin 1993] discussed in section 4.3.3.

Numerical simulations have also been used to examine whether atomic carbon could plausibly act as a growth species. Atomic carbon is not detected experimentally at the substrate in low-pressure hot-filament or microwave systems, but as discussed above is predicted to be abundant in the gas for large H/H_2 ratios. In some high-speed plasma systems with thin boundary layers, simulations predict that C may survive the boundary layer, and be present in significant amounts at the surface [Goodwin 1991, Yu and Girshick 1994, Coltrin and Dandy 1993, Dandy and Coltrin 1995].

Yu and Girshick [1994] have presented a theoretical analysis which suggests that C may be the dominant growth species in thermal plasma systems with very thin boundary layers ($<$ 0.6 mm in their calculations). This analysis is based on a modification of the methyl radical growth mechanism used in the calculations shown in Figure 9 to allow C as a reactant in addition to CH_3. They also predict from this analysis that CH_2 may contribute as much as 17% to the overall growth rate under some conditions.

The modeling and theoretical studies suggest that C may play a role in diamond growth

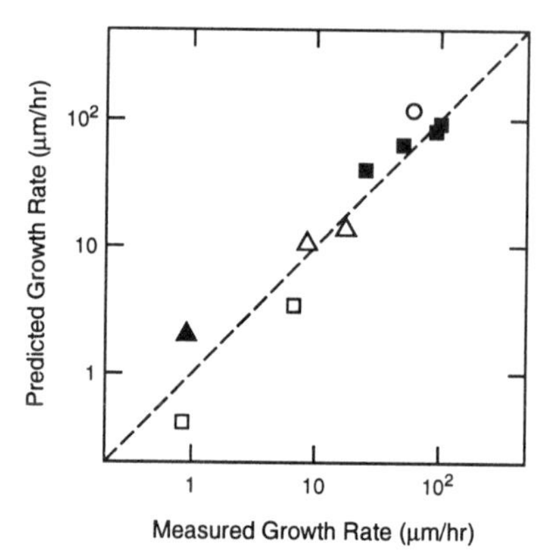

Figure 9. Comparison of measured growth rates with growth rates from numerical simulations [Goodwin 1991, Glumac and Goodwin 1992, Yu and Girshick 1994] using a growth mechanism which is first-order in CH_3 [Harris 1990, Harris and Goodwin 1993]. Experiments simulated: 1 atm oxygen-acetylene flame, Matsui et al. [1989], solid squares; 220 Torr DC arcjet, Stalder and Sharpless [1990], open circle; 1 atm RF torch, Owano et al. [1991], open squares; 1 atm RF torch, Girshick, Yu, Li and Han [1993], open triangles; 40 Torr oxygen-acetylene flame, Glumac and Goodwin [1992], solid triangle.

under some conditions. However, given the large uncertainty in the chemistry of atomic carbon on diamond, it is very likely that there are other kinetic pathways not yet considered. For example, the high reactivity of C might allow it to easily contribute to sp^2 carbon formation, unlike CH_3. Definitive conclusions about the importance of C to diamond growth will have to await experimental evidence.

In summary, a wide variety of experiments point to CH_3 as the principal growth species in conventional diamond CVD environments, although other species (particularly acetylene) may contribute small amounts – of the order of 10% or less. However, under high-speed, low-pressure conditions where acetylene is abundant and the methyl concentration is suppressed, then growth appears to be primarily due to acetylene. The role of atomic carbon in systems with a high degree of hydrogen dissociation has been predicted by models, but has not yet been verified experimentally.

4. Diamond Surface Chemistry

While gas-phase chemistry and identification of the growth species are important parts of diamond CVD research, the fundamental goal is to identify the chemistry occurring *on* the surface during CVD growth. This is a difficult experimental problem, since the electron spectroscopies used in surface science are not directly useful at the high gas pressures encountered in chemical vapor deposition.

In this section, several aspects of diamond surface chemistry are reviewed. First, the important topic of hydrogen on diamond will be discussed, followed by a summary of what is known experimentally about diamond growth kinetics. Finally, several proposed growth mechanisms will be discussed.

4.1 HYDROGEN CHEMISTRY

One of the most important aspects of diamond surface chemistry is the reaction between atomic hydrogen and the diamond surface. Because of the continual bombardment of the diamond surface with reactive atomic hydrogen during growth, most of the diamond surface is hydrogenated, and is therefore unreactive with incoming hydrocarbon species. The fraction of surface sites which are *not* hydrogenated (the open site fraction f^*) is determined by a dynamic equilibrium between the two reactions

$$\mathrm{C_dH + H \rightarrow C_d^* + H_2} \tag{38}$$

and

$$\mathrm{C_d^* + H \rightarrow C_dH}, \tag{39}$$

where $\mathrm{C_dH}$ represents a hydrogen-terminated surface site and $\mathrm{C_d^*}$ an equivalent site without the hydrogen.

The reverse of reaction (39) (thermal desorption of H) may always be neglected. The reverse rate for reaction (38) may be calculated from the forward rate and the equilibrium constant K_p ($\approx$ 800 at 1200 K). Goodwin [1993a] has shown that the dominant path for converting C_d^* to C_dH is reaction (39), rather than reaction (-38), as long as the H mole fraction is greater than 10^{-4}. Since the H mole fraction at the substrate during diamond CVD is typically 10^{-3} or greater (Figure 4), we are justified in treating reaction (38) as irreversible.

In ultra-high vacuum, hydrogen desorbs from diamond as H_2 above about 800 °C (ca. 75 kcal/mol activation energy). However, for diamond CVD conditions, this process does not appreciably contribute to the open site fraction, since the desorption rate only becomes comparable to atomic hydrogen abstraction and adsorption reactions at significantly higher temperatures ($\gg$ 1100 °C) [Brenner et al. 1992].

By balancing $\mathrm{C_d^*}$ production via reaction (38) and destruction by reaction (39), the value of f^* which results is

$$f^* = \frac{\gamma_1}{\gamma_1 + \gamma_2} \tag{40}$$

where γ_1 is the abstraction probability [reaction (38)] and γ_2 the adsorption probability [reaction (39)].

The values of γ_1 and γ_2 have not been measured directly, but have been estimated in several different ways. Brenner et al. [1992] have carried out molecular dynamics simulations of H bombarding a (111) diamond surface for temperatures from 1200 to 1800 K. At 1200 K, their simulations predict $\gamma_1 = 0.04$ and $\gamma_2 = 0.43$.

The H recombination coefficient measurements discussed in Section 2.2 also shed some light on the values of γ_1 and γ_2, since if the loss of H on the surface is primarily through reactions (38) and (39) then [Goodwin 1993a]

$$\frac{1}{\gamma_H} = \frac{1}{2}\left(\frac{1}{\gamma_1} + \frac{1}{\gamma_2}\right). \tag{41}$$

Assuming γ_2 is between the value calculated by Brenner et al. and 1.0, the value of γ_H measured by Krasnoperov et al. [1993] implies a value of γ_1 at 1200 K in the range from 0.085 – 0.096. The activation energy for γ_1 should be close to that of γ_H, which Krasnoperov et al. [1993] found to be 6 kcal/mol.

Thoms et al. [1994] have used HREELS to directly measure the *ratio* of γ_1/γ_2 on polycrystalline diamond. Measurements were made at the two substrate temperatures of 80 and 600 °C, and in both cases the ratio γ_1/γ_2 was found to be 0.05. The lack of dependence on substrate temperature indicates the reaction mechanism is an Eley-Rideal one, and that the kinetic energy of the incoming H atom is more important in determining the abstraction or adsorption rate than the substrate temperature. It should be also noted that the HREELS experiments were conducted in vacuum, and the incident H atoms arrived at the substrate directly after dissociation on an 1800 °C filament; they are therefore on average more energetic than typical H atoms in thermal equilibrium with a 1200 K substrate.

Koleske et al. [1995] have carried out measurements similar to Thoms et al. using time-of-flight recoil spectrometry. By comparing their value for γ_1/γ_2 of 0.03 at a filament temperature of 1560 °C to that of Thoms et al., Koleske et al. estimate an activation energy for abstraction of 16 kcal/mol. They also point out that theoretical estimates for the activation energy for this reaction range from 0 to 25.6 kcal/mol, depending on the calculation technique used (although many of the techniques used cannot be expected to be accurate for this reaction).

These studies all agree about the qualitative magnitudes of γ_1 and γ_2: γ_1 is of order a few percent, while γ_2 is much larger. However, there is still substantial uncertainty about the actual values and the temperature dependence.

Surface science studies have also been carried out to identify the nature of C-H bonding on hydrogenated diamond surfaces [Thoms and Butler 1995, Struck and D'Evelyn 1993, McGonigal et al. 1995]. These studies show that the (100), (110), and (111) surfaces are mainly covered by the monohydride (CH) species.

4.2 EXPERIMENTAL DIAMOND GROWTH KINETICS

The temperature dependence of the diamond growth rate can give critical information regarding the rate-limiting steps during diamond growth. Several studies have reported temperature-dependence measurements, either for growth of polycrystalline or single-crystal homoepitaxial films. The homoepitaxial measurements are easier to interpret theoretically, since polycrystalline film growth involves nucleation, coalescence, and growth competition among individual grains. These additional effects result in changing film texture and surface morphology as the film grows, which complicates the interpretation of the results.

4.2.1 *Polycrystalline Films*

Kondoh et al. [1991, 1993] have grown polycrystalline diamond films on silicon at temperatures from 620 – 1130 °C using a hot-filament reactor at 30 Torr. The growth rate was determined afterward from the film thickness measured by cross-sectional scanning electron microscopy. (Snail and Marks [1992] have pointed out that this technique may be subject to errors due to the induction time before growth begins.)

In the temperature range from 620 – 900 °C, Kondoh et al. [1993] find the growth rate exhibits Arrhenius behavior, with an activation energy of 23 kcal/mol, which is independent of filament-to-substrate distance and carbon source gas. The growth rate peaks between 900 and 1000 °C, and decreases at higher temperature. Other studies have reported similar activation energies (20 – 30 kcal/mole) for polycrystalline diamond growth [Spitysn et al. 1981, Mitsuda et al. 1989].

Yamaguchi et al. [1994] have reported activation energies in a hot-filament system for diamond particle growth at very low temperatures (210 – 700 °C). They find very low apparent activation energies – decreasing from 5 kcal/mol at the upper end of the temperature range to 1 kcal/mol at the lower end.

Snail and Marks [1992] measured polycrystalline growth rates in an atmospheric pressure turbulent flame *in situ* by monitoring the oscillations in apparent temperature recorded by a two-color pyrometer due to thin-film interference effects. This technique yields a true, nearly instantaneous growth rate, independent of any nucleation induction time. Above about 800 °C, the growth rate exhibited an activation energy in the range 16 – 23 kcal/mol, depending on flame conditions. The growth rate continued to rise with increasing temperature up to nearly 1200 °C (in contrast to the kinetics in low-pressure systems [Kondoh et al. 1993]), although the graphitic content of the film appeared to increase above 1100 °C. Other studies have also shown that atmospheric-pressure reactors can grow diamond at higher substrate temperatures (up to $\sim$ 1400 °C) than is possible in low-pressure systems [Weimer et al. 1995, Owano and Kruger 1993, Baldwin et al. 1994].

4.2.2 *Homoepitaxial Films*

Homoepitaxial growth kinetics for the (100), (110), and (111) crystal faces have been reported by Chu et al. [1992]. The results, reproduced in Figure 10, show that the (110) face has the highest growth rate, which is consistent with the absence of (110) facets in polycrystalline films. The dependence of the growth rate on CH_4 fraction was different for each orientation; it was linear for (100), slightly sublinear for (110), and sigmoidal on (111).

The temperature dependence differed as well for each orientation. In the temperature range 735 – 970 °C, the kinetics could be described by activation energies of 8 $\pm$ 3, 18 $\pm$ 2, and 12 $\pm$ 4 kcal/mol for (100), (110), and (111) growth, respectively. All three orientations showed a larger activation energy (of order 50 kcal/mol) in the temperature range 675 – 735 °C. This conclusion – that the activation energy is higher at lower temperatures – is precisely opposite that of Yamaguchi et al. [1994].

Maeda et al. [1995] measured homoepitaxial growth rates by first growing cubo-octahedral crystals on a silicon substrate in a microwave plasma reactor and then observing their change in shape with additional diamond growth. The (100) growth rate is linear in CH_4 fraction, while the (111) growth rate is much less sensitive to CH_4 fraction. Also, the (111) growth rate is more temperature-dependent than the (100) growth rate. These trends are consistent with the results of Chu et al. [1992]. Some differences are reported also: the sharp fall-off in growth rate at low temperature reported by Chu et al. is not observed, and a maximum growth rate is found between 800 – 900 °C.

The activation energies measured for homoepitaxial growth in the temperature range $\approx$

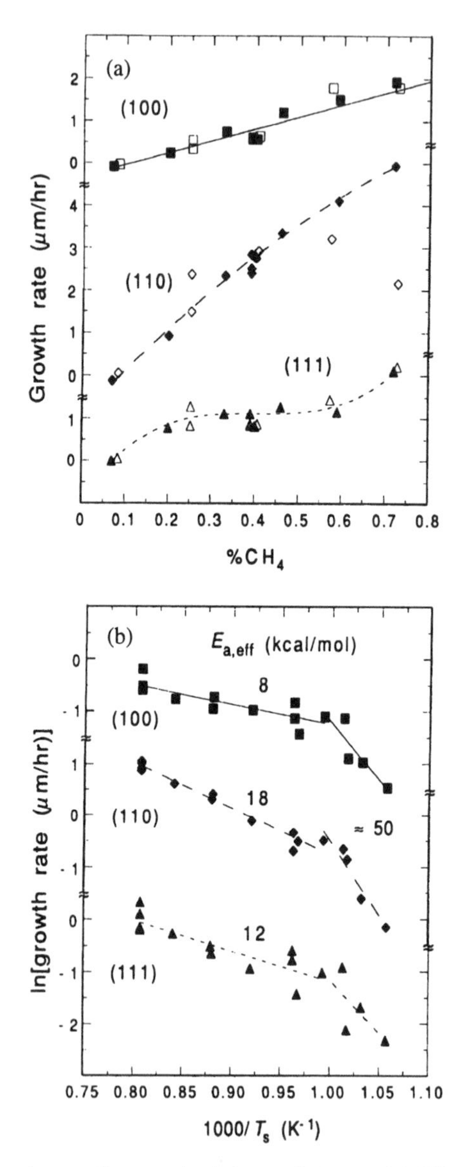

Figure 10. Measured dependence of homoepitaxial growth rate on (a) CH_4 fraction and (b) temperature [Chu et al. 1992].

700– 900 °C are all significantly lower than those measured for polycrystalline films. The reasons for this are not entirely clear, since it would be expected that the polycrystalline results should be an average of the homoepitaxial results.

The measured homoepitaxial kinetics show that the (100), (111), and (110) surfaces have growth rates which depend differently on temperature and gas composition. This feature of the kinetics can be used to control the morphology of polycrystalline films, which is determined by the ratio of the (100) and (111) growth rates [Wild et al. 1994, Clausing et al. 1992]. A parameter α is commonly defined as

$$\alpha = \sqrt{3}\frac{V_{100}}{V_{111}}. \tag{42}$$

The α parameter determines the shape of single crystals, which are cubes for $\alpha = 1$, octahedra for $\alpha = 3$, and cubo-octahedra for intermediate α values. For polycrystalline films, α controls the film texture and stability with respect to twinning. By carefully controlling α, smooth polycrystalline films with (100) texture and (100) faceting have been grown [Wild et al. 1994].

4.3 GROWTH MECHANISMS

4.3.1 *General Considerations*

The basic theory of crystal growth from the vapor phase is well developed [Zangwill 1988]. In physical vapor deposition processes such as molecular beam epitaxy, a growth monomer from the vapor phase adsorbs on the crystal surface at a random location, and then migrates laterally on the surface. Eventually, it may encounter an atomic step, where it becomes incorporated due to the increase in coordination number (and therefore in binding energy). Alternatively, it may thermally desorb from the crystal, or encounter another monomer, which may lead to nucleation of a new atomic layer (an island). Depending on factors such as the adsorbate mobility and monomer arrival rate, crystal growth may proceed by steps "flowing" across the surface, or formation of individual islands, which later coalesce.

Crystal growth by chemical vapor deposition is more complicated than the simple picture given above, since the species which adsorb from the vapor phase may have a different composition than the growing crystal (e.g. adsorption of SiH_4 to grow Si). Surface *chemistry* therefore plays a critical role in crystal growth by CVD.

This is particularly true for diamond CVD, since the conditions under which it takes place are such that the physical process of thermally-activated migration may only occur to a limited extent. During diamond growth, the diamond surface is nearly fully saturated with hydrogen. This hydrogen coverage limits the availability of sites where hydrocarbon species may chemisorb, and blocks migration sites once they are adsorbed. In addition, typical substrate temperatures during growth (600 – 1200 °C) are well above the temperatures at which physisorbed species desorb. Also, the substrate temperature during diamond CVD is well below the Debye temperature of diamond (1860 K), a useful parameter for scaling thermally driven processes in crystals. This contrasts with virtually all other crystal growth processes, which take place at roughly *twice* the Debye temperature of the crystal. This observation suggests that adspecies mobility will not be large, since they do not have enough thermal energy to overcome the energy barrier to migration.

Nevertheless, images of diamond surfaces taken after CVD growth often show distinct steps [Okada et al. 1991, Tsuno et al. 1991, Ravi and Joshi 1991, Clausing et al. 1992], which implies growth occurs by lateral motion of steps across the surface. There must either be chemical mechanisms – likely involving atomic hydrogen – which impart mobility to surface species, or else the growth sites at the step edges are much more reactive than isolated ledge sites.

Much work is still required before we will have a satisfactory understanding of the surface processes occurring during diamond CVD. At present, most theoretical studies have only considered the initial bonding of a gas-phase hydrocarbon to the diamond surface. Other questions – for example how faceted, crystalline films can be grown under conditions of apparently limited adsorbate mobility, or the mechanisms of defect formation – are only now beginning to be addressed theoretically.

4.3.2 *Steric Constraints and Kinetic Stability*

Due to the small lattice constant of diamond and to the presence of a large flux of atomic hydrogen at the surface, there are several difficulties one is faced with in attempting to develop a chemical mechanism for diamond CVD. Two of the most significant problems are those of steric constraints, due to the large size of incoming hydrocarbons relative to the carbon-carbon lattice spacings, and kinetic stability, which refers to the tendency of H to attack and remove hydrocarbons which are bonded on the surface.

If an incoming hydrocarbon species is to bond to the surface, it generally must do so at one of the open, radical sites (unless it can insert directly into a surface C-H bond, as some species such as C_2 may be able to do [Horner et al. 1995, Pitts et al. 1982]). At low values of f^*, most open sites are isolated – that is, neighboring surface carbons are hydrogenated. Since a typical C-H bond length is 1.1 Å, while the diamond C-C bond length is only 1.54 Å, the neighboring hydrogens seriously crowd the open site and may make it difficult or impossible for some incoming hydrocarbon species to bond to this site, due to the repulsion between hydrogens on the surface and those on the incoming hydrocarbon.

Even if bonding occurs, the bond may be significantly strained and weakened by the steric environment, making the chemisorbed species susceptible to thermal desorption. As pointed out by Skokov, Weimer and Frenklach [1994], this problem makes many otherwise-plausible hydrocarbon addition mechanisms unworkable. The problem is particularly acute for mechanisms which require *two* hydrocarbons to bond at adjacent positions on the surface before stabilizing cross-linking bonds can be formed, as several do. If the lifetime on the surface is shorter than the mean arrival rate per site, then the first hydrocarbon to bond will not last long enough to complete the process.

The large flux of atomic hydrogen bombarding the surface places further constraints on possible growth mechanisms. Many types of hydrocarbon adsorbates are subject to attack by atomic hydrogen, leading to removal through the process known as β-scission [Butler and Woodin 1993]. As an example of the removal of surface species by β-scission, two possible ways to attack a surface-bonded ethyl group are shown in Figure 11. An H atom is first abstracted by gas-phase H to create a radical site (an unpaired electron). In (a), a methyl hydrogen is abstracted, while in (b) the hydrogen on the diamond surface carbon is abstracted. Once this has occurred, the bond once-removed (β) from the radical site may

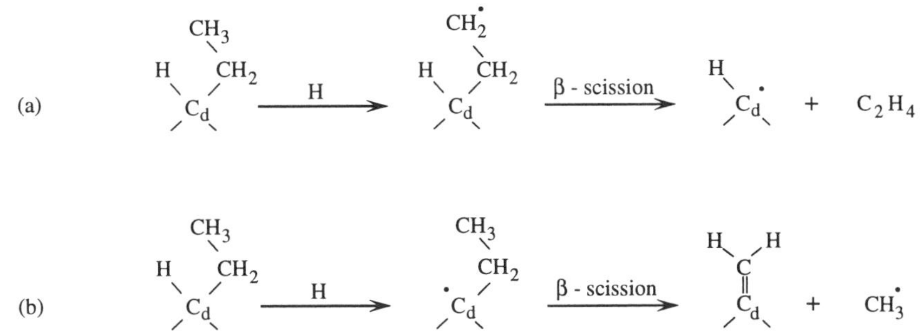

(a) $H{-}C_d{-}CH_2{-}CH_3 \xrightarrow{H} H{-}C_d{-}CH_2{-}\dot{C}H_2 \xrightarrow{\beta\text{ - scission}} H{-}\dot{C}_d + C_2H_4$

(b) $H{-}C_d{-}CH_2{-}CH_3 \xrightarrow{H} \cdot C_d{-}CH_2{-}CH_3 \xrightarrow{\beta\text{ - scission}} C_d{=}CH_2 + \dot{C}H_3$

Figure 11. Illustration of two β-scission processes to attack a surface-bonded ethyl group.

now break, allowing double bond formation to the carbon from which the hydrogen was abstracted. This process results in scission of all or part of the adsorbate from the lattice. Similar β-scission processes will occur for other multi-carbon surface species.

The β-scission process plays an important role in diamond growth, since it is believed to be an important mechanism by which H is able to etch non-diamond sp^2 carbon [Butler and Woodin 1993]. Skokov, Weimer and Frenklach [1994] have persuasively argued that this problem is a severe one for many proposed growth mechanisms which proceed through intermediate surface species susceptible to removal by β-scission. As with thermal desorption, the problem is particularly severe for mechanisms which require *two* hydrocarbons to adsorb nearby and then react. The steady-state surface coverage of these vulnerable intermediates may be so small that the probability of two being simultaneously adjacent is negligible.

In summary, these arguments suggest that the possible chemical pathways through which carbon is added to the lattice during diamond CVD are constrained in ways not encountered in other, "milder" CVD environments, or in film growth by MBE. The small open site fraction and the problem of steric hindrance limit the number of sites at which hydrocarbon species may bond. The fact that many surface species are kinetically unstable with respect to β-scission means that these species must be incorporated into the lattice very quickly, through paths which do not require waiting for a second species to bond nearby. Only relatively infrequent events, such as nucleating a new atomic layer, could conceivably proceed through surface species subject to removal by β-scission.

4.3.3 *C_1 Mechanisms*

Several mechanisms have been proposed for addition of single-carbon species to the diamond lattice. The specific growth species has usually been assumed to be CH_3, but in most cases small modifications would make the mechanisms equally applicable for CH_2 or even atomic C.

Addition to the (111) surface. One of the first diamond growth mechanisms proposed was the methyl mechanism of Tsuda et al. [1986]. In this mechanism, the diamond (111) surface is assumed to first become covered with a complete monolayer of adsorbed CH_3 groups; a methyl cation CH_3^+ then attacks three adjacent surface methyl groups, and, in a sequence of steps, binds the methyl groups together, with the evolution of 3 H_2 molecules.

One difficulty with this mechanism is the very low concentration of ionized species such as CH_3^+ found in thermal (hot-filament or combustion) environments. Also, Valone [1989] has calculated the energetics of bonding 3 methyl groups at adjacent positions on the (111) surface and has shown that this structure is energetically unstable, due to the severe steric hindrance between adjacent groups. Consequently, the mechanism of Tsuda must be regarded as improbable.

No other mechanisms to add methyl (or any other C_1 species) have been proposed for the flat (111) surface. The essential difficulty is that addition of 3 carbons is required on this surface before a second, stabilizing bond to the surface can be made. In any case, (111) surfaces of CVD diamond appear to be rough on an atomic scale [Geis 1990, Sutcu, Thomson, Chu, Hauge, Margrave and D'Evelyn 1992], and (111) growth sectors are highly defective [Wang et al. 1994]. This may indicate that growth on (111) occurs locally at sites with (100) or (110) character, or that growth on (111) may be defect-mediated.

Addition to the (100) surface. The (100) surface is unique in that addition of only one carbon is required to form two bonds to the surface. The (100) faces of CVD diamond often appear atomically smooth [Geis 1990, Sutcu, Thomson, Chu, Hauge, Margrave and D'Evelyn 1992], and (100) growth sectors have low defect density [Wang et al. 1994]. For these reasons, growth on the (100) surface has received the most theoretical attention.

The structure of the hydrogenated diamond (100) surface during diamond growth has been the subject of some debate. Early studies reported that the surface had the bulk structure, with two hydrogens terminating each surface carbon [Hamza et al. 1990]. (This surface will be denoted here as the (100)-(1x1):2H surface.) However, theoretical studies [Yang and D'Evelyn 1992, Mehandru and Anderson 1991] showed that this surface would be much more strained than any known hydrocarbon molecule, due to the strong steric repulsion between the hydrogen atoms on neighboring surface carbons.

More stable surfaces may be obtained by removing hydrogens, allowing neighboring carbon atoms to form "dimer" bonds. Yang and D'Evelyn [1992] find using the MM3 molecular mechanics force field that the (100)-(2x1):1H surface, on which every carbon has one hydrogen and one dimer bond, is thermodynamically most stable (lowest free energy of formation) above 400 K. An alternative surface reconstruction, in which rows of dimers alternate with rows of dihydride terminated carbons [the (100)-(3x1):1.33H surface] is predicted to be lower in *energy* than the (2x1):1H surface (in agreement with calculations of Skokov, Carmer, Weiner and Frenklach [1994]) but when entropy effects are included it is slightly less stable at 298 K and is significantly less stable at high temperature [Yang and D'Evelyn 1992]. Scanning-tunneling-microscopy (STM) and atomic-force-microscopy (AFM) images of as-grown (100) CVD diamond surfaces clearly show a (2x1) pattern of dimer rows, consistent with the (2x1):1H surface [Tsuno et al. 1991, Busmann et al. 1991, Sutcu, Chu, Thompson, Hauge, Margrave and D'Evelyn 1992]. LEED and HREELS measurements also indicate a (2x1) reconstruction [Thoms and Butler 1995].

The first mechanism proposed for methyl addition to the (100) surface was that of Harris [1990]. This mechanism took as a model structure for the (100) surface the molecule bicyclo[3.3.1] nonane (BCN), shown in Figure 12. A set of reactions were proposed to add a methyl group to the central opposing-hydrogen site (the "HH site"), creating the

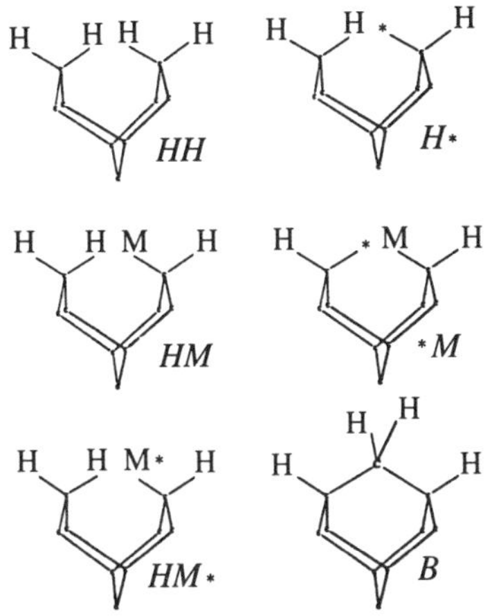

Figure 12. Model clusters for the original HH mechanism [Harris 1990]. The upper-left cluster is bicyclononane, and the lower right is adamantane. The asterisk denotes a radical site, and M a methyl group.

molecule adamantane. The mechanism consists of abstraction of one of the hydrogens of the HH site, methyl addition, abstraction of either a methyl hydrogen or the other HH hydrogen, and finally one more H-abstraction to creating the bridging methylene group.

Rate constants for each step were estimated based on known values for similar gas-phase reactions, and thermochemistry was estimated using group additivity methods. It was shown that this mechanism, with no adjustable parameters, resulted in a conversion from BCN to adamantane at a rate which, if carried out simultaneously everywhere on the (100) surface, would result in a diamond growth rate of 0.06 – 0.6 μm/hr, for typical hot-filament conditions which experimentally result in growth at 0.1 – 1.0 μm/hr. Harris states, however, that the near-perfect agreement is fortuitous, and estimates an uncertainty in growth rate of two orders of magnitude.

In this work, no account was taken of the extreme strain which would result if this mechanism were actually carried out on the (100)-(1x1):2H surface, rather than on BCN. Huang and Frenklach [1992] estimated a prohibitive barrier of 80 kcal/mole for this mechanism on the (100)-(1x1):2H surface. However, as discussed below, the basic HH site topology occurs commonly at steps on other surfaces or between dimer rows on the (100)-(2x1):1H surface. All of these HH-like sites have much lower strain than on the (1x1):2H surface.

Once it became clear that the relevant (100) surface was the (2x1):1H surface, rather than the (1x1):2H surface, several groups focused attention on developing mechanisms for methyl addition to this surface. Since the dimer bonds are part of 5-membered rings, a mechanism is required to open these rings and insert a carbon, in order to form the 6-membered rings of sp^3 diamond.

Garrison et al. [1992] ("GDSB") proposed a low-barrier mechanism to do this, as shown schematically in Figure 13. The dimer is opened by abstraction of a surface hydrogen, addi-

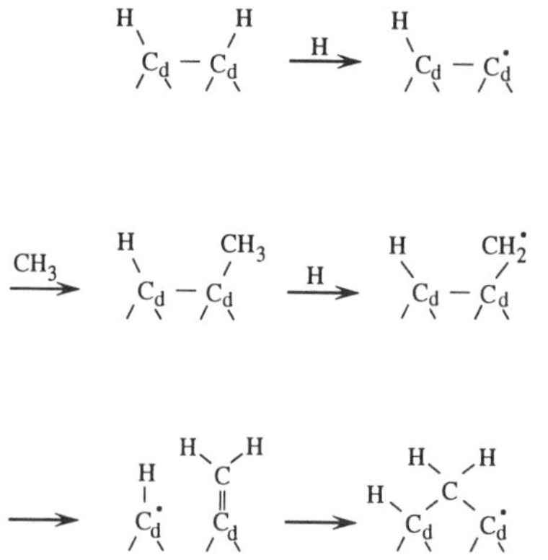

H H → H H
$C_d - C_d$ → $C_d - C_d^{\bullet}$

$\xrightarrow{CH_3}$ H CH_3 $C_d - C_d$ $\xrightarrow{H}$ H $CH_2^{\bullet}$ $C_d - C_d$

→ H $C_d^{\bullet}$ H H C = C_d → H H H C C_d $C_d^{\bullet}$

Figure 13. The GDSB mechanism for dimer opening and carbon insertion [Garrison et al. 1992].

tion of CH_3, abstraction of a methyl hydrogen, followed by a β-scission reaction leading to a methylidene intermediate and a radical site. This intermediate then reacts rapidly [Musgrave et al. 1995] with the adjacent radical, creating a bridging methylene group and propagating the radical site. More recently, Dawnkaski et al. [1995] suggested that hydrogen-free dimer sites (which are predicted to form highly strained, reactive π-bonds) may also be an important site type where CH_3 may chemisorb. In this case, an H migration would lead to the fourth structure in Figure 13.

Huang and Frenklach [1992] have proposed a similar mechanism, except that the fourth structure in Figure 13 is transformed directly into the final bridging-carbon structure though a triangular transition state. However, the activation energy for this transition state is calculated to be 55 kcal/mole, indicating it will be a less important pathway that the β-scission pathway in the GDSB mechanism. The GDSB mechanism is now widely believed to be a principal means for opening and inserting carbon into dimer bonds, and has been incorporated as a submechanism into more extensive mechanisms discussed below.

If a carbon is inserted into every dimer on the (2x1) surface, 1/2 of a monolayer results. To complete the new surface layer, carbon must be inserted between every pair of dimers on adjacent rows, thus bridging the "troughs" between dimer rows. Harris and Goodwin [1993] proposed alternating the GDSB mechanism with a modified HH mechanism to insert into dimers and to bridge between dimers, respectively. A detailed thermochemical analysis of this proposal was conducted, using the MM3 molecular mechanics program to obtain accurate estimates of the free energy change for each step.

Several points emerged from this analysis. It was found that once the GDSB mechanism inserts a methylene group into one dimer, it cannot insert one into the dimer on the adjacent row, due to steric hindrance. Also, the distance between two *dimerized* carbons on adjacent rows is too great for a methylene group to bridge them. However, once the GDSB mechanism acts on one dimer, the distance to the nearest carbon atom on the adjacent dimer is decreased to the point that a methylene group may bridge the trough with little strain. Thus, once the GDSB mechanism acts, the HH mechanism becomes enabled to bridge the trough.

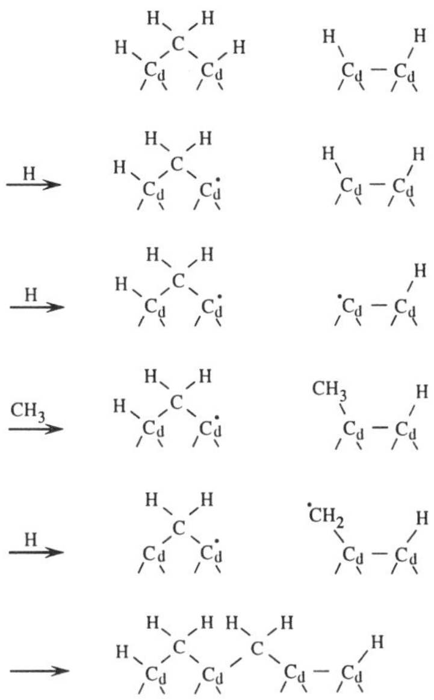

Figure 14. A simplified representation of the trough-bridging (HH) portion of the mechanism of Harris and Goodwin [1993]. Only the path proceeding through the diradical site is shown.

An additional result from this study was the prediction that the (modified) HH mechanism proceeds preferentially through a path in which *both* "HH" hydrogens are missing (a diradical site). This path is shown in Figure 14. This general point that diradical sites are important chemisorption sites is in accord with the conclusions of Dawnkaski et al. [1995], Huang and Frenklach [1992] and Skokov, Weimer and Frenklach [1994].

The thermochemical analysis of Harris and Goodwin predicted that the trough-bridging (HH) portion of the mechanism is rate-limiting, and the resulting growth rate is found to be very close to that predicted by the original HH mechanism of Harris [1990].

The simulation results shown in Figure 9 in the previous section were obtained using the HH mechanism (in either the original or modified form). These results show that this mechanism is able to reproduce the kinetics of diamond growth over a wide range. To some extent this good absolute agreement must be fortuitous, since the uncertainties in reaction rates and thermochemistry are substantial, and the experimental films are polycrystalline. Perhaps most significantly, it shows that the assumption of CH_3 as the sole growth species can account for the widely varying growth rates found in different systems.

Neither Garrison et al. [1992] nor Harris and Goodwin [1993] explicitly addressed the question of why (100) surfaces have a smooth morphology. If carbon is able to incorporate everywhere on the surface through these mechanisms, a rough surface would result unless some form of surface migration were operative. Harris and Goodwin [1993] discussed this point, and suggested that the incorporation may occur preferentially at steps, due to lower steric hindrance at these locations.

Zhu et al. [1993] have presented a mechanism for growth from methyl radicals which explicitly addresses the issue of smooth morphology. In this mechanism, preferential addition of methyl groups occurs at the end of dimer rows on a (100)-(2x1) monohydride surface with a type B single step. Their mechanism makes use of the GDSB submechanism to open the dimer on the lower level immediately in front of the propagating dimer row; this is followed by a dimer shift reaction, which allows formation of a methylene bridge at the end of the dimer row. Zhu et al. [1993] used MM3 to calculate the enthalpy of reaction for each step in their mechanism, but did not evaluate free energy changes. The most endothermic step (+6 kcal/mole) was the dimer shift step. No estimates of growth rate were made in this study.

Proposals have also been advanced for incorporation of other C_1 species on the (100) surface. Since the above methyl mechanisms all involve an H abstraction to form CH_2, direct addition of CH_2 could clearly occur through these mechanisms for any environment in which significant CH_2 is present. Also, atomic carbon would be expected to be quite reactive with the surface, and could be expected to add via mechanisms similar to these or by more direct routes [Huang and Frenklach 1992, Yu and Girshick 1994].

Huang and Frenklach [1992] have proposed a mechanism to add CO to (100), but find a large barrier (51 kcal/mole). In any case, there is no experimental evidence for thermal CO contributing to diamond growth. For example, the diamond growth rate in oxyacetylene torches at 1 atm tracks the hydrocarbon concentration at the surface (which increases dramatically for $C_2H_2/O_2 > 1$), not the CO concentration, which is substantial even for $C_2H_2/O_2 < 1$ [Goodwin 1991].

Addition to the (110) surface. On the flat (110) surface, 2 carbons must be added to form a stable structure. For this reason, like (111), it has received less attention theoretically than the (100) surface for methyl addition. The (110) surface never appears in CVD diamond films, indicating it has a high growth rate and therefore grows out. Indeed, the growth rate measured on single-crystal diamond is (slightly) higher for the (110) face than for either the (111) or (100) face. Nevertheless, (110) faces are typically rough [Chu et al. 1991], and therefore a nominally (110) surface may be microfaceted into (111) and (100) regions, and the high growth rate a reflection of a high step density of (100)-like sites.

Some suggestions have been presented for methyl addition to (110). Butler and Woodin [1993] presented a mechanism which begins with sequential addition of 2 methyl radicals to form an ethyl (C_2H_5) species. This species may be removed from the surface by thermal desorption or H-abstraction followed by β-elimination of ethylene. However, an alternative path is abstraction of a neighboring H, allowing formation of a bridging C_2H_4 structure. Once this "growth site" is formed, growth may be continued by adding *single* methyl radicals to this site. The local topology around the growth site is that of the HH site (or a dimer site if the two hydrogens are abstracted). The propagation can therefore occur through either the HH or GDSB mechanism, or variations on these.

4.3.4 C_2 *Growth Mechanisms*

Several mechanisms have been proposed for addition of C_2 species (primarily acetylene) to the diamond surface.

Addition to the (110) surface. The first detailed mechanism for addition of acetylene to diamond was that of Frenklach and Spear [1988]. While this study considered addition at a step on the (111) surface, the local topology is identical to a flat (110) surface. The mechanism begins with an H-abstraction to form a radical site, followed by acetylene addition to form a bridging C_2H_3 surface species. This step may be thought of as nucleating a new layer, similar to the Butler-Woodin methyl mechanism above. Once this structure is formed, subsequent acetylene additions propagate the (110) row. A subsequent paper [Huang et al. 1988] modified this mechanism, suggesting that high energy barriers in the original proposal could be eliminated if certain steps proceed through concerted transition states, in which multiple bonds simultaneously form and break.

Harris and Belton [1991] carried out a thermochemical analysis of this revised mechanism, and argued that the acetylene addition steps are highly reversible, leading to acetylene desorption under typical diamond CVD conditions. They point out that acetylene addition to a single radical site is only slightly exothermic, but the entropy loss is substantial, resulting in a free energy change which is not favorable for the reaction to proceed. Belton and Harris [1992] propose adding acetylene at diradical sites to overcome this difficulty, since this process is highly exothermic.

Recently, Skokov, Weimer and Frenklach [1994] have argued that all of the above acetylene-addition mechanisms are unlikely, due to either to the reversibility problem pointed out by Belton and Harris, or by kinetic instability problems. Due to the low coverage of diradical sites (at best of order 1%), they estimate that the acetylene must have a lifetime on the surface of at least 10 ms, which is much longer than the time required to remove it by β-scission.

Addition to the (100) surface. Skokov, Weimer and Frenklach [1994] argued that there is, however, a viable mechanism to add acetylene to the (100)-(2x1):1H surface. This mechanism, shown in Figure 15, begins by chemisorbing acetylene at a diradical site spanning a trough between dimer rows. H addition then creates the chemisorbed vinyl group discussed above, but now there is no need to wait for a second acetylene to chemisorb nearby. Instead, an H-abstraction followed by a β-scission opens the dimer and allows a bridging group to span it, in a manner very similar to the GDSB mechanism for C_1 species. The activation energies for this process are reported to be low.

Gruen et al. [1995] have proposed that diamond growth may occur from C_2 radicals. While simulations suggest that C_2 will have a very low concentration near the substrate for most conventional diamond CVD reactors, there may be some unconventional systems with little hydrogen present where C_2 concentrations can be significant (for example, a hydrogen-free C_{60}/argon microwave plasma [Gruen et al. 1994])

Gruen et al. [1995] suggested that C_2 insertion into dimer bonds on the (100)-(2x1):1H surface will be energetically favorable with a low barrier, although no calculations are presented to support this. If this process can occur, it will result in a new dimer, one level up and perpendicular to the dimer rows on the surface. Once this has occurred on neighboring dimer rows, additional C_2 can insert into the sites over the troughs to form a complete new (2x1):1H surface. If this process can occur with a reasonable rate, it would be novel in that no H-abstraction reactions (and therefore no hydrogen) would be required, other than for surface stabilization.

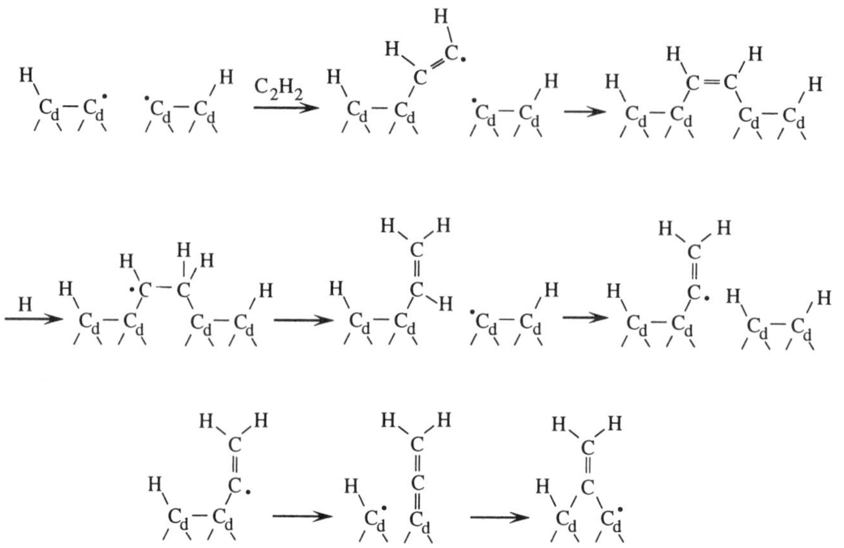

Figure 15. Mechanism of Skokov, Weimer and Frenklach [1994] for addition of acetylene to the (100)-(2x1):1H surface.

5. Diamond Quality and Defect Formation

To date, most models of diamond growth have focused on the formation mechanisms of diamond from gaseous species. Very little discussion has addressed competing processes which lead to the formation of non-diamond carbon and defects. There is clearly a spectrum of qualities in CVD diamond materials and a wealth of evidence for various types of defects. We shall first classify and summarize the various types of defects observed, and then discuss the few growth models for defects which have been proposed.

5.1 INTRINSIC DEFECTS

The most common structural defect observed in a CVD diamond grain is twinning on the (111) plane. Diamond has an ABCABC stacking sequence of planes along the 111 direction, while "hexagonal" diamond, Lonsdalaite, has an AB'AB'AB' sequence, in which B' is related to A by the mirror plane. Thus, high densities of parallel twin planes can give rise to local regions of 'hexagonal' diamond. Twinning on multiple, non-parallel (111) planes can lead to higher order grain boundaries and complicated morphological development of the macroscopic grain. Such multiple twinning is responsible for the near five-fold symmetry frequently observed in many CVD crystals, and for the parallel, re-entrant corners often seen in 110 textured polycrystalline films. Stacking faults and dislocations are also observed within the grain. Multiple nucleation events give rise to complex grain boundaries as the growing grains collide in polycrystalline material. Since carbon materials can exist in several forms of electronic hybridization (sp, sp^2, sp^3), local regions of non-diamond may exist within grains and at grain boundaries, often as non-crystalline or amorphous material. Point defects, such as isolated vacancies and interstitial carbon seem to be rare or non-existent in as-grown material.

5.2 EXTRINSIC DEFECTS

Extrinsic (impurity) defects exist both within individual grains of CVD diamond and at the grain boundaries. Defects incorporated within the grain often show a growth sector dependence in which, for example, the rate of nitrogen or boron incorporation depends on the specific growth face (site), which gives rise to different concentrations of nitrogen (boron) in the (111) vs. (100) growth sectors. Impurity incorporation can alter the growth rate of specific faces (sites) which can lead to variations in crystal morphology, polycrystalline texture, strain, and overall growth rate. The common impurities found in CVD diamond materials are hydrogen, nitrogen, boron, silicon, tungsten, tantalum, and phosphorus. While there is strong evidence for defects at grain boundaries, as of early 1996, there seems to be little detailed knowledge about the nature of these.

5.3 DEFECT GROWTH MODELS

5.3.1 *Competitive Deposition and Etching*

Probably the earliest model of defect growth in diamond is the attempt to rationalize the growth of diamond as opposed to graphite by the competition between etching and deposition of sp^2 vs. sp^3 carbon [Deryagin and Fedoseev 1977]. In this model, carbon is depositied on the surface as both sp^3 (diamond) and sp^2 (graphite/defects) material in competition with the simultaneous etching of these materials by atomic and molecular hydrogen. Since graphite is known to etch much faster in an atomic hydrogen environment, conditions can be achieved for which the deposition rate of graphite/defects is low while still having a measurable rate of diamond deposition. One of the latest variations on this model attributed the energy depositied by the atomic hydrogen flux at the surface as capable of converting a thin amorphous diamondlike carbon phase into the diamond [Singh 1994]. No atomic or molecular level mechanisms are proposed in these models.

5.3.2 *A Generic Picture of Diamond Growth and Defect Formation*

Models of defect formation have not yet been formulated at the level of detail of the diamond growth mechanisms discussed in Section 4. Instead, models which employ a "generic" hydrogen terminated surface have been used, and have worked remarkably well at explaining many of the observed trends [Frenklach and Wang 1991, Butler and Woodin 1993, Goodwin 1993a, Spear and Frenklach 1994, Angus and Evans 1994]. In these models, the surface is viewed as bulk, diamond carbon, C_d, with surface bonds terminated with hydrogen, C_dH, and gaseous atomic hydrogen drives most of the surface chemistry. While these models differ in some significant ways, the qualitative conclusions are all similar. Here we present a summary of the basic features common to most models, following largely the analysis of Butler and Woodin [1993].

Let us consider the incorporation of a generic C_xH_y species into the diamond lattice. This process begins by creation of active surface sites C_d^* by reaction with atomic hydrogen [reactions (38) and (39) discussed in Section 4.1]. The hydrocarbon may then attach at an open site:

$$C_d^* + C_xH_y \rightleftharpoons C_d - C_xH_y. \tag{43}$$

For the purposes of this analysis, the identity of C_xH_y is not critical, and could be either a radical (e.g. CH_3) or an unsaturated molecule (e.g. C_2H_2).

Under steady-state film growth conditions, the surface species coverages may be taken to be constant. Let us assume (as discussed in Section 4.1) that reaction (-38) may be neglected and that the H abstraction rate $k_{38}[C_dH][H]$ is much greater than the unimolecular desorption rate
$k_{-43}[C_d - C_xH_y]$. Then applying steady-state to $[C_d^*]$ results in an expression for the fraction of open sites:

$$\frac{[C_d^*]}{[C_dH]} = \frac{k_{38}[H]}{k_{39}[H] + k_{43}[C_xH_y]}. \tag{44}$$

The concentration of reactive hydrocarbons C_xH_y is usually small enough that $k_{43}[C_xH_y]$ is less than $k_{39}[H]$. Thus, equation 44 reduces to a form equivalent to equation 40 (only differing in notation):

$$\frac{[C_d^*]}{[C_dH]} = \frac{k_{38}}{k_{39}}. \tag{45}$$

Thus, for the small reactive hydrocarbon concentrations used in diamond CVD, the fraction of open sites is the same as found in pure H/H_2, and depends only on temperature.

Upon addition of a hydrocarbon radical or molecule to the surface, the newly added carbon has one bond to the lattice. Once added, it may unimolecularly desorb from the lattice through the the reverse of the hydrocarbon addition, reaction (-43). Since this requires enough energy to break a C-C bond, unimolecular desorption typically has a large activation energy and becomes important at high temperature. For typical hot-filament diamond CVD conditions, the desorption rate becomes significant above about 1200 K.

As discussed earlier, another way surface species may removed from the surface (either completely or as a fragment) is through attack by atomic hydrogen, via reaction sequences involving H abstraction and β-scission:

$$C_d - C_xH_y + H \Rightarrow C_d^* + C_xH_{y-1} + H_2. \tag{46}$$

As written, this reaction assumes all x carbons are removed, but actually the number detached will depend on which hydrogen atom on the surface species or on the diamond carbon is abstracted.

Clearly, addition of C_xH_y by itself is not sufficient to grow the diamond lattice. First, the newly added carbon must be stabilized against desorption and/or removal by β-scission. This will occur when the carbon atom attached to the diamond carbon makes a *second* bond to the lattice (stabilizing at least this carbon atom). To be fully incorporated, at least two bonds to neighboring carbon atoms must be made after the initial attachment bond, assuming the fourth bond will be to the next layer to be deposited. Since the hydrocarbons tend to be hydrogenated, incorporation must involve removal of hydrogen. Once again, atomic hydrogen is the primary reagent for this process, analogous to reaction (38). We shall define *growth* of the lattice as the formation of at least two bonds between a single carbon atom of the hydrocarbon adsorbate and the diamond lattice, as opposed to the process we have considered thus far, namely the *addition* of a carbon species to the surface.

Let us consider the process of bond formation between a chemisorbed hydrocarbon species and an adjacent hydrogen terminated surface site.

$$CH \cdots HC_d \rightleftharpoons C^* \cdots HC_d + H_2, \tag{47}$$

$$C^* \cdots HC_d + H \rightarrow CH \cdots HC_d, \tag{48}$$

$$C^* \cdots HC_d + H \rightarrow C^* \cdots C_d^*, \tag{49}$$

$$C^* \cdots C_d^* \rightarrow C - C_d. \tag{50}$$

Incorporation of carbon into the diamond lattice is represented by the sum of equations 49 and 50, and shows that the rate of lattice incorporation depends directly on the atomic hydrogen concentration:

$$C^* \cdots HC_d + H \Rightarrow C - C_d + H_2. \tag{51}$$

Competing with incorporation, intermediate surface species may unimolecularly decompose (possibly with desorption of a hydrocarbon) to form sp^2 structures:

$$C^* \cdots HC_d \Rightarrow C_{defect} \cdots HC_d. \tag{52}$$

Unimolecular decomposition of these species to create sp^2 lattice bonds will form a type of defect with the characteristic sp^2 Raman spectrum observed at elevated temperatures. Once formed, sp^2 bonds on the surface may be converted back into sp^3 carbon by insertion of atomic hydrogen into the bonds:

$$C_{defect} \cdots C_d + H \Rightarrow C^* \cdots HC_d. \tag{53}$$

Growth as we have defined it includes both the growth of the sp^3 diamond lattice as well as the potential for the incorporation of sp^2 defects. The sp^2 fraction will depend on the balance between sp^3 incorporation, unimolecular decomposition, and hydrogenation of sp^2 bonds.

Defects with sp^3 character may also be formed if the C_xH_y flux is high enough that the kinetics of bond formation to adjacent carbons (dependent on atomic hydrogen) cannot keep up with the rate at which additional layers of carbon deposit. In this case, partially hydrogen-terminated carbon, not yet bonded to all four neighbors, will be buried and represent a defect. The sp^3 defects will exhibit a characteristic CH stretching infrared absorption spectrum (2800–3000 cm^{-1}) [Willingham et al. 1991]. A completely missing carbon atom in the lattice will give rise to a similar hydrogen-terminated defect.

5.3.3 *Models for Diamond Quality*

Most of the proposed descriptions of diamond growth and defect formation make use of at least some of the ideas discussed in the previous section. Although in each case the specific assumptions differ – sometimes greatly – the resulting expressions for film quality share many similarities, and are in qualitative accord with experiment.

Butler and Woodin [1993] have presented a model very similar to that described above, and derived an expression for the quality, defined as the fraction of all bonds (sp^3 diamond, plus sp^2 and sp^3 defects) which are sp^3 diamond:

$$\text{quality} = \frac{k_{38}[H]}{k_{38}[H] + k_{43}C_xH_y + k_{defect}}, \tag{54}$$

where k_{defect} is the rate constant for formation of sp^2 defects by unimolecular decomposition. They estimate values for the rate constants in this expression by analogy with gas-phase reactions. In units of moles, cm^3, s, and kcal/mole, the values they use are

$$k_{38} = 7 \times 10^{13} e^{-7.3/RT}, \quad (55)$$
$$k_{43} = 1 \times 10^{13}, \quad (56)$$
$$k_{defect} = 1.12 \times 10^{16} T^{-1} e^{-48/RT}. \quad (57)$$

With these rate constants and typical concentration values, they find the quality factor (equation 54) to be near 1 between 600 K and 1200 K, falling rapidly to zero for higher or lower temperatures.

Frenklach and Wang [1991] have presented a detailed kinetic analysis of diamond growth, with generic surface site types and rate constants estimated from analogous gas-phase reactions. This analysis assumes diamond growth is primarily from acetylene, and that sp^2 carbon is formed due to condensation of aromatics and H-mediated interconversion of sp^2 and sp^3 carbon phases. While no simple expression is given, plots of sp^2 fraction show a minimum sp^2 content near 800 °C.

Angus and Evans [1994] postulated the existence of a graphitic layer on the surface, which is continuously being transformed into diamond by reaction with atomic hydrogen. A simple model of this process results in an expression for the surface concentration of sp^2 material (which is assumed to be incorporated as the film grows):

$$[\mathrm{sp}^2] = \frac{G}{k_d[\mathrm{H}]}\left[1 + \frac{k_g[\mathrm{CH_x}]}{k_h[\mathrm{H}]}\right]. \quad (58)$$

Here G is the diamond growth rate, and k_d, k_g, and k_h, are rate constants for sp^2 to sp^3 conversion, growth of sp^2 surface species, and hydrogenation of sp^2 species, respectively.

Goodwin [1993a] has proposed a generic model in which defects form due to reactions or steric interactions between nearby surface species (for example, a species attaching and blocking a neighboring hydrogen from being able to be removed, burying an sp^3 defect). In its original form, the defect formation rate is assumed second order in surface species concentration C_dA and zero order in H, while sp^3 incorporation is assumed to be first order in both C_dA and H. These assumptions lead to the result that

$$\text{defect fraction} \propto \frac{G}{[\mathrm{H}]^2}. \quad (59)$$

Due to the extreme simplifications present in the model, Goodwin suggests that the exponent 2 in the denominator is uncertain, and should be replaced with n, to be determined from experiment.

Although the starting assumptions differ, all of these models produce similar qualitative conclusions, which are generally in accord with experiment. The film quality is predicted to increase with [H] and decrease with growth rate or growth species concentration, as is found experimentally. The models which account for temperature dependence [Frenklach and Wang 1991, Butler and Woodin 1993] reproduce qualitatively the experimental temperature region for diamond growth.

TABLE IV. Growth rates, estimated H fluxes, and N parameter for several different diamond CVD techniques.

Experiment	Pressure	H flux (cm^{-2})	G (μm/hr)	$\log_{10} N$
Stagnation-point flame [Murayama and Uchida 1992, Meeks et al. 1993]	1 atm	9×10^{20}	40	3.67
Flat Flame [Glumac and Goodwin 1992]	52 Torr	6×10^{19}	1	4.1
Hot Filament [Hsu 1991b]	20 Torr	4×10^{19}	0.5	4.2
Microwave plasma	120 Torr	2×10^{21}	13.5	4.6
RF Torch [Girshick, Li, Yu and Han 1993]	1 atm	8×10^{20}	17	4.0
DC Arcjet [Stalder and Sharpless 1990]	220 Torr	7×10^{21}	60	4.4

5.3.4 *An Empirical Relationship Between G, [H], and Quality*

There is some indirect evidence that a simpler empirical approach provides at least an approximate relationship between quality, growth rate, and atomic hydrogen concentration. It was argued in Section 5.3.2 that sp^3 defects are incorporated due to growth of carbon layers at a rate too high to allow all carbon atoms to be fully bonded to their neighbors before being overgrown. Since atomic hydrogen is required for the process of bonding new carbon atoms to the surrounding ones, the rate of this process will depend on the rate at which H strikes the surface. The ratio of this rate to the rate at which diamond carbon layers grow is likely to be a significant parameter. Let us define a dimensionless parameter N by

$$N = \frac{\text{H collision rate with surface (mol/cm}^2\text{/s)}}{\text{C incorporation rate into lattice (mol/cm}^2\text{/s)}}. \tag{60}$$

The utility of the N parameter is illustrated by the data summarized in Table IV. In this table, experimental growth rates reported in the literature for many different reactors are compared to estimates of the H flux at the surface. For the hot-filament case, the H concentration at the substrate was measured [McMaster et al. 1994]. In all others, it was estimated based on the stated growth conditions. In some cases, these estimates are the results of detailed numerical simulations, while in others (e.g. microwave) a simple power balance on the plasma was used to estimate the H production rate, which balances the loss rate on the substrate.

The results in Table IV show that widely-varying diamond growth methods all require a value of $N \sim 10^4$ in order to grow diamond. Based on the models presented above, it would be expected that the diamond quality would increase for higher N, and degrade for lower N, although at present there is little experimental data to confirm this. In most of the experiments cited in Table IV, diamond with a sharp 1332 cm^{-1} Raman line was grown, but its quality (as measured, for example, by thermal conductivity) was not extensively analyzed. It is reasonable to take the value of $N \approx 10^4$ as the *minimum* N value necessary for diamond growth.

This analysis is also supported by experiments in which diamond was grown by exposing a substrate alternately to a carbon flux and then to an atomic hydrogen flux [Kelly et al. 1992, Olson et al. 1994, Kapoor 1994]. These experiments clearly showed that diamond formation did not require an particular gas-phase hydrocarbon precursor – the action of atomic hydrogen on sputtered carbon on the surface is capable of producing diamond.

While these conditions of course differ from standard CVD conditions, they provide a degree of control not possible with standard methods. Kelly et al. [1992] reported good diamond film growth when the H exposure results in 1700 H collisions per surface carbon atom per exposure; this led to 15% of the carbon being converted to diamond (most of the rest is gasified). This implies an N parameter of $1700/0.15 = 1.1 \times 10^4$ [Kelly 1994]. For conditions which produced a poor film (not all non-diamond carbon is removed), the N parameter is $800/0.12 = 6.6 \times 10^3$. In later experiments, good films were grown with $N = 7.2 \times 10^3$ [Kapoor 1994]. These results are perhaps the strongest evidence that diamond growth requires a threshold H flux per incorporated carbon atom of about 10^4.

The results discussed in this section have shown that simple models can be quite successful in describing *some* of the observed relationships between diamond growth rate, atomic hydrogen, and film quality. However, it is also clear that these models are highly simplified. In general, they do not even address the specific nature of the defect formed. Also, little theoretical attention has been paid to incorporation of extrinsic defects. Clearly, there is much work left to do before a satisfactory picture of CVD diamond film quality and the relationship to growth conditions will be in hand.

5.4 A RELATIONSHIP BETWEEN DEFECTS AND THERMAL CONDUCTIVITY

In the analysis presented above, diamond quality was only defined in the most general of terms. In practice, the appropriate measure of film quality is not the sp^3 fraction but is determined by the properties required for the application. Depending on the intended use of the diamond film, it may be most appropriate to define quality in terms of the mechanical, electrical, optical, or thermal properties of the film, or a combination of these.

The relationship of atomic level defects to macroscopic measures of quality is still only partly understood. Recently, Coltrin and Dandy [1996] have presented a simple empirical model which relates thermal conductivity to point defect concentration ρ_D (cm^{-3}) as measured by electron paramagnetic resonance (EPR) spectroscopy. The relationship they give is

$$\lambda = 9.203 \times 10^6 \frac{\ell_i \rho_D^{-1/3}}{\ell_i + \rho_D^{-1/3}}, \tag{61}$$

where λ is the thermal conductivity in units of W/cm^2/K, and $\ell_i = 4.274 \times 10^{-6}$ cm. The constants in this equation were obtained by fitting to unpublished experimental data of Newton et al. [1994].

6. Summary

In this chapter, we have tried to give an overview of the current understanding of diamond chemical vapor deposition as of 1996. We have come far in the last 10 years. Questions which stirred much debate at diamond CVD technical meetings in the late 80's and early 90's are now essentially resolved. We have a good picture of the gas-phase environment at the substrate and the factors (chemistry and transport) which conspire to create it; we understand the multiple roles of atomic hydrogen; we now have strong evidence for CH_3 as the principal growth species in most systems; and we understand the basic outlines of the diamond growth mechanism, at least for the "classical" CVD methods discussed in this chapter.

However, there is much left to do. The chemical mechanisms of step-flow growth must be explained, and reconciled with the apparent low mobility of chemisorbed surface species. Clarification of the atomic-level nature of both intrinsic and extrinsic defects is sorely needed, along with chemical mechanisms for their formation. Grain boundaries need to be understood in much greater detail. The effects of impurities such as boron and nitrogen on diamond growth have been observed, but not adequately explained. Of course, there is still much which is unknown about diamond nucleation. And finally, entirely new diamond synthesis techniques may be discovered, with new sets of questions. These issues and others will occupy researchers for the next decade and beyond, keeping diamond CVD a fascinating and rich field.

7. References

Angus, J. C., Buck, F. A., Sunkara, M., Groth, T. F., Hayman, C. C. and Gat, R. [1989]. Diamond growth at low pressures, *MRS Bulletin* **14**: 38–47.

Angus, J. C. and Evans, E. A. [1994]. Nucleation and growth processes during the chemical vapor deposition of diamond, *in* C. L. Renschler, D. M. Cox, J. J. Pouch and Y. Achiba (eds), *Novel forms of carbon II*, Mat. Res. Soc., Pittsburgh, PA.

Angus, J. C. and Hayman, C. C. [1988]. Low-pressure, metastable growth of diamond and "diamondlike" phases, *Science* **241**: 913–921.

Angus, J. C., Wang, Y. and Sunkara, M. [1991]. Metastable growth of diamond and diamond-like phases, *Ann. Rev. Mater. Sci.* **21**: 221–48.

Bachmann, P. K., Leers, D. and Lydtin, H. [1991]. Towards a general concept of diamond chemical vapor deposition, *Diamond and Related Materials* **1**: 1–12.

Baldwin, Jr., S. K., Owano, T. G. and Kruger, C. H. [1994]. Growth rate studies of CVD diamond in an RF plasma torch, *Plasma Chem. Plasma Proc.* **14**: 383–406.

Belton, D. N. and Harris, S. J. [1992]. A mechanism for growth on diamond (110) from acetylene, *J. Chem. Phys.* **96**: 2371–2377.

Brenner, D. W., Robertson, D. H., Carty, R. J., Srivastava, D. and Garrison, B. J. [1992]. Combining molecular dynamics and Monte Carlo simulations to model chemical vapor deposition: Application to diamond, *in* J. E. Mark, M. E. Glicksman and S. P. Marsh (eds), *Computational Methods in Materials Science*, Materials Research Society, Pittsburgh, PA, pp. 255–260.

Busmann, H. G., Sprang, H., Hertel, I. V., Zimmermann-Edling, W. and Güntherodt, H. J. [1991]. Scanning tunneling microscopy on chemical vapor deposited diamond films, *Appl. Phys. Lett.* **59**: 295–297.

Butler, J. E. and Celii, F. G. [1989]. Vapor phase diagnostics in CVD diamond deposition, *in* Purdes et al. [1989], pp. 317–329.

Butler, J. E. and Woodin, R. L. [1993]. Thin film diamond growth mechanisms, *Philos. Trans. R. Soc. London* **342**: 209–224.

Cappelli, M. A. and Loh, M. H. [1994]. *In-situ* mass sampling during supersonic arcjet synthesis of diamond, *Diamond and Related Materials* **3**: 417–421.

Cappelli, M. A. and Paul, P. H. [1990]. An investigation of diamond film deposition in a premixed oxyacetylene flame, *J. Appl. Phys.* **67**: 2596–2602.

Celii, F. G. and Butler, J. E. [1989]. Hydrogen atom detection in the filament-assisted diamond deposition environment, *Appl. Phys. Lett.* **54**: 1031–1033.

Celii, F. G. and Butler, J. E. [1991]. Diamond chemical vapor deposition, *Annu. Rev. Phys. Chem.* **42**: 643–84.

Celii, F. G. and Butler, J. E. [1992]. Direct monitoring of CH_3 in a filament-assisted diamond chemical vapor deposition reactor, *J. Appl. Phys.* **71**: 2877–2833.

Celii, F. G., Pehrsson, P. E., t. Wang, H. and Butler, J. E. [1988]. Infrared detection of gaseous species during the filament-assisted growth of diamond, *Appl. Phys. Lett.* **52**: 2043–2045.

Celii, F. G., Thorsheim, H. R., Butler, J. E., Plano, L. S. and Pinneo, J. M. [1990]. Detection of ground-state atomic-hydrogen in a DC-plasma using 3rd-harmonic generation, *J. Appl. Phys.* **68**: 3814–3817.

Chang, C. P., Flamm, D. L., Ibbotson, D. E. and Mucha, J. A. [1988]. Diamond crystal-growth by plasma chemical vapor deposition, *J. Appl. Phys.* **63**: 1744–1748.

Chu, C. J., D'Evelyn, M. P., Hauge, R. H. and Margrave, J. L. [1990]. Mechanism of diamond film growth by hot-filament CVD: Carbon-13 studies, *J. Mater. Res.* **5**: 2405–2413.

Chu, C. J., D'Evelyn, M. P., Hauge, R. H. and Margrave, J. L. [1991]. Mechanism of diamond growth by chemical vapor-deposition on diamond (100), (111), and (110) surfaces: Carbon-13 studies, *J. Appl. Phys.* **70**: 1695–1705.

Chu, C. J., Hauge, R. H., Margrave, J. L. and D'Evelyn, M. P. [1992]. Growth kinetics of (100), (110), and (111) homoepitaxial diamond films, *Appl. Phys. Lett.* **61**: 1393.

Clausing, R. E., Heatherly, L., Horton, L. L., Specht, E. D., Begun, G. M. and Wang, Z. L. [1992]. Textures and morphologies of chemical vapor deposited (CVD) diamond, *Diamond and Related Materials* **1**: 411–415.

Coltrin, M. E. and Dandy, D. S. [1993]. Analysis of diamond growth in subatmospheric DC plasma-gun reactors, *J. Appl. Phys.* **74**: 5803–5820.

Coltrin, M. E. and Dandy, D. S. [1996]. Simplified models of growth, defect formation, and thermal conductivity in diamond chemical vapor deposition, *Technical Report SAND96-0883*, Sandia National Laboratories.

Connell, L. L., Fleming, J. W., Chu, H. N., Vestyck, Jr., D. J., Jensen, E. and Butler, J. E. [1995]. Spatially resolved atomic hydrogen concentrations and molecular hydrogen temperature profiles in the chemical-vapor deposition of diamond, *J. Appl. Phys.* **78**: 3622–3634.

Corat, E. J. and Goodwin, D. G. [1993]. Temperature dependence of species concentrations near the substrate during diamond chemical vapor deposition, *J. Appl. Phys.* **74**: 2021–2029.

Dandy, D. S. and Coltrin, M. E. [1995]. Relationship between diamond growth rate and hydrocarbon injector location in direct-current arcjet reactors, *Appl. Phys. Lett.* **66**: 391–393.

Dawnkaski, E. J., Srivastava, D. and Garrison, B. J. [1995]. Time-dependent Monte-Carlo simulations of radical densities and distributions on the diamond (001)(2x1)/H surface, *Chem. Phys. Lett.* **232**: 524–530.

DebRoy, T., Tankala, K., Yarbrough, W. A. and Messier, R. [1990]. Role of heat transfer and fluid flow in the chemical vapor deposition of diamond, *J. Appl. Phys.* **68**: 2424–2432.

Deryagin, B. V. and Fedoseev, D. V. [1977]. *Growth of Diamond and Graphite from the Gas Phase*, Nauka, Moscow. In Russion. English translation in *Surf. Coatings Technol.* **38**, 131 (1989).

D'Evelyn, M. P., Chu, C. J., Hauge, R. H. and Margrave, J. L. [1992]. Mechanisms of diamond growth by chemical vapor-deposition – C-13 studies, *J. Appl. Phys.* **71**: 1528–1530.

DeVries, R. C. [1987]. Synthesis of diamond under metastable conditions, *Ann. Rev. Mater. Sci.* **17**: 161–187.

Donnelly, V. M. and Pasternack, L. [1979]. Reactions of C_2(a $^3\Pi_u$) with CH_4, C_2H_2, C_2H_4, C_2H_6, and O_2 using multiphoton UV excimer laser photolysis, *Chem. Phys.* **39**: 427–432.

Evans, E. A. and Angus, J. C. [1996]. Microbalance studies of diamond nucleation and growth rates, *Diamond and Related Materials* **5**: 200–205.

Frenklach, M. [1989]. The role of hydrogen in vapor deposition of diamond, *J. Appl. Phys.* **65**: 5142–5149.

Frenklach, M. and Spear, K. E. [1988]. Growth mechanism of vapor-deposited diamond, *J. Mater. Res.* **3**: 133–140.

Frenklach, M. and Wang, H. [1991]. Detailed surface and gas-phase chemical kinetics of diamond deposition, *Phys. Rev. B* **43**: 1520–1544.

Garrison, B. J., Dawnkaski, E. J., Srivastava, D. and Brenner, D. W. [1992]. Molecular-dynamics simulations of dimer opening on a diamond (001)(2x1) surface, *Science* **255**: 835–838.

Gat, R. and Angus, J. C. [1993]. Hydrogen atom recombination on tungsten and diamond in hot-filament assisted deposition of diamond, *J. Appl. Phys.* **74**: 5981–5989.

Geis, M. W. [1990]. Growth of device-quality homoepitaxial diamond thin films, *in* J. T. Glass, R. Messier and N. Fujimori (eds), *Diamond, Silicon Carbide and Related Wide Bandgap Semiconductors*, Materials Research Society, pp. 15–22.

Girshick, S. L., Li, C., Yu, B. W. and Han, H. [1993]. Fluid boundary layer effects in atmospheric-pressure plasma diamond film deposition, *Plasma Chem. Plasma Proc.* **13**: 169–187.

Girshick, S. L., Yu, B. W., Li, C. and Han, H. [1993]. Diamond deposition by atmospheric-pressure induction plasma: Effects of impinging jet fluid mechanics on film formation, *Diamond and Related Materials* **2**: 1090–1095.

Glumac, N. G. and Goodwin, D. G. [1992]. Diamond synthesis in a low-pressure flat flame, *Thin Solid Films* **212**: 122–126.

Goodwin, D. G. [1991]. Simulations of high-rate diamond synthesis: Methyl as growth species, *Appl. Phys. Lett.* **59**: 277–279.

Goodwin, D. G. [1993a]. Scaling laws for diamond chemical vapor deposition. I: Diamond surface chemistry, *J. Appl. Phys.* **74**: 6888–6894.

Goodwin, D. G. [1993b]. Scaling laws for diamond chemical vapor deposition. II: Atomic hydrogen transport, *J. Appl. Phys.* **74**: 6895–6906.

Goodwin, D. G. and Gavillet, G. G. [1990]. Numerical modeling of the filament-assisted diamond growth environment, *J. Appl. Phys.* **68**: 6393–6400.

Goodwin, D. G. and Gavillet, G. G. [1991]. Numerical modeling of filament-assisted diamond growth, *in* Messier et al. [1991], pp. 335–340.

Green, D. S., Owano, T. G., Williams, S., Goodwin, D. G., Zare, R. N. and Kruger, C. H. [1993]. Boundary layer profiles in plasma chemical vapor deposition, *Science* **259**: 1726–1728.

Gruen, D. M., Liu, S., Krauss, A. R., Luo, J. and Pan, X. [1994]. Fullerenes as precursors for diamond film growth without hydrogen or oxygen additions, *Appl. Phys. Lett.* **64**: 1502.

Gruen, D. M., Zuiker, C. D., Krauss, A. R. and Pan, X. [1995]. Carbon dimer, C_2, as a growth species for diamond films from methane/hydrogen/argon microwave plasmas, *J. Vac. Sci. Technol. A* **13**: 1628–1632.

Hamza, A. V., Kubiak, G. D. and Stulen, R. H. [1990]. Hydrogen chemisorption and the structure of the diamond C(100)-(2x1) surface, *Surf. Sci.* **237**: 35–52.

Harris, S. J. [1990]. Mechanism for diamond growth from methyl radicals, *Appl. Phys. Lett.* **56**: 2298–2300.

Harris, S. J. and Belton, D. N. [1991]. Thermochemical kinetics of a proposed mechanism for diamond growth from acetylene, *Jpn. J. Appl. Phys.* **30**: 2615–2618.

Harris, S. J., Belton, D. N., Weiner, A. M. and Schmieg, S. J. [1989]. Diamond formation on platinum, *J. Appl. Phys.* **66**: 5353–5359.

Harris, S. J. and Goodwin, D. G. [1993]. Growth on the reconstructed diamond (100) surface, *J. Phys. Chem.* **97**: 23–28.

Harris, S. J. and Martin, L. R. [1990]. Methyl versus acetylene as diamond growth species, *J. Mater. Res.* **5**: 2313–2319.

Harris, S. J. and Weiner, A. M. [1989]. Effects of oxygen on diamond growth, *Appl. Phys. Lett.* **55**: 2179–2181.

Harris, S. J. and Weiner, A. M. [1990]. Methyl radical and H atom concentrations during diamond growth, *J. Appl. Phys.* **67**: 6520–6526.

Harris, S. J. and Weiner, A. M. [1991]. Filament-assisted diamond growth kinetics, *J. Appl. Phys.* **70**: 1385–1391.

Harris, S. J. and Weiner, A. M. [1992]. Diamond growth rates vs. acetylene concentrations, *Thin Solid Films* **212**: 201–205.

Harris, S. J. and Weiner, A. M. [1993]. Reaction kinetics on diamond: Measurement of H atom destruction rates, *J. Appl. Phys.* **74**: 1022–1026.

Harris, S. J., Weiner, A. M. and Perry, T. A. [1988]. Measurement of stable species present during filament-assisted diamond growth, *Appl. Phys. Lett.* **53**: 1605–1607.

Horner, D. A., Curtiss, L. A. and Gruen, D. M. [1995]. A theoretical study of the energetics of insertion of dicarbon (C_2) and vinylidene into methane C-H bonds, *Chem. Phys. Lett.* **233**: 243–248.

Hsu, W. L. [1991a]. Mole fractions of H, CH_3 and other species during filament-assisted diamond growth, *Appl. Phys. Lett.* **59**: 1427–1429.

Hsu, W. L. [1991b]. Quantitative analysis of the gaseous composition during filament-assisted diamond growth, *in* A. J. Purdes, K. E. Spear, B. S. Meyerson, M. Yoder, R. Davis and J. C. Angus (eds), *Proc. 2nd Intl. Symp. Diamond and Related Materials*, The Electrochemical Society, Pennington, NJ, pp. 217–223.

Hsu, W. L. [1992]. Gas-phase kinetics during microwave plasma-assisted diamond deposition: Is the hydrocarbon product distribution dictated by neutral-neutral interactions?, *J. Appl. Phys.* **72**: 3102–3109.

Huang, D. and Frenklach, M. [1992]. Energetics of surface reactions on (100) diamond plane, *J. Phys. Chem.* **96**: 1868–1875.

Huang, D., Frenklach, M. and Maroncelli, M. [1988]. Energetics of acetylene-addition mechanism of diamond growth, *J. Phys. Chem.* **92**: 6379–6381.

Janssen, G., van Enckevort, W. J. P. and Giling, L. J. [1989]. CVD growth of diamond: The multiple role of atomic hydrogen, *in* Purdes et al. [1989], pp. 508–523.

Johnson, C. E., Weimer, W. A. and Cerio, F. M. [1992]. Efficiency of methane and acetylene in forming diamond by microwave plasma assisted chemical vapor deposition, *J. Mater. Res.* **7**: 1427.

Kapoor, S. [1994]. *Growth of diamond films by cyclic exposure to different chemistries*, PhD thesis, Stanford University.

Kapoor, S., Kelly, M. A. and Hagstrom, S. B. [1995]. Synthesis of diamond films using sequential chemistry: Enhanced growth rate by atomic oxygen, *J. Appl. Phys.* **77**: 6267–6272.

Kawato, T. and Kondo, K. [1987]. Effects of oxygen on CVD diamond synthesis, *Jpn. J. Appl. Phys.* **26**: 1429–1432.

Kays, W. M. and Crawford, M. E. [1980]. *Convective Heat and Mass Transfer*, 2nd edn, McGraw-Hill, New York.

Kelly, M. A. [1994]. personal communication.

Kelly, M. A., Olson, D. S., Kapoor, S. and Hagstrom, S. B. [1992]. Diamond growth by a new method based upon sequential exposure to atomic carbon and hydrogen, *Appl. Phys. Lett.* **60**: 2502–2504.

Kim, J. S. and Cappelli, M. A. [1992]. A model of diamond growth in low pressure premixed flames, *J. Appl. Phys.* **72**: 5461–5466.

Koleske, D. D., Gates, S. M., Thoms, B. D., Russell, Jr., J. N. and Butler, J. E. [1995]. Hydrogen on polycrystalline diamond films: Studies of isothermal desorption and atomic deuterium abstraction, *J. Chem. Phys.* **102**: 992–1002.

Komaki, K., Yanagisawa, M. and Yamamoto, I. [1993]. Synthesis of diamond in combustion flame under low pressures, *Jpn. J. Appl. Phys.* **32**: 1814–1817.

Kondoh, E., Ohta, T., Mitomo, T. and Ohtsuka, K. [1991]. Determination of activation energies for diamond growth by an advanced hot filament chemical vapor deposition method, *Appl. Phys. Lett.* **59**: 488–490.

Kondoh, E., Ohta, T., Mitomo, T. and Ohtsuka, K. [1993]. Surface reaction kinetics of gas-phase diamond growth, *J. Appl. Phys.* **73**: 3041–3046.

Krasnoperov, L. N., Kalinovski, I. J., Chu, H. N. and Gutman, D. [1993]. Heterogeneous reactions of H-atoms and CH_3 radicals with a diamond surface in the 300 – 1133 k temperature-range, *J. Phys. Chem.* **97**(45): 11787–11796.

Lee, S. S., Minsek, D. W., Vestyck, D. J. and Chen, P. [1994]. Growth of diamond from atomic hydrogen and a supersonic free jet of methyl radicals, *Science* **263**: 1596–1598.

Liou, Y., Inspektor, A., Weimer, R., Knight, D. and Messier, R. [1990]. The effect of oxygen in diamond deposition by microwave plasma enhanced chemical vapor-deposition, *J. Mater. Res.* **5**: 2305–2312.

Liu, H. and Dandy, D. S. [1995]. *Diamond Chemical Vapor Deposition: Nucleation and Early Growth Stages*, Noyes Publications, Park Ridge, New Jersey.

Loh, M. H. and Cappelli, M. A. [1993]. Study of precursor transport during diamond synthesis in a supersonic flow, *in* Purdes et al. [1993].

Loh, M. H. and Cappelli, M. A. [1997]. CH_3 detection in a low-density supersonic arcjet plasma during diamond synthesis, *Appl. Phys. Lett.* **70**: 1052–1054.

Maeda, H., Ohtsubo, K., Irie, M., Ohya, N., Kusakabe, K. and Morooka, S. [1995]. Determination of diamond [100] and [111] growth rate and formation of highly oriented diamond film by microwave plasma-assisted chemical vapor deposition, *J. Mater. Res.* **10**: 3115–3123.

Martin, L. R. [1993]. High quality diamonds from an acetylene mechanism, *J. Mater. Sci. Let.* **12**: 246–248.

Martin, L. R. and Harris, S. J. [1991]. Does benzene inhibit diamond film growth?, *Appl. Phys. Lett.* **59**: 1911–1913.

Martin, L. R. and Hill, M. W. [1990]. A flowtube study of diamond film growth: Methane versus acetylene, *J. Mater. Sci. Lett.* **9**: 621.

Matsui, Y., Yuuki, A., Sahara, M. and Hirose, Y. [1989]. Flame structure and diamond growth mechanism of acetylene torch, *Jpn. J. Appl. Phys.* **28**: 1718.

McGonigal, M., Russell, J. N., Pehrsson, P. E., Maguire, H. G. and Butler, J. E. [1995]. Multiple internal reflection infrared spectroscopy of hydrogen adsorbed on diamond (110), *J. Appl. Phys.* **77**: 4049–4053.

McMaster, M. C., Hsu, W. L., Coltrin, M. E. and Dandy, D. S. [1994]. Experimental measurements and numerical simulations of the gas composition in a hot-filament-assisted diamond chemical-vapor-deposition reactor, *J. Appl. Phys.* **76**: 7567–7577.

McMaster, M. C., Hsu, W. L., Coltrin, M. E., Dandy, D. S. and Fox, C. [1995]. Dependence of the gas composition in a microwave plasma-assisted diamond chemical vapor deposition reactor on the inlet carbon source, *Diamond and Related Materials* **4**: 1000–1008.

Meeks, E., Kee, R. J., Dandy, D. S. and Coltrin, M. E. [1993]. Computational simulation of diamond chemical vapor deposition in premixed $C_2H_2/O_2/H_2$ and CH_4/O_2 strained flames, *Combustion and Flame* **92**: 144–160.

Mehandru, S. P. and Anderson, A. B. [1991]. Adsorption of H, CH_3, CH_2, and C_2H_2 on 2×1 restructured diamond (100), *Surf. Sci.* **248**: 369–381.

Menningen, K. L., Childs, M. A., Chevako, P., Toyoda, H., Anderson, L. W. and Lawler, J. E. [1993]. Methyl radical production in a hot filament CVD system, *Chem. Phys. Lett.* **204**: 573–577.

Menningen, K. L., Childs, M. A., Toyoda, H., Ueda, Y., Anderson, L. W. and Lawler, J. E. [1995]. CH_3 and CH densities in a diamond growth dc discharge, *Contrib. Plasma Phys.* **35**: 359–373.

Messier, R., Glass, J. T., Butler, J. E. and Roy, R. (eds) [1991]. *Proc. 2nd Intl. Conf. New Diamond Sci. Tech.*, Materials Research Society, Pittsburgh, PA.

Mitsuda, Y., Yoshida, T. and Akashi, K. [1989]. Development of a new microwave plasma torch and its application to diamond synthesis, *Rev. Sci. Inst.* **60**: 249–252.

Mucha, J. A., Flamm, D. L. and Ibbotson, D. E. [1989]. On the role of oxygen and hydrogen in diamond-forming discharges, *J. Appl. Phys.* **65**: 3448–3452.

Murayama, M. and Uchida, K. [1992]. Synthesis of uniform diamond films by flat flame combustion of acetylene / hydrogen / oxygen mixtures, *Combustion and Flame* **91**: 239–245.

Musgrave, C. B., Harris, S. J. and III, W. A. G. [1995]. The surface-radical-surface-olefin recombination step for CVD growth of diamond. calculation of rate constant from first principles., *Chem. Phys. Lett.* **247**: 359–365.

Nemanich, R. J. [1991]. Growth and characterization of diamond thin films, *Annu. Rev. Mater. Sci.* **21**: 535–558.

Newton, M. E., Cox, A. and Baker, J. M. [1994]. unpublished.

Okada, K., Komatsu, S., Matsumoto, S. and Moriyoshi, Y. [1991]. Growth steps and etch pits appearing on {100} planes of diamond prepared by combustion-flame deposition method, *J. Cryst. Growth* **108**: 416–420.

Olson, D. S., Kelly, M. A., Kapoor, S. and Hagstrom, S. B. [1994]. A mechanism of CVD diamond film growth deduced from the sequential deposition from sputtered carbon and atomic hydrogen, *J. Mater. Res.* **9**: 1546–1551.

Owano, T. G., Goodwin, D. G., Kruger, C. H. and Cappelli, M. A. [1991]. Diamond synthesis in a 50 kW inductively coupled atmospheric pressure plasma torch, *in* Messier et al. [1991], pp. 497–502.

Owano, T. G. and Kruger, C. H. [1993]. Parametric study of atmospheric-pressure diamond synthesis with an inductively coupled plasma torch, *Plasma Chem. Plasma Proc.* **13**: 433–446.

Pehrsson, P. E., Celii, F. G. and Butler, J. E. [1993]. Chemical mechanisms of diamond CVD, *in* R. Davis (ed.), *Diamond Films and Coatings*, Noyes, Park Ridge, NJ, chapter 3.

Pitts, W. M., Pasternack, L. and McDonald, J. R. [1982]. Temperature dependence of the $C_2(X\ ^1\Sigma_g^+)$ reaction with H_2 and CH_4 and $C_2(X\ ^1\Sigma_g^+$ and a $^3\Pi_u$ equilibrated states) with O_2, *Chem. Phys.* **68**: 417–422.

Proudfit, J. A. and Cappelli, M. A. [1993]. Atomic hydrogen recombination on CVD diamond, *in* Purdes et al. [1993].

Purdes, A. J., Spear, K. E., Meyerson, B. S., Ravi, K. V., Moustakis, T. D. and Yoder, M. (eds) [1989]. *Proc. 1st Intl. Symp. Diamond and Related Materials*, The Electrochemical Society, Pennington, NJ.

Purdes, A. J., Spear, K. E., Meyerson, B. S., Yoder, M., Davis, R. and Angus, J. C. (eds) [1993]. *Proc. 3rd Intl. Symp. Diamond Materials*, The Electrochemical Society, Pennington, NJ.

Raiche, G. A. and Jeffries, J. B. [1993]. Laser-induced fluorescence temperature measurements in a dc arcjet used for diamond deposition, *Appl. Optics* **32**: 4629–4635.

Rau, H. and Picht, F. [1992]. Rate limitation in low pressure diamond growth, *J. Mater. Res.* **7**: 934–939.

Ravi, K. V. and Joshi, A. [1991]. Evidence for ledge growth and lateral epitaxy of diamond single-crystals synthesized by the combustion flame technique, *Appl. Phys. Lett.* **58**: 246–248.

Reeve, S. W. and Weimer, W. A. [1994]. Plasma diagnostics of a direct-current arcjet diamond reactor. 1. electrostatic-probe analysis, *J. Vac. Sci. Technol. A* **12**: 3131–3136.

Reeve, S. W. and Weimer, W. A. [1995]. Plasma diagnostics of a direct-current arcjet diamond reactor. 2. optical-emission spectroscopy, *J. Vac. Sci. Technol. A* **13**: 359–367.

Rye, R. R. [1977]. Reaction of thermal atomic hydrogen with carbon, *Surf. Sci.* **69**: 653–667.

Schäfer, L., Klages, C. P., Meier, U. and Kohse-Höinghaus, K. [1991]. Atomic hydrogen concentration profiles at filaments used for chemical vapor deposition of diamond, *Appl. Phys. Lett.* **58**: 571–573.

Singh, J. [1994]. Nucleation and growth mechanism of diamond during hot-filament chemical vapour deposition, *J. Mat. Sci.* **29**: 2761–2766.

Skokov, S., Carmer, C. S., Weiner, B. and Frenklach, M. [1994]. Reconstruction of (100) diamond surfaces using molecular mechanics with combined quantum and empirical forces, *Phys. Rev. B* **49**: 5662–5671.

Skokov, S., Weimer, B. and Frenklach, M. [1994]. Elementary reaction mechanism of diamond growth from acetylene, *J. Phys. Chem.* **98**: 8–11.

Snail, K. A. and Marks, C. M. [1992]. *In situ* diamond growth rate measurement using emission interferometry, *Appl. Phys. Lett.* **60**: 3135–3137.

Sommer, M. and Smith, F. W. [1990]. Activity of tungsten and rhenium filaments in CH_4/H_2 and C_2H_2/H_2 mixtures: Importance for diamond CVD, *J. Mater. Res.* **5**: 2433–2440.

Spear, K. E. [1989]. Diamond - ceramic coating of the future, *J. Am. Ceram. Soc.* **72**: 171–91.

Spear, K. E. and Frenklach, M. [1994]. High-temperature chemistry of CVD (chemical-vapor-deposition) diamond growth, *Pure Appl. Chem.* **66**: 1773–1782.

Spitsyn, B. V., Bouilov, L. L. and Derjaguin, B. V. [1988]. Diamond and diamond-like films: Deposition from the vapour phase, structure and properties, *Prog. Crystal Growth and Charact.* **17**: 79–170.

Spitysn, B. V., Bouilov, L. L. and Derjaguin, B. V. [1981]. *J. Cryst. Growth* **52**: 219.

Stalder, K. R. and Sharpless, R. L. [1990]. Plasma properties of a hydrocarbon arcjet used in the plasma deposition of diamond thin films, *J. Appl. Phys.* **68**: 6187.

Struck, L. M. and D'Evelyn, M. P. [1993]. Interaction of hydrogen and water with diamond (100): infrared spectroscopy, *J. Vac. Sci. Technol. A* pp. 1992–1997.

Sutcu, L. F., Chu, C. J., Thompson, M. S., Hauge, R. H., Margrave, J. L. and D'Evelyn, M. P. [1992]. Atomic force microscopy of (100), (110), and (111) homoepitaxial diamond films, *J. Appl. Phys.* **71**: 5930–5940.

Sutcu, L. F., Thomson, M. S., Chu, C. J., Hauge, R. H., Margrave, J. L. and D'Evelyn, M. P. [1992]. Nanometer-scale morphology of homoepitaxial diamond films by atomic force microscopy, *Appl. Phys. Lett.* **60**: 1685–1687.

Thoms, B. D. and Butler, J. E. [1995]. HREELS and LEED of H/C(100) - the 2x1 monohydride dimer row reconstruction, *Surf. Sci.* **328**: 291–301.

Thoms, B. D., Russell, Jr., J. N., Pehrsson, P. E. and Butler, J. E. [1994]. Adsorption and abstraction of hydrogen on polycrystalline diamond, *J. Chem. Phys.* **100**: 8425–8431.

Tsuda, M., Nakajima, M. and Oikawa, S. [1986]. Epitaxial growth mechanism of diamond crystal in CH_4-H_2 plasma, *J. Am. Chem. Soc.* **108**: 5780–5783.

Tsuno, T., Imai, T., Nishibayashi, Y., Hamada, K. and Fujimori, N. [1991]. Epitaxially grown diamond (001) 2x1/1x2 surface investigated by scanning tunneling microscopy in air, *Jpn. J. Appl. Phys.* **30**: 1063–1066.

Valone, S. M. [1989]. Barriers to the nucleation of methyl groups on the diamond (111) surface, *in* I. A. Aksay, G. L. McVay and D. R. Ulrich (eds), *Processing Science of Advanced Ceramics*, Materials Research Society, pp. 227–234.

Wahl, E. H., Owano, T. G., Kruger, C. H., Zalicki, P., Ma, Y. and Zare, R. N. [1996]. Measurement of absolute CH_3 concentration in a hot-filament reactor using cavity ring-down spectroscopy, *Diamond and Related Materials* **5**: 373–377.

Wang, Y. and Angus, J. C. [1993]. Microbalance studies of the kinetics of diamond growth, *in* Purdes et al. [1993], pp. 249–255.

Wang, Z. L., Bentley, J., Clausing, R. E., Heatherly, L. and Horton, L. L. [1994]. Direct correlation of microtwin distribution with growth face morphology of cvd diamond films by a novel tem technique, *J. Mater. Res.* **9**: 1552–1565.

Weimer, R. A., Thorpe, T. P. and Snail, K. A. [1995]. Growth rate and quality variation of homoepitaxial diamond grown at elevated temperatures, *J. Appl. Phys.* **77**: 641.

Wild, C., Kohl, R., Herres, N., Müller-Sebert, W. and Koidl, P. [1994]. Oriented CVD diamond films: Twin formation, structure, and morphology, *Diamond and Related Materials* **3**: 373–381.

Willingham, C., Hartnett, T., Robinson, C. and Klein, C. [1991]. Polycrystalline diamond for infrared optical applications prepared by the microwave plasma and hot filament chemical vapor deposition techniques, *in* Y. Tzeng, M. Yoshikawa, M. Murakawa and A. Feldman (eds), *Applications of diamond films and related materials*, Elsevier, Amsterdam, pp. 157–162.

Yamaguchi, A., Ihara, M. and Komiyama, H. [1994]. Temperature dependence of growth rate for diamonds grown using a hot filament assisted chemical vapor deposition method at low substrate temperatures, *Appl. Phys. Lett.* **64**: 1306–1308.

Yang, Y. L. and D'Evelyn, M. P. [1992]. Theoretical studies of clean and hydrogenated diamond (100) by molecular mechanics, *J. Vac. Sci. Technol. A* **10**: 978–984.

Yarbrough, W. A. [1992]. Vapor-phase-deposited diamond — problems and potential, *J. Am. Ceram. Soc.* **75**: 3179–200.

Yarbrough, W. A. and Messier, R. [1990]. Current issues and problems in the chemical vapor deposition of diamond, *Science* **247**: 688–696.

Yu, B. W. and Girshick, S. L. [1994]. Atomic carbon vapor as a diamond growth precursor in thermal plasmas, *J. Appl. Phys.* **75**: 3914–3923.

Yu, B. W., Han, H. and Girshick, S. L. [1993]. Chemical vapor deposition of diamond film with an atmospheric pressure plasma: Boundary-layer chemistry, *Proc. 11th Int. Symp. Plasma Chem.*, Vol. 4, p. 1674.

Zalicki, P., Ma, Y., Zare, R. N., Wahl, E. H., Owano, T. G. and Kruger, C. H. [1995]. Measurement of the methyl radical concentration profile in a hot-filament reactor, *Appl. Phys. Lett.* **67**: 144–146.

Zangwill, A. [1988]. *Physics at Surfaces*, Cambridge University Press, Cambridge.

Zhu, M., Hauge, R. H., Margrave, J. L. and D'Evelyn, M. P. [1993]. Mechanism for diamond growth on flat and stepped diamond (100) surfaces, *in* Purdes et al. [1993].

MODELING AND DIAGNOSTICS OF PLASMA REACTORS

Chapter 12
SECTION 1: INTRODUCTION

A. Gicquel
Laboratoire d'Ingénierie des Matériaux et des Hautes Pressions
CNRS-UPR 1311 -Université Paris-Nord
Avenue J. B. Clement 93430, Villetaneuse, France

The chapter "Modeling and Diagnostics of Plasma Reactors" focuses on the description of the physical characteristics of H_2 - CH_4 microwave plasma reactors under conditions typical for diamond deposition. The goal is a better understanding of the phenomena leading to increased control and an optimization of these types of reactors. This chapter is a compilation of sections written by several authors. Some repetitions might be found;

they have been kept since it seems to us that different perspectives often provide an increased understanding of a given phenomenon.

The first two sections are very general and are related to the gross characteristics of a microwave hydrogen plasma under conditions typical for diamond deposition and of the plasma / surface interface. They have been written by Alix Gicquel from the Centre National de la Recherche Scientifique and University of Paris, Nord. A description of the main phenomena responsible for the creation and loss of the different species in a plasma are described in section 2. In section 3, the dynamics of surface reactions leading to energy and mass transfers from the plasma to the surface are analyzed.

Section 4 is written by Matt Gordon from the University of Arkansas. It is focused on the Boltzmann equation for electrons: the formulation and possible simplications are presented before the calculations are made for a methane - hydrogen plasma. This chapter allows us to understand how the electrons exchange energy with others species and how their energy depends on these processes.

Section 5, written by Khaled Hassouni from Centre National de la Recherche Scientifique, Carl Scott from National Aeronautics and Space Administration and Samir Farhat from the University of Paris, Nord, addresses the description of a H_2 plasma in terms of thermochemical non equilibrium plasma flow. A three temperature chemical kinetics model coupled to the transport equations is described, and the boundary conditions are discussed. The main physico-chemical processes are described.

Section 6 is written by Timothy Grotjohn from Michigan State University. The electromagnetic field modeling of a microwave cavity plasma reactor is presented, and a discussion on limiting cases is given. From this analysis, the microwave power density distribution in the plasma is calculated.

Section 7 is devoted to the analysis of a microwave bell jar diamond deposition reactor built at the University of Paris, Nord; two parts are presented. The main contributions to the first part are due to Khaled Hassouni, Carl Scott and Samir Farhat. It describes, step by step, the chemical kinetics model coupled to the diffusive transport equation in one dimension. In this model, the electron energy distribution function is assumed to be Maxwellian and the power density distribution is only simply computed. A discussion on these assumptions is given. Alix Gicquel is gratefully thanked for very helpful discussions. The second part of this section is devoted to the experimental analysis of a plasma interacting with a diamond surface in growth. The main contributions of this are due to Alix Gicquel, Marc Chenevier from the Joseph Fourier University at Grenoble, and Michel Lefebvre from Office National d'Etudes et de Recherches Aérospatiales (ONERA). Spatial analysis giving distributions of gas temperature and of relative H-atom mole fractions are presented. The variations of the gas temperature and of the absolute H-atom mole fractions as a function of the average power density and the methane percentage introduced in the feed gas are also shown. The spectroscopic diagnostics performed in the microwave reactor were not possible without the financial support of DRET (DGA-France) and BRITE EURAM Program (E.U.). The authors are grateful to them. Michel Pealat and P. Bouchardy from ONERA are thanked from their participation in the CARS measurements. Khaled Hassouni and the students who have participated to spectroscopic measurements are deeply thanked: Yves Breton, Jean Christophe Cubertafon, Florence Souverville, Maxime Petiau and A. Tserepi. Also Eric Anger, Christophe Jany and Christophe Vivensang are thanked for their constructive technical help. Nader Sadeghi and Jean Paul Booth from the Laboratoire de Spectrométrie Physique of Grenoble are kindly thanked for the helpful discussions. Samir Farhat has performed the calculations allowing the calibration of the H-atom mole fraction measurements and Khaled Hassouni the calculations of the excitation electron-species collisional reactions rates; they are very gratefully thanked.

Chapter 13

SECTION 2: BASIC PROCESSES IN PLASMAS UNDER CONDITIONS TYPICAL FOR DIAMOND DEPOSITION

A. Gicquel

Laboratoire d'Ingéniérie des Matériaux et des Hautes Pressions
CNRS-UPR 1311- Université Paris-Nord
Avenue J. B. Clément 93430, Villetaneuse France

Beyond the phenomena induced during any deposition in thermal CVD reactors [Gicquel et al. 1991], in plasma enhanced chemical vapor deposition reactors (PECVD), the processes related to the nature of the plasma must be considered. On the contrary to the reactive medium of the hot filament reactors, the plasma is characterized by the presence of charged particles (electrons, ions), electronically excited neutrals and ground electronic state ro-vibrationally excited neutrals which can participate in the chemical processes either in the volume or at the plasma / diamond interface. The plasmas used for diamond deposition belong to the domain of weakly ionized plasmas, where typically, the electrons density is less that 10^{-4} of the total density.

Modeling both the volume reactive plasma and the plasma/surface interface is rather complex due to the number of phenomena involved. The aim of this section is to discuss the main processes of creation and loss of the species and the processes by which they gain or loose energy in volume. A review of the main phenomena is presented. After discussing the behavior of the electrons in a plasma (see also section 4), we present the main processes occurring in volume which are generated through electron-neutral collisions (primary processes) and through neutral-neutral collisions (secondary processes).

The processes occurring at the plasma/surface interface are discussed in section 4. Finally, the transport equations are presented in section 5.

2.1 PLASMA CHARACTERISTICS

In general, low pressure plasmas are typically composed of electrons, ions, molecules in a variety of excited states (electronically, vibrationally and rotationally), radicals and atoms [Gicquel et al 1991] and [Elenbasa 1951]. The electrons' energy and mole fraction range respectively between 1 to 20 eV and 10^{-8} to 10^{-4}. The plasma is characterized by a macroscopic quasi-neutrality. At the steady state, three kinetic temperatures can generally be defined: T_i, T_N, and T_e, corresponding respectively to the ions, neutrals (heavy particles) and the free electron temperatures. In the case where a Boltzmann distribution is assumed for each of the distribution concerned, they are given by the following expression:

$$1/2\, m_j \langle w_j^2 \rangle = 3/2\, kT_j$$

where m_j is the mass of the jth species, $<w_j^2>$ its average kinetic velocity, k the Boltzmann constant and T_j the jth species temperature.

The characteristics of the plasma such as electron density (n_e), electron temperature, difference in temperature between electrons and neutrals (non equilibrium state of the plasma) strongly depend on the pressure, the excitation frequency and the reduced electric field (E/N) (figures 1, 2 and 3) [Pointu 1991] and [Delcroix]. Due to the high mobility of the electrons compared to that of the heavy particles, in low pressure ($P< 10^{-1}$ hPa) and moderate pressure (10^{-1} hPa $<$ P$<$ 100 hPa) plasmas, the electron temperature is much higher than the heavy particle temperature (figure 1). The ions' temperature is equal to the heavy particles' temperature, and even at 1 to 100 hPa, the vibration modes are not necessarily in equilibrium with the rotation-translation modes. The rotational and vibrational "temperatures," defined assuming Boltzmann distributions, are then not necessary equal. As the pressure is increased, the electron temperature becomes closer to that of the heavy particle (4000-4500 K) [Girshick et al. 1993] and [Lu et al. 1992], and the difference between the vibrational and the rotational "temperatures" is decreased. Chemistry takes place in such non equilibrium plasmas [Gicquel et al. 1991].

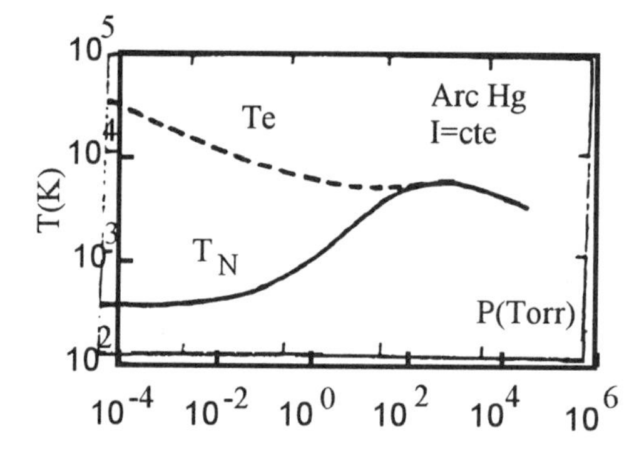

Figure 1: Variation of the electron temperature and the heavy particle temperature in a mercury plasma as a function of the pressure. From references 2 and 3.

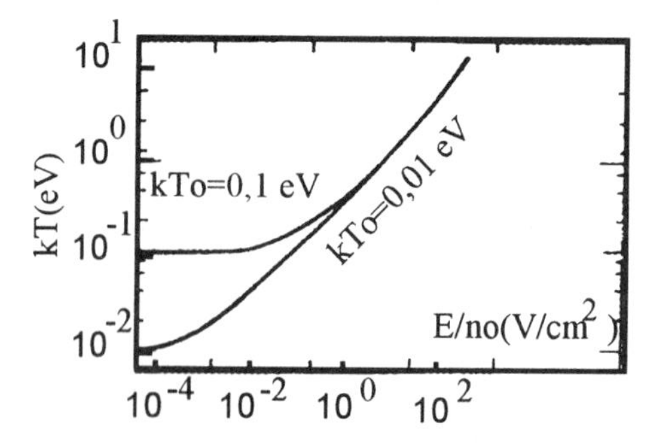

Figure 2: Relation between the electron temperature and the reduced electric field, according to Delcroix[4] and from reference 3.

2.1.1Characteristics

The plasmas considered here are weakly ionized. The electrons possess a thermal energy kT_e, and move within the plasma. As charged particles, they are surrounded by species of the opposite charge which compensate the electrostatic field. The thermal energy has to be compared with the electric energy due to charge separation $((e^2 n_e/\varepsilon_0)\lambda^2)$. At a distance λ_D (the Debye length), there is an equality between the electric energy and the thermal energy, so:

$$\lambda_D \text{ (cm)} = (k\, T_e \text{ (eV)}\, \varepsilon_0 / e^2\, n_e \text{ (cm}^{-3}))^{1/2} = 744\, (T_e/n_e)^{1/2}$$

For distances less than λ_D, the thermal energy (kT_e) is higher than the electric energy, allowing the separation of charges; for distances larger than the Debye length (λ_D), the thermal energy is lower than the electric field, and the neutrality of the gas is observed [Pointu 1991], [Delcroix] and [Arnal]. The Debye sphere defines then, the distance from which the macroscopic effects occur and where the gas neutrality can be considered. Its magnitude depends on the operating conditions (figure 3) [Pointu 1991]. If the dimensions of the reactor are much larger than λ_D, the plasma processes are dominated by macroscopic effects induced by the space charge, and individual behavior is not considered. The quasi-neutrality of the plasma is a rather good assumption. For the typical microwave diamond deposition conditions (P= 50 hPa, T_e = 1.2 eV, n_e = 10^{11} to 10^{12} cm^{-3}) [Scott et al. 1993], λ_D = 8 to 25 µm.

If an external electrical perturbation is applied to the system, the electrons react and move, resulting in a charge separation. Due to the high mobility of the electrons relative to the ions, the charge separation creates an electric field given by Poisson's law: $E = (e\, n_e/\varepsilon_0)\, \Delta x$, where ε_0 is the permitivity and e the electronic charge. The resulting potential is: $V = E\, \Delta x = (e\, n_e/\varepsilon_0)\, \Delta x^2$, the resulting electric energy, $(e^2\, n_e/\varepsilon_0)\, \Delta x^2$, and

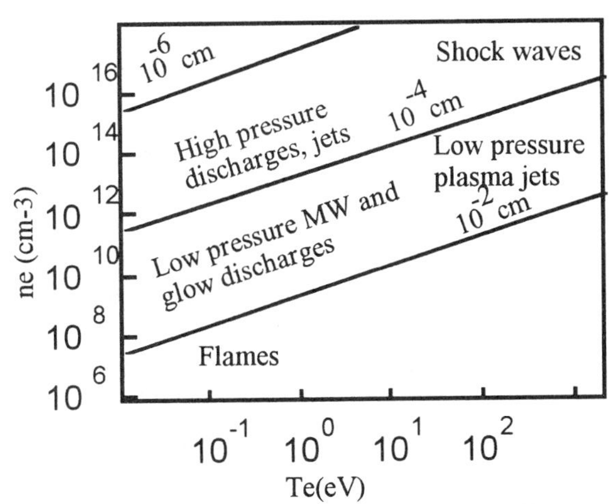

Figure 3: Averages of electron density and electron temperature for different types of discharges for λ_D = cst, from reference 3.

the force which tends to put the system back in equilibrium, $-e\,E = -(e^2\, n_e/\varepsilon_0)\,\Delta x$. In the absence of electron-molecules collisions, the electron motion is governed by [Pointu 1991], [Delcroix] and [Arnal]

$$m_e d^2\Delta x/dt^2 + (e^2\, n_e/\varepsilon_0)\,\Delta x = 0$$

The solution is a sine function with a angular momentum $\omega_{pe} = (e^2\, n_e/\, m_e\varepsilon_0)^{1/2}$. A frequency $f_{pe} = \omega_{pe} / 2\pi$ is defined as the electronic plasma frequency and a simplified expression for it, is $f_{pe} \cong 9000\,(n_e)^{1/2}$, where f_{pe} is in MHz and n_e in cm^{-3}. For typical microwave diamond deposition plasma conditions, P= 50 hPa, $n_e = 10^{11}$ to 10^{12} cm^{-3}, f_{pe}= 2.846 to 9 GHz. For a H_2 spraying R F atmospheric plasma [Girshick et al. 1993], [Lu et al. 1992] and [Fauchais 1994] at T_e = 5000 K and $n_e = 10^{12}$ cm^{-3}, f_{pe}= 9 GHz, at T_e = 10000 K and $n_e = 10^{16}$ cm^{-3}, f_{pe}= 900 GHz.

The maximum of the electron density far from the walls (on the axis of the reactor) should not exceed much more than the one corresponding to the oscillation frequency, that is 10^{11} cm^{-3}, in the case of microwave plasmas operating at 2.45 GHz. After this limit (the cut off frequency), the incident microwave would not efficiently penetrate the plasma [Delcroix] and [Veprek 1989]; the wave frequency should be lowered in order to penetrate the plasma (evanescent wave). For very high electron densities, incident waves may even be reflected back by the plasma. For 2.45 GHz microwave diamond deposition plasmas, 10^{12} cm will be considered as an upper limit for the electron density.

An ion plasma frequency is also defined: $f_{pi} = (e^2\, n_i/\, m_i\varepsilon_0)^{1/2} / 2\pi$. Obviously the ion plasma frequency is much lower than the electronic one. For the microwave diamond deposition reactor, f_{pi} = 33 to 100 MHz, indicating that the ions are not moving under the influence of the excitation frequency.

The elastic collisions between the electrons and the particles, characterized by a collision frequency $<\nu_{eN}>$, greatly slow the electron oscillations. In the case where $\omega_{pe} \cong <\nu_{eN}>$, the macroscopic effects of the perturbation may no longer be observed as a function of time. In the typical bell jar diamond deposition conditions $<\nu_{eN}>$ = 20 GHz $\cong \omega_{pe}$ (12 to 52 GHz) (see later).

2.1.2 Collision frequencies

The transport properties of the plasma species are linked to their interaction with other particles. The collision frequency between a particle i and particles j is defined by the average number of collisions occurring in one second [Arnal] and [Biberman et al.].

$$<\nu_{ij}> = n_j <\sigma_{ij}(w_{ij}) \cdot w_{ij}>$$

where n_j is the density of the jth particles, w_{ij} the relative velocity, and $\sigma_{ij}(w_{ij})$ the collisional cross section of the particle i with the particles j. The mean free path is given by:

$$\lambda_{ij} = w_i / (n_j <\sigma_{ij}(w_{ij}) . w_{ij}>)= w_i / <\nu_{ij}>$$

The cross sections of the collisions are defined by the interaction potentials of the species i and j. At large distances, the interaction potential between an electron and an ion varies as r^{-1} and between an electron and a neutral as a r^{-4}, and varies as r^{-6} for neutral-neutral collisions [Arnal]. For elastic collisions, where there is a deviation of the angle and momentum exchange, the deviation angle can be calculated using the interaction potential. If the species are assumed to rigid (hard) spheres, the interaction occurs only for distances between i and j smaller than the sum of their radii of influence: $r_i + r_j$ [Hirschfelder et al. 1964]. The collision cross section is given by: $\sigma_{ij} = \pi(r_i + r_j)^2$, and has the dimension of an area.

For the electron-neutral collisions, due to the difference of mass, the electron -neutral collision frequency is given by:

$$<\nu_{eN}> = n <w_e> \sigma_{eN} = 6\ 10^3\ n\ (T_e\ (K))^{1/2}\ \sigma_{eN}$$
$$= 4.34\ 10^{26}\ (P(Pa) / T_N(K))\ (T_e(K))^{1/2}\ \sigma_{eN}.$$

For the microwave diamond deposition reactor (P= 50 hPa, T_e = 1.2 eV, T_g = 2200 K) [Scott et al. 1993] and [Gicquel et al. 1994], the mean velocity for the electrons $<w_e> = (8\ kT_e / \pi\ m_e)^{1/2} \cong 7\cdot10^5$ m/s, $<\nu_{eN}> \cong 10^{24}\ \sigma_{eN}$ (in Hz) ≅ 20 GHz[15], and the mean free path for electron-neutral collisions, $\lambda_{eN} = 10^{-19}\ T_N / (P(Torr)\ \sigma_{eN}) \cong 30\ \mu m$. The mean velocity for the molecules $<w_{H_{21}}> \cong 5\cdot10^3$ m/s, $\sigma_{H_2} / H_2{}^{16} \cong 55\cdot10^{-16}\ cm^2$ and the magnitude for the neutral - neutral collisions mean free path is 20 µm.

2.1.3 Microscopic description of the plasma

A plasma is weakly ionized if $<\nu_{ee}> << <\nu_{eN}>^3$. For an electron temperature around 1 eV, this condition is realized for ionization degree ($n_e/(n_e + n_N) \cong n_e/n_N$) lower than 10^{-4}. For the plasmas considered here (microwave bell jar), $<\nu_{ee}> \cong 1.5\ 10^6$ Hz << $<\nu_{eN}> \cong 2.0\cdot10^{10}$ Hz =20 GHz.

In general, a plasma should be studied using the velocity distributions of each species f(r,v,t). For a given species, the number of species in the elementary volume dr^3 having their velocities in the range dv^3, is given by [Pointu 1991] and [Arnal].

$$dn= f(r,w)\ dr^3\ dw^3 \qquad \text{or} \qquad dn / dr^3 = f(r,w)\ dw^3$$

where dn/dr^3 represents the density of the species. The study of the function f as a function of the diffusion processes, the electromagnetic field, the inelastic, elastic and superelastic collisions is realized by solving the Boltzmann equation [Capitelli et al

1977], [Cacciatore et al 1986], [Capitelli et al. 1980] and[Capitelli et al. 1986]. The knowledge of the function f is enough for describing the properties of the plasma. In particular, the electron velocity distribution function characterizes the mean value of any function $a(w_e)$ with the help of the relation [Pointu 1991].

$$\langle a(w_e) \rangle = 1/n_e \int a(w_e) f(r_e, w_e) dw_e^3$$

A special section (section 4) is devoted to the resolution of the Boltzmann equation for electrons in general cases where the electron energy distribution function (EEDF) is non Maxwellian.

As described by C. M. Ferreira, M. Moisan and Z. Zakrzemski [Ferreira et al. 1992], in the presence of a harmonic electric field E exp(iωt), the electron velocity distribution function depends on the electron velocity and on the angle χ between the electron velocity direction and that of the electric field. Using a first order spherical harmonic expansion in the velocity space and a first order Fourier expansion in time, the electron velocity distribution can be defined as: $f(w) = f_0(w) + f_1(w) \cos\chi \, \exp(i\omega t)$, where $f_0(w)$ is the isotropic distribution, related to the average value of the electron energy, and $f_1(w)$ is the anisotropic term which determines the average value of the electron velocity under the influence of the electric field. $f_1(w)$ is related to the collisional momentum dissipation and is a complex quantity.
For a harmonic field:

$$f_1(w) = [e E m_e / (\nu(w) + i\omega)] \, \partial f_0(w)/\partial w$$

where $\nu(w)$ is the electron-neutral collision frequency for the momentum transfer. According to Winkler [Winkler 1992], in hydrogen the lumped momentum dissipation frequency is slightly higher than that of the lumped energy dissipation (figure 4).

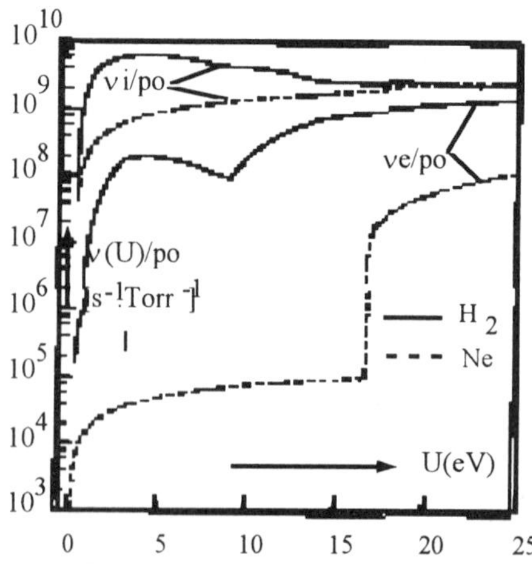

Figure 4: The lumped energy (ν_e/P_0) and momentum (ν_i/P_0) dissipation frequencies in H_2 as a function of the electron energy, after Winkler[18].

At thermodynamic equilibrium, the electron energy distribution function is isotropic, Maxwellian [Pointu 1991] and [Arnal], and is expressed as follows:

$$f_0(w) = f(w_e) = n_e\,(m_e / 2\pi\, kT_e)^{3/2} \exp(-\, m_e\, w_e^2 / 2\, kT_e)$$

The averaged velocity is given by : $<w_e> = (8\ kTe/\pi\ m_e)^{1/2}$. In the case of an isotropic distribution of the velocities, $dn_e/\,dr^3 = 4\pi\, w_e^2\, f(w_e)\, dw_e$, then:

$$dn_e/\,dr^3 = 4\pi\, n_e\, m_e^{1/2}\,(2\pi\, kT_e)^{-3/2}\,(2E/m_e)^{1/2} \exp(-E_k / kTe)\, dE_k,$$

where E_k is the kinetic energy, $dn_e/\,dr^3 = f(E_k)\,(2E_k/m_e)^{1/2}\,dE_k$, and $f(E_k)$ is called electron energy distribution function (EEDF).

For an isotropic distribution, the average of the velocity vector is zero. In a discharge where gradients and electromagnetic field are present, the real distribution of the velocity cannot be completely Maxwellian. In general, the distribution function is not Maxwellian and even the isotropic part f_0 is difficult to calculate (see section 4).

The electron energy distribution function (and therefore the electron temperature in the case of a Maxwellian distribution) might be sensitive to the excitation frequency, the pressure, the metastable and the radiative species concentrations, and the atom fraction. In addition, in the case of the diamond deposition plasmas, it should be sensitive to the fraction of methane introduced and to its degree of dissociation.

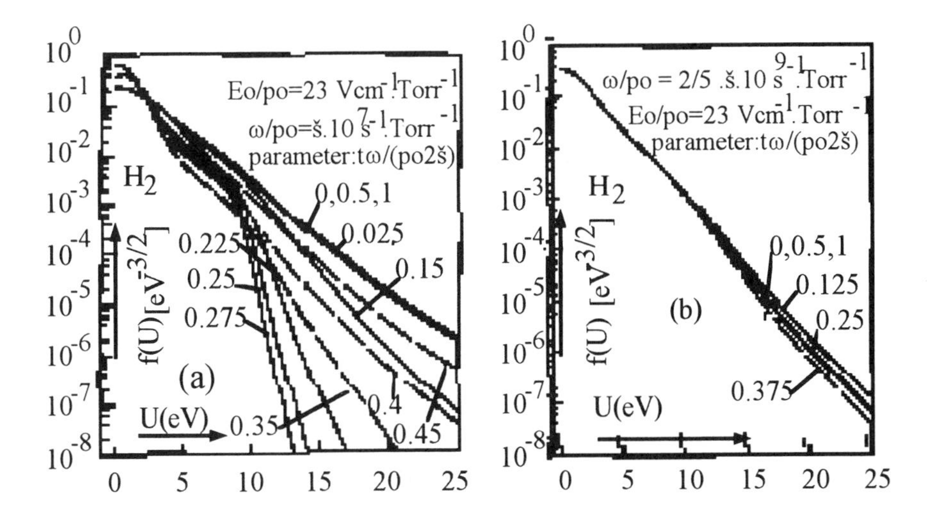

Figure 5: The periodic alteration of the isotropic distribution (a) at ' $10^7\ s^{-1}\ Torr^{-1}$, (b) 2/5 ' $10^9\ s^{-1} Torr^{-1}$, after Winkler[18].

The variations of the isotropic and the anisotropic parts of the EEDF as a function of time (frequency) have been studied by R. Winkler in the case of pure hydrogen [Winkler et al. 1992]. For values of frequencies which are very low when compared to the electron-molecule collisional frequency, both isotropic and anisotropic parts have a large periodic alteration following the alternating field (figure 5a). For values of frequencies close to the collision frequency, the isotropic part does not experience a periodic alteration and can be considered as time independent (figure 5b). In this case, the anisotropic part continues to have a periodic variation, but the higher the frequency, the lower the amplitude of this alteration.

In typical bell jar diamond deposition reactor, the collision frequency is close to the field frequency, and the isotropic part of the EEDF can then be considered as time independent.

2.1.4 Macroscopic properties of a weakly ionized gas in the collisional regime

In a weakly ionized gas, the transport of the charged particles is described independently from the neutral particles. When the dimension of analysis is large compared to the mean free path, the evolution of the density and the species' velocity is governed by the hydrodynamic equations [Pointu 1991].

For electrons, the conservation of the charge number density gives:

$$\partial n_e/\partial t + \nabla \, . \, n_e \, w_e = S$$

where S is the net source production term, and $\nabla \, . \, n_e \, w_e$ represents the gradient of the electrons flux; conservation of the momentum gives:

$$n_e m_e(\partial w_e/\partial t + w_e.\nabla.w_e) = n_e q_e \,(E + w_e \times B) - \nabla n_e k T_e - n_e m_e \, \nu_{eN'} . \, w_e$$

where B and E are the applied magnetic and electric fields. $\nabla \, n_e \, kT_e$ is the force due to pressure gradient, and $n_e \, m_e \, \nu_{eN'}$ is the friction force due to neutrals- an effective collision frequency. In the absence of a magnetic field, and for $\omega^2 >> \nu_{eN}^2$ (microwave conditions) [Ferreira et al. 1992]

$$\nu_{eN'} = \left[-4\pi/3 \int_0^\infty (w_e^3 / \nu_{eN}) \cdot (\partial f_0(w) / \partial w) dw_e \right]^{-1}$$

Conversely, for $\omega^2 << \nu_{eN}^2$ (DC conditions)[17]:

$$\nu_{eN'} = -{}^{4\pi}\!/_{3} \int_0^\infty w_e^3 \nu_{eN} (\partial f_0(w) / \partial w) dw_e$$

From the momentum equation, two coefficients are derived. The electron mobility, μ_e, is defined by $w_e = v_{ed} = \mu_e \, E$ for $B = 0$, (v_{ed} is the drift velocity), which characterizes the effect of the electrostatic field; and the free diffusion coefficient D_e, defined by $n_e w_e = - D_e \nabla \cdot n_e$, which represents the motion of the electrons due to their gradients.

$$\mu_e = (q_e / m_e) \nu_{eN'}$$
$$D_e = (kT_e / m_e) \nu_{eN'}$$

$\sigma_e = -i\, n_e q^2 / [m_e (\omega + i\ \nu_{eN'})]$ is the electron conductivity.

Ion mobility and ion free diffusion coefficients are also defined. Both electron and ion drift velocities are strong functions of the reduced electric field E/N [Pointu 1991].

In cases where the pressure effect and the diffusion of the electrons are negligible, the momentum balance equation applied to a microwave electromagnetic field ($E = E_0 \sin 2\pi\, ft$) is [Veprek 1989]:

$$n_e m_e (\partial w_e / \partial t) + n_e m_e \nu_{eN'} . w_e = n_e q_e E$$

with the solution:

$$w_e(t) = eE_0/m_e\ [(\nu_{eN'} - i2\pi f) / (\nu_{eN'}^2 + (2\pi f)^2)] \exp (i (2\pi f t)$$

For $2\pi f >> \nu_{eN'}$ (collision frequency negligible), the momentum balance equation becomes:

$n_e m_e (\partial w_e / \partial t) = - n_e q_e E$, and $w_e(t) = (e E_0 / 2\pi f m_e) \cos (2\pi f t)$.

There is a $\pi/2$ phase shift between the applied field and the current density:

$$j(t) = - (n_e e^2 E_0 / 2\pi f m_e) \cos (2\pi f t)$$

The averaged power density dissipated in the plasma is given by:

$$2\pi < dP_{abs} / dV >= \int_0^{2\pi} j(t)E(t)dt = 0$$

showing that there is no dissipation of power.

In the case where the friction term is not negligible (which is actually the case for diamond deposition conditions), there is a phase shift between the applied field and the current density less than $\pi/2$. The averaged power density dissipated in the plasma, given by:

$$2\pi < dP_{abs}/dV >= \int_0^{2\pi} j(t)E(t)dt = n_e e^2 E_0^2/2m_e\left[\nu_{eN'}/(\nu_{eN'}^2 + (2\pi f)^2)\right]$$

is positive. Thus, the electron-neutrals collisions causing the randomization of the momentum vector are responsible for the absorption of the applied electric field power [Veprek 1989]. The averaged power density must compensate for the lost power due to ionization, elastic and inelastic collisions, and diffusion processes.

2.1.5 Role of the Walls

At the laboratory scale, the plasma is limited by the reactor walls [Papoular 1963], [Arnal], [Pointu 1991] and [Delcroix]. The charged particles created in the volume can be lost in volume, at the reactor walls or at the growing diamond surface. The losses of the charged particles in volume can be considered as negligible for plasmas with electron density lower than 10^{12} cm^{-3}, and the charged particles are then mostly lost at the walls (figure 6) [Delcroix]. The consumption of charged particles at the surfaces is generally high (see later).

In the vicinity of the walls, there is a loss of the plasma neutrality owing to the formation of a space charge zone called the electrostatic sheath. An isolated piece of solid, immersed into the plasma, initially acquires the fastest charged particles (the electrons) and begins to become negatively charged. The ions are then attracted, and drag the electrons in their motion. After a while, a steady state is reached: the ion flux compensates for the electron flux. The surface is at a floating potential.

The order of magnitude of the difference between the plasma potential and the floating potential is around 10 eV, and the length of the floating potential sheath is some Debye lengths [Pointu 1991] and [Arnal]. In the case of a non collisional sheath, the Bohm theory enables the calculation of the floating potential. For high pressure (P > 10^{-1} hPa), the sheath is collisional, and the diffusion processes of the ions cannot be neglected. For pressure higher than 1 Torr, ionization phenomena occur in the sheath, the result being a decrease of the ion energy and a decrease of the sheath thickness.

Thus, owing to the high mobility of the electrons towards the surface compared with that of the ions, an electric field (the ambipolar field) is formed. It slows down the electrons and accelerates the ions. A drift velocity, v_d, governing the diffusion of both the charged particles (ions and electrons due to the motion of the ions) results. The process is named ambipolar diffusion, and an ambipolar diffusion coefficient D_a is defined. The drift velocity is calculated from the transport equations for the electrons and the ions. For a

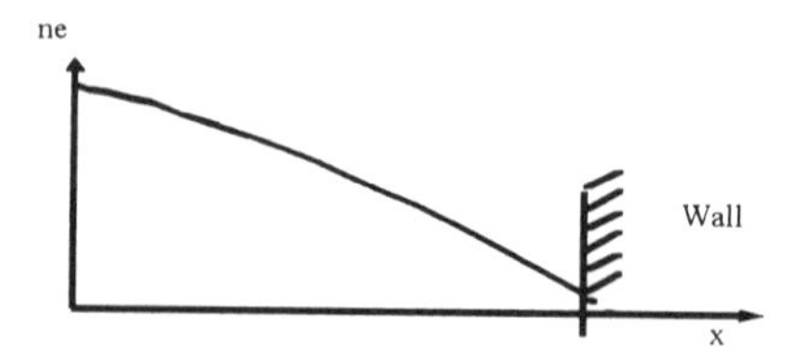

Figure 6: Radial distribution of the electron density at a wall vicinity[4,12].

species j, the diffusion flux $n_j v_d$ is equal to the sum of the diffusion term $D_i \nabla_j$ and of the term $\mu_j n_j E_s$, related to the action of the ambipolar electric field. Two equations are then obtained for the ions and the electrons, and the drift velocity is given by [Pointu 1991], [Delcroix], [Arnal], [Papoular 1963] and [Veprek 1989]:

$$v_d = - (D_e \mu_i + D_i \mu_e) / (\mu_e + \mu_i) \ \nabla n_e/n_e = D_a \nabla n_e/n_e$$

where $\mu_e >> \mu_i$. D_e and D_i are the free diffusion coefficients for the electrons and the ions. For dimensions of the reactor which are long compared to the mean free paths, for conditions where the volume loss is negligible and a Maxwellian distribution is assumed for the electrons energy, D_e and D_i are given by simplified relations:

$$D_e = \lambda_e N (kT_e/m_e)^{1/2} \text{ and } D_i = \lambda_i N (kT_i/m_i)^{1/2}$$

With the general Einstein relation: $D_j = \mu_j kT_j/e$, the drift velocity comes:

$$v_d = - (\mu_i kT_e/e) \ . \ \nabla n_e/n_e$$

The drift velocity is opposite to the electron density gradient and is oriented towards the wall. In the moderate pressure diamond deposition conditions, it is much lower than the electron average velocity. With $D_a = 100$ cm^2/s,

$$v_{ed} \cong 1 \text{ m s}^{-1} << <w_e> = 7\ 10^5 \text{ m s}^{-1}.$$

Although, this simplified approach gives a picture of the main physical processes occurring in a weakly ionized plasma, solutions of the Boltzmann equations in real media are necessary for taking into account the real collisions which are almost all strongly dependent on the electron velocity [Ferreira and Moisan 1992].

2.2 ELECTRON-NEUTRAL COLLISIONS

The electron-molecule collisions and the molecule-molecule (or ions) collisions are characterized by the type of energy transfer [Gicquel et al. 1991]. Elastic, inelastic, reactive or superelastic collisions are different kinds of collisions. These processes lead to the creation of a non equilibrium state of the plasma, with a given chemical composition. In this sub-section, the primary collisions involving collisions between electrons and molecular species, and the secondary ones involving collisions between neutrals and neutrals, or between ions and neutrals are discussed.

The efficiency of a collision between a particle i and a particle j is characterized by its cross section and the energies of the incoming particles. A chemical kinetics model would imply a number of coupled reactive and non-reactive collisions. The accuracy of the cross sections is instrumental to the accuracy of the proposed chemical kinetics model. Owing to the complexity of the processes involved, to the lack and to the dispersion of the data, serious difficulties are found in establishing realistic chemical kinetic models. Concentration of data can be found in data bases such as NIST-JILA: Boulder US, ORNL: Oak Ridge US, GAPHIOR: Orsay, France, NIFS: Nagoya Japan and JAERI : Takaimura, Japan [Delcroix]. When experimental determination of the cross sections is not possible, they can be calculated by different methods knowing the interaction potentials for the reactions or the scattering processes [Biberman et al.]. In all cases, and whatever the complexity of the models, these latter need experimental validation. Consequently, in the field of plasma physico-chemistry and plasma process engineering, the analysis of the plasma must be performed by both means of modeling and optical diagnostics.

2.2.1 Elastic collisions

During elastic collisions between electrons and neutral particles, the incident electron and the neutral species undergo changes in their direction, and the electron kinetic energy is changed while the internal energy of the species remains unchanged. At large distances [Arnal], the interaction potential between an electron and an ion varies as r^{-1} and between an electron and a neutral as r^{-4}.

The elastic collisions are responsible for momentum transfer (kinetic energy exchange). The elastic cross sections depend on the velocity of the incoming particles. Usually, there is a distribution of the particles over velocities and angles [Biberman et al.]. The total averaged cross section is given by:

$$<\sigma>=\int_0^\infty w^2\sigma(w)f(w)dw$$

where f(w) is the velocity distribution function, which is Maxwellian at equilibrium and w is the relative velocity of the particles. The cross section has the dimensions of an area and $\sigma(w)$ is already integrated over the angles. For hydrogen $<\sigma_{Ne}>$ is maximum at 1 eV and is equal to 1.8 10^{-16} cm^2 (figure 7) [Biberman et al.], [Gibson 1970] and [Crompton 1969].

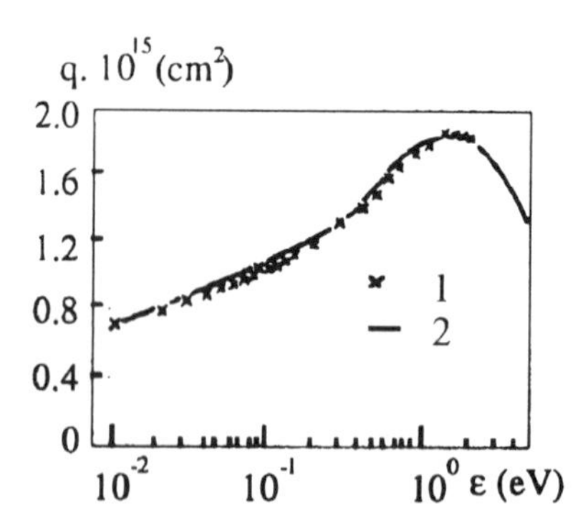

Figure 7: Transport cross section for the scattering of electrons by molecules of hydrogen, after 14 and 15, and from reference 13. 1- experiment, 2- calculated.

The average collision frequency is given by $\langle\nu(w)\rangle = \langle w\,\sigma(w)\rangle\, n$. Since $w_e \gg w_N$, the mean free path is equal to [Pointu 1991] and [Arnal]:

$$\lambda_{eN} = w / \langle\nu(w)\rangle \cong 10^{-19}\, T_N(K) / (P\,(Torr)\, \sigma_{Ne})$$

For diamond deposition conditions $\lambda_{eN} \cong 30\ \mu m$.

2.2.2 Inelastic collisions

Inelastic collisions are accompanied by a change in the internal state of the colliding molecular species. The different processes are ionization, association, electronic excitation, vibrational and rotational excitation and fractional dissociation of the molecules, and reverse reactions. The potential energy curves for the hydrogen molecule are given in figure 8 [Von Engel 1983]. Inelastic collisions are important in plasmas and change the energetic state of the particles. The cross sections are often characterized by the fact that they have a threshold (figure 9).

Electronic excitation: $A_2 + e \rightarrow A_2^* + e$ (k_e)

The transition probability is given by: $\langle\nu_{kn}\rangle = n_e \langle w_e\, \sigma_{kn}(w_e)\rangle$, where $\sigma_{kn}(w_e)$ is the cross section for the transition k to n.

$$\langle \nu_{kn} \rangle = \int_{thresh}^{\infty} w^3 \sigma_{kn}(w_e) f(w)\, dw = \int_{thresh}^{\infty} (2\varepsilon_e/m_e)^{1/2} \sigma_{kn}(\varepsilon_e) f(\varepsilon_e)\, d\varepsilon_e$$

where $f(w_e)$ is the velocity distribution function and $f(\varepsilon_e)$, the electron energy distribution function.

The reverse reaction of excitation is the quenching (de-excitation process) [Biberman et al.], owing to the microreversibility principle: $n_k\langle\nu_{kn}\rangle = n_n \langle\nu_{nk}\rangle$. Cross sections for allowed transitions of the H atom can be found in the literature [Biberman et al.], [Vainshtein et al. 1979] and [Janev et al.]. For forbidden transitions, the cross sections are around one tenth of the allowed transitions cross sections.

Ionization: $A_2 + e \rightarrow A_2^+ + e + e$ (k_i)

The ionization frequency has the same shape as the excitation frequency. The reverse reaction is a ternary process involving two electrons and an ion. We have:

$$n_k\langle\nu_{ke}\rangle = n_e\, n_+ \langle\nu_{ek}\rangle.$$

Calculated and experimental ionization cross sections are given in figure 10 for hydrogen atoms and hydrogen molecules [Biberman et al.], [Fite and Brackman 1958] and [Boyd and Green 1958]. Compared values of cross sections or rate constants for the ionization of different molecules are given in tables 1 and 2 [Gicquel 1987], [Engelhardt

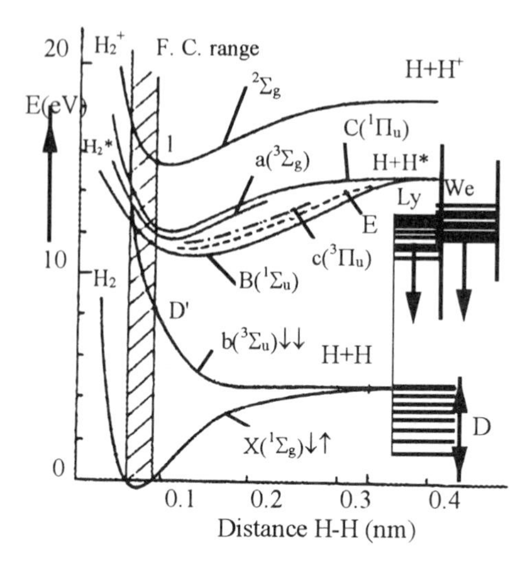

Figure 8: Energy potentials for the hydrogen molecule, after Von Engel[11].

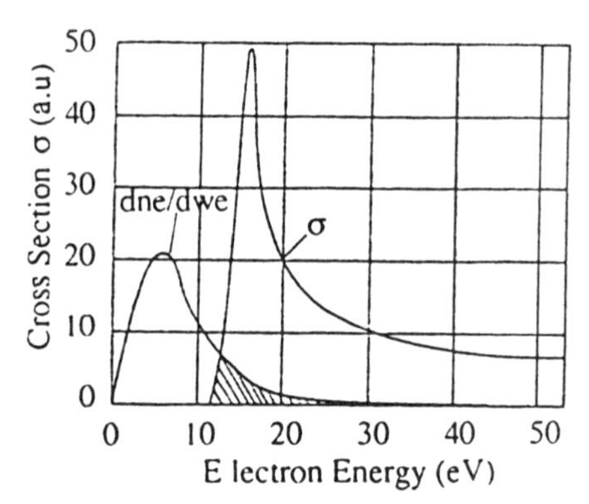

Figure 9: Overlapping of an electron velocity distribution function $dn_e/dw_e = 4'w_e^2\ f(w_e)$ and of the cross section for a excitation process by the electrons.

et al. 1964], [Hake and Phelps 1975], [Kaufman 1969], [Kaufman and Kelso 1960], [Corrigan 1966], [Myers 1969] and [Capitelli and Dilonardo 1978].

Vibrational excitation: $A_2 + e \rightarrow A_2(v) + e$ (k_{vib})

The most efficient mechanism for excitation of vibrational levels is the electron attachment process [Biberman et al.] where the auto-ionization state decays into a free electron and a molecule in a vibrationally excited state. The cross section for this process depends on the ratio of the lifetime of the resonance level to the period of the vibrations. Tables 1 and 2 present rate constants and cross sections of the electron-molecule reactions leading to vibrational excitation of hydrogen and other molecules [Biberman et al.], [Gilmore et al. 1969], [Biberman and Mnatsakanyan 1966] and [Mnatsakanyan 1970].

Table 1: Electron cross sections in H_2, O_2 and N_2 (cm^2)

	H_2	O_2	N_2
ionization at 50 eV	$1\ 10^{-16}$	$2\ 10^{-16}$	$2\ 10^{-16}$
Electronic	$< 8\ 10^{-17}$ at 14 eV ($^3\Sigma_u^+$)	$8\ 10^{-17}$ at 6eV ($^1\Delta_g$)	$6\ 10^{-17}$ at 11 eV ($A^3\Sigma_g$)
Vibration	10^{-17} at 1eV $8\ 10^{-17}$ at 2.5 eV	10^{-17} 0.2-8 eV	$5.5\ 10^{-17}$ at 2.5 eV
Dissociation (electron attach.)	$4.5\ 10^{-17}$ (16-17 eV) Phelps[26,27]	$1.5\ 10^{-18}$ at 6.5 eV	10^{-16} at 20 eV
thresh. 8.8 eV	$9\ 10^{-17}$ at 16.5 eV Corrigan[30] $5\ 10^{-17}$ at 10 eV		

Table 2: Electron rate constants (molecules/cm^3), in H_2, O_2 and N_2

	H_2	O_2	N_2
ionization	$7\ 10^{-11}$ (3 eV) $6\ 10^{-12}$ (2 eV)	$1.3\ 10^{-10}$ (3eV)	$6\ 10^{-11}$ (3 eV) $3\ 10^{-12}$ (2 eV)
Dissociation (Kaufman[28,29])	2-11 10^{-10}	$2\ 10^{-9}$ (2 eV)	$3\ 10^{-10}$ (3eV)
Dissociation (Phelps[26,27])	$1.2\ 10^{-9}$ (3 eV)	$1.9\ 10^{-9}$	$3\ 10^{-11}$ (2.2 eV)

Dissociation: $A_2(v) + e \rightarrow A + A + e$ (k_d)

The dissociation processes generally involve a step where an excited state (electronic or vibrational) is formed. In the case of the hydrogen molecule, different ways of dissociation through electron collisional reactions are possible: the dissociative attachment and the dissociation through the excitation in the repulsive state $b(^3\Sigma_u)$ are the two main channels. The role of the vibrational initial level on the dissociation rate constants of the hydrogen molecule for these two processes has been studied by M. Capitelli et al. [Capitelli et al. 1977].

2.2.3 Superelastic collisions

$$A_2^*(v) + e_{cold} \rightarrow A_2(v') + e_{hot} \qquad (v'<v)$$

The superelastic collisions lead to the heating of the electrons through energy transfers from electronic or vibrationally excited molecules to cold electrons [Cacciatore et al. 1986], [Capitelli et al. 1980], [Capitelli et al. 1986] and [Capitelli et al. 1994].

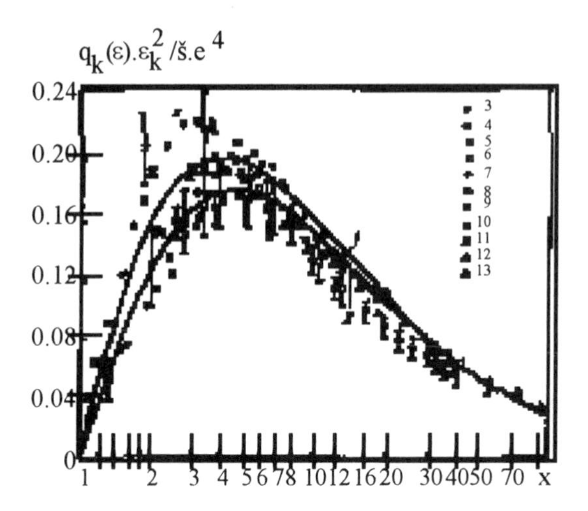

Figure 10: Calculated (1,2) and experimental (3-13) ionization cross sections, reduced to units of $q_k(\varepsilon)\ E_k{}^2/{}^{\prime}e^4$ and plotted versus $x = \varepsilon/E_k$ for various species: 1) $f(x) = 10\ (x-1)/x(x+8)$; 2) $f(x) = 10(x-1)/(x+0.5)\ (x+8)$; 3) hydrogen according to 23, 13) hydrogen molecules, according to 24. E_k is the k level energy, and ε the electron energy.

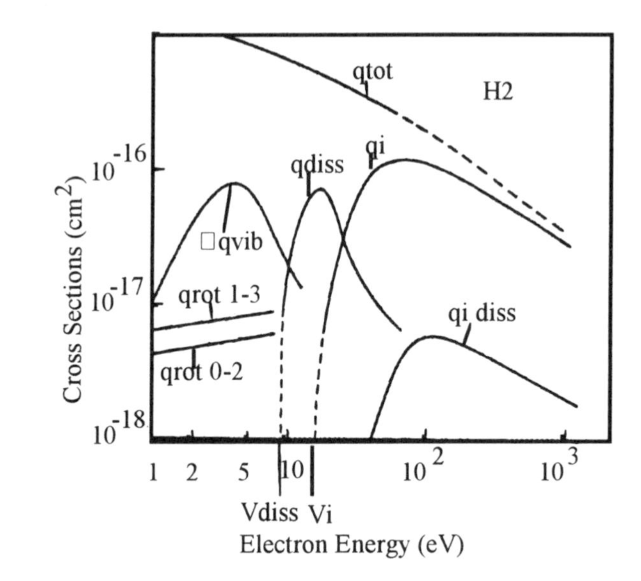

Figure 11: Excitation (vibration, rotation), ionization and dissociation cross sections for H_2, according to Von Engel[11].

The magnitude for these different electron-molecule collision cross sections in the case of H_2 are given in figure 11 [Von Engel 1983], other values can be found in the literature [Engelhardt et al 1964], [Hake and Phelps 1975], [Kaufman 1969], [Kaufman and Kelso 1960], [Corrigan 1966], [Janev et al 1980], [Capitelli 1980] and [Phelps 1990].

2.2.4 Rate constants

In a perfectly mixed reactor, the reaction rate of production of the state j in a reaction is:

$$dn_j/dt = n_j^o n_e 4\pi \int \sigma_{ej} w_e^3 f(w_e) dw_e = n_j n_e 4\pi < \nu_{0j} >$$

where n_j^0 is the population of the initial state, n_e the electron density, and $<\nu_{0j}>$ the reaction frequency. For a Maxwellian distribution of the electron velocities, we have:

$$dn_j/dt = n_j^0 n_e \left[m_e/2\pi kT_e\right]^{1/2} 4\pi \int_{thresh}^{\infty} \sigma_{ej} w_e^3 \exp(-m_e w_e/2kT_e) dw_e$$

2.3 NEUTRAL-NEUTRAL COLLISIONS

The neutral-neutral collisions involve elastic, inelastic and reactive processes. Compared with the collisions with electrons, they lead to a larger number of processes. The incoming molecules may have internal energy, which would make a complete analysis of the phenomena very complex. The inelastic collisions are responsible for internal energy changes of the molecules and for chemical reactions.

The molecule-molecule collision processes generally have a threshold, and are functions of the particle energy. The cross sections of the collisions are determined by the interaction potential of the species i and j. At large distances, the interaction potential between neutral and neutral varies as r^{-6}.

In inelastic collisions, particles undergo transfer of energy between modes such as vibration-vibration (V-V), vibration-rotation (V-R) and vibration-translation (V-T) and electronic-electronic (quenching). Likewise, chemical reactions lead to the heating of the gas and to the formation of a reactive non- equilibrium medium.

The main processes are the following:

Quenching: $A^* + M \rightarrow M^* + A$ (k_Q)

The quenching reaction leads to the destruction of metastable or radiative species (under conditions of moderate to high pressure) through electronic energy exchange.

Charge transfer: $A^+ + M \rightarrow M^+ + A$ (k_{ch})

The charge transfer becomes resonant in the case where the 'M' species is an excited state of species 'A.'

Ionization: $A + M \rightarrow A^+ + M + e$

Penning effect: $A^* + M \rightarrow A + M^+ + e$ (k_{Pen})

Table 3: Rate constants for energy transfer in H_2, O_2 and N_2 (in cm^3s^{-1}) (Tg = 500 K, v_r =1 eV, v=0) (according to Capitelli et al. [32, 36, 51])

	H_2	O_2	N_2
V-V Transfer	$2\ 10^{-12}$	$6.3\ 10^{-13}$	$2.5\ 10^{-13}$
V-T Transfer	$1\ 10^{-15}$	$3\ 10^{-17}$	$1.7\ 10^{-21}$
e-V Transfer	$2\ 10^{-9}$	$1.2\ 10^{-10}$	$1.5\ 10^{-8}$

The ionization process from the ground state is not efficient, however it becomes efficient when the particle A is in an excited state. If A is a metastable state, this process is called Penning effect. The three-particle recombination is the reverse reaction.

Associative ionization : $A + M \rightarrow (AM)^+ + e$

In many cases, the associative ionization and the dissociative recombination (reverse reaction) determine the rate of formation of the charged particles in high temperature gases and plasmas. The dissociative recombination is a fast process (10^{-6} cm^3 s^{-1}).

The formation of negative ions can also occur by different processes. The most important one is the *dissociative attachment:*

$$e + A_2 \rightarrow A_2^-(v) \rightarrow A^- + A.$$

V-V energy transfer: $A_2(v-1) + A_2(w+1) \rightarrow A_2(v+1) + A_2(w-1)$

V-R and V-T energy transfers: $A_2(v+1) + A \rightarrow A_2(v) + A(hot)$

V-V, V-R and V-R energy transfers without a chemical reaction are very important in low pressure plasmas (table 3). When a molecule has a low anharmonicity (such as nitrogen for instance), V-V energy transfers can even lead to the dissociation of the molecule (vibrational ladder) [Capitelli and Molinari 1980] and [Capitelli and Dilonardo 1978]. In nitrogen plasmas at low pressure, the vibrational temperature can reached 5 to 10 times the translational temperature [Gicquel 1987]. The magnitude of V-V, V-T and e-V energy transfer rate constants is given in table 3 for different molecules.

Chemistry: $A + BC\ (v, j) \rightarrow AB\ (v', j') + C.$

where v, j, v', and j' represent the vibrational and rotational quantum numbers of the particles respectively in the initial and the final states.

Chemical reactions involving molecules, atoms and radicals obviously occur in such media. In particular, in diamond deposition plasmas, owing to the high gas temperatures reported [Gicquel et al. 1994] and Girshick et al. 1993], thermal dissociation of the molecular hydrogen cannot be neglected.

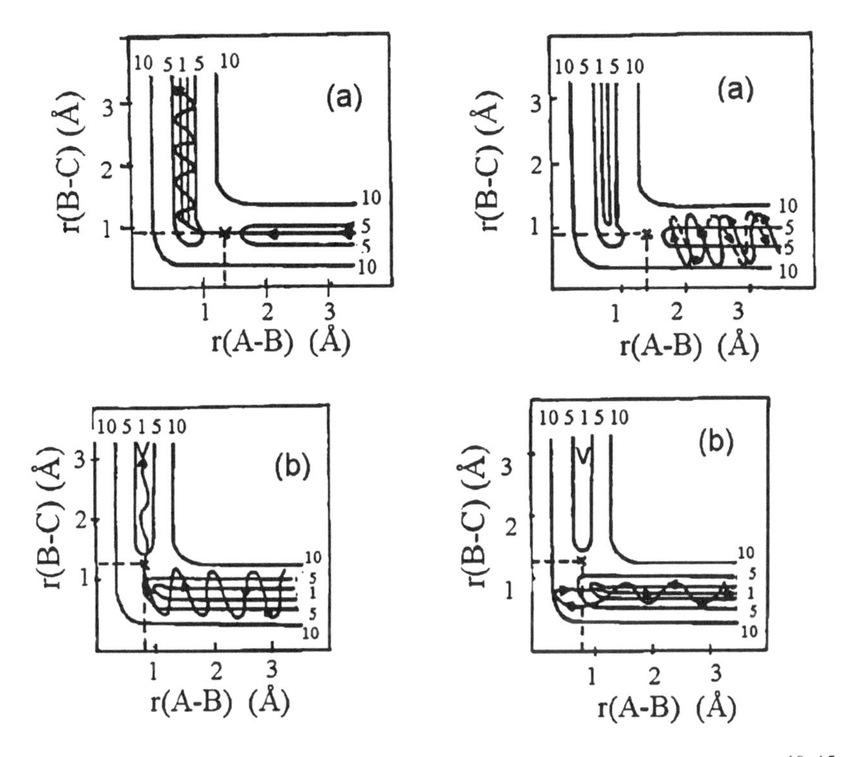

Figure 12: Dynamic trajectory on a potential-energy surface, according to J. C. Polanyi[40-45]. Reaction A + BC → AB + C (a) Type I hypersurface (attractive surface): the activation barrier is located in the entry valley; (b) (a) Type II hypersurface: the activation barrier is located in the exit valley (repulsive surface)

Although they are still strongly dependent on the gas temperature (as in a classical situation), in a plasma, the neutral-neutral collisional chemical reactions (A + BC --> AB + C) can be strongly enhanced by the vibrational or the rotational excitation contained in the incident species. A thorough study of the reaction dynamics was performed by Polanyi [Polanyi 1973], [Polanyi 1972], [Kuntz et al. 1966], [Monk and Polanyi 1969], [Polanyi and Wong 1969] and [Anlauf et al. 1969]. According to the interaction potentials (attractive or repulsive behavior) of the incoming molecules, the reaction can be favored either by vibrational excitation or by translational-rotational energy. For the case of an attractive interaction potential (figure 12a) [Polanyi 1973], [Polanyi 1972], [Kuntz et al. 1966], [Monk and Polanyi 1969] and [Polanyi and Wong 1969], the activation barrier is passed while C is still close to B (in the entry valley). Ro-translational energy on the incoming molecule (BC) is necessary while vibrational energy is inefficient. The products of such a reaction are vibrationally excited. In the case of a repulsive interaction potential (figure 12b) [Polanyi 1973], [Polanyi 1972], [Kunze et al. 1966], [Monk and Polanyi 1969] and [Polanyi and Wong 1969], the activation barrier is crossed in the exit valley while C begins to be separated from B. Vibrational energy of the incoming molecule (BC) is necessary while ro-translational energy is inefficient. The products of such a reaction possess ro-translational energy.

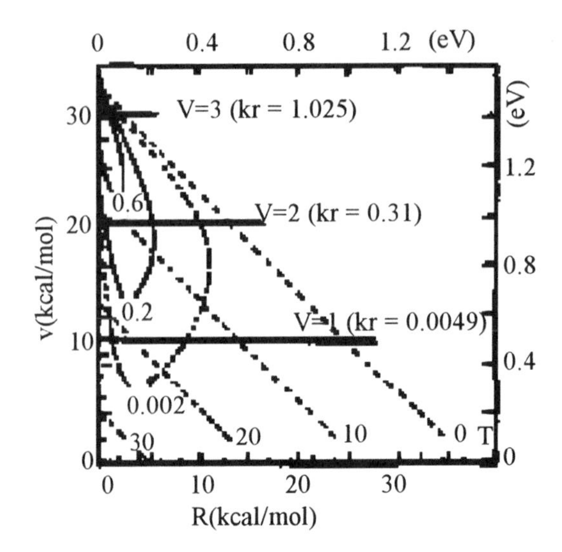

Figure 13: Role of the vibrational excitation on the rate constant of an endothermic reaction: HF + H $\rightarrow$ F + H_2. Bold lines indicated the rate constants for different vibrational energies (V) and rotational energies (R), while the dashed lines indicated the variations in the translational energy (T) of the incoming molecule, according to J. C. Polanyi[45].

The rate constants of these processes are, then, not only functions of the gas temperature, but also of the vibration and the rotational excitation of the reactants: k = k (v, v', j, j', Tg), and the reaction frequency becomes:

$$\langle \nu \rangle = n_j \langle \sigma_{ij}(v, j, w_{ij}) \cdot w_{ij} \rangle$$

$$= \int_{thresh}^{\infty} (2\varepsilon_e / m_e)^{1/2} \sigma_{kn}(v, j, v', j' m \varepsilon_e) f(\varepsilon_e) d\varepsilon_e$$

An example concerning an endothermic limited step reaction HF + H --> F + H_2 is given on figure 13 [Anlauf et al. 1969]. We can observe the strong effect of the vibrational quantum number on the rate constant; while the initial vibrational quantum number varies from 1 to 3, the rate constant is increased by a factor 200.

2.3.1 Loss of excited species

Under low pressure conditions, the main loss of radiative excited states is through photon emission owing to high transition probabilities (10^6-10^8 s^{-1}) [Ricard 1988]. Under higher pressure, radiative species may be lost through photon emission or by quenching processes in volume. Non-radiative metastable species can be lost by diffusion, followed by their destruction at the walls, by chemical reactions or by quenching processes in volume.

2.4 CONCLUSION

Processes occurring in plasmas are rather complex. They strongly depend on the discharge characteristics such as electron energy, electron density, pressure, gas temperature and ro-vibrational energy contained on the molecules. Modeling of the volume plasma is always difficult, and the difficulty is still enhanced by the lack of data. When studying low pressure plasmas for surface treatment for instance, the first step should involve determination of the average characteristics of the plasma in volume and of the main physico-chemical parameters responsible for them. Once a rather simple chemical kinetic simulation is running with a reasonable computation time, experimental validation would become the main preoccupation. In-situ diagnostics must be developed. Furthermore, once the chemistry understood, physical transport equations can be solved including chemical source terms.

2.5 REFERENCES

Anlauf, K.G., Maylotte, D.H. and Polanyi, J.C. *J. Chem. Phys.* vol 51, n° 12 5716 (1969)

Arnal, Y. "*Traitements de surface par plasmas-Initiation aux plasmas*", Formation Continue, Institut National Polytechnique de Grenoble, Fascicule I: plasmas, pp.1-31

Biberman, L.M. and Mnatsakanyan, Kh. "On the exchange of energy between electron and molecular gases" *Teplofiz. Vys. Temp.* 4, 491 (1966)

Biberman, L.M., Vorob'ev, V.S., Yakubov, I.T. *"Kinetics of non equilibrium low-temperature plasmas",* consultants Bureau. New York and London,

Boyd, R.L. and Green, G.W. *"Measurement of effective cross section for ionization by electron impact"* Proc. Phys. Soc. London, 71, 351 (1958)

Cacciatore, M. Capitelli, M., de Benedictis, S., Dilonardo, M., Gorse, C. *Topics in current Physics* vol 39 (1986)

Capitelli, M. *Pure and Appl. Chemistry* 52, 1707 (1980)

Capitelli, M. and Molinari, E. *Topics in Current Chemistry*, n° 90, p 60-109 (1980)

Capitelli, M., Colonna, G., Hassouni, K., Gicquel, A. *Chem. Phys. Letter* 228, 687-94 (1994)

Capitelli, M., Dilonardo, M. *Rev. Phys. Appl. Tome* 13, 115 (1978)

Capitelli, M., Dilonardo, M., Molinari, E. *Chem. Phys. 20,* 417-429 (1977)

Capitelli, M., Gorse, C., Ricard, A. *Topics in current Physics* vol 39 (1986)

Chapman, B. "*Glow discharge process*" Wiley- New York (1980)

Cherrington, B.E. "*Gaseous Electronics and Gas Lasers*", Pergamon Press (1979)

Corrigan, S.J.B. *J. Chem. Phys.* 43, 4381 (1966)

Crompton, R.W., Gibson, D.K., McIntosh, A.I. "The cross sections for the J=0 --> 2 rotational excitation of hydrogen by slow electrons" *Aust. J. Phys.* 22, 715 (1969)

Delcroix, J.L., Bes, A. *Physique des Plasmas*, InterEditions, CNRS Editions, vol I and II

Elenbasa, W. *The high pressure mercury vapor discharge,* Interscience, New York (1951)

Engelhardt, A.G., Phelps,A.V. and Risk, G.C. *Phys. Rev.* 135, A1566 (1964)

Fauchais, P. Course *"Overview of Thermal Spraying Techniques"*, University of Limoges (France) (1994)

Ferreira, C.M., Moisan, M. "Microwave Excited Plasma", *Plasma Technology*, editors: M. Moisan and J. Pelletier, Elsevier (1992) chapter 3: Kinetic Modeling of microwave discharges: influence of the discharge stimulating frequency, pp 53-91

Ferreira, C.M., Moisan, M., Zakrzemski, Z. "Microwave Excited Plasma", *Plasma Technology,* editors: M. Moisan and J. Pelletier, Elsevier (1992) chapter 2: Physical principles of microwave plasma generation, pages 11 - 52

Fite, W.L. and Brackman, R.T. *"Collisions of electrons with hydrogen atoms. II. Excitation of Lyman-alpha radiation"* Phys. Rev. 112, 1151 (1958)

Gibson, D.K. *"The cross section for rotational excitation of H2 and D2 by low energy electrons"* Aust. J. Phys., 23, 683 (1970)

Gicquel, A. *Thèse d'Etat,* Université Pierre et Marie Curie (1987)

Gicquel, A., Hassouni, K., Farhat, S., Breton, Y., Scott, C.D., Lefebvre, M., Pealat, M. *Diamond and Related Materials* , vol 3, n° 4-6, 581-586 (1994)

Gicquel, A., Hassouni, K., Farhat, S., Breton, Y., Scott, C.D., Lefevbre, M., Péalat, M. *Diamond and Related Materials* , vol 3, n° 4-6, 581-586 (1994)

Gicquel, A.. Catherine, Y. *Journal de Physique IV*, C2-343-356, Colloque C2, suppl. Journal de Physique II, (1991)

Gilmore, F.R., Baer, E. and McGowan, I.W. "A review of atomic and molecular excitation mechanisms in non equilibrium gases up to 2000 K". *J. Quant. Spectrosc. Radiat. Transfer.,* 9, 157 (1969)

Girshick, S.L., Li, C., Yu, B.W., Han, H. *Plasma Chemistry and Plasma Processing*, vol 13, n° 2 (1993)

Hake, R.D. and Phelps, A.V. *Phys Rev*. 158, A 70 (1975)

Hirschfelder, J.O., Curtis, C.F., Bird, R.B. *Molecular Theory of Gases and liquids* John Wiley and Sons, Inc., New York, London, Sydney (1964)

Janev, R.K., Langer, W.D., Evans, K. Jr., Post, D.E. Jr. *"Elementary Process in hydrogen plasmas"* Springer Verlag

Kaufman, F. *"Chemical Reactions in Electrical Discharges. The production of atoms and simple radicals in glow discharges"* Adv. in Chemistry, Series 80, 29, ed R.F. Gould (1969)

Kaufman, F. and Kelso, J.R. *J. Chem. Phys.*, 32, 301 (1960)

Kuntz, P.J., Nemeth, E.M., Polanyi, J.C., Rosner, S.D., Young, C.E. *J. Chem. Phys.* vol 44, n° 5, 1168 (1966)

Lu, Z.P., Heberlein, J., Pfender, E. *Plasma Chemistry and Plasma Processing*, vol 12, n° 1 (1992)

Mnatsakanyan, Kh. "Electron energy balance in a plasma containing molecular impurities" *Teplofiz. Vys. Temp.* 8, 1149 (1970)

Monk., M.H. and Polanyi, J.C. *J. Chem. Phys.* vol 51, n°4, 1452 (1969)

Myers, H. *J. Chem. Phys.* B2, 393 (1969)

Papoular, R. *Phénomènes électriques dans les gaz*, Monographies Dunod (1963)

Phelps, A.V. *J. Chem. Ref. data, vol 19*, n°3, 653-675 (1990)

Pointu, A.M. *Introduction aux plasmas de décharges*, Journées de Formation CIP91- "Dépôt et Gravure Chimiques par plasma", ed. SFV-CNRS Supplément à la Revue "Le Vide, les couches minces" Mars-Avril 1991, n° 256 pp1-56.

Polanyi, J.C. *Acc. Chem. Res.*, 5, 161 (1972)

Polanyi, J.C. *J. Chem. Soc.* Faraday disc. p. 389-409 (1973)

Polanyi, J.C. and Wong, W.H. *J. Chem. Phys.* vol 51, n° 51, 1439 (1969)

Ricard, A. *"Basic Physics of Plasma Discharges: Production of active species"*, Nato-ASI, Alicante (1988)

Scott, C.D., Farhat, S., Gicquel, A., Hassouni, K., Lefebvre,M. *AIAA Plasmadynamics and Laser Conference,* Florida Orlando AIAA 3226 (1993)

Vainshtein, L.A., Sobel'man, I.I., Yukov, E.A. *"Excitation of Atoms and Spectral Line Broadening"* (in Russian), Nauka, Moscow (1979)

Veprek, S. "Characterization of Glow discharges for plasma processing and fundamentals of plasma diagnostics" in short course on *"Low Temperature (non equilibrium) plasmas"* International Summer School on Plasma Chemistry, editors G. Bruno, G. K. Herb (1989) pp 263-288

Von Engel, A. *"Electric plasmas: their nature and uses"* Ed Taylor and Francis, Ldt London and New York (1983)

Von Engel, A. *"Ionized gases"* Ed Oxford at the Clarendon Press (1965)

Winkler, R. *Microwaves Discharges: Fundamental and Applications,* ed C. M. Ferreira, M. Moisan, Nato ASI Series, Séries B physics, vol 302, pp 339-357 (1992)

References subsection 1:

1- A. Gicquel, Y. Catherine, *Journal de Physique IV*, C2-343-356, Colloque C2, suppl. Journal de Physique II, (1991)

2- W. Elenbasa, *The High Pressure Mercury Vapor Discharge*, Interscience, New York (1951)

3- A. M. Pointu, Introduction aux plasmas de décharges, *Journées de Formation* CIP91- "Dépôt et Gravure Chimiques par plasma", ed. SFV-CNRS Supplément à la Revue "Le Vide, les couches minces" Mars-Avril 1991, n° 256, pp1-56.

4- J. L. Delcroix, A. Bes, *Physique des Plasmas*, InterEditions, CNRS Editions, vol I and II

5- S. L. Girshick, C. Li, B. W. Yu, H. Han, *Plasma Chemistry and Plasma Processing*, vol **13**, n° 2 (1993)

6- Z. P. Lu, J. Heberlein, E. Pfender, *Plasma Chemistry and Plasma Processing*, vol **12**, n° 1 (1992)

7- Y. Arnal, *"Traitements de surface par plasmas-Initiation aux plasmas"*, Formation Continue, Institut National Polytechnique de Grenoble, Fascicule I: plasmas, pp.1-31

8- P. Fauchais, Course *"Overview of Thermal Spraying Techniques"*, University of Limoges (France) (1994)

9- B. Chapman *"Glow discharge process"* Wiley- New York (1980)

10- B. E. Cherrington, *"Gaseous Electronics and Gas Lasers"*, Pergamon Press (1979)

11- A. Von Engel, *"Ionized gases"* Ed Oxford at the Clarendon Press (1965)

12- R. Papoular, *Phénomènes électriques dans les gaz*, Monographies Dunod (1963)

13- L. M. Biberman, V. S. Vorob'ev, I. T. Yakubov, *"Kinetics of non equilibrium low-temperature plasmas"*, consultants Bureau. New York and London,

14- D. K. Gibson, "The cross section for rotational excitation of H2 and D2 by low energy electrons" *Aust. J. Phys.*, **23**, 683 (1970)

15- R. W. Crompton, D. K. Gibson, A. I. McIntosh, "The cross sections for the J=0 --> 2 rotational excitation of hydrogen by slow electrons" *Aust. J. Phys.* **22**, 715 (1969)

16- J. O. Hirschfelder, C. F. Curtis, R. B. Bird, *Molecular Theory of Gases and liquids* John Wiley and Sons, INC., New York, London, Sydney (1964)

17- C. M. Ferreira, M. Moisan, Z. Zakrzemski, "Microwave Excited Plasma", *Plasma Technology*, editors: M. Moisan and J. Pelletier, Elsevier (1992) chapter 2: Physical principles of microwave plasma generation, pages 11 - 52

18- R. Winkler, *Microwaves Discharges: Fundamental and Applications*, ed C. M. Ferreira, M. Moisan, Nato ASI Series, Séries B physics, vol **302**, pp 339-357 (1992)

19- S. Veprek, "Characterization of Glow discharges for plasma processing and fundamentals of plasma diagnostics" in short course on "Low Temperature (non equilibrium) plasmas" *International Summer School on Plasma Chemistry"*, editors G. Bruno, G. K. Herb (1989) pp 263-288

20- C. M. Ferreira, M. Moisan, "Microwave Excited Plasma", *Plasma Technology*, editors: M. Moisan and J. Pelletier, Elsevier (1992) chapter 3: Kinetic Modeling of microwave discharges: influence of the discharge stimulating frequency, pp 53-91

21- A. Von Engel, *"Electric plasmas: their nature and uses"* Ed Taylor and Francis, Ldt London and New York (1983)

22- L. A. Vainshtein, I. I. Sobel'man, E. A. Yukov, *"Excitation of Atoms and Spectral Line Broadening"* (in Russian), Nauka, Moscow (1979)

23- W. L. Fite and R. T. Brackman, "Collisions of electrons with hydrogen atoms. II. Excitation of Lyman-alpha radiation" *Phys. Rev.* **112**, 1151 (1958)

24- R. L. Boyd an G. W. Green, "Measurement of effective cross section for ionization by electron impact" *Proc. Phys. Soc. London*, **71**, 351 (1958)

25- A. Gicquel, *thèse d'Etat*, Université Pierre et Marie Curie (1987)

26- A. G. Engelhardt, A. V. Phelps and G. C. Risk, *Phys. Rev.* **135**, A1566 (1964)

27- R. D. Hake and A. V. Phelps, *Phys Rev.* **158**, A 70 (1975)

28- F. Kaufman, "Chemical Reactions in Electrical Discharges. The production of atoms and simple radicals in glow discharges" *Adv. in Chemistry, Series 80*, 29, ed R.F. Gould (1969)

29- F. Kaufman, and J. R. Kelso, *J. Chem. Phys.*, **32**, 301 (1960)

30- S. J. B. Corrigan, *J. Chem. Phys.* **43**, 4381 (1966)

31- H. Myers, *J. Chem. Phys.* **B2**, 393 (1969)

32- M. Capitelli, M. Dilonardo, *Rev. Phys. Appl.* Tome **13**, 115 (1978)

33- F. R. Gilmore, E. Baer, and I. W. McGowan "A review of atomic and molecular excitation mechanisms in non equilibrium gases up to 2000 K". *J. Quant. Spectrosc. Radiat. Transfer.*, **9**, 157 (1969)

34- L. M. Biberman and Kh. Mnatsakanyan "On the exchange of energy between electron and molecular gases" *Teplofiz. Vys. Temp.* **4**, 491 (1966)

35- Kh. Mnatsakanyan, "Electron energy balance in a plasma containing molecular impurities" *Teplofiz. Vys. Temp.* **8**, 1149 (1970)

36- M. Capitelli, M. Dilonardo, E. Molinari, *Chem. Phys.* **20**, 417-429 (1977)

37- M. Cacciatore, M. Capitelli, S. de Benedictis, M. Dilonardo, C. Gorse, *Topics in current Physics* vol 39 (1986)

38- M. Capitelli, and E. Molinari, *Topics in Current Chemistry,* n° 90, p 60-109 (1980)

39- M. Capitelli, C. Gorse, A. Ricard, *Topics in current Physics* vol **39** (1986)

40- J. C. Polanyi, *J. Chem. Soc.* Faraday disc. p. 389-409 (1973)

41- J. C. Polanyi, *Acc. Chem. Res.*, **5**, 161 (1972)

42- P. J. Kuntz, E. M. Nemeth, J. C. Polanyi, S. D. Rosner, C. E. Young, *J. Chem. Phys.* vol **44**, n° 5, 1168 (1966)

43- M. H. Monk., and J. C. Polanyi, *J. Chem. Phys.* vol **51**, n°4, 1452 (1969)

44- J. C. Polanyi, and W. H. Wong, *J. Chem. Phys.* vol **51**, n° 51, 1439 (1969)

45- K. G. Anlauf, D. H. Maylotte, and J. C. Polanyi, *J. Chem. Phys.* vol **51**, n° 12 5716 (1969)

46- M. Capitelli, G. Colonna, K. Hassouni, A. Gicquel, *Chem. Phys. Letter* **228**, 687-94 (1994)

47- A. Ricard, "Basic Physics of Plasma Discharges: Production of active species", *Nato-ASI*, Alicante (1988)

48- C. D. Scott, S. Farhat, A. Gicquel, K. Hassouni, M. Lefebvre, *AIAA Plasma dynamics and Laser Conference*, Florida Orlando AIAA 3226 (1993)

49- A. Gicquel, K. Hassouni, S. Farhat, Y. Breton, C. D. Scott, M. Lefebvre, M. Pealat, *Diamond and Related Materials* , vol **3**, n° 4-6, 581-586 (1994)

50- R. K. Janev, W. D. Langer, K. Evans Jr., D. E. Post Jr. "*Elementary Process in hydrogen plasmas*" Springer Verlag

51- M. Capitelli, *Pure and Appl. Chemistry* **52**, 1707 (1980)

52- A. V. Phelps, *J. Chem. Ref. data*, vol **19**, n°3, 653-675 (1990)

53- A. Gicquel, K. Hassouni, S. Farhat, Y. Breton, C. D. Scott, M. Lefevbre, M. Péalat, *Diamond and Related Materials* , vol **3**, n° 4-6, 581-586 (1994)

Chapter 14

SECTION 3: BASIC PROCESSES: PLASMA / SURFACE INTERACTIONS

A. Gicquel
Laboratoire d'Ingéniérie des Matériaux et des Hautes Pressions
CNRS-UPR 1311- Université Paris-Nord
Avenue J. B. Clément 93430, Villetaneuse France

The description of the interaction of plasma particles with a surface would need to detail each chemical component of the plasma (in its different states) interacting with the surface, which is characterized by its distribution of states. The microscopic description of the whole system is too complex, and a macroscopic approach must be considered. In addition, we will considered each of the species created in the plasma interacting independently with the surface atoms.
The plasma considered here is mainly composed of neutrals (molecules and radicals) in their ground electronic state containing some ro-vibrational excitation, and atoms. The interaction of the neutrals with a surface are presented in terms of adsorption and recombination - desorption processes. The influence of the ro-vibration energy contained in the incoming species on the efficiency of the reactions is discussed. The distribution of energy of the different modes (vibration, rotation and translation) of the desorbed species is also discussed.
Owing to their low population, the interaction of metastable and radiative species with surfaces is only briefly described. Also, owing to the low kinetic energy of the ions under the conditions considered here, their interaction with surfaces is restricted to that of low energy ions.
Although microscopic mechanisms of interaction can be introduced, the energy and mass transfer resulting from such an interaction are described in terms of macroscopic coefficients.

3.1 INTERACTION OF NEUTRALS WITH SURFACES: DYNAMICS OF ADSORPTION AND RECOMBINATION-DESORPTION

The interaction potential of an atom approaching a surface can be considered as the sum of the potentials of interaction of this atom with each atom on the surface. Each interaction is the sum of an attractive potential (large distances), and of a repulsive one (small distances) (figure 1, curve B) [Wood and Wise 1967]. The minimum of the global curve corresponds to the chemisorption state. It provides the adsorption potential energy, which is a characteristic for the atom/surface system. Owing to the presence of different adsorption sites, the electrostatic field on the surface is not uniform. Once chemisorbed, the adatom is able to move on the surface from one adsorption site to another (superficial mobility), the efficiency of which depends on the surface temperature and on the depth of the energy potential wells.

In the case where an atom flow is interacting with a surface, the surface will become more and more covered, until a monolayer is formed (10^{14} to 10^{15} atoms cm^{-2}).

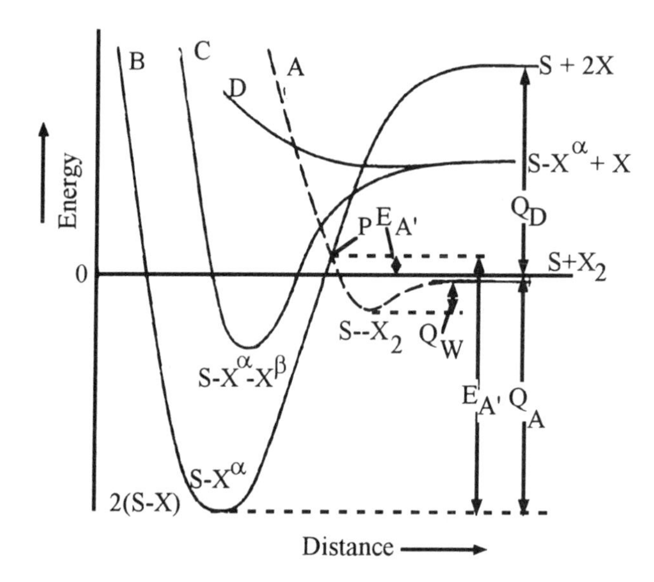

Figure 1. Energy diagram for atom-surface and molecules-surface interactions (after Wood et Wise[1])

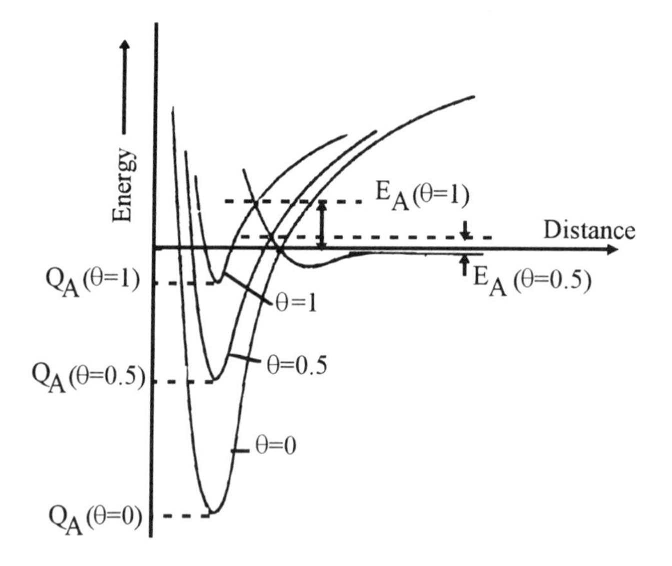

Figure 2. Formal representation of the dependence of the activation barrier on coverage (after Bond, 1962[2]).

During this time, there is a modification of the adsorption well potential (figure 2) [Wood and Wise 1967] and [Bond 1962], which decreases [Bond 1962], [Pomot] and [Boudart and Djéga - Mariadassou 1984].

3.1.1 Interaction of a molecule or a radical with a surface

Wolken has studied molecule / surface reaction dynamics to a depth similar to Polanyi's study of homogeneous phase reactions. The molecule is considered to interact with a solid surface, considered as an atom network. Only solid atoms in the vicinity of the incident molecule are assumed to participate to the interaction [McCreery and Wolken 1976]. Wolken contributed to the understanding of the effect of the internal energy modes of the incoming molecules on adsorption processes. In addition, he has studied the distribution of the energy of molecules formed by a recombination-desorption mechanism.

Adsorption process. For adsorption processes, the surface interaction potential strongly depends on the species/surface system. Let us first consider the case of an exothermic dissociative chemisorption step, which presents no activation barrier (figure 1) [Wood and Wise 1967]. In that case, the efficiency of the reaction is very high, and the distribution of energy contained in the molecule (kinetic, vibration, rotation) does not have any influence on the efficiency of the process.

As an example, let us now analyze the exothermic dissociative chemisorption of nitrogen on iron, which presents physisorption sites (figure 1, union of curves A and B [Wood and Wise 1967] and figure 3[Boszo et al. 1977]). An activation energy barrier is now present for the dissociative chemisorption mechanism.

The position of the activation barrier as well as its height (at the crossing between the atom/surface adsorption potential curve (curve A) and that of the molecule /surface physisorption potential (curve B)), as a function of the atom - surface distance coordinate are characteristics of the reaction dynamics. Wolken's calculations realized, when considering three different adsorption energy potential curves, that the higher this barrier, the higher the translational energy necessary in the incoming molecule to overcome the activation barrier. Once the threshold of translational energy is passed, the vibrational energy contained in the molecule can be efficient (table 1)[Gregory et al. 1978], [McCreery and Wolken 1975], [McCreery and Wolken 1977a], and [Wolken 1978]. The three cases of energy-potential considered are: surface I represents a system without any activation barrier; surface II, a system with an activation energy of 0.06 eV; and surface III, a system with an activation barrier of 0.15 eV. The results given in table 1 and table 2 clearly show (i) the necessity of a threshold in translational energy on the incident molecule for overcoming the activation barrier, and (ii) the effect of the vibrational excitation for overcoming the activation barrier, once the threshold in translational is reached.

The effect of surface contamination on the adsorption process can be drastically important [Gicquel 1987]. For instance, the contamination of the iron by sulfur displaces the position of the activation barrier towards smaller surface / molecule distance in the N_2 / iron system. It corresponds to a later position in the exit valley of the chemical path for the adsorption process, and to an earlier position in the entry valley for the recombination-desorption process (reverse reaction) (figure 3). The effect of such a displacement is a complete change in the reaction dynamics and in the heat transfer (see later).

Table 1: Adsorption probability of H_2 as a function of the distribution of the energy modes (v and j represent respectively the vibrational and the rotational quantum numbers). The total energy is kept constant, and equal to 0,26 eV. (after Wolken et al. [7,8,9,10])

v	j	Translational energy (eV)	adsorption probability (%) Surf. I	Surf. II (0.06 eV)	Surf. III (0.15 eV)
0	0	0.260	92	92	74
0	1	0.245	93	100	72
0	2	0.216	95	99	54
0	3	0.170	95	97	16
0	4	0.110	97	84	0
0	5	0.040	95	0	0

Table 2: Adsorption probability of H_2 as a function of the vibrational and the rotational quantum numbers (v,j). The kinetic energy is kept constant at 0,26 eV. (after Wolken et al [7,8,9,10])

v	j	Total energy (eV)	adsorption probability (%) Surf. I	Surf. II (0.06 eV)	Surf. III (0.15 eV)
0	0	0.26	92	92	74
0	3	0.35	93	96	77
0	6	0.56	90	97	85
1	0	0.77	93	100	95
1	6	1.06	92	98	95
2	0	1.26	91	94	100
2	6	1.53	91	97	99

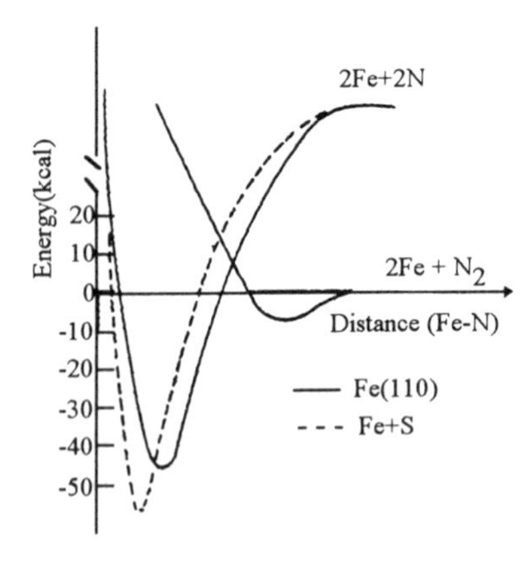

Figure 3: Energy potential curve for the adsorption of N_2 on iron. (after Boszo, 1977[5])

3.2 RECOMBINATION - DESORPTION PROCESSES

3.2.1 The recombination-desorption process

The two steps of recombination and desorption are inseparable in the way that the recombination process (formation of the new molecule) furnishes the energy needed for the desorption process (break of the surface-atom bonds). In addition, part of the energy is given to the surface, enabling a stabilization of the molecule which desorbs. The reaction dynamics depend on the energy potential shapes.

Let us consider a reaction consisting of going from the initial state, where two atoms are adsorbed on a surface to the final state where the molecule has desorbed from the surface (with the symmetry axis parallel to the surface), that is occurring through a Langmuir-Hinshelwood mechanism. Let us follow G. Wolken considering first that the whole step (recombination-desorption) is endothermic (figure 4a) [McCreery and Wolken 1977b]. In that case, calculations (table 3) clearly indicate that no molecule is formed whatever the kinetic energy of the adatoms (even if the kinetic energy is larger than the activation barrier).

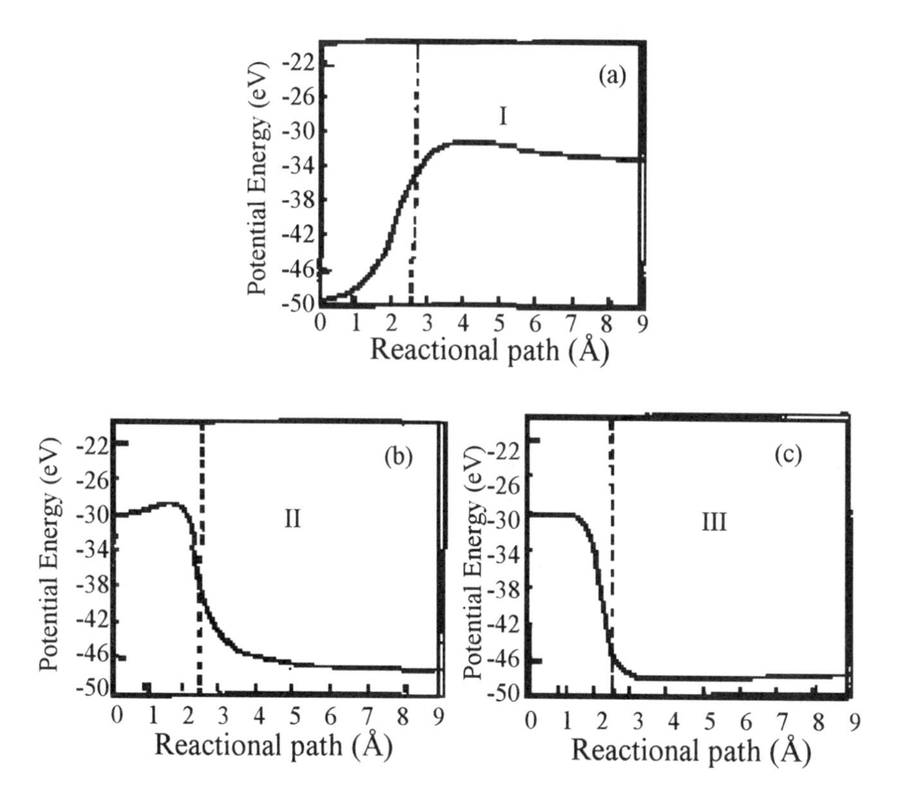

Figure 4: Energetic profiles of chemical paths (after Wolken et al[11]). The dash lines indicate the transition between the entry valley and the exit valley. The two atoms are initially adsorbed on adjacent 1CN sites. (a) endothermic reaction, (b) exothermic reaction, the energy is released partly in the entry valley and partly in the exit valley; (c) exothermic reaction, the energy is release in the entry valley.

Table 3: Energy distribution on products desorbing from surfaces, with different energy potential curves (I, II, III): (a) for H_2; (b) for a diatomic molecule where one of the atoms is very heavy compared to the other one. (after Wolken et al[11])

Potential	% molecules formed	% E_{vib}	% E_{rot}	% E_{trans}	% $E_{transferred}$
I	0				
II	(a) 65	57	8	30	5.24
	(b) 69	47	17	32	5.89
III	(a) 69	62	18	13	7.04
	(b) 70	46	27	23	4.26

Let us now consider the case of a global exothermic step. The reaction dynamics are then influenced by the shape of the energy potential. Two energy potential profiles can be considered [McCreery and Woken 1977b]: curve II (figure 4b) shows an energy diagram where the reaction energy is released partly in the entry valley of the system (the new molecule is still not formed) and partly in the exit valley (the new molecule is leaving the surface), and curve III (figure 4c) representing the situation where the whole energy is released in the entry valley, while the new molecule is still on the surface. Wolken's calculations show that the efficiency of the reaction is very similar in both cases. However, the analysis of the distribution of the energy in the different modes of the formed molecules shows differences. The vibrational energy contained on the desorbed molecules is much higher in the case of the potential III than in the case of potential II (table 3).

These calculations has been confirmed by Bernasek [Bernasek 1980], who compared the distribution of energy in N_2 when desorbing from a non-contaminated surface of iron (T_R = 400 K , T_V = 1200 K, T_{Fe}= 1050 K), and from an iron surface contaminated by sulfur, an electronegative component, (T_R = 400 K , T_V= 2600 K) (see figure 3).

3.2.2 Dynamics of a chemical surface reaction

The consequence of the previous results is that a global chemical surface reaction will be efficient if (i) the adsorption step is exothermic and occurs with a high efficiency, (this implies that the incoming molecules possess a threshold in ro-translational energy, and possibly some vibrational energy), and (ii) the recombination-desorption step is exothermic [Purvis III et al. 1979].

3.2.3 Catalytic recombination of atoms on a surface

When impinging on a surface, an atom falls to an energy potential well, and it is adsorbed. The recombination process leading to the formation of a molecule (A_2 or AB molecule) can occur according to different mechanisms: either two adspecies react and desorb as a molecule through a Langmuir-Hinshelwood mechanism, or one adspecies reacts with an atom in the gaseous phase and both desorb as a molecule through a Eley-Rideal mechanism [Wood and Wise 1967] and [Gicquel 1987].

The probability of recombination of atoms on a surface, defined by the coefficient γ, depends strongly on the mechanism involved, which is related to the gas / surface system, the surface temperature, and the type of solid lattice. For instance, the higher the temperature, the higher the surface mobility, and the more efficient the Langmuir-Hinshelwood mechanism.

Important studies of H atom recombination processes are due to B. Wood, H. Wise, M. Sancier [Wood and Wise 1967], [Wood and Wise 1962] and [Sancier and Wise 1969] and to Boudart and Kim [Kim and Boudart 1991]. For a quartz surface [Rousseau et al. 1994] at 300 K, γ_H ranges between 10^{-5} and 10^{-4}, while at 1200 K, it ranges between 10^{-3} and $5\ 10^{-2}$. A value of 0.1 for the recombination of H atoms on a diamond surface at 900 °C is reported in the literature [Goodwin 1993] and [Harris and Weiner 1993], which seems consistent with our measurements (see section 7) [Hassouni et al. 1994], [Gicquel et al. in preparation], and [Chenevier et al. 1995].

3.3 ENERGY TRANSFER TO A SURFACE

During collisions between particles and a surface, energy transfer from the gas to the surface or from the surface to the gas can occur [Gicquel 1987] and [Hassouni 1992]. In a plasma, if the surface is not submitted to external heating, the plasma particles transfer part of their energy to the surface. Each species arriving at the surface possesses its own internal energy. Each mode is able to be partly transferred to the surface, and the fraction transferred, β, depends on the symmetry of the species, the nature of the degree of freedom under consideration (vibration, rotation, translation) and the residence time of the species on the surface. The energy transfer depends also on the nature of the surface, its cleanliness, its temperature and its Debye temperature.

Interaction of neutrals with surfaces involves transfer of:
vibrational energy
translational energy
rotational energy
adsorption potential energy
recombination - desorption potential energy.

3.3.1 Vibrational energy [Hassouni 1992]

The energy transfer resulting from the interaction of vibrationally excited molecules with a surface can occur through a direct interaction with the phonons of the solid [Zdhanov and Kamrev 1982], through a direct electronic excitation of the surface (plasmons) [Miswich et al. 1985], through a dipolar interaction over a long distance [Zdhanov and Kamrev 1982] or through the creation of an electron-hole pair, that is through an adsorption process.

The first three processes are of less importance with respect to the fourth. As a matter of fact, the vibration frequencies of the molecules are usually different from the Debye frequencies of the solids, as well as from the surface electronic frequencies. In addition, in the case of H_2, no permanent dipole exists and there is a very low probability that the induced dipolar moment will interact with a diamond surface. On the contrary, the adsorption or physisorption steps allow a molecule to remain for some time on the surface or on the adspecies layer, which can lead to an efficient energy transfer. Consequently, the energy transfer efficiency depends on the efficiency of the adsorption step. The vibrational accommodation coefficient is defined as:

$$\beta_V = (E_V^{In} - E_V^{f}) / (E_V^{f} - E_S),$$

where E_V^{In} and E_V^{f} are respectively the vibrational energy of the molecule before and after the collision with the surface and E_S, the vibrational energy associated to the surface temperature; E_V^{f} and then β_v can be calculated as a function of the chemisorption probability [Hassouni 1992].

3.3.2 Translational energy transfer

Elastic collisions do not lead to kinetic energy transfer with surfaces, while inelastic collisions do, the interaction potential of which is controlled by the adsorption - desorption mechanism. As for the vibrational energy transfer, a translational accommodation coefficient is defined by:

$$\beta_t = (E_t^{In} - E_t^{f}) / (E_t^{f} - E_S).$$

E_t^{In} and E_t^{f} are respectively the translational energy of the molecule before and after the collision with the surface and E_S the translational energy corresponding to the surface temperature; β_t can be calculated knowing the probability of physisorption and chemisorption of the molecule [Hassouni 1992].

3.3.3 Rotational energy transfer [Hassouni 1992]

The rotational energy transfer mechanism is not known. However, significant energy transfer has been measured experimentally [Draper and Rosenblatt 1978], [Rosenblatt et al. 1977] and [Rosenblatt 1981]. In diamond deposition conditions, where the rotational excitation mode is in equilibrium with the translational mode, we will consider the apparent rotational energy transfer coefficient to be identical to the translational coefficient.

3.3.4 Adsorption step

During any adsorption, chemisorption, dissociative chemisorption or physisorption process, if the species remains an adspecies long enough, all of its potential energy is transferred to the substrate. If it is quickly desorbed, the energy transfer is low. Frankel [Frankel 1924] has calculated the residence time of a species chemisorbed (or physisorbed) on a surface. It is given by:

$$t_S = t_0 \exp (E / (k\, T_S))$$

where t_o represents the vibration characteristic time of the surface phonons, E the activation barrier, T_S the surface temperature, and k the Boltzmann constant. For the chemisorption of a species on a surface, t_S has the magnitude of 10^{-13} second.

3.3.5 Recombination-desorption step

During recombination-desorption processes, the global energy transfer depends both on the recombination coefficient γ and on the accommodation coefficient β. The accommodation coefficient β is defined as the part of energy transferred to the substrate per event of recombination. β and γ depend on the nature of the potential interaction of the species-substrate system and on the mechanism involved: Langmuir-Hinshelwood (L. H.) or Eley-Rideal (E. R.).

According to the process, the amount of energy available after the formation of the desorbed molecule is different: the L. H. mechanism needs two C-H bonds broken, while the E. R. mechanism requires only one. Consequently, the energy available after an E. R. process is higher than for L. H.. The reaction time is also different according to the process. The process involving an E. R. mechanism is faster than the one involving a L. H. mechanism (except for very high surface temperatures).

Thus, although the amount of energy available during a E. R. process is higher, owing to the much lower reaction time, the energy transfer to the surface is lower. As shown by Halpern and Rosner [Halpern and Rosner 1978] in the case of the system N / tungsten, at low temperature, the energy transfer to the surface (product $\beta\gamma$) is low as the E.R. mechanism is responsible for the recombination process, while the energy transfer is higher, at high temperatures because the L. H. mechanism is predominant.

As already discussed, the molecules desorb containing vibrational energy. Calculations made by Tully [Tully 1980, 1975] show that the molecules possess less vibrational energy in the case of a L.H. mechanism than in case of an E.R. mechanism. This is consistent with the fact that the desorbed molecule leaves the surface with a larger energy in the case of a E.R. mechanism. By a mathematical model simulating the reaction of atomic oxygen with carbon chemisorbed on platinum, Tully has shown that the E. R. mechanism leads to desorbed molecules possessing almost only vibrational energy, while only 10 % of the available energy is kept by the metal. The calculation done assuming a L. H. mechanism showed that the molecules possess much less vibrational energy. Simultaneously, the energy transfer to the metal has increased.

In conclusion, even if a reaction is very exothermic, the energy transfer to the surface resulting from the recombination-desorption might remain low in the case where the reaction takes place in a very short time through a E. R. mechanism. The formed molecule may desorb with a high vibrational excitation.

An L. H. mechanism mainly occurs for weak surface - atom bonds, and the resulting energy transfer is high. On the other hand, for strong surface - atom bonds, the E. R. mechanism is responsible for the recombination-desorption process, and the resulting energy transfer to the surface remains low. The desorbed molecule contains a large vibrational energy [Hall et al. 1988].

3.3.6 Recombination of ions on surfaces

Although minor species in weakly ionized plasmas, ions interact with surfaces through specific processes. In addition, ions, submitted to an acceleration in a cathodic sheath, may have gained kinetic energy and might consequently interact not only with the surface, but also within the volume of the solid. Sputtering of the surface, implantation of the ions into the substrate or activated migration of the adspecies might be decisive for the characteristics of the treated surface [Gicquel and Catherine 1991].

The kinetic energy transferred during a collision of an ion with a surface is given by [Hassouni 1992]:

$$DE = (4\, M_1\, M_2\, E_{kin} \sin^2 q) / (M_1 + M_2)^2$$

where M_1 and M_2 represent respectively the mass of the ion and the surface atom, q the incident angle, and E_{kin} the kinetic energy of the incoming particle.

Under microwave diamond deposition conditions, sheaths are strongly collisional and the substrate potential can be isolated from the cavity. Only low energy ions are allowed to hit the surface; highly energetic ions would destroy the H-atom terminated dangling bonds, which are necessary for diamond growth. Consequently, low kinetic energy transfer is expected under these conditions.

However, beyond kinetic energy, ions are able to transfer energy to the surface during their neutralization at the surface. As they approach a metallic surface, ions are neutralized, and the emission of secondary electrons results. Three main processes, well described by Hagstrum [Hagstrum 1975], are responsible for such phenomena:

- de-excitation process involving the emission of a radiation, which occurs in 10^{-8} s,
- de-excitation Auger process involving 2 electrons,
- resonant process where an electron is transferred from the metal to the ion in an identical energy level or an electron is transferred from the ion to the metal.

Schekhter [Schekhter 1937] has shown that the last two processes are more efficient than the first. The duration of these processes are short, 10^{-14} s, when compared to either radiation emission (10^{-8} s) or incorporation of a particle into the metal (10^{-12} s). In addition, most of the time, the Auger neutralization process is more efficient than the resonant process.

The yield of such a process is very high, and the probability of an ion hitting the surface to be reflected back as an ion is very low. For He^+ ions impinging on silicon with an energy of 100 eV, the probability of reflection [Parilis and Kishinerskii 1960] is $2.4 \cdot 10^{-5}$.

Owing to the efficiency of these processes, the interaction of low energy ions with a metallic surface is reduced to that of neutral species containing a few eV interacting with the surface. However, a surface electron has been transferred to the ion, and secondary electrons are emitted from the surface. The flux is given by J = A T exp(-

ef/kT), and the maximum energy of the emitted electrons is equal to E_i-2ef, where E_i is the ionization potential, ef is the work function of the metal, T the surface temperature and A is a constant. The energy transferred to the surface during the neutralization process is given by [Hassouni 1992]:

$$DE = E_i - ef - G_i (ef+Ec)$$

where G_i is the yield in secondary electrons and Ec is the kinetic energy of the secondary electrons [Hassouni 1992].

In the case where the ions are slow enough, they are reflected back to the gaseous phase as neutrals with part of their kinetic energy. In the case of molecular ions, the neutral molecules formed might have gained vibrational energy [Gicquel 1987].

As a conclusion, although the interaction of ions with surfaces may lead to the production of secondary electrons and of vibrationally excited molecules, in diamond deposition conditions, owing to their low population and to the electrical configuration of the substrate, their contribution to energy transfer will be neglected with respect to the contribution of the neutrals particles. However, the probability of neutralization of ions, as on metallic surfaces, will be considered as equal to unity.

3.3.7 *De-excitation of metastable and radiative species on surfaces*

Like the ions, electronically excited species are minor species in the plasmas considered. The radiative species, owing to their very short life time, are de-excited far from the surface, and do not have any influence on the surface reactions [Winters 1965, 1980].

In the case of metastable species, characterized by long life times, the de-excitation processes involve Auger or resonant transition mechanisms. The result is a reflection back of the species which have gained some vibrational excitation. However, as shown by Winters [Winters 1965, 1980] and [Winters et al. 1964], the potential energy contained in the metastable incoming species does not play any role in surface reactions such as chemisorption processes. The energy transfer resulting from the metastable de-excitation is difficult to evaluate, and calculations would imply the use of models where the molecule is simulated by a dipole interacting with the surface. To a first approximation, this type of interaction can reasonably be neglected when studying energy and mass transfers during diamond deposition in MWPECVD (microwave plasma enhanced CVD) reactors.

3.4 CONCLUSION

Plasma/surface interaction processes are numerous and complex. They are of extreme importance for the balance of mass and energy transfer, and can affect the volume, as well as influencing the control of the plasma characteristics. For instance, a strong consumption or production on surfaces of an easily ionized species can lead to a major change in the electrical characteristics of the plasma followed by changes in the energetic and chemical composition balances.

Under diamond deposition conditions, the main processes are limited to those resulting from the interaction of ground electronic state neutral species (molecules and radicals) containing rotational, translational and vibrational energy, and atoms with the growing diamond surface and the reactor walls. The consumption and production of species by these surfaces must be evaluated. They have to be taken into account in chemical kinetics models.

A good knowledge of the chemical reaction mechanisms and of their corresponding temperature dependent rate constants, as well as knowledge of the energy transfer are necessary for optimizing reactors. However, owing to the lack of microscopic data, macroscopic coefficients for mass and energy transfer (γ and β coefficients) will be introduced to allow modeling of the plasma and of the plasma / surface interface. Macroscopic coefficients can take into account the possible effects of internal energy contained in the incoming molecules on the surface reactions and on the energy transfers. Likewise, contamination of the diamond surface (non diamond phases, impurities such as B, N, O,...) and reactor walls might also influence the macroscopic coefficients.

3.5 REFERENCES

Asscher, M., Pollak, E., Somorjai, G.A. *Surf. Sci.* vol 149, p 146 (1989)

Bernasek, S.L. "Heterogeneous Reaction Dynamics", *Adv. Chem. Phys.*, 41, 477-513 (1980)

Bond, G.C. *Catalysis by metals* p 106 Academic Press, New York (1962)

Boszo, F. Ertl, G. and Weiss, M. *J. Catal.* 50, 519 (1977)

Boudart, M. and Djéga-Mariadassou, G. "Kinetics of heterogeneous catalytic reactions", *Physical Chemistry: Science and Engineering*, Ed. J. M. Prausnitz and L. Brewer. Princeton University Press / Princeton, N. J., (1984)

Chenevier, M., Gicquel, A., Cubertafon, J.C. *Proceedings of the 3rd International Conference on the Applications of Diamond Films and Related Materials. ADC'95* pp. 305-312. (1995). ed. Y. Tzeng, M. Yoshikawa, M. Murakawa and A. Feldman. Elsever Sciences publishers.

Draper, C.W. and Rosenblatt, G.M. *J. Chem. Phys.* vol 69, (4), 1465-72 (1978)

Frankel, J. *Zeitstschrift für Physik*, vol 26, 117 (1924)

Gicquel, A. *Doctorat d'Etat*, Université Pierre et Marie Curie, Paris VI (1987)

Goodwin, D.G. *J. Appl. Phys.* 74 (11) 6888-94 (1993)

Gregory, A.R., Gelb, A. and Silbey, R. *Surface Science* 74, 497-523 (1978)

Hagstrum, H.D. Phys. Rev. 96, 336 (1954) and *J. Vac. Sci. Technol.*, vol 12 no. 1, 7-16 (1975)

Hall, R.I., Cadez, I., Landau, M., Pichou, F., Schermann, C. *Phys. Rev. Lett,* vol 60, no. 4, 337-340 (1988)

Halpern, B.L., and Rosner, D.E. *J. Chem. Soc. Faraday Trans.* 160, 1883-1912 (1978)

Harris, J., Weiner, A.M. *J. Appl. Phys.* 74, 1022 (1993)

Hassouni, K. *Thèse de l'Université Pierre et Marie Curie* (1992)

Hassouni, K., Farhat, S. Scott, C.D., Gicquel, A. *Fourth Int. Conf. on New Diamond Science an Technology*, ed S. Saito, N. Fulimori, O. Fukunaga, M. Kamo, K. Kobashi, M Yoshikawa. MYU, Tokyo, 131-134, (1994)

Kim, Y.C., and. Boudart, M. *Langmuir*, 7, 2999 (1991)

Mc Creery, J.H. and Wolken, G. Jr. *J. Chem Phys.* Vol 66, no. 6, 2316-21 (1977a)

Mc Creery, J.H. and Wolken, G. Jr. *J. Chem Phys.* Vol 67, no. 6, 2551-59 (1977b)

Mc Creery, J.H. and Wolken, G. Jr. *J. Chem. Phys.,* Vol 65, no. 4, 1310-16 (1976)

Mc Creery, J.H. and Wolken, G. Jr. *J. Chem Phys.* Vol 63, no. 6, 2340-49 (1975)

Miswich, J., Houston,, P., Merill, R.P. *J. Chem. Phys.* vol 82, no.3, pp 1577-84 (1985)

Parilis, E.S., Kishinerskii, L.M. *Fiz. Tverd. Tela.* 3, 1219 (1960) (Sov. Phys. Solid State 3, 885 (1960)

Pomot, C. "Traitements de surface par plasmas-Initiation aux plasmas", *Formation Continue, Institut National Polytechnique de Grenoble, Fascicule II: Particules-surface*, pp.1-20

Purvis III, G.D., Redmon, M.J., Wolken, G.Jr. *J. Chem. Phys.* vol 83, no. 8, 1027-33 (1979)

Rosenblatt, G.M. *Acc. Chem. Res.* vol 14, 42-48 (1981)

Rosenblatt, G.M., Lemons, R.S., Draper, C.W. *J. Chem. Phys.* vol 67, (3), 1099-1107 (1977)

Rousseau, A., Granier, A., Gousset, G., Leprince, P. *J. Appl. Phys.* (1994) 1412-22 (1994)

Sancier, K.M., and Wise, H. *J. Chem. Phys.* 51, 1434 (1969)

Schekhter, S.S. *J. Exptl. Theoret. Phys.* (URSS) 7, 650 (1937)

Smith, W.V. *J. Chem. Phys.* 11, 110, (1943)

Tully, J.C. *J. Chem. Phys.* 73 (12) 6333-6342 (1980) and J. Chem. Phys. 62, (5) 1893 (1975)

Winters, H.F. *Topics in current chemistry*, no. 94, 69-124 (1980), J. Chem. Phys., 43, 926 (1965)

Winters, H.F., Horne, D.E., Donaldson, E.E. *J. Chem. Phys.* 41, 2766 (1964)

Wolken, G.. Jr. *J. Chem Phys.* Vol 68, no. 10, 4338-42 (1978)

Wood, H and Wise, B.J. "Reactive Collisions between gas and surface atoms" *Adv. At. Mo. Phys.* 3 291-353 (1967)

Wood, B.J., and Wise, H. *J. Chem. Phys.* 65, 1976 (1961) and J. Chem. Phys. 66, 1049 (1962)

Zdhanov, V.P. and Kamrev, K.I. *Cat. Rev. Sci. Eng.* Vol 24, no. 3, pp373-413 (1982)

References subsection 2:

1. H. Wood and B. J. Wise, "Reactive Collisions between gas and surface atoms" *Adv. At. Mo. Phys.* **3** 291-353 (1967)

2. G. C. Bond, "*Catalysis by metals*" p 106 Academic Press, New York (1962)

3. C. Pomot, "*Traitements de surface par plasmas-Initiation aux plasmas*", Formation Continue, Institut National Polytechnique de Grenoble, Fascicule II: Particules-surface, pp.1-20

4. M. Boudart and G. Djéga-Mariadassou, "Kinetics of heterogeneous catalytic reactions", *Physical Chemistry: Science and Engineering*, Ed. J. M. Prausnitz and L. Brewer, Princeton University Press / Princeton, N. J., (1984)

5. F. Boszo, G. Ertl and M. Weiss, *J. Catal.* **50**, 519 (1977)

6. J. H. Mc Creery and G. Jr. Wolken, *J. Chem. Phys.*, Vol **65**, no. 4, 1310-16 (1976)

7. A. R. Gregory, A. Gelb, R. Silbey, *Surface Science* **74**, 497-523 (1978)

8. J. H. Mc Creery and G. Jr. Wolken, *J. Chem Phys.* Vol **63**, no. 6, 2340-49 (1975)

9. J. H. Mc Creery and G. Jr. Wolken, *J. Chem Phys.* Vol **66**, no. 6, 2316-21 (1977)

10. G. Jr. Wolken, *J. Chem Phys.* Vol **68**, no. 10, 4338-42 (1978)

11. J. H. Mc Creery and G. Jr. Wolken, *J. Chem Phys.* Vol **67**, no. 6, 2551-59 (1977)

12. A. Gicquel, *Doctorat d'Etat*, Université Pierre et Marie Curie, Paris VI (1987)

13. S. L. Bernasek, "Heterogeneous Reaction Dynamics", *Adv. Chem. Phys.*, **41**, 477-513 (1980)

14. A. Rousseau, A. Granier, G. Gousset, P. Leprince *J. Appl. Phys.* (1994) 1412-22 (1994)

15. W. V. Smith, *J. Chem. Phys.* **11**, 110, (1943)

16. B. J. Wood and H. Wise, *J. Chem. Phys.* **65**, 1976 (1961) and *J. Chem. Phys.* **66**, 1049 (1962)

17. K. M. Sancier and H. Wise *J. Chem. Phys.* **51**, 1434 (1969)

18. Y. C. Kim and M. Boudart, *Langmuir*, 7, 2999 (1991)

19. D. G. Goodwin, *J. Appl. Phys.* **74** (11) 6888-94 (1993)

20. K. Hassouni, S. Farhat, C. D. Scott, A. Gicquel *Fourth Int. Conf. on New Diamond Science an Technology*, ed S. Saito, N. Fulimori, O. Fukunaga, M. Kamo, K. Kobashi, M Yoshikawa, MYU, Tokyo, 131-134, (1994)

21. K. Hassouni, *thèse de l'Université Pierre et Marie Curie* (1992)

22. V. P. Zdhanov, K. I. Kamrev, *Cat. Rev. Sci. Eng.* Vol **24**, no. 3, pp373-413 (1982)

23. J. Miswich, P. Houston, R.P. Merill, *J. Chem. Phys.* vol **82**, no.3, pp 1577-84 (1985)

24. C. W. Draper, G. M. Rosenblatt, *J. Chem. Phys.* vol **69**, (4), 1465-72 (1978)

25. G. M. Rosenblatt, R. S. Lemons, C. W. Draper, *J. Chem. Phys.* vol **67**, (3), 1099-1107 (1977)

26. G. M. Rosenblatt, *Acc. Chem. Res.* vol **14**, 42-48 (1981)

27. J. Frankel, *Zeitstschrift für Physik*, vol **26**, 117 (1924)

28. M. Asscher, E. Pollak, G. A. Somorjai, *Surf. Sci.* vol **149**, p 146 (1989)

29. R. I. Hall, I. Cadez, M. Landau, F. Pichou, C. Schermann, *Phys. Rev. Lett*, vol **60**, no. 4, 337-340 (1988)

30. J. Harris, A. M. Weiner, *J. Appl. Phys.* **74**, 1022 (1993)

31. J. C. Tully, *J. Chem. Phys.* **73** (12) 6333-6342 (1980) and *J. Chem. Phys.* **62**, (5) 1893 (1975)

32. H. D. Hagstrum, *Phys. Rev.* **96**, 336 (1954) and *J. Vac. Sci. Technol.*, vol **12** no. 1, 7-16 (1975)

33. S. S. Schekhter, *J. Exptl. Theoret. Phys.* (URSS) **7**, 650 (1937)

34. E. S. Parilis, L. M. Kishinerskii, *Fiz. Tverd. Tela.* **3**, 1219 (1960) (*Sov. Phys. Solid State*) **3**, 885 (1960)

35. H. F. Winters, *Topics in current chemistry*, no. 94, 69-124 (1980), *J. Chem. Phys.*, **43**, 926 (1965)

36. H. F. Winters, D. E Horne, E. E. Donaldson, *J. Chem. Phys.* **41**, 2766 (1964)

37. A. Gicquel, Y. Catherine, *Journal de Physique IV*, C2-343-356, Colloque C2, suppl. Journal de Physique II, (1991)

38. G. D. Purvis III, M. J. Redmon, G. Wolken Jr., *J. Chem. Phys.* vol **83**, no. 8, 1027-33 (1979)

39. A. Gicquel, M. Chenevier, K. Hassouni, J. C. Cubertafon, Y. Breton, A. Tserepi, in preparation

40. B. L. Halpern and D. E. Rosner, *J. Chem. Soc.* Faraday Trans. **160**, 1883-1912 (1978)

Chapter 15

SECTION 4: THE BOLTZMANN EQUATION

Matt Gordon

University of Arkansas

Fayetteville, Arkansas

4.1 FORMULATION AND SIMPLIFICATION

Perhaps the most difficult aspect of modeling microwave CVD reactors is the inherent nonequilibrium. This nonequilibrium is initiated by the coupling of the microwave energy to the charged particles. Because electrons are much lighter than ions, the electrons absorb the majority of the microwave energy. The other plasma constituents gain energy through collisions with the electrons. In the presence of the electromagnetic fields, electrons will always have more energy than will the neutral species. If the collision frequency, which scales with pressure, were sufficiently high, however, this difference in energy would be small. Under typical microwave CVD operating conditions, collisions frequencies are relatively low, and the difference in energy is appreciable. The electrons attain energies around 1.2-eV or roughly 4 times greater than those attained by the neutrals. Plasmas exhibiting this type of nonequilibrium are known as two-temperature plasmas.

Although more difficult to model than equilibrium plasmas, two-temperature plasmas are routinely modeled if both the electrons' and the neutrals' distribution of energies are Maxwellian. Even if the collision frequency between the electrons and the neutrals is insufficient to equilibrate their energies, the collision frequency may still be sufficient to Maxwellianize the distribution of energy within each particle group. Although the neutrals can often be described by a Maxwellian energy distribution function, the electrons cannot for the majority of microwave CVD reactors. This form of nonequilibrium hinders the development of accurate models because transport properties and reaction rates are very sensitive to the precise distribution of energy among electrons. Calculation of the electron energy distribution function (EEDF) requires the solution of the Boltzmann equation and is the focus of this chapter.

In its most general form the Boltzmann equation, Equation 1, simply represents a balance between all events which produce an electron with a given velocity against those which eliminate them (note that we will derive information about electron energy from electron velocity).

$$\frac{\partial}{\partial t}\left(n_e f_e\right) + \vec{c}\cdot\nabla n_e f_e + \frac{\vec{F}_e}{m_e}\cdot n_e \nabla_c f_e = \sum R_{er} \tag{1}$$

The EVDF is normalized such that:

$$\int_{-\infty}^{\infty}\int_{-\infty}^{\infty}\int_{-\infty}^{\infty} f_e d_{cx} d_{cy} d_{cz} = \int_{-\infty}^{\infty} f_e d_c^3 = 1 \tag{2}$$

In Equation 1, n_e represents the total electron number density, f_e represents the electron velocity distribution function (EVDF) and is a function of both spatial location and velocity class, **c** represents the electron velocity, $\mathbf{F}_e$ represents the external forces, m_e represents the electron mass, and R_{er} represents collisional rates between electrons in a particular velocity class and species r. The first term symbolizes the change with time of electrons with a particular velocity in a specific volume in space. The second term symbolizes the spatial divergence of electrons with a particular velocity due to convection in and out of a specific volume in space. The third term symbolizes the velocity divergence of electrons due to external forces which accelerate and decelerate electrons in and out of a particular velocity class in a specific volume in space. The term on the right hand side of Equation 1 symbolizes the net rate of increase or decrease of electrons in a particular velocity class due to collisions- elastic and inelastic- with species r. This last term is the most difficult to quantify and therefore presents the greatest hindrance to solving- either analytically or numerically- the Boltzmann equation. Fortunately, there are several simplifying assumptions which enable tractable solutions. The validity of these assumptions will be discussed throughout this chapter.

Perhaps the most well known solution to Equation 1 results from the assumption of equilibrium. Steady state equilibrium eliminates the first (transient) term. Spatial equilibrium- or no gradients- eliminates the second term. Mechanical equilibrium- or no external forces- eliminates the third term. If only elastic collisions among electrons are considered, then equilibrium is attained when every collision which produces an electron in a given velocity class is balanced by a collision which consumes an electron in a given velocity class. Under these conditions of equilibrium, the solution to Equation 1 is the Maxwell-Boltzmann distribution:

$$f_M = \left(\frac{m_e}{2\pi k\,T_e}\right)^{3/2} e^{-m_e/2kT_e\left(\vec{c}_x^{\,2} + \vec{c}_y^{\,2} + \vec{c}_z^{\,2}\right)} \tag{3}$$

Although, in practice, true equilibrium is never attained, if elastic collisions among electrons dominate all other terms in Equation 1, then the EVDF given in Equation 3 is valid and depends on only the electron temperature.

In the more general case of nonequilibrium, when additional terms in Equation 1 must be retained, other simplifications are implemented. To begin with, we assume that the EVDF is nearly isotropic such that it can be replaced by the first two terms of a spherical harmonic expansion (MacDonald, 1966, pp. 39-42):

$$f_e(\vec{c}) = f_o(C) + \vec{f}_1(C)\cdot\frac{\vec{c}}{C} + \cdots \tag{4}$$

where C represents the magnitude of velocity. The first term, f_o, in this expansion represents the isotropic part, and the second term, f_1, represents small deviations due to external forces (e.g. electromagnetic fields). Note that this second term, which contains directional information, is a vector. Although not demonstrated here, this derivation implicitly requires that convective speeds be much less than the average electron speed. The expanded EVDF is still normalized according to Equation 2. It is useful, however, to replace the differential velocity space term, d^3c, with its spherical coordinate equivalent, $C^2 \sin\theta\, d\phi\, d\theta\, dC$ (Figure 1). For the isotropic term, f_o, the spherical differential volume can be integrated over the two directional angles to obtain $4\pi C^2 dC$.

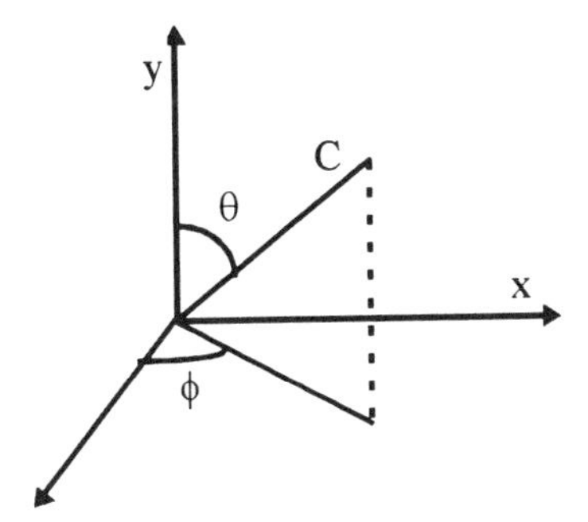

Figure 1. Spherical coordinate schematic.

The next assumptions affect the right side of Equation 1. As was done with the EVDF, the collision terms are expanded using spherical harmonics. Further, for elastic collisions between electrons and heavy- either neutral or charged- particles, we take advantage of the large mass disparity between collision partners such that the change in energy for the light electron is small. Using these assumptions, Equations 5 and 6 can be derived (Mitchner and Kruger, 1973, pp. 321-368):

$$\frac{D}{Dt}\left(n_e f_o\right) + \frac{C}{3}\nabla\cdot\left(n_e \vec{f}_1\right) - \frac{e n_e}{3 m_e C^2}\vec{E}\cdot\frac{\partial}{\partial C}\left(C^2 \vec{f}_1\right) - \frac{C}{3}\nabla\cdot\vec{u}\, n_e \frac{\partial f_o}{\partial C} \tag{5}$$
$$= n_e \sum_h \frac{m_e}{m_h}\frac{1}{C^2}\frac{\partial}{\partial C}\left[C^3\left(f_o + \frac{kT_h}{m_e C}\frac{\partial f_o}{\partial C}\right)\nu_m\right] + R^o_{ee} + \sum_r R^o_{ier}$$

$$\frac{\partial}{\partial t}\left(n_e \vec{f}_1\right) + C\nabla\left(n_e f_o\right) + \vec{a}\, n_e \frac{\partial f_o}{\partial C} - \frac{e}{m_e}\vec{B}\times n_e \vec{f}_1 = -n_e \sum_h \nu_m\left(\vec{f}_1 + \vec{U}_h \frac{\partial f_o}{\partial C}\right) + R^1_{ee} \tag{6}$$

where D/dt represents the total derivative, e represents the charge of the electron, **E** and **B** represent the electric and magnetic fields, respectively, **a** represents accelerations, **u** represents the convective velocity, **U** represents the diffusion velocity, m_h represents the mass of a heavy particle, k represents Boltzmann's constant, T_h represents the temperature of the heavy specie, ν_m represents the momentum collision frequency between the electron and the heavy specie, R_{ee}^o and R_{ee}^1 represent the isotropic and first anisotropic terms, respectively, of the spherical expansion of the electron-electron collision term, and R_{ier}^o represents the isotropic term for all inelastic processes with species r (the corresponding term R_{eir}^1 is assumed negligible compared to the anisotropic elastic loss term).

Expressions for the electron-electron collision terms are quite involved. Their derivation relies on the assumption that the resulting deflections and thus the resulting velocity changes are small. This assumption leads to the Fokker-Planck collision term (Mitchner and Kruger, 1973, pp. 338-344).

Although Equations 5 and 6 which govern f_o and f_l are coupled, typical simplifications for microwave CVD reactor conditions allow Equation 6 to be solved for the anisotropic term which is then substituted into Equation 5. The resulting equation is the starting point for most solutions (numerical and analytical) to the Boltzmann equation. These simplifications to Equation 6 include steady state (eliminating the first term) and spatial homogeneity (eliminating both the second term and the diffusive velocities) assumptions. For weakly ionized plasmas, electron-electron collisions can be neglected. If there are no external magnetic fields, and if the acceleration term is due to only a DC electric field with just one directional component, then Equation 6 can be expressed as follows:

$$f_1 = \frac{eE}{m_e \sum_h \nu_m} \frac{\partial f_o}{\partial C} \tag{7}$$

Note that most, if not all, EEDF solvers utilize Equation 7 even when the above assumptions are not strictly valid. For example, when electron-electron collisions are included, the anisotropic term, R_{ee}^1, is still neglected on the assumption that its contribution is negligible. Similarly, when a transient solution is desired, the transient term is still neglected in Equation 6 on the assumption that the temporal changes to $\mathbf{f_l}$ are small compared to the other terms retained in Equation 6. Caution should therefore be exercised if conditions to be modeled strongly contradict the implied assumptions noted above.

In Equation 7, note that, because the electric field has only one directional component, f_l also has only one directional component and thus neither are expressed as vectors. Using the same assumptions which led to Equation 7, we can derive Equation 8 by substituting Equation 7 into Equation 5:

$$-\frac{en_e}{3m_eC^2} E \frac{\partial}{\partial C}\left(C^2 \frac{eE}{m_e \sum_h \nu_m} \frac{\partial f_o}{\partial C} \right) = n_e \sum_h \frac{m_e}{m_h} \frac{1}{C^2} \frac{\partial}{\partial C}\left[C^3 \left(f_o + \frac{kT_h}{m_eC} \frac{\partial f_o}{\partial C} \right) \nu_m \right] \tag{8}$$

As noted, these equations are valid only for a constant or DC electric field. For AC electric fields, there are two simplifying cases of interest for which the same equations are still valid depending on the applied field frequency and on the electron energy loss frequency (Jiang and Economou, 1993). The electron energy loss frequency is related to the electron momentum frequency and is used to determine the characteristic time required for the energetic electrons to collisionally equilibrate with the heavy particles. The amount of energy exchanged elastically per electron-heavy collision is proportional to twice the ratio of the electron to heavy masses. If inelastic energy losses are relevant, the amount of energy exchanged is larger by a factor known as the energy loss factor (Mitchner and Kruger, 1973, pp. 51-52). Since the number of collisions is characterized by the electron momentum frequency, the electron energy loss, or relaxation, frequency is typically 2 to 3 orders of magnitude lower than the electron momentum frequency.

If the frequency of the applied field is much less than the electron energy loss collision frequency, then the plasma will adapt to the changing field quickly and will be quasi-stationary. The EEDF at any time t will therefore be the steady state solution of the Boltzmann equation using the magnitude of the electric filed at time t. If the frequency of the applied field is much greater than the electron loss collision frequency, then the electrons cannot respond to the changing field and so the EEDF does not change in time. In this case, accurate predictions from the DC model are possible by using an effective electric field defined by:

$$E_{eff} = \frac{E_o/\sqrt{2}}{\sqrt{\left(\nu / \sum_h \nu_m\right)^2 + 1}} \tag{9}$$

where ν is the frequency of the applied field, $E=E_o\sin(2\pi\nu t)$ and ν_m is the electron momentum frequency. Note that if the applied frequency is much less than the momentum frequency, then the effective field is simply the root mean square (rms) of the applied oscillating electric field. For typical microwave reactors, n is equal to $2.45\text{x}10^9\ s^{-1}$. For a plasma at 40 Torr, a gas temperature near 2,500 K, and an average electron temperature of 10,000 K, the summed total electron-heavy momentum collision frequency is about $1.6\text{x}10^{10}\ s^{-1}$. For molecular hydrogen, twice the ratios of the masses is roughly 1/1800 and the energy loss factor is near 10. Thus the electron energy loss collision frequency is about 1/180 n_m, or $8.5\text{x}10^7\ s^{-1}$. Thus, for these conditions, an effective electric field which is about 99% of E_{rms} is appropriate.

For true AC conditions, programs such as ELENDIF still rely on the assumption that $\delta f_1/\delta t$ can be neglected such that Equation 7 is valid. From there, numerical schemes rely on choosing appropriate time steps to solve for the time-dependent EVDF without stability problems.

Often, it is desirable to work with an EEDF rather than an EVDF by using the relation $\varepsilon=mC^2/2$. Making this substitution, we obtain:

$$-\frac{2en_e}{3m_e}\frac{E}{\varepsilon^{1/2}}\frac{\partial}{\partial\varepsilon}\left(\frac{eE}{\sum_h \nu_m}\varepsilon^{3/2}\frac{\partial f_o}{\partial\varepsilon}\right) = n_e\sum_h\frac{m_e}{m_h}\frac{2}{\varepsilon^{1/2}}\frac{\partial}{\partial\varepsilon}\left[\varepsilon^{3/2}\left(f_o + kT_h\frac{\partial f_o}{\partial\varepsilon}\right)\nu_m\right] \tag{10}$$

where the EEDF is normalized such that:

$$\int_0^\infty \varepsilon^{1/2} f_0(\varepsilon)d\varepsilon = 1 \tag{11}$$

To convert between the EVDF and the EEDF, the following expression is convenient:

$$f_o(C) = \frac{\sqrt{2}\,m^{3/2}}{8\pi}f_o(\varepsilon) \tag{12}$$

Equation 12 is derived using the fact that both $4\pi C^2 f(C)dC$ and $f(\varepsilon)\varepsilon^{1/2}d\varepsilon$ symbolize the fraction of electrons in a differential speed or energy volume. The defining relation expressed in Equation 11 must be known before such a transformation can be derived. In some cases, for example, the EEDF will be normalized such that $\int f_o(\varepsilon)d\varepsilon=1$. In this case, Equation 12 is not valid.

In anticipation of numerical solutions which will incorporate collisional cross section data, we will replace the electron-heavy momentum collision frequency, ν_m, with the product $\chi_h n_o \sigma_m C$ where χ_h is the mol fraction of species h, n_o is the total neutral number density, and σ_m is the electron-heavy momentum cross section. Substitution of this expression into Equation 10 yields, after some rearrangement:

$$-\frac{\sqrt{2}n_o n_e}{3m_e^{1/2}\varepsilon^{1/2}}\left(\frac{eE}{n_o}\right)^2\frac{\partial}{\partial\varepsilon}\left(\frac{\varepsilon}{\sum_h \chi_h\sigma_{mh}}\frac{\partial f_o}{\partial\varepsilon}\right) = n_e n_o\sum_h \chi_h\frac{m_e^{1/2}}{m_h}\frac{2\sqrt{2}}{\varepsilon^{1/2}}\frac{\partial}{\partial\varepsilon}\left[\varepsilon^2\left(f_o + kT_h\frac{\partial f_o}{\partial\varepsilon}\right)\sigma_m\right] \tag{13}$$

If there is more than one heavy specie, then the summation in the denominator of the left hand term can be thought of as representing the average cross section for momentum exchange between an electron and all heavies. For singly ionized plasmas, the total neutral number density is the total number density (P/kT) less twice the electron number density (see Equation 44). For weakly ionized gases, then, n_o is approximated well by the total number density. Compared to similar steady state derivations elsewhere, Equation 13 will likely differ by a factor of $(2/\varepsilon m_e)^{1/2}$ which appears on both sides. Though this extra factor can be divided out, it is important to remember that, in general, there will be other terms resulting from transients, electron-electron collision, inelastic collisions, etc. which must then also be appropriately modified. The approach here is to develop each term independently from the others so that any may be combined as necessary back in Equation 5.

Inelastic collisions often strongly affect the EEDF and therefore must be included. This inelastic term is a subset of the term R_{er} in Equation 1. We have already discussed elastic collisions between electrons and heavies, and between electrons and electrons. In an inelastic process, there is a change of the internal energy of an atom or molecule. This change of internal energy can be rotational, vibrational, or electronic. Since electrons have no internal energy, this inelastic term is only applicable to electron-neutral collisions (electron-ion collisions, though also relevant, are neglected because there are significantly fewer ions than neutrals). There are two reactions which involve an electron of energy e:

$$\begin{aligned} e(\varepsilon) + A &\Leftrightarrow e(\varepsilon - \varepsilon^*) + A^* \\ e(\varepsilon+\varepsilon^*) + A &\Leftrightarrow e(\varepsilon) + A^* \end{aligned} \tag{14}$$

where A^* represents an excited state (rotational, vibrational, or electronic) of an atom or molecule A, and ε^* represents the energy required to excite the atom or molecule. In the reverse direction (right to left as written in Equation 14), the reactions represent superelastic reactions and are often neglected since the populations of the excited states are typically negligible. In any case, each reaction either produces or depletes an electron in a given energy class depending on the direction each reaction proceeds. The development of R_{ie}^{o} begins by considering the net rate of energy gain per unit volume as a result of inelastic collisions for electrons in the energy range de. For every reaction which occurs, an energy equal to $1/2m_eC^2$ is gained or lost. The differential volume in velocity space containing the number of electrons involved is $4\pi C^2 dC$. Converting from speed to energy and multiplying these terms by R_{ie}^{o}, we obtain the desired rate of energy gain (or loss). A similar development for the pair of reactions in Equation 14 leads to (Mitchner and Kruger, 1973, p. 471):

$$
\begin{aligned}
R_{ie}^{o}\varepsilon\frac{4\pi}{m_e}\sqrt{\frac{2\varepsilon}{m_e}}d\varepsilon = & -n_e n_A f_o(\varepsilon)\sqrt{\frac{2\varepsilon}{m_e}}\sigma^{A\to A^*}(\varepsilon)\varepsilon\frac{4\pi}{m_e}\sqrt{\frac{2\varepsilon}{m_e}}d\varepsilon - n_e n_{A^*} f_o(\varepsilon)\sqrt{\frac{2\varepsilon}{m_e}}\sigma^{A^*\to A}(\varepsilon)\varepsilon\frac{4\pi}{m_e}\sqrt{\frac{2\varepsilon}{m_e}}d\varepsilon \\
& + n_e n_{A^*} f_o(\varepsilon-\varepsilon^*)\sqrt{\frac{2}{m_e}(\varepsilon-\varepsilon^*)}\sigma^{A^*\to A}(\varepsilon-\varepsilon^*)\varepsilon\frac{4\pi}{m_e}\sqrt{\frac{2}{m_e}(\varepsilon-\varepsilon^*)}d(\varepsilon-\varepsilon^*) \\
& + n_e n_A f_o(\varepsilon+\varepsilon^*)\sqrt{\frac{2}{m_e}(\varepsilon+\varepsilon^*)}\sigma^{A\to A^*}(\varepsilon+\varepsilon^*)\varepsilon\frac{4\pi}{m_e}\sqrt{\frac{2}{m_e}(\varepsilon+\varepsilon^*)}d(\varepsilon+\varepsilon^*)
\end{aligned}
\tag{15}
$$

The first term on the right hand side represents the loss of energy due to the forward direction of the first reaction in Equation 14, the second term represents the loss of energy due to the reverse direction of the second reaction in Equation 14, the third term represent the gain of energy due to the reverse direction of the first reaction in Equation 14, and the fourth term represents the gain of energy due to the forward reaction of the second reaction in Equation 14. If the populations of the excited states are negligible, which is often the case, the second and the third terms can be omitted. Note that the first reaction in Equation 14 is valid only for electrons with energy greater than ε*. Thus, the forward direction of the second reaction in Equation 14 can often be neglected since the number of electrons with energy greater than twice ε* is very small. In any case, numerically, if the cross sections are known, all reactions can be included very easily. Equation 15 can be simplified to obtain:

$$
\begin{aligned}
R_{ie}^{o} = & \sum_j -n_e n_A f_o(\varepsilon)\sqrt{\frac{2\varepsilon}{m_e}}\sigma_j^{A\to A^*}(\varepsilon) - n_e n_{A^*} f_o(\varepsilon)\sqrt{\frac{2\varepsilon}{m_e}}\sigma_j^{A^*\to A}(\varepsilon) \\
& + n_e n_{A^*}(\varepsilon-\varepsilon^*) f_o(\varepsilon-\varepsilon^*)\sqrt{\frac{2}{\varepsilon m_e}}\sigma_j^{A^*\to A}(\varepsilon-\varepsilon^*) \\
& + n_e n_A(\varepsilon+\varepsilon^*) f_o(\varepsilon+\varepsilon^*)\sqrt{\frac{2}{m_e\varepsilon}}\sigma_j^{A\to A^*}(\varepsilon+\varepsilon^*)
\end{aligned}
\tag{16}
$$

Although not developed here, similar terms for ionization, including the secondary electrons produced in this process, attachment and recombination, photon-electron (free-free) processes, an external source of electrons such as an electron beam, and diffusion have been included in commercially available EEDF solvers such as ELENDIF (Morgan and Penetrante, 1990).

4.2 TRANSPORT COEFFICIENTS AND RATE EQUATIONS

Once the Boltzmann equation has been solved, one can infer much information about the plasma under study. Rate equations for reactions involving electrons can be calculated from the isotropic part of the EVDF. For example, consider the following electron-atom ionization process:

$$e + a \rightarrow e + a^{+} + e \tag{17}$$

The energy required to ionize the atom comes from the kinetic energy of the collision partners. Because of the mass disparity between electrons and neutrals, the majority of this kinetic energy is delivered by the electron. Thus, only electrons with an energy (or velocity) greater than the atom's ionization energy will contribute to the ionization process. In general, then, the rate coefficient for this process is obtained from the following expression:

$$k^{i}_{ea}(C) = k^{i}_{ea}(\varepsilon) = \int_0^\infty f_o(C)C\sigma(C)4\pi c^2 dC = \sqrt{\frac{2}{m_e}}\int_0^\infty f_o(\varepsilon)\varepsilon\sigma(\varepsilon)d\varepsilon \tag{18}$$

where $\sigma(C)/\sigma(e)$ is the speed/energy dependent cross section for a particular reaction. As demonstrated in Section 4.4, these rate coefficients are required for the development of a collisional-radiative model.

In contrast to the evaluation of rate coefficients which depend on the isotropic portion of the EVDF, the determination of transport properties depends on the anisotropic portion of the EVDF. If the EVDF were entirely isotropic, by definition there would be no net motion of electrons and therefore no contribution to the transport properties. Before calculating the transport properties from the anisotropic part of the EVDF, it is instructive to investigate electron drift and mobility, and electron conductivity using the Langevin equation. This equation is derived for a gas in which the electrons are free to move in a stationary uniform background of ions and neutrals which provide a viscous damping force. Writing Newton's second law, we obtain

$$m_e \frac{dU_e}{dt} = -eE + F_e \tag{19}$$

where U_e is the electron's drift velocity in the direction of the electric field, E, and F_e is the average force experienced by the electron as a result of collisions with other particles. This average force can be approximated as the average rate at which an electron loses momentum in collisions, or

$$F_e \approx -m_e \sum_h \nu_m U_e \tag{20}$$

This derivation assumes that, on the average, ν_m times per second the electron loses all of its momentum in the direction of the electric field. If we next assume that the electric field is oscillatory with a frequency ωw such that $E(t)=E_o e^{j\omega t}$, and look for a solution for drift velocity with the same form ($U_e(t)=U_o e^{j\omega t}$), we obtain the following:

$$j\omega\, m_e U_o = -eE_o - m_e U_o \sum_h \nu_m$$

$$\therefore U_o = \frac{eE_o}{m_e\left(\sum_h \nu_m + j\omega\right)} \tag{21}$$

We see that the drift velocity is linearly dependent on the electric field. The proportionality constant is defined as the electron mobility. In equation form,

$$U_o \equiv -\mu_e E_o$$

$$\mu_e = \frac{e}{m_e\left(\sum_h \nu_m + j\omega\right)} \tag{22}$$

The electrical conductivity due to the electrons, σ_e, can then be related to the mobility by equating two separate expressions for the conduction current density, J, as follows:

$$J_o = \sigma_e E_o = -n_e e U_o = n_e e \mu_e E_o = \frac{n_e e^2 E_o}{m_e\left(\sum_h \nu_m + j\omega\right)} \tag{23}$$

Thus, the conductivity of the plasma medium due to the electrons is:

$$\sigma_e = n_e e \mu_e = \frac{n_e e^2}{m_e\left(\sum_h \nu_m + j\omega\right)} \tag{24}$$

Since the drift velocity of the electrons is much greater than that of the ions, the electrical conductivity derived for the electrons is also valid for the plasma as a whole.

The above derivations are independent of the actual EVDF assuming that the average momentum collision frequency is known and independent of velocity. In general, however, transport properties are dependent on the anisotropic portion of the EVDF as described at the beginning of this section. If elastic collisions predominate, then mobility and conductivity coefficients can be determined accurately using the first two terms of the spherical harmonic expansion as described in Section 4.1. We assume that the externally applied electric field is one dimensional in space and sinusoidal in time (e.g. $E(z,t)=E_z e^{j\omega t}$). The expected current density will then also be one dimensional in space with the same time dependence. If we again let U_z represent the electron drift velocity due to the applied electric field, then

$$J_z = -n_e e U_z = -n_e e \int c_z f(\vec{c}) d^3 c \tag{25}$$

where, by definition, the z-drift velocity is the average velocity component in the z-direction and is obtained by integrating the z-velocity over the entire distribution function. We see that, for an isotropic EVDF, the average drift velocity and thus the current density will be zero. For a one dimensional electric field, we need only expand the EVDF in one direction. Thus, we can rewrite Equation 25 as:

$$J_z = -n_e e \int_0^\infty \int_0^\pi (C\cos\theta_l)(f_o + f_l\cos\theta_l) 2\pi C^2 \sin\theta_l \, d\theta_l \, dC \tag{26}$$

Note that we used the relations $c_z = C\cos\theta$ and $d^3c = C^2 \sin\theta d\theta d\phi dC$ (Figure 1) and have integrated over ϕ. To evaluate this integral, we need to solve Equation 6 for f_l in terms of f_o. If we assume that f_l has the same temporal dependence as does the electric field, that f_l is uniform in space, that we can neglect the heavy diffusion velocity, and that we can neglect electron-electron elastic collisions, then

$$f_l = \frac{eE_z}{m_e} \frac{\partial f_o / \partial C}{\left(\sum_h \nu_m + j\omega\right)} \tag{27}$$

Substituting this expression in Equation 26 and noting that the integral over the isotropic f_o is zero, we obtain the following equation for the current density:

$$\begin{aligned} J_z = \sigma_e E_z &= \frac{-n_e e^2 E_z}{m} 2\pi \int_0^\infty \int_0^\pi \frac{\partial f_o / \partial C}{\sum_h \nu_m + j\omega} C^3 \cos^2\theta_l \sin\theta_l \, d\theta_l \, dC \\ &= \frac{-4\pi n_e e^2}{3 m_e} E_z \int_0^\infty \frac{C^3 \partial f_o / \partial C}{\sum_h \nu_m + j\omega} dC \end{aligned} \tag{28}$$

If $\Sigma\nu_m$ is independent of speed, then, independent of the distribution function shape,

$$\sigma_e = \frac{n_e e^2}{m\left(\sum_h \nu_m + j\omega\right)} \tag{29}$$

which is the same expression as Equation 24 which we derived from the Langevin equation. Equation 28 is valid for a non-Maxwellian f_o. If the EVDF is Maxwellian, however, then the DC electrical conductivity is

$$\sigma_e = \frac{n_e e^2}{3kT_e} \left\langle \frac{C_e^2}{\sum_h \nu_m} \right\rangle \tag{30}$$

where <> signifies a value averaged over the distribution function.

In all cases, mobility can be obtained directly from the electrical conductivity according using the relation $\mu_e=\sigma_e/\nu_e e$ (Equation 24).

Although the electron drift velocity is analogous to diffusion, these effects arise from different causes. Drift velocity results from an applied electric field. Diffusion arises from spatial electron density concentration densities. In any case, both are derived from the EVDF. To determine the diffusion coefficient, we first must determine the diffusion flux ($\Gamma_z=n_z\langle v_z\rangle$) where v_z is the one-dimensional diffusion velocity of electrons due to spatial gradients. We can then express the diffusion flux as follows:

$$\Gamma_z = n_e \int c_z\left(f_o + f_1 \cos\theta_1\right) d^3c = n_e \int c_z f_1 \cos\theta_1 d^3c \tag{31}$$

since the isotropic part of the EVDF integrates to zero. Again, using Equation 6, we can express f_1 in terms of f_o. We assume steady state, a one-dimensional spatial gradient, no external electric fields, a negligible neutral diffusion velocity, and negligible electron-electron collisions to show

$$n_e f_1 = \frac{-C}{\sum_h \nu_m} \frac{\partial(n_e f_o)}{\partial z} \tag{32}$$

and then

$$\Gamma_z = -\int \frac{c_z}{\sum_h \nu_m} C \frac{\partial n_e f_o}{\partial z} \cos\theta_1 d^3c = -\int_0^\infty 4\pi C^2 \frac{C^2}{3\sum_h \nu_m} \frac{\partial n_e f_o}{\partial z} dC \tag{33}$$

Note that we have integrated over the two directional angles. Since only n_e is a function of spatial location, Equation 33 can be written as follows:

$$\vec{\Gamma} = -\nabla \int_0^\infty 4\pi C^2 \frac{C^2}{3\sum_h \nu_m} n_e f_o dC = -\nabla(D n_e) \tag{34}$$

where D is the diffusion coefficient. In integral form,

$$D = \int_0^\infty \frac{C^2}{3\sum_h \nu_m} f_o 4\pi C^2 dC = \left\langle \frac{C^2}{3\sum_h \nu_m} \right\rangle \tag{35}$$

If the momentum collision frequency is independent of velocity, and if the EVDF is Maxwellian, then $D=kT_e/m_e\nu_m$. Independent of the momentum collision frequency's dependence on velocity, if the EVDF is Maxwellian, then the ratio of diffusion to mobility, D/μ, is equal to 2/3 kT_e. In the presence of an electric field, the above derivation is valid for the diffusion coefficient transverse to the field. Parallel to the field, both drift velocity due to the electric field and diffusion due to possible spatial concentration gradients require consideration.

4.3 SOLUTION TECHNIQUES

Although there are a few analytical solutions to the Boltzmann equation- such as the Maxwell-Boltzmann distribution function for equilibrium conditions- a numerical solution is required for most practical situations. Typically, Equation 5 is solved numerically after first solving for f_1 in Equation 6. Any other reaction which affects the population of electrons in a given velocity class can be added to Equation 5. In Section 4.2, expressions for superelastic processes and electron-heavy collisions were developed. Other processes such as photon-electron processes, electron attachment and recombination, ionization reactions which account for the secondary electron produced during the process, and an external source of electrons can be handled (Morgan and Penetrante, 1990). Whichever terms are included, the result is a second order non-linear (due to the electron-electron term) partial differential equation. Under steady state conditions, the equation to be solved becomes a second order ordinary differential equation which can be transformed into 2 first order equations. The known boundary conditions are that the EEDF have a value of 0 and a slope of 0 at an infinite energy. A possible solution method thus involves starting at a large energy and then marching back towards zero energy. To ensure that the starting energy value was sufficiently large, the solution should be repeated for a larger starting energy and compared for convergence to the first solution.

More generally, Equation 5 is solved with the transient term as a partial differential equation. If a steady state solution is desired, the solution at a large time can be chosen. The solution procedure begins by discretizing the energy domain between 0 and some finite energy. Ideally, this upper energy boundary will be large enough such that imposing the zero value and zero slope boundary conditions is justified. Again, to ensure accuracy, the solution should be re-solved with a larger upper boundary. The first and second order partial derivatives are next finite differenced as follows:

$$\frac{\partial(n_e f_o)}{\partial \varepsilon} = \frac{(n_e f_o)_{j+1} - (n_e f_o)_{j-1}}{2\Delta\varepsilon} + O(\Delta\varepsilon)^2 \tag{36}$$

$$\frac{\partial^2(n_e f_o)}{\partial \varepsilon^2} = \frac{(n_e f_o)_{j+1} - 2(n_e f_o)_j + (n_e f_o)_{j-1}}{(\Delta\varepsilon)^2} + O(\Delta\varepsilon)^2$$

The above relations are second order accurate in energy and are typically, but not necessarily, used. To advance in time, the following relation is utilized.

$$n_e f_o(\varepsilon_j, t+\Delta t) = n_e f_o(\varepsilon_j, t) + \Delta t \frac{\partial[n_e f_o(\varepsilon_j, t)]}{\partial t} + O(\Delta t) \tag{37}$$

where the partial derivative is evaluated using Equation 5. As written, Equation 37 is first order accurate in time and explicit- the right hand side of Equation 5 is evaluated at

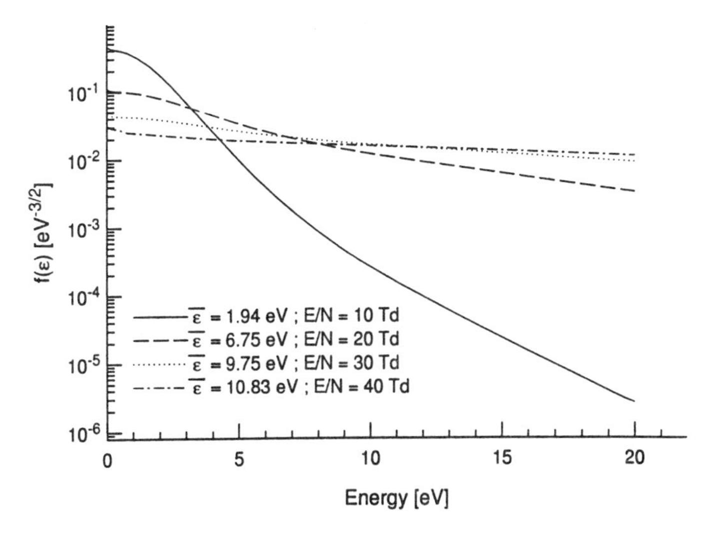

Figure 2. Numerically predicted EEDFs for varying E/N for electron-heavy elastic losses only.

time t which is known. Note that an initial condition- either user specified or arbitrarily chosen by the program- is needed to start this procedure. In practice, this approach is possible, but the maximum time step allowed is limited by stability issues (Ferziger, 1981). Alternatively, the partial derivative can be evaluated at time $t+\Delta t$ resulting in a first order accurate in time implicit formulation which has no stability restrictions on the maximum time step. However, implicit methods are difficult to apply to nonlinear equations. Because Equation 5 has several terms, most numerical solvers use a semi-implicit method in which some of terms are treated explicitly and others implicitly. This approach is conditionally stable and thus restricts the maximum time step.

If there are n points on the energy domain, then an algebraic equation, according to Equation 37, can be written for the n-2 interior points. To solve for the n unknown values of the product $n_e f_o$, the n-2 algebraic equations are combined with the two boundary conditions. The system of equations to be solved is tri-diagonal and thus very tractable. For more complete details, the interested reader is referred to Morgan and Penetrante (1990) where the solution procedure for his program ELENDIF is described. ELENDIF is a very fast a user friendly program which solves the Boltzmann equation for the EEDF. Results from this code for a methane-hydrogen plasma are presented in the following section.

4.4 APPLICATION TO A METHANE-HYDROGEN PLASMA

Having derived the governing equations and described the numerical solution techniques, we are now ready to simulate microwave plasma conditions particularly relevant for diamond CVD. Initially, a microwave reactor has virtually no free electrons. Once the microwave power is turned on, electromagnetic fields build until breakdown conditions are reached (MacDonald, 1966). At this point, atoms/molecules are ionized and free electrons are generated. These free electrons are then accelerated by the electromagnetic fields. If the energy is increased sufficiently, each electron can then further ionize other atoms/molecules. When the rate of ionization is equal to the rate of electron recombination (predominantly at the reactor walls and substrate), a steady-state electron density is attained. The distribution of energy among the electrons is then determined by the collision rates with various plasma constituents as described previously. Electron collisions with other electrons do not affect the total electron energy, and electron collisions with atoms/molecules typically result in a net electron energy loss. Recall that when electron-electron collisions dominate, the EEDF is Maxwellian and a function of only the electron temperature. In general, however, Equations 5 and 6 must be solved to determine the EEDF.

To show the effect of increasing electric field strengths, Figure 2 shows several EEDFs for varying values of E/N and the corresponding average electron energy. The EEDFs in Figure 2 do not include the effects of electron-electron collisions nor any inelastic losses. Only electron-heavy elastic processes are included. We see that the average electron energy increases with increasing E/N. Since the only electron energy loss mechanism is elastic collisions with the heavies, the following can be derived analytically from the electron energy equation for a DC field (Cherrington, 1979, p. 65):

$$T_e = T_h + \frac{2e^2 E^2}{3k\, m_e \delta\, \nu_m^2} \tag{38}$$

where δ is the twice the ratio of the electron mass to the heavy mass. Equation 38 is valid only for constant electron-heavy collision frequency, ν_m. Under this assumption, an analytical solution to the Boltzmann equation is possible and shows that the EEDF is Maxwellian at an electron temperature given by Equation 38. Further, for ν_m held constant, Equation 38 shows that T_e is proportional to the square of the electric field (for $T_e >> T_h$). However, Figure 2 shows a weaker dependence because the actual collision frequency increases with increasing applied electric field.

Before considering the effects of electron-electron collisions, there is another solution of interest when only electron-heavy elastic collisions are considered. When the electron-heavy collision frequency is constant, there is an analytical solution to the Boltzmann equation as described in the preceding paragraph. If, instead, the electron-heavy cross-section is constant- resulting in the constant mean free path assumption- then another analytical solution exists. In this case, the EEDF is Druyvesteyn and varies exponentially with the negative of C^4 (or ε^2) rather than C^2 (or ε) as for a Maxwellian EEDF. Thus, a Druyvesteyn EEDF drops off more rapidly at high energies and is demonstrated shortly.

Electron-electron collisions drive the EEDF toward a Maxwellian one independent of any other effects. Typically, as we will soon demonstrate, the largest deviations from a Maxwellian EEDF occur at high energies. Thus, the inclusion of electron-electron collisions often Maxwellianize the tail. Ginzburg and Gurevich (1960) compared the relative importance of electron-electron and electron-heavy elastic collisions on the EEDF under a constant mean free path assumption. They found that the following parameter, P, determined whether a Maxwellian or a Druyvesteyn EEDF would be present:

$$P = \frac{\nu_{ee}(v_o)}{\delta\, \nu_m(v_o)}$$

$$\text{where } v_o = \sqrt{\frac{2k\,T_e}{m_e}} \text{ and } T_e = \frac{eEl}{\sqrt{6\delta}} \tag{39}$$

where l is the assumed constant mean free path. When P>>5, then the EEDF is Maxwellian, and when P<<5, the EEDF is Druyvesteyn. Recall that the collision frequency can be expresses as $nC\sigma$. Thus, P is proportional to the ratio of the electron and heavy number densities which is often near 10^{-6}, the ratio of the electron-electron and electron-atom relative speeds which is about $\sqrt{2}$, and the ratio of the electron-electron and electron-atom momentum cross-sections which is typically about 100. For atomic hydrogen, δ is near 1/900, so that P is about 10^{-1} and thus electron-electron effects are minimized. This conclusion is particularly true for molecular plasmas because of the additional energy exchange between an electron and a molecule through inelastic processes (Mitchner and Kruger, 1973, pp. 51-52) which effectively increases the electron-heavy collision frequency. However, if the degree of ionization is large enough, then electron-electron collisions will drive the EEDF toward a Maxwellian one. In any case, programs such as ELENDIF readily include electron-electron collision provided that an electron density is specified. Figure 3 shows two cases of interest from ELENDIF when only electron-electron and electron-heavy elastic collisions are considered for a purely molecular hydrogen plasma. Further, the momentum cross-section is set equal to 1.6×10^{-19} m^2 and held constant over all energies. In one solution, the plasma is 0.1% ionized (P=180) and the EEDF is nearly Maxwellian (note the straight line appearance). In the other solution, the plasma is 10^{-5}% ionized (P=0.018) and the EEDF is nearly Druyvesteyn (note the parabolic shape). Although both cases were run at the same applied E/N (15 Td), they do not have the same average energy. The simulation with more electrons has a higher average energy because of the stronger coupling with the electric field.

Of great significance for methane-hydrogen plasmas are inelastic processes. Electrons of varying energy will rotationally, vibrationally, and electronically excite atoms and molecules. Figure 4 demonstrates the effects of rotational, vibrational, and electronic inelastic processes on a 99% H_2 1% CH_4 plasma (we will soon examine the role of atomic hydrogen) for E/N of 35 Td, for a gas temperature of 2800K, for a pressure of 40 Torr, and for an electron density of 4×10^{18} m^{-3}. Note that the average energy is strongly affected by the addition of vibrational losses (recall that electrons and molecules can efficiently exchange energy through such inelastic processes), and negligibly affected by the rotational processes. The effect of the electronic inelastic process impacts the high energy tail. For molecular hydrogen, 8.9 eV separates the ground and first excited states. Thus, inelastic processes will deplete electrons with energy larger than 8.9 eV. There will also be inelastic losses at the ionization energy. Elastic collisions will then balance these depletions in steady state. For these conditions, we see that the depletion rate is faster so that the high energy tail is substantially depressed relative to the EEDF without inelastic electronic excitation.

Clearly, to form diamond, atomic hydrogen must be present. Atomic hydrogen has its first excited state at 10.19 eV and therefore can deplete high energy electrons. However, the addition of atomic hydrogen displaces some molecular hydrogen. Thus the effect of electronic excitation processes is about the same. Because of the strong role of vibrational losses, in contrast, the displacement of molecular hydrogen with atomic hydrogen actually causes the average energy of the EEDF to increase (Koemtzopoulos et al., 1993) as shown in Figure 5.

In the above analyses, effects due to the excited states have been neglected. Their impact on the solution of the Boltzmann equation primarily results from superelastic collisions. For example, an electron can collide with hydrogen atom or molecule in its first excited state causing the excited state to transition to the ground state. The excess

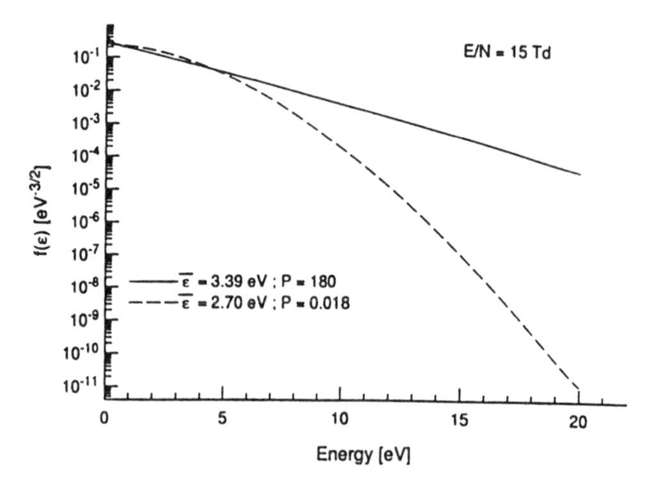

Figure 3. The effect of electron-electron collisions on the EEDF for a constant electron-heavy momentum cross section.

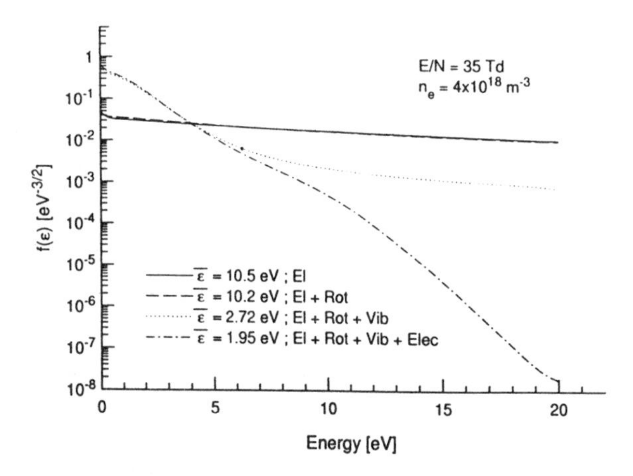

Figure 4. The effect of inelastic loss mechanisms on the EEDF.

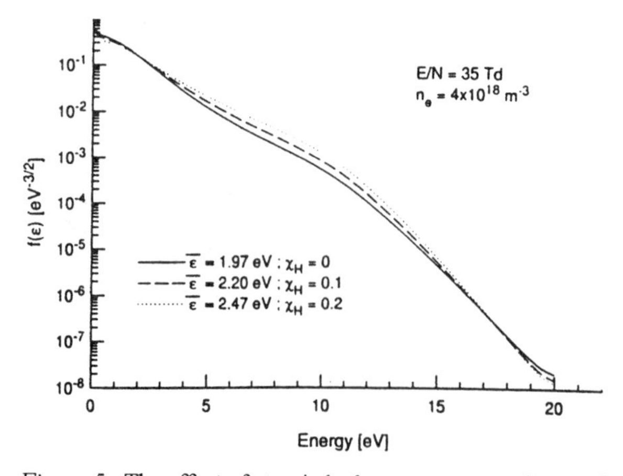

Figure 5. The effect of atomic hydrogen concentration on the EEDF.

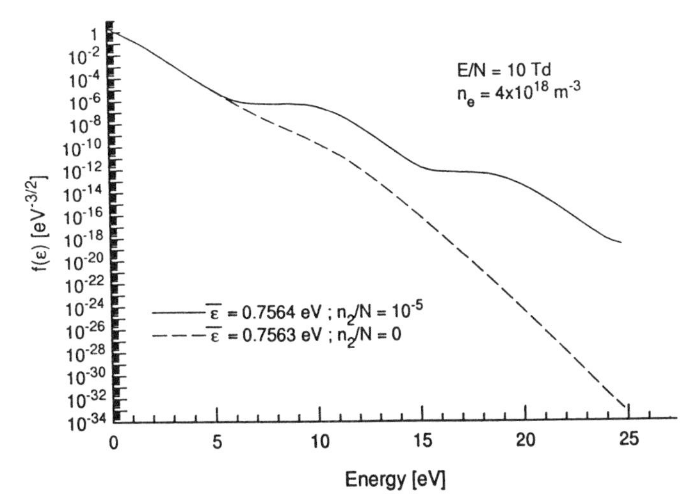

Figure 6. The effect of superelastic collisions with molecular hydrogen's first electronically excited state on the EEDF.

energy is then carried off by the electron which can selectively increase a portion of the EEDF. The most difficult aspect of modeling this effect is that the number density of the excited states must be known. Typically, the populations of the excited states are several orders of magnitude lower that the population of the ground state and so collisions involving these states are neglected. However, superelastic processes can be initiated with electrons of all energies in contrast to inelastic processes which require an electron with a minimum energy. Thus, there are several orders of magnitude more electrons which can initiate a superelastic reaction then those which can initiate the complementary inelastic reaction. Using the principle of detailed balance (Mitchner and Kruger, 1973, pp. 80-86), for a Maxwellian EEDF, the inelastic and superelastic rates are equal when

$$\frac{n_2}{g_2} = \frac{n_1}{g_1} e^{-\varepsilon_{12}/kT_e} \tag{40}$$

where n_1 and n_2 represent the populations of the ground and excited states, respectively, and ε_{12} represents the energy separating the two states. For non-Maxwellian EEDFs, an expression similar to Equation 40 would give the equilibrium condition. In any case, by neglecting the excited state populations, the reverse superelastic reaction rate is set to zero. At an excited state population equal to the equilibrium value (given by Equation 40 for a Maxwellian EEDF), the superelastic reaction rate is equal to the inelastic rate which means neither reaction affects the EEDF. Capitelli et al. (1994) have examined the effect of superelastic collisions involving electronically excited states near 11 eV in a molecular hydrogen discharge. Increasing the relative population of the excited states increases the EEDF at multiples of 11 eV. Low energy electrons collide with the excited state H_2 molecule to produce an abundance (relative to a zero population of excited states) of electrons near 11 eV. Some of these 11 eV electrons then can collide with the excited state H_2 molecule to generate an abundance of 22 eV electrons and so on. These results are shown in Figure 6.

Because the solution of the EEDF can be very sensitive to the population of the excited states, we must consider how to determine these populations. Such a determination involves a collisional-radiative model (CRM). As an example, consider a system of only atomic hydrogen in which we want to determine the populations of hydrogen's ground and first 19 excited states and of the ion and electron density. Because of the relatively low energy, the ion density and electron density can be assumed to be equal. The following electron-atom and atom-atom inelastic reactions and radiative reactions are then possible:

$$
\begin{aligned}
H_m + e &\Leftrightarrow H_n + e \quad n > m \\
H_m + H_1 &\Leftrightarrow H_n + H_1 \quad n > m \\
H_m + e &\Leftrightarrow H^+ + e + e \\
H_m + H_1 &\Leftrightarrow H^+ + e + H_1 \\
H_n &\rightarrow H_m + h\nu \quad n > m \\
H^+ + e &\rightarrow H_m + h\nu
\end{aligned}
\tag{41}
$$

The first two reactions represent electron-atom and atom-atom excitation reactions, respectively. The next two reactions represent electron-atom and atom-atom ionization reactions, respectively. The last two reactions represent radiative de-excitation and radiative electron recombination, respectively. For each unknown number density, a rate equation can be generated based on the above reaction set as follows:

$$
\begin{aligned}
\frac{dn_m}{dt} = &- n_m \sum_{j\_m} \left(n_e k_{ea}^{m\to j} + n_1 k_{aa}^{m\to j} \right) + \sum_{j\_m} n_j \left(n_e k_{ea}^{j\to m} + n_1 k_{aa}^{j\to m} \right) - n_m \left(n_e k_{ea}^{m\to i} + n_1 k_{aa}^{m\to i} \right) \\
&+ n_e n_i \left(n_e k_{ea}^{i\to m} + n_1 k_{aa}^{i\to m} \right) - \sum_{j<m} n_m k_v^{m\to j} + \sum_{j>m} n_j k_v^{j\to m} + n_i n_e k_v^{i\to m} = 0
\end{aligned}
\tag{42}
$$

where n_m represents the number density of a hydrogen excited state (m>1) or the ground state (m=1), n_e represents the electron number density, n_i represents the ion number density, and k represents the appropriate reaction rate coefficient. The first two terms account for electron-atom and atom-atom excitation and de-excitation, the third term accounts for electron-atom and atom-atom ionization, the fourth term accounts for electron-atom and atom-atom recombination, the fifth and sixth terms account for radiative excited state transitions, and the last term accounts for radiative recombination. The evaluation of the needed electron-atom collisional rate coefficients requires integrating the appropriate cross-sections (σ) over the chosen EEDF as described in Section 4.2. The rate coefficient for an electron-atom excitation reaction, for example, would be determined by

$$
k_{ea}^{m\to n}(\varepsilon) = \sqrt{\frac{2}{m_e}} \int_0^\infty f(\varepsilon)\varepsilon\, \sigma_{ea}^{m\to n}(\varepsilon)\, d\varepsilon
\tag{43}
$$

ELENDIF is used to determine the EEDF. Because the solution of the EEDF depends on the excited state populations which in turn depend on the EEDF, the process is iterative. A self-consistent solution is obtained once the chosen EEDF returns, from the CRM, the same number densities which were input into ELENDIF. In addition, both a mass and energy balance must be satisfied. The mass balance requires that the total number density as expressed by the sum of all the individual number densities and by the pressure and temperature be equal. For a two temperature plasma, the mass balance would appear as follows:

$$n_T = n_e + n_i + \sum_{j=1}^{\infty} n_j = \frac{P_e}{kT_e} + \frac{P_i + P_h}{kT_h} \tag{44}$$

For most applications, the number densities and partial pressures of the ions and electrons can be neglected. Further, the summation can generally be replaced by just the ground state number density, n_1.

The nonelastic cross sections for electron-atom ionization and excitation are taken in their analytical form from Drawin (1961) and Drawin (1963), respectively. Since the lower-energy portion of the distribution function is often nearly Maxwellian, the inverse reaction rates can be obtained through the principle of detailed balancing using forward reaction rates given by a Maxwellian EEDF (Shaw et al., 1970). This procedure is implemented to determine the three-body electron-atom recombination rate coefficients. For electron-atom de-excitation reaction rate coefficients, we can use the principle of detailed balancing to relate directly the de-excitation cross sections to the excitation cross sections (Mitchner and Kruger, 1973, pp 80-86). With these cross- sections, the rate coefficients are calculated using Equation 43.

The evaluation of the atom-atom collisional rate coefficients is more direct under the assumption that the atom energy distribution function is Maxwellian at the gas temperature (assumed to be 2800 K). Accordingly, excitation and ionization rates can be taken from Kunc (1987) and the inverse reaction rates can be evaluated through detailed balancing.

For radiative recombination, the needed cross sections can be calculated according to Mitchner and Kruger (1973, p. 85). For radiative transitions, the rate constants do not require integration over the EEDF and are simply the Einstein transition probabilities which can be taken from Wiese *et al.*(1966). As calculated, these radiative rates reflect the assumption that the emission is optically thin and escapes the plasma. However, in some cases the radiation is reabsorbed in the inverse absorption process before reaching the plasma's edge. Holstein (1951) introduced the idea of an escape factor which directly multiplies the optically thin emission rate to account for this trapping. For a cylindrically symmetric plasma of radius r and of infinite length, Holstein developed the following expressions for a radiative absorption coefficient, α_0, and an escape factor, g_{ef} for a Doppler-broadened resonance line:

$$\alpha_0 = \frac{\lambda_{21}^3 N_1 g_2 A_{21}}{8\pi^{3/2} g_1 v_0} \qquad g_{ef} = \frac{1.6}{\alpha_0 r\sqrt{\pi \log(\alpha_0 r)}} \tag{45}$$

where λ_{21} is the transition's wavelength, N_1 is the absorbing state's population, g_2 and g_1 are the emitting and absorbing state's degeneracies, respectively, A_{21} is the Einstein transition probability, and v_0 is the absorbing atom's most probable speed and equal to $(2kT_g/M)^{1/2}$ where M is the mass of the absorbing species. An escape factor of 1 represents the optically thin assumption, and an escape factor of 0 represents an

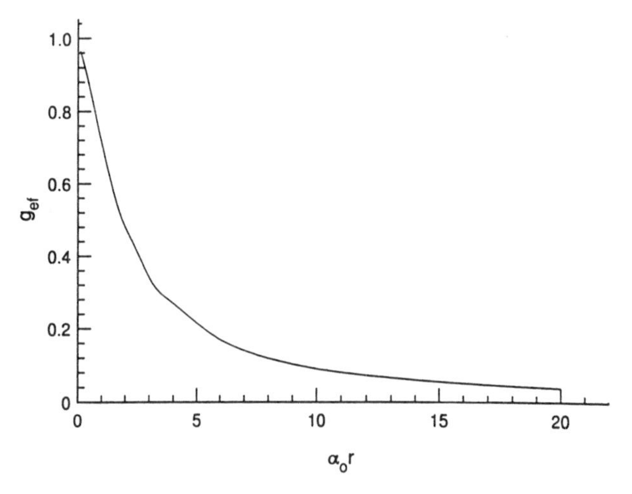

Figure 7. Radiation escape factors for an infinite homogenous plasma of radius r as a function of the absorption coefficient when Gaussian line broadening predominates.

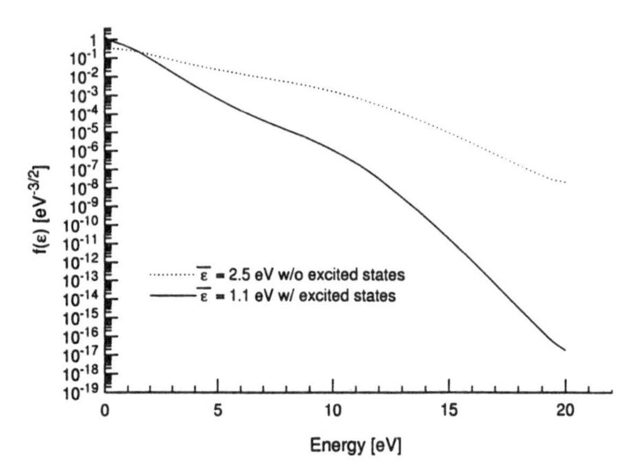

Figure 8. The effect of atomic hydrogen excited states on the EEDF.

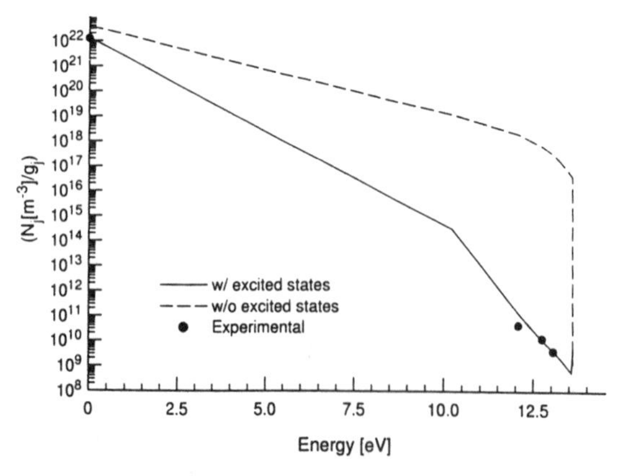

Figure 9. Comparison of numerical and experimental data to show the effect of atomic hydrogen excited states.

optically thick assumption. Note that as the absorbing population goes to zero- the optically thin assumption- g_{ef} (as defined in Equation 45) becomes undefined. Once the product a_0r becomes less than about 3, it is recommended that an escape factor according to Jolly and Touzeau (1975) be used. For a homogenous plasma where the predominant line broadening mechanism is Gaussian (e.g. Doppler broadening), Figure 7 shows the appropriate escape factor as a function of the product $\alpha_0 r$. It is sometimes convenient to note that the reciprocal of α_0 represents the characteristic length for radiative escape. Thus, when this characteristic length greatly exceeds the plasma radius, the optically thin assumption ($g_{ef}=1$) is valid.

Gordon and Kelkar (1995) have used such a CRM to determine the important role the excited states have on EEDF predictions. In particular, they showed that ionization from the excited states (as compared to ionization from the ground state) produces more than 99% of the total electrons. Thus, if the excited states are not modeled, one will infer an electric field which is much larger than actually present to generate the required ionization rate. This overestimated electric field then leads to an over-prediction in the average electron energy and a significantly different EEDF. Figure 8 shows both an EEDF for typical microwave diamond CVD conditions- 40 Torr pressure, 2800 K gas temperature, $4x10^{18}$ m^{-3} electron number density, 300 sccm H_2, 3 sccm CH_4- when the excited states are neglected and when the excited states are included. Note that the sustaining electric field and the average energy are lower by more than a factor of two when the excited states are self-consistently included. When these EEDFs are used as input to the above described CRM, a Boltzmann plot of the excited state populations is generated and shown in Figure 9. Note that each curve in Figure 9 is produced by configuring the CRM to either include or exclude all reactions involving the excited states to be consistent with the assumptions made in determining the particular EEDF. We see that, by neglecting ionization from the excited states, the resulting EEDF with a large average energy over-predicts by orders of magnitude the excited state populations. By self-consistently accounting for the excited states, in contrast, accurate predictions are possible.

4.5 REFERENCES

Capitelli, M., Colonna, G., Hassouni, K., and Gicquel, A., *Chem. Phys. Letters* **228**, 687 (1994).

Cherrington, B.E., in *Gaseous Electronics and Gas Lasers* (Permagon Press, Oxford, 1973).

Drawin, H.W., *Z. Phys.* **164**, 513 (1961).

Drawin, H.W., *Atomic Cross-Sections for Inelastic Electronic Collisions*, Report EUR-CEA-FC 236 (1963).

Ferziger, J.H., in *Numerical Methods for Engineering Application* (John Wiley & Sons, New York, 1981).

Ginzburg, V.L. and Gurevich, A.C., *Sov. Phys. Uspekhi* **3**, 115 (1960).

Gordon, M.H. and Kelkar, U.M., submitted to *Phys. of Plasmas* (1995).

Holstein, T., *Physical Review* **83**, 1159 (1951).

Jiang , P. andEconomou, D.J., *J. Appl. Phys.* **73**, 8151 (1993).

Jolly , J. and Touzeau, M., *J. Quant. Spectrosc. Radiat. Transfer* **15**, 863 (1975).

Koemtzopoulos, C.R., Economou, D.J., and Pollard, R., *Diamond and Related Materials* **2**, 25 (1993).

Kunc, J.A., *Phys. Fluids* **30**, 2255 (1987).

MacDonald, A.D., in *Microwave Breakdown in Gases* (Wiley, New York, 1966).

Mitchner , M. and Kruger, C.H., in *Partially Ionized Gases* (Wiley, New York, 1973).

Morgan , W.L. and Penetrante, B.M., *Comp. Phys. Commun.* **58**, 127 (1990).

Shaw, J.F., Mitchner, M., and Kruger, C.H., *Phys. Fluids* **13**, 325 (1970).

Wiese, W.L., Smith, M.W., and Glenon, B.M., *Atomic Transition Probabilities*, U.S. National Bureau of Standards National Standard Reference Series-4 (U.S. Government Printing Office, Washington, DC, vol. 1, 1966).

Chapter 16

SECTION 5: FLOW MODELING FOR A PLASMA ASSISTED DIAMOND DEPOSITION REACTOR

K. Hassouni[a], C. D. Scott[b], and S. Farhat[a]

[a]*Laboratoire d'Ingénierie des Matériaux et des Hautes Pressions*
CNRS-UPN, Av. J. B. Clément 93300 Villetaneuse, France
[b]*NASA, NASA Lyndon B. Johnson Space Center, EG3, Houston, TX 77058, USA*

5.1 INTRODUCTION

The knowledge of plasma flow characteristics is of prime interest for the control, the optimization and the scale-up of microwave plasma assisted diamond deposition reactors. Indeed, the growth rate and the quality of the deposited films are strongly dependent on the reactive species fluxes at the substrate surface. The estimation of these fluxes is also necessary for determining the growth and precursor species and understanding the diamond growth mechanism.

The species fluxes at the substrate surface are controlled by transport in the plasma (diffusion, convection) and by surface reactions kinetics. Therefore, to estimate these fluxes one has to build a model taking both these phenomena into account.

In this section we will consider the modeling of thermochemically non equilibrium plasma flows. After a first phenomenological description of the basic phenomena which characterize moderate pressure plasmas, we will express the transport equations with a three temperature flow assumption. We will also report the details of the fluxes and source terms involved in these equations and the expressions of the associated transport coefficients. Finally, the thermodynamic relations for a three temperature flow and boundary conditions will be discussed.

5.2 BASIC PHENOMENA IN PLASMA FLOWS

5.2.1 Introduction

In a plasma discharge, the electrons are heated by a high frequency electric field. They undergo elastic and inelastic collisions which result in momentum and energy transfer from electron kinetic energy mode to internal and translational heavy particles energy modes. Reactive collisions between electrons and feed gas species also take place, resulting in dissociation, ionization and electron attachment processes. Besides these electron impact processes, heavy particle-heavy particle collisions result in chemical reactions and energy redistribution on the different energy modes of the heavy species.

All these collisions lead to a very complex medium whose complete description requires taking into account the electronic, vibrational and rotational energy states of all the plasma species. Fortunately, for moderate pressure plasmas this complex problem may be simplified first by neglecting the population of the radiative species which usually have very short lifetimes, and second by a dynamic analysis of the collisions dealing with energy transfer between the electrons and the electronic ground state of heavy species.

5.2.2 Translational modes of heavy particles

In moderate pressure plasmas (P>10 mbar) the heavy particle-heavy particle elastic collisions frequency is very high and the translation energy transfer during a collision is very efficient [Bird, 1976]. The translational modes energy distributions functions of the different heavy species are then Maxwellian with the same temperature T_{trans}.

5.2.3 Rotational modes of heavy particles

Energy transfer between the translational and rotational modes of the heavy particles is very fast and leads to, in moderate pressure plasmas, an equilibrium between these modes [Bird, 1976] and [Brun, 1986]. As a consequence, the rotational distribution functions of all the molecular species are Boltzmann distributions with a rotational temperature equal to the translational temperature. The equilibrium between the translational and rotational modes of the heavy particles leads to a description of these modes in terms of one temperature which is usually called the gas temperature T_g.

5.2.4 Vibrational modes of molecules

The vibrational energy distribution function (VEDF) which quantifies the population of the different vibrational energy levels depends on the kinetics of vibrational excitation by electrons, vibration-vibration quanta exchange and translation-vibration energy exchange [Capitelli et al., 1990] and [Gorse et al., 1989]. Numerous studies were carried out on these kinetics in the case of diatomic molecules such as N_2, H_2 and CO [Keck and Carrier, 1965], [Cacciatore et al., 1982], [Lourreiro and Ferreira., 1986], [Gorse et al., 1992], [Lourreiro and Ferreira, 1989] and [Gorse et al., 1987]. Results show that the VEDF is generally not a Boltzmann function, but may be characterized by a Treanor distribution behavior at low energies, a large plateau for intermediate energies, and a Maxwellian distribution at high energies. However, since the major part of the energy is carried by the few first vibrational levels, the mean vibrational energy for a given species may be well estimated from the approximation of a Boltzmann distribution function for the few first levels (v<5) [Maronne and Treanor, 1963] and [Ricard, 1983]. This approximation enables the description of the vibrational distribution in terms of one variable, the vibrational temperature T_v. However, no information is available for the high energy levels. In the case of diamond deposition plasmas, the major part of the vibrational energy is carried by H_2 molecules which represent the major molecular species of the plasma [Gicquel et al., 1994] and [Koemtzopoulos et al., 1993]. As a consequence, only one vibrational energy mode has to be considered.

5.2.5 Kinetic modes of electrons

In diamond deposition plasmas the electron energy distribution function (EEDF) is not Maxwellian [Koemtzopoulos et al., 1993] and [Capitelli et al., 1994]. In cases corresponding to high reduced electric field (E/N >20 Td) and low metastables concentration ($c^*<10^{-4}$), this function is bimodal and may be characterized by a low energy electron temperature T_e and a high energy electron temperature $T_{e\text{-}h}$. A first approach for the description of the electron energy relaxation and transport could be carried out by distinguishing the low electron energy relaxation processes, controlled by T_e, such as electron-heavy particles translation and electron-vibration energy transfers and high energy processes, controlled by $T_{e\text{-}h}$, such as dissociation and ionization.

5.2.6 Chemistry in plasma flows

Two kinds of chemical kinetics processes may be distinguished in a plasma flow. The first one consists of reactions which involve the electrons. Generally, the rate constants of these reactions depend on the EEDF and on the collision cross sections. In some cases, such as the dissociative attachment of H_2, this rate also depends on the internal energy of the heavy particles [Wadhera and Bradsley, 1978] (rotational or vibrational modes). The second kind of processes include heavy particle-heavy particle reactions whose rate constants depend on the gas temperature and sometimes on the vibrational temperature (see for example the dissociation of N_2, O_2 and CO in low pressure plasma) [Lourreiro and Ferreira., 1986], [Marrone and Treanor, 1963], [Lourreiro and Ferreira, 1990] and [Park, 1987].

5.3 MODERATE PRESSURE PLASMA FLOW DESCRIPTION WITH FLUID THEORY

5.3.1 Introduction

In the moderate pressure plasma flows, the Knudsen number, defined as the ratio of the mean free path to a characteristic length of the flow (plasma diameter for example) : $Kn = \lambda/d$, is lower than 10^{-1}. In this case the continuum description of the flow is valid, [Bird, 1976] and [Brun, 1986], and the plasma species and energy transport are governed by the fluid dynamic equations which express the conservation of mass, momentum and energy in the plasma flow [Brun, 1986].

For thermochemically non equilibrium flow, a balance equation has to be written for each species, each energy mode and each momentum component. The conservative variables usually used for the description of the chemical species populations, the momentum components and the excitation modes energies are the mass densities ρ_s, the momentum per unit volume $\rho\mathbf{U}$ and the energy per unit volume for each mode E_{mode} [Lee, 1984].

If ϕ is a conserved variable (mass, energy, ...), the conservation equation for ϕ may be derived for a two dimensional geometry by considering a small volume, $\Delta v=\Delta x.\Delta y$, in the plasma flow (figure 1), and computing the variation of ϕ during the time Δt :

$$\Delta\phi = F_{ix}\,\Delta y\,dt + F_{iy}\,\Delta y\,\Delta t - F_{ox}\,\Delta x\,\Delta t - F_{oy}\,\Delta x\,\Delta t + \omega_\phi\,\Delta t \qquad (1)$$

$F_{ix}\,\Delta y\,\Delta t$ and $F_{iy}\,\Delta x\,\Delta t$ represent the amount of ϕ transported in the volume $\Delta x\Delta y$. $F_{ox}\,\Delta y\,\Delta t$ and $F_{oy}\,\Delta x\,\Delta t$ represents the amount of ϕ transported out of the volume $\Delta x\Delta y$. The term $\omega_\phi\Delta t$ is the quantity of ϕ created in the volume $\Delta x\,\Delta y$ during Δt by source mechanisms (chemistry for species, intermode transfer for energy, etc. ...).

Dividing equation (1) by $\Delta x\Delta y\Delta t$, one obtains :

$$\frac{\Delta\phi/\Delta x\Delta y}{\Delta t} = \frac{(F_{ix} - F_{ox})}{\Delta x} + \frac{(F_{iy} - F_{oy})}{\Delta y} + \frac{\omega_\phi}{\Delta x\Delta y} \qquad (2)$$

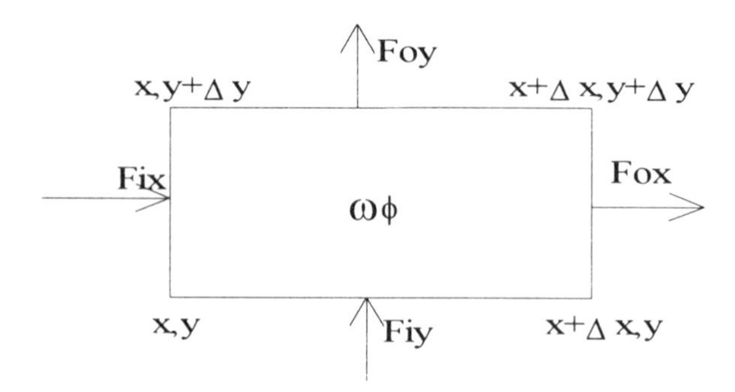

Figure 1 : Schematic of ϕ variation in a volume element $\Delta x\Delta y$.

for Δx, Δy and Δt very small the finite difference formulation (2) is equivalent to the following continuous formulation :

$$\frac{\partial \psi}{\partial t} = -\frac{\partial F_x}{\partial x} - \frac{\partial F_y}{\partial y} + W_\psi \tag{3}$$

which may also be written :

$$\frac{\partial \psi}{\partial t} = -\mathrm{div}(\mathbf{F}) + W_\psi \tag{4}$$

with

$$\psi = \frac{d\phi}{dv} \tag{5}$$

F is the flux density vector whose components are F_x and F_y, W_ψ is the rate of variation of ψ by source mechanism.

To determine the conservation equation for each conserved variable of the plasma flow, one has to express the fluxes and the source terms for each of these variables.

5.3.2 Continuity equations for chemical species

The conserved variables in these cases are the species masses and the associated conservative variables are the species mass densities.

For a given finite volume $\Delta x\Delta y$, a species "s" may be transported inside this volume by convection or diffusion, the associated flux vector components are :

$$\mathbf{F}_{s\text{-}x} = \rho_s u + \rho_s u_s \tag{6}$$

and

$$\mathbf{F}_{s\text{-}y} = \rho_s v + \rho_s v_s \tag{7}$$

Where ρ_s is the mass density of species "s", u and v are the components of the plasma species averaged velocity, u_s and v_s are the species "s" diffusion velocity components relative to the average velocity.

The source term corresponding to the species transport equation is the rate of chemical production (or consumption) W_s, which depends on the chemical kinetics of the plasma. The resulting species equation is :

$$\frac{\partial \rho_s}{\partial t} = -\frac{\partial(\rho_s u + \rho_s u_s)}{\partial x} - \frac{\partial(\rho_s u + \rho_s u_s)}{\partial y} + W_s \tag{8}$$

5.3.3 Momentum component equations

The axial momentum component of the plasma flow is transported by convection fluxes. It may also change from pressure and viscous forces exerted on the flow elements [Bird et al., 1960]. This results in the following flux components expressions :

$$F_x = (\rho u)\, u + P - \tau_{xx} \tag{9}$$

$$F_y = (\rho u)\, v - \tau_{xy} \tag{10}$$

where P is the total pressure, ρ is the plasma total mass density and τ_{xx} and τ_{xy} are the plasma stress tensor components.

The source term associated with the momentum axial component is due to the electric forces exerted on the plasma charged particles. However, in the case of a moderate pressure plasma, the Debye length is very small ($\lambda_d < 10^{-2}$ mm) [Delcroix, 1958] and [Pointu, 1983], so for the volume elements which have characteristic length greater than λ_d the electrical neutrality is satisfied and the electrical force source term for the momentum equation vanishes [Lee, 1984] and [Delcroix, 1958] : $W_{\rho u} = 0$. The resulting transport equation of the momentum axial component is :

$$\frac{\partial \rho u}{\partial t} = -\frac{\partial((\rho u)\, u + P - \tau_{xx})}{\partial x} - \frac{\partial((\rho u)\, v - \tau_{xy})}{\partial y} \tag{11}$$

The transport equation for the radial momentum component may be obtained in a similar way.

5.3.4 H_2 vibration energy equation

The modeling of the H_2 vibrational energy transport may be carried out using two different methods, depending on the assumption of a Boltzmann VEDF or not. If this assumption is not used then one has to investigate the transport of each of the vibrational levels, which is considered as a particular chemical species. This method has the advantage of being more realistic for the plasma description. Its main disadvantage consists of resulting in a huge chemical kinetics model with a high number of species and reactions. The implementation of such kinetics in flow description leads to a large number of differential equations whose numerical solutions are very difficult, or impossible, for a two dimensional geometry.

The assumption of a Boltzmann VEDF leads to a substantial simplification of the mathematical problem, since only one conservative variable, which is the mean vibrational energy per unit volume E_v, is needed for the description of the vibrational mode. The models based on this assumption are only valid for the description of the few first vibrational levels (v<5) [Ricard, 1983].

The vibrational energy may be transported by convection, conduction and diffusion fluxes. It may also change by intermodal energy transfer source terms. The resulting transport fluxes components are :

$$F_x = u.E_v + u_{H2}.E_v + q_{v-x} \quad (12)$$

and

$$F_y = v.E_v + v_{H2}.E_v + q_{v-y} \quad (13)$$

where q_{v-x} and q_{v-y} are the energy conduction flux density vector components. u_{H2} and v_{H2} are the H_2 diffusion velocity components.

The source terms account for the energy transfer rates from the vibrational mode of H_2 to the electrons kinetic mode $Q_{v->e}$ and to the translation-rotation mode $Q_{v->t}$ [Lee, 1984].

The vibration energy equation is then :

$$\frac{\partial E_v}{\partial t} = -\frac{\partial(uE_v + u_{H2}E_v + q_{v-x})}{\partial x} - \frac{\partial(vE_v + v_{H2}Ev + q_{v-y})}{\partial y} - Q_{v-e} - Q_{v->t} \quad (14)$$

5.3.5 Electron energy equation

Since the EEDF is generally not Maxwellian in moderate pressure plasma flows, a rigorous modeling of the electron energy transport requires the solution of the electron Boltzmann equation in phase space (**r**,**v**) [Winkler, 1992]. The treatment of this equation in this case may be very complex, but could be simplified by the use of two approximations. The first one, which is called the local approximation and which is valid for moderate pressure plasmas, enables one to neglect the space derivative of the normalized EEDF in the Boltzmann equation [Delcroix, 1963]. The second consists of neglecting the anisotropic parts of the EEDF [Winkler, 1992] and [Delcroix, 1963]. The problem is then reduced to the solution of the two term expansion of the homogeneous Boltzmann equation which only deals with energy space [Gorse et al., 1989]. In the particular case of a diamond deposition plasma, and for typical discharge conditions of deposition experiments the EEDF is bimodal [Koemtzopoulos et al., 1993] and [Capitelli et al., 1994]. In this case the major part of the total electron kinetic energy is carried by low energy electrons. The high energy electrons only control the reactive processes such as dissociation and ionization [Capitelli et al., 1994]. In this case the electron energy equation may be written as in the case of Maxwellian distribution, except for the source terms describing the loss of energy by dissociation and ionization electron impact processes. The electron energy may be transported by convection, diffusion and conduction. It may also change by electron energy viscous dissipation and pressure forces [Lee, 1984]. The resulting fluxes are :

$$F_{ex} = (u + u_e).(E_e + P_e - \tau_{e\text{-}xx}) - (v + v_e).\tau_{e\text{-}xy} + q_{e\text{-}x} \quad (15)$$

and

$$F_{ey} = (v + v_e).(E_e + P_e - \tau_{e\text{-}yy}) - (u + u_e).\tau_{e\text{-}xy} + q_{e\text{-}y} \quad (16)$$

where E_e is the electron energy per unit volume, $q_{e\text{-}x}$ and $q_{e\text{-}y}$ are the electron energy conduction flux vector components. P_e is the electron partial pressure and $\tau_{e\text{-}xx}$, $\tau_{e\text{-}yy}$ and $\tau_{e\text{-}xy}$ are the electron shear stress tensor components.

The source terms account for the energy transfer rates from the electron kinetic mode to the vibration and translation-rotation modes of molecular species ($Q_{e\text{-}v}$, $Q_{e\text{-}t}$), for the electron energy loss during the activation of electron-heavy particle chemical processes ($Q_{e\text{-}chem}$), [Scott et al., 1995], and for the energy gain from the high frequency electric field (PMW) [Zakrzewski et al., 1992]. The resulting electron energy equation is then:

$$\frac{\partial E_e}{\partial t} = -\frac{\partial((u+u_e)(E_e+P_e-\tau_{exx}) - (v+v_e)\,\tau_{exy} + q_{e\text{-}x})}{\partial x} - \frac{\partial((v+v_e)\,(E_e+P_e-\tau_{eyy}) - (u+u_e)\,\tau_{exy} + q_{e\text{-}y})}{\partial y} - Q_{e\text{->}chem} - Q_{e\text{->}tr} - Q_{e\text{->}v} + PMW \quad (17)$$

5.3.6 Total energy equation

The total energy of the plasma is the sum of the translation-rotation mode, vibration mode and electron mode energies. The total energy is transported by convection, conduction and enthalpic species diffusion. It may also change under the effect of pressure and viscous dissipation. The resulting flux vector components are :

$$F_{Et,x} = u.(E + P - \tau_{xx}) - v.\tau_{xy} + q_{c\text{-}x} + q_{h\text{-}x} \quad (18)$$

$$F_{Et,y} = v.(E + P - \tau_{yy}) - u.\tau_{yx} + q_{c\text{-}y} + q_{h\text{-}y} \quad (19)$$

where E is the total energy per unit volume. τ_{xx}, τ_{xy} and τ_{yy} are the plasma shear stress tensor components. $q_{c\text{-}x}$ and $q_{c\text{-}y}$ are the total energy conduction flux vector components. $q_{h\text{-}x}$ and $q_{h\text{-}y}$ are the components of the total energy flux vector due to the diffusion of enthalpic species.

The total plasma energy may also change due to source terms which are the absorbed electric power density and the energy radiated by the plasma. The resulting transport equation is the following :

$$\frac{\partial E}{\partial t} = -\frac{\partial(u(E+P-\tau_{xx}) - v.\tau_{xy} + q_{c\text{-}x} + q_{h\text{-}x})}{\partial x} - \frac{\partial(v(E+P-\tau_{yy}) - u\tau_{xy} + q_{c\text{-}y} + q_{h\text{-}y})}{\partial y} + PMW - Q_{rad} \quad (20)$$

The solution of the plasma flow transport equations requires the expression of the different fluxes and source terms as functions of the plasma characteristic variables such as species densities and energy mode temperatures (T_g, T_v, T_e). This is the object of the next section.

5.4 FLUX AND SOURCE TERM EXPRESSIONS FOR THE TRANSPORT EQUATIONS

5.4.1 Species diffusion velocity

The rigorous expressions of species diffusion velocities in a multicomponent flow are quite complex and their use for practical flow modeling is very difficult [Hirschfelder et al., 1954]. However, these velocities may be well approximated using an expression which is similar to Fick's law with an effective species diffusion coefficient in the plasma mixture $D_{s\text{-}M}$.[Curtiss et al., 1949] and [Lee, 1984] :

$$\mathbf{U}_s = -\frac{\rho . D_{s\text{-}M}}{\rho_s} . \mathbf{grad}(x_s) \tag{21}$$

where $D_{s\text{-}M}$ is the diffusion coefficient of the species 's' in the plasma mixture and x_s is the molar fraction of the species 's'.

5.4.2 Shear stress tensor components

Expressions for the shear stress tensor components are derived for Newtonian fluids as the following [Bird et al. 1960]:

$$\tau_{xx} = 2\mu \frac{\partial u}{\partial x} - \frac{2}{3}\mu . \left(\frac{\partial u}{\partial x} + \frac{\partial v}{\partial y}\right) \tag{22}$$

$$\tau_{xy} = \tau_{yx} = \mu \left(\frac{\partial v}{\partial x} + \frac{\partial u}{\partial y}\right) \tag{23}$$

where μ is the plasma species averaged viscosity.

5.4.3 Energy modes conduction fluxes

The conduction fluxes of the different energy modes are given by Fourrier laws [Lee, 1984] :

$$\mathbf{q}_v = -\lambda_v \, \mathbf{grad} \, T_v \tag{24}$$

$$\mathbf{q}_e = -\lambda_e \, \mathbf{grad} \, T_e \tag{25}$$

$$\mathbf{q}_c = -\lambda_v \, \mathbf{grad} \, T_v - \lambda_e \, \mathbf{grad} \, T_e - \lambda_{t\text{-}r} \, \mathbf{grad} \, T_g \tag{26}$$

$\lambda_{t\text{-}r}$ is the translation-rotation energy conductivity, and λ_v and λ_e are the vibration and electron energies conduction coefficients resulting from the collisions with all the plasma species.

The total energy flux density vector resulting from the diffusion of enthalpic species is given by the following expression [Bird et al., 1960] :

$$\mathbf{q}_h = \sum_s \rho_s . \mathbf{U}_s \, h_s \tag{27}$$

where, respectively, $\mathbf{U}_s$ and h_s are the diffusion velocity vector and the enthalpy of species 's'.

5.4.4 Electron-translation energy transfer source term

This term is derived from classical kinetic theory, its expression is the following [Vincenti and Kruger, 1967], [Delcroix, 1963] and [Lee, 1984] :

$$Q_{e\text{-}t} = 3\,R\,\rho_e\,(T_e - T_g)\,\sqrt{\frac{8RTe}{\pi Me}}\,\sum_{s\neq e^-}\frac{\rho_s.n}{M_s^2}\,\sigma_{e\text{-}s} \tag{28}$$

where M_e and M_s are the electron and 's' species molar masses, $\sigma_{e\text{-}s}$ is the 'e-s' momentum transfer collision cross section and n the plasma total density.

5.4.5 Electron-H_2 vibration energy transfer source term

The source term is derived from the kinetics of vibrational excitation by electron impact which consists of collisions of the type [Gorse et al., 1992], [Lourreiro and Ferreira, 1989] and [Lee, 1985]:

$$H_2(v) + e^- \rightarrow H_2(w) + e^- \quad (v\neq w)$$

For a Boltzmann VEDF and a Maxwell EEDF this term is given by a Landau-Teller type equation [Lee, 1984] :

$$Q_{e\text{-}v} = \frac{ev^{**}(T_e) - ev\,(T_v)}{\tau_{e\text{-}v}} \tag{29}$$

where $ev(T_v)$ and $ev^{**}(T_e)$ are the values of the vibrational energy per unit mass corresponding to T_v and T_e. $\tau_{e\text{-}v}$ is the electron-vibration energy transfer relaxation time, given by the following expression [Lee, 1984] :

$$\tau_{e\text{-}v} = \frac{1}{n_e.1/2\,.(1-\exp(-\theta_v/T_e))^2\,.\,\int_1^{v_l} k_{0\text{-}j}^2.dj} \tag{30}$$

n_e is the electron density, θ_v is the vibration characteristic temperature and corresponds to the energy difference between the first two vibrational levels. $k_{0\text{-}j}$ is the rate constant of the process:

$$e^- + H_2(v=0) \rightarrow e^- + H_2(v=j).$$

5.4.6 Translation-H_2 vibration energy transfer source term

Like the previous one, this source term is derived from a kinetic investigation of the following vibrational quanta evolution processes :

$H_2(v) + M \rightarrow H_2(w) + M$, where M is a heavy species.

For a Boltzmann VEDF, it has a similar expression to $Q_{e\text{-}v}$ [Brun, 1986], [Lee, 1984], [Vincenti and Kruger, 1967] and [Millikan and White, 1963] :

$$Q_{v\text{-}t} = \frac{ev(T_v)\text{-}ev^*(T_g)}{\tau_{v\text{-}t}} \tag{31}$$

where $ev^*(T_g)$ is the value of the vibrational energy per unit mass corresponding to T_g.

$\tau_{v\text{-}t}$ is the species averaged vibration-translation energy transfer relaxation time, it is given by [Millikan and White, 1963] :

$$\tau_{v\text{-}t} = \sum_{s \neq e^-} x_s \Big/ \sum_{r \neq e^-} x_r / \tau_{H2-r} \tag{32}$$

where $\tau_{s\text{-}r}$ is the relaxation time for the energy transfer between the translation mode of the species 'r' and the H_2 vibration mode. $\tau_{H2\text{-}r}$ may be estimated from two expressions, the first one is derived from a kinetic analysis of the processes : $H_2(v)$ + 'r' $\rightarrow H_2(w)$ +'r', and leads to the following relation [Vincenti and Kruger, 1967] :

$$\tau_{H2\text{-}r} = \frac{1}{n.\ k_{0\text{-}1}.(1\text{-}\exp(-\theta_v/T_g))} \tag{33}$$

$k_{0\text{-}1}$ is the rate constant of the process : $H_2(v{=}0)$ + 'r' $\rightarrow H_2(v{=}1)$ + 'r'.

The second expression, due to Milikan and White [Millikan and White, 1963], is an empirical one :

$$\tau_{H2\text{-}r} = \frac{1}{P}\exp[1.16\ 10^{-3}\ \mu_{H2}^{0.5}\ \theta_v^{4/3}\ \ (T_g^{-1/3}\text{-}0.015\ \ \mu_{H2r}^{1/4})\ \text{-}18.42\ ] \tag{34}$$

where $\mu_{H2\text{-}r}$ is the reduced mass of H_2 and 'r'.

5.4.7 Electron energy loss by chemical activation

This term depends on the plasma chemical kinetics, and is described by the following general expression [Scott et al. 1995]:

$$Q_{e\text{-}chem} = \sum_{\text{Reactions}} \alpha_{R\text{-}e}\ v_R\ .\ E_{actR} \tag{35}$$

where $\alpha_{R\text{-}e}$, is the electron stochiometric coefficient in the reaction 'R'. v_R and E_{actR} are respectively the rate and the activation energy of the reaction 'R'.

5.5 TRANSPORT COEFFICIENTS

The rigorous expressions for the viscosity and the thermal conductivity of multi-component flow in thermal equilibrium were derived by Yos [Yos, 1963] and were adapted to the case of thermally non equilibrium flow by Lee [Lee, 1984]. They depend on the first and second order collisions integrals $\Omega^{(1,1)}_{s\text{-}r}$ and $\Omega^{(2,2)}_{s\text{-}r}$ of all the species pairs in the plasma.

5.5.1 Viscosity

The viscosity of a non equilibrium multicomponent flow depends on the second order collision integrals, it is given by the following expression [Lee, 1984] :

$$\mu = \sum_{se}(m_s c_s / \sum_{re}\left(c_r \Delta^{(2)}_{s-r}(T_g)\right) + c_e \Delta^{(2)}_{s-e}(T_e)) + m_e c_e / \sum_{r}\left(c_r \Delta^{(2)}_{r-e}(T_e)\right) \tag{36}$$

where

$$\Delta^{(2)}_{s\text{-}r} = 16/5\left(\frac{2m_s m_r}{\pi.kT(m_s+m_r)}\right).\ \pi\,\Omega^{(2,2)}_{s\text{-}r} \tag{37}$$

The use of equation (36) to compute the viscosity of the plasma may be very time consuming if the species number is high. For a plasma containing one major species, such as a diamond deposition plasma, the viscosity may be well estimated from the following Wilke equation, which is simpler for practical use [Wilke, 1950] :

$$\mu = \sum_{s}\left(x_s \mu_s / \sum_{k \neq s} x_k \Phi_{sk}\right) \tag{38}$$

where x_s is the molar fraction of species 's', Φ_{sk} is given by the following expression :

$$\Phi_{sk} = \frac{1}{\sqrt{8}}\left(1+\frac{M_s}{M_k}\right)^{-1/2}\left[1+\left(\frac{\mu_s}{\mu_k}\right)^{1/2}\left(\frac{M_k}{M_s}\right)^{1/4}\right] \tag{39}$$

μ_s is the viscosity of the pure gas 's', given by [Bird et al., 1960], [Wilke, 1950] and [Hirchfelder et al., 1954] :

$$\mu_s = \frac{5}{16\pi\,\Omega_s^{(2,2)}}\sqrt{\pi.m_s\,k\,T_g} \tag{40}$$

5.5.2 Thermal conductivity

The conduction coefficient of the translation-rotation energy mode is obtained by summing the conduction coefficients of the translation and rotation modes. The translation mode conductivity, λ_t, depends on the second order collisions integrals, and is given by :

$$\lambda_t = 15/4\ k \sum_{s\neq e} \left(x_s \Big/ \sum_{r\neq e} \{ x_r\, a_{sr}\, \Delta^{(2)}_{s-r}(T_g) \} + 3.54\ x_e\, \Delta^{(2);s-e}(T_e) \right) \quad (41)$$

where

$$a_{sr} = 1 + \frac{(1-m_s/m_r)\,(0.45-2.54\ m_s/m_r)}{(1+m_s/m_r)^2} \quad (42)$$

The conduction coefficients of the rotational and vibrational mode are equal; they depend on the first order collisions integrals and are given by the following expression :

$$\lambda_r = \lambda_v = k \sum_{s=molecules} \left(x_s \Big/ \sum_{r\neq e} \left(x_r \Delta^{(1)}_{s-r}(T_g) \right) + x_e \Delta^{(1);s-e}(T_e) \right) \quad (43)$$

where

$$D^{(1)}_{s-r} = 8/3 \left(\frac{2 m_s m_r}{p.kT(m_s + m_r)} \right) .\, p\, W^{(1,1)^*}_{s-r} \quad (44)$$

The conduction coefficient of electron energy λ_e is given by :

$$\lambda_e = 15/4\ k \left(x_e \Big/ \sum_r (x_r\ 1.45\ \Delta^{(2)}_{e-r}(T_e)) \right) \quad (45)$$

As in the case of the viscosity, the estimation of the conduction coefficients using equations (41)-(45) is very time consuming, especially for a high species number mixture. In the case of a plasma with one major species, The thermal conductivities may be well estimated from the conductivities corresponding to the pure gases components. This is performed by substituting the conductivities λ_s to the viscosities μ_s in the Wilke's equation (38) [Bird et al., 1960].

The conductivity coefficients λ_s for each species are given by Eucken relations [Bird et al., 1960] :

$$\lambda_{s,tr} = \mu_s \left(\frac{5}{2}\ C_{trans} + C_{rot}\right) \quad (46)$$

$$\lambda_{s,v} = \mu_s\, C_{vs} \quad (47)$$

where C_{trans}, C_{rot} and C_{vs} are the specific heat for the translational, rotational and vibrational energy modes.

The electron thermal conductivity calculation may also be simplified with the help of the following formulae [Jaffrin, 1965] :

$$\lambda_e = \frac{75\ k}{64\ (1+\sqrt{2})\ Q_{ee}} \sqrt{\frac{\pi\, k\, T_e}{m_e}} \quad (48)$$

where Q_{ee} is the electron-electron collision cross section and is given by the same author. k is the Boltzmann constant.

5.5.3 Species diffusion coefficients

The diffusion coefficient $D_{s\text{-}M}$ for a neutral species 's' in a gas mixture is given by [Curtiss et al., 1949] :

$$D_{s\text{-}M} = \frac{M_s}{M} \frac{(1\text{-}x_s\, M_s/M)}{\sum\limits_{r \neq s} x_r/D_{s\text{-}r}} \tag{49}$$

The binary diffusion coefficient $D_{s\text{-}r}$ of a species 's' in a species 'r' is expressed as a function of the 's-r' first order collision integral [Hirchfelder et al., 1954] and [Yos, 1963]:

$$D_{s\text{-}r} = \frac{3}{16}\sqrt{\frac{2}{\pi}}\left(\frac{1}{m_s}+\frac{1}{m_r}\right)^{1/2} \frac{(k\,T_g)^{3/2};P\,\Omega^1}{{}^{1}_{s\text{-}r}} \tag{50}$$

For the charged species, one have to account for the ambipolar diffusion which occurs in moderate pressure plasmas and insures the electrical neutrality for length scale greater than Debye length [Delcroix, 1958] and [Pointu, 1983]. In this case the diffusion coefficient of the ions in the plasma are given by [Delcroix, 1958] and [Camac and Kemp, 1963] :

$$D_{i\text{-}amb} = D_{i\text{-}mix}\,(1+\frac{T_e}{T_g}) \tag{51}$$

Equation (51) is only valid for a single ion plasma. However, for low ionization degree plasmas, where ions-ions collisions are negligible, one can assume that the diffusion of a given ion is not influenced by the other charged particles. In this case equation (51) may be valid even if the plasma contains several ionic species.

The diffusion coefficient of the electrons is derived from the same assumption of electrical neutrality and zero electric current, the resulting expression is [Delcroix, 1958] :

$$D_{e\text{-}amb} = \sum_{ions} z_s\, M_e\, x_s D_{s\text{-}amb} \Big/ \sum_{ions} z_e M_s\, x_s \tag{52}$$

where z_s is the electrical charge of the ions 's'.

5.6 THERMODYNAMIC AND STATE EQUATIONS

The thermodynamic and state equations enable linking the transport equations conservative variables (ρ_s, E_v, E_e, E) with more commonly used and measured plasma characteristic variables (x_s, T_g, T_v, T_e).

The state equation for a three temperature flow is given by the following expression [Lee, 1984] :

$$P = \sum_{s \neq e^-} c_s R\, T_g + c_e\, R\, T_e \tag{53}$$

where c_s is the molar concentration of species 's' and R the gas constant.

The electron energy is the sum of the thermal energy and the mean kinetic energy, it is related to electron temperature by :

$$E_e = 3/2\; c_e\; R\, T_e + 1/2\; M_e\; c_e\; ((u+u_e)^2 + (v+v_e)^2) \tag{54}$$

where c_e is the electron molar concentration.

For a harmonic oscillator approximation, the vibrational energy of species 's' is related to its vibrational temperature $T_{v\text{-}s}$ by [Vincenti and Kruger, 1967] :

$$E_{v\text{-}s} = R\; c_s\; \theta_{v\text{-}s}\; \frac{1}{\exp(\theta_{v\text{-}s}/T_{v\text{-}s})-1} \tag{55}$$

The total energy expression is obtained by summing the electron energy, the vibration energy, the heavy species translation energy, the molecules rotation energy, the chemical species formation energy and the heavy species averaged flow kinetic energy. This leads to the following expression :

$$E = E_e + E_v + \sum_{s \neq e}(3/2\;\; c_t\, x_s\, R\, T_g + E_{0s}) + . \;\; \sum_{\text{molecules}} c_t\, x_s\, R\, T_g \; + \sum_{s \neq e} ½\;\; M_s\, c_t\, x_s\, ((u+u_s)^2 + (v+v_s)^2) \tag{56}$$

where E_{0s} is the formation energy of the species 's'.

5.7 BOUNDARY CONDITIONS

The numerical solution of the plasma flow equations for a given deposition reactor requires the specification of the flow characteristics at the computation domain boundaries. The boundary conditions may be given by experimental measurements or extracted from conservation laws.

For a cylindrical symmetric deposition reactor which contains a deposition substrate in a stagnation point configuration, one has to specify the boundary conditions at the symmetry axis, the reactor wall, the reactor inlet, the reactor outlet and the substrate surface.

At the symmetry axis, zero gradient boundary conditions are specified :

$$\frac{\partial \phi}{\partial r} = 0 \tag{57}$$

$$\phi = x_{s,s=1,ns},\ u,\ v,\ T_g,\ T_v \text{ and } T_e$$

At the reactor wall and at the substrate surfaces the velocity components and the pressure gradient vanish, the gas temperature is in equilibrium with the wall temperature and the accommodation of electrons energy at the wall is almost very weak [Scott, 1993]. The resulting boundary conditions are :

At the reactor wall :

$$u = v = 0 \tag{58}$$

$$\left.\frac{\partial P}{\partial r}\right|_{wall} = 0 \tag{59}$$

$$T_{g\text{-}wall} = T_{wall} \tag{60}$$

$$\left.\frac{\partial T_e}{\partial r}\right|_{wall} \approx 0 \tag{61}$$

At the substrate surface :

$$u = v = 0 \tag{62}$$

$$\left.\frac{\partial P}{\partial x}\right|_{substrate} = 0 \tag{63}$$

$$T_g = T_{substrate} \tag{64}$$

and

$$\left.\frac{\partial T_e}{\partial x}\right|_{substrate} \approx 0 \tag{65}$$

The boundary conditions for the vibration temperature requires either the experimental measurement of T_v at the surfaces or the knowledge of the vibrational energy mode accommodation coefficient β_v [Scott, 1993]. As a consequence two kinds of boundary conditions may be specified. However, the experimental determination of $T_{v\text{-}wall}$ or β_v may be very complex and subject to important experimental uncertainties which could affect the solution of the transport equations.

The boundary conditions on the chemical species concentration at the substrate and the wall reactor surfaces may be deduced from the modeling of the surface chemical kinetics. Such models have been already derived, for diamond deposition plasmas [Coltrin and Dandy, 1993] . They lead to an estimation of the consumption or the production coefficients of the different species at the substrate surface. The resulting boundary conditions are of the form [Scott, 1993] and [McMaster et al., 1994] :

$$-D_{s\text{-mix}} \frac{\partial c_s}{\partial x} = W_{s\text{-surface}}(c_1, \quad ,c_{ns}, T_s, k_r) \tag{66}$$

where $W_{s\text{-surface}}$ is the production or consumption rate of the species 's' by surface reactions. It depends on the species molar fractions at the surface, on the surface temperature and on the rate constants of the surface reactions.

The species concentration spatial distributions in the plasma flow are very sensitive to the surface reactions kinetics which investigation is of prime importance for plasma flow modeling.

The specification of inlet and outlet boundary conditions may be more or less difficult depending on the choice of the computation domain. Substantial simplification occurs when this domain is chosen such that its inlet and outlet boundaries are far from the discharge zone in a region where the flow behavior are very simple and well known.

5.8 CONCLUSION

In this section the transport equations for a three temperature non equilibrium plasma flow have been given. The practical use of these equations for simulating diamond deposition plasma reactors requires :

- a plasma chemical kinetic model,
- a set of transport data (collision integrals) for all the chemical species,
- the data concerning the intermode energy transfer relaxation times,
- the modeling of the surface kinetics for providing the boundary conditions at the substrate,
- the knowledge of the absorbed electrical power density in the deposition reactor.

The complexity of the numerical solution of the set of the partial differential equations describing the plasma flow increases with the number of chemical species required for the modeling of the plasma kinetics. For a two dimensional flow and a large number of species, the numerical solution of the flow equation may be practically impossible. As a consequence it is very important, when possible, to carried out a chemical kinetics sensitivity investigation which could lead to a substantial reduction of the number of species and reactions for given discharge conditions.

For a given diamond deposition setup, the numerical solution of the transport equations may be substantially simplified, according to the experimental discharge conditions and to the object of the transport investigation. Indeed, if one is only interested in the study of the effects of some discharge parameters on the plasma behavior, the governing equations may be solved in one dimension. This leads to an important reduction in the computational effort. The numerical problem may also be simplified, by neglecting the convection fluxes and ignoring the momentum components equations, if the plasma Peclet number is low. However, for process scale-up, such simplifications are not possible and the transport equations have to be solved for two dimensions ; and all the fluxes and source terms described in this section have to be considered.

The solution of the flow equations may be very sensitive to the adopted boundary conditions, especially at the substrate surface where the surface kinetics scheme are very difficult to specify and to validate. A sensitivity investigation of the solution to the adopted surface kinetic scheme is then necessary.

In addition, non equilibrium plasma flow models require the knowledge of a large amount of data whose measurements or calculations may be subject to experimental errors or theoretical assumptions. As a consequence the experimental validation of these models and the investigation of the sensitivity of the results to the different data are very important for the determination of the key parameters which governs the species production and transport.

The spatial distribution of the absorbed electrical power is an important parameter for the flow modeling [Hassouni et al., 1995]. It depends on the electrical configuration of the deposition setup, the discharge nature and the plasma composition. A strong coupling exists between the absorbed power and the plasma temperatures and composition spatial distributions [Ferreira, 1992]. The absorbed power distribution may be extracted from experimental measurements (electrical or spectroscopic) [Zhang et al., 1990]. It may be also extracted from a self consistent modeling of the plasma flow by coupling the flow equations to the Maxwell equations [Tan and Grotjhon , 1994]. This selfconsistent modeling is numerically very complex to perform, especially for plasma flows with a large number of species such as diamond deposition plasmas.

The investigation of the evolution of the calculated plasma local parameters (temperatures, concentrations) and of the diamond films quality and growth rates as functions of the experimental and modeling input parameters (input power, gas flow rate, feed gas composition, pressure...) may lead to the determination of the precursor species and of the major phenomena governing the diamond deposition [Gicquel et al. , 1994].

5.9 REFERENCES

Bird, G. A. (1976), 'Molecular Gas Dynamic,' Clarendon Press Oxford, pp 171-173.

Bird, R. B., Stewart, W. E. and Lightfoot, E. N. (1960) 'Transport Phenomena,' Eds by John Wiley & sons Inc.

Brun, R. (1986), 'Transport et Relaxation dans les milieux gazeux,' Eds. Masson Paris, pp. 112-114

Cacciatore, M., Capitelli, M. and and Gorse, C. (1982) 'Non equilibrium dissociation and ionization of nitrogen in electrical discharges. The role of the electronic collisions from vibrationaly excited molecules,' Chem. Phys. 66 : 141

Camac, N., and Kemp, N. (1963) ' A Multi-Temperatures Boundary Layer' Conferences on Physics of Entry into Planetary Atmospheres,' Massachusetts, No. 63-460.

Capitelli, M., and Gorse, C. (1990) 'Non equilibrium plasma kinetics,' in 'non equilibrium kinetics in partially ionized gases' Eds Capitelli, M. and Bradsley, J. N., Plenum, NY.

Capitelli, M., Colonna, G., Hassouni, K. and Gicquel, A, (1994), 'Electron Energy Distribution Function in Non Equilibrium H_2 Discharges : The Role of Superelastic Collisions from Electronically Excited States'. Chem. Phys. Lett., 228, pp. 687-694.

Coltrin, M. E. and Dandy, D. S. (1993), J. Appl. Phys., **74**, 5803.

Curtiss, C. F., and Hirchfelder, J. O. (1949), 'Transport Properties of Multi-Component Gas Mixture,' J. Chem. Phys., Vol. 17, No. 6, pp. 550-555.

Delcroix, J. L. (1958), 'Introduction à la Théorie des Gaz ionisés,'Eds Monographie Dunod, Paris.

Delcroix, J. L. (1963), 'Physique des Plasmas,' Eds Monographie Dunod, Paris,

Ferreira, C. M. (1992), 'Kinetic Modeling of Microwave Discharge,' in Microwave Discharges Fundamentals and Applications, edited by Ferreira, C. M. and Moisan, M., in NATO ASI series, Series B : Physics Vol. 302., pp 313-337.

Gicquel, A., Hassouni, K., Farhat, S., Breton, Y., Scott, C. D., Lefebvre, M. and Pealat, M. (1994), 'Spectroscopic Analysis and Chemical Kinetic Modeling of a Diamond Deposition Plasma Reactor,' Diam. and Rel. Mat. , Vol. 3, n° 4-6, pp 581-586.

Gicquel, A., Scott, C. D., Lefebvre, M., Breton, Y., Lamendola, R., Anger, E. and Pealat, M. (1994), 'Optimizing parameters in a Diamond Deposition Reactor,' Vuotto XXIII, No. 2, pp. 5-9.

Gorse, G., Billing, G., Caccioatore, M., Capitelli, M. and De Benedctis, S. (1987), chem phys, 111 : 357.

Gorse, C., Cacciatore, M., Celiberto, R., Cives, P. and Capitelli, M. (1989), 9th International Symposium on Plasma Chemistry. Pugnochinso, Italy, Sept. 4-8, Proceedings V. 1, ed. D'Agostino, R., pp 229-234.

Gorse, C., Celiberto, R., Cacciatore, M., Lagana, A. and Capitelli, M. (1992), 'from dynamics to modeling of plasma complex systems : negative ion (H^-) sources, Chem. Phys. 161 : 211.

Hassouni, K., Farhat, S., Scott, C. D. and Gicquel, A. (1995), 'Mass and energy transport in an H_2 microwave plasma obtained under diamond deposition discharge conditions. The role of the spatial distribution of the microwave power density,' To be published in the Proceeding of Applied Diamond Conference, August 21-24.

Hirschfelder, J. O., Curtis, C. F., Bird, R. B. (1954), 'The Molecular Theory of Gases and Liquids,' Eds J. Wiley and Sons, New York.

Jaffrin, H. J. (1965), 'Shock Structure in Partially Ionized Gases,' Phys. Fluids, Vol. 8, No 4, pp 605-625, April 1965.

Keck, J. and Carrier, G., J. Chem. Phys., **43** 2284-98, 1965.

Koemtzopoulos, C. R., Economou, D. J. and Pollard, R. (1993), 'Hydrogen Dissociation in a Microwave Discharge for Diamond Deposition', Diamond and Related Materials, Vol. 2, pp. 25-35.

Lee, J. H. (1984), 'Basic Governing Equations for the Flight Regimes of Aeroassisted Orbital Transfer Vehicles,' AIAA paper No 84-1729.

Lee, J. H.(1985), 'Electron Impact Vibrational Excitation Rates in the Flowfield of Aeroassisted Orbital Transfer Vehicles,' AIAA paper No 85-1035.

Levin, E., Partridge, H. and Stallcop, J. R. (1988), 'High Temperature Transport Properties of Air : N-O Interaction Energies and Collision Integrals,' AIAA paper No 88-2660.

Loureiro, J. and Ferreira, C. M. (1986), J. Phys. D. : Appl. Phys., Vol. 19,. 17-35.

Loureiro, J. and Ferreira, C. M. (1989), 'Electron and Vibrational Kinetics in the Hydrogen Positive column,' J. Phys. D : Appl. Phys., Vol. 22, pp. 1680-1691.

Loureiro, J. and Ferreira, C. M., Capitelli, M., Gorse, C. and Cacciatore, M. (1990), 'Non Equilibrium Kinetics in Nitrogen Discharges : A Comparative analysis of Two Theoritical Approaches'. J. Phys. D : Appl. Phys. **23** pp. 1371-1383.

Marrone, P. V. and Treanor, C. E.(1963), 'Chemical Relaxation with Preferencial Dissociation from Excited Vibrational Levels', Phys. Fluids, Vol. 6, no. 9, p 1215.

McMaster, M. C., Hsu, W. L., Coltrin, M. E. and Dandy, D. S. (1994), 'Experimental Measurements and Numerical Simulation of the Gas Composition in a Hot Filament Assisted Diamond Chemical Vapor Deposition Reactor,' J. Appl. Phys. **76** (11).

Millikan, M. C. and White, D. R. (1963), 'Systematics of Vibrational Relaxation', J. Chem. Phys. Vol. 39, No 12.

Park, C. (1987), 'Assessment of Two Temperature Kinetic Model for Ionizing Air,' AIAA paper No. 87-1292.

Pointu, A. M. (1983), 'Physique des décharges', in Ecole d'Ete Réactivite dans les Plasmas, Eds Société Française de Physique, Aussois.

Ricard, A. (1983), ' Réactivité des Etats Excités,' in Ecole d'Ete Réactivite dans les Plasmas, Aussois.

Scott, C. D. (1993), 'A Non Equilibrium Model for a Moderate Pressure Hydrogen Microwave Discharge Plasma,' NASA Technical Memorandum 104765.

Scott, C. D., Farhat, S., Gicquel, A., Hassouni, K. and Lefebvre, M. (1995), 'Finding Electron Temperature and Density in a Hydrogen Microwave Discharge Plasma,' (accepted in Thermophysics and Heat Transfer, to be published in 1995)

Tan, W. and Grotjhon, T. A. (1994), 'Modeling the Electromagnetic Excitation of a Microwave cavity Plasma Reactor,' J. Vac. Sci. Technol. A 12(4).

Vincenti, W. G. and Kruger, C. H. Jr. (1967), 'Introduction to Molecular Gas Dynamics,' John Wiley and Sons, New York.

Wadhera, J. M. and Bradsley, J. N. (1978), 'Vibrational and Rotational State Dependence of Dissociative Attachement in e^{-}-H_2 collisions', Phys. Rev. Lett., Vol. 41, pp. 1795-1798.

Wilke, C. R. (1950), ' A Viscosity Equation for Gas Mixtures,' J. Chem. Phys., Vol. 18, No 4.

Winkler, R. (1992), 'Collision Dominated Electron Kinetics in Low and High Frequency Fields,' in Microwave Discharges Fundamentals and Applications, edited by C. M. Ferreira and M. Moisan in NATO ASI series, Series B : Physics Vol. 302., pp 339-358.

Yos, J. M. (1963), 'Transport properties of Nitrogen, Hydrogen, Oxygen and Air to 30000 K,' Technical Memorandum RAD TM-63-7, AVCO-RAD, Wilmington, Mass.

Zakrzewski, Z., Moisan, M. and Sauvé, G. (1992), 'Surface wave Plasma sources in Low and High Frequency Fields,' in Microwave Discharges Fundamentals and Applications, edited by C. M. Ferreira and M. Moisan in NATO ASI series, Series B : Physics Vol. 302., pp 117-140.

Zhang, J., Huang, B., Reinhard, D. K. and Asmussen, J. (1990), 'An investigation of Electromagnetic field Patterns during Microwave Plasma Diamond Thin Film Deposition,' J. Vac. Sci. Technol. A 8(3).

Chapter 17
SECTION 6: ELECTROMAGNETIC FIELD MODELING OF DIAMOND CVD REACTORS

Timothy A. Grotjohn
Department of Electrical Engineering
Michigan State University
East Lansing, MI 48824

6.1 ELECTROMAGNETIC HEATING OF CVD PLASMAS

The generation of diamond deposition discharges can be accomplished by direct thermal heating (as occurs in hot filament reactors), by chemical reactions (as occurs in oxy-acetylene flame deposition reactors), and by electromagnetic/electric field excitation of plasma discharges (as occurs in dc, rf, and microwave plasma source reactors). This later mechanism dealing with the electromagnetic/electric field excitation is the topic of this section. Basically, the electromagnetic/electric fields first couple energy into the free electrons in the plasma discharge. The electrons absorb this energy from the electric fields and then through subsequent collisions with the neutral gas the electron gas energy is transferred to the neutrals by elastic and inelastic collision including excitation, dissociation and ionization collisions.

The key electromagnetic/electric field excitation components in microwave, rf and dc discharge reactors are shown in Fig. 1. The basic components of the microwave system are an input power coupling port or probe, an electromagnetic field boundary to contain and focus the fields, a plasma discharge containment boundary (quartz in this figure), and the plasma discharge. The rf system shown uses a rf power supply to produce current flow in the coil and the associated rf magnetic fields, which couple inductively to the plasma resulting in heating of the electron gas. The dc plasma source applies a voltage of several hundred volts across the electrodes which accelerates and heats the electron gas. The behavior of the electromagnetic energy in each of these systems and how it couples to and excites the plasma discharge depends on the geometry of the structure, the input power coupling structure, and the size, density, pressure, and composition of the CVD discharge. In the next two sections the basic equations for solving the electromagnetic/electric fields and their interaction with plasma discharges for various types of systems are formulated. Then a specific example of solving for the electromagnetic excitation of a microwave diamond CVD reactor is presented.

6.2 MAXWELL'S EQUATIONS AND THEIR FORMULATION FOR PLASMA DISCHARGE MODELING

The Maxwell equations are given by:

$$\nabla \times \mathrm{E} = -\mu_o \frac{\partial \mathrm{H}}{\partial t} \tag{1}$$

$$\nabla \times \mathrm{H} = \varepsilon \frac{\partial \mathrm{E}}{\partial t} + \mathrm{J} \tag{2}$$

$$\nabla \cdot \mathrm{E} = \frac{\rho}{\varepsilon} \tag{3}$$

$$\nabla \cdot \mathrm{B} = 0 \tag{4}$$

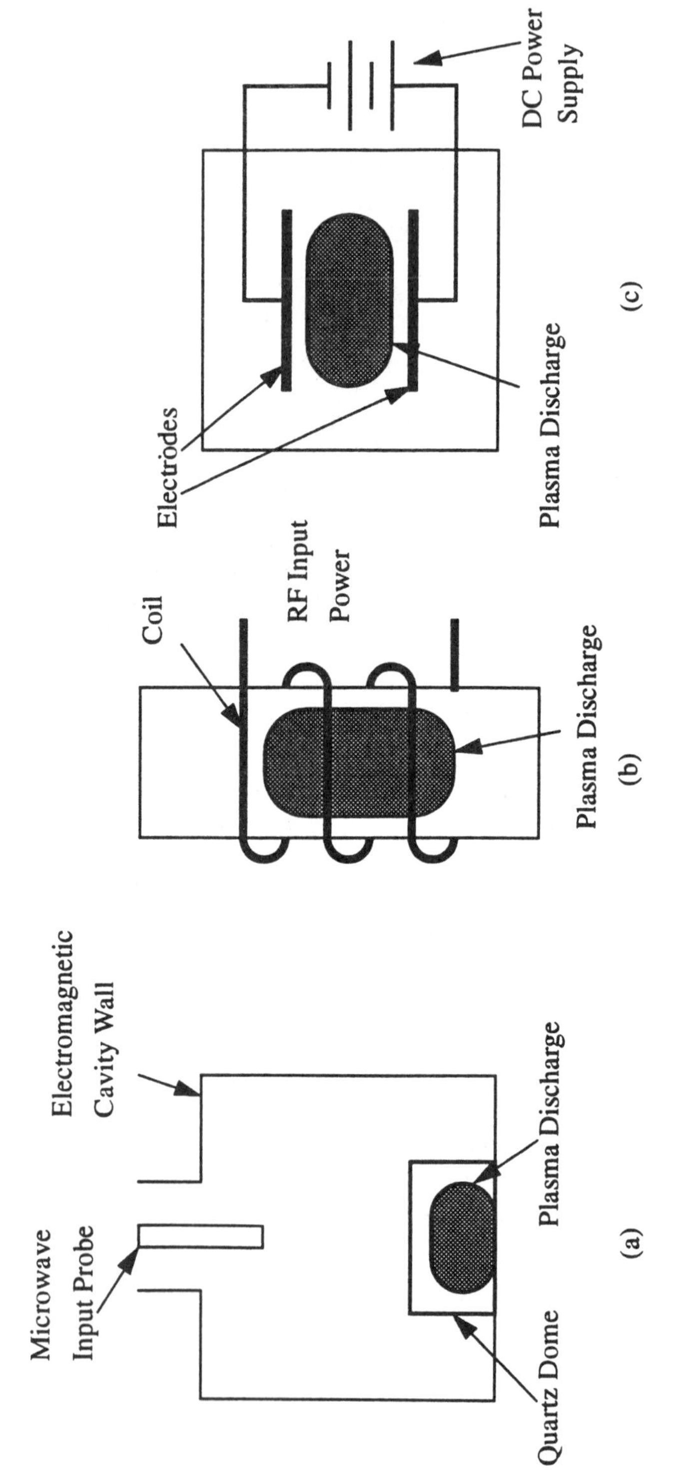

Figure 1 Microwave, rf inductively-coupled, and dc plasma discharge reactors are shown in (a), (b), and (c), respectively.

where $\boldsymbol{E}$ is the electric field, $\boldsymbol{H}$ is the magnetic field, $\boldsymbol{B}$ is the magnetic flux density, $\boldsymbol{J}$ is the current density, ε is the permittivity, μ_o is the permeability, and ρ is the charge density. The plasma discharge contributes to these four equations through the current density and the charge density terms.

Equation (4) can be treated as an initial condition for the time domain solution of Maxwell's equation because from (1) we can write

$$\nabla\cdot(\nabla\cdot \mathrm{E}) = \frac{d}{dt}(-\nabla\cdot \mathrm{B}) \tag{5}$$

and since $\nabla\cdot(\nabla \times \mathrm{E})= 0$ it can be concluded that $\nabla\cdot\mathrm{B}$ = constant. So the solution of (4) is an initial condition for the time domain solution of (1). Therefore, the solution of Maxwell equations can be done by solving equations (1), (2) and (3) in the time domain. A simplification often made for this solution is that of solving the high-frequency behavior, given primarily by (1) and (2) which have time derivative terms, separately from the low frequency or dc behavior. The time domain method used to solve (1) and (2) is called the finite-difference time-domain method (FDTD). The FDTD technique was first proposed by Yee [Yee 1966] in 1966 to solve the interaction of electromagnetic waves with two-dimensional, isotropic material. Recent studies used the FDTD method for the solution of three-dimensional scattering problems, lossy dielectric materials, resonant cavities, thin plates and wires, and bioelectromagnetic dosimetry problems [Taflove and Brodwin 1975], [Mur 1981] and [Gothard et al. 1994]. The FDTD technique has also been applied to study the electromagnetic field interactions with plasma discharges [Grotjohn 1992], [Grotjohn et al. 1994], [Tan and Grotjohn 1994], [Tan and Grotjohn 1995] and [Hunsberger et al. 1992]. Overall, the FDTD method has the advantage of ease of implementation for complicated geometries, because spatially varying dielectric and conductivity parameters can be assigned using grid points.

Another method of solving the Maxwell equations is by combining equations (1) and (2) and obtaining the double curl equation as

$$\nabla\times\nabla\times \mathrm{E} = -\mu_o\varepsilon\frac{\partial^2}{\partial t^2}\mathrm{E} - \mu_o\frac{\partial \mathrm{J}}{\partial t} \tag{6}$$

or by applying a vector identity, (6) becomes

$$\nabla(\nabla\cdot \mathrm{E}) - \nabla^2\mathrm{E} = -\mu_o\varepsilon\frac{\partial^2}{\partial t^2}\mathrm{E} - \mu_o\frac{\partial \mathrm{J}}{\partial t} \tag{7}$$

By using equation (3) for the first term on the left side of (7), it becomes

$$\underbrace{\nabla^2\mathrm{E}}_{L} = \underbrace{\nabla\left(\frac{\rho}{\varepsilon}\right)}_{R1} + \underbrace{\mu_o\varepsilon\frac{\partial^2}{\partial t^2}\mathrm{E}}_{R2} + \underbrace{\mu_o\frac{\partial \mathrm{J}}{\partial t}}_{R3} \tag{8}$$

where the symbols L, R1, R2, and R3 denote the four terms. The formulation of the electromagnetic/electric fields as shown in (8) allows the significance of spatial charge density variations, time-varying fields, and time varying current densities to be assessed and compared to each other. At high frequencies in the neutral portion of the plasma (i.e., away from the sheaths) the terms R2 and R3 dominate R1. At low frequencies or dc the term R1 often dominates R2 and R3.

Equation (7) has been solved to determine the electromagnetic fields in various applications for electromagnetic wave propagation [Paulson et al. 1988]. Unfortunately, the direct solution of (7) is prone to the generation of extraneous or parasitic solutions

[Lynch and Paulsen 1991]. However, if additional assumptions are made as indicated below, then a unique solution without any extraneous or parasitic solutions is possible. Special cases are as follows.

6.2.1 Low Frequency and/or Large Charge Density Variations

For low frequency where the variations in charge density (term R1) dominate over the time varying terms (R2 and R3), equation (8) simplifies to

$$\nabla^2 E = \nabla\left(\frac{\rho}{\varepsilon}\right) \tag{9}$$

or for one coordinate direction

$$\frac{\partial^2 E_x}{\partial x^2} + \frac{\partial^2 E_x}{\partial y^2} + \frac{\partial^2 E_x}{\partial z^2} = \frac{\partial}{\partial x}\left(\frac{\rho}{\varepsilon}\right) \tag{10}$$

In plasma sheath-like regions where

$$\frac{\partial^2 E_x}{\partial x^2} >> \frac{\partial^2 E_x}{\partial y^2} + \frac{\partial^2 E_x}{\partial z^2} \tag{11}$$

the equation (10) simplifies to

$$\frac{\partial E}{\partial x} = \frac{\rho}{\varepsilon} \tag{12}$$

or more generally the Poisson equation as given by (3). Therefore, for dc/low frequency discharges only one of the four Maxwell equations (i.e., the Poisson equation) is required to determine the fields. An example of such a solution for the diamond deposition process is the work on self-consistent dc glow-discharge simulations by Surendra, Graves and Plano [Surendra et al. 1992]. The solution of (3) alone is also the technique used for the solution of rf capacitively-coupled discharges (also called rf parallel-plate discharges.)

6.2.2 High Frequency and Negligible Charge Density Variations

For high frequencies where the charge density variations are small, R2 and R3 dominates with respect to the R1 term and (8) simplifies to

$$\nabla^2 E = \mu_o \varepsilon \frac{\partial^2}{\partial t^2} E + \mu_o \frac{\partial J}{\partial t} \tag{13}$$

For a sinusoidal solution where $E = Ee^{j\omega t}$ and for the current density written as $J = \sigma E$, equation (13) becomes

$$\nabla^2 E + \omega^2 \mu_o \varepsilon E - j\omega \mu_o \sigma E = 0 \tag{14}$$

In a collisional plasma the conductivity σ is

$$\sigma = \frac{n_e e^2 (\nu - j\omega)}{m_e(\nu^2 + \omega^2)} \tag{15}$$

where ν is the electron collision frequency, e is the electron charge, and m_e is the electron mass.

Low Pressure Plasma Discharges. The application of (14) to the solution of electromagnetic fields in a low pressure plasma discharge where $\omega >> \nu$ allows the conductivity to be written as

$$\sigma_1 = \frac{-jn_e e^2}{m_e \omega} \tag{16}$$

giving

$$\nabla^2 E + \omega^2 \mu_o \varepsilon_p E = 0 \tag{17}$$

where the plasma dielectric constant is defined as

$$\varepsilon_p = \varepsilon_o \left(1 - \frac{\omega^2 pe}{\omega^2} \right) \tag{18}$$

with ω_{pe} being the electron plasma frequency. This formulation has been used by Hopwood et al. [Hopwood et al. 1993] to describe the electromagnetic fields in a low pressure inductively-coupled plasma source. Such plasma sources have been investigated for their potential in depositing diamond-like carbon films [Pappas and Hopwood 1994].

High Pressure Plasma Discharges. The application of (14) to the solution of electromagnetic fields in a high-pressure collisional plasma discharge where $\omega << \nu$ allows the conductivity to be written as

$$\sigma_2 = \frac{n_e e^2}{m_e \nu} \tag{19}$$

giving

$$\nabla^2 E - j\omega \mu_o \sigma_2 E = 0 \tag{20}$$

This formulation, where the collision frequency is greater than the excitation frequency, has been applied to high pressure inductively-coupled rf plasma discharge [Yu and Girshick 1991] as sketched in Fig. 1b. This type of discharge reactor has also been used to deposit CVD diamond [Girshick et al. 1993].

The basic formulations for the solution of the electromagnetic/electric fields can be classified as:

1) High frequency (microwave): Solve (1) and (2) in the time domain.
2) Low-frequency or dc: Solve (3).
3) Rf highly collisional plasmas: Solve (20).
4) Rf low-pressure plasmas: Solve (17).

6.3 ELECTROMAGNETIC PROPERTIES OF PLASMA DISCHARGES

For collisional, non-magnetized plasma discharges, such as often used for diamond CVD deposition, the frequency of excitation is generally less than the plasma frequency. For this condition the rf and microwave electromagnetic fields do not propagate through the plasma discharge. Rather, the electromagnetic waves penetrate into the plasma discharge a distance on the order of a skin depth. For the region of the discharge within approximately a skin depth of the plasma surface, the electron gas absorbs energy from the rf or microwave electric fields. The general expression for the skin depth d of transverse electromagnetic wave penetration into a plasma with a plasma density n and an electron collision frequency u is [Bittencourt 1986]:

$$\frac{1}{\delta} = -Imag\left[\left(\frac{1}{c}\left(\omega^2 - \frac{\omega_{pe}^2}{1+\left(\frac{\nu}{\omega}\right)^2} - \frac{j\,\omega_{pe}^2\frac{\nu}{\omega}}{1+\left(\frac{\nu}{\omega}\right)^2}\right)\right)^{1/2}\right] \tag{21}$$

where c is the speed of light and ω_{pe} is the electron plasma frequency given by

$$\omega_{pe}^2 = \frac{e^2 n}{\varepsilon_o m_e} \tag{22}$$

In the limit of $\nu >> \omega$, the skin depth expression simplifies to

$$\delta = c\left(\frac{2\nu}{\omega_{pe}^2\omega}\right)^{1/2} \tag{23}$$

or

$$\delta = \left(\frac{2}{\omega\,\mu_o\sigma_2}\right)^{1/2} \tag{24}$$

where σ_2 is give by (19).

As an example, consider a diamond deposition microwave excited discharge with a plasma density of $n=5x10^{11}$ cm^{-3}, a microwave frequency of 2.45 GHz, and a collision frequency of $\nu=5x10^{10}$ sec^{-1} which corresponds approximately to that found in a 50 Torr hydrogen plasma discharge with a neutral gas temperature of 1000-2000 K. The skin depth for this example according to (21) is approximately 1.5-2.5 cm.

For the case of $\omega >> \nu$ and $\omega_{pe}>\omega$, the electromagnetic waves penetrate the plasma with a skin depth given by

$$\delta = \left(\frac{m_e}{e^2\mu_o n}\right)^{1/2} \tag{25}$$

which is called the collisionless skin depth.

For the case of longitudinal electric fields applied normal to the plasma discharge boundary, the skin depth is generally much smaller and is given approximately by [Cohen 1989]

$$\delta = \frac{3}{2}\lambda_D \tag{26}$$

where λ_D is the Debye length given by

$$\lambda_D = \left(\frac{\varepsilon k\,T_e}{e^2 n}\right)^{1/2} \tag{27}$$

where k is the Boltzmann constant.

In plasma reactors the electromagnetic fields are seldom pure transverse waves or pure longitudinal waves, rather the plasma discharge exists in an electromagnetic boundary that forms TM and TE modes. For the case of TM and TE modes the skin depth expressions still provide estimates of the electromagnetic field penetration in the plasma. However, the complete solution of the plasma discharge reactor behavior requires a self-consistent solution of the plasma kinetics (continuity equation, momentum transport equation, the energy equation) together with the electromagnetic equations.

6.4 NUMERICAL SIMULATIONS OF A MICROWAVE DIAMOND CVD REACTOR

In the next two sections, a numerical model is described that self-consistently solves the electromagnetic fields, as well as, the excitation mechanism and characteristics of the discharges inside a cylindrical microwave cavity plasma reactor. The electromagnetic fields are solved by using the finite-difference time-domain (FDTD) [Yee 1966] method. The characteristics of the discharge are simulated by a fluid plasma model which solves the electron and ion continuity equations, the electron energy balance equation, and the Poisson equation. A final electromagnetic field and plasma discharge solution is obtained by iteratively solving the electromagnetic field model and the fluid plasma model.

In particular, a TM_{013} mode, 17.78 cm i.d. microwave cavity plasma reactor [Zhang 1990] loaded with a hydrogen discharge is simulated using this numerical model. The TM_{013} mode is an axi-symmetric mode. The hydrogen plasma example was chosen because diamond film deposition processes often consist of high percentages of hydrogen in the discharge. And, as shown by Koemtzopoulos et al. [Koemtzopoulos et al. 1993], adding small percentages (e.g., 1%) of methane to a hydrogen discharge has only a minimal effect on the electron energy distribution function. The method used to excite electromagnetic waves inside the cavity and the boundary condition assignments for both the electromagnetic field model and fluid plasma model are discussed. The methods to evaluate the cavity quality factor Q are presented. Additionally, the spatial electric field patterns, power absorption patterns, and quality factor of the cavity loaded with a hydrogen discharge are investigated in the moderate pressure range of 40 - 60 Torr. The physical behavior of the hydrogen discharge, such as plasma density, electron temperature, and plasma potential, are also simulated and analyzed for various input conditions. And lastly, the simulated results are compared with experimental data.

6.4.1 FDTD Electromagnetic Field Model

The electromagnetic fields inside the cavity can be described by the time-dependent Maxwell's equations (1) and (2), which in three-dimensional cylindrical coordinates are written as six scalar equations given by

$$\frac{\partial H_r}{\partial t} = \frac{1}{\mu}\left(\frac{\partial E_\phi}{\partial z} - \frac{1^{\partial E_z}}{r\partial\phi}\right) \tag{28}$$

$$\frac{\partial H_\phi}{\partial t} = \frac{1}{\mu}\left(\frac{\partial E_z}{\partial r} - \frac{\partial E_r}{\partial z}\right) \tag{29}$$

$$\frac{\partial H_z}{\partial t} = \frac{1}{\mu}\left[\frac{1^{\partial E_r}}{r\partial\phi} - \frac{1}{r}\frac{\partial}{r\partial r}(rE_\phi)\right] \tag{30}$$

$$\frac{\partial E_r}{\partial t} = \frac{1}{\varepsilon}\left(\frac{1}{r}\frac{\partial H_z}{\partial \phi} - \frac{\partial H_\phi}{\partial z}\right) - \frac{J_r}{\varepsilon} \tag{31}$$

$$\frac{\partial E_\phi}{\partial t} = \frac{1}{\varepsilon}\left(\frac{\partial H_r}{\partial z} - \frac{\partial H_z}{\partial r}\right) - \frac{J_\phi}{\varepsilon} \tag{32}$$

$$\frac{\partial E_z}{\partial t} = \frac{1}{\varepsilon}\left[\frac{1}{r}\frac{\partial}{\partial r}(rH_\phi) - \frac{1}{r}\frac{\partial H_r}{\partial \phi}\right] - \frac{J_z}{\varepsilon} \tag{33}$$

The FDTD method is formulated by discretizing the partial differential equations (28)-(33) with a centered difference approximation in both the time and space domain. The six field locations are assigned to be interleaved in space to satisfy the continuity of tangential field components. A FDTD unit cell in cylindrical coordinates as shown in Figure 2 is drawn in a way similar to Yee's rectangular coordinate cube cell [Yee 1966] and [Taflove and Brodwin 1975]. The dimensions of the unit cell in the ϕ and z directions are $r\Delta\phi$ and Δz respectively. If the permittivity and permeability are space dependent, appropriate values are assigned to lattice points for each field component. The E and H values are calculated at alternating half-time steps in order to achieve centered differences for the time derivatives. Grid spaces along each direction and Δt are chosen appropriately to satisfy both the accuracy and stability conditions of the finite difference equations [Holland et al. 1980].

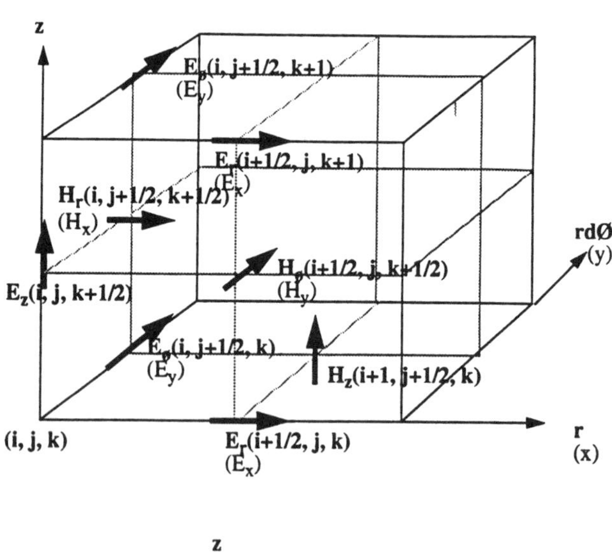

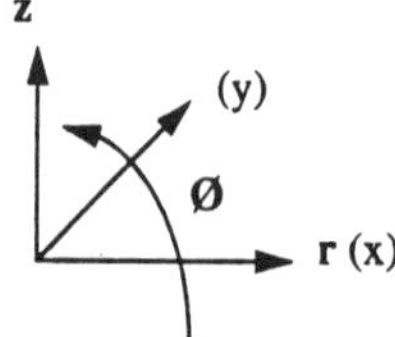

Figure 2 Finite-difference time-domain method grid cell.

The electromagnetic field energy U stored in the microwave cavity plasma reactor is

$$U(t) = 0.5\int_V (\varepsilon(\mathrm{r})(\mathrm{E}(r,t))^2 + \mu(\mathrm{r})(\mathrm{H}(\mathrm{r},t))^2)dV \tag{34}$$

where the integration is performed for the volume of the cavity, V. The power dissipation density, P(r,t), with a power absorbing load present is

$$P(\mathrm{r},t) = \mathrm{J}(\mathrm{r},t)\cdot \mathrm{E}(\mathrm{r},t) \tag{35}$$

To model the electrical behavior of the plasma in the time domain, ε and μ in equations (28) to (33) are set equal to ε_0 and μ_0 (dielectric parameters in vacuum). It is important to note that the dielectric parameter used is ε_0 and not $\varepsilon_r\varepsilon_0$ where ε_r is the nonmagnetized plasma dielectric parameter. The use of ε_r is appropriate for sinusoidal steady state solutions where the two terms on the right side of (2) have been combined. In the time domain analysis of this model the two terms are kept separated and the current density in (2), (31-33), or (35) is evaluated using an appropriate momentum transport equation. For example, if a uniform, isotropic, homogeneous plasma is assumed, and ion motion is ignored, the momentum transport equation can be written as

$$m_e\frac{d\mathrm{v}}{dt} = -q\mathrm{E} - m_e\, \nu_{eff}\, \mathrm{v} \tag{36}$$

and the current density as

$$\mathrm{J}_e = -qn_e\mathrm{v} \tag{37}$$

where **v** is the average electron velocity, q is the electron charge, m_e is the electron mass, ν_{eff} is the effective collision frequency, and n_e is the electron density. The electric field **E** used in (36) is the high frequency microwave or rf electric field found by solving (28)-(33).

Using the finite difference method to discretize this first order differential equation (36) in time, the momentum transport equation for electrons becomes

$$v^n_{r,\phi,z}(i,j,k) = (1.0 - \Delta t\cdot \nu_{eff})\cdot v^{n-1}_{r,\phi,z}(i,j,k) - \frac{q\Delta t}{m_e}E^{n-1}_{r,\phi,z}(i,j,k) \tag{38}$$

and the current density is

$$J^n_{e(r,\phi,z)}(i,j,k) = -qn_e v^n_{r,\phi,z}(i,j,k) \tag{39}$$

where n denotes the time iteration count, Δt is the time step and i, j, k denote the grid location. Equation (38) is solved simultaneously with (28)-(33).

6.4.2 Fluid Plasma Model

The behavior of the charged particles in a weakly ionized gas can be described by the particle, momentum, and energy balance equations for electrons and ions, which are obtained from the moments of the Boltzmann equation. Moreover, these equations are often combined with Poisson's equation to provide a self-consistent space charge field. Under moderate pressures (1-150 Torr) of this study, the momentum balance equations of electrons and ions at low frequency/dc are given by (43) and (44) according to [Passchier and Goeheer 1993]. In steady state, these equations are written as [Surendra et al. 1992]

$$\nabla^2 \psi = \frac{q}{\varepsilon_o}(n_e - n_i) \tag{40}$$

$$\nabla \cdot \mathbf{J}_{e(dc)} = n_e n_n k_{ion} - \alpha_r n_i n_e \tag{41}$$

$$\nabla \cdot \mathbf{J}_i = n_e n_n k_{ion} - \alpha_r n_i n_e \tag{42}$$

$$\mathbf{J}_{e(dc)} = -n_e \mu_e \mathbf{E}_{dc} - D_e \nabla n_e \tag{43}$$

$$\mathbf{J}_i = n_i \mu_i \mathbf{E}_{dc} - D_i \nabla n_i \tag{44}$$

$$\nabla \cdot \mathbf{q}_e = -q \mathbf{J}_e \cdot \mathbf{E} \tag{45}$$

$$-n_e n_n (\varepsilon_{ion} k_{ion} + \varepsilon_{ext} k_{ext} + \varepsilon_{dis} k_{dis})$$

$$-\frac{2 m_e}{m_n}\left(\frac{3}{2} k_B T_e - \frac{3}{2} k_B T_n\right) \nu_m n_e - \frac{3}{2} k_B T_e \alpha_r n_i n_e$$

$$\mathbf{q}_e = \frac{5}{2} k_B T_e \mathbf{J}_{e(dc)} - \left(\frac{5}{2} k_B D_e n_e\right) \nabla T_e \tag{46}$$

Eq. (40) is the Poisson equation, where Ψ is the potential $(\mathbf{E}_{dc} = \nabla \psi)$ and where $\mathbf{E}_{dc}$ is the low frequency or dc electric field. Eqs. (41) and (42) are the electron and ion continuity equations, respectively. Eqs. (43) and (44) are in essence the momentum balance for electrons and ions, respectively. Eq. (45) is the electron energy balance, with the total electron energy flux given by Eq. (46) where the thermal energy is assumed to be greater than the kinetic energy. In the above equations, n_e, n_i, and n_n are the electron, ion, and neutral densities, respectively. $\mathbf{J}_{e(dc)}$ and $\mathbf{J}_i$ are the electron and ion flux, respectively. Note that the low frequency/dc current density $\mathbf{J}_{e(dc)}$ is a different quantity than $\mathbf{J}_e$ defined in (37) for high frequencies. T_e is the electron temperature, and q_e is the electron energy flux. k_{ion}, k_{ext}, and k_{dis} are the inelastic rate constants, with threshold energies ε_{ion}, ε_{ext}, and ε_{dis} for ionization, excitation, and dissociation, respectively. α_r is the recombination-rate constant. Note that elastic losses have been included in Eq. (45), where ν_m is the electron-neutral momentum transfer frequency, and m_n is the neutral species mass. Moreover, $D_{e,i}$ and $\mu_{e,i}$ are the electron and ion diffusivity and mobility, respectively.

The fluid plasma model describes the characteristics of discharges by solving Eqs. (40) to (46). Finite difference methods and staggered-mesh techniques [Barnes et al. 1988] are used to discretize these equations in cylindrical coordinates. Since the resonant electromagnetic modes of the microwave plasma sources used for diamond thin film deposition in this study are normally TM_{01n} modes, the electric field spatial distributions are ϕ symmetric in nature and the E_ϕ component is zero. Therefore the discharge behavior can be assumed to be ϕ symmetric and the discretization of the equations

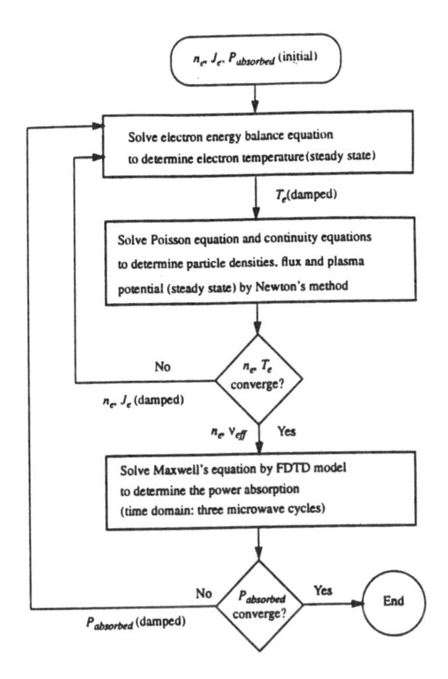

Figure 3 Flow chart for microwave cavity plasma reactor simulation.

reduced to a two dimensional problem. Thus, the simulation region remains in the r-z plane only. Note that the heating term $q\mathbf{J}_e \cdot \mathbf{E}$ in Eq. (45) is the microwave power absorbed by the plasma, which is determined by the microwave electric field and the microwave induced current density using the FDTD electromagnetic field model described earlier. Therefore, for simplicity, this heating term given in Eq. (35) as P can be viewed as an input parameter for the electron energy balance equation (45).

The flow chart of the microwave cavity plasma simulation is shown in Fig. 3 [Tan and Grotjohn 1994]. The solution of the discretized fluid plasma equations, Eqs. (40) to (46), on a grid of N nodes involves a total of 4N unknowns for the fluid model since at each point the unknowns are Ψ, n_e, n_i, and T_e. To solve the system of 4N non-linear type, discretized equations, both direct method (Newton's method) and iterative method techniques are used. In particular, the Poisson equation, electron continuity equation and ion continuity equation are tightly coupled and solved by Newton's method. The unknowns solved at each grid point are the electron density, ion density and plasma potential, and the electron and ion flux (steady state DC flux) can be determined by Eqs. (43) and (44) using the solution of these unknowns. Then, the electron density and flux are coupled into the discretized electron energy balance equation, Eq. (45), to solve the electron temperature. The calculated electron temperature is then coupled back to the continuity equations and Poisson equation to modify the reaction rates and parameters and update the electron density, ion density, and plasma potential. The final stable solution of density and temperature is achieved by iteratively solving the continuity/Poisson equations and electron energy balance equation.

As described above an input term in the electron energy balance equation (45) is the absorbed microwave power density P, which comes from the FDTD electromagnetic field model. The FDTD model which solves the electromagnetic fields inside the discharges provides output information such as the microwave power absorption of the discharges by simultaneously solving the microwave electron momentum transport equation, Eq. (36). The discharge characteristics such as the plasma density and electron temperature are determined in steady state under this power absorption condition. The discharge characteristics information is then coupled to the FDTD model to modify the plasma conductivity and calculate a new discharge power absorption. Therefore, in this iterative manner, the power absorbed by the discharges will converge to a stable value, and the electromagnetic fields inside the reactor source and the plasma characteristics are solved self-consistently.

6.5 MODELING CASE STUDY: A MICROWAVE CAVITY PLASMA REACTOR

6.5.1 Microwave Cavity Plasma Reactor Model

The numerical model developed in Section 4 is now applied to simulate a microwave cavity plasma reactor used for diamond thin film deposition as shown in figure 4. The cavity is a cylindrical, 17.78 cm i.d. microwave cavity in which the microwave energy is coupled into the cavity through a coaxial input probe. The specific resonant mode is determined by the geometrical size of the cylindrical cavity and it can be adjusted by the movable sliding short. In this study, the height of the cavity is adjusted for TM_{013} mode excitation with a 2.45 GHz input microwave frequency. The coaxial input probe and movable short position are adjusted for maximum power transfer to the plasma load as indicated by minimum reflected power from the cavity. A hydrogen dominated discharge is excited and sustained inside the quartz dome by the input microwave power. The volume and the temperature of the discharge are dependent on the input power and pressure. The substrate and substrate holder are placed on the top of a quartz tube which is used to adjust the height of the substrate. A metal screen is placed at the bottom of the cavity to prevent the microwave field from propagating into the downward region.

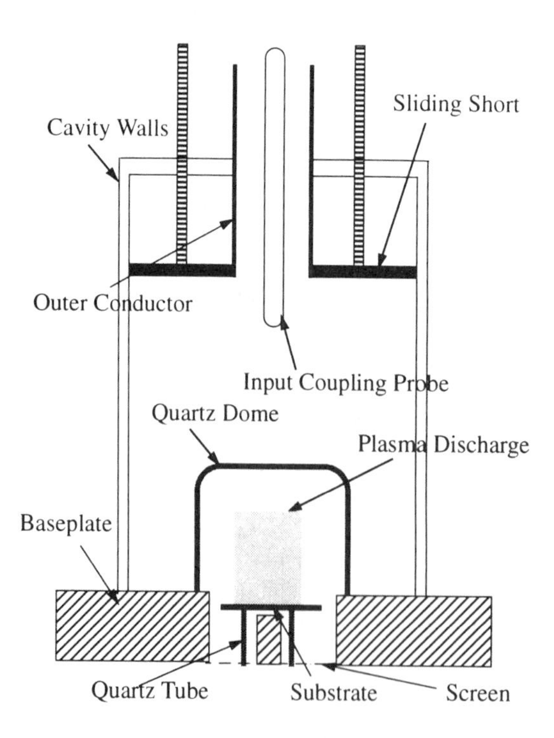

Figure 4 Configuration of a microwave cavity plasma reactor used for diamond film deposition. The cavity diameter is 17.78 cm.

The simulation structure for the FDTD model and the plasma fluid model is shown in Fig. 5. The sliding short, the base plate, the substrate holder, the coaxial probe (both the inner and outer conductor), and the cavity sidewalls form the boundary of the FDTD simulation region. The boundary conditions for the electric fields on these surfaces are that only the normal components of the electric fields exist and the tangential electric fields on these surfaces are zero. This is based on assuming these boundaries are formed by perfect conductors. At the input end of the coaxial probe is an open boundary where the region in which the field has to be computed is unbounded. At this open boundary a truncation method is used to prevent any artificial reflection of outgoing waves. This non-reflecting boundary condition has been investigated extensively in solving electromagnetic wave scattering problems [Mur 1981]. Basically, this boundary condition allows electromagnetic waves arriving at the open boundary of the coaxial probe to leave the simulation region without any artificial reflections at the termination of the grid structure. The non-reflecting boundary condition which is applied in this model can be written as:

$$E_r^{n+1}(i,k_o) = E_r^n(i,(k_o-1)) + \left(\frac{\Delta t}{\sqrt{\varepsilon\mu}} - \Delta z\right) \Big/ \left(\frac{\Delta t}{\sqrt{\varepsilon\mu}} + \Delta z\right) \cdot \left(E_r^{n+1}(i,(k_o-1)) - E_r^n(i,k_o)\right) \tag{47}$$

and

$$E_\phi^{n+1}(i,k_o) = E_\phi^n(i,(k_o-1)) + \left(\frac{\Delta t}{\sqrt{\varepsilon\mu}} - \Delta z\right) \Big/ \left(\frac{\Delta t}{\sqrt{\varepsilon\mu}} + \Delta z\right) \cdot \left(E_\phi^{n+1}(i,(k_o-1)) - E_\phi^n(i,k_o)\right) \tag{48}$$

where i denotes the grid location in the r direction, Δt is the time step, and k_0 is the grid terminating point in the z direction. Additionally, the quartz dome and the quartz tube, where the dielectric constant is equal to 3.75 ε_0, are included in this model.

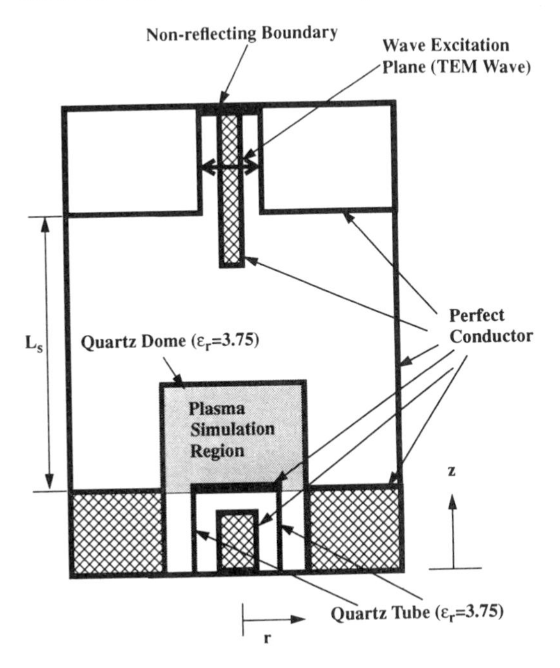

Figure 5 Simulation structure cross-section for microwave cavity plasma reactor with quartz tube.

Table 1: H_2 Discharge Reaction Rates

Reaction	Expression	Threshold energy	pre-exponential factor or rate coefficient (m^3/sec)
Ionization	$e + H_2 \rightarrow e + H_2^+ + e$	15.4 eV	1.0×10^{-14}
	$H_2^+ + H_2 \rightarrow H_3^+ + H$	0 eV	H_3^+ is the dominant ion species
Excitation	$e + H_2 \rightarrow H_2^* + e$	12.0 eV	6.5×10^{-15}
Dissociation	$e + H_2 \rightarrow e + H + H$	10.0 eV	1.0×10^{-14}
Recombination	$e + ion \rightarrow neutral$	0 eV	1.0×10^{-14}

The boundaries of the fluid plasma simulation are the quartz dome and the substrate holder. The diameter of the substrate holder is assumed to be the same as that of the quartz tube. The boundary conditions of the fluid plasma model are:

at the substrate:

$$n_e = n_i = 0, \quad T_e = T_n, \quad \psi = 0 \tag{49}$$

and at the quartz wall:

$$n_e = n_i = 0, \quad T_e = T_n, \quad \psi = \psi_f. \tag{50}$$

The boundary conditions at the quartz wall and substrate assume wall recombination occurs when electrons and ions reach the walls. The substrate and the substrate holder can be either grounded and floating. In the simulations of this paper the substrate potential is assumed grounded. The potential is assumed to be at the floating potential Ψ_f on the quartz disk.

The grid structure used in the plasma simulation is in the r and z directions. In the z direction a uniform grid spacing is used, and in the r direction the grids are constructed such that all the unit cell areas (or volumes, $\Delta z r \Delta r 2\pi$) are approximately equal. For the FDTD electromagnetic field model, the number of grids used in the simulation is 30 x 83 in the r and z direction respectively, and the time step is 0.5 psec. The grid spacing for the FDTD model is uniform in both r and z directions. Since the grid structure spacing for the FDTD model is different from that of the fluid plasma model, a linear interpolation technique is used to interpolate the absorbed power to each grid point of the plasma model, and the plasma density and electron temperature to each grid point of the FDTD model.

For the partially ionized and partially dissociated H_2 plasma discharge simulation, the major particle interaction processes are the electron-H_2 molecule inelastic collisions, electron-H_2 molecule elastic collisions and electron-hydrogen ion recombination collisions. The electron-H_2 inelastic collisions include the H_2 molecule ionization, excitation and dissociation processes. The rate coefficients in Equations (41), (42) and (45) for these collision processes can be expressed using the Arrhenius relationship [Surendra et al. 1992] as

$$\begin{aligned} k_{ion} &= A_{ion} \exp(-\varepsilon_{ion} / k T_e) \\ k_{ext} &= A_{ext} \exp(-\varepsilon_{ext} / k T_e) \\ k_{dis} &= A_{dis} \exp(-\varepsilon_{dis} / k T_e) \end{aligned} \quad (51)$$

where ε_{ion}, ε_{ext} and ε_{dis} are the threshold energy for H_2 molecule ionization, excitation and dissociation; and A_{ion}, A_{ext} and A_{dis} are the pre-exponential factors, which are obtained by approximating the rate constant data at low electron temperatures to these relationships [Surendra et al. 1992] and [Janev et al. 1987]. For simplicity, only the reactions with higher rate coefficients are considered in this study for H_2 discharges. The types of inelastic collisions and their corresponding rate parameters are summarized in Table 1. The collision frequency for electron-H_2 molecule momentum transfer is relatively independent of the electron temperature, so it can be written as [Cherrington 1979]

$$\nu_m(H_2) \approx 1.44x\,10^{12} \frac{Pressure(Torr)}{T_n(^\circ K)} \quad (52)$$

where T_n is the neutral temperature which can be represented by the translational temperature of H_2 gas. Since ν_m is relatively independent of the electron temperature, the effective collision frequency ν_{eff} in (45) is set equal to ν_m. It should be noted that the only neutral species considered in the ion and electron simulations was the H_2 species and that the dominant ionic species in the plasma is H_3^+ [Surendra et al. 1992] and [Koentzopoulos et al. 1993]. Hence these simulations assume a low hydrogen dissociation percentage.

6.5.2 Microwave Cavity Plasma Reactor Experimental Measurements

In order to verify the cavity resonant mode, the radial component of electric field strength at the cavity outer wall was measured by calibrated micro-coax electrical probes inserted through the cavity wall versus z and ϕ [Zhang et al. 1990]. The center conductor of the probe is inserted 2 mm beyond the inside surface of the cavity side walls, and the other end is connected to a microwave power meter. Since the probe power reading is proportional to the square of the rms electric field strength normal to the inside cavity, the cavity resonant mode and the cavity stored energy can be determined using this technique [Zhang et al. 1990]. Probe experiments were done by measuring the electric field strength at various z positions along the cylindrical cavity wall of the microwave plasma reactor during diamond thin film deposition. The microwave excitation frequency was 2.45 GHz, and the cavity height was adjusted to a

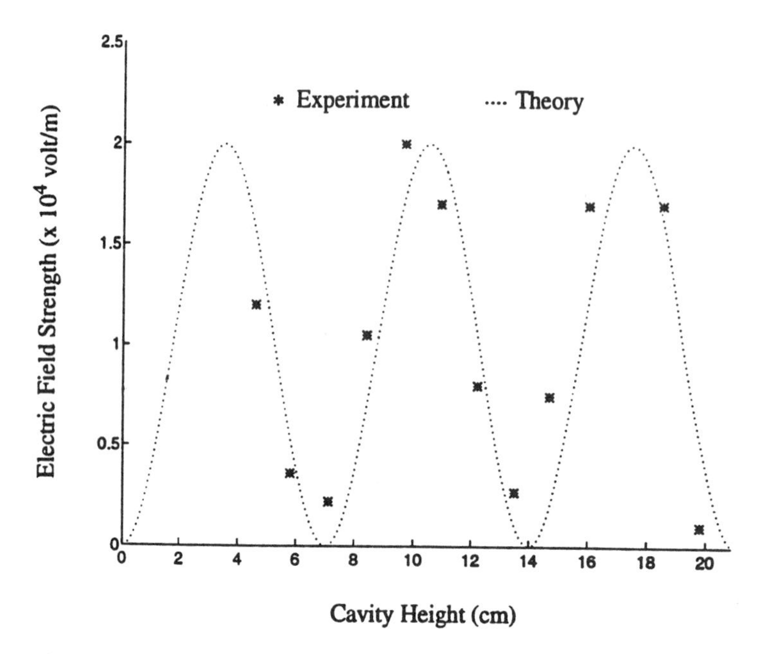

Figure 6 Microwave electric field strength (radial directed) versus cavity height for microwave cavity plasma reactor during the diamond deposition process.

position near 20.4 cm for a minimum reflected power condition. The result is shown in Fig. 6 where the resonant condition for the microwave cavity during a deposition process matches closely to the ideal TM_{013} mode. Additionally, the electric field was probed circumferential around the cavity. The field strength showed no significant variation along the ϕ direction as expected for the TM_{013} mode.

Once the electromagnetic field strength and mode is determined inside the cavity, the cavity quality factor Q of the plasma discharge loaded cavity reactor can be found by calculating the electromagnetic energy U stored in the cavity and the power absorbed in the cavity P_{abs}. The energy stored in the cavity U is expressed as

$$U = \varepsilon \int |E|^2 dV \tag{53}$$

and the Q factor is given by,

$$Q = \frac{\omega U}{P_{abs}} \tag{54}$$

where ω is the excitation frequency. The measured Q value at a pressure of 50 Torr and an absorbed microwave power of approximately 1500 watts was 100. A Q value of 100 indicates that the microwave energy pattern with the plasma load present is largely influenced by the resonant cavity structure.

6.5.3 Microwave Cavity Plasma Reactor Simulation Results

The microwave cavity plasma reactor loaded with a H_2 discharge for diamond film deposition was simulated and investigated by the self-consistent numerical model as described above. The input parameters for this model include the pressure and input microwave power. A set of experimentally determined empirical equations developed by G. King [King 1994] are used to establish the neutral temperature for this simulation.

These empirical equations were obtained by doing discharge diagnostics across a parameter space including pressure (35-65 Torr) and input microwave power (1.6-2.6 kWatt) variations. The empirical equations used for the translational temperature of H_2 gas and the discharge volume are [King 1994]:

Translational Temperature (°K) = 228.6 + 374.3 × Incident Power (kW) + 16.5 × Pressure (Torr) ± 94.2 (55)

Plasma Volume (cm³) = 449.7 + 116.2 X Incident Power (kW) -18.1 × Pressure (Torr) + 57.1 × [Incident Power (kW)]² + 0.25 × Pressure (Torr)]² - 5.4 × Pressure (Torr) × Incident Power (kW) ± 15.4 (56)

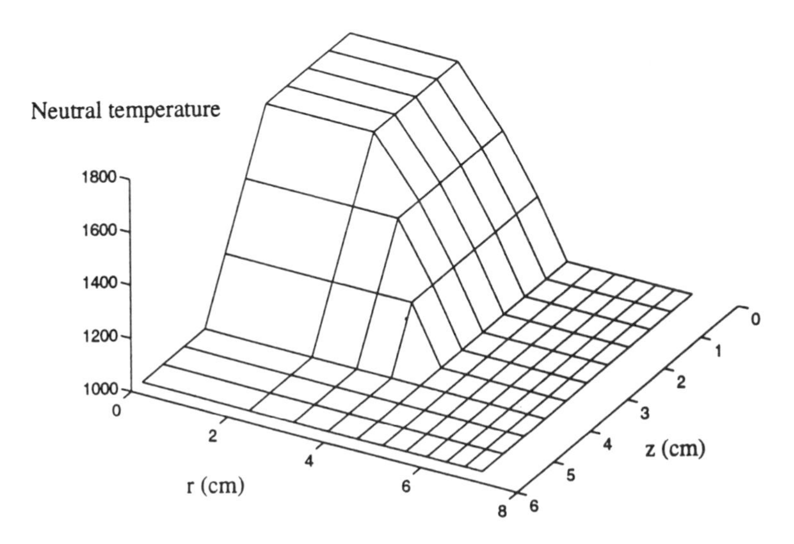

Figure 7 Neutral temperature distribution in the plasma simulation region.

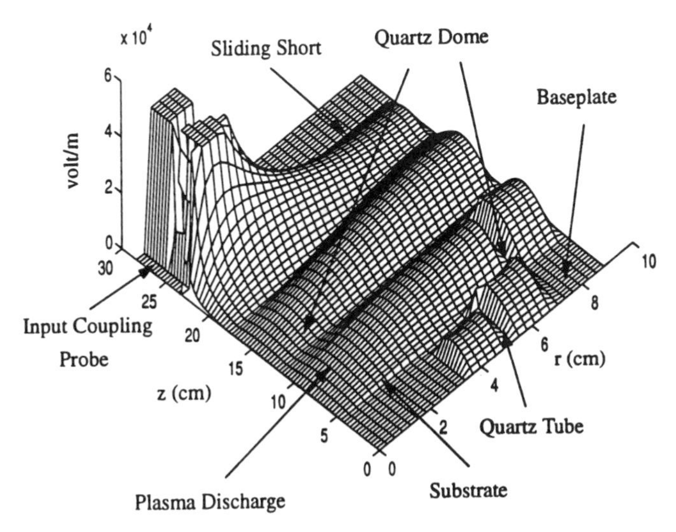

Figure 8 E_r field distribution at a pressure of 50 Torr and an input microwave power of 1500 W.

The translational temperature was determined using doppler broadening measurements of the H atom emission and the plasma discharge volume was determined using photographs of the plasma.

The plasma discharge volume serves as a boundary for the portion of the plasma simulation region where the neutral temperature (T_n) is high. Inside the discharge volume region, the translational temperature given by Eq. (55) is the neutral temperature. Outside the volume, the neutral temperature is equal to the temperature assigned on the boundaries (T_n = 1000 °K). Between these two regions, a linear temperature change profile is used to prevent an abrupt change of neutral temperature. This is done by assigning the boundary temperature (T_n = 1000 °K) to the region several grids, namely three grids, away from the plasma volume in both the r and z directions. The temperature assigned on the grids between these two regions are linearly interpolated from the neutral temperature given by Eq. (55) and boundary temperature (T_n = 1000 °K) as shown in Fig. 7.

For the FDTD electromagnetic field model, the technique used to excite the electromagnetic field in this study is to select grid points on a cross section plane of the coaxial probe as source points (see Fig. 5) and assign the time-varying electric field component at these points based at theoretical transverse electromagnetic (TEM) wave solutions in a coaxial structure. The electromagnetic wave then propagates down into the cavity region where power is absorbed by the discharge. Any reflected electromagnetic wave will propagate to the end of the open boundary of the input power probe and be terminated by the non-reflecting boundary.

The simulation results for the microwave electric field distribution are shown in Fig. 8 and Fig. 9 for E_r and E_z components, respectively. The input condition for this simulation is 50 Torr in pressure and 1500 Watts absorbed microwave power. The electric field patterns basically follow the TM_{013} mode electromagnetic field distribution. The effect of the electric field caused by the presence of discharge can be observed. It reduces the amplitude of the radial directed electric field component, E_r, in the plasma discharge region since electromagnetic power is absorbed in the discharge region. The discontinuity of the electric field resulting from the difference of dielectric constant between the quartz and air can also be seen. It can also be observed that in the plasma discharge simulation region the E_z field has the highest magnitude just above the substrate. The conductivity of the plasma can be determined using the solved density and collision frequency according to (19). Fig. 10 shows the maximum conductivity of the discharge is around 0.4 mho-m. Fig. 11 shows the power absorption pattern versus r and z and indicates that the microwave power is highly absorbed in the central portion of the plasma region close to the substrate.

The simulated electric field strength in Figs. 8 and 9, as well as, the field measured in Section 5b, indicate that the microwave electric field strengths are on the order of 10^4V/m. If this electric field strength is assumed to contribute to the heating of the substrate through direct electromagnetic energy absorption, then the amount of this direct electromagnetic heating can be estimated. Specifically, if a conductivity is assumed for the substrate holder similar to that of graphite, the absorbed power in the substrate and substrate holder is calculated to be less than one percent of the power entering the plasma source. Hence, in the resonant cavity microwave system described in this paper, the microwave electric fields are not strong enough to contribute substantially to the substrate heating, rather, the substrate is heat by the plasma.

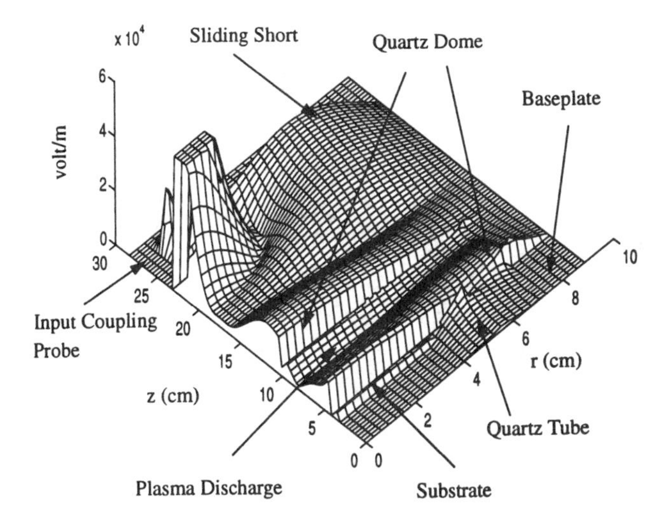

Figure 9 E_z field distribution.

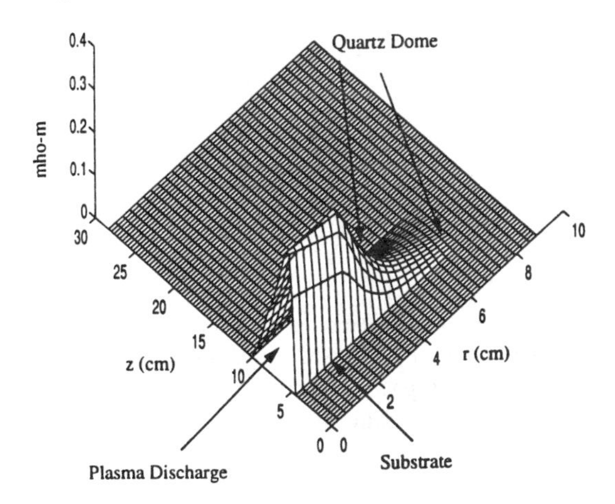

Figure 10 Conductivity versus r and z.

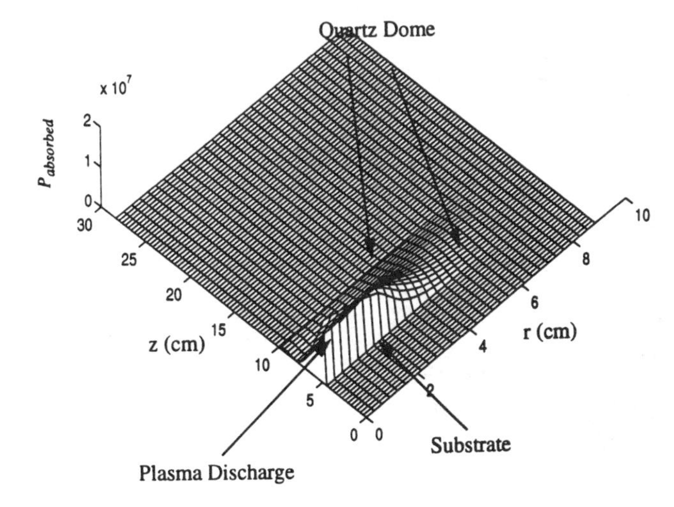

Figure 11 Absorbed microwave power in units of watts/m^3 versus r and z.

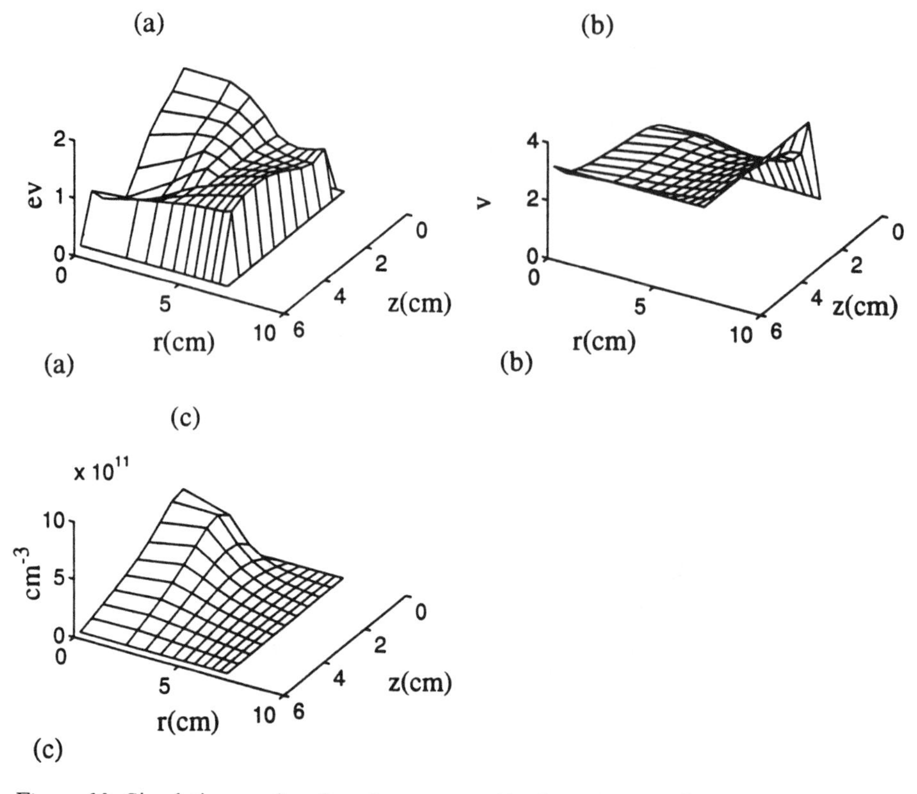

Figure 12 Simulation results of a microwave cavity plasma reactor for pressure = 50 Torr and input microwave power = 1500 watts. (a) Electron density, (b) plasma potential, (c) electron temperature. Note that z=0 is located at the substrate position for the plasma simulations. (This same z location is at 4.5 cm for the electromagnetic field simulations.)

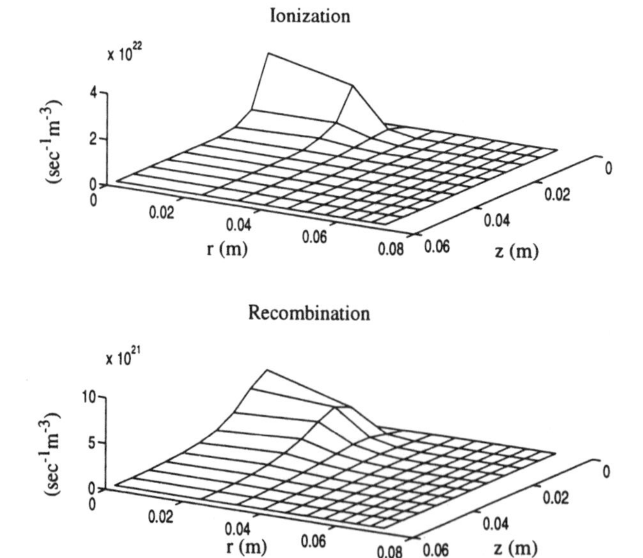

Figure 13 Spatial variation of ionization rate and recombination rate for pressure = 50 Torr and input microwave power = 1500 watts.

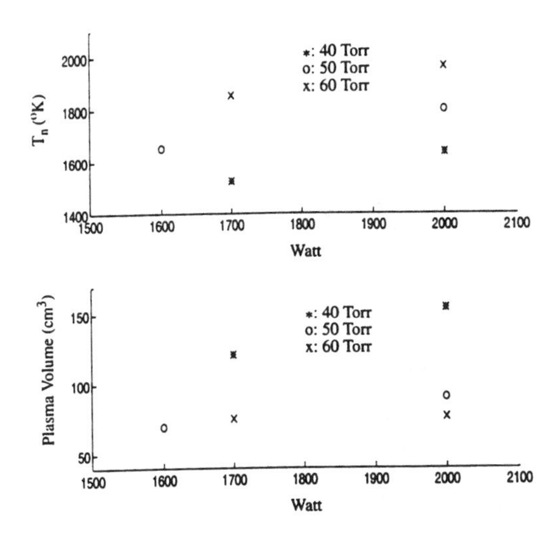

Figure 14 Experimentally measured neutral temperature and plasma volume versus incident power for various pressures.

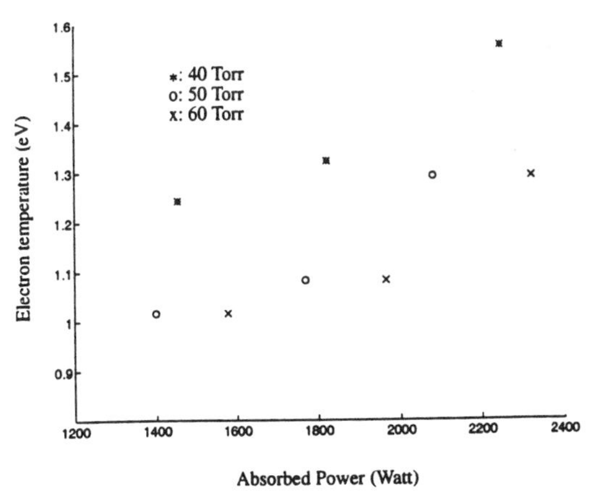

Figure 15 Simulated electron temperature versus absorbed power for three different pressures.

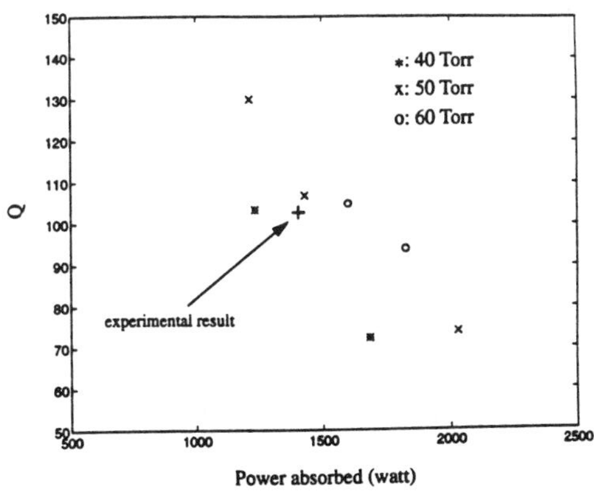

Figure 16 Simulated Q value vs. absorbed power for three different pressures. The experimentally measured Q at a pressure of 50 Torr and an input microwave power of 1500 Watts is also indicated.

The simulation result of the plasma characteristics of a H_2 discharge at 50 Torr and 1500 Watts input power is shown in Fig. 12. The plasma density is highest near the center of the discharge. The electron temperature is also highest at the center of the discharge. The spatial variation of ionization and recombination rate are shown in Fig. 13. The ionization rate is higher near the substrate due to the electron temperature being higher at that region. The recombination rate profile basically follows the electron density variation.

The characteristics of H_2 discharges and cavity quality factor Q were analyzed for various pressures and absorbed powers by this numerical model. The neutral temperature and plasma volume determined by Eqs. (55) and (56) are the input data for simulations and their variations for different incident power and pressure are shown Fig. 14. The simulation results of average electron temperature vs. absorbed microwave power are shown in Fig. 15. These average values are calculated by averaging the electron temperature across the plasma volume region. The average electron temperature is generally about 1.0-1.5 eV. The electron temperature increases with higher absorbed microwave power. This occurs because the measured neutral gas temperature increases with microwave power as indicated in Eq. (55). The higher neutral gas temperature leads to a reduced neutral density and, hence, to a reduced electron-neutral collisions rate at a given electron temperature. Also, the electron temperature drops with increases of pressure. This can be understood since higher pressure causes higher collision frequency for electrons with neutrals. Therefore, electrons more easily transfer their energy to neutral particles and reduce their kinetic energy. The results for Q versus absorbed microwave power are shown in Fig. 16. It shows that when the input microwave power is increased the Q factor decreases. The pressure has only a small effect on the Q factor. At the 50 Torr pressure and 1500 microwave power condition, the Q factor of the microwave cavity plasma reactor with a H_2 discharge was experimental determined by using the technique described earlier Section 5b. The Q value determined experimentally is about 100 which is close to the simulation result ($Q = 107$). This close agreement is an indication of the validity of the model.

6.5.4 Microwave Cavity Plasma Reactor Summary

The numerical modeling example presented here allows the calculation of the power absorption profile in microwave discharges which has often been a difficulty in plasma simulations. The understanding of the interactions between the microwave fields and the plasma discharges allows the reactor design to be analyzed for improvement and optimization of such quantities as uniformity of microwave power absorption and uniformity of the plasma species. Additionally, the simulation and understanding of the electromagnetic excitation of the discharge is expected to be important for microwave discharge diamond deposition machine control.

6.6 REFERENCES

Barnes, M.S., Cotler, T.J. and Elta, M.E. "A staggered-mesh finite difference numerical method for solving the transport equations in low pressure RF glow discharges," *J. Comp. Phys.*, 77, pp. 53-72, 1988.

Bittencourt, J.A. *Fundamentals of Plasma Physics*, Pergamon Press, Oxford, pp. 425-427, 1986.

Cherrington, B.E. *Gaseous Electronic and Gas Lasers*, Pergamon, New York, 1979.

Cohen, S.A. "An introduction to plasma physics for materials processing," in *Plasma Etching: An Introduction*, Edited by D. M. Manos and D. L. Flamm, Academic Press, Boston, pp. 213-217, 1989.

Girshick, S.L. Li, C., Yu, B.W. and Han, H. "Fluid boundary layer effects in atmospheric-pressure plasma diamond film deposition," *Plasma Chem. and Plasma Processing*, 13, pp. 169-187, 1993.

Gothard, G.K., Vechinski, D.A. and Rao, S.M. "Computational methods in transient electromagnetics: A selective survey," *IEEE Computational Science and Eng.*, pp. 50-59, Summer, 1994.

Grotjohn, T.A. "Numerical modeling of a compact ECR ion source," *Review of Scientific Instruments*, vol. 63, pp. 2535-2537, 1992.

Grotjohn, T.A., Tan, W., Gopinath, V., Srivastava, A.K. and Asmussen, J. "Modeling the electromagnetic excitation of a compact ECR ion/free radical source," *Rev. Scientific Instruments*, vol. 65, pp. 1761-1765, 1994.

Holland, R., Simpson, L. and Kunz, K.S. "Finite-difference analysis of EMP coupling to lossy dielectric structures," *IEEE Trans. Electromagnetic Compat.*, EC-22, pp. 203-209, 1980.

Hopwood, J., Guarnieri, C.R., Whitehair, S.J. and Cuomo, J.J. "Electromagnetic fields in a radio-frequency induction plasma," *J. Vac. Sci. Technol.* A, 11, pp. 147-151, 1993.

Hunsberger, F., Luebbers, R. and Kunz, K. "Finite-difference time-domain analysis of gyrotropic media-I: Magnetized plasmas," *IEEE Trans. on Antennas and Propagation*, 40, pp. 1489-1495, 1992.

Janev, R.K. et al. *Elementary Processes in Hydrogen-Helium Plasmas*, Springer, Berlin, 1987.

King, G. "Temperature and concentration of ionic and neutral species in resonant microwave cavity plasma discharges," Ph.D. Dissertation, Michigan State University, 1994.

Koemtzopoulos, C.R. et al. *Diamond and Related Materials*, 2, p. 25, 1993.

Lynch, D.R. and Paulsen, K.D. "Origin of vector parasites in numerical Maxwell solutions," *IEEE Trans. Microwave Theory and Tech.*, 39, pp. 383-394, 1991.

Mur, G. "Absorbing boundary conditions for the finite-difference approximation of the time-domain electromagnetic-field equations," *IEEE Trans. Electromagnetic Compat.*, EC-23, pp. 377-382, 1981.

Pappas D.L. and Hopwood, J. "Deposition of diamondlike carbon using a planar radio frequency induction plasma," *J. Vac. Sci. Technol.* A, 12, pp. 1576-1582, 1994.

Passchier, J.D.P. and Goeheer, W.J. "A two-dimensional fluid model for an argon rf discharge," *J. Appl. Phys.*, 74, pp. 3744-3751, 1993.

Paulson, K.D., Lynch, D.R. and Strohbehn, J.W. "Three-dimensional finite, boundary, and hybrid element solutions of the Maxwell equations for lossy dielectric media," *IEEE Trans. Microwave Theory Tech.*, 36, pp. 682-693, 1988.

Surendra, M., Graves, D.B. and Plano, L.S. "Self-consistent dc glow-discharge simulations applied to diamond film deposition reactors," *J. Appl. Phys.*, 71, pp. 5189-5198, 1992.

Taflove, A. and Brodwin, M.E. "Numerical solution of steady-state electromagnetic scattering problems using the time-dependent Maxwell's equations," *IEEE Trans. Microwave Theory Tech.*, MTT-23, pp. 623-630, 1975.

Tan, W. and Grotjohn, T.A. "Modeling the electromagnetic field and plasma discharge in a microwave plsma diamond deposition reactor," *Diamond and Related Materials*, vol. 4, pp. 1145-1154, 1995.

Tan, W. and Grotjohn, T.A. "Modeling the electromagnetic excitation of a microwave cavity plasma reactor," *J. of Vacuum Sci. and Technol.*, vol. A-12, pp. 1216-1220, 1994.

Yee, K.S. "Numerical solution of initial boundary value problems involving Maxwell's equation in isotropic media," *IEEE Trans. Antennas Propagat.*, 14, pp. 302-307, 1966.

Yu, B.W. and Girshick, S.L. "Modeling inductively coupled plasmas: The coil current boundary condition," *J. Appl. Phys.*, 69, pp. 656-661, 1991.

Zhang, J., Huang, B., Reinhard, D.K. and Asmussen, J. "An investigation of electromagnetic field patterns during microwave plasma diamond thin film deposition," *J. Vac. Sci. Technol.*, A, 8, pp. 2124-2128, 1990.

Chapter 18

SECTION 7.1: MODELING OF THE DIFFUSIONAL TRANSPORT OF AN H_2 PLASMA OBTAINED UNDER DIAMOND DEPOSITION DISCHARGE CONDITIONS

K. Hassouni[a], C. D. Scott[b], S. Farhat[a]

[a]*Laboratoire d'Ingénierie des Matériaux et des Hautes Pressions CNRS-UPN, Av. J. B. Clément 93300 Villetaneuse, France*
[b]*NASA,NASA Lyndon B. Johnson Space Center, EG3, Houston, TX 77058, USA*

7.1.1 Introduction

As indicated in Section 5 of this chapter, the understanding of diamond growth process in MPACVD devices requires the development of satisfactory H_2/CH_4 plasma transport models which provide the estimation of species and temperatures spatial distributions in the deposition reactor. Such models have to take into account the numerous physical phenomena and chemical processes which occur in the plasma and at the plasma/substrate interface. This requires the investigation of the following points :

1. Excitation wave propagation in the reactor and its interaction with the plasma.
2. Electron energy distribution function (EEDF) in the plasma
3. Chemical kinetics, including electron-heavy species collisions (e-hs) and heavy species-heavy species collisions (hs-hs), in the discharge
4. Energy transfer between the different energy modes of the plasma
5. Transport of chemical species and energy in the plasma
6. Plasma species/substrate surface interactions.

In a microwave plasma reactor the high frequency electric field and the plasma temperatures and species concentrations are strongly coupled. In addition, they are linked to the heterogeneous gas-surface chemical kinetics at the reactor wall and at the substrate surface. As a consequence, the modeling of the deposition reactor requires the solution of the transport equations for chemical species and energy modes and of the wave equations. Furthermore, a surface chemical kinetics model must specify the boundary conditions of the species equations.

All these requirements show that a complete modeling of the deposition reactor may be very complex, especially when considering the large number of species in H_2/CH_4 plasmas.

Since hydrogen represents the major part of the deposition plasma feed gas, modeling of the deposition reactor requires first a good description of H_2 plasmas obtained under diamond deposition discharge conditions. This will be the scope of the present section where we will especially focus on an investigation of species and energy transport in the MPACVD deposition reactor described in figure 7.1.1.

7.1.2 Physical model of H_2 plasma. chemical kinetics and energy exchange

Introduction. The hydrogen plasma considered in this study is obtained under moderate pressure discharge conditions (25 mbar $< P <$ 100 mbar) with an injected microwave

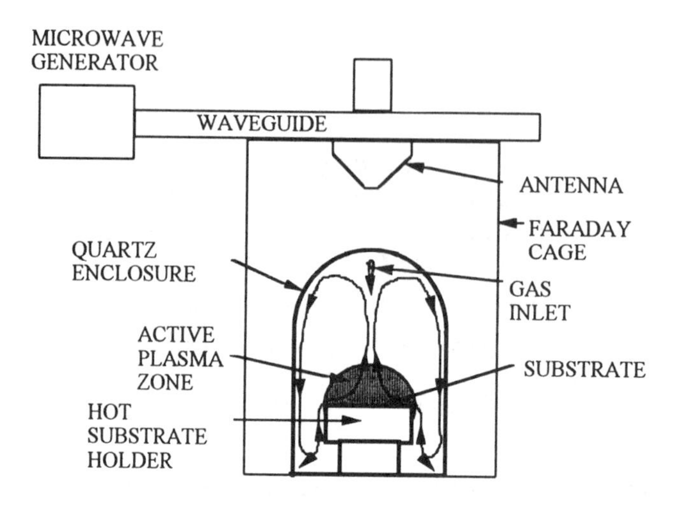

Figure 7.1.1 Schematic of hydrogen microwave plasma diamond deposition apparatus.

power in the range 600 W to 2000 W. The plasma reactor is an axisymetrical quartz bell jar containing a 50 mm diameter deposition substrate. The plasmas obtained in this reactor are in thermal and chemical non equilibrium. The major chemical species are H_2, H, H^+, H_2^+, H_3^+, H^- and e^- [Gicquel at al. 1994]. These species may exist in the electronic ground state or in electronically excited, radiative or metastable states. Due to their very short life-time and to the typical average electron energy in the considered discharge, the radiative states have very low concentrations with respect to the ground state. The metastable state concentrations are also very low because of the relatively discharge pressures, which induce important quenching rates for these states. For the typical discharge conditions considered in this study (average electron energy < 3 eV and Pressure > 10 mbar), an upper limit of the ratio between the electronically excited states and the ground state concentrations is 10^{-4} [Gicquel et al., 1996]. Therefore, only the populations of the electronic ground states of the chemical species will be investigated in this model. The electronically excited states will be considered only when they are involved in the chemical kinetics of the ground state species.

Thermal non equilibrium leads to the distinction between the rotational-translational mode of heavy species, the vibrational mode of the molecules and the translational mode of electrons. For the conditions studied in this work, the translational-rotational and vibrational modes of heavy species are described by Maxwell-Boltzmann distribution functions characterized by two different temperatures T_g and T_v. The electron energy distribution function (EEDF) presents non Maxwellian behavior [Capitelli et al., 1994]; this description requires the solution of the two term expansion of the electron Boltzmann equation. An extensive investigation of the electron kinetics and energy distribution function in spatially homogenous H_2/CH_4 plasmas is presented by M. Gordon in Section 4 of this chapter. The scope of the present section is very different from that of section 4; our objective is to investigate the coupling between energy transport, chemical species transport, chemical kinetics and energy exchange kinetics. Therefore, for taking into account the non Maxwellian behavior of the EEDF we should solve the electron Boltzmann equation in the whole discharge volume. From the practical point of view, the solution of this equation in the whole flow field, makes the numerical treatment of the transport equations very time consuming. We have then chosen, as a first approximation, to assume a Maxwellian EEDF characterized by an electron temperature T_e. With this assumption the H_2 plasma may be described using the three temperature chemically non equilibrium flow model of section 5.

The description of H_2 plasma flows by a three temperature chemically non equilibrium model requires the implementation of a chemical reactions set which describes the plasma kinetics and provides the chemical source terms appearing in the species transport equations. This is the scope of the next subsections where the energy source terms which control the energy exchange between the different modes and the energy dissipation in the plasma will also be investigated.

Chemical kinetics in H_2 plasmas under moderate pressure discharge conditions [Scott et al., 1996]. Under moderate pressure and power discharge conditions (10 mbar < P < 200 mbar and 100 W < MWP < 2000 W), the average electron energy is in the range 1eV to 4eV and the ionization degree of H_2 plasma is less than 10^{-4}[Gicquel et al. 1996]. In such conditions, H_2 plasmas are composed of 7 major species which are H_2, H, H^+, H_2^+, H_3^+, H^- and e^-. There are at least 27 chemical reactions that describe the chemical kinetics of a H_2 nonequilibrium plasma at conditions associated with the present experiments. They fall into two basic categories. First are those that involve reactions with electrons, e.g., dissociation and ionization, while the other category includes heavy particles, including charge exchange and recombination of ions and atoms. This complete set of reactions is given in Table 2, along with their corresponding rate coefficients. This set of chemical reactions is written in general forward form as

$$\sum_i \nu'_{ir} A_i \rightarrow \sum_i \nu''_{ir} A_i \tag{1}$$

where n'_{ir} and n''_{ir} are the stoichiometric coefficients for species A_i. Reverse reactions are written explicitly as forward reactions. Thus the general species production rates are written

$$w_i = \sum_r (n''_{ir} - n'_{ir}) k_{fr} \prod c_i^{n'_{ir}} \tag{2}$$

The reaction rates for ionization of H_2 and H by electrons given in Table 2 are calculated from cross section data from [Buckman and Phelps, 1985 A] and [Buckman and Phelps, 1985 B] by integration over the electron energy distribution function that is assumed to be Boltzmannian at an electron temperature T_e. A plot and curve fit of the ionization rate coefficient for H_2 are given in Figure 7.1.2, where the cross sections were obtained from [Buckman and Phelps, 1985 A] and [Buckman and Phelps, 1985 B]. Even though the conditions under which these cross sections were obtained may not correspond to a Maxwellian EEDF, we have assumed a Boltzmann distribution for the electrons here; and we have used the cross sections of [Buckman and Phelps, 1985 B] directly. These

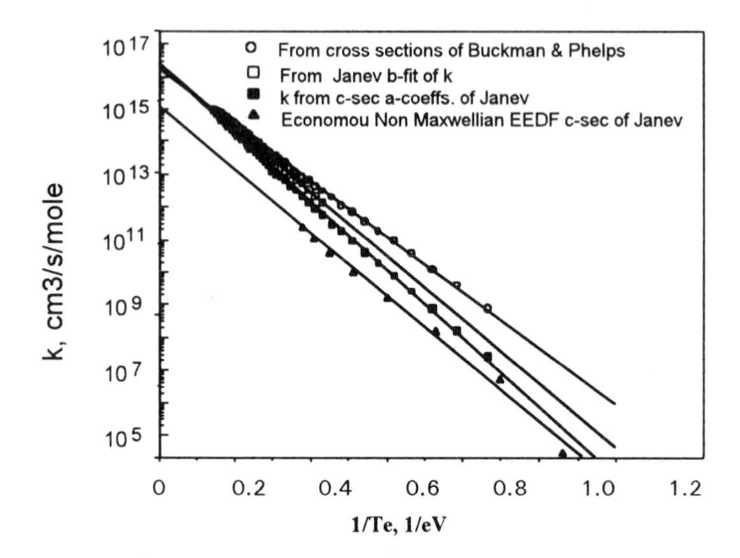

Figure 7.1.2 Comparison of rate coefficients for ionization of H_2 by electrons: cross sections from [Buckman and Phelps, 1985] and from [Janev et al, 1987] using Maxwellian EEDF and from [Janev et al, 1987] using [Koemtzopoulos et al. 1993] non-Maxwellian EEDF.

rates are compared with rates computed from cross section curve fits given in [Janev et al. 1987] and with rates computed from [Janev et al. 1987] cross sections using a non-Maxwellian EEDF [Koemtzopoulos et al., 1993] There is a wide difference in the rates, depending on which distribution function is used and depending on which source of cross section data. It is believed that the Buckman and Phelps cross sections are preferred. The curve fits of Janev et al. are not very good at low energies due to the fact that they fit the data over a very wide range of energies and near threshold the cross section rises steeply with energy.

The rate of deionization is evaluated at the *electron* temperature T_e. One of the dominant ions has been estimated to be the H_3^+ ion; therefore, reactions involving the production and depletion of H_3^+ are important for electron recombination. Electron detachment and dissociative attachment also contribute to the charge balance. H^+ is the other dominant positive ion. Recombination of atomic hydrogen ions can follow two paths; one is three body recombination, whereas the other is radiative recombination as indicated in Table 2. The rate of three body recombination depends on the energy level in which the H-atom is found, thus on the mechanism of the collision and the electron density and temperature. For sufficiently high electron temperature and low electron density the recombination coefficient is fairly insensitive to the electron density because it is dominated by radiative recombination. Radiation trapping of the Lyman-α line can affect the rate of recombination since the number density of the ground and excited states of the atoms may be affected. Drawin and Emard have solved the master equations, taking into account both three body recombination and radiative recombination with radiative absorption [Drawin and Emard, 1977]. Their results show

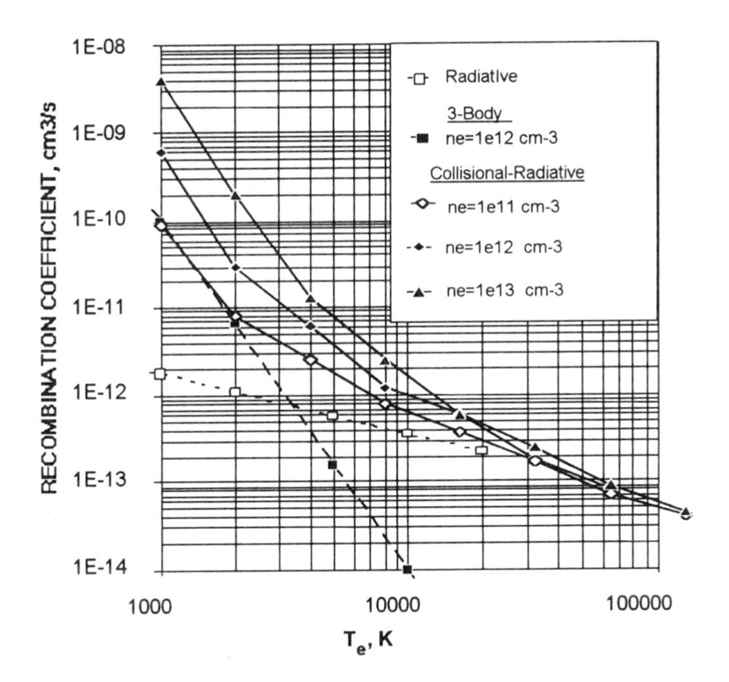

Figure 7.1.3 Recombination rate for $H^+ + e^-$ in various T_e, n_e domains given by 3-body, radiative and collisional-radiative recombination paths.

a dependence on electron density in lower temperature and higher density regimes. Their results are shown in Figure 7.1.3 along with the purely radiative recombination rate from [Massey, 1969] and the pure three body recombination rate for relatively high electron density by [Johnson and Hinnov, 1973] who curve fit measurements of [Johnson and Hinnov, 1969]. Both these paths to H^+ recombination are slow for the conditions here.

Energy exchange kinetics in H_2 plasma [Scott et al., 1996]. In a plasma reactor, the energy gained by the electrons from the electromagnetic field is transferred to the different modes of heavy particles and dissipated for the activation of some chemical reactions. Energy transfer also occurs between the different modes of heavy particles. At steady state the energy gained by the plasma is balanced by the energy lost by radiation and at the reactor wall. Because of the relatively high discharge pressure values, the rotational and the translational modes are in equilibrium. Since in the present plasmas the ionization degree is rather low, the vibrational energy carried by the molecular ions may be neglected and only the H_2 vibrational energy has to be considered. As a consequence, the following energy exchange channels may be distinguished:

- Electron-vibration energy exchange
- Electron-(translation-rotation) energy exchange
- (translation-rotation)-Vibration relaxation with molecules and atoms
- Energy loss by H_2 dissociation

- Energy loss by H_2 and H ionization

The source term corresponding to the electron-translation energy exchange may be simply derived from equation (28) of section 5. Since H_2 and H are the major heavy species in the plasma investigated in this work, one could neglect the participation of the other species to the e-T energy transfer and only the momentum transfer collision cross sections relative to H and H_2 have to be considered in the expression of the source term $Q_{e\text{-}T}$.

The energy loss by electron for the activation of the dissociation and ionization processes may be derived from equation (35) of section 5, where the rate constants given in Table 2 have to be used for computing the involved reaction rates. The H_2 dissociation by electron proceeds by way of the excitation of the three triplet states $a^3\Sigma$, $b^3\Sigma$ and $c^3\Sigma$. Therefore, the dissociation energy threshold which has to be considered in equation (35) is an average of those of the triplet states excitation. For the source terms corresponding to H_2 and H ionization, the energy thresholds are 15.5 eV and 13.6 eV respectively.

The electron-vibration and vibration-translation energy exchanges are due to several kinds of interactions and mechanisms which will be analyzed in the next subsections where the source terms associated with these energy exchanges will be given.

Electron-Vibration Exchange. There are two paths to vibrational energy exchange with electrons. The first is direct vibrational excitation and de-excitation in a given electronic state—in particular, the ground state. The second is vibrational excitation via electronic excitation of excited electronic singlet states followed by a radiative transition to a vibrationally excited ground state. Cross sections for these reactions are given by [Buckman and Phelps, 1985 A and 1985 B].

* *Direct Ground Electronic State Excitation of $H_2(v)$*

The rate coefficients in cm^3/s/atom for these reactions are calculated by integrating the cross sections over the Maxwellian electron energy distribution function.

$$k_{oj}(T_e) = 6.70\times10^7\, T_e^{-3/2} \int_{E_o}^{\infty} \sigma_{oj}(E)\, E \exp(-E/T_e)\, dE \tag{3}$$

where the energy E and the temperature are in electron-volts, E_o is the threshold energy of the exchange, and cross section σ_{oj} is in cm^2.

The relaxation time, τ_{ev} is determined following the procedure developed in the appendix of [Lee, 1985].

$$\tau_{ev} = \frac{1}{n_e(1 - \exp(-\theta_v/T_e))^2 \frac{1}{2}\int k_{oj}\, j^2\, dj} \tag{4}$$

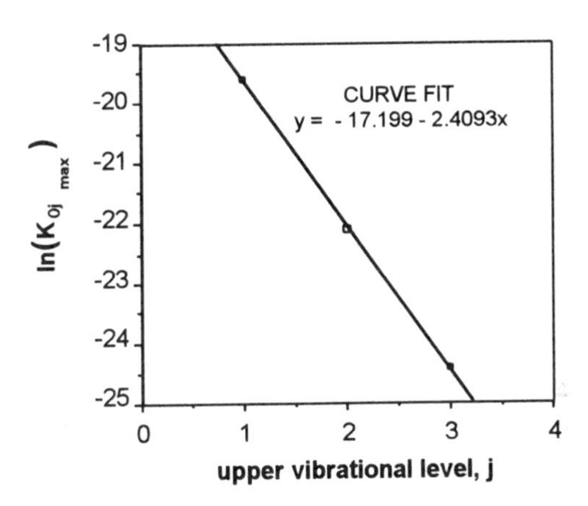

Figure 7.1.4 Curve fit of peak values of electron-H_2-vibration rate coefficients

It has been assumed that the vibrational levels are closely spaced so the integral in (4) well represents a sum over levels. This may not be the case for H_2, but we will use it until the error is determined to be too large. The form of k_{oj} for the first three vibrational levels (j=1-3) are very similar in shape. Therefore, an approximation was made that the integral in (4) could be done semi-analytically by fitting the maxima of the k_{oj} to a simple function of the form

$$k_{oj}^{max} = 3.39x10^{-8}\ e^{-2.401j} \tag{5}$$

which can be integrated analytically. The peak values are shown in Figure 7. 1.4. A polynomial curve fit of the result is shown in the figure. The relative rate coefficient function was integrated numerically to obtain a constant factor. The resulting relaxation time τ_{ev} as a function of electron energy is given in Figure 7. 1. 5.

The energy loss rate is obtained from the relation time by the relation

$$Q_{eV} = \rho_{H_2}\frac{de_v}{dt} = \rho_{H_2}\frac{e_v^{**}(T_e) - e_v}{\tau_{ev}} \tag{6}$$

where $e_V^{**}(T_e)$ is the equilibrium vibrational energy evaluated at the electron temperature (the driving energy) and e_v is the vibrational energy per unit mass to be determined.

* *Vibrational Excitation of $H_2(v)$ via Excitation of H_2 Singlets*

Electrons can excite the $B^1\Sigma$ and $C^1\Pi$ states which then radiate to the $H_2(X,v)$ ground electronic state. It requires about 10.5 eV to excite the singlets, but the molecules end up with only about 1 eV in vibration. Thus there is a large electron energy loss, but only a modest gain of vibrational energy. The reminder of the energy is lost from the plasma via radiation. The rate coefficients may be calculated from total cross sections of [Buckman and Phelps, 1985 A and 1985 B] with the probabilities for specific vibrational

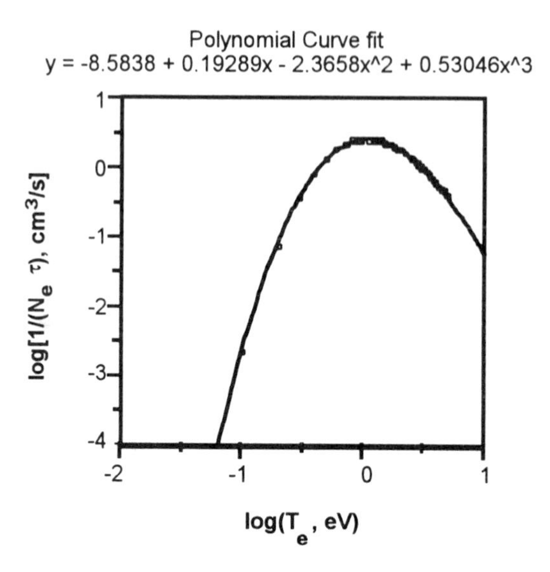

Figure 7.1.5 Curve fit of electron-H_2-vibration relaxation time based on cross sections.

states obtained from [Hiskes, 1991] who has calculated F(B,v,v") and F(C,v,v"), the probabilities of exciting an $X^1\Sigma(v")$ state from an $X^1\Sigma(v)$ state via excitations of the $B^1\Sigma$ and $C^1\Pi$ states, respectively. We have used these probabilities to compute the energy transferred from electrons to vibration using the relation

$$Q_{ev}^{1v} = n_{H_2} \; R\theta_v \left[k_{ev}^{B1}(T_e)\, S_B(T_v) + k_{ev}^{C1}(T_e)\, S_C(T_v) \right] \tag{7}$$

where

$$S_B(T_V) = \sum_{v=0}^{14} \exp(-\varepsilon_v/T_v)/Z_v(T_v) \sum_{v"=0}^{14} F(B,v,v")(\varepsilon_v - \varepsilon_{v"}) \tag{8}$$

$$S_C(T_V) = \sum_{v=0}^{14} \exp(-\varepsilon_v/T_v)/Z_v(T_v) \sum_{v"=0}^{14} F(C,v,v")(\varepsilon_v - \varepsilon_{v"}) \tag{9}$$

where $\varepsilon_v = E_v/R\theta_v$ is the dimensionless vibrational energy of level v; and $k_{ev}^{B1}(T_e)$ and $k_{ev}^{C1}(T_e)$ are the rate coefficients for electron excitation of the singlets calculated from the total cross sections. These cross sections could have been taken to be equal as indicated by [Hiskes, 1991], however, here they were calculated individually from the cross sections given in [Buckman and Phelps, 1985 B]. Curve fits of the results are given in Table 1. The initial ground vibrational quantum number is v and the final one is v". Here, the term $\exp(-E_v/RT_v)/Z_v(T_v)$ is the weighting factor for the number of H_2 molecules in state v and $Z_v(T_v)$ is the vibrational partition function. The factors $S(T_v)$ were evaluated as functions of T_v and were then curve fit using third degree polynomials. The results are given in Table 1. The vibrational energy levels are calculated from the spectroscopic constants ω_e=4401.21 cm^{-1} and $\omega_e\chi_e$=121.336 cm^{-1}. The S functions are calculated assuming anharmonic oscillator energy levels given by

$$E_v = hc\left[\omega_e(v+\frac{1}{2}) - \omega_e\chi_e(v+\frac{1}{2})^2\right] \tag{10}$$

Also shown in Table 1 are rates obtained assuming an harmonic oscillator. The values for an anharmonic oscillator were used in the solving the present model.

* *Electron Energy Loss by Excitation of Singlet H_2 States*
The loss of energy by electrons in exciting the singlet states of H_2 is given by the rate of excitation of those states times the energy lost per excitation ε.

$$Q_{ev}^{1e} = n_e \, n_{H_2} \, (k_{ev}^{B1} \varepsilon_B + k_{ev}^{C1} \varepsilon_C) \tag{11}$$

Translation-Vibration Relaxation with Molecules and Atoms. One could calculate the T-v exchange for molecules by summing over all rate probabilities as shown in [Cacciatore et al, 1978] and basing the rates on the data of [Billing and Fisher, 1976] and the method of [Audibert et al, 1974]. This requires keeping track of the population of all vibrational states. Instead, in keeping with the assumption of Boltzmann distributions we shall adopt some of the modeling methods of [Lee, 1985], who applied Landau-Teller type relations for vibration-translation relaxation to mixtures of nitrogen and oxygen. The energy exchange rate is given by

$$Q_{Tv} = \rho_{H_2} \frac{e_v^*(T) - e_v}{\tau_{v_{H_2}}} + \rho_{H_2} \frac{e_v^*(T) - e_v}{\tau_{v_H}} \tag{12}$$

where $e_v^*(T)$ is the specific vibrational energy at the translational temperature, e_v is the vibrational energy of the flow, and τ_{v_H} is the relaxation time for T-v relaxation due to collisions of H_2 with H and $\tau_{v_{H_2}}$ is the relaxation time for collisions of H_2 with H_2. [Lee, 1985] has given the general form for relaxation of molecular species l in a mixture.

$$\tau_{vl} = \frac{\sum_{i \neq e-} X_i}{\sum_{i \neq e-} X_i / \tau_{v_{li}}} \tag{13}$$

The $\tau_{v_{H2}}$ values can be determined from either the [Millikan and White, 1963] or [Kiefer and Lutz, 1966] correlations. [Audibert et al., 1974] have shown that the [Kiefer and Lutz, 1966] correlations agree with their higher temperature results over a fairly broad temperature range of interest here. The [Kiefer and Lutz, 1966] correlation, which we have used in the present calculations, is

$$\tau v_{H_2H_2}\ p = 3.9\times10^{-10}\exp\left(\frac{100}{\sqrt[3]{T}}\right)\ \text{atm-s} \tag{14}$$

The relaxation scheme for atoms exchanging translational energy with molecules involves significant quenching of hydrogen molecule vibrational states. The rates are given in the appendix of [Gorse et al. 1987] and we have determined τv_{H2H} in a fashion similar to that for electron-vibration exchange. In particular the relaxation time was based on the rate k_{01} for excitation of $H_2(v=0)$ to $H_2(v=1)$ considering both reactive and non reactive collisions. [Gorse et al., 1987] obtained the rate coefficients in Arrhenius form by a curve fit from which the relaxation time was derived. The curve fit coefficients are given in Table 1. The relaxation times are obtained from the rate coefficients using

$$n_H\ \tau v_{H_2H} = \frac{1}{k_{10}^{nr}+k_{10}^{r}} \tag{15}$$

where the explicit vibrational excitation by collisions with atoms does not appear.

7.1.3 Plasma reactor model

Even with the assumption of Maxwellian EEDF, the modeling of H_2 plasma transport in the whole deposition reactor remains complex. Indeed, the transport equations have to be coupled with the Maxwell equations which gives the absorbed microwave power density appearing in the energy equations (17) and (20) of section 5. The simulation of the plasma transport requires then the solution of the coupled sets of these equations for an axisymetrical geometry.

The set of Maxwell equation coupled to charged species momentum and continuity equations as well as Poisson equation, have been extensively investigated by T. Grothjohn in section 6 of this chapter, where the spatial distributions of the electromagnetic field and the microwave power absorbed by the electron in the reactor have been determined.

The objective of this section is different from that of section 6, since we are mainly interested in the study of the coupling between energy transport, chemical species transport, chemical kinetics and energy exchange kinetics. Therefore, in the model presented here the spatial distribution of the absorbed microwave power density will be treated as a model parameter and estimated from experimental measurements. Such treatment makes the model non-self consistent but leads to an important simplification of the mathematical description of the plasma reactor.

The axisymetric geometry of the plasma reactor enables another simplification since on the reactor axis, where all the radial gradients are zero, the transport equations may be reduced and expressed in one-dimension. This simplification leads to an important reduction of the numerical complexity of the transport equations, but prevents the study of the transport along the radial coordinate.

For the deposition setup considered in this work, the feed gas flow rates are usually very low and in the range of [100 sccm , 300 sccm] [Gicquel et al., 1996]. We have then assumed that convection fluxes are negligible with respect to diffusion ones. If the convection transport is mainly forced, this assumption is valid near the substrate surface where species and temperatures gradients are very high. However, for high power density plasmas, the thermal gradients in the boundary layer may be high enough to induce an important free convection. In this case the assumption of negligible convection fluxes is no longer valid even in the high gradients region.

7.1.4 Description of the final model

The simplifications and assumptions given in the two last subsections lead to an H_2 plasma one-dimensional diffusion model which presents the following characteristics :

- The plasma is described as a three temperature chemically non equilibrium medium.
- The absorbed microwave power density spatial distribution is an input parameter for the model
- The model is one-dimensional and describes the plasma transport on the reactor axis and especially at the plasma/substrate boundary layer (figure 7. 1. 1).
- The convection fluxes are neglected

7.1.5 Mathematical formulation

For negligible convection fluxes, the reduction of the transport equations described in section 5 to a one dimensional geometry leads to the following plasma diffusion equations[1] [Scott et al., 1993] :

- Species continuity equations for H_2, H, H^+, H_2^+, H_3^+, H^- and e^- :

$$-\frac{d}{dz}\left[\rho . D_s \frac{dx_s}{dz}\right] = W_s \quad (s=1, 7) \tag{16}$$

- Electron energy equation :

$$-\frac{d}{dz}\left[\lambda_{e} . \frac{dT_e}{dz}\right] - \frac{d}{dz}\left[\rho . D_{e} . h_e \frac{dx_e}{dz}\right] = PMW - Q_{e\text{-}v} - Q_{e\text{-}t} - Q_{e\text{-}chem} \tag{17}$$

- H_2 vibration energy equation :

$$-\frac{d}{dz}\left[\lambda_{v} . \frac{dT_v}{dz}\right] - \frac{d}{dz}\left[\rho . D_{H2} . E_{v\text{-}H2} \frac{dx_{H2}}{dz}\right] = -Q_{v\text{-}e} - Q_{v\text{-}t} \tag{18}$$

- Total energy equation :

---

[1] Since the convection fluxes are neglected and since we are only interested in species and energy transport, the momentum equations have not to be considered.

$$-\frac{d}{dz}\left[\lambda_{e}.\frac{dT_e}{dz}+\lambda_{v}.\frac{dT_v}{dz}+\lambda_{t\text{-}r}.\frac{dT_g}{dz}\right]-\sum_{s}\frac{d}{dz}\left[\rho.D_s.\ h_s\frac{dx_s}{dz}\right] \tag{19}$$

$$= PMW - Q_{rad}$$

The different symbols appearing in these equations are described in sections 5 and 7.1.2, where the expressions which give the transport coefficients and source terms were also reported. Because of the low average electron energy the electronic states excitation rates are relatively low and the energy loss by radiation is expected to be negligible with respect to the energy loss at the substrate surface and at the reactor wall (conduction transport). The source term Q_{rad} was then neglected in the present work. In addition, for the typical average electron energy the e-v energy exchange is mainly governed by the direct mechanism and the source term corresponding to the e-v exchange by way of singlet excitation was neglected in $Q_{e\text{-}v}$ estimation.

The use of this set of equations to describe the plasma diffusion in the one dimensional domain of figure 7.1.1, requires the specification of the boundary conditions at the substrate surface and at the discharge inlet and of the absorbed microwave power spatial distribution.

7.1.6 Boundary conditions and spatial distribution of the absorbed microwave power density

The visible region of the plasma volume usually obtained in the typical deposition discharge conditions is a 5 cm diameter hemisphere located just above the deposition substrate. The 1D simulation domain length was chosen larger than the plasma radius (figure 7.1.6). In this way, the computation domain inlet is out of the discharge region, and reasonable inlet boundary conditions may be easily specified. It should be noted that there actually exists free convection in the reactor due to the heated sample holder and the plasma heating. The stagnation point at the center of the substrate is therefore in a region where the gas flows upwards in the z-direction and away from the surface and radially towards the stagnation point. See Fig. 7.1.1.

Boundary conditions

* *At the computation domain inlet : $z=z_{inlet}$.* Because of the relatively high pressure in the reactor (P>10 mbar) the electron temperature is expected to be very low in the post-discharge region(z>3 cm), where the absorbed microwave power density is zero. As a consequence, most of the electron-heavy species collisions lead to recombination reactions in this region. The ionic species involved in the model may be divided in two groups:

- H_2^+, H_3^+ which rapidly recombine in the post-discharge
- H^+ and H^- which slowly recombine in the post-discharge

For the first two species the concentration far from the discharge may be expected to be very low and almost constant. For the second two species the very low recombination

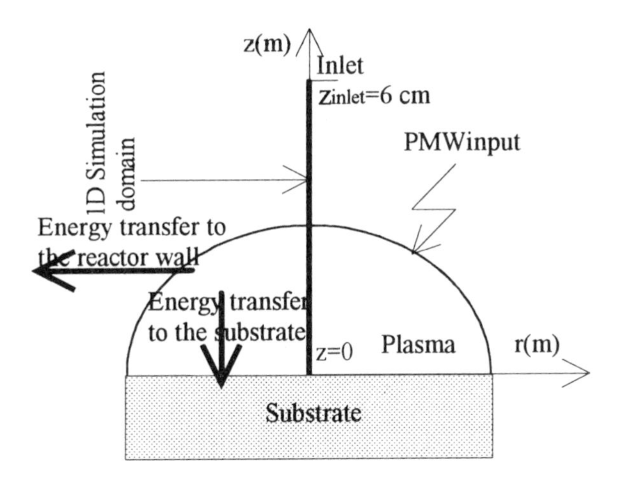

Figure 7.1.6 Modeling principle.

kinetic may result in low concentration gradients far from the discharge area. It seems reasonable to specify zero concentration gradient boundary conditions for the charged species.

For H-atom and H_2-molecule the chemical kinetics outside the discharge region consists of the thermal recombination and dissociation reactions which depend on the gas temperature. If this latter is low enough to insure very slow dissociation and recombination rates of H_2 and H, the concentrations of these species may be expected to be approximately constant, which leads to approximately zero concentration gradient outside the discharge area ($z>3$).

As far as the gas and H_2 vibrational temperatures are concerned, Coherent Antistokes Raman Spectroscopy (CARS) measurements previously carried out on the studied reactor, [Gicquel, et al., 1994] showed that the gradients of these temperatures are almost very weak at $z \approx 3$ cm. Since, the H_2/H mixture has quite high conductivity at high temperature, the temperature may be expected to be quite constant far from the reactor wall and from the substrate surface. Therefore, since the simulation domain inlet is far from the reactor surface, we have used zero gas and H_2 vibration temperature gradient boundary conditions at z_{inlet}.

Due to the absence of any electric field and to the high pressure, the electron temperature may be expected to be in quasi-equilibrium with the gas temperature in the post-discharge region. As a consequence, zero gradient boundary condition may also be used for T_e.

From the above analysis the boundary conditions at the computation domain inlet will be of the form :

$$\left(\frac{d\phi}{dz}\right)_{z=z_{inlet}} = 0 \quad (\phi = x_1, x_2, ..., x_7, Tg, T_v, T_e) \tag{20}$$

Before closing this subsection, we have to point out that the boundary conditions used here are only approximate ones. Their validity depends on the chosen value of z_{inlet} and has to be checked. This may be performed by studying the sensitivity of the transport equations solution to the value of z_{inlet}. Furthermore, the use of these kind of conditions is equivalent to assuming that there are no mass and energy exchanges between the system involving the plasma and the deposition substrate and the reminder of the deposition setup. This is physically equivalent to considering that the plasma and the substrate surface are isolated from the rest of the reactor (adiabatic assumption).

* *At the substrate surface: z=0*

For the pressure conditions considered here the gas temperature at the substrate surface is in equilibrium with the substrate temperature which is experimentally measured with the help of a bichromatic pyrometer. This leads to the following boundary condition :

$$T_g\,|_{substrate} = T_{substrate} \tag{21}$$

The H_2 vibrational temperature at the substrate surface may be different from the gas temperature and, as indicated in section 5, either the value of T_v or of the vibrational energy accommodation coefficient β_v must be given. For the investigated reactor, CARS measurements showed that, at 1 mm from the substrate surface, T_v is slightly higher than T_g, the difference T_v-T_g being of approximately 200 K. We have then to specify the following T_v boundary condition :

$$T_v\,|_{substrate} = T_{substrate} + 200 \tag{22}$$

The accommodation of the electron kinetic energy is also very weak [Scott, 1993] and the electron temperature at the substrate surface may be extracted from the following relation :

$$\left(\frac{dT_e}{dz}\right)_{substrate} = 0 \tag{23}$$

The chemical species concentration at the substrate are governed by a balance between the species quantity arriving to the surface by diffusion and produced at the substrate surface by heterogeneous reactions. The heterogeneous chemical model for the seven species H_2 plasma was reported by [Scott, 1993], it consists in the following reactions, where "s" denotes a surface site:

$$H + H\text{-}s \longrightarrow \tfrac{1}{2} H_2 + s \tag{R1}$$

$$H^+ + e^-\text{-}s \longrightarrow H + s \tag{R2}$$

$$H_3^+ + e^-\text{-}s \longrightarrow H + H_2 + s \tag{R3}$$

$$H_2^+ + e^-\text{-}s \longrightarrow H_2 + s \tag{R4}$$

$$H^- + s \longrightarrow H + e^- + s \tag{R5}$$

The rate of each of these reactions may be simply expressed as the product of the reacting gas species collision frequency with the substrate surface and a recombination coefficient which represents the recombination probability once the collisions occurs. As an example, the rate of reaction R1 may be expressed as :

$$v_{R1} = \frac{1}{4}\sqrt{\frac{8RT_g}{\pi\, M_H}}\; c_{H\text{-}sub}\; \gamma_{R1} \tag{24}$$

where M_H and $c_{H\text{-}sub}$ are H-atom molar mass and concentration at the substrate surface, R the gas constant and γ_{R1} the recombination coefficient of H-atom on diamond surfaces.

The charged species usually totally recombine at a surface and a value of 1 can be adopted for the recombination coefficients of reactions (R2)-(R5). For H-atom, the catalytic recombination is partial and no measurements of the recombination coefficient of this species on diamond surfaces were performed. We have then chosen to use the value of 0.1 suggested by [Goodwin et al., 1993]. The mathematical formulation of the species boundary conditions obeys to equation (66) of section 5 where the source terms $w_{s\text{-}surface}$ must be calculated using the set of heterogeneous reactions (R1)-R(5).

Spatial distribution of the absorbed microwave power density

The spatial distribution of the absorbed microwave power density (SDMWPD) depends on the plasma composition. Its exact determination requires the solution of the Maxwell equations which have to be coupled to the plasma transport equations. The SDMWPD was determined experimentally by [Zhang, et al.] and theoretically by [Tan, et al.] who solved the Maxwell equations. Although these experimental measurements and theoretical calculations were performed for slightly different discharge conditions and deposition setup they result in similar SDMWPD shapes. However, the plasma reactor investigated in this work is slightly different from both the deposition setup considered by these authors.

In this study, we have estimated the behavior of the SDMWPD on the reactor axis from the solution of the one-dimensional wave equation, where the wave is assumed to be injected at the top of the reactor and reflected back on a plate located at the bottom end of the reactor. The plasma electrical permitivity was assumed constant over all the plasma volume, and we have also assumed that the radial behavior of the SDMWPD is similar to the axial distribution. The absorbed microwave power density is then given by the following expressions :

for $0 < z < z_0 + \lambda/4$ and $0 < r < (\lambda^2/16 - z^2)^{1/2}$

$$MWPD(z,r) = MWPD_0 \,.\, \cos^2\!\left(\frac{2\pi(z-z_0)}{\lambda}\right) \,.\, \cos^2\!\left(\frac{2\pi r}{\sqrt{\lambda^2/16 - z^2}}\right) \tag{25}$$

and for $z > z_0 + \lambda/4$ or $r > (\lambda^2/16 - z^2)^{1/2}$

$$MWPD(z,r) = 0 \tag{26}$$

λ is the wavelength of the high frequency electric field (λ=12.24 cm) and z_0 is the position of the maximum power density. The parameter $MWPD_0$ is directly linked to the total absorbed microwave power by the following energy balance equation :

$$MWP_{total} = \iint_{hemisphere} [MWPD(z,r)\ 2\pi.r.dr.dz] \qquad (27)$$

For given discharge conditions, the exact determination of the absorbed microwave power density in the whole plasma volume is now equivalent to the estimation of its maximum location (z_0) and of the total absorbed power MWP_{total} which gives the parameter $MWPD_0$.

7.1.7 Use of the 1D model for the investigation of the H_2 plasma obtained at 600 W and 25 mbar

Introduction

Extensive experimental investigations have already been performed on the H_2 plasma obtained in the considered reactor under discharge conditions corresponding to : MWP_{input}=600 W, P=25 mbar, $T_{substrate}$=900 °C and Flw=300 sccm. The axial profiles of gas and H_2 vibration mode temperatures were determined by CARS and axial and radial profiles of several emission lines intensities (H_α (656.5 nm), H_β (486.1 nm), Ar(750 nm), ...) were obtained from Optical Emission Spectroscopy measurements (OES) (see section 7.2).

Using the actinometry technique, the axial profile of H-atom molar fraction was deduced from that of the ratio $IH_\alpha/IAr(750\ nm)$. The validity of this technique was checked by comparing the resulting axial profiles with those determined by Two Photons Allowed Laser Induced Fluorescence (TALIF) [Gicquel et al., 1996].

The large number of experimental data obtained for these discharge conditions makes possible the estimation of the spatial distribution of the absorbed microwave power density and the validation of the model.

In the investigated reactor, a direct experimental determination of the SDPMW parameters is very difficult since only the total input power is known in our reactor. We propose in this section a method based on the simultaneous uses of the 1D diffusion model and of CARS and OES measurements for determining these parameters.

In the investigated reactor the major part of the input microwave power is absorbed by the plasma and according to Grotjohn, the power directly transferred to the substrate surface and the reactor walls is negligible (see section 6). At the steady state regime, the microwave power absorbed by the plasma is transferred to the substrate and to the reactor wall (figure 7.1.6). Since in this model the plasma and the substrate are considered to be isolated from the rest of the reactor (see subsection 7.1.6, boundary conditions), only the fraction of the total power absorbed by the plasma which is transferred to the substrate surface has to be considered in the model.

The energy transfer from the plasma to the substrate occurs by three mechanisms :

- the energy flux resulting from the heat conduction of the translational mode which can be estimated from the axial profile of T_g (figure 7.1.7) and using the following relation :

$$q_{t-r} = -\lambda_{t-r} \left[\frac{dT_g}{dz} \right]_{substrate} \tag{28}$$

- the energy flux resulting from the heat conduction of the H_2 vibrational mode which can be estimated from the axial profile of T_v (figure 7. 1. 7) and using the following relation :

$$q_v = -\lambda_v \left[\frac{dT_v}{dz} \right]_{substrate} \tag{29}$$

- the energy flux resulting from the catalytic recombination of H, H^+, H_2^+, H_3^+ and H^-. These fluxes may be deduced from the axial profiles of the species molar fractions and using the following relation :

$$q_s = -\left[\frac{\rho D_s}{M_s} \right]_{substrate} \left[\frac{dx_s}{dz} \right]_{substrate} \beta_{rec-s} E_{r-s} \quad (s= H, H^+, H_2^+, H_3^+ \text{ and } H^-). \tag{30}$$

where E_{r-s} and β_{rec-s} are, respectively, the energy and the accommodation coefficients of species 's' catalytic recombination reaction.

From T_g and T_v axial profiles (figure 7.1.7) and using equations (28) and (29), we can calculate a thermal conduction energy flux of 20 W/cm^2 at the substrate center. Due to the high conductivity of H_2 plasma these fluxes density are expected to be approximately constant over all the substrate section. The resulting value of the power transferred to the substrate by conduction mechanism is PW_{cond} = 390 W.

Neglecting the power transferred to the reactor walls, the total power absorbed by the plasma must be equal to the power transferred to the substrate. This may be written :

$$MWP_{total} = PW_{cond} + PW_{rec} \tag{31}$$

The determination of the total absorbed power is now equivalent to that of the power transferred by recombination reactions to the substrate. The latter may be estimated by using the value of 390 W as a rough estimation of PMW_{total}, solving the 1D equations and calculating the ratio $\alpha_p = PW_{rec}/PW_{cond}$ from the computed profiles of the temperatures and the chemical species molar fractions. The actual power transferred to the substrate may then be extracted from the following equation :

$$PMW_{total} = Pw_{cond} (1+\alpha_p) \tag{32}$$

To determine the values of z_0, we will compute the plasma temperatures and species concentrations for three SDMPWD corresponding to MWP_{total} = 390 W and to three values of z_0 : 0, 5 mm and 10 mm (figure 7. 1. 8). The calculated axial profiles of T_g, T_e and x_e will be used for determining the axial distributions of the 750 nm argon emission line intensity corresponding to the different values of z_0. These calculated distributions will be compared with that measured by OES. From this comparison we will determine the value of z_0 which corresponds to the discharge conditions considered here.

Once the value of z_0 is estimated, we will use the axial profiles of T_g, T_v and x_s to calculate the power fraction transferred to the substrate by recombination.

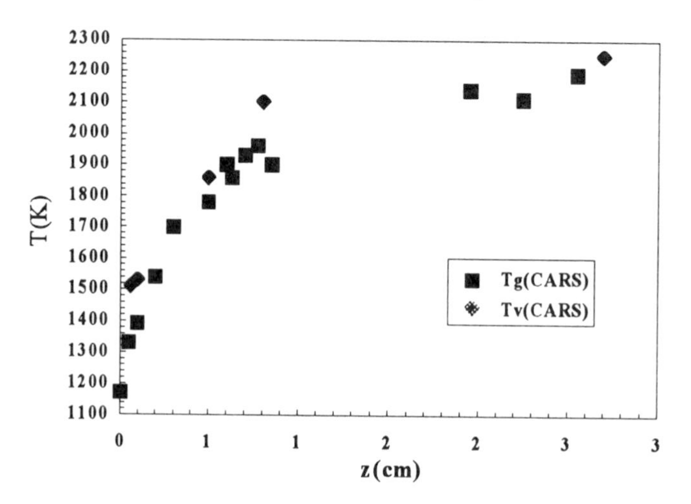

Figure 7.1.7 T_g and T_v axial profiles determined by CARS.

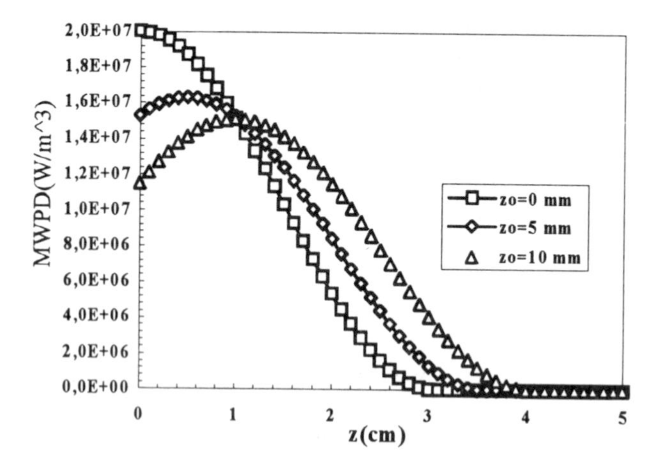

Figure 7.1.8 Axial profiles of the investigated microwave power density distributions. MWP_{total}=390 W

Preliminary simulations

The calculated axial profiles of gas temperature, electron temperature and electron molar fraction corresponding to the SDMWPD of figure 7.1.8 are presented in figures 7.1.9, 7.1.10 and 7.1.11. The gas temperature in the plasma bulk (z=2 cm) is very sensitive to the value of z_0. It increases from 2000 K to 2500 K when z_0 increases from 0 to 10 mm. The thickness of the high gradients layer presents the same behavior, since it increases from 2 cm to 3 cm when z_0 increases.

The electron temperature axial profiles are also very sensitive to the value of z_0, and we can distinguish a first domain corresponding to z<1 cm where T_e is approximately constant (18000 K < T_e < 20000 K) and slightly decreases with z_0 and a second domain

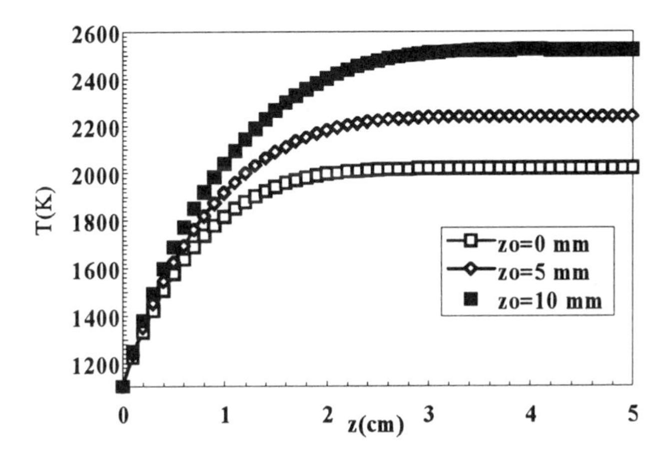

Figure 7.1.9 Calculated axial profiles of T_g for z_0 = 0, 5 and 10 mm.

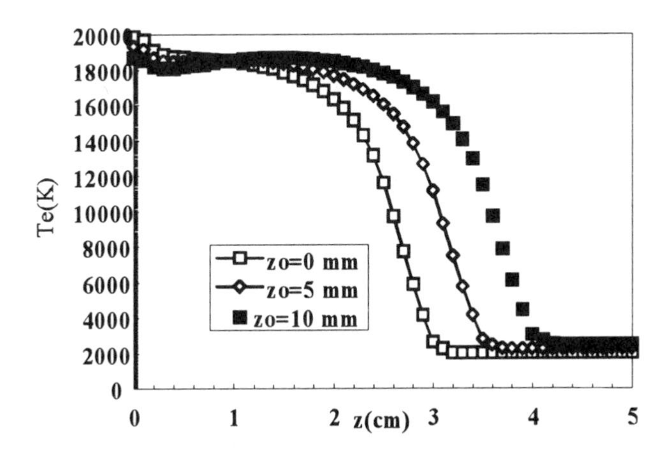

Figure 7.1.10 Calculated axial profiles of T_e for z_0 = 0, 5 and 10 mm.

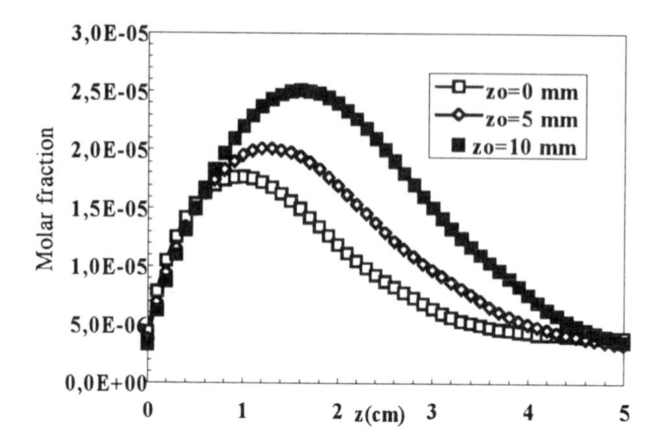

Figure 7.1.11 Calculated axial profiles of electron molar fraction for z_0 = 0, 5 and 10 mm.

corresponding to z>1 cm where the axial profile of T_e shows a sharp decrease (from 18000 K to 20000 K) and where T_e strongly increases with z_0.

The electron molar fraction, which approximately represents the ionization degree of the plasma, shows a maximum, the value and location of which depend on z_0. The maximum ionization degree is about 1.7 10^{-5} for z_0=0 and increases with z_0. The distance between the position of this maximum and of the substrate surface increases from 7 mm to 15 mm when z_0 increases from 0 to 1 cm.

Estimation of z_0

Now suppose that we introduce a small amount of argon in the plasma (1% for example). Because of its low concentration and chemical inertness, the argon does not perturb the discharge composition and temperatures. This has been observed experimentally by investigating the variation of several emission lines intensities of the plasma as functions of argon mole fraction in the plasma feed gas. In addition, due to the low averaged electron energy, most of the argon atoms will be in their electronic ground state. The transport of argon may be then described by the diffusion equation (16) where the source term has to be set equal to zero. The mass density and the diffusion coefficient axial profiles required for solving this equation, may be estimated from the calculated gas temperature profile. Since the major species of the plasma is H_2, the argon diffusion coefficient may be well approximated by the Ar-H_2 binary diffusion coefficient estimated from equation (50) of section 5.

The solution of this diffusion equation, with boundary conditions corresponding to a given argon mole fraction at the plasma inlet and zero gradient at the substrate surface (no surface reaction), leads to a constant argon mole fraction in the plasma.

The populations of the radiative excited states of argon in the discharge are mainly governed by the excitation-deexcitation kinetics, the characteristic times of which are much smaller than that of diffusion.

As a consequence, the Ar (4p'[1/2]) excited state population , is governed by the following balance between the excitation and deexcitation rates (for convenience the Ar (4p'[1/2]) will be written Ar*) :

$$k_{exc}\, c^2\, x_e\, x_{Ar} = k_{rad} \,.\, c\, x_{Ar^*} + c^2\, x_{Ar^*}\, (k_{q\text{-}H2}\, x_{H2} + k_{q\text{-}H}\, x_H + k_{q\text{-}Ar}\, x_{Ar}) \qquad (33)$$

where c is the total concentration of the plasma and k_{exc}, k_{rad} and $k_{q\text{-}s}$ denote the rate constants of the following processes :

$e^- + Ar$ -------> $e^- + Ar^*$ (Electron excitation) (R5)
$s^- + Ar^*$ -------> $s + Ar$ (quenching s= H_2, Ar or H) (R6)
Ar^* -------> $Ar + h\nu$ (radiative deexcitation) (R7)

k_{exc} depends on the electron temperature, k_{rad} is constant and k_q is function of the gas temperature and the different quenching cross sections.

The Ar(750 nm) emission line intensity at a given point of the plasma is proportional to Ar* population and the axial profile of this emission line intensity may be calculated from the following relation :

$$I_{Ar} = K \frac{k_{exc}\, x_e\, x_{Ar}}{k_{rad}/\, c + (k_{q\text{-}H2}\, x_{H2} + k_{q\text{-}H}\, x_H + k_{q\text{-}Ar}\, x_{Ar})} \qquad (34)$$

where K is a constant, the determination of its exact value is not necessary since we are only interested in the shape, especially the maximum location, of the emission line intensity axial profile.

In figure 7.1.12 we have presented the axial profiles of $I_{Ar}/I_{Ar\text{-}max}$ determined from equation (34). The comparison of these profiles with that determined by OES shows that the more realistic value of z_0 is the one corresponding to a maximum of the SDMWPD on the substrate surface (z_0=0). This result qualitatively agrees with those of [Zhang et al.] and [Tan et al.] who obtained similar SDMWPD shapes.

Estimation of MWP_{total} (or $MWPD_0$)

As mentioned in section 6.1 the energy gained by the plasma from the high frequency electric field is transferred to the substrate surface.

The energy flux due to the recombination of the charged species is negligible. Assuming a total surface accommodation of the energy due to recombination, the energy fluxes resulting from heavy species translation and vibration modes and H-atom recombination are respectively $1.733\cdot10^5$ W/m^3, $2.56\cdot10^4$ W/m^3 and $7.5\cdot10^4$ W/m^3. The resulting value of α_p is 0.38.

The actual total absorbed power can be obtained by multiplying the value of 390 W by the factor 1.38 which takes into account the energy transferred by the recombination of H-atom. The resulting value of the total power is then MWP_{total} = 530 W.

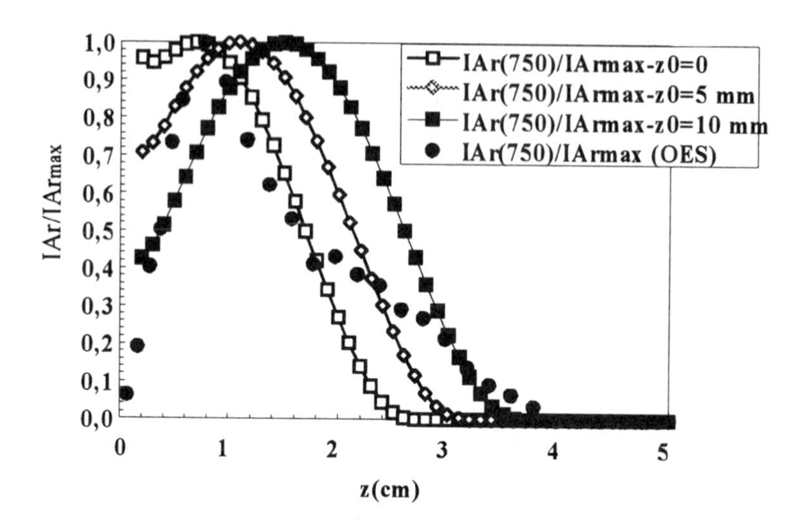

Figure 7.1.12 Calculated axial profiles of the 750 nm argon emission line intensity for the different values of z_0 (0, 5 and 10 mm). MWP_{total}=390 W and P=2500 Pa.

Summary and conclusion

The method used here to determine the SDMWPD parameters for the discharge conditions corresponding to MWP_{input} = 600 W and P =25 mbar can be summarized as follow :

- MWP_{total} was roughly estimated from conduction heat flux measurements using CARS results.
- The 1D transport equations were then solved for three values of z_0 and for the estimated value of MWP_{total}.
- The resulting electron temperature and mole fraction profiles were used to calculate the axial profiles of the 750 nm argon emission line intensity corresponding to the three values of z_0.
- The comparison of these profiles with the experimental one show that the value of z_0=0 is the more realistic one.
- For z_0=0, the ratio of the energy flux due to the recombination of H-atom to the energy flux due to conduction effect was estimated to 0.38. This enables a better estimation of the total absorbed power :MWP_{total} = 530 W.

This method makes use of the following assumptions :

- The SDMWPD is described by equation (10)
- The accommodation coefficient of H-atom catalytic recombination is 1
- The power fraction transferred to the substrate by the accommodation of the recombination energy is approximately constant for MWP_{total} in the range [400 W, 600 W].

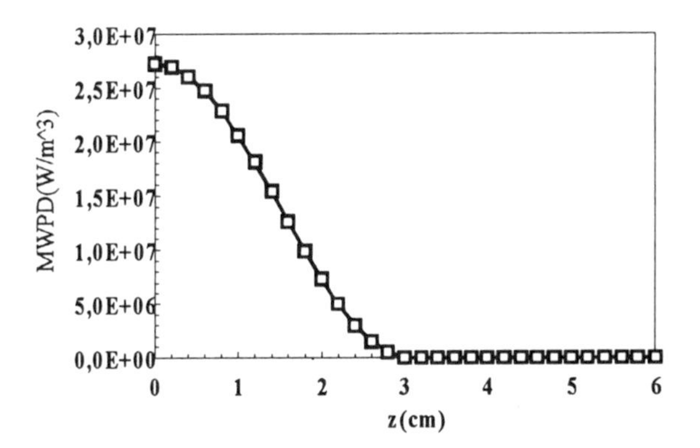

Figure 7.1.13 Axial profile of the microwave power density on the reactor axis. MWP_{total} = 530 W, P=2500 Pa and $MWP_{averaged}$ = 9 W/cm^3

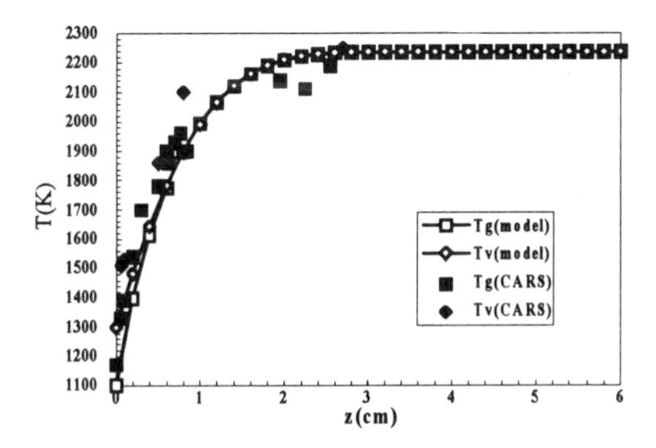

Figure 7.1.14 Calculated and measured axial profile of gas and vibration temperatures. MWP_{total} = 530 W, P=2500 Pa and $MWP_{averaged}$ = 9 W/cm^3

7.1.8 Comparison with experimental results

For the absorbed microwave power density axial profile presented in fig. 7.1.13 and corresponding to MWP_{total} = 530 W and z_0=0, the computed gas and H_2 vibration temperatures are nearly in equilibrium in the plasma bulk, the small non equilibrium observed in the boundary layer near the substrate surface is mainly due to the boundary condition used in this calculation. The calculated axial profiles are in good agreement with those determined by CARS (Tg_{exp}-Tg_{model} < 100 K) even if the small vibrational non equilibrium (Tv-Tg ≈ 200 K) observed experimentally [Gicquel et al., 1994] is not predicted by the model (fig. 7. 1. 14). They show a sharp decrease of T_g from a value of 2300 K in the plasma bulk (z>2cm) to 1173 K at the substrate surface. The value of the thickness of the high gradient layer obtained from both the experiment and the model is about 2 cm.

For the investigated discharge conditions, the calculated H-atom mole fraction in the plasma bulk is 6.1 10^{-2}. The calculated axial profile of the H-atom molar fraction is

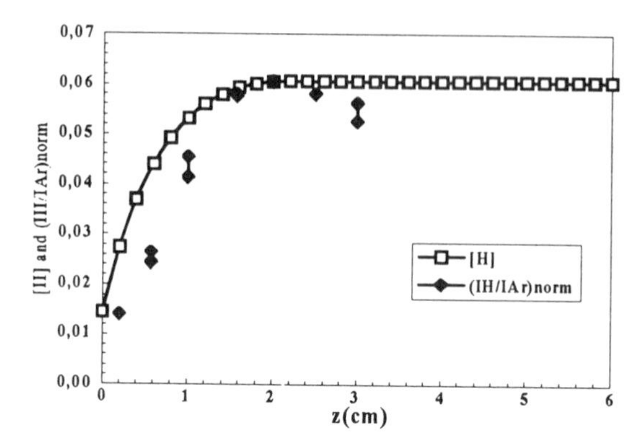

Figure 7.1.15 Calculated axial profile of H-atom mole fraction and measured axial profiles of the normalized IH/IAr for the reference experiment. MWP_{total} = 530 W, P=2500 Pa and $MWP_{averaged}$ = 9 W/cm^3

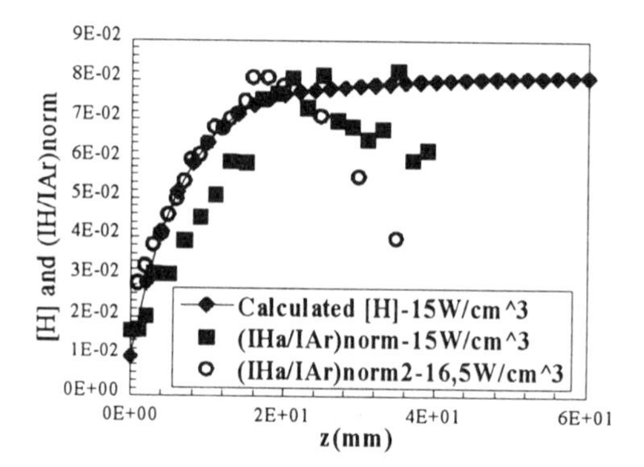

Figure 7.1.16 Calculated axial profile of H-atom mole fraction and measured axial profiles of the normalized IH/IAr at high power density

given in figure 7.1.15, where we have also reported the normalized ratio $(IH/IAr)_{norm}$ given by the following equation :

$$(IH/IAr)_{norm} = [(IH/IAr)/(IH/IAr)_{max}].[H]_{bulk} \tag{35}$$

where IH and IAr are the H_α and Ar(750 nm) emission lines intensities, and $[H]_{bulk}$ denotes the H-atom molar fraction in the plasma bulk. The ratio IH/IAr is a linear function of the H-atom mole fraction. [Gicquel et al. 1996]. The comparison between the measured and calculated profiles shows that the thickness of the high H-atom mole fraction gradient is well predicted by the model (δ=2 cm). The calculated values of $[H]/[H]_{bulk}$ near the substrate (0.1 cm $<z<$ 1 cm) are however much higher than those measured by actinometry (figure 7.1.15). This discrepancy may be due to the assumption of a Maxwellian EEDF, which leads to an overestimation of the rate constants of high energy electron-heavy species collisions (dissociation and ionization). The calculated

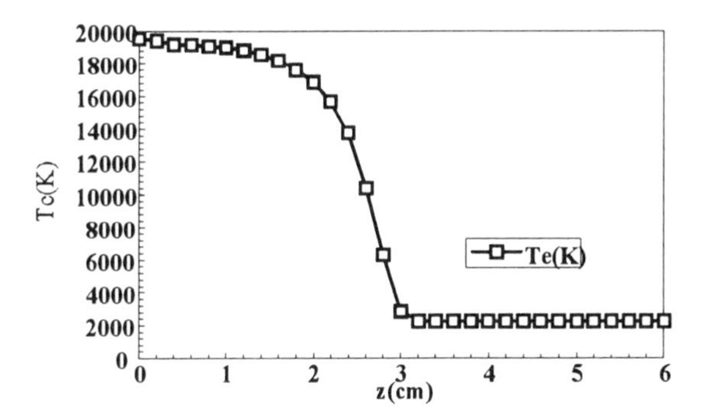

Figure 7.1.17 Calculated axial profile of electron temperature. MWP_{total} = 530 W, P=2500 Pa and $MWP_{averaged}$ = 9 W/cm^3

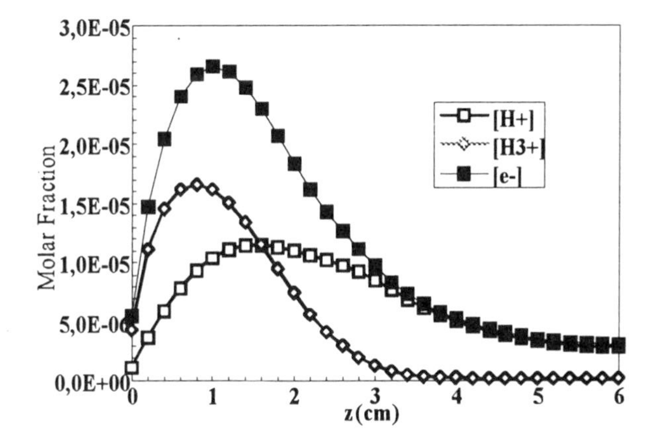

Figure 7.1.18 Calculated axial profiles of the mole fraction of the major charged species H^+, H_3^+ and e^-.

and measured H-atom axial profiles were also compared at an averaged power density of 15 W/cm^3 (figure 7.1.16). The results show that the calculated values are greater than those measured in the high gradient region. However one can note that the calculated axial profile obtained at 15 W/cm^3 and the measured profile at 16.5 W/cm^3 agree well. The difference between the calculated and the measured H-atom mole fraction at the plasma inlet (for $z > 3$ cm) may be explained by the fact that convection is not taken into account in the present model. In fact, since the model is only diffusional, it does not take into account the molecular hydrogen flow entering in the discharge zone. This should lead to the underestimation of the molar fraction of H_2-molecule and to the overestimation of that of H-atom at the discharge inlet where convection flux is not negligible with respect to the diffusion one.

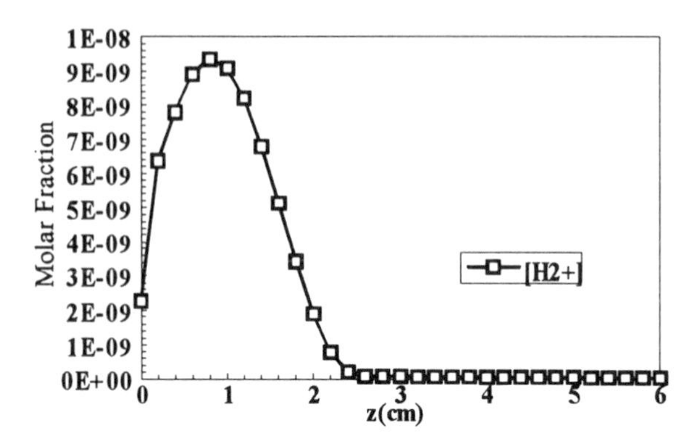

Figure 7.1.19 Calculated axial profiles of H_2^+ mole fraction. MWP_{total} = 530 W, P=2500 Pa and $MWP_{averaged}$ = 9 W/cm^3

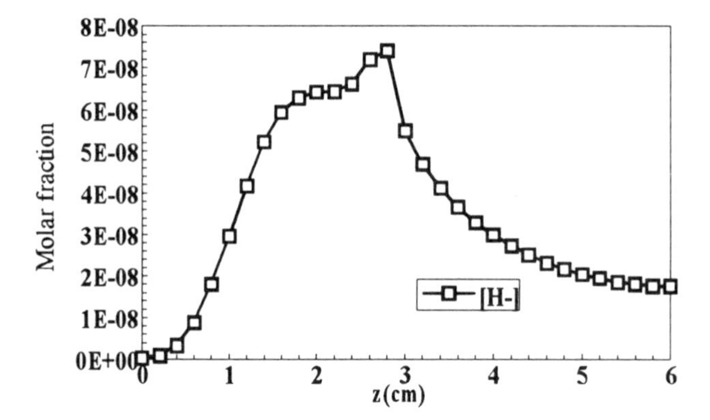

Figure 7.1.20 Calculated axial profiles of H^- mole fraction. MWP_{total} = 530 W, P=2500 Pa and $MWP_{averaged}$ = 9 W/cm^3

The calculated electron temperature (fig. 7.1.17) is about 18000 K in the plasma bulk (0.5cm<z<2cm) and increases in the vicinity of the substrate surface where it reaches a value of 20000 K (0<z<0.3 cm).

The major ionic species are the electrons, H_3^+ and H^+, the mole fraction axial distributions of which are presented in figure 7.1.18. In the high power density discharge zone (z<2 cm), the major ion is H_3^+, its profile shows a maximum value of 1.7 10^{-5} at 8 mm from the substrate surface. This ion rapidly recombines out of the discharge zone where the major ion is H^+. The mole fraction axial profile of the latter shows a maximum value of 1.2 10^{-5} at 16 mm from the substrate surface. The mole fraction of electron may be deduced by summing those of H^+ and H_3^+, its axial profile shows a maximum value of $2.8.10^{-5}$ at 11 mm from the substrate surface.

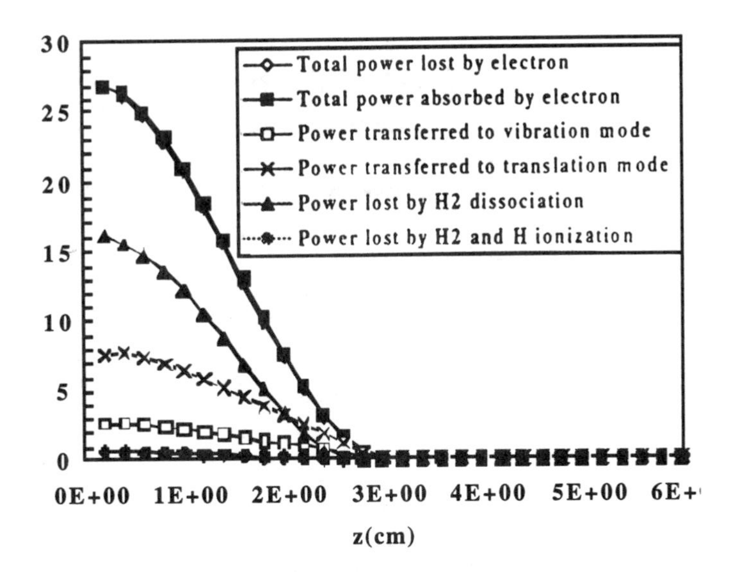

Figure 7.1.21 Power dissipation channels in the plasma MWP_{total} = 530 W, P = 2500 Pa et $MWP_{averaged}$ = 9 W/cm^3

The H_2^+ and H^- ions, even very important for the kinetic modeling of this kind of plasma, have very low mole fractions which are characterized by maximum values of 5.10^{-9} for H_2^+ and 2.10^{-8} for H^- (figures 7. 1. 19 and 7. 1. 20).

7. 1. 9. Energy dissipation in the discharge

It may be very interesting to investigate the relative magnitudes of the different energy transfer channels. In figure 7.1.21 we have reported the axial profiles of the source terms corresponding to the power densities gained and lost by electrons, those transferred from electrons to the other modes, and those lost by electrons for the activation of the dissociation and ionization processes. The total power densities gained and lost by the electron are approximately equal ; the slight difference between these densities is due to the electron enthalpic flux gradient which also participates in the electron energy equation (17). This weak difference shows that the electron energy transport may be well described by neglecting the enthalpic diffusion. This simplification should lead to an important simplification of the mathematical problem, since the resulting electron energy equation is algebraic and not differential. Figure 7.1.21 shows that most of the energy gained by electron from the electromagnetic field is lost by activation of the H_2 dissociation which constitutes 58 % of the total lost power at 1 cm from the substrate surface. The second channel for the electron energy dissipation consists of the transfer to the translational mode of the heavy particles, e. g. heating of the gas. It represents 30% of the total lost power at 1 cm from the substrate surface. Then we have the energy transfer to the H_2 vibration mode which presents 10% of the total power. The power dissipated for the ionization of H and H_2 is very weak and almost negligible compared to the other dissipation channels (2%).

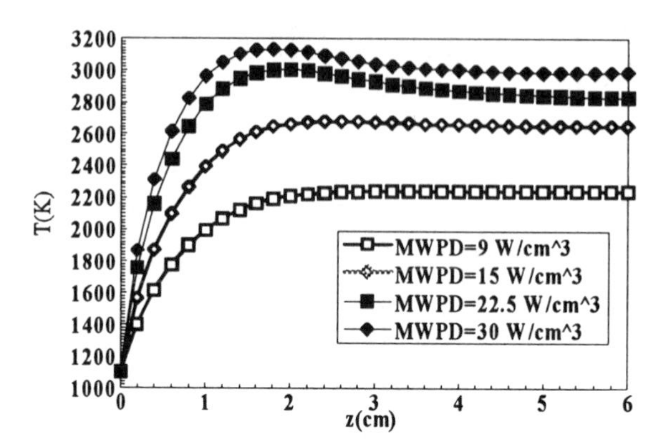

Figure 7.1.22 Evolution of T_g axial profile with the averaged microwave power density

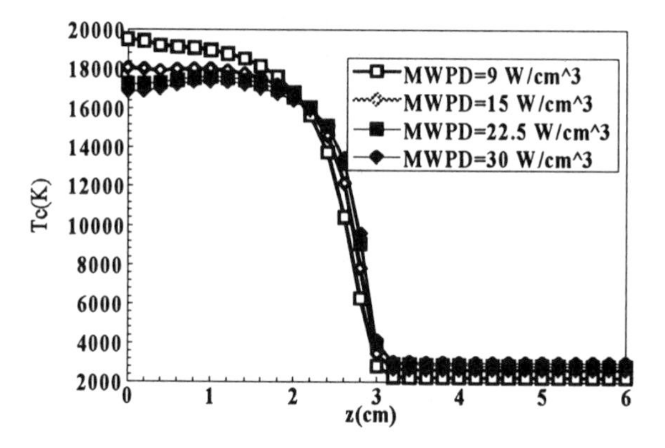

Figure 7.1.23 Evolution of T_e axial profile with the averaged microwave power density

7.1.10 Effect of the averaged microwave power density

Deposition experiments have previously shown that an increase of the averaged power density (MWP_{av}) from 9 W/cm^3 to 18 W/cm^3, while keeping the plasma volume constant, has a strong effect on the diamond growth rate and quality [Gicquel et al. 1994]. To investigate the effect of such a parameter on the plasma properties, the transport equations were solved for four couples (MWP_{total}, P) corresponding to a constant discharge volume V≈32.7 cm^3 and to MWP_{av} = 9W/cm^3, 15 W/cm^3, 22.5 W/cm^3 and 30 W/cm^3.

The increase of the averaged power density from 9W/cm^3 to 30 W/ cm^3 leads to a strong increase of the gas temperature which varies from 2300 K to 3100 K in the bulk of the plasma (z>3 cm) (figure 7.1.22). The variation of T_g with MWP_{av} is non linear and is attenuated at high power density ($\Delta T_g/\Delta PMW$ decreases with MWP_{av}).

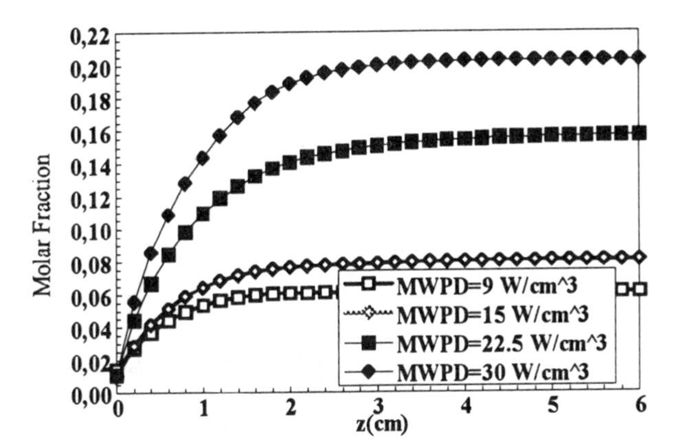

Figure 7.1.24 Evolution of H-atom mole fraction axial profile with the averaged microwave power density.

The increase of T_g with MWP_{av} results in a sharp increase of H-atom mole fraction which varies from 0.06 to 0.21 in the investigated domain of the averaged power density (figure 7.1.24). The variation of H-atom mole fraction with the power density is also non linear and is more important at high power density. At low MWP_{av} values the H-atom is mainly produced by electron-H_2 collision which is controlled by electron temperature. This is not the case at high MWP_{av}, where the dissociation of H_2 is mainly due to H_2-H_2 and H_2-H collisions which are controlled by gas temperature (thermal dissociation).

These results show that at high power density the energy transferred to the heavy species is dissipated in the activation of the H_2 thermal dissociation which leads to a higher increase of H-atom molar fraction and slower increase of T_g. They are in a good qualitative agreement with CARS and actinometry measurements. They also explain the high quality of the diamond films obtained under high power density discharge conditions, since the high H-atom concentration obtained for these conditions insures a high etching rate of the non-diamond phase.

In the discharge zone ($z < 3$ cm), the electron temperature decreases when increasing the averaged microwave power density. It varies from 20000 K to 17000 K in the investigated absorbed power domain (figure 7.1.23). This behavior is due to the increase of the pressure which leads to higher energy transfer rates from the electron kinetic mode to the other modes.

As far as the major charged species are concerned, lower values of the ionization degree are obtained at high microwave power density. Furthermore the increase of the power density leads to a sharp decrease of the charged species concentrations out of the discharge zone. This is due to the recombination reactions kinetic which is accelerated at the high pressure values characterizing the high power density discharge conditions. In addition the variations of the charged species mole fraction with MWP_{av} are also non linear and highly attenuated at high MWP_{av} values (figures 7.1.25, 7.1.26 and 7.1.27).

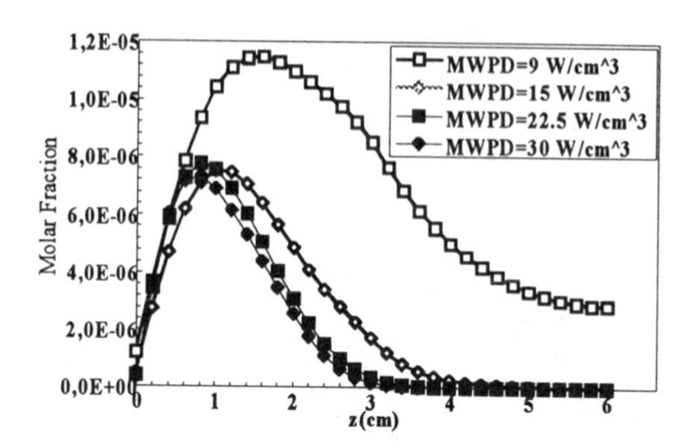

Figure 7.1.25 Evolution of H^+ mole fraction axial profile with the averaged microwave power density.

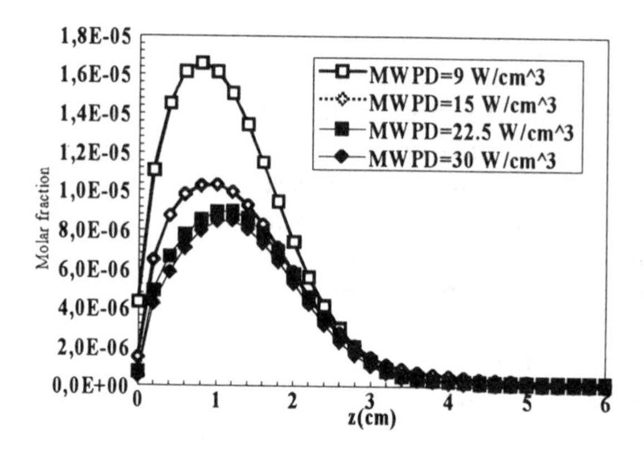

Figure 7.1.26 Evolution of H_3^+ mole fraction axial profile with the averaged microwave power density.

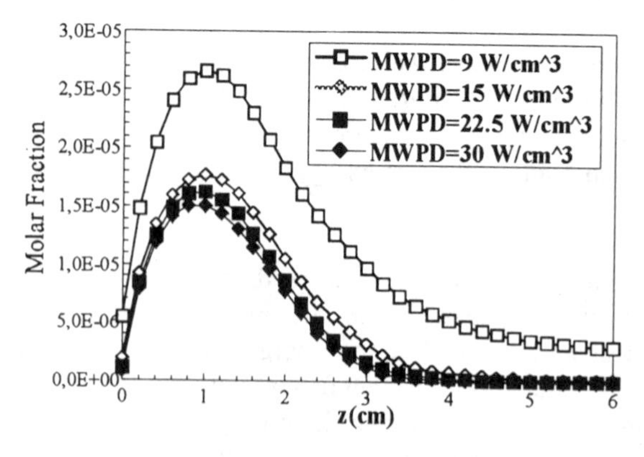

Figure 7.1.27 Evolution of H^+ mole fraction axial profile with the averaged microwave power density.

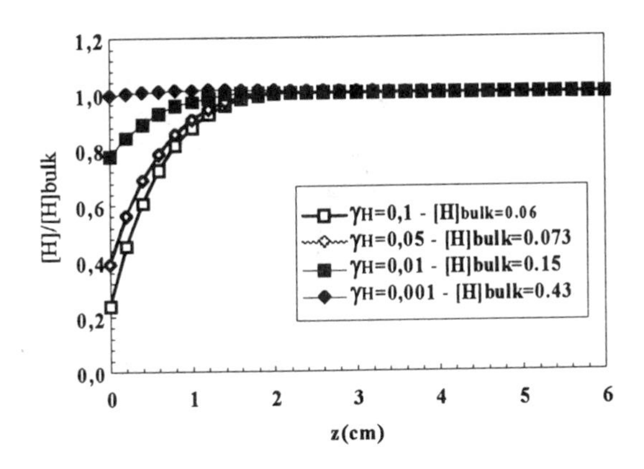

Figure 7.1.28 Evolution of the axial profile of H-atom mole fraction with its recombination coefficient. MWP_{total} = 530 W, P=2500 Pa and $MWP_{averaged}$ = 9 W/cm^3.

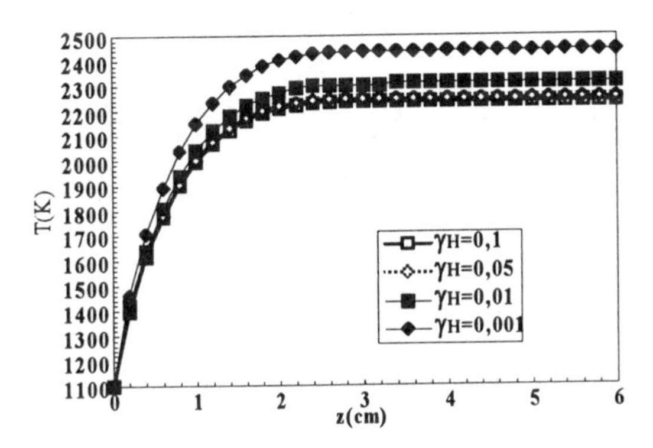

Figure 7.1.29 Evolution of the axial profile of T_g with H-atom recombination coefficient. MWP_{total} = 530 W, P=2500 Pa and $MWP_{averaged}$ = 9 W/cm^3.

7.1.11 Effect of the H-atom recombination coefficient

On of the most important species for diamond deposition kinetics is H-atom, the concentration of which is very sensitive to its recombination coefficient at the substrate surface. This latter depends not only on the chemical nature of the substrate but also on its physical and crystallographic states (roughness, grain boundary density, texture, previous treatments ...). it seems therefore very interesting to investigate the effect of this parameter on the plasma composition and temperatures.

We have then performed simulations of the H_2 plasma, corresponding to discharge conditions such as P=2500 Pa, MWP_{input}=600 W, for four values of γ_H : 10^{-1}, 5 10^{-2}, 10^{-2} and 10^{-3}.

As we could expect, the decrease of γ_H from 0.1 to 10^{-3} results in a sharp increase of H-atom mole fraction in the plasma bulk from 6.6 10^{-2} to 0.42 (figure 7.1.28). The shape

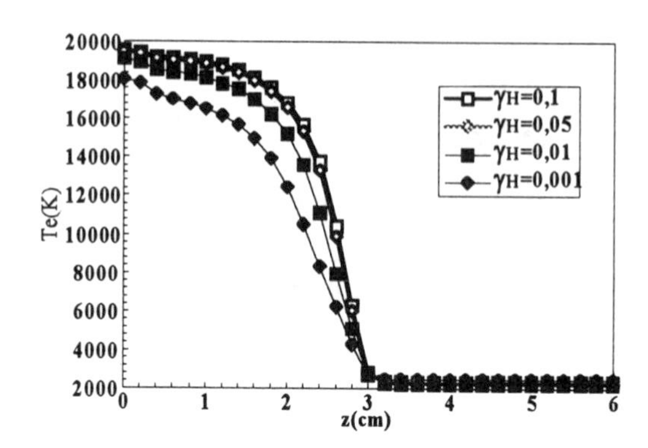

Figure 7.1.30 Evolution of the axial profile of T_e with H-atom recombination coefficient. MWP_{total} = 530 W, P=2500 Pa and $MWP_{averaged}$ = 9 W/cm^3.

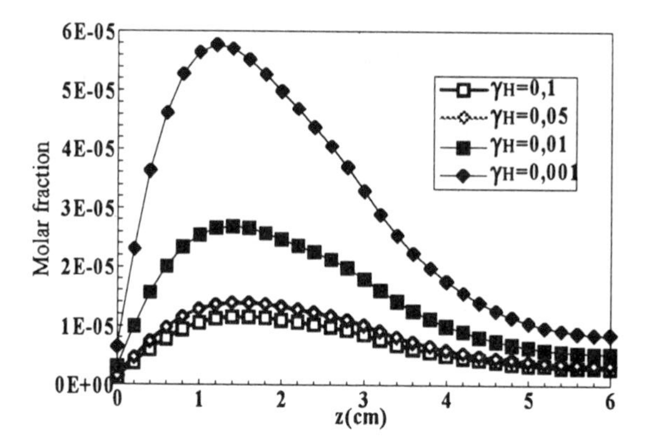

Figure 7.1.31 Evolution of the axial profile of H^+ mole fraction with H-atom recombination coefficient. MWP_{total} = 530 W, P=2500 Pa and $MWP_{averaged}$ = 9 W/cm^3.

of the axial profile of this molar fraction also changes with γ_H. Especially the ratio $[H]_{bulk}/[H]_{substrate}$ increases with γ_H and the H-atom mole fraction axial profile is practically flat at low recombination coefficient values. The gas temperature is slightly sensitive to γ_H, and its value in the bulk of the plasma is higher for a non catalytic substrate. The variation magnitude of T_g is less than 200 K for the investigated range of the recombination coefficient (figure 7.1.29).

The variation of γ_H results in a strong change of the axial profile of the electron temperature (figure 7.1.30). As a matter of fact, the decrease of T_e in the discharge zone (z<2 cm) is more important at low recombination coefficient. In addition, the maximum value of T_e (at the substrate surface) decreases from 20000 K to 16000 K when γ_H decreases from 0.1 to 10^{-3}.

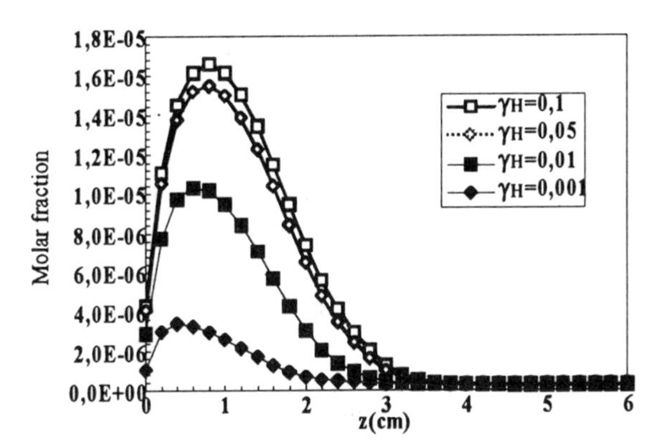

Figure 7.1.32 Evolution of the axial profile of H_3^+ mole fraction with H-atom recombination coefficient. MWP_{total} = 530 W, P=2500 Pa and $MWP_{averaged}$ = 9 W/cm^3.

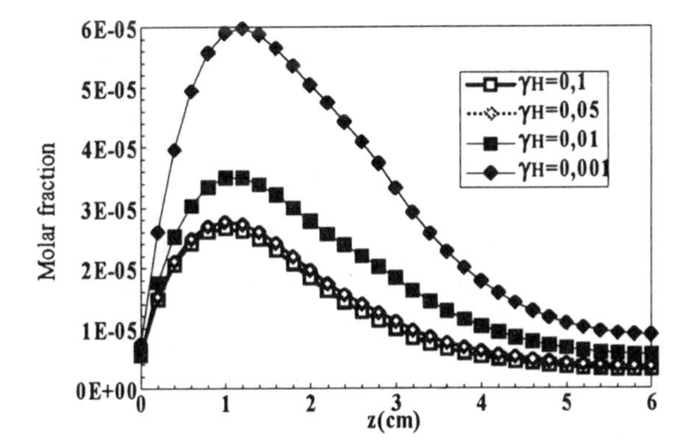

Figure 7.1.33 Evolution of the axial profile of e^- mole fraction with H-atom recombination coefficient. MWP_{total} = 530 W, P=2500 Pa and $MWP_{averaged}$ = 9 W/cm^3.

The variation of T_e with γ_H is linked to a strong effect of γ_H on the major charged species mole fractions. The increase of γ_H leads to the increase H_3^+ and to the decrease of H^+ and e^- (figures 7.1.31, 7.1.32 and 7.1.33).

The effect of γ_H on the temperatures and the chemical species mole fraction may be explained as follows : the decrease of γ_H leads to the increase of H-atom mole fraction and since the ionization of H is faster than that of H_2, the ionization degree of the plasma increases. The increase of the electron mole fraction leads to the increase of the rates of ionization, dissociation and of energy transfer from the electron mode to the other modes ; and the electron temperature is lower.

Before closing this section we have to point out that the actual sensitivity of the plasma parameters to the H-atom recombination coefficient should not be as strong as predicted by the present diffusion model. In fact, the convection transport would reduce the sensitivity of the plasma parameters to the boundary conditions. This would be especially the case far from the substrate surface ($z > 2$ cm).

7.1.12 Conclusion

In this section we have used the three temperature non equilibrium model of section 5 to investigate the H_2 plasma obtained under diamond deposition discharge conditions. The use of this kind of model requires the assumption of a Maxwellian EEDF. Since a self consistent modeling of the reactor is very complex, we have simplified the mathematical description of the plasma by treating the spatial distribution of the microwave power density as a model parameter, neglecting the convection fluxes and reducing the transport equation to one-dimensional geometry.

The shape of the spatial distribution of the absorbed microwave power distribution was simply extracted from the solution of the one dimensional wave equation. This enables its description in term of two parameters : the maximum position and the value of the total absorbed power.

For discharge conditions corresponding to a total input power of 600 W and a pressure of 25 mbar, these parameters were determined by the simultaneous use of the 1D diffusion model, the experimental values of the heat fluxes at the substrate surface and the axial profile of the Ar (750 nm) emission line intensity determined by OES.

Once the spatial distribution of the absorbed microwave power density was determined the plasma diffusion equations were solved and the calculated axial profiles of gas temperature, H_2 vibration temperature and H-atom mole fraction were compared with the measured ones.

This comparison shows that the axial profiles of the temperatures are well predicted by the model, while a slight discrepancy was observed concerning the H-atom mole fraction which is overestimated by the model even if the calculated and measured profiles are quite similar. This discrepancy may be due to the assumption of a Maxwellian EEDF which leads to an overestimation of the rate constant of the H_2 dissociation by electron-H_2 collision.

The model also enabled the estimation of the electron temperature and the charged species mole fractions which are difficult to determine experimentally. Results show that T_e is in the range [16000 K , 20000 K] and reaches its maximum value at the substrate surface. The calculated axial profiles show that the major charged species are H^+, H_3^+ and e^-. The ionization degree shows a strong spatial variation and reaches a maximum value of 2.75 10^{-5}. However, this value must be an overestimation of the actual ionization degree since the assumption of a Maxwellian EEDF leads to the overestimation of the rate constants of high energy electron-heavy species collisions.

The model was then used to investigate the effect of the averaged absorbed microwave power density and the H-atom recombination coefficient on the plasma composition and temperatures. The increase of the averaged absorbed power results in the increase of the gas temperature and H-atom mole fraction and in the decrease of the charged species mole fractions and electron temperature. The variation of the gas temperature with this parameter is non linear and less important at high absorbed microwave power values.

This model also shows that, when the convection fluxes are neglected, the H-atom catalytic recombination coefficient has a strong effect on all the plasma local parameters except the gas and H_2 vibration temperatures. Its decrease results not only in the increase of H-atom mole fraction but also in a sharp increase of the ionization degree and in a decrease of the electron temperature. This parameter is then very important for the investigation of H_2 plasma and its experimental determination is necessary for further studies.

These results show that it is possible to perform a rather satisfying description of the H_2 plasma obtained under diamond deposition conditions even with a simplified and non self consistent model. Of course, due to the assumption used in this model the conclusion concerning the chemical species molar fractions must remain qualitative and only the order of magnitude and the variations of these molar fractions with the investigated parameters may be used. In fact, more quantitative results on the chemical species molar fraction requires to take into account the non Maxwellian behavior of the EEDF by solving the electron Boltzmann equation.

A better description of both the plasma and the absorbed microwave power spatial distribution may be obtained by solving the transport equation in such a way to match the calculated and measured Ar(750 nm) emission line intensity profile. In this case the spatial distribution of the absorbed microwave density may be accurately calculated without coupling the Maxwell equations to the transport equations.

Finally, the high gas temperature gradients obtained from the solution of the 1D diffusion equations, show that it should be better to take into account the convection fluxes, at least at high averaged absorbed microwave power.

Table 1. Energy Exchange Rate Formulas and Constants

Interaction	Constants, a_i			
Formula	a_0	a_1	a_2	a_3
Singlet electron/vibration				
$500<T_v<4700$ K				
$S_B(T_v)=\sum_{i=0}^{2} a_i T_v^i$ HO**	4.5608	1.4915e-5	1.3416e-8	
AHO**	3.2179	1.9571e-5	-1.4701e-8	
$S_C(T_v)=\sum_{i=0}^{2} a_i T_v^i$ HO**	4.2096	8.1733e-6	2.3308e-8	
AHO**	3.3031	-2.0333e-6	-2.4737e-8	
Arrhenius†				
$0.1<T_e<5.0$ eV				
$k_{ev}^{B1}(T_e)$, cm^3/s/molecule	9.441e-9	0	133270 K	
$\varepsilon_B = 1.776 \times 10^{-18}$ J/molecule				
$k_{ev}^{C1}(T_e)$, cm^3/s/molecule	1.0563e-8	0	150900 K	
$\varepsilon_C = 1.952 \times 10^{-18}$ J/molecule				

Direct electron/vibration† $0.1<T_e<5.0$ eV $\log\left(\frac{1}{n_e \tau_{ev}}\right) = \sum_{i=0}^{3} a_i(\log(T_e, eV))^i$ cm^3/s/molecule	-9.7748	0.0073464	-2.2301	0.54929
	Arrhenius			
Vibration-Translation $(H+H_2^* \rightarrow H+H_2)$ quenching§ k_{10}, cm^3/s/molecule Reactive Non reactive $300<T<4000$ K	7.682e-11 3.707e-11	0 0	2192 K 2140 K	
Vibration-Translation (H2)* atm-s	$p\tau_{H2-H2} = 3.9 \times 10^{-10} \exp\left(\frac{-100}{T^{1/3}}\right)$			

†Calculated from cross sections in Buckman and Phelps

§Gorse, et al.

*Kiefer and Lutz

**HO Harmonic Oscillator energy levels. AHO Anharmonic Oscillator

Table 2. Reactions and rate constants for a moderate pressure hydrogen plasma

Reaction	Constants $k = AT^{\eta}e^{-T_a/T}$			Temp. Range	Gov T	Data
	A, $cm^3/s/mole$	η	T_a, K			
$e + H_2(v) \rightarrow e + H_2^+ + e^-$	$1.18x10^{16}$	0	191500	$<10^4$ - $5x10^4$ K	T_e	from cross sections, Phelps
$e + H \rightarrow e + H^+ + e^-$	$1.08x10^{16}$	0	178210	1-6 eV	T_e	from cross sections[2]
$e + H_2(v) \rightarrow e + H_2^* \rightarrow e + 2H$	$1.2x10^{16}$	0	113500	$<10^4$ - $5x10^4$ K	T_e	from cross sections, Phelps
$e + H_3^+ \rightarrow 3H$ $e + H_3^+ \rightarrow H_2^*(v>5) + H^*(n=2)$	$8.0x10^{17}$ $3.2x10^{17}$ 2.5:1	-0.404	0	0.1-1 eV	T_e	Janev, from cross sections 2.2.15a,b
$e + H_3^+ \rightarrow e + H^+ + 2H$	$1.22x10^{17}$	0	179380	5000-50000 K	T_e	Janev, from curve fit 2.2.16
$e + H_2(v) \rightarrow H + H^-$ v^3 4	$2.24x10^{22}$	-1.45	9592	5000-50000 K	T_e, T_v	Janev, 2.2.17¶
$e + H_2^+ \rightarrow e + H^+ + H$	$1.46x10^{17}$	0	37460	5000-50000 K	T_e	Janev, 2.2.12
$e + H_2^+ \rightarrow H + H(n)$	$9.44x10^{18}$	-0.604	0	5000-50000 K	T_e	Janev, 2.2.14
$2e + H^+ \rightarrow e + H$	$3.63x10^{37}$ $cm^6/s/mole^2$	-4.0	0	300-5500 K	T_e	Johnson and Hinnov

[2]Takayanage, K. and Suzuki, H. "Cross Sections for Atomic Processes," Vol. 1, Research Information Center, Instiitute of Plasma Physics, Nagoya University, Chikusa-Ku, Nagoya, 464, Japan, November 1978.

$e + H^- \rightarrow 2e + H$	$1.34x10^{13}$	0.9	22700.	5000-50000 K		Janev, 7.1.1
$e + H^+ \rightarrow h\nu + H$	$1.46x10^{14}$	-0.699	0	250-64000 K	T_e	Massey
$2H + H_2 \rightarrow H_2 + H_2$	$1.0x10^{17}$ $cm^6/s/mole^2$	-0.6	0	50-5000 K	T	Cohen & Westburg[3]
$2H + H \rightarrow H_2 + H$	$3.2x10^{15}$ $cm^6/s/mole^2$	0	0	50-5000 K	T	Cohen & Westburg
$H_2^+ + H \rightarrow H^+ + H_2$	$3.85x10^{14}$	0	0	not given	T	Karpas et al.[4]
$H_2^+ + H_2 \rightarrow H_3^+ + H$	$1.27x10^{15}$	0	0	not given	T	Karpas et al.
$H_2 + H_2 \rightarrow H + H + H_2$	$8.61x10^{17}$	-0.7	52530	600-5000 K	T†	Cohen & Westburg
$H_2 + H \rightarrow H + H + H$	$2.7x10^{16}$	-0.1	52530	600-5000 K	T†	Cohen & Westburg
$H + H^- \rightarrow e + 2H$	$4.5x10^8$	1.5	698	1100-11000 K	T	Janev 7.3.2a
$H + H^- \rightarrow e + H_2(v)$	$1.43x10^{15}$	-0.146	815.	1100-11000 K	T	Janev 7.3.2b
$H^+ + H_2 \rightarrow H + H_2^+$	$1.90x10^{14}$	0	21902.	1100-11000 K	T	Phelps[5]
$H^+ + H^- \rightarrow H(n=3) + H(1s)$	$1.78 x 10^{17}$	0	1768.	1100-11000 K	T	Janev, 7.2.3
$2e + H_2^+ \rightarrow \varepsilon + 2H$	$3.17 x 10^{21}$	-4.5	0		Te	Raizer[6]

[3]Cohen, N. and Westburg, K. R., "Chemical Kinetic Data Sheets for High Temperature Chemical Reactions," *J. Phys. Chem. Ref. Data*, Vol. 2, 1983, pp. 531-564.
[4]Karpas, Z., Anicich, V., and Huntress, W. T., Jr., "An Ion Cyclotron Resonance Study of Ions with Hydrogen Atoms,", *J. Chem. Phys.*, Vol. 70, 1979, pp. 2877-2881.
[5]Phelps, A. V., "Cross Sections and Swarm Coefficients for H^+, H_2^+, H_3^+, H, H_2, and H^- in H_2 for Energies from 0.1 eV to 10 keV," *J. Phys. Chem. Ref. Data*, Vol. 19, 1990, pp. 653-674.
[6]Raizer, Yu. P., *Physics of Gas Discharge*, Nauka., Moscow, 1987.

$2e + H_3^+ \rightarrow \varepsilon + H + H_2$	3.17×10^{21}	-4.5	0		T_e	Raizer
$H^+ + 2H_2 \rightarrow H_3^+ + H_2$	1.95×10^{20}	-0.5	0		T	Matveyev & Silakov[7]
$H^- + H_2^+ \rightarrow H_2 + H$	2.08×10^{18}	-0.5	0		T	Matveyev & Silakov
$H^- + H_3^+ \rightarrow H_2 + H_2$	2.08×10^{18}	-0.5	0		T	Matveyev & Silakov

†These reactions actually depend on T_V as well as T, but cross section data that depend on vibrational excitation state are not available. Rate must be reduced by Boltzmann factor for v³ 4.

7.1.13 References

Audibert, M. M., Joffrin, C., and Ducuing, J., "Vibrational Relaxation of H_2 in the range 500-40 K," *Chem. Phys. Letters*, V. 25, 1974, pp. 158-163.

Billing, G. D. and Fisher, V. V., "VV and VT Rate Coefficients in H_2 by a Quantum-Classical Model," *Chem. Phys.*, V. 18, 1976, pp. 225-232.

Buckman, S. J., and Phelps, A. V., "Vibrational Excitation of D_2 by Low Energy Electrons," *J. Chem. Phys.*, V. 82, 1985, pp. 4999-5011.

Buckman, S. J., and Phelps, A. V., "Tabulations of Collision Cross Sections and Calculated Transport and Reaction Coefficients for Electrons in H_2 and D_2," JILA Information Center Report No. 27, University of Colorado, Boulder, May 1, 1985.

Cacciatore, M., Capitelli, M., and Dilonardo, M., "A Joint Vibro-electronic Mechanism in the Dissociation of Molecular Hydrogen in Nonequilibrium Plasmas," *Chem. Phys.* V. 34, 1978, pp. 193-204.

Capitelli, M., Colonna, G., Hassouni, K. and Gicquel, A. Chem. Phys. Lett., 228, pp. 687-694, 1994.

Drawin, H. W. and Emard, F., "Instantaneous Population Densities of the Excited Levels of Hydrogen Atoms and Hydrogen-like Ions in Plasmas," *Physica C*, Vol. 85C, 1977, pp. 333-356.

[7]Matveyev, A. A. and Silakov, V. P., "Non-equilibrium Kinetic Processes in Low-Temperature Hydrogen Plasma," Preprint 8, IOFAN,Russian Academy of Sciences General Physics Institute, Moscow, 1994.

Gicquel, A., Hassouni, K., Farhat, S., Breton, Y., Scott, C. D., Lefebvre, M., and Péalat, M., "Spectroscopic Analysis and Chemical Kinetics Modeling of a Diamond Deposition Plasma Reactor," *Diamond and Related Materials*, Vol. 3, 1994, pp. 581-586.

Gicquel, A., Chenevier, M., Hassouni, K., Breton, Y., Cubertafon, J. C. abstract submitted to 95 Diamond Films Conference, (Barcelona, 1995)

Gicquel, A., (1996). Section 4 of the present Chapter.

Gorse, C., Capitelli, M., Bacal, M., Bretagne, J., and Lagana, A., "Progress in the Non-equilibrium Vibrational Kinetics of Hydrogen in Magnetic Multicusp H^- Ion Sources," *Chemical Physics*, V. 117, 1987, pp. 177-195.

Hiskes, J. R., "Cross Sections for the Vibrational Excitation of the H_2 $X^1\Sigma^+{}_g(v)$ Levels Generated by Electron Collisional Excitation of the Higher Singlet States," *J. Appl. Phys.*, Vol. 70, 1991, pp. 3409-3417.

Janev, R. K., Langer, W. D., Evans, K., Jr., Post, D. E., Jr., *Elementary Processes in Hydrogen-Helium Plasmas*, Springer-Verlag, Berlin, 1987.

Johnson, L. C. and Hinnov, E., "Ionization, Recombination, and Population of Excited Levels in Hydrogen Plasmas," *J. Quant. Spectros. Radiat. Trans.*, Vol. 13, 1973, pp. 333-358.

Johnson, L. C. and Hinnov, E., "Rates of Electron-Impact Transitions between Excited States of Helium," *Phys. Rev.*, Vol. 187, 1969, pp. 143-152.

Kiefer, J. H. and Lutz, R. W., "Vibrational Relaxation of Hydrogen," *J. Chem. Phys., V. 44, 1966, pp. 668-672.*

Koemtzopoulos, C. R., Economou, D. J., and Pollard, R., "Hydrogen Dissociation in a Microwave Discharge for Diamond Deposition," *Diamond and Related Materials*, Vol. 2, 1993, pp. 25-35.

Lee, J.-H., "Basic Governing Equations for the Flight Regimes of Aeroassisted Orbital Transfer Vehicles," in *Thermal Design of Aeroassisted Orbital Transfer Vehicles*, edited by H. F. Nelson, Vol. 96 of Progress in Astronautics and Aeronautics, 1985, pp. 3-

Millikan, R. C. and White, D. R., "Systematics of Vibrational Relaxation," *J. Chem. Phys.*, V. 39, 1963, pp. 3209-3213.

Massey, H. S. W., *Electronic and Ionic Impact Phenomena*, Vol. II, Electron Collisions with Molecules and Photo-ionization, Oxford, 1969.

Scott, C. D., Farhat, S., Gicquel, A., Hassouni, K. , Lefebvre, M., Péalat, M. AIAA 93-3226, 1993

Tan, W. and Grotjohn, T. A. (1994), 'Modeling the Electromagnetic Excitation of a Microwave cavity Plasma Reactor,' J. Vac. Sci. Technol. A 12(4).

Zhang, J., Huang, B., Reinhard, D. K. and Asmussen, J. (1990), 'An investigation of Electromagnetic field Patterns during Microwave Plasma Diamond Thin Film Deposition,' J. Vac. Sci. Technol. A 8(3).

Chapter 19
SECTION 7.2: SPATIALLY RESOLVED SPECTROSCOPIC ANALYSIS OF THE PLASMA

A. Gicquel
Laboratoire d'Ingéniérie des Matériaux et des Hautes Pressions
CNRS-UPR 1311- Université Paris-Nord
Avenue J. B. Clément 93430, Villetaneuse, FRANCE
M. Chenevier
Laboratoire de spectrométrie Physique
CNRS-URA 008, Université Joseph Fourier de Grenoble
B. P. 87, 38402, Saint Martin d'Hères Cedex, FRANCE
M. Lefebvre
Office National d'Etudes et de Recherches Aérospatiales
B. P. 72, 92322 Châtillon Cedex, FRANCE

7.2.1 Introduction
The spatially resolved spectroscopic analysis of plasmas provides measurements of local conditions in the plasma and at the plasma/surface interface (temperatures, concentrations). Knowledge of these conditions allows the validation of models leading to an increased understanding of the phenomena. Spectroscopic analysis also provides a means for monitoring the reactors. However, since the control of industrial reactors by laser spectroscopy is unrealistic, the development of measurements by optical emission spectroscopy, associated with their calibration by laser diagnostics techniques, is needed.

In low pressure diamond deposition reactors (hot-filament and microwave-assisted) operating with H_2 + CH_4 mixtures, the role of the H atoms and of the carbon species radicals (in particular CH_3 and C_2H_2) have been reported by several authors[1-12,29,36]. As a consequence, these species, as well as the parameters responsible for their production and destruction must be analyzed carefully.

H-atom net production
Owing to the important role of H atoms in diamond deposition, the control of the parameters governing their concentration, production and consumption is crucial. The main elementary processes responsible for their production and destruction in the plasma and at the plasma / surface interface have already been discussed above (section 7.1 and references 12 to 15). The H-atoms production rate is a function, in plasma reactors, of both the electron temperature and the gas temperature, which respectively control the electron impact dissociation and the thermal dissociation of molecular hydrogen. The atoms consumption is partly due to the dissociation in the volume of the CH_4 molecules and the CH3 radicals via H + CHx $\Leftrightarrow$ CHx-1 + H_2(x=3,4) or H + CH3 + M $\rightarrow$ CH_4 + M collisional reactions[6,7,16], which occur at relatively high temperatures. It is also partly due to surface reactions such as etching and catalytic recombination processes[12,13,16,19]. At high pressure, volume recombination reactions might become important.

Experiments for measuring relative H-atom concentrations have already been reported in hot filament and plasma reactors using resonance enhanced multiphoton ionization (REMPI), third harmonic generation[21], coherent anti-Stokes Raman spectroscopy (CARS)[22], two photon allowed transitions laser induced fluorescence (TALIF)[23,24,39,53] and optical emission spectroscopy (OES)[12,26,34,37]. Several absolute measurements of the H-atom mole fraction were performed in hot filament reactors[21-23,27-31] but only one value has been reported in moderate pressure plasma reactors[7]. However, in this latter case, the measurements were carried out at the substrate and the H-atom mole fraction in the plasma volume was deduced from calculations.

Gas temperatures have been already measured in both types of reactors, using CARS[12,22,29,], TALIF[53,39], T_gH (Third Harmonic Generation)[29], degenerate four-wave mixing (DFWM)[43,29], and OES[12,34,35,51].

Methane dissociation and carbon containing species net production

Owing to the relatively low electron density in moderate to high pressure (10 hPa to 1 atm.) H_2 + CH_4 plasma systems, the dissociation of methane, introduced in small amounts (<5 %), is mainly controlled by the gas temperature and, under low power density, by the H-atom density (see later). Its destruction is due to gas phase reactions (effect of pressure and temperature) and surface reactions[12-16,19,]. The production of carbon containing species can then in principle be controlled by the gas temperature, the electron temperature and the H-atom concentration. However, the quality of the diamond films under hot filament and moderate pressure plasma systems seems to depend, beyond the substrate temperature and the H-atom concentration, on the nature of the carbon containing radicals, in particular by the ratio $[CH_3]$ / $[C_2H_2]$[41,17]. Consequently, the direct measurement of the different concentrations of the carbon containing species is necessary for validation of models and for understanding the growth process. However, owing to the diversity of carbon containing species, it is almost impossible to detect and to measure all of them. Laser diagnostics such as laser induced fluorescence (LIF), laser or UV absorption spectroscopy, DFWM[43], cavity ring-down spectroscopy[42] and mass spectrometry[7], have been used for measuring some possible major species for diamond growth (CH_3, CH, C_2, C_2H_2)[42-48]. Optical emission spectroscopy[40,41] can also provide plasma controlling parameters via some of the radiative excited state species. However, again, calibrations are needed for obtaining information concerning the growth species, that is on ground electronic state species.

Contents

In this section, we report measurements of vibrational and rotational temperatures of the ground state molecular hydrogen, of ground and excited state H-atom temperatures and of H-atom relative concentrations using different techniques. Spatial distributions of these parameters as well as their variations as a function of the plasma parameters (power density and percentage of methane) and the substrate temperature are presented.

Ground state H-atom temperature (T_H) are measured by two-photon allowed transition laser induced fluorescence (TALIF) (Doppler broadening), temperature of H-atoms in the n=3 excited state ($T_{H\alpha}$) by OES (Doppler broadening), vibrational and rotational temperatures of the molecular hydrogen in the ground electronic state by coherent anti-

Stokes Raman spectroscopy[1] and rotational temperatures of the $G^1\Sigma_g^+$ electronically excited state of molecular hydrogen by OES. All were measured in identical microwave plasma reactors used for diamond deposition, and can be compared. The spatially resolved ground state H-atom relative concentrations are measured by TALIF.

We also discuss the use of optical emission spectroscopy for measuring relative concentrations of ground state H-atoms. On one hand, optical emission spectroscopy (OES) only gives access to the properties of species in electronic excited states (not involved in the diamond growth process), the concentration of which is less than 10^{-4} relatively to species in the electronic ground state. On the other hand, diamond growth depends only on the characteristics of species in the electronic ground state. Consequently, correlations between the concentrations and the temperatures of the species in the electronic excited state and in the electronic ground state must be established. They allow the proposition of spectroscopic parameters for the use of OES as a controlling and monitoring technique. Validation of the method of actinometry (OES) used for acquiring relative H-atom mole fraction from OES measurements is largely discussed. Emission transition line intensity ratios of H-atom (transition n=3 to n=2) over that of argon (4p to 4s transition) (with argon introduced at 1 % in the gas mixture) are compared to the H-atom relative mole fraction measured by TALIF. Measurements of absolute H-atom mole fractions in microwave diamond deposition plasma reactors as a function of operating parameters are reported for the first time.

7.2.2 Experimental set up and diagnostics

The reactor, made of a silica bell jar low pressure chamber, is presented in figure 1. The feed gas is a mixture of CH_4(0 - 5 %) diluted in hydrogen. The mixture is activated by a

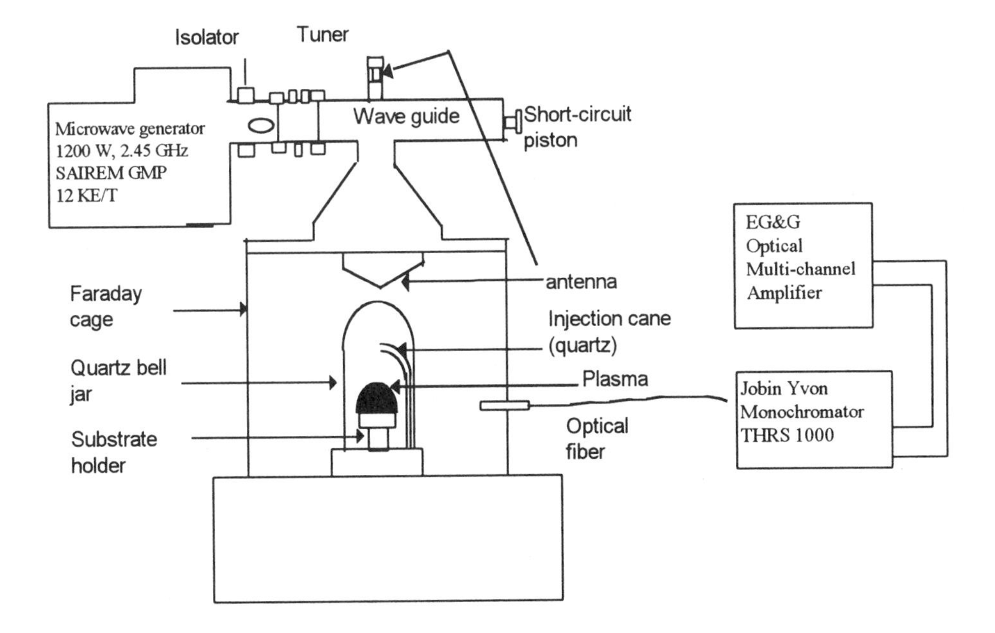

Figure 1: The bell jar microwave reactor.

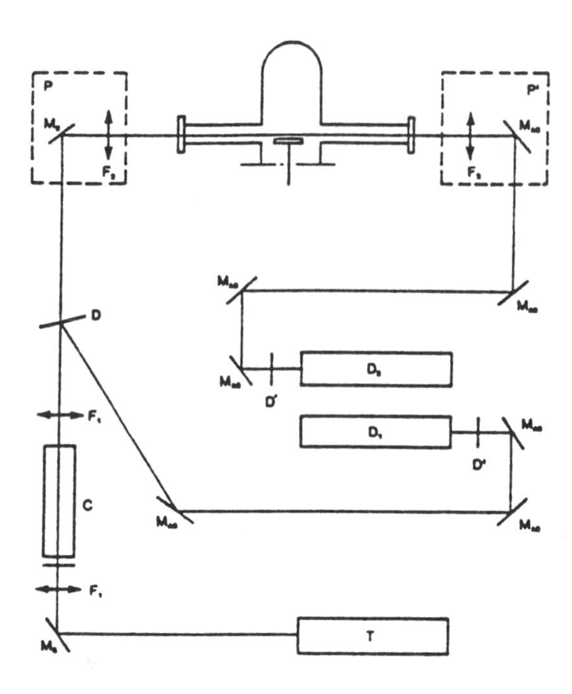

Figure 2: Coherent anti-Stokes Raman spectroscopy experimental block.

1200 W or a 6 kW, 2.45 GHz SAIREM microwave generator. The averaged input power density (in W cm^{-3}), defined as the input power over the volume of the plasma in absence of the substrate holder (plasma ball), was changed by a simultaneous variation of the pressure and the microwave power, keeping the plasma volume constant. The input power over the total density (W cm^{-3}) was kept almost constant. The power was varied from 400 to 2000 W, and the pressure from 14 hPa to 100 hPa. 5-cm-diameter silicon wafers covered by a polycrystalline untextured diamond film were placed inside the plasma ball. Depending on the plasma conditions, in particular the averaged microwave power density, heating or cooling of the substrate holder is used to maintain the substrate temperature constant at the chosen temperature (measured by a bichromatic pyrometer). The substrate temperature was varied from 700 °C to 900 °C, independently from the plasma conditions.

Three techniques were used for analyzing the plasma local parameters: coherent anti-Stokes Raman spectroscopy (CARS), two-photon allowed transition laser induced fluorescence (TALIF), and optical emission spectroscopy (OES).

Coherent Anti-Stokes Raman Spectroscopy (CARS)

Coherent anti-Stokes Raman spectroscopy (CARS)[49,50] has been performed in the diamond bell jar plasma reactor using a modified version of the bell jar, adapted for the laser beam paths. Three sets of 2 silica optical windows of 20 mm diameter were mounted on the silica bell jar allowing an axial analysis. The measurements were performed in a planar boxcar geometry. The probe volume is 15 mm long and 0.2 mm in diameter. The axis of the probe volume is parallel to the diamond substrate, and its location can be moved in the vertical direction (at 90° to the substrate) using micrometric tables.

Rotational and vibrational temperatures of the ground electronic state of molecular hydrogen have been measured by probing Q lines. First, the rotational temperature was deduced from a Boltzmann plot of successive Q lines of the fundamental vibrational level (J= 1 to 6, v=0). Second, the vibrational temperature was obtained by comparing two Q lines (J=3, v=0) and (J=3, v =1). Variations of flow rate, pressure, gas mixture (H_2 + CH_4), microwave power density, and substrate temperature were investigated.

The optical set up, shown in figure 2, is composed of:

- the CARS bench including a Nd:YAG laser and a dye laser, which delivered, at the exit of the bench, respectively 60 mJ, and 1 to 2 mJ. The pulse duration is 12 ns (full width at half maximum), and the repetition rate is 12.5 Hz,
- the reference and the measurement cells mounted in series,
- dichroic plates placed on the laser beams in order to separate the Anti-Stokes signal from the input laser beams (D and D'),
- two identical blocks for the detection of the two CARS signals.

Rotational temperature. Rotational temperature profiles where recorded using the broad band technique. In this situation, each detector block is composed of a spectrometer and a photodiode multichannel analyzer. The spectrometer has a focal length of 0.8 m; it is equipped with a 2100 groves per mm grating and with an internal set up providing a magnification of a factor of 4 before imaging the spectrum on the sensitive surface of the detector (model EGG 1456). The detection system has a resolution of approximately 0.7 cm^{-1} (full width at half maximum) and a dispersion of 0.166 cm^{-1} per diode (1 diode = 25 μm).

For each position of the probe volume, 500 spectra are averaged, and the recording time is only around 1 minute. The accuracy of the measurements of the intensity is around 3 % for the most intense lines. For the sets of rotational levels ranging from J=1 to J=6, the populations are distributed according to a Boltzmann distribution. The measurement accuracy is deduced from the standard deviation associated to the linear regression of the Boltzmann straight line. It is estimated at $\pm$ 25 K at 1400 K, and is twice as good than if spectral scanning of the dye laser were used.

Vibrational temperature. For the vibrational temperature measurements, the acquisition was performed by scanning the dye laser wavelength ($5 \cdot 10^{-3}$ Å) step by step providing a higher detectivity than the broad band at which sensitivity is not enough to probe the rotational levels of v=1. Each detection block is now composed of a monochromator and a photomultiplier (RTC XP 2012 B). In this configuration, the spectral resolution is due to the dye laser width (0.06 cm^{-1}). In order to avoid saturation effects, the beams are attenuated at the entrance of the reactor by a factor of 2.2. A vibrational temperature measurement needs around ten minutes. The rotational lines J= 3 from v=0 and v=1 are recorded successively and, for each spectral position, 5 signal/reference ratios are averaged. The accuracy is deduced from the standard deviation associated to a set of measurements performed under the same experimental conditions, it is estimated at $\pm$ 100 K at 2200 K.

Two photon L.I.F. technique (TALIF)

Two photon allowed transition laser induced fluorescence (TALIF)[51-53] measurements were conducted in a reactor equipped with two U.V. grade silica windows allowing transmission of 205 nm light. The whole reactor was accurately translated vertically and horizontally with respect to the laser beam in order to obtain axial and radial profiles. The spatial resolution is estimated at 0.5 mm.

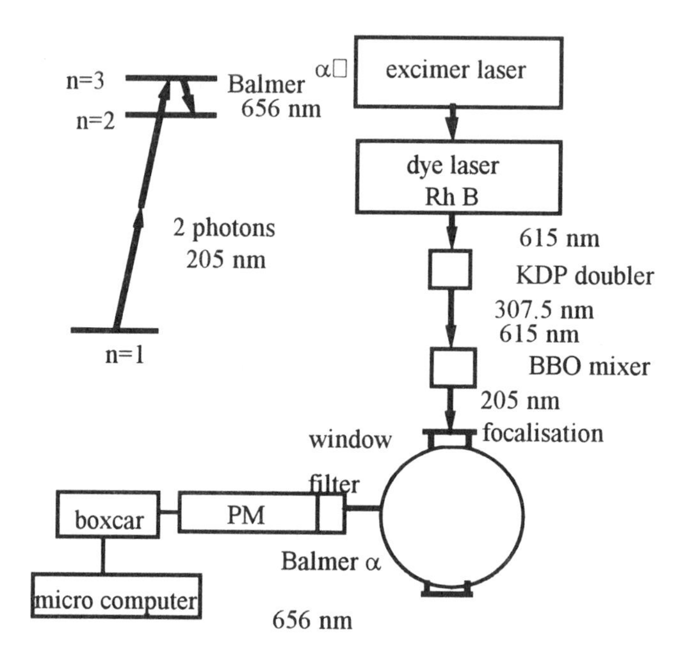

Figure 3: Two photons allowed transition laser induced fluorescence (TALIF) experimental block.

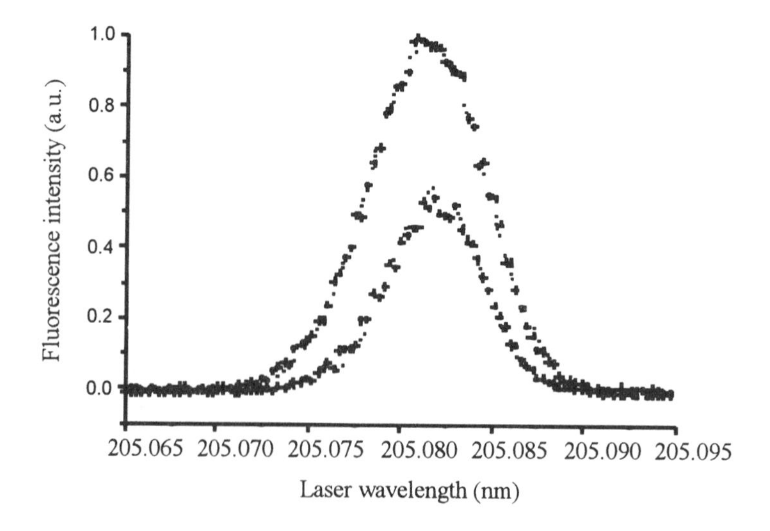

Figure 4: : Doppler broadened TALIF fluorescence signals.

The system used to generate the light at about 205 nm consists of a pulsed excimer laser (XeCl) emitting at 308 nm. This excimer pumps a dye laser composed of a tunable oscillator and an amplifier. With Rhodamine B, an intense beam is obtained at 615 nm which is frequency doubled in a KDP crystal to give 307.5 nm. Mixing with the residual beam at 615 nm in a BBO crystal, produces the 205 nm beam, with a repetition rate of 10 Hz and a pulse duration of about 25 ns; the energy and line width are typically 50 μJ per pulse and 0.0025 nm respectively. The whole laser system is computer controlled and allows wavelength scanning over a few nm. In our experiment a scan of 0.030 nm is sufficient to cover the line profile with 90 steps. The profile line is obtained with a laser wavelength varying from 205.065 nm to 205.095 nm. The fluorescence light is collected at 90° to the laser beam by two lenses and detected directly by a photomultiplier in front of which an interference filter centered at 656.5 nm, which is used to eliminate scattered laser light. The resulting signal is processed by a boxcar integrator and sent to a personal computer. The fluorescence signal is averaged over 20 laser shots for each wavelength step, thus a scan takes about 3 minutes. A block diagram of the experimental set up is presented in figure 3[14,15]. Under the experimental conditions, the observed two-photon line profile (figure 4) is Doppler broadening dominated. Its full-width at half-maximum is directly related to the translational temperature of the atoms, and its area to their concentration.

Width and Temperature. As atomic hydrogen is the lightest atom, the Doppler broadening line is particularly large, and this is therefore advantageous for temperature measurements from line profiles. By measuring the Doppler line width $\Delta\lambda_D$ (full width at half maximum, FWHM) of the fluorescence excitation profile, the H-atom temperature can be determined using the formula :

$$\frac{\Delta\lambda_D}{\lambda_o} = \frac{2}{c}\sqrt{\frac{2kT\ln 2}{m}} = 7.16\cdot 10^{-7}\sqrt{\frac{T}{M}} \tag{1}$$

where c is the speed of light, k the Boltzmann constant, m is the mass of the atom, T is the temperature of the atom and M is the mass of the atom in amu.

We estimate the magnitude of $\Delta\lambda_D$ by applying Eq.(1) to the H atom at 1500 K for the doubled $1\,^2S \rightarrow 3\,^2S\ (^2D)$ H atom transition, i.e. for $\lambda_0 = 205.14$ nm in vacuum (205.082 in air). A value of 0.0057 nm is obtained for $\Delta\lambda_D$. For our experimental conditions, i.e. pressure of about 25 mbar, the atomic fine structure and the collisional broadening, which are of the order of 0.00015 nm for the transition involved, can be neglected. Therefore, the line profile can be considered as a pure Gaussian. By fitting the experimental profile with a Gaussian profile, this becomes:

$$f(\lambda) = a + b\exp\left[-4\ln 2\left(\frac{\lambda - \lambda_o}{\Delta\lambda_R}\right)^2\right] \tag{2}$$

which depends on four parameters (a, b, λ_0, $\Delta\lambda_R$) , where $\Delta\lambda_R$ is the resulting LIF signal FWHM. $\Delta\lambda_R$ is directly related to the Doppler broadening by the relation:

$$\Delta\lambda_R^2 = \Delta\lambda_D^2 + \Delta\lambda_L^2 \quad (3)$$

By deconvolving $\Delta\lambda_R$, we can obtain $\Delta\lambda_D$ and estimate the H atom temperature. For the determination of $\Delta\lambda_L$ (laser width) we performed measurements with a room temperature atomic hydrogen source and a value of 0.0025 nm was obtained.

Owing to experimental difficulties, the error in the measurements from the long-term reproducibility is quite large, it is estimated at $\pm$ 250 K, at 2000K.

Optical Emission Spectroscopy (OES)
A Jobin Yvon THR 1000 mounted with an 1800 groves per mm grating, blazed at 450 nm and equipped with a photomultiplier (Hamamatsu R 3896) and a red region intensified OMA III (EGG-Princeton Instrument -1460) was used to perform emission spectroscopy. The light emitted from the plasma was collected by a one millimeter optical collimator and transported via an optical fiber to the entrance slit of the monochromator. This device enables a spatial resolution in the plasma emissive cylinder of approximately 2 mm in diameter. The optical system was mounted on a computer controlled moving table, allowing axial and radial measurements. Emission intensities profiles, averaged on the line-of-sight, were measured at 90° to the axis of the reactor. An Abel Inversion procedure was applied on these profiles in order to obtain radial distributions of the emission intensity of the excited species[12,35,41].

Concentrations. The emission intensities of H_α (λ= 656.5 nm) and H_β (λ= 486.1 nm) were recorded systematically using the optical multi-channel analyzer. Argon was introduced at 1 % in the plasma and the 4p $\rightarrow$ 4s argon transitions (λ=750.3 nm (2p1 $\rightarrow$ 1s2) and λ=811.5 nm (2p9 $\rightarrow$ 1s5) (in Paschen notation) were measured.

Argon was chosen as an actinometer for the H-atom. To be a good actinometer, argon must induce no perturbation in the plasma, and, like the H-atom, must be excited only by a direct electron impact process. A discussion on the validity of actinometry for measuring the H-atom relative mole fraction, based on theory and experiments, under these experimental conditions, is presented below.

Temperature of the G electronically excited state of molecular hydrogen. When performing rotational temperature measurements, the resolution of the optical system was around 0.6 nm. The optical device was not calibrated, however owing to the reduced spectral domain [450 nm, 470 nm] involved for the G state rotational temperature measurements, we have neglected any variation as a function of the wavelength in the detection system sensitivity.

Under typical operating conditions (25 hPa, gas temperature > 2000 K), the plasma with a neutral gas density of around 10^{17} cm^{-3}, is far from local thermodynamic equilibrium. However, we can expect that at this pressure, the quenching rate is high enough to equilibrate the rotational modes of the long lived electronic excited states of H_2 with the heavy particles kinetic mode. This is the case for the electronic ground state. If it is the case for the electronic excited sates, the plasma heavy particle temperature should be determined from the rotational bands of the radiative excited state of H_2. Under diamond growth conditions, two strong emission bands of H_2 corresponding to $G^1\Sigma_g^+ \rightarrow B^1\Sigma_u^+$ (0,0) and $d^3\pi_g \rightarrow a^3\pi_g^+$ (0,0) transitions can be used. No direct relationship between the rotational temperature measured in the d state (Fulcher-α system) and that of the ground electronic state was found. Then, only the G state was extensively studied. The life time of the G state is 33.3 ns, and, for the range of pressure considered here, should be high enough to allow the rotational distribution to reflect the gas kinetic temperature[38]. A discussion on that is given later.

The bands are supposed to be optically thin and most of their lines are not blended with other systems[38]. The temperature was determined using the R branch of the $G^1\Sigma_g^+ \rightarrow B^1\Sigma_u^+$ (0,0) electronic transition. The $B^1\Sigma_u^+$ level is well described by Hund's case b approximation. This is not the case for the $G^1\Sigma_g^+$ level which is intermediate between the Hund's limiting cases b and d[56,57], so the rotational energy is no longer a linear function of K(K+1). So, one has to consider the exact numerical values of the rotational energy levels in the temperature Boltzmann plot[51-60]. Although these levels are perturbed by the high vibrational levels of the EF $^1\Sigma_g^+$, the strength of most of the R branch rotational lines can be well described by the Höln-London formulae, i. e. $S_k = (K+1)/2$. Considering the nuclear spin degeneracy $(T_k)^{12}$, the intensity of the Kth rotational emission line can be expressed by : $I_k = (K+1)(2T_k+1)\exp[-E_{K+1}/kT_{rot}]$.

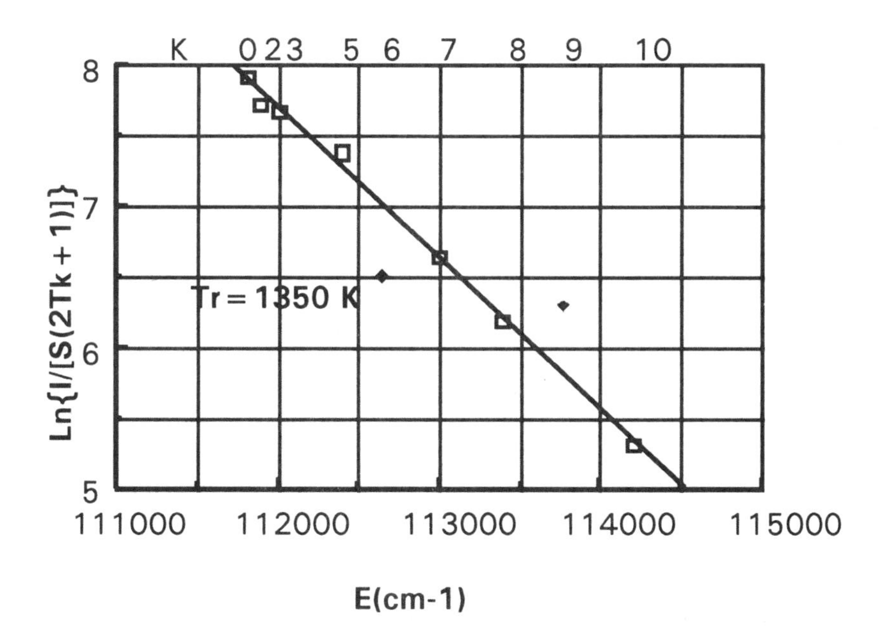

Figure 5: Boltzmann diagram for the determination of the rotational temperature of the $G^1\Sigma_g^+$ electronically excited state of molecular hydrogen. Plasma conditions: 600 W, 25 mbars, 9W cm^{-3}, axial position 20 mm from the substrate, radial position at the edge of the plasma (22.5 mm from the axis).

Only ten emission lines (R_0 - R_{10}) were identified. The R_1 and R_4 lines were not resolved and were not used in the Boltzmann plot. The R_6 and R_9 lines show a large deviation from the Boltzmann plot obtained with the ten first lines. This phenomenon was observed by other authors[38] and was attributed to the perturbation by the nearby high vibrational levels of the $EF^1\Sigma_g^+$ electronic state. Neither the R_6 or R_9 lines were considered in $H_2(G)$ rotational temperature measurements. Figure 5 shows a typical Boltzmann plot of $\ln(I / [Sk\ (2T_k + 1)])$ versus E_{k+1} for the R branch. The accuracy on these measurements, mainly due to reproducibility, is estimated at $\pm$ 200 K, at 1500K.

Abel inversion procedure. The volume of the plasma created in the bell jar reactor is axi-symmetric, thus allowing the use of the Abel inversion method, which gives the radial distribution of each line intensity. The emission intensities of the band are strong enough to allow the determination of the radial distribution of the rotational temperatures of $H_2(G)$. Values of the temperature on the axis of the reactor can be compared to those of T_R H_2 (X) obtained by CARS, and to T_H measured by Doppler broadening from TALIF measurements.

For a given axial position, we determined line-of-sight averaged intensity profile I(y)[61] (y represents the position at 90° to the axis) of all the lines involved in the Boltzmann plot of the rotational temperature. The profile is then separated into two parts. The first part corresponds to the central region of the plasma, and was fitted with a fourth order polynomial function. The second part which corresponds to the edge of the plasma was fitted with a third order polynomial function. The intensity value at the edge of the plasma was imposed equal to zero by subtracting the residual intensity due to reflection and diffraction phenomena. The radial profile of the intensity is then deduced using the following Abel-Olsen transformation:

$$I(r) = -1/^{1} \int_{r}^{R} \frac{dI(y)/dy}{(x2-r2)}\, dx \tag{4}$$

Measurements of the temperature of H-atom in the n=3 excited state (H_α-atom temperature). The use of the THR 1000 equipped with the photomultiplier and a Pelletier effect cooling system allowed us, with spectrometer slit apertures of 1.5 μm, to reach a resolution of around 5 pm, making possible measurements of Doppler broadening on H atoms during the radiative decay from the n=3 to n=2 fluorescence line (H_α).

As explained before, the Doppler broadened peak is a Gaussian, its FWHM is given by equation (1). Owing to the finite resolution of the spectrometer, the peak is a convolution of the line with the spectrometer response function. This latter function has been approximated to a Gaussian function, and was determined using a low pressure mercury lamp (the natural broadening is known). At λ = 5790 Å, the broadening generated by the optical system is given by:

$$\Delta\lambda^2_{opt.syst.} = \Delta\lambda^2_{measur.} - \Delta\lambda^2_{natural} \tag{5}$$

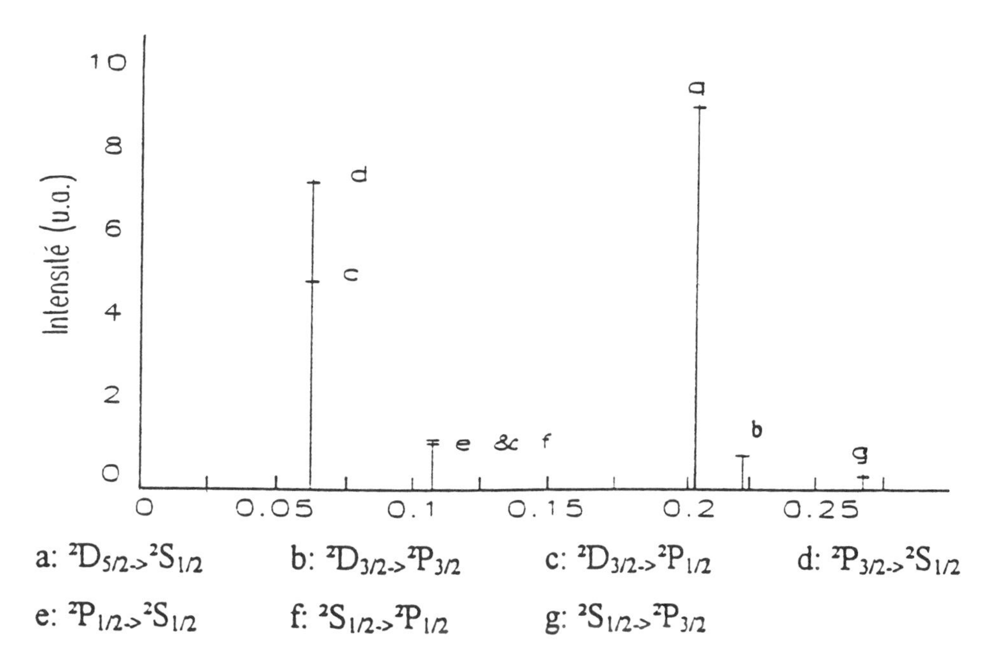

Figure 6: Fine structure of H_α(according to 34).

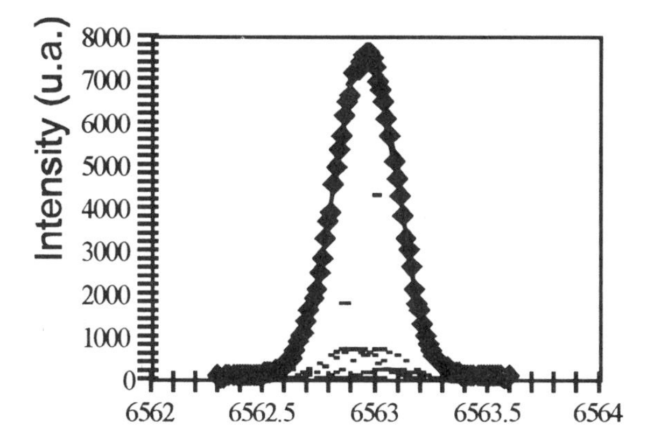

Figure 7: Experimental Doppler broadened spectrum of H_α transition and the decomposition in seven features.

We found $\Delta\lambda_{opt.syst.}$ = 6.5 pm. This value has been confirmed later by a measurement using a He-Ne laser. Owing to the fine structure of the H_α line (figure 6), a systematic decomposition of the spectrum must be done. According to Röpcke et al.[34, 35] and E. U. Condon et al.[100, 36], the spectrum has been decomposed into seven Gaussian functions. For each of the components, the FWHM was obtained after subtracting the optical system broadening.

Finally, the FWHM is the result of the Gaussian Doppler broadening, the Lorentzian pressure broadening and the Lorentzian Stark broadening. The contribution of the pressure broadening as well as the Stark broadening have been estimated. According to[62,102,103], the pressure broadening is given by:

$$\Delta\lambda_p = n\, \sigma_{H\alpha/H2}\, \lambda_0^2\, v_{H\alpha/H2} / \Pi = 1.54 \cdot 10^{-7}\, \lambda_0^2\, n\, \sigma_{H\alpha/H2}\, (T/\mu_H)^{1/2} \quad (6)$$

where n is the total density (in m^{-3}), $v_{H\alpha/H2}$ the relative mean velocity, the quenching cross section of H_α by H_2 molecules (in m^2), and the reduced mass (in g). $\Delta\lambda_p$ is given in m. It has been evaluated at less than 1 pm over the operating conditions used here. The contribution of the Stark broadening has been estimated from expressions given by Wiese[104]:

$$\Delta\lambda_S = 5 \cdot 10^{-9}\, \alpha_{1/2}\, ne^{2/3} \quad (7)$$

where ne is the electron density (in cm^{-3}), $\alpha_{1/2}$ depends on the principal quantum number (we took $\alpha_{1/2} = 0.018$ for the H_α line)[104] and $\Delta\lambda_S$ is given in Angström. Over the conditions used here, $\Delta\lambda_S$ has been estimated at up to 2 pm.

An experimental broadened peak corresponding to the H_α emission is given in figure 7, where the decomposition of the spectrum with the seven features is shown. After subtraction of the approximately Gaussian contribution of the optical system, the Lorentzian contributions were subtracted in order to determine the H_α temperature (from Doppler broadening). For each measurement of the H_α -atom temperature, nine spectra were recorded (each in 4 minutes), and an averaged value for the temperature was calculated. The error on the absolute measurement has been estimated at around 10 %[107]. Note that only line-of-sight averaged H_α-atom temperature are available to date. The determination of real axial temperature would need mathematical treatment of the line-of-sight signals (amplitude and width).

7.2.3 Data Analysis: Temperature Measurement Results

Spatial distributions of the rotational and vibrational temperatures of molecular hydrogen in the ground electronic state $H_2(X)$. The $H_2(X)$ rotational and vibrational temperatures (T_RH_2 (X) and T_VH_2 (X)) were measured by CARS along the axis of the reactor from the plasma bulk to the diamond substrate for typical conditions of diamond deposition except the absence of CH_4 in the gas.

Typical axial distributions are given in figure 8 for the following experimental conditions: power 600 W, pressure 25 hPa, power density of 9 W cm^{-3}, flow rate =300 sccm, substrate temperature = 900°C. The rotational temperature reaches 2150 K $\pm$ 50 K and the vibrational temperature 2250 K $\pm$ 100 K in the bulk of the plasma. A quasi-equilibrium between these two temperatures is observed: the vibrational temperature is higher than the rotational by not more than 100 K.

In the vicinity of the substrate, the rotational temperatures decrease and approach the substrate temperature, while at 0.6 mm from the diamond substrate the vibrational temperature remains around 200 K to 250 K higher than the rotational temperature. The accommodation coefficient for the rotational-translational energy is then equal to 1, while that of the vibrational mode appears to be lower. The rotational - translational

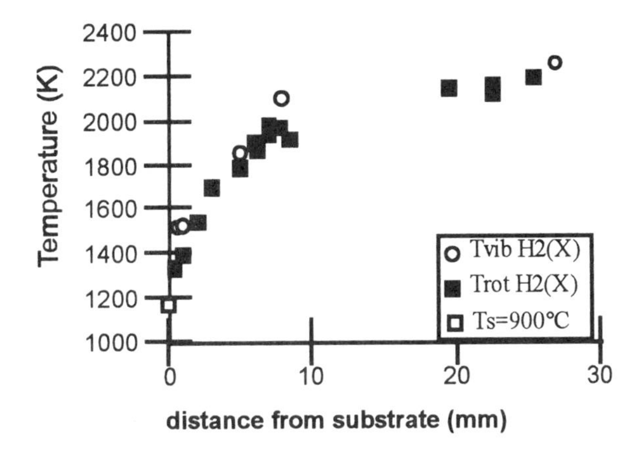

Figure 8: Axial $H_2(X)$ vibrational and rotational temperature distributions. (600 W, 25 hPa, 9 W cm^{-3}, 300 sccm, Ts= 900 °C) in pure hydrogen plasma. Interaction with an untextured diamond surface.

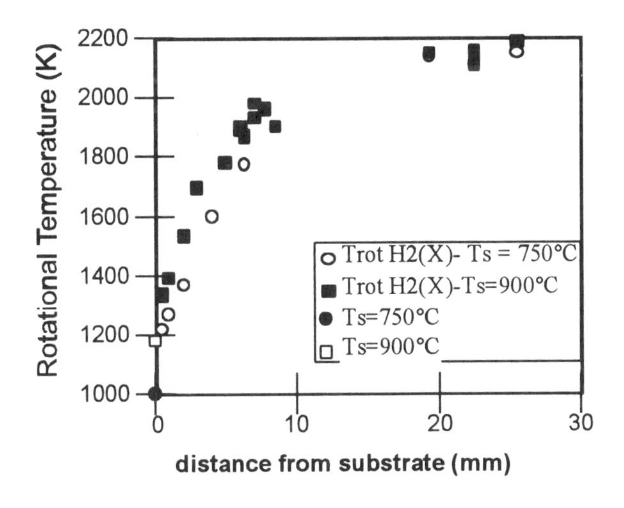

Figure 9: Axial distributions of $H_2(X)$ rotational temperatures for the same plasma conditions (600 W, 25 hPa, 9 W cm^{-3}, 300 sccm) in pure hydrogen plasma and two substrate temperatures. Interaction with an untextured diamond surface.

energy transfer from the plasma to the surface is estimated at 18 W cm^{-2}, taking $5 \cdot 10^{-3}$ W cm^{-1} K^{-1} as a value for the thermal conductivity.

Influence of the substrate temperature: Thermal boundary layer thickness. The thickness of the boundary layer has been estimated through measurements of the axial distribution of the $H_2(X)$ rotational temperature when applying the same plasma conditions and two different substrate temperatures (1173 K and 1000 K) (figure 9). At a distance more than 15 mm from the surface, the plasma characteristics are no longer

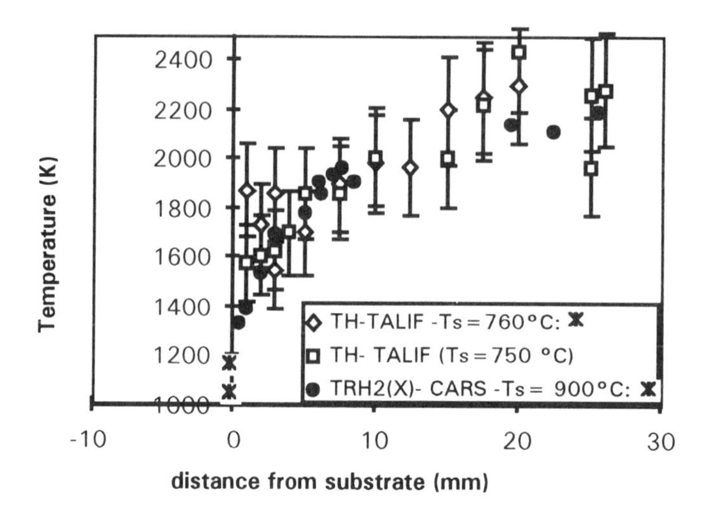

Figure 10: Axial distributions of the rotational temperature of ground electronic state molecular hydrogen ($T_RH_2(X)$) and of the ground state H-atom temperature (T_H). Plasma conditions: 9 W cm^{-3}, 600 W, 25 mbar, 300 sccm, H2, interaction with an untextured diamond surface, positioned at 0 mm.

influenced by its presence. Taking $\delta_T = T_\infty - T_{surf}/(\partial T/\partial x)_{surf}$ as a definition for the thermal boundary layer thickness[16], we found an estimate of 6 mm.

Comparison between H-atom temperature and the rotational temperature of the ground electronic state of molecular hydrogen. In the presence of a substrate holder, axial distributions of the H-atom temperature (T_H) measured from Doppler broadening obtained by TALIF are reported in figure 10 for the same set of conditions as before (9 W cm^{-3}). They are compared to T_RH_2 (X) axial distribution, measured by CARS. In the plasma volume, T_H is equal to T_RH_2 (X). The H-atom temperature is then in equilibrium with the ground electronic state rotational temperature of molecular hydrogen, which is equal to the gas temperature. Both T_RH_2 (X) and T_H reflect the gas kinetic temperature.

At the interface, although the error bars are large, the gradient of T_H appears much lower than that of T_RH_2 (X). This might indicate a lower energy accommodation for the translational mode of the H-atoms than for the translational-rotational mode of the H_2 molecules.

Influence of the diamond substrate - holder position on the spatial distributions. Figure 11 shows the measurements of the H-atom temperature (T_H), in the presence and in the absence of the substrate holder maintained at 900 °C. In the absence of the substrate holder, T_H is slows down less than in its presence. The effect of the presence of the substrate on the translational temperature is given by the difference between the axial distribution of T_H and the axial distribution of $T_{H2}(X)$. The fact that the plasma characteristics are not influenced by the substrate beyond 15 mm from the substrate is

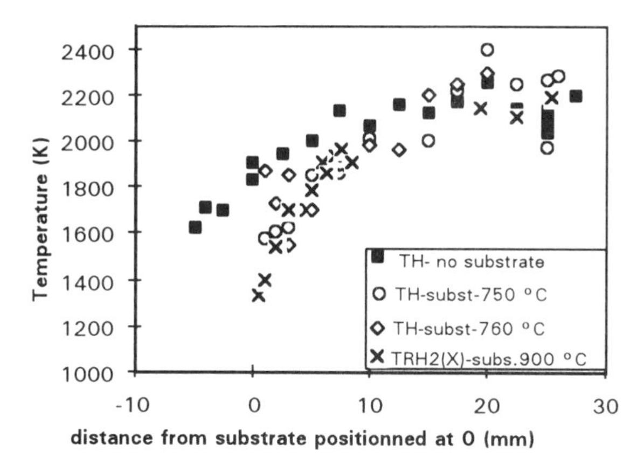

Figure 11: Axial distributions of the rotational temperature of the electronic ground state molecular hydrogen ($T_RH_2(X)$) and of the ground state H atoms temperature (T_H), in presence and in the absence of the untextured diamond substrate positioned at 0 mm. Plasma conditions: 9 W/cm^3, 25 mbar, 300 sccm, pure H_2.

confirmed here. This feature shows that the substrate holder does not actively participate in the microwave plasma cavity. The results also indicate that convective transport phenomena in the bulk of the plasma might not be completely negligible.

Comparison of the radial distributions of Abel inverted G electronically excited state of molecular hydrogen and H-atom temperature. The radial distribution in the bulk of the plasma of the Abel inverted G electronically excited state of molecular hydrogen ($T_RH_2(G)$) is compared to that of H-atom temperature (figure 12). Abel inverted $T_RH_2(G)$ is less than the H-atom temperature by 550 K (or $T_RH_2(G) = 0.75\ T_RH_2(X)$), however, despite large error bars, the shapes of the radial profiles are rather similar. An important non uniformity in the radial distribution of temperatures is shown in the bulk of the plasma. At the interface (5 mm from the substrate), the non uniformity is strongly reduced.

Variations of $T_RH_2(X)$, $T_RH_2(G)$ and T_H as a function of the plasma parameters. The $H_2(X)$ rotational and vibrational temperatures measured in the bulk of the plasma (20-25 mm from the substrate) were not influenced by the flow rate when varying from 50 to 300 cm3 / min, indicating that even at 300 cm3 min-1, the residence time is high enough to reach the equilibrium regime of vibrational and rotational excitation. They were also not influenced by the substrate temperature, probably indicating that free convection transport phenomena cannot be completely neglected is the bulk of the plasma.

Variations as a function of the methane percentage. The $H_2(X)$ rotational temperature and the H-atom temperature were measured in the bulk of the plasma (at a distance of 20 to 25 mm from the diamond substrate), as a function of the CH_4 percentage. As this varied from 0 to 5 %, the temperatures remain constant at around 2150 K (figure 13).

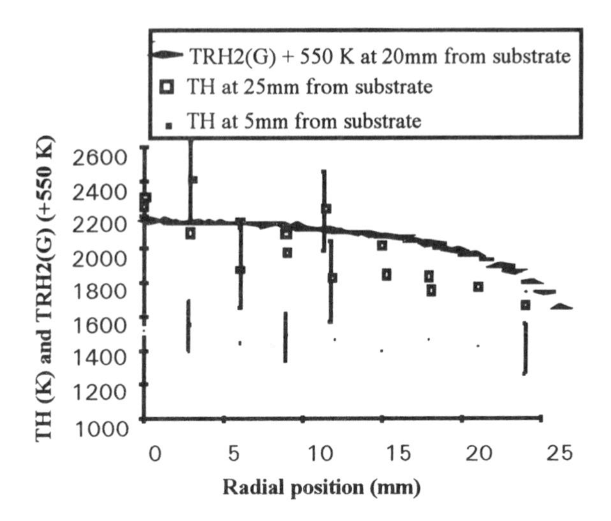

Figure 12: Radial distributions of the ground state H-atom temperature (T_H), at two axial positions from the diamond substrate (25 mm and 5 mm from the substrate), and of the rotational temperature G electronically excited state of molecular hydrogen increased by 550 K, measured at 20 mm from the substrate. Plasma conditions: 9 W/cm3, 600 W, 25 mbar, 300 sccm, H2.

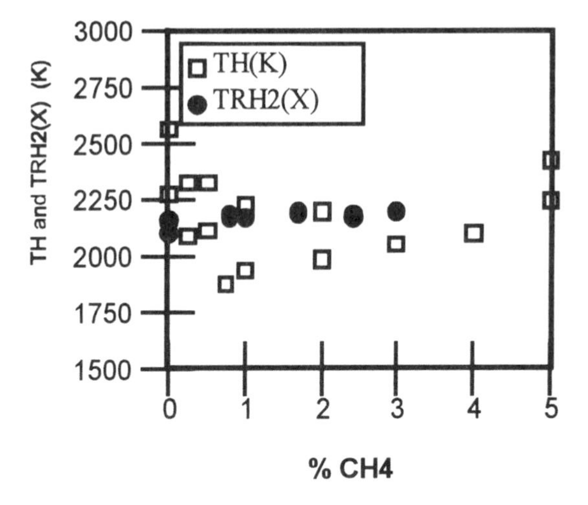

Figure 13: Variations of the H2(X) rotational temperature (CARS) and H-atom temperature T_H (TALIF) as a function of the percentage of methane in the feed gas.

Note that the H-atom temperatures measured by Doppler broadening (TALIF) are more scattered than that of $T_RH_2(X)$.

Variations as a function of the microwave power density. The variations of T_H, measured at 25 mm from the substrate (bulk of the plasma), as a function of the averaged microwave power density are presented in figure 14. T_H increases from 1550 $\pm$ 180 K to 2600 $\pm$ 300 K as the power density increases from 6 to 15 W cm^{-3}. This figure reports measurements of the $H_2(X)$ rotational temperature measured by CARS at 22.5 mm from the substrate (bulk plasma) (only one value of 2150 K) and that of $H_2(X)$ rotational temperature performed at 6 mm from the substrate (in the thermal boundary layer at the interface). Note that between 20 and 25 mm from substrate, i. e. in the bulk plasma, there is no effect of the position on the plasma characteristics (see figures 9-11). Although lower than that in the bulk of the plasma, the values at the interface follow exactly the variations of T_H. The values of the gas temperature in the plasma volume, calculated using the 1D diffusive model presented in sub-section 7.1, are also reported in figure 14. There is very good agreement between the measured and the calculated gas temperatures, for the range of power densities where $T_R H_2(X)$ and T_H were measured ($<$ 15 W cm^{-3}). At 30 W cm^{-3}, the calculated gas temperature in the bulk of the plasma reaches 3100 K.

The line-of-sight averaged H_α-atom temperature can be deduced first from the experimental H_α peak FWHM (measured by OES) assuming that the broadening is Doppler dominated. In the bulk of the plasma (at 20 mm from the substrate holder), it ($T_{H\alpha}$ without corrections) varies from 1680 $\pm$ 170 K to 3900 $\pm$ 400 K as the power density varies from 4.5 to 35 W cm^{-3} (figure 15)[101]. The higher the power density (the pressure), the higher the difference between $T_{H\alpha}$ without corrections and the gas temperature, indicating that pressure and Stark broadening effects are probably not negligible. Using relations (6) and (7), we can estimate the contributions of the pressure and Stark broadenings to the H_α-FWHM as a function of the power density (pressure and temperature variations). The temperatures ($T_{H\alpha}$ pressure and Stark effects corrected) are reported on figure 15. It is equal to T_H=T_g at power density less than 16 W cm^{-3}, and seems slightly higher than the calculated T_g at higher power density (by around 300 K), although this difference still ranges within the error bar. It varies from 1540 K $\pm$ 150 K to around 3500 K $\pm$ 350 K as the power density increases from 4.5 to 35 W cm^{-3}. An error of around 10 % in the power density must be considered owing to the difficulty in measuring the exact volume of the plasma.

The fact that $T_{H\alpha}$ seems higher than T_g might be due to underestimated Stark effect. Production of very hot H_α atoms needs strongly energetic repulsive dissociative excitation processes (threshold at around 27 eV)[35,105] which are unrealistic in such plasmas. In addition, the higher the pressure the less their contribution, owing to the decrease in the electron temperature (calculations in sub section 7.1). However, the increase of the electron density (and then of the Stark effect) as a function of the power density (increase of pressure) should remain limited in microwave plasmas owing to the cut off frequency (see subsection 1).

The variations of Abel inverted $T_R H_2(G)$, measured at 20 mm from the substrate (bulk plasma), as a function of the averaged microwave power density are presented in figure 16. The Abel inverted G state rotational temperature follows rather well the variations of T_g, although its value is always lower by 550 K (T_g remain around 3/4 T_g), for the range of experimental conditions tested here. This result disagrees with those of Chu et al.[38] who found an equilibrium between the $T_R H_2(G)$ and the rotational temperature of the CN(B) system measured by laser induced fluorescence, which is in principle equal to the gas kinetic temperature.

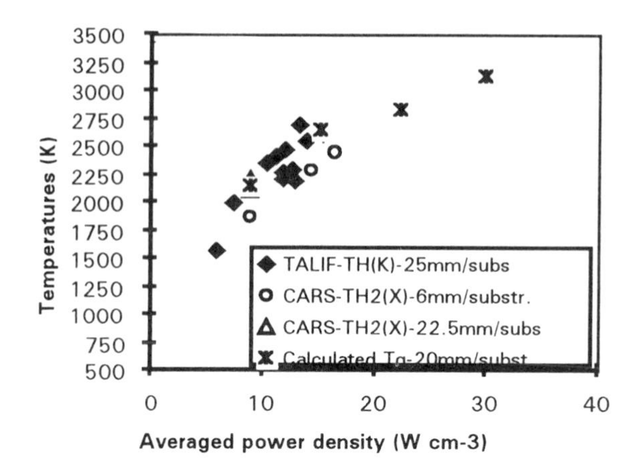

Figure 14: $T_R(H2(X))$ at 6 mm from the substrate (CARS), T_H (Doppler broadening: TALIF) measured at 25 mm from the substrate, and calculated gas temperature (1 D diffusive model) as a function of the averaged input microwave power density. One value of $T_R(H2(X))$ measured at 22.5 mm from the substrate (CARS) is reported for P_{MW}= 9 W cm^{-3}. The temperatures are measured on the axis of the plasma at distances of 20 to 25 mm from the substrate, i;e. in the bulk of the plasma where there is no dependence on the position.

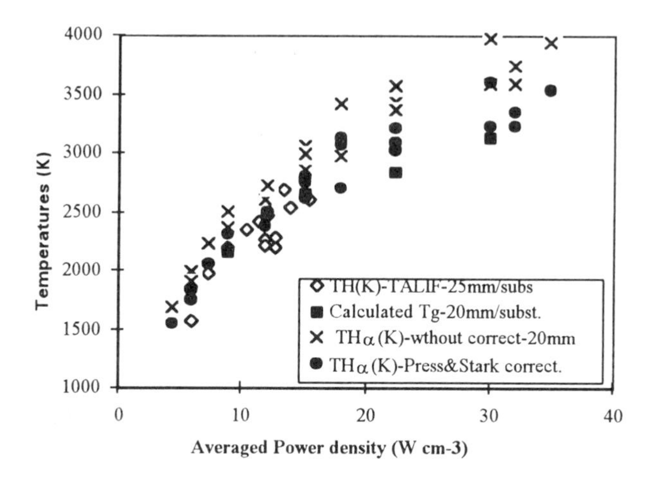

Figure 15: T_H (TALIF) and line-of-sight $T_{H\alpha}$ (OES) (both from Doppler broadening) measured at distances of 20 to 25 mm from the substrate in the bulk of the plasma, where there is no dependence of temperature on the location.

Discussion of the G State Rotational Temperature Measurement

The determination of the rotational temperature of an excited state requires measurements of the line emission intensity for different K rotational numbers. For a given K number, the emission intensity is a result of excitation processes (electron impact, other processes), processes leading to energy transfers between molecules (Vibration-Rotation (V-R), Rotation-Rotation (R-R) or Rotation-Translation (R-T)), and

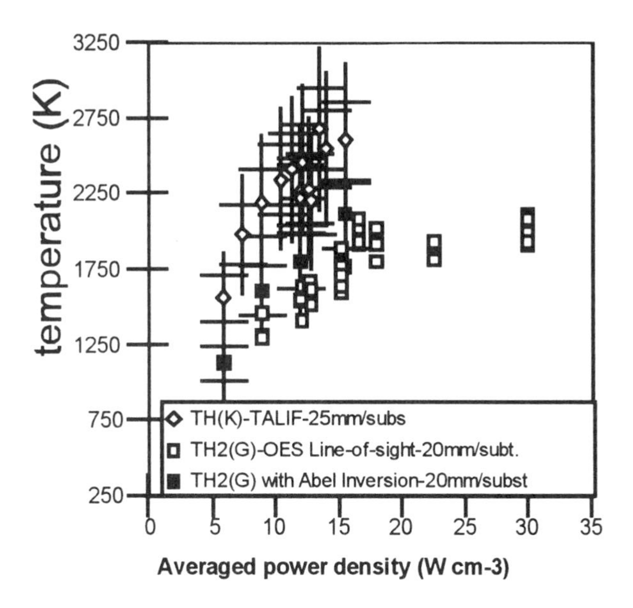

Figure 16: T_H (Doppler broadening) measured at 25 mm from the substrate, and Abel inverted $T_{H2}(G)$ and line-of-sight $T_{H2}(G)$ (OES) measured at 20 mm from the substrate as a function of the averaged input microwave power density. The temperatures are, except for the line-of-sight averaged $T_{H2}(G)$, measured on the axis of the plasma in the bulk of the plasma.

radiative de-excitation mechanisms. Auto-absorption process could also be involved, however due to the small radiative lifetime of the B electronic state (1 ns)[62], we will consider it negligible here.

Excitation processes. Otorbaev et al.[63] have studied the electron-impact excitation levels of the rotational levels of molecular hydrogen in detail. They demonstrated that during the electron-impact excitation of the G state, there is a dependence of the cross section of the molecule on the rotational quantum numbers, owing to momentum transfer from the electron to the molecule. However, for high gas kinetic temperature (kT_g > rotational constant of the ground electronic state B^0; that is here for $T_g > 800$ K), they claim that the rotational distribution of the excited state is near a Boltzmann distribution and have demonstrated it for the $d^3\pi_u$ state of molecular hydrogen. Thus, at low pressure (no thermalization) and high temperature, the rotational distribution of the G state should reproduce that of the ground state. In that case, the excited state rotational temperature is linked to that of the ground state by the relation $B^0\, T_{rot} = B_v\, T_{rot}^0$, where B_v and B^0 are respectively the rotational constants of the excited state (for the G state, $B_v = 28.4$ cm^{-1}) and of the ground state ($B^0 = 60.8$ cm^{-1}), and T_{rot} and T_{rot}^0 respectively the rotational temperatures of the excited state and the ground state.

De-excitation. The radiative life time of the rotational levels can also depend on the rotational quantum number. For instance, the radiative life time of $G^1\Sigma_g^+$ (v=0, K=1) is 27 ns, while that of $G^1\Sigma_g^+$ (v=0, K=2, 3) is 39 ns[62]. This (independently of the upper state rotational distribution) can lead to a systematic error in the determination of the slope of the Boltzmann plot, that is, on the G-state rotational temperature. Further information on the oscillator strength of this radiative state as a function of the K number are needed to take this effect into account.

Thermalization. At high pressure, molecule/molecule collisions involve, besides momentum transfers, V-T, R-R and R-T energy transfers, which may lead to the thermalization of the rotational distribution of the excited state. In the case of a complete thermalization, the rotational Boltzmann distribution of the excited state is equal to that of the gas kinetic temperature, and $T_RH_2(G) = T_RH_2(X) = T_g$.

Under diamond deposition conditions, a difference of 550 K between $T_RH_2(G)$ and $T_RH_2(X)$ is observed, in other words $T_RH_2(G) = 0.75\ T_RH_2(X)$ indicating probably that the thermalization is efficient even if not still completed.

Let us compare the R-R and T-R collision frequencies to the decay rate of the G-state, owing to the radiative decay ($3 \cdot 10^7\ s^{-1}$) and to the quenching decay rate. The quenching decay rate is not known for the G state, but it should be of the same magnitude as the elastic collision frequency ($\nu = 6.8 \cdot 10^7\ s^{-1}$ at $T_g = 2200K$, P= 25 hPa, with a cross section[54] of $17 \cdot 10^{-16}\ cm^2$). Thus, under diamond deposition conditions, the lifetime of the G state should be mainly controlled by the quenching process, and not by the radiative process. Its is estimated at around 14 ns instead of 33 ns (radiative life time). The T-R energy transfer cross sections ($3 \cdot 10^{-17}\ cm^2$ at around 2000 K[54]) are around 60 times lower than that of the elastic collisions[54]. Thus, the time needed for R-R energy transfer is around 60 times longer than the time needed for momentum exchange. The same magnitude is found for R-R energy transfers (factor of 100 instead of 60)[54,55].

Consequently, due to the weak life-time of the G-state owed to quenching, thermalization is not efficient at all. In that case, $T_RH_2(G)$ should be half of $T_RH_2(X)$. The experimental results show that we are in an intermediate situation. This indicates that either, as reported by Otorbaev, the excitation cross sections depend on the rotational quantum number, or the de-excitation probabilities depend on the rotation quantum number. For going further, cross sections of the electron-impact excitation and the radiative probabilities, both function of K, are needed.

Conclusion. The determination of the gas kinetic temperature through measurements of the G state rotational temperature of molecular hydrogen appears to be difficult. The difference of 550 K between $T_RH_2(G)$ and $T_RH_2(X)$ or the value of the proportionality factor B between $T_RH_2(G)$ and $T_RH_2(X)$) of around 0.75 (instead of being equal to 0.5) cannot be attributed to an efficient thermalization occurring during molecule / molecule collisions, but either to the variation of the excitation cross sections or to that of the radiative probabilities as a function of the rotational quantum number

Nevertheless, our experimental results show that even if the G state rotational temperature measurement does not provide the gas kinetic temperature, its evolution as a function of the averaged power density is rather similar to that of the gas temperature. The proportionality factor B remains almost constant (0.72 to 0.77) as the pressure increases from 14 to 52 hPa (the power density varies from 6 to 15 W cm^{-3}). For the range of conditions tested here, we propose to use $T_RH_2(G)$ as a "spectroscopic parameter" able to represent the evolution of gas temperature as a function of experimental conditions, under similar experimental conditions. It can be used for analyzing the plasma when CARS or TALIF measurements are not available, and when the resolution of the spectrometer does not allow the determination of $T_{H\alpha}$.

For the bell jar reactor characterized by an axi-symmetric behavior, a comparison of these temperatures with line-of-sight averaged G state rotational temperatures can be interesting since the Abel inversion procedure is long and difficult. Line-of-sight averaged G state rotational temperatures are compared to Abel inverted ones in figure 16; they are lower than the Abel inverted temperature by around 250 K, but again they follow the evolution of the ground state molecular hydrogen temperature and that of the H atoms. Line-of-sight averaged $T_RH_2(G)$ is around 800 K lower than the gas temperature. Note that its evolution also follows the calculated gas kinetic temperature rather well. We propose to use line-of-sight averaged G state rotational temperatures for estimating the evolution of the gas temperature as a function of the macroscopic parameters in the bell jar reactor. This technique can be used for optimizing the bell jar reactor and for its in-situ long-time monitoring.

Discussion and Conclusions on the Temperature Measurements

Under typical conditions used for diamond deposition, we have determined vibrational and rotational temperatures of the molecular hydrogen in ground electronic state and in the $G^1\Sigma_g^+$ excited state by CARS and OES respectively. We have measured the H-atom temperature in the ground state and in the n=3 excited state (H_α) from the Doppler broadening, using TALIF and OES respectively. All of these have been compared, and we have proposed two spectroscopic parameters that allow the control of the gas temperature as well as its long-time in-situ monitoring in a microwave bell jar reactor.

First, a comparison of the vibrational and the rotational temperatures show that they are nearly in equilibrium, in particular in the bulk of the plasma. At the interface, the vibrational temperature is a little higher than the rotational temperature. At the surface, the translational temperature is in complete equilibrium with the substrate temperature, while the vibrational temperature is not.

A comparison between the H-atom temperature measured by TALIF (T_H) and the rotational temperature of the ground electronic state of molecular hydrogen ($T_RH_2(X)$) shows that these two temperatures are in equilibrium at least in the bulk of the plasma. Both of them represent the gas kinetic temperature. Diamond deposition reactors are characterized by very high temperatures. Temperatures are very sensitive to the averaged power density, and increase from 1550 K to 2600 K as the power density increases from 6 to 15 W cm^{-3}. Calculations show that the gas temperature reaches 3100 K at 30 W cm^{-3}.

Line-of-sight averaged H_α-atom temperatures ($T_{H\alpha}$) were measured by OES. Under diamond deposition conditions, Doppler, pressure and Stark broadenings contribute to the FWHM, the pressure broadening remaining still low at pressures up to 100 mbars.

$T_{H\alpha}$ was seen to be close to the gas temperature, although slightly higher at high power densities. It varies from 1540 $\pm$ 150 K to 3500 $\pm$ 350 K as the power density varies from 4.5 to 35 W cm^{-3}. The difference might be due to an underestimation of the Stark effect. Very high temperatures were confirmed at 30 to 35 W cm^{-3}.

The gas kinetic temperatures are comparable to that determined by C. Kaminski et al.[106], who found a value of 2100 K ± 200 K when measuring the rotational temperature by LIF from the $d^3\pi_g$-$a^3\pi_u$ (1,0) Swan C_2 band spectrum, at 12 mm above the substrate in plasma discharges similar to ours, but containing argon (800 W and 8500 Pa). Also Prepperneau et al.[39] reported similar values.

A comparison of the gas temperatures found in microwave plasma reactors with those reported in hot filament reactors is interesting since the pressure conditions are similar for both reactors. Also the growth rate and diamond film quality are comparable. In hot filament reactors, typical filaments temperatures ranging from 2000 K to 2800 K according to the power supplied to the filaments (0.5 kW to 2.7 kW)[22,29], are reported. However, measurements by CARS[29] very close to the filament showed that the gas temperature is lower by 200 K than the filament temperature. The gas temperature decreases slightly from this initial value until that of the substrate temperature (diffusional transport). The typical gas temperatures obtained in hot filament reactors are then similar to those measured in microwave reactors, although slightly lower. However one of the main difference between these reactors is the temperature gradient at the diamond substrate. In plasmas reactors, much higher temperature gradients at the diamond substrate are found, owing to plasma processes which maintain high temperatures closer to the surface. In reactors such as arc-jet or plasma torch, higher temperatures ranging from 3000 K to 5000 K are reported[48,95].

Measurements of the radial distribution of the G electronically excited state of molecular hydrogen rotational temperature T_R H_2 (G)) were performed using Abel inversion for each of the rotational line of the $G^1\Sigma_g^+ \rightarrow B^1\Sigma_u^+$ (0,0) band. $T_RH_2(G)$ measured on the axis of the plasma was compared to the axial T_H and $T_RH_2(X)$, i.e. to the gas temperature. $T_RH_2(G)$ is lower than the gas temperature and equal to 3/4 of $T_RH_2(X)$. Thus, T_R $H_2(G)$ is not in equilibrium with the gas temperature even at pressures as high as 20 to 50 hPa. This result disagrees with those of Chu et al[13]. Owing to the very high efficiency of the quenching processes compared to that of thermalization, $T_RH_2(G)$ should be equal to 1/2 $T_RH_2(X)$. The proportionality factor B= 0.75 instead of 0.5 has been attributed to the G state excitation cross sections variations as a function of the rotation quantum number K, as suggested by Otorbaev or to the variations of the radiative probabilities of the G state as a function of K.

Nevertheless, experimental results show that the G state rotational temperature follows the variations of the ground state temperature as a function of the operating conditions, in particular the averaged microwave power density. Under the experimental conditions tested here, the G state rotational temperature has been proposed as an interesting spectroscopic parameter allowing optimization and monitoring of the local plasma conditions using optical emission spectroscopy. The measurements of the H_α temperature can be also realized, however a very high resolution of the spectrometer is required. In bell jar reactors, line-of-sight averaged $T_RH_2(G)$ temperature is lower by around 800 K than the gas temperature, but again follows the gas temperature as a function of the power density. It can be used as well for monitoring plasma conditions for long deposition time.

7.2.4 Data Analysis: H-atom Relative Mole Fractions and Absolute Mole Fractions
Introduction
This subsection is devoted to measurements of relative and absolute H-atom mole fractions by means of Optical Emission Spectroscopy (OES) and Two-photon Allowed transitions Laser Induced Fluorescence (TALIF).

First, the spectroscopic method called actinometry is presented. We carefully analyze the validity of this method in typical conditions for diamond deposition. In order to define the domain of validity of actinometry for the H-atom, we first give a theoretical analysis based on data from the literature. It allows us not only to find its main domain of validity, but also to determine the parameters which may drastically change as a function of the operating conditions, provoking a possible loss of the validity of actinometry. In order to prove experimentally the validity of actinometry, a comparative study of the variations of the H-atom relative mole fraction as a function of the operating conditions obtained respectively by TALIF and by OES is shown.

Second, an estimate of the quenching cross section of the radiative electronically excited 4p state of argon (2p1) by molecular hydrogen, which was unknown and necessary for our study is found.

Third, a calibration of the TALIF measurements based on the comparison between an experimental H-atom depletion observed as methane is introduced into the feed gas and the calculated H-atom depletion using chemical kinetics calculations is shown. This allowed us to estimate the H-atom mole fraction for a set of specific conditions. Then, for the first time, we present variations of the absolute H-atom mole fraction as a function of some operating conditions (power density and methane percentage) in a microwave diamond deposition reactor.

Finally, spatial distributions (axial and radial) of H-atom mole fractions are shown. They are compared with distributions calculated using the 1-D diffusive model presented in section 7.1.

Actinometry Method and Validity: Theoretical Approach
Actinometry is a method allowing one to estimate the relative concentrations or mole fractions of a ground electronic state species X from measurements provided by OES. It was introduced by Coburn et al[64], in the 1980's. The principle of the method is based on the excitation processes of two species (the species X and the actinometer): in the case where (i) the actinometer does not perturb the plasma, (ii) both species are excited to their respective radiative excited state by a direct electron impact process, and (iii) the excitation processes cross-sections (as a function of the electron energy) have, for both species, the same shape and a similar threshold, then there is a very simple relationship linking the emission intensity ratio to the concentration ratios of species X and the actinometer: $[X]/[Act] = k\, I_X/I_{Act}$. Furthermore, if the quenching processes as well as all

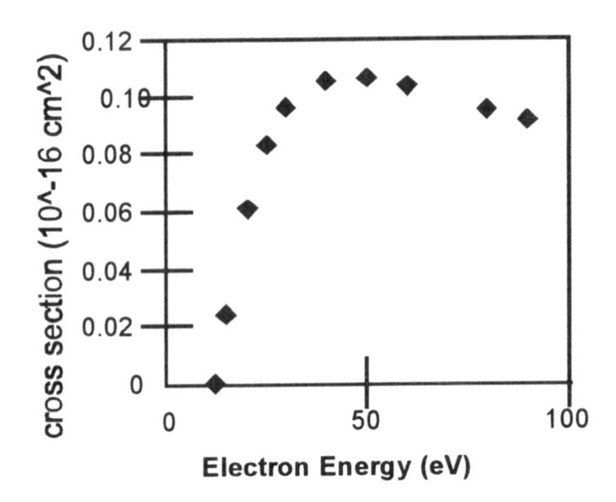

Figure 17a: Electron impact excitation cross sections for H(n=3) as a function of the electron energy (according to 68).

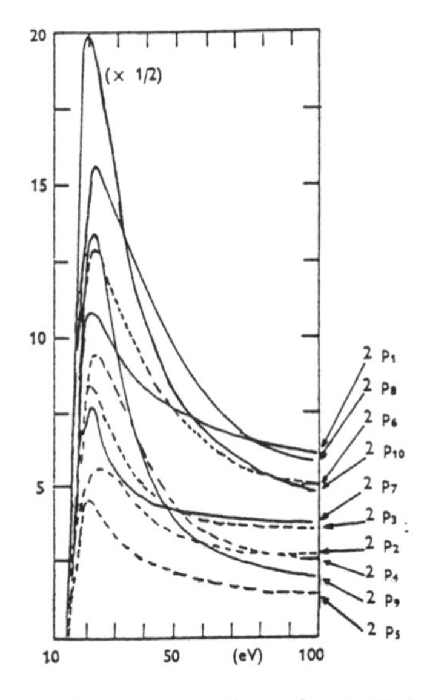

Figure 17b: Electron impact excitation cross sections for Ar(4p) as a function of the electron energy (according to 69).

the others processes of loss of the respective excited state species are negligible, the relationship is linear (k = constant). Very often, but not necessarily, a rare gas is chosen as actinometer for a given species X.

The use of actinometry for studying plasma discharges has been reported for different species[65-67]. Comparison of the results obtained by laser induced techniques and by actinometry led to the conclusion that the validity of actinometry must be, every time, analyzed carefully under the specific set of conditions used.

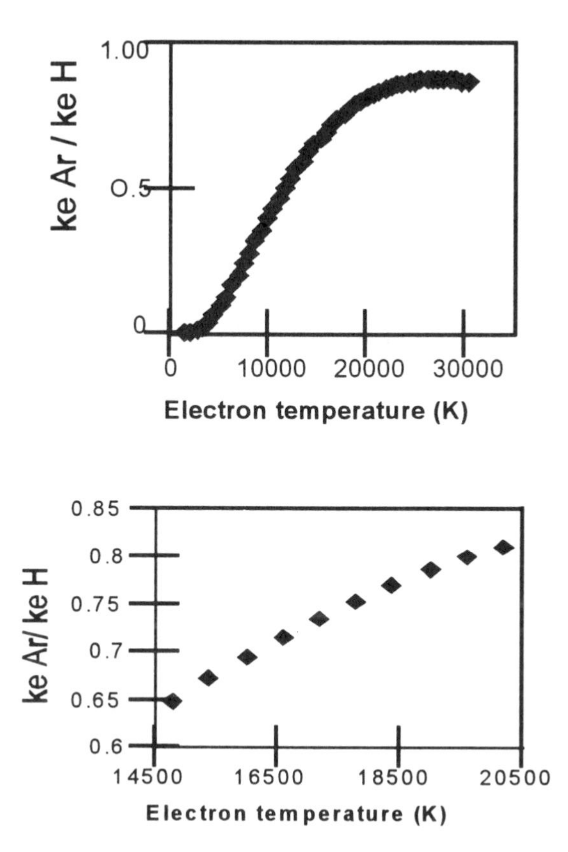

Figure 18: Ratio of the rate constants of excitation of Ar(4p) and H(n=3) by electron impact as a function of the electron temperature (assumption of a Maxwell distribution for the electron's energy).

Theoretical approach: excitation processes in the plasma. Argon has been chosen as the actinometer, since it presents radiative states with excitation thresholds close to that of the H-atom in the n=3 level. For instance, the 2p1 radiative state of argon has a threshold at 13.48 eV which is very close to that of the H-atom in the n=3 level (threshold 12.09 eV). The electron impact excitation cross sections of H(n=3) and Ar(2p1) as a function of the electron energy are given in figure 17[68-70]. The variation of the ratio of the excitation rate constants as a function of the electron temperature in the range 1000 to 30000 K, with the assumption of a Maxwellian distribution function for the electron's energy, is given in figures 18. Although very sensitive to T_e at low electron temperatures, its variations are less important for T_e ranging from 15000 K to 20000 K, corresponding to our conditions (calculations), making realistic the use of argon as an actinometer for H atoms.

First, we have to verify that both excited species (Ar(4p) and H(n=3)) are mainly excited directly from their corresponding ground electronic states through direct electron-ground electronic state species collisional reactions. Calculations based on kinetics allow us to estimate the relative importance of the different mechanisms in the production of the H and Ar atoms in their excited states.

The main processes involved in the production and consumption of the H(n=3) electronic state of H-atom are the following:

Production of H(n=3)

electron excitation from ground state $H(n=1) + e \rightarrow H(n=3\ (s,p,d)) + e$ (1)
electron excitation from the n=2 state: $H(2s) + e \rightarrow H(n=3\ (s,p,d)) + e$ (2)
dissociative excitation: $H_2 + e \rightarrow H(n=3) + H(n=1) + e$ (3)

Destruction

radiative de-excitation: $H(3(s,p,d)) \rightarrow H(2(s,p)) + h\nu$ (4)
radiative de-excitation: $H(3p) \rightarrow H(1s) + h\nu$ (5)
quenching processes: $H(3) + M_i \rightarrow H(n=1) + M_i^*$ (6)
ionization process from H(3) state: $H(3) + e \rightarrow H^+ + 2\,e$ (7)

Since H(n=2) may be a step in the excitation of H(n=3), its production and loss terms must be analyzed:

Production and destruction of H(n=2)

electron excitation from ground state: $H(n=1) + e \rightarrow H(n=2\ (s,p)) + e$ (8)
electron excitation to the n=3 state: $H(2s) + e \rightarrow H(n=3\ (s,p,d)) + e$ (9)
$Ar((3p^5 4s^1)$: (metastable $^3p_0$ or $^3p_2$) $+ H(n=1) \rightarrow H(n=2) + Ar$ (10)
mixing of the 2s / 2p H-atom (electric field): $H(2s) \rightarrow H(n=2\ (p,d))$ (11)
radiative de-excitation: $H(2p) \rightarrow H(1s) + h\nu$ (12)
quenching processes: $H(2s) + M_i \rightarrow H(n=1) + M_i^*$ (13)
Collisional mixing processes: $H(2s) + Ar \rightarrow H(2p) + Ar$ (14)

The main processes for the production and the destruction of the Ar-atom are the following:

Production

direct electronic impact excitation: $Ar(3p) + e \rightarrow Ar(4p) + e$ (15)
excitation to a metastable: $Ar(3p) + e \rightarrow Ar\ (^3P_2$ or $^3P_0) + e$ (16)

Excitation from metastable:

$Ar\ (^3P_2$ or $^3P_0)$ (11.5 or 11.72 eV) $+ e \rightarrow Ar(4p) + e$ (17)

Destruction

radiative de-excitation: $Ar(4p) \rightarrow Ar(4s) + h\nu$ (18)
quenching: $Ar\ (4p) + M_i \rightarrow Ar\ (3s, 4p) + M_i^*$ (19)
collisional mixing: $Ar\ (4p) + M_i \rightarrow Ar\ (4p$ (other states)$) + M_i^*$ (20)

We have neglected all the processes of production of the states H(n=3) and H(n=2) through radiative cascades due to the high pressure considered here (high quenching efficiency) and the relatively low electron temperature (the higher the level the less efficient the electron impact excitation process).

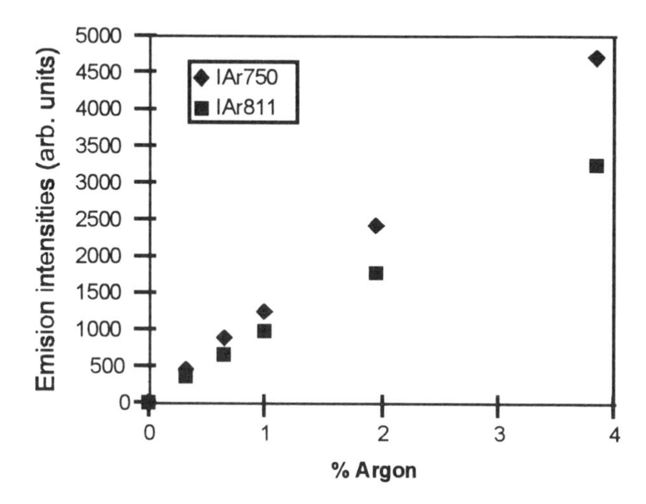

Figure 19: Emission intensities of Ar(2p1→1s2) (750.3 nm) and Ar(2p9→1s5) (811.5 nm) transition lines as a function of the percentage of argon introduced; (H_2 Plasma: 600 W, 25 hPa, 9 W cm^{-3}).

In principle, actinometry can be used only if the excitation processes (1) and (15), and de-excitation processes (4) and (18) are predominant. Then, a very simple relationship is obtained:

$$[H] / [Ar] = x_H / x_{Ar} = k\, I_{H\alpha} / I_{Ar}$$

where k depends (only slightly) on the electron energy distribution function (EEDF) due to the fact that the shapes and thresholds of the excitation processes cross sections for the two excited species are not strictly equal (see figure 18). [H] and [Ar] represent the H-atom and Ar-atom concentrations, x_H and x_{Ar} their mole fractions, and $I_{H\alpha}$ and I_{Ar} their associated emission intensities. In cases where quenching processes are not negligible, the constant k depends on the quenching cross sections, the pressure and the gas temperature.

At pressures as high as those used for growing diamond under microwave plasmas (20 to 100 hPa), the quenching rate of metastable argon atoms is higher than its excitation to upper excited state[72]. For confirming that argon is produced in the 4p level through a direct electron impact, we have measured the emission intensity ratio of lines corresponding to two transitions of argon issued from different excited sublevels (2p1 and 2p9) (in Paschen notation) of the Ar(4p) state as a function of the percentage of argon introduced (0 to 4 %) (figure 19). The emission intensity $I_{750.3\,nm}$ corresponding to the transition 2p1 (excitation threshold 13.48 eV) to 1s2, is proportional to the amount of argon introduced while $I_{811.5\,nm}$ corresponds to the transition 2p9 (excitation threshold 13.05 eV) to 1s5 (metastable $^3p_2$) which is not. The 2p1 state is then very probably excited directly from the ground state of argon by collision with an electron, while for the 2p9, other processes seems to contribute to its excitation[71-73]. For the Ar(4p) (2p1 state), the main process of excitation is most probably the direct excitation from the ground state of argon.

The rate of the mechanism of production of H(3) involving the H(2s) state (non radiative state) (channel (2)), depends on the rate constant of the process as well as on the H(2s) and the electron densities. The rate must be compared to the rate of the direct electron excitation mechanism (channel (1)). The density of the H(2s) depends on its production and on its consumption rates.

The H(n=2) state can be produced through reaction (8) or through reaction (16) followed by reaction (10). The process of production of H(n=2) involving the metastable $Ar(^3p_2)$ is one order of magnitude higher than the one involving $Ar(^3p_0)$[74], and the process is rather efficient (σ = 10 $Å^2$)[75]. The reaction rate depends on the density of $Ar(^3p_2)$, which is governed by the direct electron impact process for the production of $Ar(^3p_2)$. This latter process has the same magnitude of σ as the electron impact excitation leading to H(n=2) (σ = 0.6 $Å^2$ for reaction (16) and σ = 0.8 $Å^2$ for reaction (8)). The consequence is that the contribution of reactions (16) + (10) to the population of H(n=2s) is as important as reaction (8).

H(n=2s) consumption is mainly due to the quenching processes[80] (reactions (13)) (destruction frequency at 25 hPa and 2000 K, $\nu_Q = 3.5 \cdot 10^8\ s^{-1}$) and to the mixing with H(2p) state (radiative state) (reaction (11)) due to the action of the electric field[80, 81, 98] ($\nu E \cong 7 \cdot 10^6\ s^{-1}$) and to the collisional mixing reaction with argon atoms (reaction (14) (cross section 200 $Å^2$ and collision frequency $1.3 \cdot 10^7\ s^{-1}$ for 1 % argon at 25 hPa)[77,82,83]. Trapping processes may also contribute to maintaining a relatively high population of H(n=2s) species[84]. Due to the high pressure considered here, the quenching might be more important than this process, and trapping processes are neglected. Taking into account these remarks, the corresponding lifetime of the H(n=2s) state is less than $3 \cdot 10^{-9}$ s. For an electron temperature of 1.5 eV (reasonable for conditions considered here as seen from calculations), the density of H(n=2) is estimated at $4 \cdot 10^8\ cm^{-3}$. For a density of H(n=1) of $10^{15}\ cm^{-3}$ (1 % of H atoms), the ratio of the rate of channel (1) over that of channel (2) is larger than 25. Consequently, the channel (2) can be neglected compared to channel (1).

The fact that the contribution of argon to the excitation of H(n=3) is negligible has been confirmed experimentally by the measurement of the emission intensity owing to the H_α transition as a function of the percentage of argon. It remains unchanged as the percentage of argon is increased from 0 to 4 % (as also does the emission intensity due to the H_β transition) (figure 20).

The last process of production of H(3s) which may compete with the channel (1) process is the dissociative excitation mechanism (channel (3)). This can strongly prohibit the use of actinometry under some conditions. The term $k_e^{H\alpha}\ n_H n_e$ leading to the excitation to the state n=3 (channel (1)) have to be compared to the $k_{diss}\ n_{H2} n_e$ term of channel (3). The ratio of $k_{diss} / k_e^{H\alpha}$ can be estimated from the cross sections reported in reference 68 as a function of the principal quantum number n and the electron temperature, with the assumption of a Boltzmann distribution for the electron energy[72]. For the state n=3, and

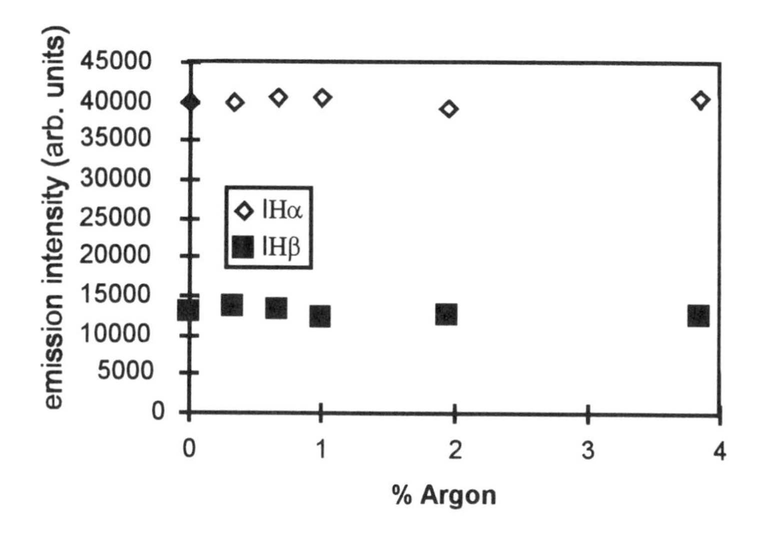

Figure 20: Emission intensities of H_α and H_β line as a function of the percentage of argon introduced; (H_2 Plasma: 600 W, 25 hPa, 9 W cm^{-3}).

an electron temperature varying from 1 to 2 eV, the variations of $k_e^{H\alpha}.n_H/\ k_{diss}.n_{H2}$ (r_{dir}/r_{diss}) are presented in figure 21. r_{dir}/r_{diss} varies drastically as a function of the electron temperature and the H-atom mole fraction. For electron temperatures less than 19000 K (typical in our reactor), the direct electron impact excitation rate is more than 3 times the dissociative excitation process rate once the H-atom mole fraction is higher than 5 %. Similar results are reported by A. Rousseau[71] when assuming a non Maxwellian distribution for the electrons. In our case, the dissociative excitation mechanism (channel (3)) will become as important as the direct excitation process (channel (1)) for conditions where the percentage of H atoms in the plasma is less than 1 % and the electron temperature more than 20000 K (power density < 9 W cm^{-3}). The lower the electron temperature, the lower the atomic hydrogen mole fraction, allowing the possibility of neglecting the dissociative excitation process contribution.

In addition to these requirements, the use of actinometry for mapping the H atom concentration in the reactor or for studying the effect of the operating conditions (% CH_4, power density) requires no drastic evolution in the electron energy distribution function (EEDF) as a function of the coordinates[85] or operating conditions. As a matter of fact, owing to the difference in the excitation thresholds for H(n=3) and Ar(4p) and in the shape of the cross sections, changes in the rate constants ratios would occur.

The variations of the ratio of the two excitation lines $I_{H\alpha}/I_{H\beta}$ (owing to the H_α and H_β transitions with the respective energies of 12.06 and 12.75 eV (H(n=3) and H(n=4) levels)) and of $I_{811.5\ nm}\ /\ I_{750.3\ nm}$ were measured as a function of the percentage of methane and the microwave power density. No variations were observed as the methane was varied from 0 to almost 6 % and as the power density was varied from 9 W cm^{-3} to

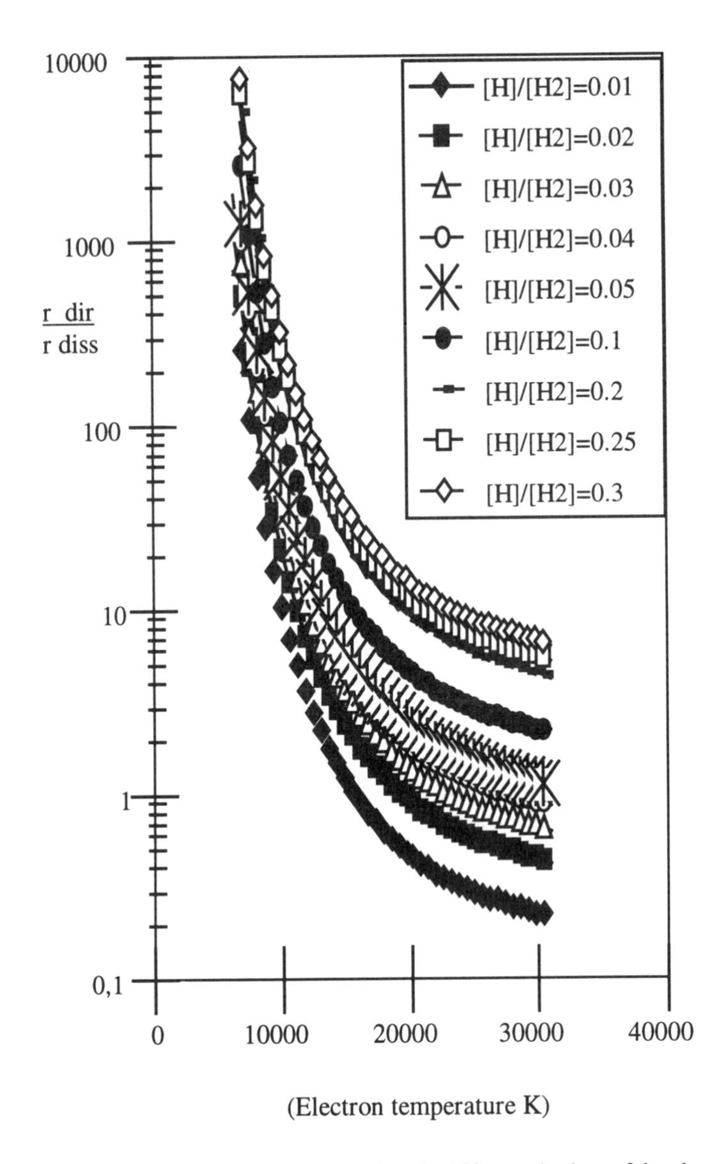

Figure 21: Ratio of reaction rates of production of excited H-atom in the n=3 level as a function of the electron temperature: rate of direct electronic impact excitation over that of dissociative excitation reaction.(cross section after reference 68).

15 W cm^{-3}, indicating that there is no drastic change in the EEDF. The same observations were obtained by Milne et al. when varying the percentage of methane[26]. These results also indicate that the mechanisms leading to the excited states are not drastically changed as the methane percentage or the power density are varied. Under these conditions, calculations have shown a decrease in the electron temperature from 19500 K to 18000 K as the power density is varied from 9 W cm^{-3} to 15 W cm^{-3}.

Moreover, it is reported in the literature[28] that the electron energy distribution function (then the electron temperature) remains constant as a function of the percentage of methane. Experimentally the gas temperature is seen to be unchanged, which is an indication that the electron temperature is not drastically changed as a function of the methane percentage introduced in the feed gas.

In addition to the processes of production of H(3), we may have to consider the processes of its destruction. Under the conditions tested here, owing to the high pressure (> 5 hPa) we can consider that the ionization process (reaction (7)) is negligible compared to the quenching processes (reaction (6)). The quenching of excited argon atoms must also be taken into account (reactions (19) and (20)).

Once the conditions of validity of actinometry realized, that is the H-atom mole fraction is higher than 1 to 5 % (according to the electron temperature), the electron temperature remains in the range 1 to 2 eV, the pressure is higher than 5 hPa and there is no drastic change of the EEDF in the reactor (spatially and as a function of the operating conditions), the proportionality constant k linking the ground state H atoms concentration to the ratio $I_{H\alpha}$ / I_{Ar}, depends on the electron temperature (see figure18), and on the collisional quenching terms for the H(3) and for the Ar 4p'[1/2] excited states by H_2 molecules and H atoms[23,39]. Consequently, as a function of the experimental conditions, a linear variation or a monotone function of $I_{H\alpha}$ / I_{Ar} versus [H] / [Ar] is expected.

$$\frac{[H]}{[Ar]} = \frac{xH}{xAr} = k(\text{EEDF, quenching terms})\,\frac{IH\alpha}{IAr}$$

$$= \frac{keAr^*}{keH\alpha}\left[\frac{\nu Ar^*}{\nu H\alpha}\right] Q_T\,\frac{IH\alpha}{IAr^*}\cdot F$$

where Q_T is the de-excitation term taking into account all the processes of radiative de-excitation and for the collisional quenching processes. $k_e^{Ar^*}$ and $k_e^{H\alpha}$ are the excitation rate constants for the transitions Ar (3p) $\rightarrow$ Ar (4p)[69] and H(n=1) $\rightarrow$ H(n=3)[68] respectively, ν_{Ar^*} and $\nu_{H\alpha}$ the de-excitation frequencies for H_α and Ar* transitions, and x_H and x_{Ar} the H-atom and Ar-atom mole fractions respectively. The reduced chemical scheme that we can consider, under the domain of conditions discussed here, is now:

electron excitation from the ground state:

H(n=1) + e $\rightarrow$ H(n=3 (s,p,d)) + e	($k_e^{H\alpha}$)	(1)
radiative de-excitation: H(3) $\rightarrow$ H(2) + hν	(A_{32})	(4)
radiative de-excitation: H(3p) $\rightarrow$ H(1s) + hν	(A_{31})	(5)
dissociative excitation: H_2 + e $\rightarrow$ H(n=3) + H(n=1) + e	(k_{diss})	(3)
quenching processes: H(3) + M_i $\rightarrow$ H(n=1) + M_i^*	($k_Q^{H^*}$)	(6)
direct electronic impact excitation:		
Ar(3p) + e $\rightarrow$ Ar(4p) + e	($k_e^{Ar^*}$)	(15)
de-excitation: Ar(4p) $\rightarrow$ Ar(4s) + hν	(A_{44})	(18)
quenching: Ar (4p) + M_i $\rightarrow$ Ar (3s,4p) + M_i^*	($k_Q^{Ar^*}$)	(19–20)

We will assume that the distribution of the 3s, 3p and 3d excited states levels owing to electron impact excitation is the statistical weighting given by the number of sublevel degeneracies of each angular momentum L-state, that is 11.8 % for the 3s state, 33.3 % for the 3p state and 55.5 % for the 3d state[39].

The emission intensity of H_α is equal to:

$$I_{H\alpha} = K(\nu_{H\alpha})\ A_{32}\ \nu_{H\alpha}\ v_{emiss.}\ \frac{[H(n=1)]\ keH\alpha\ ne + k_{diss}\ [H2]\ ne}{[H]\ k_{QH}^{H} + [H_2]\ k_{QH}^{H2} + k_R}$$

with $k_R = (A_{32} + A_{31})$, and $A_{32} = ([3s]/\ \tau_s) + ([3p]*f_p/\ \tau_p) + ([3d]/\ \tau_d)$, where τ_s, τ_p and τ_d are respectively the lifetimes of the 3s, 3p and 3d levels, in nanoseconds. f_p represents the branching term of de-excitation of the 3p level towards the 2s level relative to the 1s level.

$$A_{32} = [(0.118/159) + (0.333*0.1183/5.4) + (0.555/15.6)]\cdot 10^9 = 4.36\cdot 10^7\ s^{-1}$$
$$\text{and } A_{31} = [3p]*(1-f_p) / \tau_p) = 5.39\cdot 10^7\ s^{-1}$$

which lead to a total radiative de-excitation rate constant $A_{32} + A_{31} = 9.8\cdot 10^7\ s^{-1}$.

$K(\nu_{H\alpha})$ is a constant taking into account the optical device response, k_{QH}^{H} and k_{QH}^{H2} represent respectively the quenching rate constants of H_α by the H_2 molecules and the H atoms, they are equal respectively to $(P / RT)\ v_{H/H2}\ \sigma_{Ar*/H2}\ x_{H2}$ and $(P / RT)\ v_{H/H}\ \sigma_{H\alpha/H}\ x_H$, where $v_{H/H2}$ and $v_{H/H}$ are the respective relative mean velocities of H_α -atom and H_2-molecules and H_α and H atoms and $\sigma_{Ar*/H2}$ and $\sigma_{H\alpha/H}$ the quenching cross sections of H_α by the H_2 molecules and by the H atoms.

Neglecting the dissociative excitation mechanism lead to:

$$I_{H\alpha} = K(\nu_{H\alpha})\ \nu_{H\alpha}\ A_{32}\ v_{emss.}\ \frac{[H(n=1)]\ keH\alpha\ ne}{[H]\ k_{QH}^{H} + [H_2]\ k_{QH}^{H2} + k_R}$$

$$= (K(\nu_{H\alpha})\ \nu_{H\alpha}\ A_{32}\ v_{emss.} / k_R)\ \frac{\{[H(n=1)]\ keH\alpha\ ne\}}{\{[H]\ (k_{QH}^{H} / k_R) + [H_2]\ (k_{QH}^{H2} / k_R) + 1\}}$$

Concerning the excitation of argon, we have:

$$I_{Ar*} = K(\nu_{Ar*})\ \nu_{Ar*} A_{44}\ v_{emss.}\ \frac{[Ar(3p)]\ keAr*\ ne}{[H]\ k_{QAr}^{H} + [H_2]\ k_{QAr}^{H2} + k_{RAr}}$$

where $A_{44}$is the Einstein coefficient for the transition $Ar(4p) \rightarrow Ar(4s)$ ($k_{RAr} = A_{44}$), k_{QAr}^{H} and k_{QAr}^{H2} are the quenching terms of Ar* by molecular hydrogen and atomic hydrogen. $[H_2]\ k_{QAr}^{H} = (P / RT)\ v_{Ar/H2}\ \sigma_{Ar*/H2}\ x_{H2}$ and $k_{QAr}^{H2} = (P / RT)\ v_{Ar/H}\ \sigma_{Ar*/H}\ x_H$, where $v_{Ar/H2}$ and $v_{Ar/H}$ are the relative mean velocities of Ar by respectively H_2 and H, and $\sigma_{Ar*/H2}$ and $\sigma_{Ar*/H}$ the associated quenching cross sections. $K(\nu_{Ar*})$ is a constant taking into account for the optical device response.

$$I_{Ar*} = K(\nu_{Ar*})\, \nu_{Ar*} A_{44}\, v_{emss} / k_{RAr}\,) \frac{[Ar(3p)]\; keAr^*\; ne}{[H]\; k_{QAr}{}^{H} / k_{RAr} + [H_2]\; k_{QAr}{}^{H2} / k_{RAr} + 1}$$

Then :

$$\frac{I_{H\alpha}}{I_{Ar*}} = \frac{K(\nu_{H\alpha})\nu_{H\alpha} A_{32} v_{emiss.} / k_R \dfrac{[H(n=1)]k_e^{H\alpha} ne}{[H](k_{QH}^{H} / k_R) + [H_2](k_{QH}^{H2} / k_R) + 1}}{K(\nu_{Ar*})\nu_{Ar*} A_{44} v_{emiss.} / k_{RAr} \dfrac{[Ar(3p)]k_e^{Ar*} ne}{[H](k_{QAr}^{H} / k_{RAr}) + [H_2](k_{QAr}^{H2} / k_{RAr}) + 1}}$$

Then,

$$[H(n=1)] / [Ar(3p)] = F\, k_e^{Ar*} / k_e^{H\alpha}\; x\; (\nu_{Ar*} A_{44} / k_{RAr}) / (\nu_{H\alpha}\, A_{32} / (A_{32} + A_{31}))\; Q_T\, I_{H\alpha} / I_{Ar}$$

with

$$Q_T = \frac{1 + (P / RT\, (A_{32} + A_{31}))\; [v_{H/H2}\, \sigma H\alpha/H2\; xH2 + v_{H/H}\, \sigma H\alpha/H\; (1 - xH2)]}{1 + (P / RT\, A_{44})\; [v_{Ar/H2}\, \sigma Ar^*/H2\; xH2 + v_{Ar/H}\, \sigma Ar^*/H\; (1 - xH2)]}$$

and $F = \dfrac{K(\nu Ar^*)}{K(\nu H\alpha)}$. Giving the cross sections in $Å^2$, the gas temperature in K and the pressure in hPa, we have:

$$Q_T = \frac{1 + 0.10755 PT^{-1/2}(\sqrt{3/2}\sigma_{H\alpha/H2} x_{H2} + \sqrt{2}\sigma_{H\alpha/H}(1 - x_{H2}))}{1 + 0.223 PT^{-1/2}(0.72\sigma_{Ar*/H2} x_{H2} + 1.012\sigma_{Ar*/H}(1 - x_{H2}))}$$

or

$$Q_T = \frac{1 + PT^{-1/2}(0.132\sigma_{H\alpha/H2} x_{H2} + 0.152\sigma_{H\alpha/H}(1 - x_{H2}))}{1 + PT^{-1/2}(0.162\sigma_{Ar*/H2} x_{H2} + 0.226\sigma_{Ar*/H}(1 - x_{H2}))}$$

The quenching of H_α by argon atoms has been neglected due to the small amount of argon introduced and to the small cross section of the process ($\sigma_{H\alpha/Ar}/\sigma_{H\alpha/H2}=0.3$)[23,39]. As well, the quenching of Ar^* by Ar atoms has been neglected since its cross section is very small ($\sigma_{Ar*/Ar}/\sigma_{H\alpha/H2}=0.027$) as well. The cross section for the quenching of H (n=3) by atomic hydrogen ($\sigma_{H\alpha/H}$) is not known, nor is that of Ar* by atomic hydrogen. They can be evaluated using the hard sphere model or using energy potential functions when available. Using the hard sphere model and taking the diametric cross section from literature[87,88], we can obtain upper values for the cross sections. We found 18 $Å^2$ and 50 $Å^2$ respectively for $\sigma_{H\alpha/H}$ and $\sigma_{Ar*/H}$. Using the energy potential functions from reference 78 and reference 74 respectively for the interaction of H(n=3) with H_{1s} and Ar* with H_{1s} lead to values of around 2 $Å^2$ and 17 $Å^2$ respectively for $\sigma_{H\alpha/H}$ and $\sigma_{Ar*/H}$. These latter will be probably the lowest values. The cross section for the quenching of H (n=3) by molecular hydrogen ($\sigma_{H\alpha/H2}$) has been found by Bittner et al. and Prepperneau et al. (respectively 65 and 58 Å^2)[23,39] while the cross section for Ar(4p) by molecular hydrogen is unknown. It is estimated below.

For low dissociation yield of molecular hydrogen, the contribution of the quenching by H atoms will be neglected. The expression of k is reduced to:

$$k = \frac{(keAr^*/keH\alpha)\ F'\ [1+0.132\ \sigma H\alpha/H2\ P\ xH2\ T\text{-}1/2]}{[1+0.162\ \sigma Ar^*/H2\ P\ xH2\ T\text{-}1/2]}$$

and then:

$$x_H = \frac{keAr^*}{keH\alpha}\ F'\ .\ Q_{H2}.\ xAr\left(\frac{IH\alpha}{IAr}\right)_{act}$$

with $$Q_{H2} = \frac{[1+0.132\ xH2\ \sigma H\alpha/H2\ P\ T\text{-}1/2]}{[1+0.162\ xH2\ \sigma Ar^*/H2\ P\ T\text{-}1/2]}$$

and $$F' = F\ \frac{(\nu Ar^* A_{44}/\ k_{RAr})}{(\nu H\alpha\ A_{32}/(A_{32}+A_{31}))} = \frac{F\ \nu Ar^*\ (A_{32}+A_{31})}{\nu H\alpha\ A_{32}}$$

As a conclusion, a relationship linking the ratio of concentrations of H-atom and Ar-atom to their emission intensities ratio has been established. It can reasonably be considered as valid for conditions characterized by an electron temperature in the range 1 to 2 eV, a H-atom mole fraction higher than 1 to 5 %, a pressure higher than 5 hPa, and a very slight evolution of the EEDF as a function of the spatial coordinates, or as a function of the operating conditions. Depending on the H-atom mole fraction, the contribution of the quenching of the excited species by atomic hydrogen must be taken into account or not. An experimental determination of the cross sections of H_α and Ar(4p) by H atoms and of that of Ar(4p) by molecular hydrogen should be important for our study.

Two-photon allowed transition LIF excitation (TALIF). The H-atom concentration is proportional to the product of the fluorescence amplitude (parameter b in equation 2, subsection 7.2.2) and the width $\Delta\lambda_D$, which must be corrected due to the quenching effect. This effect is due to the collisional de-excitation of the H atoms before they fluoresce. Absolute values of the concentration could be directly obtained after an experimental calibration procedure[89], which was not performed here. Our calibration procedure is based on a comparison between measurement of H-atom depletion observed experimentally as the methane is introduced at low power density and the H-atom depletion calculated at low power density. A 0-D chemical kinetics model operating in a mixture $H_2 + CH_4$ was used for simulating the experimental conditions (see below).

The processes of excitation and de-excitation of the H(n=3) excited state of H-atom are the following:

Laser excitation: H(n=1) + 2 hν → H(n=3s, 3d) (21)

electron excitation: H(n=1) + e → H(n=3 (s, p, d)) + e (1)

collisional mixing: $H(n=3\ s,d) + M_i \rightarrow H(n=3\ s,p,d)$ (22)
radiative de-excitation: $H(n=3s,3d) \rightarrow H(n=2p) + h\nu$ (4b)
radiative de-excitation: $H(n=3p) \rightarrow H(n=2s,\ 1s) + h\nu$ (5b)
quenching processes: $H(n=3) + M_i \rightarrow H(n=1) + M_i^*$ (6)
3 photon ionization process: $H(n=3) + h\nu \rightarrow H^+$ (7)

If we assume[90] that (i) the excitation by the laser is more powerful than by the electrons, (ii) the 3 photon ionization is negligible compared to the quenching of the H(n=3) atoms, then, the fluorescence signal intensity including all the radiative processes leading to the H(n=2) level is given by:

$$I_{H\alpha} = \frac{K'(\nu H\alpha)\ v_{emiss.}\ A'_{32}\ \nu_{H\alpha}\ W^2_{13}/\ k_R}{1 + k_Q\ [M]/k_R}\ [H(n=1)]$$

where $k_Q\ [M] = (P/RT)\ (v_{H/H2}\ \sigma_{H\alpha/H2}\ x_{H2} + v_{H/H}\ \sigma_{H\alpha/H}\ (1 - x_{H2}))$

and $k_R = A'_{32} + A'_{31}$, where $A'_{32} + A'_{31}$ are calculated from the Einstein coefficients for the spontaneous emissions (H(n=3 (s,p,d)) $\rightarrow$ H(n=2s,p, 1s))[39]. W_{13} is the two photon excitation rate coefficient, $v_{emiss.}$, the emissive volume, and $K'(\nu_{H\alpha})$ an optical device function.

For laser excitation, the n=3 distribution is not given by the statistical distribution as in the case for electron excitation. The distribution depends on the electric field polarization of the laser light in the direction of propagation which interacts with the medium. In our case, the distribution is the same as that reported by Prepperneau et al.[39] and Tung et al.[86]: 12 % for the 3s state and 88 % for the 3d state. Under conditions of high quenching efficiency, the lifetime of the excited species is small (0.5 to 2.5 ns, according to the experimental conditions), so the mixing between the 3s, 3p and 3d levels of the H atoms is not complete[39]. The total mixing would be expected for a lifetime of 4 ns. The consequence is that the population of the 3p level is not contributing to radiative decay of the n=3 level, in particular through the H(3p) $\rightarrow$ H(1s) process which is very efficient, as in the case of the electron excitation. The radiative lifetime of the n=3 level is then higher than in the case of excitation by electrons. If we assume that no mixing is occurring at all, the overall lifetime of the n=3 level becomes 17.4 ns instead of 10.2 ns and the radiative decay rate $k_R = A'_{32} + A'_{31}$ is equal to $A'_{32} = 5.73 \cdot 10^7\ s^{-1}$ (or $4.78 \cdot 10^7\ s^{-1}$) (according to Prepperneau or Bittner), instead of $9.8 \cdot 10^7\ s^{-1}$.

In a pure hydrogen plasma, we have:

$$k_Q\ [M] = (x_{H2}\sigma_{H\alpha/H2}\ v_{H\alpha/H2} + x_H\ v_{H\alpha/H}\ \sigma_{H\alpha/H})\ P\ /\ R\ T$$
$$k_Q\ [M]\ /k_R = \{(\ x_{H2}\ \sigma_{H\alpha/H2}\ v_{H\alpha/H2} + x_H\ v_{H\alpha/H}\ \sigma_{H\alpha/H})\ P\ /\ R\ T\}/\ k_R$$
$$k_Q\ [M]\ /k_R = \{(\ x_{H2}\ \sigma_{H\alpha/H2}\ v_{H\alpha/H2} + x_H\ v_{H\alpha/H}\ \sigma_{H\alpha/H})\ P\ /\ R\ T\}/\ 5.73 \cdot 10^7$$
$$k_Q\ [M]\ /k_R = (x_{H2}\ \sigma_{H\alpha/H2}\ (3/2)^{1/2} + x_H(2)^{1/2}\ \sigma_{H\alpha/H})\ 0.186\ P\ T^{-1/2}$$
$$k_Q\ [M]\ /k_R = (0.228\ x_{H2}\sigma_{H\alpha/H2} + 0.263\ x_H\ \sigma_{H\alpha/H}\ )\ P\ T^{-1/2}$$

where P is given in hPa, T in K and the cross sections in $Å^2$.

$$(I_H)_{fluo} = \frac{K'(\nu_{H\alpha})\nu_{H\alpha}\nu_{emiss.}W_{13}^2[H(n=1)]}{0.228x_{H2}\sigma_{H\alpha/H2} + 0.263x_H\sigma_{H\alpha/H}}$$

For low H-atom mole fraction, the fluorescence signal is given by:

$$(IH)_{fluo} = \frac{K'(\nu H\alpha)\{\nu H\alpha \text{ vemiss. } W^2 13 \; [H(n=1)]\}}{1 + 0.228 \text{ xH2 } \sigma H\alpha/H2 \text{ P T-1/2}}$$

$$x_H = RT\,[H(n=1)] / P$$

$$x_H = \frac{(IH\alpha \text{ T/P})\text{fluo R } [1 + 0.228 \text{ xH2 } \sigma H\alpha/H2 \text{ P T-1/2}]}{\{\nu H\alpha \text{ vemiss. } W^2 13.\ K'(\nu H\alpha)\}}$$

$$= (I_{H\alpha} \text{ T/P})_{fluo} \{ B/ W^2 13.\} [1 + 0.228\ x_{H2}\sigma_{Ar^*/H2} \text{ P T-1/2}]$$

With $B = R /\{ \nu_{H\alpha}\, \nu_{emiss}.\ K'(\nu_{H\alpha})\}$

Estimation of the cross section for quenching of Ar(4p) by H_2 molecules ($\sigma_{Ar^*/H2}$). At low H-atom mole fraction, dividing the H-atoms mole fraction provided by OES by that provided by TALIF, it comes:

$$\frac{(IH\alpha/IAr)act}{(IH\alpha \text{ T/P})fluo} = \frac{\dfrac{[1 + 0.228 \text{ xH2 } \sigma H\alpha/H2 \text{ P T-1/2}]}{Q_{H2}}\ B\,\dfrac{keH\alpha}{keAr.W^2 13}}{xAr\ F'}$$

$$= \frac{A}{xAr}\ \frac{[1+0.228 \text{ xH2 } \sigma H\alpha/H2 \text{ P T-1/2}]\ [1+0.162\sigma Ar^*/H2 \text{ xH2 P T-1/2}]}{1+0.132 \text{ xH2}\sigma H\alpha/H2PT\text{-}1/2}$$

$$\text{with } A = A(EEDF, W^2 13) = \frac{\dfrac{B\ keH\alpha}{(keAr^*\ W^2 13)}}{F'},$$

$$\text{and } F' = F\,\frac{(\nu Ar^* A_{44}/\ k_{RAr})}{(\nu H\alpha\ A_{32}/(A_{32} + A_{31}))} = F\,\nu_{Ar^*}\frac{(A_{32} + A_{31})}{(\nu H\alpha\ A_{32})}$$

In order to limit the dissociation of the molecular hydrogen, and to consider x_{H2} and x_{Ar} as constant (1 and 0.01 respectively), we have carried out experiments in a region of the plasma where the dissociation is still low due to the low residence time (entry of the plasma). In that region, the gas temperature, I_H/I_{Ar} and the fluorescence signal were measured as the microwave power and the pressure were varied. Experimentally, the variations of I_H/I_{Ar} and of the fluorescence area remained small.

Considering $k_e^{H\alpha}/\ k_e^{Ar^*}$ as a constant (see figure 18) and $\sigma_{Ar^*/H2} = 58$ Å^2 (Prepperneau value), we have:

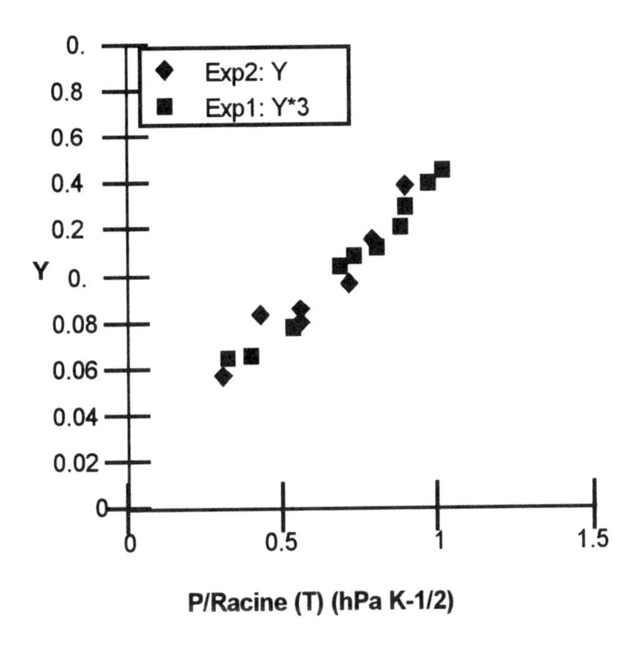

Figure 22: Y= $(1+0.132\ \sigma_{H\alpha/H2}PT^{-1/2})/(1+0.228\ \sigma_{H\alpha/H2}PT^{-1/2})$ (IHa/IAr)act/(IHa T/P)fluo as a function of $PT^{-1/2}$. The slope multiplied by origin lead to an estimate of the cross section for the quenching of Ar(4p) by molecular hydrogen.

$$\frac{(I_{H\alpha}/I_{Ar})_{act}}{(I_{H\alpha}TP)_{fluo}} = \frac{A}{x_{Ar}} \frac{[1+13.22PT^{-1/2}]\cdot[1+0.162\sigma_{Ar^*/H2}x_{H2}PT^{-1/2}]}{[1.766PT^{-1/2}]} \frac{[1.766PT^{-1/2}]}{[1+13.22PT^{-1/2}]}$$

$$\frac{(IH\alpha/IAr)act}{(IH\alpha\ T/P)fluo} = \frac{A}{xAr}\ [1+0.162\ \sigma_{Ar^*/H2}\ x_{H2}\ PT\text{-}1/2]$$

With the Bittner et al. value ($\sigma_{Ar^*/H2}$= 65 Å^2), we have:

$$\frac{1+8.58PT^{-1/2}}{1+14.8PT^{-1/2}} \frac{(I_{H\alpha}/I_{Ar})_{act}}{(I_{H\alpha}T/P)_{fluo}} = \left(\frac{A}{x_A}\right)(1+0.162\sigma_{Ar^*/H2}x_{H2}PT^{-1/2})$$

The plots of {[1+7.66PT-1/2] /[1+13.22P T-1/2]} $(I_{H\alpha}/I_{Ar})$act/($I_{H\alpha}$ T/P)$_{fluo}$ as a function of P $T^{-1/2}$ are presented on figure 22 (x_{H2} is taken equal to 1), for two experiments performed under similar experimental conditions and different laser power (W_{13}). Straight lines are obtained indicating that the assumptions made previously (for actinometry, and for x_{H2} and $k_e^{H\alpha}/k_e^{Ar^*}$ considered as rather constant) are reasonable for the domain of pressure and temperature considered. In particular, the dissociative excitation process does not appear important under the conditions considered here (even at this low dissociation yield), otherwise straight lines would not be found.

From the slopes and origins, an estimate of the cross section for the quenching of Ar(4p) by molecular hydrogen is obtained. Using Prepperneau's or Bittner's cross section, the mean square line provides a value of 42 $Å^2$. However, owing to the uncertainties on the origin and on the variations of $k_e^{H\alpha}/k_e^{Ar*}$, and to the unknown dissociation yield, a rather large error in the value given for $\sigma_{Ar*/H2}$ is probable. 42 $Å^2$ constitutes a magnitude for $\sigma_{Ar*/H2}$, which is consistent with our experiments. This value can be compared with the cross section calculated on the basis of the hard sphere model (58 $Å^2$).

Note: The value of $\sigma_{Ar*/H2}$ is directly related to the radiative decay of the H(n=3) state in TALIF experiments. We have considered here that no mixing had occurred at all. If we consider that on the contrary a complete mixing between the 3s, 3p and 3d sublevels is effective (the lifetime of the H(n=3) is higher than 4ns), then a cross section value of 70 $Å^2$ is found.

Using $\sigma_{Ar*/H2}$= 58 $Å^2$, $\sigma_{Ar*/H2}$ = 42 $Å^2$, and the two sets of calculated cross sections for the quenching processes by the H atoms and the variations of the factors Q_{H2} and Q_T as a function of the dissociation of H_2, the pressure and the gas temperature can be calculated. This makes it possible to estimate the variations of the relative H-atom mole fraction as a function of the operating conditions, in particular as a function of the power density, once a calibration of the technique is performed. Larger errors will be found as the H-atom mole fraction increases due to the large uncertainty of the quenching terms by atomic hydrogen.

Calibration of the H-Atom Mole Fraction

The principle of the calibration is based on a comparison between measurements (TALIF) of the depletion of the H-atom relative mole fraction as a function of the percentage of methane in the feed gas observed at low power density (figure 23), and a calculated depletion[97].

A depletion of the H atoms is observed experimentally (by TALIF and OES) as a function of the percentage of methane, when the H-atom mole fraction is comparable to that of methane introduced in the feed gas (figures 23 and 24). At higher H-atom mole fractions obtained under more energetic conditions (15 W cm^{-3}), the relative H-atom mole fraction measured by OES remains unchanged as methane is introduced (figure 24). The depletion of H atoms is then directly related to the H-atom mole fraction and to the efficiency of the reactions involving carbon containing species. This can certainly be estimated using chemical kinetics models. The higher the H-atom mole fraction, the lower the depletion.

The model used is a 0-D chemical kinetics model operating in H_2 + CH_4 mixture. We did not used the 1-D diffusive model presented before owing to the increased complexity generated by the introduction of methane. To the H_2 plasma chemical kinetics scheme, already presented in sub-section 6, we have added chemical reactions involving the carbon containing species. A complete description of a chemical kinetics model including methane in the hydrogen plasma should take into account the electron-neutral collisions and the neutral-neutral collisions. However, owing to the small mole fraction of electrons in the plasma (10^{-6}) and to the small amount of methane introduced,

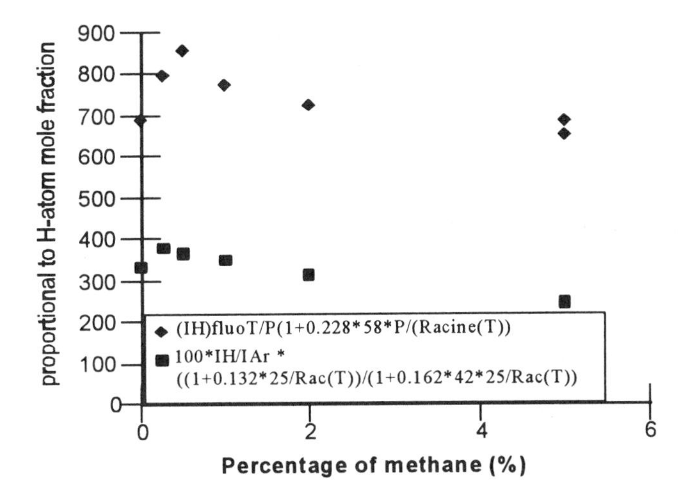

Figure 23: Relative H-atom mole fraction measured by TALIF and OES as a function of the percentage of methane introduced in the discharge. Plasma conditions: 9 W/cm3, 600 W, 25 mbar, 300 sccm.

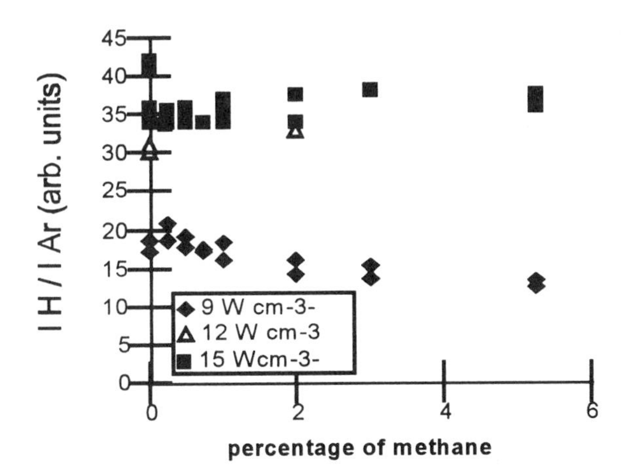

Figure 24: Emission intensity ration $I_{H\alpha}/I_{Ar}$ measured by actinometry as a function of the percentage of methane introduced in the discharge for two plasma conditions: 9 W/cm^3: (600 W, 25 mbar) and 15 W cm^{-3}:1000W, 52 hPa). The optical system used for OES is not the same as that used for figure 23.

collisions between CH_4 and electrons are less probable than those with H_2(or H). Nevertheless owing to the high rate constant of the reactions involving electrons, the rate of dissociation of CH_4 through the H-atom - CH_4 collisions must be compared to that through electron - CH_4 collisions. For electron temperatures lower than 20000 K and gas temperatures higher than 2000 K, the ratio is greater than 1000. Consequently, the thermal dissociation is much more important than the one occurring through electron-CH_4 collisions. For higher gas temperatures, the thermal effect will be reinforced. For higher electron temperatures, the electronic effect will be reinforced. Owing to the fact that the electron temperature is typically lower than 20000 K (see calculations) and that an increase of the methane percentage or of the power density (by a simultaneous increase of the power and pressure) should not increase the electron temperature, the assumption of considering only the thermal dissociation of CH_4 appears reasonable for the experimental conditions studied here. This analysis is in agreement with Hsu[7].

The chemical kinetics scheme for the thermal dissociation of CH_4 in mixtures H_2 + CH_4 has been taken from Harris,[17, 91, 92] J.A. Miller[93], M. Frenklach[94] and S. Girshick[48]. The chemical reactions[97] added to the previous model are the following:

Chemical reactions involving methane:

$2\ CH_3 + M \rightarrow C_2H_6 + M$ M= H_2 enhanced
$CH_3 + H + M = CH_4 + M$ M= H_2 enhanced
$CH_4 + H \rightarrow CH_3 + H_2$
$CH_3 + H \rightarrow CH_2 + H_2$
$C + H_2 \rightarrow CH + H$
$CH + C_2H_2 \rightarrow C_3H_2 + H$
$CH + CH_2 \rightarrow C_2H_2 + H$
$CH + CH_3 \rightarrow C_2H_3 + H$
$CH + CH_4 \rightarrow C_2H_4 + H$
$C + CH_3 \rightarrow C_2H_2 + H$
$C + CH_2 \rightarrow C_2H + H$
$C_2H_6 + CH_3 \rightarrow C_2H_5 + CH_4$
$C_2H_6 + H \rightarrow C_2H_5 + H_2$
$C_2H_4 + H \rightarrow C_2H_3 + H_2$
$CH_2 + CH_3 \rightarrow C_2H_4 + H$
$H + C_2H_4 + M \rightarrow C_2H_5 + M$ M=H_2 enhanced
$C_2H_5 + H \rightarrow 2\ CH_3$
$H_2 + C_2H \rightarrow C_2H_2 + H$
$C_2H_3 + H \rightarrow C_2H_2 + H_2$
$C_2H_3 + CH_2 = C_2H_2 + CH_3$
$C_2H_3 + C_2H \rightarrow 2\ C_2H_2$
$C_2H_3 + CH \rightarrow C_2H_2 + CH_2$
$C_2H_2 + C_2H \rightarrow C_4H_2 + H$
$CH_2(*) + M = CH_2 + M$ M = H enhanced
$CH_2(*) + CH_4 \rightarrow 2\ CH_3$
$CH_2 + CH_4 \rightarrow 2\ CH_3$
$CH_2(*) + C_2H_6 \rightarrow CH_3 + C_2H_5$
$CH_2(*) + H_2 \rightarrow CH_3 + H$
$CH_2(*) + H \rightarrow CH_2 + H$
$2\ CH_2 \rightarrow C_2H_2 + H_2$
$C_2H_2 + M \rightarrow C_2H + H + M$

$C_2H_4 + M \rightarrow C_2H_2 + H_2 + M$
$C_2H_4 + M \rightarrow C_2H_3 + H + M$
$2\,H + M \rightarrow H_2 + M$ H_2 enhanced
$C + CH_4 \rightarrow CH_3 + CH$
$C_2 + M = 2\,C + M$
$C_2 + H_2 + M = C_2H_2 + M$
$CH + M = C + H + M$
$C_2H + M = C_2 + H + M$
$C_2H + H = C_2 + H_2$
$C_3H_4 + H = CH_3 + C_2H_2$
$C_2H_3 + CH_3 = C_2H_2 + CH_4$
$C_2H_4 + CH_3 = CH_4 + C_2H_3$
$C_2H_5 + CH_3 = C_2H_4 + CH_4$
$CH_3 + M = CH_2 + H + M$
$CH_3 + CH_3 = C_2H_4 + H_2$
$C_2H_5 + C_2H_3 = C_4H_8$
$C_4H_6 + H = C_2H_4 + C_2H_3$
$C_4H_6 + C_2H_3 = C_2H_4 + C_2H_2 + C_2H_3$
$C_2H_3 + C_2H_3 = C_4H_6$
$C_4H_2 + M = C_4H + H + M$
$C_3H_4 + CH_2(*) = C_4H_6$
$C_4H + H_2 = C_4H_2 + H$
$CH_2(*) + H = CH + H_2$
$CH_2(*) + CH_3 = C_2H_4 + H$
$CH_2 + H = CH + H_2$
$H + C_2H_2 + M = C_2H_3 + M$ $M = H_2$ enhanced

In conditions where the electron temperature is not influenced by the introduction of methane, the depletion (or variation) of the H -atom mole fraction is only related to the chemistry involving thermal dissociation of the carbon containing species. It then depends only on the gas temperature and the initial H-atom mole fraction calculated in the absence of methane.

For a given set of experimental conditions, the initial H-atom mole fraction (in absence of methane) is the result of energy and mass balance and of the surface reactions. The 1-D diffusive model allows such calculations, the convective transport being neglected to date. The determination of the initial H-atom mole fraction when using the 0-D model can be done in a pure hydrogen plasma by solving the energy equations and by introducing losses of species (simulating the surface reactions). The determination of the H-atom mole fraction in H_2 + CH_4 plasma, using the 0-D model would require the solution of the energy equations, including all the electron-CH_4 processes such as vibrational excitation. It would also require the introduction of carbon containing species losses for simulating surface reactions. This work has not yet been done in detail, and should be somewhat difficult. Consequently, we have chosen to use a simpler model where no losses due to surface reactions are included. The variation of the initial H-atom mole fraction has been compensated for by varying the electron temperature, while the gas temperature and pressure are those provided by the experiment. The

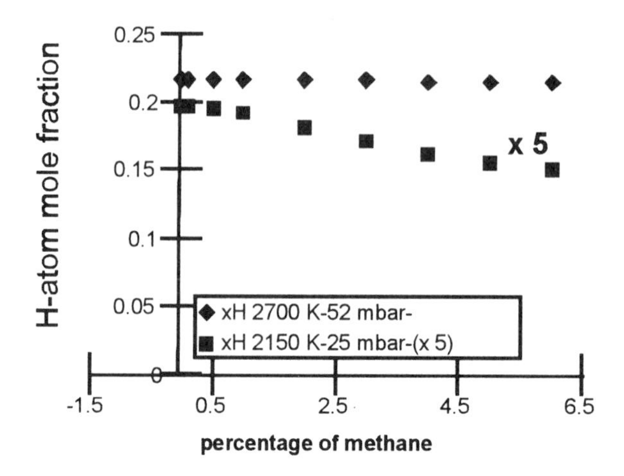

Figure 25: Calculated H-atom mole fraction as a function of the methane percentage, for two conditions of gas temperature and pressure: T_g =2150 K, 25 hPa simulating 9 W cm^{-3}; T_g =2700 K, 52 hPa simulating power of around 20 W cm^{-3}. Electron temperature = 15000 K, steady state simulation. x_H has been multiplied by 5 for the conditions: T_g=2150 K, P=25 hPa.

variation of the H atoms as a function of the methane percentage is then calculated keeping this electron temperature constant, and taking the measured values for the gas temperature and pressure. The calculations (variations of the H-atom mole fraction as a function of the electron temperature) are performed until the calculated variation of the H-atom mole fraction fits the measured variation. This simple method provides only a magnitude of the H-atom mole fraction for a given set of operating conditions.

Simulations were performed at steady state since high residence time are expected in this reactor due to recirculation of the gas (high thermal gradients). The experimental gas temperature is given (T_g = 2150 K at 9 W cm^{-3}) and the electron temperature is considered as a parameter for the simulation allowing variation of the initial H-atom mole fraction (absence of methane). Within the approximations, the results provide a magnitude of 4 % for the H-atom mole fraction in the absence of methane, for the set of conditions corresponding approximately to a power density of 9 W cm^{-3} (figure 25). This value is compatible with the amount of methane introduced (up to 5 %). The error is difficult to estimate owing to the approximations made. The 4 % H-atom mole fraction must be compared to the value of 6 % provided by the 1-D diffusive pure H_2 plasma model presented in sub-section 6 for the same experimental conditions, where energy balance equations were solved and surface reactions considered. We can conclude to a rather satisfactory agreement (around 50 %).

The decrease of 25 % of the H-atoms observed at 9 W cm^{-3} as 5 % of methane is introduced into the feed gas can be compared to the decrease of atomic hydrogen which has been observed in hot filament reactors by different authors. For example, Shäfer et al.[27] reported a decrease of 30 % of the atomic hydrogen at 2 mm from the filament when introducing 5 % of methane into the feed gas. The filament temperature was 2540

K and the pressure 30 hPa. Under these conditions the H-atom mole fraction is around 2 %, which is compatible with our results. On the other hand, Celii et al.[28] reported a decrease of almost 50 % when adding 1 % of methane in the feed gas with a filament at 2000 °C. Also, Tankala et al.[30] observed a decrease of 37 % of the atomic hydrogen when adding 1 % of methane at 10 mm from the filament maintained at 2473 K and 30 hPa.

While Goodwin[16] attributes the main contribution of the atomic hydrogen decrease to the reactions of H with the CH_4 molecules and the CH_3 radicals through $CH_3 + H + M \rightarrow CH_4 + M$ and $CH_4 + H \Leftrightarrow CH_3 + H_2$, Butler et al. as well as Tankala et al. attribute it to the positioning of the filament. Although their arguments are convincing, our results found in microwave plasma experiments show the strong effect of the homogeneous chemical reactions of the H atoms with the hydrocarbon species on the H-atom mole fraction when this latter is of the same magnitude as the percentage of methane which is added in the feed gas.

At higher gas temperatures (2700 K) corresponding to a higher power density (around 20 W cm^{-3}), the variation of the H-atom mole fraction as a function of the methane percentage is calculated assuming the same electron temperature as for the previous case (figure 25). As also observed experimentally by actinometry (figure 24), there is no depletion of the H-atom mole fraction. The H-atom mole fraction is overestimated here relative to the previous case since the electron temperature is kept constant. However, the purpose here is only to show that when the H-atom mole fraction is higher than the percentage of methane added, no depletion is observed.

Note: From 0 % to 0.5 % methane, an increase of the fluorescence signal as well as of the ratio $I_{H\alpha}/I_{Ar}$ (OES) is observed (figure 23). This has been attributed to an increase in the electron density owing to an increased ionization processes in presence of carbon containing species, which have lower ionization thresholds than H.

Variation of the H-Atom Mole Fraction as a Function of the Power Density

From the previous sub-sections, the variations of the Q_T and Q_{H2} quenching factors of the electron temperature (calculations with the 1-D diffusive model), and of the rate constant ratio k_{Ar^*}/k_{H^*} as a function of the electron temperature are known. Furthermore, using the magnitude for the H-atom mole fraction determined for a given set of experimental conditions allows us to determine the variations of the absolute H-atom mole fraction as a function of the power density using actinometric measurements. The variation of the argon-atom mole fraction as a function of the H_2 dissociation yield has been calculated. No effect of diffusional transport was seen on the argon-atom mole fraction.

The variation of the $I_{H\alpha}$ / I_{Ar} ratios has been systematically recorded as a function of the power density at a distance of 20 mm from the substrate (bulk of the plasma) (figure 26). The results show a strong increase of the $I_{H\alpha}$ / I_{Ar} ratio, particularly above a power density of 15 W cm^{-3}. At 30 W cm^{-3}(2000 W and 100 hPa), $I_{H\alpha}$ / I_{Ar} has increased by a factor of 16.5 compared that at 9 W cm^{-3}. Owing to the difficulties in measuring the dimensions of the plasma volume, an error of 10% on the power density is estimated.

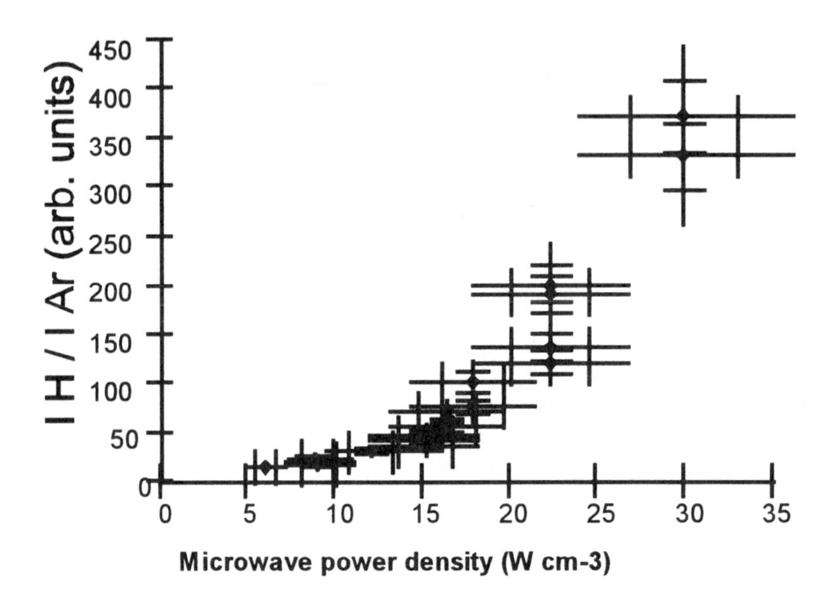

Figure 26: Emission intensity ratio I_H/I_{Ar} as a function of the averaged microwave power density. The microwave power density is varied by changing simultaneously the power and pressure, while the volume is kept constant as well as the ratio of the power by the total density.

The factor Q_{H2}, which takes into account the quenching by only molecular hydrogen, is almost constant for the range of conditions tested here (within 3 %). While Q_T, which takes into account the quenching by both atomic and molecular hydrogen, varies by around 60 % as the power density varies from 6 to 30 W cm^{-3} for both sets of cross sections calculated for the quenching by atomic hydrogen. From the variation of the electron temperature calculated with the 1-D diffusive model (decrease from 19500 K to 17000 K for an increase of the power density from 9 W cm^{-3} to 30 W cm^{-3}), the variation of the ratio k_{Ar*}/k_H has been estimated (decrease from 0.8 to 0.7).

Using the calibration of 4 % obtained for the H-atom mole fraction at 9 W cm^{-3}, the variation of the H-atom mole fraction as a function of the power density is reported in figure 27.

If we neglect the quenching by atomic hydrogen, we find that x_H increases from 0.01 to 0.45 as the power density increases from 6 W cm^{-3} to 30 W cm^{-3}. However, when considering the quenching by atomic hydrogen, x_H varies from 0.01 to 0.32. The same result is obtained using both sets of cross sections calculated (hard sphere model and Lennard-Jones potential curves) for the quenching by atomic hydrogen. The error on the H-atom mole fraction is quite large owing to the many steps we used for reaching the final result. It can even reach almost 50 %.

The H-atom mole fractions deduced from experiments can be compared to that calculated with the 1-D diffusive model (figure 27). Considering the experimental errors as well as the assumptions made for elaborating the model, a rather satisfactory

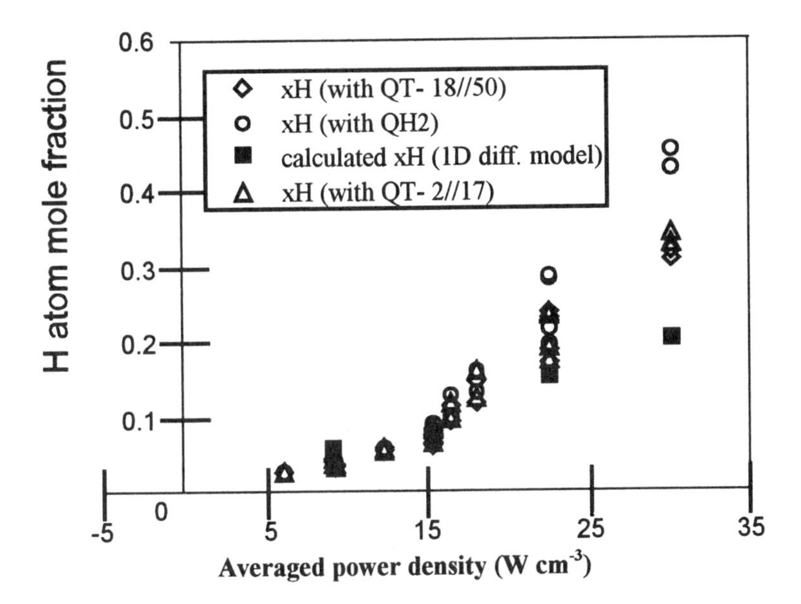

Figure 27: H-atom mole fraction as a function of the power density deduced from experiments presented in figure 26. Presented values were obtained when considering only the quenching by molecular hydrogen (Q_{H2}), and those when considering the quenching by molecular hydrogen and atomic hydrogen (Q_T). For this latter case, two sets for the quenching of Ar(4p) and H(n=3) by atomic hydrogen were used: respectively (50 Å^2 and 18 Å^2) and (17 Å^2 and 2 Å^2). Values calculated with the 1D diffusive model are also given on the figure.

agreement between experiment and simulation is found. On the one hand, the calculated values enter into the experimental error bar, which is quite large (within 50 %). On the other hand, the convection transport has been neglected in the model, and a Maxwell distribution has been assumed for the electron energy. Both assumptions may lead to some errors.

Over the range of conditions tested here, the main mechanisms of dissociation might be different. At low power density (6-12 W cm^{-3}), the electron-molecule collisions are probably more efficient (T_e high, T_g moderated) than at high power density (> 20 W cm^{-3}), where the thermal processes become rather important (T_e is lower, and T_g higher).

Note: The variations of the experimental H-atom mole fraction as a function of the power density has been deduced on the basis of four important assumptions: (i) the cross sections of the quenching of Hα and Ar* by H atoms are approximated, (ii) the H-atom mole fraction is equal to 4 % at 9 W cm^{-3} in a pure H_2 plasma. This value has been found assuming that there is no variation in the electron temperature as a function of the methane percentage, (iii) actinometry is valid for H-atom mole fraction higher than 1 to 5 %. Consequently, these results must be taken with precautions. In particular, in the range 1 to 5 %, we have not taken any correction factor owing to the possible contribution of the dissociative excitation mechanism. Nevertheless, the rather good agreement between the experimental and the modeling approaches suggests that the magnitude for x_H can be considered as reasonable, (iv) the variations of k_{Ar*}/k_H are well enough described by the cross sections from reference 68.

The values of the H-atom mole fraction obtained as a function of the power density can be compared to the H-atom mole fraction reported in the literature. Concerning the microwave plasma reactor, to our knowledge, only Hsu[7] reported a value of $1.2 \cdot 10^{-3}$ for plasma conditions of 800 W and 26 hPa. Strictly, we cannot compare our measurements to his, since Hsu had measured the H-atom mole fraction at the surface. However, Hsu calculated the corresponding H-atom mole fraction in the bulk of the plasma (at the entry of the diffusive boundary layer) and found a value of around 0.08 which is consistent with our results. Concerning the hot filament reactor, Shäfer et al.[27] reported a value of around 2 % at 2 mm from the filament at a filament temperature of 2540 K and a pressure of 30 hPa. L. Connell et al.[29] reported values of 2.8 % at 1 mm from the filament maintained at 2373 K at a pressure of 35 hPa. Chen et al.[22] have studied the variation of the H-atom mole fraction as a function of the power applied to the filament. As the power increases from 0.5 kW to 2.7 kW, the filament temperature increases from 2000 K to 2820 K, and the H-atom mole fraction at 1 mm from the filament increases from less than 5 % to 50 %.

This comparison shows that hot filament and microwave plasma reactors produce the same magnitude of H atoms, and that, for both reactors, the H-atom mole fraction is very sensitive to the experimental conditions. The diamond growth rate and quality are known to be very sensitive to the experimental conditions.

Spatial Distributions of the H-Atom Mole Fraction

As discussed above, for a given set of discharge conditions, the spatial evolution of the H-atom mole fraction can be reached by the means of TALIF. In addition, if the conditions required for using actinometry are satisfied, in particular if there is no drastic change in the spatial distribution of the electron energy distribution function, the emission intensity ratio $I_{H\alpha}/I_{Ar}$ should be proportional to the H-atom mole fraction. The profiles obtained by TALIF and actinometry are compared for a given set of experimental conditions.

Axial distributions. In figure 28a, the axial distributions of the H-atom relative mole fraction obtained by TALIF $[(IH)_{fluo}*(T/P)*(1+0.228\ \sigma_{Ar*/H2}\ PT^{-1/2})]$ in the presence and in the absence of the diamond substrate holder are presented (power density (9 W cm^{-3})). The H-atom mole fraction measured in the bulk of the plasma remained unchanged as the substrate holder is removed, confirming that the surface does not influence the plasma characteristics in the bulk of the plasma (at distance greater than 15 mm). Using the definition given by Goodwin[16], a boundary layer thickness of 13 mm is found, which is twice that of the thermal boundary layer. This results is very surprising since the diffusion coefficient is only around 1.4 times the thermal diffusivity[16]. The deposition process is then completely controlled by diffusion in a layer of around 13 mm thick at the plasma/surface interface.

Again, the fact that both the gas temperature and the H-atom mole fraction remained unchanged as the substrate holder is removed, would indicate that in the bulk of the

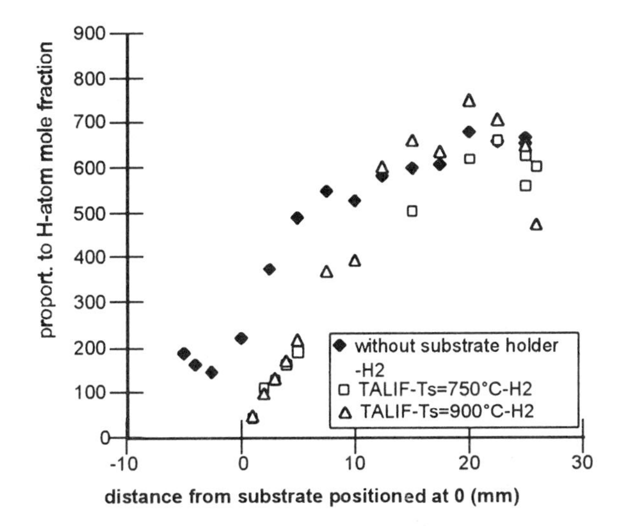

Figure 28a: Axial distribution of the relative H-atom concentration measured by TALIF in presence and in the absence of an untextured diamond substrate with its holder. Plasma conditions: 9 W/cm3, 25 mbar, 300 sccm,H2.

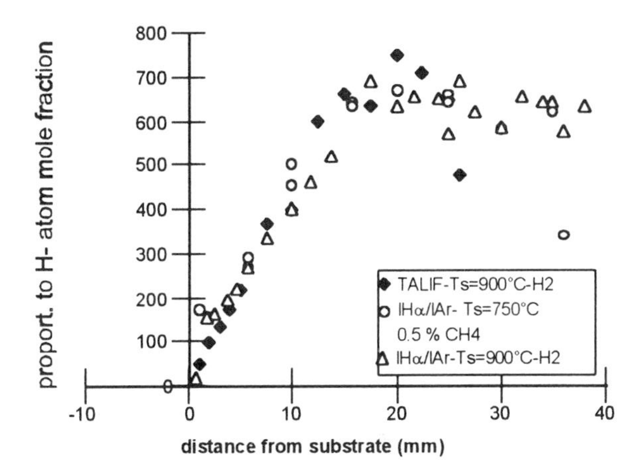

Figure 28b: Axial distribution of the relative H-atom concentration measured by OES and TALIF in presence of an untextured diamond substrate. Plasma conditions: 9 W/cm3, 25 mbar, 300 sccm, H2.

plasma, far from the diffusive and thermal layers, the species and energy fluxes are probably partly controlled by convective transport.

A decrease of the H-atom mole fraction at the border of the plasma is observed in the experiment performed in the absence of the substrate. It shows the strong non uniformity of these types of plasmas, controlled by the distribution of the microwave power density (see the modeling section).

For the same experimental conditions, line-of-sight averaged $I_{H\alpha}/I_{Ar}$, recorded along the reactor are presented on figure 28b. Although $I_{H\alpha}/I_{Ar}$ were line-of-sight averaged and the spatial variations of the ratio $k_e^{Ar^*}/ k_e^{H\alpha}$ (owing to the spatial distribution of the electron temperature) is not considered, a very good agreement with TALIF measurements is obtained. The profiles reveal a region of production of the H atoms, and a region of loss owing to the very efficient surface reactions such as H-atom recombination reactions (pure H_2 plasma). In between, a plateau is observed corresponding to a region where the rates of production and loss are balanced.

Note: The error made by considering line-of-sight averaged instead of Abel inverted measurements is estimated at 10 % on the axis of the reactor (in the bulk of the plasma) for this power density (see later). Owing to the decrease in the electron temperature calculated at the borders of the plasma, the values of $I_{H\alpha} / I_{Ar}$ are probably slightly overestimated in that region (entry of the plasma).

Under these conditions, a magnitude of the H-atom mole fraction of 4 % was found in the bulk of the plasma. The agreement between TALIF and OES results suggest again that the effect of the contribution of the dissociative excitation is less than the experimental uncertainty. This contributed to the belief that dissociative excitation does not appear to be as crucial under these experimental conditions, although the H-atom mole fraction is low.

The relative measurements can be compared to the spatial distribution of the H-atom mole fraction calculated using the 1-D diffusive model in the presence of the substrate (figure 29a). When considering the same power density (9 W cm^{-3}), a discrepancy is observed in the values of the relative H-atom mole fraction both in the bulk of the plasma and in the diffusive layer. This can be attributed to three main possible causes: (i) the surface H-atom recombination coefficient is not equal to 0.1 as found in the literature[16] (error in the gradient of x_H at the surface), (ii) the Maxwell distribution assumption made for the electron energy is not realistic (error of x_H in the plasma volume), (iii) the convection transport cannot be neglected (error of x_H in the plasma volume). However, it is interesting to compare the axial distributions calculated at 9 W cm^{-3} to an experimental distribution obtained at a slightly higher power density (12 ± 3 W cm^{-3}). This time, a pretty good agreement is observed in both the bulk of the plasma and in the diffusive layer (figure 29a). A diffusive layer thickness of 10 mm is estimated, which is closer to the calculated thickness (8.5 mm= 1.4 x 6 (thermal boundary layer)). Also, at a higher power density (15 W cm^{-3}) (figure 29b), very good agreement is observed between the calculated H-atom mole fraction distribution and the experimental distributions. The discrepancy observed at the lower power densities might be related to the actually unknown real power density injected into the plasma.

Owing to the number of assumptions made and to the unknown exact absorbed power density, we can conclude a very satisfactory agreement between calculated and measured distributions of atomic hydrogen. A rather high level of confidence in the model can be accepted now, allowing us to increase its complexity (coupling to the Navier Stokes equations, coupling to the Boltzmann equation for the electrons).

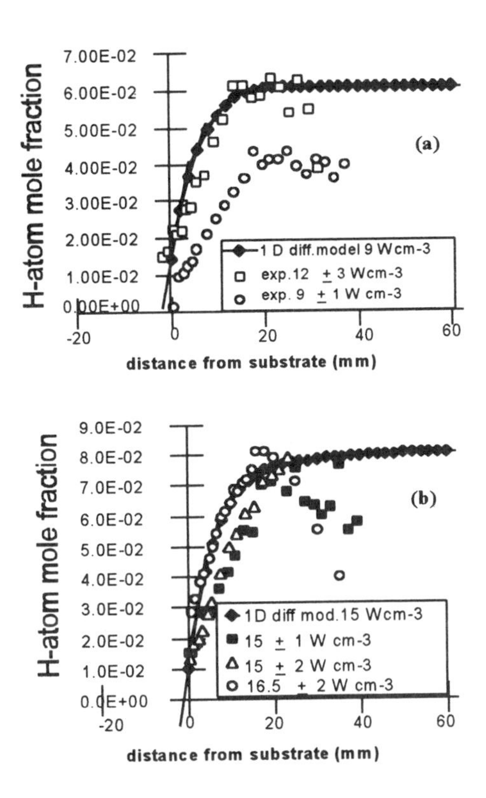

Figure 29: Comparison between experiments and simulations. Plasma conditions: (a) 9-12 W/cm3 corresponding to (600 W, 25 hPa) and (800 W, 40 hPa), 300 sccm,H2, Ts =900 °C; (b) 15-16.5 W cm^{-3} corresponding to (1000W, 52 hPa) and (1100 W, 60 hPa), 300 sccm, H_2 , Ts =900 °C. The errors indicated for the averaged power density depend on the microwave generator used (1,2 kW or 6 kW).

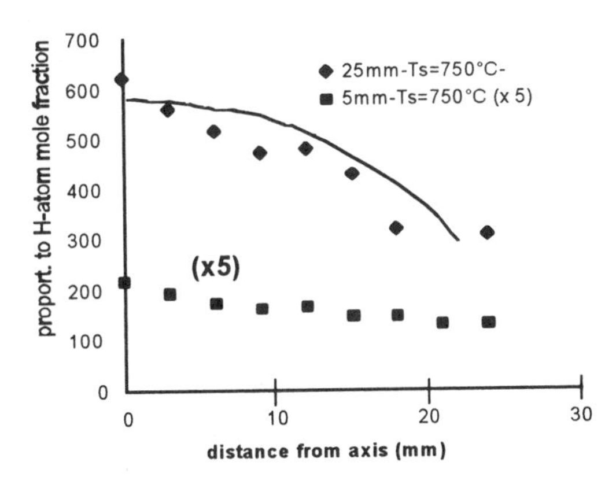

Figure 30: Radial distributions of the (IH*)fluo*(T/P)*QH2 obtained by TALIF, proportional to H-atom mole fractions. Plasma conditions: 600 W, 25 hPa, 9 W cm-3, 300 sccm, diamond temperature =750°C, H2. The values corresponding to the profile at 5 mm from the substrate are multiplied by 5.

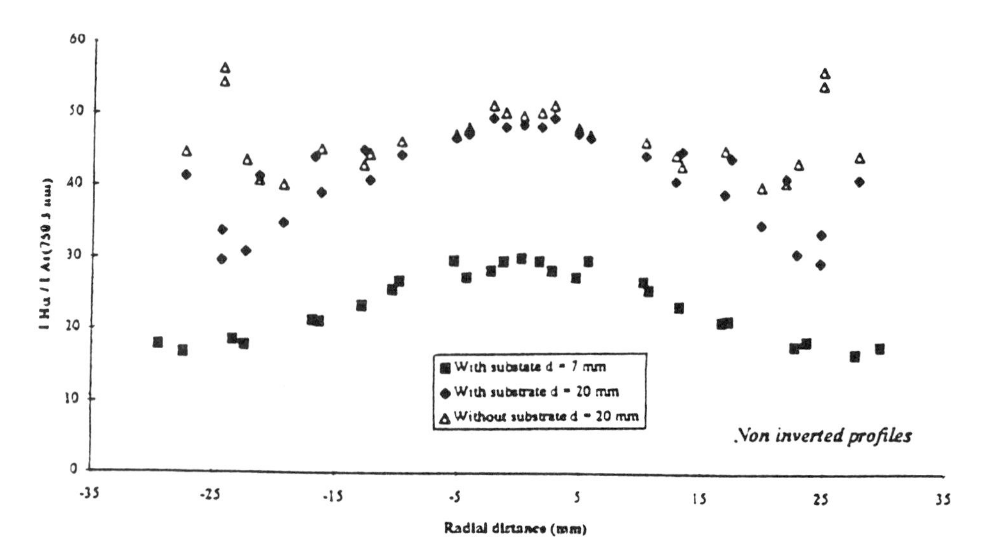

Figure 31: Line-of-sight averaged radial distributions of $I_{H\alpha}/I_{Ar}$ at two axial positions (bulk of the plasma at 20 mm from the substrate and plasma/surface interface at 7 mm from the substrate). At 20 mm from the substrate, two profiles are shown monitored respectively in presence and in absence of the diamond surface. Plasma conditions: 600 W, 25 hPa, 9 W cm^{-3}, 300 sccm, Ts = 740°C.

Radial distributions. For a power density of 9 W cm^{-3}, radial distributions of H-atom relative mole fractions have been measured by TALIF in the presence of a diamond substrate maintained at 900 °C. Radial distributions are given in figure 30, corresponding to two axial positions (respectively 25 and 5 mm from the substrate). In the bulk of the plasma, we can observe a strong non uniformity of the plasma which has already been discussed. It is strongly reduced at the interface (5 mm from the substrate). A rather high non uniformity in the radial H-atom fluxes (and energy) is a consequence of these results.

Under the same experimental conditions, radial distributions of $I_{H\alpha}$ / I_{Ar} have been measured both in the bulk of the plasma and at the plasma/ surface interface (respectively at 20 mm and 7 mm from the substrate). Line-of-sight averaged profiles and Abel inverted profiles are presented respectively in figures 31 and 32, without corrections, owing to the radial variation of $k_e^{Ar^*}/ k_e^{H\alpha}$, corresponding to the radial distribution of the electron temperature. In the bulk of the plasma, the error made in considering line-of-sight averaged measurements instead of Abel inverted profiles is around 10 % on the axis of the reactor.

Within the experimental errors, a rather good agreement is found between TALIF and actinometric measurements, even without considering any radial variation for $k_e^{Ar^*}/ k_e^{H\alpha}$.

The radial decrease of the H-atom mole fraction from the center to 20 mm from the axis is around 33 % using actinometry and around 38 % using TALIF.

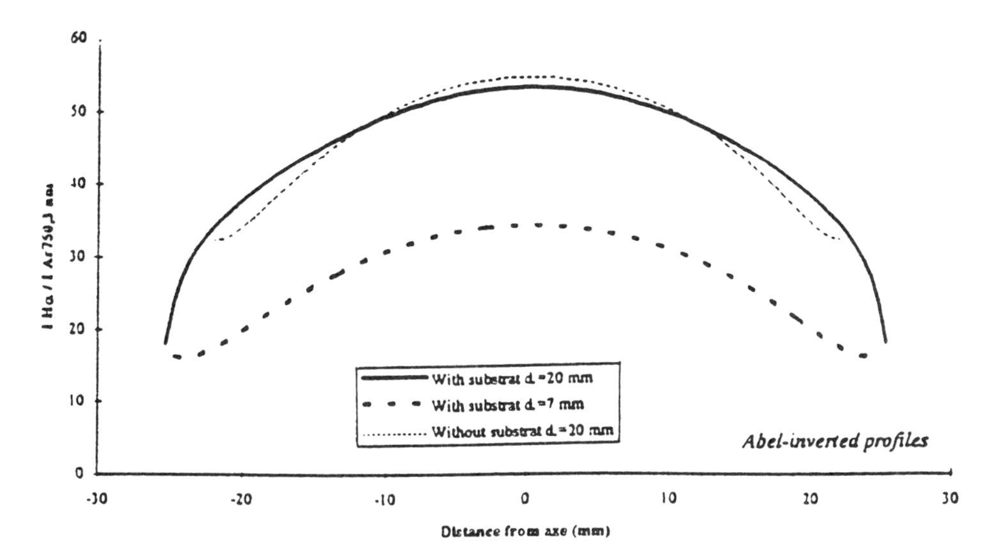

Figure 32: Abel inverted radial distributions of I_{Ha}/I_{Ar} at two axial positions (bulk of the plasma at 20 mm from the substrate and plasma/surface interface at 7 mm from the substrate). At 20 mm from the substrate, two profiles are shown monitored respectively in presence and in absence of the diamond surface. Plasma conditions: 600 W, 25 hPa, 9 W cm^{-3}, 300 sccm, Ts = 740°C.

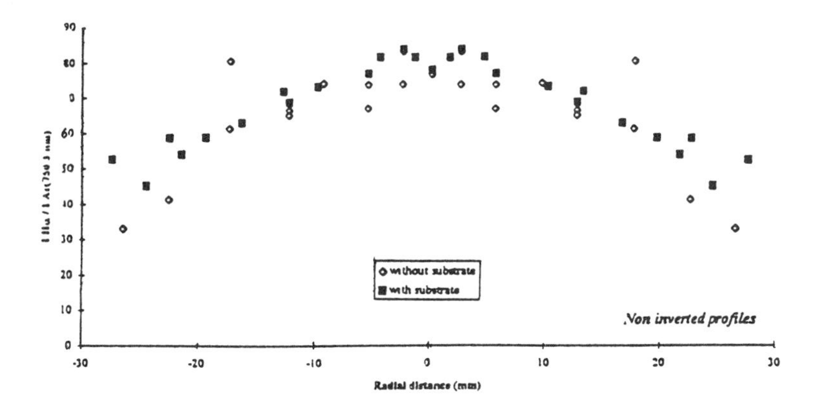

Figure 33: Line-of-sight averaged radial distributions of I_{Ha}/I_{Ar} at 20 mm from the substrate. Two profiles are shown monitored respectively in presence and in absence of the diamond surface plasma conditions: 1000 W, 52 hPa, 15 W cm^{-3}, 300 sccm.

At the same power density, the radial distribution of $I_{H\alpha}$ / I_{Ar}, measured in the bulk of the plasma in the presence of the substrate holder (hemispherical plasma) is compared to that obtained in its absence (spherical plasma) (figures 31 and 32). There is no variation of $I_{H\alpha}$ / I_{Ar} in the plasma volume as a function of the position of the substrate holder, showing again that the substrate is not influencing the plasma volume characteristics.

Radial variations of the $I_{H\alpha}$ / I_{Ar} at 20 mm from the substrate has been measured also in the presence and the absence of the substrate holder, at higher averaged power density (15 W cm^{-3}). Line-of-sight averaged profiles are presented in figure 33, again, at this power density, there is no variation of $I_{H\alpha}/I_{Ar}$ as the substrate holder is removed.

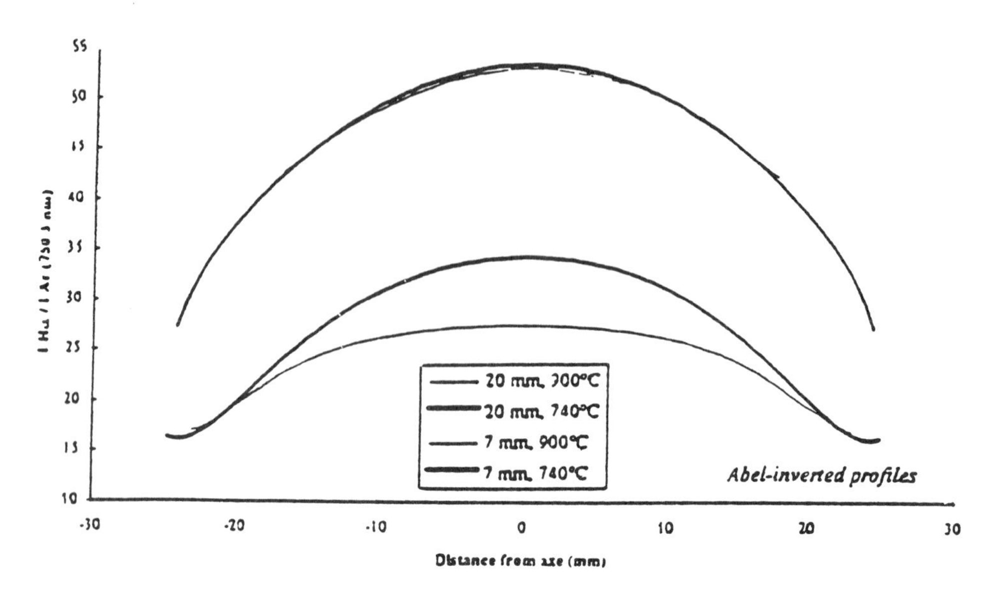

Figure 34: Radial Abel inverted distributions of $I_{H\alpha}/I_{Ar}$ at two positions, in the bulk plasma at 20 mm from the substrate and at the plasma/surface interface at 7 mm from the substrate. Two substrate temperatures (900°C and 750°C) are applied while the plasma conditions are identical: power=600 W, pressure = 25 hPa, power density = 9 W cm^{-3}, flow rate =300 sccm.

Influence of the substrate temperature. Keeping constant the plasma parameters (power density of 9 W cm^{-3}), the effect of the substrate temperature on the H-atom mole fraction measured at the plasma / surface interface has been studied by actinometry.

Two substrate temperatures, respectively 740 °C and 900 °C, were tested. The effect of the substrate temperature on the $I_{H\alpha}$ / I_{Ar}, measured at 7 mm from the substrate, is rather important: the higher the substrate temperature, the lower the H-atom mole fraction, showing the increased efficiency of the surface reactions as the substrate temperature increases (figure 34). The ratios of $(x_H)_{7\ mm\ from\ surface}/(x_H)_{volume}$ decreases from 0.65 at 750 °C to 0.5 at 900 °C.

Conclusion

This sub-section has been dedicated to the quantitative estimation of the H-atom mole fraction in a microwave bell jar plasma reactor. Two photon allowed transition laser induced fluorescence and optical emission spectroscopy have been used for (i) analyzing the variations of the H-atom mole fraction as a function of the operating conditions (percentage of methane in the feed gas and power density), and (ii) spatially (axially and radially) analyzing the plasma.

The validity of the spectroscopic method of actinometry has been carefully analyzed as a function of the operating conditions. It can be used under conditions where the electron temperature is less that 20000 K, and the H-atom mole fraction more than few percent. Simultaneous measurements of TALIF and OES showed that at 9 W cm^{-3}, the actinometry method is valid. For using actinometry at higher power densities, we needed to determine correction factors owing to elementary processes, which were unknown.

Simultaneous measurements by TALIF and OES allowed us to estimate the quenching cross section of the excited 4p (2p1) argon state by molecular hydrogen. A value of 42 $\pm$ 30 $Å^2$ was found. The quenching cross sections of excited 4p (2p1) argon state and of excited H(n=3) state by atomic hydrogen have been estimated on the basis of the hard sphere model and the energy Lennard-Jones potential curves found in the literature.

Then, a calibration of the TALIF measurements has been performed by comparing the experimental depletion of the H atoms observed as the methane is introduced into the discharge at low H-atom mole fraction, to that calculated with a 0-D chemical kinetics model working with H_2 + CH_4. A H-atom mole fraction of 0.04 was estimated for an averaged power density of 9 W cm^{-3} (gas temperature of the bulk of the plasma was 2150 K). The decrease of the H-atom mole fraction is only observed in conditions where the H-atom mole fraction has the same magnitude as the methane percentage introduced in the feed gas. At high power density, no decrease is observed.

With these data, the variations of the H-atom mole fraction as a function of the power density was obtained. It increases from 0.01 $\pm$ 0.005 to around 0.32 $\pm$ 0.15 as the power density increases from 6 W cm^{-3} to 30 W cm^{-3}. Under these conditions, the gas temperature increases from 1550 K to 3100 K, and the electron temperature decreases from 19500K to 17000 K. The dissociation mechanism of molecular hydrogen is mainly attributed, at low power density (6-12 W cm^{-3}), to the efficiency of the electron-molecule collisions (T_e high, T_g moderated), while, at high power density (> 20 W cm^{-3}), to the thermal processes efficiency (T_e is lower, and T_g higher).

Owing to the large experimental error and to the approximations made for establishing the model, we have concluded to a rather satisfactory agreement between the experimental H-atom mole fraction and those calculated with a 1 dimension diffusive H_2 plasma where energy and mass balance equations are solved and the surface reactions considered.

The H-atom mole fractions obtained in our microwave bell jar reactor were compared to those measured in hot filament reactors. The same magnitude for the H-atom mole fraction was found for both reactors. It was seen to be very sensitive to the experimental conditions, in particular to the power density for the microwave reactors and to the power supplied to the filament for the hot filament reactors. As it has already been stated, the main difference between hot filament reactors and microwave plasma reactors concerns the concentration and temperature gradients (fluxes) at the diamond surface in growth. Higher H-atom fluxes can be obtained using microwave plasma reactors than using conventional hot filament reactors.

With this work, we have shown that actinometry can be used for optimizing the plasma conditions used for diamond deposition. In addition to measurements of $T(H_2(G))$ and $T_{H\alpha}$, the emission intensity ratio, $I_{H\alpha}/I_{ar}$, can be used for controlling the plasma reactivity. Correction factors owing to quenching processes need to be used for estimating the variation of H-atom mole fraction. For in-situ monitoring, $I_{H\alpha}/I_{Ar}$ can be used with confidence.

Acknowledgments
The spectroscopic diagnostics performed in the microwave reactor were not possible without the financial support of DRET (DGA-France) and of BRITE EURAM Program. The authors are grateful to them. The students who have participated to spectroscopic measurements are deeply thanked: Yves Breton, Jean Christophe Cubertafon, Florence Souverville, Maxime Petiau and A. Tserepi. Also Christophe Jany and Christophe Vivensang are thanked for their constructive help. Nader Sadeghi and J. P. Booth from the Laboratoire de Spectrometrie Physique of Grenoble are kindly thanked for the very helpful discussions.

References
1. E. Vietzke, V. Philipps, K. Flaskamp, J. Winter, S. Veprek, P. Koidl, in U. Ehlemann, H. G. Lergon, K. Wiesmann (eds), *Proc. 10th Int. Symp. Plasma Chemistry*, Bochum, vol **3**, 1991, p 3.-1.
2. *Status on Applications of diamond and diamond like materials: an emerging technology.* National Materials Advisory board. Committee on superhard materials, Chairman J. D. Venables (1990)
3. N. Setaka, *J. Mater. Res.* **4**, 664 (1989)
4. B. B. Pate, *Surf. Sci. Lett.* **165**, 83 (1986)
5. F. G. Celii, J. E. Butler *Appl. Phys. Lett.* **54** (11), 1031-1033 (1989)
6. S. O. Hay, W. Roman, M. B. Colket III, *J. Mater. Res.*, Vol **5**, no. 11, (1990) 2387-97
7. W. L. Hsu, *J. Appl. Phys.* **72** (7) 3102-3109 (1992)
8. M. Tsuda, M. Nakajima, S. Oikawa, *J. Am. Chem. Soc.* **108**, 5780 (1986)
9. M. Tsuda, M. Nakajima, S. Oikawa, *ISPC-8 .eding*, vol **4**, 2452-2457 (1987)
10. M. Frenklach, K. E. Spear, *J. Mat. Science*, **3** (1) 133-140 (1987)
11. J. Angus et al., *Diamond and Related Materials* (1996) to be published
12. A. Gicquel, K. Hassouni, S. Farhat, Y. Breton, C. D. Scott, M. Lefebvre and M. Pealat. *Diamond and Related Materials*, vol **3**, no. 4-6, (1994) 581- 586.
13. K. Hassouni, S. Farhat, C. D. Scott, A. Gicquel, *Advances in New Diamond Science and Technology*, ed K. Kobashi, N. Fujimori, O. Fukunaga, M. Kamo, 131-134 (1994)
14. A. Gicquel, K. Hassouni, S. Farhat, C.D. Scott, M. Lefebvre, *Proceedings of the workshop on Microwave Plasma and Applications* (Sept 5-8, 1994, Zvenigorod, Russie), ed. Y Lebedev *The Moscow Physical Society*, pp. 105-120 (1995);
15. C. D. Scott, S. Farhat, A. Gicquel, K. Hassouni, M. Lefebvre, *AIAA Journal* 93-3226, 24th Plasmadynamics and Lasers Conference. July 6-9 (1993) Orlando, Florida, USA; C. D. Scott, S. Farhat, A. Gicquel, K. Hassouni, M. Lefebvre, J. Heat Transfers (1996) submitted
16. D. G. Goodwin, *J. Appl. Phys.* **74** (11) 6895-6906 (1993);
17. J. Harris and A. M. Weiner, *J. Appl. Phys.* **74**, 1022 (1993);. S. J. Harris, A. M. Weiner, R. J. Blint, Combust. Flame, **65**, 177 (1986)

18. B. J. Wood, H. Wise, *J. Phys. Chem.* **65**, 1976 (1961); **66**, 1049 (1962); H. Wise, B. J. Wood, *Adv. At. Mol. Phys.* **3**, 291 (1967); J. Wood and H. Wise, *J. Chem. Phys.* **66**, 1049 (1962).
19. B. Halpern, D. E. Rosner, *J. Chem. Soc. Faraday Trans.* **160**, 1883-1912 (1978)
20. F. G. Celii and J. E. Butler, *Appl. Phys. Lett.* **54**, 1031 (1989)
21. F. G. Celii, H. R. Thorsheim, J. E. Butler, L. S. Plano, J. M. Pinneo, *J. Appl. Phys.* **68**, 3814 (1990)
22. K. H. Chen, M. C. Chuang, C. M. Penney, W. F. Banhollzer, *J. Appl. Phys.* **71** (3) (1992), 1485-1493
23. J. Bittner, K. Kohse-Höinghaus, U. Meier, *Th. Just. Chem. Phys. Letters*, Vol. **143** n° 6 (1988).
24. M. Chenevier, J. C. Cubertafon, A. Campargue, J. P. Booth, *Diam. Rel. Mat.* **3**, 587-592 (1994).
25. A. Mucha, D.L. Flamm, D. E. Ibbotson, *J. Appl. Phys.* **65** (9), 3448-3452 (1989)
26. D.K. Milne, P. John, I.C. Drummond, P.G. Roberts, M. G. Jubber, J. I. B. Wilson, *Diamond and Related Materials* (1994) ?
27. L. Schäfer, U. Bringmann, C. P. Kalges, U. Meier, K. Koshe-Höinghaus, *Appl. Phys. Lett.* **58** (6) 571-573 (1991)
28. C. R. Koemtzopoulos et al *Diamond and Related Materials* **2**, 25 (1993)
29. L. L. Connell, J. W. Fleming, H. N. Chu, D. J. Vestryck, E. Jansen, J. E. Butler, *J. Appl. Phys.* **78** (6) (1995) 1-13
30. K. Tankala, T. DebRoy, *J. Appl. Phys.* **72** (2) (1992) 712-718
31. D. G. Goodwin, *J. Appl. Phys.* **74** (11) 6895-6906 (1993)
32. S. O. Hay, W. Roman, M. B. Colket III, *J. Mater. Res.*, Vol **5**, no. 11, (1990) 2387-97
33. T. G. Owano, E. H. Wahl, C. H. Kruger, R. N. Zare, *26th AIAA Plasmadynamics and Lasers Conference* (1995), Proceed. ISPC Minneapolis
34. J. Ropcke, A. Ohl, *Contributed Plasma Physics*, vol **34** (4) (1994) 575.586, K. Ito, N. Oda, Y. Hatano, T. Tsuboi,, Chem. Phys. vol **17** (1976) 35-43
35. G. Sultan G. Baravian, M. Gantois, G. Henrion, H. Michel, A. Ricard; *Chem. Phys.* vol **123** (1988) 423-429
36. J. Vetterhösser, A. Campargue, F. Stoeckel, M. Chenevier, *Diamond and Related Materials* **2** (1993) 481-485
37. A. Gicquel, M. Chenevier, K. Hassouni, Y. Breton, J. C. Cubertafon, *Diamond and Related Materials* (1996) to be published
38. H. N. Chu, E.A. Den Hartog, A. R. Lefkow, J. Jacobs, L. W. Anderson, M. G. Legally, J. E. Lawler, *Phys. Rev. A*, Vol **44**, n° 6 Sept (1991)
39. B. L. Prepperneau, K. Pearce, A. Tserepi, E. Wurzberg and T. A. Miller, Chemical Physics (1995); B. L. Prepperneau, *Optical diagnostics in a diamond deposition reactor. Ecole Diagnostics plasma*, Les Houches, Janv. 1995
40. S. W. Reeve, W. A. Weimer, *J. Vac. Sci. Techn.* **A 13** (2) (1995) 359-367
41. A. Gicquel et al- *Brite-Euram Report- contract* n° BRE 2 CT 920147- June 1995
42. M. Chenevier, *Optical diagnostics in a diamond deposition reactor. Ecole Diagnostics plasma*, Les Houches, Janv. 1995
43. P. Ewart, *Optical diagnostics in a diamond deposition reactor. Ecole Diagnostics plasma,* Les Houches, Janv. 1995

44. P. Zalicki, Y. Ma, R. N. Zare, E. H. Wahl, J. R. Dadamio, T. G. Owano, C. H. Kruger, *Chem. Phys. Lett.* **234** (1995) 269-274; P. Zalicki, R. N. Zare, *J. Chem. Phys.* (1995) vol. **102** (7) 2708-17.
45.M. A Childs, K. L. Menningen, H. Toyoda, L. W Anderson, J. E. Lawler, *EuroPhys. Lett.* **(25)** (9) (1994) 729-734
46. M. Cappelli , Ph. Paul, *Thin Solid film* (1991)
47. M. A. Cappelli, M. H. Loh; *Diamond and Related Mat.* **3** (1994) 417-421
48. S. L. Girshick, C. Li, W. Yu, H. Han, *Plasma Chem. Plasma Process.* **13** (1993); C. Li, B. W. Yu, S. L. Girshick, Proc. 10th Intl. Symp. Plasma Chem., Bochum, vol **3**, paper 3.1.9 (1991)
49. A. C. Eckberth, *Appl. Phys. Lett.*, Vol **32**, 421 (1978)
50. N. Herlin, M. Pealat, M. Lefebvre, P. Alnot, J. Perrin, *Surface Science* 258,381 (1991); B. Attal-Trétout, P. Bouchardy, P. Magre, M. Pealat, J. P. Taran, *Appl. Phys. B* **51**, 17 (1990)
51. T. Lang et al, *Diamond and Related Materials* (1996) to be published
52. A. Gicquel, M. Chenevier, J. C. Cubertafon, C. Jany, A. Tserepi, *Optical diagnostics in a diamond deposition reactor. Ecole Diagnostics plasma*, Les Houches, Janv. 1995
53. M. Chenevier, A. Gicquel, J. C. Cubertafon, *Proceedings of the third International Conference on the Applications of Diamond films and related Materials- ADC'95-* , 305-312 (1995) ed Y. Tzeng, M. Yoshikawa, M. Murakawa, A. Feldman, Elsevier Sciences publishers.)
54. A. Phelps, *J. Phys. Chem. Ref Data*, **19**, 653 (1990)
55. S. Green *J. Chem. Phys.* vol **62**, n°6 (1974) 2271-77
56. K. Dressler, L. Wolniewicz, *J. Mol. Spect.* **67,** 416 (1977)
57. K. Dressler, R. Galluser, P. Quadrelli, and L. Wolniewicz, *J. Mol. Spect.* **75**, 205, (1977)
58. I. Kovacs, *Rotational Structure in the Spectra of Diatomic molecules*, Adam Hilger LTD, London, 1969.
59. G. H. Dieke, *J. Mol. Spectrosc.* **2**, 494 (1958)
60. C. Barbeau " *Etude de la région cathodique d'une décharge luminescente d'hydrogène: diagnostics spectroscopique*" Université de Paris Sud 13 Decembre 1991.
61. J. M. Mermet, J. P. Robin, *Rev. Int. des Hautes Temp. et Refract.*, 1973, tome **10**, pp 133-139.
62. Jayr de Amorim Filho, *PhD thesis, Université d'Orsay* (Paris XI) (1994)
63. D. K. Otorbaev, V. N. Ochkin, P. L. Rubin, S. Yu Savinov, N. N. Sobolev, S. N. Tskhai, *Electron-excited molecules in non equilibrium plasma, Technical report of the Lebedev Physics Institute*, Academy of Sciences of the USSR, seres editor N. G. Basov, vol. **179**, supplemental volume , edited by Sobolev, translated by Kevin S. Hendzel, Nova Science Publishers, Commack, pp 121-173 (1989).
64. J. W Coburn, M. Chen*J. Appl. Phys.* **51** (6) (1980) 3134-36
65. V. M. Donelly, D. L. Flamm, W. C. Dautremont-Smith, D. J. Werder, *J. Appl. Phys.* (1) (1984) 242-252
66. R. E. Walkup, K. L. Saenger, G. S. Selwyn, *J. Chem. Phys.* **84** (5) (1986) 2668-74
67. R. d'Agostino, F. Cramarossa, S. de Benedicti, G. Ferraro, *J. Appl. Phys.* **52** (3) (1981) 1259-65

68. S.J. Buckmann and A. V. Phelps, *JILA Informations Center Report n°27*, University of Colorado Boulder (1985)
69. P. Laborie, J. M. Rocard, J. A. Rees, *Tables de sections efficaces électroniques et coefficients macroscopiques 1 Hydrogène et gaz rares ed Dunod Paris* 1968; I. P. Zapesochnyi, P. V. Feltsan, Opt. and Spectrosc., U.R.S.S., english transl. 20, 291 (1966)
70. R. K. Janev, W. D. Langer, K. Evans Jr, D. E. Post Jr, *"Elemantary Process in Hydrogen Plasma"*, Springer Verlag 1987.
71. A. Rousseau, A. Granier, G. Gousset, P. Leprince; *J. Phys.D: Appl. Phys* (1994) 1412-1422
72. L. St. Onge, M. Moisan, *Plasma Chemistry and Plasma Processing*, vol **14**, n° 2 (1994)
73. O. Krogh, T. Wicker, B. Chapmann *J. Vac. Sci. Techn.* **B4** 1292 (1986)
74. N. Sadeghi, D. W. Setser, *Chem. Phys.* **95** (1985) 305-311
75. M. A. A Clyne, P. B. Monkhouse, D. W. Setser, *Chem. Phys.* **28** (1978) 447
76. J. Lorenzen, H. Hotop, M. W. Morgner, *Z. Physik A* **297** (1980) 19
77. P. Ségur, M. C. Bordage, *Proc. XIXth ICPIG*, Invited paper page 86 (1989) ed. U. J. Zigman Belgrade
78. H. Tawara, Y. Itikawa, H. Nishimura, M. Yoshino, *J. Phys Chem ref Data*, Vol **19**, n° 3, (1990) 617-636
79. D. M. Cox and S. J. Smith, *Phys. Rev.* **A5**, 2428 (1972) (H2--H(2s)
80. W. L. Fite, R. T. Brackmann, D. G. Hummer, R. F. Stebbings, *Phys. Rev.* Vol **116**, 363-367 (1959)
81. W. E. Lamb. Jr., R. C. Retherford, *Phys. Rev.* **79**, 549 (1950), M. Glass-Maujean Phys. Rev. Let (1989) vol **62** 144
82. S. R. Ryan, S. J. Czuchlewski, M. V. McCusker, *Phys. Rev.* **A16** (1977) 1892
83. V. Dose, A. Richard, J. Phys. **B14** (1981) 63
84. Y. Lebedev, *Proceedings of the workshop on Microwave Plasma and Applications* (Sept 5-8, 1994, Zvenigorod, Russie), ed. Y Lebedev The Moscow Physical Society (1995); Holstein, *Phys. Rev.* (1951) vol **83**, 1159; L.L. Alves, G. Gousset, C. M. Ferreira, *J. Phys. D: Appl. Phys.* (1992) vol **25** 1713-32
85. A. Gicquel, P. Saillard, S. Piétré, M. Cappelli, J. Amouroux, *Proceedings of the 10th International Symposium on Plasma Chemistry ISPC10*, vol **2**, 2.1.1-2.1.6 (1991)
86. J. Tung, A. Tang, G. Salamo, F. Chan, *J. Opt. Soc. Amer.* **B3**, 837 (1986)
87. J. E. Velazco, J. H. Kolts, D. W. Setser, *J. Chem. Phys.* **69** (10) (1978) 4357-4373; J. O. Hirschfelder, C. F. Curtis, R. B. Bird, Molecular theory of gases and liquids, Wiley, New York (1954)
88. Y. Matsui, A. Yuuki, N. Morita, K. Tachibana, *Japn. J. Appl. Phys.*, **26**, 1575 (1987); M. J. Kushner, *J. Appl. Phys.*, **63**, 2532 (1988)
89. R. C. Cheshire, V. Kornas, H. F. Döbele, K. Donnelly, D. Dowling, W. G. Graham, T. Morrow, T. O'Brien, *Proceed. ESCAMPIG 94, Twelfth European Sectional Conference on the Atomic and Molecular Physics of Ionized Gases*, vol **18** E, 474 (1994)
90. J. Bittner, K. Kohse-Höinghaus, U. Meier, S. Kelm, *Th. Just, Combustion and Flame* **71**, 41-50 (1988)
91. S. Harris, *J. Appl. Phys.* **65**, (8), 3044-48 (1989)

92. S. J. Harris, A. M. Weiner, R. J. Blint, *Combust. Flame*, **65**, 177 (1986)
93. J. A. Miller, C. T. Bowman, Prog. *Energy Combust. Sci.* **15**, 287 (1989)
94. M. Frenklach, H. Wang, *Phys. Rev. B* **43**, 1520 (1991)
95. S. W. Reeve, W. A. Weimer, F. M. Cerio, *J. Appl. Phys.* **74** (12) (1993) 7521-30
96. S. W. Reeve, W. A. Weimer, *Thin Solid Films* **253** (1994) 103-108
97. S. Farhat, K. Hassouni, C. Scott, A. Gicquel, in preparation
98. T. Grotjohn, *Handbook on Industrials Diamond Films section 6*, ed . M. Prelas, K. Bigelow, G. Popovicci
99. K. Hassouni, S. Farhat, C. Scott, A. Gicquel, submitted to *Journal of Physic III* (1996)
100. E. U. Condon, G. H. Shortley, *Theory of atomic spectra*, Cambridge University Press, Cambs (1959)
101. A. Gicquel, M. Chenevier, Y. Breton, M. Petiau, J. P. Booth, K. Hassouni, in preparation
102. W. Demtroder. Laser Spectroscopy. *Basic concepts and Instrumentation. Volume 5 of Series in Chemical Physics.* Springer- Verlag edition (1981)
103. *"Les Applications Analytiques des Plasmas HF"* C. Tracy et J.M. Mermet; ed. Technique et Documentation, Lavoisier (1984)
104. W. L. Wiese, *Plasma Diagnostics Techniques-Pure and Applied Physics* (21) R. H. Huddlestone, S. L. Leonard Eds - Academic Press- (1965)
105. R. Ferdinand, *PhD Université d'ORSAY- Paris* (1990)
106. C. Kaminski, P. Ewart, *App. Phys. B*, **61**, 585-592 (1995)
107. H. R. Griem, *Plasma Spectroscopy*, Mac Graw Hill Book Company. (1964)

Chapter 20
HOT-FILAMENT CVD METHODS

Alberto Argoitia, Christopher S. Kovach and John C. Angus
Department of Chemical Engineering
Case Western Reserve University
Cleveland, OH 44106-7217

Contents

1. Introduction

Hot tungsten filaments were first used to generate atomic hydrogen for removal of non-diamond carbon from diamond by Chauhan et al. [1976]. Matsumoto et al. [1982[a]] gave the first description of a process that used hot filaments during growth. Virtually all contemporary hot-filament assisted processes for diamond deposition are derived from the latter publication.

The design of hot-filament reactors involves an artful choice of dimensions, materials and operating conditions that permit the generation and transport of sufficient quantities of atomic hydrogen and hydrocarbon free radicals to the growth surface. One principal difference between hot-filament and plasma assisted methods is the constraint imposed by the upper temperature limit of the filament material. Hot-filament assisted processes operate at significantly lower gas activation temperatures than plasma processes, and consequently produce less atomic hydrogen. This leads directly to the use of low pressures in hot-filament reactors to enhance atomic hydrogen production and transport. The low gas phase concentrations give relatively low growth rates compared to the plasma methods.

Despite these drawbacks, hot-filament assisted deposition has remained popular because of its low capital cost and simplicity. Also, hot-filament reactors are directly scaleable to large sizes and can be used to coat complex shapes and internal surfaces.

The typical range of operating parameters used in hot-filament reactors is given in Table I. The list of carbon sources is not meant to be exhaustive and there have been reports of diamond growth outside the parameter ranges listed. A typical laboratory version of a hot-filament reactor is shown in figure 1.

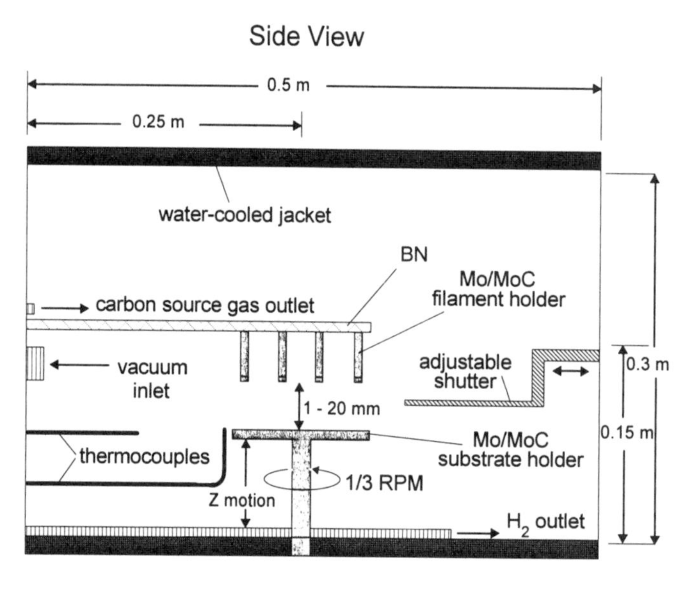

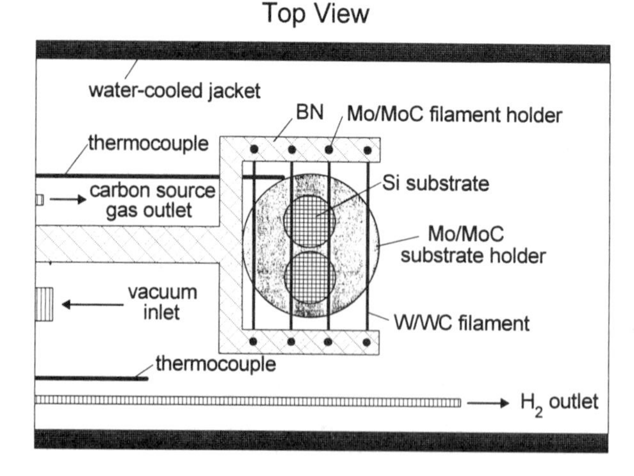

Figure 1. Side and top views of a typical hot-filament laboratory reactor. [Cassidy et al. 1995]

Table 1. Range of typical operating conditions for hot-filament CVD of diamond.

Operating Parameter	Typical Range
Pressure (torr)	1 - 80
Substrate Temperature (C)	600 - 1200
Substrate to Filament Distance (mm)	1 - 20
Filament Temperature (C)	2000 - 2600
Filament Materials	tungsten, tantalum, rhenium
Carbon Sources	CH_4, $C_2H_{(X=2,4,6)}$, CH_3OH and other alcohols, H_2CO
Feed Conditions	0.1 - 7% hydrocarbon in H_2 and 0 - 3% oxygen

2. Chemical and Transport Processes in Hot-filament Reactors

2.1 DISSOCIATION OF MOLECULAR HYDROGEN

The principal role of the hot-filaments is to dissociate molecular hydrogen to atomic hydrogen. The atomic hydrogen formed in this highly endothermic step drives the rest of the chemistry through a series of exothermic reactions [Angus et al. 1993]. The thermal dissociation of molecular hydrogen to atomic hydrogen occurs at a solid surface by direct transfer of vibrational energy, [Langmuir and Mackay 1914] and [Langmuir 1915; 1916], and requires 4.5 eV [Jansen et al. 1989].

The atomic hydrogen concentration predicted by equilibrium for reaction (1)

$$H_2 \Leftrightarrow 2H \tag{1}$$

depends on pressure and temperature and is given by (2)

$$[H]^{eq.} = \frac{\left\{\left[\left(P^\circ\right)^2 + 4K^{-1}P^\circ P_T\right]^{1/2} - P^\circ\right\}}{2K^{-1}RT} \tag{2}$$

where P° is the standard state pressure,
P_T is the total pressure,
R is the gas constant,
T is the temperature, and
K is the equilibrium constant for reaction (1).

The equilibrium constant, K, and the equilibrium atomic hydrogen concentration, $[H]^{eq}$, each increase strongly with T; $[H]^{eq}$ increases weakly with total pressure, P_T. These effects are shown in Table 2.

Table 2. Equilibrium concentrations of atomic hydrogen at various temperatures and pressures.

Temperature, (°K)	$[H]^{eq}$ (kgmol/m^3) P = 20 torr	$[H]^{eq}$ (kgmol/m^3) P = 200 torr
1600	7.62×10^{-8}	2.41×10^{-7}
1800	4.43×10^{-7}	1.40×10^{-6}
2000	1.78×10^{-6}	5.66×10^{-6}
2200	5.47×10^{-6}	1.75×10^{-5}
2400	1.35×10^{-5}	4.43×10^{-5}
2600	2.75×10^{-5}	9.50×10^{-5}
2800	4.69×10^{-5}	1.77×10^{-4}
3000	6.65×10^{-5}	2.92×10^{-4}

Bare metal filaments are more efficient at generating atomic hydrogen than carbide filaments [Chevenier et al. 1994] and generation of atomic hydrogen essentially stops when elemental carbon is deposited on the filament. Rapid diffusion causes the atomic hydrogen concentration near the filament to be lower than the equilibrium value; the atomic hydrogen concentration near the substrate is much greater than the equilibrium concentration at the local gas temperature. In pure hydrogen, smaller diameter wires generated more atomic hydrogen per unit surface area than larger diameter wires; flat rhenium sheets were found to be less effective than wires [Jansen et al. 1989].

Two-photon laser-induced fluorescence (LIF) was used to measure the atomic hydrogen concentration near tantalum filaments in 5% CH_4 in H_2 at 30 mbar [Schafer et al. 1991]. The atomic hydrogen concentration increased by a factor of three when the filament diameter was increased from 0.3 to 2.0 mm. The gas temperature and atomic hydrogen concentration distribution measured by coherent anti-Stokes Raman scattering (CARS) [Chen et al. 1992] deviated from the distribution predicted by thermal equilibrium except very near the filament.

2.2 ATOMIC HYDROGEN DESTRUCTION

Several processes act to destroy atomic hydrogen. H atom recombination takes place by reaction (3).

$$2H + M \rightarrow H_2 + M \qquad (3)$$

Recombination requires the presence of a third body to remove the energy of reaction. This third body, M, is either another gas phase molecule (homogeneous recombination) or a solid surface (heterogeneous recombination). Because it is a three-body reaction, the rate of homogeneous recombination depends strongly on the total concentration of gas phase molecules.

Reaction with hydrocarbons in the gas phase also destroys atomic hydrogen, e.g.,

$$H + CH_4 \rightarrow CH_3 + H_2 \qquad (4)$$

Celli and Butler, [1989], using resonance enhanced multiphoton ionization (REMPI) found that the atomic hydrogen REMPI signal 8 mm from a tungsten filament at 2500 °C decreased by an order of magnitude as the methane concentration increased from one to three percent. Similar results were obtained using gas chromatography [Harris et al. 1988].

Meier et al. [1990], and Schafer et al. [1991], using LIF found that the atomic hydrogen concentration decreased by 30% when the methane concentration increased by 5 percent. Intracavity laser absorption spectroscopy (ICLAS) showed that the atomic hydrogen concentration decreases sharply with small incremental increases in methane concentration at low methane concentrations [Vetterhoffer et al. 1993]; at higher methane concentrations, the atomic hydrogen concentration decreases less strongly as the methane concentration is increased. The LIF technique gave similar results [Chevenier et al. 1994]. The atomic hydrogen concentration at the substrate measured by molecular beam mass spectrometry (MBMS) decreased by more than an order of magnitude when the methane concentration increased from 0.4% to 7.2% [Hsu 1991].

Atomic hydrogen concentration profiles have been determined using the temperature difference between two thermocouples, one of which was passivated to minimize atomic hydrogen recombination [Tankala and DebRoy 1992]. A 37% decrease in atomic hydrogen concentration was reported when moving 10 mm from a 2000 °C tantalum hoop filament in 1% CH_4 in H_2 at 30 torr. Thermocouple probes were also used in a dual hot-filament reactor [Gat and Angus 1993] in which the substrate was a thin wire whose temperature was independently controlled by resistive heating. The power required to maintain a constant substrate temperature at different methane concentrations was measured. Addition of up to 7% methane lowered the atomic hydrogen recombination rate at the substrate by a factor of three.

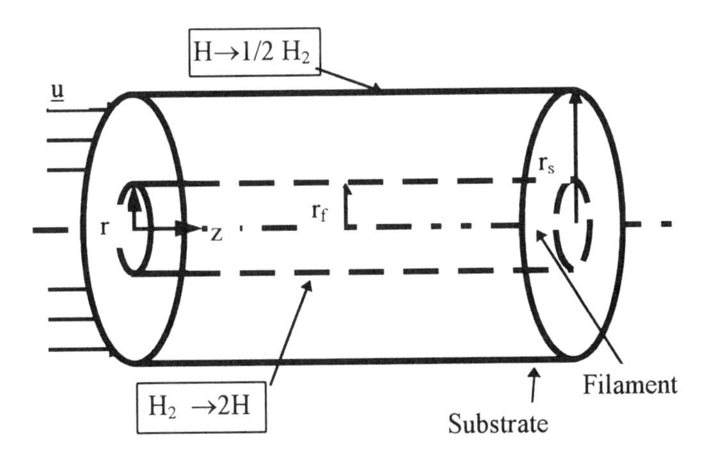

Figure 2. Idealized cylindrically symmetric reactor used to determine an analytical solution for atomic hydrogen concentration [Zeatoun, 1996].

The importance of heterogeneous recombination is illustrated by an analytical solution for the atomic hydrogen concentration in an idealized, cylindrically symmetric reactor (Fig. 2) [Zeatoun, 1996; Kovach et al. 1994]. A uniform flow of H_2 enters at one end of the reactor. Atomic hydrogen is generated on the filament and recombines at the wall but not in the gas phase. The fully developed concentration profiles are plotted as a function of the filament and wall Damkohler numbers (Figs. 3 and 4). In these solutions, the Damkohler number at the filament is defined as $Da_F = k_F r_W/D$ and at the wall as $Da_w = k_W r_W/D$, where k_F is the rate constant for molecular hydrogen dissociation at the filament, k_W is the rate constant for atomic hydrogen recombination at the reactor walls, r_W is the reactor radius and D is the diffusion coefficient for atomic hydrogen. The recombination rate constant on the reactor walls, k_W, was estimated using the collision rate at 20 torr from gas kinetic theory. This leads to $k_W = 36.4\beta T^{0.5}$ (m/sec) where β is the probability of a collision leading to recombination. Figures 3 and 4 can be used for other conditions by using appropriate values of Da_F and Da_W.

2.3 GAS PHASE CARBON CHEMISTRY

The conditions leading to carbon deposition on the filaments can be estimated by calculating the equilibrium gas phase composition adjacent to the filament. Sommer et al. [1989] used a quasi-equilibrium thermodynamic analysis to obtain a phase diagram for the C-H system. They introduced kinetic factors to account for the enhanced etching of graphite by hydrogen [Matsumoto et al. 1982[b]]. The results show that the gas phase carbon solubility passes through a minimum at about 1600 °K. Below 1600 °K, CH_4 is the dominant gas phase species, while above 1600 °K, C_2H_2 is dominant. The solubility minimum is shifted towards lower temperatures when the pressure is decreased and is slightly higher for diamond than for graphite. These theoretical predictions agree with experimental results from Matsumoto et al. [1982[b]] and other thermodynamic analyses [Piekarczk et al. 1989; Anthony, 1990].

Prijaya et al. [1993] performed constrained chemical equilibrium calculations for the ternary C-H-O and C-H-F systems. A dramatic change in equilibrium concentrations was found at an O/C ratio around one, i.e., near the H_2-CO tie line where diamond growth occurs [Bachmann et al. 1991]. Small deviations in source gas composition near the H_2-CO tie line produce large deviations in gas phase species that can influence diamond growth rates and quality.

Methyl radicals, CH_3, and acetylene, C_2H_2, are believed to be the most likely growth species in hot-filament reactors. Infrared diode laser absorption spectroscopy (IR) was used by Celli et al. [1991], for detecting these species in a hot-filament reactor. Vibrational spectra of C_2H_2, C_2H_4 and CH_3 were measured for 0.5% CH_4 in H_2 at 25 torr. Acetylene was the most abundant gas phase species; the concentrations of other stable hydrocarbons, e.g., C_2H_6, C_3H_4 and C_3H_8, were below the limits of detection.

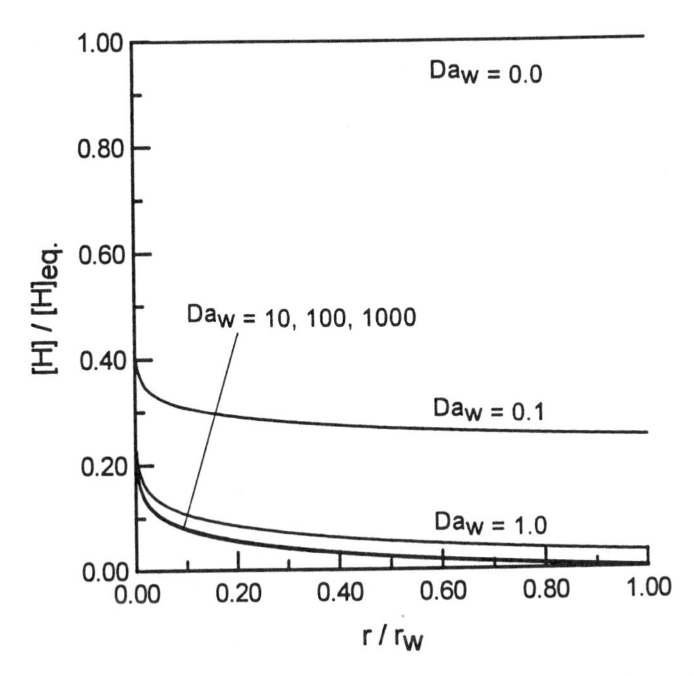

Figure 3. Fully developed radial atomic hydrogen concentration profiles for a $Da_F = 10^{-4}$ in the cylindrically symmetric reactor shown in Fig. 2.

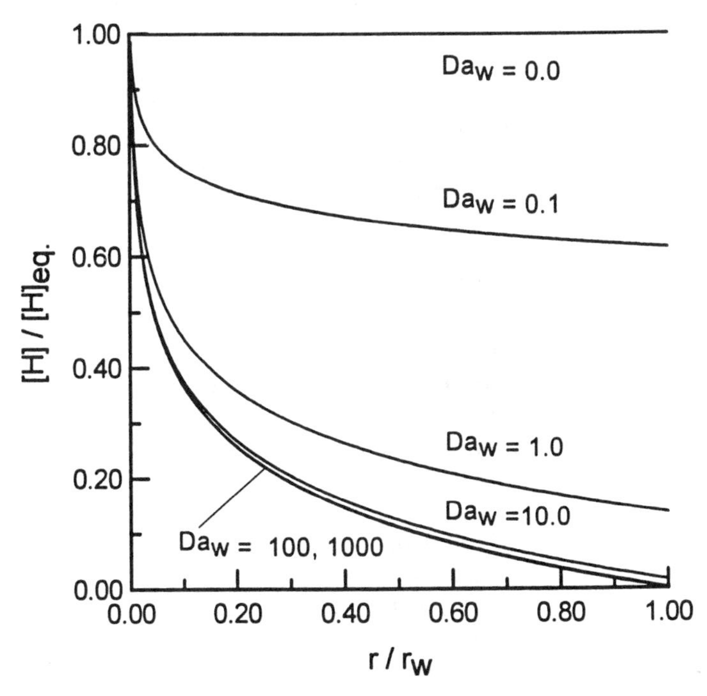

Figure 4. Fully developed radial atomic hydrogen concentration profiles for a $Da_F = 1.0$ in the cylindrically symmetric reactor shown in Fig. 2.

Using a radical-scavenging method with a microprobe operating at low pressure, stable species and methyl radical concentrations were detected by mass-spectrometry [Harris et al. 1988] and gas-chromatography [Harris and Weiner 1990]. The results are in good agreement with those obtained by Celli et al. [1991]. Molecular beam mass spectrometry was used to detect stable species and free radicals near the substrate surface [Hsu 1991].

Oxygen reduces the partial pressure of all hydrocarbons [Sommer and Smith 1991], favors etching of non-diamond carbon by OH, and increases radical concentrations [Harris and Weiner 1989]. Z. Li Tolt et al. [1995], reported that the addition of 1000 ppm of oxygen to a CH_4/H_2 mixture allows diamond film deposition at filament and substrate temperatures below 1750 °C and 600 °C, respectively. Hirose and Terasawa [1986] obtained growth rates more than 10 times faster using ethanol, methanol, acetone, diethyl-ether and trimethylamine than with methane. Cassidy et al. [1995], found that the diamond quality was lowered as the growth rate increased; however, the decrease in quality was less when using alcohol carbon sources compared to methane or acetone. In contrast to Hirose and Terasawa [1986], diamond growth from methanol was not observed. No differences in the growth rates were found for equivalent carbon concentrations using CH_4/H_2 and C_2H_2/H_2 gas mixtures [Wu et al. 1990; Wang and Angus 1993]. At equal molar concentrations, higher growth rates were obtained using acetone rather than methane [Okoli et al. 1989]; however, similar growth rates and diamond quality were obtained for equivalent CH_3 concentrations. Lower growth rates were found using acetone, methanol and dimethyl-propanol than with methane at the same C/H ratio [Beckmann et al. 1992].

The required substrate temperatures are lower when CH_2Cl_2 and $CHCl_3$ are used [Hong et al. 1993]. Substrate temperatures as low as 380 °C have been reported using CCl_4 [Hong et al. 1993] and to 390°C using $CF_4/CH_4/H_2$ gas mixtures[Corat et al. 1994]. Homoepitaxial diamond films have been deposited from C_3F_8/H_2 mixtures [Huangfu et al. 1994].

2.4 ENERGY, MASS AND MOMENTUM TRANSPORT

The relative importance of the various transport mechanisms for energy, mass and momentum can be described by dimensional analysis. Dimensionless groups are defined and their values given in Table 3 for typical hot-filament deposition conditions. The physical properties were evaluated at two pressures, 20 and 200 torr; the characteristic length is taken as 1 cm.

The Knudsen number indicates that, for typical dimensions and pressures, hot-filament reactors operate in a continuum regime. The Grashof number is a measure of the ratio of buoyant to frictional forces. On a length scale of L = 1 cm, the Grashof number is less than one, indicating convective roll cells between the filament and substrate are suppressed. On a larger scale of L = 25 cm, the Grashof number is greater than one, indicating that convective roll cells of this scale are present in the reactor. These conclusions are supported by computer simulations [Wang et al. 1993[a]].

Table 3. Estimated values of dimensionless numbers that describe transport and reaction processes in hot-filament reactors.

Dimensionless Group	Definition	Physical Interpretation	Value (20 torr)	Value (200 torr)
Knudsen	$\frac{\lambda}{L}$	$\frac{\text{Mean Free Path}}{\text{Characteristic Length}}$	3.5×10^{-4}	9.2×10^{-6}
Grashof (Thermal)	$\frac{\beta g L^3 \Delta T}{v^2}$	$\frac{\text{Momentum Flux by Free Convection}}{\text{Momentum Flux by Diffusion}}$	2.4×10^{-3}	2.6×10^{-1}
Peclet (Thermal)	$\frac{VL}{\alpha}$	$\frac{\text{Thermal Flux by Forced Convection}}{\text{Thermal Flux by Conduction}}$	1.3×10^{-2}	1.4×10^{-1}
Rayleigh (Thermal)	$\frac{\beta g L^3 \Delta T}{\alpha v}$	$\frac{\text{Thermal Flux by Free Convection}}{\text{Thermal Flux by Conduction}}$	1.9×10^{-3}	2.0×10^{-1}
Stefan	$\frac{\eta_s A_r T^4}{k A_g \left(\frac{\delta T}{\delta L}\right)}$	$\frac{\text{Thermal Flux by Radiation}}{\text{Thermal Flux by Conduction}}$	0.84	0.84
Damkohler Group IV	$\frac{\left(\frac{\Delta H}{m}\right) r_a L}{hT}$	$\frac{\text{Heat Released by Chemical Reaction}}{\text{Thermal Flux by Conduction}}$	2.7	14
Peclet (Mass)	$\frac{VL}{D}$	$\frac{\text{Mass Flux by Convection}}{\text{Mass Flux by Diffusion}}$	2.3×10^{-2}	2.3×10^{-1}
Theile Modulus (Surface Reaction)	$\frac{k_s C_a^{n-1} L}{D}$	$\frac{\text{Characteristic Time for Diffusion to Surface}}{\text{Characteristic Time for Surface Reaction}}$	1	10

λ is the mean free path, L is a characteristic length, β is the coefficient of volume expansion ($\beta \equiv 1/T$ for an ideal gas where T is a temperature), g is the gravitational constant, ΔT is a temperature difference, v is the kinematic viscosity ($v \equiv \mu/\rho$ where μ is the viscosity and ρ is the density), V is the bulk gas velocity, α is the thermal diffusivity ($\alpha \equiv k/\rho C_p$ where k is the thermal conductivity and C_p is the heat capacity), η_s is the Stefan-Boltzmann constant, A_r is the radiating area, A_g is the conducting area, ΔH is the heat of reaction (3), r_a is the rate of reaction (3), D is the diffusion coefficient, k_i is the reaction rate constant, C_a is the concentration of species a, and n is the reaction order. The Theile Moduli are estimated values for the simple attachment of methyl radicals to a bare surface site: $CH_3 + S \rightarrow CH_3 - S$.

The small thermal Peclet and Rayleigh numbers at 20 torr indicate that heat transfer by conduction dominates heat transfer by forced and free convection. The Stefan number is approximately one, indicating radiation and conduction contribute about equally to heat transfer from the filaments to the substrate. Damkohler's fourth number is approximately unity, which means that the heat liberated by recombination of H atoms is approximately the same as the heat transfer by conduction. That the Stefan number and Damkohler's Group IV number are all approximately unity shows that radiation, conduction and atomic hydrogen recombination each contribute approximately equally to heating the substrate. This is consistent with other estimates [Kuczmarski et al. 1991]. The Peclet number for mass transfer is much less than one at 20 torr and indicates that mass transfer is dominated by diffusion rather than convection. The estimated Theile modulus, Th, shows that diffusion and attachment to the surface occur on roughly the same time scale.

There have been several modeling studies of atomic hydrogen generation [Debroy et al. 1990[a,b]], gas phase and surface attachment kinetics, and fluid mechanics in hot-filament reactors [Goodwin and Gavillet, 1990; Kuczmarski et al. 1991; Kondoh et al. 1993; Wolden and Gleason, 1993; Mankelevich et al. 1995; Roozbehani, 1995].

3. Hot-filament Reactor Design

3.1 FILAMENT MATERIALS AND KINETICS OF CARBURIZATION

Considerations in choosing filament material include mechanical stability, chemical stability, and incorporation of metal into the diamond. The filaments can change structurally, e.g., by cracking, bending or creeping, due to carburization and hydrogen embrittlement. Typical filament materials are rhenium, tantalum and tungsten. The physical properties of these metals and their stable carbides are listed in several sources [Lyon 1962; Toth 1971; Lynch et al. 1979; Lynch 1989; Anthony 1990; Sommer and Smith 1990; Buckley-Golder et al. 1992; Lux 1993].

Rhenium does not form a carbide; tungsten and tantalum each form two stable carbides. Initially, W_2C or Ta_2C is formed; further carburization gives WC or TaC. Melting of the filaments through formation of low melting point solid solutions can be avoided by using precarburized filaments.

Although rhenium does not form a carbide, carbon diffuses through rhenium rapidly [Gaines et al. 1959]. After carburization, the rhenium hardness is increased, and a single phase interstitial-carbon solid solution remains. Carburized rhenium is ductile [Gaines et al. 1959] and, therefore, has a longer lifetime than the brittle tungsten and tantalum carbides; however, rhenium is more expensive.

Carbon absorption by tantalum from methane and acetylene was studied at high temperatures (1500 - 2300 °C) and low hydrocarbon partial pressures ($< 10^{-1}$ torr). These studies indicate that carburization can be described by the heterogeneous reactions

$$CH_4\,(g) \rightarrow Cs + 2H_2(g) \tag{4}$$

$$C_2H_2(g) \rightarrow 2Cs + H_2(g) \tag{5}$$

where Cs indicates carbon dissolved in the metal [Horz et al. 1974]. The reaction mechanism is believed to involve several elementary steps: 1) hydrocarbon transport to the surface, 2) physical adsorption of hydrocarbon, 3) hydrocarbon decomposition, 4) hydrogen chemisorption, 5) hydrogen atom recombination and H_2 gas liberation, 6) carbon dissolution, and 7) carbon diffusion into the metal. These elementary steps may also apply to tungsten and rhenium.

After carbon dissolves into the metal, its activity dictates which carbide will form. A tungsten filament heated to 1700 °C in a mixture of 0.5 % CH_4 in H_2 at 30 torr is only partially converted to W_2C after five hours [Moustakas et al. 1987]. Both W_2C and WC are formed after one hour at 2200 °C and 5% CH_4 [Matsumara and Sakuma 1990]. Carburization at the same conditions for an additional two hours yields only WC. The same conditions applied to tantalum filaments led to the formation of Ta_2C and TaC after one hour and full conversion to TaC after two hours [Matsumara and Sakuma 1990]. Tantalum filaments were also completely converted to TaC in three hours at 2200 °C and 5% CH_4 in H_2 followed by two hours at 2600 °C [Bruckner and Mantyla 1993]. Low hydrocarbon concentrations dramatically increase the carburization time [Okoli et al. 1991]. Tantalum filaments at 2200 °C and 0.5% CH_4 in H_2 at 10 torr are fully carburized after 60 hours; increasing the filament temperature to 2400 °C yields full carburization after 20 hours. In the same study, a tungsten filament at 2000 °C (0.5% CH_4 in H_2, 10 torr) is fully carburized after 20 hours. At the same hydrocarbon partial pressure, the carburization rate of tantalum with methane is several orders of magnitude lower than with acetylene [Horz and Lindenmaier 1974].

The filaments recrystallize during exposure to a carburizing atmosphere. The filament volume expands due to carbon incorporation. Lengthwise crack formation has been observed in cross-sectional metallographic samples of tungsten and tantalum filaments [Moustakas et al. 1987; Matsumara and Sakuma 1990].

3.2 OTHER DESIGN CONSIDERATIONS

Ferrous alloys should be avoided in the deposition region because they promote rapid formation of non-diamond carbon. Atomic hydrogen may regasify this carbon and affect the reactor carbon balance. Electrically conducting materials suitable for the deposition region include copper, molybdenum and titanium; insulating materials include alumina and boron nitride (HIP).

Coiled filaments, by their geometry, account for filament expansion. Straight wires must be allowed to expand and contract to compensate for thermal expansion and carburization. Expansion can be taken up by using springs at the end of the filament. Alternatively, one end of the filament can be allowed to move through a sleeve. In the latter case, it is essential that the sleeve be sufficiently close-fitting to provide good electrical contact. Deposition of non-diamond carbon between the sleeve and filament may aid in maintaining good electrical contact. Filament sag can be reduced for very thin horizontal filaments by attaching a weight to the end of the filament that passes through the sleeve. Vertical filaments eliminate filament sag entirely and are very useful for coating large areas. Various other design considerations for hot-filament reactors have been discussed by Joffreau et al. [1988] and Litos et al. [1989].

For laboratory scale reactors, a single filament is often used; in industrial reactors, a multi-filament array is more common. Jansen et al. [1990], found that the film thickness had a maximum at the center of the substrate for a single filament and for two filaments 9 mm apart. For two filaments 20 mm apart, the film thickness was smaller at the center and passed through a maximum directly under the filaments. A similar effect was obtained by C. P. Klages et al. [1993], in a 4-filament array, which was attributed to a nearly complete conversion of CH_4 to C_2H_2 directly beneath the filaments. Molecular beam mass spectrometry experiments for a three filament array, show that this configuration forces CH_4 to flow through regions of hot gas where it is converted to C_2H_2, in agreement with the preceding results [Hsu 1992].

Heat transfer calculations and measured substrate temperature profiles, under vacuum, H_2 and He [Wolden et al. 1992] show that the temperature uniformity was improved when both the filament number and the filament-substrate distance were increased. In addition, the calculated temperature gradients are much smaller for a rotating substrate compared to a stationary one. The radiation flux profiles agreed with the diamond growth rates obtained by Jansen et al. [1990]. The only departure from the relationship was obtained for the two filaments 20 mm apart, and is explained by a depletion of gas-phase diamond precursors.

The effect of substrate rotation on growth rates and quality was studied by Kovach et al. [1994]. At 20 torr, rotation up to 7200 rpm did not affect growth rates, diamond quality or crystal morphology; at 200 torr, the crystal morphology, quality and deposit uniformity were improved by spinning the substrate. No significant changes in growth rates were observed. Kassman and Badgewell [1995], have reported modeling studies of rotating substrate hot-filament reactors.

The gas mixture can be introduced into the reaction zone at high speed using nozzles. Klages et al. [1993], measured stable species at the substrate surface and growth rates for a single filament and a multi-filament array with gas flow velocities normal to the substrate up to 1800 m/s. The growth rates increased nearly linearly with the gas nozzle exit velocity and decreased with pressure. With the 4-filament array, the growth rates were higher at the center of the substrate and decreased towards the edge, opposite to the distribution obtained without forced convection. Diamond deposition at filament-substrate distances as great as 5 cm with a tungsten filament have been reported using convective gas flows [Dua et al. 1993].

3.3 INCORPORATION OF FILAMENT MATERIAL IN THE DIAMOND

Tungsten impurities have been detected by secondary ion mass spectroscopy (SIMS) in diamond films [Chifre et al. 1992] grown using hot-filaments. The tungsten concentration was constant in the bulk of the film and increased near the film-substrate interface; the lower the operating pressure and CH_4 concentration, the higher the tungsten concentration in the film. Films grown with a tungsten filament at 2500 °K and 0.1 - 1.2% CH_4 show a nearly constant total tungsten concentration of 10^{16} atom/cm^3 incorporated as WC_{1-x} microcrystallites in the diamond grain boundaries [Gheeraert et al. 1992]. A pure polycrystalline tungsten film 10 microns thick was reported using a non-carburized tungsten filament operating at 2400 °C with a 1% CH_4-H_2 gas feed at 2000 Pa [Venter and Neething 1993]. Tungsten incorporation may involve oxides such as WO_3 or WO_2 , which are highly volatile even at 1100°C [Nagenser-Naidu and Rama-Rao 1991]. To reduce tungsten contamination, Argoitia et al. [1993], used a movable molybdenum shutter between the filament and substrate during filament carburization. No tungsten contamination was detected by SIMS either in the bulk of the film or at the silicon-diamond film interface. Using the same procedure, no evidence of tungsten contamination was observed on cross-sectional high resolution TEM analysis of heteroepitaxial films grown on c-BN [Argoitia et al. 1994]. Griesser et al. [1994], using SIMS, Rutherford backscattering spectroscopy (RBS) and transmission electron microscopy (TEM) found tantalum concentrations between 2 and 300 ppm in the bulk of the film that increased by one to three orders of magnitude at the diamond-silicon interface. Electron diffraction and direct observation by TEM indicated the presence of TaC. More tantalum is incorporated during the induction period when diamond growth is slow. The tantalum contamination at the interface is about four orders of magnitude higher when non-carburized filaments are used. The contamination level increases with filament temperature and addition of oxygen to the gas mixture [Griesser et al. 1994]. Hinneberg et al. [1992], found tungsten concentrations of 30 ppm in the upper 0.4 microns of a film for a filament temperature of 2000 °C; the contamination decreased below the detection limit (about 5 ppm) for a filament temperature of 1900 °C. The incorporation rate for tantalum was considerably higher than for tungsten. Jansen et al. [1990], using SIMS found that the amount of incorporated rhenium increased exponentially with filament temperature from approximately 7 ppm at 1900 °C to 2000 ppm at 2350 °C.

4. Applications of Hot-filament Reactors

4.1 BIAS-ASSISTED PROCESSES

Electron assisted chemical vapor deposition (EACVD) was introduced by Sawake and Inuzuka [1985, 1986] and Hirose and Teresawa [1986]. The substrate is biased positively with respect to the filament (forward bias); electrons are emitted from the filament and flow to the substrate. High quality diamond films grown at higher rates were reported. Lee et al. [1990], studied the influence of both forward bias and reverse bias (substrate at negative potential with respect to the filament). The film quality deteriorated at 150V forward and 180V reverse bias. In contrast, for low current reverse biasing, diamond quality is enhanced. Banholzer and Kehl [1991] suggested that the increase in growth rate under forward biasing is due to an increase in the substrate temperature from electron bombardment. Okoli et al. [1992] concluded that forward bias has little or no effect on the nucleation rate, but that it influenced the growth rate of the diamond crystals, once formed. Honma et al. [1994], found that the highest growth rate and quality were obtained at an optimum value of E/p = 500 V/mtorr for forward biasing. Popovici et al. [1994], reported rapid growth of microscopically smooth, transparent diamond films under forward bias potentials (50 - 150 V) at fairly standard conditions.

4.2 *IN SITU* MEASUREMENTS

Hot-filament reactors operate in a fluid mechanical regime that permits *in situ* measurement of growth rates using a microbalance [Fedoseev and Deryagin 1979; Harris et al. 1991; Wang et al. 1993[a]; Evans 1994]. In a study using a platinum wire substrate suspended 5 - 6 mm above a tungsten filament at 2400 °K, a growth rate maximum was found at 30 torr; growth rates decreased approximately 40% between 30 and 60 torr [Harris and Weiner 1994]. Using 0.3 - 1.0% methane at both 10 and 20 torr, the deposition rate was first-order in methane; the deposition rate approaches zero-order above 1% methane [Wang and Angus 1993]. With two-carbon atom source gases, e.g., C_2H_2, C_2H_4, and C_2H_6, the growth rate is approximately half-order, which offers supporting evidence that the methyl radical is the principal growth species in hot-filament assisted diamond growth.

4.3 HETEROEPITAXY AND HOMOEPITAXY

Hot-filament assisted deposition has been successfully used to grow heteroepitaxial diamond on cubic boron nitride [Wang et al. 1993[a]; Argoitia et al. 1994] and on BeO [Argoitia et al. 1993]. The diamond-substrate interfaces were examined by high resolution transmission electron microscopy; no evidence of tungsten contamination at the interface was found.

High quality homoepitaxial films using hot-filament assisted deposition have also been reported [Geis 1990; van Enckevort et al. 1995].

4.4 GROWTH OF DOPED DIAMOND

Mort et al. [1991[a]], using a reactant gas mixture containing 70 - 900 ppm diborane relative to the methane concentration, obtained free standing boron-doped diamond films with 10^{18} - 10^{19} boron atoms/cm^3. Infrared spectroscopy confirmed the substitutional nature of the boron. These authors also reported growth of nitrogen-doped diamond films from a gas feed containing 1% NH_3, 1.5% CH_4 and 0.4 % O_2 in H_2 at 35 torr [Mort et al. 1991[b]]. Nitrogen concentrations of $\sim 2 \times 10^{18}$cm^{-3} and $\sim 4 \times 10^{19}$cm^{-3} measured by SIMS were obtained for different NH_3/H_2 feed ratios. Okano et al. [1988, 1989], introduced boron with boron trioxide (B_2O_3) in methyl alcohol and acetone. A SIMS measurement indicated boron concentrations of 2×10^{20} cm^{-3} and 3×10^{20} cm^{-3} for silicon and diamond substrates, respectively. Growth of phosphorous containing diamond films using phosphine has also been reported [S. Bohr et al. 1995]. Homoepitaxial boron-doped p-type semiconducting diamond films have been obtained by placing natural diamond substrates on an h-BN holder [Janssen et al. 1992] Boron-doped diamond films have been grown on tungsten wires for electrochemical applications by placing the wire substrates adjacent to h-BN during deposition [Martin et al. 1995; Argoitia et al. 1995]. Boron metal [Spitsyn 1990; Gildenblat et al. 1990] and boron carbide [Masood et al. 1992] placed next to the substrate during deposition have also been used as boron sources.

4.5 INDUSTRIAL APPLICATIONS

Hot-filament assisted deposition has been applied to larger scale industrial processes. Advantages are simplicity, scalability and the ability to coat complex shapes and internal surfaces. Some representative patents are those issued to Diamonex [Garg et al. 1991], General Electric [Anthony and Fleischer 1992 and 1993], Sumitomo Electric Co. [Ikegaya and Fukimori 1992], Toyota [Taki 1992], and Matsushita Electric Industrial Co. [Nakagami et al. 1992]. Despite a large number of patents, much industrial practice remains proprietary and has not been revealed.

5. References

Angus, J. C., Argoitia, A., Gat, R., Li, Z., Sunkara, M., Wang, L. and Wang, Y. (1993) Chemical vapor deposition of diamond, *Phil Trans. Roy. Soc. Lond.- A* **342**, 195-208.

Anthony, T. R. (1990) Methods of diamond making, *Diamond and Diamond-Like Films and Coatings*, NATO ASI series, Series B: Physics Vol. **266**, pp. 555 - 577.

Anthony, T. R. and Fleischer, J. F. (1992) Transparent diamond films and method for making, *U.S. Patent* 5,110,579, May 5.

Anthony, T. R. and Fleischer, J. F. (1993) Substantially transparent free standing diamond films, *U.S. Patent* 5,273,731, December 28.

Argoitia, A., Angus, J. C., Wang, L., Ning, X. J. and Pirouz, P. (1993) Diamond grown on single-crystal beryllium oxide, *J. Appl. Phys.*, **73**(9), 4305 - 4312.

Argoitia, A., Angus, J. C., Ma, J. S., Wang, L., Pirouz, P., and Lambrecht, W. R. L., (1994) Heteroepitaxy of diamond on c-BN: Growth mechanisms and defect characterization, *J. Mat. Res.* **9**, 1849-18655.

Argoitia, A., Martin, H. B., Landau, U. and Angus, J. C. (1995) Electrochemical studies of boron-doped diamond electrodes, to be published *Symposium on Diamond for Electronic Applications,* Materials Res. Soc., Boston, MA, Nov.27 - Dec.1.

Bachmann, P. K., Leers, D. and Lydtin, H. (1991) Towards a general concept of diamond chemical vapor deposition, *Diam. Relat. Mater.* **1**, 1 - 12.

Banholzer, W. and Kehl, R. (1991) Clarification of the effect of bias in the hot filament process, *Surf. Coat. Tech.* **47**, 51 - 58.

Beckmann, R., Kulisch, W., Frenck, H. J. and Kassing, R. (1992) Influence of gas parameters on the deposition kinetics and morphology of thin diamond films deposited by HFCVD and MWCVD technique, *Diam. Relat. Mater.* **1**, 164 - 167.

Bohr, S., Haubner, R. and Lux, B. (1995) Influence of phosphorous addition on diamond CVD, *Diam. Relat. Mater.* **4**, 133 - 144.

Bruckner, J. and Mantyla, T. (1993) Diamond CVD using tantalum filaments in $H_2/CH_4/O_2$ gas mixtures, *Diam. Relat. Mater.* **2**, 373 -377.

Buckley-Golder, I. M. and Collins, A. T. (1992) Active electronic applications for diamond, *Diam. Relat. Mater.* **1**, 1083 - 1101.

Cassidy, W. D., Morrison Jr., P. W. and Angus, J. C. (1995) Growth rates and quality of diamond grown by hot-filament assisted chemical vapor deposition, *Proc. 4th Int. Symp. on Diamond Materials,* K. V. Ravi and J. P. Dismukes Eds., Reno, Nevada, Vol. **95-4**, Electrochem. Soc., Pennington, NJ, 85 - 90.

Celli, F. G. and Butler, J. E. (1989) Hydrogen atom detection in the filament-assisted diamond deposition environment, *Appl. Phys. Lett.,* **54**(11), 1031 - 1033.

Celli, F. G., Pehrsson, P. E., Wang, H. T. and Butler, J. E. (1991) Infrared detection of gaseous species during the filament-assisted growth of diamond, *Appl. Phys. Lett.,* **52**(24), 2043 - 2045.

Chauhan, S. P., Angus, J. C., and Gardner, N. C. (1976) Kinetics of carbon deposition on diamond powder, *J. Appl. Phys.,* **47**(11), 4746 - 4754.

Chen, K. H., Chuang, M. C., Penney, M. and Banholzer, W. F. (1992) Temperature and concentration distribution of H_2 and H atoms in hot-filament chemical vapor deposition of diamond, *J. Appl. Phys.,* **71**(3), 1485 - 1493.

Chevenier, M., Cubertafon, J. C., Campargue, A. and Booth, J. P. (1994) Measurement of atomic hydrogen in a hot filament reactor by two-photon laser-induced fluorescence, *Diam. Relat. Mater.* **3**, 587 - 592.

Chifre, J., Lopez, F., Morenza, J. and Esteve, L. (1992) Analysis of contamination in diamond films by secondary ion mass spectroscopy, *J. Diam. Relat. Mater.* **1**, 500 - 503.

Corat, E. J., Airoldi, V. J. T., Leite, N. F., Pena, A. F. V. and Baranauskas, V. (1994) Low temperature diamond growth with CF_4 addition in a hot filament reactor, *Mat. Res. Soc. Symp. Proc.,* Vol. **349**, 421 - 426.

Debroy, T., Tankala, K., Yarbrough, W. A. and Li, H. (1990[a]) Species convection and diffusion in the hot filament assisted chemical vapor deposition of diamond in the presence and absence of applied emf, *Proceedings Second International Conference on New Diamond Science and Technology*, Washington, DC, *Mat. Res. Soc.,* Pittsburgh PA, 359 - 364.

Debroy, T., Tankala, K. and Yarbrough, W. A. (1990[b]) Role of heat transfer and fluid flow in the chemical vapor deposition of diamond, *J. Appl. Phys.* **68**, 2424 - 2432.

Dua, A. K., George, V. C., Pruthi, D. D. and Raj, P. (1993) Large area deposition in HFCVD technique employing convective flow of gases, *Sol. State Comm.,* **86**(1), 39 - 41.

Evans, E. A. (1994) Growth kinetics during hot-filament assisted CVD of diamond, *MS Thesis*, Department of Chemical Engineering, Case Western Reserve University, Cleveland, Ohio.

Fedoseev, D. F. and Deryagin, B. V. (1979) Co-crystallization of diamond and graphite, *Zh. Fiz. Khimii.*, **53**(3),752 - 755.

Gaines, G. B., Sims, C. T. and Jaffee, R. I. (1959) The behavior of rhenium in electron tube environments, *J. Electrochem. Soc.* **106**, 881 - 885.

Garg, D., Tsai, W., Iampietro, R. L., Kimock F. M. and Kelly, C. M. "A hot-filament chemical vapor deposition reactor", *WO* 9,114,798 October 3, 1991.

Gat, R. and Angus, J. C. (1993) Hydrogen recombination on tungsten and diamond in hot-filament assisted deposition of diamond, *J. Appl. Phys.*, **74**(8), 5981 - 5991.

Geis, M. W. (1990) Growth of device-quality homoepitaxial diamond thin films. *Mat. Res. Soc. Symp. Proc.* **162**, 15 - 22.

Gheeraert, E., Deneuville, A., Brunel, M. and Oberlin, J. C. (1992) Tungsten incorporation in diamond thin films prepared by the hot-filament technique, *Diam. Relat. Mater.* **1**, 504 - 507.

Gildenblat, G. S., Grot, S. A., Hatfield, C. W., Wronski, C. R., Badzian, A. R., Badzian, T. and Messier, R. (1990) High temperature Schottky diodes with boron doped homoepitaxial diamond base, *Mater. Res. Bull.* **25**, 129 - 134.

Goodwin, D. G. and Gavillet, G. G. (1990) Numerical modeling of the filament-assisted diamond growth environment, *J. Appl. Phys.*, **68**(12), 6393 - 6400.

Griesser, M., Stingeder, G., Grasserbauer, M., Baumann, H., Link, F., Wurzinger, P., Lux, H., Haubner, R. and Lux, B. (1994) Characterization of tantalum impurities in hot-filament diamond layers, *Diam. Relat. Mater.* **3**, 638 - 644.

Harris, S. J., Weiner, A. M., and Perry, T. A. (1988) Measurement of stable species present during filament-assisted diamond growth, *Appl. Phys. Lett.*, **53**(17), 1605 - 1607.

Harris, S. J. and Weiner, A. M. (1989) Effects of oxygen on diamond growth, *Appl.Phys Lett.*, **55**(21), 2179 - 2181.

Harris S. J. and Weiner, A. M. (1990) Methyl radical and H-atom concentrations during diamond growth, *J. Appl. Phys.*, **67**(10), 6520 - 6526.

Harris, S. J., Weiner, A. M. and Perry, T. A. (1991) Filament-assisted diamond growth kinetics, *J. Appl. Phys.*, **70**(3), 1385 - 1391.

Harris, S. J. and Weiner, A. M. (1994) Pressure and temperature effects on the kinetics and quality of diamond films, *J. Appl. Phys.* **75**, 5026 - 5031.

Hinneberg, H. J., Eck, M. and Schmidt, K. (1992) Hot-filament grown diamond films on Si: Characterization of impurities, *Diam. Relat. Mater.* **1**, 810 - 816.

Hirose, Y. and Terasawa, Y. (1986) Synthesis of diamond thin films by thermal CVD using organic compounds, *Japn. J. Appl. Phys.*, **25**(6), L519 - L521.

Hong, F. C. N., Hsieh, J. C., Wu, J. J., Liang, G. T. and Hwang, J. H. (1993) Low temperature deposition of diamond using chloromethane in a hot-filament chemical vapor deposition reactor, *Diam. Relat. Mater.* **2**, 365 - 372.

Honma, S., Saitoh, H., Ishiguro, T. and Ichinose, Y. (1994) Problems of mean free path and electric field on hot filament diamond CVD, in *Proc. 4th Int. Conf. on New Diamond Science and Technology*, MYU, Tokyo, p97 - 103.

Horz, G., Lindenmaier, K. and Klaiss, R. (1974) High temperature solid solubility limit of carbon in niobium and tantalum, *J. Less Com. Met.* **35**, 97 - 105.

Horz, G. and Lindenmaier, K. (1974) The kinetics and mechanisms of the absorption of carbon by niobium and tantalum in a methane or acetylene stream, *J. Less Com. Met.* **35**, 85 - 95.

Hsu, W. L. (1991) Mole fractions of H_2, CH_3 and other species during filament-assisted diamond growth, *Appl. Phys. Lett.*, **59**(12), 1427 - 11429.

Hsu, W. L. (1992) Gas phase kinetics during MW plasma assisted diamond deposition. Is the hydrocarbon product distribution dictated by neutral-neutral interactions?, *J. Appl. Phys.* **72**, 3102 - 3108.

Huangfu, P., Jin, Z., Lu, X., Zou, G., Xiao, H., Chen, T. and Liu, M. (1994) Growth of diamond films using C_3F_8 and H_2 in a new system, *Mat. Res. Soc. Symp. Proc.* **349**, 415 - 420.

Ikegaya, A.and Fukimori, N. (1992) Manufacture of diamond by hot-filament chemical vapor deposition, *WO* 9,201,828, February 6.

Jansen, F., Chen, I. and Machonkin, M. A. (1989) On the thermal dissociation of hydrogen, *J. Appl. Phys.*, **66**(12), 5749 - 5755.

Jansen, F., Machonkin, M. A. and Kuhman, D. E. (1990) The deposition of diamond films by filament techniques, *J. Vac. Sci. Technol. A*, **8**(5), 3785 - 3790.

Janssen, G., van Enckevort, W. J. P., Vollenberg , W. and Giling, L. J. (1992) Characterization of single-crystal diamond grown by chemical vapour deposition processes, *Diam. Relat. Mater.* **1**, 789 - 800.

Joffreau, P. O., Haubner, R. and Lux, B. (1988) Low pressure diamond growth on different refractory metals, *J. Ref. Hard Metals*, **7**(4), 186 - 194.

Kassmann, D. E. and Badgewell, T. A. (1995) Modeling and scale-up of diamond chemical vapor deposition in a rotating disk reactor, in *Proceedings Diamond Films '95*, Barcelona, Spain, *to appear in Diam. Relat. Mater.*

Klages, C. P., Sattler, M. and Schafer, L. (1993) Physical chemistry of multifilament diamond CVD processes. *Proc. 3rd Int. Symp. on Diamond Materials*, Honolulu, May 16 - 21, Edited by Dismukes, J. P. and Ravi, K.V. The Electrochemical Society, Pennington, NJ, USA, p 24 - 33.

Kondoh, E., Ohta, T., Mitomo, T. and Ohtsuka, K. (1993) Surface Reaction kinetics of gas-phase diamond growth, *J. Appl. Phys.*, **73**(6), 3041- 3046.

Kovach, C. S., Zeatoun, L., Roozbehani, B., Greber, I. and Angus, J. C. (1994) Influence of transport and chemical reaction processes on diamond growth rates, morphology and quality, *Proceedings Fourth International Conference on New Diamond Science and Technology*, MYU, Tokyo, p 93 - 96.

Kuczmarski, M. A., Washlock, P. A. and Angus, J. C. (1991) Computer simulation of a hot-filament CVD reactor for diamond deposition, *Applications of Diamond Films and Related Materials.*, Materials Science Monographs **73**, Elsevier, NY, 591 - 596.

Langmuir, I. and Mackay, G. M. J. (1914) Dissociation of hydrogen into atoms. I., *J. Amer. Chem. Soc.* **36**, 1708 - 1722.

Langmuir, I. (1915) Dissociation of hydrogen into atoms. II., *J. Amer. Chem. Soc.* **37**, 417 - 458.

Langmuir, I. (1916) Dissociation of hydrogen into atoms. III., *J. Amer. Chem. Soc.* **38**, 1145 - 1156.

Lee, Y. H., Richard, P. D., Bachmann, K. J. and Glass, J. T. (1990) Bias-controlled chemical vapor deposition of diamond thin films, *Appl. Phys. Lett.*, **56**(7), 620 - 622.

Litos, R., Haubner, R., Lux, B. (1990) Untersuchung der Kinetik von Diamantabscheidungen auf Molybdan-Drahtsubstraten. (Investigation of kinetics of diamond synthesis with molybdenum wire substrates), *High Temperatures - High Pressures.* **22**, 99-118.

Lux, B. (1993) Techniques of hot-filament assisted deposition of diamond, *Diam. Relat. Mater.* **2**,1277 - 1289.

Lynch, J. F., Ruderer, G. G., Duckworth, W. H. (1979) *Engineering property data on selected ceramics volume 2, carbides*, MCIC report, MCIC-HB-07-vol. **2**.

Lynch, C. T. (1989) *Practical Handbook of Materials Science*. Ed. by C. T. Lynch, CRC Press, Inc., Boca Raton, Florida.

Lyon, T. F. (1962) Vapor pressures of several carbides and hafnium metal by the Langmuir method, *Proceedings of the International Symposium on Condensation and Evaporation of Solids,* Dayton, Ohio, Edited by Rutner, E., Goldfinger, P. and Hirth, J. P., The Gordon and Breach Science Publishers, NY, London, p. 435 - 449.

Mankelevich, Y. A., Rakhimov, A. T. and Suetin, N. V. (1995) Two-dimensional simulation of HFCVD reactor, *Proc. 4th Int. Symp. on Diamond Materials,* K. V. Ravi and J. P. Dismukes Eds., Reno, Nevada, Vol. **95-4**, Electrochem. Soc., Pennington, NJ, 687 - 692.

Martin, H. B., Argoitia, A., Angus, J. C., Anderson, A. B. and Landau, U. (1995) Boron doped diamond electrodes for electrochemical applications, *Proc. 3rd. Int. Conf. Applications of Diamond Films and Related Materials*, Gaithersburg, Maryland, 91 - 94.

Masood, A., Aslam, M., Tamor, M. A and Potter, T. J. (1992) Synthesis and electrical characterization of boron-doped thin diamond films, *Appl. Phys. Lett.*, **61**(15), 1832 - 1834.

Matsumara, H. and Sakuma, T. (1990) Diamond deposition on cemented carbide by chemical vapor deposition using a tantalum filament, *J. Mat. Sci.* **25**, 4472 - 4476.

Matsumoto, S., Sato, S. Y., Kamo, M., and Setaka, N. (1982[a]) Vapor deposition of diamond particles from methane, *Jpn. J. Appl. Phys.* **21**, L183 - L185.

Matsumoto, S., Sato, Y., Tsutsumi, M. and Setaka, N. (1982[b]) Growth of diamond particles from methane-hydrogen gas, *J. Mat. Sci.* **17**, 3106 - 3112.

Meier, U., Hoinghaus, K. K., Schafer, L. and Klages, C. P. (1990) Two-photon excited LIF determination of H-atom concentrations near a heated filament in a low pressure H_2 environment, *Appl. Opt.,* **29**(33), 4993 - 4999.

Mort, J., Machonkin, M. A. and Okumura, K. (1991[a]) Infrared absorption in boron-doped diamond thin films, *Appl. Phys. Lett.,* **58**(17), 1908 - 1910.

Mort, J., Machonkin, M. A. and Okumura, K. (1991[b]) Compensation effects in nitrogen-doped diamond thin films, *Appl. Phys. Lett.*, **59**(24), 3148 - 3150.

Moustakas, T. D., Dismukes, J. P., Tiedje, J. T., Walton, K. R. and Ye, L.(1987) Polycrystalline diamond deposition from methane-hydrogen mixtures, *J. Electrochem. Soc.* **134**, C483 - C486.

Nagenser-Naidu, S. V. and Rama-Rao, P., Eds. (1991) *Phase Diagrams of Binary Tungsten Alloys. Monograph series on alloy phase diagrams*, Calcutta, Indian Institute of Metals, p.184.

Nakagami, Y., Kurokawa, H. and Mitani, T. (1992) Hot-filament deposition of hard carbon films and apparatus therefore, Japan Kokai Tokyo Koho, 4,318,172, November 9.

Okano, K., Naruki, H., Akiba, Y., Koruso, T., Iida, M. and Hirose, Y. (1988) Synthesis of diamond thin films having semiconductive properties, *Jpn. J. Appl. Phys.*, **27**(2), L173 - L175.

Okano, K., Naruki, H., Akiba, Y., Koruso, T., Iida, M., Hirose, Y. and. Nakamura, T. (1989) Characterization of boron-doped diamond film. *Jpn. J. Appl. Phys.*, **28**(16), 1066 - 1071.

Okoli, S., Haubner, R. and Lux, B. (1989) Deposition of diamond layers by hot-filament activated CVD using acetone as a carbon source, *J. Phys. IV*, **C5**(5), 159 - 168.

Okoli, S., Haubner, R. and Lux, B. (1991) Influence of the filament material on low-pressure hot-filament CVD diamond deposition, *J. Phy. IV* **C2**, 923 - 930.

Okoli, S., Haubner, R. and Lux, B. (1992) Influence of hot filament-d.c. plasma co-enhancement on low pressure diamond synthesis, *Diam. Relat. Mater.* **1**, 955 - 962.

Piekarczk, W., Roy, R. and Messier, R. (1989) Application of thermodynamics to the examination of the diamond CVD process from hydrocarbon-hydrogen, *J. Cryst. Growth* **98**, 765 - 776.

Popovici, G., Chao, C. H., Prelas, M. A., Charlson, E. J. and Meese, J. M. (1994) Phase transformation of smooth diamond films grown by hot filament chemical vapor deposition on positively biased silicon substrates, *Mat. Res. Soc. Symp. Proc.* Vol. **339**, 325 - 328.

Prijaya, N. A., Angus, J. C. and Bachmann, P. K. (1993) Thermochemical computation of the diamond deposition domain, *Diam. Relat. Mater.* **3**, 129 - 136.

Roozbehani, B., to be published, Ph.D. Thesis, Case Western Reserve University (1995).

Sawake, A., and Inuzuka, T. (1985) Growth of diamond thin films by electron assisted chemical vapor deposition, *Appl. Phys. Lett.*, **46(2)**, 146 -147.

Sawabe, A. and Inuzuka, T. (1986) Growth of diamond thin films by electron-assisted chemical vapor deposition and their characterization, *Thin Sol. Films* **137**, 89 - 99.

Schafer, L., Klages, C. P., Meier, U. and Hoinghaus, K. K. (1991) Atomic hydrogen concentration profiles at filaments used for chemical vapor deposition of diamond, *Appl. Phys. Lett.*, **58**(6), 571 - 573.

Sommer, M., Mui, K. and Smith, F. W. (1989) Thermodynamic analysis of the chemical vapor deposition of diamond films, *Sol. State Comm.*, **69**(7), 775 - 778.

Sommer, M. and Smith, F. W. (1990) Activity of tungsten and rhenium filaments in CH_4/H_2 and C_2H_2/H_2 mixtures: Importance for CVD diamond, *J. Mat. Res.*, **5** (11), 2433 - 2444.

Sommer, M. and Smith, F. W. (1991) Effect of oxygen on filament activity in diamond chemical vapor deposition, *J. Vac. Sci. Tech. A*, **9**(3), 1134 - 1139.

Spitsyn, B. V. (1990) Chemical crystallization of diamond from the activated vapor phase, *J. Cryst. Growth* **99**, 1162 - 1167.

Taki, M. (1992) Hot-filament chemical vapor deposition of diamond films, Japan Kokai Tokyo Koho, *JP* 9,292,891, March 25.

Tankala, K. and DebRoy, T. (1992) Modeling of the role of atomic hydrogen in heat transfer during hot-filament assisted deposition of diamond, *J. Appl. Phys.* **72**, 712 - 718.

Tolt, Z. L., Heatherly, L., Clausing, R. E., Shaw, R. W. and Feigerle, C. S. (1995) HFCVD of diamond at low substrate and low filament temperatures, *Proc. 4th Int. Symp. on Diamond Materials,* K. V. Ravi and J. P. Dismukes Eds., Reno, Nevada, Vol. **95-4**, Electrochem. Soc., Pennington, NJ, 303 - 307.

Toth, L. E. (1971) *Transition Metal Carbides and Nitrides. Vol. 7, Refractories Materials.* Edited by Academic Press New York-London.

van Enckevort, W. J. P., Janssen, G., Schermer, J. J. and Giling, L. J. (1995) Step-related growth phenomena on exact and misoriented {001} surfaces of CVD-grown single crystal diamonds, *Diam. Relat. Mater.* **4**, 250 - 255.

Venter, A. and Neething, J. H. (1993) Effect of filament temperature on the growth of diamond using hot-filament chemical vapor deposition, *Diam. Relat. Mater.* **3**, 168 - 172.

Vetterhoffer, J., Campargue, A., Chevenier, M. and Stoeckel, F. (1993) Temperature measurement of atomic hydrogen produced by hot-filament dissociation in a CH_4-H_2 mixture, *Diam. Relat. Mater.* **2**, 481 - 485.

Wang, Y., Evans, E. A., Zeatoun, L. and Angus, J. C. (1993[a]) Microbalance measurements of diamond growth rates, *Proceedings of the Third IUMRS International Symposium on Diamond and Related Materials.,* Aug.31- Sept. 4, Tokyo, Japan.

Wang, L., Pirouz, P., Argoitia, A. and Angus, J. C. (1993[b]) Heteroepitaxial grown diamond on a c-BN {111} surface. *Appl. Phys. Lett.* **63**, 1336 - 1342.

Wang, Y. and Angus, J. C. (1993) Micro-balance studies of the kinetics of diamond growth, *Proc. 3rd Int. Symp. on Diamond Materials*, Honolulu, May, The Electrochemical Society, Pennington, NJ, USA, p. 249 - 256.

Wolden, C., Mitra, S. and Gleason, K. K. (1992) Radiative heat transfer in hot-filament chemical vapor deposition diamond reactors, *J. Appl. Phys.* **72**, 3750 - 3764

Wolden, C. and Gleason, K. K. (1993) Heterogeneous formation of atomic hydrogen in hot-filament diamond deposition, *Appl. Phys. Lett.*, **62**(19), 2329 - 2331.

Wu, C., Tamor, M. A., Potter, T. J. and Kaiser, E. W. (1990) A study of gas chemistry during hot-filament vapor deposition of diamond films using methane/hydrogen and acetylene/hydrogen mixtures, *J. Appl. Phys.* **68**, 4825 - 4829.

Zeatoun, L., to be published, Ph.D. Thesis, Case Western Reserve University (1996).

Chapter 21

MICROWAVE PLASMA CHEMICAL VAPOR DEPOSITION OF DIAMOND

Peter K. Bachmann
Philips Research Laboratories
Solid Films & Deposition Technologies
P.O. Box 1980
D-52021 Aachen, Germany

Contents

1. Introduction

Early approaches to grow diamond from the vapor phase used thermal decomposition of carbon-containing gases such as CBr_4, CI_4 [Spitsyn and Derjaguin 1980], CO [Eversole 1962], [Hibshman 1968], or CH_4 [Eversole 1962], [Angus et al. 1968], and were carried out at gas temperatures between 600°C and 1200°C. For these thermal low pressure CVD processes, the gas temperatures did not differ much from the surface temperature of the diamond seeds used as substrates. The resulting linear diamond growth rates of approximately 0.01 µm per hour were far too low for any industrialization of diamond CVD and, being only of scientific interest, the field went dormant for more than a decade.

Industrially applicable diamond deposition processes came within reach only after the substrate surface temperatures and the gas phase temperatures in the deposition systems were decoupled [Spitsyn et al. 1981], [Spitsyn 1991] and [Matsumoto et al. 1982].

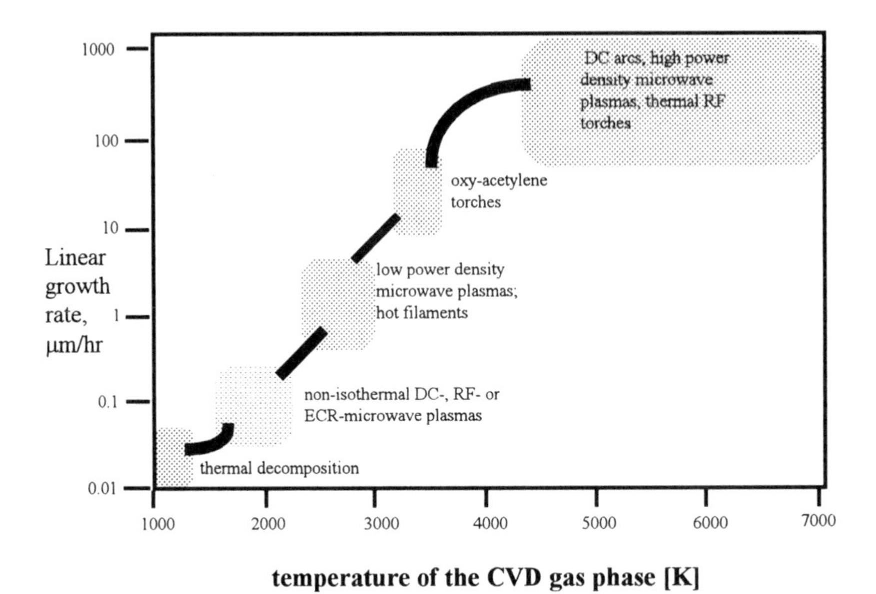

Figure 1. Correlation of gas temperatures and growth rates of diamond CVD processes. High gas temperatures combined with substrate temperatures below the graphitization limit of diamond lead to the highest rates (after [Bachmann and Lydtin 1990]).

Introduction of a hot zone in the CVD gas phase by means of activation elements such as hot graphite disks [Spitsyn 1981], [Spitsyn 1991] or hot filaments [Matsumoto 1982] led to substantially higher growth rates, revitalization of the field, and growing interest in vapor-deposited diamond by industrial corporations.

Figure 1 correlates the linear diamond growth rates with the gas temperatures for the various growth methods available today [Bachmann and Lydtin 1990]. It illustrates that the formation of diamond growth species, as well as the generation of the atomic hydrogen necessary to suppress non-diamond carbon formation, is obviously more efficient at gas temperatures that are substantially higher than the 1500 °C required for the graphitization of diamond [Field 1992]. Thus, decoupling of gas and substrate temperatures was the single most important step for the development of industrially applicable growth methods. Both the "chemical transport reaction (CTR)" method [Spitsyn et al. 1981], [Spitsyn 1991] and the now widespread "hot filament" technique [Matsumoto et al. 1982] create the desired hot zones by introducing hot solids (graphite disks or wires) into the CVD gas phase. The maximum temperature of such "activation" elements is, of course, limited by the physical properties of the materials used. In CTR, the hot graphite disk is usually at around 2000 °C [Sptisyn 1991] and the temperature of a hot filament in a diamond CVD reactor is limited to 2200-2400 °C [Matsumoto et al. 1982], [Sommer and Smith 1991] by filament burn-out. B.V. Spitsyn and his group suggested in the mid 1970's the use of electrical discharges, more specifically the application of DC plasmas, as alternative means to radicalize hydrogen and to decompose the carbon carrier gas in diamond CVD [Spitsyn et al. 1981] and [Spitsyn 1991].

Immediately after the successful introduction of the hot filament CVD method by the diamond research group at the National Institute of Research of Inorganic Materials (NIRIM), Japan [Matusmoto et al. 1982], they also published the first report on the application of a plasma discharge sustained by 2.45 GHz microwaves for the growth of diamond from the vapor phase [Kamo et al. 1983].

2. Diamonds from 2.45 GHz Microwave Plasmas

The provision of energy required to initiate chemical vapor deposition reactions by means of a plasma discharge is not unique to the vapor-growth of diamond. Variants of this deposition technology were and are being used for the fabrication of polymers, amorphous silicon, silicon nitride, silica optical fibers and other sophisticated materials and composites (see, e.g., [Bachmann et al. 1988a] for details and references). For plasma CVD, DC glow discharges, 13.45 MHz (2239 cm) radio frequency (r.f.) plasmas, and microwave plasmas excited by 915 MHz (32.8 cm) and 2.45 GHz (12.2 cm) radiation are commonly used[1] [Bachmann et al. 1988a], [Coburn and Gottscho 1986], [Plasma Chemistry and Plasma Processing], and [Janzen 1992]. The choice of very specific frequencies is mainly due to the availability of components that comply with national regulations. Introducing plasma CVD -especially the application of 2.45 GHz microwave plasmas- to the area of diamond film deposition turned out to be a very important technological advance. It is microwave plasma CVD, maybe along with hot filament CVD, that ultimately moved diamond deposition from its niche of a scientific curiosity into the area of industrially applicable technologies.

2.1 THE NIRIM TUBULAR REACTOR

In 1982, M. Kamo and collaborators [Kamo et al. 1983], NIRIM, Japan reported for the first time on the successful growth of diamond films from a methane/hydrogen 2.45 GHz microwave plasma discharge. In their paper, the authors sketch an experimental set-up that consists of a gas supply and metering system, a silica tubular reactor, a microwave source with applicator, and a vacuum pump. Fig. 2 depicts a cross-section of such a deposition system. What is today termed a NIRIM-type reactor usually consists of a 4 cm outer diameter silica tube that runs through bores in a rectangular metal microwave guide (4.30 x 2.15 inches for type WR430 wave guides) suitable for 2.45 GHz radiation (WR284, or WR340 can be used alternatively). The flow of a carbon containing gas mixture, e.g., 1 % methane in hydrogen, is maintained by a gas metering system and a vacuum pump.

Irradiating the low pressure region inside the reactor tube with microwaves of power levels of typically >200 Watt ignites a plasma discharge. At pressures below 1 mbar, such plasmas, i.e., highly reactive, radicalized, partially ionized, dissociated, quasi-neutral gases, are "non-isothermal" (see, e.g., [Janzen 1992]). That means, that within the high electric fields generated by the microwave radiation, free electrons are accelerated to energies that correspond to several thousand degrees in the case of

---

[1] See *Plasma Chemistry and Plasma Processing*, published monthly by Plenum Press, New york and London, ISSN 0272-4324 or
Plasma Processing, J.W. Coburn, R.A. Gottscho and D.W. Hess, Mat. Res. Soc. Symp. Proc. 68, Pittsburgh, PA, 1986.

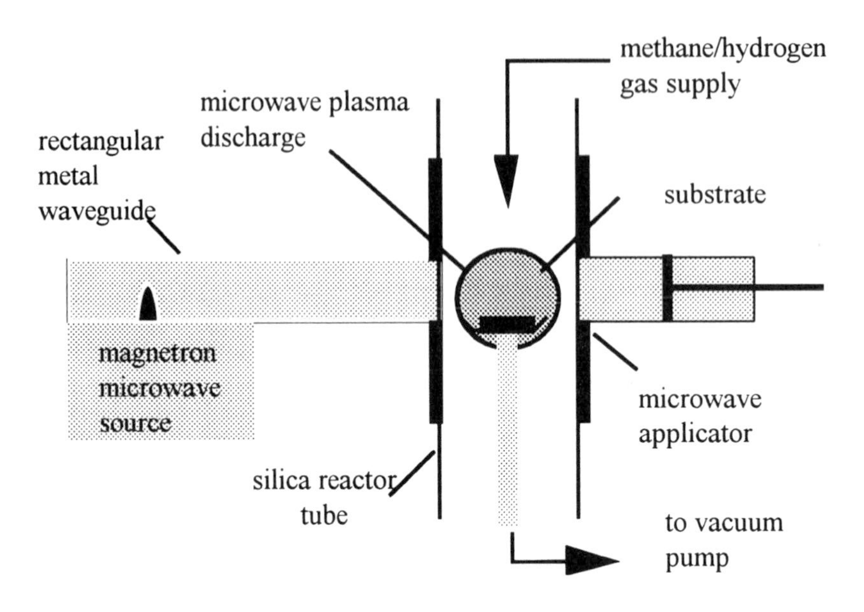

Figure 2. Reactor design originally used by M. Kamo et al [Kamo et al. 1983], NIRIM, Japan to grow diamond films on silicon from a methane/hydrogen plasma sustained by 2.45 GHz microwave radiation.

thermal activation. Neutral species and heavy ions in the gas phase are either not influenced by the alternating field or can not follow the field changes fast enough. Non-isothermal low pressures plasmas, therefore, exhibit a difference between electron temperature T_e and gas temperature T_g, with the electron temperature being substantially higher than the temperature of the almost cold heavy species (Fig. 3). Although NIRIM reactors are often started up in the non-isothermal plasma regime at pressures below 1 mbar, they are usually operated at higher pressures of 10-100 mbar during deposition. In this pressure region, however, plasmas undergo a very important change and approach "local thermal equilibrium (LTE)." In such "LTE", "thermal" or "isothermal" plasmas, not only the electrons move rapidly, but, due to frequent collisions with the hot electrons, the heavy gas species, i.e., neutrals and ions, also heat up (Fig. 3). The gas phase becomes hot and the microwave field is nothing but a convenient method to supply energy to the system and to replace gas phase activation by hot solids without the disadvantages and limitations imposed by the physical properties of such activation elements. As already mentioned in conjunction with figure 1, thermal plasmas with their higher power densities are much more efficient in producing diamond growth species. Hydrogen is also radicalized more efficiently and helps to suppress the formation of unwanted non-diamond carbon phases and with the substrate temperature stabilized below 1000°C either by plasma heating, by additional external heating or, typically at higher plasma power densities, by external cooling, diamond grows on appropriately pretreated substrates that are placed near or inside the light emitting discharge region.

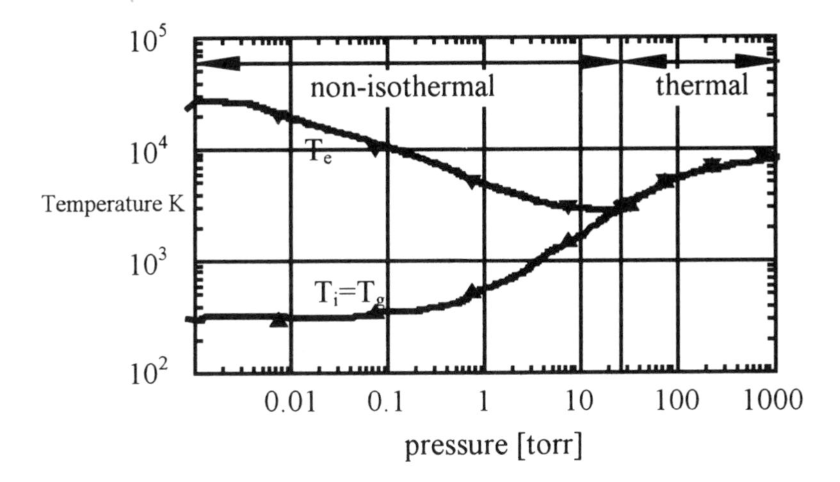

Figure 3. Electron and heavy particle (neutrals and ions) temperatures in non-isothermal and thermal plasmas (after [Janzen 1992]). Microwave plasma CVD of diamond in tubular NIRIM reactors started out in the border region between thermal and non-isothermal plasma conditions. For high rate growth, however, thermal plasma conditions turned out to be advantageous.

For NIRIM-type reactors, power levels of <1.5 kW at pressures of <100 mbar result in growth rates of 0.5-3 μm/hour for substrate temperatures around 1000 °C, with 0.5 μm/hour being more realistic for phase pure, well-faceted material.

Today, many research deposition systems are still based on the NIRIM concept and by running several tubes in parallel, Idemitsu Petrochemical Co., Japan, even scaled this reactor type to pilot production level (see [Bachmann and Messier 1989] for details). The major advantages of NIRIM-type tubular reactors are:

- simple reactor design with low set-up costs.
- easy and cheap replacement of damaged or contaminated reactor tubes.
- variation of the substrate position relative to the plasma feasible.
- plasma spectroscopy and visible observation of the substrate possible.

NIRIM-type reactors have, however, also disadvantages which limit their applicability in industrial environments. The size of substrates that can be mounted in such a unit is limited to approximately 2-3 cm^2 by the inner diameter of the silica tube and by the dimensions of the wave guide microwave applicator. For industrial applications this is often too small. The plasma is generated very close to the reactor walls. Etching and redeposition of tube material can cause severe contamination of the growing film. Implementation of an external heater or cooler can be very tedious and, in addition, the evacuated tube may get rather hot and deform.

Moreover, microwave absorbing carbonaceous films may deposit on the inner surface of the reactor tube. This is particularly problematic for the high plasma powers densities

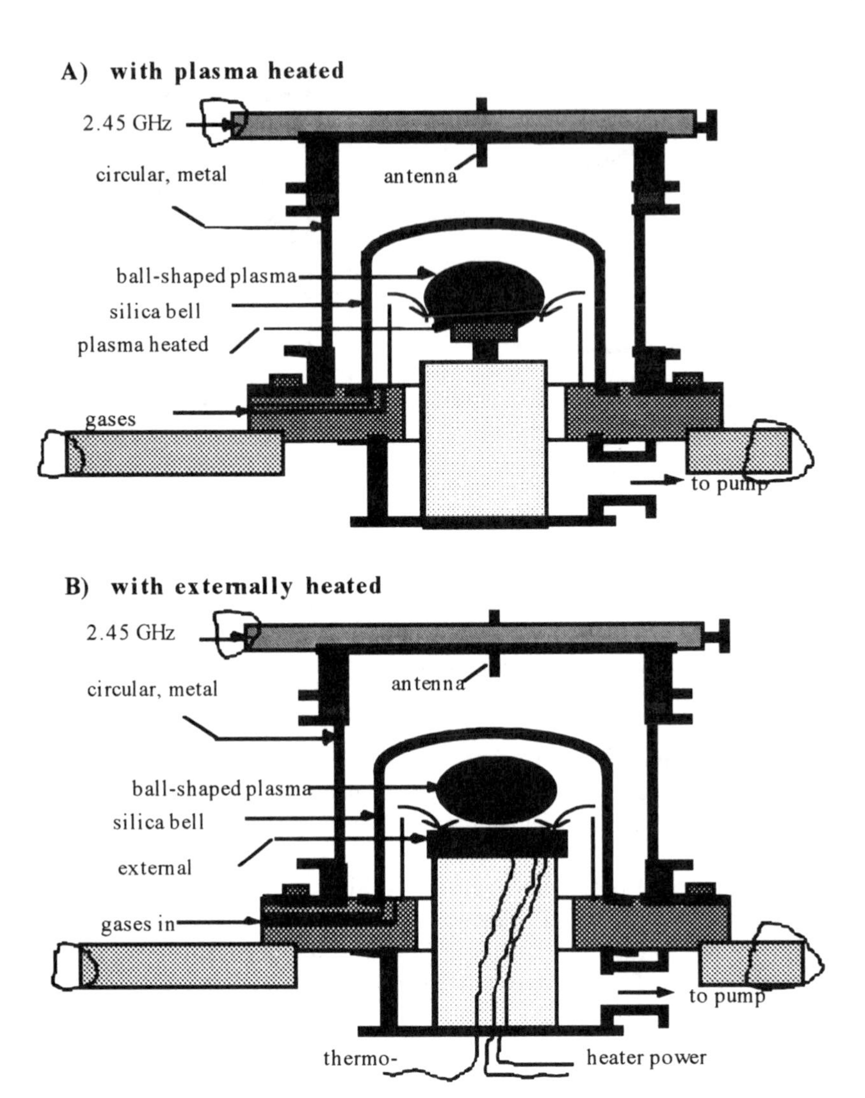

Figure 4. The "bell jar" microwave plasma CVD reactor. This 1.0 kW diamond deposition unit was jointly developed by P.K. Bachmann [Bachmann et al. 1988b] and ASTeX, Woburn, MA. It allows to deposit films onto substrates with < 3 inch diameter. Top: The plasma is used to heat substrates of < 1.5 inches up to 1000 °C. Bottom: System equipped with separate external heater for 3 inch substrates.

required to grow diamond at high rates (Fig. 1) and limits reactor lifetime, applicable power levels, material purity, and the achievable growth rates.

2.2 THE "BELL JAR" REACTOR

In 1987, P.K. Bachmann et al. [1988b] jointly with D. Smith and collaborators at ASTeX, Applied Science and Technology, Woburn, Massachusetts, developed a different set-up for 2.45 GHz microwave plasma CVD of diamond to overcome some of the limitations of the tubular reactor design. This so-called "bell jar" reactor (Fig. 4) was the first microwave plasma CVD apparatus suitable for diamond growth on substrates of up to 7.5 cm (3 inches) diameter and later became the first commercially available reactor.

A dome-shaped, 10 cm inner diameter silica bell jar replaces the silica tube of a NIRIM-type reactor. A microwave coupler picks up the 2.45 GHz radiation inside the rectangular wave guide and, via an antenna, emits it centro-symmetrically into an air-cooled circular wave guide that contains the evacuated bell jar. For convenient visual inspection during deposition, for in-situ surface analysis, and for plasma spectroscopy, the circular wave guide is perforated (hole diameter <5 mm to prevent microwave leakage). The plasma is generated at the point of highest electrical field strength inside the bell jar. Its shape, size and stability depends on gas composition, plasma power, pressure and on the reactor furniture, i.e., shape, size, material etc., of the substrate holder and other parts present inside the reactor. For gas mixtures containing more than 5% hydrogen at pressures of more than 10 mbar, a ball-shaped, stable plasma forms in the center of the bell jar, far away from any reactor walls. Wall etching and related problems of the tubular design are thus avoided. A graphite disk mounted on a moveable silica rod and heated by direct interaction with the discharge can be used as a holder for smaller substrates (Fig. 4, top). In this mode, energy released by recombination of plasma-generated radicals on the substrate surface is the main heat source. Similar to the tubular system, the substrate can be positioned anywhere in or near the plasma by means of this holder, with the plasma location hardly affected by its position. With the microwave power of the first reactor generation being limited to 1.0 kW, external heating by a resistance heater was mandatory to bring 3 inch diameter silicon wafers to temperatures of 900-1000 °C (Fig. 4, bottom). Under these conditions polycrystalline diamond films grew from 1% CH_4/H_2 at linear rates of 0.2-0.3 μm/hour. Implementation of an external heater in the system permitted, for the first time, to (at least partly) decouple plasma conditions and substrate temperature in microwave plasma CVD of diamond.

2.3 THE ASTEX HIGH PRESSURE MICROWAVE SOURCE REACTOR

Between 1988 and 1992, ASTeX improved and commercialized the bell jar reactor. Performance and user friendliness were enhanced while maintaining the basic operation principle, i.e., rectangular-to-circular coupling of microwave radiation into a circular reactor to form a centro-symmetrical, ball-shaped plasma discharge.

In the second reactor generation, the ASTeX High Pressure Microwave Source (HPMS) reactor [2] (Fig. 5), the silica bell jar is replaced by a flat silica microwave window and the once air-cooled perforated wave guide turned into a double-walled, water-cooled reactor shell equipped with windows for inspection and diagnostics. A 4.5 inch wide door allows for simple mounting of 4 inch diameter substrates. Heating of 4 inch diameter silicon wafers to approximately 900°C without the plasma being ignited is possible by a radio frequency heater incorporated into the substrate stage.

---

[2]Product details available from ASTeX, Applied Science & Technology, Inc., 35 Cabot Road, Woburn, MA 01801, U.S.A.

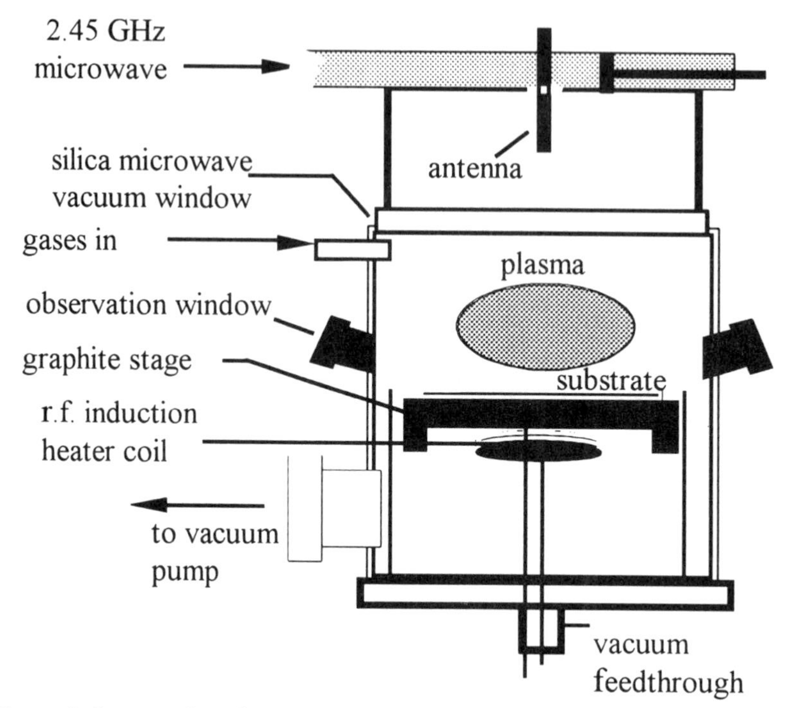

Figure 5. Cross section of a 2.45 GHz ASTeX HPMS reactor. Equipped with radio frequency induction heater and graphite substrate stage this unit is suitable for deposition onto 4 inch diameter substrates. The initial 1 kW magnetron microwave sources were later replaced by 1.5 kW and 5 kW tubes.

In 1990, a 1.5 kW microwave magnetron and power supply was introduced and with the power density vs. rate dependency (Fig. 1) in mind the system was further improved in 1992 by adding a 5 kW 2.45 GHz magnetron and power supply. The power increase resulted immediately in higher growth rates. From 5 kW microwave plasma HPMS systems it is possible to grow diamond films at linear rates of 4-14 μm/hour, depending on the gas composition (see below) and the phase purity of the deposit [Bachmann et al. 1994a]. With such high power levels implemented, additional external heating is usually obsolete. It is, on the contrary, necessary to control the substrate temperature by external cooling of the substrate stage.

Today, the 1.5 kW and 5 kW ASTeX reactors are probably the most widespread reactor types used in R&D for microwave plasma CVD of diamond. Even prototyping or small scale production is feasible with the rates and substrate sizes offered by these units. If operated under well-defined deposition conditions, these systems run extremely stably and can deposit diamond continuously and unattended for several days or even weeks. This stability is absolutely necessary in order to grow top quality, 0.5 -1 mm thick diamond disks for heat sink applications with thermal conductivities of more than 1300 W/m K within approximately one week [Bachmann et al. 1995].

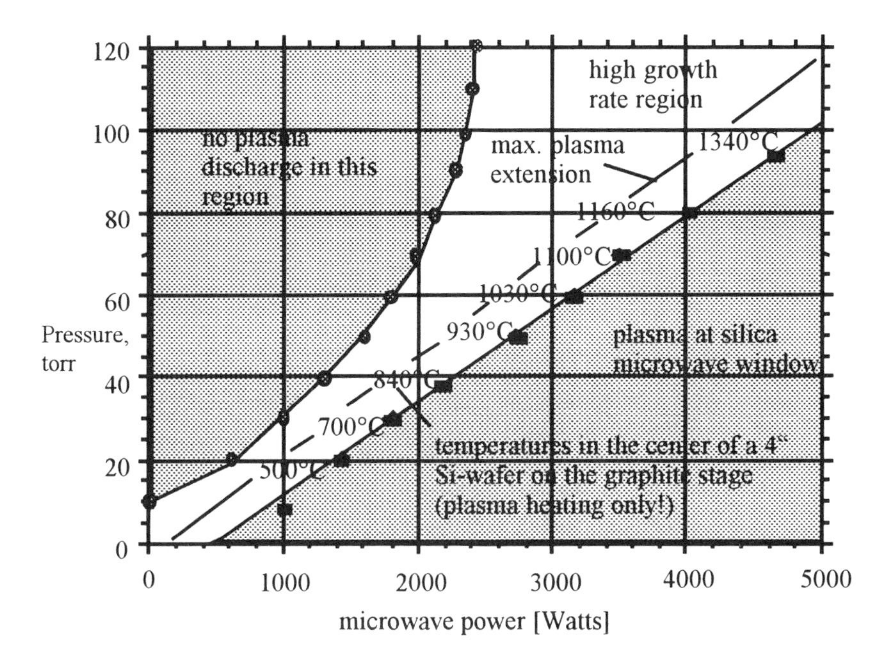

Figure 6. Operation regime of an ASTeX HPMS 5 kW microwave plasma diamond CVD reactor. Operation is limited to the central region of the diagram.

Despite its importance for the development of an emerging CVD diamond technology and for first steps towards industrialization of diamond thin films, the reactor type shown in Fig. 5 has also important drawbacks. Especially at higher plasma power levels and low reactor pressures, the plasma discharge is not stable above the substrate, but tends to jump to the silica microwave vacuum window of the reactor. Exact parameters depend on the details of the reactor configuration and vary a little from set-up to set-up and figure 6 gives approximate values for the operation regime of a HPMS source. At too low microwave power levels and/or too high pressures a plasma can not be sustained. If power levels are too high for a given reactor pressure, the plasma becomes unstable and jumps to the vacuum window. For 5 kW discharges, quick and complete destruction of the window is almost always the consequence. Stable operation is only feasible in the blank corridor (see Fig. 6) in between. This implies that reactor pressure and microwave power can not be varied completely independently for HPMS reactors. In figure 6, temperatures measured in the center of a 4 inch diameter, 0.5 mm thick silicon wafer sitting on the graphite substrate stage (Fig. 5) of the set-up are also given. For these measurements, the heater was left switched off. The data illustrate the absolute necessity of substrate cooling rather than heating for the high power density levels needed for high growth rates. Inside the bell jar as well as the HPMS microwave plasma reactor, the discharge is almost ball-shaped under common diamond growth conditions. As sketched in figure 7, this results in distance variations between the substrate surface and the center of the plasma, i.e., the source of growth species and radicals. Consequently, in HPMS reactors as well as in bell jar reactors growth rate, phase purity and transferred energy are radius-dependent.

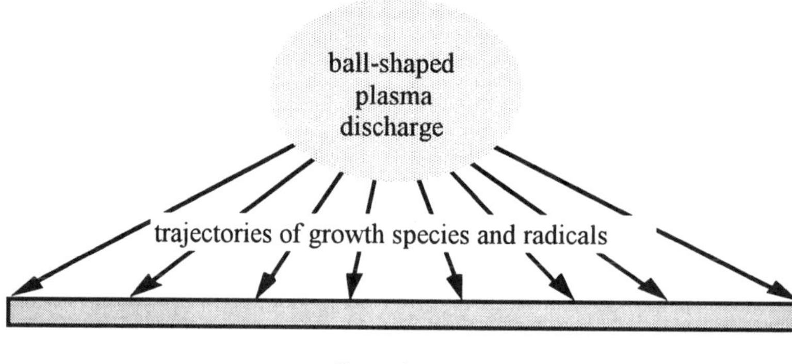

Figure 7. Schematic of trajectories of growth species, radicals and radiation emerging from the ball-shaped plasma of a bell jar or a HPMS diamond CVD microwave plasma reactor. The variation across the surface of a flat substrate results in thickness, temperature and phase purity variations.

The thickness profiles of the deposits are usually Gaussian-shaped with a 2 cm wide, fairly flat central area and less than 50% of the central thickness at diameters greater than 5 cm. The radial concentration variation of the hydrogen radicals that interact with the substrate surface leads to phase purity changes across the deposit that are easily detectable by Raman spectroscopy. In a standard set-up with the plasma above a flat 4 inch diameter substrate (Fig. 7), substantially more non-diamond carbon is being deposited at the perimeter of the Gaussian-shaped profile.

Despite its drawbacks and limitations, the bell jar and later the HPMS microwave reactor were of considerable importance for the development of an industrial CVD diamond technology. For the first time, research institutions and company laboratories were able to buy a reliable diamond CVD reactor rather than spend the time to develop there own set-up. The phase purity of the material grown from the high power density plasmas feasible in such reactors were comparable to the best natural stones and the rates were sufficient to exploit the extreme properties of diamond on a larger scale. By adding a negative bias potential to the substrate holder, J.T. Glass and colleagues at Kobe Steel Inc. North Carolina, USA [Stoner and Glass 1992] and C.P. Klages at IST, Fraunhofer Institute, Braunschweig, Germany [Jiang and Klages 1993] demonstrated nucleation and growth of highly oriented textured films with much improved electronic properties in such reactors. H. Windischmann [Windischmann et al. 1991] grew the first diamond x-ray lithography membranes with apertures of more than 6 cm in this reactor type, x-ray windows [Bachmann et al. 1993] and ultra-thin, low wear coatings [Bachmann et al. 1994b] were prepared in HPMS reactors, cutting tools were coated in such units [Lux and Haubner 1994] and top quality heat sinks with thermal conductivities of up to 2200 W/m K were grown in modified 5 kW HPMS reactors [Bachmann et al. 1995]. Full size industrialization of 2.45 GHZ microwave plasmas for diamond growth, however, still requires larger deposition areas, more homogeneous and

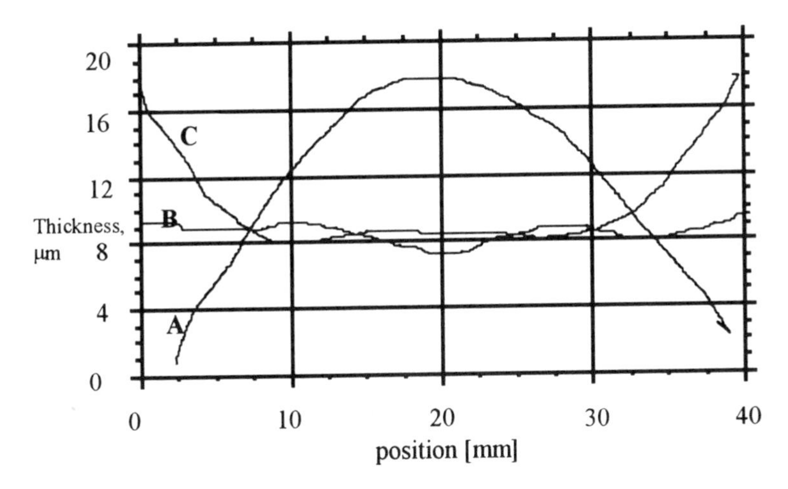

Figure 8. Measured thickness profiles of diamond films grown at Philips Research Aachen, Germany in a modified HPMS reactor. Profile variations from (A) Gaussian to (B) nearly flat to (C) U-shaped are possible by changing the electrical field distribution via the geometry of the substrate stage inside the ASTeX HPMS shell.

more uniform deposits, and higher rates than those offered by a standard 2.45 GHz HPMS unit.

Operating a HPMS reactor at the lowest possible pressure with the highest possible power (see Fig. 5) increases the flat region of the Gaussian-shaped thickness profile somewhat and homogeneous, uniform regions with 3 cm diameter are possible. If flat, uniform films with diameters less than 5 cm are sufficient for practical applications, modification of the electrical field distribution inside the reactor by changing the geometry of the substrate stage is a way to flatten out the profile and, depending on actual geometry, films with flat or even U-shaped profiles may be obtained. In these modified reactors the plasma is attracted close to the substrate surface. This results not only in thickness profile modifications, but also increase the deposition rate considerably. Fig. 8 illustrates some of the thickness distributions that are feasible by changing the field distribution in a HPMS reactor by varying the geometry of the substrate stage.

Delivering microwave radiation through the reactor bottom and using the substrate stage itself as the microwave coupling electrode inside the vacuum region of the reactor is another way to flatten the plasma and the thickness profile of the deposit. Again the plasma is closer to the substrate surface and the rate increases. With their new design, ASTeX reactors maintain centro-symmetrical coupling into the low pressure region. The flat plasma is, however, more stable and the rate increases by a factor of 1.5 [Post 1993]. With the implementation of these modifications and the addition of a 8 kW/2.45 GHz magnetron ASTeX further improved film uniformity, homogeneity and rate of diamond growth and developed their 2.45 GHz reactor line to its present status. In parallel other groups and corporations developed and tested other concepts to enlarge the deposition area and to increase the deposition rate.

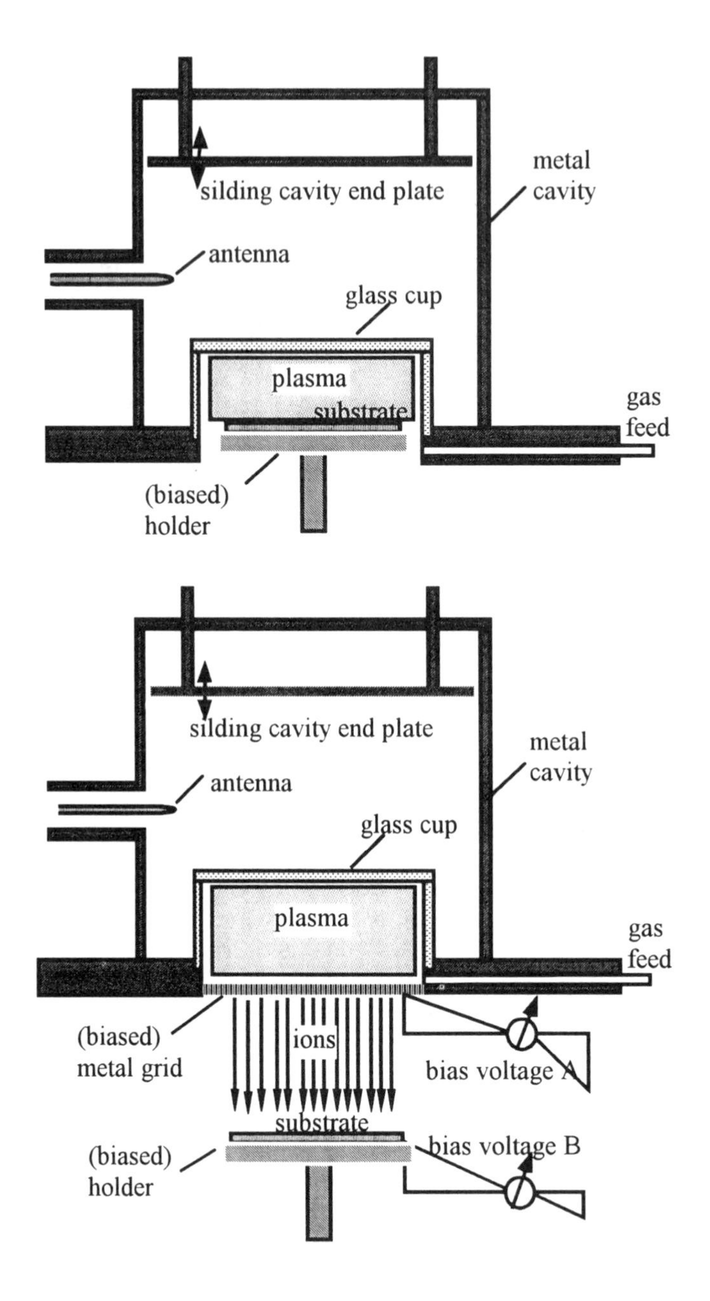

Figure 9. The Wavemat MPDR unit may be operated with the substrate immersed into the plasma (top) or with the substrate in remote position (bottom). A biased screen allows to extract and direct ions towards the substrate surface. Equipped with permanent magnets, electron cyclotron resonance (ECR) plasma operation is a third option [3].

---

[3]Product details available from Wavemat, Inc.

2.4 THE WAVEMAT REACTOR

ASTeX microwave plasma CVD reactors for diamond synthesis are probably the most successful, however, not the only commercially available reactors. Soon after the bell jar and HPMS reactors appeared on the market, Wavemat Inc. supported by Norton of Northboro, MA, USA announced the availability of their MPDR. MPDR stands for Microwave Plasma Disk Reactor with a design based on the work of J. Asmussen and collaborators at Michigan State University [Asmussen and Root 1984]. Fig. 9 depicts cross sections of two variants of the MPDR and illustrates the major differences between HPMS and MPDR. The microwave energy from a 2.45 GHz magnetron is directly coupled into the applicator. In the MPDR, the antenna is, however, not located in the center but at the perimeter of the circular wave guide. In both reactor types system tuning is done by moving the antenna in and out of the circular wave guide section and/or by adjusting the position of the substrate stage (HPMS) or a separate sliding end plate. In the MPDR, a disk-shaped plasma is generated inside a fairly small silica glass cup. The substrate can either be positioned in or close to the plasma (Fig. 9, top), or remotely, with a biased metal screen between substrate surface and the plasma itself (Fig. 9; bottom). With the substrate next to the plasma, deposition conditions resemble those of a HMPS reactor (profile differences due to the different coupling methods are possible). However, the plasma is very close to the silica glass cup. This is no problem as long as the system operates under non-isothermal plasma conditions with a cold gas phase. However, as in the NIRIM tube reactor, it may lead to wall etching, wall contamination, heat dissipation problems and deformation under the thermal plasma conditions needed for high growth rates.

2.5 MAGNETIZED AND ECR PLASMAS

Increasing the size of microwave plasmas by operating it at very low pressures in order to deposit onto large areas has its limits. Below 0.1 mbar, reduced collision probabilities result in insufficient production rates for radicals and ions. To sustain a plasma becomes increasingly difficult. Here, the use of magnetic fields that force free electrons to resonantly move in closed loops (electron cyclotron resonance, ECR) and absorb additional energy helps. For electrons in a 2.45 GHz microwave field, a magnetic field of at least 0.875 kG is required to induce ECR. The Wavemat MPDR unit offers the option of adding strong rare earth permanent magnets around the plasma containing glass cup for ECR microwave plasma CVD. Two horizontal electromagnets put around an ASTeX HPMS reactor shell (see Fig. 10) create a centro-symmetrical field along the system axis and by changing the spacing and location of the magnets distribution and strength of this field with respect to the substrate can be varied. The pressure regime for true ECR action is well below 0.1 mbar. Therefore, ECR plasmas are typical "non-isothermal" plasmas (see Fig. 3) that are often used for plasma etching or deposition in the semiconductor industry.

ECR plasmas were introduced to diamond CVD because they seemed to offer an option to coat large substrate areas [Kawarada 1987], [Suzuki 1989] and [Yuasa et al. 1991]. Unfortunately, at pressures of more than 10 mbar, where plasmas approach equilibrium and phase pure diamond can be grown, the electron collision rate in such plasmas is far too high for full size ECR. The plasma is hardly affected by the magnetic field and the

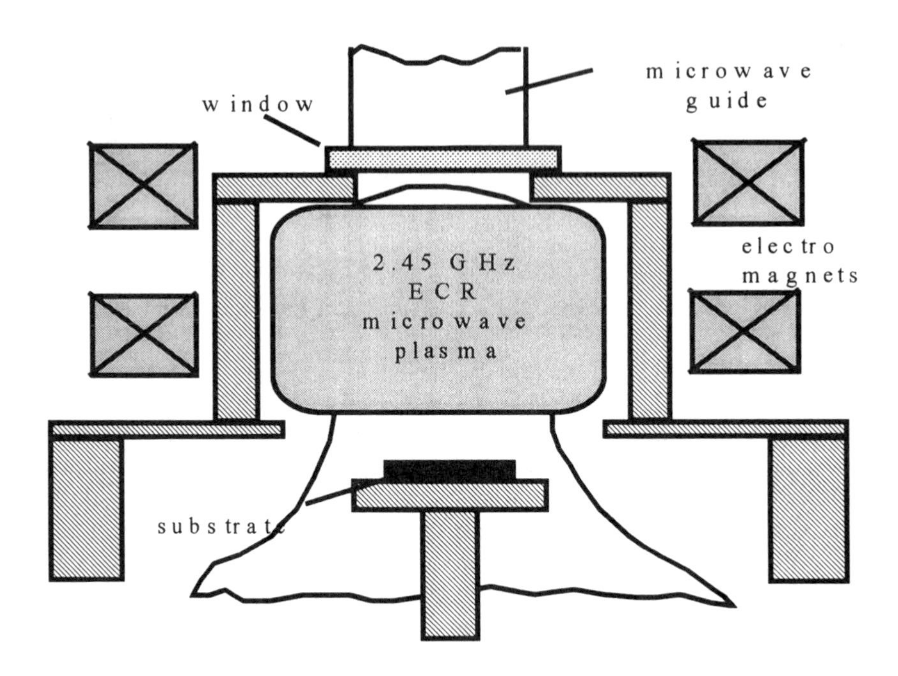

Figure 10. Electron cyclotron resonance plasmas are generated at low pressures inside a magnetic field with at least 0.875 kG for plasmas sustained by 2.45 GHz microwave radiation. The magnetic field forces free electrons to move in closed loops and absorb additional energy. With ECR it is possible create large non-isothermal plasmas between 0.1 and 10^{-5} mbar.

deposition results do not differ from those achieved with non-magnetized microwave plasmas. At pressures of less than 0.1 mbar, where ECR starts to become significant, deposition rate, crystal size, and phase purity are already markedly reduced. On the other hand, the fact that only mixed phase material is deposited from non-isothermal ECR microwave plasmas underlines quite nicely the importance of high gas temperatures for diamond formation (see Fig. 1). Although ECR is an interesting way to create large area plasmas, it turned out to be unsuitable for the growth of high quality diamond. For applications where thin, mixed phase films with small diamond crystals embedded into an amorphous carbon matrix might be sufficient, e.g., for low wear coatings [Bachmann et al. 1994b] or electron emission applications [Kumar 1993], ECR plasma CVD might still be an option.

2.6 2.45 GHZ PLASMA TORCHES

Given the correlation in figure 1 and the temperature behavior illustrated in figure 2 attempts to grow diamond from atmospheric pressure microwave plasmas nowadays seem only logical. In 1988, however, this was not so clear. Y. Mitsuda et al. [1989] were the first to develop a microwave plasma diamond deposition system capable of operating at pressures of up to 1 bar. Figure 11 depicts a cross-section of their plasma torch. While diamonds grow at linear rates of 0.5-1 μm/hour from the standard 0.5%

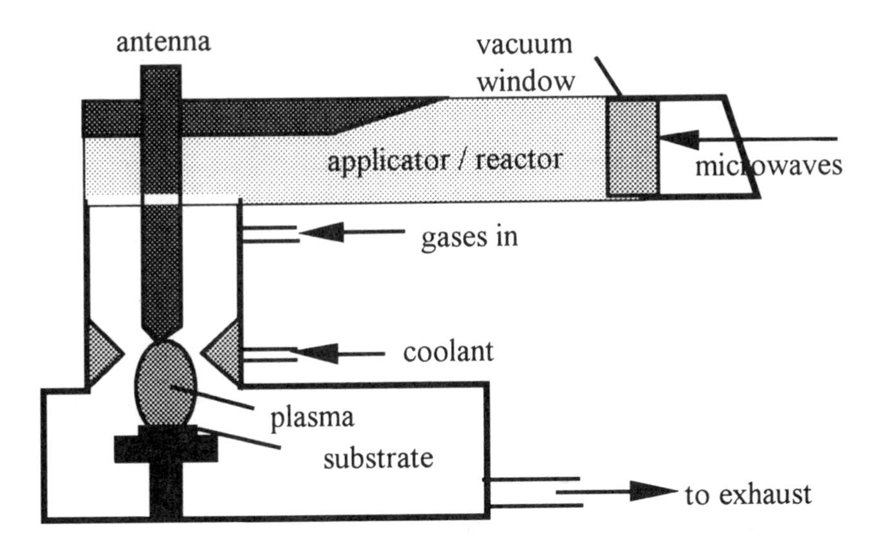

Figure 11. Atmospheric pressure microwave plasma torch developed by Y. Mitsuda et al [1989]. With this unit, high rate diamond growth from 2.45 GHz microwave plasmas were demonstrated for the first time.

CH_4/H_2 at power levels of 1-1.5 kW and pressures of 50 mbar in a NIRIM or bell jar reactor, Mitsuda's microwave 5 kW atmospheric pressure plasma produces good quality films at rates of 30 μm/hour. This early attempt to grow at higher power densities had, of course, its difficulties. The specific design of this reactor led to fairly small deposition areas of only 5 cm^2. With the metal microwave antenna in close contact with the plasma, contamination was unavoidable. Nevertheless, the work of Mitsuda et al. demonstrated that also for microwave plasma CVD of diamond, thermal plasmas have a clear advantage over non-isothermal plasmas and were important for the development of the correlation given in figure 1.

2.7 DIAMONDS FROM REMOTE MICROWAVE PLASMAS

Interaction of plasmas with substrate surfaces can be a serious problem for any plasma-based deposition process. Impinging ions, electrons or neutrals may damage the structure of the substrate, lead to material removal, change surface properties or even change bulk properties. In semiconductor processing, the negative influence of ions or photons emitted from discharges is well-known. Remote generation and subsequent transport of growth species helps to reduce these problems and can also help to distribute the material evenly over larger substrate areas [Hsu et al. 1990]. Therefore, remote plasma CVD of diamond seems to be ideal to deposit perfect films onto large substrate areas, maybe even at lower temperatures than with conventional methods. Examples for remote microwave plasma CVD already briefly described in sections 2.4 and 2.5 of this chapter. The Wavemat Inc. MPDR and also other ECR microwave plasma sources are capable of generating plasmas in one part of a reactor and to deposit onto substrates placed in a remote position. The degree of interaction between plasma

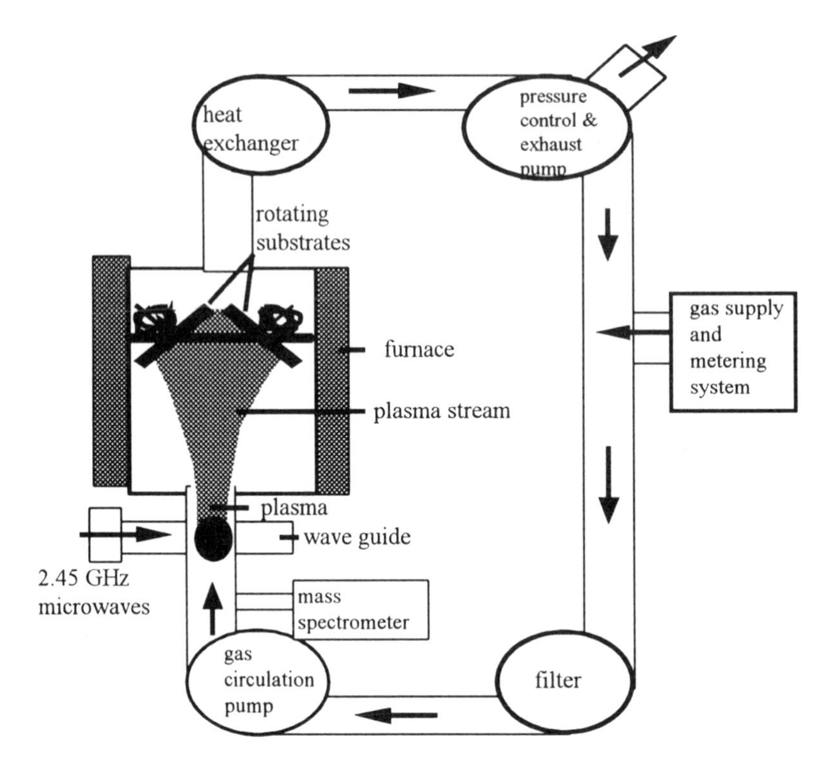

Figure 12. Schematic of the ASTeX LADS system. Diamond growth species are generated by a 2.45 GHz microwave remote plasma source and transported to the surface stationary or rotating 8 inch substrates by a high velocity gas stream. Gas recycling and mass spectrometry compositional control are additional features (after [Post 1993]).

and substrate surface can be controlled by the distance between the two regions, the gas velocity and direction and bias voltages applied to the substrate and/or to screens between plasma and substrate surface. Over the past years, however, it turned out that negative biasing is important for *in-situ* generation of nucleation sites that lead to oriented films [Stoner and Glass 1992] and [Jiang and Klages 1993], but during growth, continuous ion bombardment leads to graphitization of the growing diamond layers. Remote plasma generation and extraction of growth species by permanent biasing is, therefore, not an option to improve the film quality or enlarge the deposition area.

Remote plasma generation and transport of growth species via high velocity gas streams was investigated by several groups (see, e.g., [Rudder et al. 1988]) and in 1992 ASTeX introduced their 2.45 GHz remote microwave plasma LADS (Large Area Deposition System) reactor for up to 8 inch diameter substrates [4][Post, 1993]. Figure 12 sketches the system set-up. A microwave plasma is generated in a tubular reactor and the species

[4]See footnote 2.

are ejected into a chamber with the substrates mounted inside a furnace. Stationary or rotating substrates of up to 8 inches can be used in such a system. The high gas velocities required to produce the plasma jet would consume substantial amounts of gas. Therefore, the LADS approach includes a gas recycling system with gas circulation pump, filter and heat exchanger. The gas composition is measured by mass spectrometry and kept at the desired level by a gas supply and metering unit.

The major advantage of this remote microwave plasma CVD system is clearly the large coating area rather than the film quality. The influence of moving the plasma closer to the substrate surface on quality and rate were discussed earlier. Despite the high gas velocity applied, diamond growth species - be it C-, CH–, CH_2-, or CH_3-radicals - still have ample time between plasma generator and substrate surface to react or recombine. Outside the generator the gas phase will cool and fine grain diamond with substantial amounts of non-diamond carbon rather than phase pure material tend to form. As in low density ECR plasmas it is difficult to fabricate phase pure diamond from such a remote plasma. For low wear coatings or electron emitting layers such films may be suitable, but for, e.g., heat sinks the phase purity deposited from remote microwave plasmas is usually not sufficient.

2.8 OTHER 2.45 GHZ PLASMA APPROACHES

In addition to the various approaches already discussed in this chapter, several other configurations to grow diamond films from microwave plasmas have been tested in various laboratories. M. Kamo et al. [1990], e.g., simultaneously fed microwave radiation from two (at 180°) or three magnetrons (at 120°) into a single reactor to increase power density and size of the plasma. Unfortunately, this approach tends to generate plasma instabilities, is difficult to handle, and not practical for industrialization. Much closer to industry is the work of K. Ishibori and Y. Ohira at Denki Kogyo Ltd., Japan [Ishibori and Ohira 1990]. As early as 1988, they developed and commercialized a deposition system similar to the bell jar set-up [Bachmann et al. 1988b]. In their rather expensive system they couple microwaves centro-symmetrically into a bell jar by means of a tapered wave guide, a so-called microwave horn. Using powers of 3 kW at pressures of 40 torr they demonstrated uniform deposition of good quality diamond at linear rates of 0.3 μm/hour onto 5 inch silicon wafers. Well-known in Japan and certainly an interesting and versatile competitor for (at least) the early ASTeX units, the bulkiness of the wave guide/horn system and probably also the high price hampered the spread of these systems in the USA and Europe.

Other designs of 2.45 GHz deposition units similar to those already discussed, were tested and are used in different CVD diamond research groups [Pinneo 1993], [Rau and Trafford 1990], [Gicquel et al. 1994], [Hong et al. 1994], [Miyake et al. 1988]. Additional features include, e.g., rotating substrate stages to coat ball bearings [Pinneo et al. 1993], full glass systems to use aggressive gas compositions [Rau and Trafford 1990], [Bachmann et al. 1994a] or modifications for plasma diagnostics [Gicquel et al. 1994] and [Hong et al. 1994]. Those features of these units that are of general relevance for diamond growth in 2.45 GHz microwave plasma reactors are, however, already covered.

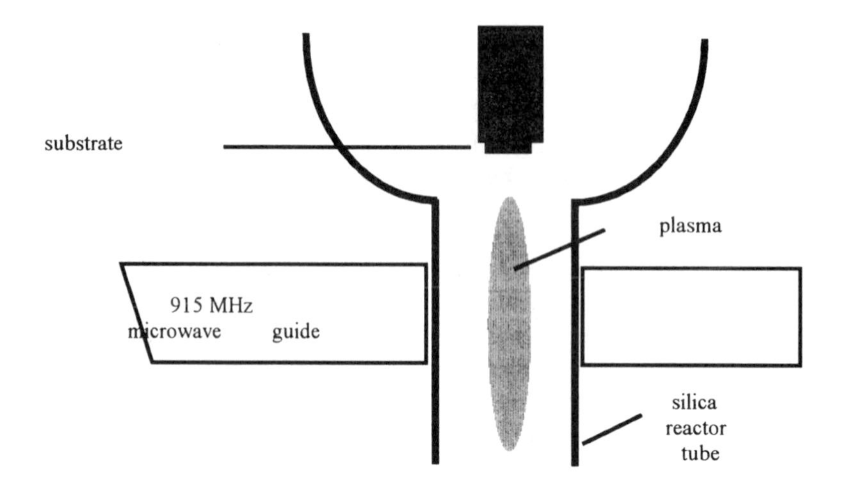

Figure 13. 915 MHz microwave plasma deposition system used by S. Miyake et al. [1988]. The reactor is similar to the tubular NIRIM set-up for 2.45 GHz, but uses larger wave guides to adapt to the longer wavelength at 915 MHz.

2.9 DRAWBACKS OF 2.45 GHZ PLASMAS

The variety of deposition systems and reactors discussed so far illustrates already a major advantage of using 2.45 GHz for plasma generation: required components such as microwave sources, wave guides, tuners, antennas, power supplies, or measurement and safety equipment are readily available. The fairly small wave guide dimension required for a wavelength of 12.2 cm allows the design of systems that are acceptable in size and weight. With magnetron sources of up to 8 kW, high rate diamond synthesis is possible and research, development and even (pilot) production diamond CVD based on 2.45 GHz plasmas is feasible. Nevertheless, there are also substantial drawbacks associated with the use of 2.45 GHz microwave plasmas for diamond deposition.

First and foremost, power density, plasma size, and achievable diamond mass deposition rate are limited. 2.45 GHz magnetron sources with more than 8 kW output power are not available. Higher powered klystron sources (30 and 50 kW) exist, but are rather complex, more expensive and not very common. None of the presently existing reactor designs include klystrons. As the size of a plasma at constant power decreases with increasing pressure, these power limitations restrict the achievable linear rates (Fig. 1) to smaller substrate areas, as already demonstrated by the plasma torch approach of Y. Mitsuda et al (section 2.6). For high rate, large area growth of phase pure diamond, higher microwave power levels than those offered by 2.45 GHz magnetrons are necessary.

In addition, plasmas sustained by 12.2 cm/2.45 GHz radiation are fairly sensitive to small objects that disturb the electrical field distribution inside the reactor. Such plasmas tend to concentrate at tips and edges. Non-uniform temperature distributions,

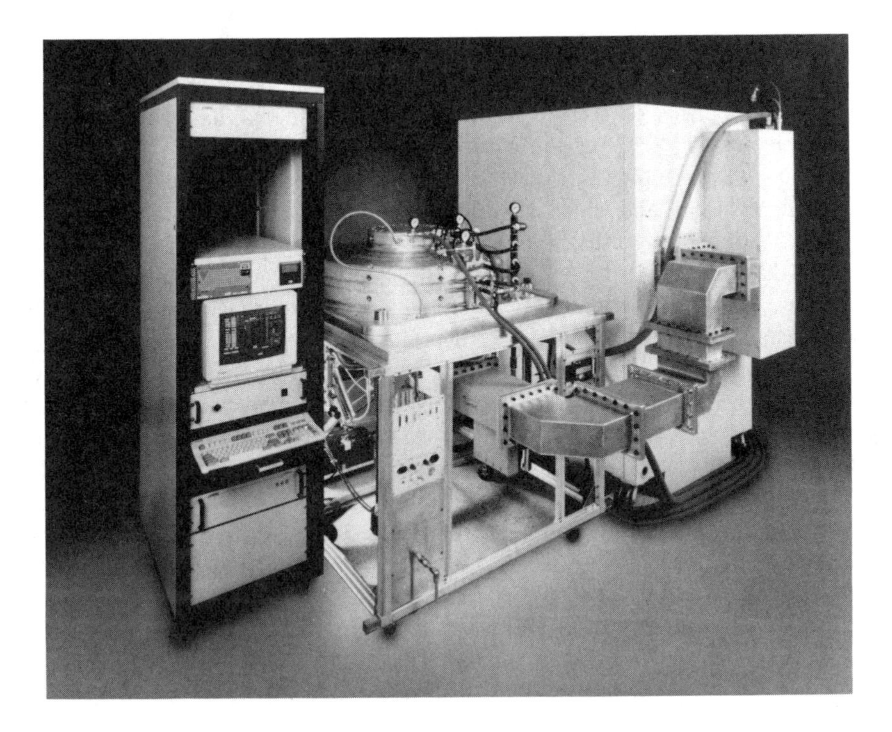

Figure 14. High rate/large area 75 kW/915 MHz microwave plasma diamond deposition system developed by ASTeX, Woburn. Like in their 5 kW/2.45 GHz "flat plasma" system the microwaves are centro-symmetrically coupled into the reactor from its bottom, using the substrate stage as coupling electrode (photo courtesy of ASTeX).

film thickness and phase purity variations are possible results. To coat complex shapes and three-dimensional objects in a 2.45 GHz reactor is, therefore, rather difficult. Fortunately, shifting from 2.45 GHz to 915 MHz microwave sources helps to overcome both, the power limitations and the geometrical sensitivity of 2.45 GHz plasmas.

3. Diamonds from 915 MHz Plasmas

S. Miyake and coworkers [Miyake et al. 1988] were the first to report on 915 MHz microwave plasma diamond deposition experiments. Their set-up, depicted in figure 13, is very similar to the NIRIM tubular reactor. However, with dimensions of 10.00 by 5.13 inches (25 by 12.5 cm), the WR975 wave guides used to adapt to the larger wavelength of 32.8 cm at 915 MHz are substantially bigger than the WR430 wave guides used at 2.45 GHz.

The authors report on experiments at 25 mbar and 1 kW input power with growth rates of approximately 1 μm/hour. At 1 bar and 7 kW, rates of up to 50 μm/hour were achieved. This corresponds well with the low and atmospheric pressure 2.45 GHz microwave torch experiments by Y. Mitsuda [Mitsuda et al. 1989] and illustrates that the excitation frequency of the plasma is not import for the deposition process. Higher plasma power densities, i.e., higher gas temperatures, again, result in higher deposition rates. Until recently Miyake's results were the only experimental data available for 915 MHz microwave plasma CVD of diamond. As higher microwave power levels became more and more important for higher rates and larger deposition areas, ASTeX Inc. Woburn, MA, developed and, since early 1995, markets a 915 MHz microwave plasma CVD system for diamond synthesis[5] [Post 1993]. Figure 14 displays a photo of such a system with the impressive dimensions of the wave guides and the reactor chamber clearly visible. Microwave radiation is fed into the reactor through the reactor bottom, using the substrate stage as coupling electrode, in much the same way as in their 2.45 GHz "flat plasma" system. At 915 MHz, the power limitations mentioned for 2.45 GHz no longer play a role. Substrates of up to 12 inches (30 cm) diameter can be mounted in such a unit. Diamond was demonstrated to grow at a rate of approximately 10 µm/hour over an entire 8 inch diameter substrates with a thickness variation of ±20% [Sevillano 1995]. With a total mass deposition rate nearly 1 g of diamond (5 ct.) per hour this unit offers the highest diamond production rate available today. This 915 MHz approach exceeds the total mass deposition rate of 5 kW / 2.45 GHz systems by a factor of 10 to 100, its coating area by a factor of 4-5 [Post 1993], [Sevillano 1995] and is clearly directed towards serious industrial use, e.g. mass production of heat sinks or tool coatings.

4. Gas Compositions used in Microwave Plasma Diamond CVD

Not only the gas temperature and, hence, a combination of gas pressure and microwave power inside a reactor are vital for rate, purity, crystallinity, uniformity, and other properties of vapor-grown diamond films. Equally important for plasma CVD (and any other diamond CVD method) is the composition of the gas phase. Early in the development of this field, experiments using C/H-mixtures, usually <1% methane in hydrogen clearly dominated. Later, a wide and sometimes confusing variety of C/H/O-, C/H/F- and C/H/Cl-compositions with and without the addition of noble gases and dopants was used. A significant amount of the available data are based on microwave plasma experiments, because for this method the gas composition is inherently much less restricted than for other methods. In hot filament CVD, high partial pressures of oxygen and other reactive gases may lead to severe damage of the filament. Flame synthesis of diamond relies on the combustion energy produced by the flame. Varying the mandatory oxy-acetylene mixtures significantly from a ratio of 1:1 or adding other gases, e.g., additional hydrogen or noble gases is difficult and usually downgrades the process performance. To avoid electrode damage and film contamination in DC arc discharge CVD of diamond from C/H-mixtures requires already sophisticated techniques. Adding O, F, Cl or other reactive components greatly increases the difficulties. Only other electrode-less plasma method of diamond growth, such as inductively coupled radio frequency plasma CVD of diamond are as flexible with respect to gas composition variations as microwave plasma CVD.

---

[5] See footnote 2.

4.1 THE C/H/O-DIAGRAM OF DIAMOND CVD

By analyzing experimental data available in the literature and subsequent careful, additional 2.45 GHz microwave plasma CVD experiments, Bachmann et al. [1991, 1994a] developed a scheme to explain and predict the dependency of diamond deposition results from the gas mixture used in the growth process. This scheme, the C/H/O gas phase compositional diagram of diamond CVD, provides a common roof for all CVD approaches to grow diamond, helps to explain experimental results, to select successful gas compositions, and to predict trends in deposition rates or film properties. It includes, but is not restricted to, the use of microwaves to initialize the CVD reactions. The principles of constructing the C/H/O-diagram are detailed in [Bachmann et al. 1991], but are relatively simple. Experimental results are categorized in "successful diamond growth" (e.g., characterized by the diamond Raman signature of the deposit), "non-diamond carbon" growth (i.e., carbon is deposited, but the diamond content is too low for Raman detection), and "no growth." In a ternary diagram, with C/H/O-atom fractions calculated from the gas mixtures used in the experiments, successful diamond growth is found to be clustered in a region, the "diamond domain," that extends along the H-CO tie line in this diagram (Fig. 15). In its first version, the "diamond domain" was wedge-shaped (dashed lines in Fig. 15) and extended fully to the hydrogen-free C-O-side line of the diagram, however, with only very little data on low H-concentration experiments. In subsequent publications [Bachmann et al. 1994a] and [Bachmann et al. 1994c], a few minor modifications were necessary. The diamond domain turned out to be lens-shaped (white area in the center of Fig. 15) rather than wedge-shaped after performing additional 1.5 kW/2.45 GHz microwave plasma CVD experiments and thermodynamic calculations [Prijaya et al 1993] and [Cassidy et al. 1994]. The domain was found to narrow down to a single point slightly off the C-O side line and diamond CVD without <u>any</u> source of hydrogen in the system is not feasible. At least 0.5% H needs to be present in the C/H/O-mixture. The straight oxygen rich border of the diamond domain is almost identical to the H-CO tie line of the diagram. Its position was not only confirmed in many growth experiments, but is also nicely illustrated by the complete disappearance of any C-related optical plasma emission when crossing over from the diamond domain into the no-growth region of the diagram [Bachmann et al. 1994a] and [Cassidy et al. 1994]. The sharpness of this border is again confirmed by thermodynamic calculations [Cassidy et al. 1994].

It is indeed surprising that, despite all possible experimental errors and differences in the way and degree the CVD gas phase is activated, the vast majority of deposition results complies with the C/H/O-diagram. Of course, the correct gas composition is not the only requirement for diamond CVD. If the energy provided to the system is not sufficient to sustain a plasma, diamond growth will not proceed, i.e., the diamond domain will disappear. With insufficient activation of the gas phase at low microwave energies (< 500 W) and pressures, the diamond domain extends a little further into the "no growth" region and diamond can grow at low rates [Bachmann et al. 1991]. At high power densities, i.e., at maximum gas phase activation, the H-rich corner of the diamond domain widens to H/H+C = 0.9, with the C-rich border approaching a thermodynamically calculated iso-potential line for carbon in the CVD gas phase over its entire length. At least under high power density conditions and for sufficiently long reactor residence times, diamond CVD is well characterized by simple thermodynamics.

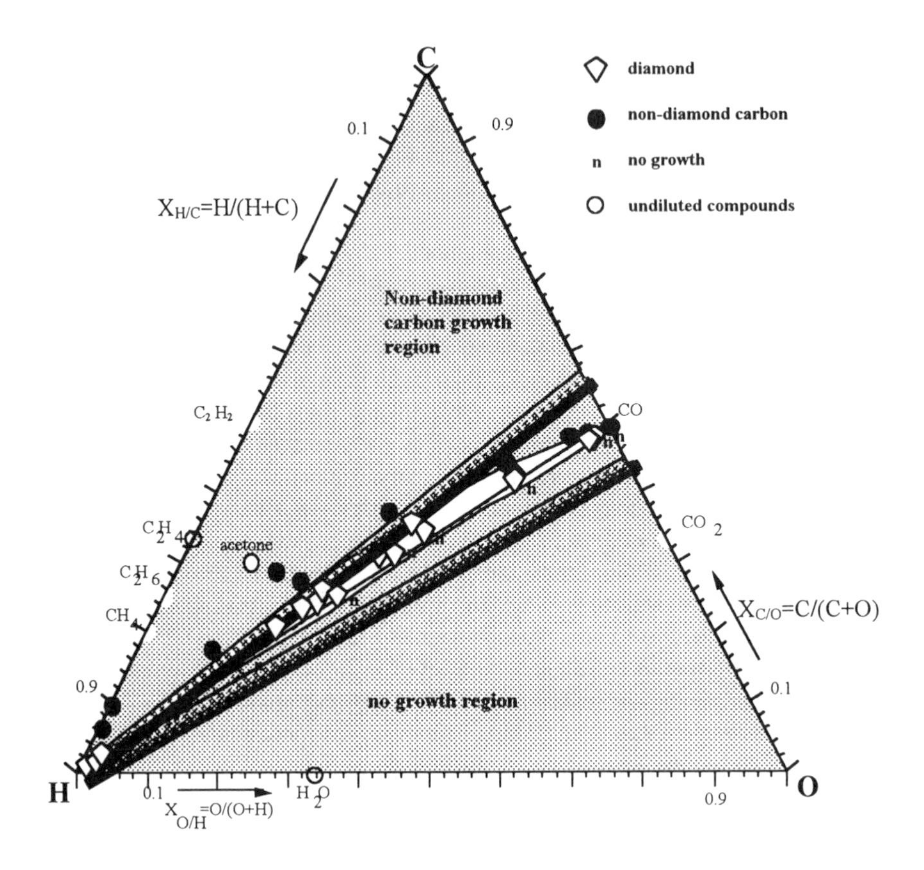

Figure 15. C/H/O-diagram of diamond CVD developed by P.K. Bachmann et al. [1991]. In the first version of this graphical representation of C/H/O-ratios used in diamond CVD experiments, the domain of successful diamond growth that extends along the H-CO tie line of the diagram is wedge-shaped (dashed lines). Later, based on additional 1-5 kW microwave plasma CVD experiments, the shape of the diamond domain changed into its lens shape that is shown here [Bachmann et al. 1994a].

The C/H/O-diagram of diamond CVD certainly has its scientific merits. In an industrial environment, however, it also turned out to be an excellent strategy tool to adapt and optimize diamond deposition to specific applications. As the diamond domain is confined between the "no growth" and the "non-diamond carbon" growth region, two trends are logical and confirmed by experiments. First, the deposition rate drops from its C- to its O-rich border and, secondly, the phase purity increases in the same direction. If, e.g., heat sinks and thermal management substrates are required, high phase purity and large grain sizes are necessary and gas compositions close to the O-rich border of the diamond domain are recommended. If a fine grain, smooth surface is required, e.g., for low wear coating applications, with phase purity being less important, the higher rates possible for mixed phase films grown from gas mixtures closer to the C-

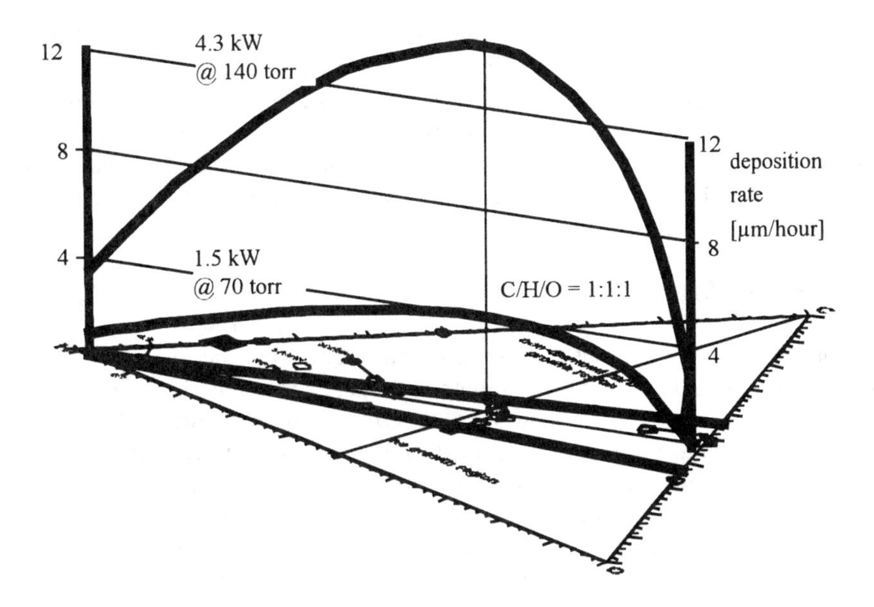

Figure 16. Diamond deposition rates along the diamond domain for different microwave power levels. The rate is highest in the center of the diagram, probably due to the additional combustion energy provided by a gas phase stochiometry of C/H/O=1:1:1.

rich domain border are advantageous. The C/H/O-concept also illustrates that specific starting materials are not required for diamond CVD and, important in industry, many patents related to specific gas phase compositions and carbon carrier gases are most likely obsolete.

Along the diamond domain, i.e., when passing from its H-rich corner (Fig. 16; left) to the C–O side line (Fig. 16; right), gas composition, deposition conditions, and film properties change considerably. Experiments to grow diamond X-ray windows and lithography membranes indicate that, e.g., the mechanical strength of polycrystalline films grown in the center of the C/H/O-diagram is worse than that of films grown from oxygen-free gas mixtures. The rate changes significantly along the domain. As shown in figure 16, it is highest in the center of the diagram and drops to zero when approaching the hydrogen-free C–O side line. For 2.45 GHz microwave plasmas and power levels of 4 to 4.5 kW, rates in the H-rich corner of the diagram are a factor of 2-3 lower than near C/H/O = 1:1:1, where, in addition to the external microwave energy, internal combustion energy helps to heat up the gas phase (cf. Fig. 1). Vice versa, similar growth rates in the center of the C/H/O-diagram require less external power and therefore tolerate lower substrate temperatures than H-rich mixtures. Adding substantial amounts of oxygen to the microwave plasma CVD gas phase, allows reduction of the substrate temperatures by 100-150 °C while retaining the diamond growth rate and quality.

The basic principles of the C/H/O-diagram have withstood the test of time. Despite its usefulness one should, however, never forget that operating a diamond CVD system inside the diamond domain does not guarantee successful diamond growth. For diamond CVD, the correct composition of the gas phase is necessary but by no means sufficient and other factors, e.g., substrate temperature, substrate material, substrate pretreatment etc. have also a significant influence on the deposition process and the properties of the deposit.

4.2 C/H/HALOGEN-MIXTURES

A major obstacle for the industrial use of diamond films are the rather high substrate temperatures of 800-1100°C, required to grow films at sufficiently high rates and phase purity. Adding oxygen to the gas phase of a microwave plasma reactor allows reduction of the substrate temperature by approximately 100 °C without sacrificing rate. Even more promising were early reports on growth rates >1 μm/hour at substrate temperatures of <300°C achieved by using C/H/halogen-mixtures in diamond CVD [Patterson et al. 1991], [Wong and Wu 1992], [Kadono et al. 1992], [Rudder et al. 1991], [Chu and Hon 1992] and [Hong et al. 1992]. To evaluate the benefits of using halogen compounds in microwave plasma CVD of diamond, P.K. Bachmann et al. [1994a, 1994c] performed a series of experiments using C/H/Cl- and C/H/F-gas mixtures. While there is no problem associated with adding (a few percent) of fluorine-containing compounds, e.g., freons, to the gas phase inside a stainless steel ASTeX HPMS shell, the use of chlorine-containing materials, e.g., chloroform, in such a system is not advisable. Decomposition of the chlorine-containing molecules results in formation of chlorine radicals and molecules that rapidly corrode stainless steel. Therefore, experiments with chlorine compounds were performed in a reactor similar to the ASTeX HPMS but cladded internally with silica glass [Rau and Trafford 1990]. Microwave plasma CVD of diamond was confirmed for both, C/H/F- and C/H/Cl-mixtures. Similar to the C/H/O-diagram it was possible to develop the first, rudimentary C/H/Cl- and C/H/F-gas phase compositional diagrams depicted in Fig. 17 and 18, respectively. For C/H/Cl-mixtures, the diamond domain seems to be confined to the H-rich corner of the diagram. High chlorine concentrations tend to promote the deposition of non-diamond carbon phases. Unfortunately, advantages like higher rates, lower deposition temperatures or higher phase purity for diamonds grown from C/H/Cl- rather than C/H/O-mixtures could not be confirmed [Bachmann et al. 1994a, 1994c].

The data for the C/H/F-system (Fig. 18) indicate that, positioned near the H-CF_4 tie line, the diamond domain extends well into the fluorine-rich region of the ternary system. Unfortunately, fluorine-assisted microwave plasma CVD of diamond had also no advantage over growth from C/H/O-mixtures. On the other hand, the use of fluorine in a plasma severely restricts the type of substrate material that can be used for diamond growth. Silicon, e.g, is always damaged, even if only small quantities of fluorine are present in the plasma-activated gas phase. Therefore, for the time being, C/H- and C/H/O-mixtures remain the gas phase compositions of choice for microwave plasma CVD of diamond.

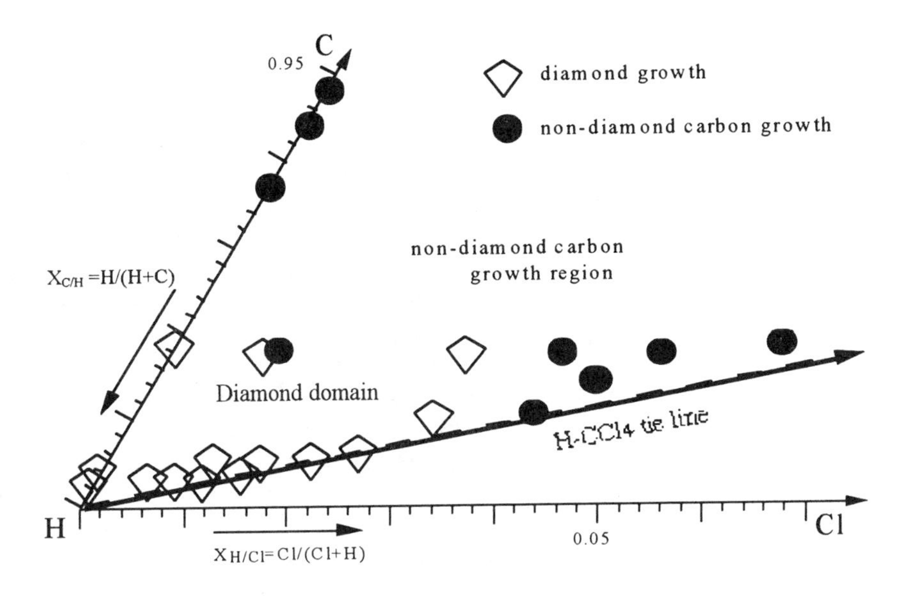

Figure 17. Diamond growth from C/H/Cl-gas compositions [Bachmann et al. 1994a]. Published data and additional experimental results indicate that the diamond domain is confined to the hydrogen-rich corner of the C/H/Cl-system.

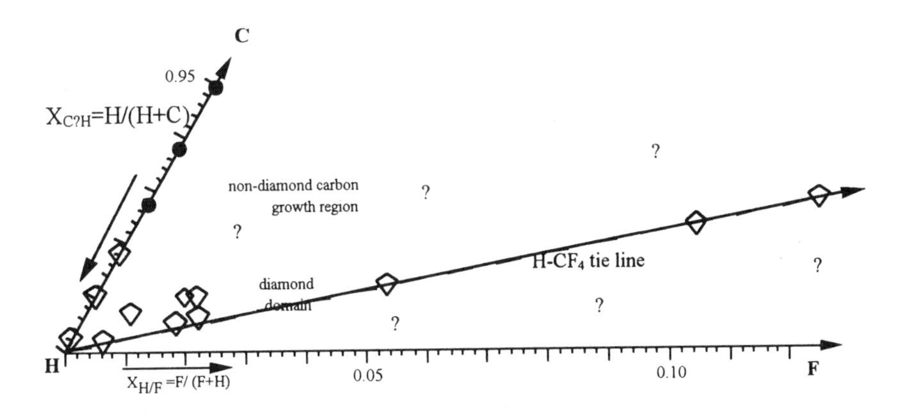

Figure 18. Diamond growth from C/H/F-gas compositions [Bachmann et al. 1994a]. Published data and additional experimental results indicate that the diamond domain extends along the H-CF_4 tie line.

5. Summary and Outlook

During the past decade, diamond thin film deposition technologies have come a long way. Deposition rates increased from less than 0.1 μm per hour to almost 1000 μm per hour. Deposition areas went from 1 cm^2 to more than 300 cm^2 and substrate temperatures dropped from 900 °C to less than 500 °C, or even lower. Of the wide variety of different methods that exist today to grow high quality diamond thin and thick films, plasma CVD methods in general and microwave plasma CVD at 2.45 GHz and 915 MHz have certainly the highest potential for large scale industrial application. If contamination-free films are required, radio frequency and microwave plasma CVD are the only suitable methods available, with the latter being much more power efficient. Deposition areas of 8-12 inches in diameter and, simultaneously, mass deposition rates of 1 g diamond per hour are feasible using newly developed 915 MHz systems. Sophisticated equipment with the stable performance mandatory in an industrial environment is, although expensive, already available on the market. Automated deposition processes are feasible and allow the production of millimeter-thick diamond wafers as well as nanometer-thin diamond coatings from a wide variety of C/H- and C/H/O-gas mixtures. Film properties comparable to those of the highest quality natural counterparts are possible and led or will soon lead to applications in the cutting tool industry, as components in passive and active electronic devices, in displays or as protective layers or low wear coatings. For certain applications, e.g., for flat panel displays, still larger deposition areas, better uniformity and homogeneity of the film composition and properties are still an issue. Deposition processes that reduce the required substrate temperatures below 300 °C while, simultaneously, maintaining good film adhesion, quality and rate would open many more application areas for diamond films.

Despite all the remaining problems and with a number of CVD diamond products already on the market and others at least demonstrated as prototypes, there is no doubt that the diamond thin film technology, and with it microwave plasma CVD as a major production method, will play an important role in the area of advanced materials.

6. References

Angus, J.C., Will, H.A., Stanko, W.S., *J. Appl. Phys.* **39**, 2915 (1968).

Asmussen, J., Root, J., *Appl. Phys. Lett.* **44** (4), 396 (1984).

Bachmann, P. K., Gärtner, G., Lydtin, H., *MRS Bulletin*, **No.12** , p. 52 (1988a).

Bachmann, P.K., Drawl, W., Knight, D., Weimer, R., Messier, R.F.in "Diamond and Diamond-like Materials" edited by A. Badzian, M. Geis, and G. Johnson, MRS Symposium Proceedings, Vol. EA-15 p. 99 (1988b).

Bachmann, Peter K., Messier, Russell F., *Chemical & Engineering News*, **Vol. 67**, No. 20, p. 24 , May 15 (1989).

Bachmann, P.K., Lydtin, H., in "Characterization of Plasma Processes" edited by G. Lukovsky, D.E. Ibbotson and D.W. Hess, *MRS Symposium Proceedings*, Vol. 165, **181** (1990).

Bachmann, P.K., Leers, D., Lydtin, H., Wiechert, D.U., *Diamond and Related Materials*, Vol.**1**, No.1, p.1 (1991).

Bachmann, P.K. in "The Business and Technical Outlook for CVD Diamond and Diamond-Like Carbon", *Proc. 2. Gorham Advanced Materials Conference*, Monterey, CA (1993).

Bachmann, P.K., Lade, H., Leers, D., Wiechert, D.U., *Diamond Films'93*, 4. European Conference on Diamond, Diamond-like and Related Material, abstract 16.2, Albufeira, Portugal, Sept. 20-24, 1993.

Bachmann, P.K., Hagemann, H.J., Lade, H., Leers, D., Picht, F., Wiechert, D.U. in "Diamond, SiC, and Nitride Wide Band Gap Semiconductors", edited by C.H. Carter, G. Gildenblatt, S. Nakamura, R.J. Nemanich MRS Symp.Proc., Vol. 339, Materials Res. Soc., Pittsburgh, PA, 267 (1994a).

Bachmann, P.K., Lade, H., Leers, , D., Wiechert, D.U., *Diamond and Related Materials*, Vol. **3**, No. 4-6, 799 (1994b).

Bachmann, P.K., Hagemann, H.J., Lade, H., Leers, D., Picht, F., Wiechert, D.U., *Proc. ADVANCED MATERIALS '94, Intl. Symposium on Advanced Materials, NIRIM*, Tsukuba, Japan, p. 115-120 (1994c).

Bachmann, P.K., Hagemann, H.J., Lade, H., Leers, D., Wiechert, D.U., Wilson, H., *Diamond and Rel. Mat.* **4**, 820 (1995).

Cassidy, W.D., Evans, E.A., Wang, Y., Angus, J.C., Bachmann, P.K., Hagemann, H.J., Leers, D., Wiechert, D.U. in "Diamond, SiC, and Nitride Wide Band Gap Semiconductors", edited by C.H. Carter, G. Gildenblatt, S. Nakamura, R.J. Nemanich MRS Symp.Proc., Vol. 339, Materials Res. Soc., Pittsburgh, PA, 285 (1994).

Chu, C.H., Hon, M.H., presented at DIAMOND 1992, Heidelberg, Germany (unpublished) (1992).

Eversole, W., US patents 3030187 and 3030188 (1962) (filed 1959).

Gicquel, A., Hassouni, K., Farhat, S., Breton, Y., Scott, C.D., Leverbre, M., Pealat, M., *Diamond and Related Materials*, Vol. **3**, No. 4-6, 581 (1994).

Hibshman, H.J., US patent 3371996 (1968) (filed 1964)

Hong, B., Wakagi, M., Collins, R.W., An, I., Engdahl, N.C., Drawl, W., Messier, R., *Diamond and Related Materials*, Vol. **3**, No. 4-6, 431 (1994).

Hong, F.C., Hsieh, J. C., Wu, J.J., presented at DIAMOND 1992, Heidelberg, Germany (unpublished) (1992).

Hsu, T., Anthony, B., Breaux, L., Banerjee, S., Tasch, A. in "Characterization of Plasma Processes" edited by G. Lukovsky, D.E. Ibbotson and D.W. Hess, *MRS Symposium Proceedings*, Vol. **165**, 139 (1990).

Ishibori , K. and Ohira, Y., in "Science and Technology of New Diamond", edited by S. Saito, O. Fukanaga, and M. Yoshikawa, KTK Science Publishers, Tokyo, Japan, 167 (1990).

Janzen, G., "Plasmatechnik", ISBN 3-7785-2086-5, Hüthig Verlag, Heidelberg (1992).

Jiang, X., Klages, C.P., *Diamond and Related Materials*, Vol. **2**, No. 5-7, 1112 (1993).

Kadono, M., Inoue, T., Miyanaga, M., Yamazaki, S., *Appl. Phys. Lett.* **61**, 7, 772 (1992).

Kamo, M., Sato, Y., Matsumoto, S., and Setaka, N., *J. of Crystal Growth*, **62**, 642 (1983).

Kamo, M., Takamura, F., and Sato, Y., in "Science and Technology of New Diamond", edited by S. Saito, O. Fukanaga, and M. Yoshikawa, KTK Science Publishers, Tokyo, Japan, **183** (1990).

Kawarada, H., Mar, K.S., and Hiraki, A., *Jpn. J. Appl.Phys.* **26**, 6, L1032 (1987).

Kumar, N., in "The Business and Technical Outlook for CVD Diamond and Diamond-Like Carbon", *Proc. 2. Gorham Advanced Materials Conference*, Monterey, CA (1993).

Lux, B., Haubner, R., in "Thin Film Diamond" edited by A. Lettington and J.W. Steeds, Chapman and Hall, Royal Society London, ISBN 0-412-49630-5, 127 (1994).

Matsumoto, S., Sato, Y., Kamo, M. and Setaka, N., *Jpn. J. of Appl. Phys.* **21**, L183 (1982).

Mitsuda, Y., Yoshida, T., and Akashi, K., *Rev. Sci. Instrum.* , **60**, 2, 249 (1989).

Miyake, S., Chen, W., Hoshino, A., and Arata, Y., *Trans. JWRI*, **17**, 2, 323 (1988).

Patterson, D.E., Bai, B.J., Chu, C.J., Hauge, R.H., and Margrave, J.L. in "New Diamond Science and Technology", edited by R.F. Messier, J.T. Glass, J.E. Butler, and R. Roy, Mater. Res. Soc. Pittsburgh, PA, 433 (1991).

Pinneo, M., Crystallume Inc., private communication (1993).

Post, R., in "The Business and Technical Outlook for CVD Diamond and Diamond-Like Carbon", *Proc. 2. Gorham Advanced Materials Conference,* Monterey, CA (1993).

Prijaya, N.A., Angus, J. C., Bachmann, P.K., *Diamond and Related Materials*, Vol. **3**, No.1-2, 129 (1993).

Field, J.E., ed. *Properties of Natural and Synthetic Diamond*, ISBN 012-255352-7, Academic Press, London (1992).

Rau, H., Trafford, B., *Physics D; Appl. Phys.* **23**, 1637 (1990).

Rudder, R.A., Hudson, G.C., LeGrice, Y., Mantini, M.J., Posthill, J.B., Nemanich, R.J., Markunasin, R.J. "Diamond and Diamond-like Materials" edited by A. Badzian, M. Geis, and G. Johnson, *MRS Symposium Proceedings*, Vol. **EA-15,** 89 (1988).

Rudder, R.A., Hudson, G.C., Malta, D.P., Posthill, J.B., Thomas, R.E., Markunas, R.J., in "Applications of Diamond and Related Materials, edited by Y. Teng, M. Yoshikawa, M. Muranaka, A. Feldman, Elsevier Science Publishers, 583 (1991).

Sevillano, E. , ASTeX, Woburn, MA, private communication (1995).

Sommer, M. and Smith., F.W., in "New Diamond Science and Technology", edited by R.F. Messier, J.T. Glass, J.E. Butler, and R. Roy, Materials Research Socienty, Pittsburgh, PA, 443 (1991).

Spitsyn, B.V. in "Applications of Diamond Films and Related Materials" edited by Y. Tzeng, M. Yoshikawa, M. Muranaka, and A. Feldman, Elsevier Science Publishers B.V. Amsterdam, 475 (1991).

Spitsyn, B.V. and Derjaguin, B.V., author's certificate July 10 (1956); USSR Patent 339 134, May 5, (1980).

Spitsyn, B.V., Bouilov, L.L., and Derjaguin, B.V., *J. of Crystal Growth*, **52**, 219 (1981).

Stoner, B.R., Glass, J.T., *Appl. Phys. Lett.* **60**, 698 (1992).

Suzuki, J., *Jpn. J. Appl.Phys.* **28**, 2, L 281 (1989).

Windischmann, H., Epps, G.F., Caesar , G.P. in "New Diamond Science and Technology", edited by R.F. Messier, J.T. Glass, J.E. Butler, and R. Roy, Materials Research Society, Pittsburgh, PA, 791 (1991).

Wong, M.S., Wu, C.H., *Diamond and Related Materials*, **1**, 369 (1992).

Yuasa, M., Kawarada, H., Wei, J., Ma, J.S., Suzuki, J., Okada, S., and Hiraki, A., *Surface and Coating Technology*, **49**, 374 (1991).

Chapter 22
DIAMOND CVD USING RADIO-FREQUENCY PLASMAS

Steven L. Girshick
University of Minnesota, Minneapolis, MN 55455, USA

Contents

1. Introduction

This chapter considers diamond CVD using radio-frequency (RF) plasmas. RF plasmas are generated by power sources with frequencies of hundreds of kHz to tens of MHz. In this sense they lie between DC and microwave plasmas. Overviews of work to about 1990 were given by Bachmann and Lydtin [Bachmann and Lydtin 1991] and by Anthony [Anthony 1991].

2. Types of RF Plasmas

Several types of RF plasmas have been used to deposit diamond. These may be distinguished primarily by the type of coupling of the plasma to the RF power source—either capacitive or inductive—and by the total gas pressure. Roughly speaking, plasmas can be separated into "glow discharge" and "thermal" plasmas, according to whether their total pressure is much less or much greater, respectively, than about 10 kPa (75 Torr). Thermal RF plasmas are almost always induction plasmas, while glow discharge plasmas of both capacitive and inductive types have been used for diamond CVD.

Glow discharges are characterized by electron temperatures which are much higher than heavy species temperatures. In this regime the heavy species may be cold—only slightly higher than room temperature—while the electron temperature commonly exceeds 2 eV (> 20,000 K). The ionization mechanism is primarily electron impact.

Thermal plasmas, in contrast, are characterized by heavy species temperatures which are roughly equilibrated with the electron temperature. Under this condition the heavy species must be quite hot—at least several thousand kelvin—to maintain a high enough degree of ionization to sustain the plasma.

3. Diamond CVD in Capacitive RF Discharges

Parallel-plate capacitive RF discharges operating at a frequency of 13.56 MHz are routinely used in the microelectronics industry for the etching and deposition processes involved in the manufacture of integrated circuits. Consequently, as Bachmann and Lydtin [Bachmann and Lydtin 1991] commented, "From the equipment point of view, this method would have been the most likely candidate for scale-up and industrialization of diamond deposition." However capacitive discharges have been notably less successful than competing technologies for depositing diamond. The major reason for this appears to be that electron energies are too low—typically about 3 eV, compared to about 10 eV for microwave plasmas—to accomplish sufficient dissociation of hydrogen [Anthony 1991; Beckman et al. 1994]. Another reason may be that parallel-plate capacitive discharges with high enough pressures to be practical for diamond CVD have plasma and sheath potentials which result in an ion energy distribution with a high-energy tail; bombardment by these energetic ions may damage films or inhibit diamond growth [Beckman et al. 1994].

The first report of a capacitive discharge (although not parallel-plate) for diamond synthesis appears to be that of Mania, Stobierski and Pampuch [Mania et al. 1981], at the Academy of Mining and Metallurgy in Cracow, Poland. They explored a wide range of discharge frequencies, from 100 Hz to 10 MHz, with reactor pressures ranging from 0.08 to 15 Torr. The reactor geometry was coaxial, and electrodes of graphite and various metals were tested. They did not produce a film, but rather discrete crystals of diamond, including needle-like crystals with a diameter of about 50 nm.

A group at Mitsubishi Metal Co. in Japan designed a reactor which utilized both a parallel-plate capacitive RF discharge and a separate hot filament [Komatsu et al. 1987]. Experiments were compared in which either the discharge, the hot filament, or both were powered. The RF-only experiments were conducted at 20 Torr pressure and with 3 kW of RF power. Six-hour depositions produced only isolated crystallites, with a significant fraction of non-diamond carbon. The superposition of RF power and a hot filament produced a higher growth rate than hot filament alone, and the growth rate increased with the magnitude of the RF power. However the growth rates were modest (less than 0.4 μm/h), and crystallinity was poor.

Lee and Gallois, at Stevens Institute of Technology, attempted diamond CVD with a parallel-plate RF plasma at a pressure of 20 Torr and with 100 W of RF power [Lee and Gallois 1992]. Ball-like deposits were obtained, with a considerable (~60%) graphitic component.

A group at Tokyo Denki University, Japan, used a 13.56-MHz parallel-plate RF reactor to deposit diamond on a room-temperature substrate consisting of either pure silicon or silicon with an iron film [Shimada et al. 1993; Shimada and Machi 1994]. In these experiments the pressure was 50 mTorr, with an RF power of 100 W. Iron was found to

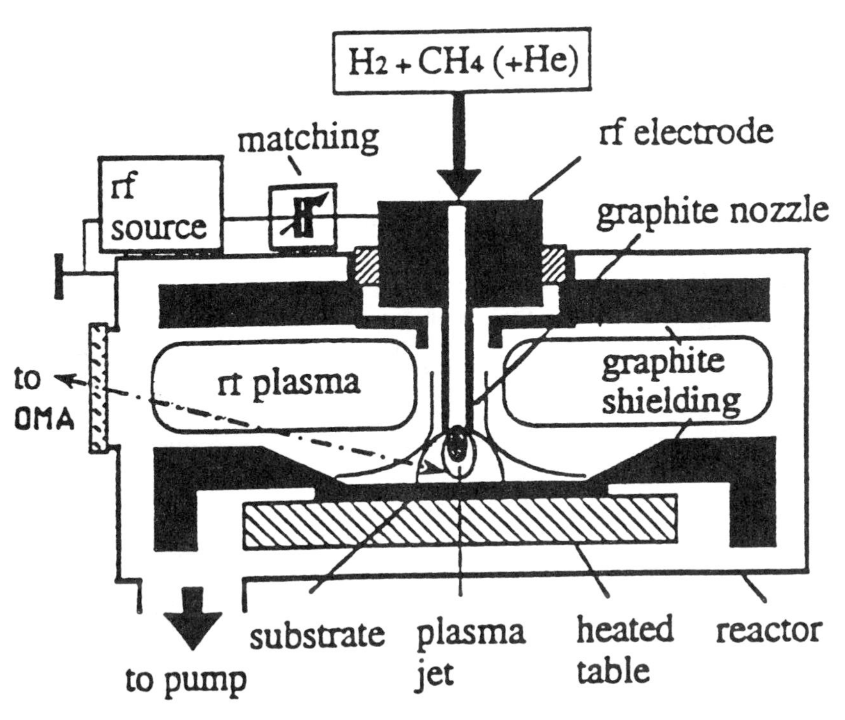

Figure 1. Schematic diagram of the RF plasma jet apparatus of Bàrdos et al. [Bàrdos et al. 1993].

enhance the diamond nucleation density by about one order of magnitude, although again only ball-like, isolated diamond particles were obtained.

Researchers at University of Cambridge [Silva et al. 1995] reported the presence of isolated, small (< 50 nm) diamond crystallites within a mostly amorphous carbon film deposited in a parallel-plate RF reactor. Conditions included 260 W RF power, 0.3 Torr pressure, and low substrate temperatures of 80–150°C.

A recent approach toward overcoming the limitations of capacitively-coupled RF for diamond CVD has been described by Jackman, Beckman and Foord, of University College London and the University of Oxford [Beckman et al. 1994; Jackman et al. 1995; Jackman et al. 1995; Jackman et al. 1995]. In their method the plasma is magnetically enhanced by an external magnetic field parallel to the electrodes. The electrodes are formed of separate rings which surround the plasma tube, and the deposition substrate is placed normal to the tube axis. The potential advantages of this approach over the parallel-plate geometry are that (1) the magnetic field suppresses the loss of electrons from the plasma, thereby promoting plasma densities as high as those in microwave plasmas, and (2) sputtering from the electrodes is avoided. Most conditions investigated produced at best isolated crystallites, but recently [Jackman et al. 1995] the authors reported success in depositing a continuous diamond film with a grain size of about 50 nm. The RF power was 300 W, pressure was in the 0.1–10 Torr range, and the substrate temperature was 1080°C. The growth rate was quite low, about 10 nm/h.

A different approach toward capacitively-coupled plasma CVD has been taken by a group at Uppsala University in Sweden [Bárdos et al. 1993; Bárdos et al. 1993]. In their process, shown in Figure 1, reactants flow through a tubular, water-cooled RF electrode, powered at 13.56 MHz, and are accelerated through a graphite nozzle to form a supersonic plasma jet, which impinges in stagnation flow upon the deposition substrate. The authors argue that the plasma generated inside the nozzle is similar to a DC hollow cathode discharge, while the plasma inside the deposition chamber has a predominantly RF character. A modified version [Bárdos et al. 1994] added a wire mesh electrode between the nozzle and the substrate, with an applied DC bias between the mesh and the substrate. A wide range of reactor pressure, from 0.8 to 100 Torr, and RF power, from 50 to 500 W, was investigated. However, although this system was previously used to deposit silicon nitride and oxynitride films at high growth rates [Bárdos and Dusek 1988], the results for diamond have been less encouraging. Diamond growth rates of about 0.2 μm/h were achieved, with Raman spectra showing significant inclusions of non-diamond carbon.

In summary, the results to date suggest that capacitive RF plasmas are unlikely to compete successfully with more established methods of diamond CVD. In spite of its equipment advantage, parallel-plate RF has yielded discouraging results, and recent work appears to be mainly directed at developing innovative methods, such as magnetic enhancement or the formation of supersonic jets, which utilize capacitive RF plasmas but which abandon the parallel-plate configuration.

4. Diamond CVD in Inductive RF Glow Discharges

Induction plasmas are often referred to as "electrodeless," because the RF current passes through a coil which surrounds the plasma tube, which has the advantage that no electrode or filament comes into contact with the reacting gases. While even less has appeared in the diamond literature regarding inductive glow discharges than for capacitive discharges, the results appear more promising, in that inductive glow discharges (unlike capacitive) have been shown to produce continuous, well-faceted diamond films, which look similar to those produced by the more widely used methods. All of the RF induction glow discharges discussed below utilized 13.56 MHz frequency.

The first report of diamond CVD using an inductive RF glow discharge appears to be by S. Matsumoto at the National Institute of Research in Inorganic Metals, Japan, in 1985 [Matsumoto 1985]. In Matsumoto's reactor the substrate was positioned parallel to the gas flow, as shown in Figure 2. Experiments were conducted with pressures ranging from about 4 to 23 Torr, with a total gas flow rate of 50 sccm, and with RF power of 0.5-1.0 kW. SEM micrographs show well-faceted, continuous diamond films. Growth rates were not reported, but Matsumoto noted that at pressures lower than about 8 Torr a small amount of silicon carbide co-deposited with the diamond, which he attributed to sputtering of silicon atoms from the quartz plasma tube.

A group at the University of Nebraska and the Solar Energy Research Institute, in Golden, Colorado, reported experiments with a similar geometry as Matsumoto's [Meyer et al. 1988; Meyer et al. 1989]. Pressure ranged from 0.5 to 40 Torr, flow rates

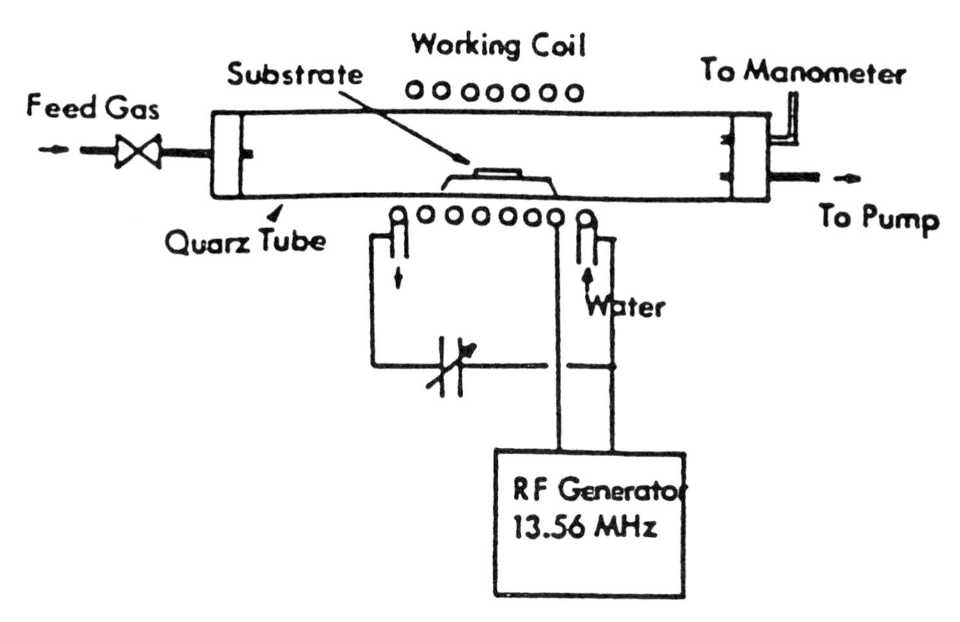

Figure 2. Schematic diagram of the reactor used by Matsumoto, Hino and Kobayashi for the first reported diamond CVD in an induction plasma [Matsumoto et al. 1985].

from 8.2 to 100 sccm, and RF power from 200 to 800 W. Isolated particles, either egg-shaped or faceted, were obtained. At the highest power levels the silicon substrate was first coated with diamond-like carbon in an attempt to increase nucleation density. Continuous films were produced, but mainly of spherical particles, apparently with a significant non-diamond component.

A different geometry, with the substrate oriented normal to the gas flow, has been used by researchers at Research Triangle Institute (RTI) and the University of North Carolina [Vitkavage et al. 1988; Rudder et al. 1991; Rudder et al. 1992; Rudder et al. 1992]. They found that a certain critical power, which increased with gas pressure, was required to achieve a high enough plasma density for diamond deposition. They obtained continuous, well-faceted films for various conditions. They achieved a growth rate of 1 μm/h using a hydrogen-methane plasma at 5 Torr pressure and 2 kW RF power, at a substrate temperature of 800°C. Their work is of particular interest because they also obtained polycrystalline films, at 0.4-0.8 μm/h, at a substrate temperature of only 300°C. To achieve this they used feed gas consisting of a mixture of water vapor and various alcohols. For these experiments the pressure was 0.5-1.0 Torr and the RF power was 800-1000 W.

Thus the work of Matsumoto and of the RTI group suggests that, within the glow discharge regime, induction plasmas are more conducive to diamond growth than are capacitive plasmas. While the highest growth rate reported was a modest 1 μm/h, the RTI results for low-temperature diamond growth are noteworthy.

5. Diamond CVD Using Inductively-Coupled Thermal Plasmas

Induction plasmas in the thermal plasma regime have much in common with DC thermal plasmas, and to a lesser extent with combustion flames. Whether the electric current input to the plasma is provided by a DC arc or by RF induction, in either case the electrical power goes primarily to heat a mostly neutral gas to very high temperature (~10,000 K) at relatively high pressure, often atmospheric. Differences between different systems may be attributable more to differences in geometry than to inherent differences between DC and RF plasmas.

On the other hand, there are some differences between RF and DC thermal plasmas which may in some cases be important. DC torches inevitably experience electrode erosion, which is a potential source of contamination, while induction plasmas are electrodeless. RF plasmas tend to have a larger, more uniform volume, which is conducive to coating large areas uniformly. However if either type of plasma is expanded through a nozzle (other than the anode of a DC arcjet), than the conditions above the substrate have more to do with the nozzle parameters (geometry and pressure ratio) than with the type of plasma upstream of the nozzle.

Typically in RF thermal plasmas the main plasma gas is argon, and hydrogen rarely exceeds about 20% of the input volumetric flow rate. To our knowledge a pure atmospheric-pressure hydrogen RF plasma has never been achieved. Argon (or other inert gas) helps to stabilize the plasma and to reduce the high heat flux to the walls associated with hydrogen. Carbon is almost always introduced as methane, with similar methane-hydrogen ratios reported as in other deposition techniques.

The first reported use of an RF thermal plasma for diamond deposition was by S. Matsumoto and coworkers in 1987 [Matsumoto et al. 1987], following Matsumoto's work with RF glow discharges discussed in the preceding section. Approximately 60 KW of RF generator plate power was supplied at 4 MHz to the plasma at 1 atm. A 20-mm-diameter deposition substrate was inserted directly into the 45-mm-diameter plasma tube, with the plasma impinging in stagnation-point flow. A ten-minute run produced a continuous polycrystalline film, with thickness ranging from 6 μm in the center to 12 μm at the edge of the 2-cm-diameter substrate.

A group at Stanford University has combined growth experiments and optical diagnostics to study diamond CVD in an RF reactor at atmospheric-pressure. The facility includes a 50-kW torch, operating at 4 MHz. Typical flow rates included about 110 slm argon and about 10 slm hydrogen. Early experiments [Cappelli et al. 1990; Owano et al. 1991] explored a geometry in which the substrate was oriented parallel to the flow exiting the plasma torch. Only isolated crystallites were obtained, with higher growth rates observed near the leading edge of the substrate, where the boundary layer is thinner. Later work focused on the more usual stagnation-point geometry, and a series of parametric studies [Owano and Kruger 1993; Baldwin et al. 1994] investigated the effects of reactant flow rates and substrate temperature on diamond growth rates. Gas phase diagnostics included emission spectroscopy of H and C_2 [Owano et al. 1993; Baldwin et al. 1994] to characterize freestream temperature; measurements of boundary-layer concentration profiles of CH [Green et al. 1993; Owano et al. 1993] and C_2

[Owano et al. 1995] using degenerative four-wave mixing (DFWM) spectroscopy; and boundary-layer temperature profiles using DFWM measurements of CH vibrational [Green et al. 1993; Owano et al. 1993] and rotational [Owano et al. 1995] temperatures.

Girshick and coworkers at the University of Minnesota have reported a number of studies of atmospheric-pressure RF plasma diamond CVD. Two reactors have been used: in one the substrate is inserted directly into a 44-mm ID plasma tube, while in the more recent configuration the plasma is accelerated through a converging nozzle before impinging on the substrate. RF generators operating at 3-4 MHz and plate powers of 10-15 kW were used. Typical flow rates in both systems included ~40 slm argon and 3-8 slm hydrogen. Early results demonstrated well-faceted polycrystalline film growth, with growth rates of about 10 μm/h, and uniform coverage over about 300 mm^2 [Li et al. 1991; Li et al. 1991]. In later work the concept of "central jet injection" was introduced. In this approach boundary layer thickness is manipulated by adding additional argon to the hydrogen-methane reactant jet, so as to vary the jet momentum. A series of parametric studies [Girshick et al. 1993; Girshick et al. 1993] demonstrated that central jet injection somewhat increased the growth rate, to about 18 μm/h, and considerably increased the range of conditions under which well-faceted, continuous films could be deposited. Diagnostics employed by the Minnesota group include gas chromatography [Lindsay et al. 1995] and mass spectrometry [Greuel et al. 1995] for the measurement of surface concentrations of several hydrocarbon species sampled through an orifice in the growth substrate, and laser-induced fluorescence (LIF) of OH [Hofeldt and Wang 1995] to measure boundary-layer temperature profiles (with a small quantity of water vapor added to the plasma to generate OH). Recently the Minnesota group conducted studies of the effect of growth conditions on the diamond morphology parameter α, which characterizes the ratio of growth rates in the <100> and <111> crystal directions [Lindsay et al. 1995]. The results for the effects on α of methane-hydrogen ratio and substrate temperature are qualitatively similar to results reported for microwave systems [Tamor and Everson 1994; Wild et al. 1994], but are systematically shifted in a manner which indicates that the gas composition at the growth surface is different for RF thermal plasmas than for microwave plasmas.

A related thrust of Girshick and coworkers has been the development of two-dimensional numerical models which describe the RF reactor temperature and flow fields, coupled to detailed one-dimensional kinetic models of the boundary layer and surface chemistry [Yu and Girshick 1989; Li et al. 1991; Girshick et al. 1993]. Figure 3 shows temperature profiles calculated for a case with central jet injection. It is seen that the peak plasma temperature exceeds 10,000 K, and that a thin boundary layer (~2 mm) exists above the substrate, across which the temperature drops from about 4000 K to the substrate temperature of about 1200 K. In this environment the injected reactants are virtually fully dissociated in the hot freestream, and recombination chemistry occurs in the boundary layer. The crucial role of boundary-layer thickness was highlighted by a numerical study [Yu and Girshick, 1994] in which Harris's growth-by-methyl mechanism [Harris 1990] was extended to include diamond growth from all C_1 radicals—monatomic carbon vapor being a potential growth species in these systems, because of the high degree of dissociation in the freestream. The model predicted that

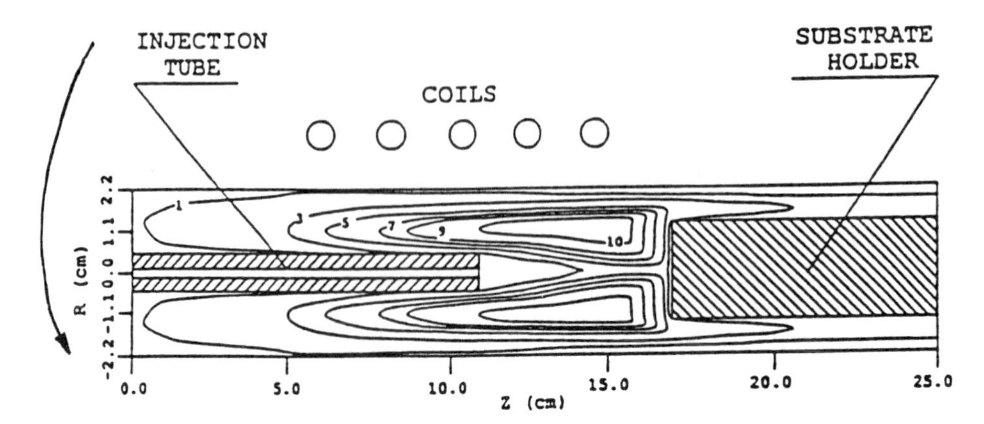

Figure 3. Temperature distribution (temperatures in kK) predicted by numerical model of Girshick et al. for an atmospheric-pressure RF plasma [Girshick et al. 1983].

linear growth rates would increase as boundary layer thickness was reduced, in accord with a growing body of experimental evidence.

Hernberg and coworkers at Tampere University of Technology, Finland, deposited diamond in an induction plasma operating at subatmospheric pressures, ranging from 20 to 60 kPa (about 150 to 460 Torr) [Hernberg et al. 1992]. Flow rates were argon, 100 slm, and hydrogen, 9.4 slm. The effect of substrate location (axial distance from the induction coil) was investigated. Continuous diamond films were obtained at ~10 μm/h, and in some cases extremely fine-grained films were observed (grain size 0.01-0.1 μm).

A group under the auspices of the Commisariat à l'Energie Atomique in France compared experiments with a DC plasma torch expanded to 5 kPa (38 Torr) and a conventional atmospheric-pressure RF plasma [Verven et al. 1993]. They obtained over 400 μm/h growth with the DC torch, compared to 50-60 μm/h for the RF plasma, but reported that the RF plasma-deposited films were more uniform over a larger area (5 cm^2) than for the DC torch.

Probably the most impressive diamond growth results reported by any RF plasma method were those obtained by Kohzaki and coworkers at Toyota Central Research and Development Laboratories, Japan [Kohzaki et al. 1993; Kohzaki et al. 1993]. Their reactor is shown schematically in Figure 4. They placed the growth substrate beneath the plasma tube, and obtained an expanded plasma by operating at a reduced pressure of 150 Torr. Input plate power was 40 kW at 3.4 MHz frequency, and flow rates included argon at 80 slm and hydrogen at 20 slm. Well-faceted, uniform films were obtained over 10-cm-diameter substrates at 30 μm/h, corresponding to a growth rate of ~40 carats/h, a mass deposition rate comparable to the highest reported rates for DC arcjets depositing diamond at several hundred μm/h but over much smaller areas.

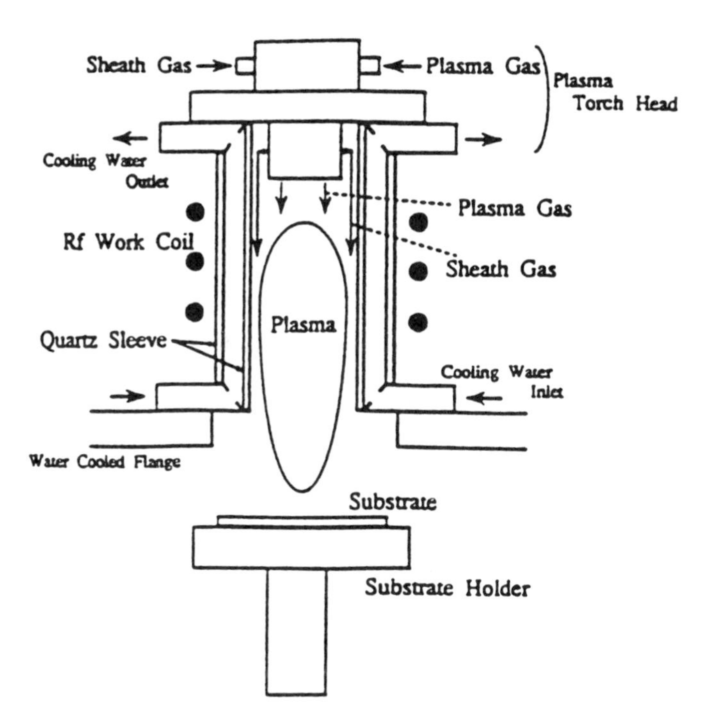

Figure 4. Schematic diagram of RF reactor used by Kohzaki et al. to deposit diamond at 30 μm/h over 10-cm diameter substrates [Kohzaki et al. 1993].

As with DC thermal plasmas and combustion flames, RF thermal plasmas expose the deposition substrate to a high heat load (typically 0.1-1.0 kW/cm^2), and the substrate must be actively cooled to maintain a low enough temperature for diamond deposition. Considering that substrate temperature is known to be one of the most important parameters for film growth, substrate temperature control must be considered a key engineering issue for the success of diamond CVD in thermal plasmas or flames. Both the Minnesota [Girshick et al. 1993] and Stanford [Owano and Kruger 1993] groups have varied substrate temperature in their RF plasma systems by inserting a thermal insulator between the substrate and the water-cooled copper base beneath the substrate. Substrate temperature was changed from run to run by using insulators of various thicknesses. Recently Bieberich and Girshick [Bieberich and Girshick 1995] developed an active temperature control system, based on a modification of an earlier design by Snail and coworkers for flame CVD [Thorpe et al. 1993; Snail and Thorpe 1994], in which an argon-helium mixture flows through small channels in an inconel insert located beneath the substrate. The substrate temperature is sensitively controlled over a range exceeding 600 K by varying the relative proportions of argon and helium.

In summary, RF thermal plasmas have produced significantly higher growth rates than the low-pressure methods, and film quality in general appears to be at least as good as for low-pressure induction CVD. Reported growth rates are generally in the tens of μm/h, and the results of the Toyota group [Kohzaki et al. 1993; Kohzaki et al. 1993] demonstrate that good uniformity can be achieved over large areas. However, while the method is promising it does not appear that there has been the level of industrial interest which is evident for microwave plasmas, DC arcjets, or hot filament CVD. This may be due to the relative unfamiliarity of industry with RF thermal plasmas, and to the relatively less developed state of RF torches for stable operation under diamond CVD conditions. Substrate temperature control is also a crucial engineering challenge, although this is shared by other high-heat-flux techniques.

6. Summary

RF plasmas of various types have received some attention as a viable method for diamond CVD, although not nearly the attention given to microwave plasmas, hot filament CVD, or DC arcjets. The widespread use of RF parallel-plate capacitive plasmas makes them ideal candidates for diamond CVD. Unfortunately this method has not proven successful, and current work with capacitive plasmas emphasizes alternative configurations. RF inductive glow discharges have been used to deposit good quality diamond at ~1 μm/h, including on low-temperature substrates. RF thermal plasmas routinely achieve growth rates in the tens of μm/h, with good crystallinity and uniformity demonstrated over substrates as large as 10 cm in diameter. Key issues are control of boundary layer thickness and control of substrate temperature in a high heat flux environment.

7. References

Anthony, T. R. (1991) Methods of diamond making, in R. E. Clausing, L. L. Horton, J. C. Angus and P. Koidl (eds), *Diamond and Diamond-like Films and Coatings*, Plenum Pr., pp. 555-577.

Bachmann, P. K. and Lydtin, H. (1991) High rate versus low rate diamond CVD methods, in R. E. Clausing, L. L. Horton, J. C. Angus and P. Koidl (eds), *Diamond and Diamond-like Films and Coatings*, Plenum Press, pp. 829-853.

Baldwin, S. K., Owano, T. G. and Kruger, C. H. (1994) Growth rate studies of CVD diamond in an rf plasma torch, *Plasma Chem. Plasma Process.*, **14**, 383-406.

Bárdos, L., Berg, S., Baránková, H. and Carlsson, J.-O. (1993) Reactive deposition of diamond and Si carbide films by hydrogen plasma etching of graphite and Si in the r.f. plasma jet, *Thin Solid Films*, **223**, 218-222.

Bárdos, L., Berg, S., Nyberg, T., Barankova, H. and Nender, C. (1993) An r.f. plasma jet applied to diamond, glassy carbon and silicon carbide film synthesis, *Diamond Relat. Mater.*, **2**, 517-522.

Bárdos, L. and Dusek, V. (1988) High rate jet plasma-assisted chemical vapour deposition, *Thin Solid Films*, **158**, 265-270.

Bárdos, L., Nyberg, T., Baráranková, H. and Berg, S. (1994) Effect of the space charge sheath on properties of carbon and diamond films in the r.f. plasma jet, *Diamond Relat. Mater.*, **3**, 528-530.

Beckman, J., Jackman, R. B. and Foord, J. S. (1994) Capacitively coupled r.f. plasma sources: a viable approach to CVD diamond growth?, *Diamond Relat. Mater.*, **3**, 602-607.

Bieberich, M. T. and Girshick, S. L. (1995) Control of substrate temperature during diamond deposition, *Plasma Chem. Plasma Process.*, submitted.

Cappelli, M. A., Owano, T. G. and Kruger, C. H. (1990) High growth rate diamond synthesis in a large area atmospheric pressure inductively coupled plasma, *J. Mater. Res.*, **5**, 2326-2333.

Girshick, S. L., Li, C., Yu, B. W. and Han, H. (1993) Fluid boundary layer effects in atmospheric-pressure plasma diamond film deposition, *Plasma Chem. Plasma Process.*, **13**, 169-187.

Girshick, S. L., Yu, B. W., Li, C. and Han, H. (1993) Diamond deposition by atmospheric pressure induction plasma: effects of impinging jet fluid mechanics on film formation, *Diamond Relat. Mater.*, **2**, 1090-1095.

Green, D. S., Owano, T. G., Williams, S., Goodwin, D. G., Zare, R. N. and Kruger, C. H. (1993) Boundary layer profiles in plasma chemical vapor deposition, *Science*, **259**, 1726-1729.

Greuel, P. G., Roberts, J. T. and Ernie, D. W. (1995) Mass spectrometric analysis of a thermal plasma assisted CVD of diamond films, to be published in *Proc. 12th Intl. Symp. Plasma Chem.*, Minneapolis, Aug. 21-25, 1995.

Harris, S. J. (1990) Mechanism for diamond growth from methyl radicals, *Appl. Phys. Lett.*, **56**, 2298-2300.

Hernberg, R., Lepistö, T., Mäntylä, T., Stenberg, T. and Vattulainen, J. (1992) Diamond film synthesis on Mo in thermal RF plasma, *Diamond Relat. Mater.*, **1**, 255-261.

Hofeldt, D. H. and Wang, Q. (1995) unpublished work.

Jackman, R. B., Beckman, J. and Foord, J. S. (1995) Chemical vapour deposition of diamond from a novel capacitively coupled r.f. plasma source, *Mater. Sci. Eng.*, **B29**, 216-219.

Jackman, R. B., Beckman, J. and Foord, J. S. (1995) Diamond chemical vapor deposition from a capacitively coupled radio frequency plasma, *Appl. Phys. Lett.*, **66**, 1018-1020.

Jackman, R. B., Beckman, J. and Foord, J. S. (1995) The growth of nucleation layers for high-quality diamond CVD from an r.f. plasma, *Diamond Relat. Mater.*, **4**, 735-739.

Kohzaki, M., Higuchi, K., Noda, S. and Uchia, K. (1993) Large-area diamond deposition and brazing of the diamond films on steel substrates for tribological applications, *Diamond Relat. Mater.*, **2**, 612-616.

Kohzaki, M., Uchida, K., Higuchi, K. and Noda, S. (1993) Large-area high-speed diamond deposition by rf induction thermal plasma chemical vapor deposition method, *Jpn. J. Appl. Phys. Part 2*, **32**, L438-L440.

Komatsu, T., Yamashita, H., Tamou, Y. and Kikuchi, N. (1987) Diamond synthesis using hot filament thermal CVD assisted by RF plasma, *Proc. 8th Intl. Symp. Plasma Chem.*, Tokyo, Aug. 31-Sep. 4, 1987, v.4, pp. 2487-2492.

Lee, S. R. and Gallois, R. (1992) A study on synthesis of diamond by capacitively coupled RF plasma-assisted chemical vapor deposition, *Diamond Relat. Mater.*, **1**, 235-238.

Li, C., Lau, Y. C. and Girshick, S. L. (1991) Parametric study of diamond film deposition in a radio-frequency thermal plasma, *2nd Intl. Symp. Diamond Mater.*, Washington, DC, May 5-10, 1991, Electrochemical Society Proc. Vol. 91-8, pp. 57-64.

Li, C., Yu, B. W. and Girshick, S. L. (1991) Diamond film deposition in a radio-frequency thermal plasma, *Proc. 10th Intl. Symp. Plasma Chem.*, Bochum, Germany, Aug. 4-9., 1991, v.3, paper 3.1-9.

Lindsay, J. W., Han, H. and Girshick, S. L. (1995) Experimental study of diamond film morphology and surface chemistry in RF induction plasma CVD, to be published in *Proc. 12th Intl. Symp. Plasma Chem.*, Minneapolis, Aug. 21-25, 1995.

Mania, R., Stobierski, L. and Pampuch, R. (1981) Diamond synthesis in cool plasma, *Cryst. Res. Technol.*, **16**, 785-788.

Matsumoto, S. (1985) Chemical vapor deposition of diamond in RF glow discharge, *J. Mater. Sci. Lett.*, **4**, 600-602.

Matsumoto, S., Hino, M. and Kobayashi, T. (1987) Synthesis of diamond films in a rf induction thermal plasma, *Appl. Phys. Lett.*, **51**, 737-739.

Meyer, D. E., Dillon, R. O. and Woollam, J. A. (1989) Radio-frequency plasma chemical vapor deposition growth of diamond, *J. Vac. Sci. Technol. A*, **7**, 2325-2327.

Meyer, D. E., Ianno, N. J., Woollam, J. A., Swartzlander, A. B. and Nelson, A. J. (1988) Growth of diamond by rf plasma-assisted chemical vapor deposition, *J. Mater. Res.*, **3**, 1397-1403.

Owano, T. G., Goodwin, D. G., Kruger, C. H. and Cappelli, M. A. (1991) Diamond synthesis in a 50 kW inductively coupled atmospheric pressure plasma torch, in R. Messier, J. T. Glass, J. E. Butler and R. Roy (eds), *New Diamond Science and Technology*, Materials Research Society, pp. 497-502.

Owano, T. G. and Kruger, C. H. (1993) Parametric study of atmospheric-pressure diamond synthesis with an inductively coupled plasma torch, *Plasma Chem. Plasma Process.*, **13**, 433-446.

Owano, T. G., Kruger, C. H., Green, D. S., Williams, S. and Zare, R. N. (1993) Degenerate four-wave mixing diagnostics of atmospheric pressure diamond deposition, *Diamond and Related Materials*, **2**, 661-666.

Owano, T.G., Wahl, E.H., Kruger, C.H. and Zare, R.N. (1995) Degenerate four-wave mixing as a spectroscopic probe of boundary layer chemistry in thermal plasma CVD, 26th AIAA Plasmadynamics and Lasers Conference, San Diego, June 19-22, 1995, AIAA paper 95-1954.

Rudder, R. A., Hudson, G. C., Hendry, R. C., Thomas, R. E., Posthill, J. B. and Markunas, R. J. (1992) Formation of diamond films from low pressure radio frequency induction discharges, *Surf. Coatings Technol.*, **54/55**, 397-402.

Rudder, R. A., Hudson, G. C., Posthill, J. B., Thomas, R. E., Hendry, R. C., Malta, D. P. and Markunas, R. J. (1992) Chemical vapor deposition of diamond films from water vapor rf-plasma discharges, *Appl. Phys. Lett.*, **60**, 329-331.

Rudder, R. A., Hudson, G. C., Posthill, J. B., Thomas, R. E. and Markunas, R. J. (1991) Direct deposition of polycrystalline films on Si(100) without surface pretreatment, *Appl. Phys. Lett.*, **59**, 791-793.

Shimada, Y. and Machi, Y. (1994) Selective growth of diamond using an iron catalyst, *Diamond Relat. Mater.*, **3**, 403-407.

Shimada, Y., Mutsukura, N. and Machi, Y. (1993) Synthesis of diamond using iron catalyst by r.f. plasma chemical vapor deposition, *Diamond Relat. Mater.*, **2**, 656-660.

Silva, S. R. P., Knowles, K. M., Amaratunga, G. A. J. and Putnis, A. (1995) The microstructure of inclusions in nanocrystalline carbon films deposited at low temperature, *Diamond Relat. Mater.*, **3**, 1048-1055.

Snail, K. A. and Thorpe, T. P. (1994) Substrate temperature control apparatus and technique for CVD reactors. *U.S. Patent No. 5,318,801*, June 7, 1994.

Tamor, M. A. and Everson, M. P. (1994) On the role of penetration twins in the morphological development of vapor-grown diamond films, *J. Mater. Res.*, **9**, 1839-1848.

Thorpe, T. P., Snail, K. A., Vardiman, R. G. and Smith, T. (1993) Advances in technique for high-temperature homoepitaxial growth of diamond using an oxygen-acetylene torch. *3rd Intl. Symp. Diamond Mater.*, Honolulu, May 1993, Electrochemical Society Proc. Vol. 93-17, 468.

Verven, G., Priem, T., Paidassi, S., Blein, F. and Bianchi, L. (1993) Production and characterization of d.c. and h.f. plasma jet diamond films, *Diamond Relat. Mater.*, **2**, 468-475.

Vitkavage, D. J., Rudder, R. A., Fountain, G. G. and Markunas, R. J. (1988) Plasma enhanced chemical vapor deposition of polycrystalline diamond and diamondlike films, *J. Vac. Sci. Technol. A*, **6**, 1812-1815.

Wild, C., Kohl, R., Herres, N., Müller-Sebert, W. and Koidl, P. (1994) Oriented CVD diamond films: twin formation, structure and morphology, *Diamond Related Mater.*, **3**, 373-381.

Yu, W. and Girshick, S. L. (1989) Diamond film deposition in a thermal plasma: effect of substrate location. *Proc. 9th Intl. Symp. Plasma Chem.*, Pugnochiuso, Italy, v. 3, pp. 1439-1444.

Yu, B. W. and Girshick, S. L. (1994) Atomic carbon vapor as a diamond growth precursor in thermal plasmas, *J. Appl. Phys.* **75**, 3914-3923.

Chapter 23
ARCJET SYNTHESIS OF DIAMOND

Mark A. Cappelli
Stanford University, Stanford, CA 94305-3032, USA

Contents

1. Introduction

The first successful demonstration of diamond synthesis in a direct-current arc plasma jet was published in 1988 [Kurihara et al. 1988]. Since then, these and similar plasma jets have been widely used in diamond synthesis research circles, and have been challenging other commercial diamond chemical vapor deposition (CVD) technologies in the manufacture of both thin and thick film diamond products [Carts 1990].

A plasma jet, or, "arcjet", is a generic expression for a high pressure direct-current (dc) plasma discharge in which convection plays a significant role in transport processes. The name arcjet stems from the fact that an electric arc discharge is used to dissipate energy, via ohmic heating, into a flowing, high pressure gas, thereby increasing the gas enthalpy and kinetic energy. In a typical arcjet, the arc is confined to a chamber that contains a nozzle from which the plasma issues. The arc chamber itself, or arcjet body, may sit in a second chamber which is maintained at a pressure lower than that of the arc chamber, forcing the issuing plasma to expand. It is this pressure difference which is primarily responsible for the gas dynamic conversion of the thermal energy in the arc to kinetic energy of the flow. Such high-enthalpy arc-heated flows found early applications in fields ranging from the study of high-speed aerodynamics to the vaporization of refractory materials [Giannini 1957]. The temperature in the core of the arc, T_{arc}, of such a plasma can reach 40,000 K, and the plasma's conductivity increases with increasing temperature. This behavior leads to a discharge that has a negative impedance, a feature characteristic of an electric arc, and one that poses a significant challenge to those designing high efficiency arc power supplies.

The increased average gas temperature due to ohmic heating surpasses that which can be generated by thermal activation methods such as hot filament CVD. It was this feature that motivated Akatsuka et. al [1988] to insert a horizontal arc discharge in place of a resistively-heated tungsten filament in a diamond CVD reactor. As a result of the high temperatures, the reactant gas decomposition process is much more vigorous, and a greater fraction of the molecular hydrogen is dissociated to produce atomic hydrogen, one of the most important chemical precursors in the diamond CVD process.

In an arcjet flow, the relatively high velocities (~1-10 km/sec) that result from the expansion or acceleration process allows for a more efficient delivery of the atomic hydrogen to a substrate because of reduced boundary layer thicknesses. The expansion to lower pressures enhances the diffusive transport of atomic hydrogen from the plasma jet to the substrate surface (the diffusion velocity scales inversely with pressure), which is driven primarily by the reaction of atomic hydrogen at the substrate surface. All these factors combined, one can derive an expression for the fraction of the plasma jet atomic hydrogen that is delivered to a substrate surface in terms of the expansion pressure (in torr), p_e, the plasma jet velocity (in cm/sec), u_o, the substrate diameter (in cm), d_s, and the reaction probability of atomic hydrogen on the substrate, γ [Goodwin 1993b]:

$$\frac{[H]_s}{[H]_o} \cong \frac{1}{\left\{1 + 1100\,\gamma\sqrt{\frac{p_e d_s}{u_o}}\right\}} \tag{1}$$

Here, $[H]_o$ and $[H]_s$ are the mole fractions of atomic hydrogen in the incident plasma jet and substrate surface respectively. It is apparent from this equation, that the maximum delivery of atomic hydrogen to a substrate from an arcjet is favored by reduced pressure and increased jet velocities.

The advantages of such high enthalpy arcjet flows in the deposition of diamond films were exemplified by the results of Ohtake and colleagues [Ohtake and Yoshikawa 1990, Ohtake et al. 1991a], who, shortly following the work of Kurihara et al. [1988], demonstrated diamond growth rates exceeding 900 mm/hr with a carbon conversion efficiency of about 8%. This growth rate is nearly three orders of magnitude higher than some of the first hot-filament and microwave CVD methods reported. However, despite the high linear growth rates, the high heat fluxes and spatially limited deposits that accompanied these early arcjet processes restricted projected applications to those involving direct use of the thick, free-standing, diamond plates that were typically produced. These applications included heat sinks for high power laser diodes [Kurihara et al. 1989], and diamond particles for incorporation into grinding wheels [Ohtake et al. 1991b]. A photograph of the free-standing thick-diamond film and cross-sectional SEM deposited in a dc arcjet first reported by Kurihara et al.[1988, 1989] is shown in Fig. 1. The discovery by Kurihara et al. was significant not only because it introduced a new diamond CVD technology, but also because it led to the first report of free-standing diamond plates synthesized by any CVD method. The Kurihara studies [Kurihara et al. 1988, 1989] spawned a plethora of activities in dc arcjet deposition in the four years that followed [Furakawa et al. 1990, Klocek et al. 1990, Li et al. 1991, Loh and Cappelli, 1992ab, Lu et al. 1991ab, 1992, Matsumoto et al. 1990, Ohtake et al. 1990, 1991ab, 1992, Stalder and Sharpless 1990].

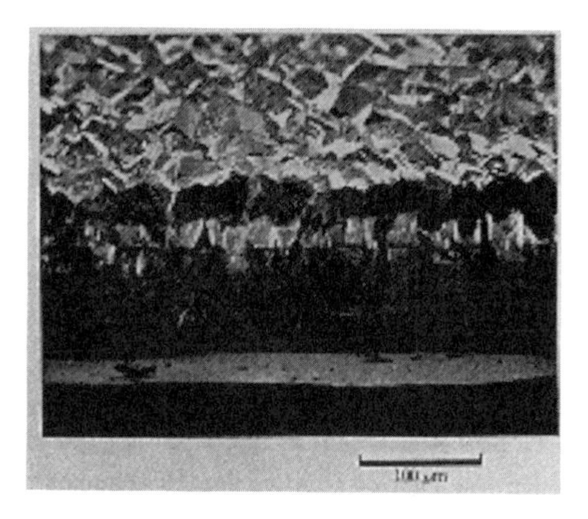

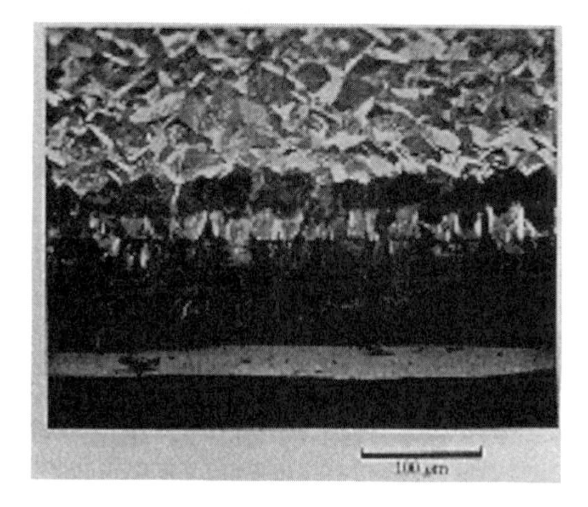

Figure 1. Photograph (top) and SEM cross-sectional image (bottom) of dc arcjet deposited diamond thick-film [Kurihara et al. 1989].

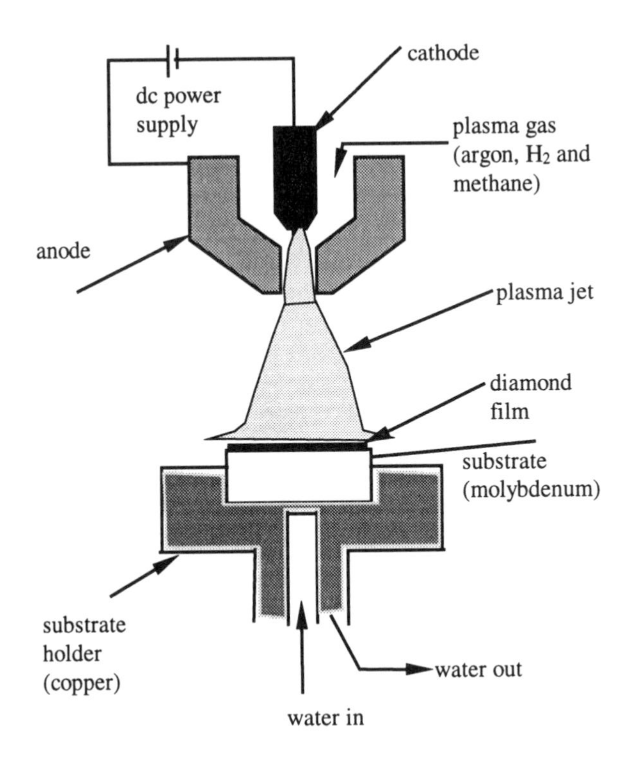

Figure 2. Schematic of typical arcjet diamond CVD process.

2. Basic dc Arcjet Processes

A schematic illustration of a typical arcjet in use as a source for diamond CVD is presented in Figure 2. Electrical energy is converted to thermal and kinetic energy of a flowing gas mixture by an electric arc discharge. Like other CVD methods, a major constituent of the gas mixture is hydrogen, which is needed to provide a source of atomic hydrogen, and methane is most often introduced into the plasma jet to provide a source of carbon. The plasma temperature is sufficiently high (1000 - 5000K on average) to partially dissociate the gas. The plasma jet containing these reactive species impinges onto a cooled substrate surface ($T_s \sim$ 1000 - 1500K). Molybdenum is often used as a substrate, although other materials have also been employed [Furakawa et al. 1990]. In arcjet deposition of diamond, pretreatment of the substrate by polishing with diamond paste is not a necessary condition for film growth, however, it greatly enhances the nucleation density [Lu et al, 1992].

In most designs, the electric discharge is sustained between a concentric cathode rod and a surrounding cylindrical anode, creating an arc column (localized path of high current density) that is nearly fully-ionized. Positive current (I) flows from the anode to the cathode, establishing the voltage drop (V) required to dissipate the total power ($P=IV$) in the arc. The current is generally concentrated in the hot region of the plasma due to the thermal pinch effect [Giannini 1957]. At sufficiently high current densities, self-induced magnetic fields further confine the motion of the charged particles within the arc column.

The arc column diameter is strongly influenced by the conduction and convection of energy away from its core to the colder surrounding gas. The conductivity of the plasma in the core of a fully-ionized arc varies as $T_{arc}^{3/2}$ [Spitzer and Harms 1953]. As a result, an increase in arc current further increases the temperature, and this results in an increased conductivity which thereby reduces the arc voltage. Except for regions very near the electrodes, there is a linear variation in the potential along the arc column from the cathode to the anode. In this respect, the arc column, or, positive column, behaves as a resistor with a negative impedance. The actual length of the arc column is also determined by electromagnetic and gasdynamic forces acting on it. Increased mass flow rates tend to increase the arc length and reduce the diameter, thereby increasing the resistance to current flow and hence raise the operating voltage. The arc voltage established for any given operating current is therefore strongly influenced by the plasma gas mixture and flow conditions such as the operating discharge chamber pressure and mass flow rates. An example of the sensitivity of the current-voltage characteristics of an arcjet used in diamond CVD to operating mixtures and mass flow rates is shown in Figure 3 [Loh and Cappelli 1992a].

2.1 ELECTRODE PROCESSES

The thermionic emission of electrons occurs at the cathode tip. Like the arc column, the arc attachment at the cathode is highly constricted, implying that electron current is emitted from a small area (characterized by a spot diameter that is usually less than the cathode diameter). The spot where the electron current is emitted is heated primarily by the concomitant ion bombardment of the cathode surface. This ion current, a consequence of the finite mobility of ions, is necessary to heat the cathode to temperatures that can support the high current densities (~10^6-10^7 A/m^2). The relationship between the cathode current density and cathode temperature, *Tcat*, is given by the Richardson-Duschman Law [Duschman 1930]:

$$J_{cat} = AT_{cat}^2 \exp-\left(\frac{\phi_w}{kT_{cat}}\right) \tag{2}$$

where A = $1.2\cdot10^6$ A/m^2/K^2, k is the Boltzmann constant, and ϕ_w is the work function of the cathode material.

Selection of a cathode material with a low work function is desirable, so as to obtain the highest current densities possible for a given cathode temperature. 2% thoriated tungsten (tungsten with an adsorbed layer of thorium when molten) has a relatively low work function of only 3.35 eV, and is often employed as a cathode material. The arc constriction and ion bombardment generally results in a cathode tip that is in a molten state. It is interesting to note that a molten pool of tungsten (melting temperature of about 3400K), 1 mm in diameter, can support an electron current of 11 A.

At extremely high current densities (~10^8 A/m^2), field enhanced thermionic emission may also be as important as thermionic emission [Schockley 1915]. The electrode heating process driven by ion bombardment also causes cathode erosion, contaminating the plasma with cathode material; a source of potential contamination of the diamond

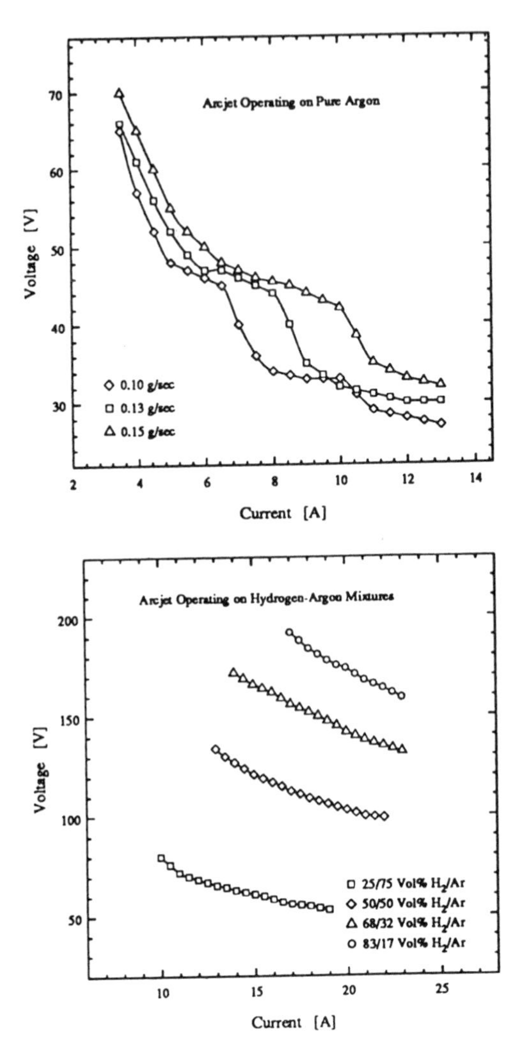

Figure 3. Typical current-voltage characteristics of a supersonic dc arcjet used in diamond CVD: (top) pure argon; (bottom) hydrogen-argon mixtures [Loh and Cappelli 1992a].

film. In some cases, the cathode may be water-cooled; however, in most arc sources, the source gas is injected upstream of the cathode, providing in itself a means of cathode cooling beyond that achieved by conduction alone.

Electron current is collected at the anode, which, in most designs, is water cooled to a temperature well below that of the arc. The strong property gradients and charged particle transport arising from these gradients that ensue near the anode due to this cooling give rise to relatively strong electric fields and a potential drop between the anode and plasma that is often greater than that determined by the plasma arc column

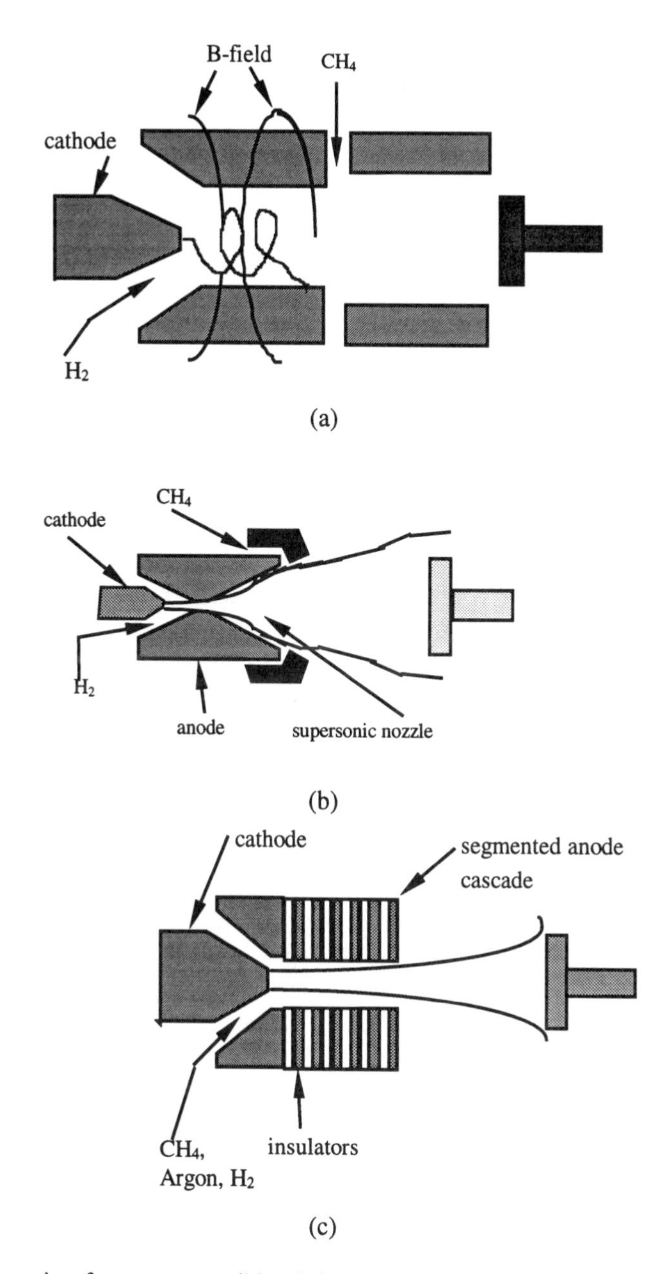

Figure 4. Schematic of some non-traditional dc arcjet diamond CVD methods: (a) magnetic stirring; (b) dc supersonic attachment; (c) segmented anode cascade.

resistance alone. This potential fall is the so-called anode fall voltage. Electrons accelerated by this fall voltage gain energy, and, when captured by the anode, transfer this energy (along with that released by the capture process-that associated with the anode work function) to the anode. This ubiquitous heating process may also lead to anode erosion and plasma contamination when the arc attachment to the anode is constricted.

Anode damage can be mitigated by methods such as magnetic stirring (applying a magnetic field to destabilize the constricted attachment mode and elongate the arc) [Cann 1984]; the use of gasdynamic designs that force the anode to attach in a more diffuse mode [Cappelli and Loh 1994a]; or reducing the current density by elongating the arc in a segmented anode discharge [Buelens 1992]. All of these methods have been incorporated into arcjet designs that have been used to deposit diamond films [Woodin et al. 1991, Loh and Cappelli, 1992a, 1992b, Beulens 1992]. Schematic illustrations of these non-traditional arcjet sources for diamond CVD are shown in Figure 4.

The plasma source gas-mixture is that gas mixture that flows through the discharge chamber with which the arcjet can be independently stably operated. The source gas mixture is the major factor that determines the operating voltage of the arc. A primary consideration in selecting the plasma gas-mixture is this operating voltage and the availability of inexpensive and efficient (i.e., operating without a resistive ballast) power supplies that support this voltage. Since arc welders conventionally operate with argon, power supplies that are designed for voltages in the 20-80 V range are readily available. As a result, argon, a relatively inexpensive and inert gas, is often the gas of choice in most arcjet diamond CVD systems. Hydrogen is often an additive to the argon as a source gas, and, the addition of even small amounts of hydrogen to the argon (< 20% by volume) can significantly affect the current-voltage characteristics. In some cases, hydrogen may be added downstream of the plasma jet, along with the carbon-containing reactants such as methane or acetylene. With the power supply constraints aside, a relatively straightforward analysis of the atomic hydrogen production efficiency in such an arcjet flow favors the use of pure hydrogen as the source gas, since there is a penalty paid for the seemingly unnecessary heating and ionization of the argon (discussed in Section 2.2 below). However, it should be noted that no comprehensive studies exist that indicate that argon is entirely inert in such CVD systems. It is well known that argon atoms and ions have metastable electronically excited states that can participate via collisions in energy transfer to molecular species, thereby affecting the reactive mixture composition.

2.2 ARCJET ENTHALPY BALANCE

Much can be learned about the basic features of most arcjet chemical vapor deposition reactors used in depositing diamond films by performing simple energy audits (measuring thermal efficiencies) and by applying straightforward energy conservation principals to simplified model systems. In doing so, it is important to define a figure of merit which represents the merit of any arcjet operating regime or arcjet design to depositing diamond films. The production of atomic hydrogen is recognized as the major function of such a source, since the growth rate and film quality improve with increasing mole fraction of atomic hydrogen [Goodwin 1993a]. For these reasons, we define a figure of merit, f_m, that characterizes the energy invested to produce atomic hydrogen:

$$f_m = \frac{\chi_H \dot{m}}{m_H P_{arc}} \tag{3}$$

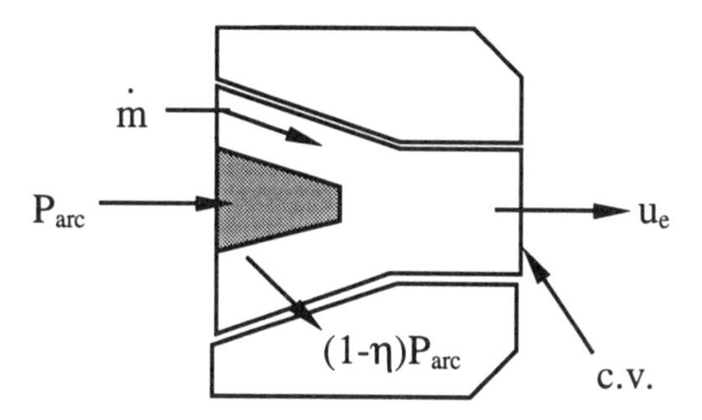

Figure 5. Schematic illustration of arcjet control volume employed for the simple model analysis.

where $\dot{m}$ is the total mass flow rate, χ_H is the atomic hydrogen mass fraction at the exit of the nozzle, and m_H is the mass of the hydrogen atom.

In the following exercise, we shall demonstrate through a simple model, how f_m depends on the operating conditions and design of the arcjet. In particular, we find that the energy invested to produce atomic hydrogen varies strongly with the arcjet thermal efficiency , η (defined here as the power deposited into the gas divided by the total discharge power, $P=IV$), and the fraction of added argon, if any. To understand this relationship, mass and energy conservation can be applied to the arcjet control volume schematically illustrated in Figure 5.

To fist order, we shall assume uniform flow of velocity, u_e, at the exit of the arcjet, and that the species comprising the plasma gas at the arcjet exit share a common temperature, T. For now, we shall assume that the gas introduced at the inlet is a mixture of molecular hydrogen and argon with a negligible kinetic energy, and that the gas exiting the nozzle is comprised mainly of atomic and molecular hydrogen (a reasonable assumption, since the plasma is only weakly ionized if averaged over the exit of the arc source) and argon. We shall neglect the impact of an added hydrocarbon on the enthalpy balance, since, in most cases, it represents a small fraction of the mass when argon is used as a plasma gas, and, in many cases, is added downstream of the discharge chamber when pure hydrogen is used as the source gas. Under these conditions, conservation of energy gives rise to the enthalpy balance:

$$\eta P_{arc} = \dot{m}^o_{H_2}\left\{\frac{u_e^2}{2} + h_{H_2} - h^o_{H_2} + \theta_D\left(h_H - h_{H_2}\right)\right\} + \dot{m}^o_{Ar}\left\{\frac{u_e^2}{2} + h_{Ar} - h^o_{Ar}\right\} \tag{4}$$

Here, h_j, and h_j^o represent the specific enthalpies of species j at the exit and inlet to the discharge chamber respectively, $\dot{m}_j^o$ is the inlet mass flow rate of species j, and θ_D is the hydrogen dissociation fraction, defined here as:

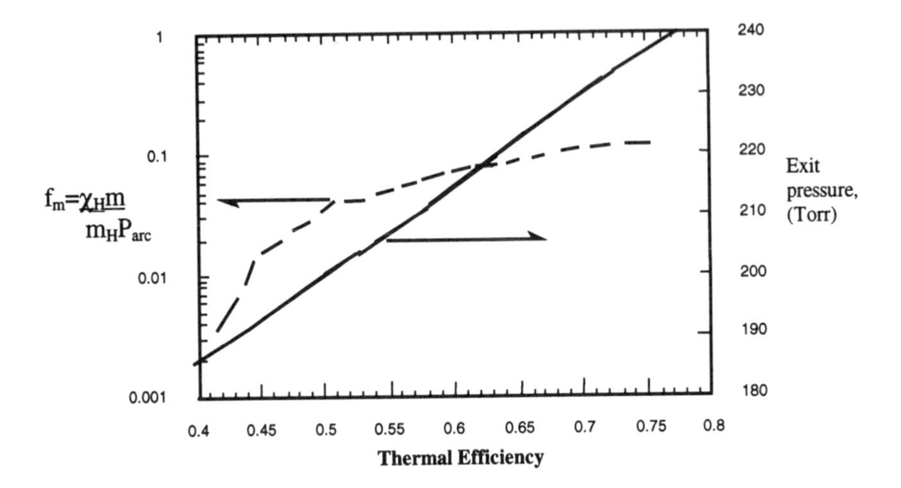

Figure 6. Computed variation in exit pressure and cost figure of merit, f_m, with changing arcjet thermal efficiency. For simulation purposes, P=10,000 W, T = 5000K, u_e=5380 m/sec (chosen to give unity Mach number at the exit), $\dot{m} = 5 \times 10^{-5}$ Kg/sec, and $A_e = 7.8 \cdot 10^{-5}$ m^2.

$$\theta_D = \frac{\chi_H}{\chi_{H_2} + \chi_H} = \frac{\dot{m}_H}{\dot{m}^o_{H_2}} \tag{5}$$

For a specified exit velocity, temperature, dissociation fraction, thermal efficiency, and inlet mass flow rates, the required discharge power can be readily obtained from Equation (4). Alternatively, the power (or specific energy, which is defined as the discharge power divided by total mass flow rate) can be specified instead of the dissociation fraction. In either case, we now have sufficient information to determine f_m. If the exit area, A_e, is specified, we can determine the mass density at the arcjet exit, from $\dot{m} = \rho u_e A_e$. In addition, we can further assume that the flow at the exit of the arcjet behaves according to the perfect gas equation of state:

$$p_e = kT \sum_j \frac{\chi_j}{m_j} \tag{6}$$

which allows us to calculate the exit pressure. The variation in the exit pressure and f_m with thermal efficiency for a pure hydrogen arcjet with operating conditions and design parameters typical of those used to deposit diamond is shown in Figure 6. Note that the exit pressure variation arises from the constraints imposed on the plasma flow conditions and changes in computed mass fractions with variations in the thermal efficiency. In the results shown in Figure 6, the exit velocity is selected to give a unity

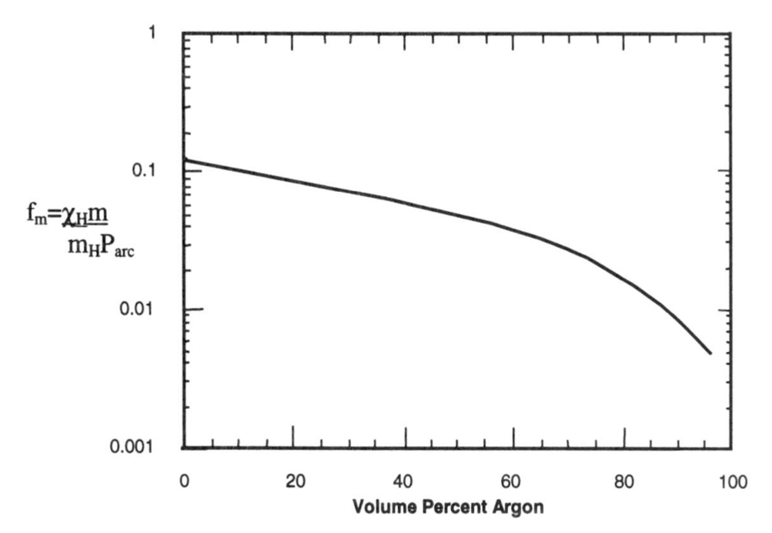

Figure 7. Computed variation in *fm* with changing volume fraction of added argon. For simulation purposes, pe = 200 torr, T = 5000K, u_e=5380 m/sec (chosen to give unity Mach number at the exit for pure hydrogen), $\dot{m}$ =1 x 10^{-4} Kg/sec, $\eta = 0.5$, $\theta_D = 0.5$, and $A_e = 7.8 \cdot 10^{-6}$ m^2.

Mach number (based on pure hydrogen at the exit temperature), i.e., a sonic jet at the nozzle exit.

Note that there is at least a decade increase in f_m for even a modest increase in the thermal efficiency (0.4 - 0.65), clearly illustrating the importance that should be placed on thermal design of arcjet reactors. In some cases, a 10-fold increase in f_m translates directly into a 10-fold decrease in deposition cost. It is this factor which motivated the use of radiation-cooled arcjets over water-cooled designs for the diamond CVD studies of Loh et al. [1993].

As mentioned earlier, most arcjets employed in diamond CVD use argon as the primary plasma gas source with only small volume flow fractions of added hydrogen. Although there are power supply benefits to operating on argon-hydrogen mixtures, and, it does appear that argon addition helps stabilize arcjet operation, it should be recognized that there is a significant penalty that is paid for the heating of argon. Argon is not believed to play a significant role in the plasma chemistry. The sensitivity of f_m to changes in added argon volume fraction is illustrated in Figure 7. For the computational results presented in Figure 7, we have used arcjet operating conditions and design parameters that closely resemble those used by Ohtake et al. [1992].

In the above figure, the dissociation fraction is held constant for illustrative purposes, and the variation in the specific energy with added volume percent argon is a consequence of the imposed constraints on the problem. In practice, the thermal

Table I. Comparison between conventional and supersonic hydrogen arcjet.

	Conventional Arcjet	Supersonic Arcjet
Specified Parameters		
Inlet temperature (K)	300	300
Exit temperature (K)	5000	1500
mass flow (Kg/sec)	$5 \cdot 10^{-5}$	$5 \cdot 10^{-5}$
thermal efficiency	0.5	0.5
Power (kW)	10	10
Mach number	1 (uo = 5380 m/sec)	2.5 (uo = 7360 m/sec)
Pressure (torr)	200 ($A_e=7.8 \cdot 10^{-6}$ m^2)	1 ($A_e= 4.0 \cdot 10^{-4}$ m^2)
Calc. Parameters		
χ_H	0.071	0.246
f_m (atom/eV)	0.034	0.118

efficiency and exit plasma properties may vary depending on the precise selection of operating conditions. Nevertheless, the trend seen in Figure 7 is representative of what would happen in practice: there would be a dramatic increase in the cost to produce atomic hydrogen with increases in added argon, a result that may translate directly into an increase in the cost to produce diamond film.

As pointed out by Goodwin [1993b], atomic hydrogen transport to the substrate from most plasma jet sources employed in diamond CVD is diffusion-limited at expansion pressures and for substrate diameters that are typically employed (200 torr and 2 cm respectively, say). We see this by considering Equation (1), with p_e = 200 torr, d_s = 2 cm, $u_o = 3 \times 10^5$ cm/sec, and $\gamma \approx 0.1$:

$$\frac{[H]_s}{[H]_o} \approx 0.2 \tag{7}$$

This represents yet another factor that limits the growth rates achievable in an arcjet reactor. More importantly however, it introduces a diffusion boundary layer that affects the interpretation of surface chemistry. The concentrations of reactive species either measured or calculated to be outside this boundary layer is not necessarily representative of what may be at the substrate surface. In order to circumvent this transport limitation or to minimize such ambiguities introduced by this diffusion layer, it may be desirable to operate at lower pressures and with higher jet velocities such that $1100\gamma\sqrt{p_e d_s / u_o} << 1$. This condition is satisfied for p_e = 1 torr, d_s = 2 cm, $u_o = 9 \cdot 10^5$ cm/sec, and $\gamma \approx 0.1$, properties common to highly expanded, supersonic, hydrogen arcjet flows [Loh et al. 1993ab]. The hydrogen atom transport to the substrate in these supersonic dc arcjets are therefore not necessarily diffusion-limited. One should also recognize that some cost benefits can be gained by supersonically expanding the flow to low temperatures. Table I compares f_m for two cases of interest: that of a conventional arcjet design expanding to unity Mach number and modest pressure (200 torr); and that for a supersonic arcjet expanding to a relatively high Mach number (M = 2.5) and very low pressure (1 torr).

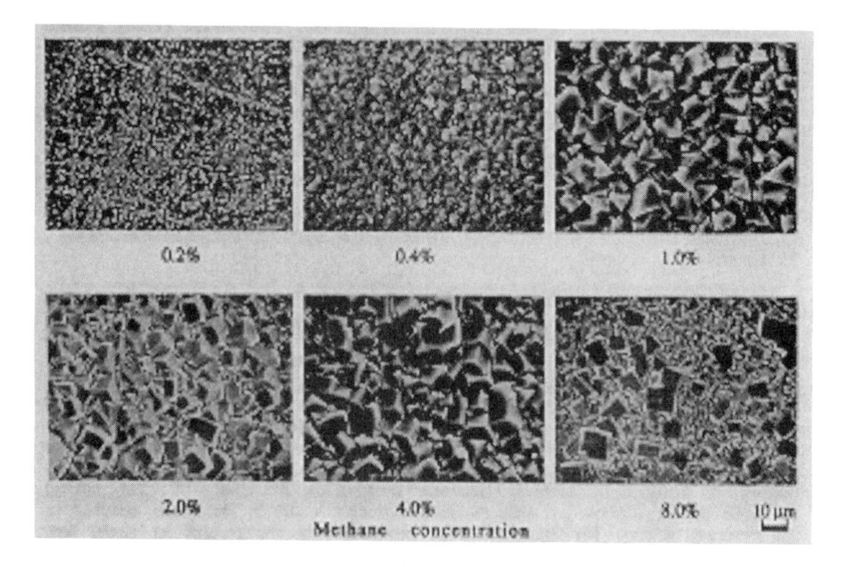

Figure 8. SEM images of the growth surface of dc arcjet deposited diamond films for various percentages of methane in hydrogen [Kurihara et al. 1990a].

The Mach number and expansion pressure is determined by the assignment of the jet velocity and exit nozzle area. The specified parameters in the table are consistent with operating and design parameters used in diamond synthesis facilities [Ohtake et al. 1992, Loh et al. 1993].

3. Progress in the Study of Arcjet Diamond CVD

3.1 BASIC RESEARCH

The relatively high growth rates afforded by dc arcjet diamond CVD provided some of the first convenient opportunities to study the morphological changes occurring during diamond crystal growth [Kurihara et al. 1990a]. Scanning electron microscope images of the growth surface of diamond films deposited by a dc arcjet under varying methane to hydrogen ratios are shown in Figure 8. The growth habit of dc arcjet deposited diamond at substrate temperatures of approximately 1000°C was found to change dramatically from a predominantly {111} surface morphology at low volume fractions of methane in hydrogen (~1%) to a {110} and {100} surface morphology at relatively high volume fraction (~ 8%). Mixed morphologies were seen in between these values.

A similar trend was observed when the substrate-to-nozzle distance was varied (increased distances favored a shift from {111} to {110} or {100} morphologies). The authors used this data as well as optical emission studies to argue that the growth of any particular crystallographic low-index plane was strongly correlated to the mole fraction ratios; [CH]/[H] and [C]/[H]. The study illustrated that there are strong variations in plasma composition along the axial dimension as one moves from the plasma jet nozzle to the substrate, in part because of the strong tendency for atomic hydrogen to diffuse towards the jet boundary more readily than heavier hydrocarbon fragments.

High resolution scanning electron microscope analyses on dc arcjet deposited diamond films were performed by Ohtake et al. [1991c] and Hirabayashi et al. [1992]. These studies clearly revealed the presence of steps on both the {100} and {111} surfaces. In comparison to the {111} surface, the {100} was more resistant to subsequent plasma etching in air, leading the authors to suggest that the {111} growth surface was more defective than the {100} growth sector; an argument that was supported by micro-Raman analysis of the deposits. A later study by the same authors revealed that addition of oxygen to the arcjet the plasma gas dramatically altered the growth morphology [Ohtake and Yoshikawa 1993]. With added oxygen, the films were smoother in texture and the {111} faces showed improved crystallinity. However, oxygen addition also reduced the overall growth rate. The authors concluded that the improved crystallinity arises from a reduction in the amount of active carbon and hydrocarbon radicals in the gas phase due to the formation of carbon-monoxide.

Consistent with the findings of many other studies of diamond CVD in plasma, combustion, and thermal activation methods, the quality of the diamond deposited in dc arcjets degrades very rapidly with increasing methane to hydrogen volume flow ratios [Ohtake and Yoshikawa 1990]. This suggests at first that the chemical mechanism controlling the synthesis of diamond over graphite in such high deposition rate arcjets is similar to that in other CVD systems, where the diamond growth rate can be orders of magnitude lower. Indeed, Goodwin [1991] has shown that such a condition is possible if the methyl radical (CH_3) is the growth species and if the surface growth mechanism follows that proposed by Harris [1989]. However, it is recognized that in thermal plasmas such as dc arcjets, atomic carbon may be as abundant as CH_3. Recent theoretical calculations by Yu and Girshick [1994] suggest that in many thermal plasma flows such as dc arcjet flows operating at modest pressures, atomic carbon can account for the observed growth rates.

In many of the first arcjet diamond CVD studies, the hydrocarbon source gas (usually methane) was mixed with the plasma carrier gas and introduced upstream of the cathode [Kurihara et al. 1988, 1989, 1990ab]. Although upstream injection was attempted by Ohtake and Yoshikawa [1990], it was found that the plasma was unstable and that the deposited films were of poor quality. As a result, their arcjet operating conditions specified downstream remote injection near the point where the plasma exited the discharge nozzle. In the supersonic dc arcjet studies of Loh and Cappelli [1992ab, 1993], diamond was deposited with either CH_4 or acetylene (C_2H_2). The point of injection was downstream of the nozzle and was selected so as to optimize the transport of either CH_3 radicals or C_2H_2 to the substrate, as it is speculated that even acetylene can act as a precursor to diamond growth [Skokov et al. 1994]. Recent experimental and modeling studies [Reeve et al. 1993, Dandy and Coltrin 1993] of diamond growth in a more conventional arcjet plasma (argon carrier gas and injected hydrogen) indicate that some increase in growth rate and quality can be obtained by optimizing the point of injection.

Extremely high diamond growth rates (~ 1mm/hour) have been achieved in dc arcjets that use liquid precursors as the carbon source [Han et al. 1991]. In that study, the liquid precursors included a variety of ketones, alcohols, halogenated, and aromatic compounds injected through the substrate opposite the direction of the impinging plasma jet. The highest growth rates were observed with ethanol and acetone.

Attempts have been made to enhance the diamond film growth rates in dc arcjets by applying a positive dc electrical bias to the substrates [Matsumoto et al. 1990, Baldwin et al. 1995]. Using a 100 kW atmospheric pressure dc arcjet facility, Baldwin et al. [1995] has shown that a six-fold increase in growth rate can be induced with only a modest secondary discharge (115 V, 3A) while maintaining uniformity over a 1 cm diameter substrate. In such atmospheric pressure discharges, there is considerable boundary-layer recombination of atomic hydrogen. The secondary discharge is believed to increase the electron temperature so as to maintain an elevated atomic hydrogen concentration in the near-surface region.

Measurements of the plasma temperature, velocity and plasma composition in arcjets under conditions of diamond growth are needed so as to better understand the energy balance in the system [Brinkman and Jeffries, 1995], and, to verify detailed arcjet CVD computational flow models that have been recently formulated [Ohtake et al. 1992, Moen and Dwyer, 1994]. Diagnostics of arcjet flows employed in diamond CVD has been limited largely to optical emission spectroscopy [Kawarada et al. 1989, Woodin et al. 1991, Cappelli and Loh, 1993]. Mass spectrometric analysis of a dc arcjet plume used in diamond cvd was performed by Stalder and Sharpless [1990], and by Loh and Cappelli [1993]. A few groups have employed laser-induced fluorescence (LIF) to characterize flow properties [Raiche et al. 1991, Loh et al. 1993, Jeffries et al. 1994, Brinkman and Jeffries 1995].

Comparison is often made of dc arcjet emission to the emission from a CVD diamond microwave plasma (see for example Kawarada et al. 1989) where it is found that considerably greater emission is seen from molecular hydrogen. Spectral emission from the atomic hydrogen Balmer series indicates that there is a departure from local thermodynamic equilibrium in these plasmas. As a result, an estimate of the plasma temperature from the relative intensity of electronic transitions in atomic hydrogen is unreliable. Atomic hydrogen laser-induced fluorescence (LIF) measurements of plasma temperature and velocity have been made in supersonic arcjets used in diamond CVD [Loh et al. 1993]. A comparison of optical emission-based measurements to LIF measurements of temperature in a similar arcjet was reported by Jeffries et al. [1994], and by Brinkman and Jeffries [1995]. In those studies, it was found that emission-based measurements of arcjet exit temperature using the rotational and vibrational state distributions in electronically excited CH and C2 radicals gave temperatures in the 3000 - 7000 K range, whereas LIF measurements of the rotational and vibrational state distributions in the electronic ground state gave temperatures in the 1200-2200 K range. The latter temperatures are consistent with those measured by Loh et al. [1993]. It is apparent that the interpretation of plasma properties based on optical emission spectroscopy is difficult because of the nonequilibrium nature of the plume.

3.2 APPLIED RESEARCH

Arcjets have a distinct advantage over other diamond CVD methods in that the deposition of diamond can proceed simultaneously with the deposition of other ceramics and metals by introducing powders of various types into the plasma stream. Using this approach, Kurihara et al. [1990b, 1991] deposited well adhered (40 mm thick) diamond films on tungsten-molybdenum substrates by spraying an interlayer (70 mm thick) composed of tungsten carbide, followed by a composite diamond-tungsten-carbide (40 mm thick) intermediate layer. Such a functional gradient material reduces the thermal stress between the diamond film and the substrate and increased the adhesion strength by an order of magnitude over that which was obtained in the absence of the tungsten-carbide interlayer.

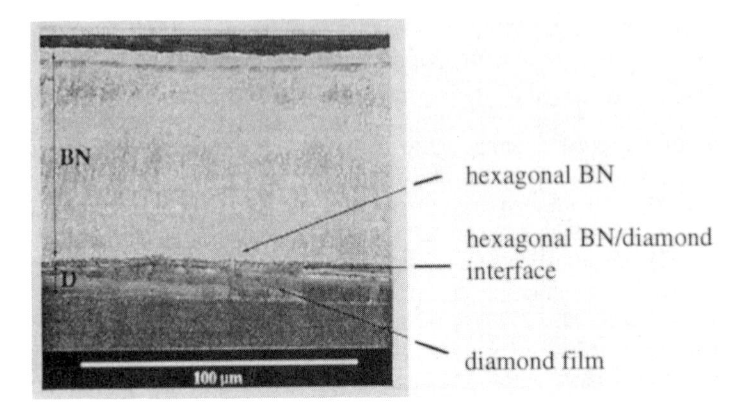

Figure 9. SEM cross-section image of a free-standing diamond/boron-nitride laminate deposited by a supersonic dc arcjet [Loh and Cappelli 1994].

Thermal stresses encountered during the cool-down period following dc arcjet diamond CVD often causes premature delamination and cracking of the diamond deposit. As a result, it is difficult to manufacture diamond films of intermediate thickness (10-50 mm). Diamond and boron-nitride laminates were deposited in a supersonic dc arcjet by first depositing a 10 mm thick diamond film on molybdenum, followed by a gradual conversion of the plasma and injected gases to nitrogen (N_2) and BCl_3 to deposit a 100 mm thick boron-nitride coating on the diamond film [Loh et al. 1994]. Care was taken during the conversion to avoid changes in substrate temperature. The laminate that resulted was easily removed from the substrate, resulting in a free-standing composite structure. A cross section SEM image of the free-standing laminate is shown in Figure 9.

The highest growth rate ever achieved in homoeptixaial diamond CVD was reported by Snail et al. [1991]. That study employed a triple-arcjet design (referred to as a "triple torch plasma reactor -TTPR") operating on argon-hydrogen mixtures as the plasma gas and with remote methane and hydrogen injection. The substrates consisted of single-crystal (0.25 mm thick) natural diamond heat sinks (type IIa) that were cut into circular cross sections and brazed onto threaded water-cooled molybdenum rods. Growth temperatures ranged from 1200-1400°C, and growth rates were in the range of 100 - 200 mm/hour, depending on the crystallographic plane. An SEM image of a {100}

Figure 10. SEM images of a diamond crystal deposited on a cylindrical {100} diamond seed in a dc-arcjet plasma: (a) 100 mm marker;(b) 1 mm marker [Snail et al. 1991].

cylindrical shaped seed crystal following deposition for 30 minutes is shown in Figure 10.

Bradley et al. [1995], have presented technical and market strategies for arcjet CVD diamond applications. Although in that paper, they discuss a wide range of potential applications, it is apparent that dc arcjet processes will be most competitive in applications that make use of either large thicknesses of diamond (so-called thick-film diamond), or, in general, large volumes. These applications include heat sinks for diode lasers, packaging for high power electronics, and brazed, thick-film tool inserts.

4. Summary

In this paper, we have summarized the past and most significant developments in arcjet diamond CVD research. In addition, a simple analysis is presented which can aid in the design of arcjet process conditions. Although there has been much activity in the CVD of diamond films, the use of arcjets as a plasma source is limited only to a handful of researchers worldwide, despite the fact that arcjets have recorded the highest deposition rates of any CVD method. The future for arcjet applications in diamond CVD will no doubt be determined by the ability to reduce variable cost such as the power required to deposit a specified mass of diamond. Although to some extent, this involves minimizing the price paid to dissociate hydrogen, some effort should be made to understand chemistries that will lead to the efficient transport and conversion of carbon precursors.

5. Acknowledgment

The author would like to acknowledge the many stimulating discussions that he has had with his graduate students, in particular, M. Loh, A. Kull, and D. Berns. The author also acknowledges S. Jaffe for discussions about arcjet performance. Research on arcjet CVD at Stanford has been supported by the ONR, NSF, Olin Aerospace Company, Norton Company, and Daimler Benz Corporation.

6. References

Akatsuka, F., Hirose, Y., and Komaki, K. [1988] Rapid growth of diamond films by arc discharge plasma CVD, *Jpn. J. Appl. Phys.* **27**(9), L1600-L1602.

Baldwin, S.K. Jr., Owano, T.G., and Kruger, C.H. [1995] Increased deposition rate of CVD diamond in a dc arcjet with a secondary discharge., *Appl. Phys. Lett.* **67**, pp. 194-196.

Beluens, J.J. [1992] Surface modification using a cascade arc plasma source, Doctoral Thesis, University of Eindhoven.

Bradley, D.S., Jaffe, S.M., and Bigelow, L.K. [1995] Technical and market strategies for arcjet cvd diamond applications, *Workshop on the Industrial Applications of Plasma Chemistry, August 25-26, Proc. Vol B*, Thermal Plasma Applications (IUPAC), pp. 77-88.

Brinkman, E.A., and Jeffries, J.B. [1995] Comparison of optical emission and laser-induced fluorescence in arcjet thruster plumes used for diamond deposition, *AIAA 95-1955*, 26th Plasmadynamics and Lasers Conference, San Diego, CA.

Cann, G.L. [1984], Magnetoplasmadynamic apparatus and process for the separation and deposition of materials, *U.S. Patent Nos.* 4,471,003 and 4,487,162.

Cappelli, M.A., and Loh, M.H. [1994a] A method and apparatus for the growing diamond films, *U.S. Patent No.* 5,358,596.

Cappelli, M.A., and Loh, M.H. [1994b] In-situ mass sampling during supersonic arcjet synthesis of diamond, *Diamond and Related Materials* **3**, pp. 417-421.

Carts, Y.A. [1990] Thin-and thick-film diamond goes commercial, *Laser Focus World*, December, p. 41.

Dandy, D.S., and Coltrin, M.E. [1994] Relationship between diamond growth rate and hydrocarbon injector location in direct-current arcjet reactors, *Appl. Phys. Lett.* **66**, pp. 391-393.

Dushman, S. [1930] Thermionic emission, *Reviews of Modern Physics* **2**, pp. 381-476.

Furakawa, R., Uyama, H., and Matsumoto, O. [1990] Diamond deposition with plasma jet at reduced pressures, *IEEE Trans. Plasma Sci.* 18 (6), pp. 930 - 933.

Giannini, G.M. [1957] The plasma jet, *Scientific American*, August, pp. 80-88.

Goodwin, D.G. [1991] Simulations of high-rate diamond synthesis: methyl as the growth species, *Appl. Phys. Lett.* **59**, pp. 277-279.

Goodwin, D.G. [1993a] Scaling laws for diamond chemical vapor deposition. I. diamond surface chemsitry, *J. Appl. Phys.* **74**, pp. 6888-6894.

Goodwin, D.G. [1993b] Scaling laws for diamond chemical vapor deposition. II. atomic hydogen transport, *J. Appl. Phys.* **74**, pp. 6895-6906.

Harris, S.J. [1990] Mechanism for diamond growth from methyl radicals, *Appl. Phys. Lett.* **56**, pp. 2298-2300.

Han, Q.Y., Or, T.W., Lu, Z.P., Heberlein, J., and Pfender, E. [1991] *Proc. 2nd International Symposium on Diamond Materials* (The Electrochemical Society), Washington D.C., Vol. 91 (8), pp. 115-122.

Hirabayashi, K., Kurihara, N.I., Ohtake, N., and Yoshikawa, M. [1992] Size dependence of morphology of diamond surfaces prepared by dc arc plasma jet chemical vapor deposition, *Jpn. J. Appl. Phys.* **31**, pp. 355-360.

Kawarada, M., Kurihara, K., Sasaki, K., Teshima, A., and Koshino, K. [1989] Diamond synthesis by dc plasma jet cvd, *SPIE Proceedings Vol. 1146*, Diamond Optics II (The International Society for Optical Engineering) pp. 28 - 36.

Klocek, P., Hoggins, J., Taborek, P., and Mckenna, T. [1990] CVD diamond by dc plasma torch, *SPIE Proceedings Vol. 1325*, Diamond Optics III (The International Society for Optical Engineering) pp. 63-72.

Kurihara, K, Sasaki, K., Kawarada M., and Koshino N. [1988] High rate synthesis of diamond by dc plasma jet chemical vapor deposition, *Appl. Phys. Lett.* **52**(6), pp. 437-438.

Kurihara, K, Sasaki, K., and Kawarada M. [1989] Diamond film synthesis using dc plasma jet cvd, *Fujitsu Scientific and Technical Journal* **25**(1), pp. 48-51.

Kurihara, K, Sasaki, K., Kawarada M., and Koshino N. [1990a] Morphology of diamond films grown by dc plasma jet cvd, *Mat. Res. Soc. Symp. Proc.* Vol. 162, pp. 115-118.

Kurihara, K, Sasaki, K., and Kawarada M. [1990b] Adhesion improvement of diamond films, *Proc. 1st International Symposium*, FGM, Sendai, pp. 65-69.

Kurihara, K, Sasaki, K., Kawarada M., and Goto Y. [1991] Formation of functionally gradient diamond films, *Applications of Diamond Films and Related Materials*, (Y. Tzeng, M. Yoshikawa, M. Murakawa, and A. Feldman, Eds.), Elsevier Science Publishers B.V. pp. 461-466.

Li, R., Shi, H., Yan, Z., Tang, S., and Zhu, H. [1991] Transparent diamond film deposited by optimized dc arc plasma jet, *Applications of Diamond Films and Related Materials*, (Y. Tzeng, M. Yoshikawa, M. Murakawa, and A. Feldman, Eds.), Elsevier Science Publishers B.V. pp. 207-212.

Lu, Z.P., Heberlein, J., and Pfender [1991a] Process study of thermal plasma chemical vapor deposition of diamond, *Plasma Chem. and Plasma Process.* 12, pp. 35-53.

Lu, Z.P., Stachowicz, L., Kong, P., Heberlein, J., and Pfender [1991b] Plasma synthesis by dc thermal plasma cvd, *Plasma Chem. Plasma Process.* **11**, pp. 387-394.

Lu, Z.P., Heberlein, J., and Pfender [1992] Process study of thermal plasma chemical vapor deposition of diamond, part II: pressure dependence and effect of substrate treatment, *Plasma Chem. and Plasma Process.* **12**, pp. 55-69.

Loh, M.H., and Cappelli, M.A. [1992a] Supersonic dc arcjet plasma at subtorr pressures as a medium for diamond film synthesis, *AIAA 92-3534, 28th Joint Propulsion Conference*, Nashville, TN.

Loh, M.H., and Cappelli, M.A. [1992b] Diamond synthesis in a supersonic direct-current arcjet plasma at subtorr pressures, *Surface and Coatings Technology* **54/55**, pp. 408 - 413.

Loh, M.H., and Cappelli, M.A. [1993] Study of precursor transport during diamond synthesis in a supersonic flow, *Proc. 3rd Intl. Symp. Diamond Materials* Vol. 93-17, The Electrochemical Society, Honolulu, HI. pp. 17-23.

Loh, M.H., Liebeskind, J.G., and Cappelli, M.A. [1993] Characterization of a supersonic hydrogen arcjet plamsa tuster employed in diamond film synthesis, *AIAA 93-2227, 29th Joint Propulsion Conference*, Monterey, CA.

Loh, M.H., Liebeskind, J.G., and Cappelli, M.A. [1994] Arcjet thrusters for the synthesis of diamond and cubic boron-nitride films, *AIAA 94-3233, 30th Joint Propulsion Conference*, Indianapolis,IN.

Matsumoto, S., Hosoya, I., and Chounan, T. [1990] Substrate bias effect on diamond deposition by dc plasma jet, *Jpn. J. Appl. Phys.* **29** (10) pp. 2082-2086.

Moen, C.D., and Dwyer, H.A. [1994] Streamline chemistry in arc heated plasma reactors for diamond chemical vapor deposition, *1994 Spring Meeting of the Western States Section/The Combustion Institute.*

Ohtake, N., and Yoshikawa, M. [1990] Diamond film preparation by arc discharge plasma jet chemical vapor deposition in the methane atmosphere, *J. Electrochem. Soc.* **137**(2), pp. 717-722.

Ohtake, N., Kuriyama, Y., Yoshikawa, M., Obana, H., Kito, M., and Saito, H. [1991a] Development of an arc-discharge plasma apparatus for the high-rate synthesis of diamond, *Int. J. Japan Soc. Prec.* Eng. **25** (1), pp. 5 -10.

Ohtake, N., Mashimo, Y., and Yoshikawa, M. [1991b] Applications of CVD diamond grains fabricated by roll milling to grinding wheels, *Proceedings of the 1991 International Conference on New Diamond Science and Technology* (Materials Research Society), Washington, D.C., pp. 173-178.

Ohtake, N., Yoshikawa, M., Suzuki, K., and Takeuchi, S. [1991c] Growth process of diamond films by arc discharge plasma jet cvd, *Applications of Diamond Films and Related Materials*, (Y. Tzeng, M. Yoshikawa, M. Murakawa, and A. Feldman, Eds.), Elsevier Science Publishers B.V., pp. 431 - 438.

Ohtake, N., Ikegami, M., and Yoshikawa, M. [1992] Numerical simulation of a flow field and temperature field of plsma jet for diamond film preparation, *Diamond Films and Technology* **2**(1), pp. 1 - 15.

Ohtake, N., and Yoshikawa, M. [1993] Effects of oxygen addition on growth of diamond film by arc discharge plasma jet chemical vapor deposition, *Jpn. J. Appl. Phys.* **32**, pp. 2067 - 2073.

Reeve, S.W., Weimer, W.A., and Dandy, D.S. (1993] Diamond growth using remote methane injection in a direct current arcjet chemical vapor deposition reactor, *Appl. Phys. Lett.* **63**, pp. 2487-2489.

Schottky, W. [1914] *Ann. d. Physik* **44**, 1011-1032.

Skokov, S., Weiner, B., and Frenklach, M. [1994] Elementary reaction mechanism of diamond growth from acetylene. *J. Phys. Chem.* **90**, pp. 8 - 11.

Snail, K.A., Marks, C.M., Lu, Z.P., Heberlein, J., and Pfender, E. [1991] High temperature, high rate homoepitaxial synthesis of diamond in a thermal plasma reactor, *Materials Letters* **12**, 301-305.

Spitzer, L., and Harm, R. [1953] Transport phenomenon in a completely ionized gas, *Phys. Rev.* **89**, p. 977.

Stalder, K.R., and Sharpless, R.L., Plasma properties of a hydrocarbon arcjet used in the plasma deposition of diamond thin films, *J. Appl. Phys.* **68**, pp. 6187-6190.

Woodin, R.L., Bigelow, L.K., and Cann, G.L. [1991] Synthesis of large area diamond films by a low pressure dc plasma jet, *Applications of Diamond Films and Related Materials*, (Y. Tzeng, M. Yoshikawa, M. Murakawa, and A. Feldman, Eds.), Elsevier Science Publishers B.V. pp. 439-444.

Yu, B.W., and Girshick, S.L. [1994] Atomic carbon vapor as a diamond growth precursor in thermal plasmas, *J. Appl. Phys.* **75**,3114-3123.

Chapter 24

Low Temperature Chemical Vapor Deposition

Akimitsu Hatta and Akio Hiraki
Electrical Engineering, Osaka University, 2-1 Yamada-oka, Suita 565, Japan

Contents

1. Introduction

Many efforts have been made to lower the substrate temperature for the diamond chemical vapor deposition process. Current processes require high temperatures typically 800 to 900 °C and work with only a limited choice of materials for substrates. Low temperature growth is an essential technique required for electronics application. It is necessary to grow diamond on a substrate without stress due to the difference in expansion rates between the diamond and the substrate material and without missing the performance of devices previously fabricated.

The trends in low temperature diamond growth research have been well reviewed by Muranaka et al. [1994]. There have been several reports of successful fabrication of diamond films at temperatures lower than 200 °C. Nakao et al. have reported fabrication of diamond films on Al substrates at 140 °C by a DC plasma chemical vapor deposition method [Nakac S. et al. 1990]. Ihara et al. [1991] have shown that diamond particles could be formed at substrate temperatures of 135 °C by hot filament assisted chemical vapor deposition by using water cooling. Muranaka et al. [1991] have reported fabrication of diamond films on silicon substrates at 130 °C by the microwave plasma chemical vapor deposition method using $CO/O_2/H_2$ gas mixtures and cooling the substrate by ventilation. Yara et al. [1994] have reported fabrication of microcrystalline diamond films at temperatures below 100 °C. In these works, however, the substrate temperature has not correctly been measured because there is a temperature gradient from the surface of the substrate to the holder with forced cooling that was not taken into account.

The following are indispensable problems for the fabrication of diamond films at low temperature, i.e., heating of the substrates by the plasma, forced cooling resulting in problems in the measurement of substrate temperature, growth and nucleation at low temperature.

2. Heating of the Substrate by the Plasma

There are several kinds of energy flow onto the substrate from the plasma, i.e., radiation in infrared, visible or ultraviolet regions, electromagnetic waves, charged or neutral species with kinetic energy, or release of chemical or electronic excitation. These energy flows will heat up the substrate in addition to their contribution to the surface reactions. In the conventional microwave plasma for diamond synthesis, the substrate temperature automatically exceeds 500 °C in the plasma. For the low temperature growth of diamond, it is important not only to cool the substrate but also to minimize the excess energy flow to the substrate during growth.

Thermal flux due to neutral gas molecules, Γ_g, is estimated from the energy loss on the surface. By means of the general kinetic theory of gas, Γ_g is expressed as follows,

$$\Gamma_g = \frac{1}{4} n_g \bar{v} \cdot \eta \cdot \frac{1}{2} m \bar{v}^2$$

$$\Gamma_g = \frac{3}{8} \eta p \sqrt{\frac{3RT_g}{M}} \tag{1}$$

Here n_g is gas density at pressure p, m and $\bar{v}$ denote mass and mean velocity of the molecule, respectively, M is the atomic mass number and η is a coefficient of energy transfer from the molecule to the substrate in a collision. The coefficient 1/4 appears in the calculation for collision frequency of gas molecules with a wall of unit area. In the case of conventional microwave discharges, $\Gamma_g \approx 11$ W/cm^2 under the condition of p=50 kPa, Tg =1000 K, M=2 (H_2 molecule) and η=0.5. The gas temperature, T_g, may be higher than 1000 K due to the high collision frequency with high energy electrons. The actual Γ_g should be determined by the gradient of temperature near the substrate. It will decrease when the substrate becomes close to the gas temperature.

Energy flux of charged particles to the substrate, Γ_c, is expressed as,

$$\Gamma_c = V_{sh} I_{is} + T_e \frac{I_e}{e} \tag{2}$$

The first and second terms of the right-hand side of equation 2 are energy flux of ions and electrons, respectively. V_{sh} and T_e denote the sheath potential and the electron temperature, respectively. Electron current, I_e is equal to the ion saturation current, I_{is}, for an electrically floating substrate, where I_{is} is limited by the electron density. By using typical values, V_{sh}=20 V, T_e=5 eV and I_{is} =10 mA/cm^2, for the conventional microwave plasma chemical vapor deposition process, Γ_c=250 mW/cm^2.

Heating by photon flux is negligible as compared with Γ_g, because a substrate that is not in the plasma column is nearly at room temperature during deposition. Dielectric heating of the substrate by microwave can be also ignored because silicon substrates are usually conductive. The dominant source of energy in heating the substrate in the conventional microwave plasmas is the thermal flux, Γ_g, due to collisions with high temperature gas. Therefore, one of the most effective ways to cool the substrate is to lower the gas pressure during deposition.

3. Forced Cooling and Measurement of Temperature

Even if the extra heating of the substrate was successfully reduced by lowering the pressure, some other kind of heat flux would be present. At minimum, the surface reaction during diamond growth in the vapor phase is an exothermic reaction itself, i.e., chemically or electronically excited species loose their energy on the surface to form the diamond structure. Forced cooling is inevitable to keep the substrate at a low temperature.

Figure 1 shows a substrate holder with forced cooling using a circulating fluid, such as water or ethylene glycol. The material of the holder should have good thermal conductivity to transfer the heat flux from the substrate to the fluid. The thickness of the wall between the fluid and the substrate should be thick enough to hold the pressure of the circulating fluid and still maintain a sufficient thermal conductance. Most of the attention should be paid to the contact between the substrate and the holder. The thermal conductivity between the substrate and the holder greatly depends on the roughness of both the backside of the substrate and the surface of the holder. Thermal contact between the substrate and the holder is not sufficient even if the surface of the holder is polished to a mirror like finish. For reliable control of the substrate temperature, paste of good thermal conductivity should be used such as silicon grease as shown in Figure 1 (a) and (b). The resulting temperature of the substrate is quite different for the substrates contact with and without silicon grease. For example, the substrate temperature without silicon grease was more than 100 °C higher than with silicon grease.

It is not easy to measure the substrate temperature accurately. With forced cooling from the backside, the temperature decreases from the surface of the substrate to the fluid in the holder. In measuring temperatures, an optical pyrometer using visible emission is not useful near room temperature. The radiation thermometer, using the infrared region, is available for measurements in a wide range of temperatures except they are not useful for semiconductor substrates. Substrates, such as Si are almost transparent in the range of the infrared wavelength. Thus the thermometer will not measure the temperature of the Si substrate but that of the holder under the substrate which will be cooler than the substrate. The thermocouple is also useful in a wide range of temperatures though attention should be paid in making the thermal contact between the substrate and the thermocouple. If the thermocouple is fixed on the backside of the substrate as shown in Figure 1, it will thermally contact the holder rather than the substrate. The temperature of the thermocouple is sometimes governed by the substrate holder. Much of the

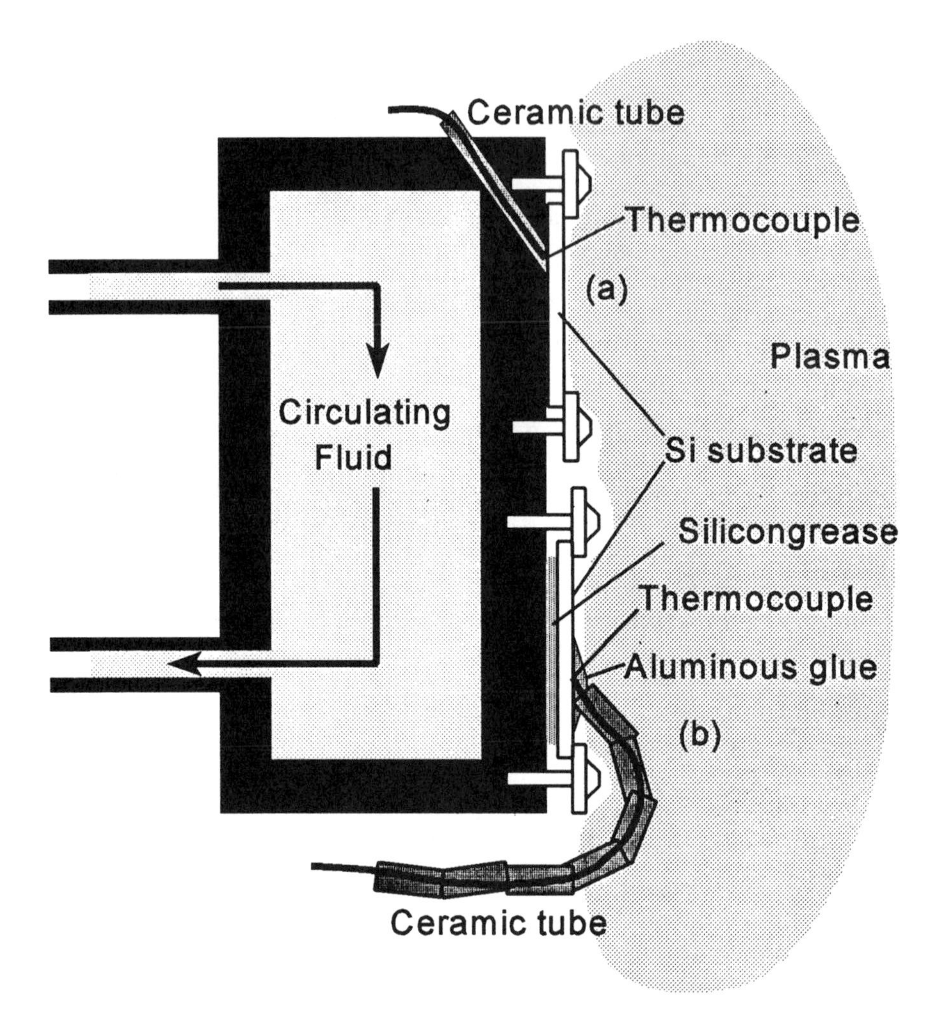

Figure 1: Substrate holder with forced cooling by circulating fluid, such as water with ethylene glycol (a) in the conventional way and (b) in the recommended way.

available literature on the topic reported temperatures of substrate holder which is lower than that of the substrate by several hundreds of degrees Celsius. The thermocouple should be attached to the surface of the substrate and should be covered with insulator to avoid detecting the plasma potential as shown in Figure 1.

4. Magnetoactive Microwave Plasma Chemical Vapor Deposition

By reducing the pressure for control of the substrate temperature, it becomes difficult to ignite or maintain a discharge due to loss of charged particles to the wall. A magnetic field is effective for low-pressure discharges because of electron trapping along magnetic lines of force. It is also advantageous to use electron cyclotron resonance (ECR) heating by microwaves and to use R-wave propagation from the higher magnetic field side [Chen, 1974], which is called an ECR discharge.

In the plasma, reaction processes are triggered by electron impact. The collision frequency of electron with neutral gas molecules governs the production rate of the reactive species. For example, the dissociative collision frequency, ν_d, is

$$\nu_d = n_e \cdot n_g < \sigma_d \cdot v_e > \tag{3}$$

where n_e and n_g denote the electron density and the gas density, respectively, σ_d is the dissociation cross section of the gas molecule, and v_e is the electron velocity.

In a conventional microwave plasma without a magnetic field, the electron density saturates at the cutoff density for the microwave frequency even if the incident power is increased. Lowering gas density by lowering pressure means a decrease in collision frequency, i.e., decrease in the production rate of reactive species and also in the ionization rate. By using a magnetic field, however, the electron density can exceed the cutoff density by additional power input through whistler mode propagation. Increase in electron density is expected to compensate the decrease in gas density. In addition, electrons will be heated by ECR resulting in a higher reaction probability $<\sigma_d \cdot v_e>$. ECR heating will contribute not only to v_e but also to σ_d which depends on the threshold energy for dissociation [Winters, 1974].

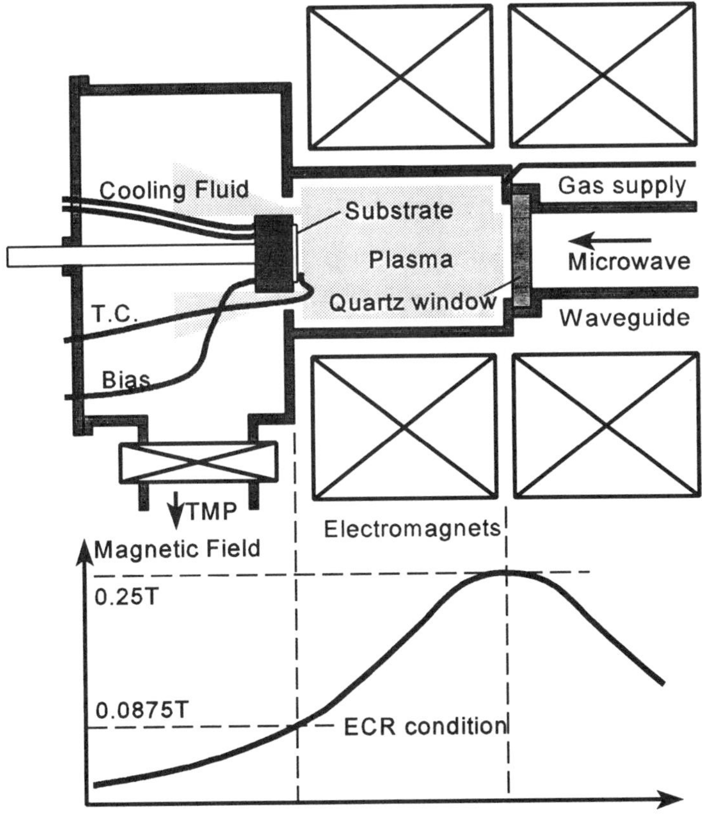

Figure 2: The magnetoactive microwave plasma chemical vapor deposition system. The substrate holder is positively biased (usually +30V) from the grounded chamber.

Table 1. Typical conditions for diamond growth by the magnetoactive microwave plasma CVD at low temperature.

Source Gas	CH_3OH 15 sccm
Dilution Gas	H_2 85 sccm
Total Pressure	10 Pa
Microwave Power	1.3 kW
Substrate	Si p+ (100) ultrasonic agitated
Substrate Temperature	200 - 600 °C
Substrate Bias	+30V

Figure 2 shows the magnetoactive microwave plasma chemical vapor deposition system. It is almost the same apparatus as ECR plasma chemical vapor deposition. For the operation condition (Table 1), higher gas pressure and higher magnetic field than a conventional ECR plasma are employed, such as l0Pa and 0.25T whereas the ECR plasma is produced at a pressure range below 0.lPa in the magnetic field about 0.1T. It requires a magnetic field several times higher than that used in the ECR condition to suppress reflection of microwave due to the higher pressures. Microwave energy is absorbed in the plasma mainly due to collisional damping in the higher magnetic field region. The operating pressure, l0Pa, is still lower by 2 orders than the conventional microwave plasma chemical vapor deposition process. Low pressure results in large deposition areas in a uniform plasma more than 6 inches wide.

In the magnetoactive microwave plasma at l0Pa, Γ_g is smaller than in the conventional microwave plasma by two orders of magnitude due to low gas pressure with the same η, T_g and M are assumed.

5. Diamond Growth at Low Temperature

Although the mechanism of diamond growth in the vapor phase has not yet been clarified, it is believed that selective etching of nondiamond phase carbon leaving partly diamond phase is essential [Kobashi, et al., 1988]. We should consider not only the growth of carbon film but also the etching of nondiamond phase carbon.

The success of diamond growth in the vapor phase by a hot filament or a microwave discharge is due to sufficient production of atomic hydrogen as an etchant. Etching by atomic hydrogen will match the growth rate of nondiamond phase carbon on the growing surface. When the substrate is cooled below 400 °C, however, atomic hydrogen can not perform sufficient removal of nondiamond phase carbon. In order to match the etching rate and growth rate of nondiamond phase carbon at lower temperature, the concentration of carbon source has been reduced to slow down the growth [Ihara et al. 1991], oxygen has been incorporated [Chang et al. 1988], or fluorine has been substituted for hydrogen to enhance the etching[Rudder R.A. et al. 1991]. Oxygen containing source gases, such as CO [Muranaka et al. 1991], CO_2 [Wei et al. 1990], and CH_3OH [Yuasa et al. 1992], is popular for diamond growth at low temperature because they can supply oxygen atoms in the plasma.

As for the growth of carbon films, it consists of production of chemically activated species, transportation to the substrates, sticking and migration on the surface. The production of reactive species is governed by plasma conditions and is almost independent of the surface. The sticking and migration on the surface is drastically changed by substrate temperature. In general discussion, the sticking coefficient will increase by decreasing the substrate temperature, and it will result in a decrease of its migration length on the surface. Without etching by sufficient hydrogen and/or oxygen, the growth rate of carbon film increases by decreasing the substrate temperature. Especially below 200 °C, polymer like films are formed very quickly. The etching of the nondiamond phase of carbon should be enhanced to match the fast growth of such a polymer film. If the nondiamond phase of carbon can be removed sufficiently by hydrogen and/or oxygen, the decrease in migration length results in a small probability for reactive species to find a proper site for diamond growth. The actual growth rate will decrease due to decrease in migration length of chemically active species on the surface. Figure 3 shows the growth rate of polycrystalline diamond films by the magnetoactive microwave plasma chemical vapor deposition process at low temperature on Si substrate seeded with nanocrystal diamond powder. A similar result in the hot filament CVD was also reported by Ihara et al. [1991].

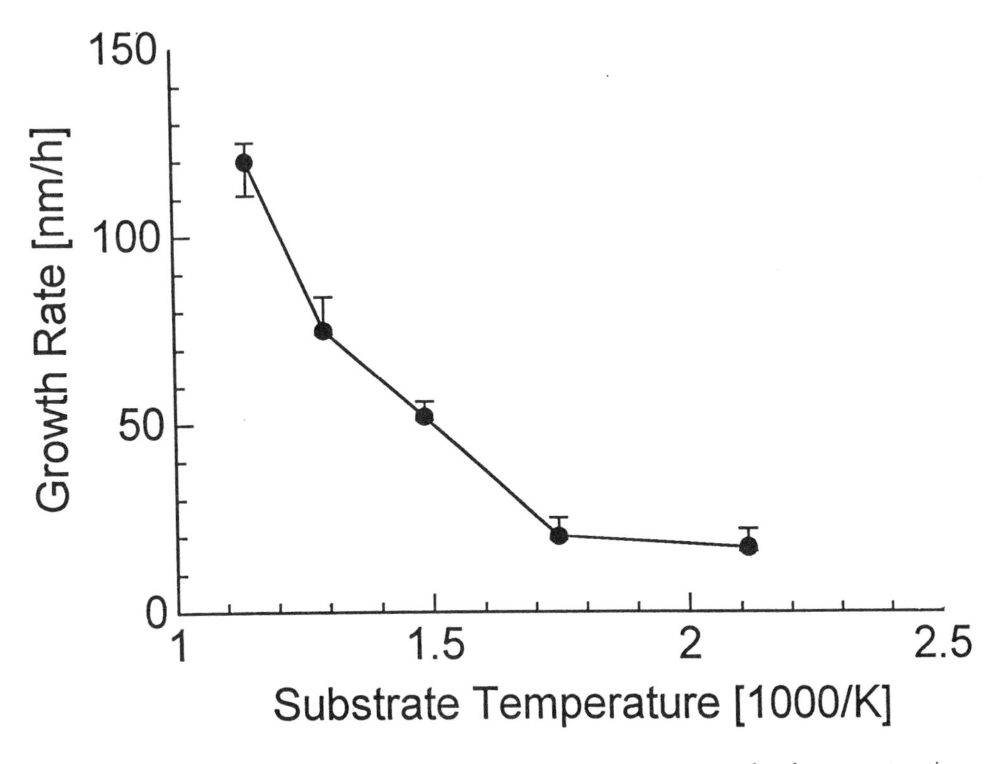

Figure 3: Dependence of the growth rate on the substrate temperature by the magnetoactive microwave plasma chemical vapor deposition at the typical condition.

At temperatures lower than 200 °C, the fabrication of diamond films becomes more difficult because of too fast of a growth rate of polymer films rather than decrease in the diamond growth rate. The authors have confirmed that diamond film can be formed also at 150 °C, which was measured on the surface, by enhancing the etching with increasing the oxygen concentration in the source gas.

6. Diamond Nucleation at Low Temperature

The proper condition for diamond nucleation is quite different from the condition for crystal growth. That is why it is necessary to enhance nucleation by seeding, biasing, scratching, or ultrasonic agitation before starting growth process. Figure 4 shows the dependence of nucleation time on growth temperature of ultrasonic agitated Si substrates at the proper condition for diamond growth in the magnetoactive microwave plasma chemical vapor deposition process. The nucleation time is the delay of growth start after stating deposition, which is evaluated by plotting the film thickness as a function of the deposition time. From this systematic study, it was revealed that the nucleation time on the conventionally pretreated Si substrate even at the proper growth condition became so long as to decrease the actual growth time when the substrate was below 400 °C [Yara et al. 1995].

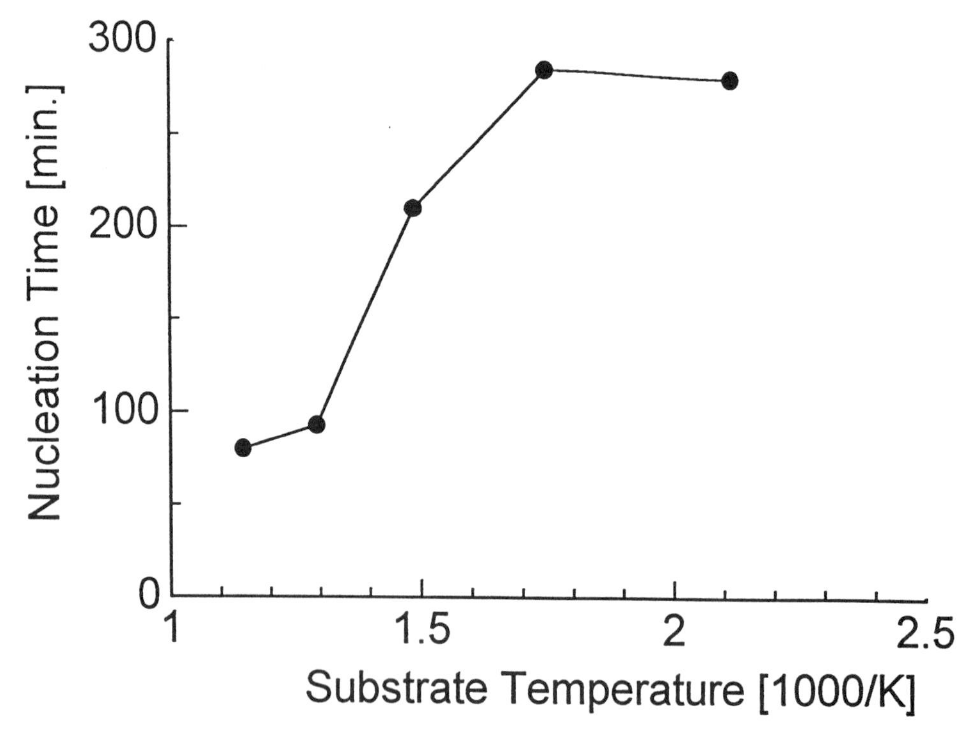

Figure 4: Dependence of the nucleation time on the substrate temperature on the silicon substrate pretreated with diamond powder by ultrasonic agitation in the conventional way. The nucleation time is the delay time from the start of growth process to the actual increasing in the film thickness.

Because of the decrease in nucleation density at low temperature, it is difficult to fabricate a continuous film. If a continuous film was fabricated by increasing the carbon concentration in the source gases, the SEM images of the resulting films show no facet and the X-ray diffraction peak for diamond of the film is broadened due to its small crystallite size. Therefore, nucleation by the conventional scratching, ultrasonic agitation, or biasing are not expected as a pretreatment for diamond film formation at low temperature.

Instead of the self-nucleation during deposition process, seeding with diamond powder on the substrate is expected as the reliable pretreatment for high quality diamond growth. Requirements for this seeding process are high seeding density so as to cover the whole surface of the substrate, and good adhesion of those diamond particle on the substrate. Finer powder will perform better for both the density and the adhesion. Nanocrystal diamond powder synthesized by an implosion process, the size of which is 3 to 5 nm, is commercially available. By using a colloidal solution of the nanocrystal diamond purified by acid, high density seeding of more than 10^{11} cm^{-2} was obtained on silicon substrates [Makita et al. 1996].

Figure 5: SEM images of the fabricated film in a same batch at 200 °C by the magnetoactive microwave plasma chemical vapor deposition on (a) Si substrate pretreated by ultrasonic agitation with diamond powder in the conventional way and (b) Si substrate seeded with nanocrystal diamond powder.

The nanocrystal seeding has been used for low temperature fabrication of diamond films [Yara et al. 1995]. With the pretreatment of the high density seeding on silicon substrate, diamond growth begins with a small delay after the deposition starts. It improved the actual growth rate by shortening the loss time for nucleation, which was more than 5 hours at 200 °C. In addition, as shown in Figure 5, the nanocrystal seeding results in well faceted polycrystalline diamond films even at 200 °C. In contrast, only the ball like diamond could be formed on the ultrasonic agitated silicon substrate after deposition in the same batch. The nanocrystal seeding improved not only the growth rate but also the crystal quality of the fabricated films.

7. Film Properties

The quality of the diamond films fabricated at low temperature is still inferior to conventionally fabricated ones, even though they show almost the same morphology in SEM observations. Figure 6 (a) shows the typical Raman spectra of a diamond film fabricated at 200 °C with an optimal condition by the magnetoactive microwave plasma chemical vapor deposition process, and Figure 6 (b) shows one of a film fabricated at 930 °C by the conventional microwave plasma CVD. The spectrum for the conventionally fabricated film has only a sharp Raman peak for diamond at 1330 cm^{-1} with the FWHM of 6 cm^{-1}. The diamond peak for the low temperature fabricated film is broadened to 21 cm^{-1} and shifted to the lower wavenumber side. In addition, the spectrum for the low temperature fabricated film includes a swell centered at 1600 cm^{-1} due to inclusion of non diamond phase carbon in the film. The film properties of the low temperature fabricated diamond films should be improved for actual applications.

8. Diamond Coating on Polymer

A diamond coated polymer sheet is expected to show enhanced performance in industrial applications. It is necessary to fabricate diamond films on polymers without losing the characteristics of the substrates. Even if they succeeded to fabricate diamond films on metal or Si at lower temperature than the heat-resistant temperature of a polymer material, those methods would not be applicable to the polymer due to the following reasons;

1. Because thermal conductivity of polymer is much smaller than metal or Si, the surface of the substrate would be heated even if it is cooled from the backside.

2. Etching of the non-diamond phase carbon by hydrogen and/or oxygen will also etched away the substrate itself, otherwise, some polymer film will grow rapidly at the low temperature.

The difficulty of diamond coating on polymer is not only the substrate temperature but also the etching of the substrate during growth.

For the cooling down of the substrate, it is effective to lower the gas pressure as mentioned before, and to use thinner film for the substrate. It is quite difficult to settle the latter problem of etching. At the lower temperature than 200 °C, growth of non-diamond phase of carbon is very fast due to the higher sticking coefficient and polymerization of hydrocarbons. For the diamond growth, carbon concentration should be decreased and/or oxygen should be incorporated. The enhancement of etching of nondiamond phase carbon will result in etching away of the polymer substrate itself. The substrate should be protected from such etching by using other precoatings on the substrate during deposition, resulting in formation of an interlayer between the substrate and diamond. Figure 7 shows diamond films fabricated by the magnetoactive microwave plasma chemical vapor deposition process on polyimide films with a-Si film as the protection layer.

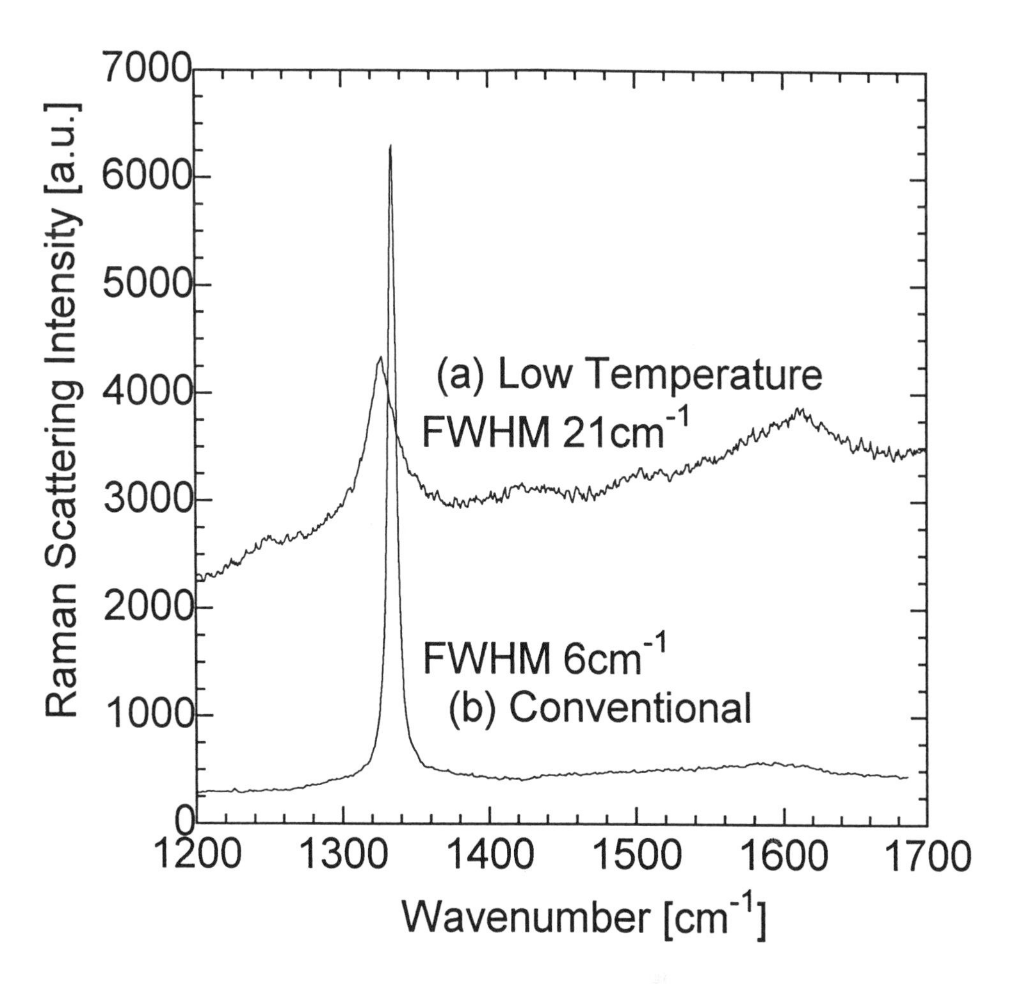

Figure 6: Raman spectra of the films fabricated (a) at 200 °C by the magnetoactive microwave plasma chemical vapor deposition at the proper condition, and (b) at 930 °C by the conventional microwave plasma CVD with gas mixture of CO/H_2 (10%) on Si substrate pretreated by ultrasonic agitation.

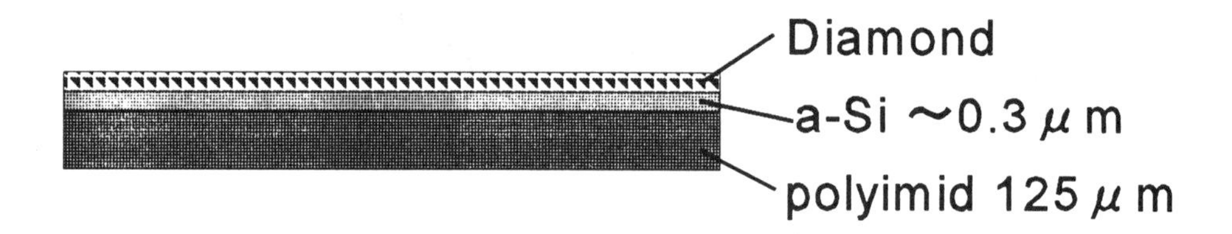

Figure 7: Polyimid films coated with diamond on the a-Si interlayer.

9. Future Prospects

The remaining challenge for low temperature growth of diamond is improvement in growth rate and film quality. The diamond growth rate at low temperature is still too slow to use for actual application. One of the possibilities to improve the growth process is pulse modulation of discharge [Hatta et al. 1995], or gas feeding.

The growth of diamond in the magnetoactive microwave plasma increases with increasing microwave power. However, it is difficult to keep the substrate at low temperature, such as 200 °C, at the high power such as 5 kW. Higher microwave power brings higher growth rate possibly because of higher production of the species necessary for diamond growth. Here, the total power should not be increased to avoid heating substrate, but selective application of the power for production of those key species for diamond growth is important. Selective and effective production of such key species at the same or lower microwave power than what is currently used may be required to improve the growth rate at low temperature. Pulse modulation of microwave power is quite interesting because of lower time averaged power and higher controllability of the reaction process. The time averaged power can be reduced by decreasing the duty ratio resulting in a lower and more easily controlled substrate temperature. In addition, the density of each species changes drastically due to its production or loss with individual time constants during each pulse discharge. Repetition of such transitional reaction stages will make it possible to control radical densities in the plasma. The control of the chemical vapor deposition process will also improve the film quality.

10. References

Chang C.P., Flamm D.L., Ibbotson D.E. and Mucha J.A.: *J. Appl. Phys.*, **63** (1988) 1744.

Chen F.F.: *Introduction to Plasma Physics*, (Plenum Press, NY, 1974) chap.4.

Hatta A., Kadota K., Mori Y., Ito T. , Odaka S., Sasaki T. and Hiraki A.: *Appl. Phys. Lett.*, **66** (1995)1602.

Ihara M., Macno H., Miyamoto K. and Komiyama H.: *Appl. Phys. Lett.*, **59** (1991) 1473.

Makita H., Nishimura K., Jiang N., Hatta A., Ito T., Hiraki A.: Thin Solid Films **281-282** (1996) 279.

Muranaka Y., Yamashita H. and Miyadera H.: *Diamond and Related Materials*, **3** (1994) 313.

MuranakaY., Yamashita H. and Miyashita H.: *J. Appl. Phys.*, **69** (1991) 8145.

Nakac S., Noda M., Kusakabe H., Shimizu H. and Maruno S.: *Jpn. J. Appl. Phys.*, **29** (1990) 1511.

Kobashi K., K. Nishimura, Y. Kawate and Y. Horiuchi: *Phys. Rev.*, **B38** (1988) 4067.

Rudder R.A., Hudson G.C., Posthill J.B., Thomas R.E. and Markunas R.J.: *Appl. Phys. Lett.*, **59** (1991) 791.

Wei J., Kawarada H., Suzuki J. and Hiraki A.: *J. Cryst. Growth*, **99** (1990)1201.

Winters H.F.: *J. Chem. Phys.*, **63** (1975) 3462.

Yara T. , Makita H., Hatta A., Ito T. and Hiraki A.: *Jpn. J. Appl. Phys.*, **34** (1995) L312.

Yara T., Yuasa M., Shimizu M., Makita H., Hatta A., Suzuki J., Ito T., and Hiraki A.: *Jpn. J. Appl. Phys.*, **33** (1994) 4404.

Yuasa M., Arakaki O., Ma J. S., Kawarada H. and Hiraki A.: *Diamond and Related Materials* **1** (1992) 168.

Chapter 25
Nucleation and Epitaxy

S.P. Bozeman
Department of Materials Science and Engineering,
North Carolina State University, Raleigh, NC, 27695-7919, USA

B.R. Stoner and J.T. Glass
Kobe Steel USA Inc., PO Box 13608,
Research Triangle Park, NC 27709, USA

Contents

1. Introduction

This chapter discusses the nucleation of diamond on nondiamond surfaces. An understanding and control of the nucleation process is important to the production of diamond films in several ways. Because of its high surface energy, diamond tends to nucleate and grow three dimensionally in a Volmer-Weber growth mode [Bachman 1995; Field 1992; Kern, LeLay et al. 1979]. This tendency leads to the production of continuous films by growing out three dimensional particles until they coalesce. If this approach is taken, a high nucleation density is necessary to produce a smooth film and thus enable the growth of thin continuous films. In light of this need for a high nucleation density, the chapter begins with a review of methods for initiating and increasing the nucleation density of diamond on a nondiamond substrate.

The need for understanding nucleation also arises in attempting the growth of single crystal heteroepitaxial films. Deposition of single crystal films is important if diamond is to be used in active electronic devices, because such films eliminate the barriers to conduction created by grain boundaries. While single crystal films can be grown on natural or synthetic diamond substrates, such substrates are small and expensive, making them undesirable for use in diamond based electronics. Thus the heteroepitaxial nucleation of diamond is an important step in the economical realization of the potential of diamond as an electronic material. Section 3 of this chapter deals with techniques leading to local epitaxy of diamond nuclei. The chapter next turns to the subject of selective nucleation, and then closes with proposed models of the diamond nucleation process.

This chapter considers only the nucleation of diamond on nondiamond surfaces. While much important research has been conducted in the area of homoepitaxial growth, the problems associated with it are distinct from the issues involved in heteroepitaxial growth. In addition, as noted above, the expense and small size of natural diamond substrates makes both heterogeneous and heteroepitaxial nucleation important for the economical realization of diamond products.

2. Nucleation Enhancement Techniques

In general, the nucleation of diamond on nondiamond substrates is difficult unless some technique is used to promote nucleation. This is believed to be caused by a combination of competition from nondiamond phases, the low sticking coefficient of the nucleation precursor, and the high surface energy of diamond. A general discussion of nucleation phenomena is given in the references [Kern, LeLay et al. 1979; Robins 1988; Venables and Price 1975]. The enhancement of diamond nucleation has been achieved technologically by a number of different pretreatments. This section surveys the different pretreatments while a detailed discussion of the nucleation mechanisms is deferred until the end of the chapter. The enhancement techniques which we will cover include abrasion with either diamond or non diamond particles, seeding or coating the substrate surface with nondiamond carbon, and the application of an electric field to the substrate (biasing).

2.1 ABRASIVE PRETREATMENTS

Scratching is probably the simplest and most commonly used of the nucleation enhancement techniques. The diamond nucleation density improves by several orders of magnitude if the substrate is scratched with diamond paste prior to placing it in the growth chamber. This improvement is regarded in the literature as common knowledge since the early part of the 1980s. One early report associated with this technique is the observation that nuclei tend to form on defects such as scratches [Spitsyn, Bouilov et al. 1981]. Diamond particles are typically used for the scratching, but other abrasive particles have also been utilized, including cubic boron nitride [Suzuki, Sawabe et al. 1987], SiC [Ravi, Koch et al. 1990; Sawabe and Inuzuka 1986] and stainless steel [Chang, Flamm et al. 1988]. Scratching with nondiamond particles improves the nucleation density, but is not as effective as scratching with diamond particles [Rebello, Straub et al. 1991]. Increases in diamond nucleation on silicon are typically from 10^5 cm^{-2} without scratching to as high as 10^{10} cm^{-2} with scratching. The density of nucleation on scratched substrates is inversely proportional to the size of the abrasive particles [Ascarelli and Fontana 1992]. Numerical estimates of nucleation densities are usually obtained by counting particles in an SEM micrograph after a fixed growth interval and are only approximate. Still, this improvement of several orders of magnitude is readily apparent.

Abrasive pretreatments have been applied to nearly every substrate used for diamond growth, including Si, Ta, Ni, Cu, and others. The methods of scratching vary in complexity and reproducibility, and include applying a diamond paste via a swab, both wet and dry abrasion using a polishing wheel, and ultrasonic agitation in a solution of the abrasive particles. Scratching pretreatments are attractive in that they are inexpensive, simple, and effective. Scratching can, however, contaminate or damage the substrate, making it unsuitable for optical and electronic applications. Furthermore, it is not easily utilized for complex shapes or three dimensional surfaces. It is generally believed that scratching promotes diamond nucleation through a combination of seeding and the creation of high energy defects which serve as nucleation sites.

2.2 SEEDING

Diamond nucleation can also be enhanced by nonabrasive pretreatments. The next of these pretreatments in order of complexity involves seeding or coating the substrate with a carbonaceous material. Materials which have been used for this purpose include carbon fibers, hydrocarbon oils, carbon clusters, and a-C:H films. This technique is more suitable than abrasion for optical coatings, but is still undesirable for electronic applications or heteroepitaxy because of the amorphous or polycrystalline layer which it usually produces at the interface. This method can be divided onto two broad categories: placing carbon compounds on the substrate before placing it in the growth chamber (*ex situ* pretreatments) or performing an initial deposition step under conditions which favor deposition of nondiamond carbon rather than diamond particles (*in situ* pretreatments). These categories are not exclusive; some researchers have combined *in situ* and *ex situ* pretreatments in an attempt to capture the advantages of both.

Ex situ pretreatments have taken many forms. Sprinkling graphite powders onto unscratched substrates has been shown to increase the diamond nucleation density [Angus, Li et al. 1991]. Diamond nucleation was noted to be higher on carbon fibers placed on silicon than on the surrounding surface [Pehrsson, Glesener et al. 1992]. Hydrocarbon oil and even fingerprints have also been demonstrated as effective coatings for nucleation enhancement [Morrish and Pehrsson 1991]. It should also be noted that one of the promising techniques for heteroepitaxy includes seeding with nondiamond carbon [Yang, Zhu et al. 1994].

In situ pretreatments using different types of nondiamond carbon films have been widely implemented using an initial deposition step at lower temperatures and higher carbon concentrations, i.e., conditions which tend to favor deposition of nondiamond carbon. This technique was reported by Ravi and co-workers [Ravi, Koch et al. 1990] in both combustion and DC growth and is commonly used in combustion growth [Tzeng, Cutshaw et al. 1990; Windheim and Glass 1992]. Mitsuda and co-workers used a high methane pretreatment to predeposit a carbon film which increased the diamond nucleation density by two orders of magnitude [Mitsuda, Moriyasu et al. 1993]. Both a-C:H and turbostratic carbon films have also been used as a nucleation enhancing coating [Dubray, Pantano et al. 1991, Oral, Ece et al. 1993].

Unfortunately, the carbonaceous films used for nucleation enhancement have generally not been well characterized. Usually the pretreatments are evaluated for their relative effectiveness at increasing diamond nucleation, but the properties of the carbonaceous films are not examined. In addition to this lack of characterization, other researchers have been unsuccessful in attempts to increase diamond nucleation using high carbon gas phase pretreatments [Hartnett, Miller et al. 1990; Menningen, Childs et al. 1994]. Furthermore, even when the pretreatment film is examined, it is difficult to ascertain the precise structure of the nucleating species. If one estimates an initial nucleus size of 100 atoms, a nucleation density of 10^{10} cm^{-2} can be obtained with only 0.01% coverage. Thus, the total surface occupied by the nucleating species can be quite low while still enhancing the nucleation density by several orders of magnitude. Because of these questions, the optimum coating for diamond nucleation has been difficult to determine.

A comparative study of several of the different pretreatments has shed some light on this issue [Feng, Brewer et al. 1994]. Researchers studied nucleation on silicon substrates coated with different carbonaceous materials including evaporated carbon, clusters of C_{60} and C_{70}, hard carbon produced by a vacuum arc, and diamond particles. These *ex situ* pretreatments were combined with a high methane *in situ* pretreatment. The highest diamond nucleation was produced by diamond seeding and the second highest was produced by coating with hard carbon films. The evaporated carbon, arc-deposited carbon, and diamond had sp^3 contents of 1, 85, and 100% and resulted in nucleation densities of 10^4, 10^8, and 10^9 cm^{-2} respectively. Thus the most effective pretreatment which did not involve diamond seeding consisted of coating with a hard carbon film via a vacuum arc and then exposing to a high methane pretreatment before diamond growth. Overall, this study indicated that the diamond nucleation density increased with the thickness and etching resistance of the carbon pretreatment. While these results are not conclusive, they do suggest that the ineffectiveness of some carbonaceous coatings can be interpreted as a sign that the predeposited films did not contain enough sp^3 bonding.

In summary, seeding and coating have included many forms of both diamond and nondiamond carbon and are often accompanied by a pretreatment in the growth reactor under conditions favorable for nondiamond carbon. The most successful of these pretreatments can increase nucleation densities by several orders of magnitude, although direct comparisons are difficult to make as reported results are usually semiquantitative at best. The implications of these results for models of nucleation mechanisms are discussed later in this chapter (§5).

2.3 BIASING

Another method for nucleation enhancement involves biasing the substrate negatively with respect to ground. This technique has been termed bias enhanced nucleation (BEN). Such an increase in nucleation upon the application of a negative bias was first reported by Yugo [Yugo, Kimura et al. 1990]. Since that time, BEN has been extensively studied and refined to allow heteroepitaxial deposition on SiC and the production of highly oriented films on Si [Jiang, Klages et al. 1993; Stoner, Ma et al. 1993; Stoner, Sahaida et al. 1993]. BEN has several advantages; it is less damaging to the surface than abrasion, it does not require the deposition of amorphous layers which would preclude the possibility of epitaxy, it is easily controllable through the bias voltage and time, and it can be performed *in situ*. BEN provides an increase in diamond nucleation on unscratched silicon from about 10^5 cm^{-2} to 10^{10} cm^{-2} and can provide a higher nucleation density than scratching (Figure 1). It is also promising for the coating of highly curved surfaces for which scratching would be impractical. For example, Liu and coworkers have applied BEN to the nucleation of diamond on silicon field emitter tips [Liu, Zhirnov et al. 1994]. Its primary disadvantage is that its effectiveness is dependent upon the specific substrate material employed.

During BEN, a negative potential of about 100-250 V is applied and methane concentrations are raised above normal growth concentrations to 4-10% in hydrogen [Jiang, Klages et al. 1993; Stoner, Ma et al. 1992; Yugo, Kimura et al. 1990]. The bias can be applied with respect to the grounded chamber walls or to a second electrode inside the chamber (Figure 2). Laser reflectance interferometry (LRI) has been used as an *in situ* growth monitor in conjunction with BEN [Stoner, Williams et al. 1991]. LRI allows one to determine the onset of nucleation and to measure film thicknesses during

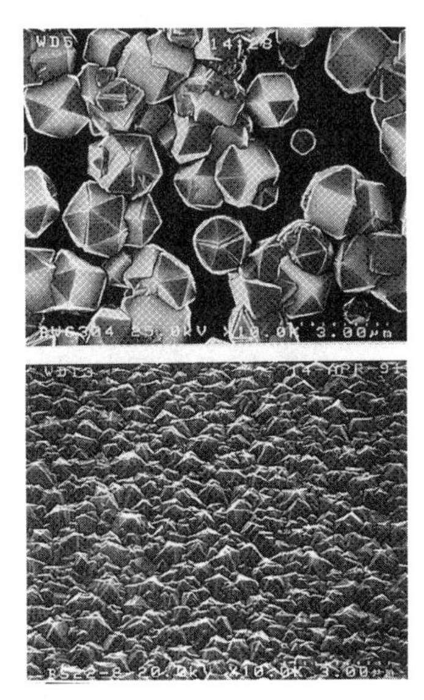

Figure 1. SEM micrographs of samples grown with scratching and biasing pretreatments: (top) silicon wafer scratched with 0.25 μm powder and (bottom) pristine silicon wafer biased in 2% CH_4 in H_2 plasma at -250V for 1 hour.

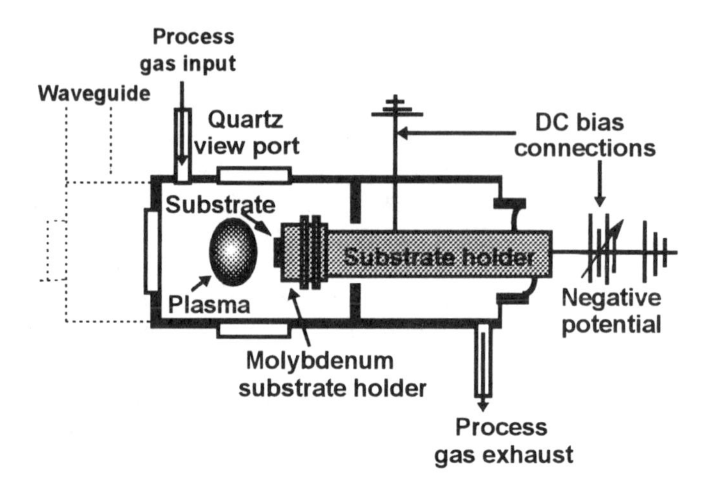

Figure 2. Schematic of microwave plasma CVD system showing one possible substrate biasing arrangement.

growth by measuring the reflected intensity of a laser from the film surface. Such an *in situ* growth monitor aids in the application of BEN by making it possible to better determine and control the onset of nucleation. Conditions effective for nucleation are often detrimental to growth, leading to a need for careful control of the transition process. BEN has been applied primarily to silicon and provides greater nucleation enhancement on Si than on any other substrate material reported to date. Other substrates which have been studied using BEN include Cu, SiC, several of the refractory metals, and their carbides [Wolter, Glass et al. 1995; Wolter, Stoner et al. 1994] . These studies clearly indicate a substrate effect, the details of which are discussed in §5.2.

BEN has been applied most commonly with a DC bias in microwave CVD reactors, but other variations have has also been used. BEN has been successfully applied to hot filament CVD systems [Zhu, Sivazlian et al. 1995]. In this case, researchers achieved BEN by negatively biasing the substrate and positively biasing the filament with respect to the reactor walls, which were grounded. Diamond nucleation enhancement occurred only at the edges of the Si wafer unless steps were taken to make the electric field more uniform. Two successful steps were raising the Si wafer above the substrate holder on a graphite pedestal and placing a wire mesh above the substrate. Other researchers have replaced the DC bias with a radio frequency bias to improve nucleation densities on insulating substrates [Joseph, Wei et al. 1993]. The radio frequency voltage induced a DC self bias of -100V and resulted in increased nucleation densities of up to 10^{10} cm^{-2} on insulating substrates such as quartz and sapphire.

Overall, BEN is one of the most promising techniques for nucleation enhancement. While it is not as simple or inexpensive as scratching, it is clean, controllable and does little damage to the interface. These attributes have led to its use as one of the primary methods for heteroepitaxial nucleation. The use of BEN as applied to heteroepitaxial nucleation will be discussed in detail in §3.4 and possible mechanisms of BEN will be discussed in §5.4.

Table I. Summary of pretreatment techniques used for nucleation enhancement. References are intended to be representative rather than exhaustive. Further details can be found in the text.

Pretreatment method	Nucleation Density (cm^{-2})	References
None	10^3-10^5	[Popovici and Prelas 1992]
Abrasion	10^6-10^{10}	[Iijima, Aikawa et al. 1990]
		[Ascarelli and Fontana 1992]
Seeding	10^4-10^{10}	[Feng, Brewer et al. 1994]
		[Pehrsson, Glesener et al. 1992]
		[Ravi, Koch et al. 1990]
Biasing	10^8-10^{11}	[Yugo, Kanai et al. 1991]
		[Stoner, Ma et al. 1992]

2.4 SUMMARY

The pretreatment techniques discussed in this section are summarized in Table 1.

3. Heteroepitaxial Nucleation

To realize the potential of diamond for electronics applications, it is desirable to deposit large single crystal films. However, diamond substrates of sufficient size for most devices are not economically available at this time. In addition, the concentrations of defects and impurities in synthetic and natural diamond substrates are generally too high for them to be suitable for electronic devices. Thus the development of heteroepitaxial diamond films on economically available substrates appears essential for the potential of diamond as an electronic material to be realized.

Although the heteroepitaxial growth of diamond on nondiamond substrates has generated much interest, only limited progress has been made in this area. The principal reason for this limited progress is the difficulty of finding a substrate with a good match to the small atomic spacing and high surface energy of diamond. Because of diamond's extreme properties in these areas, there are no materials which provide well matched substrates. Table II lists several materials which have been investigated as potential heteroepitaxial substrates, along with their lattice constant, mismatch with diamond, coefficient of thermal expansion, and surface energy. A notable exception to this lack of a suitable substrate is cubic boron nitride (c-BN). However, at present, large c-BN substrates are at least as difficult to fabricate as diamond substrates. The largest c-BN substrates reported are 3 mm size particles produced by high pressure methods [Mishima 1991]. Even this relatively small size is currently quite rare, although research is ongoing in the growth of c-BN films.

The term heteroepitaxy will be strictly used in this chapter to mean a registry between the substrate atoms and the deposited atoms. It is not intended to imply that a complete

Table II. Properties of materials used as substrates for heteroepitaxial diamond growth.

Material (space group)	Lattice Constant (Å)	Mismatch (%)	Thermal Expansion (10^{-6}/K)	Surface Energy (J/m^2) / plane
Diamond (Fd3m)	3.56 (a) (1.54) (m)	•	0.8 (b)	5.3 (111) (b) 9.2 (100) (b)
c-BN ($F\bar{4}3m$)	3.62 (a)	1.4	(o)	4.8 (111) (k)
Graphite ($P6_3mmC$)				
a-axis	2.46 (c)	4 (p)	negative (c)	2.80 ($10\bar{1}0$) (f)
c-axis	6.71 (c)		25 (c)	0.17 (0001) (f)
BeO ($P6_3mmC$)				
a-axis	1.65 (m)	7 (m)	7.1 ⊥ c (d)	4.8 (110) (j)
c-axis			6.3 ∥ c (d)	
β-SiC ($F\bar{4}3m$)	4.35 (a)	18	4.63 (d)	(o)
Si (Fd3m)	5.42 (a)	34	2.6 (i)	1.46 (111) (b)
Ni (Fm3m)	3.52 (a)	1.1	13.3 (i)	2.36 (g)
Co (Fm3m)	3.54 (a)	0.6	12.5 (i)	2.71 (g)
TiN (Fm3m)	4.24 (d)	18	9.35 (d)	(o)
TiC (Fm3m)	4.32 (d)	21	7.85 (d)	(o)
Ni_3Si (Pm3m)	3.50 (e)	1.8	9.00 (e)	1.74 (h)

Mismatch is defined as the difference in substrate and overgrowth lattice constants divided by the lattice constant of the substrate. Thermal expansion is for a temperature of 300 K, except for β-SiC, TiN, and TiC (300-800 K) and BeO (300-450 K). Surface energies are for 300 K. Notes: (a) [Weast 1987] (b) [Field 1992] (c) [Pierson 1993] (d) [Shackelford and Alexander 1992] (e) [Tucker, Sivazlian et al. 1995] (f) [Kern 1985] (g) [Mezey and Giber 1982] (h) [Keene 1987] (i) [Brandes and Brook 1992] (j) [Lambrecht and Segall 1992] (k) [Zhu, Wang et al. 1993], (m) based on Be-O nearest-neighbor distance, (o) data not available (p) based on 3:2 matching of diamond (111) to graphite (0001).

heteroepitaxial film has been deposited. To underscore this distinction, some authors have referred to oriented diamond films as epitaxially textured rather than heteroepitaxial [Wild, Kohl et al. 1994]. In this chapter, diamond particles which are oriented with respect to the underlying substrate are referred to as heteroepitaxial while it is recognized that complete heteroepitaxial films of diamond have not been deposited.

The results of investigations on the substrates described in Table II are summarized in Table III. The remainder of this section discusses in greater detail studies of heteroepitaxial diamond growth on c-BN, graphite, BeO, Si, and Ni. Three substrates (TiC, Ni_3Si, and TiN) are listed in the tables but not discussed further in the text. This is because some oriented diamond nucleation has been obtained on these substrates, but less extensive work has been reported than on the other materials which we discuss.

Table III. Summary of heteroepitaxial results. References are intended to be representative rather than exhaustive. Further discussion can be found in the text.

Substrate	Results	References
c-BN	2-D growth (Frank-van der Merwe)	[Koizumi, Murakami et al. 1990] [Maeda, Masuda et al. 1994] [Argoitia, Angus et al. 1994]
Graphite	Oriented particles on edges of HOPG sheets Diamond(111) || Graphite(0001)	[Li, Wang et al. 1993]
BeO	Oriented particles on basal plane Diamond(111) || BeO (0001)	[Argoitia, Angus et al. 1993] [Lambrecht and Segall 1992]
β-SiC	Highly oriented continuous films Diamond(100) || SiC(100) and Diamond<110> || SiC <110>	[Stoner and Glass 1992] [Zhu, Wang et al. 1993]
Si	Highly oriented continuous films Diamond(100) || Si(100) and Diamond<110> || Si<110>, twist and tilt $< 7°$	[Jiang, Klages et al. 1993] [Wolter, Stoner et al. 1993]
Ni	Oriented particles on Ni(100) and Ni(111)	[Sato, Yashima et al. 1991] [Zhu, Yang et al. 1993]
Co	Oriented particles on Co(100)	[Liu, Tucker et al. 1995]
TiN	Oriented particles on TiN(100)/MgO(100)/Si	[Chalker, Johnston et al. 1994]
TiC	Oriented particles on TiC(111)	[Wolter, McClure et al. 1995]
Ni_3Si	Oriented particles on (111) textured Ni_3Si	[Tucker, Sivazlian et al. 1995]

3.1 CUBIC BORON NITRIDE

The most promising of the materials in Table II as regards lattice match and surface energy is c-BN. These promising properties have been borne out by several independent research groups. Epitaxial diamond growth on c-BN particles has been achieved by DC plasma CVD [Koizumi, Murakami et al. 1990], microwave plasma CVD [Maeda, Masuda et al. 1994], and hot filament CVD [Wang, Pirouz et al. 1993]. Unfortunately,

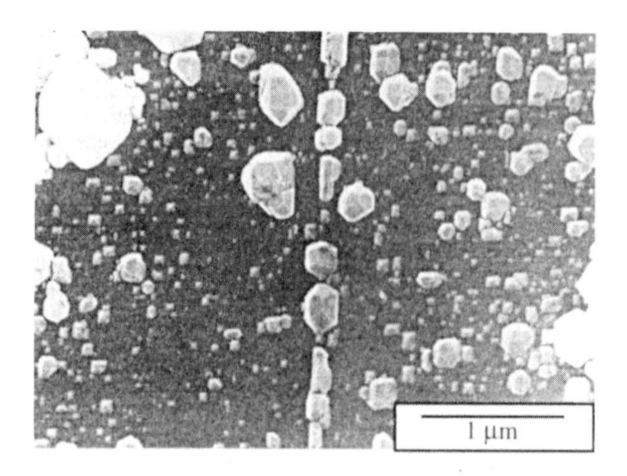

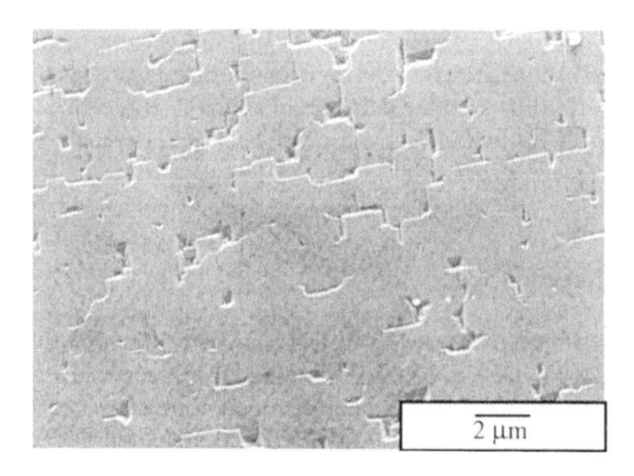

Figure 3 . SEM micrograph of epitaxial diamond nuclei (top) and film (bottom) on c-BN (100). From [Inuzuka, Koizumi et al. 1992].

as noted above, while the small lattice constant and high bond energy make c-BN an ideal substrate for diamond heteroepitaxy, large single crystal c-BN substrates are not available. At present, c-BN substrates larger than 1 mm in diameter are more difficult to obtain than diamond substrates.

Diamond thin films were deposited on 200 μm size c-BN particles and the epitaxy was verified using RHEED [Koizumi, Murakami et al. 1990]. The deposition was performed in a DC plasma CVD system at 0.5 to 2% CH_4 in H_2 with substrate temperatures of 800-1050°C and a pressure of 180 Torr. Diamond nucleated as islands several hundred angstroms in size at a density of 10^{11} cm^{-2}. The islands grew along surface steps until they coalesced to produce a smooth continuous film at a thickness of about 2 μm (Figure 3). Further studies showed that nucleation occurred more readily on the boron terminated (111) face, while nucleation occurred only at scratches or cracks on the nitrogen terminated plane [Inuzuka, Koizumi et al. 1992]. Epitaxy on the nitrogen plane could not be obtained under the conditions which gave epitaxy on the boron plane.

Growth on both the (111) and (100) planes was examined using Raman spectroscopy. No nondiamond carbon peaks were observed in the Raman spectrum. For a thin film (0.1μm) grown on a (111) substrate, a tensile stress of 2.2×10^{11} dynes/cm^2 was derived from the shift of the Raman peak of 8.2 cm^{-1}. This value agrees with the stress that would be expected from the lattice mismatch. A much lower tensile stress of 4.3×10^9 dynes/cm^2 was derived from the shift of the Raman peak of a thicker film (10 μm) grown on the (100) substrate, presumably indicating relaxation in the film as it grew thicker. Polarization sensitive Raman spectroscopy was also used to verify the epitaxial relationship on both c-BN (111) [Yoshikawa, Ishida et al. 1990] and (100) [Yoshikawa, Ishida et al. 1991]. Polarization sensitive Raman spectroscopy measures the orientational relationship between the film and substrate by observing the relative intensity of the Raman peak as the sample is rotated with respect to the polarization of the incident beam. For this Raman mode, the angular dependence is $\sin^2(2\theta)$ for the (100) plane and constant for the (111) plane. These dependencies were observed for diamond growth on both the (100) and (111) planes of c-BN. The lack of variation with polarization of the (111) planes would be satisfied by a polycrystalline film as well; however, a single crystal electron diffraction pattern was observed.

Diamond was also grown on 50 μm c-BN particles using microwave CVD and epitaxy on the (100) and (111) planes was confirmed [Maeda, Masuda et al. 1994]. The microwave results indicate that epitaxy is obtained on the boron-terminated (111) plane but not the nitrogen-terminated plane. Orientation on the (100) surface was observed in a relatively narrow parametric range of pressure greater than 80 Torr, temperature greater than 950°C, and CH_4 content of 2-3% in H_2. Epitaxy on the (111) plane occurred over a wider range of pressure as low as 20 Torr, temperature of 800-1000°C, and CH_4 content of 1-3%. These conditions are similar to those reported for homoepitaxy on the (111) plane [Kamo, Yurimoto et al. 1988].

Another study where diamond has been grown heteroepitaxially on c-BN used a hot filament CVD system and 500 μm sized particles [Argoitia, Angus et al. 1994; Wang, Pirouz et al. 1993]. Cross sectional selected area diffraction and HRTEM were used to confirm the epitaxial relationship between the diamond and c-BN. Furthermore, TEM showed an atomically sharp, clean interface with no contamination from the filament. The lattice mismatch of 1.4% would be expected to result in a misfit dislocations spaced at 15 nm; however, while misfit dislocations were observed, the spacing was irregular (Figure 4). A dislocation with a projected Burgers vector of $a/4[11\bar{2}]$ along $[1\bar{1}0]$ was also observed via TEM. This projection is consistent with an actual Burgers vector of $a/2[01\bar{1}]$ inclined at 60° with respect to the $[1\bar{1}0]$, which is a lattice dislocation in diamond.

3.2 GRAPHITE

An orientational relationship has been observed between diamond particles and the edges of graphite sheets [Angus, Li et al. 1993a; Li, Wang et al. 1993]. Diamond deposited on graphite powder and highly oriented pyrolitic graphite (HOPG) showed a strong preference for nucleation on the edges of the graphite sheets. SEM, TEM and electron diffraction images indicate that a large percentage of the diamond particles nucleate on the edges of graphite sheets in a relationship where the diamond (111) plane is parallel to the graphite (0001) plane and diamond $[1\bar{1}0]$ is parallel to graphite $[11\bar{2}0]$ (Figure 5). Of 40 particles examined on the edges of sheets of HOPG, 10 had the diamond (111) plane parallel to the graphite (0001) plane to within ±3°, a much higher percentage than would be given by a random orientation. This observed orientational relationship supports a nucleation model proposed by Angus and co-workers where the diamond growth precursor is produced by hydrogenation of the

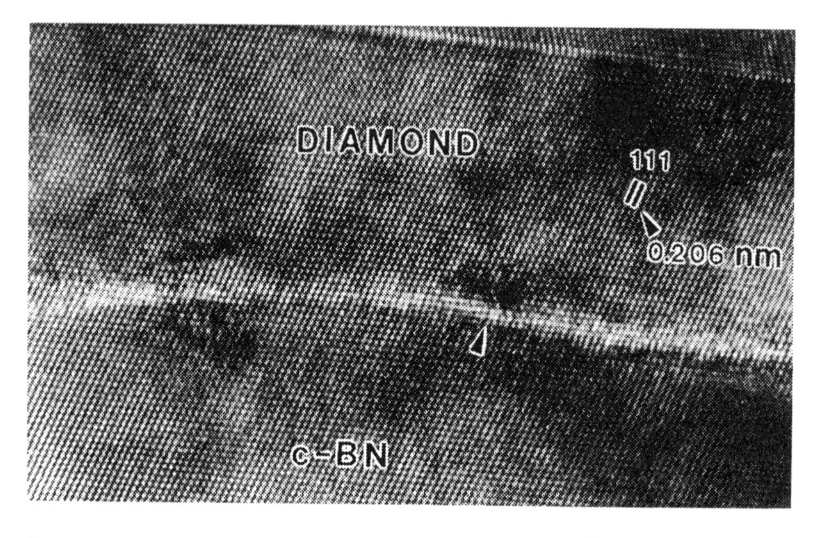

Figure 4. HRTEM of the diamond/c-BN interface. The arrow indicates a misfit dislocation at the interface. From [Argoitia, Angus et al. 1994].

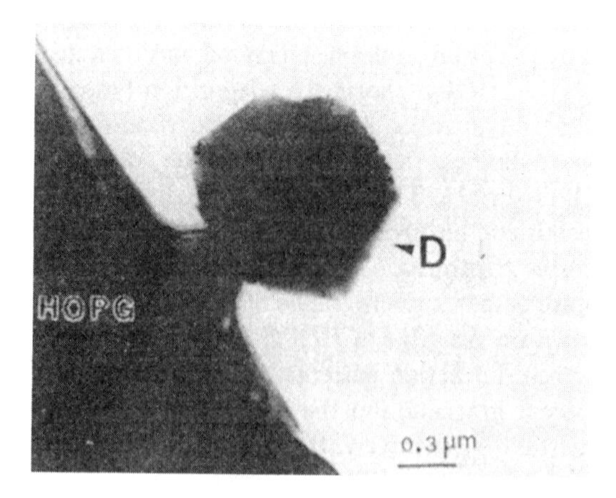

Figure 5. Transmission electron micrograph of diamond particle nucleated on the edge of highly oriented pyrolitic graphite (HOPG). View is normal to graphite (0001) basal plane. Hexagonal outline indicates view is normal to diamond (111) plane. From [Li, Wang et al. 1993].

graphite edges to form tetrahedrally coordinated sp^3 bonded edges [Angus, Hoffman et al. 1988]. In this model the (111) planes of diamond match with the (0001) planes of graphite in a 3:2 relationship. The interfacial energies were calculated to be 1.6 J/m^2 which is significantly lower than the corresponding free surface energies ($\approx$6 J/m^2)[Lambrecht, Lee et al. 1993]. Further discussion of this model is given in §5.3.

3.3 BERYLLIUM OXIDE

Beryllium oxide has been proposed as a heteroepitaxial substrate for diamond growth because of the favorable bond length and adhesion energy which it provides. The Be-O bond length (1.65Å) is only 7% greater than the C-C bond length (1.54Å) in diamond, and the BeO/diamond adhesion energy has been estimated at 4.6 J/m^2 which is nearly as high as the c-BN/diamond adhesion energy of 5.4 J/m^2 [Lambrecht and Segall 1989; Lambrecht and Segall 1992]. Diamond has been grown on both the (0001) and $(11\bar{2}0)$ surfaces of single crystal BeO by hot filament CVD. SEM, TEM and selected area electron diffraction indicate that the diamond particles grew epitaxially on the basal plane. The diamond (111) planes are nearly parallel to the (0001) planes of BeO and the diamond $\langle 1\bar{1}1 \rangle$ direction is rotated by about 6° with respect to the BeO $\langle 11\bar{2}0 \rangle$. No epitaxy was observed on the $(11\bar{2}0)$ plane, but particles of Be_2C were observed in this case which may have inhibited the diamond heteroepitaxy. An unreconstructed step with a height of three layers on the (0001) BeO surface presents the same geometrical configuration as the boat-boat bicyclodecane that the authors had previously proposed as a precursor species for the nucleation of diamond [Angus, Hoffman et al. 1988].

3.4 SILICON AND SILICON CARBIDE

The most technologically developed process for producing oriented diamond films involves bias-enhanced nucleation and textured growth on silicon [Jiang, Klages et al. 1993; Stoner, Sahaida et al. 1993]. Analytical studies of BEN on Si indicated that diamond nucleated on a SiC interfacial layer [Stoner, Ma et al. 1992]. This observation of a SiC interfacial layer led to successful attempts to grow oriented diamond particles on SiC single crystal substrates [Stoner and Glass 1992]. These growth experiments on (100) β-SiC substrates produced deposits where approximately 50% of the nuclei were oriented within $\approx$8° with respect to the SiC substrate (Figure 6). TEM performed on these samples confirmed the epitaxial relationship of diamond (100) parallel to SiC (100) and diamond $\langle 110 \rangle$ parallel to SiC $\langle 110 \rangle$ within 8° [Stoner, Ma et al. 1993; Zhu, Wang et al. 1993]. The micrographs show one dislocation every seven lattice planes, a dislocation density consistent with the observed tilt of the particles [Stoner, Ma et al. 1993] (Figure 7). The inclined nature of these dislocations was believed to cause the observed tilting and rotation of the particles in addition to accommodating the strain. Because they were inclined to the interface, their Burgers vector was made up of three components: misfit, tilt and screw (Figure 8).

This success with SiC was followed by further work on Si substrates resulting in epitaxial growth on silicon via a combination of BEN with a higher temperature *in situ* carburization treatment [Wolter, Stoner et al. 1993]. Local epitaxy of diamond on silicon was also obtained using bias enhanced nucleation without a carburization step [Jiang

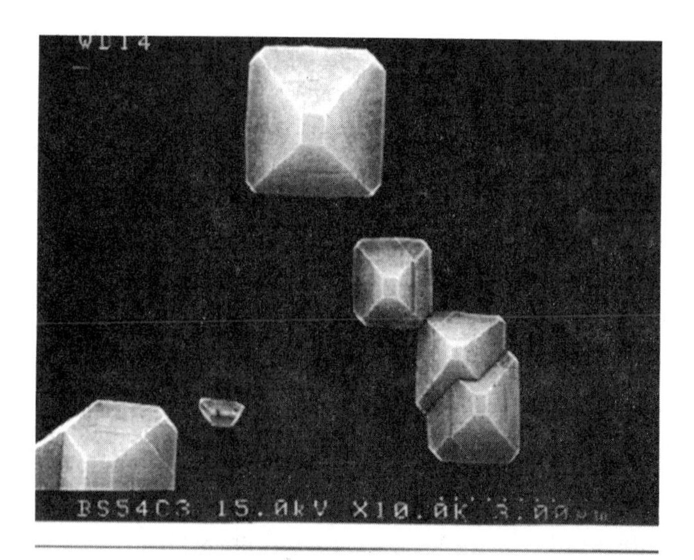

Figure 6. SEM micrographs of oriented diamond particles on SiC near the center (top) and edge (bottom) of the sample.

and Klages 1993; Jiang, Klages et al. 1993]. While SiC was not formed intentionally with a carburization step, it may have been formed during the bias-enhanced nucleation process as observed via *in vacuo* studies of BEN [Kulisch, Sobisch et al. 1995; Stoner, Ma et al. 1992]. The observation of diamond epitaxy on SiC and of SiC interlayers in the growth of polycrystalline films has led to suggestions that diamond epitaxy on Si occurs via a SiC interlayer. However, TEM investigations have found diamond/silicon interfaces both with and without the SiC interlayer so the interlayer exists in some cases but may not be essential [Tucker, Sivazlian et al. 1995], [Tucker, Sivazlian et al. 1995]. Still, the exact location of nucleation in such samples is unknown and a final conclusion cannot be made.

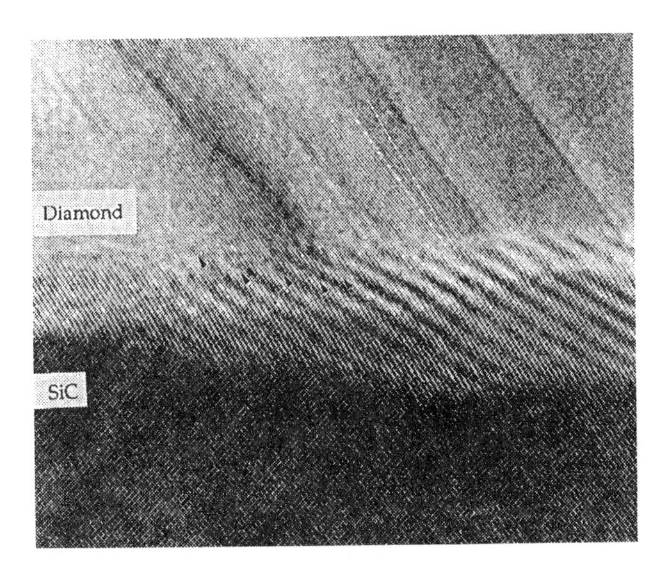

Figure 7. HRTEM micrograph of the diamond/β-SiC interface. Arrows indicate the position of misfit dislocations. From [Zhu, Wang et al. 1993].

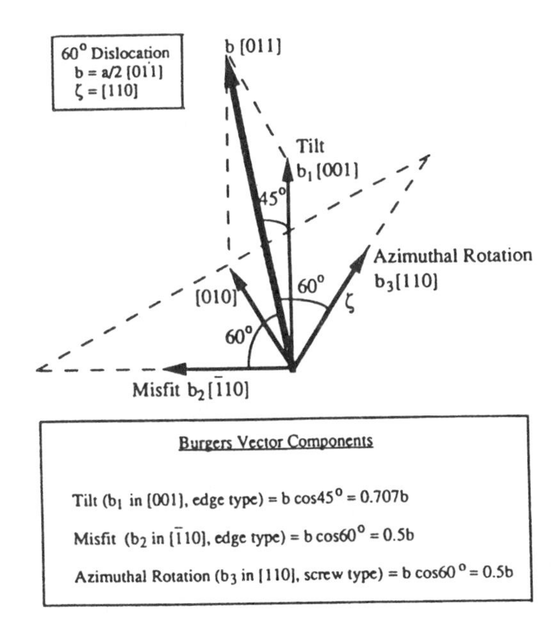

Figure 8. The decomposition of a 60° dislocation in diamond into tilt (b_1), misfit (b_2), and azimuthal rotation (b_3) components. From [Zhu, Wang et al. 1993].

The studies described above which reported oriented films of diamond utilized an evolutionary growth process first described by van der Drift [van der Drift 1967]. By growing under conditions which promote (100) texturing, the misoriented grains can be

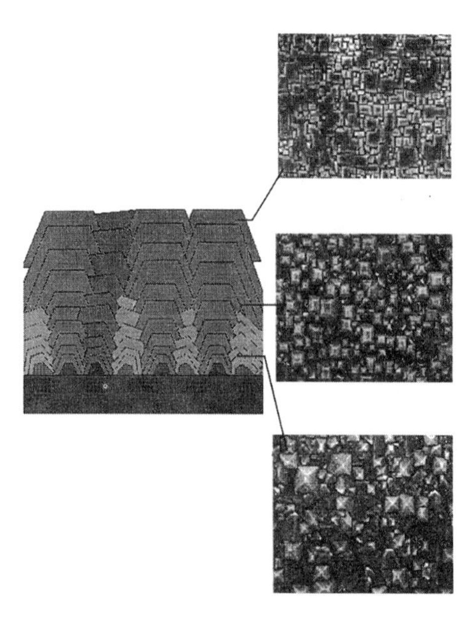

Figure 9. Evolutionary growth for the production of highly oriented diamond films. By growing under conditions which promote (100) texturing, the misoriented grains can be covered by the oriented particles. The three micrographs show the improvement in morphology with thickness.

covered by the oriented particles (Figure 9) [Stoner, Sahaida et al. 1993]. This same approach has been used by Clausing [Clausing, Heatherly et al. 1991] and Wild [Wild, Herres et al. 1990] to obtain fiber textured films from randomly oriented nuclei. Both polarization sensitive Raman spectroscopy and X-ray texture diffractometry have been used to quantify the orientational relationship between the diamond and silicon [Stoner, Kao et al. 1993]. The angular spread of the orientation was found to range from 8 to 13° in both the polar and azimuthal directions. The range of orientation angles depended on the growth conditions used to grow out the oriented nuclei and on the thickness of the film (Figure 10).

Calculations have been made of the molecular geometry and bonding of the C(100)/Si(100) and C(111)/Si(111) interfaces. Semi-empirical quantum chemical calculations have been used to model the C(100)/Si(100) interface [Verwoerd 1992; Verwoerd 1994a; Verwoerd 1994b; Verwoerd and Weimer 1991]. Interfacial structures were examined which would reduce the strain arising from the 34% mismatch between diamond and silicon. One structure that was examined involved a 45° rotation between that two materials which would reduce the mismatch to 7% [Verwoerd and Weimer 1991]. Another possibility that has been considered is a parallel 3:2 matching of diamond to Si unit cells. Such a matching would reduce the mismatch at the interface to 1.5% [Verwoerd 1994a]. These molecular geometry calculations indicated that the interface energies were lower for the 45° rotation, while the adhesive energy was more favorable for the 3:2 matching. On the other hand, independent band structure calculations indicate that the 3:2 match with like planes and directions parallel is more energetically favorable than the 45° rotation [Tucker, Sivazlian et al. 1995]. The parallel arrangement is in agreement with the polarization sensitive Raman, X-ray pole figure, and TEM measurements discussed above.

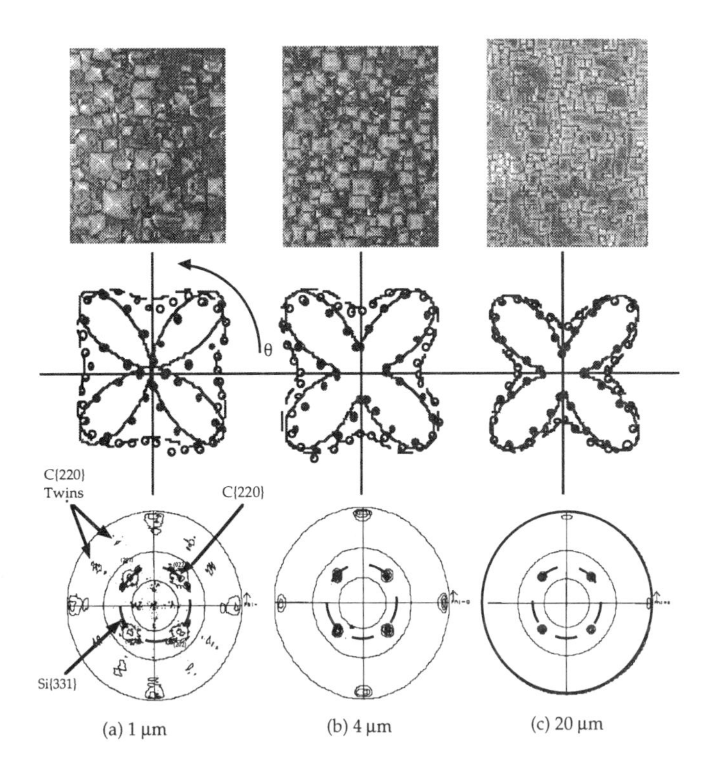

Figure 10. SEM, polar Raman and polar XRD spectra taken from (a) 1 μm, (b) 4 μm and (c) 20 μm thick highly oriented diamond films. The open circles on the polar Raman plot are from the diamond film, the filled circles are from the Si substrate. The {022} spots of (c) are asymmetric with a FWHM in the radial and transversal directions of 10 and 8 degrees, respectively. This is indicative of texturing decreasing tilting, but not azimuthal misorientation and is consistent with the van der Drift method.

The highly oriented diamond films resulting from the combination of bias-enhanced nucleation with van der Drift growth texturing exhibit improved electronic properties over polycrystalline films [Stoner, Kao et al. 1993; Stoner, Malta et al. 1994]. Typical room temperature Hall mobility data are shown in Figure 11 for boron doped films deposited on (i) polycrystalline, (ii) highly oriented, and (iii) single crystal diamond substrates. These data represent typical films doped in-situ via the gas phase (to dopant concentrations of $N_A = 10^{17}$-10^{19} cm^{-3}) as opposed to ion-implantation and HTHP growth. This figure shows clearly three separate ranges of values with little overlap. There are polycrystalline films with room temperature mobilities ranging from 2-50 cm^2/V-s at room temperature, highly oriented films falling between 135 and 278 cm^2/V-s, and single crystal films with values ranging from 250 to 1340 cm^2/V-s. These values are for films with similar room temperature carrier concentrations: $5x10^{13} < p < 5x10^{14}$ cm^{-3}. Much of the variation within each film class can be attributed to (i) variations in deposition quality, (ii) differences in boron concentration or substrate morphology, and (iii) condition prior to deposition of the p-type layer. However, it is clear in general that single crystal films have superior transport properties to highly oriented films which again have better properties than polycrystalline films deposited under similar growth conditions. These data indicate that although the highly oriented films have mobilities 5

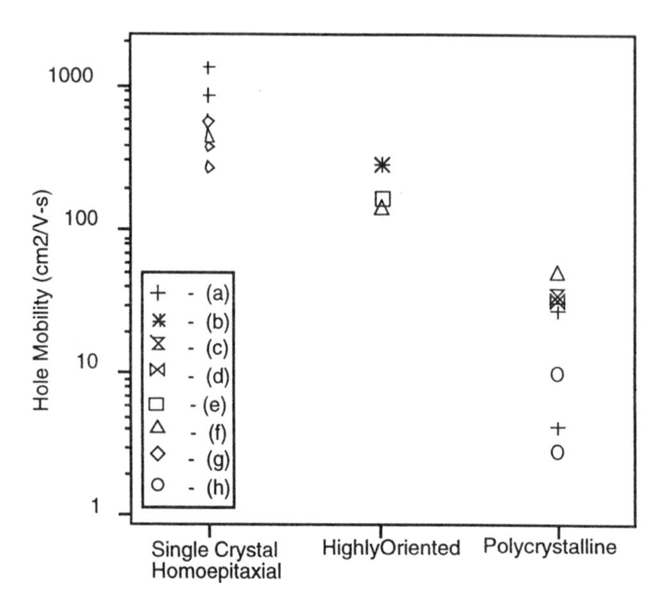

Figure 11. Hole mobility data for homoepitaxial films, highly oriented films, and polycrystalline films. Notes: (a) [Plano 1994], (b) [Stoner, Malta et al. 1994], (c) [Malta, Windheim et al. 1993], (d) [Masood, Aslam et al. 1992], (e) [Stoner, Kao et al. 1993], (f) [Fox, Stoner et al. 1994], (g) [Fujimori, Nakahata et al. 1990], (h) [Windheim, Venkatesan et al. 1993].

times higher than polycrystalline films, they are still only 1/5 that of single crystal homoepitaxial layers. These differences emphasize the importance of further reducing the film misorientation and associated defects at grain boundaries.

The results of thermal conductivity measurements on HOD films have been similar to the electrical results in that the thermal conductivity of HOD films is superior to polycrystalline films, but not as high as single crystal diamond [Käding, Rösler et al. 1994].

3.5 TRANSITION METAL SUBSTRATES

Nickel is an interesting heteroepitaxial substrate because of its good lattice match with diamond (the mismatch is less than 1.5%) and because it is a catalyst for diamond crystallization in high pressure high temperature (HPHT) diamond synthesis [Strong 1964]. Nickel differs from many of the substrates used for diamond growth in that it does not readily form a carbide and has a high carbon solubility and diffusivity. *In vacuo* analytical studies of diamond growth on Ni in a HFCVD system have indicated the formation of an intermediate graphite layer prior to diamond growth [Badzian and Badzian 1992; Belton and Schmieg 1989; Belton and Schmieg 1990]. *In vacuo* XPS, EELS, AES, and LEED were used to examine carbon deposition on Ni (100) at selected intervals in the growth process. The results of this characterization indicated that the initial deposition was of a graphitic layer which was not oriented with respect to the Ni surface. Although diamond nucleates on this graphite layer, the orientation of the substrate is not preserved and epitaxy is prevented.

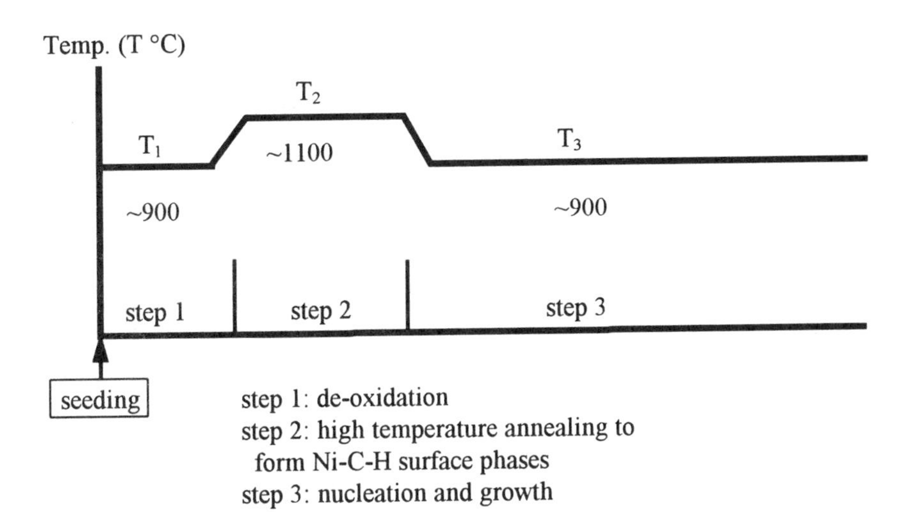

Figure 12. Schematic of multistep process used to obtain oriented diamond particles on Ni and Co. From [Yang, Kistenmacher et al. 1995].

In contrast, the possibility of diamond deposition directly on Ni has been demonstrated by Sato and coworkers who reported the production of oriented nuclei on both (100) and (111) Ni surfaces via microwave CVD [Sato, Fujita et al. 1993; Sato, Yashima et al. 1991]. In this work, Sato studied diamond growth on Ni as a function of methane percentage and substrate temperature in a microwave plasma. Local epitaxy was observed for diamond particles grown with methane concentration in hydrogen less than 0.9%, while disordered graphite formed for methane concentrations grater than 0.9%. Few details of the process were given, but it appears that graphite suppression is related to maintaining a relatively low methane concentration and high temperature.

Another process for epitaxial nucleation of diamond on nickel is multi-step hot filament CVD. This multi-step process involves seeding the nickel surface with carbon, annealing in atomic hydrogen at two different temperatures, and subsequent growth (Figure 12) [Yang, Zhu et al. 1993; Yang, Zhu et al. 1994; Zhu, Yang et al. 1993]. The suppression of graphite formation by this annealing process is similar to the observed prevention of graphite formation by nickel hydride [Badzian and Badzian 1992]. It is believed that carbon seeds dissolve to form a molten Ni-C-H phase during the annealing and that when the substrate is subsequently cooled the carbon precipitates onto the Ni substrate, possibly through Ni_4C [Yang, Kistenmacher et al. 1995]. It has been postulated in the literature that Ni-C intermediate states in which the carbon is tetrahedrally bonded precede diamond crystallization under HPHT conditions [Shelton, Patil et al. 1974] and in an atmospheric recrystallization process [Strong 1964]. Accordingly, the annealing step must be long enough to form the intermediate phase

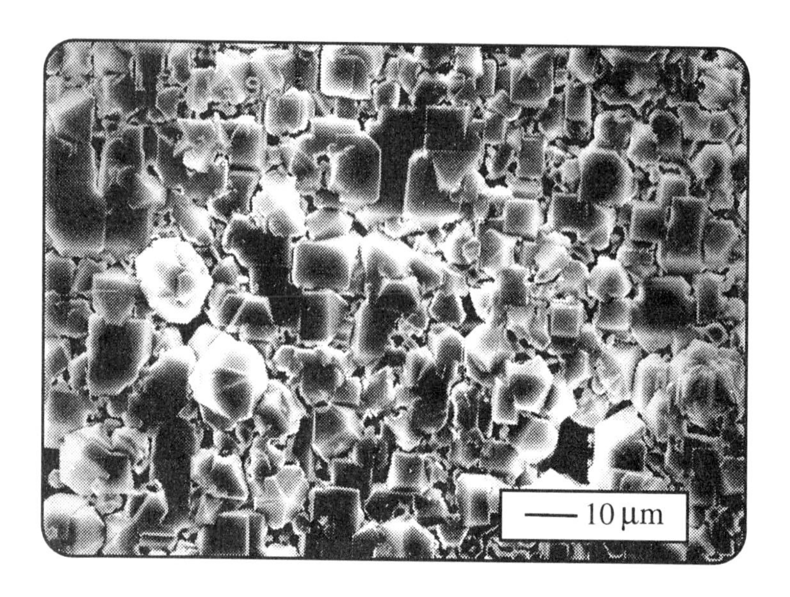

Figure 13. Scanning electron microscopy (SEM) micrographs of a diamond film grown on a (100) oriented single crystal Ni surface with (top) an overview at low magnification and (bottom) a high magnification image. From [Zhu, Yang et al. 1993].

(typically 5 to 120 minutes) but short enough to prevent the seed carbon from dissolving too far into the substrate. Effective sources of seed carbon have included diamond, graphite, and C_{60} as well as a high methane concentration in the gas phase which saturates the nickel surface. This seeding and annealing process has resulted in oriented nuclei on the Ni(111) , Ni(100) , and Co(111) surfaces (Figure 13). The technique has produced 5x5 mm^2 regions in which about 85% of the nuclei are oriented, but continuous films have not been produced. The difficulty in obtaining a complete film is probably caused by the high solubility of carbon in nickel. Diamond particles are believed to continuously dissolve into the substrate even as growth is taking place. One can, however, see grains in the SEM micrographs which have grown together with no discernible grain boundaries (Figure 13). This elimination of the grain boundaries suggests that single crystal films may be obtained by this method in the future.

4. Selected Area Nucleation

As noted previously, diamond nucleation is sparse unless one of the treatments described in §2 is used. This low nucleation density leads naturally to methods for selective nucleation. That is, selective area nucleation of diamond films is typically achieved by selective application of one of the techniques for nucleation enhancement. Since it is not possible to selectively polish areas mechanically on the micron scale, these techniques have involved scratching entire wafers and then either masking portions of the scratched wafer or etching the wafer to remove the beneficial effects of scratching.

Several researchers have achieved selective deposition by scratching the entire wafer and then masking some areas to prevent diamond growth. One procedure was to scratch an entire Si wafer, mask and pattern with Si_3N_4, oxidize the bare Si, and remove the Si_3N_4 mask [Davidson, Ellis et al. 1989]. This procedure left a surface which consisted of patterned scratched Si in some areas with the remaining areas covered by SiO_2. Diamond was then grown on the exposed Si to achieve selected area deposition with a resolution of about 2 μm. Another approach utilized amorphous silicon masking to promote growth on selected portions of scratched Si wafers [Inoue, Tachibana et al. 1990; Kobashi, Inoue et al. 1990]. In the amorphous silicon approach, scratched Si wafers were selectively masked with a-Si:H and diamond nucleated only in the areas which were scratched but not masked (Figure 14). The abrasion can also be performed after masking the surface. In this approach, a pristine Si wafer can be covered with photoresist, patterned, and then ultrasonically abraded with a diamond suspension [Hänni, Müller et al. 1993]. This procedure has the advantage that the scratching pretreatment is applied after the initial patterning steps so that only the Si areas to be diamond covered are damaged by abrasion.

An alternative to masking is the use of plasmas or ion beams to remove nucleation sites. In the plasma case, a scratched Si wafer was patterned with a photoresist and etched using a reactive ion etching system (Figure 15) [Inoue, Tachibana et al. 1990; Kobashi, Inoue et al. 1990]. Diamond grew only in areas which were scratched but not etched, yielding line widths of a few microns. Presumably the etching removed the nucleation sites which were produced by the scratching pretreatment. In the ion beam case, researchers scratched a wafer with diamond grit, covered selected areas of the wafer with photoresist, and irradiated the patterned wafer with an Ar+ ion beam [Hirabayashi, Taniguchi et al. 1988]. Diamond nucleated in the areas which were scratched but not Ar+ irradiated. The ion beam appears to remove the nucleation sites which are created

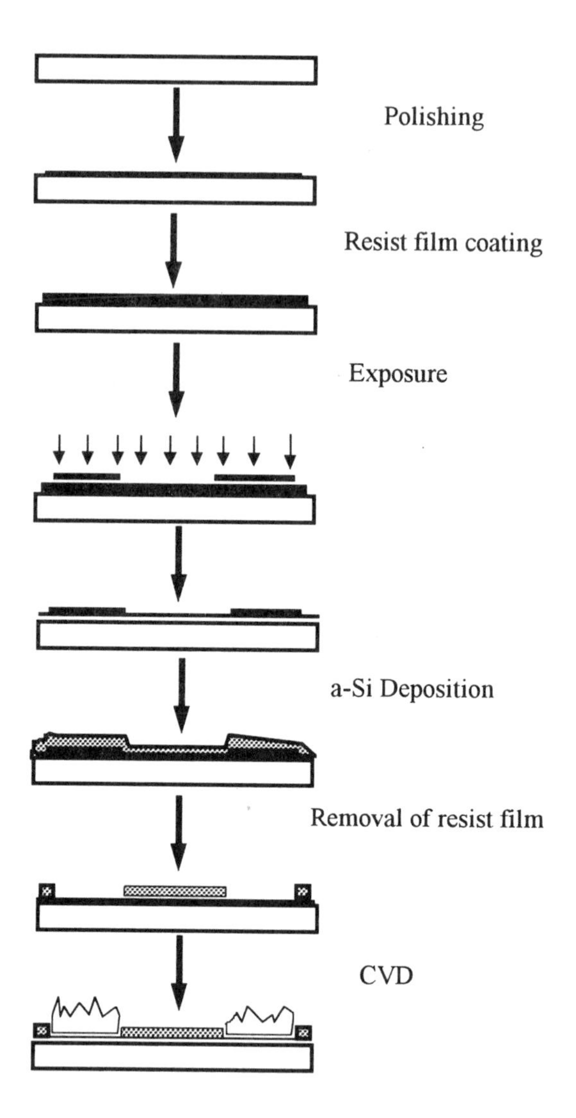

Figure 14. Process flow for selected area deposition of polycrystalline diamond on silicon using amorphous silicon masking. From [Kobashi, Inoue et al. 1990].

by the scratching process, an effect which has been seen in other studies of nucleation pretreatments [Williams 1991].

Instead of protecting portions of the wafer with a photoresist, the nucleation sites can be protected by shadowing. A 10x10 μm array of diamond particles has been produced on silicon by scratching a wafer covered with micron sized dots of Si and then irradiating the wafer with an Argon ion beam at an oblique angle [Kawarada, Ma et al. 1990]. The argon beam destroyed most of the nucleation sites, leaving only those which were shadowed by the Si dots (Figure 16). Diamond grew on the remaining sites to produce a regular array. In subsequent work, the argon ion beam was replaced by a H_2 electron

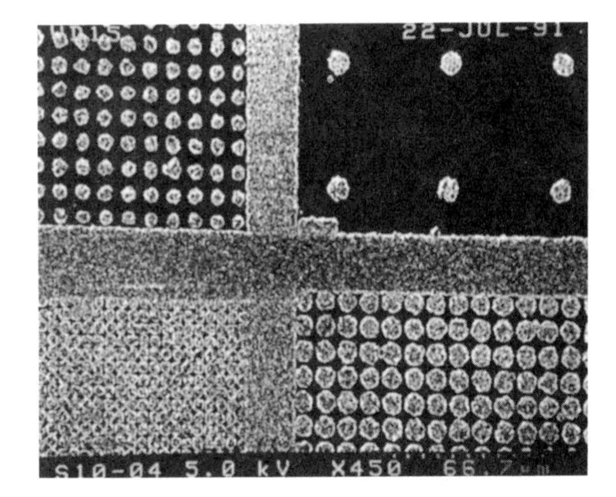

Figure 15. Selected area deposition of diamond on silicon fabricates by SiO_2 masking method. From [Windheim and Glass 1992].

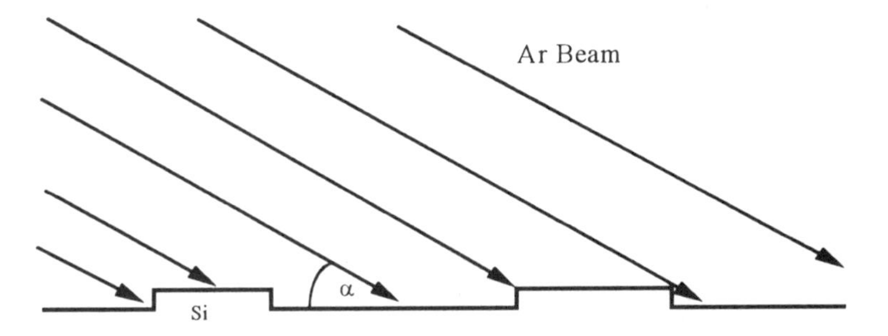

Figure 16. Schematic of oblique irradiation of Si wafer with argon ion beam. From [Kawarada, Ma et al. 1990].

cyclotron resonance (ECR) plasma [Ma, Yagyu et al. 1991]. Again, a substrate with raised Si dots was scratched and most of the nucleation sites were etched away to produce a regular array of diamond crystals. Etching via the H_2 plasma eliminated problems with sputtering by the ion beam and allowed both the etching and subsequent diamond deposition to be performed *in situ*.

The techniques described thus far have combined diamond abrasion of silicon with either etching or masking to produce patterned diamond deposits. Recently, selective deposition has been produced in conjunction with bias enhanced nucleation. Investigations of the application of BEN in a hot filament CVD system demonstrated that placing a wire mesh above the sample surface during biasing results in selective nucleation (Figure 17) [Zhu, Sivazlian et al. 1995]. This observation suggests further tailoring of the electric field configuration during biasing for selective area nucleation. Selected area nucleation has also been combined with heteroepitaxial nucleation via

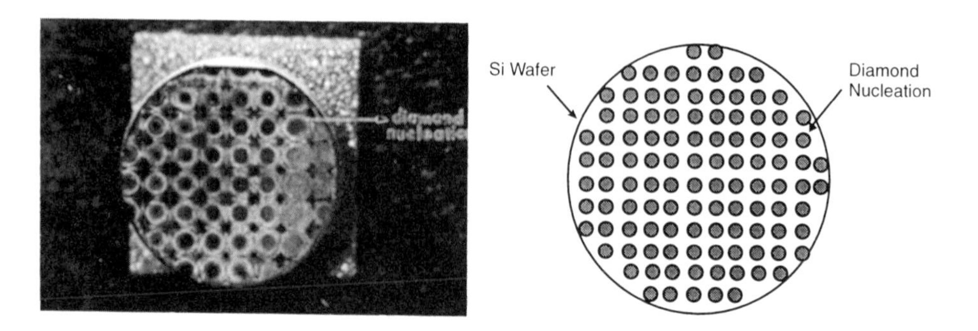

Figure 17. Selected area deposition using a wire screen during BEN. (Left) optical photo and (right) schematic pattern showing the circular selective diamond nucleation pattern. From [Zhu, Sivazlian et al. 1995].

BEN [Jiang, Boettger et al. 1994]. This combination was achieved by using SiO_2 to mask portions of an unscratched (100) Si wafer and applying BEN to produce oriented crystallites. This method produced high contrast but the smallest channels reported were 8 μm in size, larger than channels produced in earlier work involving scratching. The ultimate limit to the resolution of this method is likely to be defined by the density of oriented particles produced by the biasing process. If the misoriented particles are to be overgrown by evolutionary growth, then it is desirable to have several oriented particles within the width of a channel.

In summary, selective nucleation has been achieved in diamond and has resulted in approximately 2 μm line widths. SAD has been achieved by combining the nucleation enhancement techniques of abrasion or biasing with methods to inhibit nucleation. The methods used to inhibit nucleation include masks of SiO_2 or a-Si:H and application of reactive ion etching, Ar+ ion beams, and H_2 plasmas to etch nucleation sites. In addition to these techniques, selective application of biasing has been achieved by using a wire mesh to shadow the ion flux.

5. Nucleation Mechanisms

The preceding sections have discussed what could be considered the phenomenological side of diamond nucleation. Those sections described the technology of nucleation enhancement, heteroepitaxial nucleation, and selected area nucleation. The discussion now returns to a more fundamental examination of diamond nucleation. Several references are recommended for a general discussion of thin film nucleation and growth [Bachman 1995; Kern, LeLay et al. 1979; Robins 1988; Venables and Price 1975]. This section will address the following four sub-topics: (1) geometrical effects, (2) substrate effects such as carbide formation, (3) carbonaceous precursors, and (4) nucleation specific to bias-enhancement. It is likely that each of these four topics will highlight mechanisms which are operative to some degree, depending on the particular deposition process and conditions.

5.1 GEOMETRICAL EFFECTS

In order to examine the mechanism of nucleation enhancement via abrasion without introducing diamond seeds, diamond nucleation has been investigated on chemically etched silicon substrates and on cleaved surfaces [Dennig and Stevenson 1991a; Dennig and Stevenson 1991b; Dennig and Stevenson 1991c]. These investigations indicated that diamond nuclei tend to form on sharp convex surface features. Such features provide a large surface area per unit volume and allow more substrate vapor interface to be replaced by substrate condensate interface for a given increase in nucleus size than would be possible on a flat surface. For the appropriate values of the interfacial free energies, the total free energy of a diamond nucleus could be lowered by securing a minimum contact area with the substrate [Dennig and Stevenson 1991c; Williams 1991]. Nucleation at surface defects has also been observed in other systems [Hirth and Pound 1963]. Scratching probably enhances nucleation through a combination of these effects in addition to carbon seeding, as discussed further in §5.3.

5.2 SUBSTRATE EFFECTS AND CARBIDE FORMATION

To form diamond via chemical vapor deposition, a carbonaceous gas such as methane is utilized and substrate temperatures are generally greater than 600°C. In this high temperature process environment it is likely that carbide phases will form on the substrate during the nucleation process, under conditions where they are thermodynamically stable. An understanding of the role that carbide formation plays in the scheme of diamond nucleation is important if improvements in nucleation are to be achieved. Silicon substrates have been used extensively for diamond deposition based on this material's predominance in the microelectronics industry. Results from early diamond research on silicon indicated that a silicon carbide interlayer formed between the diamond and silicon substrate. These carbide interfacial layers have been detected via TEM [Williams and Glass 1989], XRD [Joffreau, Haubner et al. 1988] and surface sensitive analytical spectroscopies [Belton, Harris et al. 1989]. Furthermore, in work involving diamond deposition on refractory metal substrates Joffreau et al. observed a correlation between diamond nucleation and the kinetic properties of the resulting carbides [Joffreau, Haubner et al. 1988]. These results indicated that the resultant carbide is an important factor influencing diamond nucleation.

The following discussion covers work by Wolter and coworkers corresponding to their evaluation of bias-enhanced nucleation on Cu and Si [Wolter, Stoner et al. 1994] and their subsequent evaluation of nucleation on various refractory metal substrates [Wolter, Glass et al. 1995]. The BEN process was reported to have a relatively negligible effect for diamond nucleation on Cu substrates as compared to silicon, suggesting further that the carbide forming nature of the substrates strongly effects the nucleation efficiency during biasing. The variation in carbide forming nature of the refractory metals was expected to expand on this correlation and further improve the understanding of the nucleation processes during this bias-enhanced pretreatment.

Results from the above study did indicate a variation in nucleation trends for the different substrates observed. The plot of nucleation density versus bias time for

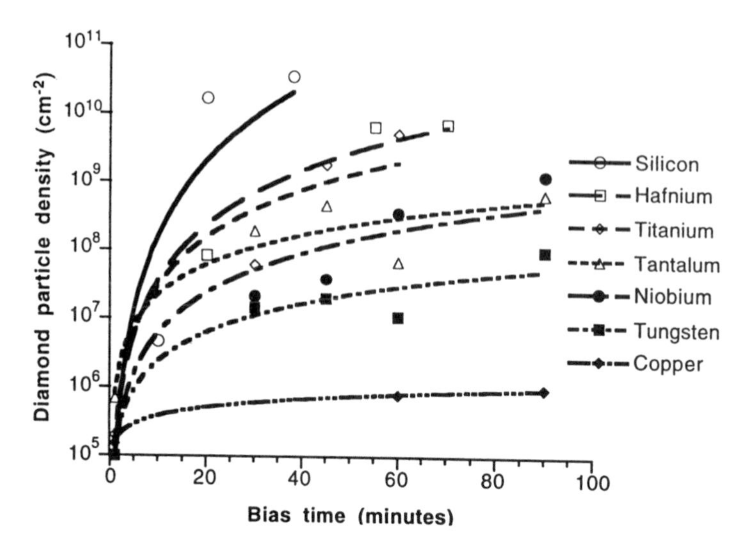

Figure 18. Nucleation density as a function of bias time for several refractory metals compared to silicon and copper. From [Wolter, Glass et al. 1995].

refractory metals compared to silicon and copper is shown in Figure 18. The diamond nucleation density on copper substrates as discussed previously is relatively unaffected by the application of a negative substrate bias. The influence of this biasing pretreatment on diamond nucleation corresponds to the positions of these substrate materials in the periodic table. Specifically, the group IVB, VB, VIB metals follow a trend of the most effective to least effective with regard to nucleation densities. This trend can be related to the carbide heat of formation, ΔH_f per metal atom, as displayed graphically in Figure 19. Although a correlation was found to exist between the carbide heat of formation and the nucleation density for the refractories, silicon did not fit the above data suggesting that additional material factors might also play a role. Silicon possesses a comparable carbide heat of formation to that of Tungsten yet was much more influenced by the BEN process. The substrate crystal structures and carbon diffusivities are possible explanations for the differences observed.

As with previous studies, the above correlation between the carbide heat of formation and the nucleation density for refractory metals supported the importance of the carbide formation on the bias-enhanced nucleation of diamond. Substrates capable of forming carbides in general have a high affinity for carbon, and substrates possessing the more negative heats of formation have the greatest affinity. Figure 20 depicts a general nucleation model through adatom addition and surface diffusion. The basic steps involve the following: (i) a flux of species to the surface leading to (ii) adatom capture followed by either (iii) reevaporation or surface diffusion. (iv) Statistical fluctuations in adatom concentrations allow clusters to form that fluctuate in size through surface migration, (v) until they eventually reach a stable size (vi) which can grow through either further surface diffusion or direct impingement. The critical transition in this model is from a metastable to a stable cluster size. It has been speculated that the strong carbide forming nature of the refractories and Si may help to stabilize the smaller carbon clusters long enough for them to become stable nuclei. Their high affinity for carbon helps to increase and stabilize the surface carbon concentrations, facilitating the nucleation process.

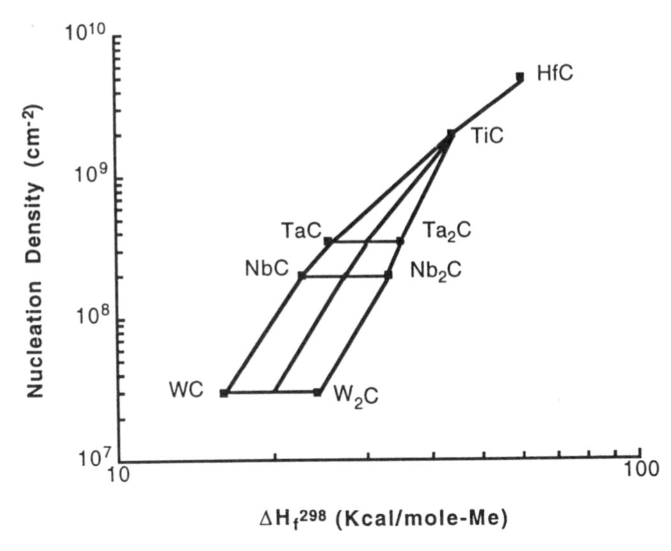

Figure 19. Correlation between nucleation density after 1 hour of BEN and carbide heat of formation. From [Wolter, Glass et al. 1995].

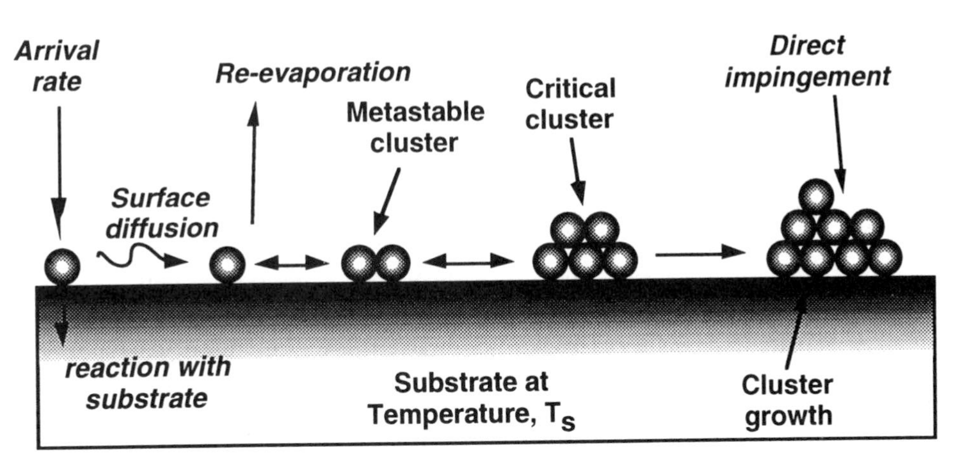

Figure 20. General model of steps involved in nucleation of three dimensional clusters.

5.3 CARBONACEOUS PRECURSORS

While there are a wide variety of surface treatments which promote nucleation, most are related to the production of higher surface carbon concentrations, as noted in §2.2 and related references [Dubray, Pantano et al. 1991; Ravi, Koch et al. 1990; Williams 1991]. For example, the deposition of an amorphous carbon layer has been shown to enhance diamond nucleation on unscratched silicon [Dubray, Pantano et al. 1991]; however the detailed mechanism for the enhancement remains unclear. In the previous section, the authors suggested that the carbide forming nature of the substrate may play a role in helping to stabilize these carbonaceous precursors. While there are several reports that propose potential mechanisms and insight into these elusive nucleation precursors, the prevailing ones suggest an intermediate precursor consisting of hydrogenated sp^2 bonded carbon structures [Angus, Li et al. 1993b]. Although this mechanism is widely supported, it is by no means unanimously accepted. The focus of this section is to review several of these proposed mechanisms allowing this information to complement the data on nucleation from previous sections.

Surface studies performed on pretreated silicon substrates at intervals during the nucleation process support a correlation between nucleation density and surface carbon concentration [Arezzo, Zacchetti et al. 1994; Williams 1991]. Some pretreatments such as biasing, scratching, and etching, have been effective at promoting nucleation while others such as scratching and hydrogen cleaning or scratching and Ar-sputtering are less effective. The results from these two studies suggested carbonaceous species or phases played active or transitional roles in the nucleation process. High diamond nucleation densities were found to correlate directly with high concentrations of surface carbon.

Carbonaceous precursors have been suggested previously as being leading factors [Celii and Butler 1991]. However, the exact nature or structure of the carbonaceous phases remains a primary question of diamond CVD. Angus and co-workers [Angus, Li et al. 1993b] have proposed a mechanism whereby diamond nucleation occurs on the edges of graphite-like sp^2 bonded intermediates. Earlier work from several of these authors reported that diamond nucleates preferentially on the edges of graphite seed crystals with the following orientations: [Angus, Li et al. 1991; Li, Wang et al. 1993]

$$(111)D \parallel (0001)G;$$

$$<1\bar{1}0>D \| <11\bar{2}0>G$$

They noted that the corrugated, saturated hexagonal rings in the diamond (111) planes have the same spatial orientation as the flat, unsaturated hexagonal rings in the graphite substrate (Figure 21). With this nucleation mechanism they went on to propose that nucleation might occur through the formation of a less stable, graphite intermediate that is subsequently hydrogenated by atomic hydrogen at its edges. Diamond nuclei form via the addition of C_xH_y species to these hydrogenated edges.

The model by Angus et al., proposed six diamond (111) planes adjoined to four graphite (0001) planes as depicted in Figure 22. The diamond (111) planes align to the graphite planes to within 4% and the mass densities of diamond and graphite are 3.515 and 2.26 g/cm^3 respectively, giving the 1.56 (or $\approx$ 3:2) ratio observed in the model. Energy minimization calculations using Tersoff semi-classical potentials were used to analyze the interfacial energies. The calculations suggested that interfacial energies were low for the above configuration as compared to surface energies, despite the dangling bonds every third diamond layer, and that hydrogen saturation of the bonds would further

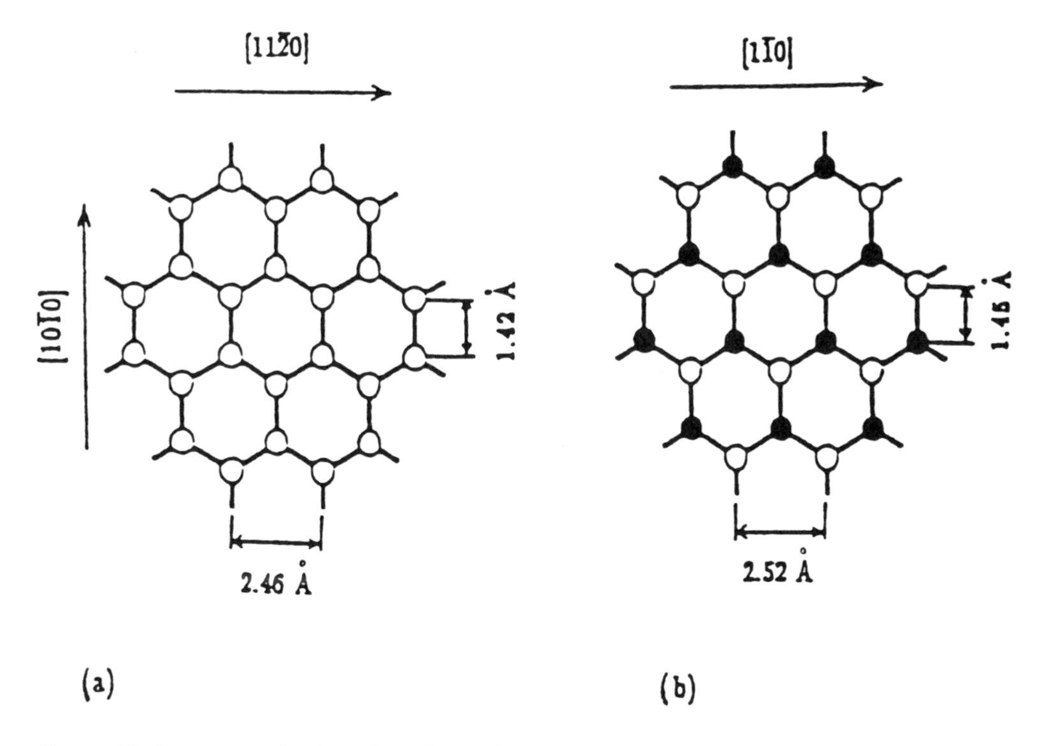

Figure 21. Structure and orientation of graphite (0001) plane and diamond (111) plane. (a) graphite (0001) plane viewed along [0001] direction; (b) diamond (111) plane viewed along [111]direction. The filled and open circles represent raised and lowered atoms. From [Angus, Li et al. 1993b].

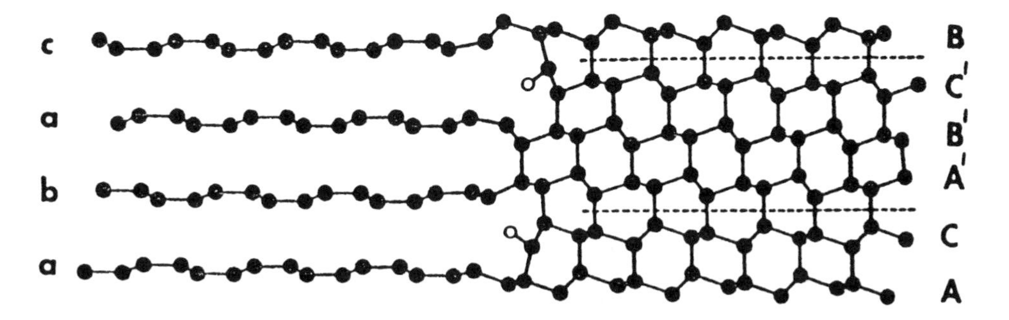

Figure 22. Diamond/graphite interface with orientation shown in Figure 21. Six diamond (111) planes are joined to four graphite (0001) planes. In the interface shown, there are stacking errors in the diamond that are indicated with short dashes.

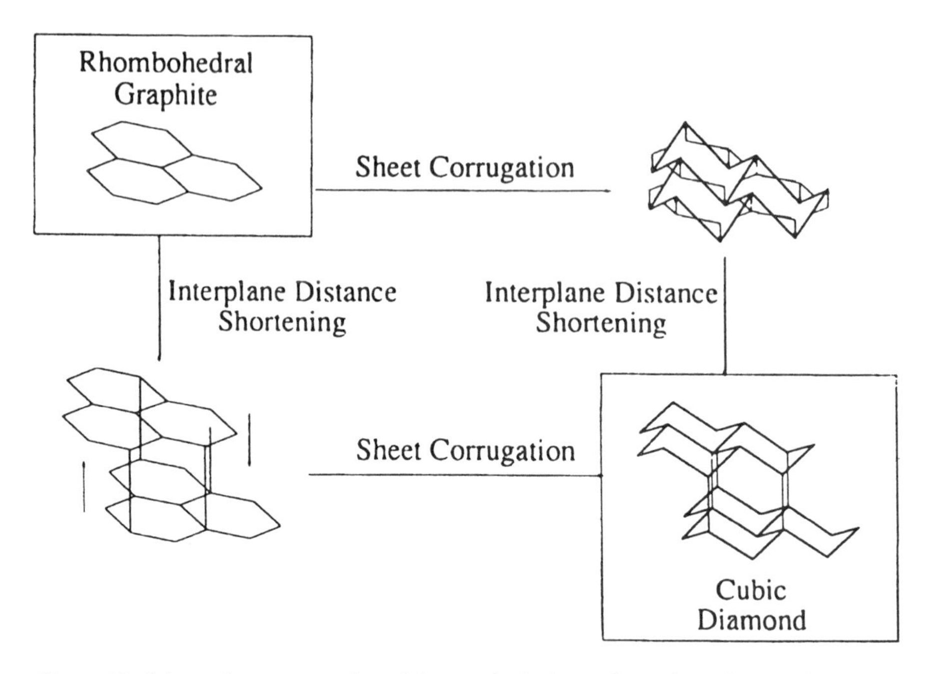

Figure 23. Schematic representation of the topological transformation of r-graphite to c-diamond. From [Sandré, Bonnot et al. 1994]

reduce the energy of the system. They also indicated that stacking faults in the graphite (i.e. abac as opposed to abab stacking) would result in more optimal interfacial bonding to the diamond.

The above model would thus proceed as follows: local fluctuations in carbon concentrations on the surface of the substrate result in the formation of graphitic or aromatic structures, which upon hydrogenation will give rise to diamond precursors. The authors note that this model does not preclude other valid nucleation mechanisms such as that resulting from the homogeneous nucleation of diamond on seeds left behind following a conventional scratching pretreatment as discussed in §2.1. The model does not explain the observed crystallographic registry across the diamond/substrate interface as in the case of highly oriented diamond films discussed in §3.4. The details of heteroepitaxial nucleation as facilitated through bias-enhanced nucleation will be discussed in more detail in §5.4.

This sp^2 to sp^3 transformation nucleation mechanism was further supported by theoretical work from Sandré and co-workers [Sandré, Bonnot et al. 1994] investigating the transformation of a graphite sheet into a (111) "diamond-like" structure. The authors reported that the transformation of a single rhombohedral graphite sheet to a corrugated sheet resembling cubic diamond can be energetically favorable (Figure 23). They further suggested that such a nucleation transformation process might be more favorable on such surfaces as amorphous silicon carbide and/or amorphous sp^3 carbon as well as exposed surfaces such as edge protrusions and scratches. The suggestion that this corrugated graphite mechanism might dominate even in the presence of a predominantly sp^3-carbon surface alleviates some of the controversy surrounding the nucleation precursor. According to the reports by both Angus et al. [Angus, Li et al. 1993b], and Sandré et al. [Sandré, Bonnot et al. 1994], an sp^2 intermediate may still dominate even where sp^3 carbon saturates the surface, such as in the case of an amorphous carbon pretreatment.

5.4 BIAS ENHANCED NUCLEATION

As discussed in §2.3, bias-enhanced nucleation was first reported by Yugo et al. [Yugo, Kanai et al. 1991] as an effective means to achieve high nucleation densities on clean silicon wafers without the need for an abrasive or seeding pretreatment. Yugo later proposed an ion-impingement based model in attempts to explain the nucleation process observed under this bias-pretreatment (Figure 24). Later work by Stoner et al. [Stoner, Ma et al. 1992] studied both the chemical and structural variations on the surface during biasing and resulted in a crude model portraying a time-lapse schematic of their analytical data (Figure 25). In terms of the precursor model described in §5.3, determining the actual mechanism(s) for BEN becomes a dual question of (i) the nature of the precursor and (ii) how biasing promotes the deposition of this relevant precursor. Several sources of a precursor have been examined or suggested in the literature as they relate specifically to BEN. There are a number of ways in which biasing may promote the formation of this precursor for diamond nucleation: (i) Biasing may provide an increased flux of positive C_xH_y+ ions to the surface, (ii) reduce the flux of electrons to the substrate surface, (iii) increase energy transfer of ions to the substrate surface, (iv) enhance dissociation reactions above the substrate surface, or (v) suppress oxide formation on the surface [Stoner, Ma et al. 1992]. Furthermore, optical emission spectroscopic measurements during biasing showed increased atomic hydrogen emission near the substrate surface when a bias was applied [Shigesato, Boekenhauer et al. 1993]. This increased atomic hydrogen might etch graphite clusters, stabilize small sp_3 clusters or aid in the dissociation of methane into more reactive carbon or hydrocarbon fragments. Identifying which of these theories are valid has been the subject of many excellent scientific studies on bias-enhanced nucleation. The following discussion attempts to highlight a subset of these studies.

As mentioned above, Yugo et al. proposed a bias-enhanced ion-impingement based nucleation mechanism. Several reports helped to support this mechanism and experimentally highlighted the importance of ion bombardment during BEN. McGinnis et al. [McGinnis, Kelly et al. 1995] reported on experiments where a small (5x5mm) silicon substrate was placed on an insulator in the center of a larger wafer and the combination was biased. The insulated sample resulted in a nucleation density several orders of magnitude lower than for the wafer on which it was placed. These data suggested that the material on which diamond is to nucleate needs to be biased and that gas phase chemistry and biasing of the area around the substrate are secondary effects. In another study by Jiang and coworkers [Jiang, Paul et al. 1995], biasing was performed on a substrate in which trenches had been cut so that surfaces were both parallel and perpendicular to the ion flux. Biasing only increased the nucleation on the surfaces bombarded by the ions. Without biasing, nucleation did not occur on either of the surfaces so it is difficult to know if another factor might have inhibited nucleation on

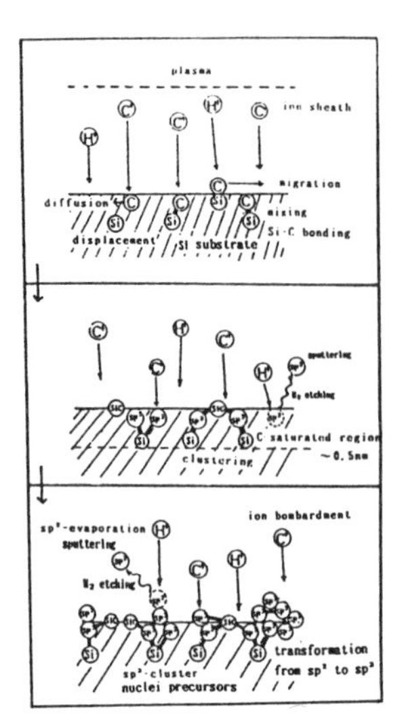

Figure 24. A model of the generation of diamond nuclei by ion impingement effects during BEN. From [Yugo, Kimura et al. 1993].

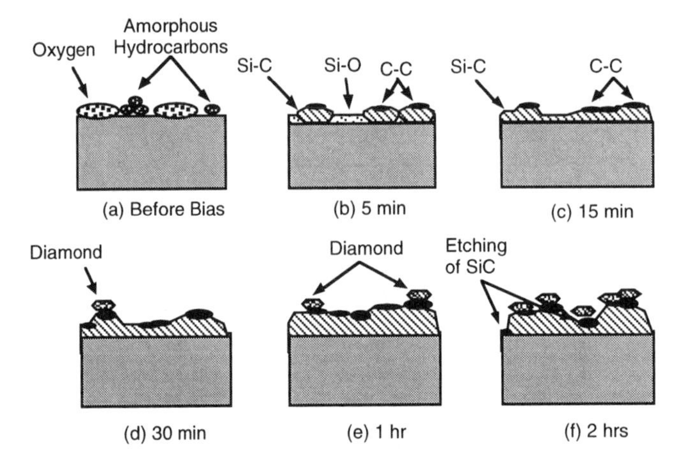

Figure 25. Model of diamond nucleation on silicon via biasing.

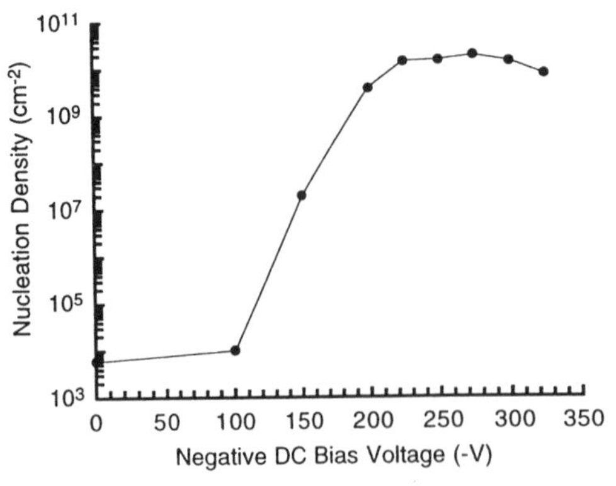

Figure 26. Nucleation density as a function of the applied bias voltage with constant time-integrated bias current to the substrate. From [McGinnis, Kelly et al. 1995].

the vertical surfaces. BEN can be a difficult process to optimize, so one must be cautious in interpreting negative results. Still, these experiments suggest strongly that BEN requires ion bombardment of the substrate.

Further evidence for the connection between BEN and ion bombardment is given by the energy distribution of ions striking the substrate. Both Monte Carlo calculations and measured ion energy distributions show an agreement between the threshold voltage for BEN and the energy required for an ion to be implanted a few layers below the substrate surface [Robertson 1995a; Robertson 1995b; Robertson, Gerber et al. 1995]. Estimates of the mean ion energy at a bias voltage of -250 V have ranged from 2 to 80 eV, while the peak in the ion energy distribution function has been measured as 80 eV for a -250V bias using a retarding voltage probe [Sheldon, Csencsits et al. 1994; McGinnis, Kelly et al. 1995; Robertson, Gerber et al. 1995]. The range of calculated ion energies arises from uncertainties in the sheath thicknesses and ion collision processes. Figure 26, from work by McGinnis and co-workers, supports such a threshold energy theory by depicting orders of magnitude increases in nucleation density as the voltage is varied from 100 to 200 volts negative and little or no variations above and below this range. Although this correlation has been investigated for only silicon it would be an informative test of the subplantation hypothesis to measure the change in bias voltage with other substrate materials and compare these to the variations in subplantation energy.

The deposition of a thin amorphous carbon layer has also been observed during BEN [Sheldon, Csencsits et al. 1994]. The thickness of this layer was independent of temperature, leading to speculation that it was produced by a nonchemical mechanism such as ion bombardment. McGinnis et al. also reported little variation in nucleation densities with temperature during biasing, further supporting a physical, non thermodynamic process. It has also been reported that SiC forms under certain conditions [Stoner, Ma et al. 1992] which was especially noteworthy in cases where local epitaxy was observed, resulting in the deposition of highly oriented films [Stoner, Sahaida et al. 1993]. With regard to heteroepitaxy, it is emphasized that the BEN effectiveness requires a balance between ion energy and flux such that carbon fluxes are high enough to promote nucleation but low enough such that amorphous material does not form [Sheldon, Csencsits et al. 1994]. Likewise (as in the model proposed by McGinnis et al.), ion energies should be sufficiently high to implant themselves beneath the surface yet low enough such that they do not cause excess damage and an amorphous transition layer.

It is likely as well that chemical reactions contribute to the stabilization of the carbon clusters deposited by the ion bombardment. This concept was suggested in the work by Wolter et al. and their evaluation of various carbide substrates; however the fact that a strong carbide former is not solely sufficient to enhance nucleation suggests that chemical reactions are a secondary effect. Others have also discussed the BEN mechanism in terms of an increased ion flux which reduces re-evaporation and diffusion, damages weakly bonded sp2 clusters, and increases bond strength via "ion mixing" [Yugo, Kimura et al. 1993]. In general, these reports indicate that BEN is strongly related to ion bombardment, but the precise nature of this connection remains unclear.

5.5 SUMMARY OF NUCLEATION MECHANISMS

In summary, the density of nuclei is often a function of the concentration of surface carbon species either present prior to growth or produced in the early stages of the deposition process. Conditions that increase the local concentrations of carbon, either through the creation of high-surface energy sites or by leaving behind residual carbonaceous compounds tend to be successful in enhancing nucleation. The exact form or forms of the ideal carbonaceous precursor is still a topic for debate, however there are both strong arguments and evidence for a mechanism involving sp^2 -type intermediates and their subsequent transformation into stable sp^3 "diamond-like" structures. A variation due to substrate material properties was also indicated. To first order, there appears to be a relationship between the substrate-carbon reactivity and the nucleation efficiency; suggesting that materials with higher carbide heat of formations are more efficient surfaces for diamond nucleation. As with the arguments for the sp^2 precursor, the stronger carbides may facilitate the increase in local carbon concentrations long enough for a stable "diamond" cluster to form more readily. In addition, bias-enhanced nucleation can be an effective technique for both enhancing diamond nucleation as well as for studying the nucleation phenomena itself. The specific mechanisms for BEN are still unknown, however increasing evidence suggests that the nucleation occurs via surface bombardment with low energy, carbonaceous ions. Further evidence and speculation suggests that carbon clusters are stabilized through subplantation in shallow sites beneath the substrate surface. As a final note, the authors would like to suggest that although there appears to be a variety of potential mechanisms depending upon the experiments performed, there does exist a common theme in all the results presented. This common theme is the initial stabilization of the diamond critical cluster which can be facilitated through one or a combination of the aforementioned techniques.

6. Acknowledgments

The authors have been assisted in the preparation of this manuscript by many insightful discussions. Participants in these discussions include D.A. Tucker, M.T. McClure, F.R. Sivazlian, and P.C. Yang at North Carolina State University. The authors would also like to thank the individuals who granted permission to reproduce figures in this chapter.

7. References

Angus, J. C., Hoffman, R. W. and Schmidt, P. H. (1988) in S. Saito, O. Fukunga and M. Yoshikawa (eds), *First International Conference on New Diamond Science and Technology,* KTK Scientific, pp. 9-16.

Angus, J. C., Li, Z., Sunkara, M., Gat, R., Anderson, A. B., Mehandru, S. P. and Geis, M. W. (1991) Nucleation and Growth Processes in Chemical Vapor Deposition of Diamond, in A. J. Purdes, J. C. Angus, R. F. Davis, B. M. Meyerson, K. E. Spear and M. Yoder (eds), *Diamond Materials*, Electrochemical Society, Inc., vol. 91-8, pp. 125.

Angus, J. C., Li, Z., Sunkara, M., Lee, C., Lambrecht, W. R. L. and Segall, B. (1993a) Diamond nucleation, in J. P. Dismukes and K. V. Ravi (eds), *Third International Symposium on Diamond Materials,* The Electrochemical Society, vol. 93-17, pp. 435-440.

Angus, J. C., Li, Z., Sunkara, M., Wang, Y., Lee, C., Lambrecht, W. R. and Segall, B. (1993b) Nucleation and Growth in the Chemical Vapor Deposition of Diamond, in M. Yoshikawa, M. Murakawa, Y. Tzeng and W. A. Yarbrough (eds), *Second International Conference on the Applications of Diamond Films and Related Materials*, MYU, Tokyo, pp. 153-158.

Arezzo, F., Zacchetti, N. and Zhu, W. (1994) X-ray photoelectron spectroscopy study of substrate surface pretreatments for diamond nucleation, *Journal of Applied Physics*, **75**, 5375.

Argoitia, A., Angus, J. C., Ma, J. S., Wang, L., Pirouz, P. and Lambrecht, W. R. L. (1994) Heteroepitaxy of diamond on c-BN: Growth mechanisms and defect characterization, *Journal of Materials Research*, **9**, 1849.

Argoitia, A., Angus, J. C., Wang, L., Ning, X. I. and Pirouz, P. (1993) Diamond grown on single crystal Beryllium Oxide, *Journal of Applied Physics*, **73**, 4305-4312.

Ascarelli, P. and Fontana, S. (1992) Dissimilar grit-size dependence of the diamond nucleation density on substrate surface pretreatments, *Applied Surface Science*, **64**, 307-311.

Bachman, K. J. (1995) *The Materials Science of Microelectronics*, New York, VCH Publishers.

Badzian, A. and Badzian, T. (1992) Routes to diamond heteroepitaxy, *Materials Research Society Symposium Proceedings*, Materials Research Society, vol. 250, pp. 339-349.

Belton, D. N., Harris, S. J., Schmieg, S. J., Weiner, A. M. and Perry, T. A. (1989) *In situ* characterization of diamond nucleation and growth, *Applied Physics Letters*, **54**, 416-417.

Belton, D. N. and Schmieg, S. J. (1989) Loss of epitaxy during diamond film growth on ordered Ni(100), *Journal of Applied Physics*, **66**, 4223.

Belton, D. N. and Schmieg, S. J. (1990) Electron spectroscopic identification of carbon species formed during diamond growth, *Journal of Vacuum Science and Technology A*, **8**, 2353-2362.

Brandes, E. A. and Brook, G. B., eds., (1992) *Smithells Metals Reference Book*, Oxford, Butterworth-Heinemann Ltd.

Celii, F. G. and Butler, J. E. (1991) Diamond Chemical Vapor Deposition, *Annual Reviews of Physical Chemistry*, **42**, 643-684.

Chalker, P. R., Johnston, C., Romani, S., Ayres, C. F. and Buckley-Golder, I. M. (1994) Nucleation and growth of CVD diamond on magnesium oxide (100) and titanium nitride-magnesium oxide (100) surfaces, *Diamond and Related Materials*, **3**, 393-397.

Chang, C.-P., Flamm, D. L., Ibbotson, D. E. and Mucha, J. A. (1988) Diamond crystal growth by plasma chemical vapor deposition, *Journal of Applied Physics*, **63**, 1744.

Clausing, R. E., Heatherly, L., Specht, E. D. and More, K. L. (1991) Texture development in diamond films grown by hot filament CVD, in R. Messier, J. T. Glass, J. E. Bulter and R. Roy (eds), *New Diamond Science and Technology*, Materials Research Society, pp. 575.

Davidson, J. L., Ellis, C. and Ramesham, R. (1989) Selective deposition of diamond films, *Journal of Electronic Materials*, **18**, 711-715.

Dennig, P. A. and Stevenson, D. A. (1991a) Influence of Substrate Preparation upon the Nucleation of Diamond Thin Films, in R. Messier, J. T. Glass, J. E. Butler and R. Roy (eds), *New Diamond Science and Technology*, Materials Research Society, pp. 403.

Dennig, P. A. and Stevenson, D. A. (1991b) Influence of substrate topography on the nucleation of diamond thin films, *Applied Physics Letters*, **59**, 1562.

Dennig, P. A. and Stevenson, D. A. (1991c) Influence of Substrate Topography on the Nucleation of Diamond Thin Films, in Y. Tzeng, M. Yoshikawa, M. Murakawa and A. Feldman (eds), *Applications of Diamond Films and Related Materials*, Elsevier Science Publishers B.V., vol. **73**, pp. 383-388.

Dubray, J. J., Pantano, C. G., Meloncelli, M. and Bertran, E. (1991) Nucleation of diamond on silicon, SiAlON, and graphite substrates coated with an a-C:H layer, *Journal of Vacuum Science and Technology A*, **9**, 3012.

Feng, Z., Brewer, M. A., Komvopoulos, K., Brown, I. G. and Bogy, D. B. (1994) Diamond nucleation on unscratched silicon substrates coated with various non-diamond carbon films by microwave plasma-enhanced chemical vapor deposition, *Journal of Materials Research*, **10**, 165.

Field, J. E. (1992) *The Properties of Natural and Synthetic Diamond*, Academic Press.

Fox, B. A., Stoner, B. R., Malta, D. M., Ellis, P. J., Glass, R. C. and Sivazlian, F. R. (1994) Epitaxial nucleation, growth and characterization of highly oriented, (100)-textured diamond films on silicon, *Diamond and Related Materials*, **3**, 382-387.

Fujimori, N., Nakahata, H. and Imai, T. (1990) Properties of Boron-Doped Epitaxial Diamond Films, *Japanese Journal of Applied Physics*, **29**, 824.

Hänni, W., Müller, C., Binggeli, M., Hintermann, H. E., Krebs, P. and Grisel, A. (1993) Selective area deposition of diamond on 4 in. Si wafers, *Thin Solid Films*, **236**, 87.

Hartnett, T., Miller, R., Montanari, D., Willingham, C. and Tustison, R. (1990) Intermediate layers for the deposition of polycrystalline diamond films, *Journal of Vacuum Science and Technology A*, **8**, 2129-2136.

Hirabayashi, K., Taniguchi, Y., Takamatsu, O., Ikeda, T., Ikoma, K. and Iwasaki-Kurihara, N. (1988) Selective deposition of diamond crystals by chemical vapor deposition using a tungsten-filament method, *Applied Physics Letters*, **53**, 1815-1817.

Hirth, J. P. and Pound, G. M. (1963) *Condensation and Evaporation: Nucleation and Growth Kinetics*, New York, Macmillan.

Iijima, S., Aikawa, Y. and Baba, K. (1990) Early formation of chemical vapor deposition diamond films, *Applied Physics Letters*, **57**, 2646.

Inoue, T., Tachibana, H., Kumagai, K., Miyata, K., Nishimura, K., Kobashi, K. and Nakaue, A. (1990) Selected-area deposition of diamond films, *Journal of Applied Physics*, **67**, 7329.

Inuzuka, T., Koizumi, S. and Suzuki, K. (1992) Epitaxial growth of diamond thin films on foreign substrates, *Diamond and Related Materials*, **1**, 175-179.

Jiang, X., Boettger, E., Paul, M. and Klages, C.-P. (1994) Approach of selective nucleation and epitaxy of diamond films on Si(100), *Applied Physics Letters*, **65**, 1519.

Jiang, X. and Klages, C.-P. (1993) Heteroepitaxial diamond growth on (100) silicon, *Diamond and Related Materials*, **2**, 1112.

Jiang, X., Klages, C.-P., Zachai, R., Hartweg, M. and Füsser, H.-J. (1993) Epitaxial diamond thin films on (001) silicon substrates, *Applied Physics Letters*, **62**, 3438.

Jiang, X., Paul, M., Klages, C.-P. and Jia, C. L. (1995) Studies of Heteroepitaxial Nucleation and Growth of Diamond on Silicon, in K. V. Ravi and J. P. Dismukes (eds), *Diamond Materials IV*, Electrochemical Society, vol. 95-4, pp. 50-55.

Joffreau, P.-O., Haubner, R. and Lux, B. (1988) Low-pressure diamond growth on refractory metals, in G. H. Johnson, A. R. Badzian and M. W. Geis (eds), *The Materials Research Society*, vol. EA-15, pp. 15-18.

Joseph, A., Wei, J., Tanger, C., Tzeng, Y., M., B. and Vohra, Y. (1993) High density diamond nucleation on unseeded substrates by a combined microwave and radio-frequency plasma, in J. P. Dismukes and K. V. Ravi (eds), *Third International Symposium on Diamond Materials*, Electrochemical Society, vol. 93-17, pp. 315-321.

Käding, O. W., Rösler, M., Zachai, R., Füβer, H.-J. and Matthias, E. (1994) Lateral thermal diffusivity of epitaxial diamond films, *Diamond and Related Materials*, **3**, 1178-1182.

Kamo, M., Yurimoto, H. and Sato, Y. (1988) Epitaxial Growth of Diamond on Diamond Substrate by Plasma Assisted CVD, *Applied Surface Science*, **33/34**, 553.

Kawarada, H., Ma, J. S., Yonehara, T. and Hiraki, A. (1990) Selective nucleation of single crystal CVD diamond and its applicability to semiconductor devices, *Materials Research Society*, vol. 162, pp. 195-200.

Keene, B. J. (1987) A Review of the Surface Tension of Silicon and its Binary Alloys with Reference to Maragoni Flow, *Surface and Interface Analysis*, **10**, 367-383.

Kern, R. (1985) Metastable Phases in the Bulk and on Substrates, in E. Kaldis (ed), *Current Topics in Materials Science*, Amsterdam, North-Holland.

Kern, R., LeLay, G. and Metios, J. J. (1979) Basic Mechanisms in the Early Stages of Epitaxy, in E. Kaldis (ed), *Current Topics in Materials Science*, North-Holland.

Kobashi, K., Inoue, T., Tachibana, H., Kumagai, K., Miyata, K., Nishimura, K. and Nakaue, A. (1990) Selected-area deposition of diamond films, *Vacuum*, **41**, 1383.

Koizumi, S., Murakami, T., Inuzuka, T. and Suzuki, K. (1990) Epitaxial growth of diamond thin films on cubic boron nitride {111} surfaces by dc plasma chemical vapor deposition, *Applied Physics Letters*, **57**, 563-565.

Kulisch, W., Sobisch, B., Kuhr, M. and Beckman, R. (1995) Characterization of the bias nucleation process, *Diamond and Related Materials*, **4**, 401-405.

Lambrecht, W. R. L., Lee, C. H., Segall, B., Angus, J. C., Li, Z. and Sunkara, M. (1993) Diamond nucleation by hydrogenation of the edges of graphitic precursors, *Nature*, **364**, 607.

Lambrecht, W. R. L. and Segall, B. (1989) Electronic structure of (diamond C)/(sphalerite BN) (110) interfaces and superlattices, *Physical Review B*, **40**, 9909.

Lambrecht, W. R. L. and Segall, B. (1992) Electronic structure and total energy of diamond/BeO interfaces, *Journal of Materials Research*, **7**, 696.

Li, Z., Wang, L., Suzuki, T., Argoita, A., Pirouz, P. and Angus, J. C. (1993) Orientation relationship between chemical vapor deposited diamond and graphite substrates, *Journal of Applied Physics*, **73**, 711.

Liu, J., et al. (1994) Electron emission from diamond coated silicon field emitters, *Applied Physics Letters*, **65**, 2842.

Liu, W., Tucker, D. A., Yang, P. C. and Glass, J. T. (1995) Nucleation of oriented diamond particles on cobalt substrates, *Journal of Applied Physics*, **78**, 1291-1296.

Ma, J. S., Yagyu, H., Hiraki, A., Kawarada, H. and Yonehara, T. (1991) Large area diamond selective nucleation based epitaxy, *Thin Solid Films*, **206**, 192-197.

Maeda, H., Masuda, S., Kusakabe, K. and Morooka, S. (1994) Heteroepitaxial growth of diamond on c-BN in a microwave plasma, *Diamond and Related Materials*, **3**, 398-402.

Malta, D. M., von Windheim, J. A. and Fox, B. A. (1993) Comparison of electron transport in boron-doped homoepitaxial, polycrystalline, and natural single crystal diamond, *Applied Physics Letters*, **62**, 2926-2928.

Masood, A., Aslam, M., Tamor, M. A. and Potter, T. J. (1992) Synthesis and electrical characterization of boron-doped thin diamond films, *Applied Physics Letters*, **61**, 1832-1834.

McGinnis, S. P., Kelly, M. A. and Hagström, S. B. (1995) Evidence of an energetic ion bombardment mechanism for bias-enhanced nucleation of diamond, *Applied Physics Letters*, **66**, 3117-3119.

Menningen, K. L., Childs, M. A., Toyoda, H., Anderson, L. W. and Lawler, J.E. (1994) Evaluation of a substrate pretreatment for hot filament CVD of diamond, *Journal of Materials Research*, **9**, 915.

Mezey, L. Z. and Giber, J. (1982) The Surface Free Energies of Solid Chemical Elements: Calculation from Internal Free Enthalpies of Atomization, Japanese *Journal of Applied Physics*, **21**, 1569-1571.

Mishima, O. (1991) Growth and Polar Properties of Cubic Boron Nitride, in Y. Tzeng, M. Yoshikawa, M. Murakawa and A. Feldman (eds), *Applications of Diamond Films and Related Materials*, Elsevier, vol. 73, pp. 647-651.

Mitsuda, Y., Moriyasu, T. and Masuko, N. (1993) Effect of high supersaturation at the initial stage on diamond nucleation phenomena, in P. K. Bachmann, A. T. Collins and M. Seal (eds), Third International Conference on the *New Diamond Science and Technology* (ICNDST-3), Elsevier Science Publishing Co., Inc., vol. 2, pp. 333-336.

Morrish, A. A. and Pehrsson, P. E. (1991) Effects of surface pretreatments on nucleation and growth of diamond films on a variety of substrates, *Applied Physics Letters*, **59**, 417.

Oral, B., Ece, M., Rogelet, T. and Yu, Z. M. (1993) Characterization of catalytically synthesized turbostratic carbon films for improved rates of diamond nucleation, *Diamond and Related Materials*, **2**, 225-228.

Pehrsson, P. E., Glesener, J. and Morrish, A. (1992) Chemical vapor deposition diamond nucleation induced by sp2 carbon on unscratched silicon, *Thin Solid Films*, **212**, 81-90.

Pierson, H. O., eds., (1993) *Handbook of carbon, diamond, graphite, and fullerenes*, New Jersey, Noyes Publications.

Plano, L. S. (1994). Measurements Performed at Kobe Steel USA Inc., Electronic Materials Center.

Popovici, G. and Prelas, M. A. (1992) Nucleation and Selective Deposition of Diamond Thin Films, *Phys. Stat. Sol. (a)*, **132**, 233.

Ravi, K. V., Koch, C. A., Hu, H. S. and Joshi, A. (1990) The nucleation and morphology of diamond crystals and films synthesized by the combustion flame technique, *Journal of Materials Research*, **5**, 2356-2366.

Rebello, J. H. D., Straub, D. L., Subramaniam, V. V., Tan, E. K., Dregia, S. A., Preppernau, B. L. and Miller, T. A. (1991) Nucleation and growth of diamond on silicon using hot filament chemical vapor deposition, in T. S. Sudarshan, D. G. Bhat and M. Jeandin (eds), *Surface Modification Technologies IV*, The Minerals, Metals, and Materials Society, pp. 569-582.

Robertson, J. (1995a) Deposition of Diamond-like Carbon, cubic Boron Nitride and Bias-enhanced Nucleation of Diamond as a Subplantation Process, in K. V. Ravi and J. P. Dismukes (eds), *Diamond IV*, Electrochemical Society, vol. 95-4, pp. 165-170.

Robertson, J. (1995b) Mechanism of bias-enhanced nucleation and heteroepitaxy of diamond on Si, *Diamond and Related Materials*, **4**, 549-552.

Robertson, J., Gerber, J., Sattel, S., Weiler, M., Jung, K. and Ehrhardt, H. (1995) Mechanism of bias-enhanced nucleation of diamond on Si, *Applied Physics Letters*, **66**, 3287-3289.

Robins, J. L. (1988) Thin Film Nucleation and Growth Kinetics, *Applied Surface Science*, **33/34**, 379-394.

Sandr□, E., Bonnot, A. and Cyrot-Lackman, F. (1994) Stabilization of diamond relative to different substrate-carbon interfaces: a nucleation model for CVD diamond growth based on a charge transfer consideration, *Diamond and Related Materials*, **3**, 448-451.

Sato, Y., Fujita, H., Ando, T., Tanaka, T. and Kamo, M. (1993) Local epitaxial growth of diamond on nickel from the vapor phase, *Philosophical Transactions of the Royal Society of London A*, **342**, 225-231.

Sato, Y., Yashima, I., Fujita, H., Ando, T. and Kamo, M. (1991) Epitaxial growth of diamond from the gas phase, in R. Messier, J. T. Glass, J. E. Butler and R. Roy (eds), *New Diamond Science and Technology*, Materials Research Society, pp. 371-376.

Sawabe, A. and Inuzuka, T. (1986) Growth of diamond thin films by electron-assisted chemical vapour deposition and their characterization, *Thin Solid Films*, **137**, 89-99.

Shackelford, J. and Alexander, W., eds., (1992) *The CRC Materials Science and Engineering Handbook*, Boca Raton, Florida, CRC Press, Inc.

Sheldon, B. W., Csencsits, R., Rankin, J., Boekenhauer, R. E. and Shigesato, Y. (1994) Bias-enhanced nucleation of diamond during microwave-assisted chemical vapor deposition, *Journal of Applied Physics*, **75**, 5001-5008.

Shelton, J. C., Patil, H. R. and J.M. Blakely (1974) *Surface Science*, **43**, 493.

Shigesato, Y., Boekenhauer, R. E. and Sheldon, B. W. (1993) Emission spectroscopy during direct-current-biased, microwave-plasma chemical vapor deposition of diamond, *Applied Physics Letters*, **63**, 314-316.

Spitsyn, B. V., Bouilov, L. L. and Derjaguin, B. V. (1981) Vapor growth of diamond on diamond and other surfaces, *Journal of Crystal Growth*, **52**, 219-226.

Stoner, B., Kao, C., Malta, D. and Glass, R. (1993) Hall effect measurements on boron-doped, highly oriented diamond films grown on silicon via microwave plasma chemical vapor deposition, *Applied Physics Letters*, **62**, 2347-2349.

Stoner, B. R. and Glass, J. T. (1992) Textured diamond growth on (100) β-SiC via microwave plasma chemical vapor deposition, *Applied Physics Letters*, **60**, 698-700.

Stoner, B. R., Ma, G.-H. M., Wolter, S. D. and Glass, J. T. (1992) Characterization of bias-enhanced nucleation of diamond on silicon by *in vacuo* surface analysis and transmission electron microscopy, *Physical Review B*, **45**, 11067-11084.

Stoner, B. R., Ma, G. H., Wolter, S. D., Zhu, W., Wang, Y.-C., Davis, R. F. and Glass, J. T. (1993) Epitaxial nucleation of diamond on β-SiC via bias-enhanced microwave plasma chemical vapor deposition, *Diamond and Related Materials*, **2**, 142-146.

Stoner, B. R., et al. (1994) Highly Oriented Diamond Films on Si: Growth, Characterization and Devices, Diamond-Film Semiconductors, *SPIE,* vol. 2151, pp. 2-10.

Stoner, B. R., Sahaida, S., Bade, J. P., Southworth, P. and Ellis, P. J. (1993) Highly oriented, textured diamond films on silicon via bias-enhanced nucleation and textured growth, *Journal of Materials Research*, **8**, 1334-1340.

Stoner, B. R., Williams, B. E., Wolter, S. D., Nishimura, K. and Glass, J. T. (1991) *In situ* growth rate measurement and nucleation enhancement for microwave plasma CVD of diamond, *Journal of Materials Research*, 7, 257-260.

Strong, H. M. (1964) Variation with pressure of the Nickel-Carbon Eutectic, *Acta Metallurgica*, **12**, 1411.

Suzuki, K., Sawabe, A., Yasuda, H. and Inuzuka, T. (1987) Growth of diamond thin films by DC plasma chemical vapor deposition, *Applied Physics Letters*, **50**, 728.

Tucker, D. A., et al. (1995) Comparison of Silicon, Nickel, and Nickel Silicide (Ni_3Si) as Substrates for Epitaxial Diamond Growth, *Surface Science*, **334**, 179-194.

Tzeng, Y., Cutshaw, C., Phillips, R., Srivinyunon, T., Ibrahim, A. and Loo, B. H. (1990) *Applied Physics Letters*, **56**, 134.

van der Drift, A. (1967) Evolutionary selection, a principle governing growth orientation in vapour-deposited layers, *Philips Research Reports*, **22**, 267-288.

Venables, J. A. and Price, G. L. (1975) *Epitaxial Growth*, New York, Academic Press.

Verwoerd, W. S. (1992) Interface structures for epitaxy of diamond on Si (100), *Diamond and Related Materials*, **1**, 195-199.

Verwoerd, W. S. (1994a) Diamond epitaxy on a Si (001) substrate: a comparison of structural models, *Surface Science,* **304**, 24-32.

Verwoerd, W. S. (1994b) Parallel versus diagonal epitaxy models of diamond and c-BN on Si (001), *Diamond and Related Materials*, **3**, 457-461.

Verwoerd, W. S. and Weimer, K. (1991) Atomic cluster study of chemisorption and epitaxial diamond films on an Si (100) substrate, *Surface and Coatings Technology*, **47**, 578-584.

Wang, L., Pirouz, P., Argoitia, A., Ma, J.-S. and Angus, J. C. (1993) Heteroepitaxially grown diamond on a c-BN {111} surface, *Applied Physics Letters*, **63**, 1336-1338.

Weast, R. C., eds., (1987) *CRC Handbook of Chemistry and Physics*, Boca Raton, Florida, CRC Press, Inc.

Wild, C., Herres, N. and Koidl, P. (1990) Texture formation in polycrystalline diamond films, *Journal of Applied Physics*, **68**, 973-978.

Wild, C., Kohl, R., Herres, N., Müller-Sebert, W. and Koidl, P. (1994) Oriented CVD diamond films: twin formation, structure and morphology, *Diamond and Related Materials*, **3**, 373-381.

Williams, B. E. (1991) Surface Analysis and Microstructural Characterization of Diamond Thin Films, Ph.D. Dissertation, North Carolina State University.

Williams, B. E. and Glass, J. T. (1989) Characterization of diamond thin films: Diamond phase identification, surface morphology, and defect structures, *Journal of Materials Research*, **4**, 373-384.

von Windheim, J. A. and Glass, J. T. (1992) Improved Uniformity and Selected Area Deposition of Diamond by the Oxy-Acetylene Flame Method, *Journal of Materials Research*, **7**, 2144.

von Windheim, J. A., Venkatesan, V., Malta, D. M. and Das, K. (1993) *Diamond and Related Materials*, **2**, 841.

Wolter, S. D., Glass, J. T. and Stoner, B. R. (1995) Bias induced diamond nucleation studies on refractory metal substrates, *Journal of Applied Physics*, **77**, 5119.

Wolter, S. D., McClure, M. T., Glass, J. T. and Stoner, B. R. (1995) Bias-enhanced nucleation of highly oriented diamond on titanium carbide (111) substrates, *Applied Physics Letters*, **66**, 2810.

Wolter, S. D., Stoner, B. R. and Glass, J. T. (1994) The effect of substrate material on bias-enhanced diamond nucleation, *Diamond and Related Materials*, **3**, 1188-1195.

Wolter, S. D., Stoner, B. R., Glass, J. T., Ellis, P. J., Buhaenko, D. S., Jenkins, C. E. and Southworth, P. (1993) Textured growth of diamond on silicon via in-situ carburization and bias-enhanced nucleation, *Applied Physics Letters*, **62**, 1215-1217.

Yang, P. C., et al. (1995) Nucleation and Growth of Oriented Diamond Films on Ni Substrates, in A. Feldman, Y. Tzeng, W. A. Yarbrough, M. Yoshikawa and M. Murakawa (eds), *Third International Conference on the Applications of Diamond Films and Related Materials, NIST*, vol. SP-885, pp. 329-332.

Yang, P. C., Zhu, W. and Glass, J. T. (1993) Nucleation of oriented diamond films on nickel substrates, *Journal of Materials Research*, **8**, 1773-1776.

Yang, P. C., Zhu, W. and Glass, J. T. (1994) Diamond nucleation on nickel substrates seeded with non-diamond carbon, *Journal of Materials Research*, **9**, 1063-1066.

Yoshikawa, M., Ishida, H., Ishitana, A., Koizumi, S. and Inuzuka, T. (1991) Study of crystallographic orientations in the diamond film on the (100) surface of cubic boron nitride using a Raman microprobe, *Applied Physics Letters*, **58**, 1387- 1388.

Yoshikawa, M., Ishida, H., Ishitani, A., Murakami, T., Koizumi, S. and Inuzuka, T. (1990) Study of crystallographic orientations in the diamond film on cubic boron nitride using Raman microprobe, *Applied Physics Letters*, **57**, 428-430.

Yugo, S., Kanai, T., Kimura, T. and Muto, T. (1991) Generation of diamond nuclei by electric field in plasma chemical vapor deposition, *Applied Physics Letters*, **58**, 1036.

Yugo, S., Kimura, T. and Kanai, T. (1993) Nucleation mechanisms of diamond in plasma chemical vapor deposition, *Diamond and Related Materials*, **2**, 328.

Yugo, S., Kimura, T. and Muto, T. (1990) Effects of electric field on the growth of diamond by microwave CVD, *Vacuum*, **41**, 1364-1367.

Zhu, W., Sivazlian, F. R., Stoner, B. R. and Glass, J. T. (1995) Nucleation and Selected Area Deposition by Biased Hot Filament Chemical Vapor Deposition, *Journal of Materials Research*, **10**, 425.

Zhu, W., Wang, X. H., Stoner, B. R., Ma, G. H. M., Kong, H. S., Braun, M. W. H. and Glass, J. T. (1993) Diamond and β-SiC heteroepitaxial interfaces: A theoretical and experimental study, *Physical Review B*, **47**, 6529-6542.

Zhu, W., Yang, P. C. and Glass, J. T. (1993) Oriented diamond films grown on nickel substrates, *Applied Physics Letters*, **63**, 1640-1642.

Chapter 26
ION IMPLANTATION OF DIAMOND AND DIAMOND FILMS

R. Kalish
Solid State Institute and Physics Department, Technion,
Haifa, Israel 32000

S. Prawer
School of Physics
University of Melbourne, Parkville, Victoria, Australia 3052

Contents

1. Introduction

Ion-implantation is a method by which energetic atoms (ions) are forced into solid targets due to their high kinetic energy. It is commonly used as the method of choice to modify many of the near-surface properties of materials. The major applications of ion-implantation are in the field of Si based micro-electronics, where ion implantation has in many cases replaced diffusion as a means of introducing dopant atoms into semiconductors. Such doping applications also apply, in principle, to diamond. However, as will be shown below, the doping of diamond by ion implantation is much more complicated than the doping of other semiconductors due to the tendency of diamond to graphitize. On the other hand, ion implantation may be the only practical method for the doping of diamond with potential donors or acceptors (apart from B) due to the negligible diffusion of impurities in diamond and due to difficulties in the incorporation of dopants into the diamond lattice during CVD diamond growth.

Ion implantation is also used in fields other than semiconductors, for example, to improve the mechanical properties of materials or to synthesize new phases. These uses have also been applied to diamond yielding interesting and technologically promising results. In addition, ion implantation has been used as a method of introducing controlled amounts of lattice damage into diamond thus allowing the study of the transformation of metastable crystalline diamond to an amorphous structure and eventually to the most stable form of carbon, namely graphite. The understanding of the effects that ion-bombardment has on diamond, whether desirable (i.e. material hardening) or undesirable (i.e. degradation of semiconductor device performances) is important. Furthermore, the novel techniques of diamond film deposition or of ion-beam deposition of amorphous carbon, all hinge on the ability to build carbon materials with predominantly sp^3 (diamond) bonding, while preferentially removing undesirable sp^2 (graphite) bonds. The role that ion impact has on the formation and removal of these bonds under various external conditions (ion energy, ion species and target temperature) needs to be understood. Information about it can be obtained from experiments in which known amounts of damage are inflicted on diamond by ion-implantation.

This review is arranged as follows: In section two, a very brief review of the principles of the slowing down processes in solids is provided. In section three, the present understanding of the changes induced in diamond under ion impact is summarized, and in section four the various methods which have been used to remove the radiation damage and activate the implanted dopants are presented and evaluated. While p-type doping of diamond has been achieved by ion implantation (and also during growth), n-type doping is still an unmet challenge. The most recent results on attempts to achieve n-type doping using ion implantation are presented in the last part of section four. The application of ion implantation techniques to the doping of polycrystalline diamond films is outlined in section five, which compares the response to ion implantation of polycrystalline CVD and natural single crystal diamond. As will be made clear in section three, ion beam irradiation induces major changes in the chemical, structural and electrical properties of diamond. Normally, these changes are detrimental to the desired end-use of diamond as an active semiconductor; however some innovative (mainly non-electronic) applications for these ion beam induced modifications have now been demonstrated and these are reviewed in section six. We summarize the state-of-the-art and indicate some future trends for research in section seven.

2. The Slowing Down of Ions in Solids - Damage Processes

2.1 BASIC PRINCIPLES

Ion-implantation is a violent process in which energetic ions (the implants) are forced into the target material. During their slowing down in the solid, large amounts of damage are inflicted upon the target structure until the implants come to rest. It is important to bear in mind that, in general, and for diamond in particular, unless the implantation related damage is annealed out, most measurable changes in the implanted layer are due to the damage and not to the direct presence of the implanted species. The volume density of the energy deposited into the stopping medium by the ion during its slowing down is the important parameter which determines the damage inflicted on the material by each implanted ion.

The general way to treat the slowing down of an ion in matter is through the "stopping power" (dE/dx), defined as the energy dE lost by an ion traversing a distance dx [Lindhard et al. 1963, Ziegler et al. 1985]. The stopping power is usually considered to be comprised of two major independent components: The "electronic stopping" denoted by $(dE/dx)_e$ is the process in which collisions with electrons in the solid (considered free) slow down the moving ions. The "nuclear stopping," denoted by $(dE/dx)_n$ is the process in which the moving ion undergoes elastic collisions with the atoms of the target material, hence loosing energy by imparting momentum to target atoms. Of interest to the present discussion is mainly the nuclear stopping as it is the one which is responsible for most of the damage which accompanies the ion implantation. Because of the crucial importance of the slowing down process to all ion-implantation and ion-beam-probing experiments, it has been extensively studied, theoretically, experimentally and phenomenologically [Zeigler et al. 1985]. Whereas the electronic slowing down process does not inflict substantial damage to the target material, the nuclear stopping, being due to ion-host atom collisions, is accompanied by the dislodging of many atoms from their regular sites in the solid.

From the total energy loss, the ion range can be calculated according to

$$R = \int_{E_0}^{0} (dE/dx) \cdot dE \tag{1}$$

where the integration limits are from the initial ion energy to zero. Since the slowing down processes of an ion moving in an amorphous solid is a statistical process, the locations at which the implants finally come to rest are also of a statistical nature. The ion range R is related to the mean track length of the ion before coming to rest, while the projected range R_p gives the mean penetration depth of the ion relative to the surface. The final ion distribution in the projected range, R_p, is assumed to the first order, to be of Gaussian shape with a standard deviation, ΔR_p, referred to as the ion

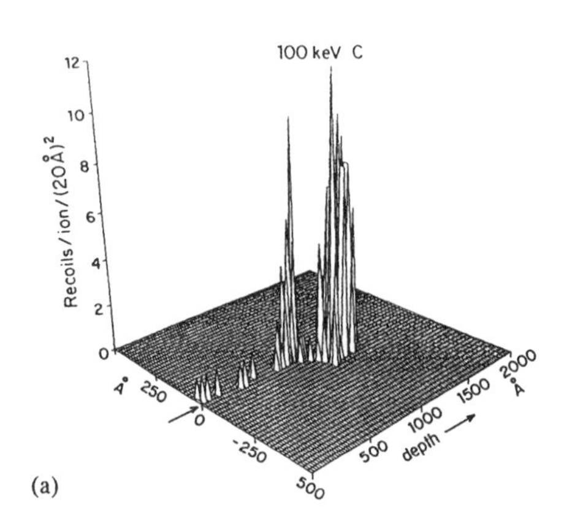

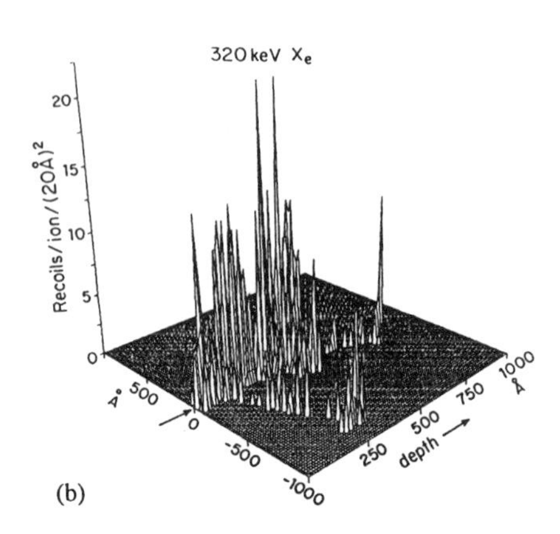

Figure 1. Collision cascade following the penetration into diamond of a single 100 keV C ion (Figure 1(a)) and a single 320 keV Xe ion (Figure 1(b)).

range straggling. It should be noted that the assumption of a Gaussian profile ignores channeling and diffusion, which may take place during the implantation as well [Feldman and Mayer, 1986].

2.2 COMPUTER SIMULATION OF ION BEAM INDUCED DAMAGE

More realistic information about the implant distribution in the target and the related damage inflicted is obtained from computer simulations, of which TRIM [Ziegler et al. 1985, and Biersack and Haggermark 1980] is the most commonly used one. The TRIM code is a Monte-Carlo program which exactly follows the collisions that individual ions undergo while in motion in the target material. The input parameters to the program are the required experimental conditions (ion type and energy, target material, composition and density) and an intrinsic parameter, E_d, which is the energy required to displace a target atom far enough from its lattice site so that it will not fall back into the vacancy that it has left behind. For diamond, E_d has been estimated both experimentally and theoretically to be 45-55 eV [Prins et al. 1986; Davies, 1994; Wu and Fahy, 1994].

The procedure followed by the TRIM program is to shoot the required ion into the target; the ion then proceeds into the solid gradually losing its energy to electrons until it collides with a target atom with a randomly selected impact parameter. The program calculates the kinematics of the collision, and then follows the trajectory of the recoiling target atom which itself undergoes collisions that may put new target atoms in motion, thus creating a cascade of recoiling target atoms. The program will return and follow the motion of the original implant, once all recoiling target atoms have attained energies below a preset cutoff energy (~2 eV). The projectile may now proceed with its new energy and direction of motion until another statistically selected collision occurs with

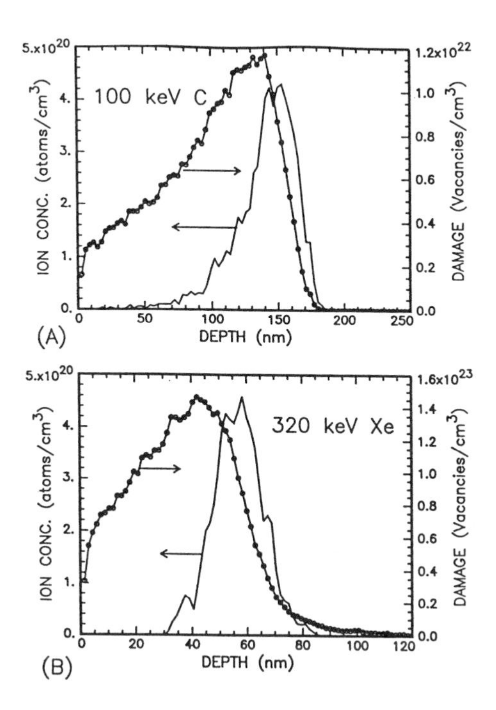

Figure 2. TRIM calculated depth profiles of implanted species and vacancy distributions for a dose of 1×10^{15} ions cm^{-2} of 100 keV C (Figure 2(a)) and 320 keV Xe implantations (Figure 2(b)).

its branch of recoils. The procedure continues until the energy of the implant itself is below the cut-off energy and it comes to rest. The program now stores the collision history of the primary ion and of all secondary events and proceeds to follow a new ion shot into the target.

As will become clear below (section 3.2), the response of diamond to 100 keV C and 320 keV Xe irradiation has been extensively investigated experimentally. Hence we present here the application of TRIM for these two cases. Typical damage cascades created by a single light (C) and a single heavy (Xe) ion implanted at 100 keV and 320 keV, respectively into diamond are shown in Figs. 1 (a) and (b). The damage cascade is best described as clusters of point defects surrounding the ion track. Note that the clusters appear to be larger for the Xe than the C case, due to the greater mass of the Xe ion.

Figures 2 (a) (100 keV C) and (b) (320 keV Xe) show the number of vacancies/cm^3 created for an ion dose of 1×10^{15} cm^{-2} as a function of depth from the surface, together with the final distribution of the implanted ions. Using a value of 45 eV for the displacement energy E_d, the TRIM calculations show that for 100 keV C ions a total of about 110 vacancies are created per incident ion, whilst about 710 vacancies are created per ion for 320 keV Xe implantations. Since the energy required to create a single vacancy is of the order of the binding energy of the solid i.e. about 2 eV/atom it is clear that only a very small fraction of the incident ion energy is in fact converted into the creation of vacancies and interstitials. In fact, most of the energy is lost via electronic energy loss processes (which create little or no permanent damage) or in nuclear knock-on events which impart an energy less than E_d to the C target atoms. In this latter case the energy imparted to the C atom is dissipated in the production of phonons; hence once again no permanent damage is produced. Nevertheless, we once again stress that despite the fact that only a small fraction of the incident ion energy results in the formation of vacancies and interstitials it is this fraction which drives the ion beam transformation of the material (see section 3.2 below).

Even though the TRIM simulation gives excellent insight into the implantation process and the damage which accompanies ion implantation, the simulation program ignores the possibility that the implanted ions undergo accidental channeling when in motion in a single crystal and the possible rearrangement of atoms in the implantation-affected volume to form new phases, such as the conversion of diamond into graphite. Dynamic annealing, i.e. the ability of the displaced atoms to regain their lattice positions during the ion irradiation is also not taken into account by TRIM. Nevertheless, TRIM and similar computer simulation programs give excellent estimates of both the implant and the damage profiles, and hence offer very valuable information on the state of a target after ion implantation. They are most applicable to the implantation conditions in which no diffusion or bond rearrangement takes place, i.e. implantations performed at low temperatures. As will be shown below for diamond, 'low temperatures' means implantations performed at or below room temperature since at these temperatures both interstitials and vacancies are believed to be immobile.

3. Ion Beam Induced Damage in Diamond

3.1 STRUCTURAL CHANGES

For covalently bonded solids high dose ion implantation usually leads to amorphization [Williams, 1986]. For diamond an additional level of complexity is involved since the bonds broken by ion impact may rearrange to form the more stable sp^2 structure. Indeed it is known that high dose implantation of diamond leads to graphitization of the irradiated volume [Vavilov, 1974]. Any attempt to dope diamond by ion implantation must first take into account the nature of this radiation damage which accompanies the implantation; therefore an understanding of the ion beam induced structural transformation of diamond and its evolution following thermal treatments is very important.

The facts that diamond is metastable with respect to graphite at room temperature and pressure and that the physical, electrical, and chemical properties of these two carbon allotropes are about as different as can be, have important consequences for the nature of damage in diamond and for its annealing. The transformation of diamond to graphite, which can be thermally driven, is much enhanced when a sufficient number of sp^3 bonds are broken by ion implantation. Such an ion beam induced transformation of diamond to graphite is accompanied by dramatic changes to the material, including large changes in density, electrical conductivity, optical transparency, hardness and chemical stability. Since the passage of energetic ions through matter is accompanied by massive bond breakage, it is expected that there should be a threshold damage level in implanted diamond, beyond which the diamond structure can not recover any more, but will, instead, "collapse" to the graphite phase. This transformation of the implantation affected volume should be easily noticeable by a large increase in density, a tremendous increase in electrical conductivity, a mechanical weakening of the material, the appearance of new optical absorption lines, a dramatic change in the Raman spectrum, and changes in chemical properties (graphite being etchable, while diamond is extremely chemically inert). Such changes have indeed been observed in implanted diamond, the main features of which have been studied by various techniques. We list these only briefly as they have been well reviewed previously [Dresselhaus and Kalish, 1992; Prins, 1992]. Rutherford Backscattering Channeling (c-RBS) [Braunstein et al. 1980] and electron diffraction [Vavilov et al. 1974] measurements show that as a result of ion beam induced damage, diamond changes from a perfect crystal to some disordered structure; EPR shows that many bonds are broken [Tiecher and Besserman, 1982, Brosius et al. 1974]. Optically, the transparent diamond turns black and new absorption lines appear [Connel et. al 1988] and Raman scattering shows that as a result of high dose implantation diamond can either transform into amorphous carbon (sp^2) or into micropolycrystalline graphite, depending on implantation temperature [Sato and Iwaki, 1988]. Chemically, the inert diamond changes to an etchable material and electrically, the conductivity changes from highly insulating to highly conducting with a conductivity approaching that of graphite [Prins, 1983]. For low dose implantations, there is some data which suggests that the diamond exhibits p-type conduction for implantations performed at [Fang et al. 1989] and below [Fang et al. 1989; Uzan-Saguy et al. 1995a] room temperature, but n-type for implantations at elevated temperatures [Fang et al. 1989]. Electron spectroscopies also show characteristic signatures for the transformation from diamond-like sp^3 bonding to graphite-like sp^2 bonding upon ion impact [Hoffman et al. 1992]. Many cathodo- and photo-luminescence lines also appear [Zaitsev 1991], one of the most notable of which is the GR1 center (1.673 eV) which is believed to be due to the presence of neutral vacancies [Davies, 1994]. The interested reader should also consult the work of Dresselhaus and Kalish [1992] and Prins [1992] for comprehensive reviews of the nature of this transformation. These changes are all characteristic of the ion beam induced conversion of diamond to graphite, or more precisely, to a form of sp^2 bonded carbon, the nature of the "end-product" at high doses depending primarily on the irradiation temperature (see below). Since the focus of this chapter is on the use of ion implantation for the doping of diamond, the next section will be devoted to a review of the changes to the electrical properties of diamond as a result of ion beam damage. Conveniently, these changes are easy to detect and may be used to gain an understanding of the effects of ion beam damage from both the macroscopic and microscopic points of view.

3.2 ION BEAM INDUCED CHANGES TO THE ELECTRICAL PROPERTIES OF DIAMOND

Since diamond (sp^3 bonding) is an insulator and graphite (sp^2bonding) is a conductor, and since ion implantation breaks many sp^3 bonds and allows their conversion to sp^2 type bonding, it is not surprising that many studies have used ion beam conductivity measurements to monitor the ion beam induced structural modification of diamond [Vavilov et al. 1974; Sato and Iwaki, 1988, Prins 1983; Kalish et al. 1980; Prawer et al. 1990; Prins 1985]. Recently an extensive set of such measurements has been performed

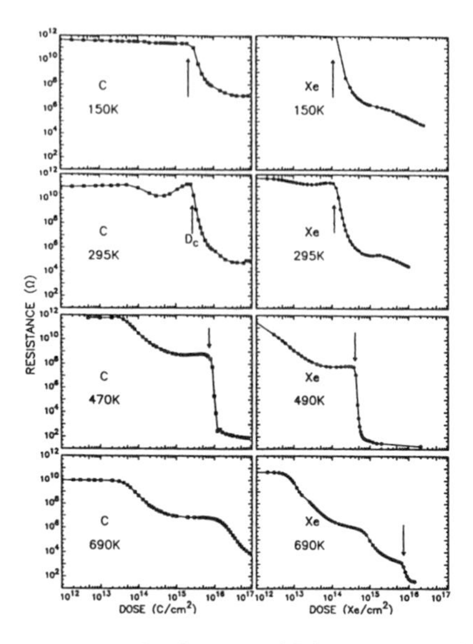

Figure 3. Resistance of Diamond as a function of ion dose for 100 keV C and 320 keV Xe implantations. The implantation temperature is shown on the figure. The arrows indicate the critical dose, D_c, at which overlap between the damage spheres occurs (see text) [Prawer and Kalish 1995a].

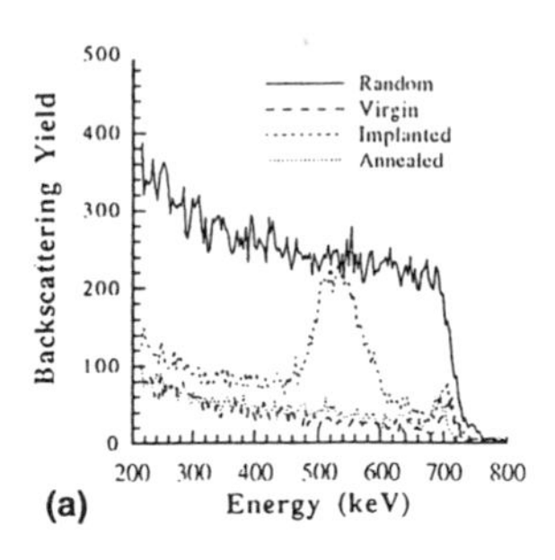

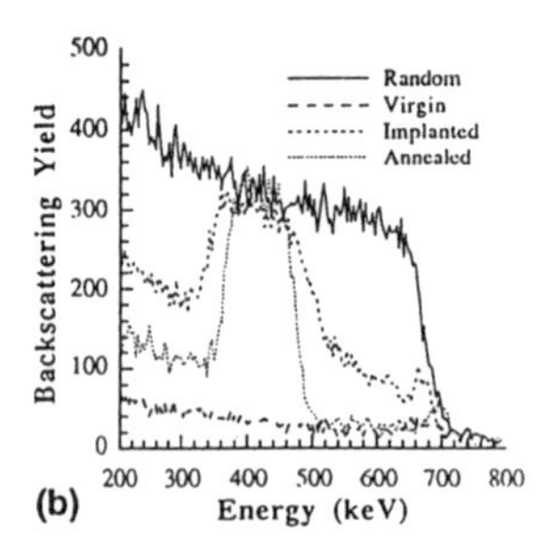

Figure 4. (a) RBS channeling of a diamond implanted with $1.5 \cdot 10^{15}$ C cm^{-2} at 100 keV and annealed at 950° C for 30 min. [Parikh et al. 1993], (b) RBS channeling of a diamond implanted with $3.0 \cdot 10^{15}$ C cm^{-2} at 100 keV and annealed at 950°C for 30 min. [Parikh et al. 1993].

over a wide variety of implantation temperatures for both light (C) and heavy (Xe) ions [Prawer and Kalish 1995a]. The results of these measurements are shown in figure 3, and serve to conveniently demonstrate the most important features of the ion beam transformation of diamond.

It is immediately obvious that for each case, there exists a critical dose D_c (marked by arrows in figure 3) at which the resistance begins to fall rapidly, and that the value of D_c depends on ion species and on implantation temperature. Examination of figure 3 reveals that the functional form of the dependence of R(D) is remarkably similar for both Xe and C implants for the same implantation temperature with the only difference being that the curves for C are displaced in dose by a factor of about 20 as compared to Xe. This factor can be entirely accounted for by taking into consideration the difference between nuclear energy deposited by the Xe and C ion in the irradiated volume. Each Xe ion deposits about 20 times more nuclear energy per target C atom than does each incident C ion. Hence it is quite understandable why the transformation takes place at a much lower Xe than C dose. In fact, D_c can be normalized to take into account the damage caused by different ions by expressing D_c in terms of the nuclear energy deposited per target C atom in the irradiated volume. The later can be extracted from TRIM calculations. When this is done for room temperature implantations, D_c has been shown to correspond to about 5.5 eV/target atom for C, Ar, Sb, and Xe irradiations [Vavilov et al. 1974, Kalish et al. 1980, Prawer et al. 1990, Hoffman et. al 1992], thus demonstrating that it is the volume density of defects which drives the transformation for both light (C) and heavy (Xe) ions. The absolute value of this threshold energy depends on the details of the TRIM calculations. Unfortunately there appears to be some inconsistency between the above value of 5.5 eV/(target atom) calculated using TRIM88 (i.e. the version of TRIM issued in 1988) and a value of about 18 eV/(target atom) calculated using TRIM91. However, it must be stressed that although the absolute value of the critical energy necessary to trigger the transformation appears to depend on the version of TRIM used in the calculation, the scaling between different ions still holds. In other words, the conclusion that it is nuclear knock-on events which drive the ion beam induced transformation is valid no matter what version of TRIM is used. Perhaps a more useful way to interpret the value of D_c is in terms of the vacancy concentration required to bring about the collapse of the diamond structure. Recent work [C. Uzan-Saguy et al. 1995b] has shown that this concentration is about 1×10^{22} vacancies cm^{-3} (for implantations at or below room temperature).

Furnace annealing experiments demonstrate that for damage levels due to implantation doses below D_c, thermal annealing can restore the diamond to a form close to its pristine state [Braunstein et al. 1980]. However, for $D > D_c$ thermal annealing leads to the conversion of the damaged diamond layer to graphite, as is shown, for example, by the susceptibility of the material to etching in hot acids. Figure 4(a) [Parikh et al. 1993] shows the channeling RBS spectrum of diamond irradiated with 1.5×10^{15} C cm^{-2} at 100 keV (i.e. $D < D_c$). Following annealing, the RBS yield is returned to the value for virgin diamond, indicating repair of the lattice. However, Figure 4(b) [Parikh et al. 1993] shows that for diamond implanted with 3×10^{15} C cm^{-2}, (i.e. $D > D_c$) thermal annealing results in the formation of a buried layer whose RBS yield reaches the random level. This buried layer is either amorphous sp^2 bonded carbon, or more likely, fine grain polycrystalline graphite. Hence, these measurements show that for thermal annealing of diamond to be able to restore the diamond structure, the dose must be kept below D_c.

3.3 THE ROLE OF IMPLANTATION TEMPERATURE, T_i.

Figure 5 [Sato and Iwaki, 1988] shows the sheet resistivity of diamond as a function of the implantation temperature for diamond irradiated with doses of 2.5 and 5 x 10^{16} Ar cm^{-2}. At these high doses, the resistivity of the diamond saturates as a function of ion dose (see Figure 3). Note the very strong dependence of the resistivity as a function of implantation temperature, which happens particularly close to "room temperature." The importance of this is that slight variations in implantation power (i.e. beam current or energy) or differences in diamond heat sinking may result in a very large spread in the properties of the ion-implanted layer. Hence accurate temperature control is important, in particular for implantations at or near room temperature.

On the basis of swelling, conductivity and other experimental data, Prins [1992] and others have concluded that there are three important implantation temperature regimes which determine the type and behavior of the resultant damage when implanting diamonds:
(i) Low temperature regime (T_i < 320 K) in which both interstitials and vacancies are immobile and the implantation induced damage is "frozen in."
(ii) Intermediate temperature regime (320 < T_i < 800 K) in which interstitials are mobile while vacancies are not.
(iii) Higher temperatures (T_i > 800 K), in which both interstitials and vacancies can diffuse.
Indeed from figure 3 it is also obvious that not only does the saturation resistivity depend on the implantation temperature, but also the critical dose itself increases with increasing implantation temperature. Prins [1985, 1992] has explained such behavior in

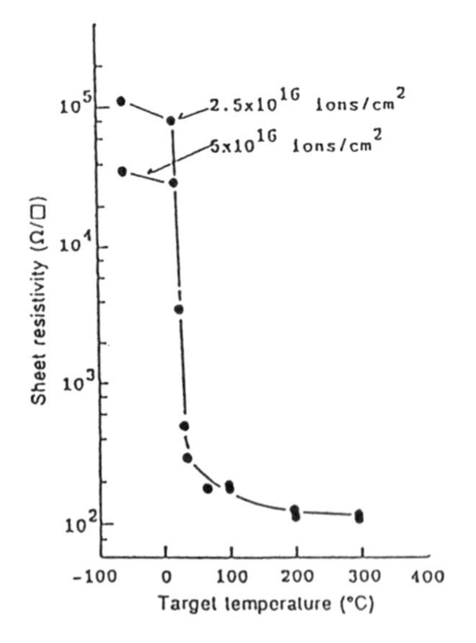

Figure 5. Sheet resistivity as a function of implantation temperature for diamond implanted with 2.5 and 5.0 x 10^{16} Ar cm^{-2} [Sato and Iwaki 1988].

terms of the average vacancy and interstitial concentrations remaining in the irradiated volume for different implantation temperatures. A different approach which attempts to explain the existence of a critical dose, D_c, for the transformation of diamond and its dependence on implantation temperature in terms of a microscopic model has recently been published [Prawer and Kalish, 1995a]. It was assumed that the passage of each ion leaves in its wake a 'trail' of damaged spheres of average radius r (see Figure 1). The conductivity and size of each of these spheres depends on the implantation temperature. When the density of these spheres reaches a sufficient concentration, a connective pathway may be formed between them giving rise to a sharp increase in the conductivity. Following the approach of Kalish et al. [1980], the critical dose, D_c, was assumed to be related to the radius, r, of the conducting sphere by

$$r = \left[\left(0.135\left(R_p + \Delta R_p\right)\right)/D_c\right]^{1/3} \tag{2}$$

where R_p and ΔR_p are the range and straggling respectively of the implanted ion.

Hence by measuring D_c under a variety of conditions, the radii of the spheres could be calculated, and these are shown as a function of implantation temperature in Figure 6. The damage spheres created by the passage of a Xe ion are about twice the radius of those produced by the C irradiation; a difference largely accounted for by the fact that Xe, being heavier than C, creates larger damage spheres in the damage cascade. Clearly r diminishes with increasing T_i due to the increased mobility of defects at elevated temperatures which results in dynamic annealing. Morehead and Crowder [1970] originally proposed a model to explain the temperature dependence of the critical dose for the ion beam amorphization of silicon which can be applied to the data in figure 6. In their original model they assumed that the radius of the damaged (i.e. amorphous) region left behind after the passage of an ion in Si decreases with increasing implantation temperature. The degree of "shrinkage" of these spheres at different T_i was considered by them to be due to dynamic annealing caused by the diffusion of point defects and their annihilation during the thermal spike which accompanies the passage of an ion through the solid. Hence the degree of shrinkage will depend on the activation energy for the diffusion of defects during the ion beam irradiation.

If we replace the amorphous spheres considered by Moorehead and Crowder [1970] by conducting spheres of graphite-like material, their model can be applied to the data in figure 6. Details of the model may be found in Kalish and Prawer [1995], but the point here is that the fit of the data of figure 6 to the functional dependence predicted my Moorehead and Crowder [1970] yields an activation energy for the defect diffusion of about 0.2 eV for both the Xe and C cases. Given the results of Prins [1992] that in the temperature range investigated up to 700 K, only interstitials are expected to be mobile, we assume that the above determined value is the activation energy for the diffusion of interstitials in diamond.

Note that the data for the size of the damage sphere extrapolates to zero radius for T_i = 805 K, i.e. at this temperature the damage introduced by the passage of the ions instantaneously anneals so that no identifiable damage region is left behind. Diamond

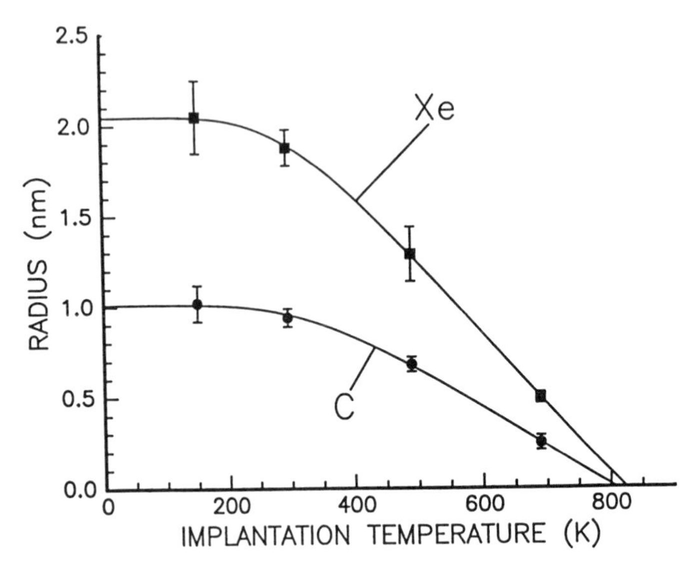

Figure 6. The radius of the damage spheres as a function of implantation temperature for 100 keV C and 320 keV Xe irradiations. The solid lines are fits to the data which allow the extraction of an activation energy of about 0.2 eV for defect diffusion during ion implantation. [Prawer and Kalish 1995a, Kalish and Prawer 1995]

implanted at or above this temperature will never graphitize, no matter how high the dose. The estimate of the 805 K as the temperature at which the radius of the damage sphere shrinks to zero is in reasonable agreement with the temperature at which the swelling which accompanies the ion beam irradiation diminishes to zero [Spits et al. 1990] and corresponds to the minimum temperature at which vacancies are believed to become mobile [Prins 1992]. This is also the temperature at which high dose C implantation can lead to the growth of new diamond layers, rather than resulting in graphitization [Freeman et al. 1978; Nelson et al. 1983].

3.4 AMORPHIZATION OF DIAMOND

The relationship between D_c as described above and the dose required to "amorphize" diamond requires a brief comment. The existence of a critical dose for the amorphization of diamond, in analogy to the case of damaged silicon has long been recognized [Dresselhaus and Kalish 1992; Prins, 1992; Kalish 1993]. It may be taken as the dose at which the c-RBS spectra first reaches the random level. An alternative and perhaps more useful definition is the dose at which furnace heating cannot anneal the damaged layer back to diamond but instead results in the formation of a thermally stable graphite-like layer [Spits et al, 1994]. A careful comparison of the available literature shows that for implantation at or below room temperature, the "amorphization dose" is always somewhat less than the dose at which the conductivity begins to increase sharply (i.e. D_c) [Hoffman et al, 1992; Uzan-Saguy et al. 1995a]. The conclusion reached, then, is that for $T_i < 320$ K, the transition from sp^3 to sp^2 bonded carbon is inhibited until the diamond lattice is amorphized, a conclusion reinforced by recent electron spectroscopic measurements of Ar irradiated diamond [Hoffman et al,

1992]. This observation may be important in explaining the success of doping methods based on cold implantations as described below.

3.5 SUMMARY

The picture to emerge of the structural changes in diamond implanted at different temperatures with different ions is as follows:
(i) For $T_i < 320$ K, implantation results in the formation of point defect clusters which at a critical dose D_c overlap to create an a-sp^3 bonded structure. For doses below D_c the diamond can be furnace annealed. This critical dose, Dc, corresponds to an ion beam induced defect concentration of about 1×10^{22} vacancies cm^{-3}, i.e. about 6% of the C atoms have been dislodged from their lattice locations. This value holds for a wide range of ion species and implantation energies. For doses exceeding D_c, annealing is no longer possible, and with further increases in ion dose a gradual increase in the conductivity occurs as sp^3 bonded material is gradually converted into sp^2 bonded micropolycrystalline graphite.
(ii) For intermediate temperatures ($320 < T_i < 800$ K), the defect mobility is sufficient to enable many of the point defects to dynamically anneal, but at the same time sufficient activation energy is available to enable those that survive to cluster into graphitic islands. When these islands overlap at D_c, a percolative transition occurs which results in a very sharp increase in the electrical conductivity.
(iii) For hot implantations ($T_i > 800$ K) the defect mobilities are sufficiently high such that dynamic annealing prevents any permanent damage zones from being created. In this temperature regime, graphitization never occurs, even for very high ion doses. However, defect clustering still takes place and diamond implanted at these elevated temperatures is usually rich in dislocations and other extended defects [Braunstein and Kalish 1983].

4. Doping by Ion Implantation: Annealing and Dopant Activation Schemes

For effective doping of diamond, the ion beam induced radiation damage must be annealed and a sufficient fraction of the dopants must occupy the appropriate sites in the repaired diamond lattice. The common way to restore the structure of an implantation damaged single crystal layer is by post implantation annealing. However, in contrast to most damaged crystals which will recover to their original structure upon heat treatment, diamond will not necessarily do so because of the metastability of the diamond phase under ambient conditions. Most of the work to date has concentrated on B implantation, as B is a well proven substitutional p-type dopant in diamond with an acceptor level 0.37 eV above the valence band.

4.1 HOT IMPLANTATION

This approach relies on the instantaneous annealing which takes place during implantation into hot ($T_i > 800$ K) diamond. As shown above, implantation into diamond held at elevated temperatures does not lead to any permanent damage (apart from possible extended defects) due to the dynamic annealing of the damage created within the cascade, i.e. the complete shrinkage of the damage spheres surrounding each

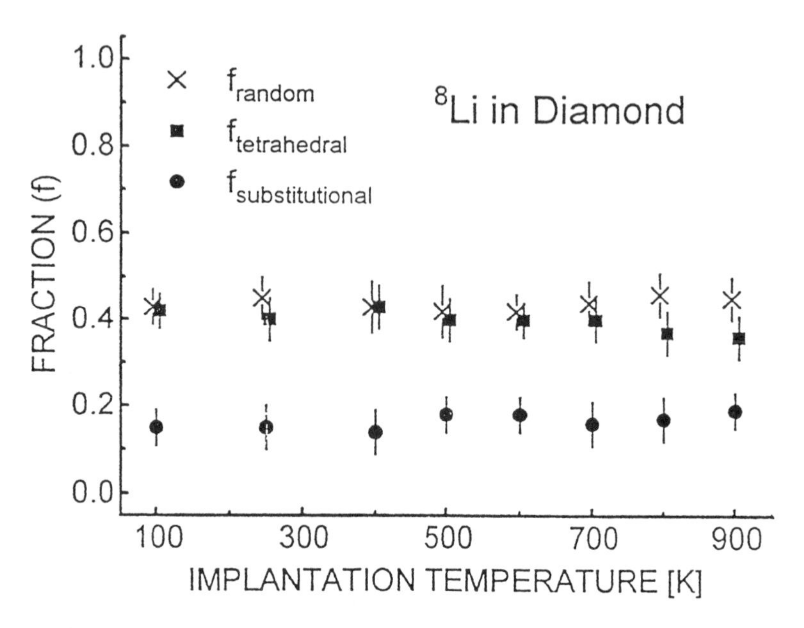

Figure 7. Fractions of Li atoms located on tetrahedral interstitial sites ($f_{tetrahedral}$), substitutional sites ($f_{substitutional}$) and random fraction (f_{random},) as a function of implantation temperature for diamond. [Restle et al. 1995].

ion track. Hence, advantage can be taken of these hot implantations to avoid the need for post-implantation annealing.

Early work by the Harwell group [Hartley and Poole 1973; Hartley 1982] showed that diamond growth can be achieved by high dose C implantations into hot diamond; hence the extension of this method to dope diamond is obvious. Indeed both the Harwell [Buckley-Golder et al, 1991] and Technion groups [Braunstein and Kalish 1981a] have employed hot implantations to dope diamond p- and n- type by B and Li implantations.

It is interesting to note that although implantations into heated diamond diminishes the accumulation of point defects and often drive the implants onto lattice sites where they are expected to be electrically active [Braunstein and Kalish, 1981a; Braunstein and R. Kalish 1981b], the overall results on activation energies and, in particular, carrier mobilities are disappointing [Zeidler et al. 1993a]. The poor carrier mobilities are probably due to defect clusters which act as traps. These defects are almost impossible to remove by post implantation annealing, even at the highest annealing temperatures employed. A similar situation is found for doping of Si; high temperature implantations can avoid amorphization but the defects left behind are very difficult to remove and degrade the electrical performance.

An interesting result, using emission channeling measurements on diamond implanted with radioactive ^{8}Li is shown in figure 7 [Restle et al. 1995], which displays the fraction of Li residing on interstitial, substitutional and random sites as a function of the implantation temperature. This figure shows that the fraction of Li occupying a particular site is virtually independent of implantation temperature. Li is expected to be an interstitial dopant in diamond, and Restle et al. [1995] suggest that the partial incorporation of Li onto substitutional sites may explain the difficulties of n-type doping of diamond using Li ion implantation (see section 4.5) because Li donors are compensated by Li acceptors. It should be noted that the behavior of substitutional type dopants may be quite different from the Li case reported here, and therefore one should not draw the conclusion that the incorporation of dopants onto lattice sites is, in general, independent of implantation temperature.

4.2 HIGH DOSE IMPLANTATION FOLLOWED BY CHEMICAL ETCHING

An alternative method for achieving implant activation in diamond relies on high dose implantations (exceeding the graphitization limit, D_c) followed by high temperature (1400° C) annealing and subsequent chemical removal of the graphitized top layer [Braunstein and Kalish 1983]. This process has been shown to leave, after graphite removal, a surface layer of intact diamond, which contains of the "tail" of the implant distribution, with the implants residing on lattice sites where they are expected to be electrically active (B on substitutional and Li on interstitial sites [Braunstein and Kalish 1981a]. Indeed, by employing this method and performing control experiments on samples implanted with C ions (i.e. an inert, non-dopant ion) at similar energies and doses and then subjected to the same annealing and etching treatment as for the B implanted samples, Braunstein and Kalish [1983] could unambiguously show that p-type activity due to B doping had been achieved.

At first glance this is an attractive and simple technique which can give some electrical activation of B dopants. However, it is clear that little or no control of the dopant concentration is possible, as this is a self limiting process which is determined primarily by D_c. Also, the carriers are located in a very shallow region close to the surface and the carrier mobilities appear to be very low. The latter is probably due to the very high concentration of B (typically 6 x 10^{21} B cm^{-3}) in the implanted layer. The technique does, however, find a major use in the fabrication of ohmic contacts to p-type diamond (see 4.6 below).

4.3 COLD IMPLANTATION FOLLOWED BY RAPID ANNEALING TECHNIQUE (CIRA)

The most successful schemes for reliable doping appear at present to involve cold implantations followed by rapid thermal annealing, the so-called CIRA techniques [Prins 1993a, Prins 1993b and references therein]. The basic idea behind the CIRA scheme relies on freezing in the point defect damage by employing low temperature implantations. The diamond is kept cold until it is ready to be rapidly thermally annealed. When the layer is annealed, a fraction of B interstitials can successfully combine with vacancies (and thus become activated) while the self-interstitials can recombine with vacancies to anneal the residual damage. This is less likely to occur for implantation temperatures in excess of 320 K, in which case the C self-interstitials can diffuse and recombine with vacancies during the implantation, thus depleting the very sites which the B implants need in order to become activated. There is also evidence to suggest that these C self-interstitials can combine to form stable defect clusters, even for doses much below D_c, further reducing the efficiency of the implantation process. For

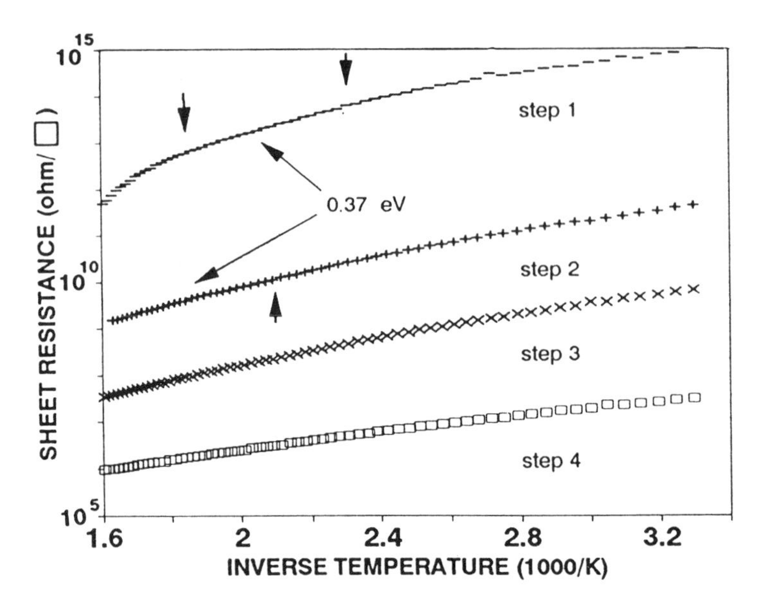

Figure 8. The temperature dependence of the sheet resistivity of a diamond subjected to multiple cold-implantation, rapid thermal annealing steps. The B dose was 3 x 10^{15} cm^{-2} spread over different energies to generate a thicker implanted region. After each implantation step, the diamond was heated to 1200°C by means of rapid thermal annealing. After the third step the acceptor density became large enough to cause impurity band conduction over the whole of the temperature range studied [Prins 1993b].

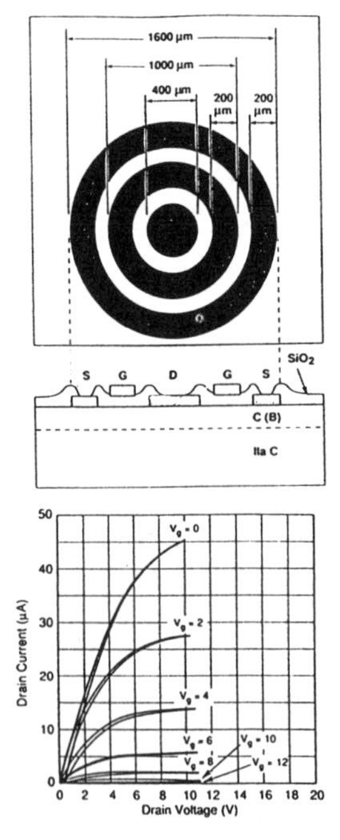

Figure 9. Schematic outline of transistor structure (a), and its room temperature transistor characteristics (b). The B implanted regions were produced by cold implantation followed by rapid thermal annealing. [Zeidler, 1993a].

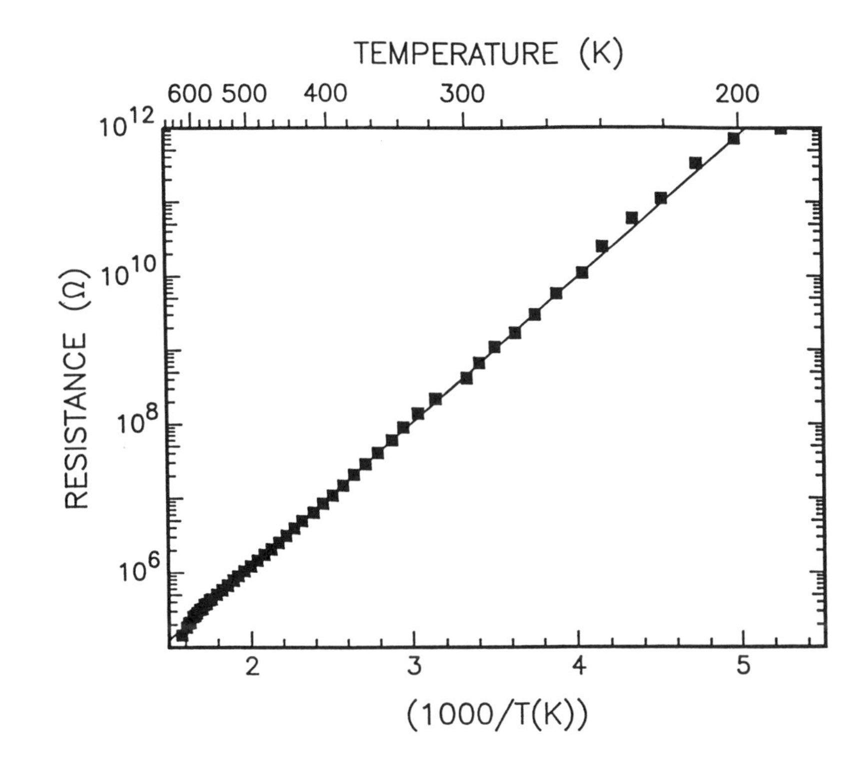

Figure 10. Sheet resistance as a function of temperature for a diamond implanted with B using the cold implantation rapid thermal annealing scheme. Note the activation energy of 0.38 eV over the whole of the temperature regime 190 to 630 K. [Fontaine et al. 1996]

effective annealing the ion dose must of course be kept below D_c, but, in addition, in order to maximize the chance for the B interstitial-vacancy combination, it is important to select implantation energies and doses such that the vacancy, interstitial, and B implant distributions overlap as much as is possible. In order to accomplish this, Prins [1988] and later Sandhu et al. [1989] have co-implanted C together with B, and have varied the C to B fraction to search for the level at which dopant activation is maximized. However, later work has indicated that this coimplantation step is probably unnecessary. Furthermore, the annealing temperature needs to be carefully selected as there is some evidence to suggest that self-interstitial-vacancy recombination rates increase with increasing temperature, whilst boron-interstitial-vacancy combinations decrease [Prins, 1991]. Finally, Prins and coworkers have recently been experimenting with a multiple step CIRA technique [Prins, 1993b]. The basic idea here is that by using many implantation/annealing cycles, with small doses at each step, it is possible to introduce a much higher level of activated B dopants onto lattice sites than is the case with a single B implant, without increasing the number of compensating donors due to residual damage. Figure 8 [Prins 1993b] shows the temperature dependence of the sheet resistance of a diamond implanted with B spread over different energies to produce a rather homogeneous, thicker doped layer with small dose increments, each followed by a rapid anneal for 5 minutes at 1200° C to remove the radiation damage. After the third step the acceptor density became large enough to cause impurity band conduction, as evidenced by the decrease in activation energy.

Zeidler et al. [1993a] have directly tested the effectiveness of the CIRA scheme. They found that the highest carrier concentration and mobility are indeed produced for implantations at 77 K followed by rapid thermal annealing (RTA). Carbon co-implantation, whilst leading to increased hole concentration, unfortunately also leads to a decreased hole mobility, (presumably due to an increased defect density), thus limiting the usefulness of the implanted layers for electronic applications. The mobility could, however, be somewhat improved by a final high temperature anneal [Hewett et al. 1993]. Although the hole concentrations and the resistivities of the diamond layers are not very different from those obtained in natural type IIa diamonds, the mobilities are still relatively low (10-50 cm^2/V-s) for the B doped films as compared to natural type IIb diamonds (> 1000 cm^2/V-s). Despite this, the p-type doped layer produced by the above CIRA scheme was used in the fabrication of an insulated-gate field-effect transistor which, for some devices, demonstrated a voltage gain of 2 [Zeidler et al. 1993b]. A schematic of the FET produced by Zeider and its transistor characteristics are shown in Figure 9. Further improvements in hole mobilities appear likely in the near future. Very recent results obtained by Fontaine et al. [1996] are shown in figure 10, which shows the log(R) vs. 1/T plot for a diamond implanted using the CIRA scheme. The diamond was implanted at -40° C with 25 keV B (1.5 x 10^{14} cm^{-2}), 50 keV B (2.1 x 10^{14} cm^{-2}) and 100 keV B (3 x 10^{14} cm^{-2}) in order to produce a thicker doped layer. Immediately following the implantation, the diamond was rapidly heated (within less than 10 s) to 900°C and kept at this temperature for 10 minutes, followed by ex-situ annealing in a vacuum furnace at 1100° C for 1 hour. Note the activation energy of 0.38 eV over the entire temperature range $190 < T < 630$ K. The mobility of the carriers was measured by the Hall effect and was found for some diamonds to be as high as 300 cm^2/V-s.

4.4 PULSED LASER ANNEALING

The annealing schemes employed by most researchers use furnace or rapid thermal heating of the samples to temperatures up to about 1800 K and anneal times ranging from seconds to many hours. A completely different method for annealing deeply buried radiation damage in diamond has recently been proposed [Prawer et al. 1992; Allen et al. 1993] using pulsed lasers. In this method, the fact that damaged diamond absorbs more laser irradiation than undamaged diamond is used. By careful selection of the laser wavelength, it is possible to heat only the damaged diamond layer. The potential advantages of pulsed laser annealing over furnace annealing include: selective heating of the damaged layer is possible by careful choice of the laser wavelength, (ii) spatial selectivity is achievable, and (iii) in-situ annealing in the implantation chamber is, in principle, possible. In addition, pulsed laser annealing offers the opportunity of studying the annealing kinetics of diamond at extremely high temperatures (up to 5000 K) and pressures and much shorter anneal times (on the order of nanoseconds) than are typical of furnace annealing.

Although pulsed laser beams have been extensively used to graphitize, ablate, and etch diamonds, the successful use of a pulsed laser (pulse time 10-100 ns) to anneal ion implanted diamond has only recently been reported [Prawer et al, 1992, Allen et al. 1993]. It was found that in order to prevent graphitization and/or ablation it was

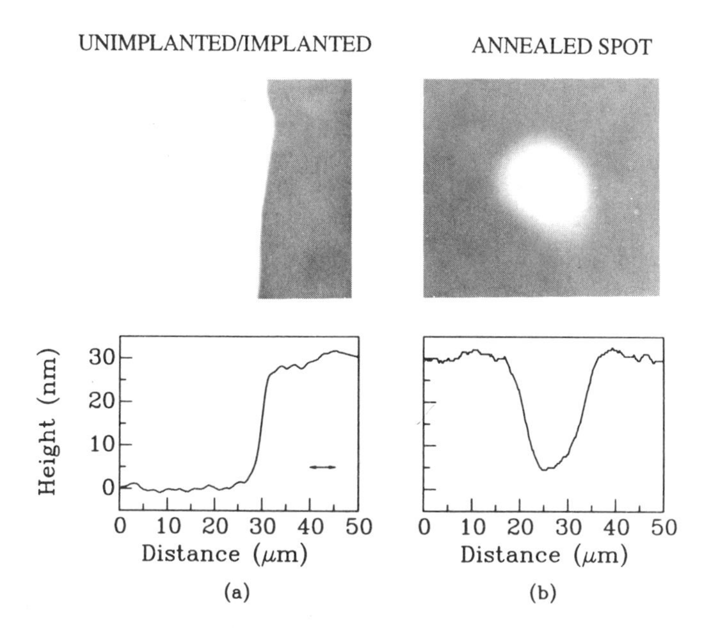

Figure 11. (a) Optical micrograph and surface profile of the boundary between implanted (1×10^{15} P cm^{-2} at 77 K, 4.0 MeV) and unimplanted portions of a diamond showing the increase in optical absorption and ion implantation induced swelling, respectively. The micrographs and profiles have the same lateral scale. (b) The optical micrograph of the annealed spot demonstrates the increased transparency upon laser treatment. The surface profile shows the compaction of the surface over the laser annealed spot [Allen et al. 1993].

necessary to use a constraining cap of undamaged diamond in order to maintain the damaged diamond under high pressure. This was accomplished by the use of deep MeV implants which create a layer of damaged diamond buried about 1 μm beneath the diamond surface. Over a small range of laser powers annealing of the radiation damage, formation of buried graphitic layers and melting of diamond followed by its conversion upon cooling into graphite were all observed [Prawer et al. 1992]. The results were explained in terms of the behavior of carbon under high internal pressures close to the diamond-graphite-liquid triple point in the carbon phase diagram.

The technique was later refined in order to anneal diamond implanted with 4 MeV P ions [Allen et al, 1993]. Optimization of the degree of annealing was found to be possible using multiple, variable energy, laser pulses. The laser power was ramped up with each successive pulse, the degree of ramping being limited by the need to avoid graphitization, which occurs if the laser power is too high. Figure 11 shows optical micrographs and step height measurements of the diamond before and after laser annealing. In Figure 11(a) the increased optical absorption and swelling of the diamond as a result of the P (1×10^{15} P cm^{-2} at 4 MeV) implantation are clearly evident. In Figure 11(b) the increased transmission and compaction of the laser annealed area are strong evidence for successful removal of the radiation damage. Figure 12 shows channeling Rutherford Backscattering (c-RBS) spectra from the as-implanted, and laser annealed sections of the sample. Figure 12(a) shows that the implantation has created a damaged buried layer in which the backscattered yield reaches 80% of the random level.

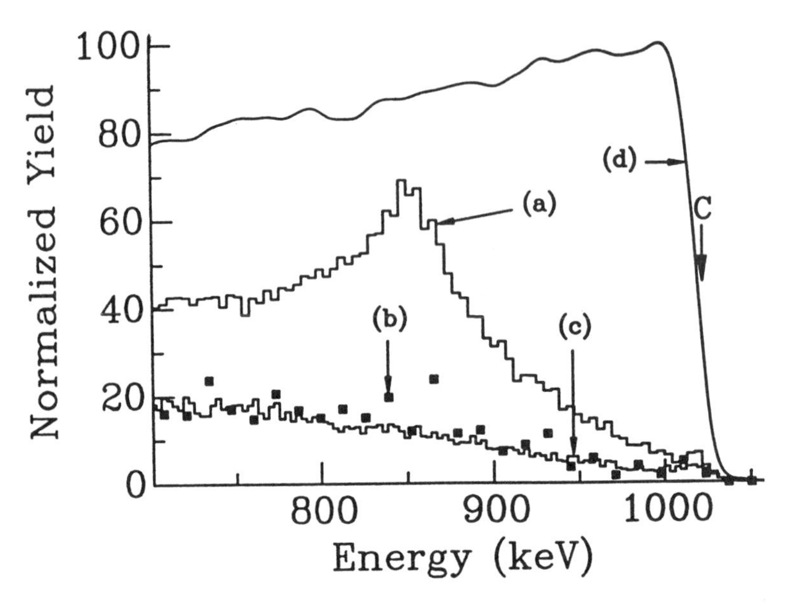

Figure 12: Channeling Rutherford Backscattering spectra of a type IIa diamond implanted with 4.0 MeV P (1 x 10^{15} cm $^{-2}$) taken with 1.4 MeV H^+ and a detector angle of 150°, displayed as the normalized yield compared to the spectrum from a randomly aligned sample (d). (a): the implanted region, showing a peak due to the damage buried at 1.3 μm, (b): the center of the annealed spot, (c): virgin diamond and (d): the spectrum in the randomly aligned orientation. The decrease in dechanneling in the laser annealed spot (c) to near its virgin value is indicative of excellent annealing of the radiation damage [Allen et al. 1993].

Figure 12(b) shows that following annealing the damage peak completely disappears and the crystal quality returns to near its pre-implantation value.

The fact that damaged diamond can be annealed by a short laser pulse is remarkable because the annealing takes place within an extremely short time, of the order of 100ns. This implies that the diffusion coefficient of defects at the annealing temperature must be very high. An estimate of the annealing temperature was obtained by solving the one-dimensional heat equation for the laser power employed; it was found that the implanted region reached a maximum temperature of about 2500 K [Allen et al. 1993]. Whether or not the short anneal times and very high annealing temperatures will result in a better dopant activation is the subject of ongoing research. In this regard, some encouraging results have been obtained recently using channeling Proton Induced X-ray Emission (c-PIXE), which demonstrates that about 50% of the P ions have occupied substitutional lattice sites following annealing [Prawer and Jamieson 1995], but the electrical characteristics of the implanted/ annealed layer have yet to be reported.

The work of Zaitsev [1991] suggests that the defect structures created by high energy ion implantation are qualitatively different from those appearing after 'conventional' ion implantation using keV energy ions. The suggestion is that under high energy implantation conditions quasi - one dimensional tracklike structures with diameters of the order of a few nm are formed in which there are zones constrained under high pressure [Zaitsev 1991; Erchak et al. 1992]. It appears that dopant diffusion may be enhanced along these tracks, and this might explain why the apparent defect mobility is so high for the laser annealed MeV implanted diamond.

Compared to furnace annealing, pulsed laser annealing of diamond is in its infancy. Whilst it is clear that the pulsed laser annealing technique is a viable alternative to furnace annealing for the removal of damage, it remains to be demonstrated whether this technique will lead to better activation of dopants. In addition, the fact that the implanted layer is buried below the surface and that the laser spot is small makes electrical characterization difficult. Nevertheless this technique certainly warrants further investigation.

4.5 N-TYPE DOPING OF DIAMOND

True n-type doping of diamond has proved to be an elusive goal. The obvious donor in diamond is nitrogen which does enter substitutionally into the diamond lattice, but unfortunately its donor level is 1.7eV below the bottom of the conduction band, too deep to be useful in device applications. Other potential n-type dopants from Group V of the periodic table have been tried without much success, (both during CVD diamond growth and by ion implantation), possibly because of the difficulty of "fitting" larger atoms into the very "tight" diamond lattice.

Recently, theoretical work has been published which investigates the possibility of achieving n-type doping of diamond with P, Na, and Li [Kajihara et al. 1991; Kajihara et al. 1993]. This important work showed that Li and Na should be interstitial dopants in diamond, occupying the tetrahedral interstitial site with donor levels 0.1 and 0.3 eV respectively below the conduction band, while P is predicted to occupy substitutional sites. The calculations showed that the solubilities of these three potential dopants are very low, so their incorporation into the diamond lattice via in-diffusion will be difficult, once again leaving ion implantation as the method of choice for their introduction into the diamond lattice.

The fact that Li and Na are interstitial dopants is important because it means that they do not have to compete with C interstitials for recombination with vacancies. The calculations also suggest that Li should be a fast diffuser at elevated temperatures, while Na should be relatively stable in its interstitial site in the diamond matrix. However, the prediction that Li should diffuse in diamond has not been verified since it has recently been shown that Li is rather immobile in ion-implanted diamond, even at high annealing temperatures [Cyterman et al. 1994]. On the other hand, Popovici et al. [1995] found that Li does diffuse into natural diamond, and make the point that in the case of ion-implanted diamond, Li (and other dopants) may be trapped at ion beam induced defects. Much more research is required to clarify the diffusion of dopants in diamond.

Recently, the electrical properties of diamond implanted with Li and Na have been experimentally determined [Prawer et al. 1993]. It is important to stress that in performing such experiments it is imperative to compare the results of the Na or Li implantation with those from control implantations with inert ions of comparable mass such as Ne (for Na) or C (for Li). Such control experiments are necessary because

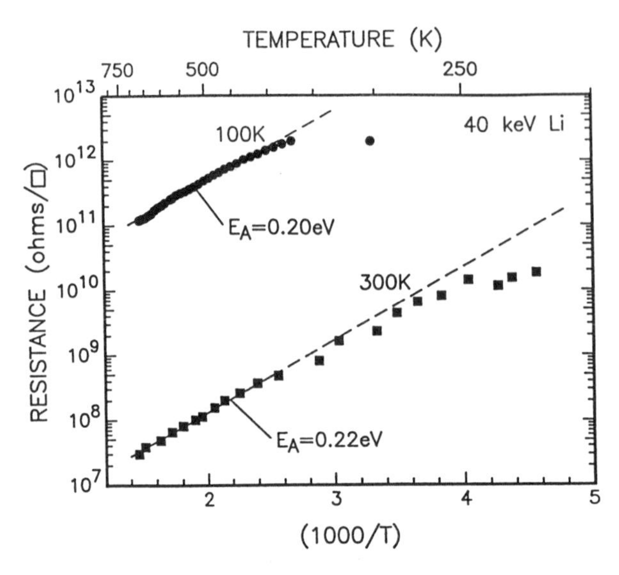

Figure 13. Temperature dependence of the resistivity of type IIa diamond implanted with 40 keV Li after furnace annealing (1400°C) and acid etching to remove the graphitized layer. The filled squares are for implantation with 1 x 10^{16} Li cm^{-2} at 300 K, and the filled circles are for implantation with 5 x 10^{16} Li cm^{-2} at 100 K [Prawer et al. 1993].

damaged diamond alone (even after thermal annealing) can display increased conductivity which may exhibit n-type features due to residual defects rather than the presence of chemical dopants [Prins 1982]. The results in figure 13 which show log(R) vs 1/T plots for diamond implanted with 40 keV Li under different conditions (see figure caption for details). Similar results were found for Na implantations. These results showed that Li and Na are indeed electrically active in the diamond matrix, with activation energies over a limited temperature range ($400 < T < 680$ K) of about 0.2 eV in each case. Interestingly, in this case the CIRA method did not yield the most encouraging results. Rather, the best results were obtained using the method outlined in section 4.2, in which the diamond was implanted with very high doses of Li and Na (well in excess of D_c). The material was then annealed, and the graphitized layer removed by etching in boiling acids. The control samples implanted with C and Ne ions and annealed and etched in a similar fashion did not display any electrical conductivity. However, as reported in the same work [Prawer et al. 1993], the R vs. T data for the Na and Li doped layers could also be well explained over a wide temperature range by variable range hopping conduction (whose signature is a linear log(R) versus $T^{-1/4}$ dependence [Mott and Davis 1979] rather than by simple activation, which would be the case if true n-type doping had been achieved. Prawer et al. [1993] also replotted some of the previously published R vs. T data to find that log(R) vs $T^{-1/4}$ plots often provided superior linear fits over the log(R) vs. T^{-1} plots. In particular, the variable range hopping dependence was found to very well describe the data for Li implantations performed at elevated temperatures by the Harwell group [Buckley-Golder et al. 1991]. However, despite the possible hopping nature of the conductivity, such Li doped layers were used successfully as the n-type part of a diamond p-n junction.

For the case of B implantations, it has taken many years of research for the right conditions to be found for effective p-type doping of diamond, and even now research continues in order to find the optimum implantation/annealing "recipe". Since Li and Na are interstitial dopants, it is likely that the optimum conditions will be different from those found for B implantations. Hence, much more research is called for in exploring Na and Li implantation of diamond, especially as there appear to be theoretical reasons why these should be effective n-type interstitial dopants in diamond. Perhaps the way forward is the application of a multiple-step CIRA technique to increase the concentration of the dopant without increasing the residual damage, or alternatively the use of low energy ion implantation as described below. Indeed, Prins [1995] has presented some interesting results using a multiple-step, CIRA technique with extremely low doses of P ($< 10^{11}$ P cm $^{-2}$ per step). Low resistivity material has been produced, and hot point measurements confirm n-type conductivity. Control experiments with similar low doses of B yielded p-type conductivity whereas C implantation did not yield any increase in the conductivity over that obtained for virgin diamond. The low doses have been used in order to minimize the residual damage; nevertheless the fact that effective doping can be obtained by such low concentrations of P, even assuming that they all occupy lattice sites and are electrically active, is somewhat surprising; it is clear that these preliminary results need to be repeated and confirmed. Nevertheless these recent developments bode well for the solution to the long-standing problem of n-type doping in diamond.

4.6 FORMATION OF OHMIC CONTACTS TO DIAMOND

One of the problems facing researchers working with diamond in general is the production of reliable and reproducible electrical contacts which are ohmic over a wide range of applied voltage. For p-type diamond, this is best achieved by producing a thin highly B doped layer. Such layers can be realized by high dose B ion implantation (exceeding D_c), followed by thermal annealing to graphitize the heavily damaged surface layer and finally removal of the graphitized layer by chemical etching. This procedure produces a relatively undamaged, but highly doped layer at the newly exposed surface which contains the tail of the ion distribution with the surface B concentration being as high as 7×10^{20} B cm $^{-3}$ [Uzan-Saguy et al. 1995b]. Indeed this is now one of the most commonly used methods for the realization of high quality ohmic contacts to diamond [Prins 1989]. Similarly, high dose B implantation followed by Ti/Au metallization has been used to fabricate ohmic contacts to polycrystalline diamond films with a specific contact resistance as low as 10^{-6} Ω-cm ^{2} [Venkatesan and Das 1992].

Despite the success of this scheme for the production of ohmic contacts to p-type diamond and diamond films, it is not clear at present what scheme should be employed in order to create ohmic contacts for n-type diamond.

4.7 SUMMARY

In summary, although some optimization still needs to be made, B doping of diamond by ion implantation now appears to have been realized. The most important feature of the successful doping scheme appears to be that the implantations must be performed at low temperatures, where the radiation damage is "frozen-in" until being annealed. There appears to be great promise in the multiple-step implant/annealing technique for selecting the degree of B doping, whilst keeping the residual damage under control. Ideally, the smallest dose steps possible should be used, as these will facilitate as near to complete removal of the radiation damage as possible. Potentially, the next step may be the combination of the multiple step CIRA technique with pulsed laser annealing. One can envisage an arrangement in which the laser could be directed onto the diamond while it is still held on the cold stage in the implantation chamber, thus ensuring that all the damage remains frozen in until the annealing step. By varying the laser power, wavelength, and pulse duration times it should be possible, at least in principle, to explore a very wide range of annealing temperatures and times. There are at least two possible reasons for the finding that these Li and Na doped layers display hopping rather than simply activated electrical behavior. The first is that it is possible that hopping occurs between the dopant sites in the crystal. Both Na and Li are conducting species and it has been recently shown that hopping conduction can result from the implantation of conducting species into an insulating matrix. The second possibility is that the presence of the dopant in some way inhibits the complete restoration of the diamond matrix, so that even though the C or Ne control specimen showed no increase in the conductivity, nevertheless some residual damage may remain for the Li and Na implanted specimens. The interested reader should also consult the work of Kalish and Prins for comprehensive reviews of the nature of this transformation.

5. Ion Implantation of CVD Diamond Films

It has been established that CVD diamond films can be doped with B to produce p-type conductivity by adding B containing compounds to the growth mixture [Nishimura et al. 1991]. However, spatial and depth control of the dopant distribution and concentration (which will be important for device fabrication) will be difficult using such a technique. In addition, despite some initial optimism, it appears that it will be difficult to incorporate n-type dopants into diamond during growth. The calculations of Kajihara et al. [1991] show that the formation energy of B acceptors in diamond is negative, which provides a thermodynamic driving force for its incorporation into the diamond lattice during growth. By contrast, the formation energies of potential n-type donors in diamond (viz., P, Li and Na) are all positive. These considerations provide strong motivation to investigate the ion beam modification and doping of CVD diamond films. One obvious question is to what extent can the extensive knowledge acquired over the years for the modification of natural single crystal diamond be applied to polycrystalline diamond films.

5.1 ION BEAM INDUCED DAMAGE IN POLYCRYSTALLINE CVD DIAMOND FILMS

Polycrystalline CVD diamond films differ from natural diamond in that they contain grain boundaries, defects, and often at least some graphitic or amorphous carbon impurities. It is by no means clear a-priori that the effects of ion beams on CVD films will be the same as for single crystalline material. For example, one might have

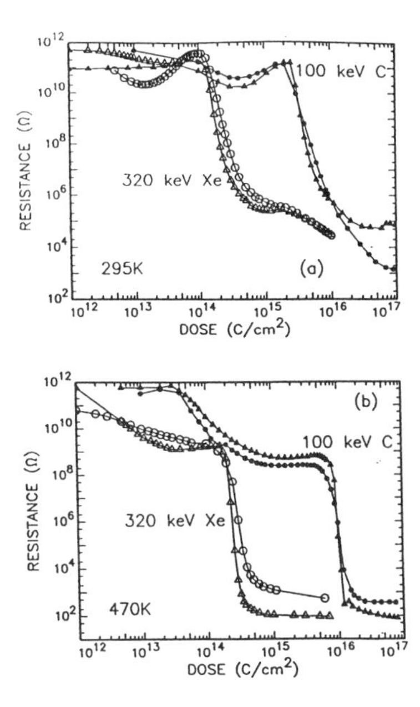

Figure 14. Resistance versus dose for irradiations carried out at room temperature and at 470 K for Type IIa diamond (triangles) and for polycrystalline CVD diamond films (circles). The open symbols are for irradiation with 320 keV Xe ions and the closed symbols are for irradiation with 100 keV C ions. [Prawer et al. 1990].

expected the material in the grain boundaries to be much more susceptible to amorphization and graphitization during ion implantation and after subsequent annealing than the bulk diamond. This, in fact, does not appear to be the case. Figure 14 [Prawer et al. 1990] compares the decrease in resistivity for natural diamond and a CVD diamond film irradiated with 100 keV C and 320 keV Xe at room temperature and at 470 K. It is clear that in each case the data for the polycrystalline films and the single crystal sample follow a very similar R vs. D dependence, even displaying the same non-monotonic behavior in the low dose regime ($D < D_c$). These results demonstrate that, at least as far as the damage related electrical conductivity is concerned, the polycrystalline nature of the film does not appear to affect the response of the material to ion beam irradiation.

5.2 BORON IMPLANTATION

The electrical properties of polycrystalline CVD films implanted with B have recently been reported [Kalish et al. 1994a; Kalish et al. 1994b]. The techniques employed were similar to those used for the implantation doping of crystalline diamond by B ion implantation, i.e. B was either implanted to a high dose (exceeding the graphitization limit), annealed and etched in acids to remove the graphite layer or to doses below D_c followed by thermal annealing. SIMS measurements on the high dose implanted and etched films showed a retention of a very high concentration of B in the remaining layer (about 10^{21} B cm^{-3}). The R vs. T curves showed low activation energies (< 0.1 eV) but this is not surprising in view of the high B concentration which would almost certainly result in an overdoped layer and impurity band conduction. However, the important point of this work is that the electrical properties of the B implanted layer are similar to those of B implanted natural diamond and of CVD diamond films heavily doped with B during growth. Hence the polycrystalline nature of the films does not appear to affect the ability of the films to be effectively p-type doped using ion implantation.

Other attempts to B dope diamond films by ion implantation have also been recently reported [Fontaine et al. 1994], using cold (77 K) implantations. A critical dose for amorphization was found to be about 3×10^{15} B cm^{-2}. Above this dose, the diamond film graphitized upon furnace annealing at 1100 K, but below it the implanted film could be rendered semiconducting after annealing. The activation energies extracted from log(R) vs. 1/T curves for T > 450 K decreased from 0.61 eV to 0.15 eV as the B dose was increased from 1×10^{13} to 2×10^{15} B cm^{-2}. This decrease in activation energy with increasing B content is also typical of the behavior found for single crystal diamond doped with high B concentrations.

Despite these similarities in response to implantation damage and to doping, some subtle differences in the electrical properties of diamond films and single crystals have been observed. For example, there is some evidence to suggest that polycrystalline films may be more sensitive to radiation damage (from 5 MeV He^{++}) than single crystal material [Han et al. 1993]. The possibility that the native defects in CVD films will act as compensation centers which will alter the apparent activation energies and hole mobilities should also be considered, as these may ultimately limit the electrical properties of the doped films. Indeed, as mentioned above, hole mobilities for B doped polycrystalline films are generally lower than for single crystals [Fox et al. 1993]. More careful measurements comparing the electrical properties of CVD films and single crystals implanted and annealed under identical conditions will need to be performed before the effects of native defects on the electrical properties of ion implanted CVD diamond can be firmly established.

5.3 ELECTRICAL CONTACTS

The methods for making ohmic contacts to p-type diamond films are the same as for natural diamond using the techniques discussed in 4.5 above. However a noteworthy result was reported by Cheng et al. [1993]. They used BF_2^+ implantation followed by furnace annealing to create a p-type layer in a CVD film, which was then metallized with Pt. For films displaying a predominantly (100) facet the Pt/diamond interface displayed Schottky characteristics, whilst for the interface with predominantly (111) facets, the Pt/diamond contact displayed ohmic behavior. This difference may be due to the difference in electron affinity between the (111) and (100) diamond surfaces as reported by Geis [1991]. This finding may point to the importance of taking crystal orientation into account when designing particular doping and contacting schemes.

5.4 LOW ENERGY ION IMPLANTATION DURING GROWTH

A novel approach to the problem of doping films has recently been published by Jamison et al. [1993]. In this method a stream of low energy dopant atoms (< 10 keV) are incident onto a diamond film during homoepitaxial growth. Typical CVD growth conditions employ gas pressures of the order of 10 Torr and hence, due to collisions with the gas molecules in the growth chamber, the dopant ions were actually incident on the growing film at energies of only 10's or 100's of eV. The result is that the ions are incorporated into the growing film, but without much of the concomitant radiation damage usually associated with an ion implantation process. The technique was successfully applied to incorporate high levels of Na, Rb and P into the films. C-RBS measurements show that the crystalline quality of the epitaxially grown doped diamond layer is high. The Na doped samples displayed high room temperature conductivities, very low activation energies (< 0.1eV) but disappointingly the carriers were determined to be p-type. This result is not in accord with the theoretical prediction [Kajihara et al. 1991; Kajihara et al. 1993] that Na should be an n-type dopant. In order to elucidate these results control experiments in which implantations with C and Ne are done during epitaxial growth would be most helpful. Nevertheless this technique is worthy of further consideration as it appears that almost any desired level of dopant can be incorporated into the growing diamond film, which remains relatively defect free.

In summary, the response of CVD diamond films and natural single crystal diamond to ion beam modification and doping appears to be very similar. This is good news indeed as the "know-how" (e.g. critical doses, dependence on implantation temperature, annealing behavior, etc.) for ion implantation of natural diamonds can now be applied to CVD films and, if single crystal stones are difficult or expensive to procure, further research into optimizing conditions for doping diamond using ion implantation can be carried out on high quality, more reproducible, and less rare polycrystalline diamond films.

6. Non-Electrical Applications for Ion Beam Modified Diamond

Most of this review has concentrated on research designed to dope diamond or diamond films using ion implantation. However, there are many potential uses for ion beam modified diamond which rely on the modification to the electrical, mechanical and chemical properties produced by ion beam damage alone. From figure 3 it is clear that practically any desired surface conductivity can be "dialed up" by selecting the right ion dose, species and implantation temperature. This has lead to interesting and innovative applications, some of which are briefly reviewed below. In particular, it is possible to produce (i) a material which is partially graphitized and hence electrically conducting while still maintaining its chemical resistance and mechanical strength or (ii) a material which is completely graphitized and is therefore susceptible to etching in hot acids.

6.1 APPLICATIONS OF PARTIALLY GRAPHITIZED DIAMOND

The recipe for the production of conducting, but chemically resistant diamond layers is moderate dose, low temperature ion implantation. This results in a damaged diamond layer which exhibits hopping conduction between broken bonds yet still maintains the diamond skeleton, i.e. the mechanical and chemical toughness of diamond. Some applications for such surface layers in diamond are briefly provided below.

6.1.1 Finger Printing

Devries et al. [1989] have proposed a scheme by which a permanent, but under normal viewing conditions, invisible, identifying pattern can be fabricated into diamond gemstones by ion implantation. The pattern only becomes visible once the implanted diamond is electrostatically charged and dusted with fine powders. This is achieved by implanting the diamond through a mask to an ion dose low enough such that no visible damage is imparted to the gemstone, but nevertheless sufficient to increase its surface conductivity and hence the ability of the irradiated diamond to hold an electrostatic charge. Various patterns were created in diamond by implantation of 350 keV C ions through suitable masks. No dust pattern was observed for doses of 1×10^{12} C cm^{-2}, whereas visible damage was observed for doses of 1×10^{14} C cm^{-2}, and dust patterns without visible damage were observed for intermediate doses of 5×10^{12} and 1×10^{13} C cm^{-2}.

6.1.2 Patterned Electrical Contacts

It is often desirable to realize conducting pathways in diamonds and diamond films, particularly when these materials are used as highly thermally conducting heat sinks for thermal management in high power electronic devices. Any desired conductivity between that of diamond and that of graphite may be induced in the diamond by the appropriate choice of the implantation conditions (see section 3.2) and hence pathways may be formed by implantations through appropriate masks. Furthermore, if metallic conductivities are required, the ion implanted regions can be used as templates for further electroplating. This has recently been demonstrated by Miller et al. [1994], who have shown the capacity of the electroplating technique to provide excellent spatial selectivity for the production of patterned Au plated contacts restricted only to those regions of the diamond which have been selectively damaged by ion implantation. Although Miller et al.[1994] report the use of hot implantation to produce highly graphitic regions suitable for electroplating, further work [Kalish 1995] has shown that this is not necessary, and that low temperature implantations, which produce a more mechanically and chemically robust layer, can be used to produce these patterned contacts.

6.1.3 Electrodes for Electrochemistry

One of the requirements in electrochemistry is the availability of chemically inert yet electrically conductive electrodes which will not degrade in chemically reactive environments. Diamond when in its conductive state is an excellent candidate for this application. It has indeed been used both in the doped (CVD grown) and damaged state as an electrode for the reduction of nitride and nitrate to ammonia [Reuben et al. 1995].

6.1.4 Conducting Tip for Atomic Force Microscopy (STM)

A diamond tip sharpened to a radius of 100 nm was rendered conductive by B ion implantation (the conductivity being most likely due to the damage accompanying the implantation). The induced conductivity was found to be sufficient to enable the tip to be used in a scanning tunneling microscope [Kanecko 1990]. Such a diamond tip was shown to yield high quality STM images with atomic scale resolution. Unlike the commonly employed W tips, the diamond tip can be used repeatedly even if it contacts the surface of the sample under investigation. In addition, if the tip is contaminated it

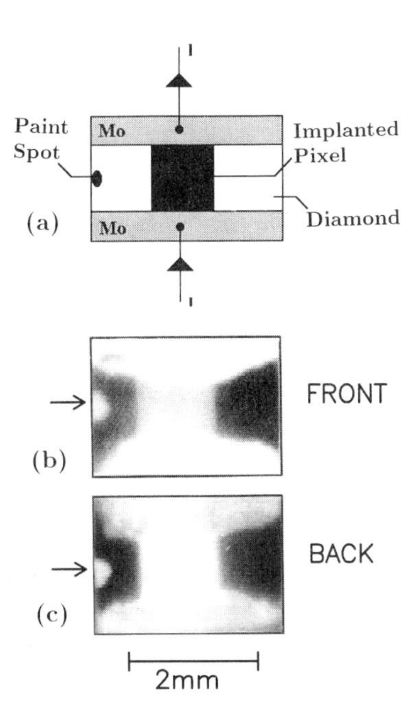

Figure 15. (a) Schematic diagram of a single infra-red emitting pixel into type IIa diamond. The spot of paint acted as a black body for calibration purposes. The emitting pixel of approximate dimension 1 x 1 mm^2 was produced by implanting the diamond through a mask to a dose of 1 x 10^{17} C cm^{-2}. Current carrying wires were bonded to the Mo strip contacts by silver epoxy. (b) and (c): Infra-red images of the emission from the device depicted in (a) as viewed from the front (i.e. implanted surface facing the detector) and the back (i.e. implanted surface facing away from the detector). Note the increase in infra-red emission in back viewing [Prawer et al. 1995b].

can be wiped clean and reused, a procedure which is not possible with a W tip which must be discarded once contaminated.

6.2 APPLICATIONS OF COMPLETE GRAPHITIZATION

Complete graphitization is usually achieved either by cold implantation to a dose exceeding D_c followed by high temperature annealing (≈1000 C), or by hot implantations to high doses. Some uses which take advantage of the relatively high electrical conductivity and/or the chemical etchability of the graphitized layer are given below.

6.2.1 Infra-red Scene Generator

Dynamic infra-red scene generators are required in a number of applications to simulate the IR signatures of moving objects. Among the major requirements for such a device are fast response times, high dynamic range, robustness, high fill factor and low cross-talk between pixels [Daehler et al. 1987]. One way of fabricating such a scene generator is by the use an electrically heated pixel array. Recently, it has been proposed that such arrays might be fabricated in diamond [Prawer et al. 1995b], taking advantage of the large difference in emissivity between diamond and graphite. In a proof-of-principle experiment, a single infra-red emitting pixel was fabricated in a type IIa diamond by implanting 1 x 10^{17} C cm^{-2} at 470 K. Once again these are the conditions for obtaining the maximum possible electrical conductivity (see Figure 3). When current was passed through the pixel the entire diamond heated up. Nevertheless, it was shown that excellent IR contrast can be obtained between the ion implanted and virgin portions of the sample due to the increased emissivity following the implantation as is shown in Figure 15. Interestingly, more infra-red radiation is emitted from the heated pixel when the diamond is viewed with the implanted side facing away from rather than towards the IR detector. This observation was explained in terms of the reflection losses for the different paths involved in viewing the device from the front and from the rear. Potentially, such a device can operate at high temperatures and very rapid switching rates.

6.2.2 Lift-off Technique for Producing Thin Diamond Plates

A new method for producing very thin, large area plates of diamond has recently been reported [Parikh et al. 1992]. For light ions, or for heavy ions implanted at MeV energies, the damage profile may be such that it is buried deep inside the diamond, leaving a relatively undamaged surface cap. In such a case the surface region will contain defects below the critical value for graphitization so that it will, following thermal treatment, regrow back to diamond, whereas the deeply buried damage region will graphitize. The basic scheme for the lift off technique therefore consists of the following steps: (i) the diamond is implanted with high doses of MeV C or O ions. The implantation leaves a relatively undamaged surface cap of diamond with a heavily damaged layer buried 1-2 μm below the surface. (ii) The diamond is then thermally annealed to repair the surface damage and to graphitize the buried end-of-range damage. (3) The damaged layer is removed either by etching in hot acids or by heating in an oxygen rich atmosphere. This results in the "lift-off" of a thin, relatively undamaged diamond plate. Sheets of up to 4 x 4 mm were lifted off in this way. The technique was taken one step further by Tzeng et al. [1993] who prior to lift-off have grown a homoepitaxial diamond layer, about 10 μm thick, on top of the implanted and annealed surface by microwave plasma enhanced chemical vapor deposition. The implanted layer with the homoepitaxial film was then heated in flowing oxygen in order to etch away the end-of-range damage so as to achieve lift-off of the underlying "seed" layer and the over-grown film, resulting in a free standing film which was flat and transparent. The front (growth) side of the film showed excellent crystal quality as measured by Raman spectroscopy, whereas the back (i.e. nucleation) side showed evidence for the presence of polycrystalline graphite which is the residue of the end of range damage due to the ion implantation.

An impressive demonstration of the capacity of this technology, combined with the fact that diamond can be cut with a laser, is the realization of diamond microcomponents, such as gears for possible use in micromachines. The processing steps to accomplish this are illustrated in figure 16 [Hunn and Christensen 1994].

Recently, Marchywka et al. [1993a] have demonstrated that instead of aggressive etching in boiling acids or etching in oxygen to dissolve the end-of-range damage and achieve "lift-off," a simple electrolytic etch in water can be used to dissolve the end of range damage. The diamond is immersed in distilled water between two Pt electrodes.

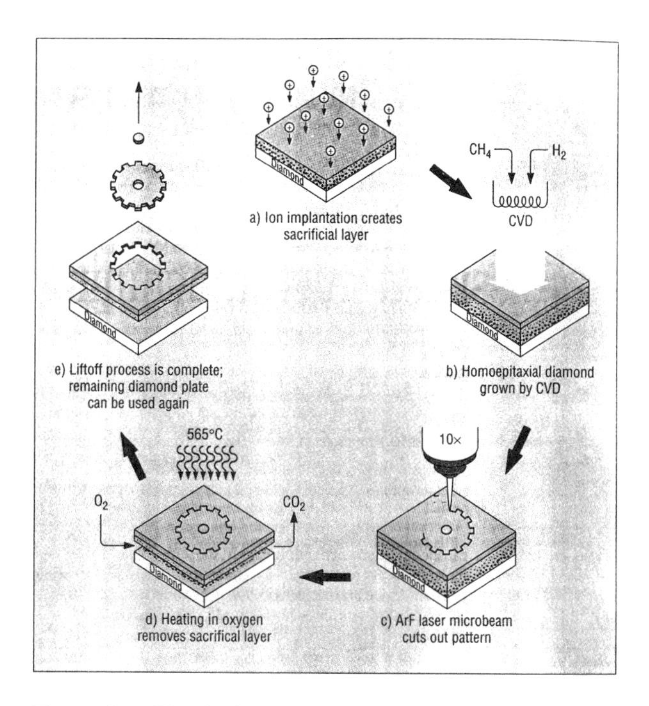

Figure 16. The basic process steps used to fabricate single crystal diamond microstructures [Hunn and Christensen 1994].

Sufficient chromic acid is added in order that a current of about 100 mA could flow with an applied bias of 200 V. The etch selectively attacks the graphitized layer because of its enhanced conductivity. The etch does not attack photoresist or wax so that patterning is also possible [Marchywka et al. 1993b].

Finally, recent work by Marchywka et al. [1993c] shows that MeV implantations may not be necessary for the successful use of the lift-off technique. They used 175 keV C implants followed by an electrochemical etch as described above, to lift off a film only 50nm thick from a thick CVD diamond film. The lifted-off film was curled due to residual stress and contained many pinholes due to the rough surface of the original polycrystalline CVD diamond film. Marchykwa et al. [1993c] also report that it is possible to grow homoepitaxial diamond layers on single crystal diamond surfaces implanted with 1×10^{16} and 4×10^{16} C cm^{-2} (175 keV) at room temperature and annealed at 1250 K in Ar. The grown films were then etched off using the electrochemical etch described above. In light of the discussion above in section three, it is somewhat surprising that homoepitaxial diamond will grow under these circumstances as the dose employed well exceeds D_c (see Figure 2), and hence after annealing one would expect the surface to be either graphitized, or at least be highly defective. Interestingly, there is some very recent evidence [Weiser et al. 1996] which suggests that the presence of defects in the implanted diamond substrate may actually improve the quality of the homoepitaxial diamond layer. In the absence of techniques for large area heteroepitaxial growth of diamond on non-diamond substrates one can expect further research and optimization of this very elegant technique for producing many CVD single crystal plates from a single diamond seed.

6.2.3 Controlled Removal of nm Thick Layers of Diamond

Finally we note that by implanting diamond with low energy heavy ions, graphitization of very thin (of the order of 10nm) near-surface regions can be achieved. Chemical etching of these allows very fine, well controlled removal of diamond. This method has been applied for the depth profiling of implants in diamond [Braunstein 1979], and may find other applications in surface treatments of diamond. It may also be used to expose deeply buried layers which have been created using the laser annealing technique described above.

7. Summary and Future Trends

The last few years have been active ones indeed in the area of ion implantation in diamond. Our understanding of the radiation damage which accompanies the ion implantation is now approaching the state at which effective defect engineering can be employed to maximize dopant activation. Indeed, research into the ion implantation of diamond for the purposes of electronic device fabrication has developed to the stage that effective recipes for p-type doping are now available. There appears to be a good chance that mobilities and carrier concentrations similar to those found in naturally B doped diamond will eventually be obtained.

However, it also appears that the best results will be limited to high quality single crystal diamond materials such as type IIa stones. Even then, the natural variability which exists between different samples will make the production of a material of consistent conductivity, carrier concentration and carrier mobility a difficult, if not impossible, task.

Therefore, the large scale application of ion implantation for the production of electronic devices in diamond will probably have to await the solution to the problem of the heteroepitaxial growth of CVD diamond films and the ability to synthesize large area single crystal diamond of consistent quality and purity.

Until then, the use of diamond as an electronic device will be of necessity limited to special applications such as, for example, radiation detectors operational under extremely harsh environments and space applications. These applications are so unique that the cost and effort involved in the selection of the highest quality natural diamond is of minor consideration. There are other electronic applications which do not require very high mobilities or very accurate control of the hole concentration (e.g. piezoresistors, and biological radiation monitoring) so that they are practical even today, using either natural diamond or the very highest purity CVD diamond films.

N-type doping of diamond remains an unmet challenge. Progress on this front has been frustratingly slow, but it seems clear that given the energetics of the incorporation of n-type dopants into the diamond lattice, ion implantation remains the most promising (and perhaps only) technique for n-type doping of diamond and diamond films. Is also appears at the moment that the multiple step Cold Implantation followed by Rapid Annealing (CIRA) is the best recipe to follow. Despite the difficulties, we believe that it is premature to give up hope in this regard and that steady and consistent research examining different dopants and activation schemes will eventually yield success. We also note, that even if n-type doping is not yet realizable, there are applications which can utilize only p-type material (e.g. the FET shown in Figure 9). It is worth mentioning that the radiation damage (which one usually tries to remove) sometimes displays n-type behavior (although with low mobility). The potential use of the n-type characteristics of this damage has not yet been fully exploited in device applications.

The non electronic uses for diamond modified by ion implantation are more obvious and are realizable using today's know-how and technology. Some of these have been outlined above, the most recent and exciting of which appears to be the use of ion implantation as a key step for the production of diamond membranes and micro components. Applications for ion beam modified diamond as an electrode in electrochemical cells also appear to be promising and practical.

An intriguing and highly speculative approach to device manufacture in diamond has recently been proposed by Zaitsev et al. [1996]. They propose the possible construction of nanometer scale devices in diamond based on the nm scale ion tracks formed during MeV ion irradiation. They note that it has been established that the ion tracks formed during high energy irradiation can be doped with impurities at temperatures which are low compared to the those required for impurity diffusion in keV implanted material. These doped regions, which can be separated by undoped insulating regions, could in principle act as active electronic devices. However, formidable (and perhaps insurmountable) problems lie ahead before such "diamond nanoelectronics" could become a reality. Nevertheless, it appears that the study of MeV irradiated diamond is likely to become a new and productive research direction for the future.

In summary, ion implantation into diamond offers a unique tool for the study of the fundamental physics of ion-solid interactions due to the conversion of damaged diamond to graphite. Furthermore, ion implantation is likely to play a key role in the future of diamond technology. The major challenges which lie ahead are n-type doping, and the improvement of the hole and electron mobilities. The knowledge base which exists today on the ion implantation of diamond bodes well for the solution to these problems.

8. References

Allen, M.G., Prawer, S., Jamieson, D.N., and Kalish, R., [1993] *Appl. Phys. Lett.*, **63**, 2062.

Biersack, J.P.,Haggermark, L.G., [1980] *Nucl. Inst. Math. Phys. Res.* **174**, 257.

Braunstein, G., Bernstein, T., Carsenti, U., and Kalish, R., [1979] *J. Appl. Phys.*, **50,** 5731.

Braunstein, G., Talmi, A., Kalish, R., Bernstein, T., Besserman, R., [1980] *Radiation Effects*, **48**, 139.

Braunstein, G. and Kalish, R., [1981a], *Nucl. Inst. Meth. Phys. Res.* **182-183,** 691.

Braunstein, G. and Kalish, R., [1981b], *Applied Physics Letters*, **38**, 416.

Braunstein, G. and Kalish, R., [1983], *J. Appl. Phys.* **54,** 2106.

Brosius, P.R., Corbett, J.W. and Bourgoin, J.C. [1974], *Phys. Stat. Solidi* **A21,** 677.

Buckley-Golder, I.M., Bullough, R., Hayns, M.R., Willis, J.R., Piller, R.C., Blamires, N.G., Gard, G., and Stephen, J., [1991] *Diamond and Related Materials*, **1**, 43.

Cheng, Y.T., Lin, S.J., Hwang, J., [1993] *Appl. Phys. Lett.*, **63**, 3344.

Connell, S., Comins, J.D. and Sellschop, J.P.F., [1988], *Radiation Effects Express*, **2,** 57.

Cyterman, C., Brener, R. and Kalish, R., [1994] *Diamond and Related Materials*, **3**, 677.

Daehler, M. [1987], *SPIE*, **765**, 94.

Burriesci, L. and Keezer, D., [1987] *SPIE*, **765**, 112.

Davies, G., [1994] *Properties and Growth Of Diamond*, INSPEC, London, 193.

DeVries, R.C., Reihl, R.F., Tuft, R.E., [1989] *Journal of Materials Science*, **24**, 505.

Dresselhaus, M.S. and Kalish, R. [1992] *Ion Implantation in Diamond, Graphite and Related Materials*, Springer-Verlag, Berlin.

Erchak, D.P., Efimov, V.G., Zaitsev, A.M., Stelmakh, V.F., Penina, M., Varichenko, V.S. and Tolstykh, V.P. [1992], *Nucl. Inst. Meth. Phys. Res.*, **B69**, 443.

Fang, F., Hewett, C.A., Fernandes, M.G., and Lau, S.S. [1989], *IEEE Transactions on Electronic Devices*, **36**, 1783.

Feldman, L.C. and Mayer, J.W., [1986], *Fundamentals of Surface and Thin Film Analysis*, North Holland, New York.

Fontaine, F., Deneuville, A., Gheeraert, E., Gonon, P., Abello, L., and Lucazeau, L., [1994], *Diamond and Related Materials*, **3**, 623.

Fox, B.A., Malta, D.M., Wynands , H.A.and von Windheim, J.A., [1993], *Proceedings of the Electrochemical Society*, Electrochemical Society, Honolulu, 759.

Freeman, H., Temple, W. and Guard, G.A., [1978] *Nature*, **275**, 634.

Geis, M.W., Gregory, J.A., and Pate, B.B., [1991] *IEEE Trans. Electron. Devices*,**38**, 619.

Han, S., Prussin, S.G., Ager, J.W. III, Plan, L.S., Kania, D.R., Lane, S.M., and Wagner, R.S., [1993], *Nucl. Inst. Meth. Phys. Res.,* **B80/81**, 1446.

Hartley, N.E.W., and Poole, M.J., [1973], *Mater. Sci.* **8**, 900.

Hartley, N.E.W. [1982], in S.T. Picraux, W.J. Choyke (eds.) *Metastable Materials Formation by Ion Implantation*, North Holland, Amsterdam, 295.

Hewett, C.A., Nguyen, R., Zeidler, J.R., and Wilson, R.G. [1993], *roceedings of the Electrochemical Society*, Electrochemical Society, Honolulu, p. 993.

Hoffman, A., Prawer, S. and Kalish, R. [1992] *Phys. Rev* **B45**, 12736.

Hunn, J.D. and Christensen, C.P., [December 1994] *Solid State Technology*, 57.

Jamison, K.D., Schmidt, H.K., Eisenmann, D., and Hellmer, R.P., [1993], *Mat. Res. Soc. Symp. Proc.* **302**, 251.

Kajihara, S.A., Antonelli, A., Bernholc, J., and Car, R., [1991], *Phys. Rev. Lett.*, **66**, 2010.

Kajihara, S.A., Antonelli, A., and Bernholc, J., [1993], *Physica B* **185**, 144.

Kalish, R., Bernstein, T., Shapiro, B.and Talmi, A. [1980], *Radiation Effects*, **52**, 153.

Kalish, R. [1993] *Diamond and Related Materials*, **2**, 621.

Kalish, R., and Prawer, S., [1995], *Nucl. Inst. Meth. Phys. Res.*, **B106**, 492.

Kalish, R., Uzan-Saguy, C., Samoioloff, A., Locher, R., Koidl, P. [1994a] *Appl. Phys. Lett.* **64**, 2532.

Kalish, R., Uzan-Saguy, C. and Samoiloff, A. [1994b] in S. Saito, N. Fujimori, O. Fukunaga, M. Kamo, K. Kobashi, and M. Yoshikawa, (eds.) *Advances in New Diamond Science and Technology*, MYU, Tokyo, 449.

Kalish, R. [1995] unpublished.

Kaneko, R. and Oguchi, S. [1990] *Japaneese Journal of Applied Physics*, **29**, 1854.

Lindhard, J., Scharff, M. and Schiott, H.E. [1963] *Nat. Fys. Medd. Dan. Vid. Selsk.* **33**, No. 14.

Marchywka, M., Pehrsson, P.E., Binari, S.C. and Moses, D. [1993a] *J. Electrochemical Society*, **140**, L19.

Marchywka, M., Pehrsson, P.E. and Moses, D. [1993b] in *Proceedings of the Electrochemical Society*, Electrochemical Society, Honolulu, **626**, [1993b]

Marychenka, M., Pehrsson, P.E., Vestyck Jr., D.J. and Moses, D. [1993c] *Appl. Phys. Lett.* **63,** 3521.

Miller, B., Kalish, R., Feldman, R.C., Katz, A., Moriya, N., Short, K. and White, A.E. [1994] *J. Electrochem. Soc.* **141**, L41.

Morehead Jr., F.F. and Crowder, B.L. [1970] *Radiation Effects*, **6**, 27.

Mott, N.F. and Davis, E.A. [1979] *Electronic Processes in Non-crystalline Solids*, Clarendon. Oxford.

Nelson, R.S., Hudson, J.A., Mazey, D.J. and Piller, R.C. [1983] *Proc. Roy. Soc. London*, **A386**, 211.

Nishimura, K., Das, K. and Glass, J.T. [1991] *J. Appl. Phys.*, **69**, 3142.

Parikh, N.R., McGucken, E., Swanson, M.L., Hunn, J.D., White, L.W. and Zhur, R.A. [1993] *1993 International Conference on Advanced Materials*, Tokyo, Japan.

Parikh, N.R., Hunn, J.D., McGuken, E., Swanson, M.L., White, C.W., Rudder, R.A., Malta, D.P., Posthill, J.B. and Markunas, R.J. [1992] *Appl. Phys. Lett*, **61**, 3124.

Popovici, G., Wilson, R.G., Sung, T., Prelas, M.A. and Khasawinah, S. [1995] *J. Appl. Phys.* **77**, 5103.

Prawer, S., Hoffman, A. and Kalish, R. [1990] *Appl. Phys. Lett.*, **57**, 2187.

Prawer, S., Jamieson, D.N. and Kalish, R. [1992] *Phys. Rev Lett.*, **69**, 2991.

Prawer, S., Jamieson, D.N., Walker, R.J., and Kalish, R., [1996] *Diamond Films and Technology*, in press.

Prawer, S., Uzan-Saguy, C., Braunstein, G. and Kalish, R. [1993] *Appl. Phys. Lett.*, **63**, 2502.

Prawer, S. and Kalish, R. [1995a] *Phys. Rev.* **B51**, 15711.

Prawer, S., Devir, A.D., Balfour, L.S. and Kalish, R. [1995b] *Applied Optics*, **34**, 636.

Prins, J.F. [1982] *Appl. Phys. Lett.* **41**, 950.

Prins, J.F. [1983] *Radiation Effects Letters*, **76**, 79.

Prins, J.F. [1985] *Phys. Rev.* **B31**, 2472.

Prins, J.F., Derry, T.E. and Sellschop, J.P.G. [1986] *Phys. Rev.* **B34**, 8870.

Prins, J.F. [1988] *Phys. Rev* **B38**, 5576.

Prins, J.F. [1989] *J. Phys. D: Applied Phys.* **22**, 1562.

Prins, J.F. [1991] *Nucl. Inst. Meth. Phys. Res.* **B59/60**, 1387.

Prins, J.F. [1992] *Materials Science Reports*, **7**, 271.

Prins, J.F. [1993a] *Physica B*, **185**, 132.

Prins, J.F. [1993a] *Nucl. Inst. Meth. Phys. Res.* **B80/81**, 1433.

Prins, J.F. [1995] *Diamond and Related Materials*, **4**, 580.

Restle, M., Bharuth-Ram, K., Quintel, H., Ronning, C., Hofsass, H., Jahn, S.G. and Wahl, U [1995] ISOLDE-Collaboration *Appl. Phys. Lett.*, **66**, 2733.

Reuben, C., Galun, E., Cohen, H., Tenne, R., Kalish, R., Muraki, I., Hashimoto, K., Fujishima, A., Butler, J.M., and Levy-Clement, C., [1995] *Wide Band Gap Electronic Materials,* M.A. Prelas, ed. Kluwar Academic Publishers, p. 137.

Sandhu, G.S., Swanson, M.L. and Chu, W.K. [1989] *Appl. Phys. Lett.*, **55**, 1397.

Sato, S. and Iwaki, M. [1988] *Nucl. Inst. Meth. Phys. Res.* **B32**, 145.

Spits, R.A., Derry, T.E., Prins, J.F. and Sellschop, J.P.F. [1990] *Nucl. Inst. Meth. Phys. Res.* **B51**, 63.

Spits, R.A., Prins, J.F. and Derry, T.E. [1994] *Nucl. Inst. Meth. Phys. Res.* **B85**, 347.

Teicher, M. and Besserman, R. [1982] *J. Appl. Phys.* **53**, 1467.

Tzeng, Y., Wei, J., Woo, J.T. and Lanford, W. [1993] *Appl. Phys. Lett.*, **63**, 2216.

Uzan-Saguy, C., Richter, V., Prawer, S., Lifshitz, Y., Grossman,E. and Kalish, R. [1995a] *Diamond and Related Materials*, **4**, 569-574.

Uzan-Saguy, C., Cytermann, C., Brener, R., Shaanan, M. and Kalish, R. [1995b] *Appl. Phys. Lett.,* **67,** 1194.

Fontaine, F., Uzan-Saguy, C., Philosoph, B. and Kalish, R. [1996] Appl. Phys. Lett. Fontaine, F., **68, 2264**.

Vavilov, V.S., Krasnopevtsev, V.V., Milijutin, Y.V., Gorodetsky, A.E. and Zakharov, A.P. [1974] *Radiation Effects*, **22**, 141.

Venkatesan, V. and Das, K. [1992] *IEEE Electron Device Letters*, **13**, 126.

Weiser, P.S., Prawer, S., Nugent, K.W., Bettiol, A.A., Kostidis, L.I. and Jamieson, D.N. [1996] *Diamond and Related Materials*, **5,** 272.

Williams, J.S. [1986] *Rep. Prog. Phys.* **49**, 491.

Wu, W. and Fahy, S. [1994] *Phys. Rev.* **B49**, 3030.

Zaitsev, A.M. [1991] *Nucl. Inst. Meth. Phys. Res.,* **B62**, 81.

Zaitsev, A.M., Fahrner, W.R., Varichenko, V.S., Melnikov, A.A., Fedotov, S.A., Denisenko, A.V., Burchard, B. and Fink, D. [1996] in J.S. Williams, R.G. Elliman, and M.C. Ridgway *Proceedings of the Ninth International Conference on Ion Beam Modification of Materials, Canberra, Australia, 5-10 February, 1995*, Elsevier, p928.

Zeidler, J.R., Hewett, C.A. and Wilson, R.G. [1993a] *Phys. Rev* **B47**, 2065.

Zeidler, J.R., Hewett, C.A., Nguyen, R., Zeisse, C.R. and Wilson, R.G. [1993b] *Diamond and Related Materials*, **2**, 1341.

Ziegler, J.F., Biersack, J.P. and Littmark, U. [1985] *The Stopping and Range of Ions in Solids*, Pergamon Press, New York.

Chapter 27
PROCESSING

Victor G. Ralchenko and Sergei M. Pimenov
General Physics Institute, ul. Vavilova 38, 117942 Moscow, Russia

Contents

1. Introduction

Although diamond is the hardest known material it can be shaped in many forms. Historically diamond processing was related to jewelry needs; cutting and polishing of diamond remained a trade secret for centuries. The fabrication of a brilliant diamond involves cleavage, sawing, bruiting, grinding and polishing, and traditionally is based on the use of hardness of diamond powder or another diamond stone counterpart in accordance with the principle "diamond cuts diamond." Historical, technical and scientific aspects of these classical methods as well as others, like drilling for making dies, are described in a number of excellent books [see for example, Grodzinski 1953, Bruton 1978, Epifanov et al. 1982, Wilks and Wilks 1991, Field 1992]. In this chapter the mechanical treatment is reviewed briefly, the major attention being focused on new techniques such as laser cutting, smoothing and patterning, ion beam and plasma etching, and polishing by hot metals. The recent progress in these methods of processing is especially important in view of appearance of polycrystalline diamond films on the market for use in tools, micromechanics, optics and electronics.

2. Cleavage

Cleaving is often used in order to produce two or more diamonds from one crystal without loss. For production of brilliants, cleaving serves to remove inclusions, flaws and other defects to obtain smaller pieces of higher quality. Cleavage is a much more rapid method of shaping diamond than sawing, but is very limiting because only (111) faces can be produced. Cleaving continues to be a skilled hand operation, unchanged from early days until now [Grodzinski 1953]. First, after careful inspection of the stone and choice of right cleavage direction, the line of cleavage is marked with India ink. A fine groove is rubbed into the stone along the mark with a sharp piece of diamond. Then the stone is fixed in cement at a support, and a sharp blow with a cleaver's stick on a blunted steel knife placed into the kerf divides the stone along the cleavage plane.

The cleaved surfaces is often found to be broken in conchoidal form. This is the characteristic cleavage of type I diamond [Tolansky 1955], while the cleavage of type II stones is almost perfect. There are almost flat regions separated by steps of a few angstroms high [Bruton 1978]. Cleavage planes other than (111), in particular nearby planes (332) and (221), were also reported but as very rare facts [Field 1992]. The dominance of (111) cleavage is explained by several reasons. First, these planes have minimum fracture surface energies, which are 5.3, 5.85 and 6.1 J m^{-2} for the (111), (332) and (221) planes, respectively [Field 1992]. However, this energy difference is not large enough to ensure the absolute preference of (111) plane cleavage, especially if one takes into account that the direction of cleaving force may not align exactly in the (111) plane. Field (1992) also noted that since the (111) plane is the growth plane, it must be more defective and therefore weaker. Ansell (1984) pointed out that there is a compressive stress in (111) plane caused by interaction of C-C bonds with neighboring non-bonded atoms. Therefore, if one bond is broken, this results in a "domino effect", whereby bond breaking (cleavage) spreads in the direction of compression, that is along the (111) plane.

3. Cutting

Cutting of diamond is one of the basic procedures in jewelry as well as in manufacturing of technical products like heat sinks, tool inserts, etc. The cutting (sawing) of diamond is documented as early as the 17th century [Bruton 1978, Epifanov et al. 1982]. At those times the sawing was carried out with an iron or bronze wire charged with diamond powder. The process was extremely slow, for instance, it took one year to cut the stone "Regent" of 410 carats. Today the sawing of diamond crystals is carried out mostly mechanically with a thin rotating bronze wheel, however, fairly new techniques such as electric discharge machine or laser cutting have been developed, which are especially important for the processing of polycrystalline diamonds.

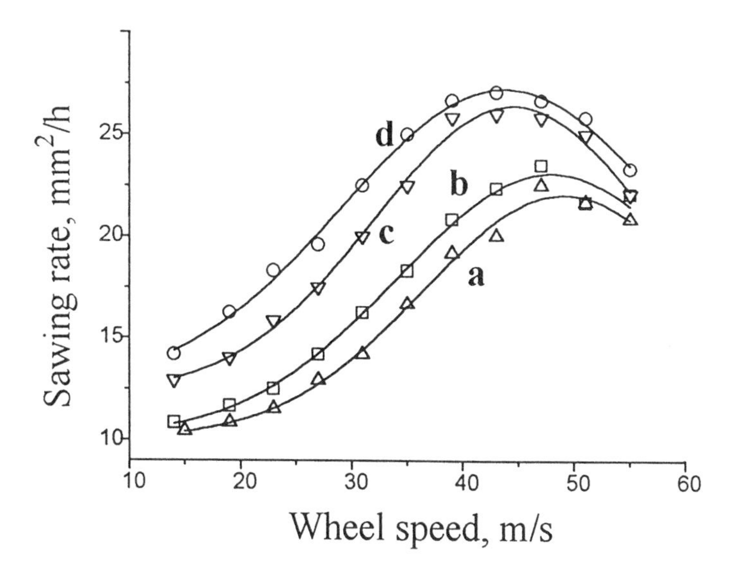

Figure 1. Sawing rate, Q, for diamond as a function of rotation speed, V, of a bronze wheel charged with diamond powder of different sizes: (a) 10/7 μm, (b) 14/10 μm, (c) 20/14 μm and (d) 28/20 μm [Epifanov et al. 1982].

3.1 MECHANICAL CUTTING

The cutting tool used for diamond sawing is a wheel made of bronze usually doped with phosphorus (0.06-0.25%). The wheel of thickness 0.04 mm to 0.09 mm, and diameter up to 76 mm, rotates at 10,000 revolutions per minute. The wheel is charged with diamond powder periodically during the work. A monolayer of the abrasive particles at the edge of the wheel creates a net of cracks at the diamond surface. A high load generated by the action of high speed (around 40 m/s) abrasive particles leads to removal of small microcrystals from the work piece, thus increasing the cut depth. The cutting rate depends on the size of abrasive powder, wheel rotation speed and load at which the wheel is pressed on the diamond [Epifanov et al. 1982]. The cutting rate Q, vs wheel velocity, V, for different sizes of abrasive powder is shown in Figure 1. Due to more frequent impacts of abrasive particles, the cutting efficiency at first increases with rotation speed, but then drops at V>45 m/s because the largest (and most useful) working particles leave the disk rapidly. So the optimal wheel speed is about 40 m/s. The extensive wear of the wheel requires changing 1-2 disks after cutting of an approximately 5 carat stone [Grodzinski 1953, Epifanov et al. 1982]. The losses in stone mass as a result of sawing is 2-7% depending on the size of a crystal. Crystals can be sawed only along (100) and (110) planes, thus polycrystalline diamonds cannot be processed in this way.

3.2 SPARK EROSION

Different shapes can be cut from single crystal, PCD or CVD diamond film using a wire electric discharge machine (EDM). The material removal proceeds via erosion by an electric spark between the work piece placed on one electrode and another metal electrode, the entire assembly being immersed in a dielectric fluid like oil, deionized water or kerosene. Generally, the material processed by EDM must be conductive, therefore, diamond compacts containing a metal binder can be treated without any problem, while the formation of a metallized or graphitized surface layer on insulating diamond samples is required prior to their EDM machining. Levitt (1968) described an apparatus for delineation of conductive marks on diamond stones prior to EDM processing. At the first stage, a graphitized strip was formed by passing the current between two needle-like electrodes contacting to the work piece. The graphitized strips, up to 0.75 mm long were produced in a paraffin oil medium at an applied voltage of 15 kV. Diamond crystals of an arbitrary shape were cut with an electrode in the form of a plate arranged along the graphitized line, while the work piece was supported on the opposite electrode at 20 kV. It took 135 min to cut a 2.5 mm thick diamond with a blade lightly loaded against the work piece. The EDM technique is very fast; it is especially useful in production of dies, molds, gear wheels and other components where great accuracy and high surface finish are needed. In the case of machining PCD with 0.2 mm diameter brass discharge wire, the optimized cutting rates were in the range of 2.8-6.5 mm/min at pulse frequencies of 10 to 15 kHz [Steinmetz et al. 1990]. For preparation of high quality sharp cutting edges of PCD blanks the sequential trimming procedure with EDM was effective, a surface roughness as small as R_a=0.3 μm being achieved [Steinmetz 1990].

The EDM method has been used to produce PCD microtools such as small drills and pin tools [Koike et al. 1993]. A part of the wire electrode in contact with the surface of the tool machines the work piece similarly to conventional grinding with a diamond wheel, but the process of material removal is realized via microsparks between the work piece and the wire. The relative positions of the work piece and the wire guide, both being submerged in a kerosene dielectric, are controlled by 5 axes. The dimensional errors caused by deformation of the work piece are eliminated as the machining force is negligible.

3.3 LASER CUTTING

Lasers applications are increasing in the diamond industry. The laser etching of a material is based on rapid heating of a work piece by a short pulse of photons which being absorbed in a thin surface layer, causes a local vaporization (sublimation). The laser beam can be focused into a spot as small as a few microns in diameter. Very accurate grooves can be produced by scanning the beam along a desired path. The laser cutting is a contactless technique, the hardness of the treated material is not significant parameter, but optical absorption and thermal conductivity are important factors, which determine the etch rate at given laser parameters.

Laser cutting is especially effective for processing CVD diamond films, PCD and single crystal diamonds with defects, i.e. for those objects for which conventional mechanical cutting is difficult or even impossible. A variety of lasers emitting in the spectral range from UV to IR are available now for cutting purposes. The use of Nd:YAG lasers operated at 1.06 μm wavelength is most common in the diamond industry. Its advantages are the very reliable construction which has been thoroughly developed for more than 30 years since the first demonstration of this laser in 1961, the ability to use simple silica glass optics to handle the laser radiation, and the ability to sharply focus the beam. The average power of a Nd:YAG laser of a few to tens of watts, and pulse repetition frequency of a few kilohertz are typical of cutting process. Although the pure diamond is transparent at wavelengths larger than 0.227 μm, the graphitization of the surface provides enough absorption of the Nd:YAG laser. The primary graphitization occurs as a result of laser absorption on defects, inclusions or intentionally created opaque marks.

In processing a diamond crystal, the groove depth becomes larger with increasing either peak or average power [Tezuka et al. 1990]. The groove can be deepened by repetitively scanning the beam along the same trace, and by lowering the focus point under the surface. The groove aspect ratio (depth/width ratio) of about 20 may be achieved at optimum conditions. The cut surface has roughness R_{max}=10-20 μm (peak-to-valley height). The amount of graphite at the sidewalls of the groove and redeposited graphite around the cut are minimized when irradiation is performed in air or pure oxygen, as a part of the converted carbon is desorbed in the form of CO and CO_2 molecules. The remaining graphitized layer can be eliminated with a hot acid mixture of HNO_3:H_2SO_4. Figure 2 illustrates the shape of a typical kerf produced by a Nd:YAG laser in 300 μm thick diamond film.

The cutting depth is lower for PCD with metal binder compared to that for diamond single crystal [Tezuka et al. 1990]. While diamond heated by laser beam is vaporized, the molten metal (Co or Fe) is exposed to the vapor pressure, which however is not sufficient to remove the metal from the groove. The molten metal solidifies inside the groove, thus limiting the further increase of the groove depth.

Recently, frequency-doubled Nd:YAG lasers (532 nm wavelength) with a high enough power have been introduced into practice. These green lasers yield a smaller beam size and higher absorption, improving the interaction with diamond. A successful cutting of a 300 μm thick diamond film has been demonstrated, the kerf width being as small as 20 μm [Golden 1992]. Excimer lasers, which are the most powerful lasers operated in ultraviolet range, were also used to treat diamonds [Rothschild et al. 1986, Ageev et al. 1988, Pimenov et al. 1993a, Harano et al. 1994]. The most important in a family of excimer lasers which have appeared in laboratories in mid-1970s are the rare-gas halides ArF, KrF and XeCl, emitting at wavelengths of 193, 248 and 308 nm respectively. Due to the high absorption in diamond among other lasers, the excimer lasers have the lowest damage (ablation) threshold fluence (J/cm^2) compared to other lasers routinely used for material processing (see Table I). Yet the excimer lasers have not found a wide application in diamond cutting. The primary reason for this are strong safety demands, the necessity to periodically refill the cavity with halogen gas, and often insufficient average power.

The damage thresholds summarized for different lasers in Table I refer to the onset of ablation of diamond, while the laser-induced surface graphitization thresholds are an order of magnitude lower [Rothschild et al. 1986, Ageev et al. 1988, Bouilov et al. 1990]. Damage thresholds generally are higher for longer wavelengths and diamonds of higher quality and polished surfaces.

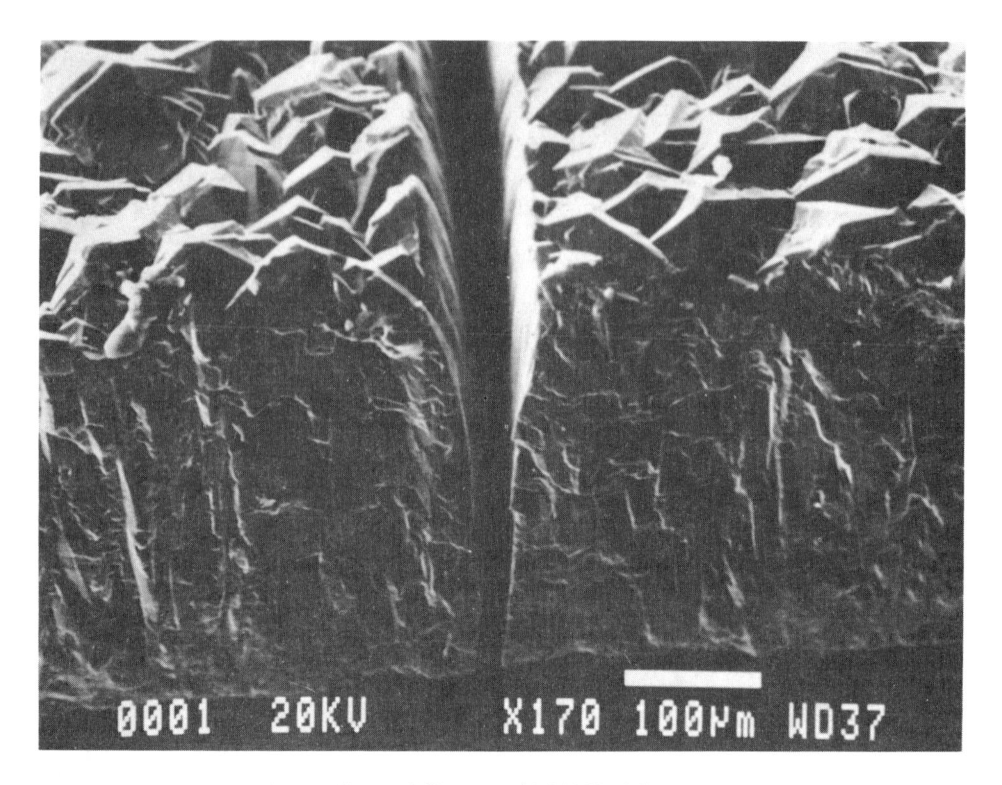

Figure 2. SEM picture of CVD diamond film cut with Nd:YAG laser.

4. Polishing

4.1 MECHANICAL POLISHING

4.1.1 *Polishing of diamond crystals*

Traditionally, single crystal diamonds are polished with a rotating cast iron scaife charged with a mixture of diamond powder and oil. The main factors influencing the polishing of diamond with this technique are well established, and the state-of-the-art of the mechanical polishing of diamond has been reviewed by Wilks and Wilks (1992) and by van Enckevort and van Halevijn (1994). The polishing regimes conventionally used in jewelry are given by Grodzinski (1953) and Epifanov et al. (1982), depending on various parameters such as orientation of polished faces and direction of abrasion, scaife speed and load, and size and concentration of diamond grit on the scaife.

It is well known that the rate of material removal during grinding and polishing of diamond crystals depends strongly on the crystal orientation and direction of abrasion. Maximum removal rates for the (110), (100) and (111) planes are in the ratio 1:0.6:0.1. The removal rate for the (110) and (100) faces in the soft directions <100> is 2-3 orders of magnitude higher than that along the hard directions <110>, while the anisotropy in the rate of material removal is not large for the (111) plane which is abrasion resistant in all directions [Epifanov et al. 1982, van Enckevort and van Halevijn 1994].

Table I. Damage threshold energy, *E*, for diamond films, DF, and single crystals for different lasers: l is wavelength, Ó is pulse duration.

Laser	l (nm)	Ó (ns)	E (J/cm^2)	Sample	Ref.
ArF	193	15	1-3	IIa type	[Rothschild et al. 1986]
KrF	248	15	1-3	DF	[Blatter et al. 1991]
			7	Ia type	[Kononenko 1993]
			28	IIa type	[Harano et al. 1994]
		0.8	0.5	DF	[Pimenov et al. 1993a]
XeCl	308	20	1	DF	[Ageev et al. 1988]
Nd:YAG	532	10	1-3	DF unpolished	[Sussman et al. 1994]
(2nd harm)		10	8-14	DF polished	[Sussman et al. 1994]
		20	10	DF	[Cropper et al. 1993]
Nd:YAG	1064	12	21-31	DF	[Sussman et al. 1994]
CO_2	10,600	50	29-66	DF	[Sussman et al. 1994]
		150	50	DF	[Bouilov et al. 1990]
		50	>93	IIa type	[Sussman et al. 1994]

The rate of material removal during diamond polishing increases with the speed of a scaife and with the load. For instance, the value of the wear rate increases by 2.5-3 times with increasing speed from 25 m/s to 35 m/s, and a scaife speed of about 50 m/s is usually used in the grinding and polishing processes [Epifanov et al. 1982]. The increase in the load leads to an increase in the wear rate while, when the load exceeds a certain value, it decreases the final surface roughness of the polished surface. It is recommended to apply a contact pressure of 2.5-6.5 MPa for diamond grinding and 1-2.5 MPa for polishing. A choice of a size of diamond grit is also determined by the value of roughness to be achieved, although larger diamond grit size provides higher wear rates.

The rate of polishing is influenced by environment conditions. Yarnitsky et al. (1988) and Epifanov et al. (1982) reported the effect of heating the scaife on the polishing rate, the maximum rates being observed at about 100°C. Reduction in the air pressure from 760 Torr to 0.1 Torr resulted in a decrease in the removal rate [Hitchiner et al. 1984], whereas polishing with a scaife covered by water (or isopropanol) provided a considerable increase in the polishing rate [Wilks and Wilks 1992].

The roughness of polished diamond crystals was reported to be about R_{rms}=1 nm for the (110) face and about R_{rms}=2 nm for the (100) face [Gaissmaier and Weis 1993]. The polishing was performed with a cast iron scaife charged with 0.25 μm diamond grit. Moreover, these authors found a "superpolishing" regime by tin sliding on the (110) and (100) diamond faces in air, for which a residual rms roughness below 0.1 nm was achieved. On the contrary, hydrogenated diamonds remained unabraded by this procedure. They suggested a mechanism of mechanochemical polishing by oxidation to interpret these results.

For a long time, the problem of the anisotropy of diamond polishing as well as the mechanism of material removal have not been completely understood. Reported recently were the investigations by Couto et al. (1994a,b) on the morphology of the polished surfaces by scanning tunneling microscopy and atomic force microscopy, giving deeper insight into the mechanisms of diamond polishing in hard and soft directions. It was observed that in the hard direction, the polishing occurred by fracture and chipping on a nanoscale, with a little evidence being found in favor of the (111) microcleaving mechanism proposed by Tolkowsky (1920). In the soft direction, the polishing occurred via the formation of numerous smooth nanogrooves due to interaction of abrasive particles with the diamond surface under high contact pressures. Considering the abrading particle-diamond interaction to proceed by direct rupture of atomic bonds rather than by plastic deformation, Couto et al. (1994b) suggested that the dependence of wear rate on crystal orientation might be explained by a relation between the directions of the atomic bonds in the diamond lattice and the direction of abrasion. These results are consistent with understanding the mechanisms of diamond-on-diamond friction discussed by Tabor and Field (1992).

4.1.2 Polishing of diamond films with a cast iron scaife

One of the reasons why conventional abrasive polishing of CVD diamond films is not systematically studied is a number of difficulties associated with the slow polishing rate, comparatively poor quality of the polished surface and problems with the film adhesion to a substrate. This is, principally, true for rather thin films. For example, Moran et al. (1991) reported that 10 μm thick films polished with a scaife for a long time (20 hours), with a final root mean square roughness of only R_{rms} =50 nm being achieved. However, acceptable polishing regimes can be realized for relatively thick diamond films which are characterized by a certain microstructure (texture) formed in the course of diamond growth.

Spitsyn et al. (1987) and Belyanin et al. (1991) studied grinding and polishing of diamond films of thickness ranging from 9 μm to 200 μm. They found a correlation between the polishing rate and the film microstructure being changed with thickness. The rate of material removal was in the range of 0.2 μm/h to 3 μm/h, and the final surface roughness was R_{max}=25 nm. The lowest polishing rate was observed for the films of thickness 30 μm. As the film thickness is increased, a resultant axial texture along the <221> and <553> directions is formed, which is characterized by high surface energy and, consequently, creates favorable conditions for polishing with acceptable polishing rates and roughness of the finished surface [Spitsyn et al. 1987].

Wilks and Wilks (1992) compared the abrasion resistance of a polycrystalline diamond film with that of an easy <100> direction on a polished (001) face and with that of an easy <112> direction on a natural (111) face of single crystal diamond. They found that the film was more resistant to polishing than an easy <100> direction on a polished (001) face but less resistant than an easy <112> direction on a natural (111) face, concluding that the abrasion resistance must depend on the conditions of deposition.

4.1.3 Polishing of diamond films with abrasive-liquid jets
Hashish and Bothell (1992) reported on a novel technique for diamond film polishing using abrasive suspension jets characterized by high flow rates and velocities of abrasive particles. An abrasive jet was obtained by means of pressurizing an abrasive suspension up to 138 MPa with the abrasive concentration (600-mesh SiC) of 20 wt.%. A nozzle system produced a radial abrasive flow between the nozzle and diamond sample with high particle velocities of 150 m/s and with near zero angles of impact. The diamond film sample was rotated at 2400 rpm and the nozzle was translated across the film at a velocity of 0.3 mm/s. Under these experimental conditions, the film surface area of 13 mm in diameter was processed by the abrasion jet during 40 minutes, resulting in a decrease in the roughness value from 3 μm to 1 μm. It was speculated that the abrasive jet polishing technique could be extended to angstrom level surface finish with a reduced size of abrasive particles.

4.2 THERMOCHEMICAL POLISHING WITH HOT METALS

A number of metals like iron, nickel or cobalt intensively interact with diamond at elevated temperatures by means of dissolving carbon. In particular, carbon solubility in iron is 0.8% at 730°C and 4.3% at the eutectic point of Fe-C phase diagram at 1153°C. Grigoriev et al. (1982, 1984) described a number of experiments in which iron, nickel and platinum templates were placed on natural diamond stones at temperatures between 600°C and 1800°C, and penetration of the metal "die" into diamond was observed. They discovered that the diamond surface becomes bright and smooth when the metal plate is rotated. Grigoriev et al. (1983) patented this method of thermochemical polishing in Russia, and utilized it for sharpening surgical diamond knives.

Later Yoshikawa (1990) developed a hot metal polishing technique which is especially effective for treatment of polycrystalline diamond. The diamond work piece is attached to a sample holder by a high temperature ceramic adhesive, then the work piece is placed on a pure iron or steel disk. The disk was heated electrically to temperature in the range of 700°C to 950°C with a nichrome wire heater. Both sample and disk are rotated, a planetary-type motion of the sample with respect to the disk being provided by a system of gears. The typical polishing velocity was 2.8 mm/s, the nominal load was 15 kPa, which is much smaller than for conventional mechanical polishing process. The type of surrounding gas is important for the rate and quality of polishing. Although the highest rate of material removal, 7 μm/h, was obtained by polishing in a vacuum at 950°C, the final polishing in hydrogen atmosphere at 750°C was necessary in order to achieve a mirror-like surface. The atomic hydrogen was generated by a hot filament positioned near the Fe disk. The role of hydrogen is a decarburization of the iron disk via methane molecule formation, thus keeping the disk in active state, i.e., able to dissolve new portions of carbon from diamond surface. Diamond film with initial roughness of about 3 μm has been polished to a final roughness of R_a=2.7 nm with this technique. Using a similar thermochemical polishing apparatus with a hot, low carbon steel disk in hydrogen gas (Fig. 3), Hickey et al. (1991) obtained a mirror finished surface on diamond films with rms roughness of 3.0-4.5 nm. The final polish is limited by etch pits of a few microns size, probably caused by the reaction of tiny iron particles detached from the disk. Importantly, no apparent difference in the polishing rate of diamond films with different orientations of grains, or preferential etching of grain boundaries, were observed. The production of specular diamond surfaces in the infrared and visible has been demonstrated, using films which were initially diffuse in these spectral ranges [Thorpe et al. 1990].

The presence of large (supermicron) diamond particulates on the film surface, which are an artifact of the growth process, have been found to be the major limitation to the rapid polishing with hot iron [Harker et al. 1990]. Before further polishing, such particulates can be selectively removed in oxygen plasma at rates of about 1 μm per minute. The etching selectivity was achieved by the following masking process. First an Au coating was deposited on the whole film, followed by removing the metal from the top of the particulates by physical polishing. Then the exposed areas were preferentially oxidized. Through repetition of the surface coating, mechanical polishing and etching procedure a surface can be obtained with a roughness below 50 nm, even if the particulates had initial dimensions of 20 μm to 100 μm.

The polishing process may be limited by the stress-induced curvature in polycrystalline films, which limits plate-diamond contact [Harker 1991]. Enhanced pressure of 40-50 kPa on the substrate is needed in this case. Under optimum conditions, the polishing rates of 2 to 5 microns/hour have been achieved with final surface roughness of about 10 nm.

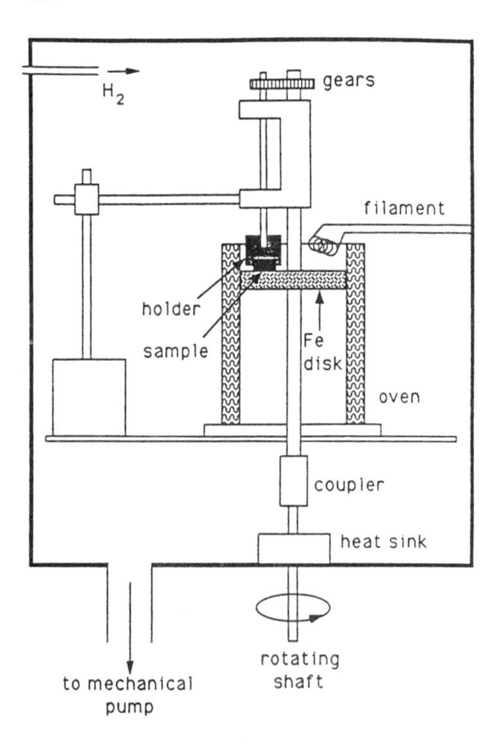

Figure 3. Experimental setup for polishing of diamond films with a hot steel disk in hydrogen atmosphere [Hickey et al. 1991].

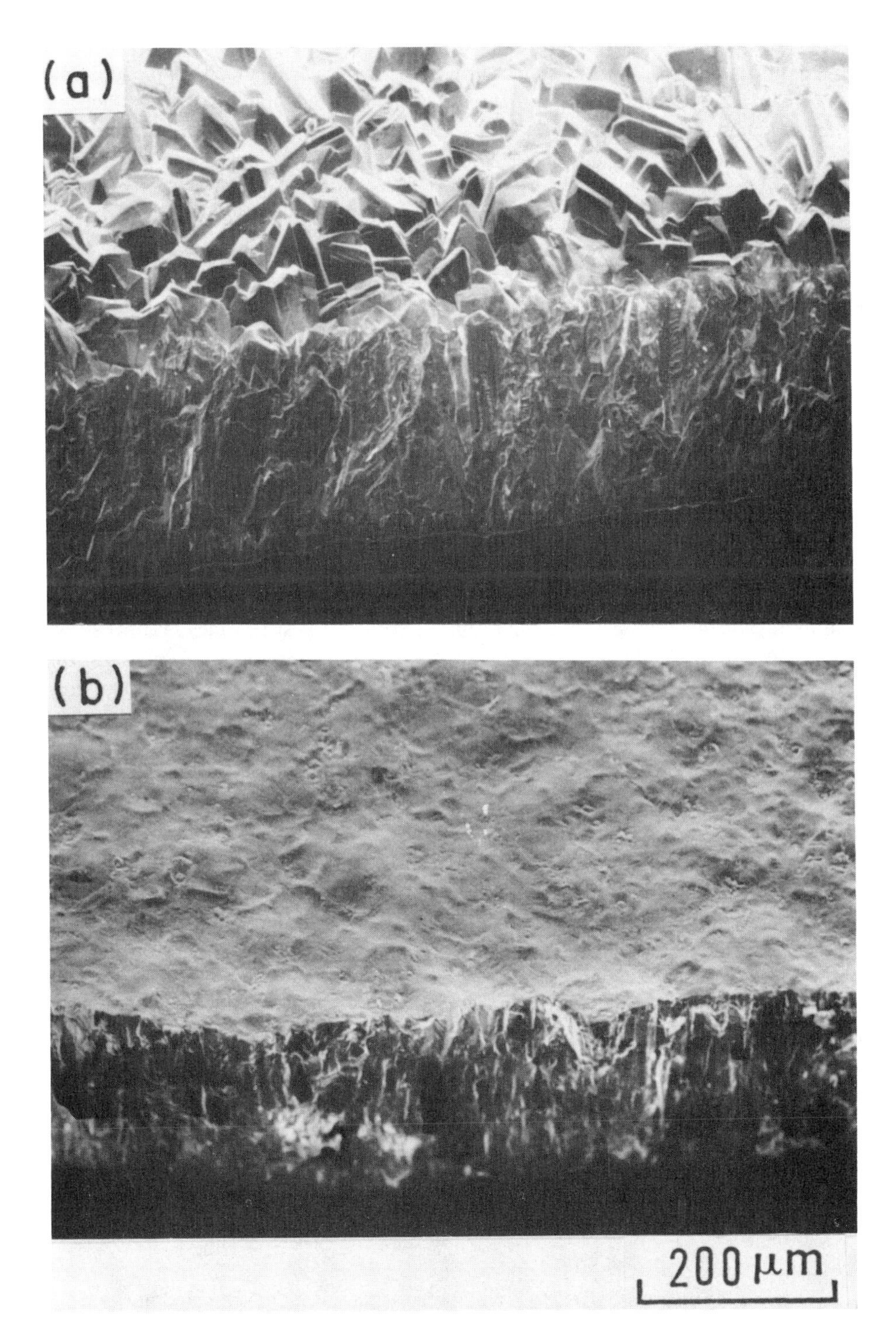

Figure 4. SEM pictures of CVD diamond films (a) before and (b) after reaction with manganese powder [Jin et al. 1992b].

Jin et al. (1992a) found that the diffusional transfer of carbon from CVD diamond film to iron foils creates relatively smooth surfaces by eliminating much of the roughness from the top faceted surface of the film. They sandwiched three ~220 μm thick free-standing diamond films between two iron foils (250 μm thickness) and heat treated them at 900°C in an argon gas atmosphere. An extensive massive thinning and smoothing of the film (100 microns thickness was removed) has been attained in 48 hours. The advantage of this smoothing technique is that it is applicable simultaneously to a large number of diamond films, avoiding the polishing of individual films. A thin reacted surface crust left on the diamond film was easily removed by polishing for 10 minutes on a diamond-bonded wheel.

Since manganese has a large solid solubility for carbon (~12 at% at 900°C) the polishing of diamond films by diffusional reaction with manganese powder appeared to be even more efficient than with iron, and it results in smoother surfaces [Jin et al. 1992b]. The Mn-treated film shown only little trace of faceted structure with an average macroscopic roughness of the order of a few microns (Fig. 4). No preferential etching of grain boundaries was observed.

Using the same technique but with molten rare-earth metals, such as Ce (melting point 798°C) and La (melting point 918°C), which can dissolve carbon up to 25 at% at 920°C, the thinning and smoothing of diamond films are possible 5 to 10 times faster than with solid metals [Jin et al. 1993]. A significant reduction in the process temperature from 900°C to 600°C has been achieved [Jin et al. 1993, McCormack et al. 1994] by using rare-earth/transition-metal alloys, such as 89%Ce/11%Ni (by weight) with a eutectic point of 477°C.

4.3 LASER POLISHING

The method of laser polishing is mainly used for treatment of CVD diamond films. It consists of selective removal of surface protrusions because of a difference in the ablation rates at “hills” and “valleys” of surface microrelief. Laser polishing technique is known as the fastest method for reducing surface roughness of diamond films due to high rates of material removal during ablative etching, and high pulse repetition rates of the lasers used. Among a variety of lasers, two types of pulsed lasers are widely used for the diamond film polishing. The first class is the excimer lasers which operate in UV range at wavelengths from 193 nm to 351 nm. One of the main advantages of the excimer lasers is a high value of light absorption in diamond, whereas the pulse repetition rate is not very high (typically up to 200 pulses per second). The second class of lasers is the lasers operating in visible and near IR regions (wavelength from 500 nm to 1000 nm) with very high repetition rate of 1-10 kHz. All the results on laser polishing to be discussed hereafter relate to these two types of lasers.

4.3.1 Excimer laser polishing

Ageev et al. (1988) were the first to have observed the laser-induced smoothing of the diamond film surface and to propose excimer lasers in diamond film polishing. In that work, ablative etching of diamond films by pulses of a XeCl excimer laser (308 nm wavelength) was investigated in ambient air and inert (xenon gas) atmosphere. It was found that the film surface inside a laser spot became much smoother than a virgin (unirradiated) surface. A decrease in surface roughness, R_a, from about 1 μm, characteristic of the as-deposited film, to 20 nm within the laser spot area was reported [Ageev et al. 1989]. A typical morphology of the laser-smoothed diamond film surface is shown in Fig. 5.

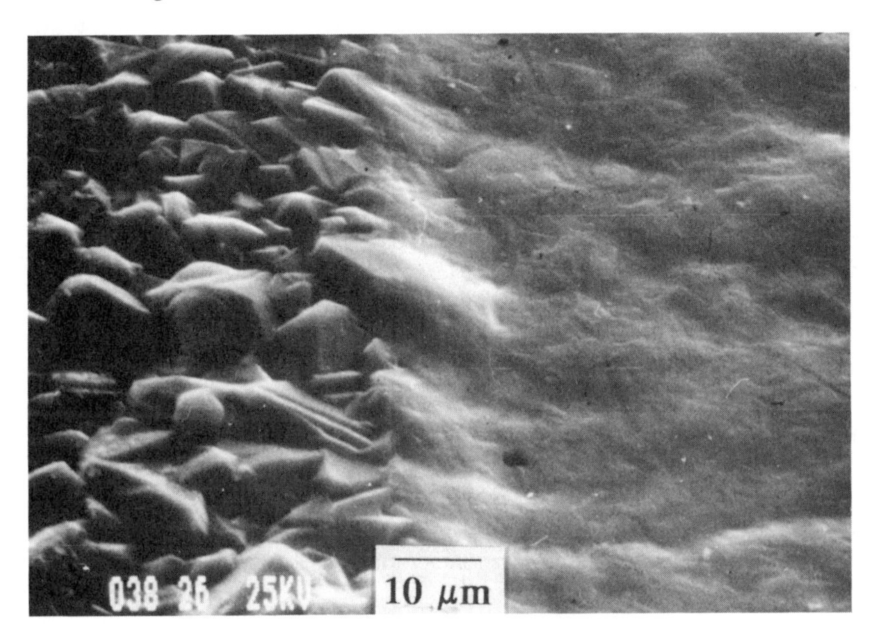

Figure 5. Smoothing of the diamond film surface as the result of KrF excimer laser etching by 100 pulses at fluence E=10 J/cm^2, [Pimenov et al. 1993a].

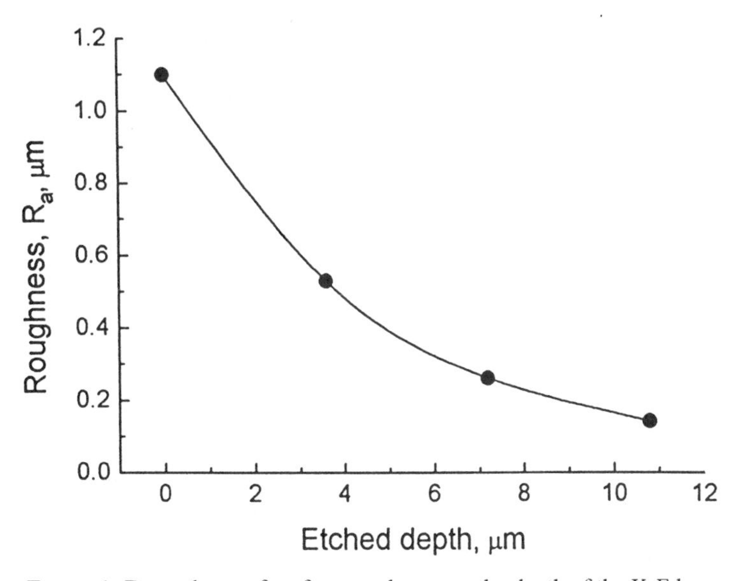

Figure 6. Dependence of surface roughness on the depth of the KrF laser-etched craters [Pimenov et al. 1993a].

The value of R_a of laser-treated diamond films was found to depend on the depth of laser-etched craters under multipulse KrF laser (l=248 nm) irradiation [Blatter et al. 1991, Pimenov et al. 1993a]. During etching down to a depth of the order of a grain size the value of R_a decreased progressively by an order of magnitude. The dependence of the surface roughness on the etched depth is shown in Fig. 6.

The final smoothness achieved was suggested to be determined by the original grain size. The polishing behavior was governed by the film microstructure. In particular, an oscillatory character of the roughness value rather than the gradual decrease in roughness, shown in Fig. 6, was observed in the case of a pronounced (100) columnar growth morphology [Blatter et al. 1991].

The first experiments on excimer laser polishing [Ageev et al. 1988, 1989, Blatter et al. 1991] were made over small laser spot areas of less than 0.5 mm x 0.5 mm, since both high laser fluences (exceeding the evaporation threshold of the diamond films) and uniform intensity distribution over the spot area were needed to quantify the etching and polishing processes. Of more practical interest is to scale up the observed polishing effect to large surface areas when a laser beam is scanned over the film surface. In the scanning mode of irradiation, the diamond samples are placed on a translation stage and are successively translated in a step of 1 μm after the action of each laser pulse (or a set of pulses). It was found [Bögli et al. 1992, Pimenov et al. 1993a,b, Boudina et al. 1993] that the large-area processing of diamond films by the scanning beam of ArF and KrF lasers resulted in uniform surface etching and smoothing over the whole irradiated area even in the case of nonuniform laser beam profiles.

Statistical analysis of the surface profiles before and after laser polishing was applied to quantify the surface finish of the diamond films etched with an ArF laser (l=193 nm) [Bögli et al. 1992]. Surface features in a wide range of spatial periods were significantly smoothed, with the smaller crystallites being smoothed more effectively than the larger crystallites. No spatial periods were observed which could be related to the discrete translation steps during laser irradiation. Consistent with the earlier results [Blatter et al. 1991], there was no polishing effect observed for the film with the columnar growth microstructure characterized by a large content of graphitic impurities at grain boundaries. Under the normal incidence of the laser beam, the efficiency of laser polishing in the scanning mode can be estimated as $(S/t) = (v \cdot a \cdot f)/Dh$, where S is the smoothed film area, t is the processing time, v is the ablation rate per pulse, a is the laser spot area, f is the pulse repetition rate and Dh is the film thickness to be ablated for achieving the best surface finish (Fig. 6). Using typical ablation and laser operation parameters, $(S/t)=(1\ cm^2/8\ min)$ was obtained for the diamond films of 20-30 μm thickness [Pimenov et al. 1993a].

From a viewpoint of transforming the rough diamond film surface into the flat one, the ablation of the surface layer with a thickness in large excess over a peak-to-valley amplitude (R_{max}) seems not to be very reasonable. What is really needed to smooth the surface is simply "to cut away" surface asperities by a laser beam. This is achieved by changing an angle of beam incidence onto the film surface. Yoshikawa (1990, 1991) was the first to apply a grazing beam of a Nd-YAG laser for smoothing diamond films (to be discussed below).

An increase in the incident angle of the KrF laser beam (with respect to the film surface normal) was found to result in more effective polishing with the best smoothing observed at the incident angles of 75-80° [Pimenov et al. 1993a,b]. At the beginning of the ablation process, when an actual angle of incidence is determined by a slope angle of surface asperities, etching of the asperities occurs with the highest rate. In the course of multipulse irradiation the actual angle is gradually increased, resulting in a gradual decrease in the ablation rate. If the value of laser energy density becomes lower than the laser evaporation threshold of a diamond film, laser etching of the film is practically stopped, the low rate oxidation in ambient air of the graphitized surface remaining the only possible mechanism for further etching.

Suzuki et al. (1994) compared the ArF laser smoothing of two films of different thickness and different initial roughness (R_{max}) of 75 μm and 3 μm. The minimum values of the final R_{max} (2 μm and 0.25 μm, respectively) were achieved at the incident angle of 80°. Recently, Tokarev et al. (1995) proposed a theoretical model for excimer laser ablating of diamond films with a grazing beam that is in agreement with the earlier experiments [Pimenov et al. 1993a,b, Suzuki et al. 1994].

The UV laser smoothing with a grazing beam gives important advantages in the large-area polishing of diamond films. First of all, it is a significant reduction in thickness of the surface layer to be ablated for achieving the lowest roughness value. It is also important, especially for thin films, that irradiation with a grazing beam attenuates an influence of substrate heating on the film adhesion in the case of low optical absorption in diamond.

Consideration of possible mechanisms of the laser-induced polishing is of great interest, and it is based on the features of laser-diamond interaction. As was reported by Rothschild et al. (1986), an excimer laser beam interacts with diamond material to form and subsequently to sublime a surface graphitic layer via evaporation and/or reaction with oxygen (or other reactive ambient gases). Tokarev and Konov (1991) proposed a theoretical model of laser-induced surface polishing under material evaporation from the solid state (without melting). In their model, the conditions were determined at which the surface microrelief is responsible for modulating the surface temperature due to a difference in the heat flow near hills and valleys at the surface (absorbed laser energy escapes easier from valleys). As the evaporation rate exponentially depends on temperature, the hills are evaporated faster than the valleys. The model is useful for qualitative interpretation of the laser-induced polishing, however, it is applicable only for small values of the surface roughness, i.e., for $(\hat{A}_g R_a) \ll 1$, where $\hat{A}_g$ is the value of light absorption (cm^{-1}) in the graphitized surface layer. Some other reasons were suggested to account for the higher evaporation rate of the hills, in particular, the vapor screen effect and electromagnetic field enhancement at the tips of crystallites [Tokarev and Konov 1991]. As pointed out by Bögli et al. (1992), all these mechanisms fade with decreasing ratio of surface height to lateral dimension, thus explaining the observed gradual decrease and eventual termination of the smoothening process as well as the dependence of final surface smoothness on the original grain size. Tokarev and Konov (1991) emphasized that by shortening the pulse duration it would be possible to polish diamond film surfaces more effectively than with longer pulses. That, however, was not confirmed experimentally [Pimenov et al. 1993a]. It should be noted that the approach proposed by Tokarev and Konov is relevant in laser-induced thermochemical etching of any material, when the surface temperature lies below the melting point and the etching rate exponentially depends on the temperature.

Bhushan et al. (1994) reported on the ArF laser polishing of diamond films in oxygen gas atmosphere, analyzing the role of chemical reactions of carbon with oxygen atoms and ozone (O_3) in the etching and polishing process. In this irradiation regime, atomic oxygen is produced by laser-induced dissociation of O_2, and ozone is formed via subsequent recombination of the O with O_2. The rms roughness of a 20 μm thick diamond film was determined to decrease from 657 nm to 114 nm after laser irradiation in a static O_2 atmosphere (50 Torr) during 5 min at a 10 Hz pulse repetition rate and intensity of about 50 MW/cm^2. It was supposed that the reaction between C and O_3 could be of major importance in the rapid film polishing. This suggestion requires further verification since the reactivity of diamond surface to O_3 is unknown. As was reported in earlier papers, the influence of reactive gases (oxygen, chlorine) on excimer laser etching of diamond single crystals [Rothschild et al. 1986] and diamond films [Ageev et al. 1988] was found, mainly, to remove a sputtered graphite-like material around the laser spot but not to change noticeably the etching rate as compared to the etching in inert gas atmosphere.

4.3.2 Polishing of diamond films with Nd-YAG lasers

Yoshikawa (1990, 1991) has demonstrated that very thick diamond films characterized by a high value of the surface roughness could be rapidly smoothed by means of Nd-YAG laser irradiation. The 0.7 mm thick films of R_{max}=250 μm were processed with a grazing beam at 83° incident angle (12 W average power and 1 kHz pulse repetition rate) to obtain a smoothed surface with R_{max}=3 μm. It took 7 min to reduce the surface roughness over the 3 mm x 2 mm film area. The Nd-YAG laser smoothing of the thick diamond films was applied as the first step in polishing followed by a high quality surface finish with a hot metal lapping technique [Yoshikawa 1990].

Ravi and Zarifis (1993) reported on the polishing of a 20 μm thick film with the Nd-YAG laser operating at 80-90 W and 840 Hz, although their attention was concentrated on the structural modification of the laser-treated diamond surface. They outlined other promising capabilities for diamond processing with the Nd-YAG laser, in particular, changing the wavelength from 1.064 μm to the second harmonic at 532 nm and using picosecond pulses with much higher peak power. The role of an oxygen environment during the Nd-YAG laser processing of diamond was found to be of great importance, which was pronounced both in an increase in the ablation rate with oxygen content and in the structure of the laser-modified diamond film surface [Yoshikawa 1990, 1991, Tezuka et al. 1990, Ravi and Zarifis 1993].

4.3.3 Surface roughness of laser-polished diamond films

The roughness data of the laser-polished surfaces are summarized in Table II, which reflects a tendency in reducing surface roughness and illustrates the limits of the roughness achieved. It follows from Table II that the laser irradiation regime with a

Table II. Reduction in surface roughness of diamond films after laser polishing. R_a^O is the roughness value of the original surface, R_a is the roughness value of the laser-irradiated surface, Â is the angle of laser beam incidence with respect to surface normal.

Laser	R_a^o, μm	R_a, μm	Â,^{o}	Refs.
Nd-YAG (l=1064 nm)	250	3	83	[Yoshikawa 1990]*
XeCl (l=308 nm)	~1	0.02	0	[Ageev et al. 1989]
KrF (l=248 nm)	0.5-0.65	0.1-0.12	0	[Blatter et al. 1991]
	0.9-1.05	0.14-0.2		
	0.65-0.8	0.13-0.17		
KrF (l=248 nm)	0.65-0.8	0.15	77	[Pimenov et al. 1993a]*
KrF (l=248 nm)	5.5	2.0	0	[Boudina et al. 1993]*
	4.5	2.0		
	7.0	4.5		
ArF (l=193 nm)	0.6	0.25	0	[Bögli et al. 1992]*
ArF (l=193 nm)	75	2	80	[Suzuki et al. 1994]*
	3	0.25		
ArF (l=193 nm)	0.65	0.11	0	[Bhushan et al. 1994]
ArF (l=193 nm)	0.45	0.12	73	[Pimenov et al. 1994]*

*Large-area processing with a scanning laser beam.
Yoshikawa (1990), Suzuki et al. (1994) and Boudina et al. (1993) used the R_{max} roughness value, and Bhushan et al. (1994) used the R_{rms} value.

grazing beam gives the best results of surface smoothing over large areas. Of interest is the laser polishing method developed by Bhushan et al. (1994). The suggestion of the dependence of final surface roughness on the grain size [Blatter et al. 1991] is confirmed by the results of Yoshikawa (1990, 1991) and Suzuki et al. (1994). The roughness data obtained with an atomic force microscope (R_a=20-40 nm, [Bhushan et al. 1994, Pimenov et al. 1994b]) demonstrated that small spatial periods of the surface microrelief were smoothed out, thus additionally indicating the roughness value to be determined by a larger spatial period dependent on the grain size.

To conclude, the laser polishing method is the fastest method for reducing roughness of the diamond film surface. The final surface roughness of the laser-polished films depends on the grain size of the original surface, which increases with the film thickness. For diamond films 20-40 μm thick, the R_a of 100 nm and the polishing rate (in terms of "polished area/processing time") of 1 cm^2/10 min have been achieved. The regime of laser ablation with a grazing beam manifests itself as the best one for technological purposes. It provides the large-area smoothing with minimum loss of diamond and reduced processing time, as well as the smoothening of complex surfaces. Laser-polished diamond films are characterized by improved optical properties in the IR region [Boudina et al. 1993] and by low friction and wear properties [Bögli et al. 1993, Bhushan et al. 1994, Pimenov et al. 1994]. Of special interest is the laser-induced structural modification of diamond surface, which is discussed in many of the referred papers (see, e.g., [Rothschild et al. 1986, Blatter et al. 1991, Bögli et al. 1992, 1993, Pimenov et al. 1993a,b, Ravi and Zarifis 1993, Bhushan et al. 1994]).

4.4 ION BEAM POLISHING

Use of ion beams for CVD diamond film polishing is attractive due to uniform etching over large areas (tens sq. cm) with a precisely controlled rate of material removal. The choice of ions determines a dominant mechanism of ion beam interaction with diamond (physical sputtering or chemical etching with reactive ions), influencing the removal rate and etching selectivity of polycrystalline material. Here we consider separately (i) the etching and polishing by reactive ions, with oxygen being mostly used, and (ii) polishing with inert gas ions.

4.4.1 Oxygen ion beam polishing

During oxygen ion beam irradiation of the diamond films both spatial and structural inhomogeneity of the film surface may prevent surface smoothing. This originates from the dependence of the sputtering yield on the angle of ion beam incidence and from a difference in etching rate of diamond and nondiamond phases. To overcome these difficulties, Zhao et al. (1990) developed a planarizing technique for diamond film polishing which made further progress in later studies [Bovard et al. 1991 and 1992, Grogan et al. 1992].

The method consists of coating the diamond surface with a planarizing solid film of a composition that matches the etching rate of the planarizing coating with the etching rate of the diamond film at a certain incident angle of the ion beam. A suitable planarizing material was found to be a mixture of the Shipley microposit photoresist 1400-17 and titanium silica emulsion [Bovard et al. 1992]. After 40 min irradiation with 500 eV oxygen ions at 47° incident angle the initial diamond film roughness of about R_{max}=1 μm was reduced to R_{max}=55 nm (R_a=4.4 nm) uniformly over the film area of 5 cm in diameter [Bovard et al. 1991, 1992]. These roughness values are well compared with the values (R_{max}=23.6 nm and R_a=2.7 nm) of the diamond films polished with a hot metal plate [Yoshikawa 1990]. The roughness data corresponding to successive stages of planarization and polishing are given in Table III. Table III illustrates the roughness measurements of (i) the original Si wafer, (ii) Si surface scratched with diamond grit before diamond deposition (to increase nucleation density), (iii) diamond film covered with a planarizing coating, and (iv) ion-polished diamond surface.

Infra-red transmission spectra confirmed the high optical quality of the ion-beam polished films [Bovard et al. 1991 and 1992, Grogan et al. 1992]. The developed planarizing technique would be promising for surface finish of free-standing films and curved surfaces, in particular, for optical applications. Hirata et al. (1991) reported on smoothing diamond films under irradiation with 1.5 keV oxygen ions produced by an ECR ion source. The diamond sample was rotated at 60 rpm and irradiated by the ion beam at 80° incidence angle for 8 hours, which resulted in a decrease in surface roughness (R_{max}) from 3 μm to 0.5 μm. For oxygen ion beams, processing with a grazing beam is a necessary requirement to avoid the nonuniform etching of diamond grains and grain boundaries that was observed at low incident angles [Alexenko et al. 1991].

TABLE III. Roughness of Si substrate and diamond film of 1 μm initial roughness at different stages of treatment by means of the planarizing polishing technique with oxygen ions [Bovard et al. 1992].

Sample	R_a, nm	R_{rms}, nm	R_{max}, nm
Polished Si	0.4	0.6	2.4
Roughened Si	5.4	7.0	87.1
with diamond grit	5.2	7.1	224.0
Planarized diamond film	3.7	4.9	70.8
Polished diamond film	4.4	5.5	55.4

4.4.2 Polishing with inert gas ions

Ar ion beams are most widely used in ion beam processing of various materials, including diamond and diamond-like carbon coatings. However, the sputtering yield of diamond for argon ions is too low to be able to match the etching rate of diamond with that of a planarizing photoresist coating [Zhao et al. 1990, Grogan et al. 1992].

The Ar ion-induced smoothing of diamond films was reported by several authors [Kobayashi et al. 1990, Hirata et al. 1991, Kyuno et al. 1994, Bögli et al. 1995]. The rate of material removal by Ar ion etching depends on ion energy, ion current density and the angle of incidence of Ar ions onto the diamond surface. Typically, etch rates up to 25 nm/min at an ion energy of 5 keV were reported [Alexenko et al. 1991]. Because of relatively low etching rates the argon ion-induced smoothing is a slow process, depending on the initial surface roughness [Hirata et al. 1991]. A typical morphology of the diamond film surface smoothed with Ar ions is shown in Fig. 7 in comparison with an unirradiated surface. For micron thick films, the primary smoothing effect is the rounding off of surface asperities, which was found to affect significantly the friction and wear properties of thin diamond coatings [Bögli et al. 1995].

Ullmann et al. (1993, 1994) studied diamond film etching with Ne ions in the range of ion energies from 100 eV to 1600 eV. They found that the film surface topography was modified from only rounding and flattening of grain edges to a distinctly smoothed surface, depending on the ion energy and on the film microstructure.

5. Patterning

An important step in the fabrication of diamond-based electronic and micromechanical devices is diamond patterning by a controllable etching process. Because of chemical inertness of diamond the etching methods are mostly dry ones, such as ion beam, plasma or laser etching, while liquid etchants, like molten KNO_3, are used to reveal defects in diamond crystals [Tolansky 1955, Seal 1965]. Marchiwka et al. (1993) reported also on a "wet" selective removal of a nondiamond carbon layer created by ion implantation into (100) oriented natural and synthetic diamonds. The damaged layer was eliminated by electrochemical etching in deionized water and various water

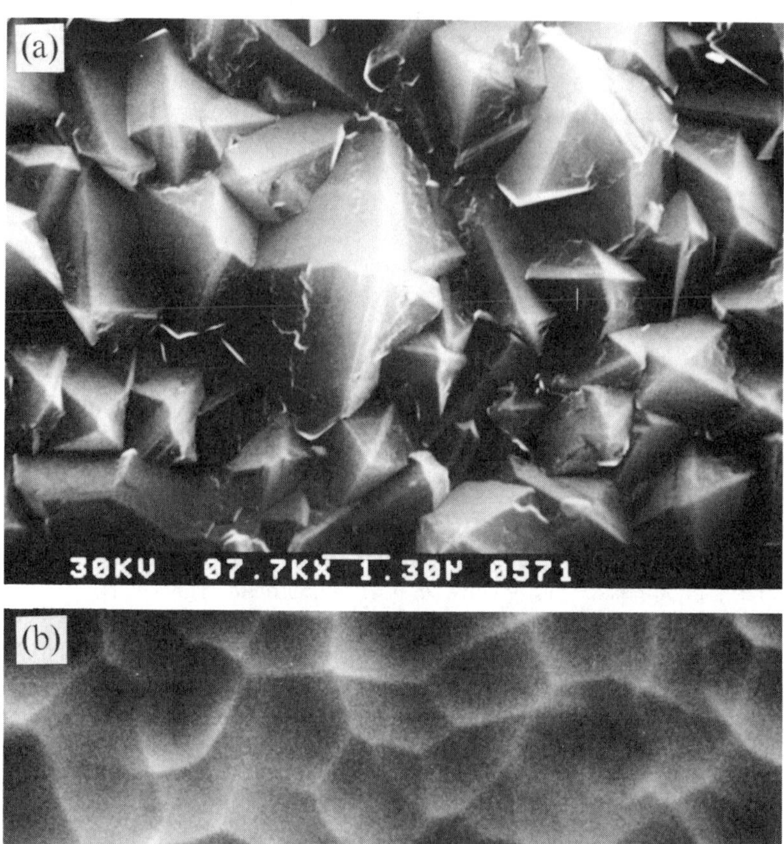

Figure 7. Surface morphology of a 10 μm thick diamond film: (a) original surface, R_a=185 nm, (b) surface smoothed with Ar ions in ECR microwave plasma, R_a=129 nm (bias voltage -500 V, etching time 100 min) [Bögli et al. 1995].

solutions. The implanted diamond substrates were covered with a photoresist, patterned by standard lithographic technique and then immersed in etch medium between two electrodes biased to provide a field of about 100 V/cm. The method is simple, selective, clean, safe and can be used for patterning of various forms of non-diamond carbon near the surface of a diamond.

5.1 ION SPUTTERING

Etching of diamond with inert ions occurs via physical sputtering of carbon atoms by energetic ions bombarding the surface. Whetten et al. (1984) studied sputtering of (100) single crystal diamond by argon and oxygen ion beams. Ion flux with energy of 500 to 1000 eV was generated by a Kaufman type ion source. For Ar^+ ions a large increase in sputtering yield (number of carbon atoms removed per one ion) was observed as the angle of incidence increased from the normal, with maximum yield of 1.1 atoms per ion occurring at 60° from the normal. This was explained by increased deposited energy (and increased damage) which becomes concentrated near the surface at oblique incidence angles. Later, Miyamoto (1990) used 1 keV argon ion beam machining at tilt angles to sharpen diamond knives and styli to final edge radius of the order of few tens nanometers.

Ullmann et al. (1994) studied the etching of diamond films by low energy Ne ions. The etch rate increased from 2.3 nm/min to 4.5 nm/min as ion energy increased from 400 eV to 1600 eV, corresponding to the sputtering yield variation from 0.29 to 0.58. The etch rates of fine and coarse grained films were comparable. The 400 eV ions, which have a penetration depth in diamond of about 1.3 nm, caused only rounding of the grain edges, while the 1600 eV ions resulted in a surface smoothing. An amorphous carbon layer 1-7 nm thick was formed at the etched surface.

A Penning discharge in Ar gas was applied to pattern polycrystalline diamond films using a stainless steel mask [Sugita et al. 1990, Funamoto et al. 1991]. Etch rates of 70-120 nm/min was achieved under a pressure $1.5 \cdot 10^{-2}$ Pa, discharge current of 6-7 mA and anode potential of 0.8-1.5 keV. The same etch rate has been obtained for synthetic diamond crystals. Surface smoothing from R_a=200 nm to 110 nm was obtained after removal of a 3 μm thick top layer. The etched film demonstrated an increased optical transmission.

Kobayashi et al. (1990a,b) have found that the etching rate of polycrystalline diamond in an Ar RF plasma is three times lower (2 nm/min) than that of hard amorphous carbon. They suggested that the etch rate of these carbon materials is inversely proportional to their hardness. The diamond is modified by the Ar plasma, the amorphous and graphite phases appearing at the surface. The etched surface becomes smooth in the Ar plasma, while the etching in CF_4 RF plasma at the same conditions causes some surface roughening.

5.2 REACTIVE ION ETCHING

For reactive ion etching (RIE) processes the ion energies used are typically a few hundred to a few thousand eV, with the ion penetration range being a few to tens of angstroms. Due to simplicity of the etching process and high selectivity towards oxide and metallic masks, the RIE in O_2 plasma is suitable for patterning of diamond in a large area. The reactive species are activated in radio-frequency plasma at typical pressures of 100-300 mTorr. The etch rates of 10 nm/min to 60 nm/min were measured for polycrystalline films [Dorsch et al. 1992]. The etch rate is higher in oxygen than in hydrogen, CVD diamond and IIa type crystals being sputtered more slowly than amorphous a-C films [Sandhu and Chu 1989]. Due to preferential etching of non-diamond carbon located at grain boundaries, the diamond films exposed to microwave plasma in air or O_2 may exhibit a columnar structure [Sato et al. 1990].

Shikata et al. (1993) fabricated up to 1 μm line and space patterns in synthetic single crystal diamond by RIE in O_2/Ar mixtures. High etching selectivity of up to 40:1 was obtained using a SiO_2 mask. The optimal O_2 content in Ar was about 1%, being the result of a trade off between demands of high etch rate, high etch selectivity and smooth morphology of the etched surface.

Other types of reactive plasmas different from RF discharges were also used for diamond patterning. Miyauchi et al. (1991) investigated etching of polycrystalline diamond films in an electron-beam assisted plasma using H_2, H_2-O_2 and O_2-He mixtures. Patterns 10-20 μm wide were obtained with a gold mask. During plasma treatment the temperature of specimen was rather high, about 700°C.

An electron cyclotron resonance (ECR) plasma was used to produce damage-free etched surfaces in diamond single crystals [Beetz et al. 1990] and in boron-doped diamond films [Grot et al. 1992]. The ECR plasma provides ion energies <50 eV, which are well below the 150 eV reported [Marsch and Farnsworth 1964] to be the threshold energy for ion bombardment damage of diamond. An etch rate of 9 nm/min has been measured for a 2.7 mTorr Ar : 0.4 mTorr O_2 reactant mixture, while no etching was detected with pure Ar gas. Fine patterns with a smooth surface (roughness ~10 nm) have been fabricated using an oxidation resistant SiO_2 mask. Damage- and graphite-free surfaces can be obtained also by bombardment of ion-implanted diamond with oxygen ions of low energy (200 eV) generated by an ion gun [Beerling and Helms 1994]. However, when the oxygen ion energy is increased a steady-state damaged layer is created. This demonstrates the importance of using lower energy ion energies to produce non-degraded etched diamond surface.

Stoner et al. (1993) used a hydrogen microwave plasma at a pressure of 10-15 Torr for negative bias-assisted etching of diamond films in order to obtain micron-scale features with vertical walls. It was suggested that the etching mechanism involves intensive electron emission from the diamond followed by surface degradation (graphitization) and subsequent etching of nondiamond component. The advantage of this technique is that the etching can be performed in the same reactor where the diamond film was grown, thus reducing the cost of the patterning process.

Diamond films were anisotropically etched in a DC glow discharge in pure H_2 or in H_2 with a few percent of O_2 at pressures of a few hundreds of mTorr [Tessmer et al. 1993]. Etch rates of about 40 nm/min were achieved when the samples were negatively biased. Patterns of minimum size of 2 microns were obtained with Au masks. The diamond temperature was below 400°C during the plasma treatment. However the etched surface was rough possibly due to local increases in surface temperature as a result of enhanced electron emission from sharp crystallite edges.

Efremow et al. (1985) used a combination of 2 keV Xe^+ ion beams and a flux of NO_2 molecules to etch a single crystal diamond. The NO_2 molecules caused the surface oxidation in presence of bombarding Xe^+ ions, with an etch rate of about 200 nm/min. Using an Al mask they fabricated narrow trenches for high-frequency traveling wave tubes. The selective etching can be accomplished also simply by thermal oxidation of diamond as demonstrated by Masood et al. (1991). They etched 2 μm thick diamond films with a 0.9 μm thick Si_3N_4 mask using a rapid thermal process in oxygen at 700°C. It was suggested that the pattern resolution can be as high as 1 micron in the case of small enough grain size in the film.

5.3 LASER ETCHING

A laser pulse of sufficient energy being absorbed by a diamond surface layer causes a rapid temperature rise up to the sublimation point of carbon, resulting in the ejection of the film upper layer. The optical band gap of pure diamond is 5.5 eV, and generally speaking, only the lasers operated at wavelengths shorter than the diamond absorption edge (227 nm), e.g., ArF excimer laser (193 nm wavelength), might be expected to be suitable for efficient diamond treatment [Rothschild et al. 1986, Bogli et al. 1992, Johnston et al. 1993, Harano et al. 1994]. However, various defects, amorphous and graphite phase inclusions in CVD diamond, concentrated predominantly at grain boundaries, provide an initial absorption even at longer wavelengths. This may cause the formation of an opaque graphitized surface layer under the action of one or the first few laser pulses, so the next pulses are absorbed automatically by that self-sustained modified layer [Rothschild et al. 1986]. Once such a graphitized layer is formed the choice of type of laser becomes less critical. To date, a variety of lasers operated in a broad spectral range from UV to IR have been used for diamond etching, including KrF excimer lasers (248 nm wavelength) [Blatter et al. 1991, Pimenov et al. 1993a,b, Harano et al. 1994] and XeCl excimer lasers (308 nm) [Ageev et al. 1988], Cu vapor lasers (510 nm) [Obraztsov and Pirogov 1993], Nd:YAG lasers (1.06 μm) [Tezuka et al. 1990] and CO_2 lasers (10.6 μm) [Bouilov et al. 1990].

In the ablation process, a part of the previously graphitized layer is removed, but simultaneously a deeper layer of diamond is converted to graphite, so the laser etching occurs in the form of pulse-by-pulse penetration of the graphitic "piston" into the diamond. Due to high optical absorption the ultraviolet excimer lasers are suitable for dry etching of diamond by ablation (vaporization). In addition, short wavelength light allows one to obtain better spatial resolution of the pattern produced. Type IIa diamond crystals have been etched with an ArF excimer laser emitting photons at 193 nm wavelength (6.4 eV) [Rothschild et al. 1986], at which the absorption length is ~10 μm. Using an optical projection scheme to image a transmission mask at the substrate surface, the well-defined gratings ~100 nm deep with a period as small as 250 nm were obtained in single 15 ns long pulses. Deeper structures can be produced with multiple pulsing. In another optical geometry (direct writing mode) the laser beam was focused to a 20 μm spot size and scanned over the surface to ablate 15 μm deep trenches. An etching rate of 200 nm/pulse was achieved. The etch efficiency, defined as the ratio of the number of removed carbon atoms to the number of incident photons was 0.12. The graphite layer which is redeposited inside and around the etched pattern can be strongly

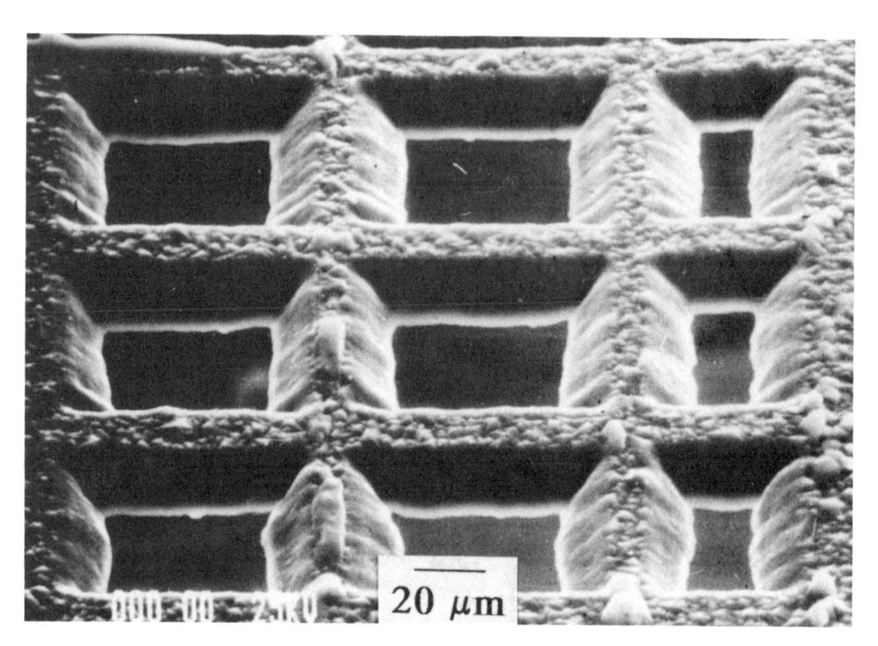

Figure 8. An array of microholes in 20 μm thick free-standing CVD diamond film etched-through with a KrF excimer laser.

reduced when the irradiation is carried out in reactive gas like Cl_2, O_2, NO_2. The laser-induced graphitic phase can be completely removed by post-etching annealing in air at 450°C [Pimenov et al. 1993b] or by H_2 plasma treatment [Harano et al. 1994]. The treatment in hot H_2SO_4-CrO_3 acid mixture removes the graphitic layer, however a thin 40-60 nm modified conductive bottom layer may still remain [Geis et al. 1989]. This layer, composed of diamond and graphite-like material can withstand heating up to 1800°C, and forms ohmic contacts to type IIb diamond. Other excimer lasers operated on KrF and XeCl molecules were also successfully used for etching of circular and rectangular pits and holes, as well as gears in diamond films and single crystals [Ageev et al. 1988, Konov et al. 1993, Pimenov et al. 1993a, Harano et al. 1994, *Cutting* 1994]. An array of holes drilled in a few seconds in 20 μm thick free-standing CVD diamond film with a KrF excimer laser is shown in Fig. 8. The film was patterned using an optical projection technique with 6X demagnification and a grid with a period of 500 μm as the mask.

Generally, the materials with lower thermal conductivity, *K*, are heated and vaporized by laser radiation more easily than those with high *K*, therefore it is not surprising, that an increase in ablation threshold fluence and etch rate with a decrease in the quality of diamond films was reported [Blatter et al. 1992, Johnston et al. 1993]. In the extreme case of nanocrystalline diamond films, which may have high optical absorption and thermal conductivity as low as $K \sim 0.4$ W $cm^{-1}K^{-1}$ [Plamann et al. 1994], the diamond can be heated to high temperatures by a continuous wave laser even at powers as low as 1-2 watts. Ralchenko et al. (1994) engraved a fine-grained diamond film with a sharply focused beam of a green Ar^+ ion laser scanned over the film surface under computer control. The diamond heated within the laser spot rapidly oxidizes (burns) with etch rates up to 50 μm/s when the irradiation is performed in oxygen or simply in air. The direct writing technique is very flexible, the dimensions of grooves can be precisely controlled by laser power, beam spot size, scan velocity, and the number of beam passes along the trace. Micron-sized holes, grooves and gratings were produced by this technique.

A version of indirect laser etching of grooves in diamond by irradiation of the crystal immersed into liquid (deionized water or KOH solution) has been described [Miyazawa et al. 1994]. The beam of a Nd:YAG laser passed vertically through the transparent diamond crystal fixed on a stainless steel underplate and heated the metal support. The authors believed that the lower surface of the diamond interacted with active species generated in the vaporized liquid resulting in etching. Another possible etching mechanism may be a mechanical removal of small pieces of diamond by a cavitation effect which induces the formation of high concentration of defects (dislocations) near surface. In this case an accumulation of defects under multipulsed irradiation results in material fatigue and surface erosion, as was observed for laser-induced silicon damage under a layer of liquid [Simakin and Shafeev 1995].

5.4 ETCHING WITH CATALYTICALLY ACTIVE METALS

An interesting method of diamond engraving with a hot metal die has been developed by Grigoriev et al. (1982). The diamond etching mechanism involves carbon dissolution in metal, diffusional transport to the metal-gas interface, and carbon desorption in the form of methane. Templates 0.05-2 mm thick made of Fe, Pt or Ni-based alloys were brought into contact with natural diamond workpieces and heated to temperatures of 600-1800°C in a vacuum, argon or hydrogen atmosphere. Metal dies in the form of a square, wire or gear dissolved the diamond leaving a deep image of the figure. The highest etch rates, up to 0.3 mm/hour, have been achieved for iron in a H_2 atmosphere. Hydrogen extracts the dissolved carbon by a methanation reaction, preventing a saturation of the metal with carbon and keeping the catalyst active for a long period of time. Uegami et al. (1992) found that the efficiency of working the diamond with a piece of iron or steel in H_2 strongly increased if the diamond was heated to temperatures above the eutectic point of the Fe-C equilibrium diagram (1153°C). In the first stage of the metal-diamond interaction, the carbon diffuses into solid iron. As the content of the dissolved carbon increases the iron melts, rapidly accelerating the carbon diffusion. No difference in etch rate for different crystal planes was observed, possibly since the etch rate was limited by the rate of the decarburization reaction rather than the dissolution process. The use of other metal films which intensively dissolve carbon, for instance, manganese, has been also reported for patterning purposes [Jin et al. 1992b].

In order to achieve better pattern resolution and higher etch rates, very thin, (0.1-12 μm), films of Fe, Ni and Pt were deposited onto polished CVD diamond films [Armeyev et al. 1991, Ralchenko et al. 1993]. First, the metal strips 10 μm wide were delineated by conventional photolithography, or selective-area laser ablation, or laser-assisted decomposition of an organometallic compound. Then, the samples were heated to 850-950°C in flowing H_2 to produce trenches in the diamond. Iron showed the highest catalytic activity of the metals examined, providing an etch rate up to 8 μm/min. In a particular experiment an iron film was transferred into diamond to a depth about 100 times the metal thickness. Thus, potentially, patterns with very high aspect ratios could be produced by this technique. However, too thin metal films may undergo mechanical rupture during the diffusion process, resulting in the loss of sharpness of pattern edges and a rough relief of the etched surface. In a similar way, Glesener and Tonucci (1993) fabricated a wear resistant diamond press mold by etching an array of 1 μm deep circular pits into a single crystal HPHT diamond with a patterned iron film. When the diamond mold was placed against gold and aluminum foils the metal surfaces accurately reproduced the inverse pattern from the diamond surface after the compression.

5.5 SELECTIVE-AREA DIAMOND DEPOSITION

Selective-area deposition of diamond (SAD) is an alternative method for diamond patterning, which may help to produce structures with less surface damage and faster (as compared to etching) techniques. The SAD process is based on the fact that CVD diamond growth occurs only on seeded areas, thus two surfaces on which the nucleation density differs by several orders of magnitude are required for fabrication of patterns with good selectivity. The most common approach involves a uniform seeding of the substrate by abrasion or ultrasonic pre-treatment followed by selective removal of the seeds (for an early review see [Popovici and Prelas 1992]). Mostly Si substrates were used for SAD, a few exclusions being diamond [Fujimori et al. 1992] and copper substrates [Narayan and Chen 1992, Konov et al. 1994]. Hirabayashi et al. (1988) seeded Si substrates by ultrasonic pre-treatment in a diamond powder suspension in ethyl alcohol, which is a common procedure, then the resist pattern was drawn by a standard lithographic method. At the next step the uncovered surface was etched 50-60 nm in depth by an Ar^+ ion beam. After resist removal the diamond film grew only on unetched area producing patterns with a resolution of a few microns.

In addition to Ar^+ ion beam etching [Hirabayashi et al. (1988), Higuchi and Noda (1992), Ma et al. 1989], "wet" chemical etching [Ramesham et al. 1991] and RIE [Kobashi et al. 1990] of Si substrates were used for annihilation of nucleation centers. A photoresist [Hirabayashi et al. 1988, Higuchi and Noda 1992, Kobashi et al. 1990], SiO_2 [Masood et al. 1991, Ma et al. 1989, Lin et al. 1992] and Si_3N_4 [Masood et al. 1991, Ramesham et al. 1991] were used as mask materials to save the seeds for successive diamond growth. Ramesham et al. (1991) used abrasion seeding and photolithographically patterned silicon nitride masks to protect seeds under the masks upon chemical etching off of grooves in a Si substrate. After removal of the mask, the diamond was deposited on the formerly masked area to form microchannels between diamond and substrate, which could be used for effective cooling of high power electronic devices mounted on the top of the diamond film.

Mask-less treatments of substrates were performed with lasers [Chapliev et al. 1991, Narayan and Chen 1992, Ralchenko et al. 1993, Pimenov et al. 1993c, Ralchenko et al. 1994, Konov et al. 1994]. Laser irradiation of seeded substrates to suppress diamond nucleation is performed either by passing a powerful laser pulse through an optical mask to melt the surface, or by a beam scan over the surface to ablate or oxidize the seeds in selected areas.

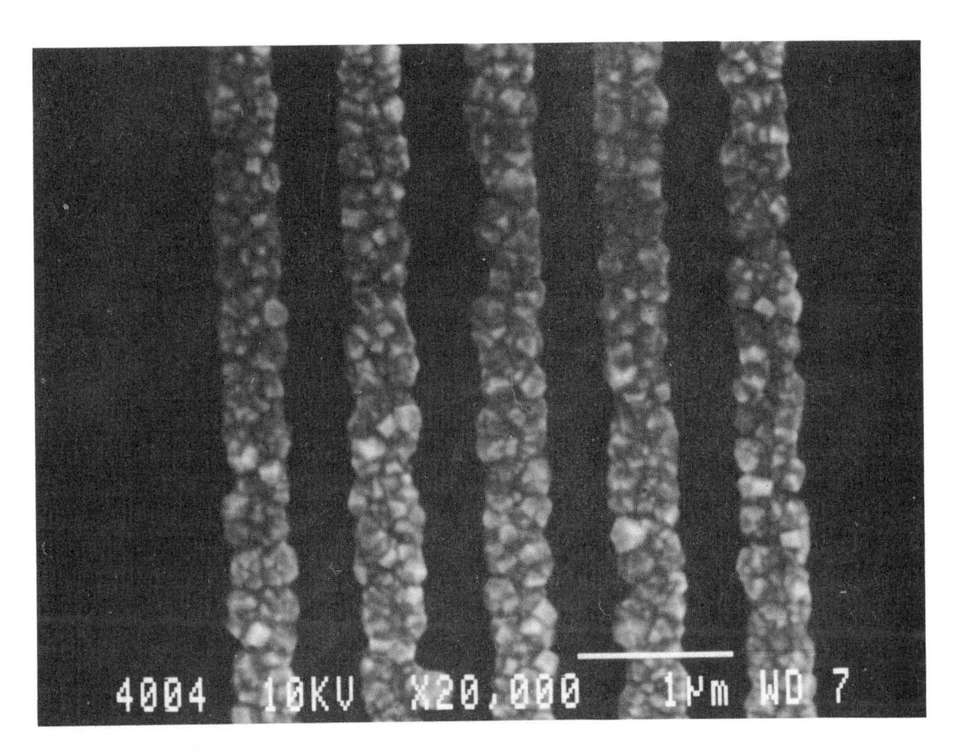

Figure 9. Selective area deposited 450 nm wide diamond lines on Si substrate. SiO_2 mask strips are between diamond lines [Katsumata and Yugo 1994].

The abrasive pre-treatment may be unsuitable for fabrication of diamond structures for electronic applications since microdamage of the substrate subsurface layer may cause a degradation of device performance. This is avoided by a spin coating of the Si substrate with a mixture of a photoresist and fine diamond powder [Masood et al. 1991, Ralchenko et al. 1993]. The resist is then patterned by a standard lithography technique, and the diamond particles are selectively removed together with the resist carrier during the resist development step. The advantage of this method is a compatibility with the standard technology of microelectronic devices fabrication.

There is another group of SAD methods which relies on selective enhancement rather than suppression of diamond nucleation. Higuchi and Noda (1992) used a die dipped into diamond powder for direct printing of seed patterns. Implantation of P^+ and As^+ ions into silicon masked with a SiO_2 film increased nucleation density by nearly 10^3 times [Lin et al. 1992], this effect being ascribed to a strain development under ion bombardment. Dots of C_{70} carbon clusters have been deposited onto Si, Mo and SiO_2 substrates to promote enhanced diamond deposition, while no diamond grew beyond the cluster spots [Meilunas et al. 1991]. Valdes et al. (1991) electrophoretically deposited submicrometer diamond seeds from a diamond powder suspension in water onto Si substrate with a patterned SiO_2 mask. The regions masked with oxide developed a negative surface charge, which blocked the deposition of negatively charged diamond particles by electrostatic repulsion. Deposition of colloidal diamond particles at biased positive potential remained unhindered, thus seeded patterns were formed only at unmasked areas. A laser technique for SAD was also reported [Pimenov et al. 1994a], in which local regions with high nucleation density were formed at the substrate surface. The method is based on the laser-induced forward transfer of a thin solid layer containing ultrafine diamond particles from a transparent quartz substrate onto a Si substrate.

Diamond patterns with submicron resolution have been grown on silicon without conventional scratching or ultrasonic pre-treatment of the substrate [Katsumata and Yugo 1994]. First, the substrate was thermally oxidized in oxygen and patterned lithographically. Then, generation of diamond nuclei and growth of diamond were carried out in a microwave methane-hydrogen plasma with an electric field applied to the substrate. Nucleation selectivity is believed to be caused by a difference in potentials near naked Si and the insulating oxide. Under the DC negative bias, cations accelerated towards the substrate penetrate into the Si to a depth of a few angstroms to form nuclei, while the SiO_2 area remains at a floating potential insufficient to accelerate the ions to enough energy. The 0.45 μm wide diamond lines were deposited between narrow SiO_2 strips (Fig. 9).

A witty method to plant diamond seeds into special small pits on a substrate for successive SAD process has been demonstrated by Kirkpatrick and Ward (1991). Regular craters of 100 nm in diameter and depth were produced in a Si substrate by focused ion milling. Then fine diamond powder suspended in alcohol was allowed to flow over the substrate to carry the diamond particles into the craters. After the CVD process, nearly the every pit held one diamond crystal. It could be possible to use large fields of such structures, for example, for fabrication of arrays of discrete optical or electronic devices. A similar approach to SAD based on substrate pitting, but using anisotropic chemical etching of Si and relatively large seeds (<100 μm) was applied to growth of mosaic diamond films with improved electronic properties [Geis 1992].

6. Conclusions

The treatment of diamond single crystals for jewelry and, to a lesser extent for industrial applications still relies on well-established mechanical approaches and on the use of the superior hardness of the diamond instrument itself. However, the traditional mechanical methods like sawing with a bronze wheel or polishing with a cast iron scaife charged with diamond powder appeared to be unsuitable for the processing of PCD and CVD diamond coatings. In recent years considerable progress has been achieved in the development of new techniques for diamond shaping. In particular, laser cutting and spark erosion machining became routine procedures to obtain these materials in a desired form and size for many applications such as cutting tool inserts, dies and heat sinks. Excimer lasers and Nd:YAG lasers can be used for fast smoothing of rough diamond surfaces over large areas, however, they do not provide an optical grade surface finish. On the other hand, lapping with a hot iron disk and ion polishing, which remove surface material at much slower rates, achieve final surface roughness of the order of few a nanometers. Miniature patterns (even with micron scale resolution) can be etched in diamond with a number of dry techniques, including laser ablation and burning, plasma and ion beam etching using an appropriate mask material, and diamond dissolution by pre-patterned thin films of catalytically active metals (iron or nickel).

Alternatively, diverse diamond patterns can be produced by selective area diamond deposition from the vapor phase, for which many methods have been suggested.

7. References

Ageev, V.P., Bouilov, L.L., Konov, V.I., Kuzmichov, A.V., Pimenov, S.M., Prokhorov, A.M., Ralchenko, V.G., Spitsyn, B.V. and Chapliev, N.I. (1988) Interaction of laser light with diamond films, *Soviet Physics-Doklady*, **33**, 840-842.

Ageev, V.P., Chapliev, N.I., Konov, V.I., Kuzmichov, A.V., Pimenov, S.M., and Ralchenko, V.G. (1989) Etching and patterning of polycrystalline diamond films by pulsed UV and IR laser radiation, *Proc. 3rd Int. Conf. on Energy Pulse and Particle Beam Modification of Materials*, Dresden, GDR, September 4-8, K. Hohmuth and E. Richter (eds), Physical Research, Akademie-Verlag Berlin, Vol. 13, pp. 318-320.

Alexenko, A.E., Belyanin, A.F., Bouilov, L.L., Semenov, A.P., and Spitsyn, B.V. (1991) Application of gas-discharge ion source for etching polycrystalline diamond films, *Proc. 2nd All-Union Workshop on Thin Films in Electronics*, Moscow-Izhevsk, pp. 74-79 (in Russian).

Ansell, M.F. (1984 Diamond cleavage, *Chemistry in Britain*, **20**, 1017-1021.

Armeyev, V.Yu., Ralchenko, V.G., Smolin, A.A. and Spitsyn, B.V. (1991) Thermochemical etching of diamond films via catalytical action of laser-deposited metal microstructures", in *Abstracts of the 2nd Europ. Conf. on Diamond, Diamond-like and Related Coatings, Nice, France,* Paper 7.37.

Beerling, T.E. and Helms, C.R. (1994) Direct observation of the etching of damaged surface layers from natural diamond by low-energy oxygen ion bombardment, *Appl. Phys. Lett.*, **64**, 288-290.

Beetz, C.P., Lincoln, B.A., Lin, B.Y. and Tan, S.H. (1991) ECR plasma etching of natural type IIa and synthetic diamonds, in R. Messier, J.T. Glass, J.E. Butler and R. Roy (eds), *Proceedings of the 2nd Int. Conf. on New Diamond Science and Technology*, Mat. Res. Soc., Pittsburgh, PA, pp. 833-838.

Belyanin, A.F., Pashchenko, P.V., Semyonov, A.P., Smirnyagina, N.N., Spitsyn, B.V., Bouilov, L.L. and Alexenko, A.E. (1991) Deposition and etching of diamond films and diamond-like carbon by ionic beam, in *Proc. 1st Int. Seminar on Diamond Films (June 30 - July 6, 1991)*, Tekhnika sredstv svyazi, Moscow, Vypusk 4, pp. 55-68 (in Russian).

Bhushan, B., Subramaniam, V.V. and Gupta, B.K. (1994) Polishing of diamond films, *Diamond Films and Technology*, **4**(2), 71-97.

Blatter, A., Bögli, U., Bouilov, L.L., Chapliev, N.I., Konov, V.I., Pimenov, S.M., Smolin, A.A. and Spitsyn, B.V. (1991) Excimer laser etching and polishing of diamond films, in *Proc. of the 2nd Int. Symp. on Diamond Materials, Washington D.C., May 5-10, 1991*, The Electrochem. Soc., Pennington, pp. 357-364.

Bögli, U., Blatter, A., Pimenov, S.M., Smolin, A.A. and Konov, V.I. (1992) Smoothening of diamond films with an ArF laser, *Diamond and Related Materials*, 7, 782-788.

Bögli, U., Blatter, A., Bächli, A., Lüthi, R., and Meyer, E. (1993) Characterization of laser-irradiated surfaces of a polycrystalline diamond film with an atomic force microscope, *Diamond and Related Materials*, **2**, 924-927.

Bögli, U., Blatter, A., Pimenov, S.M., Obraztsova, E.D., Smolin, A.A., Maillat, M., Leijala, A., Burger, J., Hintermann, H.E. and Loubnin, E.N. (1995) Tribological properties of smooth polycrystalline diamond films, *Diamond and Related Materials*, in press.

Boudina, A., Fitzer, E., Wahl, G. and Esrom, H. (1993) Improvement in IR properties of chemically vapour-deposited diamond films by smoothening with KrF excimer radiation, *Diamond and Related Materials*, **2**, 678-682.

Bouilov, L.L., Chapliev, N.I., Konov, V.I., Pimenov, S.M., Ralchenko, V.G. and Spitsyn, B.V. (1990) CO_2 laser processing of diamond films, in T.S. Sudarshan and D.G. Bhat (eds.), *Surface Modification Technologies III*, The Minerals, Metals & Materials Society, pp. 27-35.

Bovard, B.G., Zhao, T. and Macleod, H.A. (1991) Smooth diamond films by reactive ion beam polishing, *Diamond Optics IV, SPIE Proc.*, **1534**, 216-222.

Bovard, B.G., Zhao, T. and Macleod, H.A. (1992) Oxygen-ion beam polishing of a 5-cm-diameter diamond film, *Applied Optics*, **31**(13), 2366-2369.

Bruton, E. (1978), *Diamonds*, NAG Press Ltd, London, 2nd edition.

Chapliev, N.I., Konov, V.I., Pimenov, S.M., Prokhorov, A.M. and Smolin, A.A. (1991) Laser-assisted selective area deposition of diamond films, in Y. Tzeng, M. Yoshikawa, M. Murakawa and A. Feldman (eds), *Applications of Diamond Films and Related Materials*, Elsevier, Amsterdam, pp. 417-421.

Couto, M.S., van Enckevort, W.J.P. and Seal, M. (1994a) Diamond polishing mechanisms: an investigation by scanning tunneling microscopy, *Philosophical Magazine* B, **69**(4), 621-641.

Couto, M.S., van Enckevort, W.J.P. and Seal, M. (1994b) On the mechanism of diamond polishing in the soft directions, *J. Hard Mater.*, in press.

Cropper, A.D., Moore, D.J. and White, C. (1993) Damage theshold chaacterization of thin film diamond, in J.P. Dismukes, K.V. Ravi, K.E. Spear, B. Lux and N. Setaka (eds), *Proc. 3rd Int. Symp. on Diamond Materials*, Vol. 93-17, The Electrochemical Soc., Pennington, NJ, pp. 682-688.

Cutting, trimming, polishing, forming and machining the hardest material, (1994) *Diamond Depositions: Science & Tecnology*, **4**(9), pp. 1-9.

Dorsch, O., Werner, M., Obermeier, E., Harper, R.E., Johnson, C. and Buckley-Golder, I.M. (1992) Etching of polycrystalline diamond and amorphous carbon films by RIE, *Diamond and Related Materials*, **1**, 277-280.

Efremow, N.N., Geis, M.W., Flanders, D.C., Lincoln, G.A. and Economou, N.P. (1985) Ion-beam-assisted etching of diamond, *J.Vac. Sci. Technol.* **B3**, 416-418.

Epifanov, V.I., Pesina, A.Ya. and Zykov, L.V., (1982), *Technology of Diamond Processing*, Vysshaya shkola, Moscow (in Russian).

Field, J.E. (ed) (1992), *The Properties of Natural and Synthetic Diamond*, Academic Press, London.

Fujimori, N. and Nishibayashi, Y. (1992) Diamond devices made of epitaxial diamond films, *Diamond and Related Materials*, **1**, 665-668.

Funamoto, H., Koseki, O. and Sugita, T. (1991) Planning of synthetic diamond film by the Penning discharge microsputtering method, *Surface and Coating Technology*, **47**, 474-480.

Gaissmaier, K. and Weis, O. (1993) Superpolishing of diamond, *Diamond and Related Materials*, **2**, 943-948.

Geis, M.W., Rothschild, M., Kunz, R.R., Aggarwal, R.L., Wall, K.F., Parker, C.D., McIntosh, K.A., Efremow, N.N., Zayhovski, J.J., Ehrlich, D.J. and Butler, J.E. (1989) Electrical, crystallographic, and optical properties of ArF laser modified diamond surfaces, *Appl. Phys. Lett.*, **55**, 2295-2297.

Geis, M.W. (1992) Device quality diamond substrates, *Diamond and Related Materials*, **1**, 684-687.

Glesener, J.W. and Tonucci, R.J. (1993) Micropatterned diamond substrates, *J.Appl.Phys.*, **74**, 5280-5281.

Golden, J. (1992) Green lasers score good marks in semiconductor material processing, *Laser Focus World*, **28**(6), 75-88.

Grigoriev, A.P., Lifshits, S.K. and Shamaev, P.P. (1982) Method of treating diamond, *US Patent* 4,339,304.

Grigoriev, A.P., Lifshits, S.K. and Shamaev, P.P. (1983) Method of treating diamond, *USSR Sertificate of Invention*, No. 1,081,922.

Grigoriev, A.P. and Kovalsky, V.V. (1984) Working of diamond with metals, *INDIAQUA*, 39 (3) 47.

Grodzinski, P. (1953), *Diamond Technology*, NAG Press Ltd, London.

Grogan, D.F., Zhao, T., Bovard, B.G. and Macleod, H.A. (1992) Planarizing technique for ion-beam polishing of diamond films, *Applied Optics*, **31**(10), 1483-1487.

Grot, S.A., Ditizio, R.A., Gildenblatt, G.Sh., Badzian, A.R. and Fonash, S.J. (1992) Oxygen-based electron cyclotron resonance etching of semiconducting homoepitaxial diamond films, *Appl.Phys.Lett.*, **61**, 2326-2328.

Harano, K., Ota, N. and Fujimori, N. (1994) Diamond processing by excimer laser ablation, in S. Saito, N. Fujimory, O. Fukunaga, M. Kamo, K. Kobashi and M. Yoshikawa (eds), *Advances in New Diamond Science and Technology*, MYU, Tokyo, pp. 493-500.

Harker, A.B. (1991) Reactive polishing of polycrystalline diamond and measured spectroscopic properties, in Tzeng, M. Yoshikawa, M. Murakawa and A. Feldman (eds), *Applications of Diamond Films and Related Materials*, Elsevier, pp. 223-225.

Harker, A.B., Flintoff, J. and DeNatale, J.F. (1990) The polishing of polycrystalline diamond films, in *Diamond Optics III, SPIE Proc.*, **1325**, 222-229.

Hashish, M. and Bothell, D.H. (1992) Polishing of CVD diamond films with abrasive-liquidjets, *Diamond Optics V, SPIE Proc.*, **1759**, 97-105.

Hickey, C.F., Thorpe, T.P., Morrish, A.A., Butler, J.E., Vold, C. and Snail, K.A. (1991) Polishing of filament assisted CVD diamond film, in *Diamond Optics IV, SPIE Proc.*, **1534**, 67-76.

Higuchi, K. and Noda, S. (1992) Selected area diamond deposition by control the nucleation sites, *Diamond and Related Materials*, **1**, 220-229.

Hirabayashi, K., Taniguchi, Y., Takamatsu, O., Ikeda, T., Ikoma, K. and Iwasaki-Kurihara, N. (1988) Selective deposition of diamond crystals by chemical vapor deposition using a tungsten-filament method, *Appl.Phys.Lett.*, **53**, 1815-1817.

Hirata, A., Tokura, H. and Yoshikawa, M., (1991) Smoothing of diamond films by ion beam irradiation, in Y. Tzeng, M. Yoshikawa, M. Murakawa and A. Feldman (eds.), *Applications of Diamond Films and Related Materials*, Elsevier, Amsterdam, pp. 227-232.

Hitchiner, M.P., Wilks, E.M. and Wilks, J. (1984) *Wear*, **94**, 103-120.

Jin, S., Graebner, J.E., Kammlott, G.W., Tiefel, T.H., Kosinsky, S.G., Chen, L.H. and Fastnacht, R.A. (1992a) Massive thinning of diamond films by a diffusion process, *Appl.Phys.Lett.*, **60**, 1948-1950.

Jin, S., Graebner, J.E., Tiefel, T.H., Kammlott, G.W. and Zydzik, G.J. (1992b) Polishing of CVD diamond by diffusional reaction with manganese powder, *Diamond and Related Materials*, **1**, 949-953.

Jin, S., Graebner, J.E., McCormack, M., Tiefel, T.H., Katz, A. and Dautremont-Smith, W.C., (1993) Shaping of diamond films by etching with molten rare-earth metals, *Nature*, **362**, 822-824.

Johnston, C., Chalker, P.R., Buckley-Golder, I.M., Marsden, P.J. and Williams, S.W. (1993) Diamond device deliniation via excimer laser patterning, *Diamond and Related Materials*, **2**, 829-834.

Katsumata, S. and Yugo, S. (1994) Submicron selective growth of diamonds by microwave plasma chemical vapor deposition, *Diamond Films and Technology*, **3**, 199-207.

Kirkpatrick, A.R. and Ward, B.W. (1991) Control of diamond film microstructure by use of seeded focused ion beam crater arrays, *J.Vac.Sci.Technol.* **B9**, 3095-3098.

Kobashi, K., Inoue, T., Tashibana, H., Kumagai, K., Miyata, K., Nishimura, K. and Nakaue, A. (1990) Selected-area deposition of diamond films, *Vacuum*, **41**, 1383-1386.

Kobayashi, K., Shimada, Y., Mutsukura, N. and Machi, Y. (1990a) Etching characteristics of carbon films by RF plasma, in S. Saito, O. Fukunaga and M. Yoshikawa (eds), *Science and Technology of New Diamond*, KTK Scientific Publishers, Tokyo, pp. 333-338.

Kobayashi, K., Yamamoto, K., Mutsukura, N., and Machi, Y. (1990b) Sputtering characteristics of diamond and hydrogenated amorphous carbon films by r.f. plasma, *Thin Solid Films*, **185**, 71-78.

Koike, Y., Tsuzaka, H. and Masuzawa, T. (1993) Micromachining of PCD and carbide tools by WEDG method, in M. Yoshikawa, M. Murakawa, Y. Tzeng and W.A. Yarbrough (eds), *2nd Int. Conf. on the Applications of Diamond Films and Related Materials*, MYU, Tokyo, pp. 603-608.

Kononenko T.V. (1993) *Internal report*, General Physics Institute.

Konov, V.I., Ralchenko, V.G., Pimenov, S.M., Smolin, A.A., and Kononenko, T.V. (1993) Laser microprocessing of diamond and diamond-like films, in *Laser-Assisted Fabrication of Thin Films and Microstructures, SPIE Proc.*, **2045**, 184-192.

Konov, V.I., Pimenov, S.M., Ralchenko, V.G. and Smolin, A.A. (1994) Selective deposition of diamond films, in V.B. Kvaskov (ed.), *Physics and Technology of Diamond Materials*, Polaron Publishers, Moscow, pp. 42-51.

Kyuno, T., Saitoh, H. and Urao, R. (1994) Sputtering rate of polycrystalline diamond film using argon ion beam, in S. Saito, N. Fujimory, O. Fukunaga, M. Kamo, K. Kobashi and M. Yoshikawa (eds), *Advances in New Diamond Science and Technology*, MYU, Tokyo, pp. 489-492.

Levitt, C.M. (1968) Spark erosion of diamond, *Rev.Sci.Instrum*, **39**, 752-754.

Lin, S.J., Lee, S.L., Hwang, J. and Lin, T.S. (1992) Selective deposition of diamond films on ion-implanted Si (100) by microwave plasma chemical vapor deposition, *J.Electrochem.Soc.*, **139**, 3255-3258.

Ma, J.S., Kawarada, H., Yonehara, T., Suzuki, J.-I., Wei, J., Yokota, Y. and Hiraki, A. (1989) Selective nucleation and growth of diamond particles by plasma-assisted chemical vapor depostion, *Appl.Phys.Lett.*, **55**, 1071-1073.

Marchywka, M., Pehrsson, P.E., Binari, S.C. and Moses, D. (1993) Electrochemical patterning of amorphous carbon on diamond, *J.Electrochem.Soc.*, **140**, L19-L24.

Marsh, J.B. and Farnsworth, H.E. (1964) Low-energy electron diffraction studies of (100) and (111) surfaces of semiconducting diamond, *Surf. Sci.*, **1**, 3-21.

Masood, A., Aslam, M., Tamor, M.A. and Potter, T.J. (1991) Techniques for patterning of CVD diamond films on non-diamond substrates, *J.Electrochem.Soc.* **138**, L67-L68.

McCormack, M., Jin, S., Graebner, J.E., Tiefel, T.H. and Kammlott, G.W. (1994) Low temperature thinning of thick chemically vapor-deposited diamond films with a molten Ce-Ni alloy, *Diamond and Related Materials*, **3**, 254-258.

Meilunas, R.J., Chang, R.P.H., Liu, S. and Kappes, M.M. (1991) Nucleation of diamond films on surfaces using carbon clusters, *Appl.Phys.Lett.* **59**, 3461-3463.

Miyamoto, I. (1990) Ion beam forming and sharpening of diamond tools having a small apex angle, in S. Saito, O. Fukunaga and M. Yoshikawa (eds), *Science and Technology of New Diamond*, KTK Scientific Publishers, Tokyo, pp. 395-398.

Miyauchi, S., Kumagai, K., Miyata, K., Nishimura, K., Kobashi, K., Nakaue, A., Glass, J.T. and Buckley-Golder, I.M. (1991) Microfabrication of diamond films: selective deposition and ecthing, *Surface and Coating Technology*, **47**, 465-473.

Miyazawa, H., Miyake, S., Watanabe, S., Murakawa, M. and Miyazaki, T. (1994) Laser-assisted thermochemical processing of diamond, *Appl.Phys.Lett.*, **64**, 387-389.

Moran, M.B., Klemm, K.A., and Johnson, L.F. (1991) Antireflection/antioxidation coatings and polishing results for PECVD optics, in Y. Tzeng, M. Yoshikawa, M. Murakawa and A. Feldman (eds), *Applications of Diamond Films and Related Materials*, Elsevier, Amsterdam, pp. 233-240.

Narayan, J. and Chen, X. (1992) Laser patterning of diamond films, *J.Appl.Phys.*, 71, 7795-3801.

Obraztsov, A.N. and Pirogov, V.G. (1993) Direct laser writing on natural diamond monocrystals and polycrystalline diamond films, in M. Yoshikawa, M. Murakawa, Y. Tzeng and W.A. Yarbrough (eds), *2nd Int. Conf. on the Applications of Diamond Films and Related Materials*, MYU, Tokyo, pp. 579-582.

Pimenov, S.M., Smolin, A.A., Ralchenko, V.G. and Konov, V.I. (1993a) Excimer laser processing of diamond films, *Diamond Films and Technology*, **2**, 201-214 .

Pimenov, S.M., Smolin, A.A., Ralchenko, V.G., Konov, V.I., Sokolina, G.A., Bantsekov, S.V. and Spitsyn, B.V. (1993b) UV laser processing of diamond films: effects of irradiation conditions on the properties of laser-treated diamond film surface, *Diamond and Related Materials*, **2**, 291-297.

Pimenov, S.M., Smolin, A.A., Ralchenko, V.G. Konov, V.I., Pypkin, B.N., and Loubnin, E.N. (1993c) Selective area deposition of diamond on tungsten-alloyed amorphous carbon films, *Proc. 3rd Int. Symp. on Diamond Materials*, The Electrochem. Soc., Pennington, NJ, Vol. **93-17**, 559-565.

Pimenov, S.M., Shafeev, G.A., Smolin, A.A., Konov, V.I. and Vodolaga, B.K. (1994a) Laser-induced transfer of ultrafine diamond particles for selective deposition of diamond films, in S. Saito, N. Fujimory, O. Fukunaga, M. Kamo, K. Kobashi and M. Yoshikawa (eds), *Advances in New Diamond Science and Technology*, MYU, Tokyo, pp. 109-112.

Pimenov, S.M., Smolin, A.A., Obraztsova, E.D., Konov, V.I., Bögli, U., Blatter, A., Loubnin, E.N., Maillat, M., Leijala, A., Burger, J. and Hintermann, H.E. (1994b) Tribological properties of smooth diamond films, *Paper presented at the 7th Int. Conf. Solid Films and Surfaces*, Taiwan, December 12-16, pap. P2-30.

Plamann, K., Fournier, D., Anger, E. and Gicquel, A. (1994) Photothermal examination of the heat diffusion inhomogeniety in diamond films of sub-micron thickness, *Diamond and Related Materials*, **3**, 742-756.

Popovici G. and Prelas M.A. (1992) Nucleation and selective deposition of diamond thin films, *Phys.Stat.Sol. (a)*, **132**, 233-252.

Ralchenko, V.G., Kononenko, T.V., Pimenov, S.M., Chernenko, N.V., Loubnin, E.N., Armeyev, V.Yu. and Zlobin, A.Yu. (1993a) Catalytic interaction of Fe, Ni, and Pt with diamond films: patterning applications, *Diamond and Related Materials,* **2**, 904-909.

Ralchenko, V.G., Smolin, A.A., Korotoushenko, K.G., Nounouparov, M.S., Pimenov, S.M. and Vodolaga, B.K. (1993b) A technique for controllable seeding of ultrafine diamond particles for growth and selective-area deposition of diamond films, in M. Yoshikawa, M. Murakawa, Y. Tzeng and W.A. Yarbrough (eds), *Proceedings of the 2nd International Conference on the Applications of Diamond Films and Related Materials*, MIU, Tokyo, pp. 475-480.

Ralchenko, V.G., Korotoushenko, K.G., Smolin, A.A. and Konov, V.I. (1994) Patterning of diamond films by direct laser writing: Selective-area deposition, chemical etching and surface smoothing, in S. Saito, N. Fujimory, O. Fukunaga, M. Kamo, K. Kobashi and M. Yoshikawa (eds), *Advances in New Diamond Science and Technology*, MYU, Tokyo, pp. 493-496.

Ramesham, R., Roppel, T., Ellis, C. and Rose, M.F. (1991) Fabrication of microchannels in synthetic polycrystalline diamond thin films for heat sinking applications, *J.Electrochem.Soc.*, **138**, 1706-1709.

Ravi, K.V. and Zarifis, V.G. (1993) Laser polishing of diamond, *Proc. 3rd Int. Symp. on Diamond Materials*, The Electrochem. Soc., Pennington, NJ, Vol. **93-17**, 861-867.

Rothschild, M., Arnone, C. and Ehrlich, D.J. (1986) Excimer-laser etching of diamond and hard carbon films by direct writing and optical projection, *J.Vac.Sci.Technol.* **B4**, 310-314.

Sandhu, G.S. and Chu, W.K. (1989) Reactive ion etching of diamond, *Appl.Phys.Lett.*, **55**, 438-438.

Sato, Y., Hatta, C. and Kamo, M. and (1990) Formation and structural features of needle-like diamond, in S. Saito, O. Fukunaga and M. Yoshikawa (eds), *Science and Technology of New Diamond*, KTK Scientific Publishers, Tokyo, pp. 83-88.

Seal, M. (1965) Structure in diamonds as revealed by etching, *American Mineralogist*, **50**, 105-123.

Shikata, S., Nishibayashi, Y., Tomikawa, T., Toda, N. and Fujimori, N. (1993) Microfabrication technique for diamond devices, in M. Yoshikawa, M. Murakawa, Y. Tzeng and W.A. Yarbrough (eds), *2nd Int. Conf. on the Applications of Diamond Films and Related Materials*, MYU, Tokyo, pp. 377-380.

Shimada, Y. and Machi, Y. (1993) Selective growth of diamond using an iron catalyst, *Diamond and Related Materials*, **3**, 403-407.

Simakin A.V. and Shafeev G.A. (1995) Laser-assisted etching-like damage of Si, *Appl. Surf. Sci.*, **86**, 422-427.

Spitsyn, B.V., Belyanin, A.F., Bulyonkov, N.A. and Rivilis, V.M. (1987) Texture and mechanical treatment of diamond layers grown from gas phase, *Tekhnika sredstv svyazi*, Moscow, Vypusk 1, pp. 61-70 (in Russian).

Steinmetz, K., Nimmich, A. and Friedrich, P. (1990) Edging syndite by wire EDM, *Industrial Diamond Review*, **50**, 317-327.

Stoner, B.R., Tessmer, G.J. and Dreifus, D.L. (1993) Bias assisted etching of diamond in a conventional chemical vapor deposition reactor, *Appl.Phys.Lett.*, **62**, 1803-1805.

Sugita, T., Nishikawa, E., Yoshida, Y., Funamoto, H. and Koseki, O. (1990) Pattern marking on synthetic diamond film by Penning discharge sputtering, *Vacuum*, **41**, 1371-1373.

Sussmann, R.S., Scarsbrook, G.A., Wort, C.J.H. and Wood, R.M. (1994) Laser damage testing of CVD-grown diamond windows, *Diamond and Related Materials*, **3**, 1173-1177.

Suzuki, H., Yoshikawa, M. and Tokura, H. (1994) Excimer laser processing of diamond films, in S. Saito, N. Fujimory, O. Fukunaga, M. Kamo, K. Kobashi and M. Yoshikawa (eds), *Advances in New Diamond Science and Technology*, MYU, Tokyo, pp. 501-504.

Tabor, D. and Field, J.E. (1992) Friction of diamond, in J.E. Field (ed.), *The Properties of Natural and Synthetic Diamond*, Academic Press, pp. 547-571.

Tessmer, G.J., Stoner, B.R., Dreifus, D.L. and Plano, L.S. (1993) Biased etching of diamond films via DC glow discharges of H_2 and H_2O_2, in J.P. Dismukes, K.V. Ravi, K.E. Spear, B. Lux and N. Setaka (eds), *Proc. 3rd Int. Symp. on Diamond Materials*, Vol. 93-17, The Electrochemical Soc., Pennington NJ, pp. 640-646.

Tezuka, S., Tokura, H. and Yoshikawa, M. (1990) Cutting of diamond grit with YAG laser, in S. Saito, O. Fukunaga and M. Yoshikawa (eds), *Science and Technology of New Diamond*, KTK Scientific Publishers, Tokyo, pp. 469-473.

Thorpe, T.P., Morrish, A.A., Hansen, L.M., Butler, J.E. and Snail, K.A. (1990) Growth, polishing, and optical scatter of diamond thin films, in *Diamond Optics III, SPIE Proc.*, **1325**, 230-237.

Tokarev, V.N. and Konov, V.I. (1991) Light-induced polishing of diamond films, in Y. Tzeng, M. Yoshikawa, M. Murakawa and A. Feldman (eds.), *Applications of Diamond Films and Related Materials*, Elsevier, Amsterdam, pp. 249-255.

Tokarev, V.N., Wilson, J.I.B., Jubber, M.G., John, P. and Milne, D.K. (1995) Modelling of self-limiting laser ablation of rough surfaces: application to the polishing of diamond films, in *Proc. 8th CIMTEC-World Ceramics Congress & Forum on New Materials* (Florence, Italy, 1994), to be published.

Tolansky, S. (1955), *The Microstructures of Diamond Surfaces*, NAG Press Ltd, London.

Tolkowsky, M. (1920) *Ph.D. Thesis*, University of London.

Uegami, K, Tamamura, K. and Mineji, S. (1992) Working of diamond by heating ferrous metal, *Memoirs of the Faculty of Engineering, Osaka City University*, **33**, 7-12.

Ullmann, J., Delan, A. and Schmidt, G. (1993) Ion etching behaviour and surface binding energies of hard diamond-like carbon and microwave chemical vapour deposition diamond films, *Diamond and Related Materials*, **2**, 266-271.

Ullmann, J., Weber, A., Mainz, B., Stiegler, J. and Schuhrke, T. (1994) Low energy ion-induced damage of polycrystalline diamond films, *Diamond and Related Materials*, **3**, 663-671.

Valdes, J.L., Mitchell, J.W., Mucha, J.A., Seibles, L. and Huggins, H. (1991) Electrophoretic deposition of colloidal diamond for selective CVD growth of diamond films, in Y. Tzeng, M. Yoshikawa, M. Murakawa and A. Feldman (eds), *Applications of Diamond Films and Related Materials*, Elsevier, Amsterdam, pp. 423-428.

van Enckevort, W.J.P. and van Halevijn, H.J. (1994) Shaping of diamond, in G. Davies (ed), *Properties and Growth of Diamond*, INSPEC, the Institution of Electrical Engineers, London, UK, pp. 293-300.

Whetten, T.J., Armstead, A.A., Grzybowski, T.A. and Ruoff, A.L. (1984) Etching of diamond with argon and oxygen ion beams, *J.Vac.Sci.Tecnol.* **A2**, 477-480.

Wilks, J. and Wilks, E.M. (1991), *Properties and Applications of Diamond*, Butterworth-Heinemann.

Wilks, J. and Wilks, E.M. (1992) Wear and polishing of diamond, in J.E. Field (ed.), *The Properties of Natural and Synthetic Diamond*, Academic Press, pp. 573-604.

Yarnitsky, Y., Sellschop, J.P.F., Rebak, M. and Luyckx, S.B. (1988) The dynamometer as a qualitative gauge in diamond polishing, *Mat. Sci. Eng.* **A105/106**, 565-569.

Yoshikawa, M. (1990) Development and perfomance of a diamond film polishing apparatus with hot metals, in *Diamond Optics III, SPIE Proc.*, **1325**, 210-221.

Yoshikawa, M. (1991) Application of CVD diamond to tools and machine components, *Diamond Films and Technology*, **1**(1), 1-29.

Zhao, T., Grogan, D.F., Bovard, B.G., and Macleod, H.A. (1990) Diamond film polishing with argon and oxygen ion beams, in *Diamond Optics III, SPIE Proc.*, **1325**, 142-151.

Chapter 28
ABRASIVE APPLICATIONS OF DIAMOND

K. Subramanian and V. R. Shanbhag
World Grinding Technology Center
Norton Company
Worcester, MA USA

Contents

1. Introduction

Diamond and cubic boron nitride abrasives used in grinding applications are collectively referred to as superabrasives. Diamond is the hardest material known, and cubic boron nitride is the second hardest. Because of their high hardness, abrasion resistance, and other unique properties, these materials find extensive use in a wide variety of abrasive or cutting applications ranging from the rough cutting of stone and concrete to the ultra-precision machining of electronic ceramics.

The invention of the synthetic diamond caused the rapid growth of diamond in a wide variety of abrasive applications. In addition to its high hardness, diamond is an excellent heat conductor and has an extremely low coefficient of friction. However, at temperatures above 800°C (1475°F), diamond tends to graphitize, that is, convert back to carbon. Thus, it loses its value as a wear-resistant abrasive. Under such conditions, diamond suffers rapid wear and chemical dissolution (erosion) when abraded against iron. Therefore, it is not used as an abrasive against ferrous material except in special machining operations such as honing.

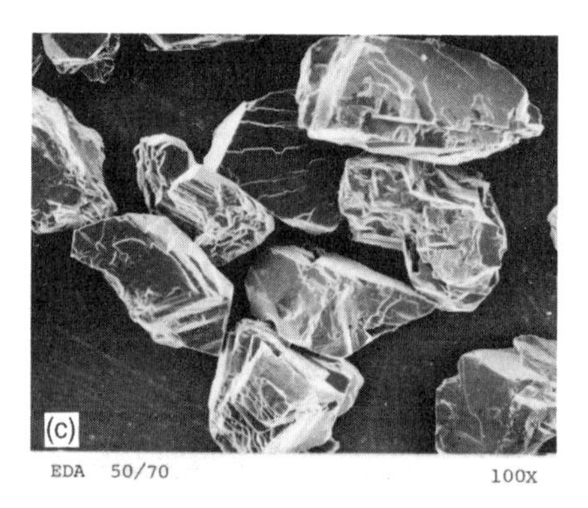

Figure 1. Typical shapes of diamond abrasive grains. (a) Strong and blocky. (b) Intermediate strength. (c) Weak and friable.

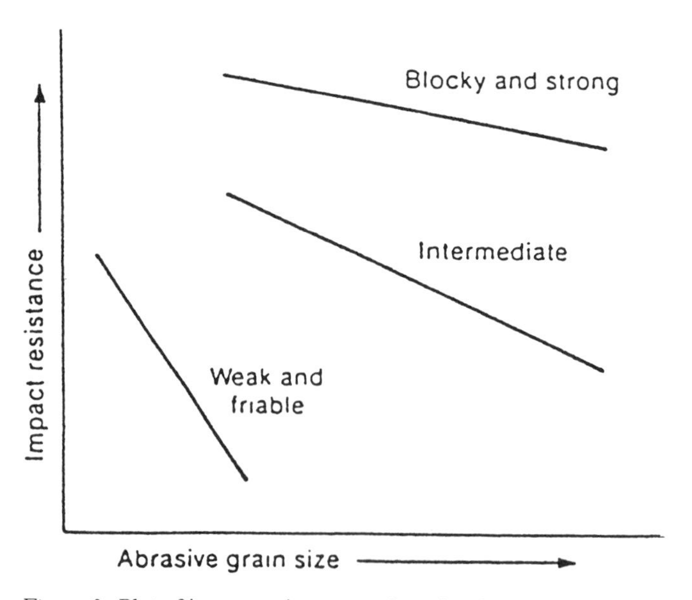

Figure 2. Plot of impact resistance against abrasive grain size showing relative toughness of synthetic diamond grains.

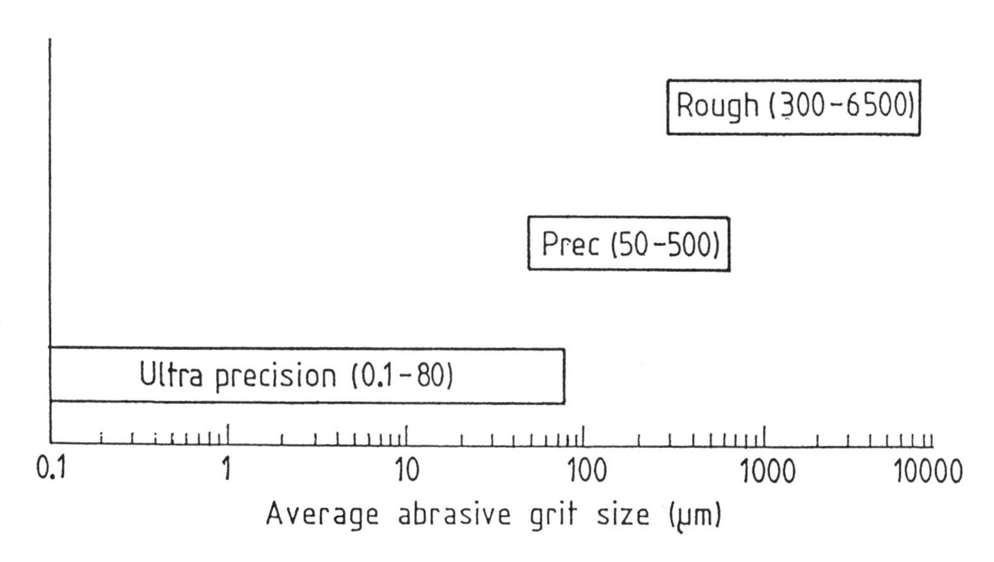

Figure 3. Abrasive grain size versus application type. Rough: cutting, sawing, mining, etc. Precision: grinding applications. Ultraprecision: fine grinding applications for electronic and optical components and lapping and polishing applications.

The shape, size, and distribution of the diamond particles or abrasive grains are strictly controlled to achieve desirable performance characteristics. Figures 1 (a), (b), and (c) show strong and blocky diamond grains, intermediate-strength grains, and weak and friable grains, respectively. The impact resistance of the abrasive grain also changes with the particle size or mesh size, as shown in Figure 2. Very small quantities (part per million) of impurities such as nitrogen, iron, and boron in diamond contribute to the color, strength, and other property variations in diamond abrasives.

The size of the diamond abrasive grain used varies with the application as noted in Figure 3. Very large and blocky grains are used for cutting, sawing and mining applications. Intermediate sizes are used for a variety of precision grinding applications. Extremely fine sized grains are used for ultra precision grinding processes as well as lapping and polishing applications.

2. Overview of Abrasive Machining Processes

2.1 DIAMOND ABRASIVE PRODUCTS

The superabrasive products can be grouped in three broad categories: coated abrasives; grinding wheels and powders; slurries; and compounds as shown in Figure 4.

The coated abrasives are those which have a layer of abrasives coated on a backing which may be cloth, film or paper. To retain the superabrasive for a long time, it becomes important to have a very long lasting backing.

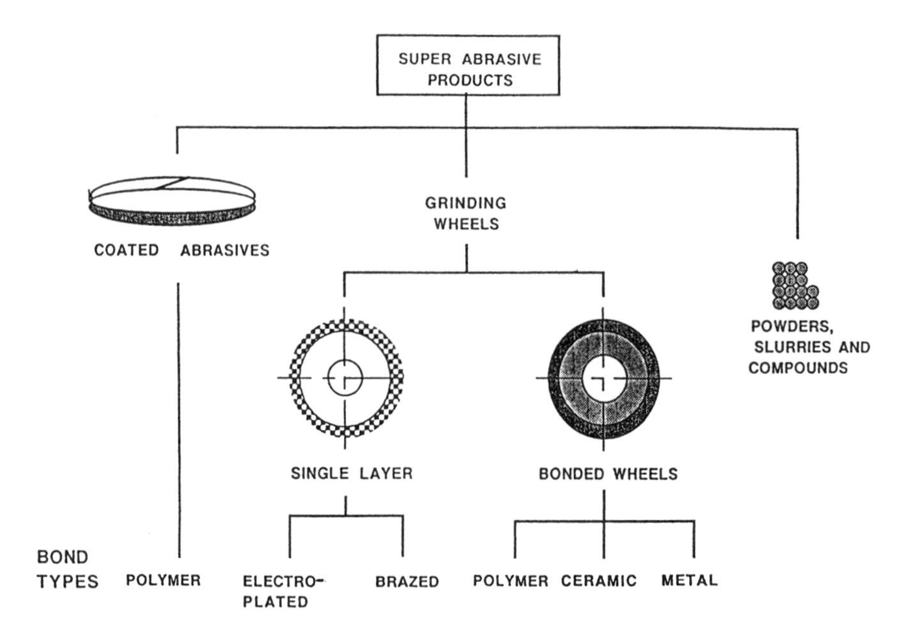

Figure 4. Superabrasive product categories.

The grinding wheels may be either single layer products or bonded wheels. The single layer products may be either an electroplated product or a brazed product. The bonded abrasives are so named because the abrasive is firmly retained in a matrix called bond. The wheels can be made in various shapes and sizes. The bonds for these products may either be polymer, ceramic or metal.

Powders, slurries and compounds refer to those products which are either abrasives in powder form or held in a liquid or semi-solid carrier of controlled viscosity.

The coated abrasive products and grinding wheels are used in rough, precision and ultra-precision grinding applications. Powders and slurries are more commonly used for ultra-precision applications like lapping and polishing.

2.2 APPLICATIONS OF DIAMOND ABRASIVE PRODUCTS

The surface generating processes are governed by two requirements: the surface generated and the effectiveness or efficiency with which it is achieved. These two requirements are related to each other as shown in Figure 5.

Coated abrasives are a single/multiple layer of abrasive products that adhere to a backing made of cloth, paper or film. The use of super abrasives on these kinds of products is still at its infancy. Diamond abrasive belts are used in finishing of carbides, ceramics and glass, frequently in the ultra-precision finishing of electronic and ceramic components.

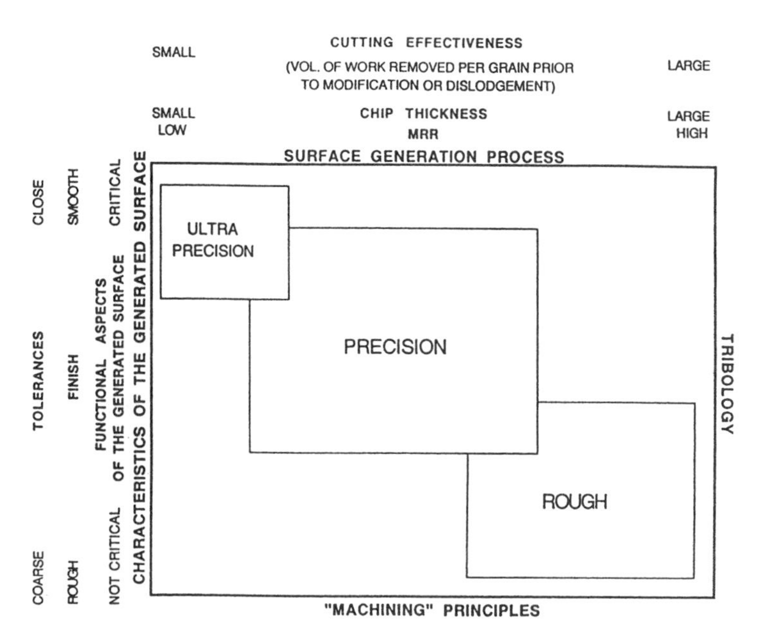

Figure 5. Overview of abrasive finishing processes.

As mentioned earlier, grinding wheels are widely used on rough, precision and ultra precision grinding applications. Table 1 lists the wide range of applications for diamond and CBN abrasives. Applications for diamond are much more extensive than those for CBN. Table 2 lists the advantages of diamond abrasives in precision production grinding.

Diamond abrasives are used in a wide variety of applications in the construction industry. These applications include the cutting and dimensioning of stones, quarry operations, polishing and shaping of stone and marbles, and cutting of concrete or asphalt to repair building, highways, and airport runways. Diamond abrasives are also used as mining bits in rock drilling and oil-well drilling operations.

The use of diamond abrasives in cutoff operations, and other surface generation needs is the basis for a variety of precision grinding applications. Carbide tools, parts, and drill bits are ground and shaped with diamond wheels. Diamond grinding wheels are used to grind the glass used in optics, the flat glass used in furniture and automotive applications, and crystal glass having intricate designs. A variety of low and high density ceramics used in kiln furniture, magnet, capacitor, spark plug, and similar applications are ground with diamond grinding wheels. Electronic ceramics such as silicon wafers, magnetic heads, optical fibers, and sensors are machined to tight tolerances and fine surface finishes with diamond grinding wheels. The potential use of engineering, technical, or fine ceramics for a wide variety of thermal and mechanical applications will call for more extensive use of diamond grinding wheels in the future. Typical products machined with diamond abrasives are shown in Figures 6 to 10.

Table 1 Typical Applications of Superabrasives

Superabrasive	Work Material	End Use/Operation
Diamond		
Construction products	Stone Concrete and asphalt (for construction and repair)	Quarrying, cutting, polishing Buildings, airports, highways
Mining products	Stone, shale, rock	Drilling and mining
Precision grinding wheels	Carbide	Tool production and re-sharpening; wear parts and others
	Glass	Optical, furniture, automotive, crystal glass
	Industrial ceramics	Electrical parts and insulators, magnets, kiln furniture
	Rubber, plastics and composites	Fiber-reinforced plastics, friction materials, optical lenses
	Electronic ceramics	Integrated circuit chip fabrication, magnetic heads, sensors, optical fibers
	Engineering ceramics	Internal combustion engine parts, wear-resistant parts and bearings, bioceramics
Precision grinding hones	Ferrous alloys	Automotive components
Powder, slurries, and compounds	Structural ceramics, ferrous alloys, electronics ceramics	Wear-resistant parts, metallurgical samples, magnetic heads, molds, dies and punches.
Cubic boron nitride		
Precision production grinding wheels	Alloy steels	Automotive parts, gears, pump parts, appliance parts, cam shafts
	Bearing steels	Bearing components
	High-nickel alloys Tool steels	Aerospace parts, Cutting tool production, die grinding, re-sharpening
Precision production honing operations	Alloy steels	Hydraulic cylinder, automotive parts

Table 2. Advantages of Diamond Abrasives in Grinding Applications

Properties	Technology (Research & Development)	End Results (factory)	Economic Results (corporate)
High hardness Thermal stability (Diamond is chemically active with ferrous materials) at elevated temperatures Chemical inertness High thermal conductivity	Wear resistant abrasive for long wheel life High grinding efficiency resulting in: Lower grinding power Lower grinding forces Better surface quality Close tolerance/ geometry Better component performance Processing of difficult-to-grind parts, difficult materials, and difficult component geometries High unit width metal removal rate New process	Fewer wheel changes (30 minutes versus days); (1 shift versus 3 months) Consistent part quality Better geometry Lower inspection Lower rejections Lower in process inventory Shorter grinding cycle New products Less labor intensive More automation	Lower total cost per part (if the system is optimized) Better asset utilization More value added products Factory automation

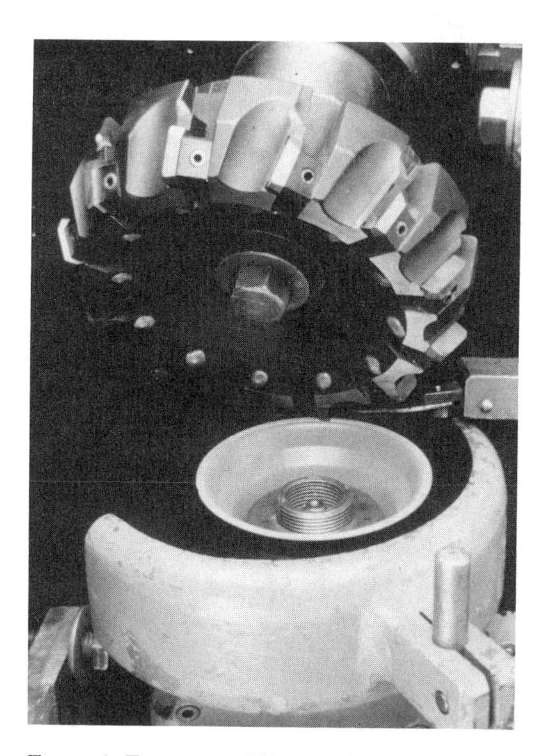

Figure 6. Tungsten carbide components having surfaces finished with diamond abrasives. Figure shows a milling cutter being reground using a diamond wheel.

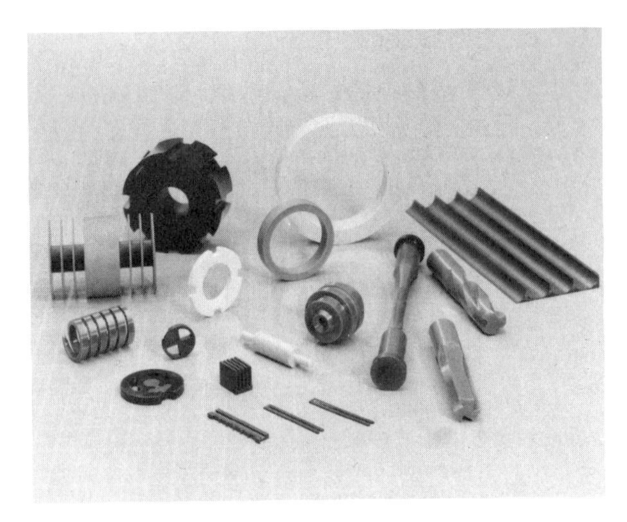

Figure 7. A variety of ceramic parts ground using diamond wheels.

Figure 8. A variety of refractory materials which are cut or sized using wheels.

Figure 9. In the manufacturing of IC chips, the silicon wafers undergo a variety of grinding applications using diamond abrasive products, ranging from lapping with slurries to back grinding wheels and dicing wheels.

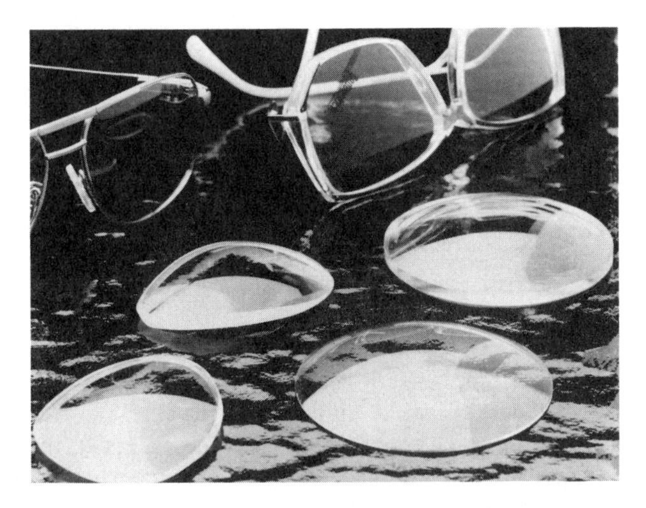

Figure 10. Diamond wheels are used to generate the surface of optical lenses from the lens blank.

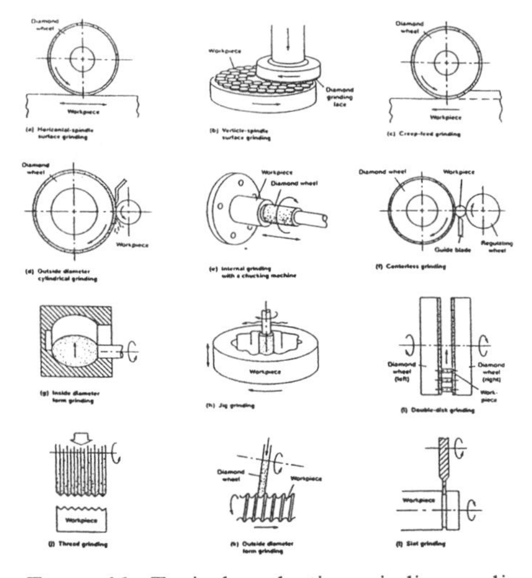

Figure 11. Typical production grinding application of diamond grinding wheels.

Grinding. A variety of process and wheel/work arrangements are used to generate surfaces. Typical processes that utilize diamond grinding wheels are illustrated in Figure 11.

A variety of high-precision processes have been in use, and new processes are constantly being developed to achieve extremely close geometric tolerances or to improve surface finish. The objective of all high-precision processes is to achieve geometrically precise components or surfaces of controlled texture or surface finish. Typical examples of these are described below:

Honing is a low-velocity abrading process that uses bonded abrasive sticks to remove stock from metallic and nonmetallic surfaces. As one of the last operations performed on the surface of a part, honing generates functional characteristics specified for a surface, such as geometric accuracy, dimensional accuracy, and surface features (roughness, lay pattern, and integrity). It also reduces or corrects geometric errors resulting from previous operations.

The most common application of honing is on internal cylindrical surfaces [Figure 12(a)]. However, honing is also used to generate functional characteristics on external cylindrical surfaces, flat surfaces, truncated spherical surfaces, and torroidal surfaces (both internal and external). A characteristic common to all these shapes is that they can be generated by a simple combination of motions.

Superfinishing is a low-velocity abrading process very similar to honing. However, unlike honing, superfinishing processes focus primarily on the improvement of surface finish and much less on correction of geometric errors [Figure 12(b)]. As a result, the pressures and amplitude of oscillation applied during superfinishing are extremely small. This process is also referred to as *microhoning, microsurfacing,* and *microstoning.*

Microgrinding is akin to the precision grinding processes described in Figure 11, except that extremely fine abrasives are used (50μm and finer). The cutting velocities in microgrinding range from very low (500 sfpm) to as large as those used for grinding (6,000 to 12,000 sfpm). This process is also called *fine grinding* or *microfinish grinding.*

Flat honing is a low-velocity abrading process, similar to honing except that a large flat honing surface is used to simultaneously finish a large number of flat parts. The predecessor to the flat honing process was *hyper lap*, in which the lapping plate was simply replaced by an abrasive product such as a grinding wheel.

Each of these high-precision processes is an emerging technology, and many advancements can be expected, driven by functional or performance improvements of industrial components and their systems. For example:

- Honing and microgrinding are used to achieve closer-fitting cylinders and pistons that reduce leakage past the piston and improve performance efficiency in hydraulic cylinders and automotive engines.
- Flat honing and microgrinding improve the quality of flat and parallel surfaces that are used to align, join, or seal other surfaces, while improving process efficiency and economics.
- Microfinishing can be used to modify surfaces ground to a certain feature to improve the bearing area, which in turn improves load-bearing area, which in turn improves load-bearing capacity (in the case of bearings) or signal processing capability (in the case of the magnetic heads used in computers).

2.3 OVERVIEW OF WORK MATERIALS FINISHED BY DIAMOND ABRASIVES

Diamond is used as an abrasive for grinding a wide variety of materials. The materials range from commonly used glass to new ceramic materials. The relative usage of

Table 3. Overview of Abrasive Machining Processes

Work Material	Drilling or Mining	Cutting, Sawing or Hole Making	Grinding and Honing	Lapping and Polishing
Plastics	None	Low	Low	Low
Composites	None	Low	Low	Low
Metals-Non ferrous	None	Low	High	Low
Metals-Ferrous	None	Low	Low	High
Glass	None	High	High	High
Carbides	None	Low	High	High
Ceramics	None	High	High	High
Wood	None	Low	Low	None
Stone	High	High	Low	Low
Minerals	High	High	None	None
Concrete	None	High	Low	None

Relative Usage: High, Low, None

diamond as an abrasive is significantly higher in grinding applications of ceramics, glass and carbides. Table 3 shows a range of work materials and their relative usage of different diamond abrasives for finishing these work materials.

2.4 SURFACE GENERATION USING DIAMOND ABRASIVES AND EMERGING TECHNOLOGIES

As mentioned earlier [Figure 5], all the surface generating processes using diamond abrasives are governed by two requirements; the generated surface and the effectiveness or efficiency with which it is achieved.

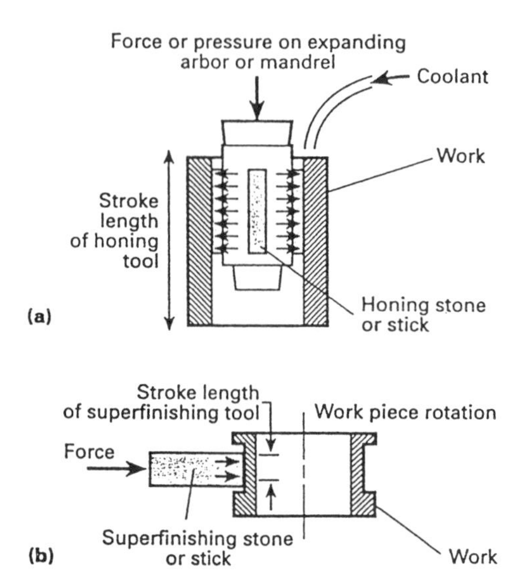

Figure 12. (a) Schematic representation of honing, (b) schematic representation of superfinishing.

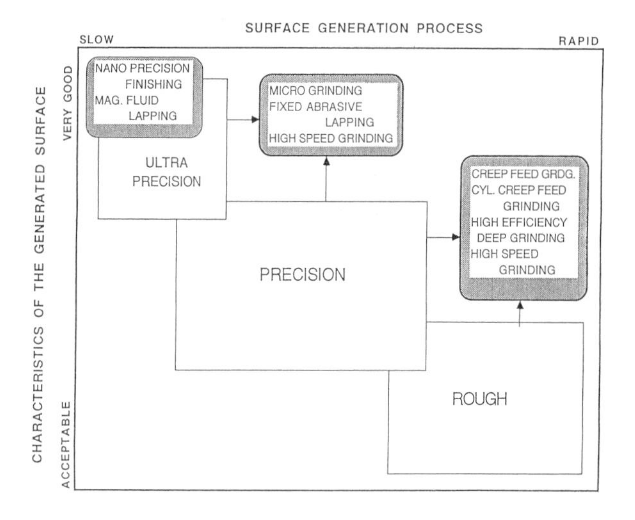

Figure 13. Overview of abrasive machining processes and emerging technologies.

The rough grinding processes are those where the rate at which the surface is generated is more critical than the feature and tolerance of the surface. Typical examples of these types of processes are the concrete cutting operation on runways, stone cutting, etc.

In precision processes, the rate of surface generation is as critical as the surface characteristics like form, geometry, tolerances, finish, surface integrity, etc. Hence these processes constantly strive to achieve a balance between these two requirements. Most of the production applications fall in the precision process category. Typical examples are finishing of various components used in automotive parts, aerospace parts, cutting tools, bearings, optical and ophthalmic components, etc.

In the ultra precision applications, the surface characteristics get the maximum attention sometimes even at the expense of speed or rate of surface generation. These processes are generally used to alter the surface characteristics such as finish, waviness, roundness, etc. without involving substantial removal of work material. Typical examples are the lapping and polishing of optical lenses, computer chips or magnetic heads, honing of cylinder liners, microfinishing of bearing races, etc.

Fine surface finishing processes such as lapping, honing, and polishing also use fine-grit diamond abrasives in loose abrasive form, such as powders, compounds and slurries. Constant effort is being made by researchers to improve productivity, surface quality or both. As these results come into use, new abrasive finishing processes are established. These emerging technologies are shown in Figure 13.

2.5 SELECTED FEATURES OF DIAMOND ABRASIVE PRODUCTS

2.5.1 Diamond Wheels

Superabrasive (Diamond and CBN) wheels are available in a variety of sizes and shapes and are classified according to construction, abrasive content and its size and type and bond system.

Size and Shape. The superabrasive wheels used in the grinding applications vary greatly in size and shape as shown in Figure 14 and 15. Precision wheels ranging in thickness from 0.025 mm (0.001 in.) or less to 255 mm (10 in.) or more are used, depending on the application. Similarly, wheels ranging from 1 to 760 mm (0.040 to 30 in.) in diameter are used. Because of their lower wear rate and longer life, smaller superabrasive wheels can do the work of larger conventional abrasive wheels, provided the machining system can accommodate smaller wheels and higher spindle speeds. Typical sizes of diamond abrasive products vs. their application are shown in Figure 16.

Bond Systems. Four bond systems are typically used in superabrasive wheels: Resin, Vitrified, Metal and layered abrasive products.

The details of the constituents of the bond and the processing techniques result in specific bond types or bond designation and determine the response of the bond to the applications. The manufacturer should be consulted for optimum bond selection. Each bond system has its own unique properties, as shown in Table 4.

Resin bond wheels provide good resilience and vibration-absorbing characteristics, which reduce chatter at the grinding zone. Wheels with resin bonds are easy to true and dress and are commonly selected for a wide range of applications.

Vitrified bond wheels offer controlled porosity, which facilitates chip removal and coolant flow to the grinding zone. They generally last longer than resin bond wheels and are suitable for producing accurate and complex forms.

Metal bond wheels, although difficult to true and dress, offer long life, good form holding or geometry retention characteristics, and good thermal conductivity. They are excellent for simple form grinding, but usually require greater grinding forces and more

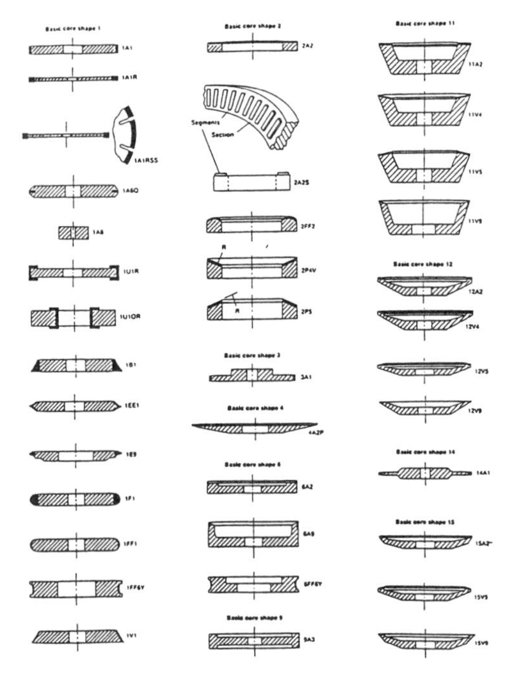

Figure 14. Superabrasive wheel configurations and their designations.

Figure 15. A range of super abrasive wheels in different sizes and shapes.

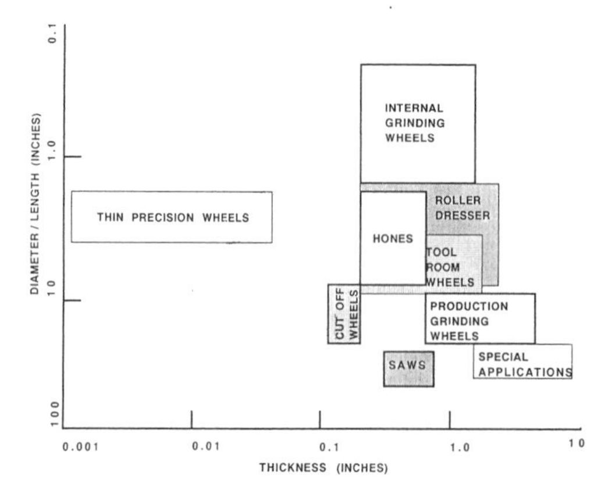

Figure 16. Typical sizes of diamond abrasive products vs. their applications.

Table 4. Advantages of Diamond Abrasive Bond Types

Bond Type and Advantages
Resin Bond Readily available Easy to true and dress Moderate freeness of cut Applicable for a range of operations First selection for learning the use of diamond wheels
Vitrified Bond Free cutting Easy to true Does not need dressing (if selected and trued properly) Controlled porosity to enable coolant flow to the grinding zone and chip removal Intricate forms can be crush formed on the wheels Suitable for creep-feed or deep grinding, inside diameter grinding, or high-conformity grinding Potential for longer wheel life than resin bond Excellent under oil as coolant
Metal Bond Very durable Excellent for thin slot, groove, cutoff, simple form, or slot grinding High stiffness Good form holding Good thermal conductivity Potential for high-speed operation Generally requires high grinding forces and power Difficult to true and dress
Layered Products Single abrasive layer plated on a pre-machined steel preform Extremely free cutting High unit-width metal removal rates Form wheels, easily produced Form accuracy dependent on preform and plating brazing accuracy High abrasive density Generally not truable Generally poorer surface finish than bonded abrasive wheels

power than resin or vitrified bond wheels. Metal bond wheels are frequently used in many applications in the construction industry, optical and ophthalmic industry as well as drilling and mining applications. Metal bond wheels constitute the largest fraction of diamond wheels for industrial applications.

Layered product wheels utilize a single abrasive layer plated or brazed to a pre-machined preform. They usually produce a poorer surface finish than bonded abrasive wheels. Layered product wheels are used for small production runs or where tolerance and surface finish are not very critical.

Figure 17. Typical diamond tools used for truing and dressing.

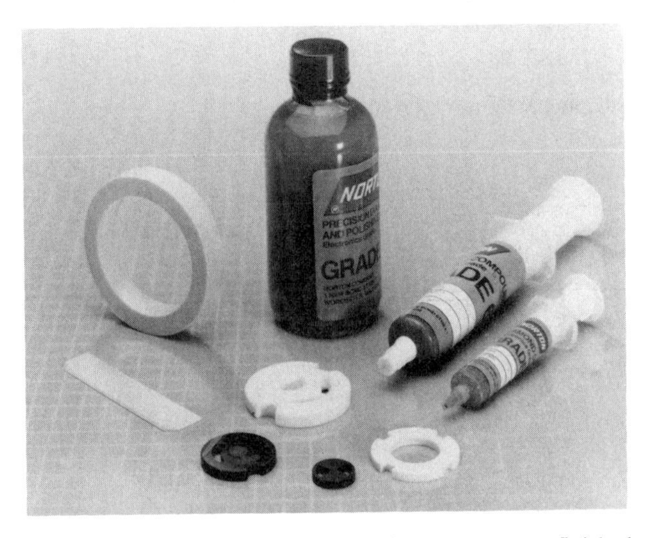

Figure 18. Lapping compounds and various components finished using these compounds.

2.5.2 Diamond Dressing Tools

Diamond abrasives are also commonly used in the truing and dressing of conventional as well as superabrasive wheels. Dressing tools can be in the form of shank tools, diamond dressing blocks and rotary diamond cutters. The shank tools can be in the form of single point tools, cluster tools, nibs (impregnated tools), etc. The diamond type and size used may vary depending on the application. Metal bonds are commonly used in all diamond tools. Typical diamond tools for truing are shown in Figure 17.

2.5.3 Slurries and Lapping Compounds

Abrasive products of this type are typically supplied as suspensions of fine grit diamond of know weight per cubic centimeter, with various colors typically representing the abrasive grain size. Typical slurries and compounds along with a few components are shown in Figure 18.

The following are some descriptions of typical processes used in these applications:

Polishing and *buffing* improve edge and surface conditions of a product for decorative or functional purposes. These techniques are abrading operations, although some plastic working of surfaces may occur, particularly in buffing. Buffing is the use of abrasives or abrasive compounds that adhere loosely to a flexible backing, such as a wheel. Polishing is the use of abrasives that are firmly attached to a flexible backing, such as a wheel or belt.

Polishing operations usually follow grinding and precede buffing. In general, polishing permits more aggressive abrading action than buffing. It has greater capability to modify the shape of a component than buffing. On the other hand buffing achieves finer finishes, has greater flexibility, and follows the contours of components.

Polishing and buffing processes are used on most metals and many nonmetals for refining edges and surfaces of castings, forging, machined and stamped components, and molded and fabricated parts. Traditionally, these processes have also been considered a means of developing attractive, decorative surfaces and generate surfaces suitable for preplate and prepaint surfaces.

Lapping is the process of finishing work materials by applying a loose abrasive slurry between a work material and a closely fitting surface, called a lapping plate. When loose abrasive is used to machine the work material, it may slide, roll, become embedded, or do all three, depending on the shape of the abrasive grain and the composition of the backup surface.

3. Systems Approach

Successful applications of diamond abrasive products require careful attention to all aspects of the "grinding system" - an approach that addresses four key inputs to the grinding process (viz.) machine tool, abrasive product selection, work material properties and operational factors. Each of these four primary input variables consist of a number of factors as shown in Table 5. Inadequate attention to details in any one of these system input variables can influence the grinding process results. Figure 19 shows an input/output model to be considered when pursuing the systems approach.

In essence, the systems approach requires an understanding of the influence of the input factors on the grinding process through cutting, plowing and the frictional interactions occurring at the grinding zone and their relationship to the resultant process variables of grinding forces and power.

These process interactions in turn influence the output characteristics such as the production of components of required geometry, surface quality, and production economics. It is nearly impossible to characterize each of these factors and their interactions. Besides, by its nature, any manufacturing process has to deal with stochastic variations and changes within control limits. Grinding, being a manufacturing process, suffers with these limitations, as well. However, it is possible to begin with a set of basic information and experimental results on most of the system input factors. Such information combined with the fine tuning of selected factors can achieve the desired output from the grinding system. This approach coupled with experience based rules or guidelines have been successfully employed in achieving desired performance in a number of grinding systems using diamond abrasives.

Table 5. Selected Variables Influencing the Superabrasive Grinding System

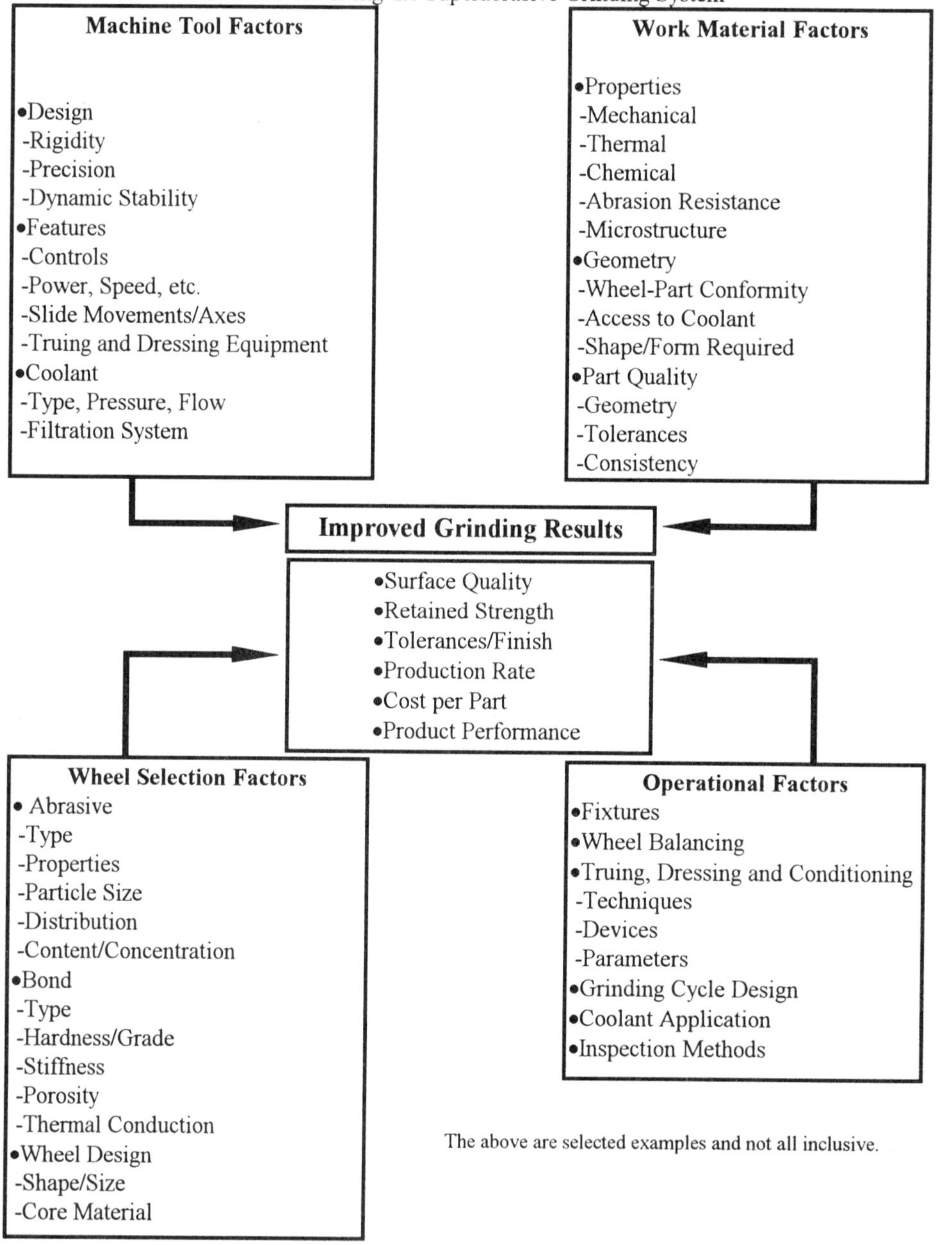

Machine Tool Factors

- Design
  - -Rigidity
  - -Precision
  - -Dynamic Stability
- Features
  - -Controls
  - -Power, Speed, etc.
  - -Slide Movements/Axes
  - -Truing and Dressing Equipment
- Coolant
  - -Type, Pressure, Flow
  - -Filtration System

Work Material Factors

- Properties
  - -Mechanical
  - -Thermal
  - -Chemical
  - -Abrasion Resistance
  - -Microstructure
- Geometry
  - -Wheel-Part Conformity
  - -Access to Coolant
  - -Shape/Form Required
- Part Quality
  - -Geometry
  - -Tolerances
  - -Consistency

Improved Grinding Results

- Surface Quality
- Retained Strength
- Tolerances/Finish
- Production Rate
- Cost per Part
- Product Performance

Wheel Selection Factors

- Abrasive
  - -Type
  - -Properties
  - -Particle Size
  - -Distribution
  - -Content/Concentration
- Bond
  - -Type
  - -Hardness/Grade
  - -Stiffness
  - -Porosity
  - -Thermal Conduction
- Wheel Design
  - -Shape/Size
  - -Core Material

Operational Factors

- Fixtures
- Wheel Balancing
- Truing, Dressing and Conditioning
  - -Techniques
  - -Devices
  - -Parameters
- Grinding Cycle Design
- Coolant Application
- Inspection Methods

The above are selected examples and not all inclusive.

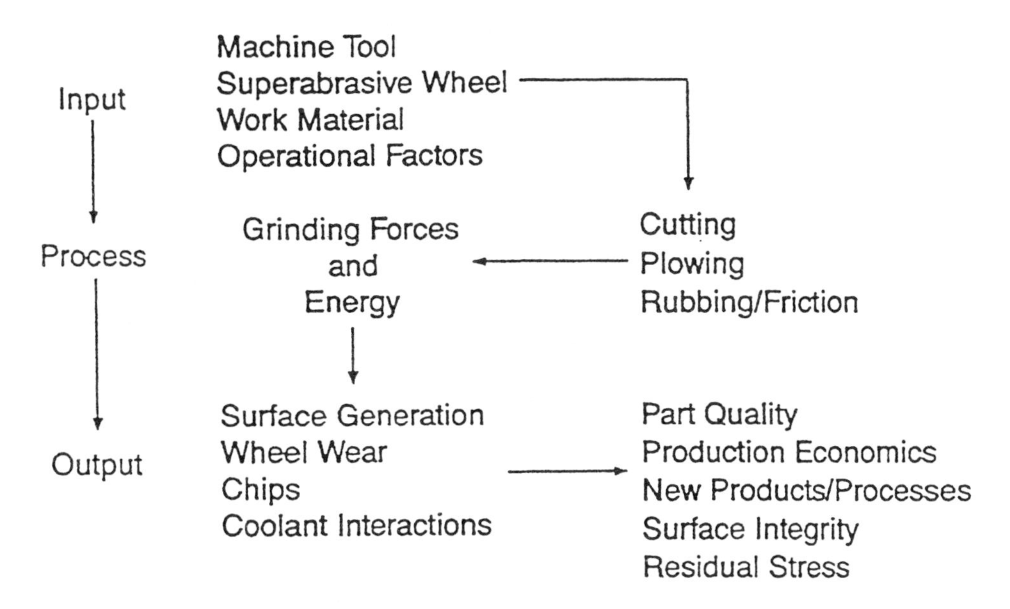

Figure 19. Input/output model of the precision grinding process.

4. Acknowledgments

The authors thank their colleagues at Norton Company for their input and valuable discussions. The help of Ms. Maryann Mekelski, Ms. Janice Vasalofsky and Mr. Frank Esposito in the preparation of this manuscript is gratefully acknowledged.

5. References

ANSI B74.3-1986, "American National Standard Specification for Shapes and Sizes of Diamond and CBN Abrasive Products". Grinding Wheel Institute 1986.

"*Advances in Ultrahard Materials Application Technology*", Edited by Barrett, Chris, Vol. 4, DeBeers Industrial Diamond Division.

Ault, William N., "Grinding Equipment and Processes", ASM Metals Handbook, Volume 16, Page 431-452.

Rohr, Norman E., 1995, "Superfinishing with Diamond Micron Powder", Supergrind 1995.

Subramanian, K.; Lindsay, R.P., "A System Approach for the Use of Vitrified Bonded Superabrasive Wheels for Precision Production Grinding", Journal of Engineering of Industry, February, 1992, Vol. 114, Page 41-52.

Subramanian, K., "Classification and Selection of Finishing Processes", ASM Handbook, Volume 5, Surface Engineering, Page 81-83.

Subramanian, K., "Materials Science and Technology", Volume 17, "Processing of Ceramics", Chapter 18, "Finishing", Page 215-259.

Subramanian, K., "Superabrasives", ASM Metals Handbook Vol. 16, Page 453-471.

"*Superabrasive Grinding*", Metzger, J. L., Butterworths, 1986.

Tricard, Marc, "Truing and Dressing Superabrasives - Matching Methods to Needs", Modern Machine Shop, June 1995.

Chapter 29
ACTIVE DEVICES

David L. Dreifus and Bradley A. Fox
Kobe Steel USA Inc., Electronic Materials Center
Research Triangle Park, NC 27709

Contents

1. Introduction

The large electric breakdown field, high saturated current velocity, high thermal conductivity and other extreme properties of diamond make it an excellent semiconductor material for active electronic devices. A summary of diamond's electronic properties is given in Table I. Figures of merit based on material properties have been developed to compare the potential device performance of different semiconductors. The Johnson and Keyes figures of merit demonstrate the applicability of diamond to high temperature, high frequency and high power applications. [Davis 1988]. Other figures of merit have also been evaluated and demonstrate the potential of active diamond electronics [Shenai 1989; Trew 1991].

Although the figures of merit suggest that diamond has tremendous potential for active electronics, there are some obstacles that must be overcome for diamond to achieve its potential as a material for commercial electronic devices. For example, the correlation of these figures of merit to device performance has been questioned. [Collins 1989; Collins 1992; Collins 1993] In a re-evaluation of diamond as high frequency material that addressed some of the concerns with the figures of merit, boron-doped diamond devices did not perform as well as devices made of SiC and GaAs. [Shin 1994a] The intent of this section on active electronic devices is to describe the potential for diamond

Table I. Electrical Properties of Diamond at room temperature.

Property	298 K	Units	Property	298 K	Units
thermal conductivity	20	W/cm-K	band gap	5.45	eV
thermal expansion	$1.1\cdot10^{-6}$	K^{-1}	undoped resistivity	$>10^{15}$	Ω-cm
electric breakdown field	10	MV/cm	dielectric constant	5.7	
saturated current velocity			mobility		
electron	$2.7\cdot10^{7}$	cm/s	electron	2200	$cm^2/V\cdot s$
hole	$1.0\cdot10^{7}$	cm/s	hole	2000	$cm^2/V\cdot s$

electronics devices and to identify areas that require additional development in order for diamond to achieve its potential.

2. Device Quality Diamond

For active electronic devices it is desirable to have i) insulating regions, ii) p and n-type semiconducting material and iii) Ohmic, Schottky or insulating contacts. Diamond's wide band gap allows undoped diamond to be a good insulator while doped diamond is used for the semiconducting regions. The ability to serve as both the insulator and the semiconductor make diamond a versatile material for active electronic devices. Additionally, through the use of various contact metallurgy, both Ohmic and Schottky contacts can be made to diamond. Various insulators can be used to form capacitive structures for field-effect devices.

2.1 INSULATING DIAMOND

The observed resistivity of high quality undoped diamond exceeds 10^{15}-10^{16} Ω–cm at room temperature. [Vandersande 1995] This resistivity limit is "apparatus-limited" due to leakage currents within the measurement setup. To characterize the resistivity of diamond, variable temperature resistivity measurements up to 1000-1200°C have been performed on both chemical vapor deposited (CVD) and natural diamond, as shown in Figure 1. [Vandersande 1995] In this figure, the conductivity, or reciprocal of the resistivity, is shown as a function of reciprocal temperature. The resistivity of the Crystallume 2 and Raytheon films exceeds the resistance of the natural diamond by about 100x at 1000°C and demonstrates the ability to fabricate high quality undoped CVD diamond.

From the temperature dependence of the resistivity, the observed activation energy is 1.6 eV. If this activation energy is used, the extrapolated room temperature resistivity is $\sim10^{27}$ Ω-cm. Using the band gap energy of 5.45 eV, the theoretical room temperature resistivity for intrinsic diamond is $>10^{42}$Ω–cm. The difference between the observed and theoretical activation energy and the difference in room temperature resistivity estimates indicate that intrinsic conduction does not dominate for insulating diamond. Based on the similarity of the activation energy and the energy transition of isolated substitutional nitrogen, this species has been proposed to be responsible for the conduction in insulating diamond. [Vandersande 1995]

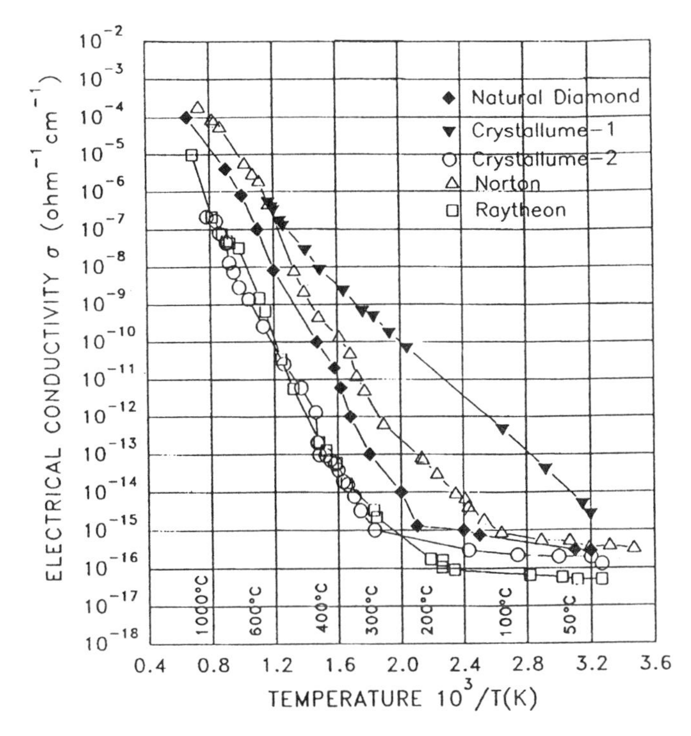

Figure 1. Temperature Dependent Electrical Conductivity of Diamond [Vandersande 1995]

For device applications, the high resistance of insulating diamond compares favorably to other semiconducting materials and to other materials used as insulators for semiconducting devices. For semi-insulating 6H-SiC, the extrapolated room temperature resistivity was estimated to be 10^{15} Ω-cm, [Hobgood 1995] while for semi-insulating GaAs, the room temperature resistivity is 10^8 Ω-cm. [Ghandi 1983] Additionally, the resistivity of insulating diamond is similar to the resistivity of other insulating materials often used for device fabrication such as silicon dioxide, 10^{14} - 10^{16} Ω-cm, or silicon nitride, ~10^{14} Ω-cm. [Sze 1981]

2.2 SEMICONDUCTING DIAMOND

For device applications it is desirable to controllably dope diamond to obtain both p and n-type material over a wide range of dopant concentrations. Doping is used to introduce a specific concentration of charge carriers in order to make the material semiconducting. As noted below, this aspect of the semiconducting properties of diamond requires additional improvement for commercial viability.

2.2.1 P-Type Doping

Boron is the primary p-type dopant in diamond[Chrenko 1973; Collins 1971] with an activation energy of 0.37 eV. [Collins 1979] For high concentrations of boron, the activation energy has been observed to decrease.[Bourgoin 1979; Nishimura 1991] In addition to valance band conduction where holes propagate in the valance band, other forms of conduction such as hopping have been observed. [Visser 1992; Williams 1970] Since carrier mobilities are significantly degraded for hopping conduction, the carrier concentration and mobility of valance band conduction will be discussed.

The temperature dependent carrier concentration is shown in Figure 2(a). [Fox 1995] The large activation energy of the boron and the presence of compensation significantly influence the carrier concentration. The 0.37 eV depth of the boron acceptor shifts the extrinsic region, where the acceptors are fully ionized, to temperatures much greater than room temperature. For example, in natural diamond samples boron-doped with concentrations of ~8 x 10^{16} cm^{-3}, the extrinsic regions begins at ~1200 K. [Collins 1979]. Since the extrinsic region is above room temperature, there is an inherent temperature variability of carrier concentration which will result in a temperature variation of the current flow of devices fabricated from diamond. Also, the lack of full ionization at temperatures less than the temperature required for the extrinsic regime reduces the total carriers available for conduction, increases the bulk resistance of the device and generates a high resistance-capacitance product that may limit the high frequency operation of diamond devices. The carrier concentration is further reduced by the presence of compensation[Fox 1995] where its influence on the hole concentration may be calculated from known semiconductor expressions. [Blakemore 1987] The net result is that compensating species trap charge carriers within the diamond, reduce the charge carriers available for conduction, and increase the bulk resistance. The lowest compensation measured in CVD diamond is 2-3 x 10^{15} cm^{-3}. [Fox 1995] For this concentration of compensation, a diamond film with an acceptor concentration of

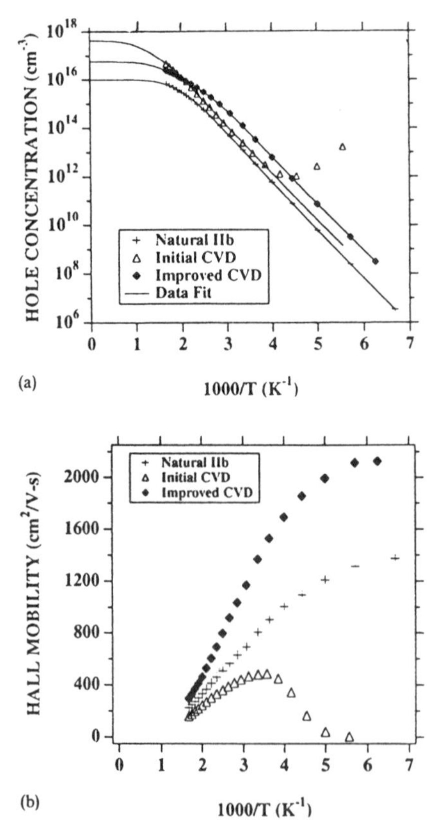

Figure 2. The Temperature Dependent a) Carrier Concentration and b) Mobility of Diamond [Fox 1995]

$1 \cdot 10^{17} cm^{-3}$ would have an additional 4x reduction in the room temperature carrier concentration due to the presence of compensation. The temperature variation of the carrier concentration and the presence of compensating species are two issue that must be addressed for diamond electronics to be viable.

The temperature dependent hole mobility is shown in Figure 2(b). [Fox 1995] The mobility of boron doped p-type diamond has been measured as high as 2010 $cm^2/V \cdot s$ in natural type IIb diamond[Williams 1971] and as high as 1590 $cm^2/V \cdot s$ in homoepitaxial CVD diamond. [Fox 1995]. The similarity of the two mobility values suggests that CVD diamond is now approaching the quality of the best type IIb natural diamond.

The mobility of diamond films is strongly affected by compensation and film morphology. Reducing the compensation of CVD diamond films to $2\text{-}3 \cdot 10^{15}$ cm^{-3} was shown to produce boron doped monocrystalline films with room temperature hole mobilities exceeding 1000 cm^2/V s. [Fox 1995] Polycrystalline films have exhibited film mobilities significantly reduced from those of monocrystalline films, typically 10x less than monocrystalline films. [Malta 1995] The carrier scattering in polycrystalline films is complex but is likely influenced by grain boundaries and intragranular defects. [Seto 1976]

2.2.2 N-Type Doping

Identification of a controllable, stable n-type dopant in diamond has been somewhat problematic. To date there are some hopeful results, but no clear confirmation of n-type diamond. The likely candidates for n-type dopants in diamond are nitrogen, phosphorus, lithium and sodium. Phosphorus and nitrogen are substitutional dopants while lithium and sodium are interstitial dopants. Theoretical calculations have been made to assess the prospects of these various candidates. [Kajihara 1990] Nitrogen is readily incorporated into the diamond lattice, but it is has an ionization energy of ~1.7 eV which produces negligible free carriers. [Collins 1979; Dyer 1965] Phosphorus has a predicted activation energy of 0.2 eV[Kajihari 1991] , but it has a high segregation energy inside diamond and its incorporation is likely to be low. Lithium and sodium are potential dopants with predicted activation energies of 0.1 eV and 0.3 eV, respectively, but neither has been demonstrated conclusively. Recent reports regarding ion implantation [Prins 1995] and biased diffusion [Popovici 1995] to produce conducting diamond appear to demonstrate some promise for achieving n-type diamond; however, further characterization of these materials is required to verify the ability to reproducibly and controllably dope diamond to make it n-type.

2.3 DIAMOND FOR ACTIVE ELECTRONICS

Diamond can be made in numerous ways, such as naturally within the earth, through high pressure-high temperature (HPHT) presses or by CVD. For active electronics, diamond will most likely be supplied by CVD due to the ability to deposited layers with controlled thickness and morphology over large areas. An additional advantage for CVD diamond is the controllable dopant concentration through gas phase incorporation of dopant species.

The availability of diamond material for active electronics is currently limited by the availability of a high quality, large diameter substrate. Large area monocrystalline diamond substrates are not available for homoepitaxial growth. Natural and HPHT synthetic diamond are typically limited to < 1 cm diameter. For the few natural diamond that exceed 1 cm in diameter, the cost is prohibitively expensive to produce commercial devices. Recently, there have been attempts at growing monolithic diamond layers on multiple aligned numbers of smaller single crystals, however, this process is still in the research phase. [Janssen 1996] Since monocrystalline diamond substrates are not currently practical, non-diamond substrates are being pursued.

There are also difficulties with finding a suitable large area nondiamond substrate for heteroepitaxy. The surface free energy of diamond is 5.3-9.2 J/m^2 on the low index planes of diamond. [Field 1979] The high surface energy makes it difficult to satisfy the free energy conditions required for monolayer-by-monolayer growth. [van der Merwe 1989] A second difficulty with heteroepitaxial growth of diamond is the small lattice constant of 0.3567 nm in the diamond lattice structure. The lattice constant of silicon is 0.5546 nm and is 60% larger than that of diamond. Heteroepitaxial diamond films have been grown on c-BN; [Koizumi 1990; Verwoerd 1994] however, large area c-BN substrates are also not available. Recently a promising technique of heteroepitaxy of {111} diamond on {111} platinum has been reported. [Tachibana 1995b] Although only preliminary investigations have been completed, this substrate appears to have some promise for diamond heteroepitaxy.

Since large area single crystal substrates for homoepitaxial or heteroepitaxial growth are not available, substantial progress has been made in the growth of highly oriented diamond films. This process involves nucleation of {100} oriented diamond grains on silicon and the preferential growth of these grains so that they coalesce into a continuous film of slightly misoriented {100} oriented grains. [Clausing 1991; Koidl 1994; Wild 1994] The advantage of this structure is that the grain boundaries have a low misorientation angle and their influence on the electrical properties is reduced. [Fox 1995] The room temperature mobility of boron doped diamond layers deposited on these highly oriented diamond substrates has been as high as 278 $cm^2/V{\cdot}s$ and is 5-10x the mobility of the best polycrystalline diamond films. [Stoner 1994]

2.4 ELECTRICAL CONTACTS TO DIAMOND

For active electronic applications, it is important to be able to make Ohmic, Schottky and insulating contacts. Complete understanding of the nature of contacts to diamond still needs to be developed, but there are some general trends that are relevant. More detailed reviews on the nature of contacts to diamond have been made by other researchers. [Das 1992; Moazed 1992; Tachibana 1995a]

2.4.1 Ohmic Contacts

Ohmic contacts have been made to diamond in a multitude of ways. There appear to be three primary structures that favor the formation of Ohmic contacts. These are the i) use of a damaged surface, ii) the use of carbide forming metals or iii) the use of heavy doping the region under the contact. [Tachibana 1995a] Each of these approaches will be briefly discussed.

Damage has been used to make reasonably good Ohmic contacts to diamond. This damage has been introduced by mechanical roughening [Collins 1970], particle bombardment [Tachibana 1992] or laser damage [Geis 1989]. Although these techniques produce reasonably good Ohmic contacts, they tend to be mechanically fragile and electrically noisy [Tachibana 1995b]. This technique is often employed along with carbide forming metals or through heavily doping the region under the contact to improve the performance of the contact.

Ohmic contacts have been explored through a variety of metallurgical combinations. There are numerous factors that affect the resulting contact. In a study of contacts to natural type IIb diamonds, carbide forming materials produced Ohmic contacts with contact resistivity of 10^{-2}-10^{-3} Ω-cm, while non-carbide formers produced rectifying contacts. [Hewett 1993] In general, it appears that carbide forming contacts, such as Ti, Mo, and Ta, provide the best contact material for producing Ohmic contacts to diamond. [Moazed 1988; Tachibana 1995a]

The nature of the contacts also depends on the dopant concentration of the diamond beneath the contact region. The increased dopant concentration may be produced by ion implantation [Kalish 1993; Prins 1989], *in situ* doping [Hewett 1993] or diffusion. [Tsai 1991] For example, epitaxial CVD films doped 10^5x higher in boron were compared to natural type IIb diamond. [Hewett 1993] Contact resistivities of 10^{-5} Ω-cm were observed on the heavily doped epitaxial films while contact resistance of 10^{-2}-10^{-3} Ω-cm were observed on the natural type IIb diamond. Even non carbide formers produced Ohmic behavior on the heavily doped boron films. Therefore, increases in boron concentration strongly facilitate the formation of Ohmic contacts.

2.4.2 Schottky Contacts

It is observed that most non-carbide forming metal make a Schottky contact to diamond. Surface preparation prior to contact formation influences the contact behavior. The effect of surface preparation is not well understood; however, recent results suggest different behavior of the contacts depending if the surface is "clean", oxygen terminated or hydrogen terminated. Kawarada, *et al.*, [Kawarada 1995] used hydrogen terminated surfaces to show a variation of the Schottky barrier height with metal electronegativity. These results suggests that the surface states do not pin the Fermi level. Baumann and Nemanich [Baumann 1995] demonstrated the ability to modify the pinned position of the Fermi level. For a clean surface, the Schottky barrier height was 1 eV while for an oxygen terminated surface the Schottky barrier height was 2 eV. The pinning of the Fermi level has been attributed to the presence of a high density of surface states. Surface preparation has also been used to minimize current leakage. [Geis 1993] The surface treatments employed appears to have two primary effects. First, a 660°C anneal removes a near surface compensation of boron atoms and second, exposure of the surface to passivating species alters the surface conductivity. Clearly, there is not a complete understanding of the effects of surface preparation on Schottky contacts to diamond; however, surface preparation appears to strongly influence the surface state of diamond which alters both the barrier height and the leakage currents.

2.4.3 Insulating Contacts

Insulating contacts to diamond often occur by either the use of an insulating diamond layer on top of a semiconducting diamond layer[Miyata 1993] or by use of a silicon dioxide layer on top of diamond. [Geis 1991a] Only a limited amount of data is available to date on these structures. In one study on type IIb diamonds, the reverse bias leakage current was $1 \cdot 10^{-12}$ A for an insulating diamond structure and $3 \cdot 10^{-11}$ A for a silicon dioxide structure. [Wynands 1994] For both structures, the density of interface traps was estimated to be in the range of $10^{-12} cm^{-2} eV^{-1}$. These results are too premature to definitively select a preferred insulating contact. The low leakage currents demonstrate the promise for each contact while the high density interface traps suggest that each contact still requires substantial development.

3. Diamond Active Electronic Device Issues

There have been many publications describing the virtues of diamond's unique combination of properties and their potential use in active electronic devices. [Geis 1991b; Grot 1992; Hewett 1991; Kobashi 1993; Shenai 1989; Trew 1991] Several theoretical analyses suggest that diamond devices could potentially exhibit performance superior to that of devices fabricated from other materials such as Si and GaAs. [Shenai 1989; Trew 1991] Hence, diamond has been regarded as a suitable material for electronic devices operating under hostile conditions such as high temperature, high power, high radiation, and high frequencies. [Geis 1991b; Gildenblat 1991b; Grot 1995; Grot 1992; Trew 1991] A great deal of excitement has been generated by reports of working transistors fabricated from single crystal diamond. [Grot 1995] The performance of diamond transistors, however, has varied greatly, depending on device structure, starting material, doping technique, and device processing technologies.

In contrast, there have been reports that refute the optimistic outlook for diamond electronic devices. [Buckley-Golder 1992; Collins 1989; Collins 1992] In these latter reports, authors indicate that many unresolved issues need to be addressed, such as the lack of low activation energy dopants, the state of diamond device process technologies, the growth of high quality doped diamond layers, and the availability of a suitable large-area substrate.

The primary issue facing the development of diamond active electronic devices is obtaining and controlling charge within a device. This is directly related to the deep B acceptor level at 0.37 eV. The concentration of charge carriers that could contribute to conduction within a device can be obtained from Poisson's equation[Blakemore 1987] :

$$N_A^- + n = N_D^+ + p, \tag{1}$$

where p is the hole concentration,
n is the electron concentration,
N_A is the ionized acceptor concentration, and
N_D is the ionized donor concentration.

Substituting for the ionized acceptor concentration and the ionized donor concentration(assumed to be completely ionized), and assuming a negligible electron concentration in p-type diamond, the following relationship can be obtained for the case of incomplete ionization of an impurity, in the presence of compensating centers: [Blakemore 1987]

$$\frac{p(p + N_D)}{(N_A - N_D - p)} = \left(\frac{N_V}{g}\right) \exp\left(\frac{E_V - E_A}{kT}\right), \tag{2}$$

where N_V is the valence band density of states,
E_V is the valence band energy level,
E_A is the acceptor energy level,
k is Boltzmann's constant,
T is the temperature, and
g is valence band degeneracy (for diamond g = 4) [Collins 1979]

Device researchers have recently been re-examining the traditional semiconductor device approximations used in device design. Existing device design software may not adequately account for the effects of incomplete ionization in the presence of compensating centers. [Dreifus 1994a; Tessmer 1995] In several design software packages such as PISCES[Silvaco-International 1988] , a simplifying approximation is made to Poisson's charge balance equation. Instead of employing the complete description, an approximation is made in which a net impurity concentration, $|N_A\text{-}N_D|$, is used for calculation of the carrier concentration. It should be noted that the carrier concentration at room temperature can be several orders of magnitude below that calculated using the above approximation. While this approximation will work for the case of a complete ionization, for the case of diamond,

$$p \neq \frac{|N_A - N_D|}{1 + g\exp\left(\frac{E_A - E_F}{kT}\right)} \tag{3}$$

Note that this has now been corrected in the most recent version of Silvaco International's modeling software. It is also important to note that the activation energy, E_A, is also a function of the doping density. Increased doping density results in a lower activation energy. [Dreifus 1996]

Another important parameter affecting the transport of charge carriers within a device are depletion regions. These regions can form due to charge re-distribution at interfaces such as metal-semiconductor junctions, heterojunctions, homojuntions, and at surfaces. The behavior of depletion layers within semiconducting diamond will depend on $|N_A - N_D|$ or the net impurity concentration. The dependence is shown for a one-sided abrupt junction is shown below[Sze 1981]

$$W_{depletion} = \sqrt{\frac{2\varepsilon_s V}{q|N_A - N_D|}}, \tag{4}$$

where e_s is the dielectric constant and
V is voltage.

Both the carrier concentration and the depletion width affect device operation. In a field-effect transistor, the extent of the channel width, and thus the potential control of charge flow depends on $|N_A\text{-}N_D|^{-0.5}$. In an FET, however, channel current is proportional to p. While the depletion layer extent is insensitive to incomplete ionization, the low carrier concentration leads to an inefficient device exhibiting poor transconductance, poor current handling capability, and unacceptable sensitivity of performance with respect to temperature. Increasing the dopant concentration will increase the number of carriers available for conduction, however, this will also limit the extent of the current controlling depletion regions.

Analyses of the effects of incomplete ionization on the performance of diamond FETs has shown a large variability in the device behavior with respect to changes in operating temperatures. [Dreifus 1994a; Grahn 1994] This has also been observed experimentally. [Dreifus 1995; Dreifus 1994b] Shin *et al.* [Shin 1994b] have re-examined the expected high frequency performance of diamond devices[Trew 1991] in light of incomplete ionization. Original performance predictions were revised such that diamond devices are now expected to exceed the performance of GaAs and SiC rf power FETs only when operated at temperatures in excess of 600°C!

Additionally, there are other obstacles facing the development of diamond active electronic devices. Specifically, the absence of other than niche applications to drive the development, high development costs, a limited number of device modeling tools that are applicable to diamond, and competition from traditional semiconductors such as Si and GaAs and more recently other wide bandgap materials such as SiC and nitrides, are issues that still must be addressed.

4. Diamond Active Electronic Devices

Despite the previously mentioned daunting hurdles, steady progress has been made in demonstrating and subsequently improving the performance of diamond active electronic devices. The state of the art has shifted from demonstrating modulation of channel conductance to the characterization of multi-device analog and digital circuits. Devices have been shown to operate at temperatures as high as 550°C. Reviews of much of the previous device work can be found in References. [Davidson 1993; Dreifus 1995; Dreifus 1995; Gildenblat 1991a; Grot 1995; Tessmer 1995; Yoder 1993] Some of this prior work is summarized in Table II. This list is not intended to be complete. As demonstrated by the number and variety of publications detailed in Table II, there are many groups engaged in the research and development of diamond active electronic devices. Innovative approaches to addressing the above diamond device issues are being investigated including modulation doping[Shiomi 1995], structural variations[Dreifus 1995; Kawarada 1994; Miyata 1995], and electron-beam injection[Lin 1995]. The remainder of this section will focus on recent developments.

In addition to the work summarized in the preceding table, several noteworthy recent advances have been reported. In particular, several approaches to engineer around the issue of the high activation energy of the B acceptor are being pursued. A few of the recent developments will be described.

Table II. Summary of Active Diamond Device Publications

Device Type [Reference]	Diamond Material	Novel Results
Bipolar Junction Transistor [Prins, 1982]	Type IIb natural p-type Single Crystal Diamond (SCD)	• C implantation used to form n-type like regions • implant depth 0.3 μm • SCD was 4 mm x 2.4 mm x 1 mm • 3.2 μm base length, 2.4 mm base width • current amplification factor of 0.11 • first diamond active electronic device
Point-Contact Transistor [Geis, 1987]	Type IIb natural p-type SCD	• modulation observed • 2 mm thick SCD, 4 mm square at base • 20 mA collector current at 90 V; base current of 4.2 mA • operated at 510°C • maximum small signal current gain up to 25 • maximum power gain up to 35 • current gain of 1.6 and power gain of 4.5 at 510°C
Point Contact Transistor [Tzeng, 1987]	natural type IIb SCD	• As implantation used to form n-type character regions • structure similar to that of Prins • operation at 350°C • current gain of 0.8
Vertical MOSFET [Geis, 1988]	Type IIb natural p-type SCD	• channel modulation • 85 μA channel current at 50 V • triode-like characteristics • 30 μS/mm transconductance • 10 μm wide vertical channels
MESFET [Shiomi, 1989]	homoepitaxial layer on SCD	• channel current modulation • 0.9 μA channel currents at 10 V • transconductance • a=2 μm, W= 1.8 mm, L=140 μm
MESFET [Tsai, 1991]	SCD	• 1400°C B-diffusion doped channel • channel current modulation • channel pinch-off • 1.5 μA channel currents at 10 V • 0.7μS./mm transconductance • Ti/Au contacts • a= 500Å, W= 1 mm, L=150 μm • 1.6 μA/mm channel current density
MiSFET [Shiomi, 1990]	homoepitaxial layer on SCD	• channel current modulation • 95 μA channel currents at 50 V • 2 μS/mm transconductance • a=0.9 μm, W~1.2 mm, L=20 μm
MOSFET [Fountain, 1991]	homoepitaxial layer on SCD	• B-doped and Li/B-doped channels • channel modulation observed • tendency towards saturation • pinch-off to parasitic channel leakage path • 5.4 μA channel current • 38 μS/mm transconductance • W= 50 μm, L = 8 μm

Table II. Summary of Active Diamond Device Publications (continued)

Device Type [Reference]	Diamond Material	Novel Results
MOSFET [Hewett, 1991]	B ion implanted SCD	• triple B ion implantation used to form channel • first pinch-off and saturation of channel current • 45 μA channel current at 10 V • 3.9 μS/mm transconductance • a=0.2 μm, W ~3.7 mm, L=400 μm
MiSFET [Nishibayashi, 1991]	homoepitaxial layer on SCD	• channel current modulation • tendency towards saturation • 1 μA channel currents at 50 V and 26°C • 0.02 μS/mm transconductance at 26°C • channel modulation observed at 300°C • 0.5 μS/mm transconductance at 300°C • a=0.8 μm, W~1.2 mm, L=30 μm
MOSFET [Kiyota. 1991]	polycrystalline layer on Si	• channel current modulation • 8 mA channel current at 45 V • high gate leakage current • triode-like characteristics • conducting Si substrate • a=10 μm, W = 1 mm, L = 100 μm
MOSFET [Gildenblat, 1991]	homoepitaxial layer on SCD	• modulation of the channel current • tendency towards saturation, no pinch-off • 200 μA channel current at 40 V at 300°C • ~5 μS/mm transconductance • a=0.1 μm, W=200 μm, L=60 μm
MOSFET [Tessmer, 1992]	polycrystalline layer on Si	• modulation of channel current • first demonstration of polycrystalline diamond FET • B ion implantation used for active channel • 8.5 mA channel current at 10 V • 121 μS/mm transconductance • a~0.15μm, W=314 μm, L=2 μm
MOSFET [Grot, 1992]	homoepitaxial layer on SCD	• modulation of the channel current • electron cyclotron plasma etching for recessed gate • tendency towards saturation, no pinch-off • 200 μA channel current at 100 V at 350°C • gate insulator leakage limited performance • 87 μS/mm transconductance • a=0.06 μm, W=30 μm, L=4-20 μm
MOSFET dc Circuit [Zeidler, 1993]	B ion implanted SCD	• triple B ion implantation used to form channel • pinch-off and saturation of channel current • enhancement-mode operation • first multi-device circuit • two devices biased to form driver/active load pair • 1.4 mA channel current at 30 V • 48 μS/mm transconductance • a=0.2 μm, W ~3.7 mm, L=400 μm • dc gain of 2

Table II. Summary of Active Diamond Device Publications (continued)

Device Type [Reference]	Diamond Material	Novel Results
MESFET [Brandes, 1993]	homoepitaxial layer on SCD	• modulation of channel current • etched diamond posts for device isolation • 116 nA channel current at 2 V • 1-2 nA of modulation
MOSFET [Tessmer, 1993]	polycrystalline layer on Si	• modulation of channel current • 900 nA channel current at 100V at 200°C • variation of performance over temperature cycling • 300 nS/mm transconductance at 250°C • operation at 285°C • a=0.5 μm, W=314 μm, L=2 μm
MiSFET [Kobashi, 1993]	polycrystalline diamond on Si_3N_4	• modulation of channel current • pinch-off of channel current • tendency towards saturation • 0.7 nA of channel current at 20 V at 22°C • 43 pS/mm at 22 °C • a= 0.03 μm. W= 1 mm, L= 10 μm • 300 °C operation • 275 nA channel current at 20 V at 300°C • 0.26 μS/mm at 300°C
MOSFET [Dreifus, 1994]	homoepitaxial layer on SCD	• modulation of channel current • pinch-off and saturation • 500°C operation • 1 mA channel current • temperature stabilized operation over 450-750K
MOSFET [Holmes, 1994]	homoepitaxial layer on SCD	• first demonstration of digital logic circuits • NAND & NOR logic operation using TTL inputs • operation at 400°C • minimal temperature variation from 350-400°C
MOSFET [Holmes, 1994]	homoepitaxial layer on SCD	• device performance uniformity data across wafer • effects of temperature on transconductance
MiSFET [Nishimura, 1994]	polycrystalline diamond on Si_3N_4	• modulation of channel current • pinch-off of channel current • tendency towards saturation • 16 nA of channel current at 20 V at 27°C • a= 0.1 μm. W=500 μm, L=5 μm • 200 °C operation • 38 nA channel current at 20 V at 200°C • modeled characteristics
MiSFET [Nishimura, 1994]	polycrystalline diamond on Si_3N_4	• modulation of channel current • triode-like characteristics • 70 μA of channel current at 10 V • a= 0.7 μm. W=1 mm, L=10 μm • 5 μS/mm transconductance

Table II. Summary of Active Diamond Device Publications (continued)

Device Type [Reference]	Diamond Material	Novel Results
MOSFET [Stoner, 1994]	HOD on Si	• modulation of channel current • first demonstration of HOD FET • saturation and pinch-off achieved • 25 nA channel current at 40 V • 6 nS/mm transconductance • W=314 μm, L=2 μm
MOSFET Amplifier [Dreifus, 1994]	homoepitaxial layer on SCD, HOD on Si, and polycrystalline layer on Si	• comparison of devices on three materials • device modeling • first demonstration of a common-source amplifier • gain of 4.8 up to 10 kHz at 250°C • W=314 μm, L=2 μm

4.1 SINGLE CRYSTAL DIAMOND FIELD-EFFECT TRANSISTORS

Plano *et al.* observed modulation for a junction field-effect transistor (JFET). [Plano 1994] Homoepitaxial layers doped with B(for the channel), and P(for the gate junction diode) were used. Although the P-doped region did not exhibit low-resistivity n-type behavior, the layer did act as an undoped region similar to an MiSFET device. Two devices were fabricated and tested. In on device, the B-doped layer was 30 μm in thickness, while the P-doped region was 0.5 μm thick. The other device had 0.2 μm and 0.5 μm B-doped and P-doped layers respectively. Channel currents of 120 μA and 1.6 mA were observed for the two devices. The devices failed due to electrostatic discharge before variable temperature measurements could be made. A maximum room temperature transconductance of 10 μS/mm was observed.

High temperature devices have been fabricated on single crystal diamond (SCD). [Dreifus 1994a] In this work, depletion-mode MOSFETs were fabricated using high mobility homoepitaxial layers grown on single crystal type IIa natural diamond substrates. The devices exhibit well-behaved characteristics, including saturation and complete pinch-off of the channel current. Variable temperature operation has been observed. These devices had gate lengths and widths of 2 μm and 314 μm respectively. Initially, elevated temperature operation was limited up to 425 °C. Non-ideal Ohmic contacts introduced parasitic channel resistance that limited channel current. In subsequent research[Dreifus 1994a; Dreifus 1994b; Holmes 1994] , improved characteristics were observed, allowing devices to be characterized up to 550°C. At temperatures in excess of 500°C, complete pinch-off of the active channel was not possible due to a parasitic conduction path. Application of larger gate biases resulted in catastrophic breakdown of the gate oxide. At intermediate temperatures of 325°C, enhancement mode operation was also observed and is shown below in Figure 3. Channel currents as large as 15 mA and transconductances in excess of 1 mS/mm were measured. Additional biasing of the device, further into enhancement-mode, does not continue to increase the channel current. From device simulations, the current in the channel appears to be limited by parasitic depletion regions that form under the gate oxide between the source and gate, and the drain and gate electrodes.

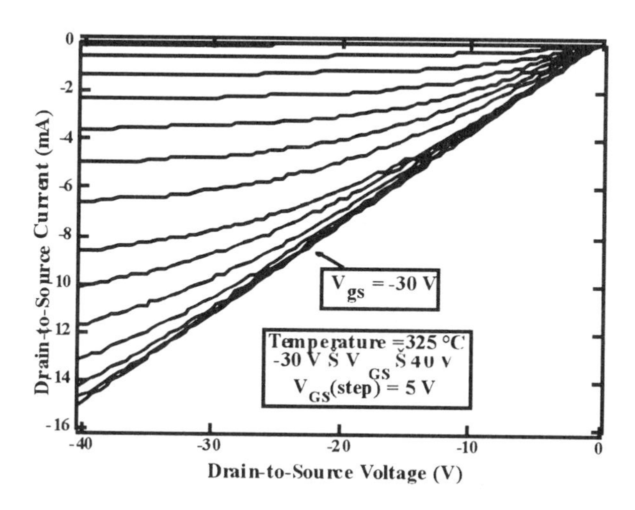

Figure 3. Drain-to-source current-voltage characteristics of a single crystal diamond field-effect transistor operating in enhancement and depletion mode at 325°C.

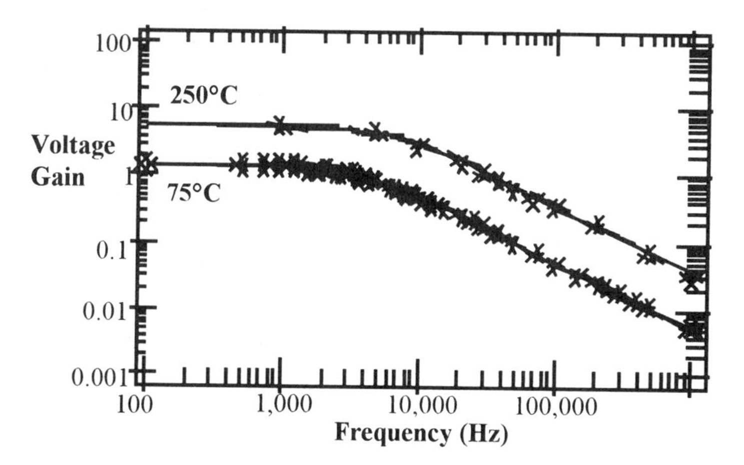

Figure 4: Frequency response for a simple common-source amplifier utilizing a diamond MOSFET. The DC voltage gain is 1.4 at 75°C and 5 at 250°C.

Individual diamond MOSFETs were biased into a simple common-source amplifier configuration. The resultant gain versus frequency plot is shown in Figure 4 for diamond amplifiers operating at 75°C and 250°C. As expected, amplifier performance improved as a function of temperature as the carrier concentration in the channel was increased. A dc gain of 5 was observed at 250°C operation. The gain versus frequency data exhibited the roll-off characteristics of a circuit dominated by a resistance-capacitance time constant. Unity gain for an amplifier operating at 250°C occurred at an input frequency of 30 kHz.

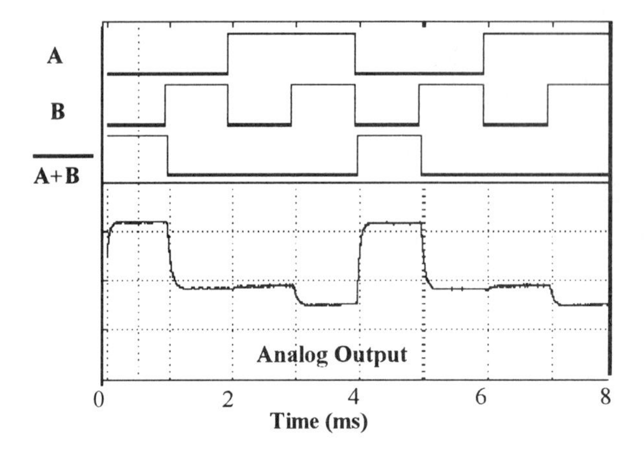

Figure 5. Plot of the input and output waveforms of a diamond NOR circuit operating at 325°C. Signals A and B shown are the TTL input signals as viewed through a logic analyzer. The input signals are followed by the logic output of the NOR circuit. The analog output is also shown as viewed by an oscilloscope.

Diamond MOSFETs were also used in simple digital logic gates. A NOR gate was constructed by connecting two diamond MOSFETs in series with an external load resistance and power supply. The input and output signals of the digital circuit operating at 400°C are shown in Figure 5 Standard TTL logic signals were used to drive the NOR circuit. Utilization of TTL inputs was facilitated by the relatively low pinch-off voltage of the constituent transistors.

This was the first demonstration of a digital circuit based solely on diamond transistors. Again, the devices were not optimized for digital circuit performance using 5 volt input logic levels. Lower pinch-off voltages of the component transistors would result in more ideal output characteristics. Also, decreased device capacitances should enhance the circuit's frequency performance.

Kawarada *et al.* have reported on enhancement-mode MESFETs fabricated using single crystal diamond with thin B-doped homoepitaxial layers. [Kawarada 1994] They employ hydrogenated diamond surfaces which exhibit a surface conduction mechanism[Landstrass 1989a; Landstrass 1989b; Nakahata 1991] . Typically this layer is eliminated through etching in chromic acid, annealing at temperatures in excess of 300 °C in an ambient containing oxygen, or electrochemical surface treatments. The motivation for this removal stems from the vulnerability of this layer to other surface effects and elevated temperatures, thereby rending it unstable during diamond device processing and subsequent characterization.

Although unstable, Kawarada *et al.* have succeeded in employing this layer to fabricate enhancement-mode diamond MESFETs. [Kawarada 1994] The thickness of the B-doped homoepitaxial layer was estimated to be 0.1 μm. The gate lengths and widths of the MESFETs were 10-40 μm and 1.0-1.2 mm respectively. Current levels of 1.8 mA have been obtained at room temperature as charge is injected from the contacts into the

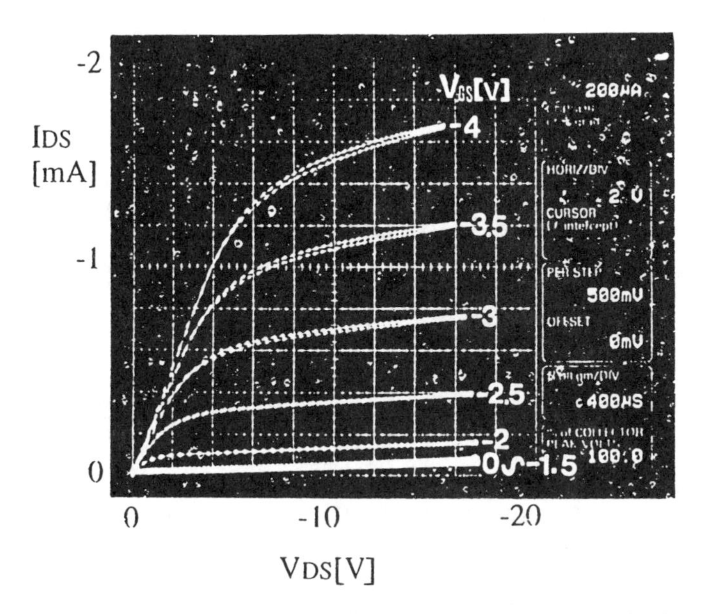

Figure 6. Current-voltage characteristics of an enhancement-mode diamond MESFET.

depleted surface layer. Saturation and pinch-off have been observed as are shown in Figure 6. Transconductances as large as 200 μS/mm have been obtained for room temperature operation of a 10 μm gate length device. More recently Kawarada and Ito[Itoh 1995] have fabricated inverters, NAND and NOR digital logic circuits using their enhancement-mode MESFET devices. The characteristics of a NAND circuit are shown in Figure 7. The authors acknowledge the unstable nature of the surface conduction layer, and indicate that the C-H bond is stable in vacuum to 800 °C and with suitable passivation they expect to extend the useful temperature range of these devices to more than 500°C.

Shiomi *et al.* employed an approach to circumventing the difficulty of obtaining complete ionization: a diamond MESFET using a d-doping or pulse-doping structure. [Shiomi 1995] The application of modulation doping to diamond active electronic devices was originally proposed by Kobayashi *et al.*[[Kobayashi 1994] as one method for improving the activation efficiency for the deep level B impurity. He describes the two primary advantages of delta-doping as 1) Fermi-level rising toward the acceptor level, leading to efficient excitation of holes from the deep level, and 2) most of the excited holes are in the surrounding unintentionally doped diamond spacer layers where holes can move with very high mobilities. In Shiomi's work, pulses of B_2H_6 containing gas were introduced into the diamond CVD reactor to form thin homoepitaxial regions of heavily B-doped diamond. A MESFET device was fabricated using a single "pulse-doped" layer. Secondary ion mass spectroscopy depth profiling indicated a 200Å-thick heavily doped region with a maximum atomic B concentration of $1 \cdot 10^{19}$ cm^{-3} as shown in Figure 8(a). The device gate lengths and widths were 2-4 μm and 39 μm respectively. Modulation of the channel current, saturation, and pinch-off were observed. A maximum channel current at room temperature was 150 μA and the device is nearly pinched-off at a gate bias of 6 V as shown in Figure 8(b). For a device with a gate length of 2 μm , a maximum transconductance of 388 μS/mm was determined from the current-voltage characteristics.

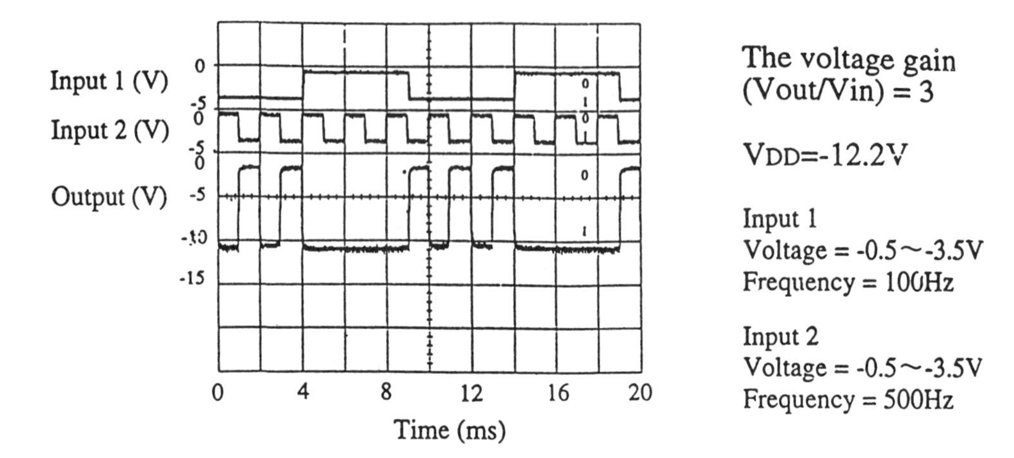

Figure 7. Input and output waveforms for a diamond NAND gate.

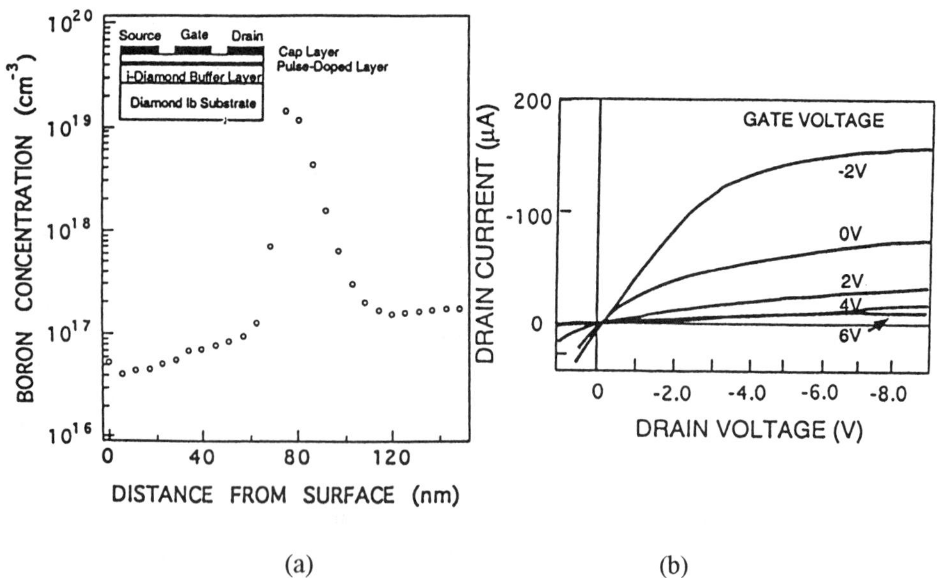

(a) (b)

Figure 8. Results for a pulse-doped diamond FET: (a)SIMS depth profile and (b) drain-to-source current voltage characteristics at room temperature.

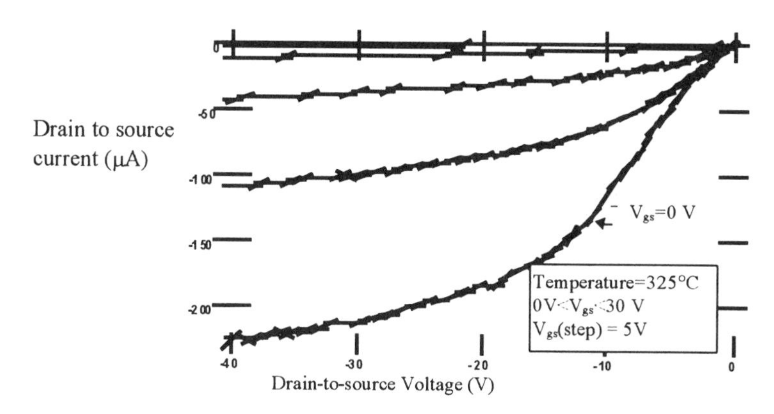

Figure 9. Drain-to-source current-voltage characteristics of a highly-oriented polycrystalline diamond field-effect transistor operating 325°C.

4.2 HIGHLY-ORIENTED DIAMOND FIELD-EFFECT TRANSISTORS

Recently, an engineered polycrystalline material, in which the individual grains are highly-oriented with respect to the substrate and adjacent grains, has been synthesized. This (100)-oriented material shows promise for becoming a potential substrate material for diamond electronic devices. [Stoner 1994; Stoner 1993] The individual grains in highly-oriented polycrystalline diamond are aligned such that only low-angle grain boundaries are present. The electrical properties of this material have already been shown to be superior to randomly-oriented polycrystalline diamond as described in §2.3 of this Chapter.

Devices fabricated from highly-oriented diamond (HOD) layers were characterized to 400°C. MOSFETs have been shown to exhibit operation consistent with depletion-mode devices and saturation and pinch-off of the channel current were observed at elevated temperatures. Current-voltage characteristics for an HOD FET operating at 325°C is shown in Figure 9. A maximum transconductance of 76 μS/mm was determined from the measured characteristics. Although the drain-to-source currents are lower than similar single crystal devices, the characteristics of highly-oriented polycrystalline diamond MOSFETs are significantly improved as compared to devices fabricated from randomly-oriented polycrystalline diamond. [Stoner 1994; Tessmer 1993] For the active devices fabricated using HOD films, the individual grain sizes are much larger than the gate length of the device. Thus the drain-to-source current is expected to traverse only a few grain boundaries.

4.3 POLYCRYSTALLINE DIAMOND FIELD-EFFECT TRANSISTORS

The economic viability and the availability of polycrystalline diamond films have many groups trying to fully characterize the electronic properties of these films. [Edwards 1993; Geis 1991b; Kobashi 1993; Malta 1993; Nishimura 1994; Werner 1992] Polycrystalline diamond films have been deposited over large areas using a variety of deposition techniques. [Nishimura 1994] Boron-doped semiconducting diamond films can be easily obtained. [Malta 1993; Malta 1995] Promisingly, the first operation and subsequent enhanced performance of FETs fabricated using polycrystalline diamond films has been reported. [Nishimura 1994; Tessmer 1995; Tessmer 1992; Tessmer 1993] The effects of grain boundaries upon electrical properties, however, has resulted in inferior performance by several orders of magnitude between polycrystalline and single crystal diamond devices.

Device performance continues to improve. Recently, Nishimura *et al.* [Nishimura 1994] reported the fabrication of polycrystalline diamond MiSFETs fabricated on Si_3N_4 substrates. Depletion-mode device operation was observed at temperatures up to 400°C. Channel currents as large as 1.2 μA at 20 V were observed. A tendency towards current saturation was observed at temperatures as high as 300°C and the devices nearly attained channel pinch-off. A transconductance of 0.26 μS/mm was observed at 300°C.

4.4 OTHER DEVICES

One method for obtaining charge carriers in diamond is to inject them from an external source. This can be accomplished using above bandgap energy light, ionizing particle radiation, or electron beam bombardment. Using this technique of external excitation of charge, many of the advantageous properties of diamond are available. The major drawback is the requirement of a high power external source.

Lin *et al.* [Lin 1993; Lin 1995] have employed electron beams for this purpose. In this work, various thicknesses of thin single crystal diamond slabs were used as anodes or targets as shown in Figure 10. Planar electrodes were placed on both surfaces. One electrode was made intentionally thin to allow for the penetration of impinging electrons into the diamond. In one experiment, the active diamond area that was illuminated was $7.1 \cdot 10^{-4}$ cm^2. The diamond thickness was 2.9 μm, and the bombarding beam voltage was 50 kV. The current voltage characteristics for electron beam currents in the range between 1 mA and 5.9 mA are shown in Figure 11.

Current gain typically ranges from 1000 to 3000. High power switching was achieved using a 10 μm thick diamond plate with an active area of 2 mm^2. An output power of 26 kW was delivered to the load. The electron beam power that controlled the switching was less than 10% of this value. Internal power dissipation within the diamond was 12 kW. Currents as large as 51 Amperes was switched using this single crystal device. This is a potentially exciting area for active diamond electronic devices, especially since Joshi *et al.* have recently demonstrated switching in polycrystalline diamond. [Joshi 1992] Additional summary information can be found in Reference [Dreifus 1995] .

Himpsel *et al.* [Himpsel 1980a; Himpsel 1980b; Himpsel 1979] proposed a number of potential applications that arise from the negative electron affinity of diamond. Field-emission from diamond at low electric fields makes diamond an ideal candidate for flat panel displays and other cold cathode applications. [Zhu 1995] One likely area for diamond devices is in the nascent field of vacuum microelectronics. Another application would be to couple a field emitter with a diamond anode. Diamond field emitters have not yet been used in active electronic devices, however, Mearini *et al.* [Mearini 1995] have employed the related property of high secondary electron emission from diamond in an electron multiplier. Diamond dynodes have been fabricated using coatings of CsI on polycrystalline diamond deposited on a Si substrate. Output measurements were made with various primary beam currents. With a primary beam current of 9.0 pA, a stable dc gain of $1.3 \cdot 10^5$ was obtained.

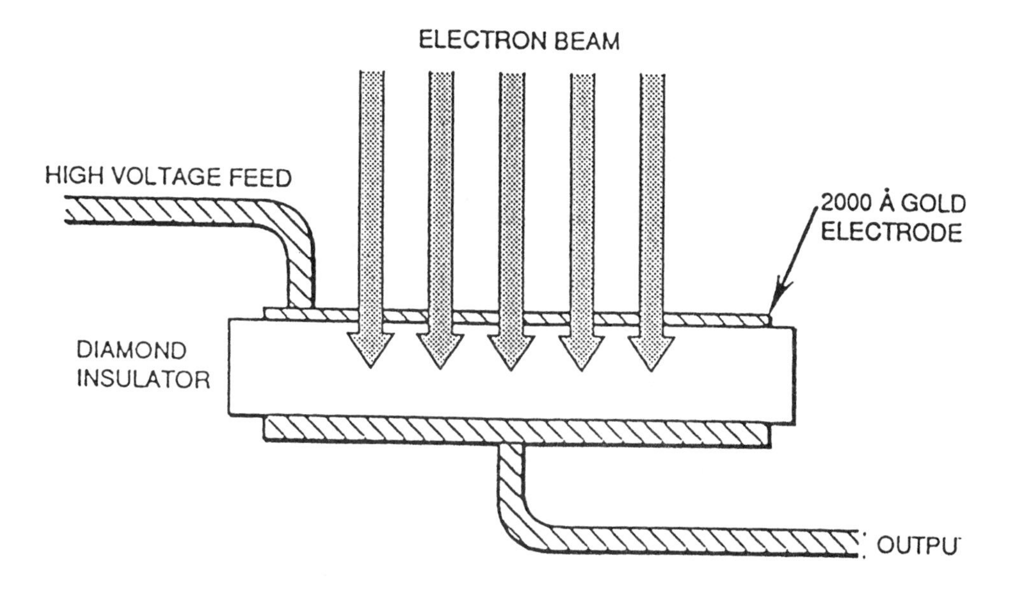

Figure 10. Schematic diagram of an electron bombarded diamond device.

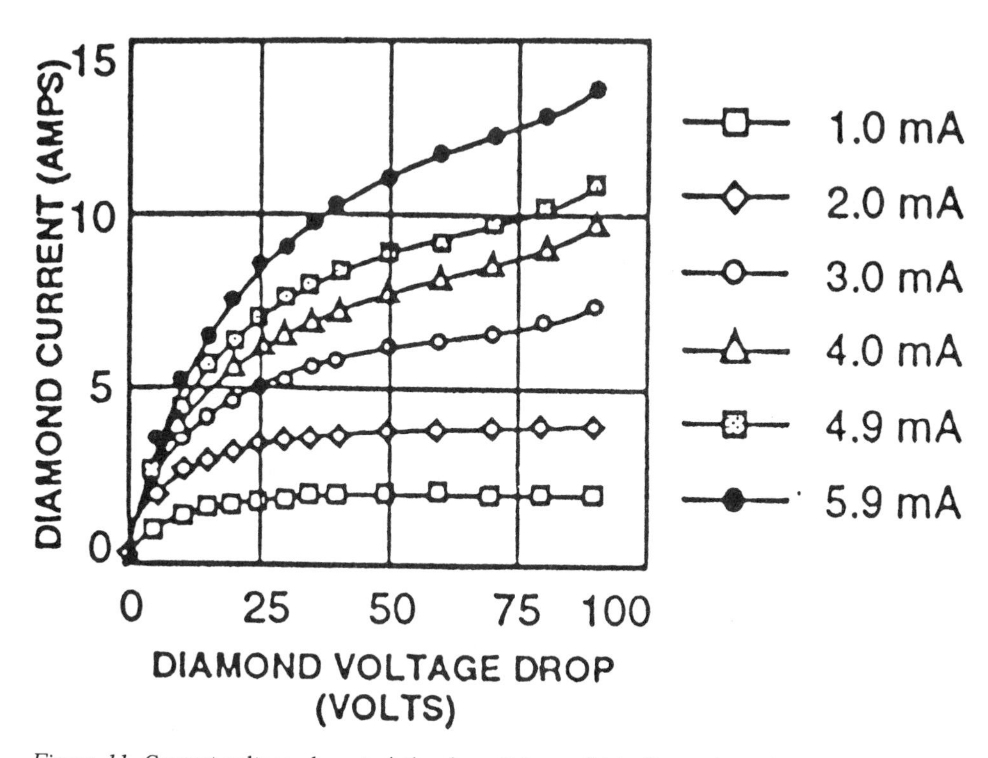

Figure 11. Current-voltage characteristics for a 2.9 μm thick diamond anode. The active area diameter was 300 μm, and the bombarding beam voltage was 50 kV.

5. Summary

The unique properties of diamond offer a tremendous potential for active electronic applications. Additionally, the preliminary device performance indicates its applicability to high temperature electronics. Substantial and continual improvements in the field of diamond electronics have been reported. Device processing technologies continue to advance. Operation of diamond MOSFETs exhibiting modulation, saturation, and pinch-off of the channel current have been demonstrated using single crystal and highly-oriented polycrystalline diamond thin films. Single crystal devices have been characterized in both enhancement and depletion-mode at elevated temperatures of 550°C. Drain-to-source currents of 15 mA and transconductances as large as 1.3 mS/mm have been reported and are similar to those first reported for SiC devices. [Davis 1991] A common-source amplifier, and NAND and NOR digital logic circuits have been operated up to 400°C. Modulation-doped FET operation has also been demonstrated. These results provide experimental evidence in support of the viability of active diamond electronic devices.

Other novel structures such as field-emission based devices, modulation doping, and electron bombarded diamond anodes have shown promise for realizing the advantages of diamond, while mitigating some of the obstacles facing diamond active electronics. New field-effect transistor structures such as a p+-i-p+ device have been proposed and simulated behavior has been shown to be comparable or in some cases superior to similar devices fabricated using SiC. [Miyata 1995] Material quality and reproducibility continue to improve with carrier mobilities approaching those of the best single crystal diamond samples. Techniques for obtaining larger-areas and less expensive device quality material, such as highly-oriented diamond or heteroepitaxial material on Pt are being developed. Applications requiring diamond's unique properties are beginning to emerge, however, progress in devices fabricated from competing materials is also continuing a rapid pace. Although substantial progress has been made, significant challenges remain for the development of commercial viable diamond active electronic devices.

6. References

Baumann, P. K. and Nemanich, R. J. (1995) Negative electron Affinity Effects and Schottky Barrier Height Measurements of Metals on Diamond (100), (110) and (111) Surfaces, *Diamond for Electronic Applications,* **416**, Materials Research Society, Pittsburgh PA, 157-162.

Blakemore, J. S. (1987) *Semiconductor Statistics,* New York, Dover Publications, Inc.

Bourgoin, J. C., Krynicki, J. and Blanchard, B. (1979) Boron Concentration and Impurity-to-Band Activation Energy in Diamond, *Phys. Stat. Sol. (a)* **52** : 293.

Buckley-Golder, I. M. and Collins, A. T. (1992) Active electronic applications for diamond, *Dia. and Rel. Mat.* **1** : 1083-1101.

Chrenko, R. M. (1973) Boron, the Dominant Acceptor in Semiconducting Diamond, *Phys. Rev. B* 7 (10): 4560.

Clausing, R. E., Heatherly, L., Specht, E. D. and Moore, K. L. (1991) Texture Development in Diamond Films Grown by Hot Filament CVD Processes, *New Diamond Science and Technology,* **NDST-2**, Materials Research Society, 575.

Collins, A. T. (1989) Diamond electronic devices-a critical appraisal, *Semicond. Sci. Technol.* **4** : 605-611.

Collins, A. T. (1992) Diamond Electronic Devices-Can They Outperform Silicon or GaAs, *Mater. Sci. Eng.* **B11** : 257.

Collins, A. T. (1993) The Optical and Electronic Properties of Semiconducting Diamond, *Phil. Trans. R. Soc. London A* **342** : 233.

Collins, A. T. and Lightowlers, E. C. (1979). "Electrical Properties," in J. E. Field, *The properties of diamond*, Academic Press, pp. 79-105.

Collins, A. T., Lightowlers, E. C. and Williams, A. W. S. (1970) Formation of Electrical Contacts on Insulating and Semiconducting Diamond, *Diamond Research* : 19.

Collins, A. T. and Williams, A. W. S. (1971) The Nature of the Acceptor Centre in Semiconducting Diamond, *J. Phys. C: Solid St. Phys.* **4** : 1789.

Das, K., Venkatesan, V., Miyata, K., Dreifus, D. L. and Glass, J. T. (1992) A Review of the Electrical Characteristics of Metal Contacts on Diamond, *Thin Solid Films* **212** (1-2): 19-24.

Davidson, J. L. (1993) Electronic Properties and Devices in Diamond, *Third International Symposium on Diamond and Related Materials,* The Electrochemical Society, extended abstract #461, **PV93-17**

Davis, R. F., Kelner, G., Shur, M., Palmour, J. W. and Edmond, J. A. (1991) Thin Film Deposition and Microelectronic and Optoelectronic Device Fabriaction and Characterization in Monocrystalline Alpha and Beta Silicon Carbide, *Proc. of IEEE* **79** (5): 677-701.

Davis, R. F., Sitar, Z., Williams, B. E., Kong, H. S., H.J., K., Palmour, J. W., Edmond, J. A., Glass, J. T. and Carter, J., C.H. (1988) *Mater. Sci. Engr.* **B1** : 77.

Dreifus, D. L. (1995). "Passive Diamond Electronic Devices," in L. S. Pan and D. R. Kania, *Diamond: Electronic Properties and Applications*, Kluwer Academic Publishers, pp. 371-442.

Dreifus, D. L., Fox, B. A. and von Windheim, J. A. (1996) *Semiconductor device with a low-doped region.* **USA:** ,

Dreifus, D. L., Holmes, J. S. and Stoner, B. R. (1995) Diamond Transistors and Circuits, *Applications of Diamond Films and Related Materials: Third International Conference,* **885**, NIST, 71-78.

Dreifus, D. L., Tessmer, A. J., Holmes, J. S., Kao, C.-t., Malta, D. M., Plano, L. S. and Stoner, B. S. (1994a) Diamond Field-Effect Transistors, *Diamond, SiC, and Nitride Wide Bandgap Semiconductors,* **339**, Materials Research Society, 109-120.

Dreifus, D. L., Tessmer, A. J., Holmes, J. S. and Plano, L. S. (1994b) High Temperature Operation of Diamond Field-Effect Transistors, *Second International High Temperature Electronics Conference,* **Vol. 1**, VI-29.

Dyer, H. B., Raal, F. A., du Preez, L. and Loubser, J. H. N. (1965) Optical absorption features associated with paramagnetic nitrogen in diamond, *Philos. Mag.* **11** : 763.

Edwards, L. M. and Davidson, J. L. (1993) Fabrication process development and characterization of polycrystalline diamond film resistors, *Dia. and Rel. Mat.* **2** : 808-811.

Field, J. E. (1979). "Strength and Fracture Properties of Diamond," in J. E. Field, *The Properties of Diamond*, Academic Press, pp. 281.

Fox, B. A., Hartsell, M. L., Malta, D. M., Wynands, H. A., Kao, C.-t., Plano, L. S., Tessmer, G. J., Henard, R. B., Holmes, J. S., Tessmer, A. J. and Dreifus, D. L. (1995) Diamond Devices and Electrical Properties, *Dia. Rel. Mater.* **4** : 622-627.

Geis, M., Gregory, J. and Pate, B. (1991a) Capacitance-Voltage Measurements on Metal-SiO_2-Diamond Structures Fabricated with (100)-and (111)-Oriented Substrates, *IEEE Trans. Elec. Dev.* **38** (8): 619-626.

Geis, M. W. (1991b) Diamond Transistor Performance and Fabrication, *Proc. of IEEE* **79** (5): 669-677.

Geis, M. W., Efremow, N. N. and von Windheim, J. A. (1993) High-Conductance, Low-Leakage Diamond Schottky Diodes, *Appl. Phys. Lett.* **63** : 952.

Geis, M. W., Rothschild, M., Kunz, R. L., Aggarwal, K. F., Wall, K. F., Parker, K. A., McIntosch, N. N., Efemow, J. J., Zayhowski, D. J., Ehrlich, D. J. and Butler, J. E. (1989) Electrical Crystallographic and Optical Properties of ArF Laser Modified Diamond Surfaces, *Appl. Phys. Lett.* **55** (22): 2295.

Ghandi, S. K. (1983) *VLSI Fabrication Principles,* New York, John Wiley & Sons.

Gildenblat, G., Grot, S. and Badzian, A. (1991a) The electrical properties and device applications of homoepitaxial and polycrystalline diamond films, *Proc. of IEEE* **79** (5): 647-668.

Gildenblat, G. S., Grot, S. A. and Badzian, A. (1991b) The Electrical Properties and Device Applications of Homoepitaxial and Polycrystalline Diamond Films, *Proc. of IEEE* **79** (5): 647-668.

Grahn, K. J., Kuivalainen, P. and Eranen, S. (1994) Effect of Partial Ionization and the Characteristics of Lateral Power Diamond MESFETs, *Physica Scripta* **T54** : 151-154.

Grot, S. A. (1995). "Active Diamond Electronic Devices," in L. S. Pan and D. R. Kania, *Diamond: Electronic Properties and Applications*, Kluwer Academic Publishers, pp. 443-461.

Grot, S. A., Gildenblat, G. S. and Badzian, A. R. (1992) Diamond Thin-Film Recessed Gate Field-Effect Transistors Fabricated by Electron Cyclotron Resonance Plasma Etching, *IEEE Elec. Dev. Lett.* **13** (9): 462-464.

Hewett, C. A. and Zeidler, J. R. (1993) Ohmic Contacts to Epitaxial and Natural Diamond, *Dia. Rel. Mat.* **2** : 1319-1321.

Hewett, C. A., Zeisse, C. R., Nguyen, R. and Zeidler, J. R. (1991) Fabrication of an Insulated Gate Diamond FET For High Temperature Applications, *1st International High Temperature Electronics Conference,* pp. 168-173.

Himpsel, F. J., Eastman, D. E. and van der Veen, J. F. (1980a) Summary Abstract: Electonic Surface Properties and Schottky Barriers for Diamond (111), *J. Vac. Sci. Technol.* **17** (5): 1085-1086.

Himpsel, F. J., Heimann, P. and Eastman, D. E. (1980b) Schottky Barriers on Diamond (111), *Solid State Comm.* **36** : 631-633.

Himpsel, F. J., Knapp, J. A., van Vechten, J. A. and Eastman, D. E. (1979) Quantum photoyield of diamond (111) — A stable negative-affinity emitter, *Phys. Rev. B* **20** (2): 624-627.

Hobgood, H. M., Glass, R. C., Augustine, G., Hopkins, R. H., Jenny, J., Skowronski, M., Mitchel, W. C. and Roth, M. (1995) Semi-Insulating 6H-SiC Grown by Physical Vapor Transport, *Appl. Phys. Lett.* **66** : 1364-1366.

Holmes, J. S. and Dreifus, D. L. (1994) Field-Effect Transistors and Circuits Fabricated for Semiconducting Diamond Thin Films, *IEDM,* IEEE, 423-426.

Itoh, M and Kawarada, H. (1995) Fabrication and Characterization of Metal-Semiconductor Field Effect Transistor Utilizing Diamond Surface Conductive Layer, *Jap. J. Appl. Phys. 1* **34** (9A): 4677-4681.

Janssen, G., Schermer, J. J. and Giling, L. J. (1996) Towards large-area diamond substrates: the mosaic process, *Diamond for Electronic Applications,* **416**. The Materials Research Society, 33-44.

Joshi, R. P., Kennedy, M. K. and Schoenbach, K. H. (1992) Studies of high field conduction in diamond for electron beam controlled switching, *J. Appl. Phys.* **72** (10): 4781-4787.

Kajihara, S. A., Antonelli, A. and Bernholc, J. (1990) N-Type doping and Diffusion of Impurities in Diamond, *Diamond, Silicon Carbide and Related Wide Bandgap Semiconductors,* **162**, The Materials Research Society, p315.

Kajihari, S. A., Antonelli, A., Bernholc, J. and Car, R. (1991) *Phys. Rev. Lett.* **66** : 2010.

Kalish, R. (1993) Ion Beam Modification of Diamond, *Dia. Rel. Mat.* **2** : 621.

Kawarada, H., Aoki, M. and Ito, M. (1994) Enhancement-mode metal-semiconductor field-effect transistors using homoepitaxial diamonds, *Appl. Phys. Lett.* **65** (12): 1563-1565.

Kawarada, H., Sasaki, H., Aoki, M. and Tsugawa, K. (1995) Homoepitaxial CVD Diamond Surfaces and Their Applications to Electron Devices,

Kobashi, K., Nishimura, K., Miyata, K., Nakamura, R., Koyama, H., Saito, K. and Dreifus, D. L. (1993) Application of Diamod Films for Electronic Devices, *2nd International Conference on the Applications of Diamond Films and Related Materials,* **ADC '93**, MYU, Japan, 35-42.

Kobayashi, T., Ariki, T., Iwabuchi, M., Maki, T., Shikama, S. and Suzuki, S. (1994) Analytical studies on multiple delta doping in diamond thin films for efficient hole excitation and conductivity enhancement, *J. Appl. Phys.* **76** (3): 1977-1979.

Koidl, P., Wild, C. and Herres, N. (1994) Structure and Morphology of Oriented Diamond Films, *Proc. NIRIM International Symposium on Advanced Materials '94,*

Koizumi, S., Murakami, T., Inuzuka, T. and Suzuki, K. (1990) Epitaxial Growth of Diamond Thin Films on Cubic Boron Nitride {111} Surfaces by DC Plasma Chemical Vapor Deposition, *Appl. Phys. Lett.* **57** : 563.

Landstrass, M. I. and Ravi, K. V. (1989a) Hydrogen passivation of electrically active defects in diamond, *Appl. Phys. Lett.* **55** (14): 1391-1393.

Landstrass, M. I. and Ravi, K. V. (1989b) Resistivity of chemical vapor deposited diamond films, *Appl. Phys Lett.* **55** (10): 975-977.

Lin, S.-H., Sverdrup, L., Garner, K., Korevaar, E., Cason, C. and C., P. (1993) Electron beam activated diamond switch experiments, *SPIE Optically Activated Switching III* **1873** : 97-109.

Lin, S. H. and Sverdrup, L. H. (1995) Electron Beam Activated Diamond Devices, *Applications of Diamond Films and Related Materials: Third International Conference,* **885**, NIST, 79-82.

Malta, D. M., von Windheim, J. A. and Fox, B. A. (1993) Comparison of electronic transport in boron-doped homoepitaxial, polycrystalline, and natural single-crystal diamond, *Appl. Phys. Lett.* **62** (23): 2926.

Malta, D. M., von Windheim, J. A., Wynands, H. A. and Fox, B. A. (1995) Comparison of the Electrical Properties of Simultaneously Deposited Homoepitaxial and Polycrystalline Diamond Films, *J. Appl Phys.* 77 : 1536.

Mearini, G. T., Krainsky, I. L. and Dayton, J. A., Jr. (1995) High Emission CsI Coated Diamond Dynodes for an Electron Multiplier, *Applications of Diamond Films and Related Materials: Third International Conference,* **885**, NIST, 13-16.

Miyata, K. and Dreifus, D. L. (1993) Metal/Intrinsic/Semiconducting Diamond Junction Diodes Fabricated from Polycrystalline Diamond Films, *J. Appl. Phys.* **73** (9): 4448.

Miyata, K., Nishimura, K. and Kobashi, K. (1995) Device Simulation of Submicrometer Gate p+-i-p+ Diamond Transistors, *IEEE Trans. on Elec. Dev.* **42** (11): 2010-2014.

Moazed, K. L. (1992) Metal Semiconductor Interfacial Reactions, *Met. Trans. A-Phys. Met. Mater. Sci.* **23** (7): 1999.

Moazed, K. L., Nguyen, R. and Zeidler, J. R. (1988) Ohmic Contacts to Semiconducting Diamond, *IEEE Elect. Dev. Lett.* **9** : 350.

Nakahata, H., Imai, T. and Fujimori, N. (1991). "Change of Resistance of Diamond Surface by Reaction with Hydrogen and Oxygen," in A. J. Purdes, J. C. Angus, R. F. Davis, B. M. Meyerson, K. E. Spear and M. Yoder, *Second International Symposium on Diamond Materials*, The Electrochemical Society, pp. 487-493.

Nishimura, K., Das, K. and Glass, J. T. (1991) Material and electrical characterization of polycrystalline boron-doped diamond films grown by microwave plasma chemical vapor deposition, *J. Appl. Phys.* **69** (5): 3142.

Nishimura, K., Kumagai, K., Nakamura, R. and Kobashi, K. (1994) Metal/intrinsic semiconductor/semiconductor field effect transistor fabricated from polycrystalline diamond films, *J. Appl. Phys.* **76** (12): 8142-8145.

Plano, M. A., Moyer, M. D. and Moreno, M. M. (1994) CVD Diamond MISFET, *Second International High Temperature Electronics Conference,* VI-23.

Popovici, G., Prelas, M. A., Sung, T., Khasawinah, S., Melinkov, A. A., Varichenko, V. S., Zaitsev, A. M., Denisenko, A. V. and Fahrner, W. R. (1995) Properties of Diffused Diamond Films with n-Type Conductivity, *Dia. Rel. Mat.* **4** : 877-881.

Prins, J. F. (1989) Preparation of Ohmic Contacts to Semiconducting Diamond, *J. Phys. D.* **22** : 1562.

Prins, J. F. (1995) Ion-Implanted n-Type Diamond: Electrical Evidence, *Dia. Rel. Mat.* **4** : 580-585.

Seto, J. Y. W. (1976) Piezoresistive properties of polycrystalline silicon, *J. Appl. Phys.* **47** (11): 4780.

Shenai, K., Scott, R. S. and Baliga, B. J. (1989) Optimum Semiconductors for High Power Electronics, *IEEE Trans. Elec. Dev.* **36** (9): 1811-1823.

Shin, M., Trew, R. J. and Bilbro, G. L. (1994a) High Temperature Operation of Diamond Digital Logic Structures, *2nd High Temperature Electronics Conference,* IV-15.

Shin, M. W., Trew, R. J. and Bilbro, G. L. (1994b) Large Signal RF and dc Performance of Wide Band-gap Semiconductor Electronic Devices for High Temperature Applications, *Second International High Temperature Electronics Conference,* **Vol. 1**, IV-15.

Shiomi, H., Nishibayashi, Y., Toda, N. and Shikata, S. (1995) Pulse-Doped Diamond P-Channel Metal Semiconductor Field-Effect Transistor, *IEEE Elec. Dev. Lett.* **16** (1): 36-38.

Silvaco-International. (1988) *PISCES IIB Manual.* May 6, 1992 .

Stoner, B. R., Malta, D. M., Tessmer, A. J., Holmes, J. S., Dreifus, D. L., Glass, R. C., Sowers, A. and Nemanich, R. J. (1994) Highly Oriented Diamond Films on Si: Growth, Characterization and Devices, *Diamond-Film Semiconductors,* **2151**, Society of Photo-Optical Instrument Engineers, 2.

Stoner, B. R., Sahaida, S. R., Bade, J. P., Southworth, P. and Ellis, P. J. (1993) Highly oriented, textured diamond films on silicon via bias-enhanced nucleation and textured growth, *J. Mater. Res.* **8** (6): 1334.

Sze, S. M. (1981) *Physics of Semiconductor Devices,* New York, John Wiley and Sons.

Tachibana, T. and Glass, J. T. (1992) Effects of Argon Sputtering on the Formation of Aluminum Contacts on Polycrystalline Diamond, *J. Appl. Phys.* **72** (12): 5912.

Tachibana, T. and Glass, J. T. (1995a). "Electrical Contacts to Diamond," in L. S. Pan and D. R. Kania, *Diamond: Electronic Properties and Applications*, Kluwer Academic Publishers, pp. 319.

Tachibana, T., Yokota, Y., Nishimura, K., Miyata, K., Kobashi, K. and Shintani, Y. (1995b) Heteroepitaxial Diamond Growth on Platinum (111) by Shitani Process, *Diamond Films 95,*

Tessmer, A. J. (1995) *Design, Fabrication, and Analysis of Diamond Field-Effect Transistors.*, Ph.D. Thesis, North Carolina State University.

Tessmer, A. J., Das, K. and Dreifus, D. (1992) Polycrystalline diamond field-effect transistors, *Dia. and Rel. Mat.* **1** : 89-92.

Tessmer, A. J., Plano, L. S. and Dreifus, D. L. (1993) High-temperature operation of polycrystalline diamond field-effect transistors, *IEEE Elec. Dev. Lett.* **14** (2): 66-68.

Trew, R. J., Yan, J.-B. and Mock, P. M. (1991) The Potential of Diamond and SiC Electronic Devices for Microwave and Millimeter-Wave Power Applications, *Proc. of IEEE* **79** (5): 598-620.

Tsai, W., Delfino, M., Hodul, D., Riazat, M., Ching, L. Y., Reynolds, G. and Cooper III, C. B. (1991) Diamond MESFET using ultrashallow RTP boron doping, *IEEE Elect. Dev. Lett.* **12** : 157-159.

van der Merwe, J. H. and Bauer, E. (1989) Influence of misfit and bonding on the mode of growth in epitaxy, *Phys, Rev. B* **39** (6): 3632-3641.

Vandersande, J. W. and Zoltan, L. D. (1995) Using High Temperature Electrical Resistivity Measurements to Determine the Quality of Diamond Films, *Dia. Rel. Mat.* **4** : 641-644.

Verwoerd, W. S. (1994) Diamond Epitaxy on a Si(001) Substrate: A Comparison of Structural Models, *Surf. Sci.* **304** : 24.

Visser, E. P., Bauhuis, G. J., Janssen, G., Vollenberg, W., van Enckevort, W. J. P. and Giling, L. J. (1992) Electrical Conduction in Homoepitaxial, Boron Doped Diamond Films, *J. Phys.: Condens. Matter* **4** : 7365.

Werner, M., Schlichting, V. and Obermeier, E. (1992) Thermistor based on doped polycrystalline diamond thin films, *Dia. and Rel. Mat.* **1** : 669-672.

Wild, C., Kohl, R., Herres, N., Müller-Sebert, W. and Koidl, P. (1994) Oriented CVD Diamond Films: Twin Formation, Structure and Morphology, *Dia. Rel. Mater.* **3** : 373.

Williams, A. W. S. (1971) *Electrical Transport Measurements in Natural and Synthetic Semiconducting Diamond.*, Ph.D. Thesis,University of London.

Williams, A. W. S., Lightowlers, E. C. and Collins, A. T. (1970) Impurity conduction in synthetic semiconducting diamond, *J Phys. C: Solid St. Phys.* **3** : 1727.

Wynands, H. A., Hartsell, M. L. and Fox, B. A. (1994) Characterization of MS, MiS, and MOS Gate Contacts fot Type IIb Diamond by Capacitance-Voltage and Current-Voltage, *Diamond, SiC, and Nitride Wide Bandgap Semiconductors,* **339**, The Materials Research Society, pp. 235-240.

Yoder, M. N. (1993). "Diamond Properties and Applications," in R. F. Davis, *Diamond Films and Coatings: Development, Properties, and Applications*, Noyes Publications, pp. 1-30.

Zhu, W., Kochanski, G. P., Jin, S., Seibles, L., Jacobson, D. C., McCormack, M. and White, A. E. (1995) Electron emission from ion-implanted diamond, *Appl. Phys. Lett.* **67** (8): 1157-1159.

Chapter 30
APPLICATIONS OF DIAMOND IN COMPUTERS

Richard C. Eden
Consultant to Norton Diamond Film
Goddard Road, Northboro, MA 01532-1545

Contents

1. Introduction

Diamond offers a unique combination of extremely high thermal conductivity and an electrically insulating nature. This combination makes diamond a very attractive material for use in computer packaging to help cope with the increasingly difficult problems of very high power densities combined with very high interconnect densities (high digital signal connectivity/bandwidth). While in many ways, diamond is an almost ideal material for such computer packaging applications, one very real constraint is cost. Most classes of computing systems have, for more than a decade, been positioned in the marketplace on a very steeply increasing performance/price curve. While achieving the radical increases in performance that have marked this industry has created the power density, etc. problems that diamond is so well positioned to help solve, the extreme cost-consciousness that characterizes most segments of the computer industry precludes using synthetic diamond materials if their cost is too high. This constraint raises two challenges to achieve market penetration for diamond in the computers: 1) to substantially lower the production costs of diamond, and 2) to identify ways of applying diamond in computer packaging which provide maximum leverage on the value that the use of diamond brings to the computer product.

2. Background

The relationship between something as seemingly tenuous or intangible as computational power and a simple concrete physical quantity like heat may not be obvious. In fact, computational throughput ("power") is, as a matter of physics, unavoidably tied to the generation of heat. Moreover, in the most efficient forms of logic circuitry, the level of generation of heat is essentially proportional to the computational throughput, albeit the exact value of the constant of proportionality is dependent on details of the integrated circuit (IC) technology used to implement the computer.

This proportionality relationship between computational throughput and the amount of dc electrical power converted to heat was not always obvious in computer circuit technologies. Early digital computing circuits (logic gates) tended to dissipate substantial levels of power, whether they were in fact engaged in computational processes (logic switching activities), or they were just standing idle waiting for signal input changes to arrive. Since a reasonable measure of computational activity is the number of such logic gates that have switched, the measure of computational throughput (rate) is simply the number of such logic gates switching per unit time, and with computational activities broken into clock periods of duration $t_c = 1/F_c$, where F_c is the clock frequency, the computing power may be expressed as proportional to the product of the number of logic gates, N_g, and F_c as:

$$\text{Computing Rate} \propto f_d \, N_g \, F_c \tag{1}$$

where f_d simply represents the average fraction of logic gates which switch per clock cycle (typically $f_d \approx 25\%$ to 30% [Bakoglu 1990]). With the earlier forms of logic circuits (from vacuum tube circuits through IC logic families like RTL, TTL, PMOS and NMOS) in which the static power dissipation was dominant, the dc power or heat dissipation was proportional to N_g, but not to F_c, so the relationship between computing power and dc power or heat was not clear. However, with the advent of CMOS IC technologies, the static power dissipation in the logic circuits may be made extremely small. While the typical CMOS logic gate dissipation is negligible with no signal changes applied ($F_c=0$), there is dynamic power dissipation to be considered. The functioning of digital electronics involves the charging and discharging of many logic circuit nodes (presumably with each driven by the output of some logic gate). Because of the capacitance, C_g, associated with each node, for operation at a voltage, V, at a minimum, an energy equal to the energy stored in this capacitance ($^1/_2 C_g V^2$) must be dissipated in each logic operation. Hence, if the total number of logic gates in the chip, board or module is N_g, the dynamic power dissipation, P_d, may be related to the clock frequency, F_c, and the average fraction of gates which switch in each clock cycle, f_d, as

$$P_d = {}^1/_2 \, f_d \, N_g \, C_g \, V_{dd}^2 \, F_c \tag{2}$$

where C_g is the average logic node capacitance and V_{dd} is the logic voltage. (More precisely, the V_{dd}^2 here is the product of the logic voltage swing, ΔV_L, times the power supply voltage, V_{dd}, but for current CMOS logic, $\Delta V_L \approx V_{dd}$, so this term is essentially equal to V_{dd}^2.) While in general, P_d is only a lower bound on the total logic power dissipation, in CMOS it is a reasonably good approximation to the total power. The implication of Eq. 2 is that (assuming effective gate utilization in the design), since the processing rate of a digital system is tied to the product of the number of gates, N_g, times the clock frequency, F_c, in any given IC logic technology (e.g., fixing C_g and ΔV_L^2), the processing rate is essentially proportional to the dynamic power dissipation, P_d (or to the total power dissipation for CMOS).

Noting that both Eqs. 1 and 2 contain the same $f_d\, N_g\, F_c$ term, we can rewrite Eq. 1 as

$$\text{Computing Rate} \propto f_d\, N_g\, F_c = P_d \,/\, [{}^1\!/_2\, C_g\, V_{dd}^2] \propto P_d \tag{3}$$

which clearly illustrates the direct proportional relationship between computational throughput ("computing power") and the dynamic power dissipation in the logic circuitry. The constant of proportionality between "computing power" and electrical power dissipation or heat generation involves logic circuit technology factors (from the $[C_g V_{dd}^2]^{-1}$term in Eq. 3) as well as factors related to the architecture of the computer. However, within any given generation of CMOS IC technology and similar family of computer architectures or functions (such as workstations), this relationship between computing power and heat dissipation applies.

The implications to be drawn from Eq. 3 for the application of diamond in computers are profound. In the first place, thermal problems are seen as an unavoidable consequence of the pursuit of higher and higher computational performance levels. Packaging materials and thermal management approaches that sufficed for earlier generations of computers can prove completely unsuitable for future generations. For example, note that the first CMOS microprocessors had less than 10^4 transistors and operated at F_c=1MHz clock rates. Current high-performance CMOS microprocessors have 10^6 to 10^7 transistors and operate at clock rates as high as several hundred MHz, which means (from Eq. 2) that their power dissipations could be expected to be well over 10,000 times higher than the earliest CMOS μPs. Indeed, CMOS microprocessor chip dissipation levels have increased from a few milliwatts in the first CMOS μPs up to nearly 50 watts in some of the fastest workstation μPs today, with dissipations up to 200 watts projected for future CMOS chips.

A further implication from Eq. 3 is that the computational performance may be degraded if the thermal performance of the packaging is inadequate. One form of this is seen directly in Eq. 2 because in CMOS, P_d is proportional to the clock rate, F_c. IC and computer packaging is usually characterized by a quantity called the junction-to-ambient thermal resistance, θ_{ja}, defined by

$$\Delta T = T_j - T_a = \theta_{ja}\, P_d \tag{4}$$

where the temperature rise, ΔT, is the difference between the IC junction temperature, T_j, and the ambient temperature, T_a, and as before, P_d is the IC chip power dissipation. There have been a number of recent examples in industry where it has not been possible to operate microprocessor chips at their (logic gate switching speed-limited) maximum clock rates (F_{cmax}) in certain computers, due to θ_{ja} values in the packaging which would give T_j values higher than the maximum allowable junction temperature, T_{jmax}, at the chip power, $P_d=P_{dmax}$, obtained (from Eq. 2) at $F_c=F_{cmax}$. Reducing F_c lowers the chip power and allows use with the given θ_{ja}, but at a corresponding sacrifice in performance. Even in cases where θ_{ja} is low enough to keep $T_j<T_{jmax}$ at $P_d=P_{dmax}$, a slightly subtler thermal mechanism still can degrade performance. Due to temperature dependencies of the electron and hole mobilities and velocities in the MOSFET channels, the switching speeds of the logic gates in the CMOS circuits degrade substantially as the IC junction temperature is increased. Hence, F_{cmax} is itself temperature dependent, even at temperatures well below the reliability-limited maximum IC junction temperature, T_{jmax}. This means that in practice, the use of a much lower thermal resistance package will, at any given power dissipation, lower the junction temperature of the ICs, increasing the switching speed of the logic, and hence allow higher operating clock frequencies and computational throughputs to be realized.

3. Use of Damond for Computer Packaging: Why and How?

The foregoing discussion clearly illustrates the benefits of, and critical need for, lowering the thermal resistance of IC packaging for computers (as well as other applications, one should add). The obvious first step toward reducing thermal resistance is to increase the thermal conductivity of the materials through which the heat must flow. In many ways diamond offers virtually an ideal material for electronic packaging applications, including both single-chip packages and multi-chip module (MCM) substrates. Of prime interest, of course, is its extremely high thermal conductivity, about k=2000 W/m°K in natural diamond, with values up to 70% to 80% of that measured in CVD synthetic diamond [Lu et al. 1993]. Table I compares the thermal conductivity of natural diamond with those of various other electronic packaging materials. Also of interest for electronic packaging applications, and included in Table I, are the coefficient of thermal expansion (CTE) values, as well as note of the electrically conducting or insulating nature of the materials. In many electronic packaging applications, a high density of insulated signal interconnections are required. With an insulating packaging material it is relatively easy to implement these isolated lines, even if a high density of interconnects must pass through the substrate itself (as required, for example, in packages with area array contacts such as ball-grid array [BGAs], as well as for 3-D packaging).

As we can see from Table 1, the metals silver and copper are the only materials to come within a factor of 5 of the thermal conductivity of diamond, but not only are they not insulators, they have CTE's much higher than that of silicon. Metals such as Kovar® and molybdenum have reasonably low CTE's, but their thermal conductivities (17 and 146 W/m°K respectively) are very low in comparison to diamond (composites such as copper-tungsten are somewhat better in that regard). The CTE of polycrystalline CVD

Table 1. Electronic Packaging Materials Comparison

Material	Thermal Conductivity W/m K	Electrical Insulator? Y/N	Thermal Expansion ppm/K
Natural Diamond	2000	Y	0.8-1
CVD Diamond	700-1700	Y	1-1.5
Beryllia, BeO	220	Y	6.4
AlN	70-230	Y	3.3
Alumina,99%	29	Y	6.3
GaAs	45	Semi	5.9
Silicon	149	N	2.6
Kovar® (FeNiCo)	17	N	5.9
Molybdenum	146	N	5.1
Aluminum	237	N	23.8
Copper	396	N	16.8
Silver	428	N	19.6
Copper-Tungsten	300	N	6
Diamond-Epoxy	8.7	Y	120
Silver-Epoxy	5.8	N	120
BCB	0.6	Y	52
Polyimide	0.2	Y	>50

diamond has been measured at about 1.5 ppm/°K [Lu et al. 1993], only about 1 ppm different from silicon (a much better match than the 3.8 ppm/°K difference between Si and the 96% alumina packages usually used for IC's). Silicon itself can be used as an inexpensive MCM substrate material (with ideal CTE match, of course), but it is not an insulator and its thermal conductivity (149 W/m°K) is only a little better than that of molybdenum, and no match to that of diamond. Most insulators tend to have poor thermal conductivities; only beryllia (very toxic and generally avoided now) and aluminum nitride come within an order of magnitude of diamond. Hence, we see that diamond, provided it can be manufactured in volume and sold at an acceptable price, is truly favored by nature as an ideal electronic packaging material.

The unique combination in diamond of extremely high thermal conductivity with an electrically insulating nature and a coefficient of thermal expansion fairly close to that of silicon provides the potential for dramatic reductions in package thermal resistance. Understanding that brings us to the question of just how to best utilize diamond in electronic packaging. One can hardly claim to have identified the "best" way to utilize a new material until the real-world competitive voice of the marketplace has been heard. Since, for the most part, diamond is at too early a stage of development and market penetration (particularly in computer-type packaging), it is unrealistic to expect to have much solid feedback from the marketplace. However, some very interesting and attractive concepts for the use of diamond for computer and other types of electronic packaging have been proposed, which have generated strong interest from potential users. These diamond application concepts range from the use of small amounts of diamond to enhance the thermal properties of conventional single-chip IC packages, to the use of diamond as a substrate material for high thermal performance multi-chip modules (MCMs), and on to the utilization of diamond as an enabling material to realize incredibly high density 3-D MCM electronic packaging. It is worthwhile to examine in some detail some of the packaging concepts which have been proposed, or are being explored, in each of these three packaging application areas.

4. Diamond Enhancements of Conventional IC Packages

In current electronic packaging practice, most IC chips are packaged into some type of single-chip package which protects the chip and allows it to be attached to the printed wiring board (PWB) that provides the electrical signal interconnections between chips with an simple, inexpensive soldering process. These so-called "first-level" packages for the IC chips frequently represent a substantial thermal resistance (θ_{jc}) in the heat flow path from junction to ambient. A common expression for the total junction-to-ambient thermal resistance, θ_{ja}, is given by

$$\theta_{ja} = \theta_{jc} + \theta_{ca} \tag{5}$$

where θ_{jc} is this "junction to case" package thermal resistance, and θ_{ca} is the case to ambient thermal resistance. Of course, θ_{ca} is strongly influenced by factors such as the use of heatsinks on the package, air velocity, etc., but regardless what a user does outside the package, θ_{ja} cannot be less than θ_{jc}. For this reason, there has always been strong interest in reducing the θ_{jc} values of IC (and other semiconductor) packages, and more recently, the use of diamond has been suggested for accomplishing this [Boudreaux et al. 1991],[Nagy and Partha],and [Fehr 1995]. We will look at just one of these in some detail.

As mentioned in the abstract above, in general the IC chip and computer businesses are extremely cost-sensitive. Choices in chip packaging are most frequently cost driven, with the package of choice being the cheapest package that will meet the electrical, thermal and environmental requirements of the application. For a fairly high percentage of ICs sold for commercial applications, including computers, plastic packaging is used because of its very low cost (≈1¢ per pin or less) and acceptable reliability, at least in reasonably well controlled environments. A serious disadvantage with the standard plastic packages is their high thermal resistance values. Figure 1 compares cross-sectional drawings and simplified thermal models of a standard 208-pin plastic quad flat pack (PQFP) package, shown at the bottom of Figure 1, with that of the "DiamaQuad" thermally enhanced PQFP package configuration under development at Norton Diamond Film [Nagy and Partha] and [Fehr 1995], shown at the top of Figure 1.

The most serious problem with the standard PQFP package is the gap of very low thermal conductivity (k=0.52W/m°K even in "enhanced" thermal conductivity filled molding plastic compounds) between the metal "die paddle" to which the IC chip is attached and the leadframe to which the 208 IC die pads are wirebonded. Most printed circuit boards (PWBs) to which the ends of the leads of packages of this type are soldered have many layers of copper signal interconnect and power planes in them, and hence can serve as good heat sinks if the heat can get from the IC chip into the board.

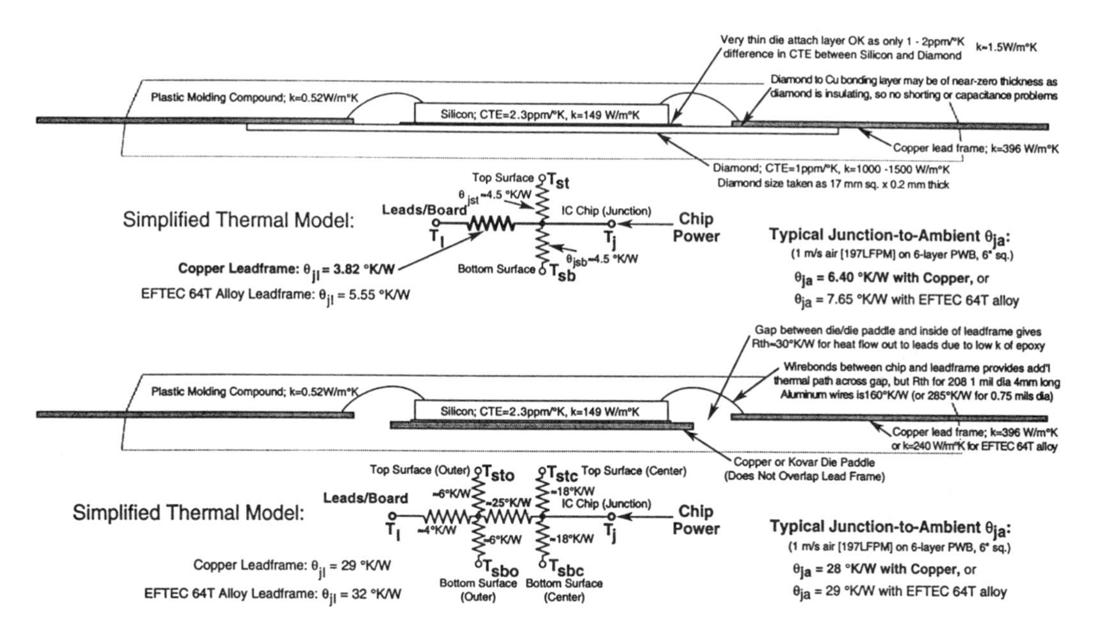

Figure 1. Comparison of standard 208 pin plastic quad flat pack IC package (bottom) with "DiamaQuad" thermally enhanced PQFP (top), in which an overlapping diamond die paddle is used to transfer heat from the die out through the lead frame into the board.

For example, for a typical 6" PWB (2 power wiring planes of "1 oz" Cu plus 2 signal wiring layers with 50% coverage of Cu), at an air velocity of 1m/s, the thermal resistance to ambient, θ_{la}, seen at the lead solder ring (due to distributed board conductance/heat transfer to ambient) is only about θ_{la}=3.6°K/W. Unfortunately, as shown in the thermal model at the bottom of Figure 1, in the standard PQFP the thermal resistance from the IC junction to the end of the leads (the point along the leads where they contact the soldered interface to the PWB), θ_{jl}, is high, about 29 °K/W even if the leadframe is pure copper. For reasons of mechanical formability and durability, the leadframes are usually made of copper alloys such as EFTEC 64T, which has only 60% of the thermal conductivity of pure copper, which raises the value of θ_{jl} to 32 °K/W for this standard PQFP package. When the additional heat transfer coefficient of about 30.5 W/m^2°K directly from the package top is considered, this gives a total θ_{ja}=28 °K/W with a copper leadframe, or θ_{ja}=29 °K/W with the 64T alloy.

The severe thermal constraints of the standard plastic PQFP package can be alleviated very nicely, without the addition of any heatsink devices or altering the package outline, by the addition of an overlapping diamond die paddle, as shown for the Norton "DiamaQuad" package at the top of Figure 1. Because of the extremely high thermal conductivity of diamond, only a thin (150μm to 200μm typical) diamond piece is required, which allows its use even in reduced-profile PQFP packages (1.4mm thick, as opposed to 3.2mm in the standard 208-pin PQFP shown in Figure 1). Because the diamond is electrically insulating, the leadframe can be directly attached to the diamond without danger of lead shorting or problems with thermal resistance of thick polymer insulating layers, as seen with metal overlapping die paddle approaches to thermally enhanced PQFPs (and without parasitic capacitance problems). For the thermal model of the DiamaQuad 208-pin PQFP at the top of Figure 1, a 17mm square piece of diamond, 200μm thick, with k=1100 W/m°K was assumed, which reduced the die-to-lead end thermal resistance to θ_{jl}=3.82 °K/W with a pure copper leadframe (7.6x better than the standard QPFP). Because the leadframe is thin in high pincount packages (6 mils typical for 208 pins), the thermal resistance of the leadframe itself (beyond the diamond die paddle) is substantial, particularly when the less thermally conductive alloys are used. For this example, θ_{jl}=5.55 °K/W is obtained with the EFTEC 64T alloy. With smaller lead count packages, thicker leadframes are used, which proportionately reduces the thermal resistance. Even with the thin leadframe metal, increasing the size of the diamond die paddle can be used to reduce θ_{jl} (e.g., increasing the diamond size to 24mm x 24mm gives θ_{jl}=2.29 W/°K with Cu, or θ_{jl}=3.01 W/°K with EFTEC 64T alloy), but this is at the expense of using twice as much diamond. Note from the thermal model in Figure 1, that even with the modest 17mm x 17mm diamond size, for the same 1 m/s air velocity and PCB characteristics as considered for the standard PQFP case above, with the DiamaQuad package, the total die-to-ambient thermal resistance is θ_{ja}=6.40 °K/W with a copper leadframe, or θ_{ja}=7.65 °K/W with the 64T alloy, respectively x4.4 and x3.8 better than the standard PQFP. With an increased diamond size to 24mm x 24mm, θ_{ja}=5.23 °K/W with Cu, or θ_{ja}=5.79 °K/W with the 64T alloy, respectively x5.35 and 5.01x better than a standard 208 PQFP. What this means is, for example, if ΔT=75°K (Eq. 4) is acceptable in an application, then the maximum power dissipation of a standard PQFP would be P_{dmax}=2.68W, whereas with a DiamaQuad 208 PQFP it would be from P_{dmax}=9.8W to 14.3W (depending on diamond size and leadframe alloy). This represents a major improvement in the thermal capability of the package. On the other hand, it is important to note that due to the extremely cost-sensitive ("next year we need twice the performance at the same cost") nature of the IC business, for the DiamaQuad package to achieve major penetration of the plastic packaging market, the cost of the diamond die paddle could not be much more than a few dollars. While this is a challenging cost goal, it is one that is expected to be realizable with diamond manufacturing technology soon.

5. Diamond Substrates for Conventional 2-D MCMs

In multi-chip module (MCM) packaging, bare IC die are crowded as closely together as possible (in order to improve performance and reduce cost and size) on a very high wiring density board. (This board is frequently termed an MCM substrate, but to avoid confusion, here we will retain use of the term substrate for the material upon which the high-density interconnects are deposited to form the MCM bare board, with the completed assembly with chips referred to as the MCM module). The MCM approach for achieving very high chip packaging densities is of particular interest for high performance, high clock rate digital systems, where signal latency due to excessive inter-chip interconnect line lengths can prove a major limitation. The problem that this brings, however, is that as IC chips are operated at higher frequencies, their power dissipations tend to increase proportionately (Eq. 2). As these chips are jammed together very closely in MCM packaging, the power density levels get very high indeed. These severe thermal management problems, difficult to cope with using conventional packaging materials, represent a serious barrier to the application of MCM technology in high performance systems. The combination of high thermal conductivity with an electrically insulating nature (which allows the easy fabrication of isolated signal interconnects), makes diamond of great interest as an ideal MCM substrate material (see [Eden 1991], [Eden 1993a], [Eden 1993b], [Bodreaux 1995], [Napolitano et al. 1993] and [Peterson et al. 1994]) , particularly in area I/O array MCM module configurations such as BGAs.

The majority of MCMs and other electronic packages currently fabricated are of the conventional 2-dimensional configuration, in which all of the IC die lie in a single plane (generally packed as closely together as their substantial chip areas and the inter-chip interconnect technology will allow). The selection of substrate material for conventional 2-dimensional packages or MCMs is generally balanced between consideration of thermal expansion coefficient (CTE) match (for which silicon is ideal), manufacturability, and cost (for which alumina is often used), and thermal performance (which often necessitates backing with copper-tungsten or employing other heat sink methods). While forced-air or other heat exchange approaches are often adequate to cope with the average power density on 2-D MCMs, a frequent problem is that of "hot spots", or single chips operating at much higher than average power levels, which, due to inadequate lateral thermal conductivity in the substrate material, operate at die temperatures which may degrade the IC reliability substantially. This case may be quantitatively illustrated by considering a circular geometry (chosen for ease in analytical solution) with a uniform power dissipation over a chip of radius R_i centered in a substrate of radius R_o with a heat sink at temperature T_o at its outer edge, assuming heat transfer only through lateral thermal conduction in the substrate. Solving the heat flow differential equation (Poisson's equation) for this cylindrical geometry is straightforward, giving the maximum temperature (at the center of the chip), T_{max}, versus total chip power, P, as [Eden 1991], [Eden 1993a] and [Eden 1993]:

$$T_{max}-T_o = (P/2\pi kz)\,[\ln(R_o/R_i) + 1/2] \qquad \text{Heat Spreading} \qquad (6)$$

where k is the substrate thermal conductivity and z its thickness, and $T_{max}-T_o$ is the maximum temperature rise above the heat sink temperature. Note that the assumption here is that the power, P, is uniformly distributed over the chip area, there are no other (e.g., convection or radiation) heat transfer mechanisms operative, and that the lateral thermal conductivity of the die is negligible in comparison to that of the substrate. For the case when the chip thickness, z_c, times chip conductivity, k_c, product is not $\ll kz$, then the 1/2 term in Eq. (6) is multiplied by $P/2\pi[kz+k_c z_c]$ instead of $P/2\pi kz$).

As a specific example, consider the case of a $2R_i$=1cm diameter chip dissipating P=25 watts at the center of a $2R_o$=10cm (4") diameter substrate, z=1mm thick, edge cooled at T_o=300°K (26.84°C). If the substrate were silicon (k=149 W/m°K), the temperature at the center of the die, T_{max}, would be 75°K above T_o, or Tj=101.84°C. Operation of the IC at this higher junction temperature may be possible, but would substantially degrade the reliability [Eden 1991]. Assuming an activation energy of 1.1 eV for the IC failure mechanism, this 75°K increase in operating temperature, in comparison to operation at the T_o=300°K heat sink temperature, would lead to a very serious x5000 degradation in failure rate (reduction in MTBF) [Eden 1991]. On the other hand, using k=2000 W/m°K

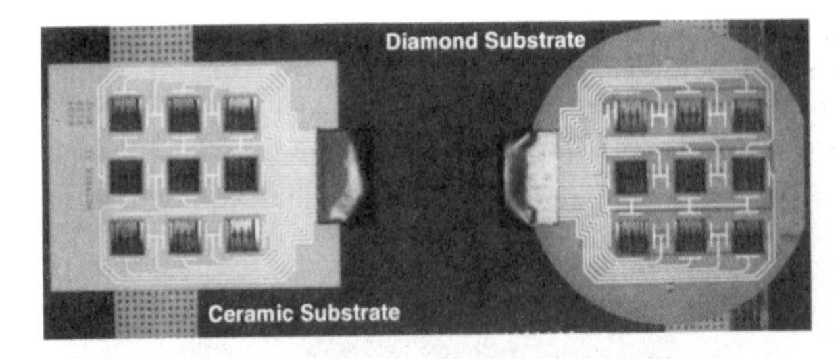

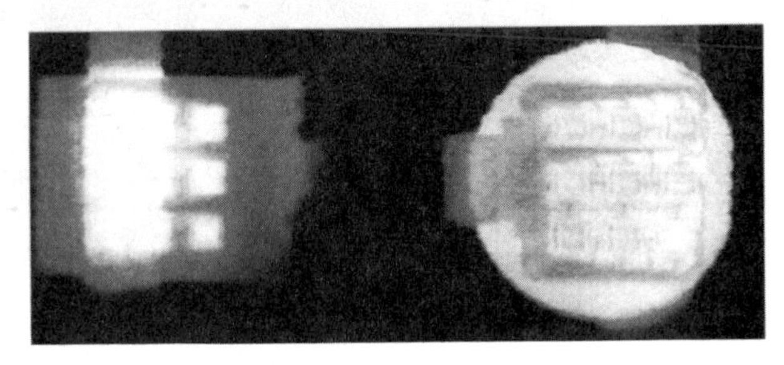

Figure 2. Thick film gold MCMs (Ref. [9]) on ceramic (rectangular, on left) and on 1 mm thick 4" diameter diamond (right) substrates, powered to 10 watts in 24°C still air. The diamond MCM is isothermal at 54°C, while the ceramic MCM has "hot spot" chip temperatures up to 87°C.

diamond would give only a $\Delta T_{max}=T_{max}-T_0=5.575°K$ temperature rise and a negligible x2.2 degradation in MTBF. The use of more normal substrate materials such as 96% to 99% alumina (k=18 to 29 W/m°K without metallization) would give catastrophic ΔT_{max} = 384°K to 619°K temperature rises. Such materials are usable only where the heat flow through the substrate is essentially normal to the substrate plane, or removed directly from the die by other means. While it is possible to devise special heat sink or active cooling approaches capable of overcoming the lateral thermal conductivity limitations of conventional substrate materials for specific applications, these generally force undesirable compromises elsewhere in the system design, degrading such factors as size, weight, cost or performance.

Figure 2, taken from [Boudreaux 1995], dramatically illustrates the capability of a diamond MCM substrate to act as a heat spreader to minimize "hot spot" temperature rises at IC die. Identical MCMs, consisting of a 3x3 array of thermal test die, were fabricated on alumina ceramic and diamond substrates. In the test shown, both MCMs were suspended by attaching them to strips of "Perfboard" (unmetallized FR-4, seen clearly in the visible light image at the top of Figure 2) and were powered to 10 watts , in 24°C still air (natural convection cooling). As seen in the IR image at the bottom of Figure 2, in the case of the alumina ceramic substrate, the very high temperature of the silicon die causes them to image as bright white, while most of the area of the ceramic remains relatively cool (meaning that it is not contributing significantly to cooling the chips). Note particularly the high temperature of the chips and substrate in the region where the back side of the ceramic substrate is covered by the Perfboard, preventing the air from reaching and cooling the back surface of the substrate. Contrast this with the case of the diamond substrate, where the entire MCM, chips and substrate, is virtually isothermal. Even the region where the back side is covered by the Perfboard appears only slightly warmer than the rest, and that could be due in part to the higher IR emissivity of the Perfboard seen through the diamond substrate. In fact, the largest contrast in the diamond substrate IR image is the dark pattern extending from the connector pattern at the left around the outside of the chips, an artifact of the low emissivity of the gold metallized lines on the substrate. In fact, with a 24°C ambient (still) air temperature, the diamond MCM was virtually isothermal at 54°C (ΔT_{ja}=30°K, determined by the heat transfer coefficient to air, not the diamond conductivity), while for the alumina ceramic case, die temperatures up to 87°C were measured (ΔT_{ja}=63°K). This example illustrates the extraordinary ability of diamond to act as a thermal spreader, minimizing the temperature rise at IC die.

It should probably be noted that Eq. 6 accounts for only the thermal resistance due to lateral heat flow through the substrate. There is additional thermal resistance due to vertical heat flow through the IC die and the bond medium between the die and the substrate. Of these, the former is usually extremely small, since IC die are only about 25 mils (0.6mm) thick and the thermal conductivity of silicon is reasonably high. As noted in Table I, the silver or diamond-loaded thermally conductive epoxy die attach materials have fairly low thermal conductivities, but are typically less than 3 mils (0.075 mm) thick, which keeps the temperature drop low. For example, at a 25 W/cm^2 chip power density, the temperature drop across the IC chip thickness is only 1.0°K, and the temperature drop across a 3 mil silver epoxy die attach layer is 3.3 °K (or 2.2 °K if diamond-filled epoxy die attach is used).

While assistance with reducing this "hot spot" problem to improve the reliability of electronic systems would alone be a major reason for interest in diamond substrates for conventional 2-D electronic packages or MCMs, there are other potential advantages as well. For example, diamond has, as noted in Table I, a very low coefficient of thermal expansion. This means that as a packaged circuit or MCM is powered up, there is very little change in dimension or alignment to high density electrical or optical interconnects, etc. (which in some cases may remain at near-ambient temperatures).

Another important advantage offered by diamond is the possibility of getting away from the undesirable and area-inefficient cavity-down package configurations. Because typical packaging materials have such poor thermal conductivity, when the IC chip power dissipation levels are fairly high, the heat sink to the package must generally be directly on the opposite side of the cavity bottom or MCM substrate from the chip mounting surface. Since usually the heat sink sits above the circuit board surface, this means that the bottom of the package or MCM substrate must be facing up (and hence the cavity facing down), away from the circuit board. The interconnect leads, however, must obviously pass from the IC (or ICs in an MCM) to the circuit board surface. In this cavity-down configuration, this necessitates considerably extending the package size in order to make room for these interconnect leads and package contacts beyond the extent of the die cavity or MCM substrate region. This not only increases board area, but degrades electrical performance due to the increased length and inductance of the leads. The extremely high thermal conductivity of diamond could make it possible, instead, to let the heat flow out to the edge of the package on its way to the heatsink while letting the electrical interconnects take the shortest and fastest possible path. This concept is

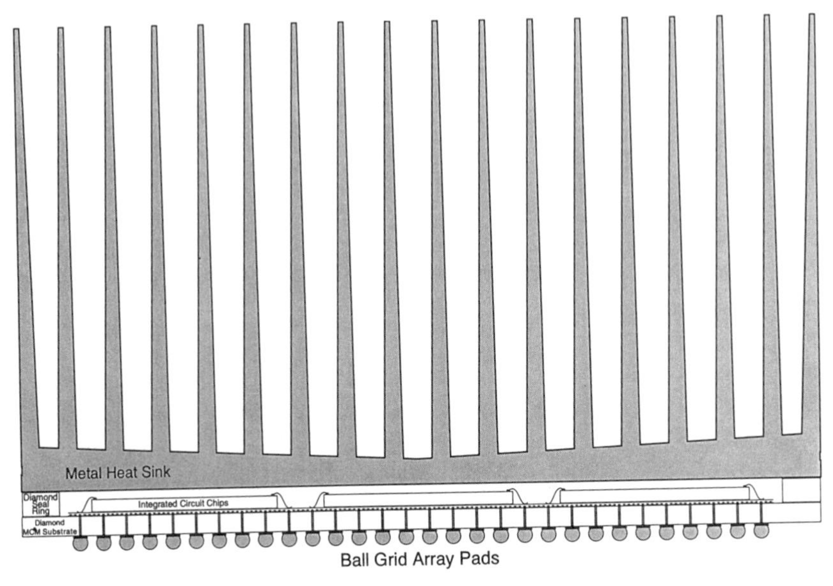

Figure 3. Illustration of very small-footprint cavity-up BGA package (single-chip or MCM) using a diamond substrate and seal ring for lateral conduction of the heat from the IC chip(s) to the topside-mounted heat sink.

illustrated for the case of a diamond BGA MCM package in Figure 3. Note that, even using normal face-up wire bonding of the IC chips, the total circuit board area required by the diamond BGA MCM is not much larger than the area of the bare IC die themselves, and lead lengths are minimal. On the other hand, the thermal path from the chip surface down into the MCM substrate, then laterally through the substrate and up through the "seal ring" at the edge of the substrate into the heat sink on top is quite long. Hence, the use of this electrically and board-area efficient cavity-down package or MCM configuration is dependent, as discussed in the next section, on the ability of diamond to maintain, by virtue of its extremely high thermal conductivity, an adequately low thermal resistance in spite of the thermally unfavorable geometry.

6. Edge- Cooled Diamond MCM Substrates

Consider, for simplicity, the case of uniform power density, P/A, applied to a package bottom or MCM substrate of thermal conductivity, k, and thickness, z. This will closely approximate the case of a large area, uniform dissipation VLSI die or an MCM essentially "tiled" with a large number of equal power dissipation chips. For the thermally most favorable case of an area heat sink on the opposite side of the substrate from the die, the temperature drop across the substrate will be given by (ignoring the thermal conductivity of the IC die and die attach layer)

$$\Delta T_{max} = T_{max} - T_o = (z/k)\,(P/A) \qquad \text{[through-substrate]} \qquad (7)$$

For a z=1.0 mm diamond substrate with k=2000 W/m°K, even with a very high power density of $P/A=50W/cm^2$, the temperature drop would be extremely small, $\Delta T=0.25°K$. Clearly, diamond is not required for this favorable ("cavity down") package geometry.

As shown in Figure 3 and discussed in the previous section, however, it would often be very desirable from an electrical standpoint to remove the heat only from one or more edges of the IC package or MCM substrate. Here the combined power dissipation from all of the die must pass laterally a long distance through the small thickness of the substrate (again, assuming only conductive heat transfer). If only one edge of the rectangular substrate were maintained at the heat sink temperature, T_o, and the length of the substrate from the cooled edge to the opposite edge is L, then the maximum temperature, T_{max}, at this far (uncooled) edge may be found for L»z by solving the one-dimensional heat flow problem to give

$$\Delta T_{max} = T_{max} - T_o = 0.500[(P/A)\,L^2]/(kz) \qquad \text{[cooled at one edge]} \qquad (8)$$

If the two opposing edges (separated by length, L) of the uniformly heated substrate are heat sunk at T_o, then the temperature distribution will also be parabolic, with a maximum temperature, T_{max}, along the center line between the heatsink edges given by

$$\Delta T_{max} = T_{max} - T_o = 0.125[(P/A)\,L^2]/(kz) \quad \text{[cooled at two opposing edges]} \qquad (9)$$

For the specific case of a square substrate (where $L^2=A$), the size of the substrate is removed from the equation if the equation is written in terms of the total power dissipation of all chips on the substrate, P. While all of the chips on such an edge-cooled MCM do not have the same temperature, so the definition used in conjunction with Eq. 5 of the "junction to case" package thermal resistance, θ_{jc}, does not precisely apply here, we can define a closely-related the "peak" thermal resistance, R_t, for the MCM as

$$R_t = \Delta T_{max}/P = (T_{max} - T_o)/P \qquad \text{[Peak Thermal Resistance]} \qquad (10)$$

From Eqs. 8 and 10, the peak thermal resistance of a uniformly heated square substrate of thickness z and thermal conductivity k is given for the case of only one edge cooled as

$$R_{t1} = 0.500 / (kz) \qquad \text{[square substrate cooled at one edge]} \qquad (11)$$

or for the case of two opposing edges cooled (from Eqs 9 and 10) as

$$R_{t2o} = 0.125 / (kz) \qquad \text{[square cooled at two opposing edges]} \qquad (12)$$

If the two adjacent sides have the heatsinks (a less favorable case than for opposing sides), or for the cases of three sides or four sides cooled, the heat flow is not 1-dimensional, but can be numerically evaluated with a 2-D solver for Poisson's equation. The results for these five edge-cooled geometry cases are summarized in the ΔT_{max} column of Figure 4.

Board Edge Heatsink Configuration	Maximum Temperature Rise ΔT_{max} =	Average Temperature Rise ΔT_{avg} =	Specific Example of Board Power, P, for $\Delta T_{max}=T_m-T_o=25°K$
1 – Edge	$0.5000\left(\frac{P}{kz}\right)$	$0.333333\left(\frac{P}{kz}\right)$	100 W
2 - Adjacent Edges	$0.29469\left(\frac{P}{kz}\right)$	$0.140899\left(\frac{P}{kz}\right)$	169.7 W
2 - Opposing Edges	$0.1250\left(\frac{P}{kz}\right)$	$0.083333\left(\frac{P}{kz}\right)$	400 W
3 – Edges	$0.113853\left(\frac{P}{kz}\right)$	$0.0572564\left(\frac{P}{kz}\right)$	439.2 W
4 – Edges	 $0.073672\left(\frac{P}{kz}\right)$	$0.0352249\left(\frac{P}{kz}\right)$	678.7 W

Figure 4. Comparison of thermal performance of laterally conductively cooled boards with heat sinks at various edges.

While the above peak thermal resistance, R_t, expressions give the peak temperature, ΔT_{max}, on the uniformly heated square substrate (divided by the total board power), also of interest is the average temperature of the board, ΔT_{avg}, which is given for these five cases in the next column of Figure 4. Note, for example, that the average board temperature, ΔT_{avg}, when all four sides have heatsinks is 58% lower than for opposing 2-sided cooling, while the peak temperature, ΔT_{max}, is only 41% lower. Because the device failure rate increases exponentially with temperature, however, the peak board temperature, ΔT_{max}, will generally be of greatest interest. The right hand column in Figure 4 compares, with a specific numerical example, these five edge cooling approaches. Tabulated here is the total board power, P (Watts), that a given board edge cooling configuration can handle and still keep the maximum temperature rise, ΔT_{max}, less than or equal to 25°K, for the specific case of a uniformly heated square board with diamond thickness of 1 mm and thermal conductivity of k=2000 W/m°K. Note that even for this very modest 25°K temperature rise, the two (opposing) edge cooled board will handle up to 400 W, as opposed to 100 W for one edge cooled, or 678.7 W for four edges cooled. The diamond BGA MCM package of Figure 3 would, of course, be an example of a four edges cooled case.

It should be noted that, while one-sided and two-opposing- sides heat sinking schemes have, with uniform chip power dissipations, essentially one-dimensional temperature distributions, and hence no thermally-generated stress in the board, the other three configurations do give thermal stresses in the boards. However, since diamond has such a low CTE (Table I), this should not cause any significant problems (such as warpage or out-of-plane distortion) unless the diamond substrate is very thin. As will be discussed later, in any edge cooling approach, the power density at the board edges is high, which means that care must be taken to insure maintenance of a low thermal resistance path all of the way to the heat sink or heat exchanger medium (the reason for the use of the diamond seal ring in Figure 3).

Use of single-side edge cooling of conventional 2-D MCMs has been demonstrated in [Peterson et al. 1994]. In this work, three identical (except for substrate material) MCMs, consisting of a 4x4 array of thermal test die, were fabricated on alumina, aluminum nitride and diamond substrates. While they did not publish the results for all four rows of chips powered (corresponding to the uniform heating case), even for the case of only the 3 rows furthest from the heat sink edge powered, Eq. 8 gives a good approximation for the measured performance (R_{t1}=0.38 °K/W calculated for uniform heating vs. 0.44 °K/W measured; slightly higher, as expected, because the heat was concentrated on the 3/4 of the chip furthest from the cooled edge). The direct comparisons between the diamond, AlN and Al_2O_3 results were dramatic. When just the last row of die (furthermost from the heat sink edge) were powered to 5 watts each (20 W total), the maximum temperature rise above heat sink for the diamond MCM was only ΔT=24°K, while for aluminum nitride it was over 5x higher, ΔT=126 °K, and for the alumina substrate MCM it was ΔT=265 °K (it would have been much higher than that, except for heat loss from wires attached to the other three edges). This is a striking illustration of the remarkable effectiveness of diamond for edge-cooled MCM substrates.

7. Attractions of 3-D Packaging of Digital Electronics

7.1 SPEED

The increasing speeds of digital integrated circuits can only be realized in increased system speeds if the signal interconnect delays can be correspondingly reduced. Such a reduction of signal propagation delays is achieved principally through the reduction in signal line lengths, such as through dense packaging of bare die in multi-chip modules (MCMs). However, due to the relatively large areas of IC die (typically 1 cm or more on a side) the interconnect distances between die in a 2-dimensional (2-D) array can be very substantial when the number of die is large. In 3-dimensional MCM packaging, use is made of the fact that IC chips are thin (typically 0.6mm or less), permitting very high volume density packaging. Figure 5 dramatically illustrates how, whereas a given die has only 8 nearest neighbors in 2-D, in 3-D (assuming a 1.5 cm X-Y pitch and a 2.5 mm Z pitch) 116 die may be reached with essentially the same signal interconnect length. For example, if the interconnect delay is 150 ps/inch, and the VLSI chips have 250,000 gates, then the 3-D geometry of Figure 5 places nearly 30 million gates within 650 ps worst case delay time. If these 117 chips had to be spread out horizontally, the worst-case delay between chips would exceed 1.7ns, which would represent a significant performance compromise for system clock rates above the 50 MHz to 100 MHz range, and virtually preclude operation at rates above 250 MHz to 350 MHz if full corner-to-corner propagation is required within the cycle. With 3-D packaging, this maximum clock rate is extended to nearly 1 GHz, a major improvement.

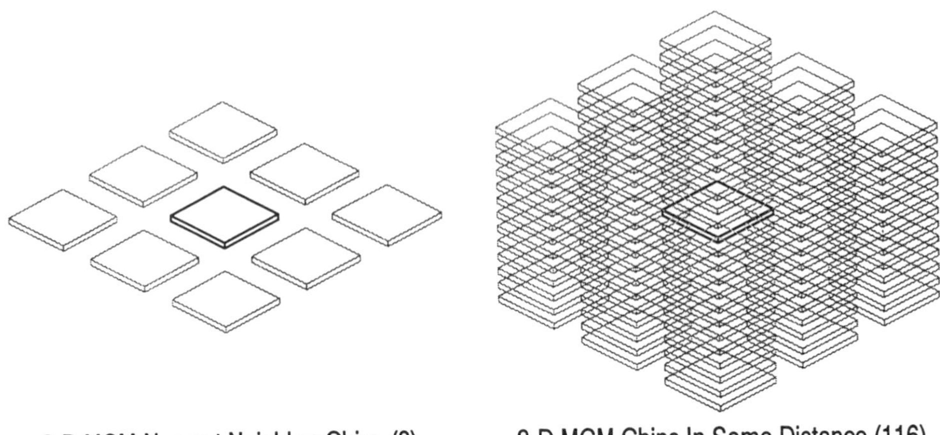

Figure 5. 2-D versus 3-D MCM architectures; comparison of the number of chips within 2-D nearest neighbor distance for 1.5 cm X and Y pitches, and 2.5 mm Z pitch in 3-D.

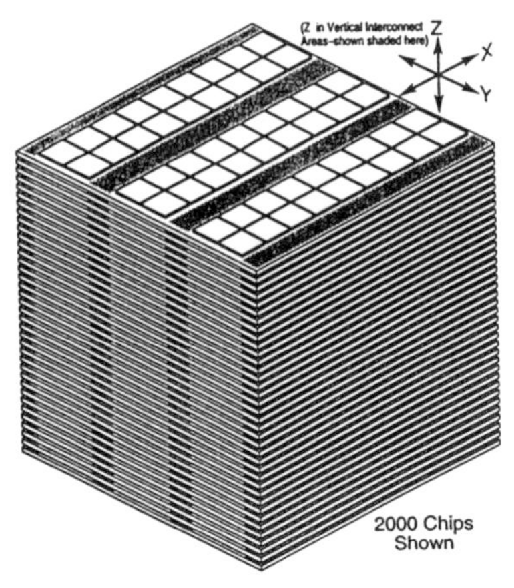

Figure 6. Area array Z-interconnected 3-D MCM configuration and its features. Typically, two opposing faces are used for heat removal and the other two faces are used for power/ground inputs and external I/O connections to each board. Volume packing density is high because of tight vertical board pitch.

7.2 SIZE, WEIGHT AND VOLUME

3-D electronic packaging offers the potential to achieve major (5x to 20x or more) reductions in the size, weight and volume of military electronic systems, while at the same time offering the potential for improved performance by virtue of the reduced latency of inter-chip interconnects noted above, as well as through increased connectivity/interconnect bandwidth. As illustrated in Figure 6, in the context used here, 3-D electronic packaging is comprised of a number (typically ≈ 4 to 50 or more) of very closely spaced 2-D MCMs which have high density through-substrate vias and interconnect pad array areas distributed over both the top and bottom surfaces of the MCMs. In addition to the usual X-Y connectivity of 2-D MCMs (generally using only edge contacts to the outside world), a high level of Z connectivity is provided between adjacent boards in the 3-D stack by means of some type of vertical interconnect technology allowing reliable, demountable contact between the area interconnect pads on the tops and bottoms of adjacent boards. The goal is to achieve high densities of high-quality controlled-impedance transmission line signal interconnects, even through long chains of inter-board Z-interconnects, so that the connectivity within the stack is truly three dimensional.

As illustrated in Figure 6, 3-D packaging offers remarkable increases in density over the conventional stacked 2-D packaging, illustrated in Figure 7 (in which stacks of 2-D boards are interconnected through some type of backplane to which the boards attach through edge connectors). The reason that the density (as measured by how many square inches of IC chip area can be supported per cubic inch of module volume) of 3-D packaging is so high is that the spacing between boards (vertical stacking pitch) in a 3-D stack can be extremely small (ultimately limited only by the thickness of the IC chips themselves plus the thickness of the 2-D (X-Y) wiring mat interconnecting them). In contrast, in stacked 2-D packaging (Fig. 7), the board stacking pitch is limited by the height of the backplane connector. This is a fundamental limitation of stacked 2-D in that there is an inherent tradeoff between board stacking pitch and connectivity (signal I/O bandwidth) to the board, because there will be a limited density of interconnections into the vertical backplane possible. In 3-D, this same interconnection density limitation exists, but it does not force any increase in the board stacking pitch; it merely uses up some (typically small) fraction of the area of the boards in the stack.

In most military systems, as well as in many non-military (portable, industrial, etc.) applications, there is a critical need to reduce the size, weight and volume of digital electronics. This basic problem of needing to reduce the size and weight and volume of digital subsystems and modules is compounded by the desire, on the system side, for fairly dramatically increasing the digital processing performance in order to meet increasing mission or application requirements. The key metric associated with packaging density, is how many square inches of silicon ICs we are able to package per cubic inch of module volume, or correspondingly, how many square inches of silicon chips we can package per pound of module weight. Since this "square inches per cubic inch" volume circuit density cannot exceed the reciprocal of the board stacking pitch, board spacing is critical. Because both IC chips and wiring mats are quite thin, board spacings under 1 mm are certainly possible, and 2 mm is quite straightforward for 3-D packaging. In a 3-D MCM module (Fig. 6), if high connectivity requirements demand $10cm^2$ of I/O interconnects be provided for, this would represent a sacrifice of 10% of the area of 10cm x 10cm boards in the 3-D stack (a 10% reduction in volume circuit density for the 3-D module). On the other hand, for stacked 2-D (Fig. 7), this $10cm^2$ area would have to be provided for on a 10cm wide backplane (assuming the same 10cm x 10cm board area) which would require a board stacking pitch of not less than 1cm. If the 3-D stacking pitch were 1mm, then the connectivity demand that gave a 10% reduction of volume circuit density for 3-D, in fact, gives a 10x reduction in density for stacked 2-D. The point here is that a given connector area in stacked 2-D must be bought with precious vertical stacking pitch, while in 3-D packaging the stacking pitch is unchanged and the connector area is paid for with "cheap" (percentage-wise) board area. Consequently, with the proper development of the Z-interconnect and other

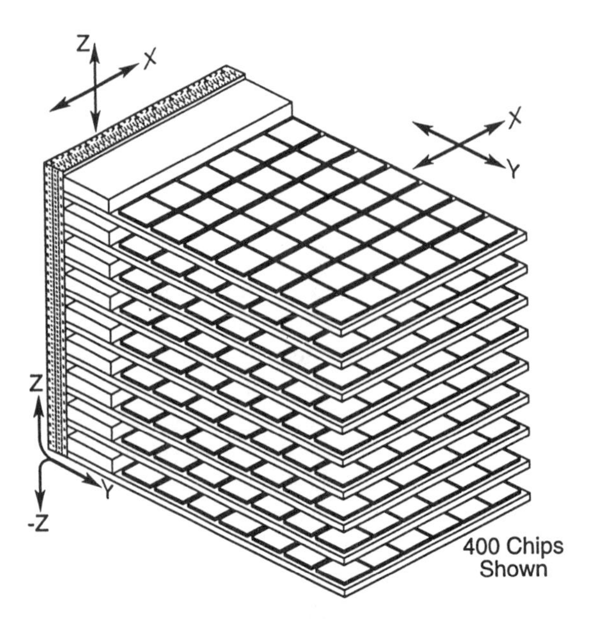

Figure 7. Conventional "stacked 2-D" packaging configuration. Volume density is limited by the requirement for connection area into the backplane, which limits vertical stacking pitch when high connectivity is required.

supporting technologies needed to realize low-cost high-density 3-D packaging, reductions in weight and volume of digital (and potentially analog) subsystems of the order of 5x to 20x can be practically realized. The implications of this, particularly for military systems, are enormous.

7.3 CONNECTIVITY / DENSITY

For many applications, both military and commercial, the density of electronics packaging is of primary concern. It is important to keep in focus, however, that the packaging density must always be judged in the context of other constraints, such as power dissipation and connectivity requirements, of the application. The implications of diamond for achieving high densities in the face of high power levels (the motivation behind applications like that described in [Napolitano et al. 1993] in which a power module was reduced from 2.25"x3" down to a 0.5"x0.5" diamond MCM, a 25x area reduction), will be further discussed later. The relationship between digital system performance and power dissipation was treated extensively earlier, in the "Background" section. There is also a corresponding relationship (called Amdahl's law) between the connectivity or internal bandwidth within a computer and its performance potential. (Bandwidth of a data path in a digital system is simply given as the product of the number of signal lines times the clock rate of the digital data, so, for a give clock rate, bandwidth is determined by connectivity.) This connectivity, upon which performance is so heavily dependent, may lie between the CPU and cache or other memory resources (or other functional elements) in single processor systems, or between processor nodes in multi-processor systems. Just how connectivity maps into performance is an issue of detailed computer architecture; that high connectivity creates the potential for high performance is a universal principal.

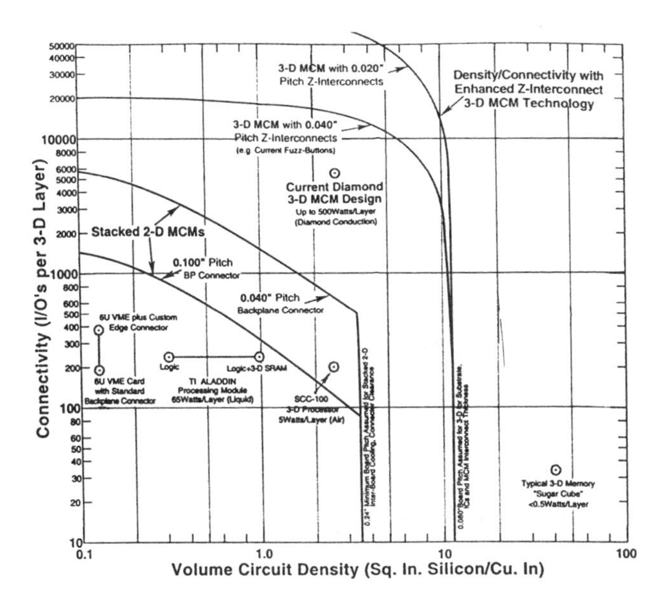

Figure 8. Comparison of calculated volume circuit density (horizontal scale) versus connectivity (number of I/O connections per layer) for area-array interconnected 3-D MCMs vs. conventional "stacked 2-D" configurations.

Understanding that connectivity is an extremely important issue in electronic packaging, it is worth examining in a bit more quantitative detail the comparisons between 3-D packaging and conventional stacked 2-D packaging, in terms of connectivity dependence of volume circuit density. Figure 8 compares the calculated volume circuit density, (number of square inches of silicon per cubic inch of volume), versus connectivity, (the number of I/Os per 2- or 3-D layer) of 3-D with conventional stacked 2-D packaging. In stacked 2-D packaging, the board stacking pitch is limited by the height of the backplane connector. This is a fundamental limitation of stacked 2-D, in that there is an inherent tradeoff between board stacking pitch and connectivity (signal I/O bandwidth) to the board, because there will be some limited density of interconnections into the vertical backplane possible. The two "Stacked 2-D MCMs" curves in Figure 8 differ in the assumed density of these backplane connector contacts, which shifts the curves vertically, but both share the same steep slope, which illustrates the sacrifice of density (horizontal scale) that connectivity increases necessitate. (The abrupt boundary at the right of these curves derives from the assumption of a practical connector-limited minimum board spacing of 0.24".) The two "Enhanced 3-D MCM Technology" curves in Figure 8 are plotted assuming a constant 2 mm board stacking pitch with two different densities of the area array Z-interconnects (the lower curve is based on the 0.040" contact pitch of current "fuzz button" vertical interconnects, while the higher curve assumes a 0.020" pitch (considered reasonable goals for enhanced 3-D technology). In contrast to the "Stacked 2-D MCMs" curves in Figure 8, the "Enhanced 3-D MCM Technology" curves show little loss of density with reasonable levels of increasing connectivity (that is, the right hand side of the curves is very steep up to about 12,000 pads per 4"x4" board for the 0.020" pitch case, or about 3000 for the 0.040" case). While in 3-D, as in stacked 2-D, an interconnection density limitation exists, the key difference is that in 3-D it does not force any increase in the board

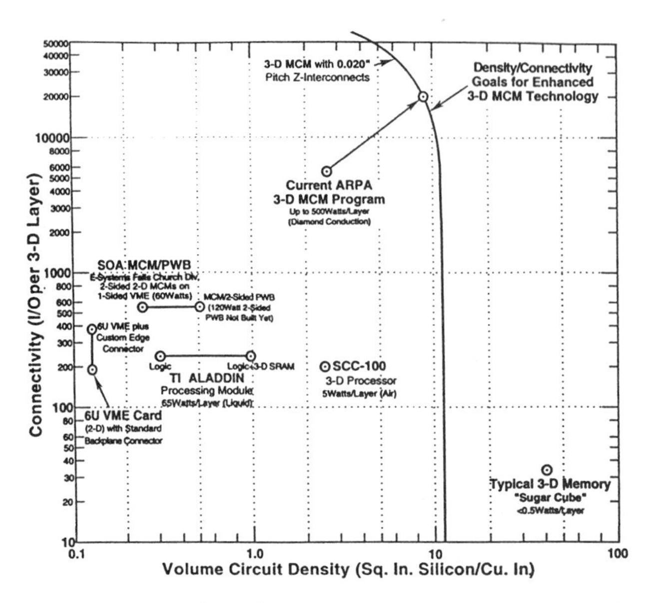

Figure 9. Comparison of present and enhanced 3-D MCM volume circuit density vs. connectivity with several examples of current state of the art and developmental packaging approaches.

stacking pitch; it merely uses up some (typically small) fraction of the area of the boards in the stack. Since the "square inches of silicon per cubic inch" volume circuit density is limited by the reciprocal of the board stacking pitch, the board spacing is the most critical factor for volume circuit density. The main point that Figure 8 illustrates is that while a given connector area in stacked 2-D must be bought with precious vertical stacking pitch, in 3-D packaging the stacking pitch is unchanged and the connector area is paid for with "cheap" (percentage-wise) board area. For example, for the 0.020" Z-interconnect pad pitch case, the maximum number of contacts to a 4"x4" board (top + bottom) would be 80,000 (off-scale in Figure 8, or 20,000 for 0.040" as seen at the left of the lower 3-D curve). Hence, supporting even 12,000 I/O pad contacts per layer would require only 15% of the board area, leaving the remaining 85% available for IC chips, which corresponds to a volume density of $\approx$10 in/in^2. As seen in Figure 8, both this 3-D level of connectivity, and this level of volume circuit density, are far beyond those achievable with conventional stacked 2-D packaging.

It is instructive to compare the calculated density vs. connectivity projections of Figure 8 with real-world packaging results. Figure 9 compares various packaging technologies on the basis of their volume circuit density, (number of square inches of silicon per cubic inch of volume), and their connectivity, (the number of I/Os per 2-D or 3-D layer). As a reference point, a 6U VME card has a 196 pin connector into the backplane, with the potential for adding a special connector on the opposite side of the card to perhaps double that connectivity. A typical density for the VME card is of the order of 0.1 in^2 of silicon per in^3 of volume. In comparison TI's ALADDIN military processor modules have somewhat similar connectivity, about 234 I/O pins per layer, but substantially higher chip densities, up to about 1.0 in^2 of silicon per in^3 (if you include the 3-D SRAM "sugar cubes" used in the design). In comparison, the Space Computer Corporation SCC-100 3-D processor modules have over 2x higher densities, about 2.4

in^2 of silicon per in^3, due to the very close spacing between adjacent boards, but the connectivity is again quite low, only about 210 pins per layer. On the other hand a typical 3-D memory "sugar cube" module with un-thinned chips offers substantially higher density, about 40 in^2 of silicon per in^3, but again with very low connectivity (typically less than 30 or 40 signal I/Os per layer).

In comparison the connectivity available in the 3-D approach of Figure 6 is much higher. As described in the next section, a program to demonstrate such a diamond 3-D "cube computer" has been funded by ARPA, headed by E-Systems Garland Division, with Norton Diamond Film supplying the via-drilled substrates and with Cray Research contributing both system and fabrication technology [Schaefer et al. 1993], [Moravec et al. 1993] and [Sienski et al. 1996]. As seen in Figure 9, in this current ARPA 3-D diamond effort [Schaefer et al. 1993], the connectivity exceeds 5000 pads per 3-D layer with a circuit density of 2.5 in^2 of silicon IC chips per in^3 of volume and a power handling capability of about 500 watts per board. This initial effort uses a face-up wire bonded chip configuration like that shown in Figure 10. Here, the through-substrate vias (connecting the peripheral pads on the IC chips to the MCM X-Y interconnects on the bottom side of the diamond substrate) are located around the periphery of the chips (instead of underneath flip-chip mounted die as assumed in Figure 6), so the area utilization is less efficient. As seen in Figure 8, if this higher density flip-chip 3-D packaging configuration is used, with an enhanced Z-interconnect technology having a pitch of 20 mils (instead of the existing 40 mil fuzz button pitch), the result would be to extend the connectivities up to an order of magnitude higher than that of the initial ARPA demonstration. In addition, the volume circuit density should be increased from about 2.5 in^2/in^3 in the current diamond 3-D MCM effort, up to about 10 in^2 of silicon per in^3 of module volume with the enhanced 3-D packaging technology.

The key point to note in Figure 9 is that with this 3-D packaging approach, we are talking about levels of both connectivity and circuit density that are far, far higher than existing standard packaging technologies, or even that have been explored in some of the special purpose military/space packaging technology development activities. The other issue noted in the chart is that, with diamond, we are talking about the capability to support much higher power levels than have been offered by the previous technologies (up to 1 to 2 orders of magnitude higher power dissipation per layer).

8. Approaches for Implementing Diamond 3-D MCM Packaging

The attractions of 3-D packaging are so great from a system size and performance standpoint that one is inclined to wonder why everybody hasn't been using it. There are two key problems involved in implementing 3-D packaging of the type illustrated in Figure 6, both of which diamond, with its remarkable combination of extremely high thermal conductivity and electrically insulating nature, and can help solve: 1) achieving the vertical interconnects between boards with high density in a practical, demountable, reworkable fashion, while at the same time, 2) getting the heat out of the "cube" (the completed 3-D interconnected system) due to the substantial power dissipation from all of the high speed IC's operating in such a small volume.

The search for practical 3-D interconnect and heat removal approaches for very high speed systems packaging is made more difficult by the fact that the two needs (dense vertical interconnects and easy heat removal) tend to be contradictory. Methods of implementing, conveniently and practically, high density arrays of vertical interconnects between adjacent boards exist, but suffer the disadvantage that they block the inter-board spaces through which cooling fluid (air usually, or possibly a liquid such as Fluorinert, as used in the CRAY2 and CRAY3 supercomputers [The 3-D packaging approach for the CRAY3 was first described in [Kiefer and Heightley 1987], and in more detail in a talk by Seymour Cray at the August 1989 Supercomputer Conference in Orlando) would ordinarily be passed. In other words, the system ends up looking much like a sealed cube, with forced-convective heat extraction approaches very difficult. As discussed in Refs. [Eden 1991], [Eden 1993a], [Eden 1993b] and [Schaefer et al. 1993], [Moravec et al. 1993], and [Sienski et al. 1996], this difficulty of extracting heat convectively from such a 3-D interconnected MCM package ("cube computer") can be solved by using an electrically insulating, super high thermal conductivity substrate material like diamond to simply conduct the heat out laterally through the substrate to heat sinks at opposing faces (as was done at much lower power densities in the CRAY1 using copper-loaded printed circuit boards). As seen from Figure 4 or Eqs. 9 or 12, with square diamond boards, 1mm thick, even a 500 watts/board design would have only a T_m-T_o =31.25°C temperature rise above the opposing-edge heat sink temperature, T_o.

For a complete 4 inch "cube computer", as illustrated in Fig. 6, consisting (assuming a conservative vertical board "stacking" pitch of 2.5 mm) of 40 diamond substrates, the cube would exhibit a thermal resistance of one fortieth of that of one board (again assuming essentially equal dissipation on all chips and boards), or, R_{th}=1.56°K/kW at the full module level. Such a 40 board module could easily handle more than 20 kW (20 kW would give a T_m-T_o=31.25°C maximum temperature rise). It is worth reflecting for a moment on just what the use of diamond accomplishes in realizing a 3-D interconnected "cube" computer. What the diamond does is to laterally conduct the heat generated by the logic out of the "battle zone" of the actively populated cube itself. This frees the designer to fully utilize all of the space between boards to implement a high area density array of vertical interconnects between all adjacent boards in the stack to minimize interconnect delays. Because there is no need to co-utilize this inter-board space to meet cooling requirements, the designer can make use of the most reliable, simple, cost effective, demountable, etc., vertical inter-connect methods available. From a performance standpoint, of course, it offers no significant speed advantage over the liquid Fluorinert cooling approach (through tens of thousands of bare wire Z-interconnects) of the CRAY3. It simply allows this same 3-D interconnect packaging advantage to be realized in a simple, straightforward way at relatively low cost for application in a wide range of high performance system applications which cannot support anything approaching high-end supercomputer price tags.

While it is not possible here to explore in great depth the details of how such a diamond-substrate 3-D MCM "cube computer" module is to be fabricated, Figure 10 shows a cross-sectional view of one approach (very similar to that being employed in the current ARPA E-Systems/Norton/Cray 3-D diamond MCM effort [Schaefer et al. 1993], [Moravec et al. 1993] and [Sienski et al. 1996]) to realizing such an assembly. In this approach, 1 mm thick synthetic diamond substrates have VLSI chips bonded to their

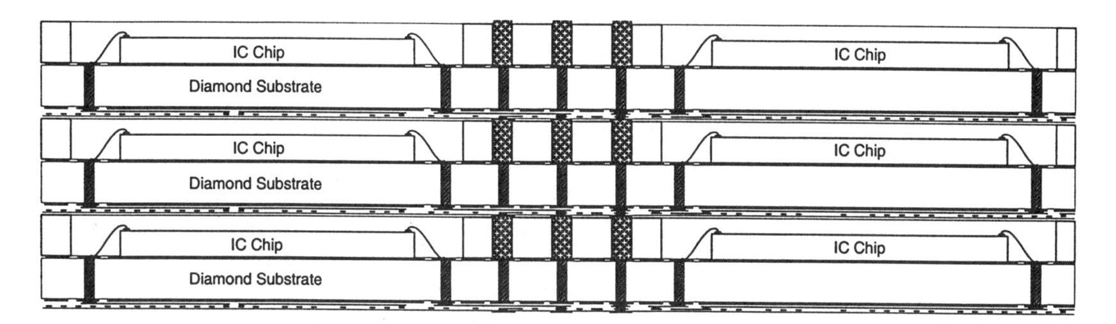

Above: Assembled (Stacked) With Mating Force Applied
(Removing Mating Force Immediately Demounts Into Separate Boards Below)

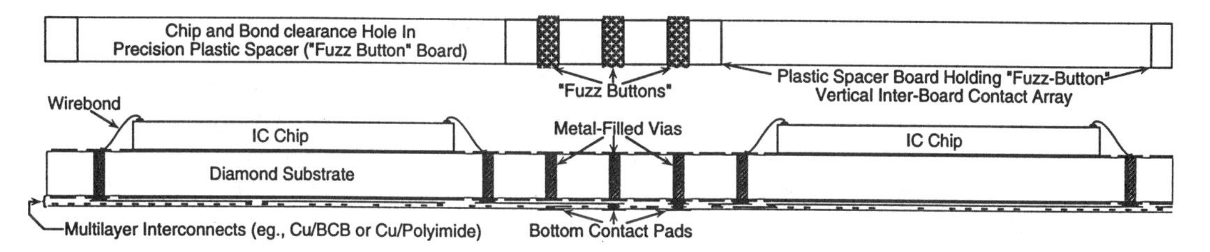

Figure 10. Cross-section view of example of approach for implementation of area array interconnected diamond 3-D MCM "cube computer" using wirebonded chips and "fuzz button" vertical interconnects. Heat is extracted conductively laterally through the diamond substrates to heat exchangers on two opposing faces of the "cube". (Higher density versions use flip-chip mounting with vias under IC die.)

upper surfaces and a high-density multi-layer Cu/BCB or Cu/Polyimide X-Y interconnect mat on the bottom side (there are also a smaller number of interconnect layers between the chips on the top side). Passing through the diamond substrate are a large number (typically ≈10,000) 4 mil diameter laser-drilled, metallized via holes for passing signals from the chips to the 50Ω controlled-impedance X-Y interconnects in the bottom, and also for handling the vertical (Z) interconnections passing through the board. These through-substrate vias are used both for completing longer vertical interconnect paths and for making connection between the edge pads on the IC chips and the X-Y MCM interconnects on the bottom of the MCM. In the current ARPA program, the minimum pitch for these 4 mil via holes through the diamond is 20 mils, and none of the holes is placed directly under the chips, allowing for the use of conductive die attach materials (e.g., silver-filled epoxy). Since there are 400 pads per die, significant area can be saved by placing these vias under the IC die (using insulating diamond-filled epoxy die attach material, of course). As noted previously, the highest area density could be achieved by flip-chip mounting area array interconnect die to matching arrays of vias through the diamond substrate. Norton Diamond Film has successfully demonstrated laser drilling arrays of 4 mil via holes on the 10 mil pitch (i.e., 10,000 vias/in^2) that would be required for mounting high pincount chips in this fashion, so this very high density configuration appears fully feasible.

In the simplest form (illustrated in Figure 10), the diamond substrate is metallized on top and bottom with fairly thick power plane metal layers in order to be able to handle the high power supply currents to the boards. Edge power contacts to the individual boards (as opposed to power contacts to the top or bottom of the stack only) are necessary because the power supply currents for the total module can be very high. For example, a 40-board 4" cube module dissipating 20 kW would have, assuming a 5V power supply voltage, a 4000 ampere supply current (clearly too high to bring in through the small "fuzz button" vertical interconnects). By bringing in the power to each board edge separately, this is reduced to 100 amperes per board (or 50 amperes per contact if two redundant power contacts are used). As long as reasonably thick (1 or 2 mils of copper or gold) power plane metallizations are used on the top and bottom diamond surfaces, neither electromigration nor IR voltage drops should cause significant problems in the power conductors.

The last metal layer on both sides of the diamond substrates in Figure 10 defines the contact pads to mate with the several thousand "fuzz button" compliant interconnects which make the signal contacts between boards. The "fuzz button" vertical contacts are tiny cylindrical wads of gold-plated silver-copper or phosphor-bronze wire. In their current standard commercial size (20 mils diameter), the buttons are spaced on ~40 mils centers, making room for about 5000 to 10,000 buttons on a 4" spacer board, even with the clearance holes for the chips accounted for. (The current ARPA demo uses ~2750 "fuzz buttons" for a 3"x3" board size.) The "fuzz buttons", which cn be formed and inserted by automated machinery, extend ~4 mils beyond the surfaces of the spacer board, which extension is compressed essentially flush (~20% compression of the button length) by the ~50 grams/button mating force applied to the entire 3-D module through pressure plates on top and bottom of the cube. The "fuzz buttons" (20 mils diameter on a 40 mil pitch in the current technology used in the ARPA diamond 3-D MCM demonstration) are retained in the spacer boards between the diamond MCM substrates. These spacer boards also have clearance holes cut in them for each IC site on the board below, and are about 1mm thick, adequate to clear the IC chips and their wirebonds. (In the current ARPA demo, the spacer boards are thicker, as flex-connect power distribution planes and signal I/Os accessed with this structure. The thicker spacer board also allows sufficient vertical clearance to place power supply bypass capacitors in the cavity above the chips.) The demountable mating of the whole 3-D module assembly is accomplished by applying mechanical force through pressure plates on the top and bottom of the stack.

9. Signal Propagation Issues

From a signal integrity standpoint, there was initially some concern expressed by people who believed that the parasitic inductances of many "fuzz button" Z-interconnect contacts in series would make it impossible to pass signals vertically through many boards (e.g., a "straight-through" vertical run as shown for the right-most of the three fuzz buttons at the top of Figure 10) without hopelessly degrading signal quality. We have shown [Schaefer et al. 1993], [Sienski et al. 1996], [Eden 1994] that, by using an alternate ground-signal-ground 2-D array arrangement of fuzz buttons, with careful design it is possible to make the sum of the distributed capacitance of the array conductors (fuzz buttons and through-diamond vias) plus the "feedthrough" lumped capacitances where these vertical conductors pass through clearance holes in the horizontal MCM interconnect "ground" planes to be of such a magnitude as to balance the distributed inductance of the conductor array to give a controlled Z_0=50Ω average impedance. As is described in [Schaefer et al. 1993], even 40-board (bottom to top in Fig. 6) "straight through" vertical runs give excellent simulated pulse propagation characteristics, with low loss and very fast signal rise times (~5Gb/s typical bandwidth). As long as both the vertical, Z interconnects and the horizontal, X-Y MCM interconnects have the same characteristic impedance, mixed vertical-horizontal runs (which would be the norm in a 3-D MCM module), should pose no signal integrity problem for this "fuzz button" vertical interconnect approach to implementing the diamond 3-D "cube computer" MCM concept.

Note that because of the ground-signal-ground pattern required to provide the controlled return current paths in the Z-interconnect arrays (necessary to insure controlled vertical transmission line impedance and signal quality), there are only half as many vertical signal paths available as there are "fuzz button" contacts. This is the same ratio as it would be if differential signal transmission (an attractive alternative to conventional single-ended transmission) were used.

10. Edge Cooling Approaches for Diamond 3-D MCMs

Inherent in the edge cooling approach is a very high power density in the substrate at the heat sink interface edges. For example, a 4 inch square board (100mm x 100mm), 1 mm thick (as described above) would have an edge power density of 50 times the average (P/A) of the chips on the board, or 250W/cm^2 for P=500 W. Obviously, even a thin layer of a relatively poor thermal conductivity material such as silver epoxy would lead to substantial temperature rise at this heat sink interface (e.g., 32.84°C for 3 mils [0.0762 mm] of silver epoxy [Table 1] at P/A=250 W/cm^2). One approach to keeping this heat sink interface temperature rise very low is to make an extension of the diamond substrate area part of the heat exchanger itself. For example, as illustrated in Figure 11, if the 100 mm x 100 mm active MCM substrate area were extended by 5 mm or so on the heat sink ends (total 110 mm x 100 mm diamond area), and if these extended ends were perforated with linear arrays of 2.5 mm long by 0.1 mm wide microchannel cooling slots laser cut through the 1.0 mm thick diamond on a 0.25 mm pitch, then the total area of the 400 slots on each end would be about 50 cm^2. This would reduce the average interface power density from 250 W/cm^2 to a very reasonable 10 W/cm^2.

Because of the thin boundary layer when cooling liquids are forced through these narrow slots, the heat transfer coefficients tend to be quite high at reasonable liquid velocities. Even with thermally "poor" liquids like Fluorinert®, the interface temperature can be kept quite low, albeit with fairly high pumping power requirements, and with water it can be made virtually negligible at these power levels (though, of course, allowing water in such close proximity to sensitive electronics would generally be considered an unacceptable hazard).

In the current ARPA 3-D diamond MCM program, the complexity, potential for leaking (because of the substantial liquid pressures) and inefficiency of the single-phase

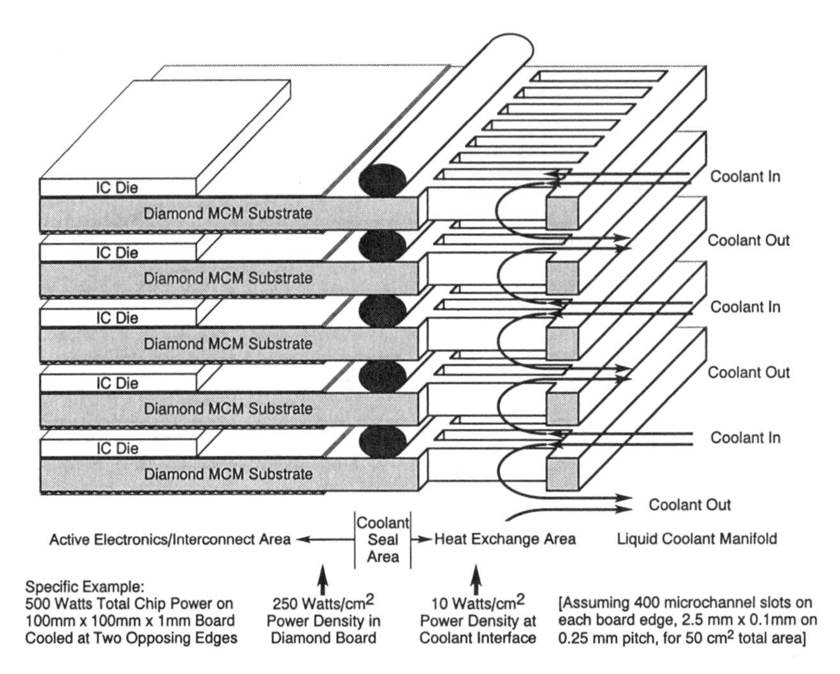

Figure 11. Example of approach to cooling of the diamond 3-D "cube computer board edges using microchannel single-phase liquid heat exchangers. (As noted in text, using ISR phase-change spray cooling is simpler, less costly and complicated, and uses much less cooling power than single-phase cooling.)

microchannel cooling approach of Figure 11 were avoided by going to a simple, but sophisticated, phase-change edge cooling approach [Schaefer et al. 1993], [Moravec et al. 1993] and [Sienski et al. 1996]. The technique, called liquid droplet impingement cooling (or spray cooling) was developed by Isothermal Systems Research. In this approach, the ends of the diamond substrates are extended through a coolant seal, as in Figure 11, with the important difference that the cooled area, like the active electronics area, is at atmospheric pressure, so there is minimal pressure difference to force leaks, and there is no need for the intricately-cut array of microchannel slots and liquid manifolding shown in Figure 11. Rather, an array of fine nozzles mist the exposed ends of the diamond substrates with a fine spray of liquid Fluorinert (recirculated through a pump and condenser [heat exchanger] system). The kinetic energy of the droplets in the spray prevents bubble formation at the diamond-liquid interface, and promotes high-rate evaporation from the surface of a thin liquid layer on the diamond, which is constantly being replenished by the spray. This leads to an extremely high heat transfer coefficient, holding the temperature of the exposed diamond fins to very near the boiling point of the particular Fluorinert liquid used. This ISR spray cooling approach appears to offer more than adequate levels of cooling power (critical heat flux and heat transfer coefficient), using safe and environmentally friendly commercial Fluorinert liquids, to meet the requirements of the diamond 3-D "cube computer". Spray cooling also offers extremely low cooling power requirements (of the order of 4% to 5% of the heat removed), with the added advantage of making the whole diamond 3-D MCM substrate and package design simpler and lower cost.

11. Power Density Comparison of Diamond 3-D MCM Packaging to Conventional Electronic Packaging Approaches

While in high heat flux engineering, the measurement standard is based on heat per area (Watts/cm^2), in electronics packaging, the real payoff is volume. The "golden triad" of electronics systems design is size, weight and cost. The "size" refers, of course, to the volume of the electronic package required to implement the function. Given reasonable bounds on density, the weight tends to follow the volume, and in mature production, the cost tends to track the weight to some degree. Hence, reducing the volume required for electronics, even over and beyond the interconnect delay issues discussed above (e.g., Figure 5), is of enormous benefit in electronic systems (and absolutely crucial in the case of military systems). It is instructive to compare the volume power density capabilities of diamond 3-D electronic packaging to more conventional packaging approaches, and, given the limiting relationship (Eq. 3) between computing power and the power dissipation of the logic circuitry, to explore the significance of this in terms of computing power density.

Figure 12 compares the volume power densities available using the diamond 3-D packaging approach discussed above with typical values from more conventional electronic packaging (circuit board or MCM) techniques. The volumes included in each case are the active electronic MCM or board itself, plus the direct heat sink or heat exchange elements attached to the module, but not remote cooling elements such as fans, heat exchangers, etc. Illustrated along with this diamond 3-D MCM "cube computer" packaging approach are examples of natural (un-forced) air convection (e.g., laptop computer), typical fan-forced air convection (e.g., VME board), more exotic ducted forced air or "air impingement" cooling (e.g., IBM RS/6000 MCM or IBM 4381 TCM), and forced water cooling (e.g., IBM 3081 TCM [thermal conduction module]). It is interesting to note that, while the volume power density level between these four "conventional" cases increases by about a factor of 3.33 between each case (from about 0.2 W/in^3 for the natural air cooled board to 7.4 W/in^3 for the water-cooled TCM), the remarkable 200 W/in^3 power density that the edge-cooled diamond 3-D configuration can handle is more than 25 times higher than the water-cooled TCM.

By making use of the relationship between processing power (computational throughput) and the power dissipation capability in power-limited processing as given by Eq. 3, the volume power densities in Figure 12 can be translated into volume computational throughput densities (e.g., millions of floating point operations [MFlops] per cubic inch). This is illustrated by the addition of the "Peak MFlops per cubic inch" scale at the right side of Figure 12. (Note that these are full 64-bit general-purpose computing MFlops depicted here, not signal processing MFlops which are much “cheaper” in terms of power and size.) This scale is technology normalized to state of the art commercial CMOS by the example of a DEC 21164-300MHz “Alpha” 64-bit processor chip which dissipates 51 watts at a clock frequency of 300 MHz, at which its peak processing throughput is about 600 MFlops, since it has two floating point units (in addition to two integer units). Note that with this technology scaling, the power-limited peak computational throughput for the edge-cooled 3-D diamond MCM at 200 W/in^3 is over 2400 MFlops/in^3, or an amazing 240 GFlops (0.24 TFlops) in a single 4" x 4" x 4" diamond “Cube Computer” module.

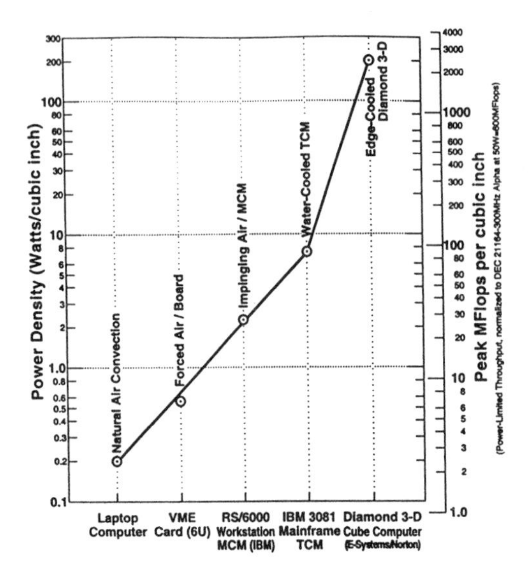

Figure 12. Comparison of the volume power density (watts per cubic inch) [left scale] and corresponding computational throughput per cubic inch [right scale] (peak Mflops of general purpose 64-bit computing power, assuming power-limited operation, normalized [Eq. 3] to DEC 21164-300Mhz "Alpha" microprocessor technology)

12. Summary

In summary, while the enormous increases in integrated circuit speeds realized over recent years offers the possibility of major enhancements of digital system speeds, the fast signal rise times and high power levels of these chips lead to formidable packaging challenges in trying to achieve these enhanced speeds at the system level. For example, high performance MCMs involve power density levels difficult to cope with using conventional packaging materials. The extremely high thermal conductivity of diamond, coupled with its electrically insulating nature and very low thermal expansion, make it an ideal choice for solving many of these packaging problems. In fact, the thermal conductivity is so high as to allow many hundreds of watts to be extracted from a 1 mm thick diamond substrate by lateral thermal conduction alone. In addition to improving normal 2-D packaging approaches, this makes possible the exciting prospect of realizing simple, practical 3-D "cube computer" MCM packaging approaches. In other words, mastering the materials science and manufacturing technology issues necessary to make available 1 mm thick diamond substrate material in 100 mm (4") or larger sizes at reasonable prices will enable electronic system designers to achieve, with a simple, cost-effective packaging approach, power densities and performance levels previously achievable with only the most exotic and expensive packaging schemes.

13. References

Bakoglu, H.B. (1990), *Circuits, Interconnections, and Packaging for VLSI*, Addison-Wesley Publishing Co., Reading, MA, 1990, ISBN 0-201-06008-6, p. 441.

Boudreaux, P.J., (1995) "Thermal Aspects of High Performance Packaging with Synthetic Diamond", *Proceedings of the Materials Research Society Third International Conference on the Application of Diamond Films and Related Materials, August 1995*, Gaithersburg, MD, pp. 603-610, NIST Special Publication 885, Edited by A. Feldman.

Boudreaux, P.J., Conner, Z., Culhane, A. and Leyendecker, A.J. (1991) "Thermal Benefits of Diamond Inserts and Diamond-Coated Substrates to IC Packages, 1991 *Digest of Papers of the Government Microcircuit Applications Conference, Volume 27,* pp. 251-256, Orlando, FL, Nov. 5-7, 1991, DTIC number GOMAC-91-B160081.

Eden, R.C., (1993b) "Application of bulk synthetic diamond for high heat flux thermal management", Paper Number 1997-11 in *the Proceedings of the SPIE Technical Conference #1997, "High Heat Flux Engineering II",* San Diego, CA, July 12, 1993.

Eden, R.C., (1993a) "Application of Diamond Substrates for Advanced High Density Packaging", *Diamond and Related Materials*, **2**, 1051-1058.

Eden, R.C., (1991) "Application of Synthetic Diamond Substrates for Thermal Management of High Performance Electronic Multi-Chip Modules", in ", Y. Tzeng, M. Yoshikawa, M. Murakawa, and A. Feldman, (eds.) *Applications of Diamond Films and Related Materials* Elsevier Science Publishers B.V., pp. 259-266.

Eden, R.C., (1994) "Capabilities of Normal Metal Electrical Interconnections for 3-D MCM Electronic Packaging", *Proceedings of 1994 OE/LASE, Los Angeles, January 26, 1994*, (SPIE Technical Conference #2153), Paper Number 2153-19.

Fehr, G.K., (1995) "The Use of Diamond Films to Create Thermally Enhanced Plastic Packages", *Proceedings of the First Annual Assembly and Packaging Foundry Conference, September 21-23, 1995.*

Kiefer, D. and Heightley, J., (1987) "CRAY-3: A GaAs Implemented Supercomputer System," *Technical Digest of the 1987 GaAs IC Symposium, Portland, Oct. 13, 1987*, pp. 3-6.

Lu, G., Gray, J., Borchelt, F., Bigelow, L.K., and Graebner, J.E., (1993) "White CVD Diamond for Thermal and Optical Applications", *Diamond and Related Materials*, **2**.

Moravec, T. J., Eden, R.C. and Schaefer, D.A., (1993) "The Use of Diamond Substrates for Implementing 3-D MCMs", *1993 Proceedings International Conf. on Multichip Modules, Denver, April 14-16, 1993*, pp. 86-91.

Nagy, Bela G. and Partha, Arjun (Norton Diamond Film), patent applied for.

Napolitano, L.M. Jr., Daily, M.R., Meeks, E., Miller, D., Norwood, D.P., Peterson, D.W., Reber,C.A., Robles J.E. and Worobey, W., (1993) "Development of a Power Electronics Multichip Module on Synthetic Diamond Substrates", *Proceedings of the 1993 International Conference on Multichip Modules, April 14-16, 1993, Denver, CO,* pp. 92-96.

Peterson, D.W., Sweet, J.N., Andaleon, D.D., Renzi, R.F. and Johnson, D.R., (1994) "Demonstration of a High Heat Removal CVD Diamond Substrate Edge-Cooled Multichip Module", *Proceedings of the 1994 International Conference on Multichip Modules, April 13-15, 1994, Denver, CO,* pp. 624-630.

Schaefer, D.A., Eden, R.C. and Moravec, T.J., (1993) "The Role of Diamond in 3-D MCMs", *Proceedings of NEPCON '93, Anaheim, February 11-14, 1993.*

Sienski, K., Eden,R.C. and Schaefer, D.A., (1996) "3-D Electronic Interconnect Packaging", *Proceedings of the 1996 IEEE Aerospace Applications Conference, Snowmass at Aspen, CO, February 3-10, Volume 1,* pp. 363-373.

Chapter 31
DIAMOND VACUUM ELECTRONICS

George R. Brandes
Advanced Technology Materials, Inc., Danbury, CT 06810

Contents

1. Introduction

Vacuum electronic systems, and diamond vacuum electronic systems in particular, are interesting to study for both fundamental and applied reasons. From a fundamental standpoint, the study of vacuum electronic systems involves investigating a complex material-vacuum system and determining how the system influences the transport of particles. From an applied standpoint, applications for vacuum electronic systems are ubiquitous, ranging from flat panel displays to high speed transistors to novel particle and photon detectors.

The significant amount of recent work in diamond vacuum electronics is in part an outgrowth of the observations that the diamond (111) surface has a negative electron affinity (NEA) [Himpsel *et al.* 1979] and that diamond readily emits electrons in the presence of electric fields as low as a few volts per micron [Wang *et al.* 1991]. Additionally, diamond has a high thermal conductivity and is resistant to damage and chemical modification, qualities that may be useful in an emitter material.

This review of diamond vacuum electronics is divided into three sections. Section 2 contains a discussion of diamond field emission and excited state electron emission from diamond.

Section 3 contains a discussion of vacuum electronic devices that capitalize upon diamond's NEA. In particular, diamond photocathodes, diamond dynodes for photomultiplier tubes, and diamond junction cathodes are reviewed.

The most frequently discussed diamond vacuum electronic application is as a cathode for flat panel displays. Section 4 contains a discussion of diamond field emission cathodes for flat panel displays. In particular, the emitter material requirements, field emission cathode design and fabrication issues, and a review of current fabrication approaches will be presented.

2. Electron Emission From Diamond

The emission of electrons from a material is critically dependent upon the surface composition, surface structure and the bulk band structure. An examination of the terms contributing to the electron work function highlights this dependence (Figure 1). The electron work function of a single crystal is defined as the minimum energy required to remove an electron from the crystal bulk to a location outside the surface. The location outside the surface must be large relative to image forces but small relative to the distance to other crystal faces. The bulk contribution to the work function is described in terms of the chemical potential μ, a quantity expressed relative to a bulk crystal reference level (ϕ_c). The surface contribution to the work function is the electronic surface dipole, $\Delta\phi$, also expressed relative to ϕ_c. The surface dipole is a consequence of the surface charge redistribution (the electrons spill out beyond the last atomic layer) and in most cases, a rearrangement of the surface ions from their ideal Bravais lattice positions. The surface dipole is further modified if the surface composition differs from the bulk, a likely occurrence if the surface is exposed to the atmosphere or other impurities.

A term related to the electron work function, the electron affinity χ, is frequently used to characterize semiconductor materials. The electron affinity of a material is defined as the energy required to remove an electron from the bottom of the conduction band to the vacuum level. As shown in Figure 1(b), if the band gap is sufficiently large or the surface is coated with a low electronegativity material, the bottom of the conduction band can lie above the vacuum level; the surface has a negative electron affinity (NEA). If a freely diffusing conduction band electron encounters a NEA surface it may be emitted into the vacuum or it may undergo quantum mechanical reflection. The electron will acquire an energy equal to the difference between the final conduction band state and the vacuum level if no inelastic collision takes place during the process of emission.

Materials with NEA surfaces are usually p-type semiconductors (activated with cesium) or insulators, both of which have few electrons, if any, in states appropriate for emission

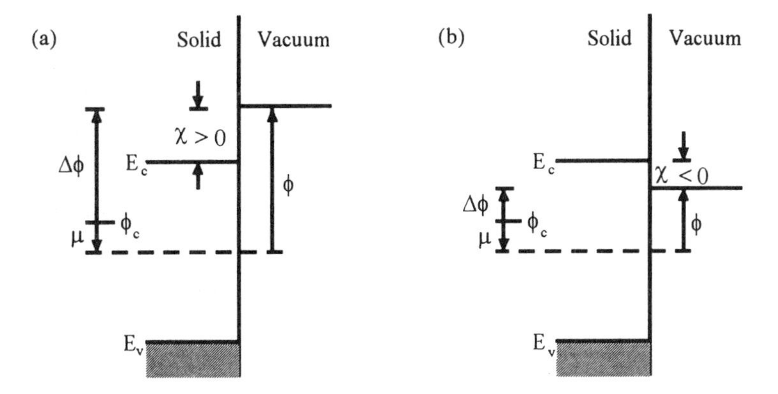

Figure 1: Energy levels at (a) a positive electron affinity surface and (b) a negative electron affinity surface.

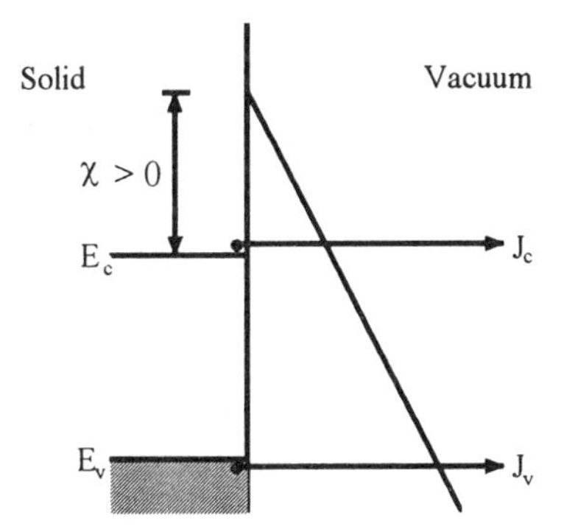

Figure 2: Schematic of surface potential energy levels in the presence of an external electric field. The width of the tunneling barrier is determined by the magnitude of the electric field as well as the electron affinity χ J_c and J_v represent energy-conserving tunneling transitions from the conduction band and valence band of the solid into the vacuum.

[Martinelli and Fisher 1974]. A means of exciting electrons into these above bandgap states must be found before NEA emission can occur. Once electrons are excited into these states, the emission probability for a material with a NEA surface is usually greater than for a material without. The energy distribution of electrons emitted from clean, ordered NEA surfaces is expected to be consistent with Maxwell-Boltzmann statistics and the angular distribution is expected to be sharply peaked about the surface normal. The high electron yield, narrow energy distribution and narrow angular distribution of NEA materials makes them excellent electron sources for many charged particle optical systems.

Electrons need not be excited into states above the vacuum level, however, for emission to occur; electrons can also be emitted into the vacuum if an electric field is present. For a suitably large electric field, the surface potential barrier will deform and electrons may tunnel into vacuum states (Figure 2). Electron emission from diamond occurs in the presence of low fields compared to metals and other semiconductors and consequently, diamond-based field emission cathodes are promising for many applications.

Excited state emission studies and field emission studies are reviewed separately to better present how the properties of diamond determine emission via each mechanism. Section 2.1 contains a discussion of electron emission following excitation by photons (photoemission) or by electron bombardment (secondary electron emission). Section 2.2 contains a discussion of diamond field emission models and the experimental evidence supporting both emission models.

2.1 EXCITED STATE EMISSION

2.1.1 Photoemission

A good way to examine energy levels in a vacuum electronic system is with photoelectron emission spectroscopy. This technique has been used extensively to characterize the electron affinity of different diamond surfaces. A peak at the conduction band minimum in the photoemission spectrum is an indication that the electron affinity is negative. Electrons excited into above-vacuum states may be emitted without energy loss, but a fraction of the electrons undergo inelastic collisions, losing energy until they occupy states at the bottom of the conduction band (thermalize). If the electron affinity is positive, the thermalized electrons will not escape into the vacuum. On the other hand, if the electron affinity is negative, electrons in low-lying conduction band states that encounter the surface may be emitted into the vacuum, giving rise to the photoemission spectrum peak at the conduction band minimum. Further evidence that the surface affinity is negative may be obtained by changing the photon energy or using sub-bandgap energy photons. If the electron affinity is negative, the position of the peak at the conduction band minimum will not change if the photon energy is increased. The observation of photoemitted electrons when the surface is illuminated with sub-bandgap energy photons is another indication of NEA, although care must be taken to insure that the electrons are originating from the valence band.

The first diamond surface shown to have a NEA was the unreconstructed, hydrogenated diamond (111) surface [Himpsel *et al.* 1979]. Himpsel *et al.* found that the photoemission spectrum showed a dominant ~0.5 eV emission peak at the conduction band minimum, consistent with NEA. The diamond had a stable quantum yield that increased linearly from photothreshold (5.5 eV) to ~20% at 9 eV, with a very large yield of 40% - 70% for 13 eV $< h\nu <$ 35 eV.

The observation of a peak at the conduction band minimum in the photoemission spectrum of the 2×1 reconstructed diamond (100) surface indicated that this surface was also NEA [van der Weide *et al.* 1994]. Electrons were emitted when the (100)-oriented, hydrogenated surface was exposed to sub-bandgap radiation (5.4 eV) [Eimori *et al.* 1995].

The NEA of the (111) oriented surface was linked to hydrogen termination [Pate 1986]. When hydrogen was driven off by heating to 1000 °C, the diamond surface reconstructed and the electron affinity became positive. Similarly, exposure to an argon plasma [van der Weide and Nemanich 1993] or an oxygen plasma [Eimori *et al.* 1994] made the diamond affinity positive, while subsequent exposure to a hydrogen plasma recovered the negative electron affinity.

In addition to hydrogen, other coatings were shown to change the dehydrogenated (111) surface affinity from positive to negative. By depositing a submonolayer of titanium onto argon-plasma cleaned diamond, the affinity changed from positive to negative [van der Weide and Nemanich 1992]. Similarly, submonolayer deposition of TiO on the (111) 2×1 diamond surface changed the affinity from positive to negative [Bandis *et al.* 1994].

The energy spread of the photoemitted electrons is larger than might be expected from a NEA material with a carefully prepared surface. Bandis and Pate showed that the low energy electrons originated not only from above vacuum-level conduction band states, but also from exciton breakup at the surface [Bandis and Pate 1995a]. Electrons originating from excitons were the dominant source of photoelectron emission from the hydrogenated diamond (111) surface for near band gap excitation and from polished diamond surfaces [Bandis and Pate 1995b].

In summary, the hydrogen-terminated surfaces of diamond possess a negative electron affinity. The electron affinity for each surface becomes positive when the hydrogen is removed. The photoyield from the NEA surfaces is high, ranging, for example, from ~ 20 % at $h\nu$ = 9 eV to 70 % at $h\nu$ = 16.5 eV for the C(111) surface. The energy and angular distribution of the emitted electrons is higher than expected from an ideal NEA material.

2.1.2 Secondary Electron Emission

An alternate means of exciting electrons is with electron bombardment. In diamond, on average, one electronic excitation occurs for every 20 eV of incident electron energy. As with photoemission, if the surface has a negative electron affinity, electrons populating the conduction band can be emitted into the vacuum.

Early secondary electron studies focused upon examining the energy distribution of secondary electrons in order to gain insight into the electronic properties of diamond. The work established a sensitivity to different carbon surfaces and crystalline order [Maguire 1976, Lurie and Wilson 1977, Hoffman 1994]. The secondary electron yield coefficient, defined as the number of secondary electrons produced ($E<50$ eV) per incident electron, ranged between 2 and 3.5 [Johnson 1953].

Bekker *et al.* first observed high secondary electron yields from diamond — typically 14 emitted electrons per 1 keV incident electron [Bekker *et al.* 1992]. Under prolonged electron bombardment the secondary electron yield dropped to around two. Presumably the electron beam removed hydrogen from the surface, changing the affinity from negative to positive. Johnson's studies of secondary electron yields were probably low because dehydrogenated surfaces were studied [Johnson 1953]. Malta *et al.* showed that diamond exposed to oxygen or annealed at high temperature had a low secondary electron yield, but exposure to atomic hydrogen increased the yield by a factor of ~30 [Malta *et al.* 1994].

One way to attain high, stable secondary electron yields has been to operate the sample in a hydrogen environment [Mearini *et al.* 1994]. A second approach has been to try to stabilize the surface; diamond coated with cesium was tried [Mearini *et al.* 1995]. The total secondary yield coefficients ranged from 25 to 50 and the surface was was observed to be stable in air, during heating in vacuum up to 120 °C, and under continuous exposure to an electron beam.

2.2 FIELD EMISSION

In field emission, the application of a large electric field deforms the surface potential barrier enough to make probable the quantum mechanical tunneling of electrons from the solid into the space outside. The size of the field required for emission to occur is determined by the bulk electronic structure and the surface structure and composition, which in turn determines the size and shape of the surface potential barrier ($\Delta\phi$).

External field emission from semiconductors is treated using an approach similar to that developed by Fowler and Nordheim for metals [Fowler and Nordheim 1928]. Fowler and Nordheim showed that the emission current density J from a material with work function ϕ was

$$J = \frac{AE}{\phi} \exp\left(\frac{-B\phi^{3/2}}{|E|} \right)$$

where E is the electric field and A and B are constants. If the geometry of the emitter is a pointed needle of length l and radius r, the electric field at the tip is $E=\beta V$, where $\beta=2/(r \ln(2l/r))$. From these equations it is clear that the work function and/or the tip radius must be made smaller if low field, high current density emission is desired.

The equation describing emission from semiconductors has a similar functional form to that for metals, but modifications are made to take into account electric field penetration into the semiconductor, the presence of surface states, and whether electrons are emitted from valence, conduction or midgap (impurity) states (Figure 2). For example, instead of using the work function ϕ, the electron affinity χ must be used if electrons are field-emitted from the conduction band. Similarly, $\chi+(E_c - E_v)$ must be used in place of ϕ if electrons are field-emitted from the top of the valence band. A detailed derivation and discussion of the equations describing emission from semiconductors may be found in books by Gomer and Modinos [Gomer 1961, Modinos 1984].

Diamond emits electrons in the presence of modest (macroscopic) electric fields, typically less than 20 V μm^{-1}, as many researchers have shown. [The earliest published results were authored by Wang *et al.* 1991, Xu *et al.* 1993, Kordesch 1993.] Figure 3 shows a typical emission current scan from a polycrystalline diamond film and the accompanying Fowler-Nordheim plot.

The low-field emission current from diamond cannot be readily explained; there appear to be no populated states close enough to the vacuum level for emission to occur, even with diamond's field-enhancing surface morphology. Two electron emission mechanisms, *external* field emission and *internal* field emission, may explain diamond's low field emission. There is evidence to support both models of emission.

With *external* field emission, electrons originate from states below the vacuum level and emission is determined by the surface characteristics. The electron states may arise from non-diamond carbon that forms on the diamond surface, gap surface states, or when the material is heavily doped, impurity band states. Changes in emission consistent with changes in the surface characteristics provides evidence supporting the external field emission mechanism. For example, by enhancing the surface morphology beyond what was obtained from as-grown polycrystalline diamond, greater emission currents were obtained for the same macroscopic field [Twichell *et al.* 1993, Okano *et al.* 1994, Asano *et al.* 1995]. Emission performance similar to that obtained from diamond was obtained from non-diamond carbon, indicating that some forms of non-diamond carbon have a low work function or field enhancing morphology [Bajic and Latham 1988, Humphreys and Khachan 1995]. Non-diamond carbon was observed on some natural and CVD diamond surfaces that also had a positive affinity [Shovlin and Kordesch 1995]. Diamond doped with a high activation energy donor impurity (P) emitted at lower field than a diamond doped with an acceptor impurity (B) [Okano and Gleason 1995]. Changing the surface structure and surface chemistry by coating the diamond with cesium yielded an exceedingly low-field emitter ($E<0.2$ V μm^{-1}) [Geis *et al.* 1995]. A theoretical study found an oxygenated diamond surface converted from a 2.45 eV positive electron affinity to a 0.85 eV negative affinity upon cesiation [Pickett *et al.* 1994]. The resulting surface is metallic due to the Cs-O-C surface states and thereby also provides a low work function (1.25 eV).

Huang *et al.* calculated emission current density from diamond as a function of applied field, taking into account band-bending, the image interaction and the presence of surface states [Huang *et al.* 1994, 1995]. They found that tunneling from the conduction and valence bands of intrinsic or p-type diamond was negligible in the presence of modest electric fields. If a suitable low activation energy donor impurity existed, significant tunneling current density could be obtained. Emission from surface states located ~ 1 eV below the conduction band could also account for the observed results.

Internal field emission might also explain diamond's low field emission. When an external electric field is applied, the field will penetrate into the diamond to a degree determined by the densities and types of dopants present. Small, non-diamond-carbon particles embedded in the diamond will facilitate the formation of electrically conducting channels in the diamond crystal [Bajic and Latham 1988, Xu *et al.* 1993].

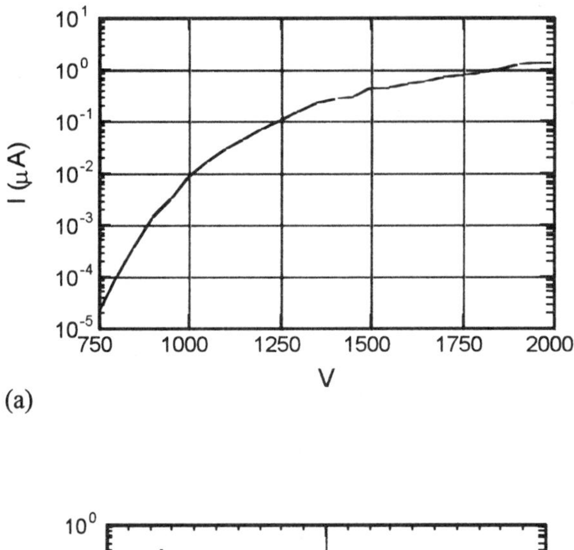

(a)

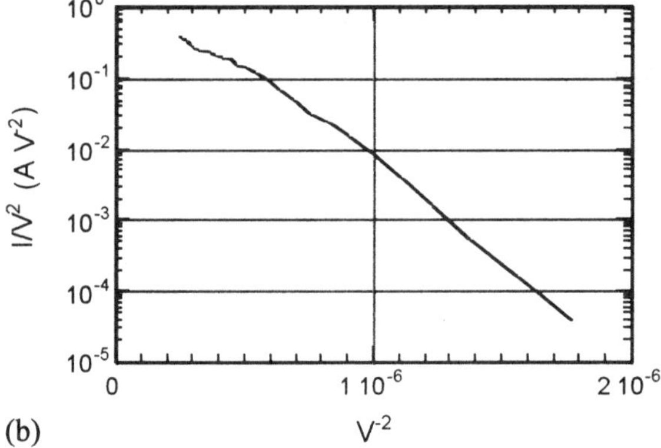

(b)

Figure 3: (a) Plot of the emission current from a P-doped polycrystalline diamond film as a function of probe voltage (measured relative to the diamond film potential). The 50 μm radius probe was placed 50 μm above the surface of the diamond film for this measurement. The emission site density for this sample was in excess of 10^4 cm^{-2}. (b) Fowler-Nordheim plot ($\ln(IV^{-2})$ versus V^{-2}) of the data shown in Figure 3(a).

The conducting channels will bias the carbon particles at or near the substrate potential. Consequently, the field at the surface of the small, embedded carbon particles will be very large, and the probability that electrons from the embedded particle will be field emitted into the conduction band will be high. The internal field will also act to enhance the transport of electrons to the surface. At the surface, the electrons will either be heated to a point where they can escape over the potential barrier [Latham and Xu 1991 and references therein] or if the surface has a NEA, the electrons will be emitted even if they occupy the lowest conduction band states.

Evidence to support this emission mechanism can be found in the field and photo emission microscope study of hydrogen-rich, chemical vapor deposited polycrystalline diamond films [Wang *et al.* 1991]. Electron emission occurred at low field strength (3 V μm^{-1}), but good field emission sites did not always coincide with the tips of the diamond facets. If external field emission were responsible for the emission current results, the likely emission sites would be the facet tips. Further work showed that previously nonemitting chemical vapor deposited diamond films were observed to emit electrons in an applied field of 3-5 V μm^{-1} after coating with a discontinuous, nonreactive sputtered gold layer [Shovlin and Kordesch 1994]. The gold overlayer acts in a manner similar to the embedded carbon particles, locally increasing the field strength. The consequent dielectric breakdown of the film provides conductive channels for the observed electron emission. Zhu *et al.* introduced defects into a diamond film by ion implantation and found the field requirements dropped by a factor of four [Zhu *et al.* 1995].

A NEA surface is not required for external low-field electron emission. A negative affinity indicates that the separation between the valence band and the vacuum level is less than the bandgap of diamond. Some researchers have suggested that diamond's negative electron affinity is responsible for the low field emission. While diamond's NEA surfaces might play a role, there must be a field-dependent means of promoting electrons to the conduction band. Once promoted to the conduction band, no additional energy is required for emission if the surface has a negative electron affinity.

A NEA surface can play a role in the model proposed by Xu *et al.*, increasing the probability that electrons produced with *internal* field emission are emitted into the vacuum. However, the surface need not be NEA for emission to occur. The conduction band electrons could be `heated' by the internal electric field and acquire enough energy to overcome the surface potential barrier. The energy spread of electrons emitted from diamond was 1 eV compared to 0.23 eV typically obtained from a clean tungsten tip [Xu *et al.* 1994]. The large energy spread may indicate electron heating, but the spread may also arise from excitonic emission or a disordered surface.

Even if the diamond surfaces are initially NEA surfaces, they are unlikely to remain so over an extended period of operation. Positive ions will be produced in the electric field region and accelerated into the diamond, causing dehydrogenization of the surface and a change in the affinity. A variety of natural and chemical vapor deposited diamond surfaces were imaged using a photoelectron emission microscope [Kordesch 1993, Shovlin *et al.* 1995]. The data showed that diamonds with positive NEA surfaces had graphitic surfaces.

Many of the studies of electron emission have focused upon the low-field emission of diamond cathodes without discussing the emission site density. As will be discussed later, if a material is to be used in a field emission cathode, it must possess a high density (>10^5 cm^{-2}) of emission sites that all turn-on at roughly the same voltage. Emission site density can be measured with a small diameter probe that can be moved across the surface of a sample, or by imaging with a field emission microscope. The few published measurements indicate that the site density is low. For example, substantial emission was obtained from a diamond coated Mo sample at 10 V μm^{-1} but the emission site density was only ~ 60 cm^{-2} [Xu *et al.* 1993].

In conclusion, many interesting experiments have been conducted to investigate electron emission from diamond. The results, as discussed above, revealed much about electron emission and how the diamond-vacuum system influences the transport of particles. Many areas are not well understood, however, most notably the emission mechanism(s) responsible for diamond's low field emission.

3. Negative Electron Affinity Devices

Diamond photocathodes, diamond dynodes and diamond junction transistors are three NEA vacuum electronic devices that have been studied. The devices were designed to capitalize upon diamond's unique properties, but their development has been hampered by unanticipated technical challenges. The design, applications, performance and problems for each device will be presented in the following three subsections. Diamond field emission cathodes, which capitalize upon diamond's low field emission properties, will be discussed in Section 4.

3.1 PHOTOCATHODES

Photocathodes composed of a NEA material are valuable not only because the emission current density can be high, but also because the cathode can have a high brightness. "Brightness" is a term used to describe the angular distribution of the emitted electrons, the current density, and the energy distribution of the emitted electrons. The degree to which geometrical, chromatic and space charge aberrations degrade an electron beam is a direct function of the cathode brightness. The geometrical aberrations are related to the angle of emission at the cathode relative to the surface normal and the emission radius at the cathode. Chromatic aberration is proportional to the kinetic energy distribution.

A useful and commonly used figure of merit for characterizing brightness is the normalized brightness-per-volt, $r_v = I/(4D^2\theta_p^2E)$. Here, I is the current, D is the cathode diameter, θ_p is the maximum emission angle at the cathode of the particle ensemble, and E is the mean energy of emission. Using this figure of merit, it is clear that a cathode with a low emission current density might perform better in certain applications than a high emission current density cathode that emits electrons with large energies and at large emission angles. Chromatic aberrations are not taken into account by r_v, but the effect of this aberration may be examined by looking at the variation of r_v with E.

NEA cathodes should be high brightness cathodes. If the magnitude of the affinity is small, the energy of the emitted electrons (E) will be small. The electron yield and consequently I/D^2, should be large from a NEA material. The spread in energy and the angular distribution (θ_p) of the particles emitted from a clean, well-ordered NEA surface should be a few kT, assuming the electrons have fully thermalized.

Himpsel and coworkers first suggested that the hydrogen terminated diamond surface had value as a stable NEA photocathode [Himpsel *et al.* 1979]. Their photoemission study showed that the photoyield from the NEA hydrogen terminated (111) diamond surface was large. Beetz and coworkers discussed the advantages of using diamond and novel optical methods to create temporal and spatially profiled electron beams [Beetz *et al.* 1991]. Several groups have produced and tested excimer-laser-driven diamond and diamond-like carbon photocathodes [Bouilov *et al.* 1991, Fischer *et al.* 1994]. A high emission current efficiency was obtained, even with sub-bandgap radiation, and the current density from diamond cathodes was higher than other carbon-based cathodes.

As discussed in the last section, the energy distribution of electrons emitted from diamond was larger than expected from a NEA cathode. The studies also showed that the NEA surface could be damaged. For these reasons, and because a high photon energy light source (>5.5 eV) is required for high yields, adoption of diamond photocathodes has been slow.

3.2 DYNODES

When light levels are low, the photomultiplier tube is the detector of choice due to its unmatched sensitivity for visible or ultraviolet wavelength photons. A photomultiplier tube consists of a photocathode, a dynode chain (continuous or discrete), and an anode. If a photon of energy greater than the photocathode work function strikes the photocathode a photoelectron is emitted a fraction of the time. The dynode chain allows one to detect this single photoelectron by amplifying the emission event. An attractive potential difference is established between the photocathode, successive dynodes and the anode. The photoelectron is accelerated to the first dynode and the collision of the electron with the dynode causes additional electrons to be emitted. Common dynode materials emit approximately three secondary electrons per incident electron. The secondary electrons are accelerated to the next dynode, collide with the dynode and produce a new, amplified burst of electrons. After passing through roughly twelve dynodes, the initial photoelectron will have grown to a packet of 10^5 to 10^6 electrons. The anode collects the electron bunch and the signal is fed into appropriate signal processing electronics.

Diamond has many characteristics that suggest it will be a good choice for dynode applications. Diamond has negative electron affinity surfaces that will facilitate electron emission and enhance charge amplification. The secondary electron yield is defined as the number of electrons emitted with energy less than 50 eV, per incident electron. High secondary electron emission from chemical vapor deposited diamond films (14-27) was observed [Mearini *et al.* 1994].

Because there are effectively no electrons in the conduction band unless they are put there (a consequence of diamond's wide bandgap and high-activation energy impurities), the probability of spurious electron emission is nearly zero. Even if the dynodes are heated to high temperatures, diamond dynodes will not be plagued with noise-causing thermionic electron emission, eliminating the need for a dynode cooling system. This feature may make diamond-dynode-based photomultiplier tubes especially useful in oil-exploration bore holes and space-based applications. Secondary electrons from NEA diamond have a relatively low emission energy (roughly 1 eV) and the energy distribution is narrow compared to other dynode materials. The low electron energy and narrow energy distribution facilitates focusing which in turn reduces noise caused by improperly focused electrons striking the glass walls.

Diamond is a relatively low cost dynode material. Because diamond can be grown over a large area and the film thickness needed is only a few microns, diamond dynodes, produced in quantity, could be competitively sold for as little as five dollars apiece, making a diamond based photomultiplier tube inexpensive. Furthermore, because diamond is air stable, assembly need not take place in vacuum and a reduction in tube vacuum integrity will not affect the non-reactive diamond surface (although the photocathode will probably be damaged).

When count rates are low or variable, dynodes made from CuBe-O and other common dynode materials can fluctuate in an unpredictable direction by up to 25%, and the settling period can be hours [Boutot *et al.* 1983, Candy 1985]. Yamashita argued that the fluctuations were due to charge build-up within the thin oxide layer on the dynode surface [Yamashita 1980]. The non-reactive, hydrogen-terminated semiconducting diamond is unlikely to suffer similar difficulties. Furthermore, boron doping makes diamond conductive to the point that the secondary electron yield is stable.

What has limited the adoption of diamond-dynodes is the drop in yield caused by extended energetic-electron-beam bombardment. The drop is due to dehydrogenization of the diamond surface [Mearini *et al.* 1994]. Without a method to extend the lifetime of diamond dynodes, diamond will not replace CuBeO dynodes except in harsh environments and for low-count rate applications, where dynode charging and unpredictable gain drift are problems.

Two methods of increasing the diamond dynode lifetime are being investigated. By operating a dynode chain in an atmosphere with a hydrogen partial pressure of 10^{6} Torr, stable dynode operation was observed, although the yield was 10 % of the predicted value [Mearini *et al.* 1994]. Diamond terminated with cesium and other alkali halides was stable and had high secondary electron yields [Mearini *et al.* 1995]. The overall performance of photomultiplier tubes containing alkali-halide-coated diamond-dynodes has not been assessed.

3.3 JUNCTION CATHODES

A compact high brightness electron source can be formed from NEA materials if a convenient means to populate the conduction band states can be found [Williams and Tietjen 1971]. One approach, first proposed by Geppert [Geppert 1966], is to form a pn junction with the p-type NEA material exposed to the vacuum (Figure 4). By forward biasing the junction, minority electrons are injected from the n-type region into the p-type region. A fraction of the electrons diffuse through the p-type material, reach the opposing surface, and are emitted into the vacuum with an energy comparable to the difference between the bulk conduction band minimum and the vacuum level.

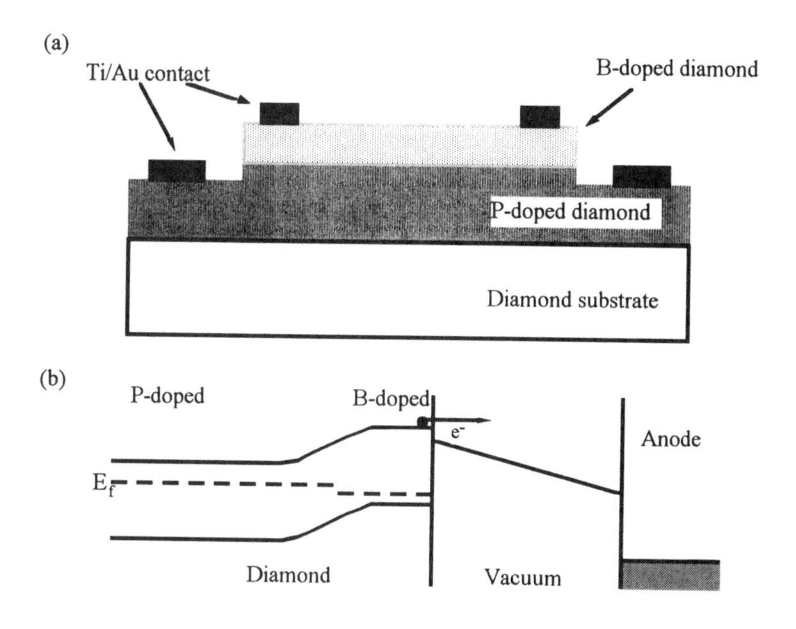

Figure 4: (a) Schematic of a diamond junction cold cathode device. (b) Energy levels in a diamond junction cold cathode. See discussion in text.

Diamond junction cathodes offer a number of advantages over other cold cathodes. Because the diamond junction cathodes can be formed with planar, well-ordered surfaces---a morphology disadvantageous to field emission cathodes---the angular and energy distribution of diamond junction cathodes should be small. The junction cathode may be less likely to form excitons because the electrons are not originally in upper level conduction band states, leading to a brighter source. The diamond surface is not as reactive as cesiated surfaces, although it is not as robust as was previously thought. Because the capacitance of diamond is low, high frequency performance is feasible.

The principal problem hindering the development of diamond junction cathodes is the lack of a low activation energy donor impurity. Geis *et al.* first fabricated a diamond junction cathode [Geis *et al.* 1991]. They implanted carbon into a type IIb diamond to produce an n-type layer, then created mesa structures by reactive ion etching in select areas through the implanted region and down into the boron doped diamond. When these diodes were forward biased, current was emitted into the vacuum. The ratio of emitted current to diode current varied from 2×10^{-4} to 1×10^{-10}. The estimated emission current density was 0.1 to 1.0 A cm^{-2} for a diode current of 10 mA.

Diamond junction cold cathodes composed of a phosphorus doped diamond layer and a boron doped diamond layer were fabricated and tested [Brandes *et al.* 1995]. Although the n-type character of P-doped single crystal diamond has yet to be convincingly established, there is evidence that heavily doped material is n-type with an activation energy of approximately 0.9 eV [Setaka 1989, Okano *et al.* 1990, Zvanut *et al.* 1994].

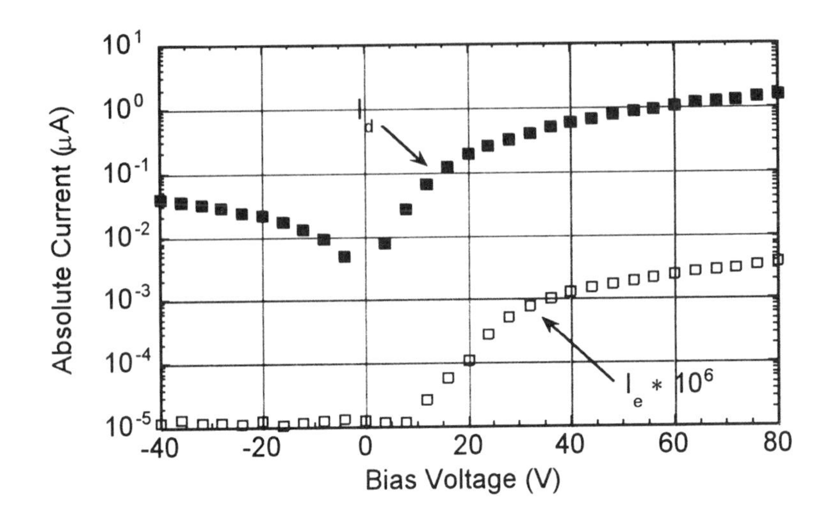

Figure 5: Plot of the absolute junction diode current and the emitted current from a 50 μm diameter device operating at 200 °C.

The junction devices were fabricated on type Ib, 4×4×0.25 mm^3, (110) oriented synthetic diamond substrates. A 2 μm thick phosphorus doped film was grown on the substrates and a 0.5 μm thick boron doped film was grown over the phosphorus doped film. Following the growth of the phosphorous and boron doped diamond epi-layers, the phosphorous doped diamond layer was exposed in select areas by reactive ion etching. Electrical contact to the P-doped diamond layer and to the B-doped diamond layer was made using 50 nm Ti and 200 nm Au. The contacts were annealed to 425 °C in a vacuum.

There were sixteen diode devices per diamond crystal. Every device tested was rectifying and had a large series resistance, most likely from the phosphorus doped diamond layer, that limited the current when the device was forward biased. The performance and temperature dependence of the device were consistent with phosphorus being a high activation energy (0.7 eV) donor impurity. The emission current from the forward biased emitter was ~10^{-8} times the diode current (Figure 5); the fraction of the diode current emitted into the vacuum was independent of temperature and nearly independent of bias voltage. Presumably when the diode was forward biased some of the electrons from the P-doped diamond were injected into the B-doped layer, diffused to the surface and were emitted into the vacuum.

Diamond junction cathodes, more so than other diamond-based cathodes, are in their infancy, and their advantages and limitations have not been fully identified. The identification of a low activation energy donor impurity, or other electrical means of populating the conduction band, would hasten development of this device.

4. Field Emission Cathodes and Displays

An alternate flat panel display technology is being sought because of the high production costs, defficient image quality, and high power requirements of liquid crystal displays (LCDs). Flat panel field emission displays (FED) are an attractive alternative to LCD displays. FEDs offer the color and brilliance of a conventional cathode ray tube (CRT) with low power requirements when high voltage phosphors are used.

Like a conventional CRT, electrons from the cathode in a FED are accelerated towards a phosphor-coated glass screen (Figure 6). But while a conventional CRT has at most three electron guns to address all pixels (one gun per color), field-emission flat panel displays use hundreds of emitters or micron-sized electron guns for each pixel [Spindt *et al.* 1976]. Because diamond emits electrons readily in the presence of modest electric fields (a few volts per micron), matrix-addressed, diamond-based field emission cathodes (FECs) are being considered for use in field-emission flat panel displays.

Although flat panel displays represent the biggest market for diamond-based field emission cathodes, there are numerous other applications. These range from pressure sensors to high frequency electronics to light bulbs. These markets can be significant, and the capital investment requirements are not as large as required for flat panel display production.

4.1 FED's AND CATHODE REQUIREMENTS

The cathode in a field emission display is just one part of a complicated and sophisticated structure. There are two basic field emission display designs: diode configuration and triode configuration (Figure 6). Both designs capitalize upon recent advances made in the microelectronics and thin-film deposition industries. With the gated cathode (triode) configuration, electrons are emitted from the cathode when a suitable potential difference is applied between mutually orthogonal row and column electrodes (also known as base and gate electrodes). A pixel is made to luminesce when the emitted electrons strike the phosphor; an image is displayed by varying the current striking each phosphor dot. With the diode configuration, the phosphor and transparent electrode are patterned to form the column electrodes. Electron emission occurs at a pixel when a suitably-large potential difference is applied between a phosphor column and an emitter row.

There are advantages and disadvantages to both approaches. The cathode for the triode display structure requires the deposition of several thin film layers and at least three patterning steps. The number of defects per cathode increases dramatically with device complexity, making yield issues a concern. The cathode in the diode display structure is simpler, but the driver electronics required to switch and regulate the large voltage between the phosphor and the emitter are expensive, even when low-voltage phosphors

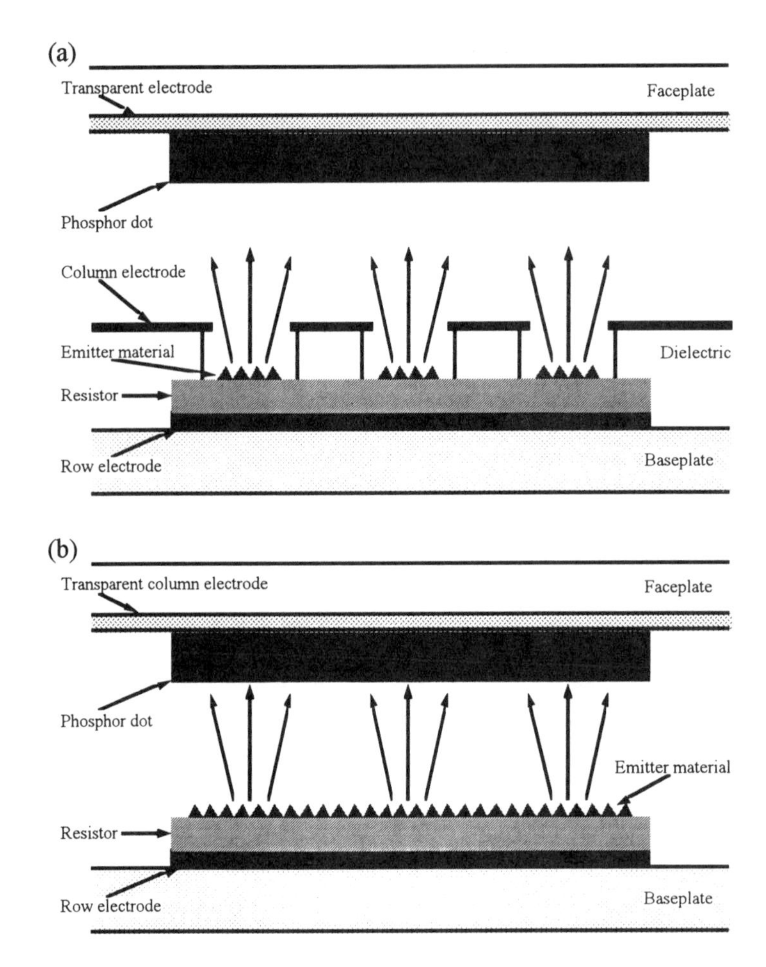

Figure 6: Schematics of diamond-based field emission displays. (a) triode structure and (b) diode structure.

are used. High-voltage phosphors are preferred from a visual-quality and reduced power requirement standpoint, but the standoff and focusing requirements are challenging.

The design of the display will determine the specifics of how the cathode must perform, but there are cathode requirements common to all display designs.

First, the display must be priced competitively and consequently, the cathode must be inexpensive. To keep pricing competitive, low cost materials and fabrication methods must be employed. Preferably the substrate will be glass which is flat, inexpensive, and easily sealed to the phosphor faceplate. Silicon or ceramic substrates have been considered, but the former is conductive and brittle and the latter is rough, or if polished, expensive. The deposition temperature of the emitter and other cathode layers may limit the choice of substrate material. Where possible, particularly for the non-emitter parts of the cathode, large area, high yield deposition systems and patterning approaches should be used to minimize the cathode cost.

The emission current per pixel must be well-regulated so that the light emitted by the phosphor can be controlled and pixel-to-pixel brightness fluctuations are minimized. A field emitter operating in 10^{-6} to 10^{-7} Torr vacuum, a pressure range common to tubes, will exhibit temporal emission current fluctuation due to surface contamination. Furthermore, the microscopic morphology of each emitter, and consequently, its performance as a function of macroscopic field, will vary from site to site. Emitter non-uniformity and surface contamination will cause undesired fluctuations in pixel brightness unless corrective measures are taken. One approach is to increase the number of emission sites per unit area, so that there are many operational emitters per pixel. Although the performance of individual emitters may vary significantly, a large number of operating emitters per pixel insures that the average variation is small. A large number of emitters per pixel also provides some redundancy, a necessary feature given the high probability of emitter failure during the display lifetime. Emitter non-uniformity can also be partially compensated for by incorporating a resistor layer into the cathode (Figure 6) [Lee 1986].

The emission current density from the emitter must be ~1 mA cm^{-2}. The angular divergence of the emitted electrons must be controlled so that there is no appreciable crosstalk between pixels. Finally the emitter must be robust so that the display lifetime is long.

4.2 DIAMOND-BASED GATED EMITTERS

A diamond-based gated cathode can be formed with processing steps that follow either a build-up or a backfill approach. In the build-up approach, the diamond emitter is deposited after the deposition and patterning of the row metal and resistor layer. The entire emitter/resistor/row-electrode structure is covered with dielectric and the gate metal. The gate metal is etched to form the column electrodes. The gate openings are patterned and the column electrode metal and dielectric are etched to form the cavity and expose the emitter material. In the backfill approach, the patterned row electrode and resistor metal are covered with dielectric. The column metal is deposited and patterned to form the column electrodes, and the gate openings are patterned. The gate openings and dielectric cavities are etched. The diamond emitter material is deposited as the final step.

Both the build-up or backfill approaches could benefit from the ability to form diamond emitters in select areas. Selective seeding for growing diamond on Si substrates was performed with conventional lithography using photoresist with fine diamond particles [Katsumata *et al.* 1994, Hong and Aslam 1995]. A selectivity up to 200 or higher was achieved with a resolution of the order of 1 μm. Good emission performance was obtained from arrays of emitters that were fabricated with the mold technique [Okano *et al.* 1995]. Other researchers found improved emission performance by forming sharp diamond structures with reactive ion etching and ion-milling [Twichell *et al.* 1993, Asano *et al.* 1995, Kang *et al.* 1995].

Ideally, to reduce fabrication costs and increase yield it would be advantageous to form the emitter without requiring fine scale lithography or a high temperature processing step. It is also important to have a large number of emission sites ($>10^5$ cm^{-2}) for uniformity, stability and redundancy requirements.

One approach to increasing the emitter site density is to use size-sorted diamond grit, purchased from commercial vendors, to form the cathode [Jaskie 1994]. If the diamond particles are uniformly sized and shaped, a similarly-sized electric field should be required for emission and the site density should scale with the number of particles. By using diamond grit particles a CVD growth step would be eliminated or made shorter, decreasing manufacturing costs. There remains a need, however, for establishing electrical and mechanical contact between the diamond particles and the underlying resistor layer [Twichell *et al.* 1994]. The contact may be obtained through diamond overgrowth or other chemical processes.

There are problems with using diamond grit and either the backfill or build-up approach. With backfilling, it is difficult to control where the emitter is deposited. If the diamond coats the dielectric walls of the cavity, conductive paths that greatly increase the row-column current (and hence the display power requirements) will develop and the electric field within the gate opening will change. With the build-up process, a substantial fraction of the diamond emitter will remain under the dielectric after the cavity regions are formed. Dielectric breakdown is likely to occur in regions where emitter material has been buried. Also, the emitter surface can be damaged when the gate holes are etched.

To circumvent site density and uniformity problems, some researchers have explored ways to capitalize upon the significant amount of work conducted to develop silicon microtip emitters. By coating the emitter cone tips with diamond or diamondlike materials, the robustness of the emitter should increase and the voltage requirements decrease. The deposition of diamond on high aspect ratio silicon tips was achieved by adjusting the growth conditions and the performance of the coated emitters was examined [Givargizov *et al.* 1993, Liu *et al.* 1994, Givargizov *et al.* 1995, Liu *et al.* 1995]. Field emission from the diamond coated field emitters exhibited significant enhancement both in total emission current and stability compared to pure silicon emitters. Complete cladding of the silicon microtips with diamond also improved the emitter performance [Cheng *et al.* 1995]. High emission currents at low voltages were observed from blunt coated tips, perhaps arising from microprotrusions at the surface [Zhirnov *et al.* 1995]. Improved stability and performance was observed from field emission array cathodes coated with carbon films, although the process was sensitive to the film deposition technique and chemistry [Djubua and Chubun 1991, Mousa 1995].

4.3 DIAMOND-BASED CATHODES FOR DIODE DISPLAYS

The diode display consists of an emitter patterned into row electrodes and a phosphor and transparent conductor patterned into orthogonal column electrodes (Figure 6). Pixels are formed at the intersection of row and phosphor-coated column electrodes. As with the triode design, the emitter site-density must be high.

A FED that uses the diode geometry is being developed; four inch mono-chrome displays have been fabricated [Kumar *et al.* 1995]. The emitter material for the FEDs is a hard, nanocrystalline (10 to 50 nm diameter) carbon thin film with an sp^3 to sp^2 ratio of approximately four. The film is deposited with a laser ablation technique that permits low-temperature deposition, but suffers from low deposition rates [Wagel and Collins 1991, Collins and Davanloo 1992]. The carbon material emits electrons at fields less than 20 V μm^{-1} and at peak current densities of 10 A cm^{-1}. An alternate, higher-rate, emitter deposition technique, pulsed ion beam ablation, was investigated [Johnston *et al.* 1994]. Initial tests revealed turn-on around 10 V μm^{-1}, as measured with a 0.2 cm diameter ball probe placed 30 μm from the substrate, and linear Fowler-Nordheim behavior.

5. Summary

Significant progress elucidating the properties of the diamond vacuum electronic system has been made in the past few years. The investigations have demonstrated that the affinity of hydrogenated diamond surfaces was negative and removal of the hydrogen made the affinity positive. Three types of devices based upon excited-state emission from NEA diamond --- photocathodes, dynodes, and junction cathodes --- have been fabricated and tested. The studies have also shown that diamond emits electrons in the presence of low electric fields; the emission mechanism has yet to be conclusively determined. Devices that capitalize upon diamond's low electric-field emissive properties, most notably matrix addressed cathodes for flat panel displays, are being developed. The Future investigations are certain to reveal more of the fundamental physics of this fascinating system, and further progress will be made in the development of diamond vacuum electronic devices.

An attempt was made when writing this chapter to include all relevant work published before the time of writing (spring 1996). In a field as active as diamond vacuum electronics it is likely that significant new material will appear in publication before this work is published, but I hope the reader will find the discussions here a useful starting point. The author is grateful for numerous informative discussions with the employees of Candescent Technology Corporation, particularly J. Macaulay and C. Spindt, as well as M. Geis, J. Twichell, B. Pate and X. Xu. This work was supported in part by a BMDO funded and ONR monitored R&D program (N00014-93-C-0264) and by the National Science Foundation (III-9361179).

6. References

Asano, T., Oobuchi, Y. and Katsumata, S. (1995) Field emission from ion milled diamond films on Si, *J. Vac. Sci. Technol. B*, **13**, 431-434.

Bajic, S. and Latham, R. V. (1988) Enhanced cold-cathode emission using composite resin-carbon coatings, *J. Phys. D: Appl. Phys.*, **21**, 200-204.

Bandis, C., Haggerty, D. and Pate, B. B. (1994) Electron emission properties of the negative electron affinity (111) 2×1 diamond-TiO interface, *Mat. Res. Soc. Symp. Proc.*, **339**, 75-80.

Bandis, C. and Pate, B. B. (1995a) Electron emission due to exciton breakup from negative electron affinity diamond, *Phys. Rev. Lett.*, **74**, 777-780.

Bandis, C. and Pate, B. B. (1995b) Photoelectron emission from negative electron affinity diamond (111) surfaces: exciton breakup versus conduction-band emission, *Phys. Rev. B*, **52**, 12056-12071.

Beetz, C. P., Lincoln, B., Segall, K., Wall, D., Vasas, M., Winn, D. R., Doering, D. and Carroll, D. (1991) Applications of diamond films to photocathode electron guns, *IEEE Part. Accel. Conf., 14th}*, **3**, 1981-1983.

Bekker, T. L., Dayton, Jr., J. A., Gilmour, Jr., A. S., Krainsky, I. L., Rose, M. F., Rameshan, R., File, D. and Mearini, G. (1992) Observations of secondary electron emission from diamond films, *Int. Electron Dev. Manuscript*, **92**, 949-952.

Bouilov, L. L., Chapliev, N. I., Kvaskov, V. B., Konov, V. I., Pimenov, S. M., Spytsin, B. V., Teremetskaya, I. G. and Zlobin, A. Yu. (1991) Photoelectron emission from diamond films, *Surf. Coatings and Tech.*, **47**, 481-486.

Boutot, J. P., Nussli, J. and Vallat, D. (1983) Recent trends in photomultipliers for nuclear physics, *Adv. Electron. Electron Phys.*, **60**, 223-305.

Brandes, G. R., Beetz, C. P., Feger, C. A. and Wright, R. L. (1995) Diamond junction cold cathode, *Diamond and Rel. Mater.*, **4**, 586-590.

Candy, B. H. (1985) Photomultiplier characteristics and practice relevant to photon counting, *Rev. Sci. Instrum.*, **56**, 183-193.

Cheng, H.-C., Ku, T.-K., Hsieh, B.-B., Chen, S.-H., Leu, S.-Y., Wang, C.-C., Chen, C.-F., Hsieh, I.-J. and Huang, J. C. M. (1995) Fabrication and characterization of diamond clad silicon field emitter arrays, *Jpn. J. Appl. Phys.*, **34**, 6926-6931.

Collins, C. B. and Davanloo, F. (1992) U. S. Patent # 5,098,737, Amorphic diamond material produced by laser plasma deposition.

Djubua, B. C. and Chubun, N. N. (1991) Emission properties of Spindt-type cold cathodes with different emission cone material, *IEEE Trans. Electron Dev.*, **38**, 2314-2316.

Eimori, N., Mori, Y., Hatta, A., Ito, T. and Hiraki, A. (1994) The electron affinity of CVD diamond with surface modifications, in I. Ohdomari, M. Oshima and A.

Hiraki (eds.), *Control Semicond. Interfaces, Proc. 1st Int. Symp.*, Elsevier Science B. V., 149-154.

Eimori, N., Mori, Y., Hatta, A., Ito, T. and Hiraki, A. (1995) Photoyield measurements of CVD diamond, *Diamond and Rel. Mater.*, **4**, 806-808.

Fischer, J., Srinivasan-Rao, T., Tsang, T. and Brandes, G. R. (1994) Photoemission from magnesium and from diamond film using high intensity laser beams, *Nucl. Instrum. Meth. in Phys. Res. A*, **340**, 190-194.

Fowler, R. H. and Nordheim, L. (1928) Electron emission in intense electric fields, *Proc. Phys. Soc. London A*, **119**, 173-181.

Geis, M. W., Efremow, N. N., Woodhouse, J. D., McAleese, M. D., Marchywka, M., Socker, D. G. and Hochedez, J. F. (1991) Diamond cold cathode, *IEEE Electron Dev. Lett.*, **12**, 456-459.

Geis, M. W., Twichell, J. C., Macaulay, J. and Okano, K. (1995) Electron field emission from diamond and other carbon materials after H_2, O_2 and Cs treatment, *Appl. Phys. Lett.*, **67**, 1328-1330.

Geppert, D. V. (1966) Research on cold cathodes, *Proc. IEEE Lett.*, **54**, 61-66.

Givargizov, E. I., Zhirnov, V. V., Kuznetsov, A. V. and Plekhanov, P. S. (1993) Growth of diamond particles on sharpened silicon tips, *Mater. Lett.*, **18**, 61-63.

Givargizov, E. I. (1995) Silicon tips with diamond particles on them: New field emitters?, *J. Vac. Sci. Technol. B*, **13**, 414-417.

Gomer, R. (1961) *Field Emission and Field Ionization*, Harvard University Press, Cambridge, MA.

Himpsel, F. J., Knapp, J. A., VanVechten, J. A. and Eastman, D. E. (1979) Quantum photoyield of diamond (111) — A stable negative-affinity emitter, *Phys. Rev. B*, **20**, 624-627.

Hoffman, A. (1994) Fine structure in the secondary electron emission spectrum as a spectroscopic tool for carbon surface characterization, *Diamond and Rel. Mater.*, **3**, 691-695, and references therein.

Hong, D. and Aslam, M. (1995) Field emission from p-type polycrystalline diamond films, *J. Vac. Sci. Technol. B*, **13**, 427-430.

Huang, Z.-H., Cutler, P. H., Miskovsky, N. M. and Sullivan, T. E. (1994) Theoretical study of field emission from diamond, *Appl. Phys. Lett.*, **65**, 2562-2564.

Huang, Z-H., Cutler, P. H., Miskovsky, N. M. and Sullivan, T. E. (1995) Calculation of electron field emission from diamond surfaces, *J. Vac. Sci. Technol. B*, **13**, 526-530.

Humphreys, V. L. and Khachan, J. (1995) Spatial correlation of electron field emission sites with non-diamond carbon content in CVD diamond, *Elec. Lett.*, **31**, 1018-1019.

Jaskie, J. E., Dworsky, L. and Kane, R. C. (1994) U. S. Patent # 5,278,475, Cathodeluminescent display apparatus and method for realization using diamond crystallites.

Johnson, J. B. (1953) Secondary electron emission from diamond, *Phys. Rev.*, **92**, 843-843.

Johnston, G. P., Tiwari, P., Rej, D. J., Davis, H. A., Waganaar, W. J., Muenchausen, E., Walter, K. C., Nastasi, M., Schmidt, H. K., Kumar, N., Lin, B., Tallant, D. R., Simpson, R. L., Williams, D. B. and Qiu, X. (1994) Preparation of diamondlike carbon films by high intensity pulsed-ion-beam deposition, *J. Appl. Phys.*, **76**, 5949-5954.

Kang, W. P., Davidson, J. L., Li, Q., Xu, J. F., Kinser, D. L. and Kerns, D. V. (1995) Patterned polycrystalline diamond microtip vacuum diode arrays, in A. Feldman, Y. Tzeng, W. A. Yarborough, M. Yoshikawa and M. Murakawa (eds.), *Proc. of the Appl. Diam. Conf. 1995.*, NIST, Gaithersburg, MD, 37-40.

Katsumata, S., Oobuchi, Y. and Asano, T. (1994) Patterning of CVD diamond films by seeding and their field emission properties, *Diamond and Rel. Mater.*, **3**, 1296-1300, and references therein.

Kordesch, M. E. (1993) Electron emission microscopy for *in situ* studies of diamond surfaces and CVD diamond nucleation and growth, in J. P. Dismukes and R. K. Ravi (eds.), *Proceedings of the 3rd International Symposium on Diamond and Related Materials*, The Electrochemical Society, Pennington, NJ, 787-793.

Kumar, N., Schmidt, H. and Xie, C. (1995) Diamond-based field emission flat panel displays, *Sol. State Tech.*, **38**, 71-74.

Latham, R. V. and Xu, N. S., (1991) 'Electron pin-holes': the limiting defect for insulating high voltages by vacuum, a basis for new cold cathode electron sources, *Vacuum*, **42**, 1175-1181.

Lee, K. J. (1986) *Current Limiting of Field Emitter Array Cathodes*, Ph.D. Dissertation, Georgia Institute of Technology, University Microfilms # 8628359.

Liu, J., Zhirnov, V. V., Wojak, G. J., Myers, A. F., Choi, W. B., Hren, J. J., Wolter, S. D., McClure, M. T., Stoner, B. R. and Glass, J. T. (1994) Electron emission from diamond coated silicon emitters, *Appl. Phys. Lett.*, **65**, 2842-2844.

Liu, J., Zhirnov, V. V., Myers, A. F., Wojak, G. J., Choi, W. B., Hren, J. J., Wolter, S. D., McClure, M. T., Stoner, B. R. and Glass, J. T. (1995) Field emission characteristics of diamond coated silicon field emitters, *J. Vac. Sci. Technol. B*, **13**, 422-426.

Lurie, P. G. and Wilson, J. M., (1977) The diamond surface II. Secondary electron emission, *Surf. Sci.*, **65**, 476-498.

Maguire, H. G. (1976) Investigation of the band structure of diamond using low energy electrons, *Phys. Status Solidi B*, **76**, 715-726.

Malta, D. P., Posthill, J. B., Humphreys, T. P., Thomas, R. E., Fountain, G. G., Rudder, R. A., Hudson, G. C., Mantini, M. J. and Markunas, R. J. (1994) Secondary electron emission enhancement and defect contrast from diamond following exposure to atomic hydrogen, *Appl. Phys. Lett.*, **64**, 1929-1931.

Martinelli, R. U. and Fisher, D. G. (1974) Application of semiconductors with negative electron affinity surfaces to electron emission devices, *Proc. of the IEEE*, **62**, 1339-1360.

Mearini, G. T., Krainsky, I. L. and Dayton, Jr., J. A. (1994) Investigation of diamond films for electronic devices, *Surface and Interface Analysis*, **21**, 138-143.

Mearini, G. T., Krainsky, I. L., Dayton, Jr., J. A., Wang, Y., Zorman, C. A., Angus, J. C., Hoffman, R. W. and Anderson, D. F. (1995) Stable secondary electron emission from chemical vapor deposited films coated with alkali-halides, *Appl. Phys. Lett.*, **66**, 242-244.

Modinos, A. (1984) *Field, Thermionic and Secondary Electron Emission Spectroscopy*, Plenum, NY.

Mousa, M. S. (1994) Investigations of *in situ* carbon coating on field emitter arrays, *Vacuum*, **45**, 241-244.

Okano, K., Kiyota, H., Iwasaki, T., Nakamura, Y., Akiba, Y., Kurosa, T., Iida, M. and Nakamura, T. (1990) Synthesis of n-type semiconducting diamond film using diphosphorus pentaoxide as the doping source, *Appl. Phys. A*, **51**, 344-346.

Okano, K., Hoshina, K., Koizumi, S. and Itoh, J. (1995) Fabrication of a miniature-size pyramidal-shape diamond field emitter array, *IEEE Electron Dev. Lett.*, **16**, 239-241, and references therein.

Okano, K. and Gleason, K. K. (1995) Electron emission from phosphorus- and boron-doped polycrystalline diamond films, *Elec. Lett.*, **31**, 74-75.

Pate, B. B. (1986) The diamond surface: atomic and electronic structure, *Surf. Sci.*, **165**, 83-142.

Pickett, W. E. (1994) Negative electron affinity and low work function surface: cesium on oxygenated diamond (100), *Phys. Rev. Lett.*, **73**, 1664-1667.

Setaka, N. (1989) A few problems in the vapor deposition of diamond, *Mat. Res. Soc. Symp. Proc.*, **152**, 3-8.

Shovlin, J. D. and Kordesch, M. E. (1994) Electron emission from chemical vapor deposited diamond and dielectric breakdown, *Appl. Phys. Lett.*, **65**, 863-865.

Shovlin, J. D., Kordesch, M. E., Dunham., D., Tonner, B. P. and Engel, W. (1995) Synchrotron radiation photoelectron emission microscopy of chemical vapor-deposited diamond electron emitters, *J. Vac. Sci. Technol. A*, **13**, 1111-1115.

Spindt, C. A., Brodie, I., Humphrey, L. and Westerberg, J. (1976) Physical properties of thin-film field emission cathodes with molybdenum cones, *J. Appl. Phys.*, **47**, 5248-5263.

Twichell, J. C., Geis, M. W., Bozler, C. O., Rathmann, D. D., Efremow, N. N., Krohn, K. E., Hollis, M. A., Uttaro, R., Lyszczarz, T. M., Kordesch, M. E. and Okano, K. (1993) Diamond field-emission cathodes, presented at Diamond Films '93, Algarve, Portugal.

Twichell, J. C., Brandes, G. R., Geis, M. W., Macaulay, J. M., Duboc, Jr. R. M. and Curtin, C. J. (1994) submitted.

Van der Weide, J. and Nemanich, R. J. (1992) Schottky barrier height and negative electron affinity of titanium on (111) diamond, *J. Vac. Sci. Technol. B*, **10**, 1940-1943.

Van der Weide, J. and Nemanich, R. J. (1993) Argon and hydrogen plasma interactions on diamond (111) surfaces: Electronic states and structure, *Appl. Phys. Lett.*, **62**, 1878-1880.

Van der Weide, J. and Nemanich, R. J. (1994) Angle-resolved photoemission of diamond (111) and (100) surfaces; negative electron affinity and band structure measurements, *J. Vac. Sci. Technol. B*, **12**, 2475-2479.

Van der Weide, J., Zhang, Z., Baumann, P. K., Wensell, M. G., Bernholc, J. and Nemanich, R. J. (1994) Negative electron affinity effects on the diamond (100) surface, *Phys. Rev. B*, **50**, 5803-5806.

Wagel, S. S. and Collins, C. B. (1991) U. S. Patent # 4,987,007, Method and apparatus for producing a layer of material from a laser ion source.

Wang, C., Garcia, A., Ingram, D. C., Lake, M. and Kordesch, M. E. (1991) Cold field emission from CVD diamond films observed in emission electron microscopy, *Electron. Lett.*, **27**, 1459-1461.

Williams, B. F. and Tietjen, J. J. (1971) Current status of negative electron affinity devices, *Proc. IEEE*, **59**, 1489-1497.

Xu, N. S., Tzeng, Y. and Latham, R. V. (1993) Similarities in the 'cold' electron emission characteristics of diamond coated molybdenum electrodes and polished bulk graphite surfaces, *J. Phys. D: Appl. Phys.*, **26**, 1776-1780.

Xu, N. S., Tzeng, Y. and Latham, R. V. (1994) A diagnostic study of the field emission characteristics of individual micro-emitters in CVD diamond films, *J. Phys. D: Appl. Phys.*, **27**, 1988-1991.

Yamashita, M. (1980) Time dependence of rate-dependent photomultiplier gain and its implications, *Rev. Sci. Instrum.*, **51**, 768-775.

Zhirnov, V. V., Givargizov, E. I. and Plekhanov, P. S. (1995) Field emission from silicon spikes with diamond coatings, *J. Vac. Sci. Technol. B*, **13**, 418-421.

Zhu, W., Kochanski, G. P., Jin, S., Seibles, L., Jacobson, D. C., McCormack, M. and White, A. E. (1995) Electron field emission from ion-implanted diamond, *Appl. Phys. Lett.*, **67**, 1157-1159.

Zvanut, M. E., Carlos, W. E., Frietas, J., and Jamison, K. (1994) Identification of phosphorus in diamond thin films using electron paramagnetic-resonance spectroscopy, *Appl. Phys. Lett.*, **39**, 2287-2289.

Chapter 32
SPECIAL APPLICATIONS

Naoji Fujimori
Sumitomo Electronic Industries, Ltd.
1-1, 1-chome Koyakita, Itami, Hyogo 664, Japan

Contents

Many attempts have been made to apply CVD diamonds in other than wear resistant parts and cutting tools. Diamond has many excellent properties but was hard to use for the following reasons:

1. Diamond was costly
2. Diamond was hard to machine and machining costs are high
3. The shape of pure diamond was limited in particles
4. Large diamond was hard to obtain and was extremely costly

Most important aspect for CVD diamond is improving conventional diamond technology solving those problems mentioned above. We can obtain pure, large and inexpensive diamonds using CVD technology. Even thin and dome shaped diamonds have already been commercialized.

1. Speaker Diaphragms

The material for speaker diaphragms has been developed utilizing materials which have high sound velocities. The development succeeded to realize advanced materials such as titanium, beryllium, boron, etc.

However, diamond is well known as the most suitable material for acoustic parts because diamond has the largest sound velocity. Table 1 shows the sound velocity of various materials including diamond.

Normally, speaker diaphragms require a stiff structure, usually in the shape of a dome. Making dome shaped diamond looks quite difficult. A diamond coated speaker diaphragm was developed as the first product of CVD diamond in 1987. Using an alumina diaphragm 30μm thick, a 1 or 2 μm thick diamond film was coated by

Table 1. Sound velocity of diamond and various materials

Material	Diamond	Beryllium	Alumina	Titanium	Aluminum
Sound Velocity, km/s	18.2	12.6	11.6	4.9	5.1

microwave plasma assisted CVD. As the adhesion between alumina and diamond was poor, intermediate layer was needed. SiC film was chosen for this film. The characteristics of diamond coated diaphragm was quite clear that diamond coating improved the highest frequency generated from 35,000 to 50,000 Hz. Although the volume percentage of diamond film was low, the effect of the film was too large. The author believes that the effect was caused by the residual stress in the diamond film.

A pure diamond diaphragm would be ideal, as evidenced by the data in table 1. The poor machinability makes it impossible to make a diaphragm of single crystal or sintered diamond. A free standing diamond diaphragm fabricated by the CVD method was developed in 1990. Diamond deposition was carried out my hot filament CVD on a silicon substrate. After deposition, silicon substrates were dissolved by a mixed acid of HF and HNO_3 to obtain free standing diamond diaphragm. Figure 1 is a photograph of this diaphragm. The diameter and height of the diaphragm is 30mm and 5mm. The thickness of diamond is 30 μm and the distribution of thickness is controlled to within 10%. Although this diaphragm is easy to break, it is quite strong against the vibration of sound generation.

The sound velocity was measured by the vibration method and was 16,500 m/s, which is a little less than the value for single crystalline diamond. The properties of the speaker which this diaphragm is used are shown in figure 2. The highest frequency of this speaker was 80,0000 Hz, which is the highest sound that a dynamic speaker can generate.

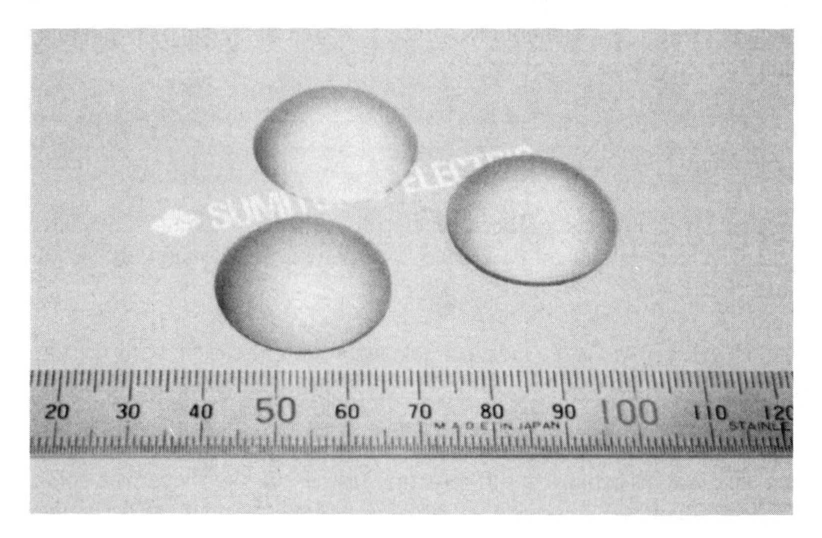

Figure 1. Diamond Speaker Diaphragm

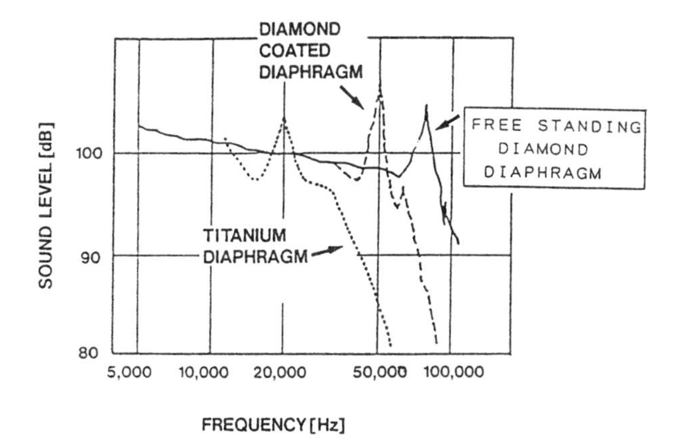

Figure 2. Characteristics of speaker using diamond coated diaphragm and diamond diaphragm

2. Surface Acoustic Wave Filter

We may expect that diamond will be applied in various passive devices and some trials have been made since CVD method of diamond was established. Stability at higher temperature, piezoresistive and photoconductivity are the candidate characteristics to be utilized. Thermistors (temperature sensor), strain gages and UV light sensors have been demonstrated.

Another expected application of diamond is surface acoustic wave (SAW) filters. SAW filters are popular devices in televisions, video recorders, mobile telephones, etc. Introduced electric signal is converted to sound signal and is reconverted to electric signal. Filtering is performed during these conversions and the higher frequency signal requires a higher sound velocity. As diamond has the highest sound velocity, we may expect high surface acoustic wave velocity. SAW devices need piezoelectricity, but diamond is not a piezoelectric material. Figure 3 shows a schematic of a diamond SAW filter.

ZnO film is used as a piezoelectric material and this structure allows the manufacture of a SAW filter. Diamond film deposited on a silicon substrate is polished by conventional methods, and an aluminum film was deposited on the diamond film. The interdigital transducer was formed by the normal lithography process. Finally, ZnO film was deposited on the transducer by sputtering. Calculated and measured SAW velocity are quite similar and the comparison of materials for SAW filters are shown in table 2.

Diamond SAW filters are essentially suitable devices for high frequency signals and the limitation of frequency for conventional materials are thought to be 2.5 GHz.

Higher than this frequency is required mostly multimedia such as satellite communication, optical communication, mobile telephones, etc. Nakahata et al. Demonstrated diamond SAW filter and one of the result is shown in figure 4.

Table 2. Characteristics of Diamond SAW filter compared with other materials

Material	Sound velocity, (m/s)	Frequency using 1μm IDT (GHZ)	Required IDT size for 2.5 GHz filter (μm)
$LiNbO_2$	3500	0.9	0.35
Quartz	3200	0.8	0.32
ZnO / Sapphire	5500	1.4	0.55
ZnO / Diamond	10000	2.5	1.00

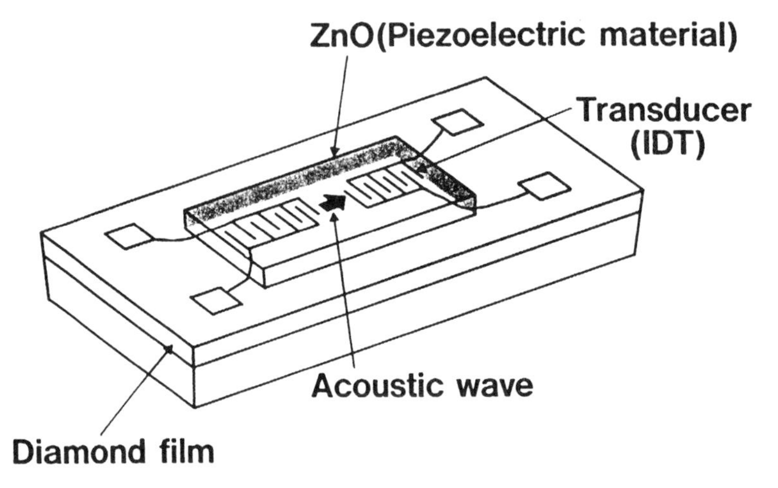

Figure 3. Schematic of diamond SAW filter

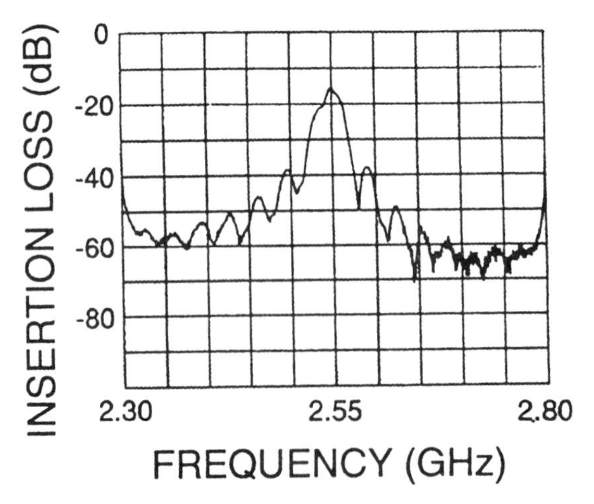

Figure 4. Characteristics of diamond SAW filter

This SAW filter was designed as 2.5 GHz and the characteristics was completely fit the simulated filter. The SAW velocity if this filter was 9500 m/s and the interdigital transducer was formed using 0.9 m lines and spaces.

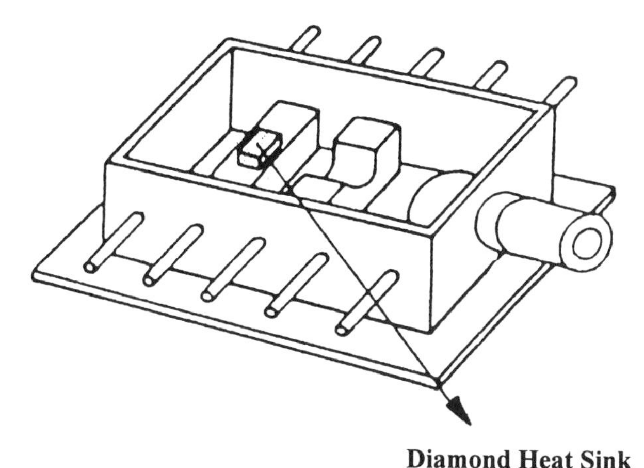

Figure 5. Schematic of a diamond heat sink in an optical communications system

ZnO is suitable material for diamond SAW filter because the strongly oriented poly-crystalline film is grown on the diamond film.

CVD methods of diamond allows the fabrication of diamond wafer and the price of this device is promised to reduce in the near future. Realizing cheap price will be key for commercialization. The author expects SAW filters will be used in various communication instruments and will contribute to life.

3. Heat Sinks

Devices using high power need heat sinks in order to avoid temperature rising because their characteristics change at higher temperatures. Optical communication requires quite stable operation of laser diode and diamond heat sinks are used in the module of optical fiber. Recently, the optical amplifying system using erbium doped optical fiber were developed and high power laser diodes are used in submarine optical fiber. This demands large heat sinks, whose length is 1 cm or longer. As large single crystalline diamonds are expensive, CVD diamonds are used.

Single-crystalline diamonds synthesized by high pressure methods are used for heat sinks for laser diodes or impat diodes which generate quite a large amount of heat in a small chip. Diamond's ultra-high heat conductivity enable to operate stably when protected by heat sinks. A schematic of diamond heat sink for laser diode is shown in figure 5.

Heat sinks are the most obvious application of CVD diamond because of the potential to be fabricated with low cost. The heat conductivity of CVD diamond depends on the conditions of deposition. The author measured the heat conductivity of 300 μm thick

CVD diamond film obtained byt hot filament CVD and result was 1600 W/m K. This value is considered to be sufficient for most applications in the same manner as single crystal diamonds. We may expect CVD diamond heat sinks to become popular because of the low cost of synthesis.

Thick diamond films (typically 300 μm) are deposited on silicon substrates and the grown surface is polished by conventional methods. Diamond is easily cut with YAG lasers, and small pieces are obtained. The width of cutting is typically 20 μm or thinner.

Various types of heat sinks made of CVD diamonds are shown in figure 6. Most of them are metallized in order to bond devices to the heat sink and the heat sink to the submount. Strong adhesion is required because bonding laser diode and submount cause misfit due to the difference in the coefficient of thermal expansion. Titanium is the best material for diamond and the multilayer structure is used for the strong bonding.

Diamond heat sinks contribute especially to optical communication systems, and we can expect it to be utilized widely.

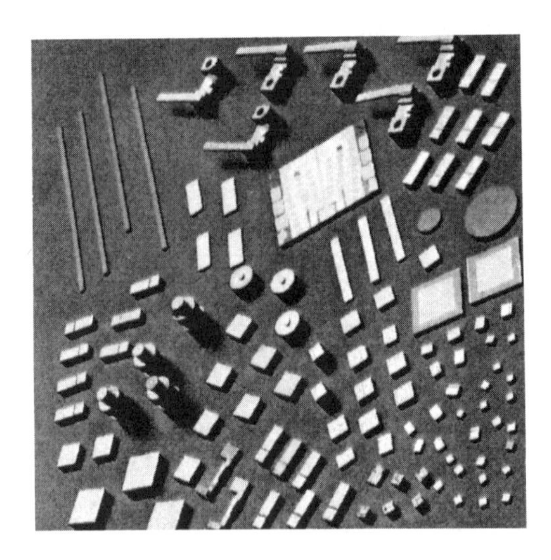

Figure 6 Various heat sinks made of diamond.

5. Reference

Nakahata, H., Hachigo, A., Shikata, S., Fujimori, N., *IEEE Ultrasonic Symposium,* 377 (1992).

Chapter 33
CUTTING AND WEAR APPLICATIONS

R.A. Hay and J. M. Galimberti
Norton Diamond Film
Goddard Road, Northboro, MA 01532-1545, USA

Contents

1. Introduction

Diamond as a cutting tool material predates diamond as a gem. From the time they were first found in Southern India, these "hard stones" were used for engraving, thus becoming the first diamond tools. They were later embedded into rods and used as tools to drill holes in stone and rock. They were also embedded into iron plates to become lapping plates. Diamond cutting tools as we know them today began when natural diamonds were attached to pieces of steel. Natural diamond cutting tools are still in use today, but on a greatly reduced scale as a result of the development of synthesized mesh-size diamonds in the mid-1950's that lead to the compacted diamond product in the early - 1970's. The cutting and wear applications of these diamonds, and of the new chemically vapor deposited diamond are discussed in this article.

2. Types of Tool Diamond

Diamond is available in five forms as a cutting tool material: single crystal natural; single crystal synthesized; high-temperature/high pressure compacted and sintered

Table I. Wear- Related Properties of Diamond

	Single Crystal Diamond	CVD Diamond	PCD
Density (g/cc)	3.52	3.51	4.10
Young's Modulus (Gpa)	1050	1180	800
Compressive Strength (Gpa)	9.0	16.0	7.4
Transverse Rupture Strength (Gpa)	2.9	1.3	1.2
Fracture Toughness (Mpa.m)	3.4	6.5	9.0
Knoop Hardness (Gpa)	50-100	85-100	50-75
Thermal Conductivity (W/mK)	1000-2000	750-1500	500
Thermal Expansion (10^{-6}/k)	2.0-5.0	3.7	4.0

Figure 1. Diamond tipped tool.

polycrystalline (PCD); chemical vapor deposited (CVD) polycrystalline "sheet"; and chemical vapor deposited (CVD) polycrystalline "film" [Hay 1994].

Single-crystal diamond tools, natural or synthetic, are still in use because extremely sharp cutting edges and very low-friction surfaces are attainable with these tools. These properties allow them to impart extremely fine surface finishes to lens in the optics industry. Single-crystal diamonds have cleavage planes allowing them to fracture rather easily along these planes. Cleavage-plane orientation relative to the cutting forces is extremely critical to a single crystal diamond tool's performance. Compacted sintered PCD, CVD diamond film, and CVD diamond sheet on the other hand are polycrystalline and are relatively isotropic, so that crystallographic orientation is a less important factor than in single crystal tools [Field and Pickles 1996].

Conventional PCD tipped inserts became commercially available in the early 1970's. The manufacture of conventional compacted polycrystalline (PCD) diamond-tipped tools such as indexable disposable cutting inserts and round tools (drills, end mills, routers, reamers, etc.) is a three-step process. The first step is to produce diamond particles in a

Table II. Tool Life Comparisons
Turning Mahle A390 Aluminum (18% Si)

Tool Type	Tool Life (minutes)
WC	<1
25 μm PCD	40
CVD "sheet/thick film" diamond	70
1 corner-CVD Thin Film diamond on WC (30μm thick)	60
3 corners - CVD Thin Film diamond on WC (30 μm thick)	180
Test Conditions	
Speed	680 m/min (2,231 sfn)
Feed Rate	0.2 mm/rev (0.008 in/rev)
Depth of cut	1.0 mm (0.040 in)
Insert style	TPG 321 with coolant
Tool life	Flank wear 0.375 mm (0.015 in)

Table III. Erosion Resistance

	Relative Volume Loss
CVD Diamond	1
PCD (Sintered High-Pressure Diamond)	4
Tungsten Carbide (6% Cobalt)	120
Alumina (99.5%)	220
Silicon Carbide	360
Silicon Nitride	920

Test Parameters: 2% SiC in Water (30 grit)
300 Ft./Sec.
45° Impingement

high-pressure, high-temperature die press. The second step is to produce a disc consisting of a compacted diamond layer of 95% diamond plus approximately 5% cobalt metallurgically bonded to a full sintered WC/Co disc, again utilizing a high-pressure, high-temperature process. The third step is to produce a finished tool by cutting the disk by wire electric discharge machining (EDM) into smaller "blanks" for silver solder brazing into a recess in the tool. This is followed by diamond wheel grinding of the exposed diamond surfaces to produce a sharp, chip-free cutting edge "tipped" tool, Figure 1.

3. Cutting and Wear Properties of Diamond

Several outstanding properties of diamond render it an ideal cutting tool material. Three of these properties are diamond's high hardness, it's high thermal conductivity, and it's low coefficient of friction. Diamond is also one of the most chemically inert substances. However, it cannot be used to machine ferrous metals, nickel-based, or titanium-based alloys because of its chemical reactivity with these materials at the high contact pressures and temperatures generated during the machining process. There is,

Figure 2. Diamond coated lathe inserts.

however, an application where these aforementioned limitations do not exist. This use is in the honing of cylinders in cast iron engine blocks. In that application diamond performs well because the heat generated is very low because the speed is very low, the contact pressures are very low, and the surface is flooded with oil.

For machining non-reactive materials such as aluminum and other nonferrous metals, plastics, wood and very abrasive composite materials, diamond has become the cost effective tool material of choice. The wear related properties, tool life, and erosion resistance of diamond and other materials are compared in Tables I, II, and III.

Compacted PCD has greater fracture toughness than CVD diamond or single crystal diamond as reported in Table I as a result of the presence of cobalt (about 5%). However, incomplete densification and the presence of cobalt reduces the wear life compared to CVD diamond. The cobalt also causes graphitization at elevated temperatures (above about 700 °C), as it catalyzes the conversion of diamond into graphite. This greatly reduces the strength and wear resistance of PCD diamond which has been exposed to elevated temperatures, or if the PCD reaches high temperatures locally in the cutting region.

Most producers of compacted polycrystalline diamond (PCD) product for the machining of nonferrous metals and nonmetals produce three "grades" based on the diamond particle size (plus 5% cobalt). These three grades contain either two micron, 10 micron, or 25 micron diamond particles. Experience has shown that the coarser particle grade of 25 micron results in a longer tool life versus the finer particle grade of two micron, whereas the grade of two micron particle size produces a smoother finish on the part surface being machined.

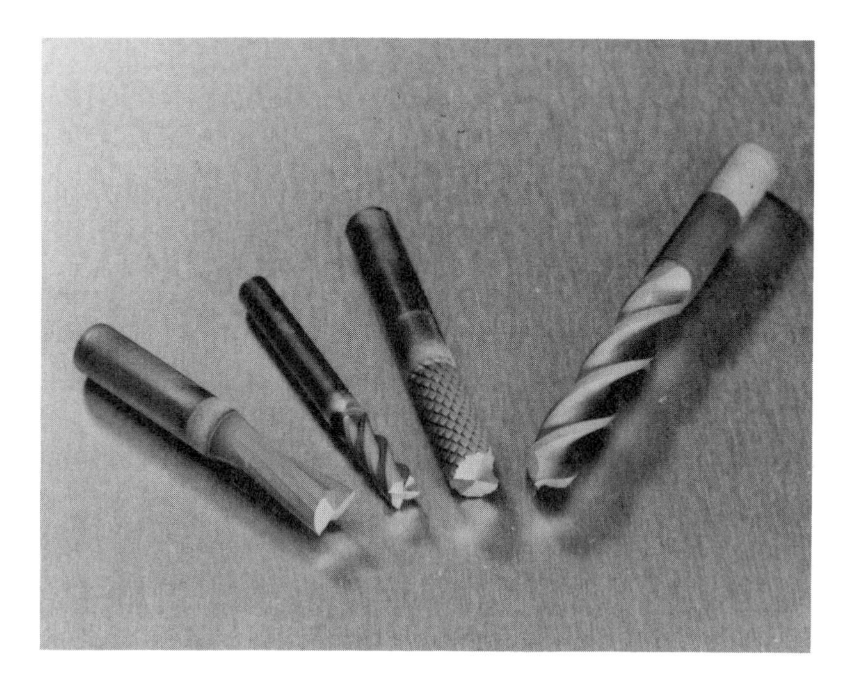

Figure 3. Diamond coated WC/Co round tools.

It is interesting to note, however, that CVD diamond film or sheet (which is 100% diamond) can be controlled to produce a variety of grain sizes from smaller than one micron up to 50 microns. Experience is showing that the smaller grains are as wear resistant as the larger grains.

Lathe inserts coated with CVD diamond became commercially available as prototypes in the early 1990's [Oles et al.1996]. The substrate material was usually silicon nitride or tungsten carbide. All diamond coated lathe inserts have the diamond on the entire face of the insert, including the cutting edges, Figure 2. The coating is 100% diamond, and has a typical thickness of 15 to 30 microns. It is frequently referred to as "thin" film diamond. As an extension of this development, solid diamond coated WC/Co round tools, Figure 3, (drills, end mills, routers, burrs, etc.) are also becoming available for machining nonferrous metals and nonmetals (fiber reinforced plastics, natural wood and manufactured wood).

CVD diamond film coated WC/Co lathe inserts can also be made with rake-face configurations to enhance chip breaking of the machined chip. Such inserts are effective in the machining of many Si-aluminum die cast parts where rather heavy cuts (depth and feed) are used, resulting in chips having considerable strength and natural resistance to deflection.

CVD diamond "thick film" cutting tool material is frequently referred to as diamond "sheet" because of its greater thickness (0.5 mm) versus "thin" film (of 5 to 30 microns). This sheet of diamond is deposited on a temporary substrate, removed as a sheet, laser cut into blanks, then high temperature brazed to a WC/Co substrate. At this point, the general appearance is that of a conventional compacted PCD blank. This composite blank can then be brazed (at a lower temperature) into a recess and further processed similarly to a conventional PCD tool.

4. Applications

4.1. METAL MACHINING

Virtually all PCD tipped lathe inserts available today have a single cutting edge. This edge consists of compacted 95% diamond/5% cobalt, having a thickness of 0.5 mm metallurgically bonded to a WC/Co substrate during the high pressure manufacturing process [Metals Handbook 1989].

All three diamond cutting tool forms - the "compacted" diamond, the CVD film diamond, and the CVD sheet diamond are polycrystalline. The application of these three types of diamond cutting tools is the same. They are used to machine nonferrous metals (aluminum, copper, brass, bronze) and nonmetals (plastics, wood, etc.)[Wilson, et al 1980]. However, the characteristics, performance (tool life), and price vary, all of which impact on cost effectiveness [Hay 1994].

Conventional compacted PCD and CVD diamond sheet lathe tools generally have a 2 to 8 microinch surface finish because these surfaces are diamond wheel ground to hold insert dimensional tolerances and to produce sharp chip-free cutting edges. In contract to this, CVD diamond film coated inserts generally have a 30 to 60 microinch surface finish because they are not usually finished after the coating process, and have the as-deposited structure [Oles et al.1996].

The effects of these surface finish differences can be quite significant. A smoother surface finish on the rake face allows freer chip flow, minimal build-up edge, and less frictional heat, taking full advantage of diamond's low coefficient of friction. A rougher surface finish, in contrast, reduces free chip flow, causes built up edge, and generates more frictional heat. However, this rougher surface finish has a benefit in that it causes more chip curling and enhanced machined chip control. A smoother surface finish on the flank face produces a smoother finish on the part machined surface whereas a rougher finish on the flank face results in a corresponding rougher finish on the part machined surface.

4.2. COMPOSITES MACHINING

Of the two families of work materials (nonferrous metals and nonmetals) for which diamond is most suited, the most easily machined relative to tool life are low alloyed aluminums, copper, brass and bronze. The most difficult of these materials to machine are the high - Si aluminums, Al_2O_3 - aluminums, SiC - aluminums and many of the fiber reinforced plastics because of their highly abrasive characteristics [Metals Handbook 1989] and [Hay 1994] and [Oles, et al 1996]. The high - Si aluminum alloys (hypereutectic) and the Al_2O_3/SiC aluminum alloys are essentially "metal bond grinding wheels"! Most of these alloys are cast with Si, SiC or Al_2O_3 abrasive particles randomly oriented and not connected within the aluminum matrix metal matrix composites.

There is also increased utilization of parts made from a skeletal ceramic preform reinforcement into which is infiltrated a molten metal alloy (also metal matrix composites). Such components are by far the most difficult to machine due to their extreme abrasive characteristics.

The use of aluminum in the automobile industry will continue to grow, at an accelerated rate, with the objective of weight reduction for improved gasoline mileage. Less than 100 pounds of aluminum was used per car in 1975 whereas almost 500 pounds of aluminum will be used per car in 2000. Most of this aluminum will be low-to-high Si aluminum chosen for strength. However, significant amounts of SiC or Al_2O_3 - aluminums will also be used.

The nonmetallic composites which are used extensively in the aircraft industry are lighter in weight than most metals, and have a higher strength-to-weight ratio. These composites include those commonly known as fiber reinforced plastics. The reinforcing fibers include carbon, glass or Aramid, or a mixture of all three. Most of these fibers are enclosed in a matrix of organic polymer resin, such as epoxy. The key to the successful machining of these composites is that the fibers must be cut and not torn. The tearing frequently leads to delamination. Clean cutting rather than tearing requires sharp, long lasting cutting edges which only diamond can provide.

4.3. WOODWORKING

The use of compacted PCD in the wood working industry is growing rapidly due to favorable cost effectiveness. Diamond tipped tools outperform WC/Co tools by 40 to 400 times depending on the type of wood (natural or manufactured), the type of operation, and the speed of the cut. Diamond tipped saw blades generally outperform carbide by 40 to 80 times, routers by 80 to 150 times, and shaper cutters by 100 to 300 times.

It is interesting to note that wood has always been a favored construction material because it was so easy to cut. Historically, wood could be easily cut using sharp stones and early bronze knives to produce canoes from tree trunks to ancient furniture. Work on cutting tools have progressed from steels to WC/Co to diamond as a result of simple economics.

The corrosive chemicals contained in both natural woods and manufactured woods are detrimental to WC/Co cutting tools. When combined with the heat generated in cutting these woods, cobalt is leached from the WC/Co cutting edge causing it to dull. Compounding this condition in manufactured wood is the large amount of abrasive particles present. For these reasons, compacted PCD is becoming more widely applied, contributing to increased production rates (three to five times) over WC/Co, improved part quality (dimensional and finish), longer tool life (40 to 400 times), increased machine uptime (10 to 20%), lower energy requirements, and lower tool costs per unit produced.

An additional wear factor when machining manufactured wood used for furniture, shelving, display panels, etc. is the laminates bonded to the wood. These laminates are very abrasive and must be machined with chip-free edges. Once again, diamond is becoming the favored tool material to meet this stringent chip-free requirement.

The process to manufacture tools for woodworking is very similar to the process to manufacture lathe inserts for machining nonferrous metals and non metals. The producer of compacted PCD provides "discs" from which a tool fabricator wire EDM cuts "blanks" which are the brazed into a recess to become a "tip". Most of these tipped tools, with the exception of circular saws and simple routers, are complex form tools containing a few to several peripheral cutting edges. These finished edges are usually an EDM'ed surface as opposed to a diamond ground surface. And, most of the compacted PCD used for these tools is the finest-grained two micron PCD available.

4.4. SCRIBES, INDENTORS AND KNIVES

One of the oldest and best known diamond tools is the glazier's diamond scribe which makes a clean scratch on a glass sheet prior to dividing it by fracture. This tool utilizes the tip of a mounted small diamond which produces a fine groove in the glass and wears very slowly. In large scale production of flat glass, a small PCD rotating wheel (scribe) is used to create the scratch on the glass replacing a tungsten carbide wheel. Diamond has also been used by engravers to produce artistic effects on materials such as glass and precious stones.

Another common use for diamond is as an indentor to measure the hardness of other softer materials. The diamond indentor is lapped to a specific geometry (Vickers & Knoop) for each type of hardness test. The diamond is an ideal material since it suffers virtually no deformation itself and very little wear or damage.

Diamond is used in knives for several applications. Diamond has been used as the cutting edge in microtone blades for producing sections of biological specimens as well as a scalpel in surgical applications [Yoder 1993]. One of the more widely used surgical applications is in eye surgery. Diamond knife edges can be produced which are one to two orders of magnitude sharper than the best steel edges. This enhanced sharpness of the diamond blade results in the advantage of greater control of the cut by the surgeon and faster healing of the incision in the patient.

Diamond knives are also being used in the telecommunications industry to cut optical fibers. The fibers are a glass material and in order to get a precision joint the two faces need to be flat, smooth, and perpendicular to the axis of the fiber. The smooth surface is achieved by indenting the fiber with the diamond knife and then cleaving the glass fiber by applying tension.

4.5. DIES

Dies for wire drawing is one of the most successful applications of diamond as a wear part. Single crystal natural and synthetic as well as PCD are used extensively as drawing dies [Mehan and Hayden 1982]. Initially, natural single crystal diamond was used, but with the advent of PCD it began to be replaced by the polycrystalline product, especially in the larger sizes . Subsequently improved synthetic single crystals have also found success for drawing fine wire in soft metals such as copper and aluminum. CVD thick film diamond and diamond coatings on carbide dies will have applications in wire drawing when grown with a fine grain size or if polished [Yoder 1993].

Dies fabricated from polycrystalline material, either PCD or CVD, offer two advantages over single crystal diamond [Field and Pickles 1996]. First, the polycrystalline material is more homogeneous and avoids the difficulty involved in orienting single crystals. Also, it is possible to obtain and coat much larger die blanks to draw larger diameter wire. The cost of large natural or synthetic diamond becomes prohibitive when drawing wire greater than approximately 2 mm in diameter.

4.6. DRILLING TOOLS

PCD drilling tools have been very successful in the drilling industry, particularly in the soft to medium hard rock formations. PCD bits are used in oil and gas drilling as well as mineral applications, surface drilling, civil engineering and construction applications [Bunting and Pope 1984] and [Glowka and Stone 1984]. PCD drilling tools are able to penetrate many types of rock far more cost effectively then other types of bit made with carbide or hardened steel.

PCD drill bits are typically manufactured by brazing the PCD segments/shapes in the cast in or machined recesses of a prefabricated tungsten carbide/matrix body. There has been a continuous evolution of PCD rock drills, and their share of the petroleum drill bit market continues to grow at a significant rate.

Another diamond application in petroleum drilling is the thrust bearing for downhole motors. In these designs, arrays of PCD discs rotate past one another and transfer the thrust load of the drill motor string to the drill bit. Diamond is far more resistant than any other material to the abrasive and erosive action of the drilling mud (Table III). Another application for diamond coated wear parts in the down hole environment is the nozzles and seals used in mud pulse telemetry.

4.7. TRIBOLOGY / WEAR PARTS

Diamond is an excellent tribological material. It's friction behavior is similar to polytetrafluroethylene in an air environment [Bowden and Taber 1964]. Natural diamond or high-pressure and high-temperature (HPHT) manufactured diamonds as well as sintered HPHT PCD are available only as simple three dimensional products [Gigl 1989]. With the advent of CVD diamond films a whole new range of products will be able to take advantage of the tribological properties of diamond. Diamond films offer tribologists and engineers the combined properties of hardness, chemical inertness as well as solid lubrication. For the manufacturer and end user this means an expansion in tribological applications [Gigl 1989].

The mechanical and tribological properties of CVD diamond films depend on several factors such as grain size and grain boundary strength, as well as the presence of non-diamond phases and the coating - substrate adhesion strength. All of these can be reasonably controlled by choice of deposition technology and process parameters [Bull and Matthews 1992]. For example, as the deposition temperature is reduced and the proportion of non-diamond phases increases, the hardness is reduced [Bull, et al 1992].

Friction in diamond is dependent on crystal orientation and rubbing direction. The mechanically hard face of diamond, (lll), is also the one with the lowest coefficient of friction; 0.05 in dry air. No significant anisotropy is observed in this coefficient. The coefficient of friction on a (100) plane, for example, varies from 0.08 to 0.12 depending on the specific crystallographic sliding direction [Tabor 1981]. This low coefficient of diamond has been attributed to strongly absorbed water molecules or hydrogen, since the friction coefficient in vacuum and inert atmospheres can be as high as 0.9 [Buckley 1981]. The wear volume on the (100) plane can be increased by two orders of magnitude by changing the sliding direction from [110] to [010] [Crompton et al. 1973].

Surface roughness of diamond films can have an appreciable influence on the initial friction of diamond films: the greater the initial surface roughness, the higher the initial coefficient of friction [Hayward and Singer 1991]. The equilibrium coefficient of friction is independent of the initial surface roughness of the diamond film in both air and vacuum. In air the initial coefficient of friction is much higher than the equilibrium coefficient of friction. But in vacuum the equilibrium coefficients of friction are greater than the initial coefficients of friction regardless of initial surface roughness of the two diamond films. In vacuum the friction is primarily due to adhesion between the sliding surface of the diamond [Dugger et al. 1992]. Therefore, the wear factors of diamond films in air are lower than those of diamond films in vacuum. The generally accepted mechanism for wear of diamond is that of small fragments chipping off the surface [Hayward 1991].

In natural diamond crystals, wear part applications remain limited to relatively small systems such as sensors, feelers, jewel bearings for clocks and watches, styli, etc. Where larger areas or parts are to be protected from wear, PCD has been used as well as the new CVD diamond either as a freestanding thick film, or as a thin coating on a material such as silicon carbide, silicon nitride, molybdenum, or tungsten carbide.

Another application where diamond can be used effectively as a wear resistant material in machining is work rest blades. PCD, thick film CVD or CVD diamond coated WC are all alternatives in diamond that can be effectively used as the contact surface in the work rest blade. These guards and supports have to have low friction and withstand high wear as a result of high contact forces and sliding speeds and also as a result of abrasive particles in the coolant used in the cylindrical grinding machine. Diamond work rest blades can result in substantial savings in machine maintenance, increased uptime, and more reliable production with improved workpiece quality.

Other excellent applications for PCD or CVD diamond are; (1) the wear resistant components in high pressure homogenizers, Figure 4, and other high fluid velocity components such as water jet nozzles; (2) vee-guides, lap stops, gauges and measuring surfaces/contact areas on micrometers and calipers; (3) coatings on mechanical seals for difficult applications where chemical inertness and a low friction coefficient are required; and (4) as the contact surface in TAB bonding tools for integrated circuit package manufacturing, as a replacement for large synthetic single crystals. In this last

Figure 4. Homogenizer parts that require wear resistant diamond coatings.

application, wear resistance, thermal conductivity and chemical inertness make diamond the best material in terms of performance. Until large area CVD diamond was available, the diamond costs were too high for wide usage, with the larger semiconductor package sizes. With the maturing of CVD diamond technology, diamond will play a much larger role in both cutting tool [Hay 1994] and wear components markets. For example, the three dimensional coatings made possible by the CVD process [Bhushan and Gupta 1991], and will create many new applications where natural or PCD diamond materials could not achieve the area coverage or shapes needed. The manufacturing costs, performance and quality of CVD diamond are steadily improving, and will open new markets and greatly expand existing ones.

5. Conclusions

In the area of tribology, the science of friction and wear, diamond has been confined until recently to several niche applications [Bhushan and Gupta 1991]. Typically diamond has been used only in environments of extreme wear. The fact that it has only been a niche player in tribological applications is due to the limited forms, shapes, and sizes available with single crystals and PCD compacts. With the development of CVD diamond film technology, the tribological applications of diamond will expand dramatically [Gigl 1989], and it will truly become an enabling technology to solve industry's tribological problems.

6. References

Bhushan, B. and Gupta, B.K., Handbook of Tribology - Materials, Coatings, and Surface Treatments, pp. 3.1-3.34, 4.52-4.63, 14.124-14.128, McGraw-Hill, Inc., New York 9 (1991).

Bowden, F.P. and Tabor, D., "The Friction and Lubrication of Solids", Pt.2, Clarendon, Oxford, 1964, pp. 159-185.

Buckley, D.H., Surface Effects in Adhesion, Friction, Wear, and Lubrication, Elsevier, Amsterdam, 1981, pp. 318-322.

Bull , S.J. and Matthews, A., Diamond Relat. Mater., 1 (1992) 1049.

Bull, S.J., Chalker , P.R. and Johnston, C., in T.S. Sudarshjan and D.G. Bhat (eds.), Proc. 5th Int. Conf. on Surface Modification Technologies, TMS, Warrendale, PA, 1992.

Bunting, J.A. and Pope, B.J., "Reduction of Thermal Failure of Sintered Diamond Drill Elements", in: Proc. ASLE, 40 (1984) 681.

Crompton, D., Hirst , W. and Howse, M.G.W., "The wear of diamond", Proc. R. Soc. London, Ser. A, Vol. 333 (1973) 435.

Dugger, M.T., Peebles, D.E., and Pope, L.E., Surface Science Investigations in Tribology, edited by Y.-W. Chung, A.M. Homola, and G.B. Street (American Chemical Society, Washington, D.C., 1992) p. 72.

Field, J.E. and Pickles, C.S.J., "Strength, Fracture and Friction Properties of Diamond" in: Diamond Films '95, (P.K. Bachmann, et al eds.) pp. 625-634, Elsevier Sciences S.A., Lausanne (1996).

Gigl, P.D., "New synthesis techniques, properties and applications for industrial diamond", IDA Ultrahard Materials Seminar, Toronto, Ontario, 1989.

Glowka, D.A. and Stone, C.M., "Effects of Thermal and Mechanical Loading on PDC Bit Life", Society of Production Engineers of AIME, Paper 13257 (1984).

Hay, R.A., "The New Diamond Technology and its Application in Cutting Tools", in: Ceramic Cutting Tools - Materials, Development and Performance, (E. Dow Whitney, ed.) pp. 305-327, Noyes Publications, Park Ridge, NJ (1994).

Hayward, I.P. and Singer, I.L., Second International Conference Proceedings (Materials Research Society, Pittsburgh, PA 1989), p. 785.

Hayward, I.P., Surf. Coat. Technol., 49, 554 (1991).

Mehan, R.L. and Hayden, S.C., "Friction and Wear of Diamond Materials and Other Ceramics Against Metals", in: Wear, 74, (1981-1982) 195.

Metals Handbook - Ninth Edition, Volume 16, Machining, pp. 71-89, 105-11, 765-770, ASM International (1989).

Oles, E.J., Inspecktor, A., and Bauer, C.E., "The New Diamond - Coated Carbide Cutting Tools", in: Diamond Films '95, (P.K. Bachmann, et al eds.) pp. 617-624, Elsevier Science S.A., Lausanne (1996).

Tabor, D., "Friction - the present state of our understanding", J. Lubr. Technical, 103, (1981) 169.

Wilson, G.F., Alworth, R., and Ramalingam, S., "Machining Sintered Carbides with Polycrystalline Diamond Tools", in: Proc. Int. Conf. on Manufacturing Engineering, Melbourne, FL., August 25-27, 1980, p. 57.

Yoder, M.N., in: Diamond Films and Coatings: Development, Properties and Applications, (R.F. Davis, ed.) p. 1, Noyes Publications, Park Ridge, NJ (1993).

Chapter 34
ECONOMICS AND COMMERCIALIZATION

Adam T. Singer and John V. Busch
IBIS Associates, Inc., Wellesley, MA 02181

Contents

1. Introduction

The international effort to manufacture and commercialize diamond has succeeded in making a commodity of diamond grit, but requires further progress to succeed in doing the same for diamond films. Diamond grit has been produced through methods developed over the past forty years, resulting in a product that has a low cost-to-performance ratio relative to competing materials. Diamond films have been produced through methods developed over only the past ten years, and progress to date has resulted in a product that has a high cost to performance ratio relative to competing materials. Table 1 lists competing materials for both diamond grit and films for different applications, along with the properties and order of magnitude prices for these materials.

Table 1. Properties and cost of diamond and several select other materials

Property	Diamond	AlN	Si	Cu	Al_2O_3	Beryllia	SiC
density (g/cm^3)	3.5	3.3	2.3	8.9	3.9	3.0	2.5 to 3.2
Young's Modulus ($x10^{10}$ N/m^2)	110	35	11	12	39	38	45
Ω-cm @25°C	10^{12} to 10^{17}	10^{14}	$3x10^5$	$2x10^{-4}$	>>10^{15}	>>10^{17}	10^2 to 10^{12}
Coefficient of Thermal Expansion ($x10^{-6}$/°C)	1 to 2	4	2.6	5.2	9	8	4
Thermal Conductivity (W/cm-K)	10-20	0.7-3	1.5	4	0.4	4	1 to 3
Dielectric Constant	5.6	9	11	N/A	9	7	40 to 100
Coefficient of Friction (on steel)	0.05 - 0.15	0.1 - 0.5	0.1 - 0.5	0.1 - 0.5	0.1 - 0.5	0.1 - 0.5	0.1 - 0.5
Vickers Hardness (GPa)	90	12	10	2	29	12 to 16	2 to 32
Order of Magnitude Price ($ /g)	100-1000 (CVD film) or 0.1 - 10 (HPHT grit)	0.1 - 10	0.1 to 1	0.1 to 1	0.1 to 1	1 to 10	0.1 to 10

Motivation for manufacturing diamond is driven by these properties listed in Table 1. Diamond's high hardness is desirable for abrasive applications: polishing slurries, bonded grit (e.g. grinding wheels), and polycrystalline film coatings (e.g. drills). With high hardness relative to the material being abraded, diamond tends to last longer while enabling higher operating speeds. Diamond's corrosion resistance enables operation with most materials and in most chemical environments.

This high resistance to wear, coupled with diamond's low coefficient of friction, makes diamond a desirable material for bearing applications. For use as a bearing, diamond would last longer than most materials while enabling higher efficiencies due to the reduced friction losses. Again, diamond's corrosion resistance provides an added attraction for its use.

Yet another quality, diamond's high thermal conductivity (low thermal resistance), is desirable for heat dissipation applications. In cases where heat generation sources such as resistive electronic circuits contact materials with high thermal resistance, thereby degrading product performance (i.e. slowing circuit operation), diamond allows lower temperatures and higher performance (i.e. faster signal transmission rates) through greater heat dissipation.

Additional benefits for electronic applications include diamond's coefficient of thermal expansion that matches well with silicon, the most widely used material in active electronic devices today, and diamond's high electrical resistance, allowing it to contact electronics directly. Lastly, the hardness and corrosion resistance enable diamond to be considered for harsh electronics operating conditions.

In another application of thermal dissipation, where heat is generated through friction, diamond's thermal conductivity can allow a lower temperature rise for a given friction heat source. This quality has utility for the sensor windows in aircraft and projectiles, where radiation is generated by the frictional heat on the sensor window surface. This radiation complicates the ability of the internal sensors to detect different sources of heat, such as other aircraft and projectiles. With diamond as part of the sensor window, less heat and therefore less radiation would be generated to reduce the ability of the sensing devices. Also, diamond is optically transparent, allowing the optical portion of the electromagnetic spectrum to pass to the sensors unimpeded. Lastly, diamond's corrosion resistance allows operation under harsh conditions for this application.

While diamond films and grit could compete for certain abrasive applications, diamond films are being studied and developed for uses where diamond materials have not been possible to date. Such applications include the curved tool coatings, heat dissipation coatings and substrates, and sensor windows mentioned above.

Overall, diamond materials have improved many products and manufacturing processes in the case of high pressure, high temperature (HPHT) diamond grit, and will improve many new and existing products in the case of the newly developing chemical vapor deposited (CVD) diamond coatings. The only barrier to widespread use for both diamond grit and film is price. To date, only diamond grit has achieved widespread use. Products for diamond grit include loose grit powder for abrasive uses and bonded grit for grinding wheels. CVD diamond films have not achieved widespread use yet, although many organizations are developing cheaper polycrystalline diamond films. This chapter does not seek to compare the economics of diamond grit technology versus the diamond film deposition technology. Instead, this chapter focuses on the economic history and potential of both HPHT and CVD diamond, and it examines how both HPHT and CVD diamond technologies could lower manufacturing costs through process scaling.

1.1 MARKET AND ECONOMIC HISTORY OF SYNTHETIC DIAMOND

HPHT diamond grit and CVD diamond films follow similar histories; CVD diamond market and pricing trends follow those created by HPHT diamond by about thirty years. Had the world developed CVD diamond technology with the same intensity as HPHT

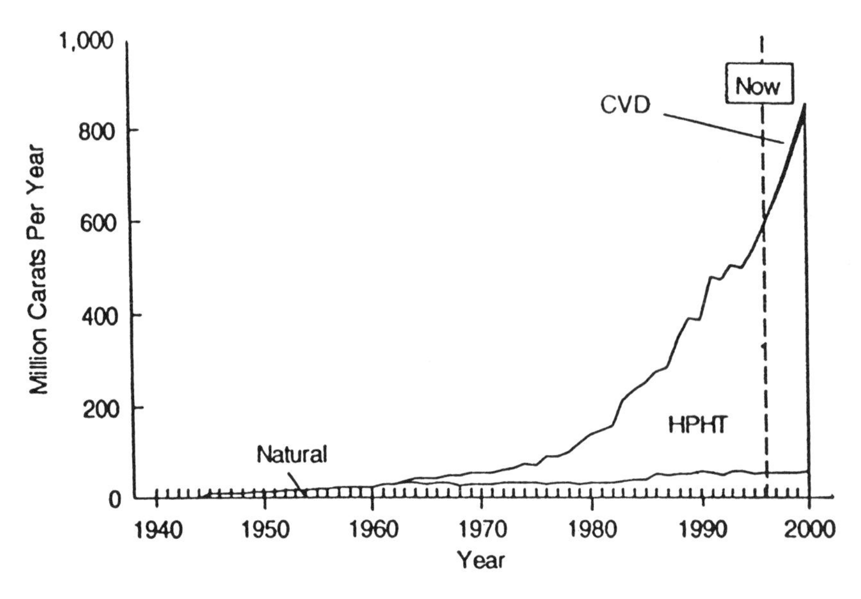

Figure 1. Growth of Synthetic Diamond Production

diamond, there might be more widespread use of CVD diamond than exists today. It is believed that HPHT diamond was chosen for commercial scale production primarily because of the following factors:

- HPHT diamond formation was understood through carbon's phase diagram, while a theory for CVD was not yet in place
- Results showed that higher mass growth rates were achieved with the HPHT method
- Characterization of thin CVD diamond films was difficult

As a result of decisions based on these observations in the mid-1950's, CVD diamond technology development lags behind that of HPHT by roughly thirty years.

The use of industrial HPHT diamond has increased steadily since its commercialization in the late 1950s. Synthetic diamond grit production surpassed that of natural diamond in about 1969, and today, reported synthetic diamond sales outnumber natural industrial diamond by about seven to one (439 million carats per year versus 59 million carats per year). Moreover, the US Bureau of Mines (USBM) forecasts continued strong demand for industrial diamond. In their 1994 Industrial Diamonds Annual Review, the USBM predicts a 10% annual growth rate for US diamond production.

General Electric pioneered the manufacture of HPHT diamond grit, supplying this material to a fast-growing market for industrial diamond in polishing, grinding, and cutting applications [Busch et al. 1991]. This dramatic growth is shown in Figure 1, where one million carats per year in 1960 has progressed to about 440 million carats per year in 1994 [Austin 1995] [Jennings 1987] [Dismukes 1990].

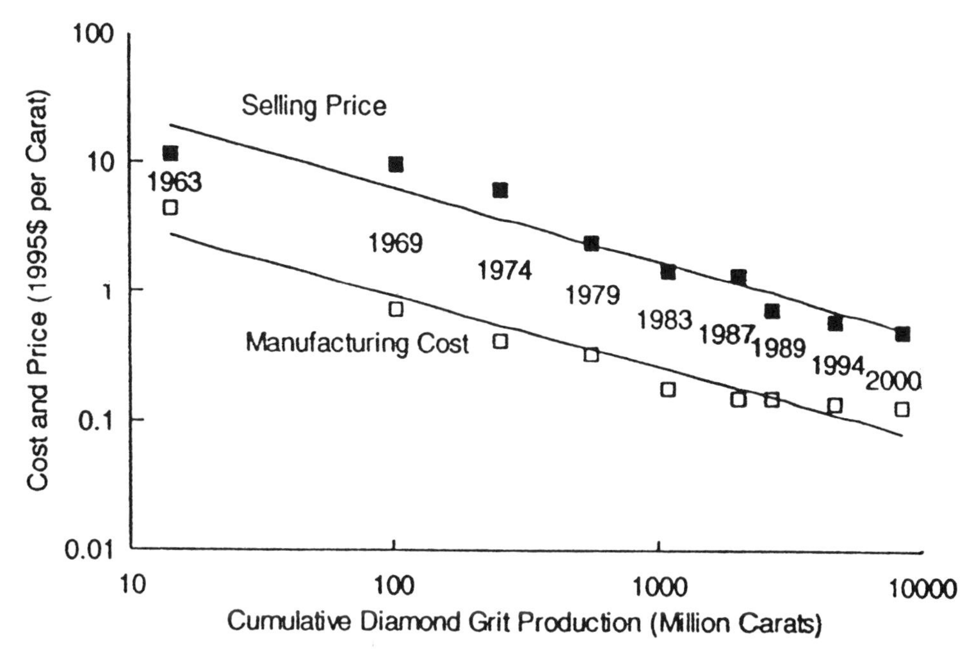

Figure 2. Diamond grit cost and price versus cumulative production and year.

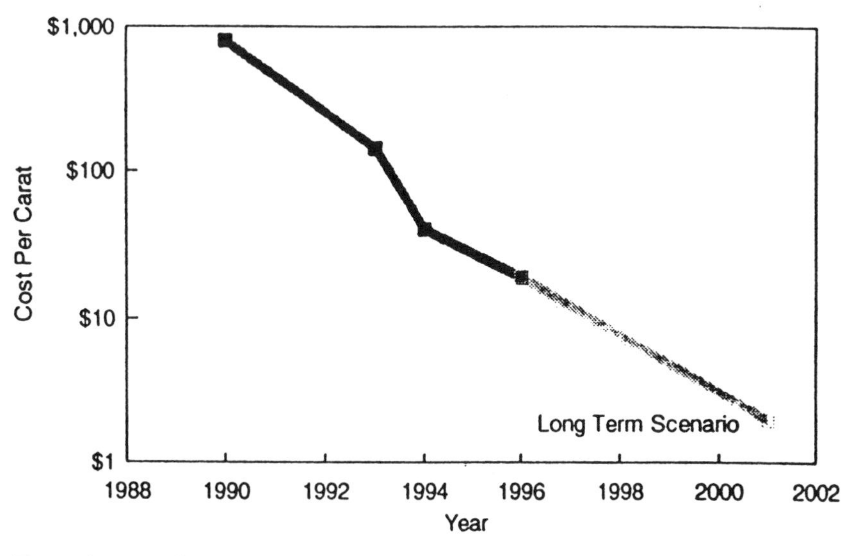

Figure 3. CVD diamond manufacturing cost versus time.

The HPHT synthesis method has thus far proved relatively limited in synthesizing diamond for use in applications other than mechanical. Science and technology advances in CVD diamond synthesis over the last twenty-five years, however, now show high potential [Dismukes 1990] for the practical, cost effective manufacture of diamond films and coatings tailored in thickness and size for acoustic, optical, electronic, corrosion resistance, as well as mechanical applications. Figure 1 also shows the growth of CVD diamond as part of the overall synthetic diamond growth trend.

The technology developments over this period of time have resulted in manufacturing cost reduction, and therefore price reduction [Busch et al. 1994]. Figure 2 follows the price trend of average quality (50 micron diameter, medium grade) HPHT diamond grit over the past thirty-five years. The two orders of magnitude price reduction over this time period reflects both technology advances for HPHT diamond grit manufacture and economies of scale reached through increased demand for this diamond product.

A similar trend exists for CVD diamond over the past five years. The deposition costs have been reduced over this period due to both the technology advances and the increase in demand, as depicted in Figure 3. The reduction in manufacturing cost, although dramatic, is hampered by the different manufacturing conditions required by the many applications of this material. Ironically, the high number of applications has slowed the pace in reaching economies of scale for CVD diamond overall: the different applications require unique deposition conditions, and even equipment in some cases, that then translate to economies of scale reached only on a product-by-product basis. Progress in CVD diamond, therefore, has been spread over many development fronts.

1.2 MOTIVATION FOR COST ANALYSIS

The commercial success of a new material depends on its cost and performance relative to its competitors. Since CVD diamond far outperforms the competition, the only barrier to its widespread use is its current cost, which is on the order of hundreds of dollars per carat. HPHT diamond outperforms all other abrasives and also would be used more widely if its price were lower than the current one dollar per carat for medium size (50 microns) and quality (opaque and roughly round). Even though potential users may be willing to pay more for synthetic diamond than other materials, the one dollar per carat level for CVD diamond and the sub-one dollar per carat level for HPHT diamond are the current goals for most organizations.

In achieving these cost goals, what technology changes must occur and to what extent? Answering this question is one of the strengths of Technical Cost Modeling. By focusing on cost as an important factor in technology development and identifying the cost "bottlenecks" for an operation (i.e., equipment throughput), companies can foresee lower costs, given the current understanding of the technology.

2. Methodology

The tool for cost prediction for this chapter is Technical Cost Modeling (TCM), a spreadsheet-based methodology for the estimation of manufacturing cost. In conjunction with TCM are predictive relationships that incorporate both empirical data and theoretical equations to provide model flexibility. This section details both the TCM methodology and the predictive relationships, and lists the inputs associated with the modeling performed in this chapter.

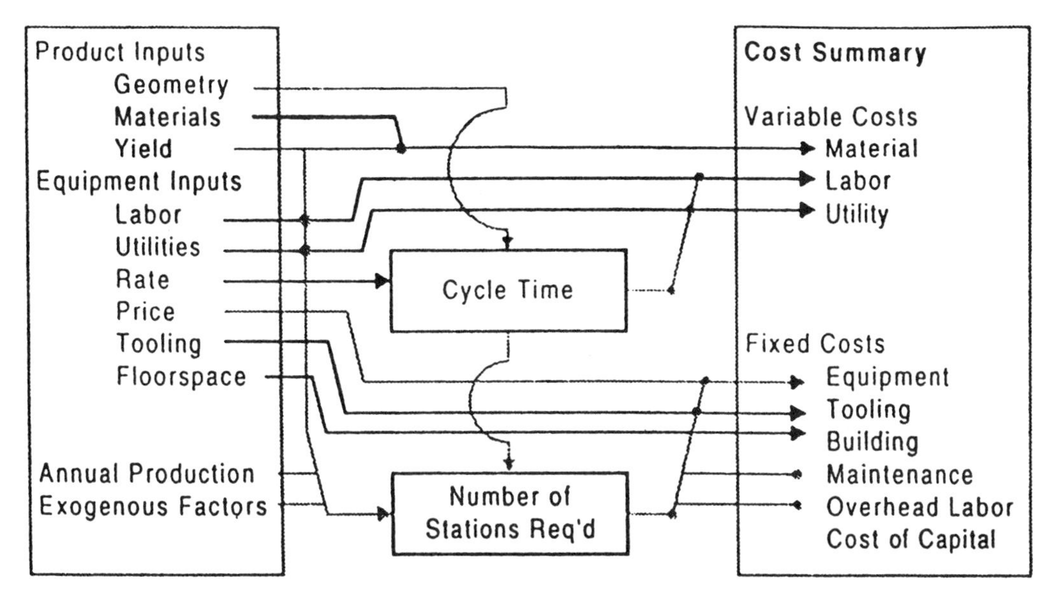

Figure 4. Typical technical cost model logic.

2.1 TECHNICAL COST MODELING

Technical Cost Modeling is a tool for estimating and simulating manufacturing costs. The technique is an extension of conventional process modeling, with particular emphasis on capturing the cost implications of material and process variables and changing economic scenarios.

In a Technical Cost Model (TCM), cost is assigned to each unit operation from a process flow diagram. For each of these unit operations, total cost is broken down into separately calculated elements:
Variable cost elements: Material, labor, and utilities (energy)
Fixed cost elements: Equipment, tooling, building, maintenance, overhead labor, cost of capital

By breaking cost down in this way, the complex task of cost estimation is reduced to a series of more simple engineering and economic calculations. A summary of the basic relationships in a Technical Cost Model is shown in Figure 4.

Technical Cost Models predictive in nature. In a predictive approach, parameters such as deposition rate and cycle time are calculated by the model as a function of the product material and geometry. These predictive functions can be derived from analysis of empirical data or through incorporation of theoretical relationships. It is this predictive nature of Technical Cost Models that enable their flexibility and subsequent utility.

Technical Cost Models can be used to accomplish tasks that include the following.

- Simulate the costs of manufacturing products
- Establish direct comparisons between material, process, and design alternatives

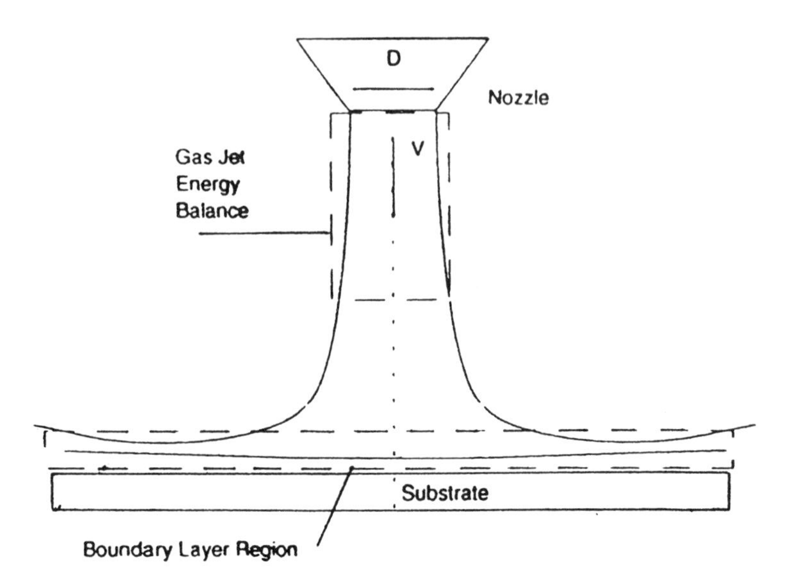

Figure 5. DC arcjet geometry assumed for simulations.

- Investigate the effect of changes in the process scenario on overall cost
- Identify limiting process steps and parameters
- Determine the merits of specific process and design improvements

2.2 PREDICTIVE RELATIONSHIPS

The incorporation of predictive relationships, in the form of first-principles theory or equations that fit empirical data, allows the cost models to predict throughput rates, equipment prices, and other parameters as functions of input variables. For the HPHT and CVD diamond technologies, industry experts were consulted to provide modeling support ranging from overall strategy to the details of the deposition rate equations. The strategy aspect included the identification of input variables, definition of process conditions, and structure of the logic of the equations; while the detailed modeling included the actual equations, chemical reaction constants, and output trend verification.

For the modeling of HPHT diamond grit manufacturing, the following intermediate outputs depend on inputs through predictive relationships:

- Cycle time = f(grit size, cell volume)
- Equipment price = f(cell volume, cell geometry)
- Material conversion = f(grit size)

All of these relationships are generated from data collected from organizations involved with HPHT diamond.

Growth rates for CVD diamond are currently creating long cycle times that lead to high material, labor, and equipment costs due to low throughput per machine. However,

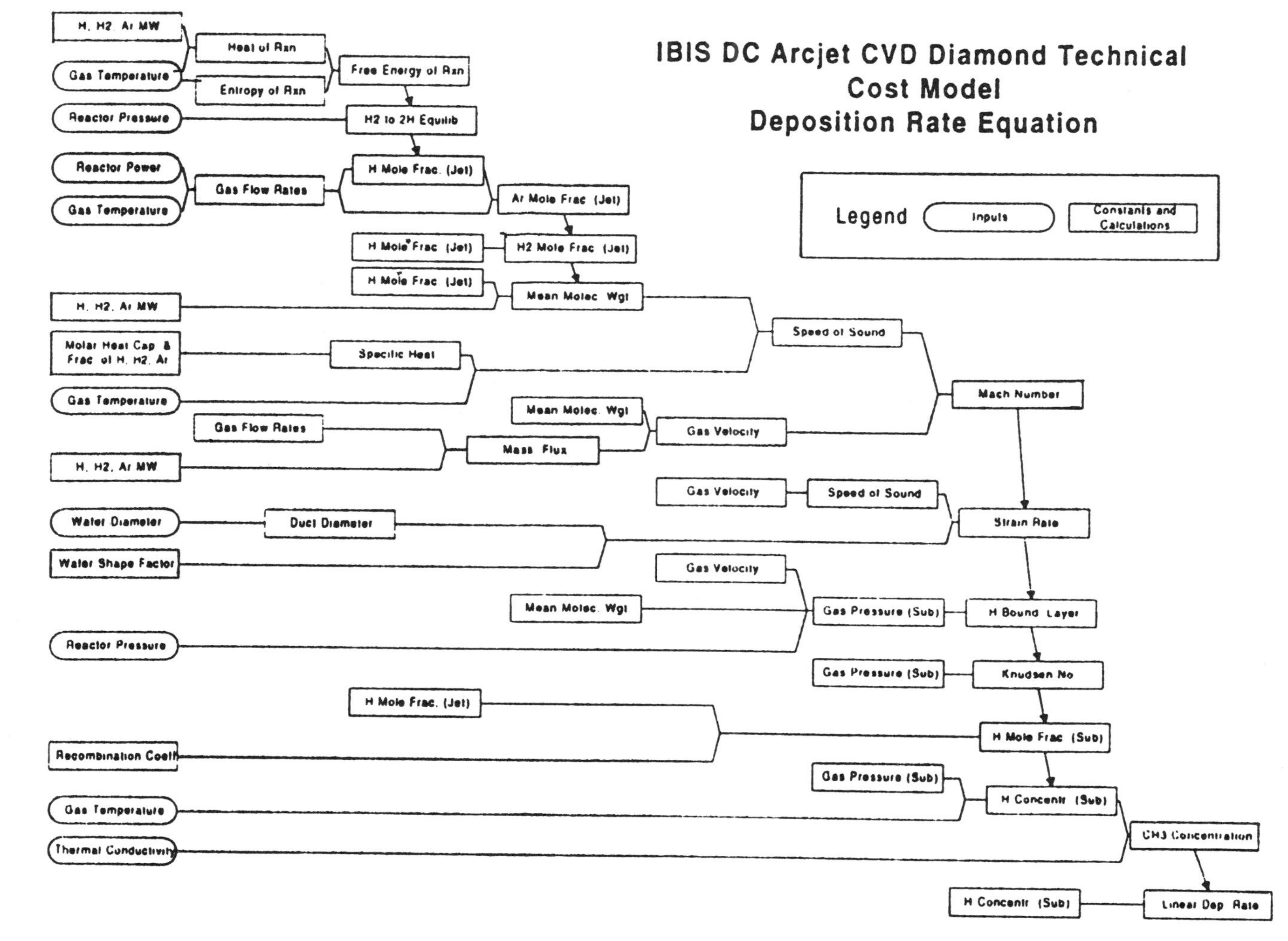

Figure 6. IBIS DC arcjet CVD diamond technical cost model; deposition rate equation.

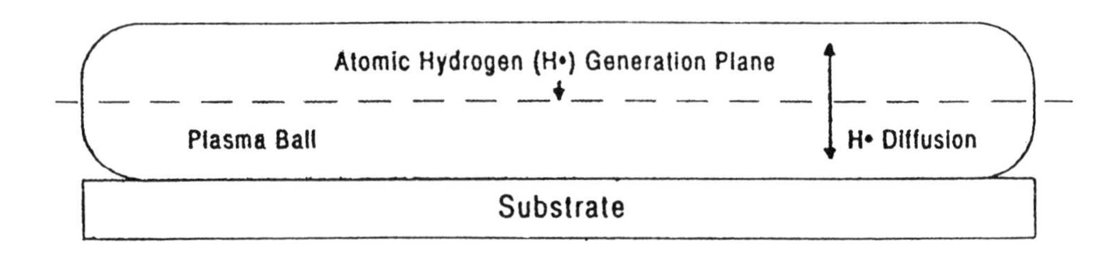

Source: Goodwin, D.G., Memo to IBIS Associates, Inc., September 1993.

Figure 7. Microwave geometry assumed for simulations.

industry data relating process parameters to growth rate are not readily available. For the modeling of commercial production a simplified deposition theory and its inherent assumptions has been incorporated into the Technical Cost Model in order to predict the scale-up costs of CVD diamond manufacture using the DC arcjet, microwave, and combustion flame technologies.

For the modeling of CVD diamond wafer manufacturing, the following intermediate outputs depend on inputs through predictive relationships:

- Deposition rate = f(reactor power, gas flow, gas mix, geometry)
- Equipment price = f(reactor power or deposition area)

The deposition rate relationship is established through first-principles diamond deposition theory, and the equipment price relationship is generated from industry data.

2.3 IMPLEMENTING CVD DIAMOND DEPOSITION THEORY

The first-principles theory for CVD diamond deposition is contained as a module inside the CVD diamond TCMs. Inputs for the theories vary by technology. The CVD diamond technologies modeled in this report are the DC arcjet, microwave, combustion flame, and hot filament methods and all of these incorporate deposition theory.

2.3.1 DC Arcjet

Figure 5 is a diagram of the overall modeling strategy for the DC arcjet model. The gas jet exiting the nozzle forms the first of two regimes; the chemistry in this region is a function of the reactor's input parameters and is assumed to be uniform. The second region is the boundary layer, where the chemistry varies with the distance from the growth surface. This goal of this approach is to calculate the atomic hydrogen

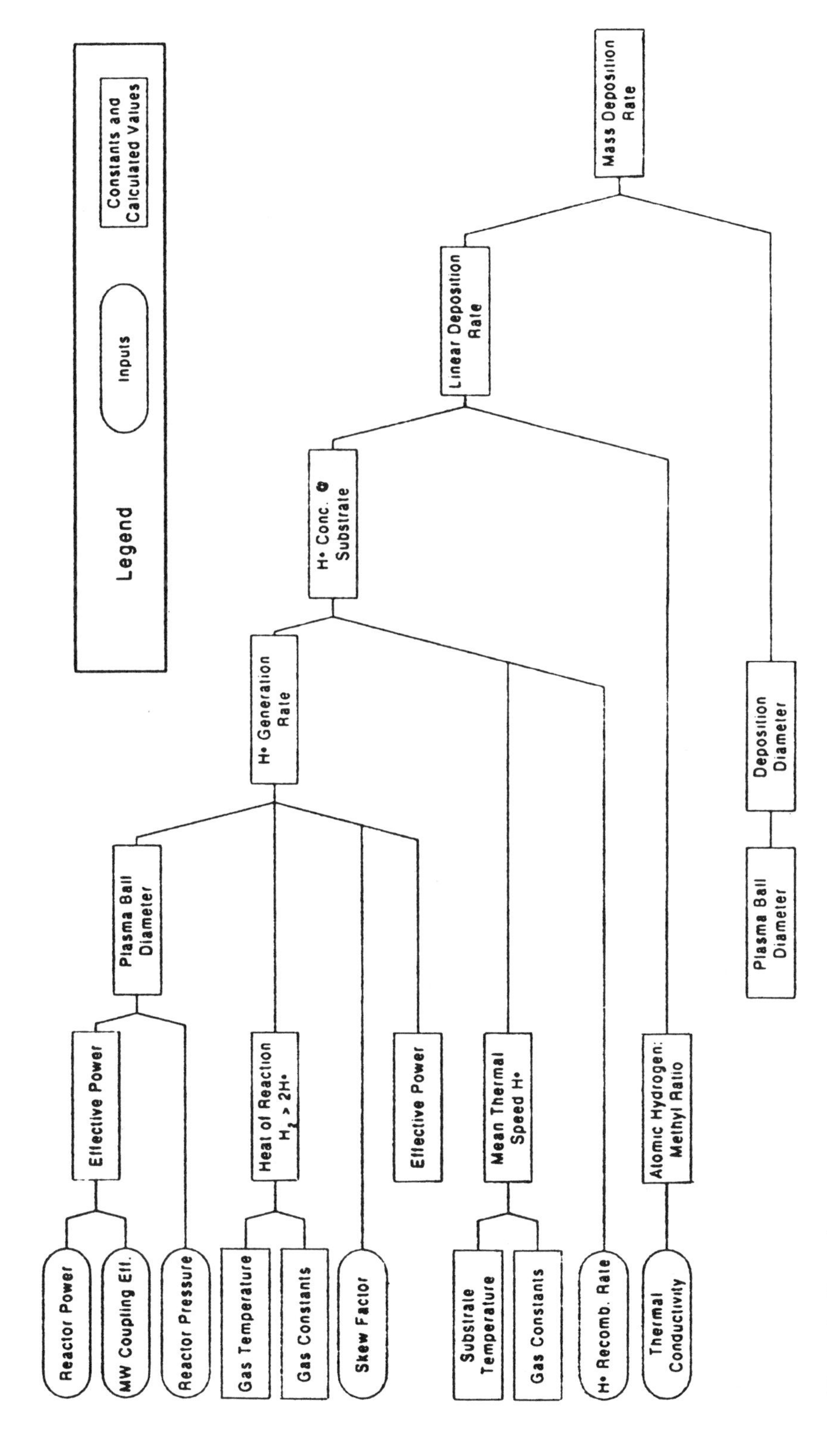

Figure 8. IBIS microwave CVD diamond technical cost model; deposition rate equation.

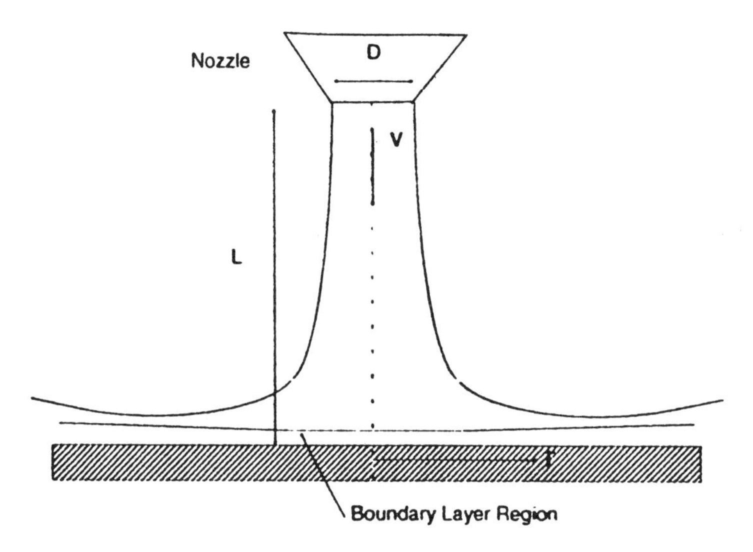

Figure 9. Combustion flame geometry assumed for simulation. Single nozzle, attached flame.

concentration at the growth surface which, along with the CH_3 (methyl radical) concentration, determines the CVD diamond growth rate. Because of the interrelationships that exist among variables such as reactor power, gas concentrations, reactor pressure, gas temperature, wafer diameter, and thermal conductivity, the calculation path to deposition rate is complex.

Figure 6 shows the equation logic flow for the deposition rate calculation in the DC arcjet model. Important calculations include the atomic hydrogen mole fraction in the gas jet (H Mole Frac. (Jet)), gas jet Mach Number (Mach Number), gas pressure at the substrate surface (Gas Pressure (Sub)), atomic hydrogen concentration at the substrate (H Concentr. (Sub)), and the linear deposition rate (Linear Dep. Rate).

2.3.2 Microwave

Figure 7 is a diagram of the overall modeling strategy for the microwave model. The model assumes atomic hydrogen is generated roughly in the middle of the plasma at a distance "L" from the substrate. This goal of this approach is to calculate the atomic hydrogen concentration at the growth surface through the characterization of both the diffusion of atomic hydrogen toward the surface and its recombination into H_2. Along with the CH_3 (methyl radical) concentration, the atomic hydrogen concentration at the surface determines the CVD diamond growth rate. Due to such variables as reactor power, reactor pressure, and thermal conductivity, the calculation path to deposition rate is fairly complex.

Figure 8 shows the equation logic flow for the deposition rate calculation in the microwave model. Important calculations include the plasma ball diameter, atomic hydrogen generation rate (H Generation Rate), atomic hydrogen concentration at the substrate (H Concentr. @ Substrate), and the mass deposition rate.

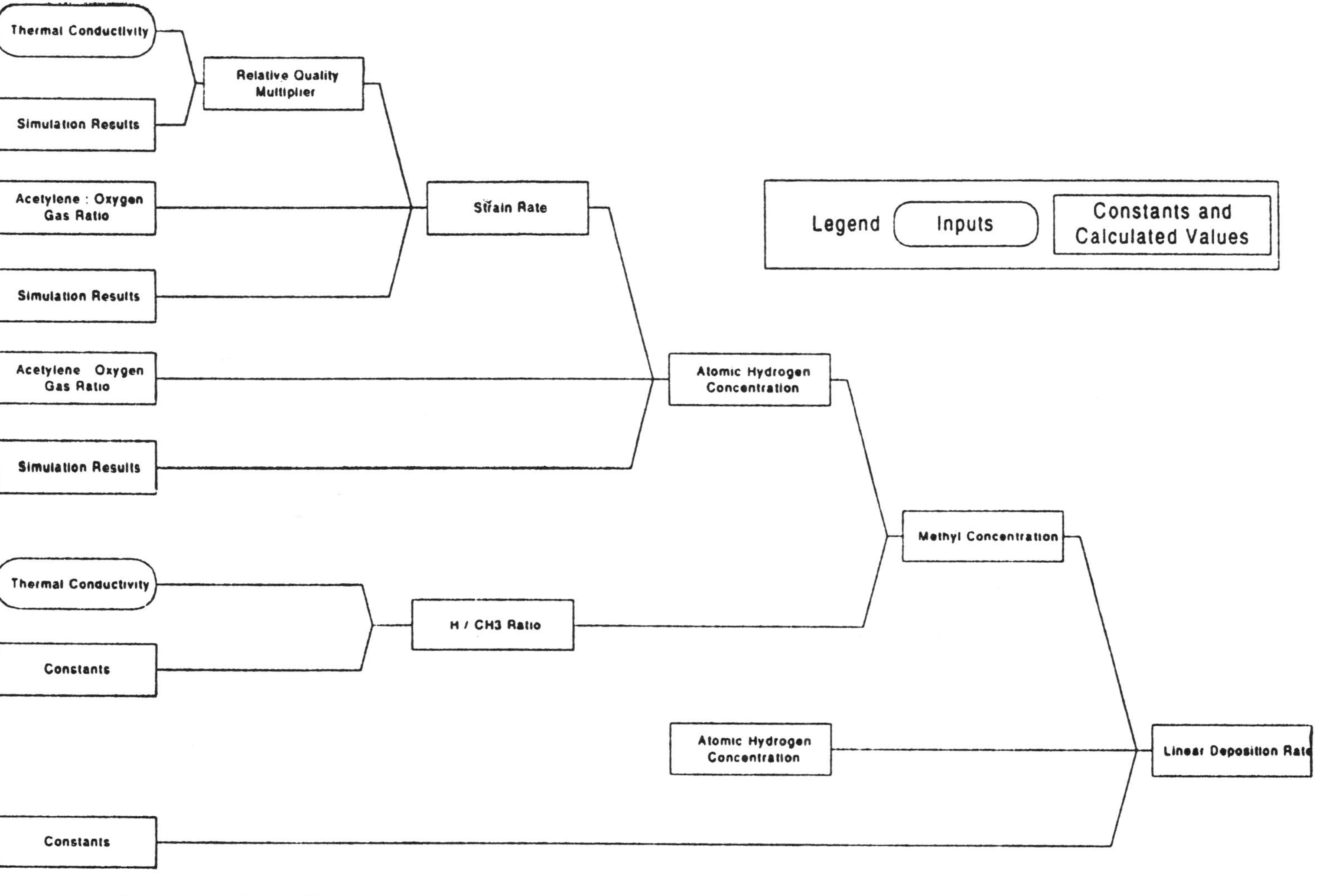

Figure 10. IBIS combustion flame CVD diamond technical cost model; deposition rate equation.

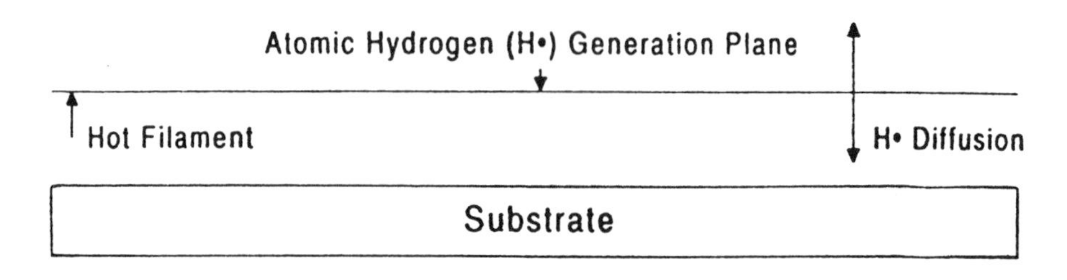

Source: Goodwin, D.G., Memo to IBIS Associates, Inc., September 1993. IBIS Associates, Inc.

Figure 11. Hot filament geometry assumed for simulations.

2.3.3 Combustion Flame

Figure 9 is a diagram of the overall modeling strategy for the combustion flame model. For numerical simulations that were generated by Professor Goodwin, it is assumed that the process gases are mixed and combust previous to accelerating through the nozzle. The resulting combustion jet is assumed to have uniform chemistry and velocity. Impinging on the substrate creates a boundary layer, through which atomic hydrogen and methyl radicals diffuse. The goal of this approach is to calculate the atomic hydrogen concentration at the growth surface which, along with the CH_3 (methyl radical) concentration, determines the CVD diamond growth rate. Due to such variables as gas concentration and thermal conductivity, the calculation path to deposition rate warrants an explanatory diagram.

Figure 10 shows the equation logic flow for the deposition rate calculation in the combustion flame model. Important calculations include the strain rate, atomic hydrogen concentration at the substrate, and the linear deposition rate.

2.3.4 Hot Filament

Figure 11 is a diagram of the overall modeling strategy for the hot filament model. The model assumes atomic hydrogen is generated at the plane of the filaments, at distance "L" from the substrate. This goal of this approach is to calculate the atomic hydrogen concentration at the growth surface through the characterization of both the diffusion of atomic hydrogen toward the surface and its recombination into H_2. Along with the CH_3 (methyl radical) concentration, the atomic hydrogen concentration at the surface determines the CVD diamond growth rate. Due to such variables as reactor power, reactor pressure, and thermal conductivity, the calculation path to deposition rate is fairly complex.

Figure 12 shows the equation logic flow for the deposition rate calculation in the hot filament model. Important calculations include radiated power, atomic hydrogen

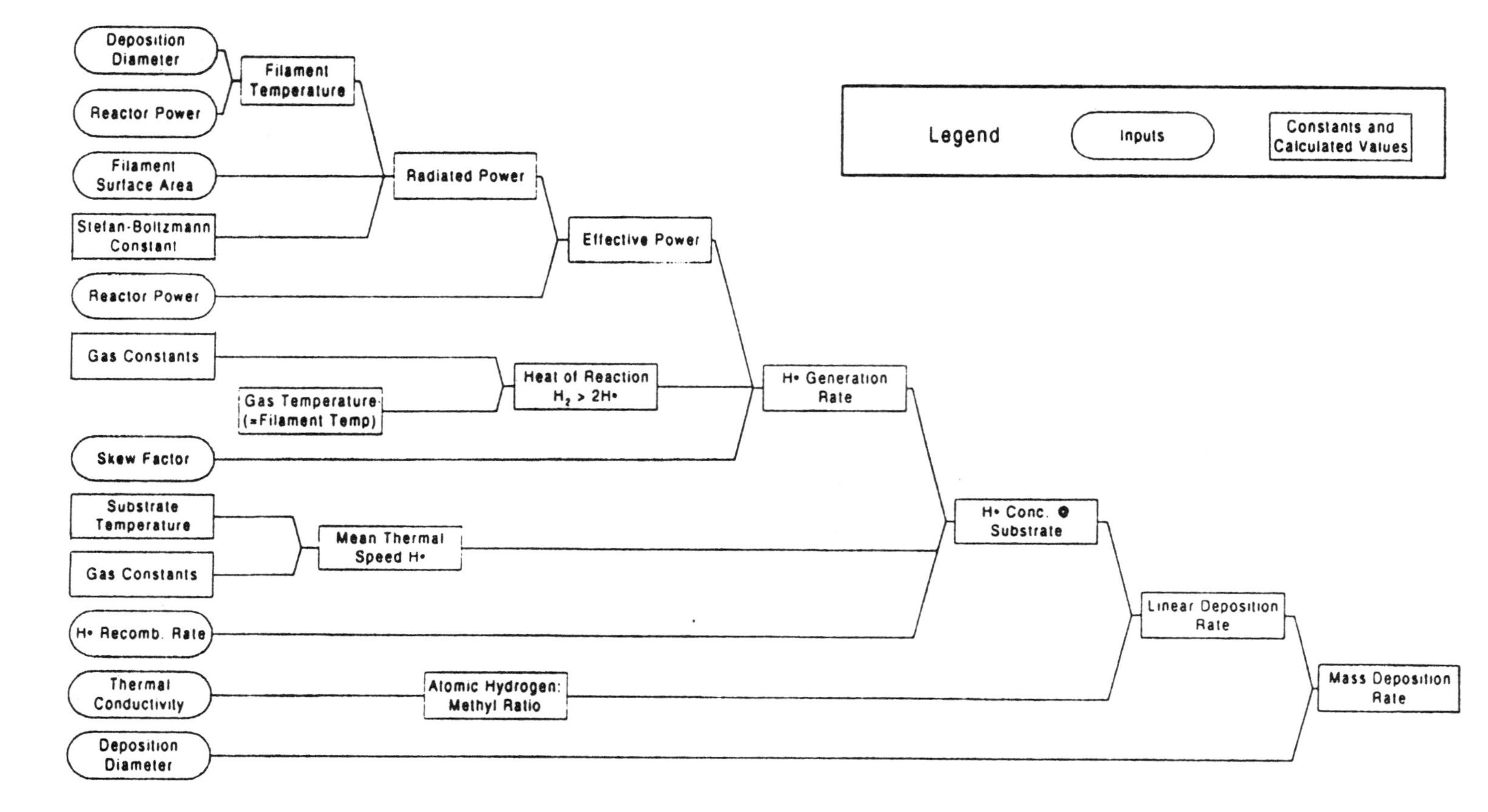

Figure 12. IBIS hot filament CVD diamond technical cost model; deposition rate equation.

Table 2. Assumptions for the time and place of manufacture.

Exogenous Cost Factors		Units
Direct Wages	$13.33	/hr
Indirect Salary	$50,000	/yr
Indirect:Direct Labor Ratio	1.00	
Benefits on Wage and Salary	35%	
Working Days per Year	360	
Working Hours per Day	24	hrs
Capital Recovery Rate	10%	
Capital Recovery Period	5	yrs
Building Recovery Life	20	yrs
Working Capital Period	3	months
Price of Electricity	$0.050	/kWh
Price of Building Space	$100	/sqft
Price of Cooling Water	$0.03	/100 gal
Auxiliary Equipment Cost	15%	
Equipment Installation Cost	35%	
Maintenance Cost	8%	

generation rate (H Generation Rate), atomic hydrogen concentration at the substrate (H Concentr. @ Substrate), and the mass deposition rate.

For an explanation of the logic and actual equations that form the foundation of this deposition theory, see an article on this subject by Professor Goodwin [Goodwin, 1993].

2.4 EXOGENOUS MODELING ASSUMPTIONS

For the modeling in this chapter, specific assumptions regarding the manufacturing operations and product specifications will be discussed in later sections. However, all of the cost models incorporate assumptions that relate to the time and place of manufacture. These inputs, such as labor wages and interest rates, are the same in all of the models and are shown in Table 2. Culled from the manufacturers in the US, these assumptions reflect an average manufacturing operation.

3. Economics of HPHT Diamond

Except for scale increases, the HPHT diamond synthesis process has changed little since its development thirty-five years ago. To manufacture HPHT diamond, cells containing graphite and catalyst mixtures are subjected to extreme pressures (730 kpsi) and temperatures (2,700 F). Specialized hydraulic presses, capable of exerting up to 10,000 tons of force, deliver both the pressure and electric current necessary to grow diamond crystals. Once pressurized, the press consumes as much as 40,000 watts of electrical power through graphite resistance heating, thereby bringing the materials to the conditions necessary for diamond growth. Grit manufacturers may control diamond shape and quality by process parameters. For example, temperature is the main variable used to control diamond shape. Synthesizing diamond at lower temperatures produces

Table 3. HPHT diamond grit manufacturing assumptions.

HPHT Manufacturing Input Parameter		Units
PRODUCT SPECIFICATIONS		
Annual Production Volume	20,000,000	carats/yr
Average Grit Diameter	50	microns
Sample Charge Volume	200	cc
Carbon to Diamond Conversion	90%	by weight
Diamond Conversion Rate	1,324	carats/cell
PROCESS RELATED FACTORS		
Pressing Equipment Price	$1,460,000	per station
Total Pressing Cycle Time	25	minutes
Press Clamping Force	5,400	tons
Tooling Life	3,000	cycles
Tooling Costs (Belt, Die and Punch)	$189,000	per set
Power Requirement	37	kW
Building Space	1,500	per station

more black or cube shaped grit, while at higher temperatures, yellow or white octahedra grit predominates.

3.1 ASSUMPTIONS

Table 3 details the assumptions of HPHT grit manufacture. The scenario being modeled reflects large scale production with current technology. For cost sensitivities in the following sections, parameters that are not varied will be set at the values in this table.

3.2 COST SENSITIVITIES

The cell, or charge volume (indicating the scale of diamond production), is an important parameter influencing the economics of grit production. Over the past thirty-five years, grit manufacturers have increased the charge volume from initial values of 25 cc to a current range of 200 - 300 cc. As shown in Figure 13, increasing charge volume from 50 cc to 200 cc drops the cost of producing 300 micron grit by nearly 45%. Furthermore, the larger charge volume triples diamond throughput. However, as charge volume increases, the press tonnage required to achieve the minimum pressure has to go up commensurably. Currently, the maximum size for presses hovers around the 10,000 ton level, limiting cell volume to about 300 cc, depending on geometry.

Also shown in Figure 13 is the impact of grit size on cost. The cost dependence on grit size is exponential, indicating significant cost increases with larger grit diameters. A doubling of the 100 micron grit size increases the cost per carat by 50% for a 200 cc cell volume manufacturing setup.

While increasing charge volume does improve equipment throughput, the effect does not correlate linearly. Because higher tonnage presses require more ramp-up time, larger charge volumes have slightly longer cycle times, thereby negatively impacting throughput. As shown in Figure 14, Technical Cost Modeling allows IBIS to account for

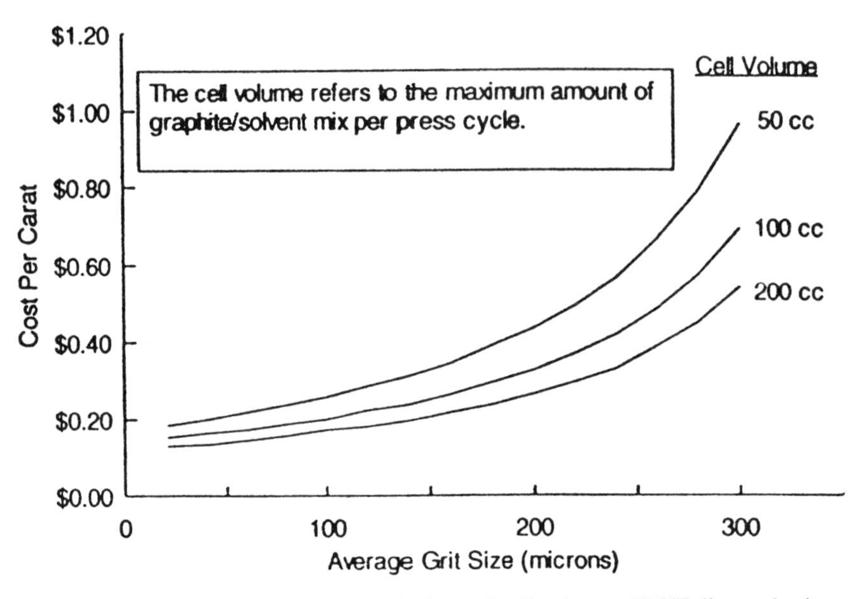

Figure 13. Cost per carat vs. average grit size and cell volume. HPHT diamond grit manufacture.

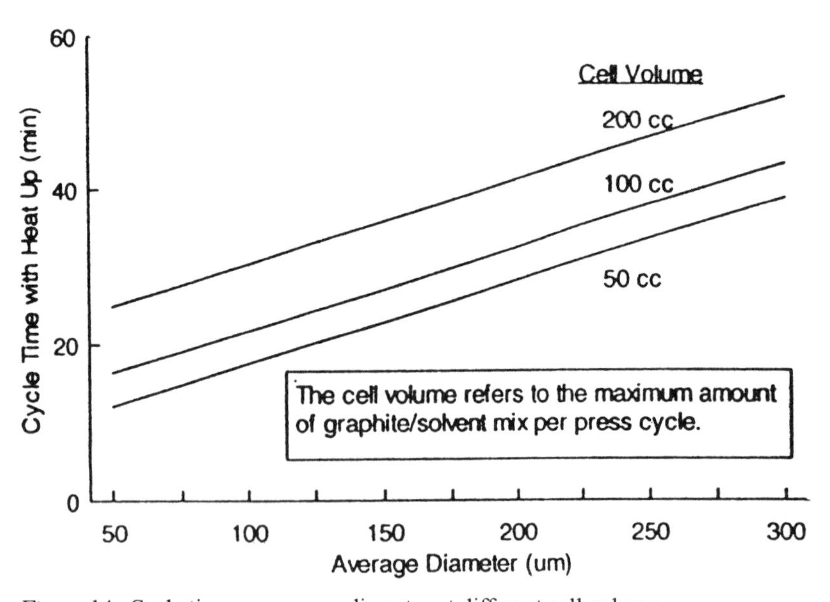

Figure 14. Cycle time vs. average diameter at different cell volume.

the effect of charge volume, as well as that of grit size, on cycle time. Increasing average grit diameter lengthens cycle time because synthesizing larger grit requires longer press dwell time. For example, to produce small grit in a small cell (50 micron average diameter, 50 cc cell), a 2,300 ton press would take roughly 12 minutes, whereas the same product in a larger cell (200 cc) would require about twice that time. Producing even larger grit, say 300 micron diameter in the 200 cc size cell, doubles cycle time again, to about 1 hour.

Table 4. CVD diamond long term inputs (Year 2001 to 2006)

	DC Arcjet	Hot Filament	Microwave	Combustion
SELECTED INPUTS				
Wafer Thickness (microns)	500	500	500	500
Thermal Conductivity (W/mK)	1,000	1,000	1,000	1,000
Machine Power (kW)	125	225	200	108
Deposition Yield	90%	90%	90%	90%
COMPUTED VALUES				
Wafer Diameter	30.5	45.7	51.1	8.9
Mass Deposition Rate (g/hr)	5.1	2.9	5.8	0.4
Linear Deposition Rate (μm/hr)	20	5.0	8.0	17
Deposition Cycle Time	25	100	63	30
Machine Cost ($/sta)	$500,000	$500,000	$1,000,000	$125,000

HPHT COST MODELING SUMMARY

Since the inception of production in the 1950s, HPHT production advances have come mainly through increasing economies of scale and yield increases, rather than through the introduction of fundamentally new technology. That is, advances in hydraulic presses and general progress in understanding grit synthesis have facilitated process improvements, resulting in lower diamond grit costs.

This progress is reflected through the IBIS Technical Cost Model, showing cost sensitivities to the average grit diameter, which determines the cycle time of the HPHT press, and the charge volume, which affects the equipment throughput.

4. Economics of CVD Diamond

This section shows results and sensitivities for the CVD diamond modeling scenarios. The following four CVD diamond technologies are analyzed: DC arcjet, microwave, combustion flame, and hot filament. The cost models for all four technologies include CVD diamond deposition theory, and the cost analysis related to these four technologies focuses on the critical deposition parameters, and cost sensitivities to these parameters.

For the purpose of these analyses it is assumed that the transport theory model that is used to estimate the diamond growth rate closely predicts actual growth rates and that input values for variables such as gas flow rate and substrate temperature are physically achievable.

4.1 ASSUMPTIONS FOR THE LONG TERM

While this chapter does provide cost model data for four CVD diamond technologies, the underlying modeling assumptions should be understood before decisions or judgments are made. For instance, it is possible that a multiple nozzle setup for the combustion flame technology would be more efficient, and therefore cost less, than the single nozzle design assumed for this analysis. Given this potential, these results should

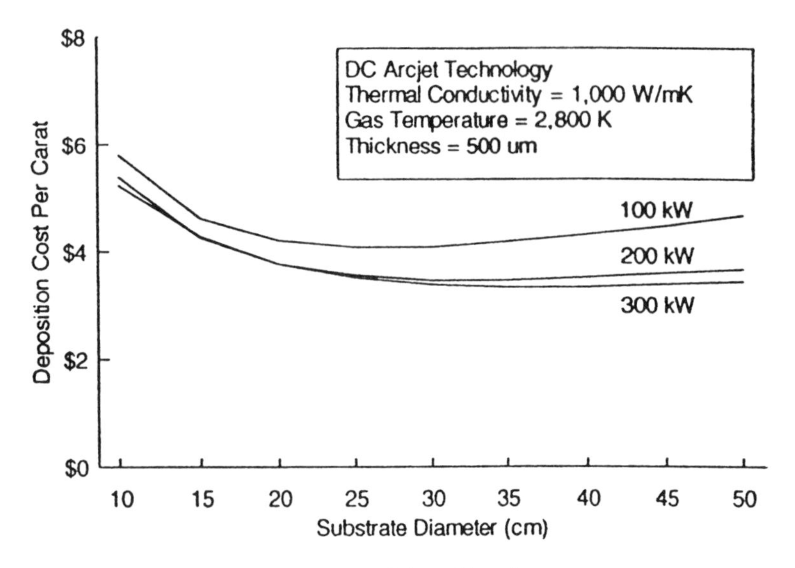

Figure 15. Cost vs. reactor power and deposition diameter.

be viewed as characterizing the costs of four specific technology implementations - not as representing four general technologies. With this understanding in mind, these specific implementations are frequently encountered throughout the CVD diamond industry.

The significant assumptions for the four CVD diamond cost models are presented in Table 4. This table reveals assumptions for the long term scenario for these technologies, where "long term" is defined as the expected state of diamond deposition five to ten years from today (i.e., in the years 2001 to 2006). With the assistance of industry experts, plausible product and process conditions have been selected to represent this long term scenario.

The combustion flame technology has the lowest cost equipment, since the machine consists of combustion-proof gas flow controls and equipment, a nozzle and nozzle cooling system, and waste gas removal system. The microwave technology possesses the highest cost equipment, due to the relatively higher cost of the power supply (versus DC arcjet and hot filament), in addition to the vacuum system, radiation-leak-proof reactor, cooling system, and gas flow controls.

Unless stated otherwise, the values in Table 4 are constant conditions. Labor wages and other exogenous (accounting-related) cost factors are held constant, and non-dedicated equipment is assumed (as if machines are rented - only the utilized percentage of annual equipment amortization is allocated to the product).

4.2 DC ARCJET

The DC arcjet technology is cost-sensitive to, among others, three process parameters: the reactor power, the upper limit on gas jet temperature (reactor limited), and the diameter of the diamond wafer being deposited. The gas flow rate is another significant

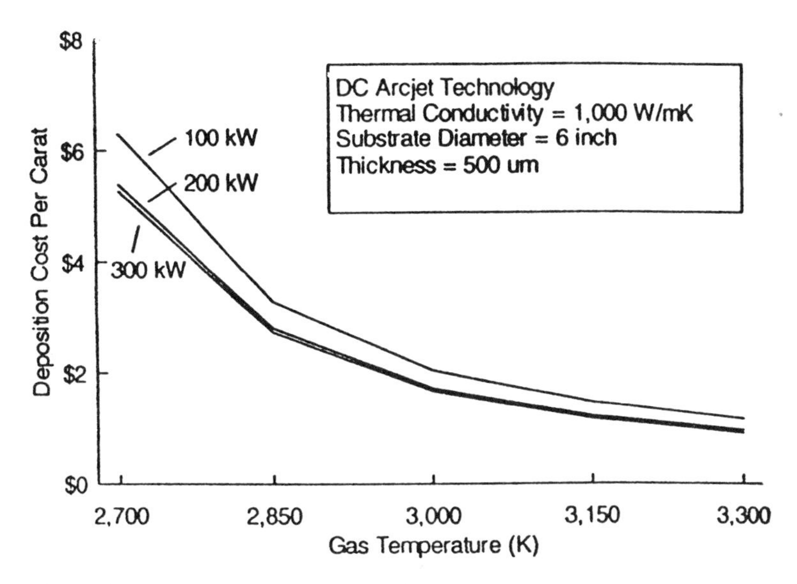

Figure 16. Cost vs. reactor power and jet temperature.

process parameter affecting the cost of diamond but is an intermediate calculation based on the process variables above. The following sections provide insight into what can reduce the cost of CVD diamond produced by this technology.

4.2.1 Cost vs Reactor Power and Substrate Diameter

Figure 15 shows that, initially, deposition cost can be reduced by increasing the deposition diameter, but that increasing the power into the range of hundreds of kilowatts does not further reduce cost. The cost optimum is due to the sum of two effects: the cost initially drops with larger areas, since the investment costs are being distributed over a greater mass; the cost then proceeds to rise as a result of a downward trend in growth rate, caused by the energy of the arcjet being distributed over higher deposition areas.

Figure 15 also indicates that there are diminishing returns on power increases. It does not, however, consider that there may be prohibitive engineering challenges when scaling up low powered reactors to high substrate diameters.

4.2.2 Cost vs Reactor Power and Gas Temperature

Figure 16 depicts the CVD diamond deposition cost as a function of both the reactor power and the temperature of the gas jet. The reigning transport theory suggests that a higher gas temperature allows more atomic hydrogen to reach the substrate, creating a higher growth rate which translates to lower cost. From interviews with industry experts, the upper limit on gas jet temperature is determined by the limitations of the DC arcjet torch nozzle. Therefore, Figure 16 indicates that the maximum allowable gas temperature must be determined in order to produce the lowest cost diamond. Fitting a curve to this data, the cost of diamond is inversely proportional to the gas temperature raised to the eighth power. This strong relationship means a gas temperature increase of just ten percent reduces the diamond cost by roughly fifty-three percent.

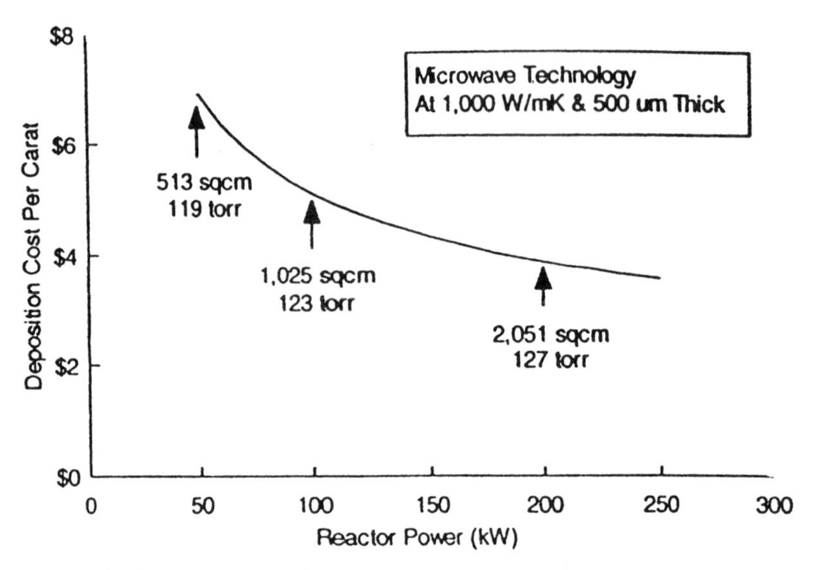

Figure 17. Cost vs. power and pressure.

In contrast to the influence of gas jet temperature is the seemingly weak effect of reactor power on deposition cost. However, this result is somewhat misleading since higher powered reactors are necessary to produce higher gas temperatures, given a wafer diameter of six inches and a thermal conductivity of 1,000 W/mK.

The same warning given for the last sensitivity must be applied in this case: both high gas temperature and large substrates require higher power reactors, therefore the lower powered reactors may not be able to realistically attain the higher gas temperatures and diameters.

4.3 MICROWAVE

The microwave technology is cost-sensitive to, among others, the reactor power. The reactor pressure and substrate diameter are other significant process parameters affecting the cost of diamond but are typically engineered to be optimized for each reactor power source. The following section provides insight into what can reduce the cost of CVD diamond produced by this technology.

For this technology, process gases are excited into a plasma through microwave radiation. A plasma ball is formed in the reactor, its area proportional to the reactor power and inversely proportional to the reactor pressure. By increasing reactor power, therefore, this technology can be scaled to larger areas.

Figure 17 shows the cost per carat of CVD diamond as a function of reactor power. As noted in the figure, both reactor pressure and deposition diameter are dependents of reactor power. Fitting a curve to the data reveals that cost is inversely proportional to reactor power to the exponent 0.43, meaning a doubling of the power allows a twenty-five percent cost reduction. This sensitivity indicates that there are cost savings and possibly new applications for scaling up this technology to higher powers and areas.

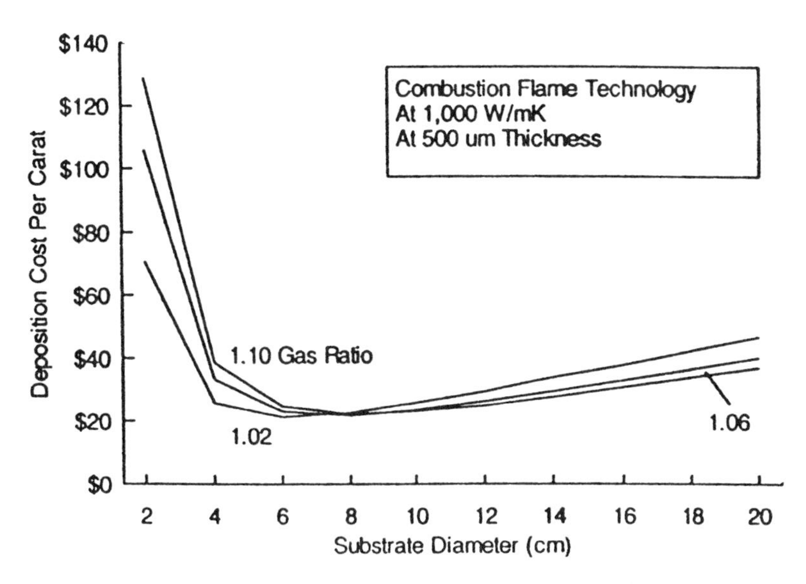

Figure 18. Cost vs. substrate diameter and acetylene:oxygen gas ratio.

4.4 COMBUSTION FLAME

The combustion flame technology is cost-sensitive to, among others, the following three parameters: the acetylene gas price, the ratio of acetylene to oxygen, and the substrate diameter. The gas flow rate is another significant process parameter affecting the cost of diamond but is dependent on the aforementioned input variables as well as the thermal conductivity. If higher quality CVD diamond is desired, the gas flow rate must be increased, assuming the gas ratios are held constant. The following section provides insight into CVD diamond cost reduction through this technology.

4.4.1 Cost vs Acetylene:Oxygen Gas Ratio and Substrate Diameter

As shown in Figure 18, there is an optimal diameter for the combustion flame technology based on the single nozzle torch design assumed in the model. There exists an optimum due to the combination of two dynamics: one where increasing substrate diameter decreases the fixed costs (i.e., equipment investment) per carat, and the dynamic where gas costs vary with the cube of substrate diameter. Depending on the ratio of incoming acetylene to oxygen, the optimal substrate diameter ranges from ten centimeters at a gas ratio of 1.02 to six centimeters at a gas ratio of 1.10. The optimal substrate diameter varies inversely with thermal conductivity; at higher thermal conductivities the flow rates must also be higher to deliver more atomic hydrogen to the growth surface. With higher flow rates, the material cost increases.

4.4.2 Cost vs Acetylene Price

Expensive acetylene gas dominates the cost of combustion flame CVD diamond. Figure 19 reveals the cost sensitivity of this technology to acetylene pricing, in the event that acetylene production becomes more efficient over time. Even acetylene pricing as low as \$0.50 per standard cubic meter cannot bring the cost per carat below the ten dollar level. This chart indicates that the single-nozzle combustion flame process as modeled for this chapter may not be the optimal for the combustion flame technology. Alternative nozzle

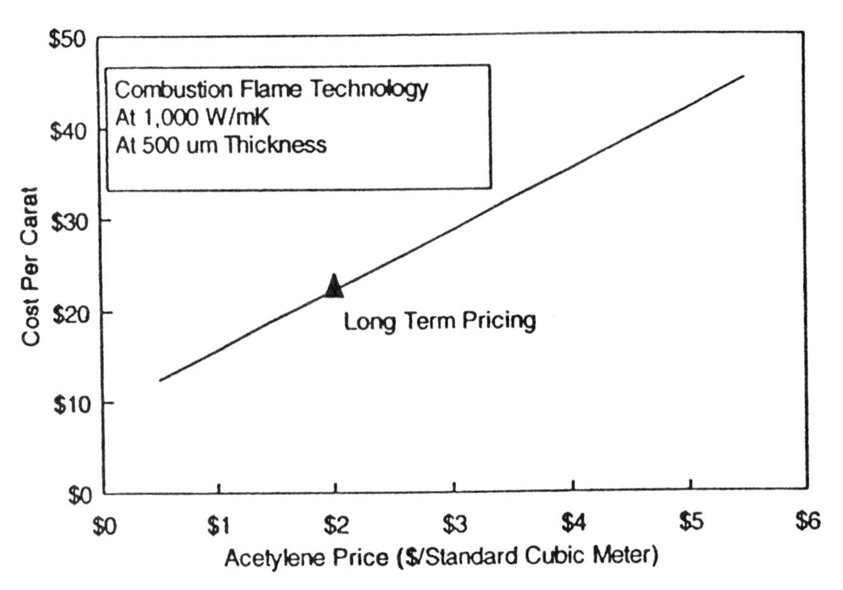

Figure 19. Cost vs. acetylene price.

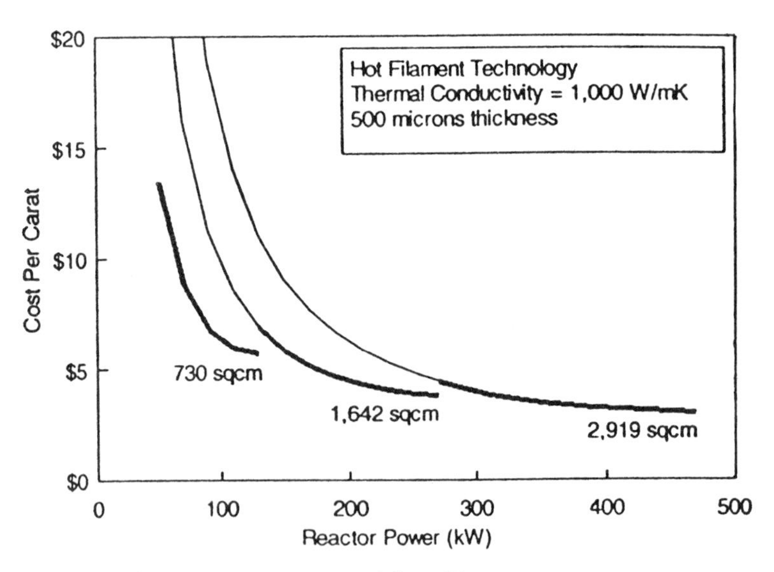

Figure 20. Cost vs. reactor power and deposition area.

geometries, cheaper process gases, or sub-atmospheric chamber pressures may boost the material utilization efficiency and therefore make this technology more competitive.

4.5 HOT FILAMENT

The hot filament technology is cost-sensitive to, among others, deposition area and operation power. The equipment cost is another significant process parameter affecting the cost of diamond but is calculated based on these input process variables. The following section provides insight into what can reduce the cost of CVD diamond produced by this technology.

4.5.1 Cost vs Deposition Area and Reactor Power

Since the hot filament equipment area and deposition power are parameters that affect the reactor cost and CVD diamond deposition rate, Figure 20 investigates the cost sensitivity to changes in these two variables. While the equipment price increases with both of these variables, economies of scale are attained that allow the cost per carat to decrease with both higher powers and deposition areas.

As shown in Figure 20, increasing the deposition area allows reduced costs per carat. However, achieving these economies of scale would require increases in reactor power in order to maintain the appropriate power density (reactor power allocated over the area of deposition).

Figure 20 shows the cost savings attained through these increases in reactor power. Fitting a curve to the data reveals that cost is inversely proportional to reactor power to the exponent 1.15, meaning a doubling of the power (at constant area) allows a fifty-five percent cost reduction. However, there are diminishing returns on reactor power, since higher power densities translate to higher filament temperatures and therefore increased radiated (unutilized) power. The maximum practical power density seems to be roughly 140 Watts per square centimeter, according to the theory implemented in this cost model. In summary, the sensitivity in Figure 20 indicates that there are cost savings, and possibly new applications, for scaling up this technology to higher powers and areas.

4.6 TECHNOLOGY SUMMARY

Due to the rapid pace of technology development to date, this chapter's presentation of long term cost modeling should not be construed as a judgment of technologies. Instead, the results should be viewed as guidance for potential improvement in each individual technology.

4.6.1 Baseline Costs in the Long Term

Figure 21 shows the relative long term costs of the DC arcjet, microwave, combustion flame, and hot filament technologies for the production of one-half millimeter thick CVD diamond wafers. The single nozzle combustion flame technology has the highest long term cost, at $22 per carat. The microwave, hot filament, and DC arcjet are significantly lower, at $3.87, $4.17, and $3.76 per carat, respectively. The combustion flame technology is dominated by the material cost due to the high consumption rate of expensive process gases. The microwave technology has a significant equipment cost per carat due to the low growth rates and relatively high equipment price per machine. The relatively high deposition area of the hot filament reactor coupled with the moderate deposition rate places the hot filament technology among the most cost effective methods. Lastly, the equipment cost is also significant for the DC arcjet technology, and its high consumption rate of process gases leads to a significant materials cost.

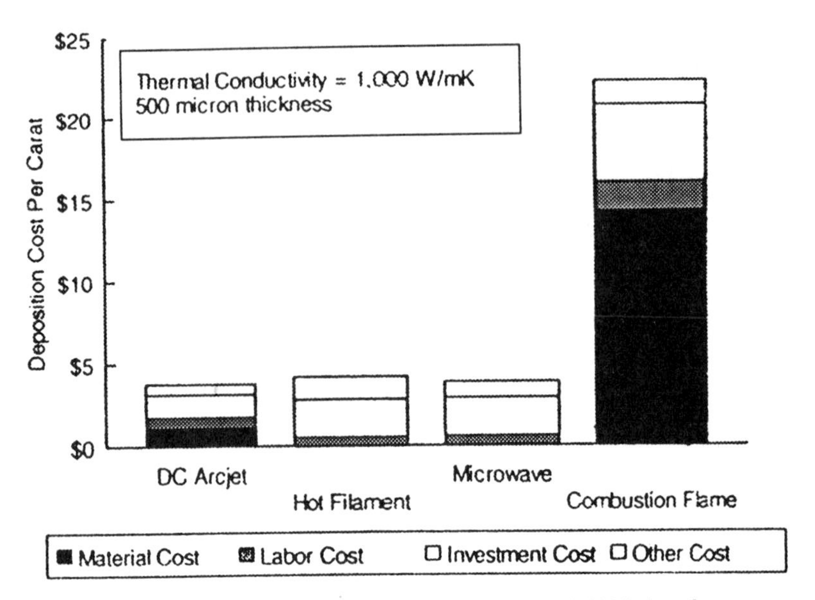

Figure 21. Deposition cost comparison. Long term- 2000-2005 time frame.

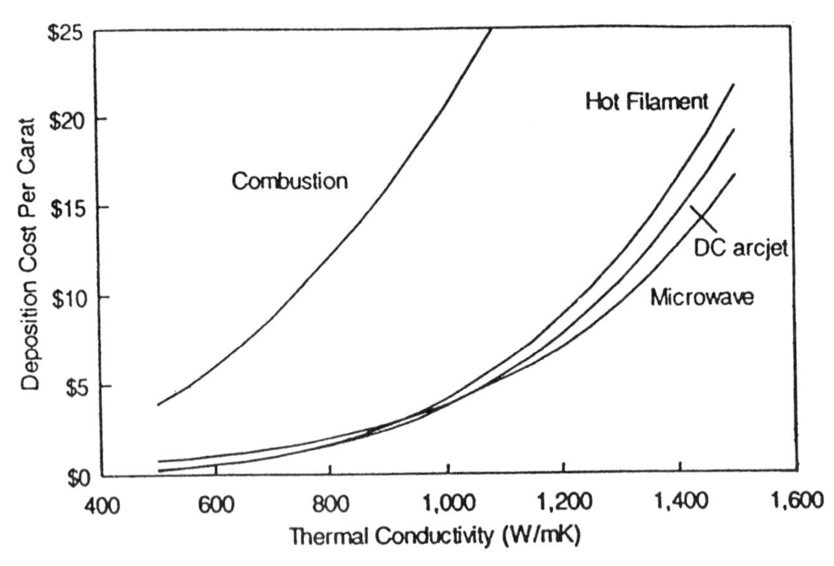

Figure 22. Deposition cost comparison versus thermal conductivity. Long term- 2001-2006 time frame.

4.6.2 Long Term Cost vs Thermal Conductivity

Competing materials for electronic thermal management applications range from as low as 200 W/mK (Aluminum) to as high as 800 W/mK (Copper/Carbon fiber composite). Industry experts believe the minimum thermal conductivity for CVD diamond would have to be higher than 800 W/mK in order to be competitive, depending on the selling price. Pure diamond has been measured at 2,000 W/mK, the upper limit for CVD diamond (although higher measurements have been made with synthetic diamond formed from pure ^{12}C.

Thermal conductivity has been implemented as an input in the four CVD diamond models. Figure 22 shows the cost per carat as a function of thermal conductivity for the these deposition models with long term input assumptions. In all cases, the cost of CVD diamond increases dramatically with thermal conductivity. The rise in cost is steepest with the hot filament technology, where a curve-fit of the model-generated data shows that cost is proportional to thermal conductivity to the exponent 4.1. In curve-fits for the microwave and DC arcjet technologies, this exponent is 2.9 and 3.8 respectively. The combustion flame technology, while having the highest cost, has the lowest sensitivity to thermal conductivity with an exponent of 2.4. The impact of this result is more apparent with the following example: if the thermal conductivity requirements of a given application can be decreased by 10%, the CVD diamond cost will consequently decrease by 23% for combustion flame diamond, 26% for microwave diamond, 33% for DC arcjet diamond, and 35% for hot filament diamond. Regardless of deposition technology, this analysis confirms that the desired thermal conductivity has a significant impact upon the cost of diamond: for a given application, the minimum acceptable value of thermal conductivity must be targeted before the CVD diamond is deposited.

4.7 CVD DIAMOND COST MODELING SUMMARY

Through the use of predictive equations in Technical Cost Models, the future (for the years 2001 to 2006) cost for CVD diamond at 1,000 W/mK is estimated in the $3 to $20 per carat range, depending on the deposition technology. With the TCM methodology predicting the cost as a function of various product and process parameters, sensitivities to critical inputs such as equipment power, gas prices, operation yield, and deposition diameter have been analyzed.

5. Summary

The extent of commercialization of man-made diamond depends on both the cost and performance of the product. The development efforts of many organizations have shown successful diamond manufacturing with respect to performance, but cost remains high for the film and coating versions of diamond.

Through the Technical Cost Modeling methodology, the paths to cost reduction are identified for both the HPHT and CVD diamond technologies. Economies of scale drive both of these diamond synthesis technologies: larger cell sizes for HPHT diamond manufacturing and larger deposition areas and higher powers for the CVD method are required for further significant cost reduction.

For HPHT diamond pressing, the cell size and average grit diameter are significant factors for the total cost per carat. A cell size of 200 cubic centimeters reflects today's technology and translates to a cost per carat of about $0.20 for an average grit size of 50 microns. Improvements are limited, but can be gained through larger cell sizes. Since the cost scales exponentially with the average grit diameter being produced, it is best to determine the minimum acceptable grit diameter for an application. From this analysis, it appears that further cost reduction is limited due to the current economies of scale.

Scaling up the deposition process reduces the cost of the CVD diamond technologies: the DC arcjet technology requires higher powers over optimal areas; the microwave method also should utilize larger areas and higher powers; the combustion flame technology demands optimum areas with cheaper fuels; and the hot filament technology should be developed over larger areas. With CVD technologies, the long term (5-10 year time frame) forecast suggests that costs per carat on the order of $1 to $2 are feasible.

In summary, the five to ten year outlook for man-made diamond entails lower cost. HPHT diamond can be expected to drop by about ten percent through cell size improvements. CVD diamond can be expected to drop by about two orders of magnitude as it completes the transition from research-size equipment to full manufacturing oriented equipment.

6. Acknowledgments

The authors would like to thank Mr. William Barker at the Advanced Research Projects Agency (ARPA) for supporting this economic research and Dr. James Butler at the Naval Research Laboratories for overseeing this project. The contributions of Professor David Goodwin at the California Institute of Technology to the models have made the theory-related analysis possible. In addition, the following individuals and organizations are recognized for their contributions to this work: Professor David Dandy at Colorado State University; Dr. Michael Coltrin, Dr. Robert Kee, and Dr. Ellen Meeks at Sandia National Laboratories; Dr. Richard Woodin (now with Crystalline Materials), Dr. Henry Windischmann, Dr. L.K. Bigelow, Mr. Edward Goss, Mr. Patrick Stephan, and many others at Norton Diamond Film; Dr. Robert Young and Dr. William Partlow at Westinghouse; Professor John Angus at Case Western Reserve University; Professor Mark Cappelli at Stanford University; Mr. Jerry White at Olin Aerospace Co.; Dr. Jeffrey Casey, Dr. Evelio Sevillano, Dr. Richard Post and many others at ASTeX; Dr. Dahimene at Wavemat; Dr. K.V. Ravi formerly at Lockheed Martin (now with Applied Materials); Dr. Charles Willingham and Dr. Robert DeKenipp at Raytheon; Dr. Roger Kidwell and Mr. Peter Santini at Diamonex; Dr. Jim Herlinger at sp3; and Dr. Vasge Shamamian at the Naval Research Laboratories.

7. References

Austin, G. (1995) US Department of the Interior Bureau of Mines Mineral Industry Surveys - Industrial Diamonds, US Bureau of Mines, pp. 1-6.

Busch, J.V., Dismukes, J.P., Nallicheri, N.V., and Walton, K.R. (1991) "Economic Assessment of HPHT Diamond Synthesis Technology," in Y. Tzeng, M. Yoshikawa, M. Murakawa, and A. Feldman (eds), *Applications of Diamond Films and Related Materials*, Elsevier Science Publishers B.V., pp.623-633.

Busch, J.V., and Dismukes, J.P. (1994) "Economics of CVD Diamond," in J. Dismukes, K. Ravi (eds), *Proceedings of the Third International Symposium on Diamond Materials*, The Electrochemical Society, Inc., pp. 880-891.

Dismukes, J.P. (1990) Technology and Manufacturing Cost Assessment of CVD Diamond Materials, *Carbon*, **28**, 789-790.

Goodwin, D.G. (1993) Scaling Laws for High Rate Diamond Growth, *J. Appl. Phys.* **74**, 6888.

Jennings, M. (1987) The Production and Uses of Industrial Diamond, *Metals and Materials*, **3**, 525-531.

Index